The names, chemical symbols, and atomic weights of the elements a[...] in Table 2.2, pages 36 and 37.

Recommended Names and Symbols for Selected Physical Quantities.

Name	Symbol
Length	l
Width	w
Height	h
Radius	r
Diameter	d
Area	A
Volume	V
Time	t
Speed	u
Speed of light in a vacuum	c
Gravitational acceleration	g
Amplitude of a wave	A
Frequency	ν
Wavelength	λ
Mass	m
Density	ρ
Force	F
Pressure	P
Work	W
Potential energy	E_p
Kinetic energy	E_k
Thermodynamic temperature	T
Electrical potential difference (voltage)	ΔV
Electromotive force	E
Faraday constant	F
Molar mass	$\mathcal{M}$
Gas constant	R
Molar volume	V_m
Amount of substance A	n_A
Concentration of substance A	c_A
Mole fraction of substance A	x_A
Mass fraction of substance A	w_A
Partial pressure of substance A	p_A
Equilibrium concentration of substance A	[A]
Quantity of heat	q
Internal energy	U
Enthalpy	H
Entropy	S
Gibbs free energy	G
Molar heat capacity	
Constant volume	C_V
Constant pressure	C_p
Specific heat capacity	c
Degree of dissociation	α
Equilibrium constant	K
Acid constant	K_a
Base constant	K_b
Self-ionization constant for water	K_w
Solubility product	K_{sp}
Standard equilibrium constant	$K^{\ominus}$
Standard increase in Gibbs free energy	$\Delta G^{\ominus}$
Standard increase in enthalpy	$\Delta H^{\ominus}$
Standard increase in entropy	$\Delta S^{\ominus}$
Electrical charge	Q
Electrical current	I
Osmotic pressure	Π
Mass number	A
Atomic number	Z
Neutron number	N
Charge of electron	$-e$
Planck constant	h
Principal quantum number	n
Half-life	$t_{1/2}$
Avogadro constant	N_A
Number of molecules	N
Molecular dipole moment	μ
Rate constant	k
Activation energy	$E^{\ddagger}$

CHEMISTRY

CHEMISTRY

John W. Moore
William G. Davies
Ronald W. Collins
Eastern Michigan University

McGraw-Hill Book Company
New York St. Louis San Francisco Auckland
Bogotá Düsseldorf Johannesburg London Madrid
Mexico Montreal New Delhi Panama Paris
São Paulo Singapore Sydney Tokyo Toronto

CHEMISTRY

1 2 3 4 5 6 7 8 9 0 VHVH 7 8 3 2 1 0 9 8

This book was set in Century Schoolbook by Progressive Typographers. The editors were Donald C. Jackson, Janet Wagner, and Anne T. Vinnicombe; the designer was Merrill Haber; the production supervisor was John F. Harte. The drawings were done by J & R Services, Inc.
Von Hoffmann Press, Inc., was printer and binder.

Library of Congress Cataloging in Publication Data

Moore, John W
Chemistry.

Includes index.
1. Chemistry.
I. Davies, William G., joint author. II. Collins, Ronald W., joint author. III. Title.
QD31.2.M62 540 77-21392
ISBN 0-07-042925-1

Cover: The front cover is an electron dot-density diagram for a carbon atom. The probability density of the electron cloud surrounding the carbon nucleus is indicated by the density of dots in the diagram. Dots corresponding to each of the six electron clouds ($1s^2 2s^2 2p^2$) have been coded by different colors. The back cover contains four smaller diagrams in which the two 1*s* electron clouds, the two 2*s* electron clouds, the two 2*p* electron clouds, and the total electron cloud are each shown. Additional electron dot-density diagrams will be found in the full-color insert in Chap. 5. All diagrams have been generated by a computer and a digital plotter using recent self-consistent field atomic wave functions. (*Copyright © 1975 by W. G. Davies and J. W. Moore.*)

CONTENTS

PREFACE

Chemistry today is a very exciting field—sufficiently complex that new discoveries are continually being made, but mature enough to be applied productively to a wide variety of other areas. This book has been designed for students who will be applying chemistry to their further studies in fields such as biology, medicine, public health, engineering, and environmental science, as well as for those who will major in chemistry and contribute to future chemical discoveries. Moreover, we believe it will be useful and interesting even to those whose career goals lie in nonscience areas. Any modern citizen who lacks understanding of chemical concepts or is not aware of their consequences in everyday life is at a considerable disadvantage.

In writing this textbook our primary aim has been to describe the most important and fundamental facts and theories of chemistry in such a way that students can attain adequate understanding and knowledge of them. Whenever we felt that applications of chemistry to other fields could be explained adequately and meshed appropriately with the principles or facts under consideration, we have included applications. Moreover, we have aimed at consistency and cohesion throughout the book so that students would be able to see how concepts already learned can be extended and applied to new situations. We agree with the remark of Joel Hildebrand that, "There's nothing that is more relevant to the problems of the world than the scientific method of attacking them," and we hope that the scientific method can be learned from the numerous examples of its use which we have provided.

Our experience with students in colleges and universities has strongly influenced decisions relating to pedagogical approach, inclusion and ordering of topics, and level of presentation. Although they are not necessarily chemistry majors, the majority of our students plan to take more advanced chemistry courses required by their academic programs. Therefore we have developed topics to a level which will give such students an adequate conceptual and factual background. On the other hand students entering the college chemistry course are often weak in the application of mathematics to chemical problems. Most have had a previous course in chemistry, but many have little or no knowledge of physics. We have taken pains in the text itself (not in an appendix, which often remains unread) to familiarize the student with the mathematics and/or physics needed for understanding a given topic. However, we have not introduced mathematics and physics for their own sakes, and we have attempted to show as clearly as possible how and why these disciplines are applicable to chemistry.

In some cases we have deferred discussion of a highly physical or mathematical topic to a time when we felt it could be better understood by the student. An example of this is the development of atomic and molecular spectroscopy and molecular orbitals in Chapter 21 rather than in an early chapter on atomic structure or chemical bonding. Understanding the significance of Bohr's interpretation of atomic spectra, the interference of waves, and the formation of molecular orbitals requires considerably more physics than most of our students have encountered. Therefore we have based our approach to atomic and molecular structure on *chemical* properties, formulas, and the periodic table. This approach is similar to the intuition which led G. N. Lewis[1] to electron pairs and octets a decade *before* the Bohr theory, and it is firmly grounded on concrete, descriptive facts about the elements and their compounds. Instructors whose students have more mathematical and physical background than we have assumed may wish to use Sections 21.1 and 21.2 at the beginning of Chapter 5 and part of Section 21.4 at the end of Chapter 7.

Our approach to the electronic structure of atoms and molecules might be termed descriptive quantum mechanics. We have described the results of wave mechanical calculations and applied them to periodic chemical properties and to chemical bonding in a pictorial rather than a mathematical way. This is possible because of a large number of highly accurate, conceptually correct computer-generated diagrams of electron density distributions and boundary surfaces. The diagrams have been plotted directly by a digital computer based on calculations involving recent self-consistent field atomic and molecular wave functions. (For examples, see Plates 4 and 5 of the color insert.) Our experience indicates that this semiquantitative, graphic display of the results of up-to-date theory is readily comprehended by students and expedites the discussion of electronic structure and bonding. Computer-generated diagrams are used in many other portions of the book as well, especially those dealing with molecular and crystal structures. Computer graphics techniques insure that each atom is in the proper location and that perspective is correct, even for complicated biopolymers like cellulose and starch (Figure 20.8). All of the diagrams in this book have been carefully thought out and designed as teaching tools. Students will often learn things from the diagrams more effectively than from the text, and we have spent considerable effort to insure that what they learn will be neither incorrect nor misleading.

Another approach which we have taken at many points throughout the text is to present at least some concrete facts *before* describing or deriving a theory to explain them. In the case of atomic structure, for example, previous knowledge of chemical formulas and periodic variation of valence raises the question of how atoms are held together in molecules and why some atoms form one bond while others form two, three, or more. When our current versions of the answers to these questions are presented, they are far more satisfying to the student than if no question had been raised in the first place. Other topics for which we have tried to avoid telling an answer before the student knows enough to ask a question are: gases, where properties are described before the kinetic molecular theory is derived to explain the properties; acids and bases, which are defined and described qualitatively before quantitative equilibrium calculations are made; dynamic equilibrium, which is first treated qualitatively, then quantitatively, and finally on a molecular basis; and kinetics, where experimental techniques are mentioned first and the results obtained are then interpreted theoretically.

Perhaps the greatest unifying concept pervading the chemist's thought processes is the quantization of matter into atomic and molecular units. The practicing chemist constantly searches for relationships between the microscopic, or molecular, world and the macroscopic, or observable, one. We introduce the terms *macroscopic* and *microscopic* in Chapter 2 and use them over and over throughout the text. From that point on, students are aware that molecules are constantly in motion and that the prop-

[1] G. N. Lewis, "Valence and the Structure of Atoms and Molecules," p. 17, Dover Publications, Inc., New York, 1966.

erties of matter depend on those movements as well as the structures and properties of the molecules themselves. This is especially evident in the treatment of equilibrium in Chapter 13 and of the second law of thermodynamics in Chapter 16. Here the crucial role of probability in determining the extent of a chemical reaction is developed in an innovative way.

We have devoted separate chapters to organic chemistry and to biochemistry. While it is clearly impossible to do justice to either of these topics in such a short space, both are far too important to be omitted from a course which may be the only chemistry course that some students will have. Organic compounds are ideal choices to illustrate principles of bonding and intermolecular forces, and so we have described them in Chapter 8 immediately following the section on bonding. Biochemistry (Chapter 20) provides a culmination (and review) of many of the principles and facts learned throughout a general chemistry course. We strongly recommend its inclusion, even by instructors whose background in this area may be minimal, because of the great research advances currently being made in understanding the molecular basis of life.

Perhaps the most controversial decision we made before setting pen to paper was to use the International System of Units (SI Units) throughout this book. Scientists and science teachers in the United States have lagged behind those abroad in adopting this updated metric system, but we feel its consistency and uniformity provide important advantages to the student in a first attempt at learning chemistry. The advantages far outweigh the slight relearning required of the instructor, and as time goes on, we are certain that SI units will become as widely accepted in the United States as they are elsewhere. (One exception to the exclusive use of SI units was so strongly urged by reviewers and our editors that we felt constrained to accept it. This is the inclusion of liters and atmospheres as well as cubic decimeters and pascals in Chapter 9 on gases. Those who wish to use SI units exclusively, however, will find that they can readily do so. Those who do not, should be equally satisfied.)

We believe that chemistry is primarily an experimental science and that students should be able to perform meaningful laboratory experiments as early as possible. Thus we have tried to introduce the descriptive material and methods of calculation needed for laboratory work in the first three chapters. In doing this we have been aided by the fact that the mole is one of the seven base units in the International System, so that it is now clearly defined as the only unit needed to measure the amount of a substance. This simplifies stoichiometric calculations considerably, and we have been able to include solution stoichiometry and titration calculations at the end of Chapter 3. This will permit a much greater variety of experiments early in the course since students can be presumed to know how to handle simple calculations involving solutions. Although we have postponed discussion of gases until Chapter 9, this chapter is relatively independent of those preceding it. If desirable, Chapter 9 could follow Chapter 3, and an even greater variety of experiments could be performed after 3 or 4 weeks of class. (A laboratory manual which takes advantage of these features of the book is available from the McGraw-Hill Book Company.)

At the end of each chapter we have provided a large number of questions and problems of various types and levels of difficulty. Answers are provided in Appendix 3 to those problems whose numbers appear in red. Worked solutions to all problems will be found in the "Student/Instructor Solution Supplement." (Some problems indicated by red numbers have multiple parts, e.g., a, b, c, etc. In such cases answers for every other part are included in Appendix 3.) Problems of greater than normal difficulty have been marked with an asterisk. A student "Study Guide" is also available from the McGraw-Hill Book Company. Overhead transparencies of many of our computer-generated illustrations are available from Science Related Materials, Inc., P.O. Box 1422, Janesville, WI, 53545.

Several of our colleagues at Eastern Michigan University have contributed to this textbook. Much of the material on stoichiometry was refined during discussions with Ms. Nina Contis. Dr. K. Rengan read and made useful comments on Chapter 19, and Drs. Stephen Schullery and Bruce West helped solidify our ideas about biochemistry. As department head, Dr. Clark Spike has been most cooperative in alleviating problems of class schedules and committee assignments which otherwise might have delayed completion of the manuscript. Students Rodney Finzel and Steve Duff have contributed computer programming expertise, and all our computer-generated diagrams were produced by Eastern Michigan's Instructional Computer Services, William Rodgers, director. Systems software for the Calcomp plotter was written by Mr. Gary Frownfelter and Dr. Jerald Griess.

The manuscript has been reviewed by Drs. David Adams, Jay Martin Anderson, Clare T. Furse, Robert R. Hubbs, James Orenberg, Theodore W. Sottery, John Stutzman, and Charles Wheeler. All have made useful comments, suggestions, and corrections, for which we are grateful. Any errors that remain are our own responsibility, of course, and we hope that readers who find them will communicate them to us. Both first and second drafts of the manuscript were typed capably by Elizabeth A. Moore, who, in addition to deciphering our handwriting, kept us organized and made numerous suggestions which have improved both style and clarity. We also thank the editors and other production staff at McGraw-Hill, especially Bob Summersgill who initiated the project, Janet Wagner who made detailed comments and improvements on the manuscript and has shepherded it to completion, and Anne Vinnicombe who has remained calm and capable in the face of numerous deadlines and crises.

Our aim in writing this textbook has been to examine the subject matter of chemistry carefully, with an eye toward improving the logic of its presentation, the comprehensibility of its development, and the efficiency with which it can be mastered. We hope that both students and teachers who use this book will agree that we have succeeded in reaching these goals.

John W. Moore
William G. Davies
Ronald W. Collins

chapter one

INTRODUCTION

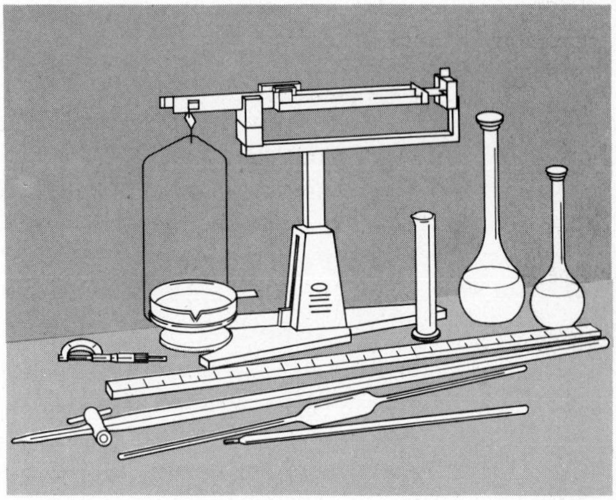

The science of chemistry is concerned with the composition, properties, and structure of matter and with the ways in which substances can change from one form to another. Since anything which has mass and occupies space can be classified as matter, this means that chemistry is involved with almost everything in the universe. Chemistry, in other words, is not just something which happens in laboratories. It happens all the time all around us, and even inside us. Chemistry is going on in places as diverse as the smallest bacterium, a field of ripening wheat, a modern manufacturing plant, the biospheres of planets such as Earth, the vast reaches of interstellar space, and even your eyes and brain as you read these words.

Because chemistry is involved in almost everything, the study of chemistry is of importance to a wide variety of persons. Biologists, for example, have examined smaller and smaller organisms, cells, and cell components, until, in the study of viruses and genes, they joined forces with chemists who were interested in larger and larger molecules. The result was a new interdisciplinary field called molecular biology, and a reinforcement of the idea that living organisms are complicated, highly organized chemical systems. Chemists interact in similar ways with scientists in areas such as chemical physics, geochemistry, pharmacology, toxicology, ecology, meteorology, oceanography, and many others. Current practice in these fields is such that a person lacking basic chemical knowledge is at a severe disadvantage.

Chemistry also underlies a great deal of modern technology. The manufacture of such basic commodities as steel, aluminum, glass, plastics, paper, textiles, and gasoline all involve chemists and chemistry. Without the abundant cheap supply of these and other substances which chemistry has helped to produce, our lives would be much less comfortable. However, as we are now beginning to discover, callous and indiscriminate use of technology can produce disadvantages as well as benefits. Automobiles, power plants, and even some aerosol sprays spew into the air harmful substances which are not always easy or cheap to eliminate. Rivers and lakes are also more easily contaminated than we once thought—substances once believed harmless now have proven to be the opposite.

Issues involving the effects of technology on the environment affect everyone—not just scientists. Decisions about them are political, at least in part, and require some chemical knowledge on the part of voters as well as their elected representatives. At the very least, a citizen needs to be able to distinguish valid and invalid arguments put forward by scientific "experts" regarding such issues. (In some cases such "expertise" may be mainly a willingness to speak out rather than a command of the scientific and political issues.) It is to be hoped also that more persons will follow the example of Russell Peterson, formerly a researcher in a large chemical company, who has served as Governor of Delaware and Chairman of the President's Council on Environmental Quality. Only by a combination of scientists willing to leave their laboratories and citizens willing to master some of the basics of science can intelligent political decisions be made in a democratic society. Indeed, such a combination may be a necessity if democracy is not to degenerate into an oligarchy ruled by those who control the experts.

Given the universality of chemistry, its central role among the sciences, and its importance in modern life, how is it possible to learn much about it in a short time? If everything is a chemical and if chemistry happens all the time and everywhere, is the field of chemistry so broad and all-encompassing that one cannot master enough to make its study worthwhile? We think the answer to this second question is a resounding *no*! This entire book has been designed to help you learn a good deal of chemistry in a short time. If it is successful, the first question will have been answered as well.

An important and valuable technique of science and scientists is that of subdividing large, seemingly unsolvable problems into smaller, simpler parts. If the latter are chosen carefully, each can be mastered. Individual small advances can then be combined to yield an important, more complicated result. Thus chemists have not acted on the assumption that because all the world is composed of chemicals and chemistry, they should try to study it all at once. Rather, many chemists do much of their work under controlled, laboratory conditions, advancing in small steps toward more general, useful, and exciting results.

Since people's process of studying and understanding chemistry is far from complete, we can narrow our area of interest considerably by redefining chemistry as those things that chemists have done and are doing. This more restricted view constitutes the main theme of this book. We hope that when you have finished with it, you will have a solid background in the facts, laws, and theories of chemistry, as well as those modes of behavior and thinking that chemists have found useful in solving the problems they have faced.

1.1 WHAT CHEMISTS DO

What are some of the things that chemists do? Like most scientists, they observe and measure components of the natural world. Based on these observations they try to place things into useful, appropriate categories and to formulate scientific laws which summarize the results of a great many observations. Indeed, it is a fundamental belief of all science that natural events do not occur in a completely unpredictable fashion. Instead, they obey natural laws. Therefore observations and measurements made on one occasion can be duplicated by the same or another person on another occasion. Communication of such results, another important activity, affords an opportunity for the entire scientific community to test an individual's work. Eventually a consensus is reached, and there is agreement on a new law.

Like other scientists, chemists try to explain their observations and laws by means of theories or models. They constantly make use of atoms, molecules, and other very small particles. Using such theories as their guides, chemists synthesize new materials. Well over 3 million compounds are now known, and more than 9000 are in large-scale commercial production. Even a backpacker going "back to nature" takes along synthetic materials such as nylon, aluminum, and aspirin.

Chemists also analyze the substances they make and those found in nature. Determining the composition of a substance is the first step in understanding its chemical properties, and detection of very small quantities of some materials in the natural world is essential in controlling air and water pollution. Another role that chemists play is in studying the processes (chemical reactions) by which one substance can be transformed into another. Will the reactions occur without prodding? If so, how quickly? Is energy given off? Can the reactions be controlled—made to occur only when we want them to?

These and many other problems that interest chemists are the subject matter of this book. We will return to each several times and in increasing detail. Do not forget, however, that chemistry is more than just what chemists do. Many persons in other sciences as well as in daily life are constantly doing chemistry, whether they call it by name or not. Indeed, each of us is an intricate combination of chemicals, and everything we do depends on chemical reactions. Although space limitations will prevent us from exploring all but a fraction of the applications of chemistry to other fields, we have included such excursions as often as seemed appropriate. From them we hope that you will be able to learn how to apply chemical facts and principles to the problems you will face in the future. Many of the problems are probably not even known yet, and we could not possibly anticipate them. If you have learned how to think "chemically" or "scientifically," however, we believe that you will be better prepared to face them.

1.2 MEASUREMENT

As an example of what we have just been saying, let us assume that you are faced with a specific problem. Then we can see how scientific thinking might help solve it. Suppose that you live near a large plant which manufactures cement. Smoke from the plant settles on your car and house, causing small pits in the paint. You would like to stop this air-pollution problem—but how?

As an individual you will probably have little influence, and even as part of a group of concerned citizens you may be ineffective, unless you can *prove* that there is a problem. Scientists have had a hand in writing most air-pollution regulations, and so you will have to employ some scientific techniques (or a scientist) to help solve your problem.

It will probably be necessary to determine *how much* air pollution the plant is producing. This might be done by comparing the smoke with a scale which ranges from white to gray to black, the assumption being that the darker the smoke, the more there is. For white cement dust, however, this is much less satisfactory than for black coal smoke. A better way to determine how much pollution there is would be to measure the mass of smoke particles which could be collected near your house or car. This could be done by using a pump (such as a vacuum cleaner) to suck the polluted gas through a filter. Weighing the filter before and after such an experiment would determine what mass of smoke had been collected.

Accurate weighing is usually done with a single-pan balance whose principle of operation is shown in Fig. 1.1. The empty pan is balanced by a counterweight. When an object is placed on the pan, gravitational attraction forces the pan downward. To restore balance, a series of weights (objects of known mass) are removed from holders above the pan. The force of gravitational attraction is proportional to mass, and when the pan is balanced, the force on it must always be the same. Therefore the mass of the object being weighed must equal that of the weights that were removed.

Numbers, Units, Quantities

If you kept a notebook or other record of your measurements, you would probably write down something like 0.0342 g to represent how much smoke had been collected. Such a result, which describes the *magnitude of some property* (in this case the magnitude of the mass), is called a **quantity.** Notice that it consists of two parts, a number and a unit. It would be ambiguous to write just 0.0342—you might not remember later whether that was measured in units of grams, ounces, pounds, or something else.

A quantity always behaves as though the number and the units are multiplied together. For example, we could write the quantity already ob-

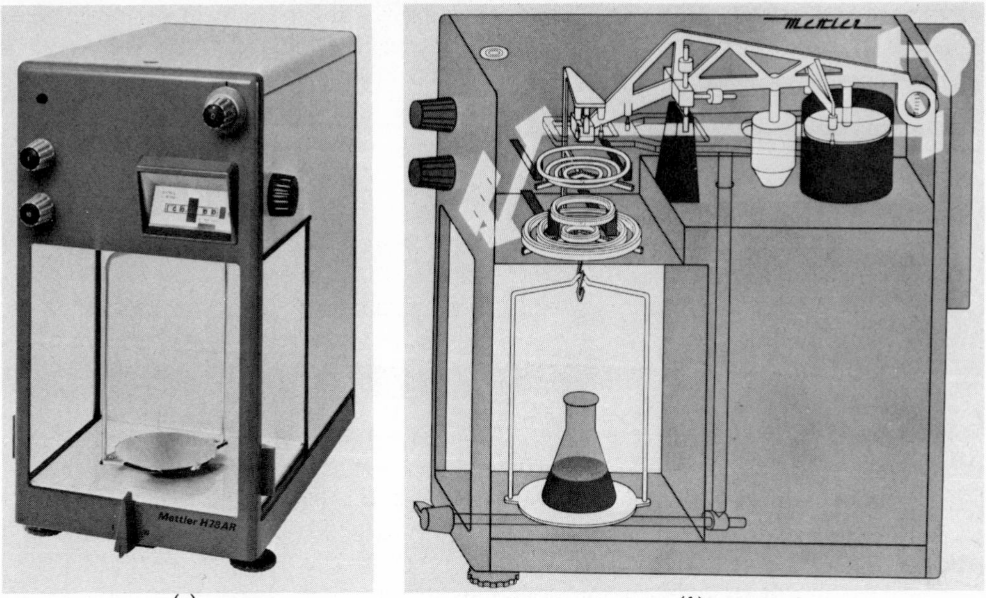

Figure 1.1 (*a*) Actual appearance of a modern substitution balance. (*b*) X-ray view showing principle of operation. When an object is placed on the balance pan, ring-shaped weights whose total mass equals that of the object are removed from holders above the pan to restore balance. (*By permission from Mettler Instrument Corporation, Princeton, New Jersey.*)

tained as 0.0342 × g. Using this simple rule of number × units, we can apply arithmetic and algebra to any quantity:

$$5\text{ g} + 2\text{ g} = (5 + 2)\text{ g} = 7\text{ g}$$

$$5\text{ g} \div 2\text{ g} = \frac{5\not{\text{g}}}{2\not{\text{g}}} = 2.5 \qquad \text{(the units cancel, and so we get a pure number)}$$

$$5\text{ in} \times 2\text{ in} = 10\text{ in}^2 \qquad \text{(10 square inches)}$$

This works perfectly well as long as we do not write equations with different kinds of quantities (i.e., those having units which measure different properties) on opposite sides of the equal sign. For example, applying algebra to the equation

$$5\text{ g} = 2\text{ in}^2$$

can lead to trouble in much the same way that dividing by zero does and should be avoided.

Notice also that whether a quantity is large or small depends on the size of the units as well as the size of the number. For example, in Fig. 1.2 the length of a metal rod has been measured with a ruler calibrated in inches and with one calibrated in centimeters. Both results (3.50 in and 8.89 cm) are the *same* quantity, the length of the rod. One involves a smaller number and larger unit (3.50 in), while the other has a larger number and smaller unit. So long as we are talking about the same quantity, it is a simple matter to adjust the number to go with any units we want.

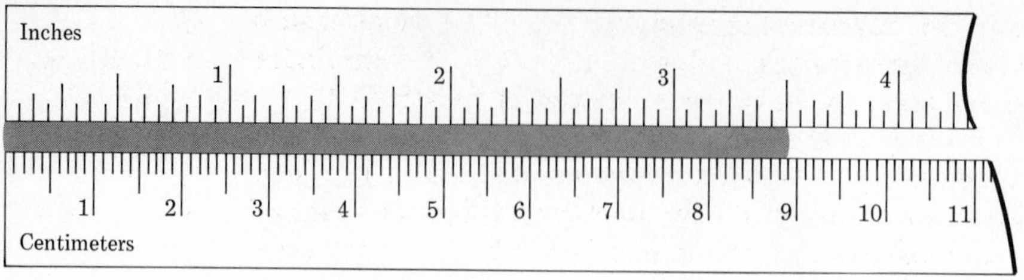

Figure 1.2 The length of a rod can be measured either in centimeters or inches. We can record the length either as 3.50 in or as 8.89 cm. In either case we are referring to the same quantity.

EXAMPLE 1.1 Express the length 8.89 cm in inches, given that 1 cm = 0.3937 in.

Solution Since 1 cm and 0.3937 in are the same quantity, we can write the equation

$$1 \text{ cm} = 0.3937 \text{ in}$$

Dividing both sides by 1 cm, we have

$$1 = \frac{0.3937 \text{ in}}{1 \text{ cm}}$$

Since the right side of this equation equals one, it is called a **unity factor.** It can be multiplied by any quantity, leaving the quantity unchanged.

$$8.89 \text{ cm} = 8.89 \text{ cm} \times 1 = 8.89 \cancel{\text{cm}} \times \frac{0.3937 \text{ in}}{1 \cancel{\text{cm}}}$$

The units *centimeter* cancel, yielding the result

$$8.89 \text{ cm} = 8.89 \times 0.3937 \text{ in} = 3.50 \text{ in}$$

This agrees with the direct observation made in Fig. 1.2.

Let us return to our air-pollution problem. It has probably already occurred to you that simply measuring the mass of smoke collected is not enough. Some other variables may affect your experiment and should also be measured if the results are to be reproducible. For example, wind direction and speed would almost certainly be important. The time of day and date when a measurement was made should be noted too. In addition you should probably specify what kind of filter you are using. Some are not fine enough to catch all the smoke particles.

Another variable which is almost always recorded is the temperature. A thermometer is easy to use, and temperature can vary a good deal outdoors, where your experiments would have to be done. In scientific work, temperatures are usually reported in degrees Celsius (°C), a scale in which the freezing point of pure water is 0°C and the normal boiling point 100°C. In the United States, however, you would be more likely to have available a thermometer calibrated in degrees Fahrenheit (°F). The relationship between these two scales of temperature is shown in Fig. 1.3. A formula for exact conversion from one scale to the other is derived in Appendix 1.

More important than any of the above variables is the fact that the more air you pump through the filter, the more smoke you will collect. Since air is a gas, it is easier to measure how much you collect in terms of volume than in terms of mass, and so you might decide to do it that way. Running your pump until it had filled a plastic weather balloon would provide a crude, inexpensive volume measurement. Assuming the balloon to be approximately spherical, you could measure its diameter and calculate its volume from the formula

$$V = \frac{4}{3}\pi r^3$$

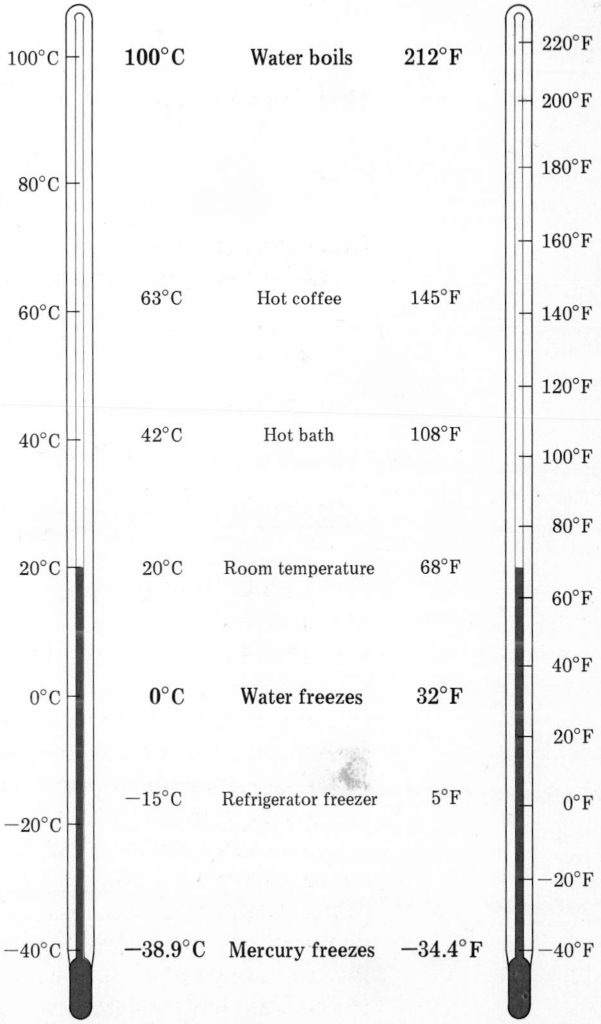

Figure 1.3 The Celsius and Fahrenheit scales compared. Temperatures in **bold face** are exact and easy to reproduce. Other temperatures are approximate and somewhat variable.

EXAMPLE 1.2 Calculate the volume of gas in a sphere whose diameter is 106 in. Express your result in cubic centimeters (cm^3).

Solution Since the radius of a sphere is half its diameter,

$$r = \frac{1}{2} \times 106 \text{ in} = 53 \text{ in}$$

We can use the same equality of quantities as in Example 1.1 to convert the radius to centimeters. When we cube the number and units, our result will be in cubic centimeters.

$$1 \text{ cm} = 0.3937 \text{ in}$$

$$\frac{1 \text{ cm}}{0.3937 \text{ in}} = 1$$

$$r = 53 \cancel{\text{in}} \times \frac{1 \text{ cm}}{0.3937 \cancel{\text{in}}} = \frac{53}{0.3937} \text{ cm}$$

Using the formula

$$V = \frac{4}{3}\pi r^3 = \frac{4}{3} \times 3.141\ 59 \times \left(\frac{53}{0.3937} \text{ cm}\right)^3$$
$$= 10\ 219\ 264 \text{ cm}^3$$

You can see from Examples 1.1 and 1.2 that *two* unity factors may be obtained from the equality

$$1 \text{ cm} = 0.3937 \text{ in}$$

We can use one of them to convert inches to centimeters and the other to convert centimeters to inches. *The correct factor is always the one which results in cancellation of the units we do not want.*

The result in Example 1.2 also shows that cubic centimeters are rather small units for expressing the volume of the balloon. If we used larger units, as shown in the following example, we would not need more than 10 million of them to report our answer.

EXAMPLE 1.3 Express the result of Example 1.2 in cubic meters, given that 1 m = 100 cm.

Solution Again we wish to use a unity factor, and since we are trying to get rid of cubic centimeters, centimeters must be in the denominator:

$$1 = \frac{1 \text{ m}}{100 \text{ cm}}$$

But this will not allow cancellation of *cubic* centimeters. However, note that $1^3 = 1$. That is, we can raise a unity factor to any power, and it remains unity. Thus

$$1 = \left(\frac{1\text{ m}}{100\text{ cm}}\right)^3 = \frac{1\text{ m}^3}{100^3\text{ cm}^3} = \frac{1\text{ m}^3}{1\ 000\ 000\text{ cm}^3}$$

and

$$10\ 219\ 264\text{ cm}^3 = 10\ 219\ 264\text{ cm}^3 \times \left(\frac{1\text{ m}}{100\text{ cm}}\right)^3$$

$$= 10\ 219\ 264\ \cancel{\text{cm}^3} \times \frac{1\text{ m}^3}{1\ 000\ 000\ \cancel{\text{cm}^3}}$$

$$= 10.219\ 264\text{ m}^3$$

Handling Large and Small Numbers and Units

Example 1.2 illustrates a common occurrence in science—results often involve very large numbers or very small fractions. The United States used 66 500 000 000 000 000 000 J (joules) of energy in 1971, and the mass of a water molecule is 0.000 000 000 000 000 000 000 029 9 g. Such numbers are inconvenient to write and hard to read correctly. (We have divided the digits into groups of three to make it easier to locate the decimal point. Spaces are used instead of commas because many countries use a comma to indicate the decimal.)

There are two ways of handling this problem. We can express a quantity in larger or smaller units, as in Example 1.3, or we can use a better way to write small and large numbers. The latter approach involves what is called **scientific notation** or **exponential notation.** The position of the decimal point is indicated by a power (or exponent) of 10. For example,

$$138 = 13.8 \times 10 = 1.38 \times 10 \times 10 = 1.38 \times 10^2$$

$$0.004\ 83 = \frac{4.83}{10 \times 10 \times 10} = 4.83 \times \frac{1}{10^3} = 4.83 \times 10^{-3}$$

A number with a *negative exponent* is simply the *reciprocal* of (one divided by) the same number with the equivalent positive exponent. Therefore decimal fractions (numbers between zero and one) may be expressed using negative powers of 10. Numbers between 1 and 10 require no exponential part, and those larger than 10 involve positive exponents. By convention the power of 10 is chosen so that there is one digit to the left of the decimal point in the ordinary number. That is, we would usually write 5280 as 5.28×10^3, not as 0.528×10^4 or 52.8×10^2.

To convert a number from ordinary to scientific notation, count how many places the decimal point must be shifted to arrive at a number between 1 and 10. If these shifts are to the *l*eft, the number was *l*arge to begin with and we multiply by a *l*arge (that is, positive) power of 10. If the shift is to the *r*ight, a *r*eciprocal (negative) power of 10 must be used.

EXAMPLE 1.4 Express the following numbers in scientific notation: (a) 7563; (b) 0.0156.

Solution

a) In this case the decimal point must be shifted *left three* places:

$$7\,.\,5\,6\,3\,.$$

Shift 3 Shift 2 Shift 1

Therefore we use an exponent of $+3$:

$$7563 = 7.563 \times 10^3$$

b) Shifting the decimal point *two* places to the *right* yields a number between 1 and 10:

$$0\,.\,0\,1\,.\,5\,6$$

Shift 1 Shift 2

Therefore the exponent is -2:

$$0.0156 = 1.56 \times 10^{-2}$$

When working with exponential notation, it is often necessary to add, subtract, multiply, or divide numbers. When multiplying and dividing, you must remember that multiplication corresponds to addition of exponents, and division to their subtraction.

Multiplication: $$10^a \times 10^b = 10^{(a+b)}$$

Division: $$\frac{10^a}{10^b} = 10^{(a-b)}$$

Hence

$$(3.0 \times 10^5) \times (5.0 \times 10^3) = 15.0 \times 10^{(5+3)} = 15.0 \times 10^8$$
$$= 1.50 \times 10^9$$

and
$$\frac{3.0 \times 10^5}{5.0 \times 10^3} = 0.6 \times 10^{(5-3)} = 0.6 \times 10^2 = 6.0 \times 10$$

EXAMPLE 1.5 Evaluate the following, giving your answer in correct exponential notation:

a) $(3.89 \times 10^5) \times (1.09 \times 10^{-3})$ **c)** $\dfrac{5.0 \times 10^6}{3.98 \times 10^8}$

b) $(6.41 \times 10^{-5}) \times (2.72 \times 10^{-2})$ **d)** $\dfrac{7.53 \times 10^{-3}}{8.57 \times 10^{-5}}$

Solution

a) $$(3.89 \times 10^5) \times (1.09 \times 10^{-3}) = 3.89 \times 1.09 \times 10^{5+(-3)}$$
$$= 4.24 \times 10^2$$

b) $(6.41 \times 10^{-5}) \times (2.72 \times 10^{-2}) = 6.41 \times 2.72 \times 10^{-5+(-2)}$

$= 17.43 \times 10^{-7} = 1.743 \times 10^{-6}$

c) $\dfrac{5.0 \times 10^6}{3.98 \times 10^8} = \dfrac{5.0}{3.98} \times 10^{6-8} = 1.26 \times 10^{-2}$

d) $\dfrac{7.53 \times 10^{-3}}{8.57 \times 10^{-5}} = \dfrac{7.53}{8.57} \times 10^{-3-(-5)} = 0.879 \times 10^2$

$= 8.79 \times 10^1$

Addition and subtraction require that all numbers be converted to the *same* power of 10. (This corresponds to lining up the decimal points.)

EXAMPLE 1.6 Evaluate the following, giving your answer in scientific notation:

a) $(6.32 \times 10^2) - (1.83 \times 10^{-1})$

b) $(3.72 \times 10^4) + (1.63 \times 10^5) - (1.7 \times 10^3)$

Solution First convert to the same power of 10; then add the ordinary numbers.

a)
$$\begin{aligned} 6.32 \times 10^2 &= 632. \\ -1.83 \times 10^{-1} &= -0.183 \\ & \overline{631.817} = 6.318\ 17 \times 10^2 \end{aligned}$$

b) Convert all powers of 10 to 10^4.

$$\begin{aligned} 3.72 \times 10^4 &= 3.72 \times 10^4 = 3.72 \times 10^4 \\ 1.63 \times 10^5 &= 1.63 \times 10 \times 10^4 = 16.3 \times 10^4 \\ -1.7 \times 10^3 &= -1.7 \times 10^{-1} \times 10^4 = -0.17 \times 10^4 \\ & \overline{19.85 \times 10^4} = 1.985 \times 10^5 \end{aligned}$$

Scientific notation is becoming more common every day. Many electronic pocket calculators use it to express numbers which otherwise would not fit into their displays. For example, an eight-digit calculator could not display the number 6 800 000 000. The decimal point would remain fixed on the right, and the 6 and the 8 would "overflow" to the left side. Such a number is often displayed as 6.8 09, which means 6.8×10^9. If you use a calculator which does *not* have scientific notation, we recommend that you express all numbers as powers of 10 before doing any arithmetic. Follow the rules in the last two examples, using your calculator to do arithmetic on the ordinary numbers. You should be able to add or subtract the powers of 10 in your head.

Computers also are prone to print results in scientific notation, and they use yet another minor modification. The printed number 2.3074 E-07 means 2.3074×10^{-7}, for example. In this case the E indicates that the number following is an exponent of 10.

Too Many Digits

Returning once more to our air-pollution experiment, we could express the mass of smoke collected as 3.42×10^{-2} g and the volume of the balloon as $1.021\,926\,4 \times 10^{7}$ cm³. There is something strange about the second quantity, though. It contains a number which was copied directly from the display of an electronic calculator and has *too many digits.*

The reliability of a quantity derived from a measurement is customarily indicated by the number of **significant figures** (or significant digits) it contains. For example, the three significant digits in the quantity 3.42×10^{-2} g tell us that a balance was used on which we could distinguish 3.42×10^{-2} g from 3.43×10^{-2} or 3.41×10^{-2} g. There might be some question about the last digit, but those to the left of it are taken as completely reliable. Another way to indicate the same thing is $(3.42 \pm 0.01) \times 10^{-2}$ g. Our measurement is somewhere between 3.41×10^{-2} and 3.43×10^{-3} g.

As another example of choosing an appropriate number of significant digits, let us read the volume of liquid in a graduated cylinder (Fig. 1.4). The bottom of the meniscus lies between graduations corresponding to 38 and 39 cm³. We can estimate that it is at 38.5 cm³, but the last digit might be off a bit—perhaps it looks like 38.4 or 38.6 cm³ to you. Since the third digit is in question, we should use three significant figures. The volume would be recorded as 38.5 cm³. Laboratory equipment is often calibrated similarly to this graduated cylinder—you should estimate to the nearest tenth of the smallest graduation.

In some ordinary numbers, for example, 0.001 23, zeros serve merely to locate the decimal point. They do not indicate the reliability of the measurement and therefore are not significant. Another advantage of scientific notation is that we can assume that *all* digits are significant. Thus if 0.001 23 is written as 1.23×10^{-3}, only the 1, 2, and 3, which indicate the reliability of the measurement, are written. The decimal point is located by the power of 10.

If the rule expressed in the previous paragraph is applied to the volume of air collected in our pollution experiment, $1.021\,926\,4 \times 10^{7}$ cm³, we find that the volume has *eight* significant digits. This implies that it was determined to ±1 cm³ out of about 10 million cm³, a reliability which corresponds to locating a grasshopper *exactly* at some point along the road from Philadelphia to New York City. For experiments as crude as ours, this is not likely. Let us see just how good the measurement was.

You will recall that we calculated the volume from the diameter of the balloon, 106 in. The three significant figures imply that this might have been as large as 107 in or as small as 105 in. We can repeat the calculation with each of these quantities to see how far off the volume would be:

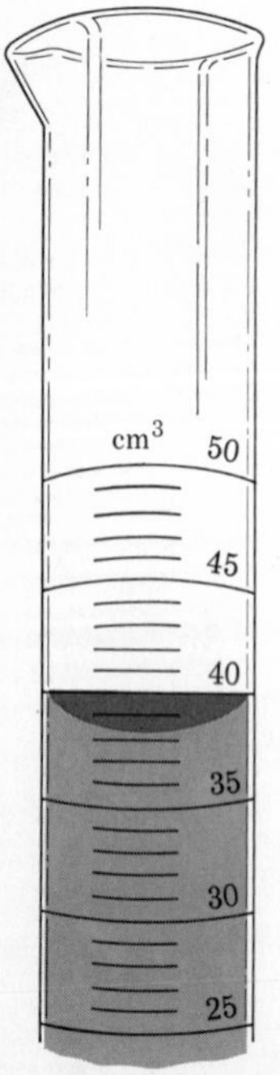

Figure 1.4 The level of a liquid in a graduated cylinder. The saucer-shaped surface of a liquid in a tube is called a **meniscus.**

$$r = \frac{1}{2} \times 107 \text{ in} = 53.5 \cancel{\text{in}} \times \frac{1 \text{ cm}}{0.3937 \cancel{\text{in}}}$$
$$= 135.890\ 27 \text{ cm}$$
$$V = \frac{4}{3} \times 3.141\ 59 \times (135.890\ 27 \text{ cm})^3$$
$$= 10\ 511\ 225 \text{ cm}^3 = 1.051\ 122\ 5 \times 10^7 \text{ cm}^3$$

or

$$V = \frac{4}{3} \times 3.141\ 59 \times \left(\frac{1}{2} \times 105 \cancel{\text{in}} \times \frac{1 \text{ cm}}{0.3937 \cancel{\text{in}}}\right)^3$$
$$= 9\ 932\ 759 \text{ cm}^3 = 0.993\ 275\ 9 \times 10^7 \text{ cm}^3$$

That is, the volume is between 0.99×10^7 and 1.05×10^7 cm^3, or $(1.02 \pm 0.03) \times 10^7$ cm^3. We should **round** our result to three significant figures, for example, 1.02×10^7 cm^3, because the last digit, namely 2, is in question.

TABLE 1.1 Rules for Rounding Numbers.

1 All digits to be rounded are removed together, not one at a time.
2 If the left-most digit to be removed is less than five, the last digit retained is not altered.
3 If the left-most digit to be removed is greater than five, the last digit retained is increased by one.
4 If the left-most digit to be removed is five and at least one of the other digits to be removed is nonzero, the last digit retained is increased by one.
5 If the left-most digit to be removed is five and all other digits to be removed are zero, the last digit retained is not altered if it is even, but is increased by one if it is odd.

The rules for rounding numbers are summarized in Table 1.1. Their application can be illustrated by an example.

EXAMPLE 1.7 Round each of the numbers below to three significant figures.

a) 34.7449 **b)** 34.864 **c)** 34.754 **d)** 34.250 **e)** 34.35

Solution

a) Apply rules 1 and 2 from Table 1.1:

Last digit retained

Digits to be removed

34.7449 ⟶ 34.7

Less than five

(Note that a different result would be obtained if the digits were incorrectly rounded one at a time from the right.)

b) Apply rules 1 and 3: 34.864 → 34.9

c) Apply rules 1 and 4: 34.754 → 34.8

d) Apply rules 1 and 5: 34.250 → 34.2

e) Apply rule 5: 34.35 → 34.4

To how many significant figures should we round our air-pollution results? We have already done a calculation involving multiplication and division to obtain the volume of our gas-collection balloon. It involved the following numbers:

106	Three significant figures
0.3937	Four significant figures
3.141 59	Six significant figures (we could obtain more if we wanted)
$\frac{4}{3}$ and $\frac{1}{2}$	An infinite number of significant figures since the integers in these fractions are exact by definition

The result of the calculation contained three significant figures—the same as the least-reliable number. This illustrates the general rule that *for multiplication and division the number of significant figures in the result is the same as in the least-reliable measurement.* Defined numbers such as π, $\frac{1}{2}$, $\frac{4}{3}$, or 100 cm/1 m are assumed to have an infinite number of significant figures.

In the case of addition and subtraction, a different rule applies. Suppose, for example, that we weighed a smoke-collection filter on a relatively inaccurate balance that could only be read to the nearest 0.01 g. After collecting a sample, the filter was reweighed on a single-pan balance to determine the mass of smoke particles.

Final mass:	2.3745 g	(colored digits are in question)
Initial mass:	−2.32 g	
Mass of smoke:	0.0545 g	

Since the initial weighing could have been anywhere from 2.31 to 2.33 g, *all three* figures in the final result are in question. (It must be between 0.0445 and 0.0645 g). Thus there is but one significant digit, and the result is 0.05 g. The rule here is that *the result of addition or subtraction cannot contain more digits to the right than there are in any of the numbers added or subtracted.* Note that subtraction can drastically reduce the number of significant digits when this rule is applied.

Rounding numbers is especially important if you use an electronic calculator, since these machines usually display a large number of digits, most of which are meaningless. The best procedure is to carry all digits to the end of the calculation (your calculator will not mind the extra work!) and then round appropriately. Answers to subsequent calculations in this book will be rounded according to the rules given. You may wish to go back to previous examples and round their answers correctly as well.

EXAMPLE 1.8 Evaluate the following expressions, rounding the answer to the appropriate number of significant figures.

a) 32.61 g + 8.446 g + 7.0 g

b) 0.136 cm^3 × 10.685 g cm^{-3}

Solution

a) 32.61 g + 8.446 g + 7.0 g = 48.056 g = 48.1 g. (7.0 has only one figure to the right of the decimal point.)

b) 0.136 cm^3 × 10.685 g cm^{-3} = 1.453 16 g = 1.45 g. (0.136 has only three significant figures.)

When we suggested filling a surplus weather balloon to measure how much gas was pumped through our air-pollution collector, we mentioned that this would be a rather crude way to determine volume. For one thing, it would not be all that simple to measure the diameter of an 8- or 9-ft sphere reliably. Using a yardstick, we would be lucky to have successive measurements agree within half an inch or so. It was for this reason that the result was reported to the nearest inch. The degree to which repeated measurements of the same quantity yield the same result is called **precision.** Repetition of a highly precise measurement would yield almost identical results, whereas low precision implies numbers would differ by a significant percentage from each other.

A highly precise measurement of the diameter of our balloon could be achieved, but it would probably not be worthwhile. We have assumed a spherical shape, but this is almost certainly not exactly correct. No matter how precisely we determine the diameter, our measurement of gas volume will be influenced by deviations from the assumed shape. When one or more of our assumptions about a measuring instrument are wrong, the **accuracy** of a result will be affected. An obvious example would be a foot rule divided into 11 equal inches. Measurements employing this instrument might agree very precisely, but they would not be very accurate.

An important point of a different kind is illustrated in the last two paragraphs. A great many common words have been adopted into the language of science. Usually such an adoption is accompanied by an unambiguous scientific definition which does not appear in a normal dictionary. *Precision and accuracy are many times treated as synonyms, but in science each has a slightly different meaning.* Another example is *quantity,* which we have defined in terms of "number × unit." Other English words like *bulk, size, amount,* and so forth, may be synonymous with quantity in everyday speech, but not in science. As you encounter other words like this, try to learn and use the scientific definition as soon as possible, and avoid confusing it with the other meanings you already know.

Even granting the crudeness of the measurements we have just described, they would be adequate to demonstrate whether or not an air-pollution problem existed. The next step would be to find a chemist or public health official who was an expert in assessing air quality, present your data, and convince that person to lend his or her skill and authority to your contention that something was wrong. Such a person would have available equipment whose precision and accuracy were adequate for highly reliable measurements and would be able to make authoritative public statements about the extent of the air-pollution problem.

1.3 THE INTERNATIONAL SYSTEM OF UNITS (SI)

The results of a scientific experiment must be communicated to be of value. This affords an opportunity for other scientists to check them. It also allows the scientific community, and sometimes the general public, to share new knowledge. Communication, however, is not always as straightforward as it might seem. Ambiguous terminology can often turn a seemingly clear statement into a morass of misunderstanding.

As an example, consider what happened when the United States first confronted the energy crisis during 1973 and 1974. Government asked various experts to estimate and compare reserves of coal, petroleum, and natural gas which could be recovered from the earth at no more than twice current costs. The result was an estimate of 350 billion tons of coal, 180 billion barrels of petroleum, and 1000 billion MCF of natural gas (1 MCF equals 1000 cubic feet) in the ground. Sounds simple, does it not?

Unfortunately these quantities have very little meaning when examined carefully because they are not accurately defined. For one thing, two different tons are used in United States commerce. These are the short ton (2000 lb) and the long ton (2240 lb). Neither should be confused with the metric ton (2204.6 lb—often written *tonne*) which most of the rest of the world uses. With regard to barrels, a U.S. barrel of petroleum contains 42 U.S. gallons (that is, 35 British or imperial gallons), but a U.S. barrel of any other liquid contains 31.5 U.S. gallons.

Even more confusing is the term MCF, where CF stands for cubic feet and M is the roman numeral for 1000 (10^3). The capital M might well be confused with the metric system prefix which means 1 000 000 (10^6). More importantly, since the measurement refers to gas volume, which varies with temperature and pressure, we need to know under what conditions it was made. To add a last confusing note, the word *billion* used in all these estimates means 1 000 000 000 (10^9) in the United States, but 1 000 000 000 000 (10^{12}—what we call a *trillion*) in Europe.

If in the midst of this hodgepodge, you asked, "Would it not be easier to have a single unit for mass, a single unit for volume, and express all masses or volumes in these units," you would not be the first person to have such an idea. The main difficulty is that it is hard to get everyone to agree on a single consistent set of units. Some units are especially convenient for some tasks. For example, a yard was originally defined as the distance from a man's nose to the end of his thumb when his arm was held horizontally to one side. This made it easy to measure cloth or ribbon by holding one end to the nose and stretching an arm's length with the other hand. Now that yardsticks, meter sticks, and other devices are readily available, the original utility of the yard is gone, but we still measure ribbon and cloth in that same unit. Many people would probably be distressed if a change were made.

Scientists are not all that different from other people—they too have favorite units which are especially suited to certain areas of research. Nevertheless, scientists have constantly pressed for improvement and uniformity in systems of measurement. The first such action occurred nearly 200 years ago when, in the aftermath of the French Revolution, the *metric system* spread over most of continental Europe and was adopted by scientists everywhere. The United States nearly followed suit, but in 1799 Thomas Jefferson was unsuccessful in persuading Congress that a system based on powers of 10 was far more convenient and would eventually become the standard of the world.

The metric system has undergone continuous evolution and improvement since its original adoption by France. Beginning in 1899, a series of international conferences have been held for the purpose of redefining and regularizing the system of units. In 1960 the Eleventh Conference on Weights and Measures proposed major changes in the metric system and suggested a new name—the **International System of Units**—for the revised metric system. (The abbreviation **SI,** from the French *Système International,* is commonly used.) Scientific bodies such as the U.S. National Bureau of Standards and the International Union of Pure and Applied Chemistry have endorsed the SI.

At the heart of the SI are the seven units listed in Table 1.2. All other units are derived from these seven so-called **base units.** For example, units for area and volume may be derived by squaring or cubing the unit for length. Some of the base units are probably familiar to you, while others, such as the mole, candela, and kelvin, may be less so. Rather than defining each of them now, we shall wait until later chapters when the less familiar units, as well as the quantities they are used to measure, can be described in

TABLE 1.2 The Seven Base Units of the SI.

Quantity Measured	Name of Unit	Symbol for Unit
Length	meter	m
Mass	kilogram	kg
Time	second	s
Electric current	ampere	A
Temperature	kelvin	K
Amount of substance	mole	mol
Luminous intensity	candela	cd

detail. The candela, which measures the intensity of light, is not used often by chemists, and so we shall pay no further attention to it.

Prefixes

The SI base units are not always of convenient size for a particular measurement. For example, the meter would be too big for reporting the thickness of this page, but rather small for the distance from Chicago to Detroit. To overcome this obstacle the SI includes a series of prefixes, each of which represents a power of 10 (Table 1.3). These allow us to reduce or enlarge the SI base units to convenient sizes. Figure 1.5 shows how these prefixes can be applied to the meter to cover almost the entire range of lengths we might wish to measure.

TABLE 1.3 Prefixes Used for Decimal Fractions and Multiples of SI Units.

Prefix	Symbol for Prefix		Scientific Notation
exa	E	1 000 000 000 000 000 000	10^{18}
peta	P	1 000 000 000 000 000	10^{15}
tera	T	1 000 000 000 000	10^{12}
giga	G	1 000 000 000	10^{9}
mega	M	1 000 000	10^{6}
kilo	k	1 000	10^{3}
hecto	h	100	10^{2}
deka	da	10	10^{1}
—	—	1	10^{0}
deci	d	0.1	10^{-1}
centi	c	0.01	10^{-2}
milli	m	0.001	10^{-3}
micro	μ	0.000 001	10^{-6}
nano	n	0.000 000 001	10^{-9}
pico	p	0.000 000 000 001	10^{-12}
femto	f	0.000 000 000 000 001	10^{-15}
atto	a	0.000 000 000 000 000 001	10^{-18}

One non-SI unit of length, the angstrom (Å), is convenient for chemists and will continue to be used for a limited time. Since 1 Å = 10^{-10} m (see Fig. 1.5), the angstrom corresponds roughly to the diameters of atoms and small molecules. Such dimensions are also conveniently expressed in picometers, 1 pm = 10^{-12} m = 0.01 Å, but the angstrom is widely used and very familiar. Therefore we will usually write atomic and molecular dimensions in both angstroms and picometers.

The SI base unit of mass, the kilogram, is unusual because it already contains a prefix. The standard kilogram is a cylinder of corrosion-resistant platinum-iridium alloy which is kept at the International Bureau of Weights and Measures near Paris. The kilogram was chosen instead of a gram because the latter would have made an inconveniently small piece of

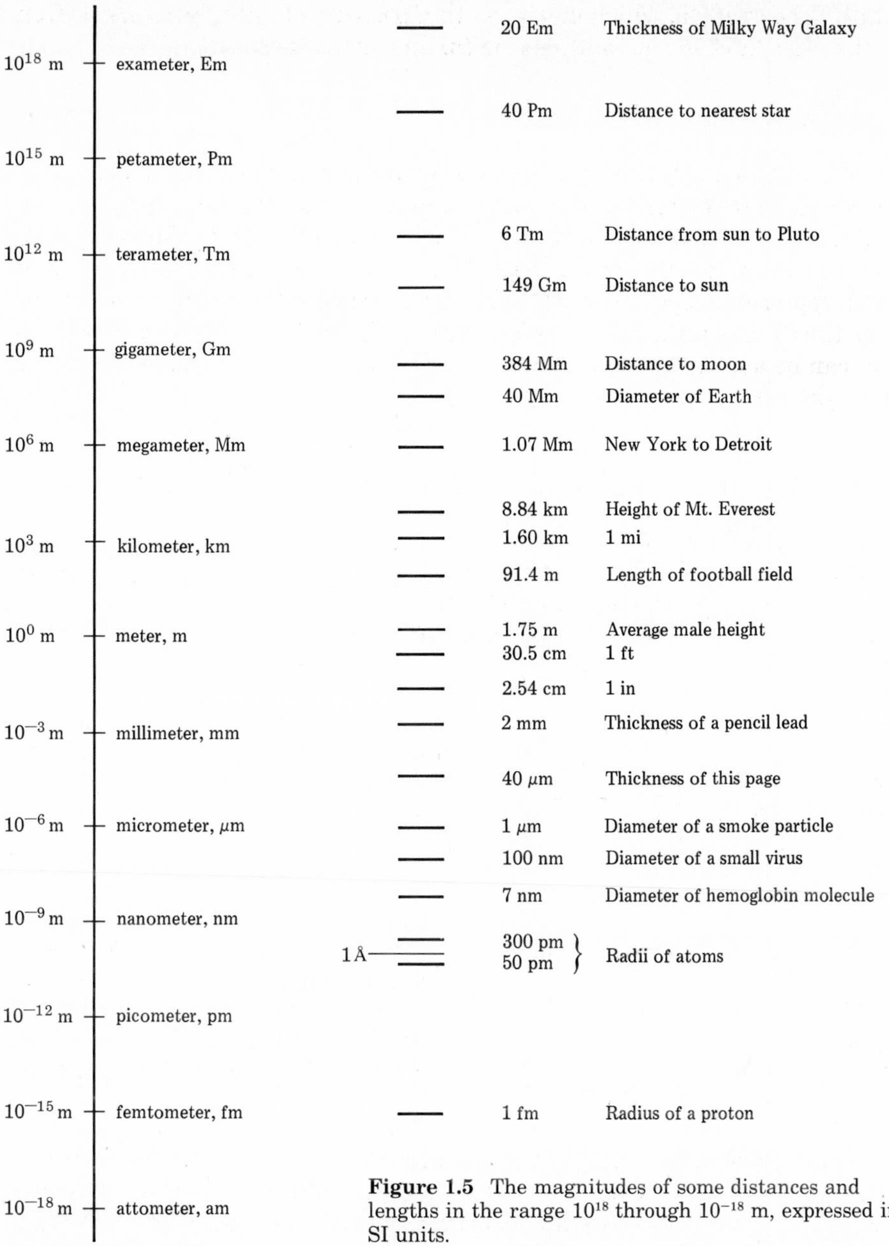

Figure 1.5 The magnitudes of some distances and lengths in the range 10^{18} through 10^{-18} m, expressed in SI units.

platinum-iridium and would have been difficult to handle. Also, units of force, pressure, energy, and power have been derived using the kilogram instead of the gram.

Despite the fact that the kilogram is the SI unit of mass, the standard prefixes are applied to the *gram* when larger or smaller mass units are needed. For example, the quantity 10^6 kg (1 million kilograms) can be written as 1 Gg (gigagram) but *not* as 1 Mkg (megakilogram). The opera-

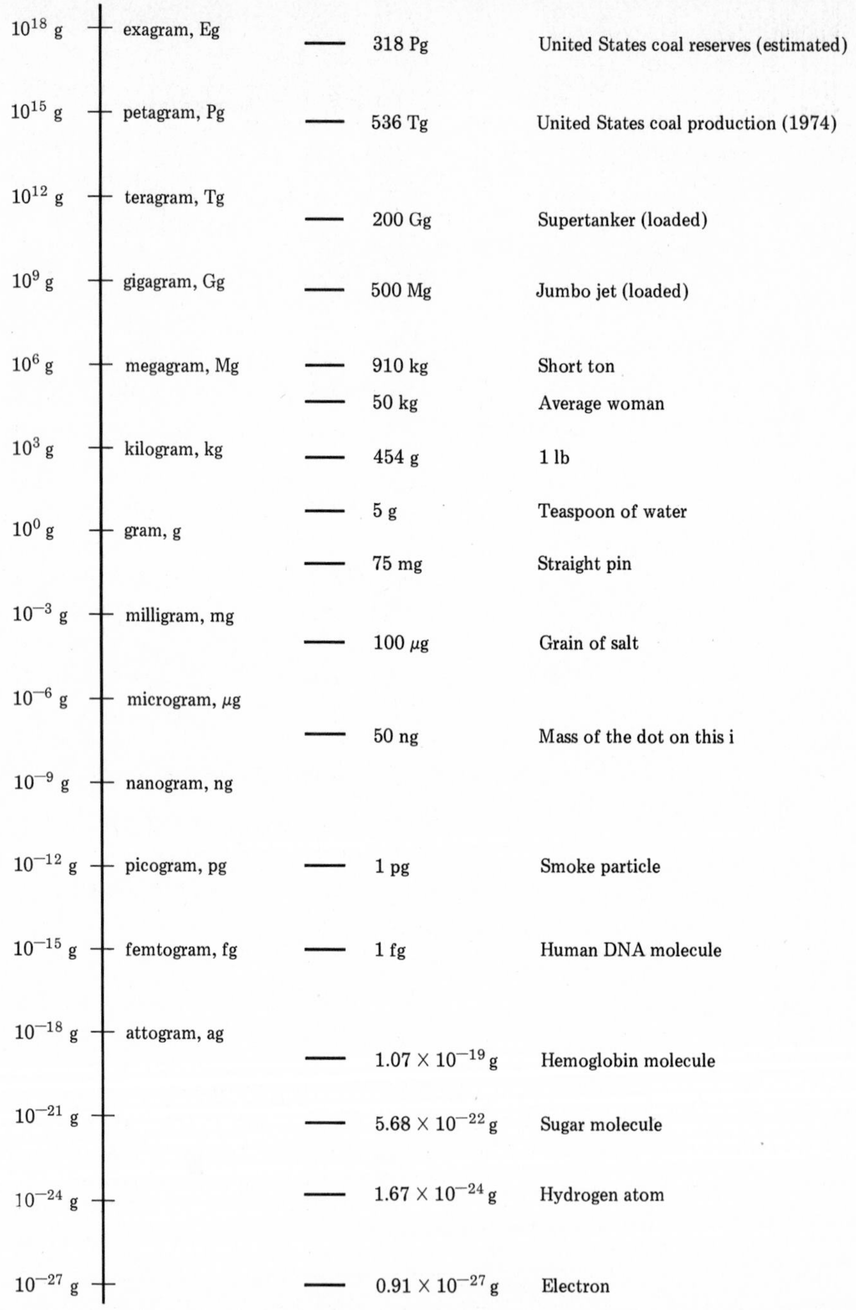

Figure 1.6 The masses of some objects in the range 10^{18} through 10^{-27} g, expressed in SI units.

tive rule here is that *one and only one prefix should be attached to the name for a unit.* Figure 1.6 illustrates the use of this rule in expressing the wide range of masses available in the universe. Note that the masses of atoms and molecules are usually so small that scientific notation must be used instead of prefixes.

Volume

The most commonly used derived units are those of volume. As we have already seen, calculation of the volume of an object requires that a length be cubed or that three lengths be multiplied together. Thus the SI unit of volume is the cubic meter (m^3). This is rather large for use in the chemical laboratory, and so the cubic decimeter (dm^3) or cubic centimeter (cm^3, formerly cc) are more commonly used. The relationship between these units and the cubic meter is easily shown:

$$1 \text{ dm} = 0.1 \text{ m} \qquad\qquad 1 \text{ cm} = 0.01 \text{ m}$$

Cubing both sides of each equation, we have

$$1 \text{ dm}^3 = 0.1^3 \text{ m}^3 \qquad\qquad 1 \text{ cm}^3 = 0.01^3 \text{ m}^3$$

$$= 0.001 \text{ m}^3 = 10^{-3} \text{ m}^3 \qquad\qquad = 0.000\ 001 \text{ m}^3 = 10^{-6} \text{ m}^3$$

Note that in the expression dm^3 *the exponent includes the prefix* as well as the base unit. A cubic decimeter is one-thousandth of a cubic meter, not one-tenth of a cubic meter.

Two other units of volume are commonly encountered in the chemical laboratory—the liter (l) and the milliliter (ml—one-thousandth of a liter). The liter was originally defined as the volume of one kilogram of pure water at the temperature of its maximum density (3.98°C), but in 1964 the definition was changed. The liter is now exactly one-thousandth of a cubic meter, that is, 1 dm^3. A milliliter is therefore exactly 1 cm^3. Because the new definition of liter altered its volume slightly, it is recommended that the results of highly accurate measurements be reported in the SI units cubic decimeters or cubic centimeters, rather than in liters or milliliters. For most situations discussed in this book, however, the units cubic decimeter and liter, and cubic centimeter and milliliter may be used interchangeably. Thus when recording a volume obtained from laboratory glassware calibrated in milliliters, you can just as well write 24.7 cm^3 as 24.7 ml.

Density

The terms *heavy* and *light* are commonly used in two different ways. We refer to weight when we say that an adult is heavier than a child. On the other hand, something else is alluded to when we say that oak is heavier than balsa wood. A small shaving of oak would obviously weigh less than a roomful of balsa wood, but oak is heavier in the sense that a piece of given size weighs more than the same-size piece of balsa.

What we are actually comparing is the *mass per unit volume,* that is, the **density.** In order to determine these densities, we might weigh a cubic centimeter of each type of wood. If the oak sample weighed 0.71 g and the balsa 0.15 g, we could describe the density of oak as 0.71 g cm^{-3} and that of balsa as 0.15 g cm^{-3}. (Note that the negative exponent in the units cubic centimeters indicates a reciprocal. Thus 1 $cm^{-3} = 1/cm^3$ and the units for our densities could be written as $\frac{\text{g}}{\text{cm}^3}$, g/$cm^3$, or g cm^{-3}. In each case the units are read as grams per cubic centimeter, the *per* indicating division.)

In general it is not necessary to weigh exactly 1 cm³ of a material in order to determine its density. We simply measure mass and volume and divide volume into mass:

$$\text{Density} = \frac{\text{mass}}{\text{volume}} \quad \text{or} \quad \rho = \frac{m}{V} \tag{1.1}$$

where ρ = density m = mass V = volume

EXAMPLE 1.9 Calculate the density of (a) a piece of aluminum whose mass is 37.42 g and which, when submerged, increases the water level in a graduated cylinder by 13.9 ml; (b) an aluminum cylinder of mass 25.07 g, radius 0.750 m, and height 5.25 cm.

Solution

a) Since the submerged metal displaces its own volume,

$$\text{Density} = \rho = \frac{m}{V} = \frac{37.42 \text{ g}}{13.9 \text{ ml}} = 2.69 \text{ g/ml or } 2.69 \text{ g ml}^{-1}$$

b) The volume of the cylinder must be calculated first, using the formula

$$\begin{aligned} V &= \pi r^2 h \\ &= 3.142 \times (0.750 \text{ cm})^2 \times 5.25 \text{ cm} \\ &= 9.278\ 718\ 8 \text{ cm}^3 \end{aligned}$$

Then

$$\rho = \frac{m}{V} = \frac{25.07 \text{ g}}{9.278\ 718\ 8 \text{ cm}^3} \left.\begin{aligned} &= 2.70 \frac{\text{g}}{\text{cm}^3} \\ &= 2.70 \text{ g cm}^{-3} \\ &= 2.70 \text{ g/cm}^3 \end{aligned}\right\} \text{all acceptable alternatives}$$

Note that unlike mass or volume, the density of a substance is independent of the size of the sample. Thus density is a property by which one substance can be distinguished from another. A sample of pure aluminum can be trimmed to any desired volume or adjusted to have any mass we choose, but its density will always be 2.70 g/cm³ at 20°C. The densities of some common pure substances are listed in Table 1.4.

Tables and graphs are designed to provide a maximum of information in a minimum of space. When a physical quantity (number × units) is involved, it is wasteful to keep repeating the same units. Therefore it is conventional to use pure numbers in a table or along the axes of a graph. A pure number can be obtained from a quantity if we divide by appropriate units. For example, when divided by the units gram per cubic centimeter, the density of aluminum becomes a pure number 2.70:

$$\frac{\text{Density of aluminum}}{1 \text{ g cm}^{-3}} = \frac{2.70 \cancel{\text{g cm}^{-3}}}{1 \cancel{\text{g cm}^{-3}}} = 2.70$$

TABLE 1.4 Density of Several Substances at 20°C.

Substance	Density/g cm^{-3}*
Helium gas	0.000 16
Dry air	0.001 185
Gasoline	0.66 → 0.69 (varies)
Kerosene	0.82
Benzene	0.880
Water	1.000
Carbon tetrachloride	1.595
Magnesium	1.74
Salt	2.16
Aluminum	2.70
Iron	7.87
Copper	8.96
Silver	10.5
Lead	11.34
Uranium	19.05
Gold	19.32

* See this section of text for a discussion of this method of labeling tables and graphs.

Therefore, a column in a table or the axis of a graph is conveniently labeled in the following form:

Quantity/units

This indicates the units that must be divided into the quantity to yield the pure number in the table or on the axis. This has been done in the second column of Table 1.4.

1.4 CONVERSION FACTORS

When we are referring to the same object or sample of material, it is often useful to be able to convert one kind of quantity into another. For example, in our discussion of fossil-fuel reserves we could define the quantities unambiguously by stating that 318 Pg (3.18×10^{17} g) of coal, 28.6 km^3 (2.86×10^{10} m^3) of petroleum, and 2.83×10^4 km^3 (2.83×10^{13} m^3) of natural gas (measured at normal atmospheric pressure and 15.5°C) are available. But none of these quantities tells us what we really want to know—how much *heat energy* could be released by burning each of these reserves? Only by converting the mass of coal and the volumes of petroleum and natural gas into their equivalent energies can we make a valid comparison. When this is done, we find that the coal could release 7.2×10^{21} J, the petroleum 1.1×10^{21} J, and the gas 1.1×10^{21} J of heat energy. Thus the reserves of coal are more than three times those of the other two fuels combined. It is for this reason that more attention is being paid to the development of new ways for using coal resources than to oil or gas.

Conversion of one kind of quantity into another is usually done with what we shall call a **conversion factor.** Since we have not yet discussed energy or the units (joules) in which it is measured, an example involving the more familiar quantities mass and volume will be used to illustrate the way conversion factors are employed. The same principles apply to finding how much energy would be released by burning a fuel, and that problem will be encountered later.

Suppose we have a rectangular solid sample of gold which measures 3.04 cm × 8.14 cm × 17.3 cm. We can easily calculate that its volume is 428 cm^3, but how much is it worth? The price of gold is about 5 dollars per gram, and so we need to know the mass rather than the volume. It is unlikely that we would have available a scale or balance which could weigh accurately such a large, heavy sample, and so we would have to determine the mass of gold equivalent to a volume of 428 cm^3. This can be done by manipulating Eq. (1.1) which defines density. If we multiply both sides by V, we obtain

$$V \times \rho = \frac{m}{\cancel{V}} \times \cancel{V} = m$$

$$m = V\rho \qquad \text{or} \qquad \text{mass} = \text{volume} \times \text{density} \tag{1.2}$$

Taking the density of gold from Table 1.4, we can now calculate

$$\text{Mass} = m = V\rho = 428\ \cancel{\text{cm}^3} \times \frac{19.32\ \text{g}}{1\ \cancel{\text{cm}^3}}$$

$$= 8.27 \times 10^3\ \text{g} = 8.27\ \text{kg}$$

This is more than 18 lb of gold. At the price quoted above, it would be worth over 40 000 dollars!

The formula which defines density can also be used to convert the mass of a sample to the corresponding volume. If both sides of Eq. (1.2) are multiplied by $1/\rho$, we have

$$\frac{1}{\rho} \times m = V\cancel{\rho} \times \frac{1}{\cancel{\rho}} = V$$

$$V = m \times \frac{1}{\rho} \tag{1.3}$$

EXAMPLE 1.10 Find the volume occupied by a 4.73-g sample of benzene.

Solution According to Table 1.4, the density of benzene is 0.880 g cm^{-3}. Using Eq. (1.3),

$$\text{Volume} = V = m \times \frac{1}{\rho} = 4.73\ \cancel{\text{g}} \times \frac{1\ \text{cm}^3}{0.880\ \cancel{\text{g}}}$$

$$= 5.38\ \text{cm}^3$$

(Note that taking the reciprocal of $\frac{0.880\ \text{g}}{1\ \text{cm}^3}$ simply inverts the fraction—1 cm^3 goes on top, and 0.880 g goes on the bottom.)

The two calculations just done show that density is a conversion factor which changes volume to mass, and the reciprocal of density is a conversion factor changing mass into volume. This can be done because of the mathematical formula, Eq. (1.1), which relates density, mass, and volume. Algebraic manipulation of this formula gave us expressions for mass and for volume [Eq. (1.2) and (1.3)], and we used them to solve our problems. In practice, however, it is unnecessary to remember all three formulas or do the algebra to derive one from another. It is much easier just to manipulate the quantities involved until the result has the correct units, as shown in the following example.

EXAMPLE 1.11 A student weighs 98.0 g of mercury. If the density of mercury is 13.6 g/cm³, what volume does the sample occupy?

Solution We know that volume is related to mass through density. Therefore

$$V = m \times \text{conversion factor}$$

Since the mass is in grams, we need to get rid of these units and replace them with volume units. This can be done if the reciprocal of the density is used as a conversion factor. This puts grams in the denominator so that these units cancel:

$$V = m \times \frac{1}{\rho}$$

$$= 98.0\ \cancel{\text{g}} \times \frac{1\ \text{cm}^3}{13.6\ \cancel{\text{g}}} = 7.21\ \text{cm}^3$$

If we had multiplied by the density instead of its reciprocal, the units of the result would immediately show our error:

$$V = 98.0\ \text{g} \times \frac{13.6\ \text{g}}{1\ \text{cm}^3} = 1333\ \text{g}^2/\text{cm}^3 \qquad \text{(no cancellation!)}$$

It is clear that square grams per cubic centimeter are not the units we want.

Using a conversion factor is very similar to using a unity factor—we know the factor is correct when units cancel appropriately. A conversion factor is *not* unity, however. Rather it is a physical quantity (or the reciprocal of a physical quantity) which is related to the two other quantities we are interconverting. The conversion factor works because of that relationship [Eqs. (1.1), (1.2), and (1.3) in the case of density, mass, and volume], *not* because it is equal to one. Once we have established that a relationship exists, it is no longer necessary to memorize a mathematical formula. The units tell us whether to use the conversion factor or its reciprocal. Without such a relationship, however, mere cancellation of units does not guarantee that we are doing the right thing.

A simple way to remember relationships among quantities and conversion factors is a "road map" of the type shown below:

$$\text{Mass} \xleftrightarrow{\text{density}} \text{volume} \qquad \text{or} \qquad m \xleftrightarrow{\rho} V$$

This indicates that the mass of a particular sample of matter is related to its volume (and the volume to its mass) through the conversion factor, density. The double arrow indicates that a conversion may be made in either direction, provided the units of the conversion factor cancel those of the quantity which was known initially. In general the road map can be written

$$\text{First quantity} \xleftrightarrow{\text{conversion factor}} \text{second quantity}$$

As we come to more complicated problems, where several steps are required to obtain a final result, such road maps will become more useful in charting a path to the solution.

EXAMPLE 1.12 Black ironwood has a density of 67.24 lb/ft³. If you had a sample whose volume was 47.3 ml, how many grams would it weigh? (1 lb = 454 g; 1 ft = 30.5 cm).

Solution The road map

$$V \xrightarrow{\rho} m$$

tells us that the mass of the sample may be obtained from its volume using the conversion factor, density. Since milliliters and cubic centimeters are the same, we use the SI units for our calculation:

$$\text{Mass} = m = 47.3\ \text{cm}^3 \times \frac{67.24\ \text{lb}}{1\ \text{ft}^3}$$

Since the volume units are different, we need a unity factor to get them to cancel:

$$m = 47.3\ \text{cm}^3 \times \left(\frac{1\ \text{ft}}{30.5\ \text{cm}}\right)^3 \times \frac{67.24\ \text{lb}}{1\ \text{ft}^3}$$

$$= 47.3\ \cancel{\text{cm}^3} \times \frac{1\ \cancel{\text{ft}^3}}{30.5^3\ \cancel{\text{cm}^3}} \times \frac{67.24\ \text{lb}}{1\ \cancel{\text{ft}^3}}$$

We now have the mass in pounds, but we want it in grams, so another unity factor is needed:

$$m = 47.3\ \cancel{\text{cm}^3} \times \frac{1\ \cancel{\text{ft}^3}}{30.5^3\ \cancel{\text{cm}^3}} \times \frac{67.24\ \cancel{\text{lb}}}{1\ \cancel{\text{ft}^3}} \times \frac{454\ \text{g}}{1\ \cancel{\text{lb}}}$$

$$m = 50.9\ \text{g}$$

In subsequent chapters we will establish a number of relationships among physical quantities. Formulas will be given which define these relationships, but we do not advocate slavish memorization and manipulation of those formulas. Instead we recommend that you remember that a relationship exists, perhaps in terms of a road map, and then adjust the quantities involved so that the units cancel appropriately. Such an approach has the advantage that you can solve a wide variety of problems by using the same technique.

SUMMARY

Chemistry is concerned with the composition, properties, and structure of matter, and with processes in which one substance changes into another. Chemistry does not just take place in laboratories but happens all the time, everywhere. Consequently this subject is very important to many persons who are not called chemists. In this book we will try to provide you with a basic foundation of the facts, laws, theories, and principles currently used by chemists, as well as some examples of how chemistry can be applied in other fields. We hope that these examples will help you learn how to think "chemically" and to apply chemistry to specific problems that face you.

The behavior of chemists and the development of science in general is based on the idea that natural events do not occur in completely unpredictable fashion. Consequently, measurements or observations made by one individual may be used by many others, provided they are communicated unambiguously. Such communication involves reporting quantities (numbers × units) in which numbers are often written in scientific notation and rounded to indicate their reliability in terms of significant digits. In many cases scientists use ordinary words like *quantity, precision,* or *accuracy,* to which highly specific scientific definitions have been attached.

Scientific communication is also facilitated when a carefully defined, consistent set of units is used. SI units were adopted for that purpose in 1960 by an international conference, and they have been endorsed by the U.S. National Bureau of Standards. In the SI there are seven base units from which others are derived, making it easier to see relationships among various quantities. The sizes of base and derived units may be adjusted using prefixes which correspond to powers of 10. Thus very small or very large measurements may be reported conveniently.

When dealing with different units (such as inches and centimeters) which measure the same quantity, unity factors are used to adjust the sizes of the numbers which multiply different-sized units. In many cases it is also useful to be able to obtain one quantity (such as mass) from another (such as volume) for the same object. If there is a known physical relationship (which can be expressed in an equation) between the quantities involved, a conversion factor (such as density or its reciprocal) may be used to interconvert the original quantities. Correct application of a unity factor or a conversion factor can be recognized because some units will cancel, leaving a final quantity with the desired units.

QUESTIONS AND PROBLEMS

When necessary, use the following equalities: 1 in = 2.540 cm; 1 lb = 453.6 g.

1.1 Modern single-pan analytical balances are often called *substitution* balances. Explain why. Why is it said that such balances have "a constant load"?

1.2 Fill in the blanks in the following calculations:

a 9.75 mm = 9.75 mm × ——— = 0.975 cm.
b 17.28 cm = 17.28 cm × ——— = ——— in.
c $253\ cm^2 = 253\ cm^2 \times (\text{———})^2$ = ——— dm^2.

1.3 Express the following quantities in the units indicated:

a 0.003 56 m in centimeters
b 415 063 g in kilograms
c 1024 cm^2 in square meters
d 8.31 m^3 in cubic centimeters

In each case use an appropriate unity factor.

1.4 Express the following quantities in the units indicated (use the data given at the beginning of the problems):

a 1 kg in pounds
b 1 ft in centimeters
c 1 mile in kilometers
d 1 cubic ft in cubic decimeters
e 1 h in seconds
f 1 mph in meters per second

1.5 Express the following temperatures in degrees Fahrenheit (see Appendix 1):

a 37°C *b* −5°C *c* 80°C

1.6 Express the following temperatures in degrees Celsius (see Appendix 1):

a 20°F *b* 120°F *c* −40°F

1.7 Which of the following equations are incorrectly written because of the omission of units or for some other reason:

a 7 g + 14 g = 21 *c* 1.0 g = 1.0 cm^3
b $(50\ cm)^2 = 50\ cm^2$ *d* 4.00 in = 10.16 cm

1.8 A cylindrical tank has a diameter of 36.0 in and a height of exactly 4 ft. Using the formulas given inside the back cover and the necessary unity factors, calculate (*a*) the height of the tank in centimeters; (*b*) the circumference of the tank in decimeters; (*c*) the cross-sectional area of the tank in square decimeters; (*d*) the volume of the tank in cubic centimeters.

1.9 Express the following numbers or quantities in scientific (exponential) notation:

a 712 473
b 0.000 37 g
c 392.68 cm
d 3119 cm × 0.41 cm
e 0.000 000 000 462 m
f $\frac{327}{0.003\ 01}$
g $\frac{1644\ g}{5.48}$
h 49 157 333 g

1.10 Rewrite each of the following numbers in a more conventional exponential notation:

a 2.795E+03
b 0.105E−05
c 5.99 06

Where would you find numbers written like this?

1.11 Evaluate each of the following expressions, giving your answer in scientific notation to the correct number of significant figures:

a $\frac{9.33 \times 10^5}{4.76 \times 10^{-5}}$

b $(32.7 \times 10^3) \times (0.006 \times 10^{-5})$

c $6.04 + (3.25 \times 10^2) - (7.67 \times 10^{-5})$

d $\frac{12.67 \times 10^9}{7.52 \times 10^6}$

1.12 How many significant figures are there, or should there be, in each of the following numbers or quantities?

a 0.000 079 9 g
b 42 000.0 mm
c 6.51×10^4 mg
d $\frac{1}{3}(12)\left(\frac{4.0}{0.7777}\right)$
e 0.199 × 47.3862
f 2.1 + 0.1234
g 37.3852 g − 3.3 g
h $\frac{120.6398}{1.2}$

1.13 Round each of the following numbers to four significant figures, in accordance with the rules given in Table 1.1:

a 78.9250
b 78.9165
c 78.9237
d 78.924 999
e 78.9150
f 78.913 762

1.14 Each of the following four sets of data represents repeated determinations of the same quantity. In each case state to how many significant figures the value of the quantity is known.

a 12.24 g, 12.25 g, 12.22 g, 12.23 g, 12.24 g
b 2.75 cm, 2.79 cm, 2.78 cm
c 0.017 25 kg, 0.019 53 kg, 0.014 49 kg
d 454 mm, 470 mm, 434 mm, 462 mm

1.15 What base SI unit accompanied by what prefix would be most convenient for describing each of the following quantities? (You might find it useful to refer to Figs. 1.5 and 1.6.)

a Distance from Washington, D.C. to Paris, France
b Distance from Minneapolis, Minnesota to St. Paul, Minnesota
c Thickness of your chemistry textbook
d Mass of this page of paper
e Diameter of the atoms found in the molecules in this page of paper
f Mass of 1 gal of water
g Mass of the earth

1.16 Using appropriate exponential notation, express the following quantities in terms of SI base units:

a 232 pm	*d* 536 Gg
b 0.107 ag	*e* 5.36 μA
c 33.4 ns	*f* 500 Mg

1.17 Express the following quantities in the units indicated:

a 0.001 31 s in microseconds
b 13.5 cm^3 in cubic decimeters
c 0.004 97 mg in grams
d 0.386 mmol in moles
e 2.36 Å in picometers
f 53 pm in nanometers
g 1.03 g cm^{-3} in kilograms per cubic meter
h 3.87 mol dm^{-3} in millimoles per cubic-centimeter

Did you use unity factors in these calculations, or did you guess? How many of your guesses were wrong?

1.18 A metal sphere has a radius of 1.00 cm and a mass of 37.6 g. (*a*) Calculate the density of the metal. (*b*) What would be the mass of a sphere of radius 2.83 cm made from the same metal? (*c*) Assuming the metal to be pure, identify it from Table 1.4. (*d*) What color is the metal?

1.19 Using Table 1.4, calculate the mass of a cube of iron which is 3.00 in on a side. Express your answer in grams.

1.20 An irregularly shaped piece of lead weighs 119.3 g. It is dropped into a graduated cylinder containing exactly 30.0 cm^3 of water. To what volume reading will the water-level rise? (Use Table 1.4.)

1.21 If you did not previously do so, write the "road map" involved in solving Probs. 1.18, 1.19, and 1.20.

1.22 The density of a metal is found to be 707.92 lb ft^{-3}. Identify this metal from Table 1.4.

1.23 Which of the following expressions are unity factors:

a 11.34 g cm^{-3}	*d* 16.4 cm^3 in^{-3}
b 2.54 cm in^{-1}	*e* 2.20 lb kg^{-1}
c 0.0936 cm^3 g^{-1}	*f* 3600 s h^{-1}

chapter two

ATOMS, MOLECULES, AND CHEMICAL REACTIONS

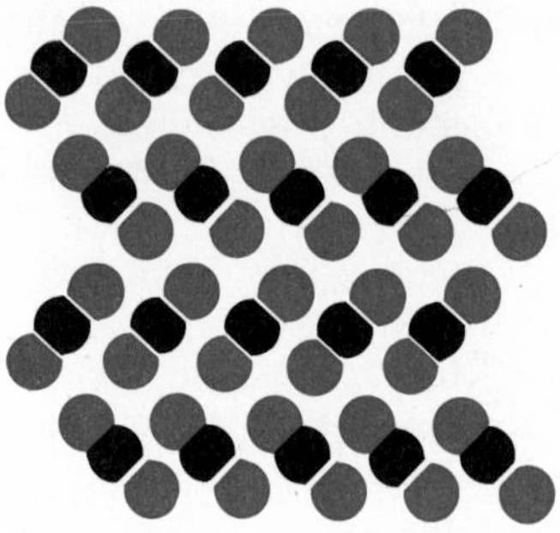

In Chap. 1 we described two very important things that chemists (and scientists in general) do. They make quantitative measurements, and they communicate the results of such experiments as clearly and unambiguously as possible. In this chapter we will deal with another important activity of chemists—the use of their imaginations to devise theories or models to interpret their observations and measurements. Such theories or models are useful in suggesting new observations or experiments which yield additional data. They also serve to summarize existing information and aid in its recall.

The **atomic theory,** first proposed in modern form by John Dalton, is extremely important to chemists. It interprets observations of the everyday world in terms of particles called atoms and molecules. **Macroscopic** events—those which humans can observe or experience with their unaided senses—are interpreted by means of **microscopic** objects—those so small that a special instrument or apparatus must be used to detect them. (Perhaps the term *submicroscopic* really ought to be used, because most atoms and molecules are much too small to be seen even under a microscope.) In any event, chemists continually try to explain the macroscopic world in microscopic terms.

2.1 MACROSCOPIC PROPERTIES AND MICROSCOPIC MODELS

As a simple example of how the macroscopic properties of a substance can be explained on a microscopic level, consider the liquid mercury. Macroscopically, mercury at ordinary temperatures is a silvery liquid which can be poured much like water—rather unusual for a metal. Mercury is also the heaviest known liquid. Its density is 13.6 g cm^{-3}, as compared with only 1.0 g cm^{-3} for water. When cooled below -38.9°C, mercury solidifies and behaves very much like more familiar solid metals such as copper and iron. Mercury frozen around the end of a wooden stick can be used to hammer nails, as long as it is kept sufficiently cold. Solid mercury has a density of 14.1 g cm^{-3}, slightly greater than that of the liquid.

When mercury is heated, it remains a liquid until quite a high temperature, finally boiling at 356.6°C to give an invisible vapor. Even at low concentrations gaseous mercury is extremely toxic if breathed into the lungs. It has been responsible for many cases of human poisoning. In other respects mercury vapor behaves much like any gas. It is easily compressible. Even when quite modest pressures are applied, the volume decreases noticeably. It is also much less dense than the liquid or the solid. At 400°C and ordinary pressures, its density is 3.6×10^{-3} g cm^{-3}, about one four-thousandth that of solid or liquid mercury.

A modern chemist would interpret these macroscopic properties in terms of a microscopic model involving atoms of mercury. As shown in Fig. 2.1, the atoms may be thought of as small, hard spheres. Like billiard balls they can move around and bounce off one another. In solid mercury the centers of adjacent atoms are separated by only 300 pm (300×10^{-12} m or 3.00 Å). Although each atom can move around a little, the others surround it so closely that it cannot escape its allotted position. Hence the solid is rigid. Very few atoms move out of position even when it strikes a nail.

As temperature increases, the atoms vibrate more violently, and eventually the solid melts. In liquid mercury, the regular, geometrically rigid structure is gone and the atoms are free to move about, but they are still rather close together and difficult to separate. This ability of the atoms to move past each other accounts for the fact that liquid mercury can flow and take the shape of its container. Note that the structure of the liquid is not as compact as that of the solid; a few gaps are present. These gaps explain why liquid mercury is less dense than the solid.

In gaseous mercury, also called mercury vapor, the atoms are very much farther apart than in the liquid and they move around quite freely and rapidly. Since there are very few atoms per unit volume, the density is considerably lower than for the liquid and solid. By moving rapidly in all directions, the atoms of mercury (or any other gas for that matter) are able to fill any container in which they are placed. When the atoms hit a wall of the container, they bounce off. This constant bombardment by atoms on the microscopic level accounts for the pressure exerted by the gas on the macroscopic level. The gas can be easily compressed because there is plenty of open space between the atoms. Reducing the volume merely reduces that

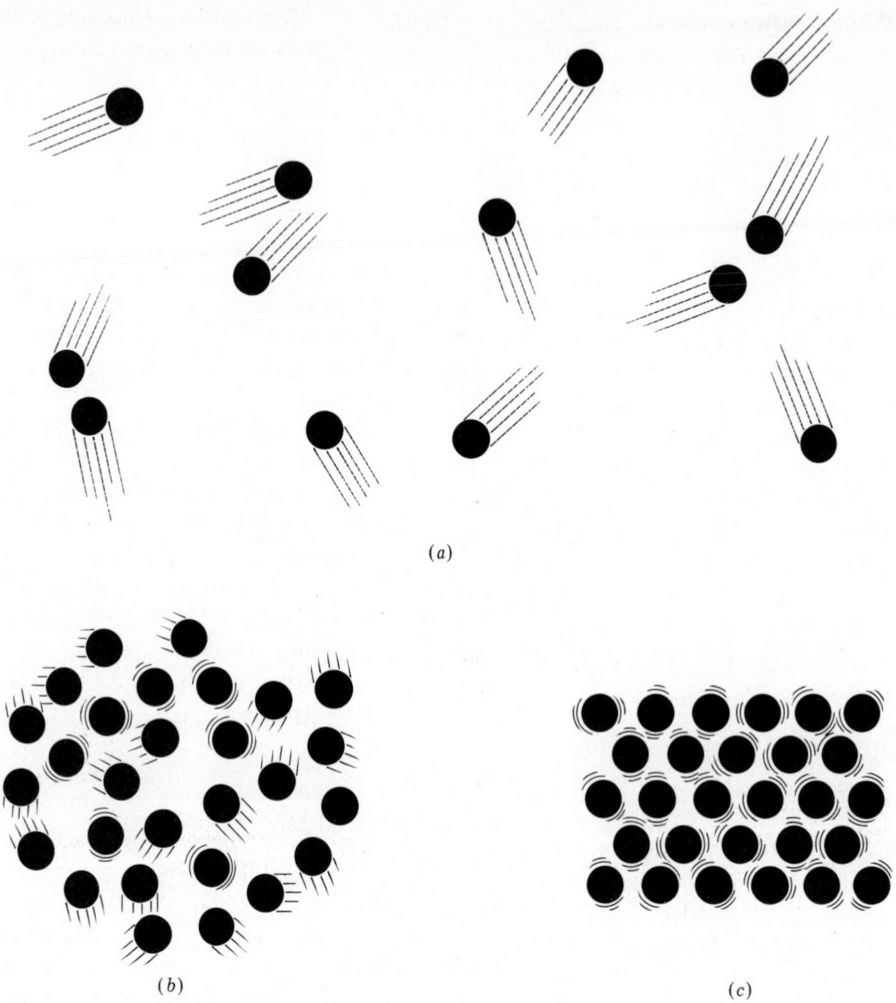

Figure 2.1 Microscopic view of the element mercury (*a*) in the gaseous state (above 356.6°C); (*b*) as a liquid (between −38.9 and 356.6°C); and (*c*) in solid form (below −38.9°C).

empty space. The liquid and the solid are not nearly so easy to compress because there is little or no empty space between the atoms.

You may have noticed that although our microscopic model can explain many of the properties of solid, liquid, and gaseous mercury, it cannot explain all of them. Mercury's silvery color and why the vapor is poisonous remain a mystery, for example. There are two approaches to such a situation. We might discard the idea of atoms in favor of a different theory which can explain more macroscopic properties. On the other hand it may be reasonable to extend the atomic theory so that it can account for more facts. The second approach has been followed by chemists. In this and the next chapter we shall discuss in more detail those facts which require only a simple atomic theory for their interpretation. Many of the subsequent chapters will describe extensions of the atomic theory which allow interpretations of far more observations.

2.2 HISTORICAL DEVELOPMENT OF THE ATOMIC THEORY

The atomic, microscopic way of looking at matter is actually a fairly new development. The United States has already celebrated its two-hundredth birthday, whereas the atomic theory is only about 175 years old. None of the Founding Fathers—not even Benjamin Franklin, the most scientific of them—thought about matter in terms of atoms.

Neither, for that matter, did the man whose experiments and ideas led directly to the theory itself. Antoine Lavoisier was born in 1743, the same year as Thomas Jefferson. The son of a wealthy French lawyer, he was well educated and became a successful businessman, gentleman farmer, economist, and social reformer, as well as the leading chemist of his day. It was Lavoisier's position as a tax collector, not his chemical research, which led to his death by guillotine in 1794, at the height of the French Revolution.

Much of Lavoisier's work as a chemist was devoted to the study of combustion. He became convinced that when a substance is burned in air, it combines with some component of the air. Eventually he realized that this component was the *dephlogisticated air* which had been discovered by Joseph Priestly (1733 to 1804) a few years earlier. Lavoisier renamed this substance *oxygen.* In an important series of experiments he showed that when mercury is heated in oxygen at a moderate temperature, a red substance, *calx of mercury,* is obtained. (A *calx* is the ash left when a substance burns in air.) At a higher temperature this calx decomposes into mercury and oxygen. Lavoisier's careful experiments also revealed that the combined masses of mercury and oxygen were exactly equal to the mass of calx of mercury. That is, there was no change in mass upon formation or decomposition of the calx. Lavoisier hypothesized that this should be true of all chemical changes, and further experiments showed that he was right. This principle is now called the **law of conservation of mass.**

As Lavoisier continued his experiments with oxygen, he noticed something else. Although oxygen combined with many other substances, it never behaved as though it were itself a combination of other substances. Lavoisier was able to decompose the red calx into mercury and oxygen, but he could find no way to break down oxygen into two or more new substances. Because of this he suggested that oxygen must be an **element**—an ultimately simple substance which could not be decomposed by chemical changes.

Lavoisier did not originate the idea that certain substances (elements) were fundamental and all others could be derived from them. This had first been proposed in Greece during the fifth century B.C. by Empedocles, who speculated that all matter consisted of combinations of earth, air, fire, and water. These ideas were further developed and taught by Aristotle and re-

ELEMENTS

TABLE OF SIMPLE SUBSTANCES.

Simple fubftances belonging to all the kingdoms of nature, which may be confidered as the elements of bodies.

New Names.	Correfpondent old Names.
Light	Light.
Caloric	Heat. Principle or element of heat. Fire. Igneous fluid. Matter of fire and of heat.
Oxygen	Dephlogifticated air. Empyreal air. Vital air, or Bafe of vital air.
Azote	Phlogifticated air or gas. Mephitis, or its bafe.
Hydrogen	Inflammable air or gas, or the bafe of inflammable air.

Oxydable and Acidifiable fimple Subftances not Metallic.

New Names.	Correfpondent old names.
Sulphur Phofphorus Charcoal	The fame names.
Muriatic radical Fluoric radical Boracic radical	Still unknown.

Oxydable and Acidifiable fimple Metallic Bodies.

New Names.		Correfpondent Old Names.
Antimony	Regulus of	Antimony.
Arfenic		Arfenic.
Bifmuth		Bifmuth.
Cobalt		Cobalt.
Copper		Copper.
Gold		Gold.
Iron		Iron.
Lead		Lead.
Manganefe		Manganefe.
Mercury		Mercury.
Molybdena		Molybdena.
Nickel		Nickel.
Platina		Platina.
Silver		Silver.
Tin		Tin.
Tungftein		Tungftein.
Zinc		Zinc.

Salifiable fimple Earthy Subftances.

New Names.	Correfpondent old Names.
Lime	Chalk, calcareous earth. Quicklime.
Magnefia	Magnefia, bafe of Epfom falt. Calcined or cauftic magnefia.
Barytes	Barytes, or heavy earth.
Argill	Clay, earth of alum.
Silex	Siliceous or vitrifiable earth.

Figure 2.2 Lavoisier's table of elements. (*From Lavoisier, "Traite de chemie," 1789. English translation, 1790.*)

mained influential for 2000 years. Lavoisier did, however, produce the first table of the elements which contained a large number of substances that modern chemists would agree should be classified as elements. The list in Fig. 2.2 is taken from the first English edition of Lavoisier's "Textbook of Chemistry" (1790). He published it with the knowledge that further research might succeed in decomposing some of the substances listed, thus showing them not to be elements. One of his objectives was to prod his contemporaries into just that kind of research. Sure enough the "earth substances" listed at the bottom were eventually shown to be combinations of certain metals with oxygen. It is also interesting to note that not even Lavoisier could entirely escape from Aristotle's influence. The second element in his list is Aristotle's "fire," which Lavoisier called "caloric," and which we now call "heat." Both heat and light, the first two items in the table, are now regarded as forms of energy rather than of matter.

Although his table of elements was incomplete, and even incorrect in some instances, Lavoisier's work represented a major step forward. By classifying certain substances as elements, he stimulated much additional chemical research and brought order and structure to the subject where none had existed before. His contemporaries accepted his ideas very readily, and he became known as the father of chemistry.

2.3 THE ATOMIC THEORY

John Dalton (1766 to 1844) was a generation younger than Lavoisier and different from him in almost every respect. Dalton came from a working class family and only attended elementary school. Apart from this, he was entirely self-taught. Even after he became famous, he never aspired beyond a modest bachelor's existence in which he supported himself by teaching mathematics to private pupils. Dalton made many contributions to science, and he seems not to have realized that his atomic theory was the most important of them. In his "New System of Chemical Philosophy" published in 1808, only the last seven pages out of a total of 168 are devoted to it!

The postulates of the atomic theory are given in Table 2.1. The first is no advance on the ancient Greek philosopher Democritus who had theorized

TABLE 2-1 The Postulates of Dalton's Atomic Theory.

1. All matter is composed of a very large number of very small particles called atoms.
2. For a given element, all atoms are identical in all respects. In particular all atoms of the same element have the same constant mass, while atoms of different elements have different masses.
3. The atoms are the units of chemical changes. Chemical reactions involve the combination, separation, or rearrangement of atoms, but atoms are neither created, destroyed, divided into parts, or converted into atoms of any other kind.
4. Atoms combine to form molecules in fixed ratios of small whole numbers.

almost 2000 years earlier that matter consists of very small particles. The second postulate, however, shows the mark of an original genius; here Dalton links the idea of *atom* to the idea of *element*. Lavoisier's criterion for an element had been essentially a macroscopic, experimental one. If a substance could not be decomposed chemically, then it was probably an element. By contrast, Dalton defines an element in theoretical, microscopic terms. *An element is an element because all its atoms are the same.* Different elements have different atoms. There are just as many different kinds of elements as there are different kinds of atoms.

Now look back a moment to Fig. 2.1, where microscopic pictures of solid, liquid, and gaseous mercury were given. Applying Dalton's second postulate to this figure, you can immediately conclude that mercury is an element, because only one kind of atom appears.

Although mercury atoms are drawn as black spheres in Fig. 2.1, it would be more common today to represent them using chemical symbols. The **chemical symbol** for an element (or an atom of that element) is a one- or two-letter abbreviation of its name. Usually, but not always, the first two letters are used. To complicate matters further, chemical symbols are

TABLE 2.2 Names, Chemical Symbols, and Atomic Weights of the Elements.

Element	Symbol	Atomic Weight	Element	Symbol	Atomic Weight
Actinium	Ac	(227)	Gallium	Ga	69.72
Aluminum	Al	26.981 5	Germanium	Ge	72.59
Americium	Am	(243)	Gold	Au	196.966 5
Antimony	Sb	121.75	Hafnium	Hf	178.49
Argon	Ar	39.948	Helium	He	4.002 60
Arsenic	As	74.921 6	Holmium	Ho	164.930 4
Astatine	At	(210)	Hydrogen	H	1.007 9
Barium	Ba	137.33	Indium	In	114.82
Berkelium	Bk	(247)	Iodine	I	126.904 5
Beryllium	Be	9.012 18	Iridium	Ir	192.22
Bismuth	Bi	208.980 4	Iron	Fe	55.847
Boron	B	10.81	Krypton	Kr	83.80
Bromine	Br	79.904	Lanthanum	La	138.905 5
Cadmium	Cd	112.41	Lawrencium	Lr	(256)
Calcium	Ca	40.08	Lead	Pb	207.2
Californium	Cf	(251)	Lithium	Li	6.941
Carbon	C	12.011	Lutetium	Lu	174.97
Cerium	Ce	140.12	Magnesium	Mg	24.305
Cesium	Cs	132.905 4	Manganese	Mn	54.938 0
Chlorine	Cl	35.453	Mendelevium	Md	(258)
Chromium	Cr	51.996	Mercury	Hg	200.59
Cobalt	Co	58.933 2	Molybdenum	Mo	95.94
Copper	Cu	63.546	Neodymium	Nd	144.24
Curium	Cm	(247)	Neon	Ne	20.179
Dysprosium	Dy	162.50	Neptunium	Np	237.048 2
Einsteinium	Es	(254)	Nickel	Ni	58.70
Erbium	Er	167.26	Niobium	Nb	92.906 4
Europium	Eu	151.96	Nitrogen	N	14.006 7
Fermium	Fm	(257)	Nobelium	No	(255)
Fluorine	F	18.998 403	Osmium	Os	190.2
Francium	Fr	(223)	Oxygen	O	15.999 4
Gadolinium	Gd	157.25	Palladium	Pd	106.4

TABLE 2.2 (*Continued*)

Element	Symbol	Atomic Weight	Element	Symbol	Atomic Weight
Phosphorus	P	30.973 8	Strontium	Sr	87.62
Platinum	Pt	195.09	Sulfur	S	32.06
Plutonium	Pu	(244)	Tantalum	Ta	180.947 9
Polonium	Po	(210)	Technetium	Tc	98.906 2
Potassium	K	39.098 3	Tellurium	Te	127.60
Praseodymium	Pr	140.907 7	Terbium	Tb	158.925 4
Promethium	Pm	(147)	Thallium	Tl	204.37
Protactinium	Pa	231.035 9	Thorium	Th	232.038 1
Radium	Ra	226.025 4	Thulium	Tm	168.934 2
Radon	Rn	(222)	Tin	Sn	118.69
Rhenium	Re	186.207	Titanium	Ti	47.90
Rhodium	Rh	102.905 5	Tungsten	W	183.85
Rubidium	Rb	85.467 8	Uranium	U	238.029
Ruthenium	Ru	101.07	Vanadium	V	50.941 4
Samarium	Sm	150.4	Xenon	Xe	131.30
Scandium	Sc	44.955 9	Ytterbium	Yb	173.04
Selenium	Se	78.96	Yttrium	Y	88.905 9
Silicon	Si	28.085 5	Zinc	Zn	65.38
Silver	Ag	107.868	Zirconium	Zr	91.22
Sodium	Na	22.989 8			

sometimes derived from a language other than English. For example, the symbol Hg for mercury comes from the first and seventh letters of the element's Latin name, *h*ydrar*g*yrum.

The chemical symbols for all the currently known elements are listed in Table 2.2. (The atomic weights in the table will be discussed later in this chapter.) These symbols are the basic vocabulary of chemistry because the atoms they represent make up all matter. You will see symbols for the more important elements over and over again, and the sooner you know what element they stand for, the easier it will be for you to learn chemistry. These more important elements have been indicated in Table 2.2 by colored shading around their names.

Dalton's fourth postulate states that atoms may combine to form molecules. An example of this is provided by bromine, the only element other than mercury which is a liquid at ordinary room temperature (20°C). Macroscopically, bromine consists of dark-colored crystals below −7.2°C and a

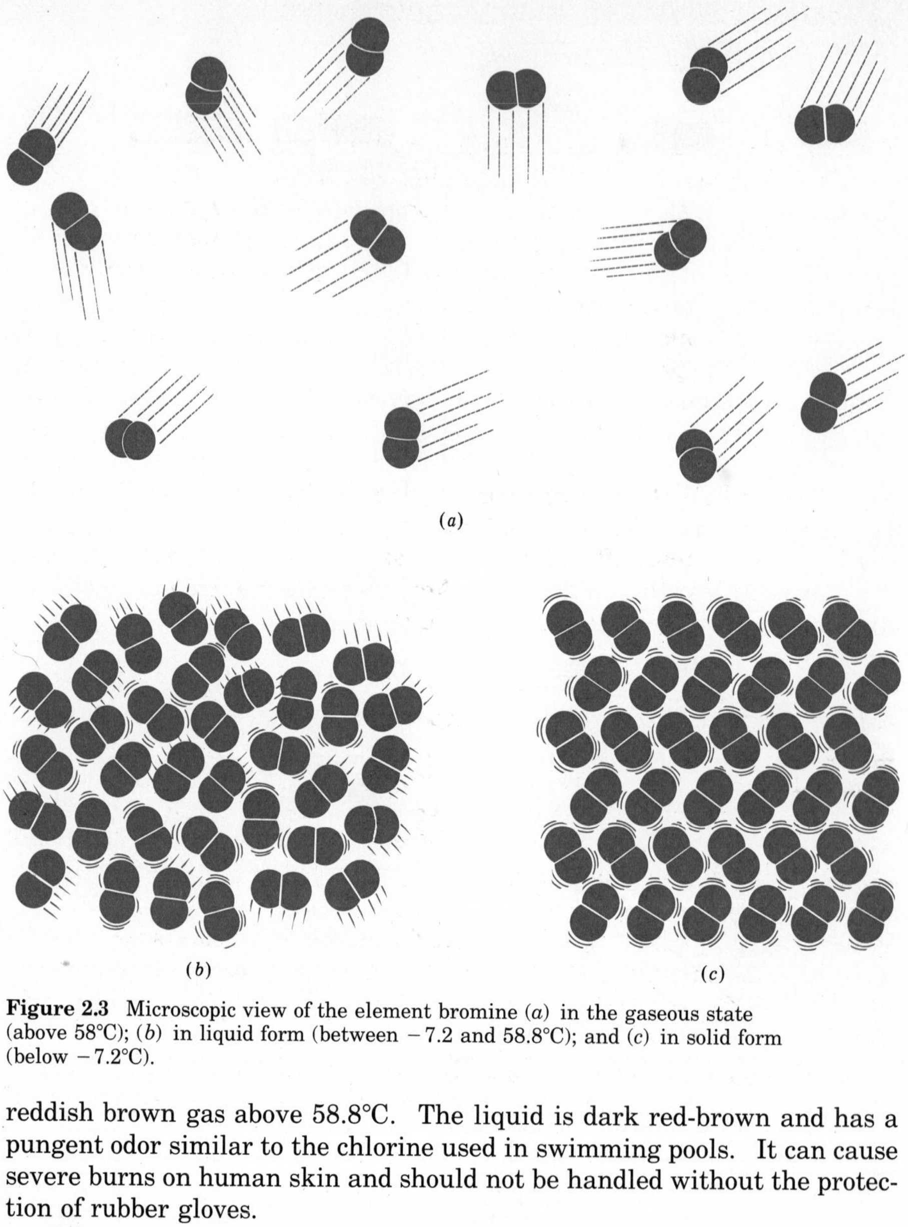

Figure 2.3 Microscopic view of the element bromine (*a*) in the gaseous state (above 58°C); (*b*) in liquid form (between −7.2 and 58.8°C); and (*c*) in solid form (below −7.2°C).

reddish brown gas above 58.8°C. The liquid is dark red-brown and has a pungent odor similar to the chlorine used in swimming pools. It can cause severe burns on human skin and should not be handled without the protection of rubber gloves.

The microscopic view of bromine in Fig. 2.3 is in agreement with its designation as an element—only one kind of atom is present. Except at very high temperatures, though, bromine atoms always double up. Whether in solid, liquid, or gas, they go around in pairs. Such a tightly held combination of two or more atoms is called a **molecule.**

The composition of a molecule is indicated by a **chemical formula.** A subscript to the right of the symbol for each element tells how many atoms of that element are in the molecule. For example, Table 2.2 gives the chemical *symbol* Br for bromine, but each molecule contains two bromine atoms, and so the chemical *formula* is Br_2. According to Dalton's fourth postulate in Table 2.1, atoms combine in the ratio of small whole numbers, and so the subscripts in a formula should be small whole numbers.

2.4 MACROSCOPIC AND MICROSCOPIC VIEWS OF A CHEMICAL REACTION

Dalton's third postulate states that atoms are the units of chemical changes. What this means can be seen in the macroscopic and microscopic views of a chemical change in Plate 3. (Color plates will be found in Chap. 5.) When macroscopic quantities of mercury and bromine are mixed at room temperature, a **chemical reaction** occurs, and a new substance, mercuric bromide, is produced. It is a white solid, quite different in appearance from the two elements from which it was formed.

A chemist's microscopic theory about what is going on is shown in Plate 3 along with the photographs of the macroscopic reaction. Soon after the two liquids are mixed together, a rearrangement of atoms begins. The two bromine atoms of each Br_2 molecule become separated and combine instead with mercury atoms. When the chemical reaction is complete, all that remains is a collection of mercuric bromide molecules, each of which contains one mercury atom and two bromine atoms. Notice that there are just as many mercury atoms after the reaction as there were before the reaction. The same applies to bromine atoms. Atoms were neither created, destroyed, divided into parts, or changed into other kinds of atoms during the chemical reaction.

The view of solid mercuric bromide shown in Plate 3*f* is our first microscopic example of a **compound.** Each molecule of a compound is made up of two (or more) different *kinds* of atoms. Since these atoms may be rearranged during a chemical reaction, the molecules may be broken apart and the compound can be decomposed into two (or more) different elements.

The formula for a compound involves at least two chemical symbols—one for each element present. In the case of mercuric bromide each molecule contains one mercury and two bromine atoms, and so the formula is $HgBr_2$. Both Plate 3*f* and the formula tell you that any sample of pure mercuric bromide contains twice as many bromine atoms as mercury atoms. This 2:1 ratio agrees with Dalton's fourth postulate that atoms combine in the ratio of small whole numbers.

Although John Dalton originally used circular symbols like those in Plate 3 to represent atoms in chemical reactions, a modern chemist would use chemical symbols and a **chemical equation** like

$$\underset{\textbf{Reactant(s)}}{Hg + Br_2} \rightarrow \underset{\textbf{Product(s)}}{HgBr_2} \tag{2.1}$$

This equation may be interpreted microscopically to mean that 1 mercury atom and 1 bromine molecule react to form 1 mercuric bromide molecule. It should also call to mind the macroscopic change shown in the photographs in Plate 3. This macroscopic interpretation is often strengthened by specifying physical states of the reactants and products:

$$Hg(l) + Br_2(l) \rightarrow HgBr_2(s) \tag{2.2}$$

Thus liquid mercury and liquid bromine react to form solid mercuric bromide. [If the bromine had been in gaseous form, $Br_2(g)$ might have been used. Occasionally (*c*) may be used instead of (*s*) to indicate a crystalline solid.] Chemical equations such as (2.1) and (2.2) summarize a great deal of information on both macroscopic and microscopic levels, for those who know how to interpret them.

2.5 TESTING THE ATOMIC THEORY

Two criteria are usually applied to any theory. First, does it agree with facts which are already known? Second, does it predict new relationships and stimulate additional observation and experimentation? Dalton's atomic theory was able to do both of these things. It was especially useful in dealing with data regarding the masses of different elements which were involved in chemical compounds or chemical reactions.

To test a theory, we first use it to make a prediction about the macroscopic world. If the prediction agrees with existing data, the theory passes the test. If it does not, the theory must be discarded or modified. If data are not available, then more research must be done. Eventually the results of new experiments can be compared with the predictions of the theory.

Several examples of this process of testing a theory against the facts are afforded by Dalton's work. For example, postulate 3 in Table 2.1 states that atoms are not created, destroyed, or changed in a chemical reaction. Postulate 2 says that atoms of a given element have a characteristic mass. By logical deduction, then, equal numbers of each type of atom must appear on left and right sides of chemical equations such as (2.1) and (2.2), and the total mass of reactants must equal the total mass of products. Dalton's atomic theory predicts Lavoisier's experimental law of conservation of mass.

A second prediction of the atomic theory is a bit more complex. A compound is made up of molecules, each of which contains a certain number of each type of atom. No matter how, when, or where a compound is made, its molecules will always be the same. Thus mercuric bromide molecules always have the formula $HgBr_2$. No matter how much we have or where the compound came from, there will always be twice as many bromine atoms as mercury atoms. Since each type of atom has a characteristic mass, the mass of one element which combines with a fixed mass of the other should always be the same. In mercuric bromide, for example, if each mercury atom is 2.510 times as heavy as a bromine atom, the ratio of masses would be

$$\frac{\text{Mass of 1 mercury atom}}{\text{Mass of 2 bromine atoms}} = \frac{2.510 \times \text{mass of 1 \sout{bromine atom}}}{2 \times \text{mass of 1 \sout{bromine atom}}} = 1.255$$

No matter how many mercuric bromide molecules we have, each has the same proportion of mercury, and so *any* sample of mercuric bromide must have that same proportion of mercury.

We have just derived the **law of constant composition,** sometimes called the **law of definite proportions.** When elements combine to form a compound, they always do so in exactly the same ratio of masses. This law had been postulated in 1799 by the French chemist Proust (1754 to 1826)

4 years before Dalton proposed the atomic theory, and its logical derivation from the theory contributed to the latter's acceptance.

The law of constant composition makes the important point that the composition and other properties of a *pure* compound are independent of who prepared it or where it came from. The carbon dioxide found on Mars, for instance, can be expected to have the same composition as that on Earth, while the natural vitamin C extracted and purified from rose hips has exactly the same composition as the synthetic vitamin C prepared by a drug company. Absolute purity is, however, an ideal limit which we can only approach, and the properties of many substances may be affected by the presence of very small quantities of impurities.

A third law of chemical composition may be deduced from the atomic theory. It involves the situation where two elements can combine in more than one way, forming more than one compound. For example, if mercuric bromide is ground and thoroughly mixed with liquid mercury, a new compound, mercurous bromide, is formed. Mercurous bromide is a white solid which is distinguishable from mercuric bromide because of its insolubility in hot or cold water. Mercurous bromide also changes directly from a solid to a gas at 345°C. From the microscopic view of solid mercurous bromide in Fig. 2.4, you can readily determine that its chemical formula is Hg_2Br_2. (Since there are *two* atoms of each kind in the molecule, it would be incorrect to write the formula as HgBr.) The chemical equation for synthesis of mercurous bromide is

$$Hg(l) + HgBr_2(s) \rightarrow Hg_2Br_2(s) \qquad (2.3)$$

From the formulas $HgBr_2$ and Hg_2Br_2 we can see that mercuric bromide has only 1 mercury atom for every 2 bromines, while mercurous bromide has 2 mercury atoms for every 2 bromines. Thus, for a given number of bromine atoms, mercurous bromide will always have twice as many mercury atoms as mercuric bromide. Again using postulate 2 in Table 2.1, the atoms have characteristic masses, and so a given number of bromine atoms corresponds to a fixed mass of bromine. Twice as many mercury atoms correspond to twice the mass of mercury. Therefore we can say that for a given

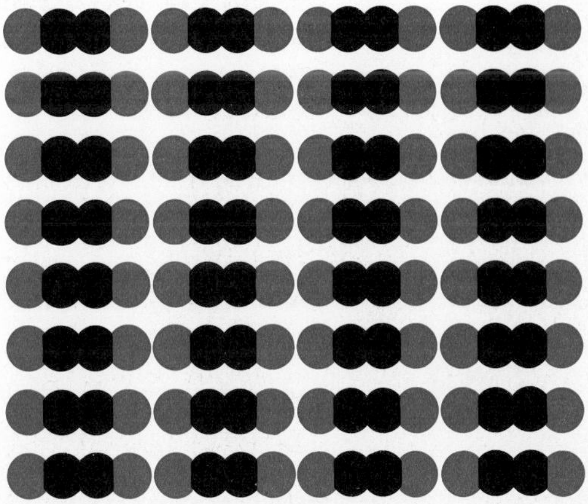

Figure 2.4 The arrangement of atoms and molecules in a crystal of mercurous bromide, Hg_2Br_2.

mass of bromine, mercurous bromide will contain twice the mass of mercury that mercuric bromide will. [The doubled mass of mercury was provided by adding liquid mercury to mercuric bromide in Eq. (2.3).]

EXAMPLE 2.1 Given that the mass of a mercury atom is 2.510 times the mass of a bromine atom, calculate the mass ratio of mercury to bromine in mercurous bromide.

Solution The formula Hg_2Br_2 tells us that there are 2 mercury atoms and 2 bromine atoms in each molecule. Thus the mass ratio is

$$\frac{\text{Mass of 2 mercury atoms}}{\text{Mass of 2 bromine atoms}}$$

$$= \frac{2 \times (2.510 \times \text{mass of 1 bromine atom})}{2 \times \text{mass of 1 bromine atom}} = 2.510$$

Note that the mass of mercury per unit mass of bromine is double that calculated earlier for mercuric bromide.

The reasoning and calculations above illustrate the **law of multiple proportions.** When two elements form several compounds, the mass ratio in one compound will be a small whole-number multiple of the mass ratio in another. In the case of mercuric bromide and mercurous bromide the mass ratios of mercury to bromine are 1.255 and 2.510, respectively. The second value is a small whole-number multiple of (2 times) the first.

Until the atomic theory was proposed, no one had expected *any* relationship to exist between mass ratios in two or more compounds containing the same elements. Because the theory predicted such relationships, Dalton and other chemists began to look for them. Before long, a great deal of experimental evidence was accumulated to show that the law of multiple proportions was valid. Thus the atomic theory was able to account for previously known facts and laws, and it also *predicted a new law.* In the process of verifying that prediction, Dalton and his contemporaries did many additional quantitative experiments. These led onward to more facts, more laws, and, eventually, new or modified theories. This characteristic of stimulating more research and thought put Dalton's postulates in the distinguished company of other good scientific theories.

2.6 ATOMIC WEIGHTS

Our discussion of the atomic theory has indicated that mass is a very important characteristic of atoms—it does not change as chemical reactions occur. Volume, on the other hand, often does change, because atoms or molecules pack together more tightly in liquids and solids or become more widely separated in gases when a reaction takes place. From the time Dalton's theory was first proposed, chemists realized the importance of the masses of atoms, and they spent much time and effort on experiments to determine how much heavier one kind of atom is than another.

Dalton, for example, studied a compound of carbon and oxygen which he called carbonic oxide. He found that a 100-g sample contained 42.9 g C and 57.1 g O. In Dalton's day there were no simple ways to determine the microscopic nature of a compound, and so he did not know the composition of the molecules (and hence the formula) of carbonic oxide. Faced with this difficulty, he did what most scientists would do—make the simplest possible assumption. This was that the molecules of carbonic oxide contained the minimum number of atoms: one of carbon and one of oxygen. Carbonic oxide was the compound we now know as carbon monoxide, CO, and so in this case Dalton was right. However, erroneous assumptions about the formulas for other compounds led to half a century of confusion about atomic weights.

Since the formula was CO, Dalton argued that the ratio of the mass of carbon to the mass of oxygen in the compound must be the same as the ratio of the mass of 1 carbon atom to the mass of 1 oxygen atom:

$$\frac{\text{Mass of 1 C atom}}{\text{Mass of 1 O atom}} = \frac{\text{mass of C in CO}}{\text{mass of O in CO}} = \frac{42.9 \text{ g}}{57.1 \text{ g}} = \frac{0.751}{1} = 0.751$$

In other words the mass of a carbon atom is about three-quarters (0.75) as great as the mass of an oxygen atom.

Notice that this method involves a *ratio* of masses and that the units grams cancel, yielding a pure number. That number (0.751, or approximately $\frac{3}{4}$) is the *relative mass* of a carbon atom compared with an oxygen atom. It tells nothing about the *actual* mass of a carbon atom or of an oxygen atom—only that carbon is three-quarters as heavy as oxygen.

The relative masses of the atoms are usually referred to as **atomic weights.** Their most recently determined values were given in Table 2.2, along with the names and symbols for the elements. The atomic-weight scale was originally based on a relative mass of 1 for the lightest atom, hydrogen. As more accurate methods for determining atomic weight were devised, it proved convenient to shift to oxygen and then carbon, but the scale was adjusted so that hydrogen's relative mass remained close to 1. Thus nitrogen's atomic weight of 14.0067 tells us that a nitrogen atom has about 14 times the mass of a hydrogen atom.

The fact that atomic weights are ratios of masses and have no units does not detract at all from their usefulness. It is very easy to determine how much heavier one kind of atom is than another.

EXAMPLE 2.2 Use the atomic weights in Table 2.2 to show that the mass of a mercury atom is 2.510 times the mass of a bromine atom, as stated earlier in this chapter.

Solution The actual masses of the atoms will be in the same proportion as their relative masses. Table 2.2 gives atomic weights of 200.59 for mercury and 79.904 for bromine. Therefore

$$\frac{\text{Mass of a Hg atom}}{\text{Mass of a Br atom}} = \frac{\text{relative mass of a Hg atom}}{\text{relative mass of a Br atom}} = \frac{200.59}{79.904} = 2.5104$$

or

$$\text{Mass of a Hg atom} = 2.5104 \times \text{mass of a Br atom}$$

The atomic-weight table also permits us to obtain the relative masses of molecules. These are called **molecular weights** and are calculated by summing the atomic weights of all atoms in the molecule.

EXAMPLE 2.3 How heavy would a mercurous bromide molecule be in comparison to a single bromine atom?

Solution First, obtain the relative mass of an Hg_2Br_2 molecule (the molecular weight):

2Hg atoms:	relative mass $= 2 \times 200.59$	$= 401.18$
2Br atoms:	relative mass $= 2 \times 79.904$	$= 159.808$
$1Hg_2Br_2$ molecule:	relative mass	$= 560.99$

Therefore

$$\frac{\text{Mass of a } Hg_2Br_2 \text{ molecule}}{\text{Mass of a Br atom}} = \frac{560.99}{79.904} = 7.0208$$

The Hg_2Br_2 molecule is about 7 times heavier than a bromine atom.

2.7 THE AMOUNT OF SUBSTANCE: MOLES

According to the atomic theory, atoms are the units of chemical reactions. The formula $HgBr_2$ indicates that each molecule of this substance contains one mercury and two bromine atoms. Therefore, if we ask how much bromine is required to make a given quantity of mercuric bromide, the answer is two bromine atoms for each mercury atom or two bromine atoms per molecule. In other words, how much substance we have depends in a very important way on how many atoms or molecules are present.

How much in the above sense of how many atoms or molecules are present is not the same thing as how much in terms of volume or mass. It takes 3.47 cm³ $Br_2(l)$ to react with a 1-cm³ sample of $Hg(l)$. That same 1 cm³ $Hg(l)$ would weigh 13.59 g, but only 10.83 g $Br_2(l)$ would be needed to react with it. In terms of volume, more bromine than mercury is needed, while in terms of mass, less bromine than mercury is required. In the atomic sense, however, there are exactly *twice* as many bromine atoms as mercury atoms and twice as much bromine as mercury.

The Mole

Because atoms and molecules are extremely small, there are a great many of them in any macroscopic sample. The 1 cm^3 of mercury referred to previously would contain 4.080×10^{22} mercury atoms, for example, and the 3.47 cm^3 of bromine would contain twice as many (8.160×10^{22}) bromine atoms. The very large numbers involved in counting microscopic particles are inconvenient to think about or to write down. Therefore chemists have chosen to count atoms and molecules using a unit called the mole. One **mole** (abbreviated mol) is 6.022×10^{23} of the microscopic particles which make up the substance in question. Thus 6.022×10^{23} Br atoms is referred to as 1 mol Br. The 8.160×10^{22} Br atoms in the sample we have been discussing would be $(8.160 \times 10^{22})/(6.022 \times 10^{23})$ mol Br, or 0.1355 mol Br.

The idea of using a large number as a unit with which to measure how many objects we have is not unique to chemists. Eggs, doughnuts, and many other things are sold by the dozen—a unit of twelve items. Smaller objects, such as pencils, may be ordered in units of 144, that is, by the gross, and paper is packaged in reams, each of which contains 500 sheets. A chemist who refers to 0.1355 mol Br is very much like a bookstore manager who orders $2\frac{1}{2}$ dozen sweat shirts, 20 gross of pencils, or 62 reams of paper.

There is a difference in degree, however, because the chemist's unit, 6.022×10^{23}, is so large. A stack of paper containing a mole of sheets would extend more than a million times the distance from the earth to the sun, and 6.022×10^{23} grains of sand would cover all the land in the world to a depth of nearly 2 ft. Obviously there are a great many particles in a mole of anything.

Why have chemists chosen such an unusual number as 6.022×10^{23} as the unit with which to count the number of atoms or molecules? Surely some nice round number would be easier to remember. The answer is that *the number of grams in the mass of 1 mol of atoms of any element is the atomic weight of that element.* For example, 1 mol of mercury atoms not only contains 6.022×10^{23} atoms, but its mass of 200.59 g is conveniently obtained by adding the unit gram to the atomic weight in Table 2.2. Some other examples are

1 mol H contains 6.022×10^{23} H atoms; its mass is 1.008 g.

1 mol C contains 6.022×10^{23} C atoms; its mass is 12.01 g.

1 mol O contains 6.022×10^{23} O atoms; its mass is 16.00 g.

1 mol Br contains 6.022×10^{23} Br atoms; its mass is 79.90 g.

(Here and in subsequent calculations atomic weights are rounded to two decimal places, unless, as in the case of H, fewer than four significant figures would remain.)

The mass of a mole of *molecules* can also be obtained from atomic weights. Just as a dozen eggs will have a dozen whites and a dozen yolks, a mole of CO molecules will contain a mole of C atoms and a mole of O atoms.

The mass of a mole of CO is thus

$$\text{Mass of 1 mol C} + \text{mass of 1 mol O} = \text{mass of 1 mol CO}$$
$$12.01 \text{ g} + 16.00 \text{ g} = 28.01 \text{ g}$$

The molecular weight of CO(28.01) expressed in grams is the mass of a mole of CO. Some other examples are

Molecule	Molecular Weight	Mass of 1 Mol of Molecules
Br_2	2(79.90) = 159.80	159.80 g
O_2	2(16.00) = 32.00	32.00 g
H_2O	2(1.008) + 16.00 = 18.02	18.02 g
$HgBr_2$	200.59 + 2(79.90) = 360.39	360.39 g
Hg_2Br_2	2(200.59) + 2(79.90) = 560.98	560.98 g

It is important to specify what kind of particle a mole refers to. A mole of Br atoms, for example, has only half as many atoms (and half as great a mass) as a mole of Br_2 molecules. It is best not to talk about a mole of bromine without specifying whether you mean 1 mol Br or 1 mol Br_2.

The Amount of Substance

Chemists use the mole so often to measure how much of a substance is present that it is convenient to have a name for the quantity which this unit measures. In the International System this quantity is called the **amount of substance** and is given the symbol n. Here again a common English word has been given a very specific scientific meaning. Although *amount* might refer to volume or mass in everyday speech, in chemistry it does not. When a chemist asks what amount of Br_2 was added to a test tube, an answer like "0.0678 mol Br_2" is expected. This would indicate that $0.0678 \times 6.022 \times 10^{23}$, or 4.08×10^{22}, Br_2 molecules had been added to the test tube.

The Avogadro Constant

Although chemists usually work with moles as units, occasionally it is helpful to refer to the actual number of atoms or molecules involved. When this is done, the symbol N is used. For example, in referring to 1 mol of mercury atoms, we could write

$$n_{\text{Hg}} = 1 \text{ mol} \qquad \text{and} \qquad N_{\text{Hg}} = 6.022 \times 10^{23}$$

Notice that N_{Hg} is a pure number, rather than a quantity. To obtain such a pure number, we need a conversion factor which involves the number of particles per unit amount of substance. The appropriate factor is given the symbol N_A and is called the **Avogadro constant.** It is defined by the equation

$$N_A = \frac{N}{n} \tag{2.4}$$

Since for any substance there are 6.022×10^{23} particles per mole, $N_A = 6.022 \times 10^{23}/1 \text{ mol} = 6.022 \times 10^{23} \text{ mol}^{-1}$.

EXAMPLE 2.4 Calculate the number of O_2 molecules in 0.189 mol O_2.

Solution Rearranging Eq. (2.4), we obtain

$$N = n \times N_A = 0.189 \text{ mol} \times 6.022 \times 10^{23} \text{ mol}^{-1}$$
$$= 1.14 \times 10^{23}$$

Alternatively, we might include the identity of the particles involved:

$$N = 0.189 \text{ mol } O_2 \times \frac{6.022 \times 10^{23} \ O_2 \text{ molecules}}{1 \text{ mol } O_2}$$
$$= 1.14 \times 10^{23} \ O_2 \text{ molecules}$$

Notice that Eq. (2.4), which defines the Avogadro constant, has the same form as Eq. (1.1), which defined density. The preceding example used the Avogadro constant as a conversion factor in the same way that density was used in Examples 1.10 and 1.11. As in those previous examples, all that is necessary is to remember that number of particles and amount of substance are related by a conversion factor, the Avogadro constant.

$$\text{Number of particles} \xleftrightarrow{\text{Avogadro constant}} \text{amount of substance} \qquad N \xleftrightarrow{N_A} n$$

As long as the units mole cancel, N_A is being used correctly.

2.8 THE MOLAR MASS

It is often convenient to express physical quantities *per unit amount of substance* (per mole), because in this way equal numbers of atoms or molecules are being compared. Such **molar quantities** often tell us something about the atoms or molecules themselves. For example, if the molar volume of one solid is larger than that of another, it is reasonable to assume that the molecules of the first substance are larger than those of the second. (Comparing the molar volumes of liquids, and especially gases, would not necessarily give the same information since the molecules would not be as tightly packed.)

A molar quantity is one which has been divided by the amount of substance. For example, an extremely useful molar quantity is the molar mass $\mathcal{M}$:

$$\text{Molar mass} = \frac{\text{mass}}{\text{amount of substance}} \qquad \mathcal{M} = \frac{m}{n} \qquad (2.5)$$

It is almost trivial to obtain the molar mass, since atomic and molecular weights expressed in grams give us the masses of 1 mol of substance.

EXAMPLE 2.5 Obtain the molar mass of (a) Hg and (b) Hg_2Br_2.

Solution

a) The atomic weight of mercury is 200.59, and so 1 mol Hg weighs 200.59 g.

$$\mathcal{M}_{Hg} = \frac{m_{Hg}}{n_{Hg}} = \frac{200.59\text{ g}}{1\text{ mol}} = 200.59\text{ g mol}^{-1}$$

b) Similarly, for Hg_2Br_2 the molecular weight is 560.98, and so

$$\mathcal{M}_{Hg_2Br_2} = \frac{m_{Hg_2Br_2}}{n_{Hg_2Br_2}} = 560.98\text{ g mol}^{-1}$$

The molar mass is numerically the same as the atomic or molecular weight, but it has units of grams per mole.

Equation (2.5), which defines the molar mass, has the same form as Eq. (1.1), which defined density, and Eq. (2.4), which defined the Avogadro constant. As in the case of density or the Avogadro constant, it is *not* necessary to memorize or manipulate a formula. Simply remember that mass and amount of substance are related *via* molar mass.

$$\text{Mass} \xleftrightarrow{\text{Molar mass}} \text{amount of substance} \qquad m \xleftrightarrow{\mathcal{M}} n$$

The molar mass is easily obtained from atomic weights and may be used as a conversion factor, provided the units cancel.

EXAMPLE 2.6 Calculate the amount of octane (C_8H_{18}) in 500 g of this liquid.

Solution Any problem involving interconversion of mass and amount of substance requires molar mass

$$\mathcal{M} = (8 \times 12.01 + 18 \times 1.008)\text{ g mol}^{-1} = 114.2\text{ g mol}^{-1}$$

The amount of substance will be the mass times a conversion factor which permits cancellation of units:

$$\begin{aligned} n &= m \times \text{conversion factor} \\ &= m \times \frac{1}{\mathcal{M}} \\ &= 500\text{ g} \times \frac{1\text{ mol}}{114.2\text{ g}} = 4.38\text{ mol} \end{aligned}$$

In this case the reciprocal of the molar mass was the appropriate conversion factor.

The Avogadro constant, molar mass, and density may be used in combination to solve more complicated problems.

EXAMPLE 2.7 How many molecules would be present in 25.0 ml of pure carbon tetrachloride (CCl_4)?

Solution In Example 2.4 we showed that the number of molecules may be obtained from the amount of substance by using the Avogadro constant. The amount of substance may be obtained from mass by using the molar mass (Example 2.6), and mass from volume by means of density (Example 1.12). A road map to the solution of this problem is

$$\text{Volume} \xrightarrow{\text{density}} \text{mass} \xrightarrow{\text{molar mass}} \text{amount} \xrightarrow{\text{Avogadro constant}} \text{number of molecules}$$

or in shorthand notation

$$V \xrightarrow{\rho} m \xrightarrow{\mathcal{M}} n \xrightarrow{N_A} N$$

The road map tells us that we must look up the density of CCl_4 in Table 1.4:

$$\rho = 1.595 \text{ g cm}^{-3}$$

The molar mass must be calculated from atomic weights:

$$\mathcal{M} = (12.01 + 4 \times 35.45) \text{ g mol}^{-1} = 153.81 \text{ g mol}^{-1}$$

and we recall that the Avogadro constant is

$$N_A = 6.022 \times 10^{23} \text{ mol}^{-1}$$

The last quantity (N) in the road map can then be obtained by starting with the first (V) and applying successive conversion factors:

$$N = 25.0 \text{ cm}^3 \times \frac{1.595 \text{ g}}{1 \text{ cm}^3} \times \frac{1 \text{ mol}}{153.81 \text{ g}} \times \frac{6.022 \times 10^{23} \text{ molecules}}{1 \text{ mol}}$$

$$= 1.56 \times 10^{23} \text{ molecules}$$

Notice that in this problem we had to *combine* techniques from three previous examples. To do this you must remember relationships among quantities. For example, a volume was given, and we knew it could be converted to the corresponding mass by means of density, and so we looked up the density in a table. By writing a road map, or at least seeing it in your mind's eye, you can keep track of such relationships, determine what conversion factors are needed, and then use them to solve the problem.

2.9 FORMULAS AND COMPOSITION

In Plate 3 we present a microscopic view of the chemical reaction between mercury and bromine. Equations (2.1) and (2.2) represent the same event in terms of chemical symbols and formulas. But how does a practicing chemist *find out* what is occurring on the microscopic scale? When a reaction is carried out for the first time, little is known about the microscopic nature of the products. It is therefore necessary to determine *experimentally* the composition and formula of a newly synthesized substance.

The first step in such a procedure is usually to separate and purify the products of a reaction. For example, although the combination of mercury with bromine produces mainly mercuric bromide, a little mercurous bromide is often formed as well. A mixture of mercurous bromide with mercuric bromide has properties which are different from a pure sample of $HgBr_2$, and so the Hg_2Br_2 must be removed. The low solubility of Hg_2Br_2 in water would permit purification by **recrystallization.** The product could be dissolved in a small quantity of hot water and filtered to remove undissolved Hg_2Br_2. Upon cooling and partial evaporation of the water, crystals of relatively pure $HgBr_2$ would form.

Once a pure product has been obtained, it may be possible to identify the substance by means of its physical and chemical properties. The reaction of mercury with bromine yields white crystals which melt at 236°C. The liquid which is formed boils at 322°C. Since it is made by combining two elements, the product is a compound. Comparing its properties with a handbook or table of data leads to the conclusion that it is mercuric bromide.

But suppose you were the *first* person who ever prepared mercuric bromide. There were no tables which listed its properties then, and so how could you determine that the formula should be $HgBr_2$? One answer involves **quantitative analysis**—the determination of the percentage by mass of each element in the compound. Such data are usually reported as the **percent composition.**

EXAMPLE 2.8 When 10.0 g mercury reacts with sufficient bromine, 18.0 g of a pure compound is formed. Calculate the percent composition from these experimental data.

Solution The percentage of mercury is the mass of mercury divided by the total mass of compound times 100 percent:

$$\%\ \text{Hg} = \frac{m_{\text{Hg}}}{m_{\text{compound}}} \times 100\% = \frac{10.0\ \cancel{\text{g}}}{18.0\ \cancel{\text{g}}} \times 100\% = 55.6\%$$

The remainder of the compound (18.0 g − 10.0 g = 8.0 g) is bromine:

$$\%\ \text{Br} = \frac{m_{\text{Br}}}{m_{\text{compound}}} \times 100\% = \frac{8.0\ \cancel{\text{g}}}{18.0\ \cancel{\text{g}}} \times 100\% = 44.4\%$$

As a check, verify that the percentages add to 100:

$$55.6\% + 44.4\% = 100\%$$

$\frac{35}{68} \times 100 = 52\%$ $\frac{68}{132} \times 100 = 52\%$

$\frac{63}{228} \times 100 = 27\%$

$\frac{340}{680} \times 100 = 50\%$

$\frac{64}{312} \times 100 = 22\%$

$\frac{81}{448} \times 100 = 18\%$ $\frac{48}{54} \times 100 = 89\%$

$\frac{12}{68} \times 100 = 17\%$ $\frac{32}{181} \times 100 = 17\%$

$\frac{47}{293} \times 100 = 16\%$ $\frac{23}{112} \times 100 = 20\%$

$\frac{29}{82} \times 100 = 35\%$

$\frac{56}{228} \times 100 = 24\%$

$\frac{79}{320} \times 100 = 25\%$ $\frac{19}{56} \times 100 = 34\%$

$\frac{48}{350} \times 100 = 14\%$

$\frac{47}{280} \times 100 = 16\%$

$\frac{32}{83} \times 100 = 38\%$

$\frac{29}{63} \times 100 = 46\%$

$\frac{72}{885} \times 100 = 8\%$ $\frac{29}{43} \times 10$

$AL_3/M = 78 g\,mol^{-1}$

$N = 3.5 \times \frac{2.7}{cm^3} \times \frac{1\,mol}{26} \times \frac{6.022 \times 10^{23}}{1\,mol} =$

$= 2 \times 10^{23}$

$\frac{48}{790} \times 100 = 6\%$ $\frac{46}{280} \times 100 =$

$AL_4/M = 104$

$N = 3.5 \times \frac{2.7}{cm^3} \times \frac{1\,mol}{26} \times \frac{6.022 \times 10^{23}}{1\,mol} = 2$

$\frac{47}{328} \times 100 = 14\%$ $\frac{32}{68} \times 100 =$

$\frac{40}{60} \times 100 = 67\%$

$\frac{48}{132} \times 100 = 36\%$ $\frac{43}{263} \times 100 = 16\%$

$\frac{69}{183} \times 100 = 38\%$

$\frac{68}{135} \times$

$6365\,g \times \frac{1\,cm^3}{58\,g} = 109\,cm^3$

$394\,g \times \frac{cm^3}{53} = 8\,cm^3$

$452\,g\ \frac{1\,cm^3}{46} = 10\,cm^3$

$681\,g \times \frac{1\,cm^3}{7\,g} = 97\,cm^3$

$683\,g \times \frac{1\,m^3}{27} = 25\,cm^3$

$3669\,g \times \frac{1\,cm^3}{69} = 54\,cm^3$

$783\,g\ \frac{1\,cm^3}{26} = 30\,cm^3$

$299\,g \times \frac{1\,cm^3}{78\,g} = 4\,cm^3$

$5253\,g \times \frac{1\,cm^3}{281\,g} = 18.6\,cm^3$

78

$711\,g \times \frac{1\,cm^3}{80\,g} = 9\,cm^3$

$Cu_3 / M = 635\,g\,mol^{-1}$

$V = m \times \frac{1}{p}$

$\%05 = 0$

$N = 3.5\,cm^3 \times \frac{9}{1\,cm^3} \times \frac{1\,mol}{63.5} \times 6.002$

~~$N = 63\,cm^3 \times \frac{2.7}{1\,cm^3}$~~ $\times \frac{1}{2}$

$= 40\,cm^3$

$N = 3.5\,g \times$

$Ag_5 / M = (5 \times 107)\,g\,mol^{-1} = 535$

$Cu_4 / M = (4 \times 63.5)\,g$

$\%L$

$N = 3.5\,cm^3 \times \frac{107}{1\,cm^3} \times \frac{1\,mol}{535} \times \frac{6.022 \times 1}{1\,mol}$

$\%9$

$N = 3.0\,cm^3 \times \frac{9}{1\,cm^3} \times \frac{1}{6}$

$= 3 \times 10^{23}$

$N = 3.5\,cm^3 \times \frac{9}{cm^3} \times \frac{1\,mol}{63.5} \times \frac{6.0}{1}$

$\%89 =$

$= 1.1 \times 10^{2}$

To obtain the formula from percent-composition data, we must find how many bromine atoms there are per mercury atom. On a macroscopic scale this corresponds to the ratio of the amount of bromine to the amount of mercury. If the formula is $HgBr_2$, it not only indicates that there are two bromine *atoms* per mercury *atom*, it also says that there are 2 *mol* of bromine atoms for each 1 *mol* of mercury atoms. That is, the *amount* of bromine is twice the *amount* of mercury. The numbers in the ratio of the amount of bromine to the amount of mercury (2:1) are the subscripts of bromine and mercury in the formula.

EXAMPLE 2.9 Determine the formula for the compound whose percent composition was calculated in the previous example.

Solution For convenience, assume that we have 100 g of the compound. Of this, 55.6 g (55.6%) is mercury and 44.4 g is bromine. Each mass can be converted to an amount of substance

$$n_{\mathrm{Hg}} = 55.6\ \cancel{\mathrm{g}} \times \frac{1\ \mathrm{mol\ Hg}}{200.59\ \cancel{\mathrm{g}}} = 0.277\ \mathrm{mol\ Hg}$$

$$n_{\mathrm{Br}} = 44.4\ \cancel{\mathrm{g}} \times \frac{1\ \mathrm{mol\ Br}}{79.90\ \cancel{\mathrm{g}}} = 0.556\ \mathrm{mol\ Br}$$

Dividing the larger amount by the smaller, we have

$$\frac{n_{\mathrm{Br}}}{n_{\mathrm{Hg}}} = \frac{0.556\ \mathrm{mol\ Br}}{0.277\ \mathrm{mol\ Hg}} = \frac{2.01\ \mathrm{mol\ Br}}{1\ \mathrm{mol\ Hg}}$$

The ratio 2.01 mol Br to 1 mol Hg also implies that there are 2.01 Br atoms per 1 Hg atom. If the atomic theory is correct, there is no such thing as 0.01 Br atom; furthermore, our numbers are only good to three significant figures. Therefore we round 2.01 to 2 and write the formula as $HgBr_2$.

EXAMPLE 2.10 A bromide of mercury has the composition 71.5% Hg, 28.5% Br. Find its formula.

Solution Again assume a 100-g sample and calculate the amount of each element:

$$n_{\mathrm{Hg}} = 71.5\ \cancel{\mathrm{g}} \times \frac{1\ \mathrm{mol\ Hg}}{200.59\ \cancel{\mathrm{g}}} = 0.356\ \mathrm{mol\ Hg}$$

$$n_{\mathrm{Br}} = 28.5\ \cancel{\mathrm{g}} \times \frac{1\ \mathrm{mol\ Br}}{79.90\ \cancel{\mathrm{g}}} = 0.357\ \mathrm{mol\ Br}$$

The ratio is

$$\frac{n_{\mathrm{Br}}}{n_{\mathrm{Hg}}} = \frac{0.357\ \mathrm{mol\ Br}}{0.356\ \mathrm{mol\ Hg}} = \frac{1.00\ \mathrm{mol\ Br}}{1\ \mathrm{mol\ Hg}}$$

We would therefore assign the formula HgBr.

The formula obtained in Example 2.10 does not correspond to either of the two mercury bromides we have already discussed. Is it a third one? The answer is no because our method can only determine the *ratio* of Br to Hg. The ratio 1:1 is the same as 2:2, and so our method will give the same result for HgBr or Hg_2Br_2 (or Hg_7Br_7, for that matter, should it exist). The formula determined by this method is called the **empirical formula** or **simplest formula.** Occasionally, as in the case of mercurous bromide, the empirical formula differs from the actual molecular composition, or the **molecular formula.** Experimental determination of the molecular weight in addition to percent composition permits calculation of the molecular formula.

EXAMPLE 2.11 A compound whose molecular weight is 28 contains 85.6% C and 14.4% H. Determine its empirical and molecular formulas.

Solution

$$n_C = 85.6\ \text{g} \times \frac{1\ \text{mol C}}{12.01\ \text{g}} = 7.13\ \text{mol C}$$

$$n_H = 14.4\ \text{g} \times \frac{1\ \text{mol H}}{1.008\ \text{g}} = 14.3\ \text{mol H}$$

$$\frac{n_H}{n_C} = \frac{14.3\ \text{mol H}}{7.13\ \text{mol C}} = \frac{2.01\ \text{mol H}}{1\ \text{mol C}}$$

The empirical formula is therefore CH_2. The molecular weight corresponding to the empirical formula is

$$12.01 + 2 \times 1.008 = 14.03$$

Since the experimental molecular weight is twice as great, all subscripts must be doubled and the molecular formula is C_2H_4.

Occasionally the ratio of amounts is not a whole number.

EXAMPLE 2.12 Aspirin contains 60.0% C, 4.48% H, and 35.5% O. What is its empirical formula?

Solution

$$n_C = 60.0\ \text{g} \times \frac{1\ \text{mol C}}{12.01\ \text{g}} = 5.00\ \text{mol C}$$

$$n_H = 4.48\ \text{g} \times \frac{1\ \text{mol H}}{1.008\ \text{g}} = 4.44\ \text{mol H}$$

$$n_O = 35.5\ \text{g} \times \frac{1\ \text{mol O}}{16.00\ \text{g}} = 2.22\ \text{mol O}$$

Divide all three by the smallest amount of substance

$$\frac{n_C}{n_O} = \frac{5.00 \text{ mol C}}{2.22 \text{ mol O}} = \frac{2.25 \text{ mol C}}{1 \text{ mol O}}$$

$$\frac{n_H}{n_O} = \frac{4.44 \text{ mol H}}{2.22 \text{ mol O}} = \frac{2.00 \text{ mol H}}{1 \text{ mol O}}$$

Clearly there are twice as many H atoms as O atoms, but the ratio of C to O is not so obvious. We must convert 2.25 to a ratio of small whole numbers. This can be done by changing the figures after the decimal point to a fraction. In this case, .25 becomes $\frac{1}{4}$. Thus $2.25 = 2\frac{1}{4} = \frac{9}{4}$, and

$$\frac{n_C}{n_O} = \frac{2.25 \text{ mol C}}{1 \text{ mol O}} = \frac{9 \text{ mol C}}{4 \text{ mol O}}$$

We can also write

$$\frac{n_H}{n_O} = \frac{2 \text{ mol H}}{1 \text{ mol O}} = \frac{8 \text{ mol H}}{4 \text{ mol O}}$$

Thus the empirical formula is $C_9H_8O_4$.

Once someone has determined a formula—empirical or molecular—it is possible for someone else to do the reverse calculation. Finding the weight-percent composition from the formula often proves quite informative, as the following example shows.

EXAMPLE 2.13 In order to replenish nitrogen removed from the soil when plants are harvested, the compounds $NaNO_3$ (sodium nitrate), NH_4NO_3 (ammonium nitrate), and NH_3 (ammonia) are used as fertilizers. If a farmer could buy each at the same cost per gram, which would be the best bargain? In other words, which compound contains the largest percentage of nitrogen?

Solution We will show the detailed calculation only for the case of NH_4NO_3.

1 mol NH_4NO_3 contains 2 mol N, 4 mol H, and 3 mol O. The molar mass is thus

$$\mathcal{M}_{NH_4NO_3} = (2 \times 14.01 + 4 \times 1.008 + 3 \times 16.00) \text{ g mol}^{-1}$$
$$= 80.05 \text{ g mol}^{-1}$$

A 1-mol sample would weigh 80.05 g. The mass of 2 mol N it contains is

$$m_N = 2 \cancel{\text{mol N}} \times \frac{14.01 \text{ g}}{1 \cancel{\text{mol N}}} = 28.02 \text{ g}$$

Therefore the percentage of N is

$$\%N = \frac{m_N}{m_{NH_4NO_3}} \times 100\% = \frac{28.02 \cancel{\text{g}}}{80.05 \cancel{\text{g}}} \times 100\% = 35.00\%$$

The percentages of H and O are easily calculated as

$$m_H = 4 \text{ mol H} \times \frac{1.008 \text{ g}}{1 \text{ mol H}} = 4.032 \text{ g}$$

$$\%H = \frac{4.032 \text{ g}}{80.05 \text{ g}} \times 100\% = 5.04\%$$

$$m_O = 3 \text{ mol O} \times \frac{16.00 \text{ g}}{1 \text{ mol O}} = 48.00 \text{ g}$$

$$\%O = \frac{48.00 \text{ g}}{80.05 \text{ g}} \times 100\% = 59.96\%$$

Though not strictly needed to answer the problem, the latter two percentages provide a check of the results. The total 35.00% + 5.04% + 59.96% = 100.00% as it should.

Similar calculations for $NaNO_3$ and NH_3 yield 16.48% and 82.24% nitrogen, respectively. The farmer who knows chemistry chooses ammonia!

2.10 BALANCING CHEMICAL EQUATIONS

We have now determined symbols and formulas for all the ingredients of our chemical equations, but one important step remains. We must be sure that the law of conservation of mass is obeyed. *The same number of atoms (or moles of atoms) of a given type must appear on each side of the equation.* This reflects our belief in Dalton's third postulate that atoms are neither created, destroyed, nor changed from one kind to another during a chemical process. When the law of conservation of mass is obeyed, the equation is said to be *balanced.*

As a simple example of how to balance an equation, let us take the reaction which occurs when a large excess of mercury combines with bromine. In this case the product is a white solid which does not melt but instead changes to a gas when heated above 345°C. It is insoluble in water. From these properties it can be identified as mercurous bromide, Hg_2Br_2. The equation for the reaction would look like this:

$$Hg + Br_2 \rightarrow Hg_2Br_2 \tag{2.6}$$

but it is not balanced because there are 2 mercury atoms (in Hg_2Br_2) on the right side of the equation and only 1 on the left.

An incorrect way of obtaining a balanced equation is to change this to

$$Hg + Br_2 \rightarrow \cancel{HgBr_2} \tag{2.7}$$

This equation is wrong because we had already determined from the properties of the product that the product was Hg_2Br_2. Equation (2.7) is balanced, but it refers to a different reaction which produces a different product. The equation might also be incorrectly written as

$$\cancel{Hg_2} + Br_2 \rightarrow Hg_2Br_2 \tag{2.8}$$

The formula Hg_2 suggests that molecules containing 2 mercury atoms each were involved, but our previous microscopic experience with this element (Figure 2.1) indicates that such molecules do not occur.

In balancing an equation you must remember that the *subscripts* in the formulas have been determined *experimentally*. Changing them indicates a change in the nature of the reactants or products. It is permissible, however, to change the amounts of reactants or products involved. For example, the equation in question is correctly balanced as follows:

$$2Hg + Br_2 \rightarrow Hg_2Br_2 \qquad (2.9)$$

The 2 written before the symbol Hg is called a **coefficient.** It indicates that on the microscopic level 2Hg atoms are required to react with the Br_2 molecule. On a macroscopic scale the coefficient 2 means that *2 mol* Hg atoms are required to react with 1 mol Br_2 molecules. Twice the amount of Hg is required to make Hg_2Br_2 as we needed for $HgBr_2$ [see Eq. (2.1)].

To summarize: Once the formulas (subscripts) have been determined, *an equation is balanced by adjusting coefficients.* Nothing else may be changed.

EXAMPLE 2.14 Balance the equation

$$Hg_2Br_2 + Cl_2 \rightarrow HgCl_2 + Br_2$$

Solution Although Br and Cl are balanced, Hg is not. A coefficient of 2 with $HgCl_2$ is needed:

$$Hg_2Br_2 + Cl_2 \rightarrow 2HgCl_2 + Br_2$$

Now Cl is not balanced. We need $2Cl_2$ molecules on the left:

$$Hg_2Br_2 + 2Cl_2 \rightarrow 2HgCl_2 + Br_2$$

We now have 2Hg atoms, 2Br atoms, and 4Cl atoms on each side, and so balancing is complete.

Most chemists use several techniques for balancing equations.[1] For example, it helps to know *which* element you should balance first. When each chemical symbol appears in a single formula on each side of the equation (as in Example 2.14), you can start wherever you want and the process will work. When a symbol appears in *three or more* formulas, however, that particular element will be more difficult to balance and should usually be left until last.

EXAMPLE 2.15 When butane (C_4H_{10}) is burned in sufficient oxygen gas (O_2), the only products are carbon dioxide (CO_2) and water. Write a balanced equation to describe this reaction.

[1] Laurence E. Strong, Balancing Chemical Equations, *Chemistry,* vol. 47, no. 1, pp. 13–16, January 1974, discusses some techniques in more detail.

Solution First write an unbalanced equation showing the correct formulas of all the reactants and products:

$$C_4H_{10} + O_2 \rightarrow CO_2 + H_2O$$

We note that O atoms appear in three formulas, one on the left and two on the right. Therefore we balance C and H first. The formula C_4H_{10} determines how many C and H atoms must remain after the reaction, and so we write coefficients of 4 for CO_2 and 5 for H_2O:

$$C_4H_{10} + O_2 \rightarrow 4CO_2 + 5H_2O$$

We now have a total of 13 O atoms on the right-hand side, and the equation can be balanced by using a coefficient of $\frac{13}{2}$ in front of O_2:

$$C_4H_{10} + \tfrac{13}{2}O_2 \rightarrow 4CO_2 + 5H_2O$$

Usually it is preferable to remove fractional coefficients since they might be interpreted to mean a fraction of a molecule. (One-half of an O_2 molecule would be an O atom, which has quite different chemical reactivity.) Therefore we multiply all coefficients on both sides of the equation by two to obtain the final result:

$$2C_4H_{10} + 13O_2 \rightarrow 8CO_2 + 10H_2O$$

(Sometimes, when we are interested in moles rather than individual molecules, it may be useful to omit this last step. Obviously the idea of half a mole of O_2 molecules, that is, 3.011×10^{23} molecules, is much more tenable than the idea of half a molecule.)

Another useful technique is illustrated in Example 2.15. When an element (such as O_2) appears by itself, it is usually best to choose its coefficient *last*. Furthermore, groups such as NO_3, SO_4, etc., often remain unchanged in a reaction and can be treated as if they consisted of a single atom. When such a group of atoms is enclosed in parentheses followed by a subscript, the subscript applies to all of them. That is, the formula $Ca(NO_3)_2$ involves 1Ca, 2N, and $2 \times 3 = 6$ O atoms.

EXAMPLE 2.16 Balance the equation

$$NaMnO_4 + H_2O_2 + H_2SO_4 \rightarrow MnSO_4 + Na_2SO_4 + O_2 + H_2O$$

Solution We note that oxygen atoms are found in every one of the seven formulas in the equation, making it especially hard to balance. However, Na appears only in two formulas:

$$2NaMnO_4 + H_2O_2 + H_2SO_4 \rightarrow MnSO_4 + Na_2SO_4 + O_2 + H_2O$$

as does manganese, Mn:

$$2NaMnO_4 + H_2O_2 + H_2SO_4 \rightarrow 2MnSO_4 + Na_2SO_4 + O_2 + H_2O$$

We now note that the element S always appears with 4 O atoms, and so we balance the SO_4 groups:

$$2NaMnO_4 + H_2O_2 + 3H_2SO_4 \rightarrow 2MnSO_4 + Na_2SO_4 + O_2 + H_2O$$

Now we are in a position to balance hydrogen:

$$2NaMnO_4 + H_2O_2 + 3H_2SO_4 \rightarrow 2MnSO_4 + Na_2SO_4 + O_2 + 4H_2O$$

and finally oxygen. (We are aided by the fact that it appears as the element.)

$$2NaMnO_4 + H_2O_2 + 3H_2SO_4 \rightarrow 2MnSO_4 + Na_2SO_4 + 3O_2 + 4H_2O$$

Notice that in this example we followed the rule of balancing first those elements whose symbols appeared in the smallest number of formulas: Na and Mn in two each, S (or SO_4) and H in three each, and finally O. Even using this rule, however, equations in which one or more elements appear in four or more formulas are difficult to balance without some additional techniques which we will develop in Chap. 11.

The balancing of chemical equations has an important environmental message for us. If atoms are conserved in a chemical reaction, then we cannot get rid of them. In other words *we cannot throw anything away*. There are only two things we can do with atoms: Move them from place to place or from compound to compound. Thus when we "dispose" of something by burning it, dumping it, or washing it down the sink, we have not really gotten rid of it at all. The atoms which constituted it are still around someplace, and it is just as well to know where they are and what kind of molecule they are in. Discarded atoms in places where we do not want them and in undesirable molecules are known as pollution.

SUMMARY

The atomic theory is one of the most important and useful ideas of chemistry. Thinking about matter in terms of microscopic particles such as atoms and molecules permits us to interpret numerous macroscopic observations. The contrasting properties of solids, liquids, and gases, for example, may be ascribed to differences in spacing between and speed of motion of the constituent atoms or molecules.

In the form originally proposed by John Dalton, the atomic theory distinguished elements from compounds and was used to explain the law of constant composition. The theory also agreed with Lavoisier's law of conservation of mass, and it was able to *predict* the law of multiple proportions. An important aspect of the atomic theory is the assignment of relative masses (atomic weights) to the elements.

Atoms and molecules are extremely small. Therefore, when calculating how much of one substance is required to react with another, chemists use a unit called the mole. One mole contains 6.022×10^{23} of whatever kind of microscopic particles one wishes to consider. Referring to 2 mol Br_2 specifies a certain number of Br_2 molecules in the same way that referring to 10 gross of pencils specifies a certain number of pencils. The quantity which is measured in the units called moles is known as the amount of substance.

The somewhat unusual number 6.022×10^{23}, which specifies how many particles are in a mole, has been chosen so that the mass of 1 mol of atoms of any element is the atomic weight of that element expressed in grams. Similarly, the mass of a mole of molecules is the molecular weight expressed in grams. The molecular weight is obtained by summing atomic weights of all atoms in the molecule. This choice for the mole makes it very convenient to obtain molar masses—simply add the units grams per mole to the atomic or molecular weight. Using molar mass and the Avogadro constant, it is possible to determine the masses of individual atoms or molecules and to find how many atoms or molecules are present in a macroscopic sample of matter.

A table of atomic weights and the molar masses which can be obtained from it can also be used to obtain the empirical formula of a substance if we know the percentage by weight of each element present. The opposite calculation, determination of weight percent from a chemical formula, is also possible. Once formulas for reactants and products are known, a balanced chemical equation can be written to describe any chemical change. Balancing an equation by adjusting the coefficients applied to each formula depends on the postulate of the atomic theory which states that atoms are neither created, destroyed, nor changed into atoms of another kind during a chemical reaction.

QUESTIONS AND PROBLEMS

(An asterisk before a number indicates the problem is more difficult.)

2.1 What are the names of the elements represented by the following symbols: Ar, Cd, Cr, Au, Fe, Pb, Mn, Hg, K, Si, Ag, Sn, and Xe?

2.2 Give a microscopic explanation of why the density of solid mercury is greater than the density of liquid mercury which in turn is greater than the density of gaseous mercury.

2.3 All mercury compounds have higher densities than the corresponding magnesium compound. For example, the density of HgO is 11.10 g cm^{-3} while that of MgO is only 3.58 g cm^{-3}. Give a microscopic explanation for this macroscopic behavior.

2.4 Explain in atomic terms why liquid mercury can adopt the shape of its container while solid mercury cannot.

2.5 The density of solid potassium is 0.86 g cm^{-3} while that of calcium is 1.55 g cm^{-3}, almost twice as much. How could you explain such a large difference in terms of differences between the atoms of K and Ca?

2.6 Differentiate between the terms *symbol, formula,* and *equation.* How does the difference between these terms correspond to the differences between the terms *atom, molecule,* and *reaction?*

2.7 A black powder is placed in a glass tube. Hydrogen is passed through the tube, and the powder is heated with a bunsen burner. The powder turns red, and water vapor can be detected emerging from the tube. Write down a clear explanation of why this behavior demonstrates that the black powder is *not* an element.

2.8 Point out all the mistakes in the following statement made by a general chemistry student: "The formula for the element potassium is P, as is clearly seen from the equation for the compound potassium chlorine, which is PCl."

2.9 The mass of a tin atom is 3.347 times the mass of a chlorine atom. (*a*) Calculate the mass ratio of Sn to Cl in the compound $SnCl_2$. (*b*) Calculate the mass ratio of Sn to Cl in the compound $SnCl_4$. (*c*) How are these two mass ratios related?

2.10 The element chromium forms three different oxides. The percentage of chromium in these compounds is 52.0, 68.4, and 76.5 percent. Show that these data conform to the law of multiple proportions.

2.11 Only one of the following correctly describes the atomic weight of calcium as defined in the text. Which is it?

a 40.08 g
b 40.08 g mol^{-1}
c 40.08 atomic mass units
d 40.08
e 6.655×10^{-23} g

2.12 Calculate the relative mass of

a An SO_2 molecule in comparison with an S atom.
b A Pb atom in comparison with a PbO_2 molecule.
c An N_2O molecule in comparison with an N atom.
d A P atom in comparison with a PCl_5 molecule.

2.13 Complete the following:

a 1.0 mol Br_2 contains ______ Br atoms.
b 3.5 mol N_2 contains ______ N_2 molecules.
c 4.014×10^{23} O atoms are contained in ______ mol O_2.
d 1.792 mol O_2 contains ______ O_2 molecules.
e 1.792 mol O_2 contains ______ O atoms.
f 5.88×10^{20} molecules of Cl_2 are contained in ______ mol Cl_2.

2.14 Which of the following quantities describes the *amount of* H_2O? (More than one may be correct.)

a 1.000 g H_2O
b 1.000 mol H_2O
c 18.01 g H_2O
d 18.01 cm^3 H_2O
e 18.01 mol H_2O

2.15 Differentiate clearly between n, the amount of substance, and N, the number of particles. How does N differ from N_A, the Avogadro constant? What is the formula relating n, N, and N_A?

2.16 Fill in the correct conversion factor and complete the following calculations:

a $N_{Si} = 1.13$ mol Si × ______ = ______.
b $N_{H_2O} = 7.3 \times 10^{-2}$ mol H_2O × ______ = ______.
c $N_{H_2} = 7.53 \times 10^{3}$ mol H_2 × ______ = ______.

2.17 Determine the molar mass ($\mathcal{M}$) of each of the following elements and compounds:

a Ba
b BaO
c $BaCl_2$
d Al
e Al_2O_3
f $AlCl_3$

2.18 If 2.384 mol of an unknown substance has a mass of 366.7 g, what is its molar mass?

2.19 2.051 mol H_2O has a mass of 36.95 g and contains 1.235×10^{24} molecules of H_2O. What symbols are conventionally used to describe these quantities:

a 2.051 mol H_2O = ______.
b 36.95 g H_2O = ______.
c 1.235×10^{24} molecules H_2O = ______.
d The molar mass of H_2O = ______.

2.20 Calculate the amount of substance in each of the following:

a 127 g Al
b 19.25 g CsCl
c 7.72×10^{-4} g SiO_2
d 5.88×10^{3} g O_2

2.21 Given exactly 45.0 cm^3 of pure liquid benzene, C_6H_6 (density = 0.88 g cm^{-3}), calculate each of the following:

a Mass of C_6H_6
b Amount of C_6H_6
c Molar mass of C_6H_6
d Number of C_6H_6 molecules
e Number of C atoms

2.22 Calculate each of the following for 3.97 mol carbon dioxide, CO_2:

a Mass
b Number of CO_2 molecules
c Amount of O — times 2
d Number of O atoms

2.23 When 50.00 g Al metal is heated with an excess of liquid Br_2, it is found that 493.9 g of a compound are formed. From this information calculate

a The amount of Al at the start of the reaction.

b The amount of Al in the product.

c The mass of Br in the product.

d The amount of Br in the product.

e The empirical formula of the product.

f The molecular formula of the product, given that its molar mass has the value 533.4 g mol^{-1}.

2.24 When 25.00 g of metallic nickel, Ni, reacts with an excess of chlorine gas, Cl_2, 55.20 g of pure nickel chloride is produced. Find the empirical formula of the nickel chloride. Why is this formula not necessarily the same as the molecular formula?

2.25 An oxide of chlorine has the composition 38.77% Cl and 61.23% O. Determine the empirical formula of this compound.

2.26 Find the empirical formulas of the three oxides of chromium whose composition is given in Prob. 2.10.

2.27 Nicotine, the toxic constituent in tobacco, has a molecular weight of 162.2 and the following percent composition: 74.07% C; 17.28% N; 8.65% H. Using these data, determine both the empirical and the molecular formulas.

2.28 Find the percent composition of the following compounds:

a CO *b* SO_2 *c* $CaCO_3$ *d* $K_2Cr_2O_7$

2.29 You are given a 100-g sample of each of the following compounds: Fe_2O_3, $FeCl_3$, $Fe(NO_3)_3$, $Fe_2(SO_4)_3$, and Fe_3O_4. Which of these samples contains the largest mass of Fe? Which contains the largest *amount* of Fe?

2.30 Balance the following equations:

a $H_2 + Br_2 \rightarrow HBr$

b $N_2 + H_2 \rightarrow NH_3$

c $SiO_2 + HF \rightarrow SiF_4 + H_2O$

d $Al_2S_3 + H_2O \rightarrow Al(OH)_3 + H_2S$

e $KClO_3 \rightarrow KClO_4 + KCl$

2.31 Write balanced equations for the following reactions:

a $Sb_2S_3 + Fe \rightarrow FeS + Sb$

b $La_2O_3 + NH_4Cl \rightarrow LaCl_3 + H_2O + NH_3$

c $NO_2 + H_2O \rightarrow HNO_3 + NO$

d $C_4H_{10} + O_2 \rightarrow CO_2 + H_2O$

e $MnO_2 + NaOH + O_2 \rightarrow Na_2MnO_4 + H_2O$

2.32 Write balanced equations for the following reactions:

a $Pb + PbO_2 + H_2SO_4 \rightarrow PbSO_4 + H_2O$

b $KNO_3 + H_2SO_4 + Hg \rightarrow K_2SO_4 + HgSO_4 + H_2O + NO$

c $Cu + HNO_3 \rightarrow Cu(NO_3)_2 + NO + H_2O$

d $Zn + H_3AsO_4 + HCl \rightarrow ZnCl_2 + AsH_3 + H_2O$

2.33 Write balanced equations to describe the reactions which occur when (*a*) hydrogen gas, H_2, is burned in oxygen to form water vapor; (*b*) methane gas, CH_4, is burned in oxygen to form water vapor and carbon dioxide; (*c*) superheated steam is passed over red-hot carbon to form a mixture of carbon monoxide and hydrogen gases; (*d*) liquid methanol is burned in oxygen to form carbon dioxide gas and water vapor; (*e*) methane and steam are passed over a catalyst at high temperature in order to produce a mixture of carbon dioxide and hydrogen gases. (The catalyst speeds up the reaction without being consumed by it.)

***2.34** There are an infinite number of ways of balancing the following equation:

$$aNH_3 + bO_2 \rightarrow cH_2O + dNO + eNO_2$$

See if you can find at least two different solutions. Also find a formula describing the general case.

chapter three

USING CHEMICAL EQUATIONS IN CALCULATIONS

$$\underset{\text{Gasoline}}{2C_8H_8} + \underset{\text{Air}}{25O_2} \longrightarrow \underset{\text{Carbon dioxide}}{16CO_2} + \underset{\text{Water}}{18H_2O}$$

There are a great many circumstances in which you may need to use a balanced equation. For example, you might want to know how much air pollution would occur when 100 metric tons of coal were burned in an electric power plant, how much heat could be obtained from a kilogram of natural gas, or how much vitamin C is really present in a 300-mg tablet. In each instance someone else would probably have determined what reaction takes place, but you would need to use the balanced equation to get the desired information.

3.1 EQUATIONS AND MASS RELATIONSHIPS

A balanced chemical equation such as

$$4NH_3(g) + 5O_2(g) \rightarrow 4NO(g) + 6H_2O(g) \qquad (3.1)$$

not only tells how many molecules of each kind are involved in a reaction, it also indicates the *amount* of each substance that is involved. Equation (3.1) says that 4 NH_3 *molecules* can react with 5 O_2 *molecules* to give 4 NO *molecules* and 6 H_2O *molecules*. It also says that 4 *mol* NH_3 would react with 5 *mol* O_2 yielding 4 *mol* NO and 6 *mol* H_2O.

The balanced equation does more than this, though. It also tells us that $2 \times 4 = 8$ mol NH_3 will react with $2 \times 5 = 10$ mol O_2, and that

$\frac{1}{2} \times 4 = 2$ mol NH_3 requires only $\frac{1}{2} \times 5 = 2.5$ mol O_2. In other words, the equation indicates that *exactly* 5 mol O_2 must react *for every* 4 mol NH_3 consumed. For the purpose of calculating how much O_2 is required to react with a certain amount of NH_3, therefore, the significant information contained in Eq. (3.1) is the *ratio*

$$\frac{5 \text{ mol } O_2}{4 \text{ mol } NH_3}$$

We shall call such a ratio derived from a balanced chemical equation a **stoichiometric ratio** and give it the symbol S. Thus, for Eq. (3.1),

$$S\left(\frac{O_2}{NH_3}\right) = \frac{5 \text{ mol } O_2}{4 \text{ mol } NH_3} \tag{3.2}$$

The word *stoichiometric* comes from the Greek words *stoicheion,* "element," and *metron,* "measure." Hence the stoichiometric ratio measures one element (or compound) against another.

EXAMPLE 3.1 Derive all possible stoichiometric ratios from Eq. (3.1).

Solution Any ratio of amounts of substance given by coefficients in the equation may be used:

$$S\left(\frac{NH_3}{O_2}\right) = \frac{4 \text{ mol } NH_3}{5 \text{ mol } O_2} \qquad S\left(\frac{O_2}{NO}\right) = \frac{5 \text{ mol } O_2}{4 \text{ mol NO}}$$

$$S\left(\frac{NH_3}{NO}\right) = \frac{4 \text{ mol } NH_3}{4 \text{ mol NO}} \qquad S\left(\frac{O_2}{H_2O}\right) = \frac{5 \text{ mol } O_2}{6 \text{ mol } H_2O}$$

$$S\left(\frac{NH_3}{H_2O}\right) = \frac{4 \text{ mol } NH_3}{6 \text{ mol } H_2O} \qquad S\left(\frac{NO}{H_2O}\right) = \frac{4 \text{ mol NO}}{6 \text{ mol } H_2O}$$

There are six more stoichiometric ratios, each of which is the reciprocal of one of these. [Eq. (3.2) gives one of them.]

When any chemical reaction occurs, the amounts of substances consumed or produced are related by the appropriate stoichiometric ratios. Using Eq. (3.1) as an example, this means that the ratio of the amount of O_2 consumed to the amount of NH_3 consumed must be the stoichiometric ratio $S(O_2/NH_3)$:

$$\frac{n_{O_2 \text{ consumed}}}{n_{NH_3 \text{ consumed}}} = S\left(\frac{O_2}{NH_3}\right) = \frac{5 \text{ mol } O_2}{4 \text{ mol } NH_3}$$

Similarly, the ratio of the amount of H_2O produced to the amount of NH_3 consumed must be $S(H_2O/NH_3)$:

$$\frac{n_{H_2O \text{ produced}}}{n_{NH_3 \text{ consumed}}} = S\left(\frac{H_2O}{NH_3}\right) = \frac{6 \text{ mol } H_2O}{4 \text{ mol } NH_3}$$

In general we can say that

$$\text{Stoichiometric ratio}\left(\frac{\text{X}}{\text{Y}}\right) = \frac{\text{amount of X consumed or produced}}{\text{amount of Y consumed or produced}} \qquad (3.3a)$$

or, in symbols,

$$S\left(\frac{\text{X}}{\text{Y}}\right) = \frac{n_{\text{X consumed or produced}}}{n_{\text{Y consumed or produced}}} \qquad (3.3b)$$

Note that in the word Eq. (3.3*a*) and the symbolic Eq. (3.3*b*), X and Y may represent *any* reactant or *any* product in the balanced chemical equation from which the stoichiometric ratio was derived. No matter how much of each reactant we have, the amounts of reactants *consumed* and the amounts of products *produced* will be in appropriate stoichiometric ratios.

EXAMPLE 3.2 Find the amount of water produced when 3.68 mol NH_3 is consumed according to Eq. (3.1).

Solution The amount of water produced must be in the stoichiometric ratio $S(H_2O/NH_3)$ to the amount of ammonia consumed:

$$S\left(\frac{H_2O}{NH_3}\right) = \frac{n_{H_2O\text{ produced}}}{n_{NH_3\text{ consumed}}}$$

Multiplying both sides by $n_{NH_3\text{ consumed}}$, we have

$$\begin{aligned} n_{H_2O\text{ produced}} &= n_{NH_3\text{ consumed}} \times S\left(\frac{H_2O}{NH_3}\right) \\ &= 3.68\ \cancel{\text{mol NH}_3} \times \frac{6\ \text{mol H}_2\text{O}}{4\ \cancel{\text{mol NH}_3}} \\ &= 5.52\ \text{mol H}_2\text{O} \end{aligned}$$

This is a typical illustration of the use of a stoichiometric ratio as a conversion factor. Example 3.2 is analogous to Examples 1.10 and 1.11, where density was employed as a conversion factor between mass and volume. Example 3.2 is also analogous to Examples 2.4 and 2.6, in which the Avogadro constant and molar mass were used as conversion factors. As in these previous cases, there is no need to memorize or do algebraic manipulations with Eq. (3.3) when using the stoichiometric ratio. Simply remember that the coefficients in a balanced chemical equation give stoichiometric ratios, and that the proper choice results in cancellation of units. In road-map form

$$\begin{matrix}\text{Amount of X consumed} \\ \text{or produced}\end{matrix} \xleftrightarrow[\text{ratio X/Y}]{\text{stoichiometric}} \begin{matrix}\text{amount of Y consumed} \\ \text{or produced}\end{matrix}$$

or symbolically:

$$n_{\text{X consumed or produced}} \xleftrightarrow{S(X/Y)} n_{\text{Y consumed or produced}}$$

When using stoichiometric ratios, be sure you *always* indicate moles *of what.* You can only cancel moles of the same substance. In other words, 1 mol NH_3 cancels 1 mol NH_3 but does *not* cancel 1 mol H_2O.

The next example shows that stoichiometric ratios are also useful in problems involving the mass of a reactant or product.

EXAMPLE 3.3 Calculate the mass of sulfur dioxide (SO_2) produced when 3.84 mol O_2 is reacted with FeS_2 according to the equation

$$4FeS_2 + 11O_2 \rightarrow 2Fe_2O_3 + 8SO_2$$

Solution The problem asks that we calculate the mass of SO_2 produced. As we learned in Sec. 2.8, Example 2.6, the molar mass can be used to convert from the amount of SO_2 to the mass of SO_2. Therefore this problem in effect is asking that we calculate the amount of SO_2 produced from the amount of O_2 consumed. This is the same problem as in Example 3.2. It requires the stoichiometric ratio

$$S\left(\frac{SO_2}{O_2}\right) = \frac{8 \text{ mol } SO_2}{11 \text{ mol } O_2}$$

The *amount* of SO_2 produced is then

$$\begin{aligned} n_{SO_2} &= n_{O_2 \text{ consumed}} \times \text{conversion factor} \\ &= 3.84 \text{ mol } O_2 \times \frac{8 \text{ mol } SO_2}{11 \text{ mol } O_2} = 2.79 \text{ mol } SO_2 \end{aligned}$$

The *mass* of SO_2 is

$$\begin{aligned} m_{SO_2} &= 2.79 \text{ mol } SO_2 \times \frac{64.06 \text{ g } SO_2}{1 \text{ mol } SO_2} \\ &= 179 \text{ g } SO_2 \end{aligned}$$

With practice this kind of problem can be solved in one step by concentrating on the units. The appropriate stoichiometric ratio will convert moles of O_2 to moles of SO_2 and the molar mass will convert moles of SO_2 to grams of SO_2. A schematic road map for the one-step calculation can be written as

$$n_{O_2} \xrightarrow{S(SO_2/O_2)} n_{SO_2} \xrightarrow{\mathcal{M}_{SO_2}} m_{SO_2}$$

Thus

$$m_{SO_2} = 3.84 \text{ mol } O_2 \times \frac{8 \text{ mol } SO_2}{11 \text{ mol } O_2} \times \frac{64.06 \text{ g}}{1 \text{ mol } SO_2} = 179 \text{ g}$$

The chemical reaction in this example is of environmental interest. Iron pyrite (FeS_2) is often an impurity in coal, and so burning this fuel in a power plant produces sulfur dioxide (SO_2), a major air pollutant.

Our next example also involves burning a fuel and its effect on the atmosphere.

EXAMPLE 3.4 What mass of oxygen would be consumed when 3.3×10^{15} g, 3.3 Pg (petagrams), of octane (C_8H_{18}) is burned to produce CO_2 and H_2O?

Solution First, write a balanced equation

$$2C_8H_{18} + 25O_2 \rightarrow 16CO_2 + 18H_2O$$

The problem gives the mass of C_8H_{18} burned and asks for the mass of O_2 required to combine with it. Thinking the problem through before trying to solve it, we realize that the molar mass of octane could be used to calculate the amount of octane consumed. Then we need a stoichiometric ratio to get the amount of O_2 consumed. Finally, the molar mass of O_2 permits calculation of the mass of O_2. Symbolically

$$m_{C_8H_{18}} \xrightarrow{\mathcal{M}_{C_8H_{18}}} n_{C_8H_{18}} \xrightarrow{S(O_2/C_8H_{18})} n_{O_2} \xrightarrow{\mathcal{M}_{O_2}} m_{O_2}$$

$$m_{O_2} = 3.3 \times 10^{15} \text{ g} \times \frac{1 \text{ mol } C_8H_{18}}{114.2 \text{ g}} \times \frac{25 \text{ mol } O_2}{2 \text{ mol } C_8H_{18}} \times \frac{32.00 \text{ g}}{1 \text{ mol } O_2}$$

$$= 1.2 \times 10^{16} \text{ g}$$

Thus 12 Pg (petagrams) of O_2 would be needed.

The large mass of oxygen obtained in this example is an estimate of how much O_2 is removed from the earth's atmosphere each year by human activities. Octane, a component of gasoline, was chosen to represent coal, gas, and other fossil fuels. Fortunately, the total mass of oxygen in the air (1.2×10^{21} g) is much larger than the yearly consumption. If we were to go on burning fuel at the present rate, it would take about 100 000 years to use up all the O_2. Actually we will consume the fossil fuels long before that! One of the least of our environmental worries is running out of atmospheric oxygen.

The Limiting Reagent

Example 3.4 also illustrates the idea that one reactant in a chemical equation may be completely consumed without using up all of another. In the laboratory as well as the environment, inexpensive reagents like atmospheric O_2 are often supplied in excess. Some portion of such a reagent will be left unchanged after the reaction. Conversely, at least one reagent is usually completely consumed. When it is gone, the other excess reactants have nothing to react with and they cannot be converted to products. The substance which is used up first is the **limiting reagent.**

EXAMPLE 3.5 When 100.0 g mercury is reacted with 100.0 g bromine to form mercuric bromide, which is the limiting reagent?

Solution The balanced equation

$$Hg + Br_2 \rightarrow HgBr_2$$

tells us that according to the atomic theory, 1 mol Hg is required for each mole of Br_2. That is, the stoichiometric ratio $S(Hg/Br_2) = 1 \text{ mol Hg}/1 \text{ mol } Br_2$. Let us see how many moles of each we actually have

$$n_{Hg} = 100.0 \text{ g} \times \frac{1 \text{ mol Hg}}{200.59 \text{ g}} = 0.4985 \text{ mol Hg}$$

$$n_{Br_2} = 100.0 \text{ g} \times \frac{1 \text{ mol } Br_2}{159.80 \text{ g}} = 0.6258 \text{ mol } Br_2$$

When the reaction ends, 0.4985 mol Hg will have reacted with 0.4985 mol Br_2 and there will be

$$(0.6258 - 0.4985) \text{ mol } Br_2 = 0.1273 \text{ mol } Br_2$$

left over. Mercury is therefore the limiting reagent.

From this example you can begin to see what needs to be done to determine which of two reagents, X or Y, is limiting. We must compare the stoichiometric ratio $S(X/Y)$ with the actual ratio of amounts of X and Y which were initially mixed together. In Example 3.5 this ratio of initial amounts

$$\frac{n_{Hg} \text{ (initial)}}{n_{Br_2} \text{ (initial)}} = \frac{0.4985 \text{ mol Hg}}{0.6258 \text{ mol } Br_2} = \frac{0.7966 \text{ mol Hg}}{1 \text{ mol } Br_2}$$

was less than the stoichiometric ratio

$$S\left(\frac{Hg}{Br_2}\right) = \frac{1 \text{ mol Hg}}{1 \text{ mol } Br_2}$$

This indicated that there was not enough Hg to react with all the Br_2, and mercury was the limiting reagent. The corresponding general rule, for any reagents X and Y, is

If $\dfrac{n_X \text{ (initial)}}{n_Y \text{ (initial)}}$ is less than $S\left(\dfrac{X}{Y}\right)$, then X is limiting.

If $\dfrac{n_X \text{ (initial)}}{n_Y \text{ (initial)}}$ is greater than $S\left(\dfrac{X}{Y}\right)$, then Y is limiting.

(Of course, when the amounts of X and Y are in exactly the stoichiometric ratio, both reagents will be completely consumed at the same time, and neither is in excess.) This general rule for determining the limiting reagent is applied in the next example.

EXAMPLE 3.6 Iron can be obtained by reacting the ore hematite (Fe_2O_3) with coke (C). The latter is converted to CO_2. As manager of a blast furnace you are told that you have 20.5 Mg (megagrams) of Fe_2O_3 and 2.84 Mg of coke on hand. (a) Which should you order first—another shipment of iron ore or one of coke? (b) How many megagrams of iron can you make with the materials you have?

Solution

a) Write a balanced equation $\quad 2Fe_2O_3 + 3C \rightarrow 3CO_2 + 4Fe$

The stoichiometric ratio connecting C and Fe_2O_3 is

$$S\left(\frac{\text{C}}{\text{Fe}_2\text{O}_3}\right) = \frac{3 \text{ mol C}}{2 \text{ mol Fe}_2\text{O}_3} = \frac{1.5 \text{ mol C}}{1 \text{ mol Fe}_2\text{O}_3}$$

The initial amounts of C and Fe_2O_3 are calculated using appropriate molar masses

$$n_{\text{C}} \text{ (initial)} = 2.84 \times 10^6 \text{ g} \times \frac{1 \text{ mol C}}{12.01 \text{ g}} = 2.36 \times 10^5 \text{ mol C}$$

$$n_{\text{Fe}_2\text{O}_3} \text{ (initial)} = 20.5 \times 10^6 \text{ g} \times \frac{1 \text{ mol Fe}_2\text{O}_3}{159.69 \text{ g}} = 1.28 \times 10^5 \text{ mol Fe}_2\text{O}_3$$

Their ratio is

$$\frac{n_{\text{C}} \text{ (initial)}}{n_{\text{Fe}_2\text{O}_3} \text{ (initial)}} = \frac{2.36 \times 10^5 \text{ mol C}}{1.28 \times 10^5 \text{ mol Fe}_2\text{O}_3} = \frac{1.84 \text{ mol C}}{1 \text{ mol Fe}_2\text{O}_3}$$

Since this ratio is larger than the stoichiometric ratio, you have more than enough C to react with all the Fe_2O_3. Fe_2O_3 is the limiting reagent, and you will want to order more of it first since it will be consumed first.

b) The amount of product formed in a reaction may be calculated via an appropriate stoichiometric ratio from the amount of a reactant which was *consumed*. Some of the excess reactant C will be left over, but *all* the initial amount of Fe_2O_3 will be consumed. Therefore we use $n_{\text{Fe}_2\text{O}_3}$ (initial) to calculate how much Fe can be obtained

$$n_{\text{Fe}_2\text{O}_3} \xrightarrow{S(\text{Fe}/\text{Fe}_2\text{O}_3)} n_{\text{Fe}} \xrightarrow{\mathcal{M}_{\text{Fe}}} m_{\text{Fe}}$$

$$m_{\text{Fe}} = 1.28 \times 10^5 \text{ mol Fe}_2\text{O}_3 \times \frac{4 \text{ mol Fe}}{2 \text{ mol Fe}_2\text{O}_3} \times \frac{55.85 \text{ g}}{\text{mol Fe}} = 1.43 \times 10^7 \text{ g Fe}$$

This is 14.3×10^6 g, or 14.3 Mg, Fe.

As you can see from the example, in a case where there is a limiting reagent, *the initial amount of the limiting reagent must be used to calculate the amount of product formed.* Using the initial amount of a reagent present in excess would be incorrect, because such a reagent is not entirely consumed.

The concept of a limiting reagent was used by the nineteenth century German chemist Justus von Liebig (1807 to 1873) to derive an important biological and ecological law. **Liebig's law of the minimum** states that the essential substance available in the smallest amount relative to some critical minimum will control growth and reproduction of any species of plant or animal life. When a group of organisms runs out of that essential limiting reagent, the chemical reactions needed for growth and reproduction must stop. Vitamins, protein, and other nutrients are essential for growth of the human body and of human populations. Similarly, the growth of algae in natural bodies of water such as Lake Erie can be inhibited by reducing the supply of nutrients such as phosphorus in the form of phosphates. It is for this reason that many states have regulated or banned the use of phosphates in detergents and are constructing treatment plants which can remove phosphates from municipal sewage before they enter lakes or streams.

Percent Yield

Not all chemical reactions are as simple as the ones we have considered so far. Quite often a mixture of two or more products containing the same element is formed. For example, when octane (or gasoline in general) burns in an excess of air, the reaction is

$$2C_8H_{18} + 25O_2 \rightarrow 16CO_2 + 18H_2O$$

If oxygen is the limiting reagent, however, the reaction does not necessarily stop short of consuming all the octane available. Instead, some carbon monoxide (CO) forms:

$$2C_8H_{18} + 24O_2 \rightarrow 14CO_2 + 2CO + 18H_2O$$

Burning gasoline in an automobile engine, where the supply of oxygen is not always as great as that demanded by the stoichiometric ratio, often produces carbon monoxide, a poisonous substance and a major source of air pollution.

In other cases, even though none of the reactants is completely consumed, no further increase in the amounts of the products occurs. We say that such a reaction does not *go to completion.* When a mixture of products is produced or a reaction does not go to completion, the effectiveness of the reaction is usually evaluated in terms of **percent yield** of the desired product. A **theoretical yield** is calculated by assuming that all the limiting reagent is converted to product. The experimentally determined mass of product is then compared to the theoretical yield and expressed as a percentage:

$$\text{Percent yield} = \frac{\text{actual yield}}{\text{theoretical yield}} \times 100 \text{ percent}$$

EXAMPLE 3.7 When 100.0 g N_2 gas and 25.0 g H_2 gas are mixed at 350°C and a high pressure, they react to form 28.96 g NH_3 (ammonia) gas. Calculate the percent yield.

Solution We must calculate the theoretical yield of NH_3, and to do this, we must first discover whether N_2 or H_2 is the limiting reagent. For the balanced equation

$$N_2 + 3H_2 \rightarrow 2NH_3$$

the stoichiometric ratio of the reactants is

$$S\left(\frac{H_2}{N_2}\right) = \frac{3 \text{ mol } H_2}{1 \text{ mol } N_2}$$

Can you do one on the board

Now, the initial amounts of the two reagents are

$$n_{H_2} \text{ (initial)} = 25.0 \text{ g } H_2 \times \frac{1 \text{ mol } H_2}{2.016 \text{ g } H_2} = 12.4 \text{ mol } H_2$$

and $$n_{N_2} \text{ (initial)} = 100.0 \text{ g } N_2 \times \frac{1 \text{ mol } N_2}{28.02 \text{ g } N_2} = 3.569 \text{ mol } N_2$$

The ratio of initial amounts is thus

$$\frac{n_{H_2} \text{ (initial)}}{n_{N_2} \text{ (initial)}} = \frac{12.4 \text{ mol } H_2}{3.569 \text{ mol } N_2} = \frac{3.47 \text{ mol } H_2}{1 \text{ mol } N_2}$$

Since this ratio is greater than $S\left(\frac{H_2}{N_2}\right)$, there is an excess of H_2. N_2 is the limiting reagent. Accordingly we must use 3.569 mol N_2 (rather than 12.4 mol H_2) to calculate the theoretical yield of NH_3. We then have

$$n_{NH_3} \text{ (theoretical)} = 3.569 \text{ mol } N_2 \times \frac{2 \text{ mol } NH_3}{1 \text{ mol } N_2}$$
$$= 7.138 \text{ mol } NH_3$$

so that

$$m_{NH_3} \text{ (theoretical)} = 7.138 \text{ mol } NH_3 \times \frac{17.03 \text{ g } NH_3}{1 \text{ mol } NH_3}$$
$$= 121.6 \text{ g } NH_3$$

The percent yield is then

$$\text{Percent yield} = \frac{\text{actual yield}}{\text{theoretical yield}} \times 100 \text{ percent}$$
$$= \frac{28.96 \text{ g}}{121.6 \text{ g}} \times 100 \text{ percent} = 23.81 \text{ percent}$$

Combination of nitrogen and hydrogen to form ammonia is a classic example of a reaction which does not go to completion. Commercial production of ammonia is accomplished using this reaction in what is called the **Haber process.** Even at the rather unusual temperatures and pressures used for this industrial synthesis, only about one-quarter of the reactants can be converted to the desired product. This is unfortunate because nearly all nitrogen fertilizers are derived from ammonia and the world has come to rely on them in order to produce enough food for its rapidly increasing population. Ammonia ranks third [after sulfuric acid (H_2SO_4) and oxygen (O_2)] in the list of most-produced chemicals, worldwide. It might rank even higher if the reaction by which it is made went to completion. Certainly ammonia and the food it helps to grow would be less expensive and would require much less energy to produce if this were the case.

3.2 ANALYSIS OF COMPOUNDS

Up to this point we have obtained all stoichiometric ratios from the coefficients of balanced chemical equations. Chemical *formulas* also indicate relative amounts of substance, however, and stoichiometric ratios may be derived from them, too. For example, the formula $HgBr_2$ tells us that no matter how large a sample of mercuric bromide we have, there will always be 2 mol of bromine atoms for each mole of mercury atoms. That is, from the formula $HgBr_2$ we have the stoichiometric ratio

$$S\left(\frac{\text{Br}}{\text{Hg}}\right) = \frac{2 \text{ mol Br}}{1 \text{ mol Hg}}$$

We could also determine that for $HgBr_2$

$$S\left(\frac{\text{Hg}}{\text{HgBr}_2}\right) = \frac{1 \text{ mol Hg}}{1 \text{ mol HgBr}_2} \qquad S\left(\frac{\text{Br}}{\text{HgBr}_2}\right) = \frac{2 \text{ mol Br}}{1 \text{ mol HgBr}_2}$$

(The reciprocals of these stoichiometric ratios are also valid for $HgBr_2$.)

Stoichiometric ratios derived from formulas instead of equations are involved in the most common procedure for determining the empirical formulas of compounds which contain only C, H, and O. A weighed quantity of the substance to be analyzed is placed in a combustion train (Fig. 3.1) and

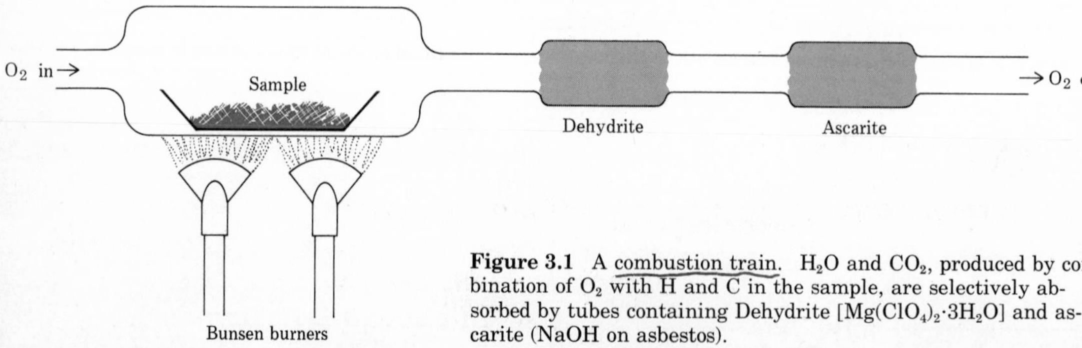

Figure 3.1 A combustion train. H_2O and CO_2, produced by combination of O_2 with H and C in the sample, are selectively absorbed by tubes containing Dehydrite [$Mg(ClO_4)_2 \cdot 3H_2O$] and ascarite (NaOH on asbestos).

heated in a stream of dry O_2. All the H in the compound is converted to $H_2O(g)$ which is trapped selectively in a previously weighed absorption tube. All the C is converted to $CO_2(g)$, and this is absorbed selectively in a second tube. The increase of mass of each tube tells, respectively, how much H_2O and CO_2 were produced by combustion of the sample.

do on the board

EXAMPLE 3.8 A 6.49-mg sample of ascorbic acid (vitamin C) was burned in a combustion train. 9.74 mg CO_2 and 2.64 mg H_2O were formed. Determine the empirical formula of ascorbic acid.

Solution We need to know the amount of C, the amount of H, and the amount of O in the sample. The ratio of these gives the subscripts in the formula. The first two may be obtained from the masses of CO_2 and H_2O, using the molar masses and the stoichiometric ratios

$$S\left(\frac{\text{C}}{\text{CO}_2}\right) = \frac{1\text{ mol C}}{1\text{ mol CO}_2} \qquad S\left(\frac{\text{H}}{\text{H}_2\text{O}}\right) = \frac{2\text{ mol H}}{1\text{ mol H}_2\text{O}}$$

221

Thus

$$n_\text{C} = 9.74 \times 10^{-3}\text{ g CO}_2 \times \frac{1\text{ mol CO}_2}{44.01\text{ g CO}_2} \times \frac{1\text{ mol C}}{1\text{ mol CO}_2}$$

$$= 2.21 \times 10^{-4}\text{ mol C}$$

$$n_\text{H} = 2.64 \times 10^{-3}\text{ g H}_2\text{O} \times \frac{1\text{ mol H}_2\text{O}}{18.02\text{ g H}_2\text{O}} \times \frac{2\text{ mol H}}{1\text{ mol H}_2\text{O}}$$

$$= 2.93 \times 10^{-4}\text{ mol H}$$

The compound may also have contained oxygen. To see if it does, calculate the masses of C and H and subtract from the total mass of sample

$$m_\text{C} = 2.21 \times 10^{-4}\text{ mol C} \times \frac{12.01\text{ g C}}{1\text{ mol C}} = 2.65 \times 10^{-3}\text{ g C}$$

$$= 2.65\text{ mg C}$$

$$m_\text{H} = 2.93 \times 10^{-4}\text{ mol H} \times \frac{1.008\text{ g H}}{1\text{ mol H}}$$

$$= 2.95 \times 10^{-4}\text{ g H}$$
$$= 0.295\text{ mg H}$$

Thus we have

$$6.49\text{ mg sample} - 2.65\text{ mg C} - 0.295\text{ mg H} = 3.54\text{ mg O}$$

and

$$n_\text{O} = 3.54 \times 10^{-3}\text{ g O} \times \frac{1\text{ mol O}}{16.00\text{ g O}} = 2.21 \times 10^{-4}\text{ mol O}$$

The ratios of the amounts of the elements in ascorbic acid are therefore

$$\frac{n_H}{n_C} = \frac{2.93 \times 10^{-4} \text{ mol H}}{2.21 \times 10^{-4} \text{ mol C}} = \frac{1.33 \text{ mol H}}{1 \text{ mol C}}$$

$$= \frac{1\frac{1}{3} \text{ mol H}}{1 \text{ mol C}} = \frac{4 \text{ mol H}}{3 \text{ mol C}}$$

$$\frac{n_O}{n_C} = \frac{2.21 \times 10^{-4} \text{ mol O}}{2.21 \times 10^{-4} \text{ mol C}} = \frac{1 \text{ mol O}}{1 \text{ mol C}} = \frac{3 \text{ mol O}}{3 \text{ mol C}}$$

Since n_C:n_H:n_O is 3 mol C:4 mol H:3 mol O, the empirical formula is $C_3H_4O_3$.

A drawing of a molecule of ascorbic acid is shown in Fig. 3.2. You can determine by counting the atoms that the molecular formula is $C_6H_8O_6$—exactly double the empirical formula. It is also evident that there is more to know about a molecule than just how many atoms of each kind are present. In ascorbic acid, as in other molecules, the way the atoms are connected together and their arrangement in three-dimensional space are quite important. A picture like Fig. 3.2, showing which atoms are connected to which, is called a **structural formula.** Empirical formulas may be obtained from percent composition or combustion-train experiments, and, if the molecular weight is known, molecular formulas may be determined from the same data. More complicated experiments are required to find structural formulas.

In Example 3.8 we obtained the mass of O by subtracting the masses of C and H from the total mass of sample. This assumed that only C, H, and O were present. Sometimes such an assumption may be incorrect. When penicillin was first isolated and analyzed, the fact that it contained sulfur

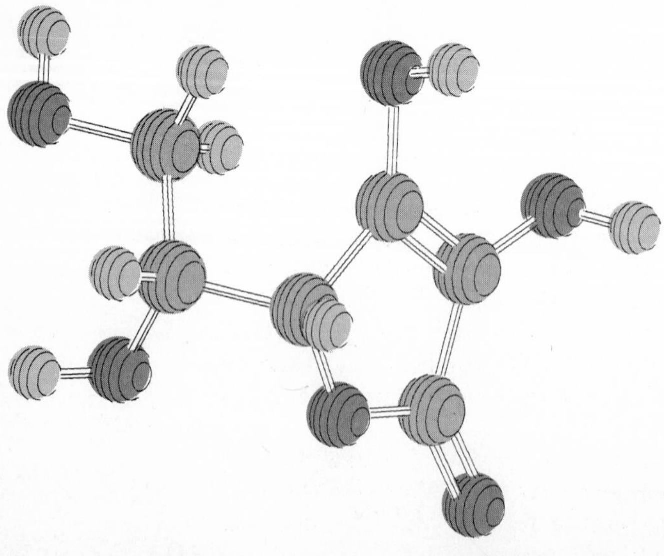

Figure 3.2 The structural formula of ascorbic acid (vitamin C), $C_6H_8O_6$. Carbon atoms are dark gray, hydrogen atoms are light red, and oxygen atoms are dark red. (Computer-generated.) *(Copyright © 1976 by W. G. Davies and J. W. Moore.)*

was missed. This mistake was not discovered for some time because the atomic weight of sulfur is almost exactly twice that of oxygen. Two atoms of oxygen were substituted in place of one sulfur atom in the formula.

3.3 THERMOCHEMISTRY

When a chemical reaction occurs, there is usually a change in temperature of the chemicals themselves and of the beaker or flask in which the reaction is carried out. If the temperature increases, the reaction is **exothermic**—energy is given off as heat when the container and its contents cool back to room temperature. (**Heat** is energy transferred from one place to another solely because of a difference in temperature.) An **endothermic** reaction produces a decrease in temperature. In this case heat is absorbed from the surroundings to return the reaction products to room temperature. **Thermochemistry,** a word derived from the Greek *thermé,* "heat," is the measurement and study of energy transferred as heat when chemical reactions take place. It is extremely important in a technological world where a great deal of work is accomplished by transforming and harnessing heat given off during combustion of coal, oil, and natural gas.

Energy

Energy is usually defined as the capability for doing work. For example, a billiard ball can collide with a second ball, changing the direction or speed of motion of the latter. In such a process the motion of the first ball would also be altered. We would say that one billiard ball did work on (transferred energy to) the other. Energy due to motion is called **kinetic energy** and is represented by E_k. For an object moving in a straight line, the kinetic energy is one-half the product of the mass and the square of the speed:

$$E_k = \tfrac{1}{2}mu^2 \qquad (3.4)$$

where m = mass of the object
u = speed of object

If the two billiard balls mentioned above were studied in outer space, where friction due to their collisions with air molecules or the surface of a pool table would be negligible, careful measurements would reveal that their total kinetic energy would be the same before and after they collided. This is an example of the **law of conservation of energy,** which states that *energy cannot be created or destroyed* under the usual conditions of everyday life. Whenever there appears to be a decrease in energy somewhere, there is a corresponding increase somewhere else.

Even when there is a great deal of friction, the law of conservation of energy still applies. If you put a milkshake on a mixer and leave it there for 10 min, you will have a warm, rather unappetizing drink. The whirling mixer blades do work on (transfer energy to) the milkshake, raising its temperature. The same effect could be produced by heating the milkshake, a fact which suggests that heating also involves a transfer of energy. The first careful experiments to determine how much work was equivalent to a given quantity of heat were done by the English physicist James Joule (1818 to 1889) in the 1840s. In an experiment very similar to our milkshake example, Joule connected falling weights through a pulley system to a paddle wheel immersed in an insulated container of water. This allowed him to compare the temperature rise which resulted from the work done by the weights with that which resulted from heating.

Units with which to measure energy may be derived from the SI base units of Table 1.2 by using Eq. (3.4).

EXAMPLE 3.9 Calculate the kinetic energy of a Volkswagen Beetle of mass 844 kg (1860 lb) which is moving at 13.4 m s^{-1} (30 miles per hour).

Solution

$$E_k = \tfrac{1}{2}mu^2$$
$$= \tfrac{1}{2} \times 844 \text{ kg} \times (13.4 \text{ m s}^{-1})^2$$
$$= 7.58 \times 10^4 \text{ kg m}^2 \text{ s}^{-2}$$

In other words the units for energy are derived from the SI base units kilogram for mass, meter for length, and second for time. A quantity of heat or any other form of energy may be expressed in kilogram meter squared per second squared. In honor of Joule's pioneering work this derived unit 1 kg m^2 s^{-2} is called the **joule,** abbreviated J. The Volkswagen in question could do nearly 76 000 J of work on anything it happened to run into.

Another unit of energy still widely used by chemists is the **calorie.** The calorie used to be defined as the energy needed to raise the temperature of one gram of water from 14.5°C to 15.5°C, but now it is defined as exactly 4.184 J. The Calorie used by dieticians and others for measuring the energy values of foods is actually a kilocalorie, i.e., 4.184 kJ. This nutritional Calorie is distinguished by a capital C in its name.

Thermochemical Equations

Energy changes which accompany chemical reactions are almost always expressed by **thermochemical equations,** such as

$$CH_4(g) + 2O_2(g) \rightarrow CO_2(g) + 2H_2O(g) \quad (25°C, 1 \text{ atm pressure})$$
$$\Delta H_m = -890.4 \text{ kJ mol}^{-1} \qquad (3.5)$$

Here the ΔH_m (delta H subscript m) tells us whether heat energy is released or absorbed when the reaction occurs and also enables us to find the actual quantity of energy involved. By convention, if ΔH_m is *positive,* heat is *absorbed* by the reaction; i.e., it is **endothermic.** More commonly, ΔH_m is *negative* as in Eq. (3.5), indicating that heat energy is *released* rather than absorbed by the reaction, and that the reaction is **exothermic.** This convention as to whether ΔH_m is positive or negative looks at the heat change in terms of the matter actually involved in the reaction rather than its surroundings. In the reaction in Eq. (3.5), the C, H, and O atoms have collectively lost energy and it is this loss which is indicated by a negative value of ΔH_m.

It is important to notice that ΔH_m is not an energy but rather an *energy per unit amount.* Its units are not kilojoules but *kilojoules per mole* (hence the m subscript in ΔH_m). This is necessary because the quantity of heat released or absorbed by a reaction is proportional to the amount of each substance consumed or produced by the reaction. Thus Eq. (3.5) tells us that 890.4 kJ of heat energy is given off *for every mole of CH_4* which is consumed. Alternatively, it tells us that 890.4 kJ is released *for every mole of $2H_2O$* produced, i.e., for every 2 mol H_2O produced. Seen in this way, ΔH_m is a conversion factor enabling us to calculate the heat absorbed when a given amount of substance is consumed or produced. If q is the quantity of heat absorbed and n is the amount of substance involved, then

$$\Delta H_m = \frac{q}{n} \tag{3.6}$$

EXAMPLE 3.10 How much heat energy is obtained when 1 kg of ethane gas, C_2H_6, is burned in oxygen according to the equation:

$$2C_2H_6(g) + 7O_2(g) \rightarrow 4CO_2(g) + 6H_2O(l)$$

$$\Delta H_m = -3083 \text{ kJ mol}^{-1} \tag{3.7}$$

Solution The mass of C_2H_6 is easily converted to the amount of C_2H_6 from which the heat energy q is easily calculated by means of Eq. (3.6). The value of ΔH_m is -3083 kJ per mole of 2 C_2H_6, i.e., per 2 mol C_2H_6. The road map is

$$m_{C_2H_6} \xrightarrow{\mathcal{M}} n_{C_2H_6} \xrightarrow{\Delta H_m} q$$

so that

$$q = 1 \times 10^3 \cancel{\text{g C}_2\text{H}_6} \times \frac{1 \cancel{\text{mol C}_2\text{H}_6}}{30.07 \cancel{\text{g C}_2\text{H}_6}} \times \frac{-3083 \text{ kJ}}{2 \cancel{\text{mol C}_2\text{H}_6}}$$

$$= -51\ 264 \text{ kJ} = -51.26 \text{ MJ}$$

Note: By convention a negative value of q corresponds to a release of heat energy by the matter involved in the reaction.

The quantity ΔH_m is usually referred to as an **enthalpy change per mole.** In this context the symbol Δ (delta) signifies "change in" while H is the symbol for the quantity being changed, namely the enthalpy. We will deal with the enthalpy in some detail in Chap. 15. For the moment we can think of it as a property of matter which increases when matter absorbs energy and decreases when matter releases energy.

It is important to realize that the value of ΔH_m given in thermochemical equations like (3.5) or (3.7) depends on the physical state of both the reactants and the products. Thus, if water were obtained as a liquid instead of a gas in the reaction in Eq. (3.5), the value of ΔH_m would be different from -890.4 kJ mol^{-1}. It is also necessary to specify both the temperature and pressure since the value of ΔH_m depends very slightly on these variables. If these are not specified [as in Eq. (3.7)] they usually refer to 25°C and to normal atmospheric pressure.

Two more characteristics of thermochemical equations arise from the law of conservation of energy. The first is that *writing an equation in the reverse direction changes the sign of the enthalpy change.* For example,

$$H_2O(l) \rightarrow H_2O(g) \qquad \Delta H_m = 44 \text{ kJ mol}^{-1} \tag{3.8a}$$

tells us that when a mole of liquid water vaporizes, 44 kJ of heat is absorbed. This corresponds to the fact that heat is absorbed from your skin when perspiration evaporates, and you cool off. Condensation of 1 mol of water vapor, on the other hand, gives off exactly the same quantity of heat.

$$H_2O(g) \rightarrow H_2O(l) \qquad \Delta H_m = -44 \text{ kJ mol}^{-1} \tag{3.8b}$$

To see why this must be true, suppose that ΔH_m [Eq. (3.8*a*)] = 44 kJ mol^{-1} while ΔH_m [Eq. (3.8*b*)] = -50.0 kJ mol^{-1}. If we took 1 mol of liquid water and allowed it to evaporate, 44 kJ would be absorbed. We could then condense the water vapor, and 50.0 kJ would be given off. We would again have 1 mol of liquid water at 25°C, but we would also have 6 kJ of heat which had been created from nowhere! This would violate the law of conservation of energy. The only way the problem can be avoided is for ΔH_m of the reverse reaction to be equal in magnitude but opposite in sign from ΔH_m of the forward reaction. That is,

$$\Delta H_{m\,\text{forward}} = -\Delta H_{m\,\text{reverse}}$$

Hess' Law

Perhaps the most useful feature of thermochemical equations is that they can be combined to determine ΔH_m values for other chemical reactions. Consider, for example, the following two-step sequence. Step 1 is reaction of 1 mol C(s) and 0.5 mol $O_2(g)$ to form 1 mol CO(g):

$$C(s) + \tfrac{1}{2}O_2(g) \rightarrow CO(g) \qquad \Delta H_m = -110.5 \text{ kJ mol}^{-1} = \Delta H_1$$

(Note that since the equation refers to moles, not molecules, fractional coefficients are permissible.) In step 2 the mole of CO reacts with an additional 0.5 mol O_2, yielding 1 mol CO_2:

$$CO(g) + \tfrac{1}{2}O_2(g) \rightarrow CO_2(g) \qquad \Delta H_m = -283.0 \text{ kJ mol}^{-1} = \Delta H_2$$

The net result of this two-step process is production of 1 mol CO_2 from the original 1 mol C and 1 mol O_2 (0.5 mol in each step). All the CO produced in step 1 is used up in step 2.

On paper this net result can be obtained by *adding* the two chemical equations as though they were algebraic equations. The CO produced is canceled by the CO consumed since it is both a reactant and a product of the overall reaction

$$\begin{array}{l} C(s) + \tfrac{1}{2}O_2(g) \rightarrow CO(g) \\ \tfrac{1}{2}O_2(g) + CO(g) \rightarrow CO_2(g) \\ \hline C(s) + O_2(g) + CO(g) \rightarrow CO(g) + CO_2(g) \end{array}$$

Experimentally it is found that the enthalpy change for the net reaction is the *sum* of the enthalpy changes for steps 1 and 2:

$$\Delta H_{net} = -110.5 \text{ kJ mol}^{-1} + (-283.0 \text{ kJ mol}^{-1}) = -393.5 \text{ kJ mol}^{-1}$$
$$= \Delta H_1 + \Delta H_2$$

That is, the thermochemical equation

$$C(s) + O_2(g) \rightarrow CO_2(g) \qquad \Delta H_m = -393.5 \text{ kJ mol}^{-1}$$

is the correct one for the overall reaction.

In the general case it is always true that *whenever two or more chemical equations can be added algebraically to give a net reaction, their enthalpy changes may also be added to give the enthalpy change of the net reaction.* This principle is known as **Hess' law.** If it were not true, it would be possible to think up a series of reactions in which energy would be created but which would end up with exactly the same substances we started with. This would contradict the law of conservation of energy.

Hess' law enables us to obtain ΔH_m values for reactions which cannot be carried out experimentally, as the next example shows.

EXAMPLE 3.11 Acetylene (C_2H_2) cannot be prepared directly from its elements according to the equation

$$2C(s) + H_2(g) \rightarrow C_2H_2(g) \qquad (3.9)$$

Calculate ΔH_m for this reaction from the following thermochemical equations, all of which can be determined experimentally:

$$C(s) + O_2(g) \rightarrow CO_2(g) \qquad \Delta H_m = -393.5 \text{ kJ mol}^{-1} \quad (3.10a)$$

$$H_2(g) + \tfrac{1}{2}O_2(g) \rightarrow H_2O(l) \qquad \Delta H_m = -285.8 \text{ kJ mol}^{-1} \quad (3.10b)$$

$$C_2H_2(g) + \tfrac{5}{2}O_2(g) \rightarrow 2CO_2(g) + H_2O(l) \qquad \Delta H_m = -1299.8 \text{ kJ mol}^{-1} \quad (3.10c)$$

Solution We use the following strategy to manipulate the three experimental equations so that when added they yield Eq. (3.9):

a) Since Eq. (3.9) has *2* mol C on the left, we multiply Eq. (3.10*a*) by 2.

b) Since Eq. (3.9) has *1* mol H_2 on the left, we leave Eq. (3.10*b*) unchanged.

c) Since Eq. (3.9) has *1* mol C_2H_2 on the *right,* whereas there is *1* mol C_2H_2 on the *left* of Eq. (3.10*c*), we write Eq. (3.10*c*) in reverse.

We then have

$$2C(s) + 2O_2(g) \rightarrow 2CO_2(g) \qquad \Delta H_m = 2(-393.5) \text{ kJ mol}^{-1}$$

$$H_2(g) + \tfrac{1}{2}O_2(g) \rightarrow H_2O(l) \qquad \Delta H_m = -285.8 \text{ kJ mol}^{-1}$$

$$2CO_2(g) + H_2O(l) \rightarrow C_2H_2(g) + \tfrac{5}{2}O_2(g) \qquad \Delta H_m = -(-1299.8) \text{ kJ mol}^{-1}$$

$$2C(s) + H_2(g) + 2\tfrac{1}{2}O_2(g) \rightarrow C_2H_2(g) + \tfrac{5}{2}O_2(g)$$

$$\Delta H_m = (-787.0 - 285.8 + 1299.8) \text{ kJ mol}^{-1} = 227.0 \text{ kJ mol}^{-1}$$

Thus the desired result is

$$2C(s) + H_2(g) \rightarrow C_2H_2(g) \qquad \Delta H_m = 227.0 \text{ kJ mol}^{-1}$$

3.4 STANDARD ENTHALPIES OF FORMATION

By now chemists have measured the enthalpy changes for so many reactions that it would take several large volumes to list all the thermochemical equations. Fortunately Hess' law makes it possible to list a single value, the **standard enthalpy of formation** ΔH_f, for each compound. The standard enthalpy of formation is the enthalpy change when 1 mol of a pure substance is formed from its elements. Each element must be in the physical and chemical form which is most stable at normal atmospheric pressure and a specified temperature (usually 25°C).

For example, if we know that $\Delta H_f[H_2O(l)] = -285.8 \text{ kJ mol}^{-1}$, we can immediately write the thermochemical equation

$$H_2(g) + \tfrac{1}{2}O_2(g) \rightarrow H_2O(l) \qquad \Delta H_m = -285.8 \text{ kJ mol}^{-1} \qquad (3.11)$$

The elements H and O appear as diatomic molecules and in gaseous form because these are their most stable chemical and physical states. Note also that 285.8 kJ are given off *per mole* of $H_2O(l)$ formed. Equation (3.11) must specify formation of *1 mol* $H_2O(l)$, and so the coefficient of O_2 must be $\tfrac{1}{2}$.

In some cases, such as that of water, the elements will react directly to form a compound, and measurement of the heat absorbed serves to determine ΔH_f. Quite often, however, elements do not react directly with each other to form the desired compound, and ΔH_f must be calculated by combining the enthalpy changes for other reactions. A case in point is the gas acetylene, C_2H_2. In Example 3.11 we used Hess' law to show that the thermochemical equation

$$2C(s) + H_2(g) \rightarrow C_2H_2(g) \qquad \Delta H_m = 227.0 \text{ kJ mol}^{-1}$$

TABLE 3.1 Some Standard Enthalpies of Formation at 25°C.

Compound	ΔH_f/kJ mol^{-1}	ΔH_f/kcal mol^{-1}	Compound	ΔH_f/kJ mol^{-1}	ΔH_f/kcal mol^{-1}
$AgCl(s)$	−127.0	−30.35	$H_2O(g)$	−241.8	−57.79
$AgN_3(s)$	+310.3	+66.78	$H_2O(l)$	−285.8	−68.31
$Ag_2O(s)$	−31.0	−7.41	$H_2O_2(l)$	−187.6	−44.84
$Al_2O_3(s)$	−1675.7	−400.50	$H_2S(g)$	−20.2	−4.83
$Br_2(l)$	0.0	0.00	$HgO(s)$	−90.8	−21.70
$Br_2(g)$	+31.0	+7.41	$I_2(s)$	0.0	0.00
$C(s)$, graphite	0.0	0.00	$I_2(g)$	+62.3	+14.89
$C(s)$, diamond	+1.9	+0.45	$KCl(s)$	−435.9	−104.18
$CH_4(g)$	−74.8	−17.88	$KBr(s)$	−392.0	−93.69
$CO(g)$	−110.5	−26.41	$MgO(s)$	−601.8	−143.83
$CO_2(g)$	−393.5	−94.05	$NH_3(g)$	−46.1	−11.02
$C_2H_2(g)$	+226.9	+54.23	$NO(g)$	+90.3	+21.58
$C_2H_4(g)$	+52.6	+12.57	$NO_2(g)$	+33.8	+8.08
$C_2H_6(g)$	−84.5	−20.20	$N_2O_4(g)$	+9.2	+2.20
$C_6H_6(l)$	+49.1	+11.74	$NF_3(g)$	−124.7	−29.80
$CaO(s)$	−635.5	−151.89	$NaBr(s)$	−359.9	−86.02
$CaCO_3(s)$	−1207.8	−288.67	$NaCl(s)$	−411.0	−98.23
$CuO(s)$	−157.3	−37.60	$O_3(g)$	+142.7	+34.11
$Fe_2O_3(s)$	−822.2	−196.51	$SO_2(g)$	−296.8	−70.94
$HBr(g)$	−36.4	−8.70	$SO_3(g)$	−395.8	−94.60
$HCl(g)$	−92.3	−22.06	$ZnO(s)$	−348.1	−83.20
$HI(g)$	+26.4	+6.31			

was valid. Since it involves 1 mol C_2H_2 and the elements are in their most stable forms, we can say that $\Delta H_f[C_2H_2(g)] = 227.0$ kJ mol^{-1}.

One further point arises from the definition of ΔH_f. *The standard enthalpy of formation for an element in its most stable state must be zero.* If we form mercury from its elements, for example, we are talking about the reaction

$$Hg(l) \rightarrow Hg(l)$$

Since the mercury is unchanged, there can be no enthalpy change, and $\Delta H_f = 0$ kJ mol^{-1}.

Standard enthalpies of formation for some common compounds are given in Table 3.1. These values may be used to calculate ΔH_m for any chemical reaction so long as all the compounds involved appear in the tables. To see how and why this may be done, consider the following example.

EXAMPLE 3.12 Use standard enthalpies of formation to calculate ΔH_m for the reaction

$$2CO(g) + O_2(g) \rightarrow 2CO_2(g)$$

Solution We can imagine that the reaction takes place in two steps, each of which involves only a standard enthalpy of formation. In the first step CO (carbon monoxide) is decomposed to its elements:

$$2CO(g) \rightarrow 2C(s) + O_2(g) \qquad \Delta H_m = \Delta H_1 \qquad (3.12)$$

Since this is the *reverse* of formation of 2 mol CO from its elements, the enthalpy change is

$$\Delta H_1 = 2 \times \{-\Delta H_f[CO(g)]\} = 2 \times [-(-110.5 \text{ kJ mol}^{-1})]$$
$$= +221.0 \text{ kJ mol}^{-1}$$

In the second step the elements are combined to give 2 mol CO_2 (carbon dioxide):

$$2C(s) + 2O_2(g) \rightarrow 2CO_2(g) \qquad \Delta H_m = \Delta H_2 \qquad (3.13)$$

In this case

$$\Delta H_2 = 2 \times \Delta H_f[CO_2(g)] = 2 \times (-393.5 \text{ kJ mol}^{-1})$$
$$= -787.0 \text{ kJ mol}^{-1}$$

You can easily verify that the sum of Eqs. (3.12) and (3.13) is

$$2CO(g) + O_2(g) \rightarrow 2CO_2(g) \qquad \Delta H_m = \Delta H_{net}$$

Therefore

$$\Delta H_{net} = \Delta H_1 + \Delta H_2$$
$$= 221.0 \text{ kJ mol}^{-1} - 393.5 \text{ kJ mol}^{-1} = -172.5 \text{ kJ mol}_{-1}$$

Note carefully how Example 3.12 was solved. In step 1 the *reactant* compound CO(*g*) was hypothetically decomposed to its elements. This equation was the reverse of formation of the compound, and so ΔH_1 was opposite in sign from ΔH_f. Step 1 also involved 2 mol CO(*g*), and so the enthalpy change had to be doubled. In step 2 we had the hypothetical formation of the *product* $CO_2(g)$ from its elements. Since 2 mol were obtained, the enthalpy change was doubled but its sign remained the same.

Any chemical reaction can be approached similarly. To calculate ΔH_m we *add* all the ΔH_f values for the products, multiplying each by the appropriate coefficient, as in step 2 above. Since the signs of ΔH_f values for the reactants had to be reversed in step 1, we *subtract* them, again multiplying by appropriate coefficients. This can be summarized by the equation

$$\Delta H_m = \Sigma\Delta H_f(\text{products}) - \Sigma\Delta H_f(\text{reactants}) \qquad (3.14)$$

The symbol Σ means "the sum of." Since ΔH_f values are given *per mole* of compound, you must be sure to multiply each ΔH_f by an appropriate coefficient derived from the equation for which ΔH_m is being calculated.

EXAMPLE 3.13 Use Table 3.1 to calculate ΔH_m for the reaction

$$4NH_3(g) + 5O_2(g) \rightarrow 6H_2O(g) + 4NO(g)$$

Solution Using Eq. (3.14), we have

$$\begin{aligned}\Delta H_m &= \Sigma\Delta H_f(\text{products}) - \Sigma\Delta H_f(\text{reactants})\\ &= [6\ \Delta H_f(H_2O) + 4\ \Delta H_f(NO)] - [4\ \Delta H_f(NH_3) + 5\ \Delta H_f(O_2)]\\ &= 6(-241.8)\text{ kJ mol}^{-1} + 4(90.3)\text{ kJ mol}^{-1}\\ &\qquad - 4(-46.1)\text{ kJ mol}^{-1} - 5 \times 0\\ &= -1450.8\text{ kJ mol}^{-1} + 361.2\text{ kJ mol}^{-1} + 184.4\text{ kJ mol}^{-1}\\ &= -905.2\text{ kJ mol}^{-1}\end{aligned}$$

Note that we were careful to use $\Delta H_f[H_2O(g)]$, not $\Delta H_f[H_2O(l)]$. Even though water vapor is not the most stable form of water at 25°C, we can still use its ΔH_f value. Also the standard enthalpy of formation of the element $O_2(g)$ is zero *by definition*. Obviously it would be a waste of space to include it in Table 3.1.

3.5 SOLUTIONS

In the laboratory, in your body, and in the outside environment, the majority of chemical reactions take place in solutions. Macroscopically a **solution** is defined as a homogeneous mixture of two or more substances, that is, a mixture which appears to be uniform throughout. On the microscopic scale a solution involves the random arrangement of one kind of atom or molecule with respect to another.

There are a number of reasons why solutions are so often encountered both in nature and in the laboratory. The most common type of solution involves a liquid **solvent** which dissolves a solid **solute.** (The term solvent usually refers to the substance present in greatest amount. There may be more than one solute dissolved in it.) Because a liquid adopts the shape of its container but does not expand to fill all space available to it, liquid solutions are convenient to handle. You can easily pour them from one container to another, and their volumes are readily measured using graduated cylinders, pipets, burets, volumetric flasks, or other laboratory glassware. Moreover, atoms or molecules of solids dissolved in a liquid are close together but still able to move past one another. They contact each other more frequently than if two solids were placed next to each other. This "intimacy" in liquid solutions often facilitates chemical reactions.

Concentration

Since solutions offer a convenient medium for carrying out chemical reactions, it is often necessary to know how much of one solution will react with a given quantity of another. Examples earlier in this chapter have shown that the amount of substance is the quantity which determines how much of one material will react with another. The ease with which solution volumes may be measured suggests that it would be very convenient to know the amount of substance dissolved per unit volume of solution. Then by measuring a certain volume of solution, we would also be measuring a certain amount of substance.

The **concentration** c of a substance in a solution (often called **molarity**) is the *amount of the substance per unit volume of solution:*

$$\text{Concentration of solute} = \frac{\text{amount of solute}}{\text{volume of solution}} \qquad c_{\text{solute}} = \frac{n_{\text{solute}}}{V_{\text{solution}}} \tag{3.15}$$

Usually the units moles per cubic decimeter (mol dm^{-3}) or moles per liter (mol $liter^{-1}$) are used to express concentration.

If a pure substance is soluble in water, it is easy to prepare a solution of known concentration. A container with a sample of the substance is weighed accurately, and an appropriate mass of sample is poured through a funnel into a volumetric flask, as shown in Fig. 3.3. The container is then reweighed. Any solid adhering to the funnel is rinsed into the flask, and water is added until the flask is about three-quarters full. After swirling the flask to dissolve the solid, water is added carefully until the bottom of the meniscus coincides with the calibration mark on the neck of the flask.

EXAMPLE 3.14 A solution of KI was prepared as described above. The initial mass of the container plus KI was 43.2874 g, and the final mass after pouring was 30.1544 g. The volume of the flask was 250.00 ml. What is the concentration of the solution?

Solution The concentration can be calculated by dividing the amount of solute by the volume of solution [Eq. (3.15)]:

$$c_{\text{KI}} = \frac{n_{\text{KI}}}{V}$$

We obtain n_{KI} from the mass of KI added to the flask:

$$m_{\text{KI}} = 43.2874 \text{ g} - 30.1544 \text{ g} = 13.1330 \text{ g}$$

$$n_{\text{KI}} = 13.1330 \text{ g} \times \frac{1 \text{ mol}}{166.00 \text{ g}} = 7.9114 \times 10^{-2} \text{ mol}$$

The volume of solution is 250.00 ml, or

$$V_{\text{solution}} = 250.00 \text{ cm}^3 \times \frac{1 \text{ dm}^3}{10^3 \text{ cm}^3} = 2.5000 \times 10^{-1} \text{ dm}^3$$

Thus

$$c_{\text{KI}} = \frac{n_{\text{KI}}}{V_{\text{solution}}} = \frac{7.9114 \times 10^{-2} \text{ mol}}{2.5000 \times 10^{-1} \text{ dm}^3} = 3.1646 \times 10^{-1} \text{ mol dm}^{-3}$$

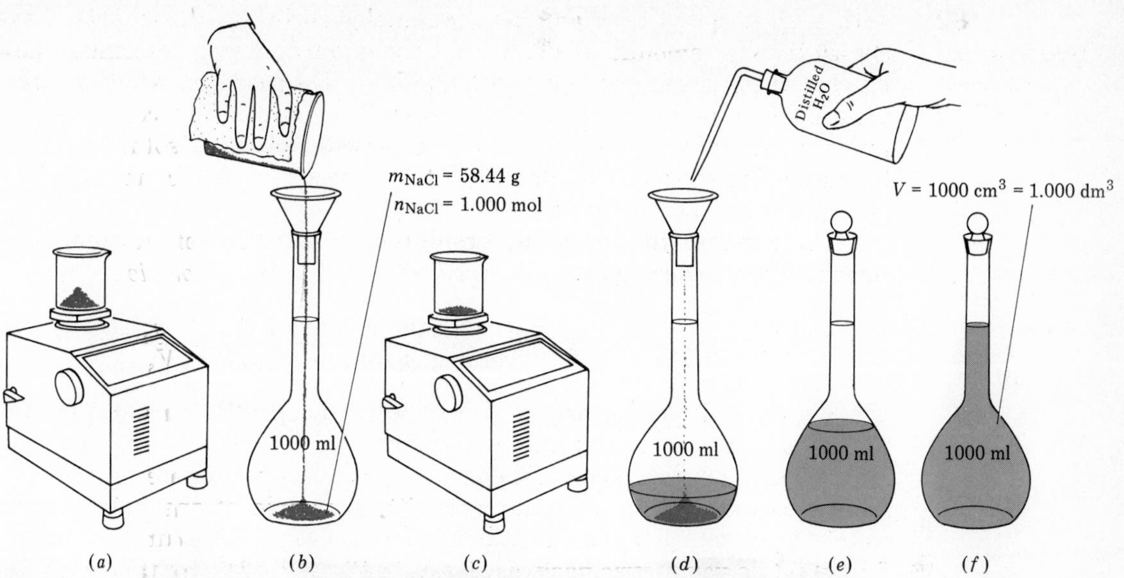

Figure 3.3 The preparation of a sodium chloride solution of concentration 1.000 mol dm^{-3}. (*a*) A sample of pure solid NaCl is first weighed with its container. (*b*) Most of the sample is now transferred into the volumetric flask. (*c*) The container and remaining solid NaCl are again weighed. Subtraction yields m_{NaCl} = 58.44 g. Thus n_{NaCl} = 1.000 mol. (*d*) Any solid remaining in the funnel is washed down into the body of the flask. (*e*) The flask is now filled to about 80 percent capacity and the solid allowed to dissolve. The flask is shaken to achieve a uniform solution. (*f*) Solvent is now added carefully until the bottom of the meniscus coincides with the mark on the flask. The flask is again shaken to achieve a uniform solution. Since the volume of the solution is 1.000 dm³, the concentration is 1.000 mol/1.000 dm³ = 1.000 mol dm^{-3}.

Note that the definition of concentration is entirely analogous to the definitions of density, molar mass, and stoichiometric ratio that we have previously encountered. Concentration will serve as a conversion factor relating the volume of solution to the amount of dissolved solute.

$$\text{Volume of solution} \xleftrightarrow{\text{concentration}} \text{amount of solute} \qquad V \overset{c}{\longleftrightarrow} n$$

Because the volume of a liquid can be measured quickly and easily, concentration is a much-used quantity. The next two examples show how this conversion factor may be applied to commonly encountered solutions in which water is the solvent (**aqueous** solutions).

EXAMPLE 3.15 An aqueous solution of HCl [represented or written HCl(*aq*)] has a concentration of 0.1396 mol dm^{-3}. If 24.71 cm³ (24.71 ml) of this solution is delivered from a buret, what amount of HCl has been delivered?

Solution Using concentration as a conversion factor, we have

$$V \xrightarrow{c} n$$

$$n_{\text{HCl}} = 24.71\ \text{cm}^3 \times \frac{0.1396\ \text{mol}}{1\ \text{dm}^3}$$

The volume units will cancel if we supply a unity factor to convert cubic centimeters to cubic decimeters:

$$\begin{aligned} n_{\text{HCl}} &= 24.71\ \text{cm}^3 \times \frac{0.1396\ \text{mol}}{1\ \text{dm}^3} \times \left(\frac{1\ \text{dm}}{10\ \text{cm}}\right)^3 \\ &= 24.71\ \cancel{\text{cm}^3} \times \frac{0.1396\ \text{mol}}{1\ \cancel{\text{dm}^3}} \times \frac{1\ \cancel{\text{dm}^3}}{10^3\ \cancel{\text{cm}^3}} \\ &= 0.003\ 450\ \text{mol} \end{aligned}$$

The concentration units of moles per cubic decimeter are often abbreviated M, pronounced *molar*. That is, a 0.1-M (one-tenth molar) solution contains 0.1 mol solute per cubic decimeter of solution. This abbreviation is very convenient for labeling laboratory bottles and for writing textbook problems; however, when doing calculations, it is difficult to see that

$$1\ \text{dm}^3 \times 1\ M = 1\ \text{mol}$$

Therefore we recommend that you *always write the units in full* when doing any calculations involving solution concentrations. That is,

$$1\ \cancel{\text{dm}^3} \times 1\ \frac{\text{mol}}{\cancel{\text{dm}^3}} = 1\ \text{mol}$$

Problems such as Example 3.15 are easier for some persons to solve if the solution concentration is expressed in millimoles per cubic centimeter (mmol cm^{-3}) instead of moles per cubic decimeter. Since the SI prefix m means 10^{-3}, 1 mmol = 10^{-3} mol, and

$$\frac{1\ \text{mol}}{1\ \text{dm}^3} = \frac{1\ \cancel{\text{mol}}}{1\ \cancel{\text{dm}^3}} \times \frac{1\ \cancel{\text{dm}^3}}{\cancel{10^3}\ \text{cm}^3} \times \frac{1\ \text{mmol}}{\cancel{10^{-3}}\ \cancel{\text{mol}}} = \frac{1\ \text{mmol}}{1\ \text{cm}^3}$$

Thus a concentration of 0.1396 mol dm^{-3} (0.1396 M) can also be expressed as 0.1396 mmol cm^{-3}. Expressing the concentration this way is very convenient when dealing with laboratory glassware calibrated in milliliters or cubic centimeters.

EXAMPLE 3.16 Exactly 25.0 ml NaOH solution whose concentration is 0.0974 M was delivered from a pipet. (a) What amount of NaOH was present? (b) What mass of NaOH would remain if all the water evaporated?

Solution

a) Since 0.0974 *M* means 0.0974 mol dm^{-3}, or 0.0974 mmol cm^{-3}, we choose the latter, more convenient quantity as a conversion factor:

$$n_{\text{NaOH}} = 25.0 \text{ cm}^3 \times \frac{0.0974 \text{ mmol}}{1 \text{ cm}^3} = 2.44 \text{ mmol}$$
$$= 2.44 \times 10^{-3} \text{ mol}$$

b) Using molar mass, we obtain

$$m_{\text{NaOH}} = 2.44 \times 10^{-3} \text{ mol} \times \frac{40.01 \text{ g}}{1 \text{ mol}} = 9.76 \times 10^{-2} \text{ g}$$

Note: The symbols n_{NaOH} and m_{NaOH} refer to the amount and mass of the *solute* NaOH, respectively. They do *not* refer to the solution. If we wanted to specify the mass of aqueous NaOH solution, the symbol $m_{\text{NaOH}(aq)}$ could be used.

Diluting and Mixing Solutions

Often it is convenient to prepare a series of solutions of known concentrations by first preparing a single **stock solution** as described in Example 3.14. **Aliquots** (carefully measured volumes) of the stock solution can then be diluted to any desired volume. In other cases it may be inconvenient to weigh accurately a small enough mass of sample to prepare a small volume of a dilute solution. Each of these situations requires that a solution be diluted to obtain the desired concentration.

EXAMPLE 3.17 A pipet is used to measure 50.0 ml of 0.1027 *M* HCl into a 250.00-ml volumetric flask. Distilled water is carefully added up to the mark on the flask. What is the concentration of the diluted solution?

Solution To calculate concentration, we first obtain the amount of HCl in the 50.0 ml (50.0 cm^3) of solution added to the volumetric flask:

$$n_{\text{HCl}} = 50.0 \text{ cm}^3 \times \frac{0.1027 \text{ mmol}}{1 \text{ cm}^3} = 5.14 \text{ mmol}$$

Dividing by the new volume gives the concentration

$$c_{\text{HCl}} = \frac{n_{\text{HCl}}}{V} = \frac{5.14 \text{ mmol}}{250.00 \text{ cm}^3} = 0.0205 \text{ mmol cm}^{-3}$$

Thus the new solution is 0.0205 *M*.

EXAMPLE 3.18 What volume of the solution prepared in Example 3.14 would be required to make 50.00 ml of 0.0500 *M* KI?

Solution Using the volume and concentration of the desired solution, we can calculate the amount of KI required. Then the concentration of the original solution (0.316 46 *M*) can be used to convert that amount of KI to the necessary volume. Schematically

$$V_{\text{new}} \xrightarrow{c_{\text{new}}} n_{\text{KI}} \xrightarrow{c_{\text{old}}} V_{\text{old}}$$

$$V_{\text{old}} = 50.00 \, \cancel{\text{cm}^3} \times \frac{0.0500 \, \cancel{\text{mmol}}}{1 \, \cancel{\text{cm}^3}} \times \frac{1 \text{ cm}^3}{0.316\ 46 \, \cancel{\text{mmol}}} = 7.90 \text{ cm}^3$$

Thus we should dilute a 7.90-ml aliquot of the stock solution to 50.00 ml. This could be done by measuring 7.90 ml from a buret into a 50.00-ml volumetric flask and adding water up to the mark.

Titrations

When solutions are used quantitatively in the laboratory, titration is usually involved. **Titration** is a technique used to determine the volume of one solution necessary to consume exactly some reactant in another solu-

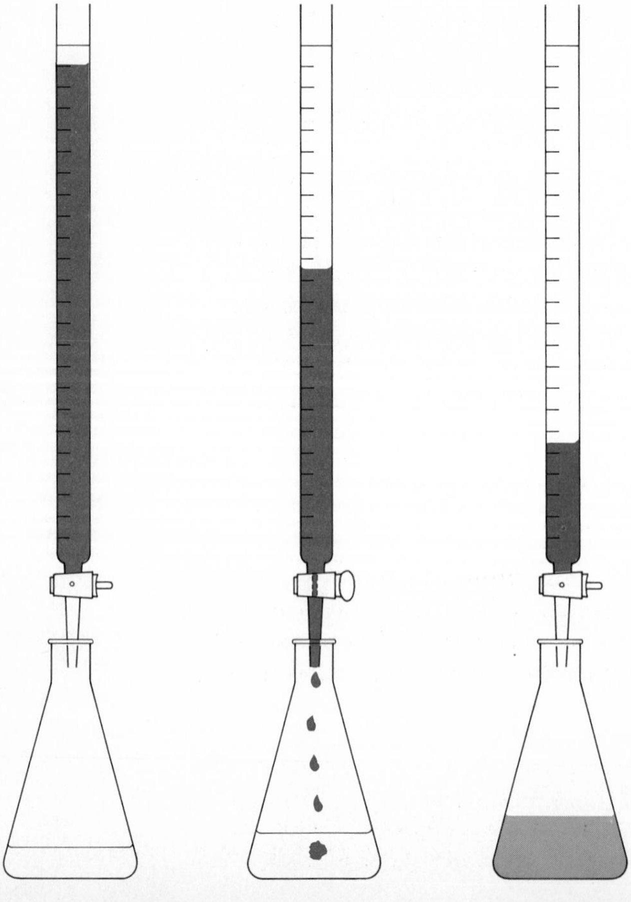

Figure 3.4 The technique of titration.

tion. As shown in Fig. 3.4, a measured volume of the solution to be titrated is placed in a flask or other container. The **titrant** (the solution to be added) is placed in a buret. The volume of titrant added can be determined by reading the level of liquid in the buret before and after titration. This reading can usually be estimated to the nearest hundredth of a milliliter, for example, 25.62 ml.

In Fig. 3.4 the solution to be titrated is colorless aqueous hydrogen peroxide, $H_2O_2(aq)$, which contains excess sulfuric acid, $H_2SO_4(aq)$. The titrant is purple-colored potassium permanganate solution, $KMnO_4(aq)$. The reaction which occurs is

$$2KMnO_4(aq) + 5H_2O_2(aq) + 3H_2SO_4(aq) \rightarrow 2MnSO_4(aq) + 5O_2(g) + K_2SO_4(aq) + 8H_2O(l) \qquad (3.16)$$

As the first few cubic centimeters of titrant flow into the flask, there is a large excess of H_2O_2. The limiting reagent $KMnO_4$ is entirely consumed, and its purple color disappears almost as soon as it is added. Eventually, though, all the H_2O_2 is consumed. Addition of just one more drop of titrant produces a lasting pink color due to unreacted $KMnO_4$ in the flask. This indicates that all the H_2O_2 has been consumed and is called the **endpoint** of the titration. If more $KMnO_4$ solution were added, there would be an excess of $KMnO_4$ and the color of the solution in the flask would get much darker. The darker color would show that we had overtitrated, or overshot the endpoint, by adding more than enough $KMnO_4$ to react with all the H_2O_2.

In the titration we have just described, the intense purple color of permanganate indicates the endpoint. Usually, however, it is necessary to add an **indicator**—a substance which combines with excess titrant to produce a visible color or to form an insoluble substance which would precipitate from solution. No matter what type of reaction occurs or how the endpoint is detected, however, the object of a titration is always to add just the amount of titrant needed to consume *exactly* the amount of substance being titrated. In the $KMnO_4$–H_2O_2 reaction [Eq. (3.16)], the endpoint occurs when exactly 2 mol $KMnO_4$ have been added from the buret for every 5 mol H_2O_2 originally in the titration flask. That is, at the endpoint the ratio of the amount of $KMnO_4$ added to the amount of H_2O_2 consumed must equal the stoichiometric ratio

$$\frac{n_{KMnO_4}\text{(added from buret)}}{n_{H_2O_2}\text{(initially in flask)}} = S\left(\frac{KMnO_4}{H_2O_2}\right) = \frac{2 \text{ mol } KMnO_4}{5 \text{ mol } H_2O_2} \qquad (3.17)$$

When the endpoint has been reached in any titration, the ratio of the amounts of substance of the two reactants is equal to the stoichiometric ratio obtained from the appropriate balanced chemical equation. Therefore we can use the stoichiometric ratio to convert from the amount of one substance to the amount of another.

EXAMPLE 3.19 What volume of 0.053 86 *M* $KMnO_4$ would be needed to reach the endpoint when titrating 25.00 ml of 0.1272 *M* H_2O_2?

Solution At the endpoint, Eq. (3.17) will apply, and we can use it to calculate the amount of $KMnO_4$ which must be added:

$$n_{KMnO_4}\text{ (added)} = n_{H_2O_2}\text{ (in flask)} \times S\left(\frac{KMnO_4}{H_2O_2}\right)$$

The amount of H_2O_2 is obtained from the volume and concentration:

$$n_{H_2O_2}\text{(in flask)} = 25.00\ \cancel{\text{cm}^3} \times 0.1272\ \frac{\text{mmol}}{\cancel{\text{cm}^3}}$$

$$= 3.180\text{ mmol } H_2O_2$$

Then

$$n_{KMnO_4}\text{(added)} = 3.180\text{ mmol } H_2O_2 \times \frac{2\text{ mol } KMnO_4}{5\text{ mol } H_2O_2} \times \frac{10^{-3}}{10^{-3}}$$

$$= 3.180\ \cancel{\text{mmol } H_2O_2} \times \frac{2\text{ mmol } KMnO_4}{5\ \cancel{\text{mmol } H_2O_2}}$$

$$= 1.272\text{ mmol } KMnO_4$$

To obtain $V_{KMnO_4(aq)}$, we use the concentration as a conversion factor:

$$V_{KMnO_4(aq)} = 1.272\ \cancel{\text{mmol } KMnO_4} \times \frac{1\text{ cm}^3}{5.386 \times 10^{-2}\ \cancel{\text{mmol } KMnO_4}}$$

$$= 23.62\text{ cm}^3$$

Note that overtitrating [adding more than 23.62 cm^3 of $KMnO_4(aq)$] would involve an excess (more than 1.272 mmol) of $KMnO_4$.

Titration is often used to determine the concentration of a solution. In many cases it is not a simple matter to obtain a pure substance, weigh it accurately, and dissolve it in a volumetric flask as was done in Example 3.14. NaOH, for example, combines rapidly with H_2O and CO_2 from the air, and so even a freshly prepared sample of solid NaOH will not be pure. Its weight would change continuously as $CO_2(g)$ and $H_2O(g)$ were absorbed. Hydrogen chloride (HCl) is a gas at ordinary temperatures and pressures, making it very difficult to handle or weigh. Aqueous solutions of both of these substances must be **standardized;** that is, their concentrations must be determined by titration.

EXAMPLE 3.20 A sample of pure potassium hydrogen phthalate ($KHC_8H_4O_4$) weighing 0.3421 g is dissolved in distilled water. Titration of the sample requires 27.03 ml $NaOH(aq)$. The titration reaction is

$$NaOH(aq) + KHC_8H_4O_4(aq) \rightarrow NaKC_8H_4O_4(aq) + H_2O$$

What is the concentration of $NaOH(aq)$?

Solution To calculate concentration, we need to know the amount of NaOH and the volume of solution in which it is dissolved. The former quantity could be obtained via a stoichiometric ratio from the amount of $KHC_8H_4O_4$, and that amount can be obtained from the mass

$$m_{KHC_8H_4O_4} \xrightarrow{M_{KHC_8H_4O_4}} n_{KHC_8H_4O_4} \xrightarrow{S(NaOH/KHC_8H_4O_4)} n_{NaOH}$$

$$n_{NaOH} = 0.3421 \text{ g} \times \frac{1 \text{ mol } KHC_8H_4O_4}{204.22 \text{ g}} \times \frac{1 \text{ mol NaOH}}{1 \text{ mol } KHC_8H_4O_4}$$

$$= 1.675 \times 10^{-3} \text{ mol NaOH} = 1.675 \text{ mmol NaOH}$$

The concentration is then

$$c_{NaOH} = \frac{n_{NaOH}}{V} = \frac{1.675 \text{ mmol NaOH}}{27.03 \text{ cm}^3} = 0.06197 \text{ mmol cm}^{-3}$$

or $0.06197\ M$.

By far the most common use of titrations is in determining unknowns, that is, in determining the concentration or amount of substance in a sample about which we initially knew nothing. The next example involves an unknown that many persons encounter every day.

EXAMPLE 3.21 Vitamin C tablets contain ascorbic acid ($C_6H_8O_6$) and a starch "filler" which holds them together. To determine how much vitamin C is present, a tablet can be dissolved in water and titrated with sodium hydroxide solution, NaOH(*aq*). The equation is

$$C_6H_8O_6(aq) + NaOH(aq) \rightarrow NaC_6H_7O_6(aq) + H_2O(l)$$

If titration of a dissolved vitamin C tablet requires 16.85 cm³ of 0.1038 *M* NaOH, how accurate is the claim on the label of the bottle that each tablet contains 300 mg of vitamin C?

Solution The known volume and concentration allow us to calculate the amount of NaOH(*aq*) which reacted with all the vitamin C. Using the stoichiometric ratio

$$S\left(\frac{C_6H_8O_6}{NaOH}\right) = \frac{1 \text{ mmol } C_6H_8O_6}{1 \text{ mmol NaOH}}$$

we can obtain the amount of $C_6H_8O_6$. The molar mass converts that amount to a mass which can be compared with the label. Schematically

$$V_{NaOH} \xrightarrow{c_{NaOH}} n_{NaOH} \xrightarrow{S(C_6H_8O_6/NaOH)} n_{C_6H_8O_6} \xrightarrow{\mathcal{M}_{C_6H_8O_6}} m_{C_6H_8O_6}$$

$$m_{C_6H_8O_6} = 16.85\ \text{cm}^3 \times \frac{0.1038\ \text{mmol NaOH}}{1\ \text{cm}^3} \times \frac{1\ \text{mmol}\ C_6H_8O_6}{1\ \text{mmolNaOH}} \times \frac{176.1\ \text{mg}}{\text{mmol}\ C_6H_8O_6} = 308.0\ \text{mg}$$

Note that the molar mass of $C_6H_8O_6$

$$\frac{176.1\ \text{g}}{1\ \text{mol}\ C_6H_8O_6} = \frac{176.1\ \text{g}}{1\ \text{mol}\ C_6H_8O_6} \times \frac{10^{-3}}{10^{-3}} = \frac{176.1 \times 10^{-3}\ \text{g}}{10^{-3}\ \text{mol}\ C_6H_8O_6} = \frac{176.1\ \text{mg}}{1\ \text{mmol}\ C_6H_8O_6}$$

can be expressed in milligrams per millimole as well as in grams per mole.

The 308.0 mg obtained in this example is in reasonably close agreement with the manufacturer's claim of 300 mg. The tablets are stamped out by machines, not weighed individually, and so some variation is expected.

SUMMARY

This chapter has been concerned with the amounts of substances which participate in chemical reactions, with the quantities of heat given off or absorbed when reactions occur, and with the volumes of solutions which react exactly with one another. These seemingly unrelated subjects have been discussed together because many of the calculations involving them are almost identical in form. The same is true of the density calculations described in Chap. 1 (Secs. 1.3 and 1.4) and of the cal-

TABLE 3.2 Summary of Related Quantities and Conversion Factors.

Related Quantities	Conversion Factor	Definition	Road Map
Volume ↔ mass	Density, ρ	$\rho = \frac{m}{V}$	$V \xleftrightarrow{\rho} m$
Amount of substance ↔ mass	Molar mass, $\mathcal{M}$	$\mathcal{M} = \frac{m}{n}$	$n \xleftrightarrow{\mathcal{M}} m$
Amount of substance ↔ number of particles	Avogadro constant, N_A	$N_A = \frac{N}{n}$	$n \xleftrightarrow{N_A} N$
Amount of X consumed or produced ↔ amount of Y consumed or produced	Stoichiometric ratio, $S(Y/X)$	$S(Y/X) = \frac{n_Y}{n_X}$	$n_X \xleftrightarrow{S(Y/X)} n_Y$
Amount of X consumed or produced ↔ quantity of heat absorbed during reaction	ΔH_m for thermochemical equation	$\Delta H_m = \frac{q}{n_X}$	$n_X \xleftrightarrow{\Delta H_m} q$
Volume of solution ↔ amount of solute	Concentration of solute, c_X	$c_X = \frac{n_X}{V}$	$V \xleftrightarrow{c_X} n_X$

culations involving molar mass and the Avogadro constant in Chap. 2 (Secs. 2.4 and 2.5). In each case one quantity is defined as the ratio of two others. The first quantity serves as a conversion factor relating the other two. A summary of the relationships and conversion factors we have encountered so far is given in Table 3.2.

An incredible variety of problems can be solved using the conversion factors in Table 3.2. Sometimes only one factor is needed, but quite often several are applied in sequence, as in Example 3.21. In solving such problems, it is necessary first to think your way through, perhaps by writing down a road map showing the relationships among the quantities given in the problem. Then you can apply conversion factors, making sure that the units cancel, and calculate the result.

The examples and end-of-chapter problems in this and the preceding two chapters should give you some indication of the broad applications of the problem-solving techniques we have developed here. Once you have mastered these techniques, you will be able to do a great many useful computations which are related to problems in the chemical laboratory, in everyday life, and in the general environment. You will find that the same type of calculations, or more complicated problems based on them, will be encountered again and again throughout your study of chemistry and other sciences.

QUESTIONS AND PROBLEMS

3.1 Complete the statements below for the following chemical reaction:

$$5A + 3B \rightarrow 2C + 4D$$

a ______ mol B reacts with 5 mol A.
b ______ mol C is formed from 6 mol B.
c The production of 0.50 mol C should be accompanied by the formation of ______ mol D.

3.2 Derive the six possible stoichiometric ratios for the following chemical reaction:

$$5I_2O_4 + 4H_2O \rightarrow 8HIO_3 + I_2$$

3.3 Using the appropriate stoichiometric ratio as a conversion factor, complete each of the statements below for the following chemical reaction:

$$7H_2 + 2NO_2 \rightarrow 2NH_3 + 4H_2O$$

a Consumption of 5.26 mol H_2 produces ______ mol H_2O.
b Reaction of 7.50 mol NO_2 yields ______ mol NH_3.
c The production of 5.00 mol NH_3 requires the consumption of ______ mol NO_2 and is accompanied by the formation of ______ mol H_2O.
d The reaction of 0.25 mol NO_2 requires ______ mol H_2.
e The formation of 15.75 mol NH_3 requires the reaction of ______ mol H_2.

3.4 (*a*) You are asked to calculate the mass of SO_2 produced when 5.38 mol MoS_2 is roasted in air according to the following equation:

$$2MoS_2 + 7O_2 \rightarrow 4SO_2 + 2MoO_3$$

Fill in the blanks in the following road map and calculation:

$$n_{MoS_2} \rightarrow n_{SO_2} \xrightarrow{\mathcal{M}_{SO_2}} \_\_\_\_\_$$

$$m_{SO_2} = 5.38 \text{ mol MoS}_2 \times \_\_\_\_\_ \times \_\_\_\_\_$$

(*b*) Construct a road map and write out the calculation similar to those in part *a* for determining the amount of MoO_3 formed by roasting 19.86 g MoS_2 in air.

3.5 Arsenic(III) oxide (As_2O_3) reacts with elemental hydrogen (H_2) to produce a mixture of the two products, AsH_3 and H_2O. After writing a balanced equation for this reaction, use the appropriate stoichiometric ratio as a conversion factor to solve each of the following:
a Consumption of 5.29 mol H_2 produces ______ mol H_2O.
b 3.73 mol AsH_3 is formed from the reaction of ______ mol As_2O_3.
c The production of 11.75 mol AsH_3 requires at least ______ mol H_2.

d The reaction of 0.475 mol As_2O_3 demands the consumption of ______ mol H_2.

e A reaction which yields 7.25 mol H_2O will also produce ______ mol AsH_3.

3.6 What mass of lithium hydride, LiH, is needed to react completely with 175 g of aluminum chloride, $AlCl_3$, according to the following reaction:

$$4LiH + AlCl_3 \rightarrow LiAlH_4 + 3LiCl$$

3.7 Calculate the mass of each product (H_3BO_3 and H_2) which is formed by the consumption of 4.18 mol B_2H_6 in the following reaction:

$$B_2H_6 + 6H_2O \rightarrow 2H_3BO_3 + 6H_2$$

3.8 The complete combustion of hydrocarbons (compounds containing only the elements carbon and hydrogen) yields the two gaseous products carbon dioxide (CO_2) and H_2O. For example,

$$2C_{12}H_{26} + 37O_2 \rightarrow 24CO_2 + 26H_2O$$

For this reaction, (*a*) calculate the amount of oxygen needed to burn 50 g $C_{12}H_{26}$; (*b*) calculate the mass of O_2 in megagrams (Mg) required to produce 1.75×10^3 kg CO_2.

3.9 Potassium chlorate ($KClO_3$) reacts as follows when heated with sulfur(s):

$$2KClO_3 + 3S \rightarrow 2KCl + 3SO_2$$

Using the stoichiometric ratio method, determine whether $KClO_3$ or S will be the limiting reagent in the following cases:

a 3 mol $KClO_3$ reacts with 3 mol S.
b 2 mol $KClO_3$ reacts with 4 mol S.
c 3.45 mol $KClO_3$ reacts with 5.00 mol S.
d 8.63 mol $KClO_3$ reacts with 13.23 mol S.

In each case also determine the amount of the other reagent which will be left unreacted.

3.10 Solid magnesium nitride (Mg_3N_2) and liquid water react on mixing according to the following equation:

$$Mg_3N_2 + 6H_2O \rightarrow 3Mg(OH)_2 + 2NH_3$$

Determine what mass of what reagent will be left in excess in the following cases:

a 10.00 g Mg_3N_2 and 60.00 g H_2O are mixed.

b 10.00 g Mg_3N_2 and 10.00 g H_2O are mixed.

c 10.00 g Mg_3N_2 and 10.71 g H_2O are mixed.

3.11 Bismuth metal can be prepared by first roasting the sulfide ore, Bi_2S_3, to an oxide, Bi_2O_3, and then treating the oxide with carbon as follows:

$$2Bi_2S_3 + 9O_2 \rightarrow 2Bi_2O_3 + 6SO_2$$
$$Bi_2O_3 + 3C \rightarrow 2Bi + 3CO$$

Assuming an excess of O_2 but only 500 kg C, calculate the mass of each of the four products (Bi_2O_3, SO_2, Bi, and CO) which would be formed from 1.00 Mg Bi_2S_3.

3.12 What mass of O_2 would be formed from the reaction of 34.0 g H_2O_2, 80.0 g $KMnO_4$, and 65.0 g H_2SO_4 according to the following equation:

$$5H_2O_2 + 2KMnO_4 + 3H_2SO_4 \rightarrow K_2SO_4 + 2MnSO_4 + 8H_2O + 5O_2$$

Also, what mass of each of the two reagents in excess would remain when the reaction was completed?

3.13 In a manned spacecraft it is necessary to remove the CO_2 gas exhaled by the crew. Here are three possible chemical reactions for this purpose:

$$2Na_2O_2 + 2CO_2 \rightarrow 2Na_2CO_3 + O_2$$
$$Mg(OH)_2 + CO_2 \rightarrow MgCO_3 + H_2O$$
$$2LiOH + CO_2 \rightarrow Li_2CO_3 + H_2O$$

Select which of the reagents, Na_2O_2, $Mg(OH)_2$, or LiOH is the most suitable for this purpose on the basis of maximum mass of CO_2 consumed per minimum mass of reactant.

3.14 Calculate the percent yield for the following reaction if 200 g phosphorus trichloride (PCl_3) in an excess of water produced 128 g hydrogen chloride (HCl):

$$PCl_3 + 3H_2O \rightarrow H_3PO_3 + 3HCl$$

3.15 Elemental chlorine (Cl_2) can undergo several different reactions in water, including the following one:

$$3Cl_2 + 3H_2O \rightarrow 5HCl + HClO_3$$

Calculate the percent yield based on this equation if 71 g Cl_2 plus 18 g H_2O produce 60 g HCl.

3.16 In the previous problem, what would the percent yield be if exactly the same quantities of reactants and products were involved, but the reaction was assumed to be

$$Cl_2 + H_2O \rightarrow HCl + HClO$$

What general conclusions can you draw regarding the percent yield concept from these two results?

3.17 Isooctane is a compound of carbon and hydrogen and also an important constituent of gasoline. When a 0.7681-g sample of this compound is burned in a combustion train similar to that shown in Fig. 3.1, 2.3717 g CO_2 and 1.0915 g H_2O are produced. What is the empirical formula of isooctane?

3.18 A compound containing only Fe and S was analyzed by roasting the compound in air and converting all the S to SO_2. If a 0.4203-g sample of the compound produced 0.4486 g SO_2, what is the empirical formula?

3.19 Calculate the kinetic energy of (*a*) a man of mass 75 kg running at 4 m s^{-1}; (*b*) a truck of mass 10 Mg moving at 25 m s^{-1}; (*c*) a woman of mass 115 lb running at 4 miles per hour.

3.20 The thermochemical equation for the fusion (melting) of sodium chloride at 808°C is

$$NaCl(s) \rightarrow NaCl(l) \qquad \Delta H_m = +33.1 \text{ kJ mol}^{-1}$$

Using a road map, calculate the quantity of heat energy needed to melt 1.00 g sodium chloride at this temperature.

3.21 From experiment it is found that 56.5 J of heat energy must be supplied to melt 1.00 g silver bromide (AgBr) at 430°C. Use this information to complete the following thermochemical equation for 430°C:

$$AgBr(l) \rightarrow AgBr(s) \qquad \Delta H_m = \_\_\_\_\_$$

3.22 What quantity of heat energy is liberated from the complete combustion of 1.000 Mg of acetaldehyde (CH_3CHO):

$$2CH_3CHO(l) + 5O_2(g) \rightarrow 4CO_2(g) + 4H_2O(g) \qquad \Delta H_m = -2716 \text{ kJ mol}^{-1}$$

3.23 Using Hess' Law and the series of thermochemical equations given below, calculate ΔH_m for the reaction

$$C(s) + 2H_2(g) \rightarrow CH_4(g) \qquad \Delta H_m = \_\_\_\_\_$$

Available equations:

$$CH_4(g) + 2O_2(g) \rightarrow CO_2(g) + 2H_2O(l) \qquad \Delta H_m = -1036 \text{ kJ mol}^{-1}$$

$$C(s) + O_2(g) \rightarrow CO_2(g) \qquad \Delta H_m = -457.8 \text{ kJ mol}^{-1}$$

$$H_2(g) + \tfrac{1}{2}O_2(g) \rightarrow H_2O(l) \qquad \Delta H_m = -332.6 \text{ kJ mol}^{-1}$$

All data refer to 25°C and normal atmospheric pressure.

3.24 Calculate ΔH_m for the reaction

$$Ca(s) + C(s) + \tfrac{3}{2}O_2(g) \rightarrow CaCO_3(s)$$

using Hess' Law in conjunction with the following thermochemical equations:

$$Ca(s) + \tfrac{1}{2}O_2(g) \rightarrow CaO(s) \qquad \Delta H_m = -739.2 \text{ kJ mol}^{-1}$$

$$C(s) + O_2(g) \rightarrow CO_2(g) \qquad \Delta H_m = -457.8 \text{ kJ mol}^{-1}$$

$$CaO(s) + CO_2(g) \rightarrow CaCO_3(s) \qquad \Delta H_m = -207.1 \text{ kJ mol}^{-1}$$

All data refer to 25°C and normal atmospheric pressure.

3.25 Using standard enthalpies of formation from Table 3.1, find ΔH_m for the following reactions at 25°C.

a $C_2H_4(g) + 3O_2(g) \rightarrow 2CO_2(g) + 2H_2O(l)$
b $SO_2(g) + 2H_2S(g) \rightarrow 2H_2O(l) + 3S(s)$
c $Fe_2O_3(s) + 3C(s) \rightarrow 2Fe(s) + 3CO(g)$
d $4NH_3(g) + 5O_2(g) \rightarrow 6H_2O(g) + 4NO(g)$

3.26 Using standard enthalpies of formation from Table 3.1, find ΔH_m for the following reactions at 25°C:

a $C_2H_2 + 2\tfrac{1}{2}O_2(g) \rightarrow 2CO_2(g) + H_2O(g)$
b $2H_2O_2(l) \rightarrow 2H_2O(l) + O_2(g)$
c $6Mg(s) + 2Fe_2O_3(s) \rightarrow 6MgO(s) + 4Fe(s)$
d $4NH_3(g) + 7O_2(g) \rightarrow 6H_2O(l) + 4NO_2(g)$

3.27 Calculate the concentration of a solution prepared by dissolving exactly 2.1735 g NaCl in sufficient H_2O to reach the calibration mark on a 500 cm^3 volumetric flask.

3.28 A small bottle containing the white crystalline salt ammonium chloride, NH_4Cl, weighs 27.8793 g. A portion of the solid was removed and transferred very carefully into a 1 liter (1 dm^3) volumetric flask. The bottle plus the remaining contents weighed 22.5293 g. An aqueous solution was then prepared by adding water up to the calibration mark in the volumetric flask. Calculate the concentration of the resultant $NH_4Cl(aq)$ solution.

3.29 If 532.6 cm^3 of a solution of potassium chloride, KCl, was evaporated and found to contain 2.9632 g of the salt, what was the concentration of KCl in the original solution?

3.30 A solution of potassium iodide, KI, was prepared by adding exactly 50.0 cm^3 of H_2O to a sample of KI weighing 0.1326 g. Why is this information not enough to find the concentration of KI? What further information would be needed?

3.31 Calculate the amount of HCl in (*a*) 2.000 dm^3 of a solution of concentration 1.863 mol HCl dm^{-3}; (*b*) 500 cm^3 of a solution of concentration 0.2486 mol HCl dm^{-3}; (*c*) 23.52 cm^3 of a solution for which c_{HCl} = 0.0986 mol dm^{-3}; (*d*) 15.86 ml of 0.1063 *M* HCl.

3.32 Calculate the new concentration when (*a*) 25.00 cm^3 of 0.1306 *M* HCl is diluted to 250 cm^3; (*b*) 28.63 cm^3 of 0.1000 *M* HCl is diluted to 1.000 dm^3; (*c*) 19.58 cm^3 of 0.0863 *M* HCl is diluted to 500 cm^3.

3.33 What volume of 0.386 *M* HCl contains 9.53 mmol HCl?

3.34 What volume of 0.1036 *M* Na_2CO_3 contains 5.00 mmol Na_2CO_3?

3.35 What volume of 1.586 *M* HNO_3 is needed to prepare 500.0 cm^3 of 0.0986 *M* HNO_3?

3.36 What volume of 0.986 *M* K_2SO_4 is needed to prepare exactly 1 liter (1 dm^3) of 0.1000 *M* K_2SO_4?

3.37 To what volume must 925.0 cm^3 of 0.1032 *M* H_2SO_4 be diluted in order to produce a solution which is exactly 0.1000 *M*?

3.38 To what volume must 732.0 cm^3 of 0.0982 *M* KCl be diluted in order to produce a solution which is exactly 0.0100 *M*?

3.39 Groundwater or river water is often found to be "hard" because of the presence of calcium sulfate ($CaSO_4$) and other calcium salts. In order to "soften" such water, the $CaSO_4$ can be removed by the addition of sodium carbonate, Na_2CO_3:

$$CaSO_4 + Na_2CO_3 \rightarrow CaCO_3 + Na_2SO_4$$

What mass of Na_2CO_3 would be required to remove all the $CaSO_4$ from 2.0×10^8 dm^3 of water containing 1.8×10^{-3} mol $CaSO_4$ dm^{-3}. (This volume of water is roughly equal to the daily consumption of a large city.)

3.40 What volume of 0.1159 *M* HCl is needed to achieve an endpoint when titrated against
a 12.50 mmol Na_2CO_3 crystals.
b 25.00 cm^3 of 0.041 64 *M* Na_2CO_3 solution.
The equation for the reaction is

$$Na_2CO_3 + 2HCl \rightarrow 2NaCl + H_2O + CO_2$$

3.41 What volume of 0.051 93 *M* H_2SO_4 is needed to achieve an endpoint when titrated against
a 0.5132 g NaOH.
b 25.00 cm^3 of 0.1034 *M* NaOH.
The equation for the reaction is

$$H_2SO_4 + 2NaOH \rightarrow Na_2SO_4 + 2H_2O$$

3.42 By using an appropriate indicator to detect the endpoint, a solution of $FeSO_4$ in H_2SO_4 can be titrated with $K_2Cr_2O_7$ according to the equation

$$6\,FeSO_4 + 7H_2SO_4 + K_2Cr_2O_7 \rightarrow 3Fe_2(SO_4)_3 + Cr_2(SO_4)_3 + K_2SO_4 + 7H_2O$$

(*a*) What volume of 0.1271 *M* $K_2Cr_2O_7$ solution would be required to titrate 25.00 cm^3 of 0.4777 *M* $FeSO_4$? (*b*) If a sample of $FeSO_4$ weighing 2.437 g was dissolved in H_2SO_4 and then titrated with 0.1271 *M* $K_2Cr_2O_7$, what volume would be required?

3.43 The concentration of aqueous NaOH solution can be determined by standardizing it against the pure solid benzoic acid, $HC_7H_5O_2$:

$$HC_7H_5O_2 + NaOH \rightarrow NaC_7H_5O_2 + H_2O$$

Calculate the concentration of an NaOH solution if 38.47 ml was required when titrated against a sample of $HC_7H_5O_2$ weighing 0.5225 g.

3.44 An impure sample of potassium hydrogen phthalate, $KHC_8H_4O_4$, weighing 1.4392 g was titrated with an NaOH solution according to the following equation:

$$KHC_8H_4O_4 + NaOH \rightarrow KNaC_8H_4O_4 + H_2O$$

25.76 ml of 0.1307 *M* NaOH was required to achieve an endpoint. Find (*a*) the mass of pure $KHC_8H_4O_4$ in the sample; (*b*) the percentage purity of the sample.

3.45 An impure sample of sodium carbonate, Na_2CO_3, weighing 0.3118 g was titrated with 0.098 73 *M* HCl solution, and 30.42 cm^3 was required to achieve an endpoint.

$$Na_2CO_3 + 2HCl \rightarrow 2NaCl + H_2O + CO_2$$

Calculate the percentage purity of the Na_2CO_3 sample.

3.46 Vinegar consists mainly of a solution of acetic acid (CH_3COOH) in H_2O. When 50.00 cm^3 of a vinegar sample is titrated with 0.1055 *M* NaOH solution, 41.09 cm^3 is required to achieve an endpoint. The reaction is

$$NaOH + CH_3COOH \rightarrow CH_3COONa + H_2O$$

(*a*) Calculate the mass of CH_3COOH in the sample. (*b*) Calculate the percent by weight of CH_3COOH in the vinegar, given that the density of vinegar is 1.052 g cm^{-3}.

chapter four

THE STRUCTURE OF ATOMS

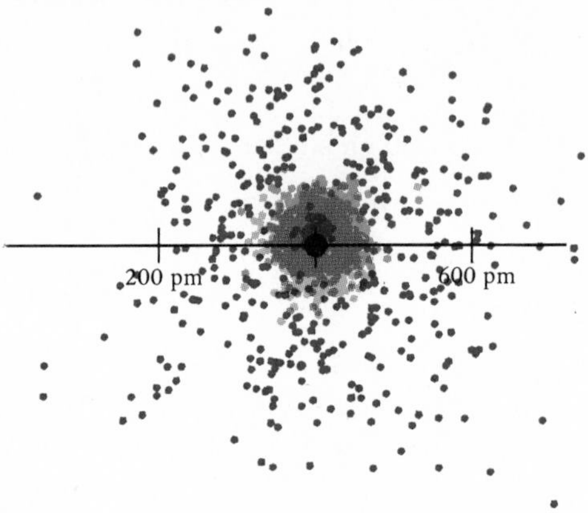

Li

We have examined the theoretical implications and practical applications of John Dalton's ideas about atoms in the preceding two chapters. Clearly the atomic theory is a powerful tool which aids our thinking about how much of one substance can combine with (or be produced from) a given quantity of another. The theory is much less helpful, however, when we try to speculate about what holds the atoms together in molecules such as Br_2, $HgBr_2$, and Hg_2Br_2. As you have seen, techniques are available for *experimental* determination of the formula of a new compound, but Dalton's theory is of little value in *predicting* formulas. Neither does it tell us which elements are likely to combine with which, nor indicate what chemical and physical properties are to be expected of the compounds which form.

The ability to make predictions about chemical reactivity and properties is very important because it guides chemists' efforts to synthesize new substances which are of value to society at large. Medicines, metals, transistors, plastics, textiles, fertilizers, and many other things that we take for granted today have been made possible by detailed knowledge of chemical and physical properties. Such knowledge also permits greater understanding of how the natural world works and what changes (favorable or detrimental) may be brought about by human activities.

Knowledge of chemical reactivity and properties may be approached on both the macroscopic and microscopic levels. Macroscopically this involves what is called **descriptive chemistry.** The person who first carries out a chemical reaction *describes* what happened, usually in terms of a balanced equation, and lists properties of any new substances. This enables other scientists to repeat the experiment if they wish. Even if the work is not carried out again, the descriptive report allows prediction of what would happen if it were repeated.

The microscopic approach uses *theory* to predict which substances will react with which. During the past century Dalton's atomic theory has been modified so that it can help us to remember the properties of elements and compounds. We now attribute *structure* to each kind of atom and expect atoms having similar structures to undergo similar reactions. The additional complication of learning about atomic structure is repaid manyfold by the increased ability of our microscopic model to predict macroscopic properties.

4.1 DESCRIPTIVE CHEMISTRY OF SOME GROUPS OF RELATED ELEMENTS

The macroscopic, descriptive approach to chemical knowledge has led to a great deal of factual information. Right now more than 3 million chemical compounds and their properties are on file at the Chemical Abstracts Service of the American Chemical Society. Anyone who wants information about these substances can look it up, although in practice it helps to have a computer do the looking! Even with a computer's memory it is hard to keep track of so many facts—no single person can remember more than a fraction of the total.

Fortunately these millions of facts are interrelated in numerous ways, and the relationships are helpful in remembering the facts. To illustrate this point, we shall present part of the descriptive chemistry of about 20 elements. Although each element has unique physical and chemical properties, it will be obvious that certain groups of elements are closely related. Members of each group are more like each other than they are like any member of another group. Because of this close relationship a special name has been assigned to each collection of elements. It is also possible to write general equations which apply to *all* members of a family of elements. Practical laboratory experience with one member gives a fairly accurate indication of how each of the others will behave. As you read the next few pages, try to concentrate on the similarities among related elements, rather than the properties of each as an individual.

Alkali Metals

The element potassium combines violently and spectacularly with water, as shown in Plate 1. The flame is due to combustion of hydrogen gas which is given off, and if the excess water is evaporated, the compound potassium hydroxide (KOH) remains behind. Thus the equation for this reaction is

$$2K(s) + 2H_2O(l) \rightarrow 2KOH(aq) + H_2(g) \qquad (4.1)$$

The elements lithium, sodium, rubidium, and cesium also combine violently with water to form hydroxides. The equations for their reactions are

$$2Li(s) + 2H_2O(l) \rightarrow 2LiOH(aq) + H_2(g)$$

$$2Na(s) + 2H_2O(l) \rightarrow 2NaOH(aq) + H_2(g)$$

$$2Rb(s) + 2H_2O(l) \rightarrow 2RbOH(aq) + H_2(g)$$

$$2Cs(s) + 2H_2O(l) \rightarrow 2CsOH(aq) + H_2(g)$$

Since potassium and these four elements all react with water in the same way, a *general equation* may be written:

$$2M(s) + 2H_2O(l) \rightarrow 2MOH(aq) + H_2(g) \qquad M = K, Li, Na, Rb, \text{ or } Cs$$

The symbol M represents any one of the five elements.

In addition to their behavior when added to water, lithium, sodium, potassium, rubidium, and cesium have a great many other properties in common. All are solids at 0°C and melt below 200°C. Each is silvery in color and has metallic properties such as good conduction of heat and electricity, malleability (the ability to be hammered into sheets), and ductility (the ability to be drawn into wires). The high thermal (heat) conductivity and the relatively low melting point (for a metal) of sodium make it an ideal heat-transfer fluid. It is used to cool certain types of nuclear reactors (liquid-metal fast breeder reactors, LMFBRs) and to cool the valves of high-powered automobile engines for this reason.

Because of their similarities, lithium, sodium, potassium, rubidium, and cesium are grouped together and called the **alkali metals.** (The term *alkali* is derived from an Arabic word meaning "ashes." Compounds of potassium as well as other alkali metals were obtained from wood ashes by early chemists.) The alkali metals all react directly with oxygen from the atmosphere, forming oxides, M_2O:

$$4M(s) + O_2(g) \rightarrow 2M_2O(s) \qquad M = Li, Na, K, Rb, \text{ or } Cs$$

(Li_2O is lithium oxide, Na_2O is sodium oxide, etc.) All except lithium react further to form peroxides, M_2O_2:

$$2M_2O(s) + O_2(g) \rightarrow 2M_2O_2(s) \qquad M = Na, K, Rb, \text{ or } Cs$$

(Na_2O_2 is sodium peroxide, etc.) Potassium, rubidium, and cesium are sufficiently reactive that superoxides (whose general formula is MO_2) can be formed:

$$M_2O_2(s) + O_2(g) \rightarrow 2MO_2(s) \qquad M = K, Rb, \text{ or } Cs$$

Unless the surface of a sample of an alkali metal is scraped clean, it will appear white instead of having a silvery metallic luster. This is due to the oxide, peroxide, or superoxide coating that forms after a few seconds of exposure to air.

The alkali metals react with most of the other chemical elements as well. For example, all combine directly with hydrogen gas to form compounds known as hydrides, MH:

$$2M(s) + H_2(g) \rightarrow 2MH(s) \qquad M = Li, Na, K, Rb, \text{ or } Cs$$

They react with sulfur to form sulfides, M_2S:

$$2M(s) + S(s) \rightarrow M_2S(s) \qquad M = \text{Li, Na, K, Rb, or Cs}$$

They also react directly with chlorine, forming chlorides, MCl:

$$2M(s) + Cl_2(g) \rightarrow 2MCl(s) \qquad M = \text{Li, Na, K, Rb, or Cs} \qquad (4.2a)$$

They react with fluorine to form fluorides, MF:

$$2M(s) + F_2(g) \rightarrow 2MF(s) \qquad M = \text{Li, Na, K, Rb, or Cs} \qquad (4.2b)$$

They react with bromine to form bromides, MBr:

$$2M(s) + Br_2(l) \rightarrow 2MBr(s) \qquad M = \text{Li, Na, K, Rb, or Cs} \qquad (4.2c)$$

Notice that each member of the chemical family of alkali metals has physical and chemical properties very similar to all the others. In most cases *all alkali metals behave the same with regard to the formulas of their compounds*. The peroxides and superoxides are exceptions to this rule, but formulas for oxides and each of the other types of compounds we have described are identical except for the chemical symbol of each alkali metal.

Halogens

The last three reactions above involve members of another important group of elements. The **halogens** include fluorine, chlorine, bromine, and iodine. Iodine combines less vigorously with alkali metals than other halogens, but its reactions would be analogous to Eqs. (4.2). Compounds of an alkali metal and a halogen, such as sodium chloride, potassium fluoride, lithium bromide, or cesium iodide, have closely related properties. (All taste salty, for example.) They belong to a general category called **salts,** all of whose members are similar to ordinary table salt, sodium chloride. The term *halogen* is derived from Greek words meaning "salt former."

The free elemental halogens all consist of diatomic molecules X_2, where X may be fluorine, chlorine, bromine, or iodine (recall the microscopic picture of bromine given in Fig. 2.3). There is somewhat more variation among their physical properties than among those of the alkali metals. Fluorine and chlorine are both gases at room temperature, the former very pale yellow, and the latter yellow-green in color. Bromine is a red-brown liquid which vaporizes rather easily (see Plate 3). Iodine forms shiny dark crystals and, when heated, sublimes (changes directly from solid to gas) to a beautiful violet vapor. All the gases produce a choking sensation when inhaled. Chlorine was used to poison soldiers on European battlefields in 1917 to 1918. Halogens are put to more humane uses such as to disinfect public water supplies by means of chlorination and to treat minor cuts by using an alcohol solution (tincture) of iodine. These applications depend on the ability of the halogens to destroy microorganisms which are harmful to humans.

All halogens are quite reactive, and in the natural world they always occur combined with other elements. Fluorine reacts so readily with almost any substance it contacts that chemists were not successful in isolating

David K. Conroy

pure fluorine until 1886, although its existence in compounds had been known for many years. Chlorine, bromine, and iodine are progressively less reactive but still form compounds with most other elements, especially metals. A good example is mercury, whose reaction with bromine was discussed in Chap. 2. Mercury reacts with other halogens in the same way

$$Hg(l) + X_2(g, l, \text{ or } s) \rightarrow HgX_2(s) \qquad X = F, Cl, Br, \text{ or } I$$

Another vigorous reaction occurs when certain compounds containing carbon and hydrogen contact the halogens. Turpentine, $C_{10}H_{16}$, reacts quite violently. In the case of fluorine and chlorine the equation is

$$C_{10}H_{16}(l) + 8X_2(g) \rightarrow 10C(s) + 16HX(g) \qquad X = F, Cl$$

but the products are different when bromine and iodine react. Before the advent of the automobile, veterinarians used solid iodine and turpentine to disinfect wounds in horses' hooves. This may have been because of the superior antiseptic qualities of the mixture. However, a more likely reason is the profound impression made on the owner of the horse by the great clouds of violet iodine vapor which sublimed as a result of the increase in temperature when the reaction occurred!

Alkaline-Earth Metals

A third family of closely related elements is the **alkaline-earth metals,** beryllium, magnesium, calcium, strontium, barium, and radium. All exhibit metallic properties and a silver or gray color. Except for beryllium, the alkaline earths react directly with hydrogen gas to form hydrides, MH_2; M = Mg, Ca, Sr, Ba, or Ra. Beryllium hydride, BeH_2, can also be prepared, but not directly from the elements.

Alkaline-earth metals combine readily with oxygen from the air to form oxides, MO. These coat the surface of the metal and prevent other substances from contacting and reacting with it. A good example of the effect of such an oxide coating is the reaction of alkaline-earth metals with water. Beryllium and magnesium react much more slowly than the others because their oxides are insoluble and prevent water from contacting the metal.

Alkaline-earth metals react directly with halogens to form salts:

$$M(s) + Cl_2(g) \rightarrow MCl_2(s) \qquad M = Be, Mg, Ca, Sr, Ba, \text{ or } Ra$$

Salt obtained by evaporating seawater (sea salt) contains a good deal of magnesium chloride and calcium chloride as well as sodium chloride. It also has small traces of iodide salts, accounting for the absence of simple goiter in communities which obtain their salt from the oceans. Simple goiter is an enlargement of the thyroid gland caused by iodine deficiency.

Other Groups of Elements

There are several other examples of related groups of elements. The **coinage metals,** copper, silver, and gold, often occur naturally as elements, not in compounds. They have been used throughout history to make coins because they do not combine rapidly with atmospheric oxygen. The reddish brown and golden colors of copper and gold are distinctive among the metals, and the electrical conductivities of the coinage metals are greater than those of any other elements. The **chalcogens** (sulfur, selenium, and tellurium) are another related group of nonmetallic elements. Their hydrogen compounds (hydrogen sulfide, hydrogen selenide, and hydrogen telluride) are all gases which have revolting odors. The familiar smell of rotten eggs is due to hydrogen sulfide and the other two are even worse. These compounds are also highly poisonous and more dense than air. Numerous cases are known where persons working in ditches or other low-lying areas have been rendered unconscious or even killed by hydrogen sulfide resulting from natural sources or from industrial activities such as petroleum refining.

One group of elements, the **noble gases** (helium, neon, argon, krypton, xenon, and radon), forms almost no chemical compounds. Although small concentrations of the noble gases are present in the earth's atmosphere, they were not discovered until 1894, largely because they underwent no reactions. Fluorine is sufficiently reactive to combine with pure samples of xenon, radon, and (under special conditions) krypton. The only other element that has been shown conclusively to occur in compounds with the noble gases is oxygen, and no more than a couple of dozen noble-gas compounds of all types are known. This group of elements is far less reactive chemically than any other.

4.2 THE PERIODIC CLASSIFICATION OF THE ELEMENTS

The similarities among macroscopic properties within each of the chemical families just described lead one to expect microscopic similarities as well. Atoms of sodium ought to be similar in some way to atoms of lithium, potassium, and the other alkali metals. This could account for the related chemical reactivities and analogous compounds of these elements.

According to Dalton's atomic theory, different kinds of atoms may be distinguished by their relative masses (atomic weights). Therefore it seems reasonable to expect some correlation between this microscopic property and macroscopic chemical behavior. You can see that such a relationship exists by listing symbols for the first dozen elements in order of increasing relative mass. Obtaining atomic weights from Table 2.2, we have

At Wt	1.0	4.0	6.9	9.0	10.8	12.0	14.0	16.0	19.0	20.2	23.0	24.3
Symbol	H	He	Li	Be	B	C	N	O	F	Ne	Na	Mg

Elements which belong to families we have already discussed are indicated by shading around their symbols. The second, third, and fourth elements on the list (He, Li, and Be) are a noble gas, an alkali metal, and an alkaline-earth metal, respectively. Exactly the same sequence is repeated eight elements later (Ne, Na, and Mg), but this time a halogen (F) precedes the noble gas. If a list were made of all elements, we would find the sequence halogen, noble gas, alkali metal, and alkaline-earth metal several more times.

The Periodic Table

In 1871 the Russian chemist Dmitri Ivanovich Mendeleev (1834 to 1907) proposed the **periodic law.** This law states that *when the elements are listed in order of increasing atomic weights, their properties vary periodically.* That is, similar elements do not have similar atomic weights. Rather, as we go down a list of elements in order of atomic weights, corresponding properties are observed at regular intervals. To emphasize this periodic repetition of similar properties, Mendeleev arranged the symbols and atomic weights of the elements in the table shown in Fig. 4.1.

Each vertical column of this **periodic table** contains a **group** or **family** of related elements. The alkali metals are in group I (*Gruppe* I), alkaline earths in group II, chalcogens in group VI, and halogens in group VII. Mendeleev was not quite sure where to put the coinage metals, and so they appear twice. Each time, however, copper, silver, and gold are arranged in a vertical column. Although the noble gases were discovered nearly a quarter century after Mendeleev's first periodic table was published, we have included them in Fig. 4.1 to indicate that they, too, fit the periodic arrangement.

In constructing his table, Mendeleev found that sometimes there were not enough elements to fill all the available spaces in each horizontal row or **period.** When this was true, he assumed that eventually someone would

Figure 4.1 Mendeleev's periodic table, redrawn from *"Annalen der Chemie,"* supplemental volume 8, 1872. The German words *Gruppe* and *Reihen* indicate, respectively, the groups and rows (or periods) in the table. Mendeleev also used the European convention of a comma instead of a period for the decimal and J instead of I for iodine. The noble gases had not yet been discovered when Mendeleev devised the periodic table, but they have been included here (in color) for completeness.

TABELLE II

Group 0	Reihen	Gruppe I	Gruppe II	Gruppe III
. . .		. . .	. . .	. . .
. . .		R^2O	RO	R^2O^3
	1	H = 1		
He = 4	2	Li = 7	Be = 9,4	B = 11
Ne = 20	3	Na = 23	Mg = 24	Al = 27,3
Ar = 40	4	K = 39	Ca = 40	______ = 44
	5	(Cu = 63)	Zn = 65	______ = 68
Kr = 84	6	Rb = 85	Sr = 87	?Yt = 88
	7	(Ag = 108)	Cd = 112	In = 113
Xe = 131	8	Cs = 133	Ba = 137	?Di = 138
	9	(______)	______	______
	10	______	______	?Er = 178
	11	(Au = 199)	Hg = 200	Tl = 204
Rn = 222	12	______	______	______

TABLE 4.1 Comparison of Mendeleev's Predictions with the Observed Properties of the Element Scandium.

	Properties Predicted for Ekaboron (Eb)* by Mendeleev in 1872	**Properties Found for Scandium after its Discovery in 1879**
Atomic weight	44	44†
Formula of oxide	Eb_2O_3	Sc_2O_3
Density of oxide	3.5	3.86
Acidity of oxide	Greater than MgO	Greater than MgO
Formula of chloride	$EbCl_3$	$ScCl_3$
Boiling point of chloride	Higher than for $AlCl_3$	Much higher than for $AlCl_3$
Color of compounds	Colorless	Colorless

* Mendeleev used the name *eka*boron because the blank space into which the element should fit was *below* boron in his periodic table.

† The modern value of the atomic weight of scandium is 44.96.

discover the element or elements needed to complete a period. Therefore he left blank spaces for undiscovered elements and predicted their properties by averaging the characteristics of other elements in the same group.

As an example of this process, look at the fourth numbered row (*Reihen*) in Fig. 4.1. Scandium (Sc) was unknown in 1872; so titanium (Ti) followed calcium (Ca) in order of atomic weights. This would have placed titanium below boron (B) in group III, but Mendeleev knew that the most common oxide of titanium, TiO_2, had a formula similar to an oxide of carbon, CO_2, rather than of boron, B_2O_3. Therefore he placed titanium below carbon in group IV. He proposed that an undiscovered element, ekaboron, would eventually be found to fit below boron. (The prefix *eka* means "below.") Properties predicted for ekaboron are shown in Table 4.1. They agreed remarkably with those measured experimentally for scandium when it was discovered 7 years later. This agreement was convincing evidence that a periodic table is a good way to summarize a great many macroscopic, experimental facts.

Gruppe IV **RH^4** **RO^2**	**Gruppe V** **RH^3** **R^2O^5**	**Gruppe VI** **RH^2** **RO^3**	**Gruppe VII** **RH** **R^2O^7**	**Gruppe VIII** . . . **RO^4**
C = 12	N = 14	O = 16	F = 19	
Si = 28	P = 31	S = 32	Cl = 35,5	
Ti = 48	V = 51	Cr = 52	Mn = 55	Fe = 56, Co = 59, Ni = 59, Cu = 63
——— = 72	As = 75	Se = 78	Br = 80	
Zr = 90	Nb = 94	Mo = 96	——— = 100	Ru = 104, Rh = 104, Pd = 106, Ag = 108
Sn = 118	Sb = 122	Te = 125	J = 127	
?Ce = 140	———	———	———	——— ——— ———
———	———	———	———	
?La = 180	Ta = 182	W = 184	———	Os = 195, Ir = 197, Pt = 198, Au = 199
Pb = 207	Bi = 208	———	———	
Th = 231	———	U = 240	———	——— ——— ———

The modern periodic table inside the front cover of this book differs in some ways from Mendeleev's original version. It contains more than 40 additional elements, and its rows are longer instead of being squeezed under one another in staggered columns. (Mendeleev's fourth and fifth rows are both contained in the fourth period of the modern table, for example.) The extremely important idea of vertical groups of related elements is still retained, as are Mendeleev's group numbers. The latter appear as roman numerals at the top of each column in the modern table.

Valence

Perhaps the most important function of the periodic table is that it helps us to predict the chemical formulas of commonly occurring compounds. At the top of each group, Mendeleev provided a general formula for oxides of the elements in the group. (See Fig. 4.1). The heading R^2O above group I, for example, means that we can expect to find compounds such as H_2O, Li_2O, Na_2O, etc. Similarly, the general formula RH^3 above group V suggests that the compounds NH_3, PH_3, VH_3, and AsH_3 (among others) should exist.

To provide a basis for checking this prediction, formulas are shown in Table 4.2 for compounds in which H, O, or Cl is combined with each of the first two dozen elements (in order of atomic weights). Even among groups of elements whose descriptive chemistry we have not discussed, you can easily confirm that most of the predicted formulas correspond to compounds which actually exist. Conversely, more than 40 percent of the formulas for known O compounds agree with Mendeleev's general formulas. (These are shaded in color in Table 4.2.)

The periodic repetition of similar formulas is even more pronounced in the case of Cl compounds. This is evident when a list is made of subscripts for Cl in combination with each of the first 24 elements. Consulting Table 4.2, we find HCl (subscript 1), no compound with He (subscript 0), LiCl (subscript 1), and so on.

Element	H	He	Li	Be	B	C	N	O	F	Ne	Na	Mg	Al	Si	P	S	Cl	*K*	*Ar*	Ca	Sc	Ti	V	Cr
Subscript of Cl	1	0	1	2	3	4	3	2	1	0	1	2	3	4	3	2	1	*1*	*0*	2	3	4	3	2

With only the two exceptions indicated in italics, at least one formula for a Cl compound of each element fits a sequence of subscripts which fluctuate regularly from 0 up to 4 and back to 0 again. (The unusual behavior of K and Ar will be discussed a bit later.) The number of Cl atoms which combines with one atom of each other element varies quite regularly as the atomic weight of the other element increases.

The experimentally determined formulas in Table 4.2 and the general formulas in Mendeleev's periodic table both imply that each element has a characteristic chemical combining capacity. This capacity is called **valence,** and it varies periodically with increasing atomic weight. The noble gases all have valences of 0 because they almost never combine with any other element. H and Cl both have the same valence. They combine with

TABLE 4.2 Molecular Formulas for Hydrogen, Oxygen, and Chlorine Compounds of the First Twenty-Four Elements in Order of Atomic Weight.*

Element	Atomic Weight	Hydrogen Compounds	Oxygen Compounds	Chlorine Compounds
Hydrogen	1.01	H_2	H_2O, H_2O_2	HCl
Helium	4.00	None formed	None formed	None formed
Lithium	6.94	LiH	Li_2O, Li_2O_2	$LiCl$
Beryllium	9.01	BeH_2	BeO	$BeCl_2$
Boron	10.81	B_2H_6	B_2O_3	BCl_3
Carbon	12.01	CH_4, C_2H_6, C_3H_8, . . .†	CO_2, CO, C_2O_3	CCl_4, C_2Cl_6
Nitrogen	14.01	NH_3, N_2H_4, HN_3	N_2O, NO, NO_2, N_2O_5	NCl_3
Oxygen	16.00	H_2O, H_2O_2	O_2, O_3	Cl_2O, ClO_2, Cl_2O_7
Fluorine	19.00	HF	OF_2, O_2F_2	ClF, ClF_3, ClF_5
Neon	20.18	None formed	None formed	None formed
Sodium	22.99	NaH	Na_2O, Na_2O_2	$NaCl$
Magnesium	24.31	MgH_2	MgO	$MgCl_2$
Aluminum	26.98	AlH_3	Al_2O_3	$AlCl_3$
Silicon	28.09	SiH_4, Si_2H_6	SiO_2	$SiCl_4$, Si_2Cl_6
Phosphorus	30.97	PH_3, P_2H_4	P_4O_{10}, P_4O_6	PCl_3, PCl_5, P_2Cl_4
Sulfur	32.06	H_2S, H_2S_2	SO_2, SO_3	S_2Cl_2, SCl_2, SCl_4
Chlorine	35.45	HCl	Cl_2O, ClO_2, Cl_2O_7	Cl_2
Potassium	39.10	KH	K_2O, K_2O_2, KO_2	KCl
Argon	39.95	None formed	None formed	None formed
Calcium	40.08	CaH_2	CaO, CaO_2	$CaCl_2$
Scandium	44.96	Relatively unstable	Sc_2O_3	$ScCl_3$
Titanium	47.90	TiH_2	TiO_2, Ti_2O_3, TiO	$TiCl_4$, $TiCl_3$, $TiCl_2$
Vanadium	50.94	VH_2	V_2O_5, V_2O_3, VO_2, VO	VCl_4, VCl_3, VCl_2
Chromium	52.00	CrH_2	Cr_2O_3, CrO_2, CrO_3	$CrCl_3$, $CrCl_2$

* For each element compounds are listed in order of decreasing stability. In some cases additional compounds are known, but these are relatively unstable.

† A great many stable compounds of carbon and hydrogen are known, but space limitations prevent listing all of them.

each other in a 1:1 ratio to form HCl, each combines with Li in the same 1:1 ratio (LiH and LiCl), each combines with Be in the same ratio (BeH_2, $BeCl_2$), and so on. Because H and Cl have the same valence, we can predict that a large number of H compounds will have formulas identical to those of Cl compounds, except, of course, that the symbol H would replace the symbol Cl. The correctness of this prediction can be verified by studying the formulas surrounded by gray shading in Table 4.2.

The combining capacity, or valence, of O is apparently twice that of H or Cl. *Two* H atoms combine with one O atom in H_2O. So do two Cl atoms or two Li atoms (Cl_2O and Li_2O). The number of atoms combining with a single O atom is usually *twice as great* as the number which combined with a single H or Cl atom. (Again, consulting the gray shaded formulas in Table 4.2 will confirm this statement.)

After careful study of the formulas in the table, it is also possible to conclude that none of the elements (except the unreactive noble gases) have smaller valences than H or Cl. Hence we assign a valence of 1 to H and to Cl. The valence of O is twice as great, and so we assign a value of 2.

EXAMPLE 4.1 Use the data in Table 4.2 to predict what formula would be expected for a compound containing (a) sodium and fluorine; (b) calcium and fluorine.

Solution

a) From the table we can obtain the following formulas for the most common sodium compounds:

$$NaH \qquad Na_2O \qquad NaCl$$

All of these would imply that sodium has a valence of 1. For fluorine compounds we have

$$HF \qquad OF_2 \qquad ClF$$

which imply that fluorine also has a valence of 1. Therefore the formula is probably

$$NaF$$

b) We already know that the valence of fluorine is 1. For calcium the formulas

$$CaH_2 \qquad CaO \qquad CaCl_2$$

argue in favor of a valence of 2. Therefore the formula is most likely

$$CaF_2$$

In some cases one element can combine in more than one way with another. For example, you have already encountered the compounds $HgBr_2$ and Hg_2Br_2. There are many other examples of such variable valence in Table 4.2. Nevertheless in its most common compounds, each element usually exhibits one characteristic valence, no matter what its partner is. Therefore it is possible to use that valence to predict formulas. Variable valence of an element may be looked upon as an exception to the rule of a specific combining capacity for each element.

The experimental observation that a given element usually has a specific valence can be explained if we assume that each of its atoms has a fixed number of valence sites. One of these sites would be required to connect with one site on another atom. In other words, a noble-gas atom such as Ar or Ne would not have any combining sites, H and Cl atoms would have one valence site each, an O atom would have two, and so on. Variable valence must involve atoms in which some valence sites are more readily used than others. In the case of the F compounds of Cl (ClF, ClF_3, ClF_5), for example, the formulas imply that at least five valence sites are available on Cl. Only one of these is used in ClF and in most of the chlorine compounds of Table 4.2. The others are apparently less readily available.

Mendeleev's inclusion of general formulas above the columns of his periodic table indicates that the table may be used to predict valences of the elements and formulas for their compounds. Two general rules may be followed:

1 In periodic groups I to IV, the group number is the most common valence.
2 In periodic groups V to VII, the most common valence is equal to 8 minus the group number, or to the group number itself.

For groups V to VII, the group number gives the valence only when the element in question is combined with oxygen, fluorine, or perhaps one of the other halogens. Otherwise 8 minus the group number is the rule.

EXAMPLE 4.2 Use the modern periodic table inside the front cover of this book to predict the formulas of compounds formed from (a) aluminum and chlorine; (b) phosphorus and chlorine. Use Table 4.2 to verify your prediction.

Solution

a) Aluminum is in group III, and so rule 1 predicts a valence of 3. Chlorine is in group VII and is not combined with oxygen or fluorine, and so its valence is $8 - 7 = 1$ by rule 2. Each aluminum has three valence sites, while each chlorine has only one, and so it requires three chlorine atoms to satisfy one aluminum, and the formula is $AlCl_3$.

b) Again chlorine has a valence of 1. Phosphorus is in group V and might have a valence of 5 or of $8 - 5 = 3$. Therefore we predict formulas PCl_5 or PCl_3. *Note:* All three predicted formulas appear in Table 4.2.

Exceptions to the Periodic Law

In the process of constructing the first periodic table, Mendeleev encountered several situations where the properties of elements were incompatible with the positions they would be forced to occupy in order of increasing atomic weight. In such a case, Mendeleev chose to emphasize the properties, because in the 1870s it was difficult to determine atomic weights accurately. He assumed that some atomic weights were in error and that ordering of elements ought to be changed to agree with chemical behavior.

We pointed out a problem of this type in the preceding section. Mendeleev did not have to contend with it because the noble gases had not been discovered in 1872, but it illustrates the difficulty nicely. There was a break in the regular sequence of valences of the first 24 elements when we came to K and Ar. The alkali metal has a *smaller* atomic weight than the noble gas and appears before the noble gas in Table 4.2. All other alkali metals immediately follow noble gases (they have slightly *larger* atomic weights). Unless we make an exception to the order of increasing atomic weight for Ar and K, the periodic table would contain a strange anomaly. One of the elements in the vertical column of noble gases would be the extremely reactive K. Likewise, the group of alkali metals would contain Ar, which is not a metal and is very unreactive.

Mendeleev's assumption that more accurate atomic weight determinations would eliminate situations such as we have just described has turned out to be incorrect. The atomic weights in Table 4.2 are modern, highly accurate values, but they still predict the wrong order for Ar and K. The same problem occurs in the case of Co and Ni and of Te and I. Apparently atomic weight, although related to chemical behavior, is not as fundamental as Mendeleev and other early developers of the periodic table thought.

Implications of Periodicity for Atomic Theory

The concept of valence implies that atoms of each element have a characteristic number of sites by which they can be connected to atoms of other elements. The number of valence sites repeats periodically as atomic weight increases, and occasionally even this regular repetition is imperfect. Atoms of similar atomic weight often have quite different properties, while some which differ widely in relative mass behave almost the same.

Dalton's atomic theory considers atoms to be indestructible spheres whose most important property is mass. This is clearly inadequate to account for the macroscopic observations described in this and the preceding section. In order to continue using the atomic theory, *we must attribute some underlying structure to atoms.* If both valence and atomic weight are determined by that structure, we should be able to account for the close but imperfect relationship between these two properties. The next section will describe some of the experiments which led to current theories about just what this atomic structure is like.

4.3 THE NUCLEAR ATOM

Radioactivity

Just prior to the turn of the twentieth century, additional observations were made which contradicted parts of Dalton's atomic theory. The French physicist Henri Becquerel (1852 to 1928) discovered by accident that compounds of uranium and thorium emitted *rays* which, like rays of sunlight, could darken photographic films. Becquerel's rays differed from light in that they could even pass through the black paper wrappings in which his

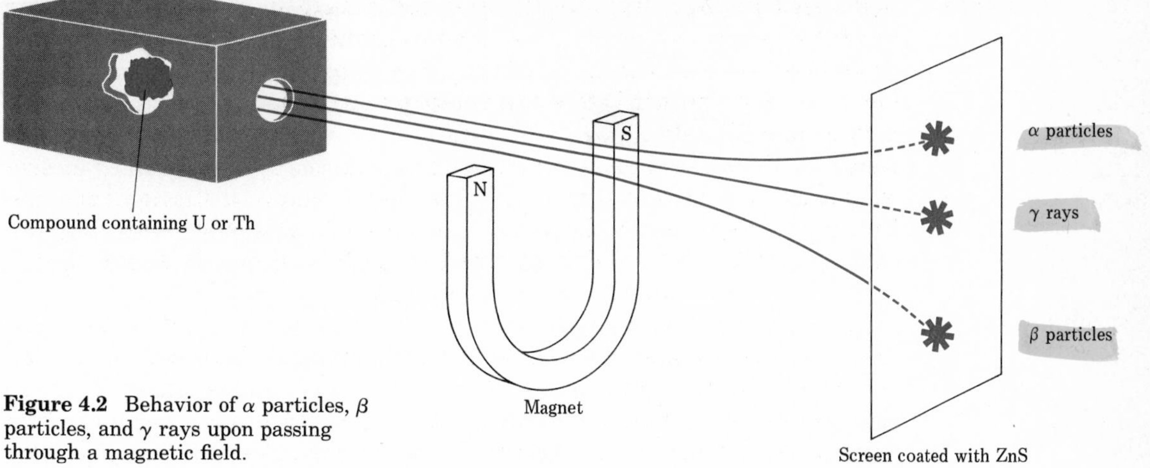

Figure 4.2 Behavior of α particles, β particles, and γ rays upon passing through a magnetic field.

films were stored. Although themselves invisible to the human eye, the rays could be detected easily because they produced visible light when they struck phosphors such as impure zinc sulfide. Such luminescence is similar to the glow of a psychedelic poster when invisible ultraviolet (black light) rays strike it.

Further experimentation showed that if the rays were allowed to pass between the poles of a magnet, they could be separated into the three groups shown in Fig. 4.2. Because little or nothing was known about these rays, they were labeled with the first three letters of the Greek alphabet. Upon passing through the magnetic field, the alpha rays (α rays) were deflected slightly in one direction, beta rays (β rays) were deflected to a much greater extent in the opposite direction, and gamma rays (γ rays) were not deflected at all.

Deflection by a magnet is a characteristic of electrically charged *particles* (as opposed to rays of light). From the direction and extent of deflection it was concluded that the β particles had a negative charge and were much less massive than the positively charged α particles. The γ rays did not behave as electrically charged particles would, and so the name *rays* was retained for them. Taken together the α particles, β particles, and γ rays were referred to as **radioactivity**, and the compounds which emitted them as **radioactive.**

Study of radioactive compounds by the French chemist Marie Curie (1867 to 1934) revealed the presence of several previously undiscovered elements (radium, polonium, actinium, and radon). These elements, and any compounds they formed, were intensely radioactive. When thorium and uranium compounds were purified to remove the newly discovered elements, the level of radioactivity decreased markedly. It increased again over a period of months or years, however. Even if the uranium or thorium compounds were carefully protected from contamination, it was possible to find small quantities of radium, polonium, actinium, or radon in them after such a time.

To chemists, who had been trained to accept Dalton's indestructible atoms, these results were intellectually distasteful. The inescapable conclusion was that some of the uranium or thorium atoms were spontaneously changing their structures and becoming atoms of the newly discovered elements. A change in atomic structure which produces a different element is called **transmutation.** Transmutation of uranium into the more radioactive elements could explain the increased emission of radiation by a carefully sealed sample of a uranium compound.

During these experiments with radioactive compounds it was observed that minerals containing uranium or thorium always contained lead as well. This lead apparently resulted from further transmutation of the highly radioactive elements radium, polonium, actinium, and radon. The lead found in uranium ores always had a significantly lower atomic weight than lead from most other sources (as low as 206.4 compared with 207.2, the accepted value). Lead associated with thorium always had an unusually high atomic weight. Nevertheless, all three forms of lead had the same chemical properties. Once mixed together, they could not be separated. Such results, as well as the reversed order of elements such as Ar and K in the periodic table, implied that atomic weight is not the fundamental determinant of chemical behavior.

The Electron

Near the middle of the nineteenth century the English chemist and physicist Michael Faraday (1791 to 1867) established a connection between electricity and chemical reactions. He already knew that an electric current flowing into certain molten compounds through metal plates called **electrodes** could cause reactions to occur. Samples of different elements would deposit on the electrodes. Faraday found that the same quantity of electric charge was required to produce 1 mol of any element whose valence was 1. Twice that quantity of charge would deposit 1 mol of an element whose valence was 2, and so on. Electric charge is measured in units called **coulombs,** abbreviated C. One coulomb is the quantity of charge which corresponds to a current of one ampere flowing for one second. It was found that 96 500 C of charge was required to deposit on an electrode 1 mol of an element whose valence is 1.

Faraday's experiments strongly suggested that electricity, like matter, consists of very small indivisible particles. The name **electron** was given to these particles, and an electric current came to be thought of as a flow of electrons from one place to another. When such a current flows into a chemical compound, one electron is required for each atom of a univalent element deposited on an electrode, two electrons for each atom of an element whose valence is 2, and so on. Thus an electric charge of 96 500 C corresponds to 1 mol of indivisible electric particles (electrons).

The relationship between electricity and atomic structure was further clarified by experiments involving **cathode-ray tubes** in the 1890s. A cathode-ray tube can be made by pumping most of the air or other gas out of a glass tube and applying a high voltage to two metal electrodes inside. If ZnS or some other phosphor is placed on the glass at the end of the tube opposite the negatively charged electrode **(cathode),** the ZnS emits light. This indicates that some kind of rays are streaming away from the cathode.

When passed between the poles of a magnet, these cathode rays behave the same way as the β particles described earlier. The fact that they were very small electrically charged particles led the English physicist J. J. Thomson (1856 to 1940) to identify them with the electrons of Faraday's experiments. Thus cathode rays are a beam of electrons which come out of the solid metal of the cathode. They behave exactly the same way no matter what the electrode is made of or what gas is in the tube. These observations allow one to conclude that *electrons must be constituents of all matter.*

In addition to being deflected by a magnet, the electron beam in a cathode-ray tube can be attracted toward a positively charged metal plate or repelled from a negative plate. By adjusting such electrodes to exactly cancel the deflection produced by a magnet of known strength, Thomson was able to determine that the ratio of charge to mass for an electron is 1.76×10^8 C/g. This is a rather large ratio. Either each electron has a very large charge, or each has a very small mass. We can see which by using Faraday's result that there are 96 500 C mol^{-1} of electrons

$$\frac{96\,500 \not{C}\ \text{mol}^{-1}}{1.76 \times 10^8 \not{C}\ \text{g}^{-1}} = 5.48 \times 10^{-4}\ \text{g mol}^{-1}$$

Thus the molar mass of an electron is 5.48×10^{-4} g mol^{-1}, and if we think of the electron as an "atom" (or indivisible particle) of electricity, its atomic weight would be 0.000 548—only $\frac{1}{1837}$ that of hydrogen, the lightest element known. In 1909 the American physicist Robert A. Millikan (1863 to 1953) was able to determine the charge on an electron independently of its mass. His value of 1.60×10^{-19} C can be combined with Thomson's charge-to-mass ratio to give an independent check on the molar mass for the electron

$$\frac{1.60 \times 10^{-19} \not{C}}{1.76 \times 10^8 \not{C}\ \text{g}^{-1}} \times 6.022 \times 10^{23}\ \text{mol}^{-1} = 5.47 \times 10^{-4}\ \text{g mol}^{-1}$$

thus confirming that the electron has much less mass than the lightest atom. (The quantity 1.60×10^{-19} C is often represented by the symbol e. Thus the charge on a single electron is $-e = -1.60 \times 10^{-19}$ C. The minus sign indicates that the electron is a *negatively* charged particle.)

The Nucleus

The results of Thomson's and other experiments implied that electrons were constituents of all matter and hence of all atoms. Since macroscopic samples of the elements are found to be electrically neutral, this meant that each atom probably contained a positively charged portion to balance the negative charge of its electrons. In an attempt to learn more about how positive and negative charges were distributed in atoms, Ernest Rutherford (1871 to 1937) and his coworkers performed numerous experiments in which α particles emitted from a radioactive element such as polonium were allowed to strike thin sheets of metals such as gold or platinum. It was

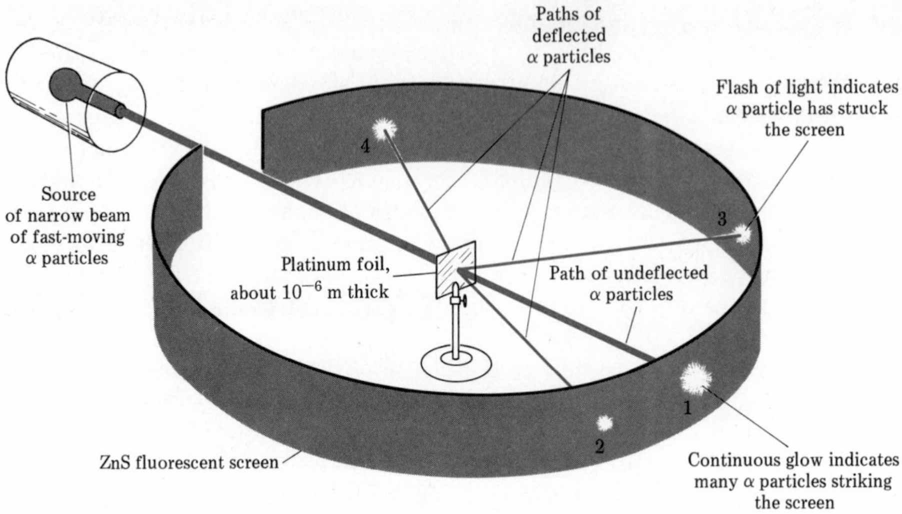

Figure 4.3 Schematic diagram of apparatus used by Geiger and Marsden to study deflection of α particles by thin metal foil. When an α particle strikes the ZnS screen, a flash of light is observed. Most of the flashes occurred at position 1, indicating the most α particles passed through the metal with little or no deflection. The few flashes at positions such as number 4 were interpreted to mean that a few α particles had struck something massive in the metal foil and hence had bounced almost straight back (see Fig. 4.4).

already known that the α particles carried a positive charge and traveled rapidly through gases in straight lines. Rutherford reasoned that in a solid, where the atoms were packed tightly together, there would be numerous collisions of α particles with electrons or with the unknown positive portions of the atoms. Since the mass of an individual electron was quite small, a great many collisions would be necessary to deflect an α particle from its original path, and Rutherford's preliminary calculations indicated that most would go right through the metal targets or be deflected very little by the electrons. In 1909, confirmation of this expected result was entrusted to Hans Geiger and a young student, Ernest Marsden, who was working on his first research project.

The results of Geiger and Marsden's work (using apparatus whose design is shown schematically in Fig. 4.3) were quite striking. Most of the α particles went straight through the sample or were deflected very little. These were observed by means of continuous luminescence of the ZnS screen at position 1 in the diagram. Observations made at greater angles from the initial path of the α particles (positions 2 and 3) revealed fewer and fewer flashes of light, but even at an angle nearly 180° from the initial path (position 4), *a few α particles were detected coming backward from the target.* This result amazed Rutherford. In his own words, "It was quite the most incredible event that has ever happened to me in my life. It was almost as incredible as if you fired a 15-inch shell at a piece of tissue paper and it came back and hit you. On consideration, I realized that this scattering backwards must be the result of a single collision, and when I made calculations I saw that it was impossible to get anything of that order of magnitude

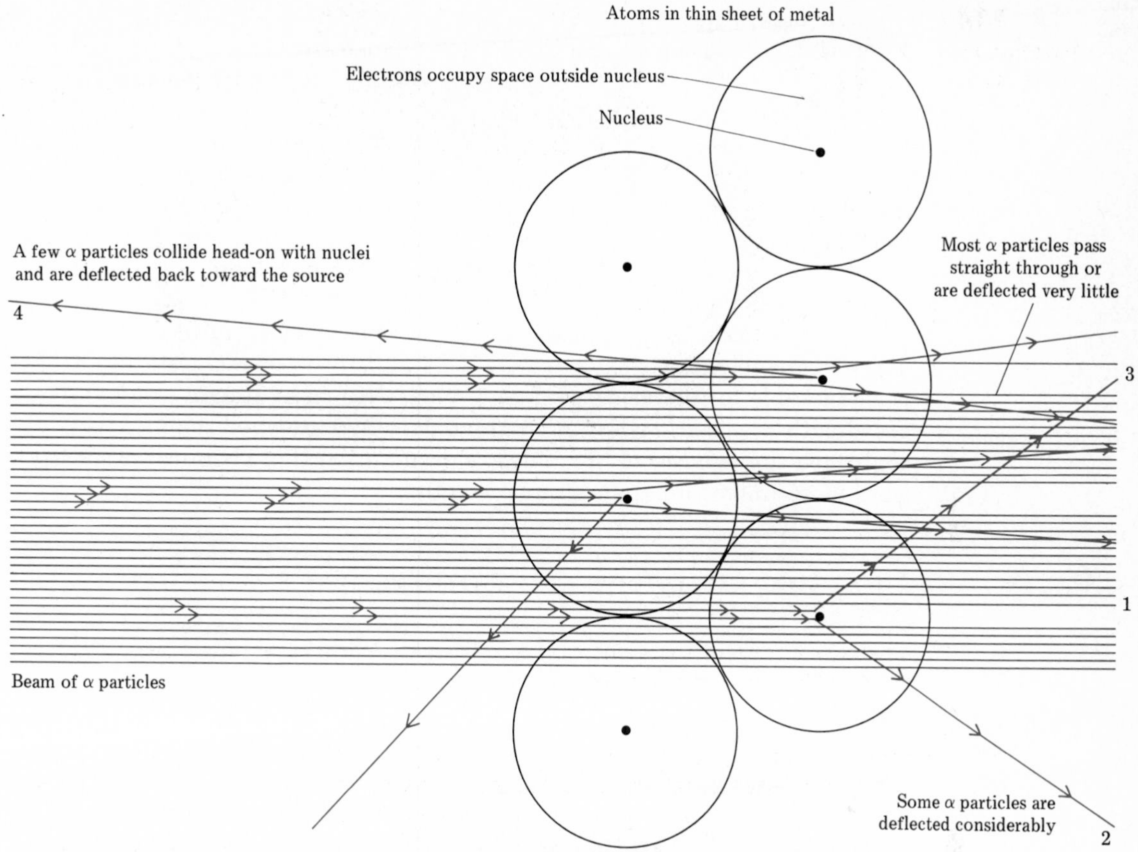

Figure 4.4 Rutherford's microscopic interpretation of the results of Geiger and Marsden's experiment.

unless you took a system in which the greater part of the mass of an atom was concentrated in a minute nucleus."[1] Rutherford's interpretation of Geiger and Marsden's experiment is shown schematically in Fig. 4.4.

Quantitative calculations using these experimental results showed that the diameter of the nucleus was about one ten-thousandth that of the atom. The positive charge on the nucleus was found to be $+Ze$, where Z is the number which indicates the position of an element in the periodic table. (For example, H is the first element and has $Z = 1$. He is the second element and has $Z = 2$. The twentieth element in Table 4.2 or Fig. 4.1 is Ca, and the nucleus of each Ca atom therefore has a charge of $+20e = 20 \times 1.60 \times 10^{-19}$ C $= 32.0 \times 10^{-19}$ C.) In order for an atom to remain electrically neutral, it must have a total of Z electrons outside the nucleus. These provide a charge of $-Ze$ to balance the positive nuclear charge. The number Z, which indicates the positive charge on the nucleus and the number of electrons in an atom, is called the **atomic number.**

[1] Ernest Rutherford, The Development of the Theory of Atomic Structure, in J. Needham and W. Pagel (eds.) "Background to Modern Science," The Macmillan Company, New York, 1938.

The significance of the atomic number was firmly established in 1914 when H. G. J. Moseley (1888 to 1915) published the results of experiments in which he bombarded a large number of different metallic elements with electrons in a cathode-ray tube. Wilhelm Roentgen (1845 to 1923) had discovered earlier that in such an experiment, rays were given off which could penetrate black paper or other materials opaque to visible light. He called this unusual radiation x-rays, the x indicating *unknown*. Moseley found that the frequency of the x-rays was unique for each different metal. It depended on the atomic number (but not on the atomic weight) of the metal. (If you are not familiar with electromagnetic radiation or the term *frequency,* read Sec. 21.1 where they are discussed more fully.) Using his x-ray frequencies, Moseley was able to establish the correct ordering in the periodic table for elements such as Co and Ni whose atomic weights disagreed with the positions to which Mendeleev had assigned them. His work confirmed the validity of Mendeleev's assumption that chemical properties were more important than atomic weights.

4.4 ATOMIC STRUCTURE AND ISOTOPES

The experimental facts described in the preceding section can be accounted for by assuming that any atom is made up of three kinds of **subatomic particles.**

1 The **electron** carries a charge of $-e$, has a mass about $\frac{1}{1837}$ that of a hydrogen atom, and occupies most of the volume of the atom.
2 The **proton** carries a charge of $+e$, has a mass about the same as a hydrogen atom, and is found within the very small volume of the nucleus.
3 The **neutron** carries no electric charge, has about the same mass as a hydrogen atom, and is found in the nucleus.

Some important properties of the three kinds of subatomic particles are listed in Table 4.3. Experimental evidence for the existence of the neutron was first correctly interpreted in 1932 by James Chadwick (1891 to), a discovery for which he was awarded the Nobel Prize in 1935.

TABLE 4.3 Important Subatomic Particles and Some of Their Properties.

Particle	Mass/kg	Electric Charge/C	Location
Electron	9.1095×10^{-31}	-1.6022×10^{-19}	Outside nucleus
Proton	1.6726×10^{-27}	$+1.6022 \times 10^{-19}$	In nucleus
Neutron	1.6750×10^{-27}	0	In nucleus

The modern picture of a helium atom, which is made up of two electrons, two protons, and two neutrons, is shown in Fig. 4.5. Because each proton and each neutron has more than 1800 times the mass of an electron, nearly all the mass of the helium atom is accounted for by the nucleus. This agrees with Rutherford's interpretation of the Geiger-Marsden experiment.

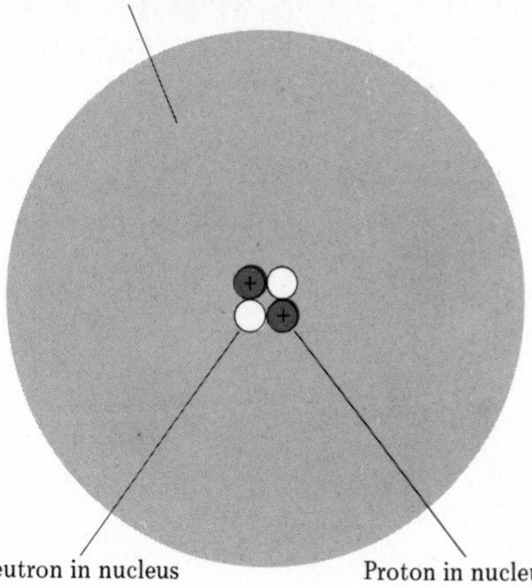

Figure 4.5 The atomic structure of a helium atom. Two electrons, two protons, and two neutrons are arranged as shown.

The number of units of positive charge on the nucleus is usually about half the number of units of mass because about half the nuclear particles are uncharged neutrons. The two electrons move about rapidly, occupying all the volume of the atom outside the nucleus. Their negative charge neutralizes the positive charge of the two protons, producing a neutral or uncharged atom.

The protons and neutrons in the nucleus of an atom such as helium are held very tightly by strong nuclear forces. It is very difficult either to separate the nuclear particles or to add extra ones. The electrons, on the other hand, are held to the atom by their electrostatic attraction for the positively charged protons in the nucleus. This force is strong, but not so strong that an atom cannot lose or gain electrons. When the number of electrons is not the same as the number of protons, an atom has a net electric charge and is called an **ion.** The α particles emitted by radioactive elements consist of two protons and two neutrons tightly bound together. Thus an α particle is the same as a helium nucleus; that is, a helium atom that has lost its two electrons or a helium ion whose charge is $+2e$. When α particles are emitted into a closed container, they slowly pick up electrons from their surroundings, and eventually the container becomes filled with helium.

The structure of any atom may be specified by indicating how many electrons, protons, and neutrons it contains. The number of protons is the same as the number of electrons and is given by the atomic number Z. Instead of directly specifying how many neutrons are present, we use the **mass number** A. This is the total number of particles in the nucleus; hence

$$A = \text{number of protons} + \text{number of neutrons}$$

$$A = Z + N$$

where N represents the number of neutrons. To symbolize a particular atom, the mass number and atomic number are written as a superscript and subscript preceding the chemical symbol (Sy) as follows:

$${}^{A}_{Z}\text{Sy}$$

The helium atom, whose structure was represented above, has 2 protons and 2 electrons ($Z = 2$) as well as 2 neutrons. Hence $A = Z + N = 2 + 2 = 4$, and the atom is represented by

$$^{4}_{2}\text{He}$$

In the case of an ion the positive or negative charge is indicated as a superscript to the right of the chemical symbol. Thus a helium atom which had lost two electrons (a helium ion with two more protons than electrons) would be written as

$$^{4}_{2}\text{He}^{2+}$$

EXAMPLE 4.3 How many electrons, protons, and neutrons are there in each of the atoms represented below?

$$^{12}_{6}\text{C} \qquad ^{40}_{20}\text{Ca} \qquad ^{206}_{82}\text{Pb} \qquad ^{40}_{20}\text{Ca}^{2+}$$

Solution For an atom the number of electrons equals the number of protons and is given by Z. For an ion the atomic number gives the number of protons, but the number of electrons must be determined from the charge. Thus

$^{12}_{6}\text{C}$ contains 6 electrons and 6 protons.

$^{40}_{20}\text{Ca}$ contains 20 electrons and 20 protons.

$^{206}_{82}\text{Pb}$ contains 82 electrons and 82 protons.

$^{40}_{20}\text{Ca}^{2+}$ has lost two electrons. Therefore it contains 18 electrons and 20 protons.

The number of neutrons can be obtained by subtracting the number of protons (Z) from the total number of particles in the nucleus (A):

$^{12}_{6}\text{C}$ $N = A - Z = 12 - 6 = 6$ neutrons

$^{40}_{20}\text{Ca}$ $N = 40 - 20 = 20$ neutrons (The same applies to $^{40}_{20}\text{Ca}^{2+}$. Only *electrons* are gained or lost when an ion forms.)

$^{206}_{82}\text{Pb}$ $N = 206 - 82 = 124$ neutrons

Isotopes

The presence of neutrons in atomic nuclei accounts for the occurrence of **isotopes**—samples of an element whose atoms contain different numbers of neutrons and hence exhibit different atomic weights. For example, naturally occurring hydrogen can be separated into two isotopes. More than 99.98 percent is "light" hydrogen, $^{1}_{1}\text{H}$. This consists of atoms each of which has one proton, one electron, and zero neutrons. The rest is "heavy" hydrogen or deuterium, $^{2}_{1}\text{H}$, which consists of atoms which contain one electron, one proton, and one neutron. Hence the atomic weight of deuterium is almost exactly twice as great as for light hydrogen. By transmutation of lithium, it is also possible to obtain a third isotope, tritium, ^3_1H. It consists of atoms whose nuclei contain two neutrons and one proton. Its atomic (or more correctly, isotopic) weight is about 3 times that of light hydrogen. The **isotopic weight** is the relative mass of an atom of a given isotope.

The discovery of isotopes and its explanation on the basis of an atomic structure built up from electrons, protons, and neutrons required a change in the ideas about atoms which John Dalton had proposed (refer to Table 2.1). For a given element all atoms are *not quite* identical in all respects —especially with regard to mass. The number of protons in the nucleus and the number of electrons which occupy most of the volume of an atom are the factors which determine its chemical behavior. *All atoms of the same element have the same atomic number,* but different isotopes have different atomic weights.

Transmutation and Radioactivity

Transmutation of one element into another requires a change in the structures of the *nuclei* of the atoms involved. For example, the first step in the spontaneous radioactive decay of uranium is emission of an α particle, $^4_2He^{2+}$, from the nucleus of $^{238}_{92}U$. Since the α particle consists of two protons and two neutrons, the atomic number must be reduced by 2 and the mass number by 4. The product of this nuclear reaction is therefore $^{234}_{90}Th$. In other words, loss of an α particle changes (transmutes) uranium into thorium.

Loss of a β particle (electron) from an atomic nucleus leaves the nucleus with an extra unit of positive charge, that is, an extra proton. This increases the atomic number by 1 and also changes one element to another. For example, the $^{234}_{90}Th$ mentioned in the previous paragraph emits β particles. Its atomic number increases by 1, but its mass number remains the same. (The β particle is an electron and has a very small mass.) In effect one neutron is converted to a proton and an electron. Thus the thorium transmutes to protactinium, $^{234}_{91}Pa$. (Note carefully that the β particle is an electron emitted from the *nucleus* of the thorium atom, not one of the electrons from outside the nucleus.)

A γ ray is not a particle, and so its emission from a nucleus does not involve a change in atomic number or mass number. Rather it involves a change in the way the same protons and neutrons are packed together in the nucleus. It is important to note, however, that radioactivity and transmutation both involve changes *within the atomic nucleus.* Such **nuclear reactions** will be discussed in more detail in Chap. 19. Because protons and neutrons are held tightly in the nucleus, nuclear reactions are much less common in everyday life than chemical reactions. The latter involve electrons surrounding the nucleus, and these are much less rigidly held.

Average Atomic Weights

Since all atoms of a given element do not necessarily have identical masses, the atomic weight must be averaged over the isotopic weights of all naturally occurring isotopes.

EXAMPLE 4.4 Naturally occurring lead is found to consist of four isotopes:

1.40% $^{204}_{82}Pb$ whose isotopic weight is 203.973.

24.10% $^{206}_{82}Pb$ whose isotopic weight is 205.974.

22.10% $^{207}_{82}Pb$ whose isotopic weight is 206.976.

52.40% $^{208}_{82}Pb$ whose isotopic weight is 207.977.

Calculate the atomic weight of an average naturally occurring sample of lead.

Solution Suppose that you had 1 mol lead. This would contain 1.40% ($\frac{1.40}{100}$ × 1 mol) $^{204}_{82}Pb$ whose molar mass is 203.973 g mol^{-1}. The mass of $^{204}_{82}Pb$ would be

$$m_{204} = n_{204} \times M_{204} = \left(\frac{1.40}{100} \times 1 \text{ mol}\right)(203.973 \text{ g mol}^{-1}) = 2.86 \text{ g}$$

Similarly for the other isotopes

$$m_{206} = n_{206} \times M_{206} = \left(\frac{24.10}{100} \times 1 \text{ mol}\right)(205.974 \text{ g mol}^{-1}) = 49.64 \text{ g}$$

$$m_{207} = n_{207} \times M_{207} = \left(\frac{22.10}{100} \times 1 \text{ mol}\right)(206.976 \text{ g mol}^{-1}) = 45.74 \text{ g}$$

$$m_{208} = n_{208} \times M_{208} = \left(\frac{52.40}{100} \times 1 \text{ mol}\right)(207.977 \text{ g mol}^{-1}) = 108.98 \text{ g}$$

Upon summing all four results, the mass of 1 mol of the mixture of isotopes is found to be

$$2.86 \text{ g} + 49.64 \text{ g} + 45.74 \text{ g} + 108.98 \text{ g} = 207.22 \text{ g}$$

Thus the atomic weight of lead is 207.2, as mentioned earlier in the discussion.

An important corollary to the existence of isotopes should be emphasized at this point. When highly accurate results are obtained, atomic weights may vary slightly depending on where a sample of an element was obtained. This can occur because the percentages of different isotopes may depend on the source of the element. For example, lead derived from transmutation of uranium contains a much larger percentage of $^{206}_{82}Pb$ than the 24.1 percent shown in Example 4.4 for the average sample. Consequently the atomic weight of lead found in uranium ores is less than 207.2 and is much closer to 205.974, the isotopic weight of $^{206}_{82}Pb$.

After the possibility of variations in the isotopic composition of the elements was recognized, it was suggested that the scale of relative masses of the atoms (the atomic weights) should use as a reference the mass of an atom of a *particular isotope* of one of the elements. The standard that was eventually chosen was $^{12}_{6}C$, and it was assigned an atomic-weight value of exactly 12.000 000. Thus the atomic weights given in Table 2.2 are the ratios of weighted averages (calculated as in Example 4.4) of the masses of atoms of all isotopes of each naturally occurring element to the mass of a single $^{12}_{6}C$ atom. Since carbon consists of two isotopes, 98.89% $^{12}_{6}C$ of isotopic weight 12.000 and 1.11% $^{13}_{6}C$ of isotopic weight 13.003, the average atomic weight of carbon is

$$\frac{98.89}{100.00} \times 12.000 + \frac{1.11}{100.00} \times 13.003 = 12.011$$

for example. Deviations from average isotopic composition are usually not large, and so the average atomic weights serve quite well for nearly all chemical calculations. In the study of nuclear reactions, however, one must be concerned about isotopic weights. This will be discussed further in Chap. 19.

The SI definition of the mole also depends on the isotope $^{12}_{6}C$ and can now be stated. One mole is defined as the amount of substance of a system which contains as many elementary entities as there are atoms in exactly 0.012 kg of $^{12}_{6}C$. The *elementary entities* may be atoms, molecules, ions, electrons, or other microscopic particles. This official definition of the mole makes possible a more accurate determination of the Avogadro constant than was reported earlier. The currently accepted value is $N_A = 6.022\ 094 \times 10^{23}\ \text{mol}^{-1}$. This is accurate to 0.0001 percent and contains three more significant figures than $6.022 \times 10^{23}\ \text{mol}^{-1}$, the number used to define the mole in Chap. 2. It is very seldom, however, that more than four significant digits are needed in the Avogadro constant. The value $6.022 \times 10^{23}\ \text{mol}^{-1}$ will certainly suffice for any calculations we shall need in this book.

4.5 MEASUREMENT OF ATOMIC WEIGHTS

You may have wondered why we have been so careful to define atomic weights and isotopic weights as *ratios* of masses. The reason will be clearer once the most important and accurate experimental technique by which isotopic weights are measured has been described. This technique, called **mass spectrometry,** has developed from the experiments with cathode-ray tubes mentioned earlier in this chapter. It depends on the fact that an electrically charged particle passing through a magnetic field of constant strength moves in a circular path. The radius r of such a path is directly proportional to the mass m and the speed u of the particle, and inversely proportional to the charge Q. Thus the greater the mass or speed of the particle, the greater the radius of its path. The greater the charge, the smaller the radius.

In a mass spectrometer (Fig. 4.6) atoms or molecules in the gaseous phase are bombarded by a beam of electrons. Occasionally one of these electrons will strike another electron in a particular atom, and both electrons

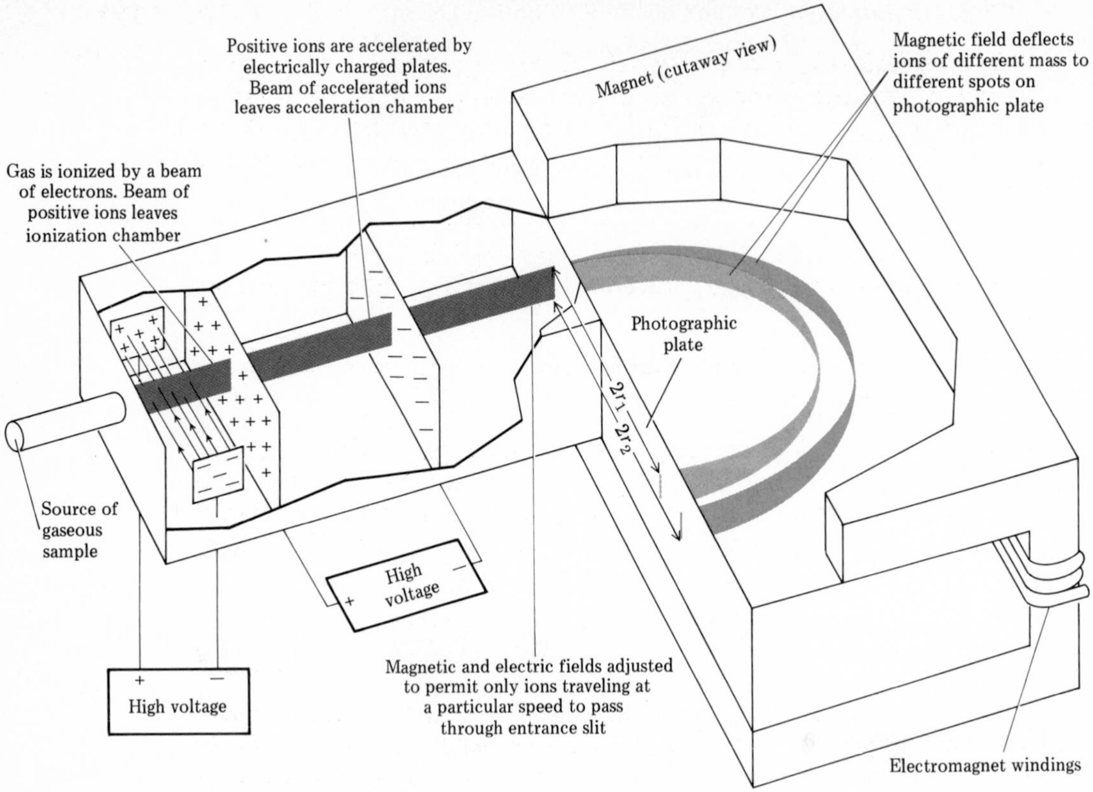
Positive ions are accelerated by electrically charged plates. Beam of accelerated ions leaves acceleration chamber
Magnet (cutaway view)
Magnetic field deflects ions of different mass to different spots on photographic plate
Gas is ionized by a beam of electrons. Beam of positive ions leaves ionization chamber
Photographic plate
2r1
2r2
Source of gaseous sample
High voltage
High voltage
Magnetic and electric fields adjusted to permit only ions traveling at a particular speed to pass through entrance slit
Electromagnet windings

Figure 4.6 Schematic diagram of a mass spectrometer.

will have enough energy to escape the attraction of the positive nucleus. This leaves behind a positive ion since the atom now has one more proton than it has electrons. For example,

$$^{12}_{6}C + e^{-}(\text{high-speed electron}) \rightarrow {}^{12}_{6}C^{+} + 2e^{-}$$

Once positive ions are produced in a mass spectrometer, they are accelerated by the attraction of a negative electrode and pass through a slit. This produces a narrow beam of ions traveling parallel to one another. The beam then passes through electric and magnetic fields. The fields deflect away all ions except those traveling at a certain speed.

The beam of ions is then passed between the poles of a large electromagnet. Since the speed and charge are the same for all ions, the radii of their paths depend only on their masses. For different ions of masses m_1 and m_2

$$\frac{r_2}{r_1} = \frac{m_2}{m_1} \tag{4.3}$$

and the ratio of masses may be obtained by measuring the ratio of radii. The paths of the ions are determined either by a photographic plate (which darkens where the ions strike it, as in the figure) or a metal plate connected to a galvanometer (a device which detects the electric current due to the beam of charged ions).

EXAMPLE 4.5 When a sample of carbon is vaporized in a mass spectrometer, two lines are observed on the photographic plate. The darker line is 27.454 cm, and the other is 29.749 cm from the entrance slit. Determine the relative atomic masses (isotopic weights) of the two isotopes of carbon.

Solution Since the distance from the entrance slit to the line on the photographic plate is twice the radius of the circular path of the ions, we have

$$\frac{m_2}{m_1} = \frac{r_2}{r_1} = \frac{2r_2}{2r_1} = \frac{29.749\ \cancel{\text{cm}}}{27.454\ \cancel{\text{cm}}} = 1.083\ 59$$

Thus $m_2 = 1.083\ 59m_1$. If we assume that the darker mark on the photographic plate is produced because there are a greater number of $^{12}_{6}C^+$ ions than of the less common $^{13}_{6}C^+$, then m_1 may be equated with the relative mass of $^{12}_{6}C$ and may be assigned a value of 12.000 000 exactly. The isotopic weight of $^{13}_{6}C$ is then

$$m_2 = (1.083\ 59)(12.000\ 000) = 13.0031$$

Notice that in mass spectrometry all that is required is that the charge and speed of the two ions whose relative masses are to be determined be *the same*. If the mass of an individual ion were to be measured accurately, its actual speed upon entering the magnetic field and the exact magnitude of its electric charge would have to be known very accurately. Therefore it is easier to measure the ratio of two masses than to determine a single absolute mass, and so atomic weights are reported as pure numbers.

SUMMARY

If you glance back through this chapter, you will see that a number of quite different kinds of experiments contributed to the extension of Dalton's atomic theory to include subatomic particles and atomic structure. The periodic variation of valence and the periodic table's successful correlation of macroscopic properties indicate that atoms must have certain specific ways of connecting to other atoms. It is reasonable to assume that valence depends on some underlying atomic structure. Atoms which are similar in structure should exhibit the same valence and have similar chemical and physical properties.

The discovery of radioactivity and transmutation implied that one kind of atom could change into another. This too can be explained if atoms have structure. A change in that structure may produce a new kind of atom. Experiments with cathode-ray tubes indicated that electrons, which are very light and carry a negative charge, are present in all atoms. Rutherford's interpretation of the Geiger-Marsden experiment suggested that electrons occupy most of the volume of the atom while most of the mass is concentrated in a small positively charged nucleus.

Moseley's x-ray spectra and the existence of isotopes made it quite clear that Dalton's emphasis on the importance of atomic weight would have to be dropped.

The chemical behavior of an atom is determined by how many protons are in the nucleus. Changing the number of neutrons changes the atomic mass but has very little effect on chemistry. The identity of an element depends on its atomic number, not on its atomic weight. If the periodic law is restated as "When the elements are listed in order of increasing atomic *number*, their properties vary periodically," there are *no* exceptions.

There is one set of observations that we have not yet dealt with from the standpoint of the theory of atomic structure. What aspect of atomic structure is responsible for the periodic repetition of valence? When two atoms approach one another, it is the electrons which make initial contact. The radius of the nucleus is only about one ten-thousandth that of the atom, and the protons and neutrons are rather inaccessible. Therefore it is reasonable to expect that electrons will determine the chemical behavior of an atom. In addition we anticipate that repeated similarities in the ways electrons are packed into atoms can account for periodic repetition of valence and other properties. Knowledge of the electronic structure of atoms should be very useful in correlating the vast number of facts available in the descriptive chemistry of the elements. In the next chapter we begin the story of electronic structure, its relationship to the periodic table, and how it applies to predicting chemical properties.

QUESTIONS AND PROBLEMS

4.1 Using the symbol M to stand for the alkali metals (that is, Li, Na, K, Rb, or Cs), write general equations for the chemical reactions of these metals with

a Water
b Oxygen
c Hydrogen
d Sulfur
e Chlorine

4.2 Using the symbol M to denote the alkaline-earth metals (that is, Be, Mg, Ca, Sr, or Ba), write general equations for the chemical reactions of these metals with

a Water
b Oxygen
c Hydrogen
d Sulfur
e Chlorine

4.3 Distinguish between a period and a group in the periodic table. What element is in group IVA of the fifth period?

4.4 What facts led Mendeleev to suspect that there were undiscovered elements? Explain how he was able to make predictions about some of the properties of these undiscovered elements which later turned out to be remarkably accurate.

4.5 In the modern form of the periodic table, not all the periods are the same length. In the first six periods, how many periods have the same length? What is the length of each period?

4.6 Remembering the answers to the previous question, see if you can write down the atomic numbers of the noble gases, He, Ne, Ar, Kr, and Xe, correctly without looking at the periodic table.

4.7 Try to sketch the modern form of the periodic table from memory. Fill in the noble gases and as many of the alkali metals, alkaline-earth metals, and halogens as you can. See if you can also fill in the first 18 elements. Keep your attempt for later reference.

4.8 Using Table 4.2, predict the probable formula of the compound formed between each of the following pairs of elements:

a Mg and Br
b As and H
c Rb and O
d Se and Cl
e Ga and F
f Sn and O
g Kr and O
h Li and F

4.9 Using the same form of reasoning as that given in Example 4.2, predict which will be the most probable valence(s) of each of the following elements:

a In
b Ba
c I
d S
e Pb
f P
g Ge
h Cs
i Sb
j Te

4.10 Discuss the term *radioactive element*, carefully distinguishing between α, β, and γ rays as to mass and charge.

4.11 What are cathode rays? How are they different from α, β, and γ rays?

4.12 What facts led J. J. Thomson to the belief that electrons are a constituent of all matter?

4.13 Using the molar mass of the electron given in the text, find the mass of a single electron.

4.14 Explain in about 15 lines of writing how the α-particle scattering experiments of Rutherford, Geiger, and Marsden led to the nuclear model of the atom.

4.15 In their scattering experiments, Rutherford, Geiger, and Marsden found that gold foil was more effective at deflecting the α particles than silver foil. Explain why this is so.

4.16 Explain how an atom can be a neutral entity even though it is made up from charged particles.

4.17 The diameter of an atom is about 10^4 times as large as the diameter of its nucleus. If the diameter of the nucleus were 1 ft, what would be the diameter of an atom in miles?

4.18 The diameter of a neutron is often given as 1 fm (femtometer). Use this value and the mass given in Table 4.3 to calculate the density of a neutron. Express your result in grams per cubic centimeter. (Astronomers believe that some stars are made up only of neutrons and have a density of this order of magnitude.)

4.19 From Table 4.3 calculate the molar mass of the proton and the neutron.

4.20 List the number of protons, neutrons, and electrons in the atoms of each of the following elements: Na, Co, Cs, Pr, Mn, and As. Note that these elements are exceptional in that they consist of only *one isotope* each. Now write out the conventional symbol for each atom, including the atomic number and mass number.

4.21 Construct a table listing the numbers of protons, neutrons, and electrons in each of the following atoms or ions:

a $^{122}_{51}Si$ *d* $^{51}_{23}V$

b $^{23}_{11}Na^+$ *e* $^{80}_{35}Br^-$

c $^{18}_{8}O$ *f* $^{79}_{34}Se^{2-}$

4.22 The most common isotope of both argon and calcium has a mass number of 40. Write out the conventional symbols for both isotopes. Is it rare to find two elements whose isotope distribution is like this? Give reasons for your answer.

4.23 Explain carefully and exactly what is meant by the phrase, *the atomic weight of mercury is 200.59.*

4.24 A naturally occurring sample of the hypothetical element X is found to consist of three isotopes with the following percent abundances and isotopic weights:

12.09% ^{113}X	113.059
40.27% ^{115}X	114.932
47.64% ^{116}X	116.326

Calculate the atomic weight of this sample of X.

4.25 Assume that naturally occurring elemental copper consists of only two isotopes $^{63}_{29}Cu$ and $^{65}_{29}Cu$, whose isotopic weights are 62.930 and 64.928, respectively. Using the known atomic weight of copper, calculate the percent abundance of each of the two isotopes.

4.26 Explain how $^{12}_{6}C$ is the standard of reference for (*a*) atomic weights; (*b*) the SI definition of the mole.

4.27 Explain why it is that the number and behavior of electrons, rather than protons or neutrons, is the major factor in determining the chemical behavior of atoms.

4.28 When a sample of carbon monoxide is introduced into a mass spectrometer, the line corresponding to $^{12}_{6}C$ appears at a distance of 27.463 cm from the slit. Three lines corresponding to oxygen isotopes appear close together but farther away from the slit. There is a strong line at 36.606 cm, a fainter one at 41.193 cm, and a just-discernible line between them at 38.904 cm. Use these figures to find the relative atomic masses of the three oxygen isotopes.

4.29 Other lines which appear in the mass spectrum of CO described above are a faint line close to 30 cm and *six* lines in the range of 87 to 94 cm from the slit. How do you account for these lines?

chapter five

THE ELECTRONIC STRUCTURE OF ATOMS

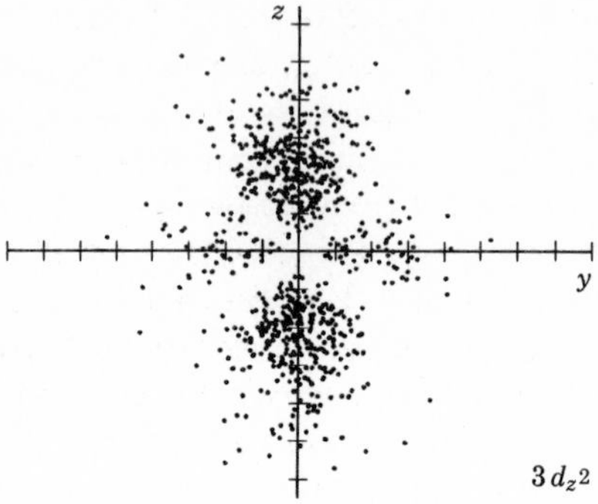

$3d_{z^2}$

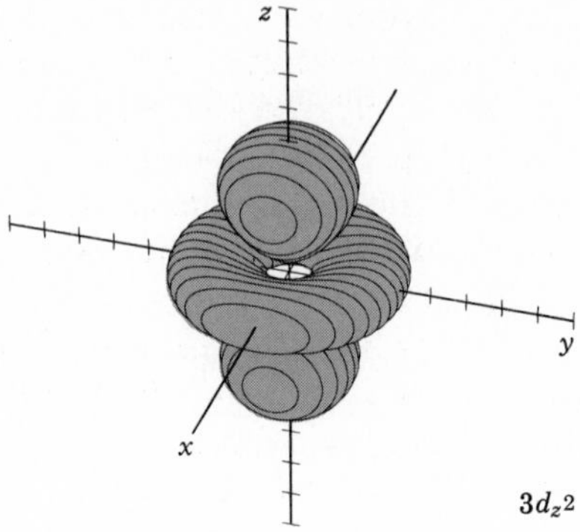

$3d_{z^2}$

Once scientists had accepted the idea that electrons were constituents of all matter, theories attempting to explain just how electrons were incorporated in the structure of the atom began to develop. This was especially true after Rutherford had discovered that most of the volume of an atom was occupied by electrons. Both chemists and physicists became interested in the electronic structure of atoms—the chemists because they wanted to explain valence and bonding, and the physicists because they wanted to explain the spectra of atoms, i.e., the light emitted when gaseous atoms were raised to a high temperature or bombarded by electrons. The chief contributors to these developments, which occurred mainly during the 15 years between 1910 and 1925, were the U.S. chemist Gilbert Newton Lewis (1875 to 1946), and the Danish physicist Niels Bohr (1885 to 1962).

5.1 ELECTRONS AND VALENCE

On a chemical level, an important clue to the unraveling of the electronic structure of atoms is the existence of noble gases, which are almost completely unable to form chemical compounds. This lack of reactivity suggests that the atoms of these elements have structures which do not permit interaction with the structures of other atoms. A second clue is the close correspondence between the valence of an element and the extent to which its atomic number differs from that of the nearest noble gas. Elements which have a valence of *1*, for instance, have atomic numbers *one more* or *one less* than that of a noble gas. Thus the atoms of the alkali metals Li, Na, K, Rb, and Cs all contain one electron more than the corresponding noble gases He, Ne, Ar, Kr, and Xe, while atoms of hydrogen H and the halogens F, Cl, Br, and I all contain one electron less. Similar remarks apply to a valence of *2*. The alkaline-earth metal atoms Be, Mg, Ca, Sr, and Ba all contain two electrons more than a noble-gas atom, while the elements O, S, Se, and Te all contain two electrons less. Exactly the same pattern of behavior also extends to elements with a valence of 3 or 4.

As early as 1902, Lewis began to suggest (in his lectures to general chemistry students, no less) that the behavior just described could be explained by assuming that the electrons in atoms were arranged in **shells,** all electrons in the same shell being approximately the same distance from the nucleus. Figure 5.1 illustrates the pictures which he eventually developed for helium, chlorine, and potassium atoms. In the helium atom the two electrons occupy only one shell, in the chlorine atom the 17 electrons are arranged in three shells, and in the potassium atom the 19 electrons occupy four shells. Lewis suggested that each shell can only accommodate so many electrons. Once this number has been reached, the shell must be regarded as *filled,* and any extra electrons are accommodated in the next shell, somewhat farther from the nucleus. Once a shell is filled, moreover, it is assumed to have a particularly stable structure which prevents the electrons in the shell from any involvement with other atoms. Thus it is only the electrons in the *outermost incompletely filled shell* (called **valence electrons**) that have any chemical importance. Furthermore, if the outermost shell is filled, then the resulting atom will have little or no tendency to react with other atoms and form compounds with them. Since this is exactly the behavior exhibited by the noble gases, Lewis concluded that the characteristic feature of an atom of a noble gas is a *filled outer shell* of electrons.

Assuming that the noble gases all contain an outermost filled shell, it is now quite simple to work out how many electrons can be accommodated in

Figure 5.1 The shell structure of atoms of He, Cl, and K, as suggested by Lewis.

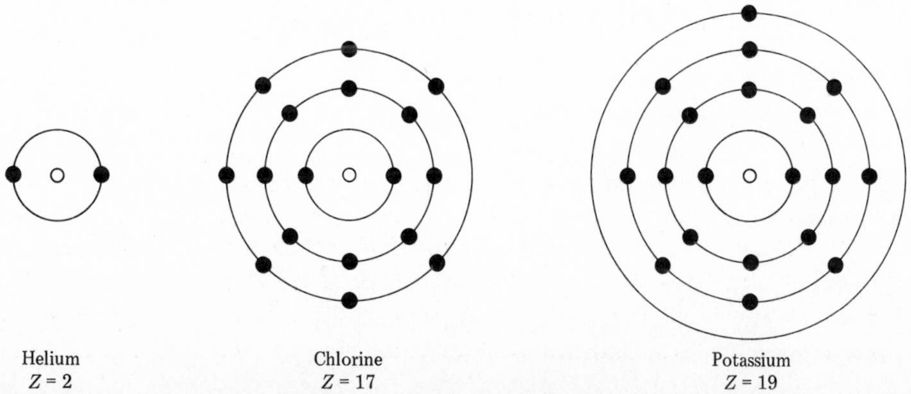

each shell. Since the first noble gas helium has two electrons, we know that only two electrons are needed to fill the first shell. A further eight electrons brings us to the next noble gas neon ($Z = 10$). Accordingly we deduce that the second shell can accommodate a maximum of eight electrons. A similar argument leads to the conclusion that the third shell also requires eight electrons to fill it and that an atom of argon has two electrons in the first shell, eight in the second, and eight in the third, a total of 18 electrons.

Lewis Diagrams

Lewis used simple diagrams (now called **Lewis diagrams**) to keep track of how many electrons were present in the outermost, or valence, shell of a given atom. The **kernel** of the atom, i.e., the nucleus together with the inner electrons, is represented by the chemical symbol, and only the valence electrons are drawn as dots surrounding the chemical symbol. Thus the three atoms shown in Fig. 5.1 can be represented by the following Lewis diagrams:

$$\mathrm{He}{:} \qquad {:}\overset{\cdot\cdot}{\underset{\cdot\cdot}{\mathrm{Cl}}}{\cdot} \qquad \mathrm{K}{\cdot}$$

If the atom is a noble-gas atom, two alternative procedures are possible. Either we can consider the atom to have zero valence electrons or we can regard the outermost filled shell as the valence shell. The first three noble gases can thus be written as

$$\mathrm{He}{:} \qquad {:}\overset{\cdot\cdot}{\underset{\cdot\cdot}{\mathrm{Ne}}}{:} \qquad {:}\overset{\cdot\cdot}{\underset{\cdot\cdot}{\mathrm{Ar}}}{:}$$

or   or   or

He   Ne   Ar

EXAMPLE 5.1 Draw Lewis diagrams for an atom of each of the following elements:

Li   N   F   Na

Solution We find from the periodic table inside the front cover that Li has an atomic number of 3. It thus contains three electrons, one more than the noble gas He. This means that the outermost, or valence, shell contains only one electron, and the Lewis diagram is

$$\mathrm{Li}{\cdot}$$

Following the same reasoning, N has seven electrons, five more than He, while F has nine electrons, seven more than He, giving

$${:}\overset{\cdot}{\underset{\cdot}{\mathrm{N}}}{\cdot} \qquad \text{and} \qquad {:}\overset{\cdot\cdot}{\underset{\cdot}{\mathrm{F}}}{:}$$

Na has nine more electrons than He, but eight of them are in the kernel, corresponding to the eight electrons in the outermost shell of Ne. Since Na has only 1 more electron than Ne, its Lewis diagram is

$$\mathrm{Na}{\cdot}$$

Notice from the preceding example that the Lewis diagrams of the alkali metals are identical except for their chemical symbols. This agrees nicely with the very similar chemical behavior of the alkali metals. Similarly, Lewis diagrams for all elements in other groups, such as the alkaline earths or halogens, look the same.

The Lewis diagrams may also be used to predict the valences of the elements. Lewis suggested that the number of valences of an atom was equal to the number of electrons in its valence shell or to the number of electrons which would have to be added to the valence shell to achieve the electronic shell structure of the next noble gas. As an example of this idea, consider the elements Be and O. Their Lewis diagrams and those of the noble gases He and Ne are

$$\text{He} \qquad \text{Be}\!: \qquad :\!\ddot{\underset{\cdot\cdot}{\text{O}}} \qquad :\!\ddot{\underset{\cdot\cdot}{\text{Ne}}}\!:$$

Comparing Be with He, we see that the former has two more electrons and therefore should have a valence of 2. The element O might be expected to have a valence of 6 or a valence of 2 since it has six valence electrons—two less than Ne.

Using rules of valence developed in this way, Lewis was able to account for the regular increase and decrease in the subscripts of the Cl compounds in Table 4.2, which we have already noted. In addition he was able to account for more than 50 percent of the formulas in the table. (Those that agree with his ideas are shaded in color or gray in Table 4.2. You may wish to refer to that table now and verify that some of the indicated formulas follow Lewis' rules.) Lewis' success in this connection gave a clear indication that electrons were the most important factor in holding atoms together when molecules formed.

Despite these successes, there are also difficulties to be found in Lewis' theories, in particular for elements beyond calcium in the periodic table. The element Br ($Z = 35$), for example, has 17 more electrons than the noble-gas Ar ($Z = 18$). This leads us to conclude that Br has 17 valence electrons, which makes it awkward to explain why Br resembles Cl and F so closely even though these two atoms have only seven valence electrons.

5.2 THE WAVE NATURE OF THE ELECTRON

At much the same time as Lewis was developing his theories of electronic structure, the physicist Niels Bohr was developing a similar, but more detailed, picture of the atom. Since Bohr was interested in light (energy) emitted by atoms under certain circumstances rather than the valence of elements, he particularly wanted to be able to calculate the energies of the electrons. To do this, he needed to know the exact path followed by each electron as it moved around the nucleus. He assumed paths similar to those of the planets around the sun. Figure 5.2, taken from a physics text of the period, illustrates Bohr's theories applied to the sodium atom. Note how the Bohr model, like that of Lewis, assumes a shell structure. There are two electrons in the innermost shell, eight electrons in the next shell, and a single electron in the outermost shell.

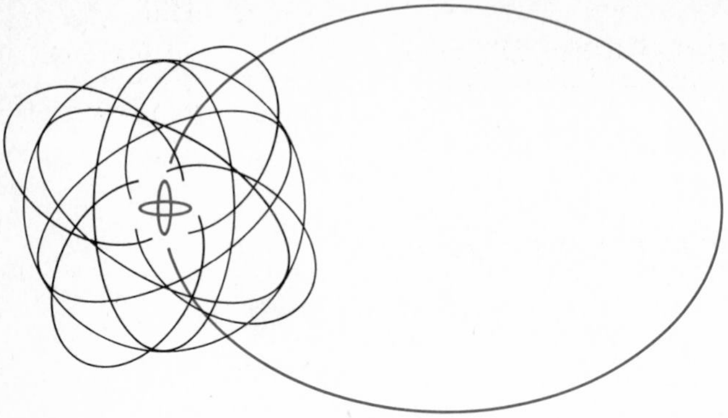

Figure 5.2 The Bohr picture of the Na atom. Two electrons orbit very close to the nucleus. Eight others move around somewhat farther away, and there is a single outermost electron in an elongated, elliptical orbit.

Like Lewis' model, Bohr's model was only partially successful. It explained some experimental results but was quite unable to account for others. In particular it failed on the quantitative mathematical level. The theory worked very well for a hydrogen atom with its single electron,[1] but calculations on atoms with more than one electron always gave the wrong answer. On a chemical level, too, certain features were inadequate. There is no evidence to suggest that atoms of sodium are ever as elongated or as flat as the one in Fig. 5.2. On the contrary, the way that sodium atoms pack together in a solid suggests that they extend out uniformly in all directions; i.e., they are spherical in shape. Another weakness in the theory was that it had to *assume* a shell structure rather than explain it. After all, there is nothing in the nature of planets moving around the sun which compels them to orbit in groups of two or eight. Bohr assumed that electrons behave much like planets; so why should they form shells in this way?

Wave Mechanics

During the middle of the nineteen-twenties some scientists began to realize that electrons must move around the nucleus in a very different way from that in which planets move around the sun. They abandoned the idea that an electron traces out a definite orbit or trajectory. Instead they adopted the point of view that it was impossible to describe the exact path of a particle whose mass was as small as that of an electron. Rather than think of the motion in planetary terms, they suggested it was much more useful to think of this motion in terms of a *wave* which could fold itself around the nucleus only in certain specific three-dimensional patterns.

This new way of approaching the behavior of electrons (and other particles too) became known as **wave mechanics** or **quantum mechanics.** In order to familiarize you with some of the concepts and terminology of wave mechanics, we shall consider the simple, though somewhat artificial, example illustrated in Fig. 5.3. This is usually referred to as a *particle in a one-dimensional box*. We consider the particle (which could be an electron) to have a mass m and to be restricted in its movement by a narrow but absolutely straight tube of length d into which it can just fit. This container, or

[1] The Bohr theory of the hydrogen atom is described in Chap. 21.

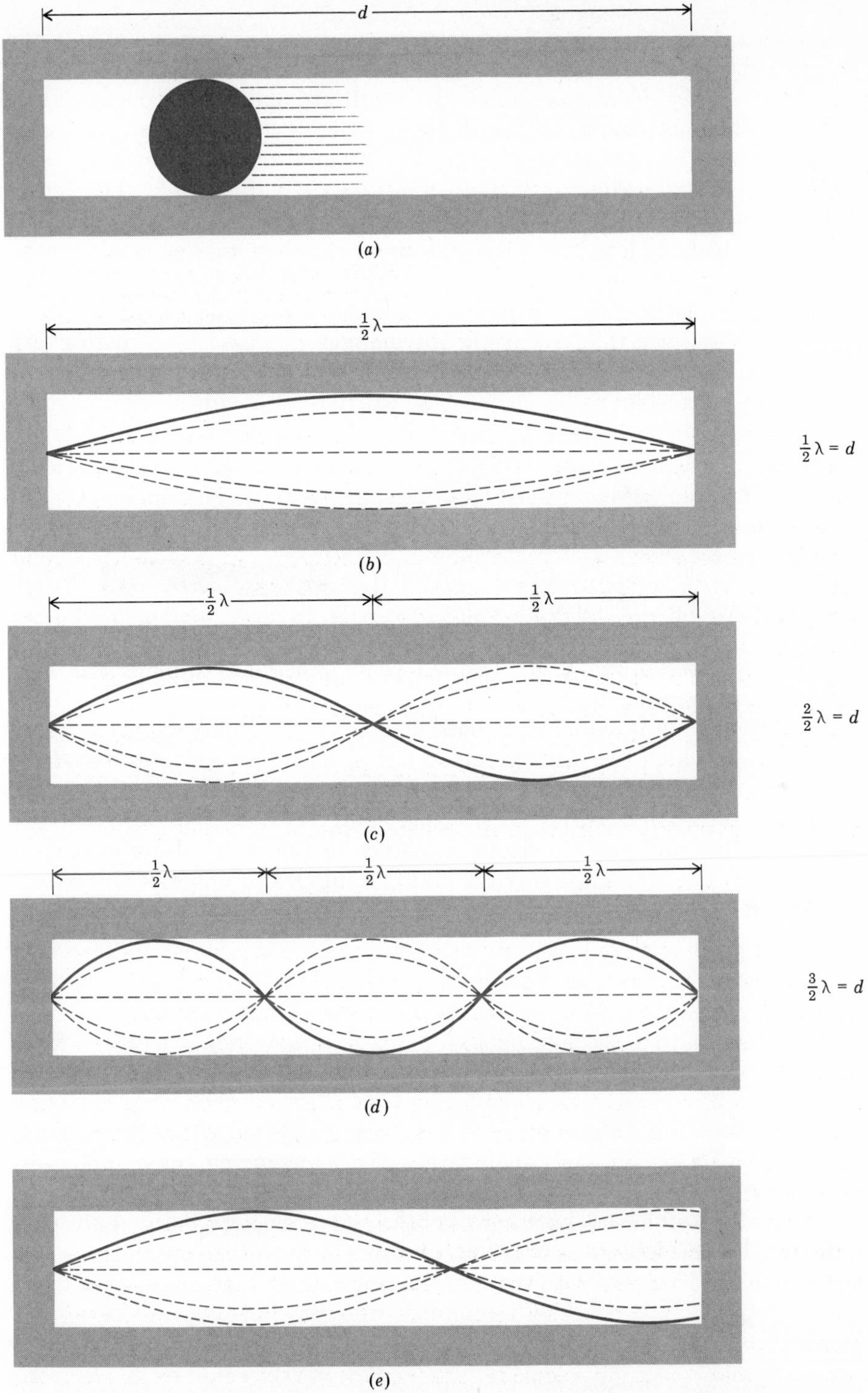

Figure 5.3 A microscopic particle in a one-dimensional box of length d. (*a*) Analogy with a macroscopic billiard ball; (*b*), (*c*), and (*d*), analogy with waves in a vibrating guitar string; (*e*) an example of a wave which would not fit the box. The right-hand end would move up and down instead of remaining fixed as in parts (*b*), (*c*), and (*d*).

box, is closed at both ends and insures that the particle can move in only one dimension within its length.

An everyday object like a marble or a billiard ball could move back and forth in this container, bouncing off either end. If there were no friction to slow it down, the particle would oscillate indefinitely, maintaining a constant speed u and a constant kinetic energy E_k of value $\frac{1}{2}mu^2$. The actual magnitude of u (and hence of E_k) would depend on how large or how small a "push" the particle was given initially to start it moving.

In order to look at this particle from a wave-mechanical point of view, we apply an idea originally suggested in 1924 by Louis de Broglie (1892 to). He proposed that a wave of wavelength λ is associated with every particle. The larger the mass of the particle and the faster it is moving, the smaller this wavelength becomes. The exact relationship is given by the formula

$$\lambda = \frac{h}{mu} \tag{5.1}$$

The constant of proportionality h is known as **Planck's constant** and has the value 6.624×10^{-34} J s.

The wave-mechanical view no longer pictures the particle as the oscillating billiard ball of Fig. 5.3*a*. Instead we must begin to think of it as having a behavior similar to that of the guitar string illustrated in Fig. 5.3*b*, *c*, or *d*. If the string is attached to both ends of the box, only those waves or vibrations in which the ends of the string do not move are possible. The length d of the box can thus correspond to a single half wavelength (Fig. 5.3*b*), to two half wavelengths (Fig. 5.3*c*), to three half wavelengths (Fig. 5.3*d*, etc., but *not* to the intermediate situation shown in Fig. 5.3*e*. In other words the length of the tube must correspond to an integral number of half wavelengths, or

$$d = n\,\frac{\lambda}{2} \tag{5.2}$$

where $n = 1, 2, 3, 4$, or some larger whole number. If $n = 1$, $d = \lambda/2$; if $n = 2$, $d = \lambda$; and so on. Rearranging Eq. (5.2), we then obtain

$$\lambda = \frac{2d}{n} \tag{5.3}$$

Since the right-hand sides of Eqs. (5.1) and (5.3) are both equal to λ, they may be set equal to each other, giving

$$\frac{2}{n}\,d = \frac{h}{mu}$$

which rearranges to give

$$u = \frac{nh}{2md} \qquad n = 1, 2, 3, 4, \ldots$$

We can now calculate the kinetic energy of our wave-particle. It is given by the formula

$$E_k = \tfrac{1}{2}mu^2 = \tfrac{1}{2}m\left(\frac{nh}{2md}\right)^2$$

or

$$E_k = n^2\left(\frac{h^2}{8md^2}\right) \qquad (5.4)$$

Since the value of n is restricted to positive whole numbers, we arrive at the interesting result that the kinetic energy of the electron can have only certain values and not others. Thus, if our particle is an electron ($m = 9.1 \times 10^{-31}$ kg) and the one-dimensional box is about the size of an atom ($d = 1 \times 10^{-10}$ m), the allowed values of the energy are given by

$$\begin{aligned} E_k &= n^2\left(\frac{h^2}{8md^2}\right) \\ &= n^2 \times \frac{(6.624 \times 10^{-34}\ \text{J s})^2}{8 \times 9.1 \times 10^{-31}\ \text{kg} \times 1 \times 10^{-20}\ \text{m}^2} \\ &= n^2 \times 6.0 \times 10^{-18}\ \frac{\text{J}^2\ \text{s}^2}{\text{kg m}^2} \\ &= n^2 \times 6.0 \times 10^{-18}\ \text{J} \qquad (\text{recall: } 1\ \text{J} = 1\ \text{kg m}^2\ \text{s}^{-2}) \end{aligned}$$

Thus if $n = 1$, $\quad E_k = 1^2 \times 6.0 \times 10^{-18}\ \text{J} = 6.0$ aJ

If $n = 2$, $\quad E_k = 2^2 \times 6.0 \times 10^{-18}\ \text{J} = 24.0$ aJ

If $n = 3$, $\quad E_k = 3^2 \times 6.0 \times 10^{-18}\ \text{J} = 54.0$ aJ

and so on.

This result means that by treating the electron as a wave, its energy is automatically restricted to certain specific values and not those in between. Although the electron can have an energy of 6.0 or 24.0 attojoules (aJ), it cannot have an intermediate energy such as 7.3 or 11.6 aJ. We describe this situation by saying that the energy of the electron is **quantized.**

We can now begin to glimpse some of the advantages of looking at the electron in wave-mechanical terms. If an electron is in some sense a wave, then only certain kinds of motion, i.e., only certain kinds of wave patterns, governed by whole numbers, are possible. As we shall shortly see, when an electron is allowed to move in three dimensions around a nucleus, this kind of behavior easily translates into a shell structure. An electron can be in shell 1 or shell 2 but not in an intermediate shell like 1.386.

One more property of the electron arises out of the wave analogy. If the electron in a box behaves like a guitar string, we can no longer state that the electron is located at a specific position within the box or is moving in one direction or the other. Indeed the electron seems to be all over the box at once! All we can say is that the wave (or the vibration of the string) has a certain *intensity* at any point along the box. This intensity is larger in some places than in others and is always zero at both ends of the box.

The Uncertainty Principle

Our inability to locate an electron exactly may seem rather strange, but it arises whether we think in terms of waves or of particles. Suppose an experiment is to be done to locate a billiard ball moving across a pool table whose surface is hidden under a black cloth. One way to do this would be to try to bounce a second billiard ball off the first one (Fig. 5.4). When a hit was made and the second ball emerged from under the cloth, we would have a pretty good idea of where the first ball was. The only trouble with the experiment is that the position and speed of the first ball would almost certainly be changed by the collision. To lessen this effect, a table-tennis ball could be substituted for the second billiard ball—its smaller mass would produce a much smaller change in the motion of the first ball. Clearly, the lighter and more delicate the "probe" we use to try to locate the first ball, the less our measurement will affect it.

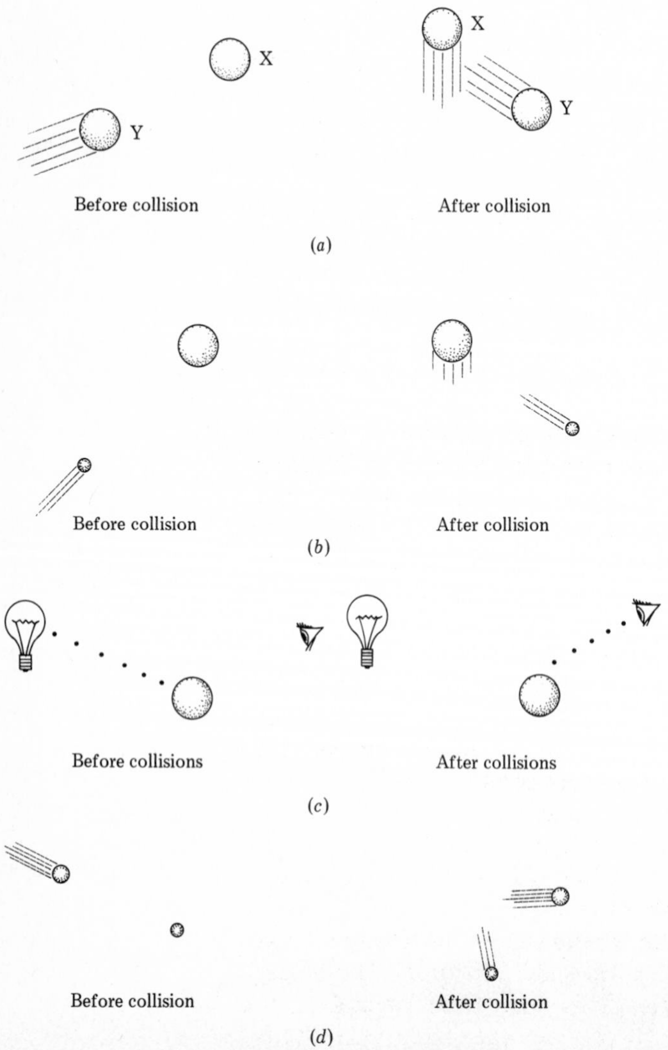

Figure 5.4 Determining the location and speed of a particle. (*a*) Locating a billiard ball (X) by bouncing a second billiard ball (Y) off it; (*b*) locating a billiard ball by bouncing a table-tennis ball off it; (*c*) locating a billiard ball by bouncing photons of visible light off it; (*d*) locating an electron by bouncing a photon off it.

The best way to locate the first billiard ball and determine its speed would be to remove the cover from the table so it could be seen. In this case, however, something is still "bouncing" off the first billiard ball (Fig. 5.4*c*). If we are to see the ball, particles of visible light, or **photons,** must strike the ball and be reflected to our eyes. Since each photon is very small and has very little energy by comparison with that needed to change the motion of the billiard ball, looking at the ball is an excellent means of observing it without changing its position or speed.

But to observe an electron is quite another story, since the mass of an electron is far smaller than that of a billiard ball. Anything (such as a photon of light) which can be bounced off an electron in such a way as to locate it precisely would have far more energy than would be required to change the path of the electron. Hence it would be impossible to predict the electron's future speed or position from the experiment. The idea that it is impossible to determine accurately *both* the location and the speed of any particle as small as an electron is called the **uncertainty principle.** It was first proposed in 1927 by Werner Heisenberg (1901 to 1976).

According to the uncertainty principle, even if we draw an analogy between the electron in a box and a billiard ball (Fig. 5.3*a*), it will be impossible to determine both the electron's exact position in the box and its exact speed. Since kinetic energy depends on speed ($\frac{1}{2}mu^2$) and Eq. (5.4) assigns exact values of kinetic energy to the electron in the box, the speed can be calculated accurately. This means that determination of the electron's position will be very inexact. It will be possible to talk about the *probability* that the electron is at a specific location, but there will also be some probability of finding it somewhere else in the box. Since it is impossible to know precisely where the electron is at a given instant, the question, "How does it get from one place to another?" is pointless. There is a finite probability that it was at the other place to begin with!

It is possible to be quantitative about the *probability* of finding a "billiard-ball" electron at a given location, however. Shortly after the uncertainty principle was proposed, the German physicist Max Born (1882 to 1969) suggested that the intensity of the electron wave at any position in the box was proportional to the probability of finding the electron (as a particle) at that same position. Thus if we can determine the shapes of the waves to be associated with an electron, we can also determine the relative probability of its being located at one point as opposed to another. The wave and particle models for the electron are thus connected to and reinforce each other. Niels Bohr suggested the term *complementary* to describe their relationship. It does no good to ask, "Is the electron a wave or a particle?" Both are ways of drawing an analogy between the microscopic world and macroscopic things whose behavior we understand. Both are useful in our thinking, and they are complementary rather than mutually exclusive.

A graphic way of indicating the probability of finding the electron at a particular location is by the density of shading or stippling along the length of the box. This has been done in Fig. 5.5 for the same three electron waves previously illustrated in Fig. 5.3. Notice that the density of dots is large wherever the electron wave is large. (This would correspond in the guitar-string analogy to places where the string was vibrating quite far

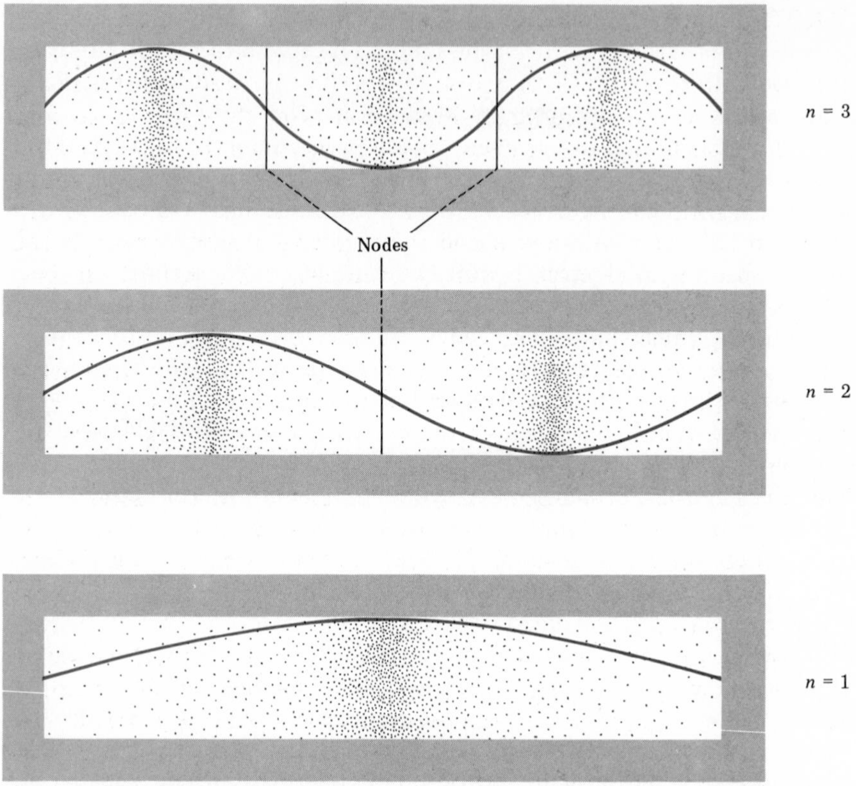

Figure 5.5 Indicating electron density or electron probability by the density of dots for the electron in a box. Nodes, where the probability of finding the electron is zero, are indicated. In the guitar-string analogy a node corresponds to a point on the string which is not vibrating up and down.

from its rest position.) Where the electron wave is small (near the ends of the box in all three cases and at the nodes indicated in the figure), there are only a small number of dots. (A **node** is a place where the intensity of the wave is zero, that is, in the guitar-string analogy, where the string has not moved from its rest position.)

If the electron is thought of as a wave occupying all parts of the box at once, we can speak of an **electron cloud** which has greater or lesser density in various parts of the box. There will be a greater quantity of negative charge in a region of high density (a region where there is a greater concentration of dots) than in one of low density. In an atom or molecule, according to the uncertainty principle, the best we can do is indicate **electron density** in various regions—we cannot locate the precise position of the electron. Therefore electron dot-density diagrams, such as the ones shown in Fig. 5.5, give a realistic and useful picture of the behavior of electrons in atoms. In such a diagram the *electron density or probability of finding the electron is indicated by the number of dots per unit area.* We will encounter electron dot-density diagrams quite often throughout this book. These have all been generated by a computer from accurate mathematical descriptions of the atom or molecule under discussion.

An electron in an atom differs in two ways from the hypothetical electron in a box that we have been discussing. First, in an atom the electron occupies all three dimensions of ordinary space. This permits the shapes of the electron waves to be more complicated. Second, the electron is not confined in an atom by the solid walls of a box. Instead, the electrostatic force of attraction between the positive nucleus and the negative electron prevents the latter from escaping. In 1926 Erwin Schrödinger (1887 to 1961) devised a mathematical procedure by which the electron waves and energies for this more complicated situation could be obtained. A solution of the **Schrödinger wave equation** is beyond the scope of a general chemistry text. However, a great many chemical phenomena can be better understood if one is familiar with Schrödinger's results, and we shall consider them in some detail.

The distribution of electron density predicted by the solution of Schrödinger's equation for a hydrogen atom which has the minimum possible quantity of energy is illustrated in Fig. 5.6. A number of general characteristics of the behavior of electrons in atoms and molecules may be observed from this figure.

First of all, the hydrogen atom does not have a well-defined boundary. The number of dots per unit area is greatest near the nucleus of the atom at the center of the diagram (where the two axes cross). Electron density decreases as distance from the nucleus increases, but there are a few dots at distances as great as 200 pm (2.00 Å) from the center. Thus as one gets closer and closer to the nucleus of an atom, electron density builds up slowly and steadily from a very small value to a large one. Another way of stating the same thing is to say that the electron cloud becomes more dense as the center of the atom is approached.

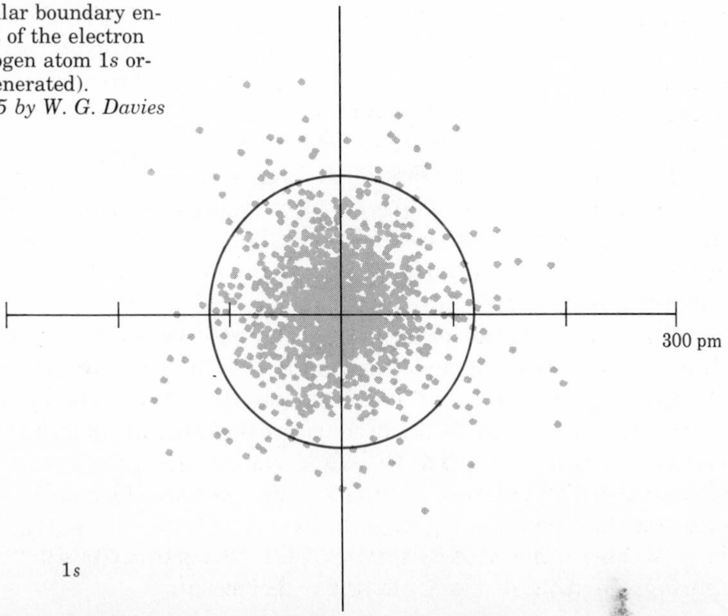

Figure 5.6 Circular boundary enclosing 90 percent of the electron density in a hydrogen atom 1*s* orbital (computer-generated). (*Copyright © 1975 by W. G. Davies and J. W. Moore.*)

A second characteristic evident from Fig. 5.6 is the *shape* of the electron cloud. In this two-dimensional diagram it appears to be approximately *circular;* in three dimensions it would be *spherical*. This can be illustrated more readily by drawing a circle (or in three dimensions, a sphere) which contains a large percentage (say 75 or 90 percent) of the dots, as has been done in the figure. Since such a sphere or circle encloses most of (but not all) the electron density, it is about as close as one can come to drawing a boundary which encloses the atom. **Boundary-surface diagrams** in two and sometimes three dimensions are easier to draw quickly than are dot-density diagrams. Therefore chemists use them a great deal.

Orbitals

A third characteristic of the diagram in Fig. 5.6 is that it has been assigned an identifying label, namely, 1*s*. This enables us to distinguish it from other wave patterns the electron could possibly adopt if it moved about the nucleus with a higher energy. Each of these three-dimensional wave patterns is different in shape, size, or orientation from all the others and is called an **orbital.** The word *orbital* is used in order to make a distinction between these wave patterns and the circular or elliptical *orbits* of the Bohr picture shown previously in Fig. 5.2. The electron density distributions for the 14 simplest orbitals of the hydrogen atom are shown in Figs. 5.7 and 5.8, both as dot-density diagrams and as boundary-surface diagrams. Note the unique label for each orbital.

At ordinary temperatures, the electron in a hydrogen atom is almost invariably found to have the lowest energy available to it. That is, the electron occupies the 1*s* orbital, and the electron cloud looks like the dot-density diagram in Fig. 5.6. At a very high temperature, though, some collisions between the atoms are sufficiently hard to provide one of the electrons with enough energy so that it can occupy one of the other orbitals, say a 2*s* orbital, but this is unusual. Nevertheless a knowledge of these higher energy orbitals is necessary since electron clouds having the same shapes as for hydrogen are found to apply to all the other atoms in the periodic table as well.

In the case of a particle in a one-dimensional box, the energy was determined by a positive whole number n. Much the same situation prevails in the case of the hydrogen atom. An integer called the **principal quantum number,** also designated by the symbol n, is used to label each orbital. The larger the value of n, the greater the energy of the electron and the larger the average distance of the electron cloud from the nucleus. In Figs. 5.7 and 5.8, $n = 1$ for the 1*s* orbital, $n = 2$ for the four orbitals 2*s*, $2p_x$, $2p_y$, and $2p_z$, while the remaining nine orbitals all correspond to $n = 3$.

Because a greater number of different shapes is available in the case of three-dimensional, as opposed to one-dimensional, waves, two other labels are used in addition to n. The first consists of one of the lowercase letters *s, p, d,* or *f*. These tell us about the *overall shapes* of the orbitals.[1] Thus all *s* orbitals, including the 1*s*, 2*s*, and 3*s* shown in Figs. 5.7 and 5.8, are *spherical.* All *p* orbitals, such as $2p_x$, $2p_y$, $2p_z$ and $3p_x$, $3p_y$, $3p_z$, have a *dumbbell*

[1] The letters *s, p, d,* and *f* originate from the words *sharp, principal, diffuse,* and *fundamental* which were used to describe certain features of spectra before wave mechanics was developed. They later became identified with orbital shapes.

shape. The *d* orbitals have more complex shapes than the *p* orbitals, while *f* orbitals (a type not shown in the figures) are even more complicated.

The third kind of label used to describe orbitals is a subscript. These subscripts distinguish between orbitals which are basically the same shape but differ in their orientation in space. In the case of *p* orbitals there are always *three* orientations possible. A *p* orbital which extends along the *x* axis is labeled a p_x orbital (for example, $2p_x$, $3p_x$, etc.). A *p* orbital along the *y* axis is labeled p_y, and one along the *z* axis is a p_z orbital. In the case of the *d* orbitals these subscripts are more difficult to follow. You can puzzle them out from Figs. 5.7 and 5.8, if you like, but we will not use them a great deal in the remainder of this book. You should, however, be aware that there are *five* possible orientations for *d* orbitals. It is also important to know that there are *seven* different orientations for *f* orbitals, since the number of orbitals of each type (*s, p, d,* etc.) is important in determining the shell structure of the atom.

Another important point is that only a limited number of orbital shapes is possible for each value of *n*. If $n = 1$, then only the spherical 1*s* orbital is possible. When *n* is increased to 2, though, *two* orbital types (2*s* and 2*p*) become possible, while if *n* equals 3, then *three* orbital types occur. The same pattern extends to $n = 4$ where *four* orbital types, namely, 4*s*, 4*p*, 4*d*, and 4*f*, are found.

5.4 THE POTENTIAL ENERGY OF ELECTRONS

As we have seen in the preceding section, the energy of an electron in an isolated hydrogen atom is determined by the principal quantum number *n*. The reason for this is not exactly the same as the reason that the energy of the particle in a box depended on a whole number *n*, however. In an atom an electron has both kinetic energy (like the particle in a box) and potential energy. The electron's potential energy is a result of the attractive force between the negatively charged electron and the positively charged nucleus.

Coulomb's Law

When unlike charges (one negative and the other positive) attract each other, or like charges (both positive or both negative) repel each other, **Coulomb's law** governs the force between them. According to this law the force of attraction or repulsion varies *inversely with the square of the distance* between the charges. Suppose two particles, one with a charge of $+1\ \mu$C (microcoulomb) and the other with a charge of $-1\ \mu$C are placed 1 cm apart. The force of attraction between these two charges is found to be 90 N (newton), about the same force as gravity exerts on a 20-lb weight. If the distance between the charges is now *multiplied* by a factor of 100 (increased to 1 m), then the force of attraction between the two charges is found to be *divided* by a factor of 100 *squared,* i.e., by a factor of 10^4. Consequently the force decreases to about the weight of a grain of sand. Another hundredfold increase in the distance (to 100 m or about 100 yd) reduces the force of attraction by a further factor of 10^4, making the force virtually undetectable.

As the above example shows, electrostatic forces of attraction and repulsion are very strong when the charges are close, but drop off fairly

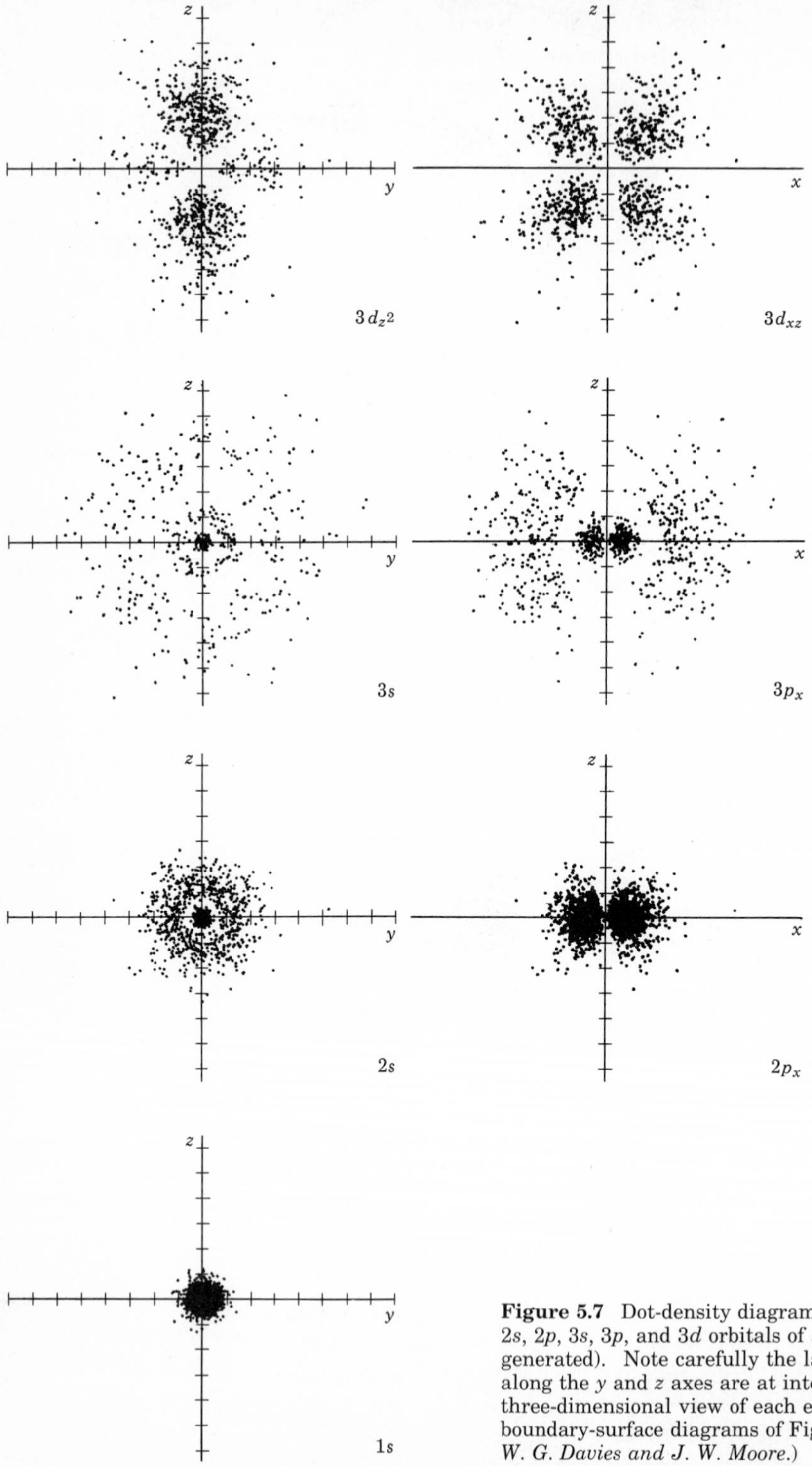

Figure 5.7 Dot-density diagrams of electron clouds for the 1*s*, 2*s*, 2*p*, 3*s*, 3*p*, and 3*d* orbitals of a hydrogen atom (computer-generated). Note carefully the labeling of the axes. Markers along the *y* and *z* axes are at intervals of 200 pm (2.00 Å). A three-dimensional view of each electron cloud is given by the boundary-surface diagrams of Fig. 5.8. (*Copyright © 1975 by W. G. Davies and J. W. Moore.*)

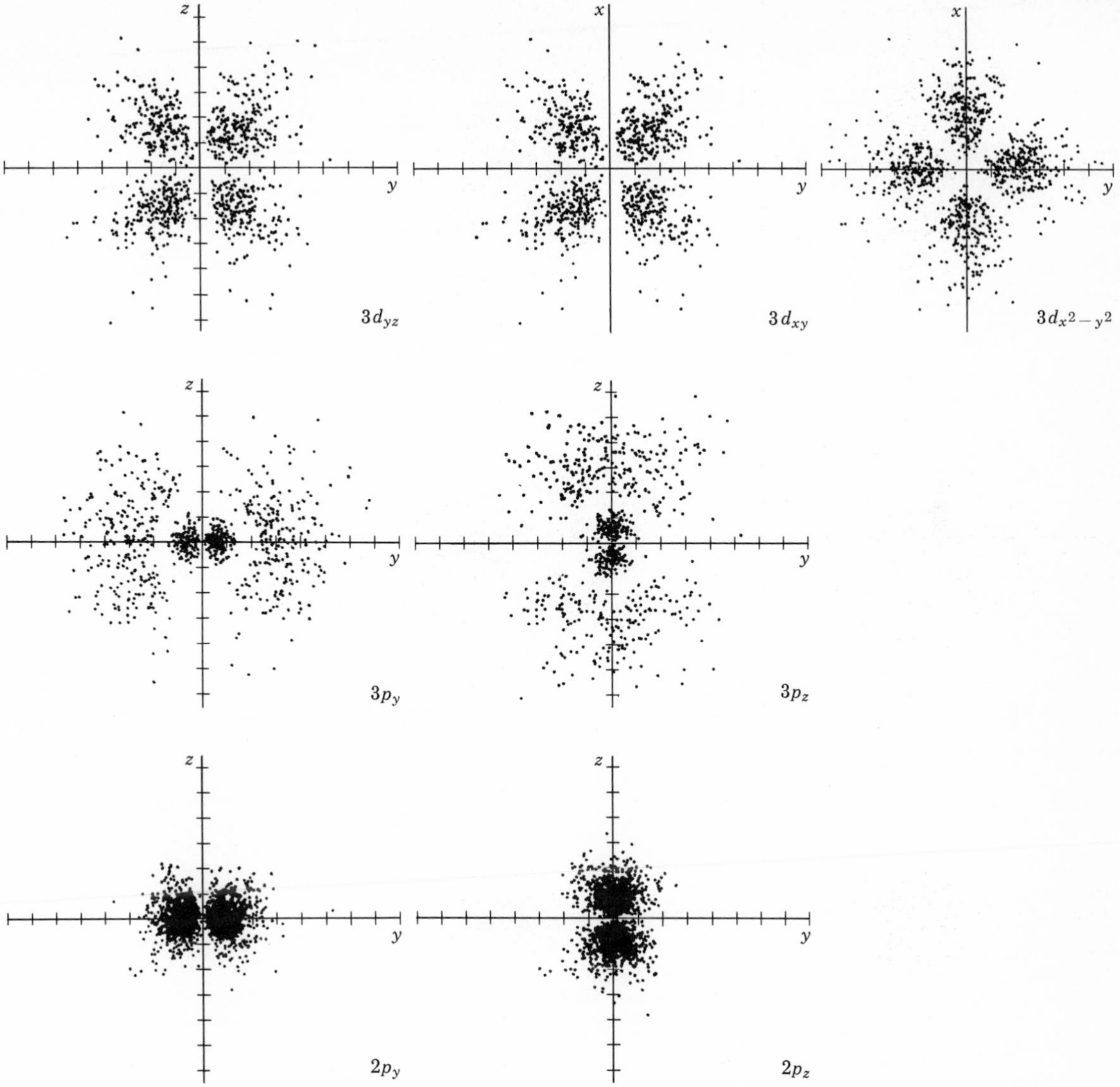
z
y
$3d_{yz}$
x
y
$3d_{xy}$
x
y
$3d_{x^2-y^2}$
z
y
$3p_y$
z
y
$3p_z$
z
y
$2p_y$
z
y
$2p_z$

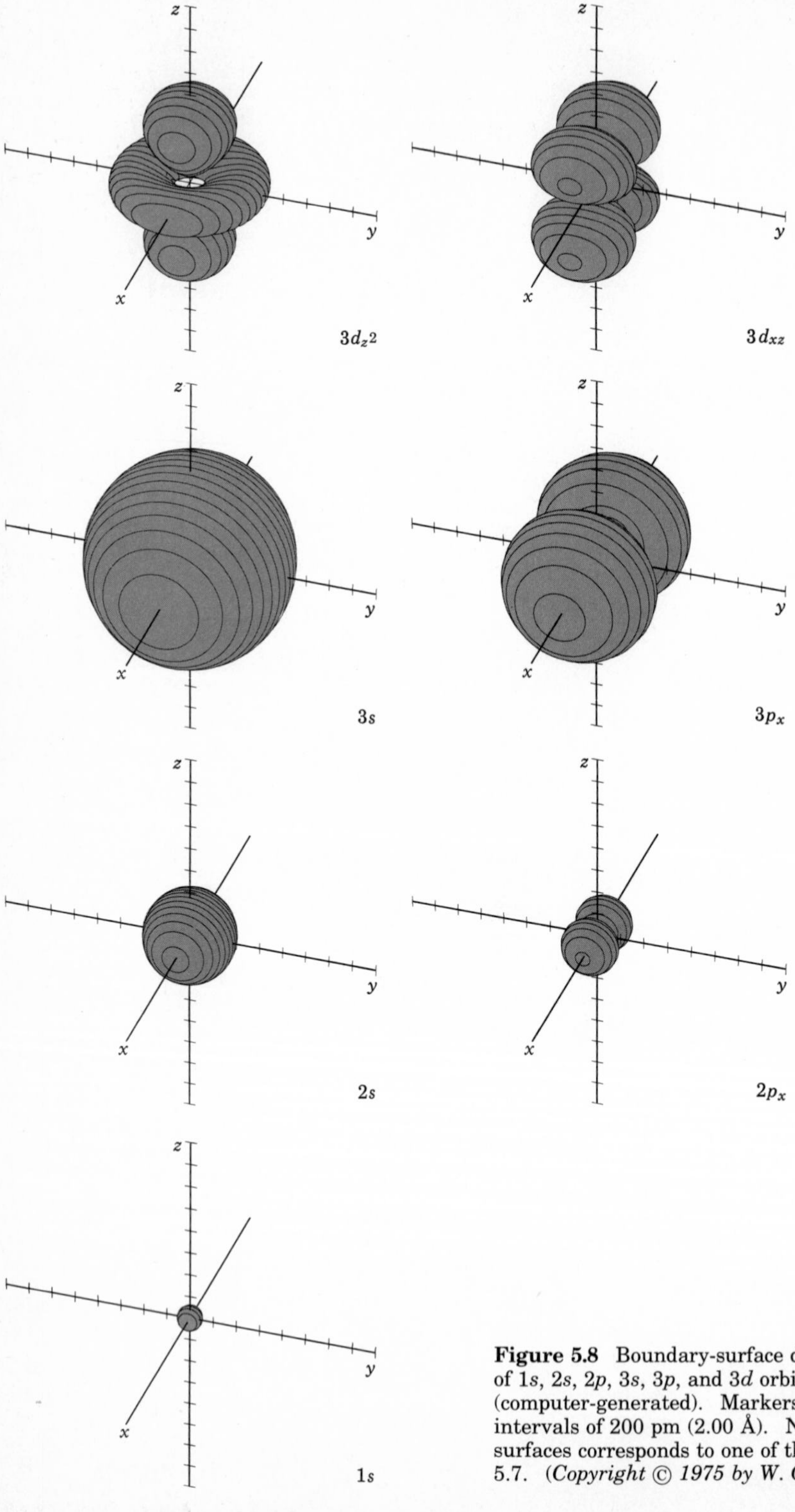

Figure 5.8 Boundary-surface diagrams for electron densities of 1*s*, 2*s*, 2*p*, 3*s*, 3*p*, and 3*d* orbitals of a hydrogen atom (computer-generated). Markers along the *y* and *z* axes are at intervals of 200 pm (2.00 Å). Note that each of these boundary surfaces corresponds to one of the dot-density diagrams in Fig. 5.7. (*Copyright © 1975 by W. G. Davies and J. W. Moore.*)

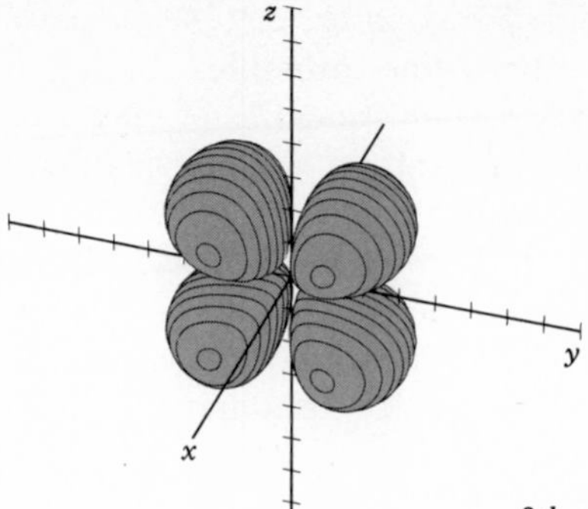

$3d_{yz}$

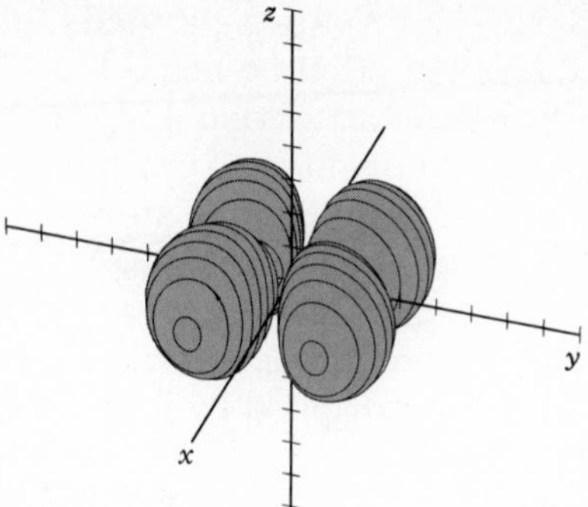

$3d_{xy}$

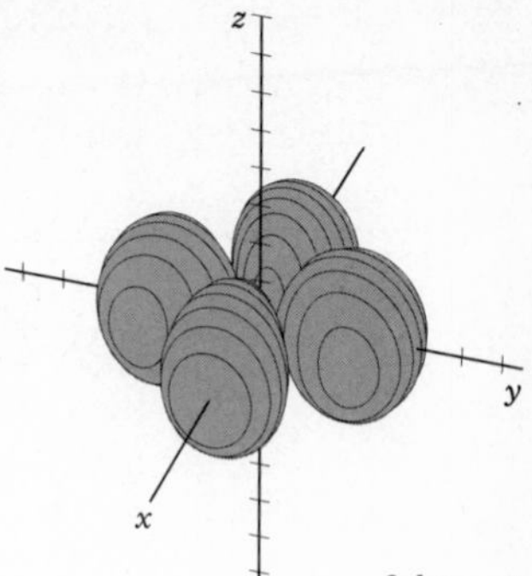

$3d_{x^2-y^2}$

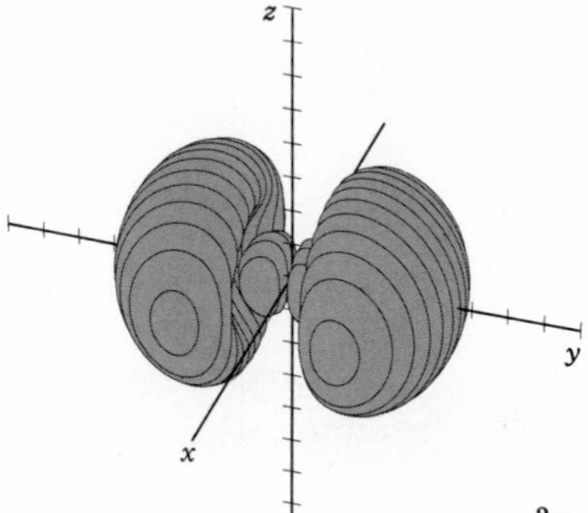

$3p_y$

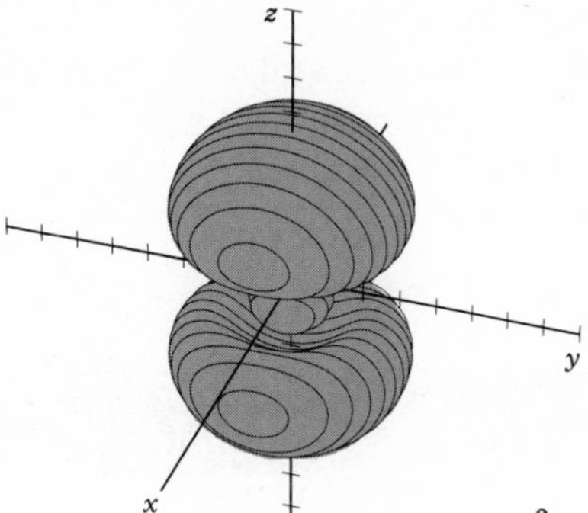

$3p_z$

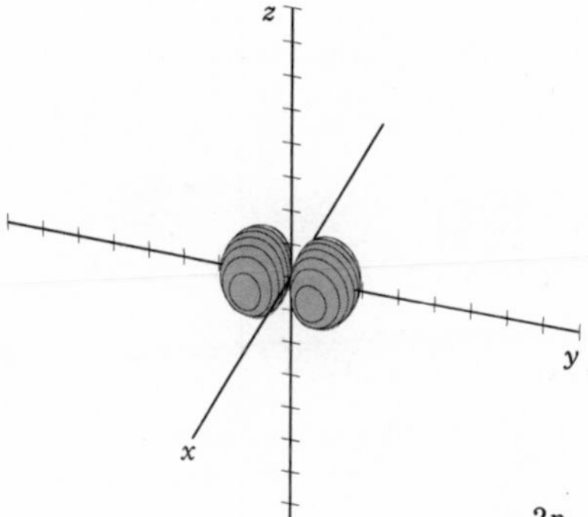

$2p_y$

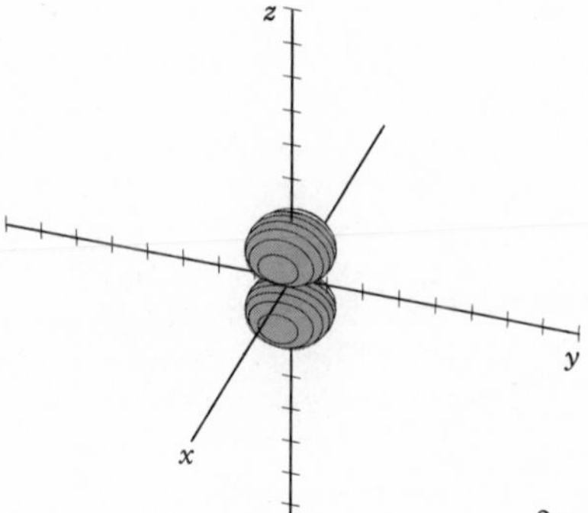

$2p_z$

quickly as the charges are separated. If the separation is very large, electrostatic forces can often be neglected. This behavior resembles the more familiar example of attraction and repulsion between the poles of magnets and also is akin to the force of gravity in the solar system. Magnetic forces and gravitational forces follow an inverse square relationship as well.

A second part of Coulomb's law states that the force is proportional to the *magnitude* of each charge. In the above example if one charge is doubled (to $\pm 2\ \mu C$), the force is likewise doubled, while if *both* charges are doubled, the force is multiplied by *four*. Coulomb's law is summarized by the equation

$$F = k\frac{Q_1 Q_2}{r^2}$$

where F is the force, Q_1 and Q_2 are the charges, and r is the distance between the charges. The constant k has the value $8.988 \times 10^9\ \text{N m}^2\ \text{C}^{-2}$.

Potential Energy

Because of the coulombic force of attraction or repulsion between them, *two charged particles will vary in energy as we alter the distance between them.* Suppose we have charges of $+1$ and $-1\ \mu C$ separated by 1 cm, for example. The charges could be separated by hand, and by the time the length of a football field lay between them, their attractive force would be negligible. Expenditure of muscle energy (0.898 J, to be exact) will be necessary to carry out such a separation. That is, because the charges attract each other, we must do work to pull them apart.

According to the law of conservation of energy (Sec. 3.3), the muscle energy expanded to pull the opposite charges apart cannot be destroyed. We say that the 0.898 J is gained by the two charges and stored as potential energy. **Potential energy** (symbol E_p) is the energy which one or more bodies have because of their *position.* We can always regain this energy by reversing the process during which it was stored. If the two opposite charges are returned to their original separation of 1 cm, their potential energy will *decrease* by 0.898 J. The energy released will appear as kinetic energy, as heat, or in some other form, but it cannot be destroyed.

If we had taken two particles both of which had a charge of $+1\ \mu C$ for our example of potential energy, work would have been required to push them together against their repulsive force. Their potential energy would increase as they were brought together from the ends of a football field, and 0.898 J would be required to move them to a distance of 1 cm apart. Because the potential energy of like-charged particles increases as they are brought closer together, while that of opposite-charged particles decreases, it is convenient to assign a value of *zero* potential energy to two charged particles which are a long distance apart. Bringing a pair of positive charges (or a pair of negative charges) closer together increases E_p to a positive value. Bringing one positive and one negative particle together decreases E_p, giving a negative value.

When we turn our attention from the potential energy of charged macroscopic particles which have a definite location in space to microscopic particles like the electron, we immediately encounter a difficulty. The electron in an atom is not a fixed distance from the nucleus but is "smeared out" in space in a wave pattern over a large range of distances. Nevertheless it is still meaningful to talk about the potential energy of such an electron cloud. Consider the 1*s* electron illustrated by the dot-density diagram in Fig. 5.6, for example. If the electron were actually positioned at one of these dots momentarily, it would have a definite potential energy at that moment. If we now add up the potential energy for each dot and divide by the number of dots, we obtain an average potential energy, which is a good approximation to the potential energy of the electron cloud. The more dots we have, the closer such an approximation is to the exact answer.

In practice we can often decide which of two electron clouds has the higher potential energy by looking at them. In Fig. 5.7, for example, it is easy to see that the potential energy of an electron in a 1*s* orbital is lower than that of a 2*s* electron. An electron in a 1*s* orbital is almost always closer than 200 pm to the nucleus, while in a 2*s* orbital it is usually farther away. In the same way we have no difficulty in estimating that a 3*s* electron is on average farther from the nucleus and hence higher in potential energy than a 2*s* electron. It is also easy to see that electron clouds which differ only in their orientation in space must have the *same* potential energy. An example would be the $2p_x$, $2p_y$, and $2p_z$ clouds.

When we compare orbitals with different basic shapes, mere inspection of the dot-density diagrams is often insufficient to tell us about the relative potential energies. It is not apparent from Fig. 5.7, for instance, whether the 2*s* or 2*p* orbital has the higher potential energy. Actually both have the same energy in a hydrogen atom, though not in other atoms. In the same way the 3*s*, the three 3*p*, and the five 3*d* orbitals are all found to have the same energy in the hydrogen atom.

Although dot-density diagrams are very informative about the potential energy of an electron in an orbital, they tell us nothing at all about its kinetic energy. It is impossible, for example, to decide from Fig. 5.6 whether the electron in a 1*s* orbital is moving faster on the whole than an electron in a 2*s* orbital, or even whether it is moving at all! Fortunately it turns out that this difficulty is unimportant. The *total* energy (kinetic + potential) of an electron in an atom or a molecule is always one-half its potential energy. Thus, for example, when an electron is shifted from a 1*s* to a 2*s* orbital, its potential energy increases by 3.27 aJ. At the same time the electron slows down and its kinetic energy drops by half this quantity, namely, 1.635 aJ. The net result is that the *total* energy (kinetic + potential) increases by *exactly half* the increase in potential energy alone; i.e., it increases by 1.635 aJ. A similar statement can be made for *any* change inflicted on *any* electron in *any* atomic or molecular system. This result is known as the **virial theorem.** Because of this theorem we can, if we want, ignore the kinetic energy of an electron and concentrate exclusively on its potential energy.

5.5 ATOMS HAVING MORE THAN ONE ELECTRON

Now that we have some familiarity with the properties of single electrons, we can move on to a discussion of atoms containing more than one electron. In order to do this, we will use the diagrams printed in full color in Plates 4 and 5 in which each electron is indicated in a different color.

Helium

The first element in the periodic table with more than one electron is helium, which has two electrons. Dot-density diagrams for both these electrons are shown in Plate 4. One electron is color coded in blue, and the other in green. Note that both electrons occupy the same orbital, namely, a 1*s* orbital. It turns out that 2 is the maximum number of electrons *any* orbital can hold. This restriction is connected with a property of the electrons not yet discussed, namely, their **spin.** Electrons can not only move about from place to place, but they can also rotate or spin about themselves. Two orientations (clockwise and counterclockwise) are possible for this spin. According to the **Pauli exclusion principle,** if two electrons occupy the same orbital, they *must* have opposite spins. Two such electrons are said to be **spin paired** and are often represented by arrows pointing in different directions, i.e., by the symbol $\downarrow\uparrow$. Two electrons spinning in the same direction are said to have their spins **parallel** and are indicated by $\downarrow\downarrow$. The Pauli principle implies that if two electrons have parallel spins, they *must* occupy different orbitals.

An obvious feature of the helium atom shown in Plate 4 is that it is somewhat *smaller* than the hydrogen atom drawn to the same scale on the same page. This contraction is caused by the increase in the charge on the nucleus from +1 in the hydrogen atom to +2 in the helium atom. This pulls both the green and the blue electron clouds in more tightly. This effect is offset, but to a lesser extent, by the mutual repulsion of the two electron clouds.

Lithium

In Plate 4 the three electrons of the lithium atom are color-coded blue, green, and red. As in the previous atom, two electrons (blue and green) occupy the 1*s* orbital. The Pauli principle prevents more than two electrons from occupying this orbital, and so the third (red) electron must occupy the next higher orbital in energy, namely, the 2*s* orbital. A convenient shorthand form for indicating this **electron configuration** is

$$1s^2 2s^1$$

The superscripts *2* and *1* indicate that there are *two* electrons in the 1*s* orbital and *one* electron in the 2*s* orbital.

As in the case of helium, the increase in nuclear charge to +3 produces a corresponding reduction in the size of the lithium 1*s* orbital. In sharp contrast to this compact inner orbital is the very large and very diffuse cloud of the outer 2*s* electron. There are two reasons why this 2*s* cloud is so large.

The first reason is that the principal quantum number n has increased from 1 to 2. As shown in Fig. 5.7, the 2s electron cloud is bigger than 1s even in the hydrogen atom with a nuclear charge of only +1. A second reason is that the two 1s electrons are usually closer to the nucleus than the 2s electron. These two 1s electrons have the effect of **screening** or **shielding** the outer electron from the full attractive force of the +3 charge on the nucleus. When the 2s electron is some distance from the nucleus, it "sees" not only the +3 charge on the nucleus but also the two negative charges close by. The overall effect is almost as though two of the three positive charges on the nucleus are canceled, leaving a net charge of +1 to hold the outer electron to the atom. This situation can also be described by saying that the **effective nuclear charge** is close to +1.

It should be clear from Plate 4 that when a lithium atom interacts with another atom, the 2s electron is far more likely to be involved than either of the two 1s electrons. In Lewis' terminology, it is a *valence electron* and occupies a *valence shell*. The pair of 1s electrons are a complete shell and form the *kernel* of the lithium atom. There is thus a close correspondence between the wave-mechanical picture and Lewis' earlier, less mathematical ideas. It is also worth noting that the wave model of lithium gives a *spherical* atom—a great advance over the elongated orbits which were needed to describe the alkali-metal atoms in the Bohr theory (see Fig. 5.2).

Beryllium

As shown in Plate 5, there are two 1s electrons and two 2s electrons in the Be atom. Its electron configuration is thus

$$1s^2 2s^2 \qquad \text{or} \qquad [\text{He}]2s^2$$

The symbol [He] denotes the inner shell of two 1s electrons which have the same configuration as the noble gas He.

The beryllium atom is noticeably smaller than the lithium atom. This is because of the increase in nuclear charge from +3 to +4. Since the two outer 2s electrons (red and orange) do not often come between each other and the nucleus, they do not screen each other from the nucleus very well. Only the two inner electrons are effective in this respect. The effective nuclear charge holding a 2s electron to the nucleus is thus nearly +2, about twice the value for lithium, and the 2s electron clouds are drawn closer to the center of the atom.

Boron

The next element after beryllium is boron. Since the 2s orbital is completely filled, a new type of orbital must be used for the fifth electron. There are three 2p orbitals available, and any of them might be used. Plate 5 shows the fifth electron (color-coded purple) occupying the $2p_x$ orbital. Note carefully the differences between the $2p_x$ and 2s electron density distributions in the boron atom. Although on the average both electron clouds extend about the same distance from the nucleus, the $2p_x$ electron wave has a node passing through the center of the atom. Thus the $2p_x$ electron cloud has a much smaller probability density very close to the nucleus than does a $2s$ cloud. This means that the $2p_x$ electron cloud is more effectively screened by the $1s$ electrons from the nuclear charge. The atom exerts a slightly smaller overall pull on the $2p$ electron than it does on the $2s$ electron. The presence of the inner electrons thus has the effect of making the $2p$ orbital somewhat *higher* in energy than the $2s$ orbital.

This difference in energy between $2s$ and $2p$ electrons in the boron atom is an example of a more general behavior. In any atom with sufficient electrons we always find that a p orbital is somewhat higher in energy than an s orbital with the same value of n. In the lithium atom, for example, the third electron occupies a $2s$ rather than a $2p$ orbital because this gives it a somewhat lower energy. Further on in the periodic table we will find a similar difference between $3s$ and $3p$ orbitals and between $4s$ and $4p$ orbitals.

Carbon

We shall examine the electron configuration of one more atom, carbon, with the aid of the color-coded diagrams. In this case six electrons must be distributed among the orbitals—four will be paired in the $1s$ and $2s$ orbitals, leaving two p-type electron clouds. These are shown color-coded purple and cyan in Plate 5 as $2p_x$ and $2p_y$, although the choice of x, y, or z directions is arbitrary. The choice of *two different p* orbitals is not arbitrary, however. It can be shown experimentally that both p electrons in the carbon atom have the same spin. Hence they cannot occupy the same orbital.

This illustrates another general rule regarding electron configurations. *When several orbitals of the same type but different orientation are available, electrons occupy them one at a time, keeping spins parallel, until forced to pair by lack of additional empty orbitals.* This is known as **Hund's rule.** Thus the electron configuration of carbon is

$$[\mathrm{He}]2s^2 2p_x^1 2p_y^1$$

This might also be written (using arrows to indicate the orientations of electron spins):

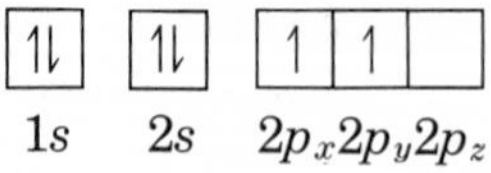

The notation

$$[\mathrm{He}]2s^2 2p^2$$

may also be found. In such a case it is assumed that the reader knows that the two $2p$ electrons are not spin paired.

It is worth noting that the arrangement of electrons in different $2p$ orbitals, necessitated by Hund's rule, produces a configuration of *lower energy.* If both $2p$ electrons could occupy the same orbital, say the $2p_x$ orbital, they would often be close to each other, and their mutual repulsion would correspond to a higher potential energy. If each is forced to occupy an orbital of different orientation, though, the electrons keep out of each other's way much more effectively. Their mutual repulsion and hence their potential energy is less.

In talking about polyelectronic atoms, the terms shell and subshell are often used. When the two electrons have the same principal quantum number, they are said to belong to the same **shell.** In the carbon atom, for example, the two $2s$ electrons and the $2p_x$ and the $2p_y$ electrons all belong to the second shell, while the two $1s$ electrons belong to the first shell. Shells defined in this manner can be further divided into **subshells** according to whether the electrons being discussed occupy *s, p, d,* or *f* orbitals. We can thus divide the second shell into $2s$ and $2p$ subshells. The third shell can similarly be divided into $3s$, $3p$, and $3d$ subshells, and so on.

5.6 ELECTRON CONFIGURATIONS

We could continue to examine the results of the wave theory of the electron for each element in the periodic table, but the examples in the previous section illustrate all the general rules that are necessary in order to predict *electron configurations* for all atoms of the elements. To review, these rules are as follows:

1 *The Aufbauprinzip (building-up principle).* The structure of an atom may be built up from that of the element preceding it in the periodic system by adding one proton (and an appropriate number of neutrons) to the nucleus and one extranuclear electron.
2 *The order of filling orbitals.* Each time an electron is added, it occupies the available subshell of *lowest energy.* The appropriate subshell may be determined from a diagram such as Fig. 5.9*a* which arranges the subshells in order of increasing energy. Once a subshell becomes filled, the subshell of the next higher energy starts to fill.
3 *The Pauli exclusion principle.* No more than *two* electrons can occupy a single orbital. When two electrons occupy the same orbital, they must be of opposite spin (*an electron pair*).
4 *Hund's rule.* When electrons are added to a subshell where more than one orbital of the same energy is available, their spins remain parallel and they occupy different orbitals. Electron pairing does not occur until it is required by lack of another empty orbital in the subshell.

The order in which the subshells are filled merits some discussion. As can be seen in Fig. 5.9*a*, within a given shell the energies of the subshells increase in the order $s < p < d < f$. When we discussed the boron atom, we saw that a *p* orbital is higher in energy than the *s* orbital in the same shell because the *p* orbital is more effectively screened from the nucleus. Similar reasoning explains why *d* orbitals are higher in energy than *p* orbitals but lower than *f* orbitals.

Not only are the energies of a given shell spread out in this way, but there is sometimes an overlap in energy between shells. As can be seen from Fig. 5.9*a*, the subshell of highest energy in the third shell, namely, $3d$,

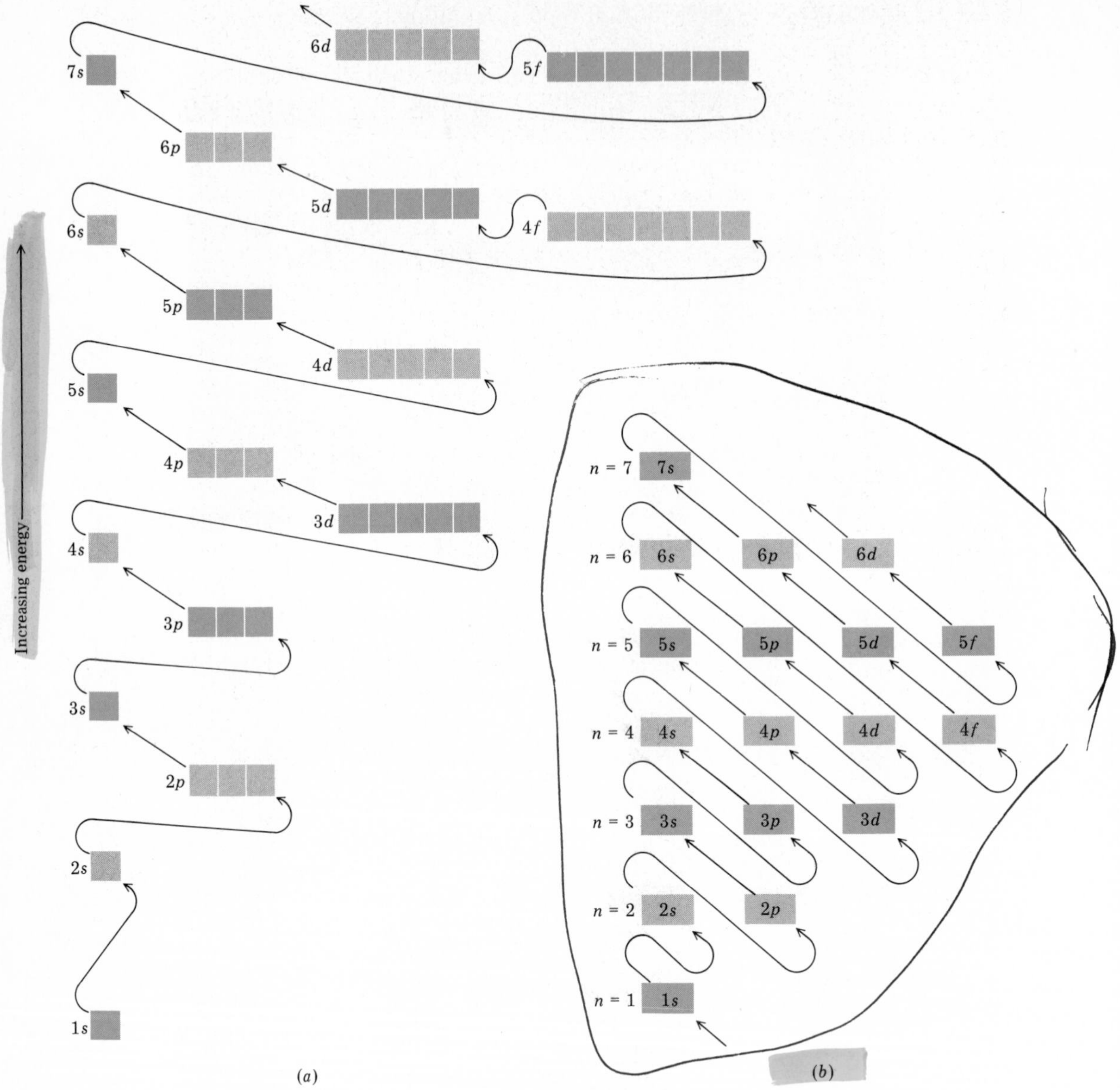

Figure 5.9 Order of filling subshells in the building-up of atomic electron configurations. (*a*) Relative energies of subshells at the time they are being filled. (*b*) Aid to remembering the order of filling subshells. All possible subshells having the same n value are written on horizontal lines. Diagonal arrows from lower right to upper left are then followed to obtain the order of filling. (The only subshells shown are those which are partially or completely occupied in atoms that have so far been discovered.)

is above the subshell of lowest energy in the fourth shell, namely, 4*s*. Similar overlaps occur among subshells of the fourth, fifth, sixth, and seventh shells. These cause exceptions to the expected order of filling subshells. The 6*s* orbital, for example, starts to fill before the 4*f*.

Although the order in which the subshells fill seems hopelessly complex at first sight, there is a very simple device available for remembering it.

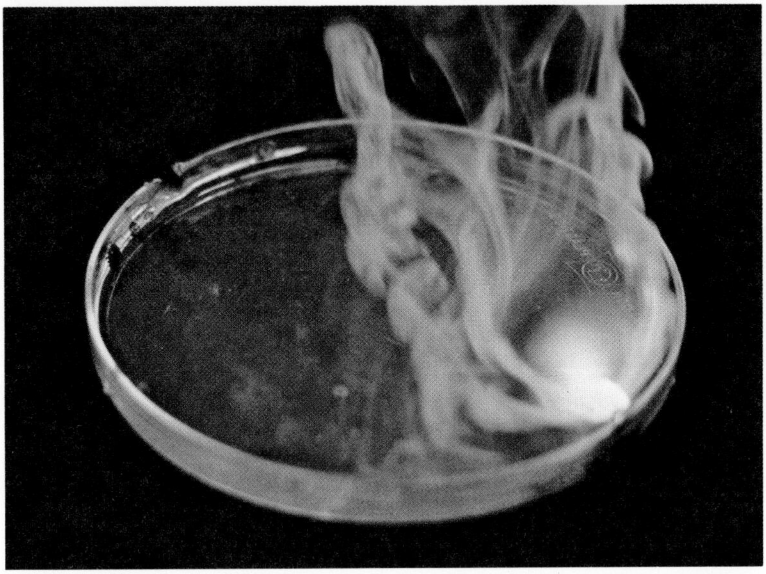

Plate 1 Reaction between potassium and water.

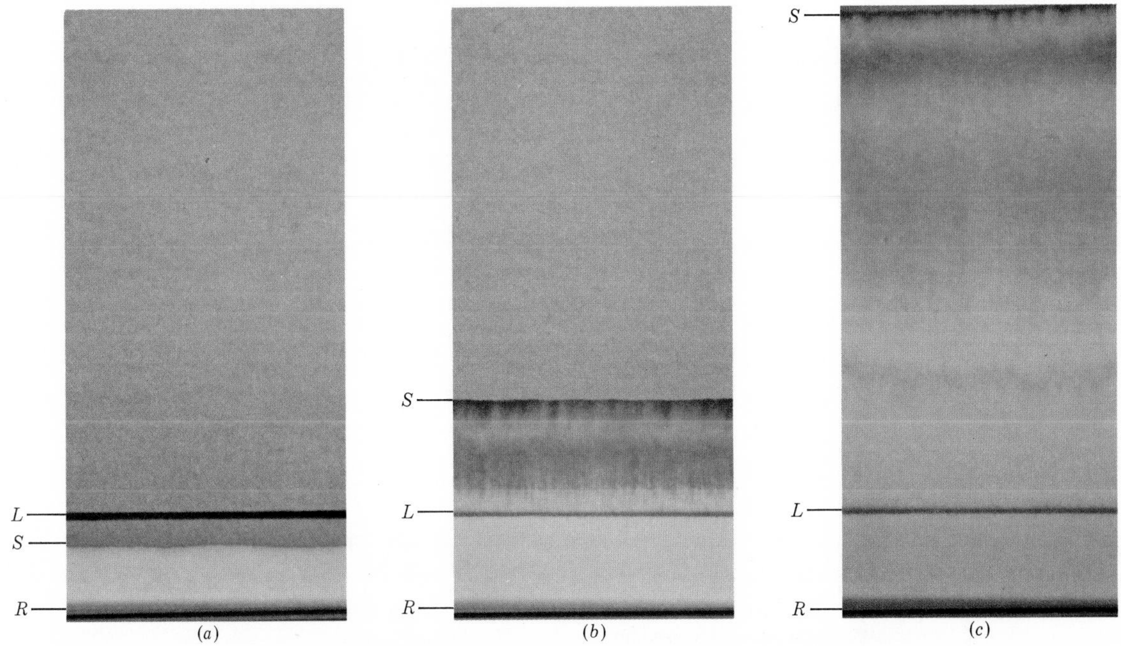

Plate 2 Separation of colored dyes in black ink by paper chromatography. (*a*) After a horizontal line has been drawn with a felt-tipped pen, the bottom of the paper is dipped in an alcohol-water mixture. (*b*) The liquid ascends the paper by capillary action, dissolving some dyes more readily than others. (*c*) The finished chromatogram, after a period of 1 hour. In each diagram the upper edge of the rising solvent is marked by *S*, the original ink line by *L*, and the reservoir of solvent at the bottom of the tank by *R*.

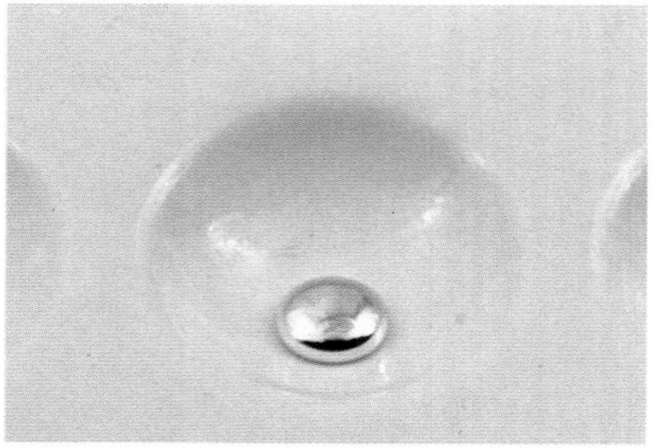

(*a*) Liquid mercury

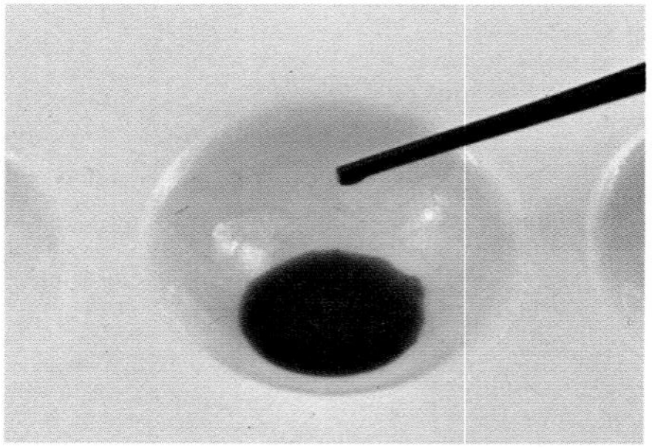

(*b*) Liquid bromine (note the bromine vapor as well)

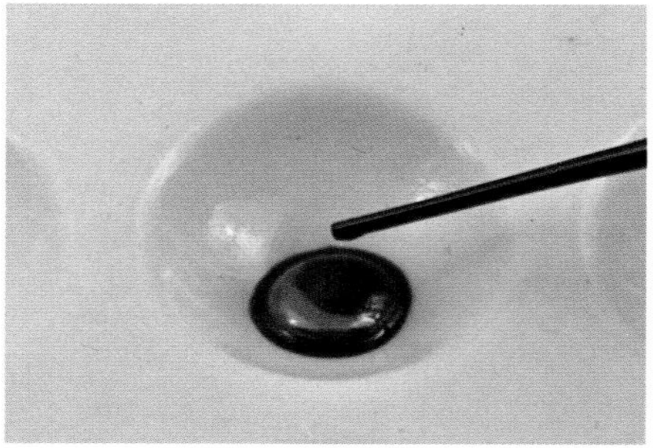

(*c*) A few drops of bromine have been added to the mercury

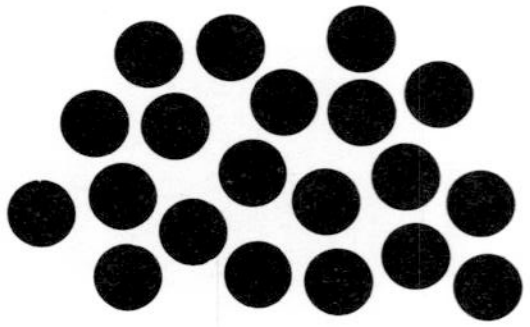

(*a*′) Hg atoms

(*b*′) Br_2 molecules

(*c*′) Hg + Br_2 before reaction

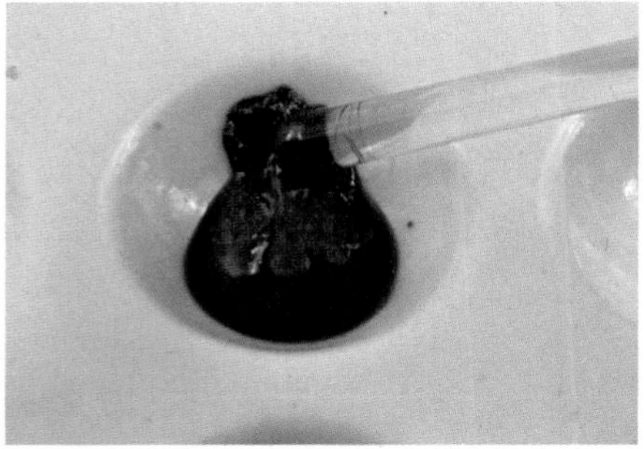

(*d*) The reactants are stirred and thoroughly mixed

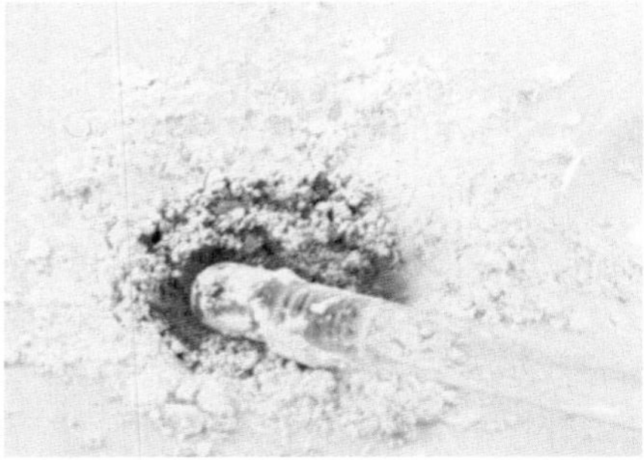

(*e*) The reaction is almost complete

(*f*) White crystals of solid mercury(II) bromide are the product

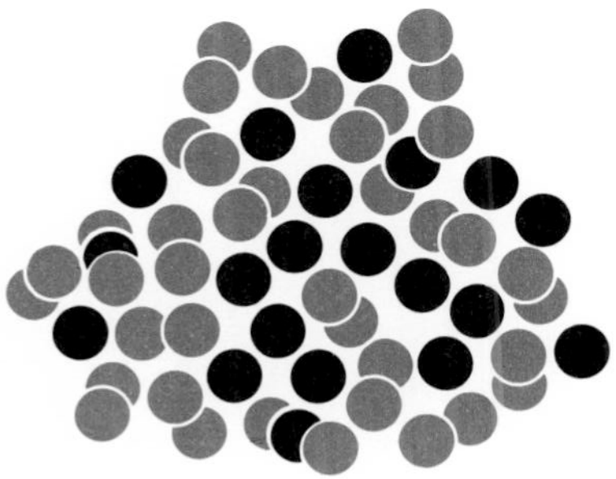

(*d'*) Hg atoms and Br_2 molecules mix together. A few $HgBr_2$ molecules form

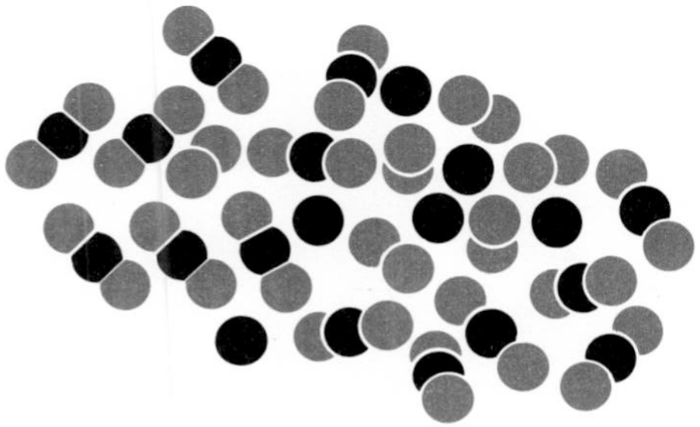

(*e'*) Only a few Br_2 molecules and Hg atoms are left

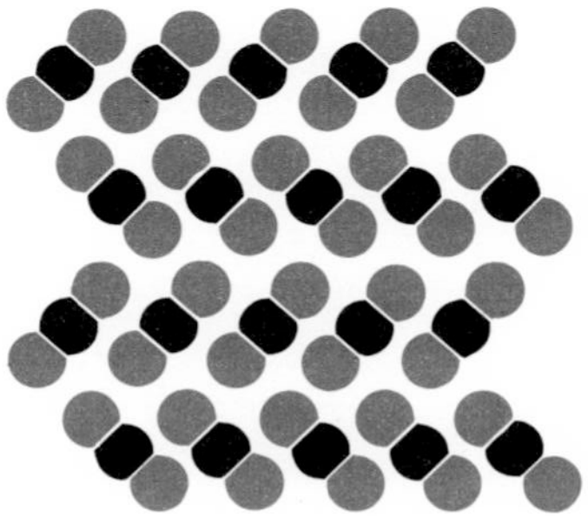

(*f'*) $HgBr_2$ molecules packed in a regular array

Plate 3 Macroscopic and microscopic descriptions of a chemical reaction. On the macroscopic level a silvery liquid, mercury, is mixed with a red-brown liquid, bromine, and white crystals are produced. On the microscopic level Hg atoms combine with Br_2 molecules to form $HgBr_2$ molecules packed together in a regular array.

Plate 4 Dot-density diagrams showing individual electrons and total electron density for H, He, and Li atoms. (Computer-generated.) *(Copyright © 1975 by W. G. Davies and J. W. Moore.)*

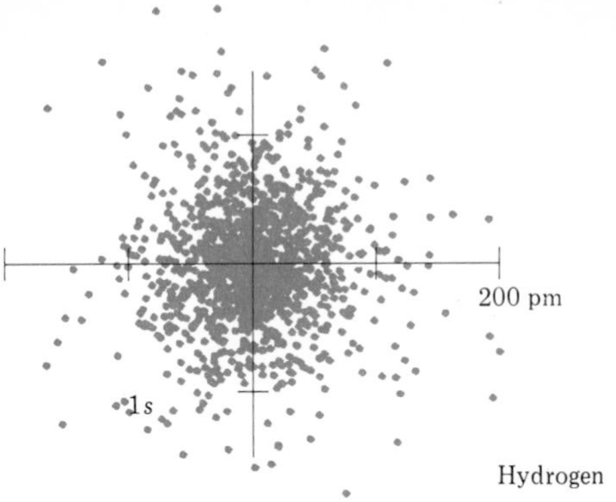

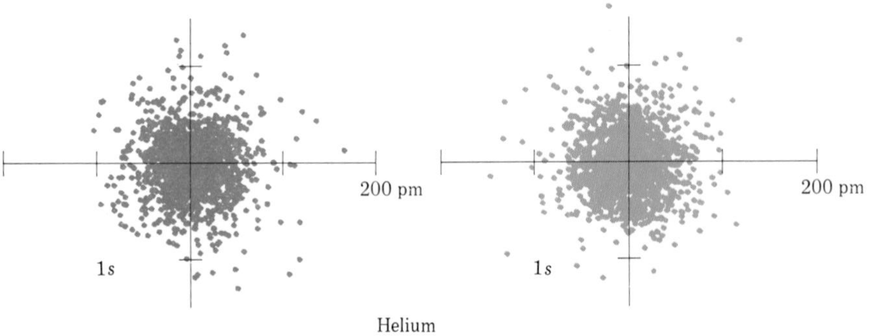

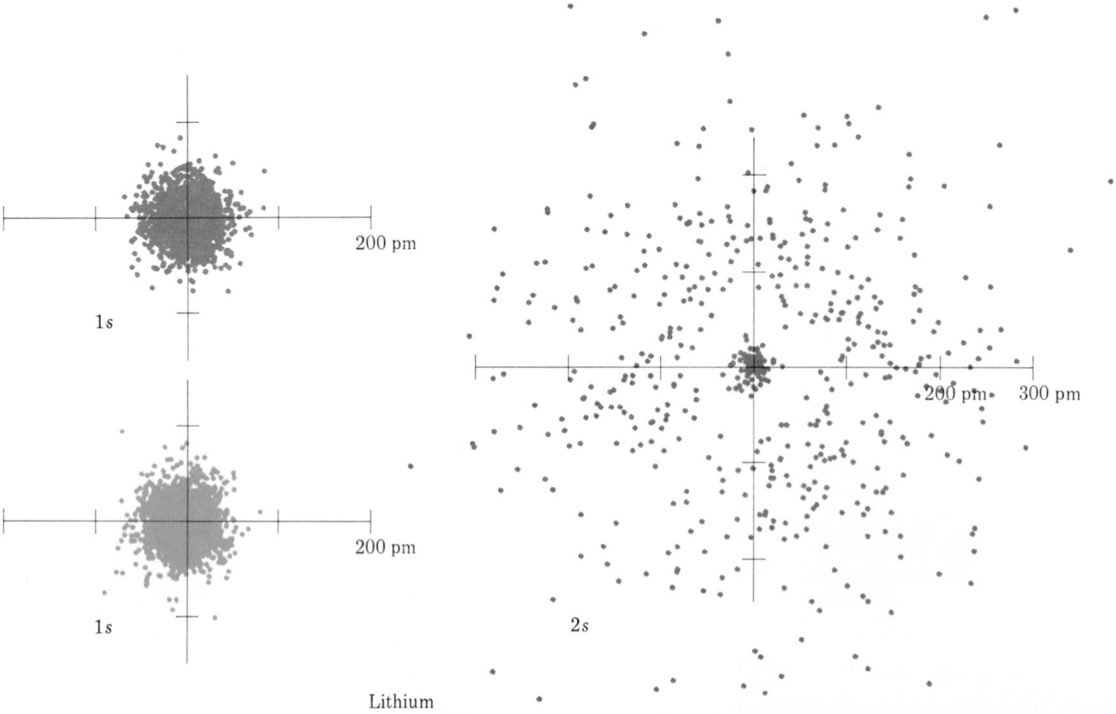

COMPLETE ATOM

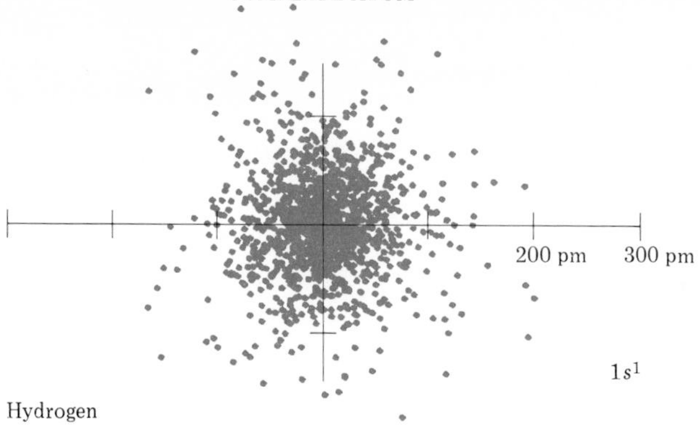

Hydrogen

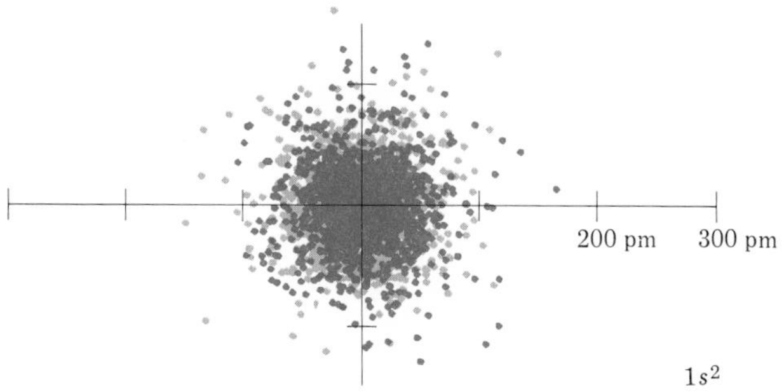

Helium

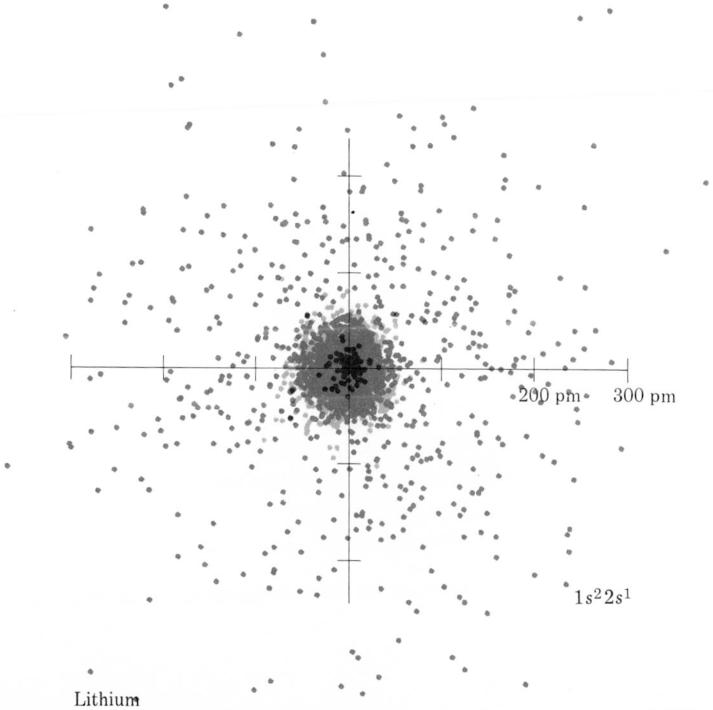

Lithium

CONSTITUENT ELECTRONS

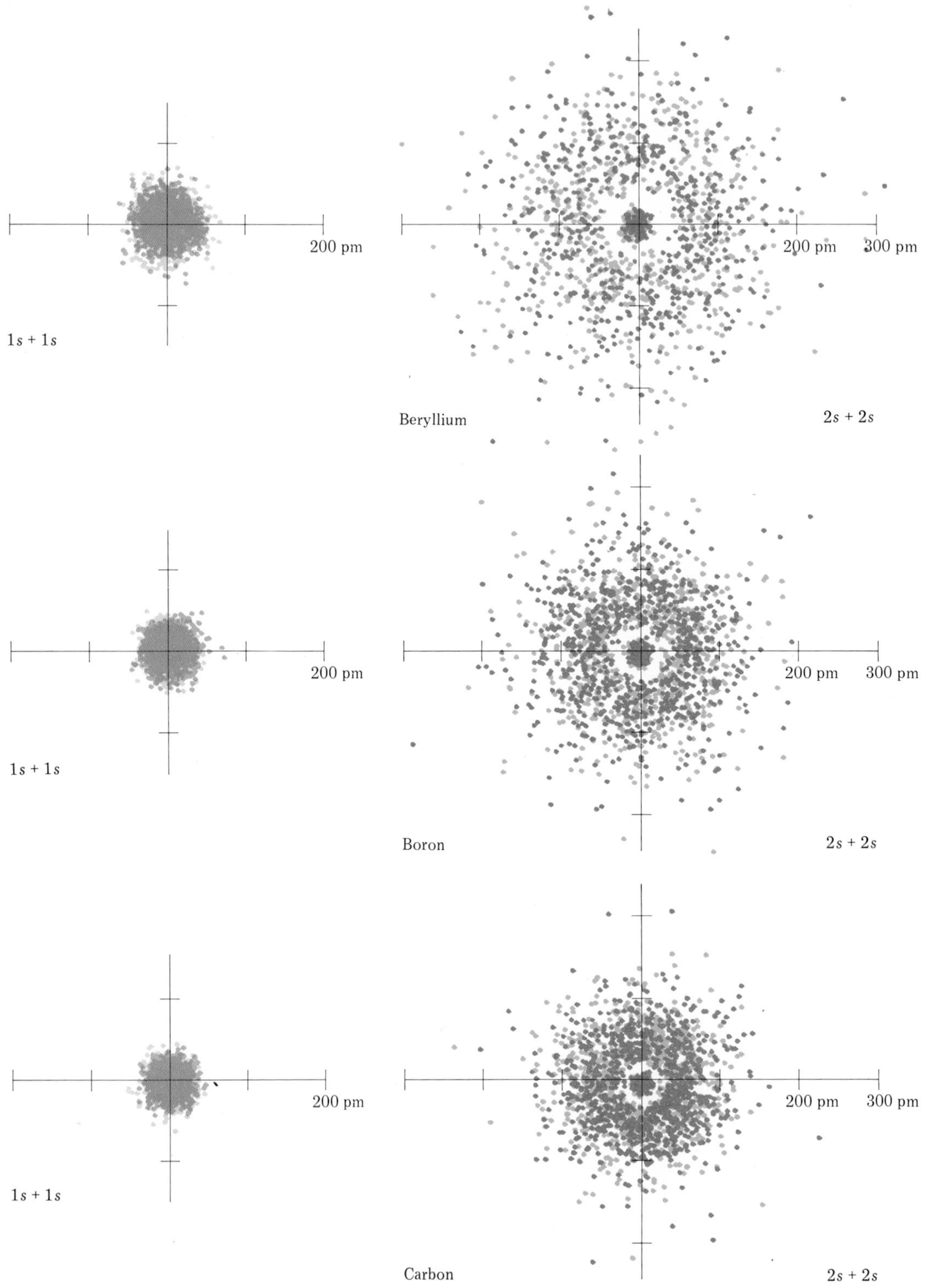

Plate 5 Dot-density diagrams showing constituent electrons and total electron density for Be, B, and C atoms. (Computer-generated.) (*Copyright © 1975 by W. G. Davies and J. W. Moore.*)

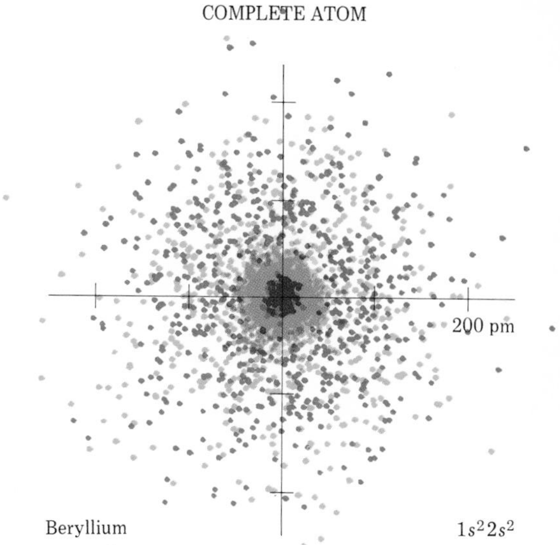

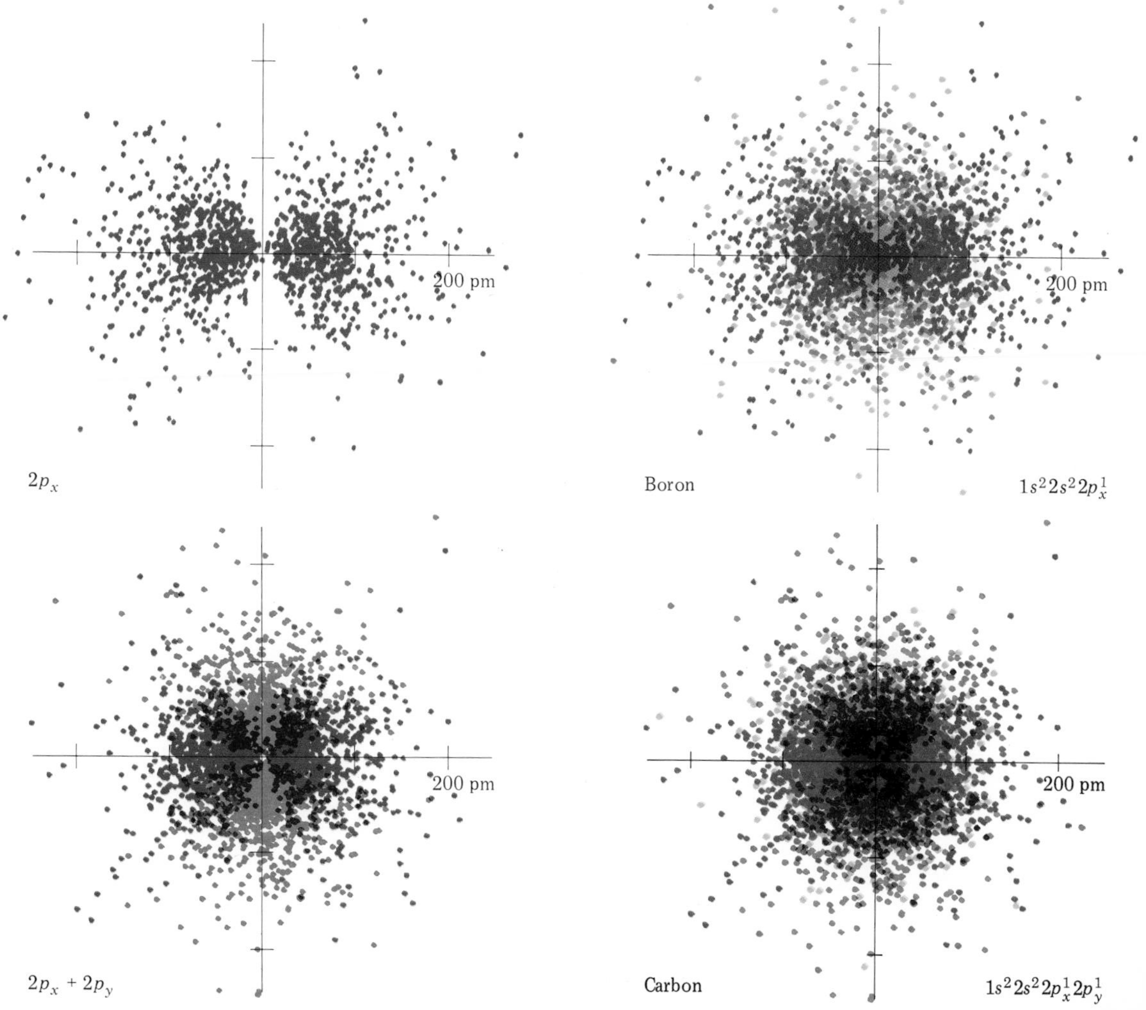

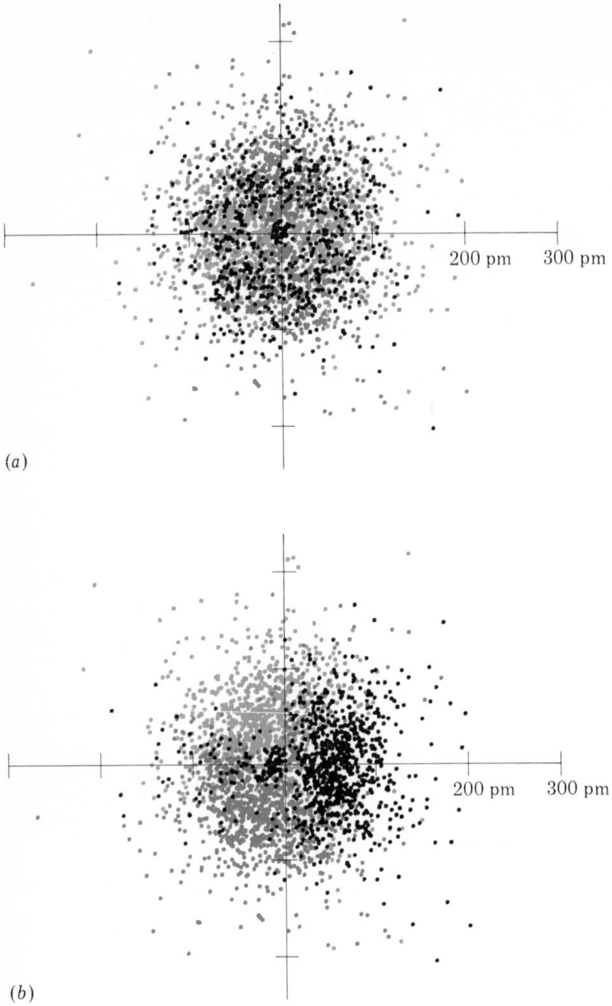

Plate 6 Electron-density distribution for the valence electron configuration $2s^2 2p_x^2 2p_y^2$. (*a*) Color coded to show $2s$ (black), $2p_x$ (green), and $2p_y$ (blue) electron densities; (*b*) color coded to show electron densities of three sp^2 hybrids at 120° angles. (Computer-generated.) (*Copyright © 1975 by W. G. Davies and J.W. Moore.*)

This is shown in Fig. 5.9*b*. The rows in this table consist of all possible subshells within each shell. For example, the second row from the bottom contains 2*s* and 2*p*, the two subshells in the second shell. Insertion of diagonal lines in the manner shown gives the right order for filling the subshells.

EXAMPLE 5.2 Predict the electron configuration for each of the following atoms: (a) $^{31}_{15}P$; (b) $^{59}_{27}Co$.

Solution In each case we follow the rules just stated.

a) For $^{31}_{15}P$ there would be 15 protons and 16 neutrons in the nucleus and 15 extranuclear electrons. Using Figure 5.9*b* to predict the order in which orbitals are filled, we have

$1s^2$	2 electrons, leaving 15 − 2 = 13 more to add
$2s^2$	2 electrons, leaving 11 more to add
$2p_x^2, 2p_y^2, 2p_z^2$ (or $2p^6$)	6 electrons, leaving 5 more to add
$3s^2$	2 electrons, leaving 3 more to add
$3p_x^1, 3p_y^1, 3p_z^1$	3 electrons

The electron configuration is thus

$$^{31}_{15}P \qquad 1s^22s^22p^63s^23p_x^13p_y^13p_z^1$$

It could also be written

$$[Ne]3s^23p_x^13p_y^13p_z^1 \qquad \text{or} \qquad [Ne]3s^23p^3$$

where [Ne] represents the neon kernel $1s^22s^22p^6$.

b) In the case of cobalt there is a total of 27 electrons to fill into the orbitals. There is no difficulty with the first 10 electrons. As in the previous example, they fill up the first and second shells:

$$1s^22s^22p^6 \qquad \text{17 more to add}$$

The third shell now begins to fill. First the 3*s* subshell then the 3*p* subshell are filled by 8 more electrons:

$$1s^22s^22p^63s^23p^6 \qquad \text{9 more to add}$$

Since this is also the structure of argon, we can use the shorthand form

$$[Ar] \qquad \text{9 more to add}$$

We now come to an energy overlap between the third and fourth shells. Because the 3*d* orbitals are so well shielded from the nucleus, they are *higher* in energy than the 4*s* orbitals. Accordingly the next orbitals to be filled are the 4*s* orbitals:

$$[Ar]4s^2 \qquad \text{7 more to add}$$

Once the 4*s* orbital is filled, the 3*d* orbitals are next in line to be filled. The 7 remaining electrons are insufficient to fill this subshell so that we have the final result

$$[\text{Ar}]3d^7 4s^2$$

Electron configurations of the atoms may be determined experimentally. Table 5.1 lists the results that have been obtained. There are some exceptions to the four rules enunciated above, but they are usually relatively minor. An obvious example of such an exception is the structure of chromium. It is found to be $[\text{Ar}]3d^5 4s^1$, whereas our rules would have predicted $[\text{Ar}]3d^4 4s^2$. Chromium adopts this structure because it allows the electrons to avoid each other more effectively. A complete discussion of this and other exceptions is beyond the scope of an introductory text.

5.7 ELECTRON CONFIGURATIONS AND THE PERIODIC TABLE

The commonly used long form of the periodic table (inside the front cover) is designed to emphasize the electron configurations that have just been described. Since it is the outermost (valence) electrons which are primarily involved in chemical interactions between atoms, the *last* electron added to an atom in the building-up process is of far more interest to a chemist than the first. This last electron is called the **distinguishing electron** because it distinguishes an atom from the one immediately preceding it in the periodic table. The type of subshell (*s, p, d, f*) into which the distinguishing electron is placed is very closely related to the chemical behavior of an element and gives rise to the classification shown by the color-coding on the periodic table inside the front cover.

The **representative elements** are those in which the distinguishing electrons enter an *s* or *p* subshell. Most of the elements whose chemistry and valence we have discussed so far fall into this category. Many of the chemical properties of the representative elements can be explained on the basis of the Lewis diagrams described in Sec. 5.1. That is, the valences of the representative elements may be predicted on the basis of the number of valence electrons they have, or from the number of electrons that would have to be added in order to attain the same electron configuration as an atom of a noble gas. For representative elements the number of valence electrons is the same as the periodic group number, and the number needed to match the next noble-gas configuration is 8 minus the group number. This agrees with the valence rules derived from the periodic table in Sec. 4.2, and results in formulas of the type shaded in Table 4.2.

The first three horizontal rows or periods in the modern periodic table consist entirely of representative elements. In the first period the distinguishing electrons for H and He are in the 1*s* subshell. Across the second period Li and Be have distinguishing electrons in the 2*s* subshell, and electrons are being added to the 2*p* subshell in the atoms from B to Ne. In the third period the $3s$ subshell is filling for Na and Mg, and the $3p$ for Al, Si, P, S, Cl, and Ar. As a general rule, in the case of the representative elements, the distinguishing electron will be in an ns or np subshell. The value of n, the principal quantum number for the distinguishing electron, can be quickly determined by counting down from the top of the periodic table. For example, iodine is a representative element in the *fifth* period. Therefore the distinguishing electron must occupy either the $5s$ or $5p$ subshell. Since I is on the right side of the table, $5p$ is the correct choice.

When the principal quantum number is three or more, d-type subshells are also possible. The **transition elements** or **transition metals** are those elements whose distinguishing electron is found in a d orbital. The first examples of transition metals (Sc, Ti, V, Cr, Mn, Fe, Co, Ni, Cu, Zn) are found in the *fourth* period even though the distinguishing electron in each case is a $3d$ electron and belongs to the *third* shell. This hiatus results, as we have already seen, because the $4s$ subshell is lower in energy than the $3d$. The $4s$ orbital thus starts to fill up, beginning the fourth period before any of the $3d$ orbitals can become occupied.

Figure 5.10 compares the probability distributions of a $4s$ and a $3d$ electron in a V atom. Although the $4s$ electron cloud lies farther from the nucleus on average than does the $3d$ cloud, a small portion of the $4s$ electron density is found very close to the nucleus where it is hardly shielded from the total nuclear charge of $+23$. It is the very strong attractive force of this small fraction of the total $4s$ electron density that lowers the energy of the $4s$ electron below that of the $3d$.

The fact that the $4s$ electron cloud is more extensive than the $3d$ has an important influence on the chemistry of the transition elements. When an atom such as V (Fig. 5.10) interacts with another atom, it is the $4s$ electrons extending farthest from the nucleus which first contact the other atom. Thus the $4s$ electrons are often more significant than the $3d$ in determining valence and the formulas of compounds. The $3d$ electrons are "buried" under the surfaces of the atoms of the transition metals. Adding one more $3d$ electron has considerably less effect on their chemical properties than adding one more $3s$ or $3p$ electron did in the case of the representative elements. Hence there is a slow but steady *transition* in properties from one transition element to another. Notice, for example, that except for Sc, all the transition metals in Table 4.2 form chlorides, MCl_2, where the metal has a valence of 2. This corresponds to the *two* $4s$ valence electrons.

Each of the transition metals also exhibits other valences where one or more of the $3d$ electrons are also involved. For example, in some compounds V (vanadium) has a valence of 2 (VO, VCl_2), in others it has a valence of 3 (V_2O_3, VCl_3), in still others it has a valence of 4 (VO_2, VCl_4), and in at least one case (V_2O_5) it has a valence of 5. The chemistry of the transition metals is more complicated and a wider variety of formulas for transition-metal compounds is possible because of this variable valence. In some cases electrons in the d subshells act as valence electrons, while in other cases they do not. Although the $3d$ electron clouds do not extend farther from the nucleus than $3s$ and $3p$ (and hence do not constitute another *shell* as the $4s$ electrons do), they are thoroughly shielded from the nuclear charge and thus often act as valence electrons. This Jekyll and Hyde

TABLE 5.1 Atomic Electron Configurations.

$_{1}$H		$1s^1$
$_{2}$He	$1s^2$	Helium core or kernel
$_{3}$Li	Helium core or kernel	$2s^1$
$_{4}$Be		$2s^2$
$_{5}$B		$2s^22p^1$
$_{6}$C		$2s^22p^2$
$_{7}$N		$2s^22p^3$
$_{8}$O		$2s^22p^4$
$_{9}$F		$2s^22p^5$
$_{10}$Ne	$1s^2\ 2s^22p^6$	Neon core or kernel
$_{11}$Na	Neon core or kernel	$3s^1$
$_{12}$Mg		$3s^2$
$_{13}$Al		$3s^23p^1$
$_{14}$Si		$3s^23p^2$
$_{15}$P		$3s^23p^3$
$_{16}$S		$3s^23p^4$
$_{17}$Cl		$3s^23p^5$
$_{18}$Ar	$1s^2\ 2s^22p^6\ 3s^23p^6$	Argon core or kernel
$_{19}$K	Argon core or kernel	$4s^1$
$_{20}$Ca		$4s^2$
$_{21}$Sc		$3d^14s^2$
$_{22}$Ti		$3d^24s^2$
$_{23}$V		$3d^34s^2$
$_{24}$Cr		$3d^54s^1$
$_{25}$Mn		$3d^54s^2$
$_{26}$Fe		$3d^64s^2$
$_{27}$Co		$3d^74s^2$
$_{28}$Ni		$3d^84s^2$
$_{29}$Cu		$3d^{10}4s^1$
$_{30}$Zn		$3d^{10}4s^2$
$_{31}$Ga		$3d^{10}4s^24p^1$
$_{32}$Ge		$3d^{10}4s^24p^2$
$_{33}$As		$3d^{10}4s^24p^3$
$_{34}$Se		$3d^{10}4s^24p^4$
$_{35}$Br		$3d^{10}4s^24p^5$
$_{36}$Kr	$1s^2\ 2s^22p^63s^23p^6\ 3d^{10}4s^24p^6$	Krypton core or kernel
$_{37}$Rb	Krypton core or kernel	$5s^1$
$_{38}$Sr		$5s^2$
$_{39}$Y		$4d^15s^2$
$_{40}$Zr		$4d^25s^2$
$_{41}$Nb		$4d^45s^1$
$_{42}$Mo		$4d^55s^1$
$_{43}$Tc		$4d^55s^2$
$_{44}$Ru		$4d^75s^1$
$_{45}$Rh		$4d^85s^1$
$_{46}$Pd		$4d^{10}$
$_{47}$Ag		$4d^{10}5s^1$
$_{48}$Cd		$4d^{10}5s^2$
$_{49}$In		$4d^{10}5s^25p^1$
$_{50}$Sn		$4d^{10}5s^25p^2$
$_{51}$Sb		$4d^{10}5s^25p^3$
$_{52}$Te		$4d^{10}5s^25p^4$
$_{53}$I		$4d^{10}5s^25p^5$

$_{54}$Xe	Krypton core or kernel	$4d^{10}$		$5s^25p^6$	Xenon core or kernel			
$_{55}$Cs		$4d^{10}$		$5s^25p^6$			$6s^1$	
$_{56}$Ba		$4d^{10}$		$5s^25p^6$			$6s^2$	
$_{57}$La		$4d^{10}$		$5s^25p^6$	$5d^1$		$6s^2$	
$_{58}$Ce		$4d^{10}$	$4f^1$	$5s^25p^6$	$5d^1$		$6s^2$	
$_{59}$Pr		$4d^{10}$	$4f^3$	$5s^25p^6$			$6s^2$	
$_{60}$Nd		$4d^{10}$	$4f^4$	$5s^25p^6$			$6s^2$	
$_{61}$Pm		$4d^{10}$	$4f^5$	$5s^25p^6$			$6s^2$	
$_{62}$Sm		$4d^{10}$	$4f^6$	$5s^25p^6$			$6s^2$	
$_{63}$Eu		$4d^{10}$	$4f^7$	$5s^25p^6$			$6s^2$	
$_{64}$Gd		$4d^{10}$	$4f^7$	$5s^25p^6$	$5d^1$		$6s^2$	
$_{65}$Tb		$4d^{10}$	$4f^9$	$5s^25p^6$			$6s^2$	
$_{66}$Dy		$4d^{10}$	$4f^{10}$	$5s^25p^6$			$6s^2$	
$_{67}$Ho		$4d^{10}$	$4f^{11}$	$5s^25p^6$			$6s^2$	
$_{68}$Er		$4d^{10}$	$4f^{12}$	$5s^25p^6$			$6s^2$	
$_{69}$Tm		$4d^{10}$	$4f^{13}$	$5s^25p^6$			$6s^2$	
$_{70}$Yb		$4d^{10}$	$4f^{14}$	$5s^25p^6$			$6s^2$	
$_{71}$Lu		$4d^{10}$	$4f^{14}$	$5s^25p^6$	$5d^1$		$6s^2$	
$_{72}$Hf		$4d^{10}$	$4f^{14}$	$5s^25p^6$	$5d^2$		$6s^2$	
$_{73}$Ta		$4d^{10}$	$4f^{14}$	$5s^25p^6$	$5d^3$		$6s^2$	
$_{74}$W		$4d^{10}$	$4f^{14}$	$5s^25p^6$	$5d^4$		$6s^2$	
$_{75}$Re		$4d^{10}$	$4f^{14}$	$5s^25p^6$	$5d^5$		$6s^2$	
$_{76}$Os		$4d^{10}$	$4f^{14}$	$5s^25p^6$	$5d^6$		$6s^2$	
$_{77}$Ir		$4d^{10}$	$4f^{14}$	$5s^25p^6$	$5d^7$		$6s^2$	
$_{78}$Pt		$4d^{10}$	$4f^{14}$	$5s^25p^6$	$5d^9$		$6s^1$	
$_{79}$Au		$4d^{10}$	$4f^{14}$	$5s^25p^6$	$5d^{10}$		$6s^1$	
$_{80}$Hg		$4d^{10}$	$4f^{14}$	$5s^25p^6$	$5d^{10}$		$6s^2$	
$_{81}$Tl		$4d^{10}$	$4f^{14}$	$5s^25p^6$	$5d^{10}$		$6s^26p^1$	
$_{82}$Pb		$4d^{10}$	$4f^{14}$	$5s^25p^6$	$5d^{10}$		$6s^26p^2$	
$_{83}$Bi		$4d^{10}$	$4f^{14}$	$5s^25p^6$	$5d^{10}$		$6s^26p^3$	
$_{84}$Po		$4d^{10}$	$4f^{14}$	$5s^25p^6$	$5d^{10}$		$6s^26p^4$	
$_{85}$At		$4d^{10}$	$4f^{14}$	$5s^25p^6$	$5d^{10}$		$6s^26p^5$	
$_{86}$Rn	Krypton core or kernel	$4d^{10}$	$4f^{14}$	$5s^25p^6$	$5d^{10}$		$6s^26p^6$	Radon core or kernel
$_{87}$Fr		$4d^{10}$	$4f^{14}$	$5s^25p^6$	$5d^{10}$		$6s^26p^6$	$7s^1$
$_{88}$Ra		$4d^{10}$	$4f^{14}$	$5s^25p^6$	$5d^{10}$		$6s^26p^6$	$7s^2$
$_{89}$Ac		$4d^{10}$	$4f^{14}$	$5s^25p^6$	$5d^{10}$		$6s^26p^6$	$6d^17s^2$
$_{90}$Th		$4d^{10}$	$4f^{14}$	$5s^25p^6$	$5d^{10}$		$6s^26p^6$	$6d^27s^2$
$_{91}$Pa		$4d^{10}$	$4f^{14}$	$5s^25p^6$	$5d^{10}$	$5f^2$	$6s^26p^6$	$6d^17s^2$
$_{92}$U		$4d^{10}$	$4f^{14}$	$5s^25p^6$	$5d^{10}$	$5f^3$	$6s^26p^6$	$6d^17s^2$
$_{93}$Np		$4d^{10}$	$4f^{14}$	$5s^25p^6$	$5d^{10}$	$5f^4$	$6s^26p^6$	$6d^17s^2$
$_{94}$Pu		$4d^{10}$	$4f^{14}$	$5s^25p^6$	$5d^{10}$	$5f^6$	$6s^26p^6$	$7s^2$
$_{95}$Am		$4d^{10}$	$4f^{14}$	$5s^25p^6$	$5d^{10}$	$5f^7$	$6s^26p^6$	$7s^2$
$_{96}$Cm		$4d^{10}$	$4f^{14}$	$5s^25p^6$	$5d^{10}$	$5f^7$	$6s^26p^6$	$6d^17s^2$
$_{97}$Bk		$4d^{10}$	$4f^{14}$	$5s^25p^6$	$5d^{10}$	$5f^9$	$6s^26p^6$	$7s^2$
$_{98}$Cf		$4d^{10}$	$4f^{14}$	$5s^25p^6$	$5d^{10}$	$5f^{10}$	$6s^26p^6$	$7s^2$
$_{99}$Es		$4d^{10}$	$4f^{14}$	$5s^25p^6$	$5d^{10}$	$5f^{11}$	$6s^26p^6$	$7s^2$
$_{100}$Fm		$4d^{10}$	$4f^{14}$	$5s^25p^6$	$5d^{10}$	$5f^{12}$	$6s^26p^6$	$7s^2$
$_{101}$Md		$4d^{10}$	$4f^{14}$	$5s^25p^6$	$5d^{10}$	$5f^{13}$	$6s^26p^6$	$7s^2$
$_{102}$No		$4d^{10}$	$4f^{14}$	$5s^25p^6$	$5d^{10}$	$5f^{14}$	$6s^26p^6$	$7s^2$
$_{103}$Lr		$4d^{10}$	$4f^{14}$	$5s^25p^6$	$5d^{10}$	$5f^{14}$	$6s^26p^6$	$6d^17s^2$
$_{104}$?		$4d^{10}$	$4f^{14}$	$5s^25p^6$	$5d^{10}$	$5f^{14}$	$6s^26p^6$	$6d^27s^2$
$_{105}$?		$4d^{10}$	$4f^{14}$	$5s^25p^6$	$5d^{10}$	$5f^{14}$	$6s^26p^6$	$6d^37s^2$
$_{106}$?		$4d^{10}$	$4f^{14}$	$5s^25p^6$	$5d^{10}$	$5f^{14}$	$6s^26p^6$	$6d^47s^2$

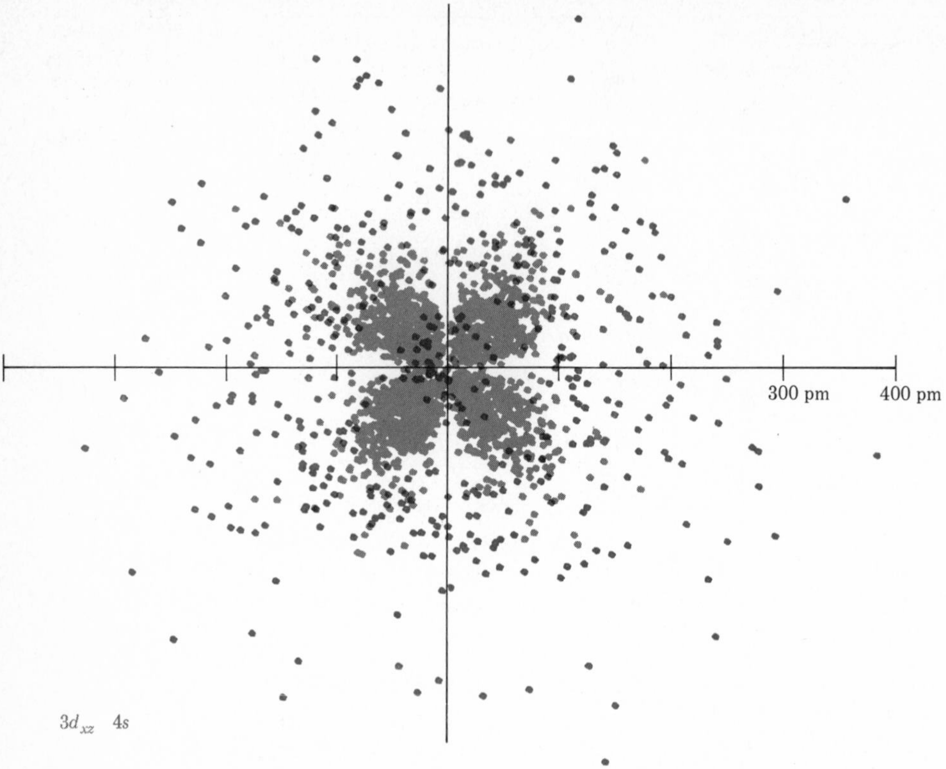

Figure 5.10 Comparison of 3*d* (gray) and 4*s* (color) electron clouds for a vanadium atom (computer-generated). (*Copyright © 1975 by W. G. Davies and J. W. Moore.*)

behavior of 3*d* electrons makes life more complicated (and often far more interesting) for chemists who study the transition elements.

The third major category of elements arises when the distinguishing electron occupies an *f* subshell. The first example occurs in the case of the **lanthanoids** (elements having atomic numbers between 57 and 71). The lanthanoids have the general electron configuration

$$[Kr]4d^{10}4f^{i}5s^{2}5p^{6}5d^{0 \text{ or } 1}6s^{2}$$

where *i* is a number between 0 and 14. Thus in the building-up process for the lanthanoids, electrons are being added to a subshell (4*f*) whose principal quantum number is two less than that of the outermost orbital (6*s*). Addition of another electron to an inner shell buried as deeply as the 4*f* has little or no effect on the chemical properties of these elements. All are quite similar to lanthanum (La) and might fit into exactly the same space in the periodic table as La. The lanthanoid elements are so similar to one another that special techniques are required to separate them. As a result, even approximately pure samples of most of them were not prepared until the 1870s.

Following the element actinium (Ac) is a series of atoms in which the 5*f* subshell is filling. The **actinoids** are somewhat less similar to Ac than the lanthanoids are to La because some exceptions to the usual order of filling orbitals occur in the case of Th, Pa, and U (see Table 5.1). Because

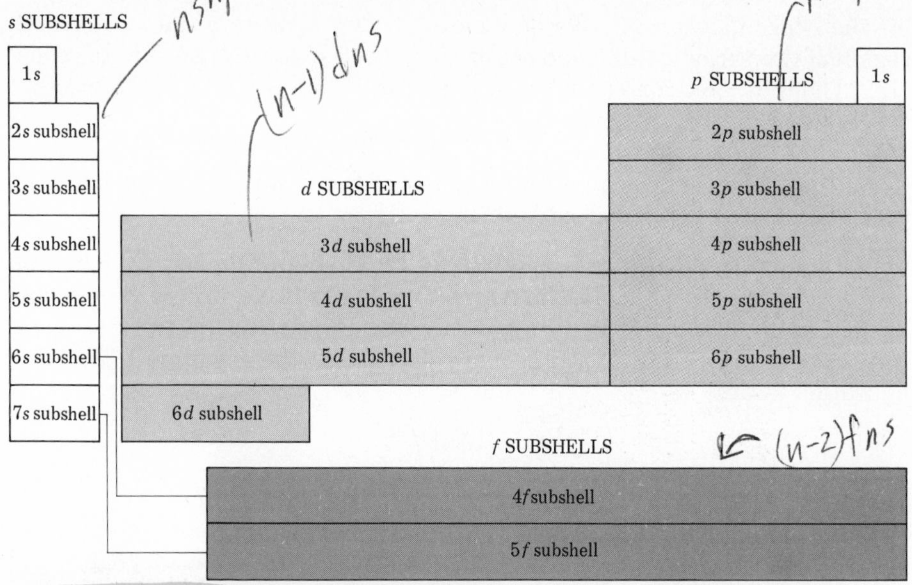

Figure 5.11 Type of subshell containing the distinguishing electron for atoms of elements in different regions of the periodic table.

the lanthanoids and most of the actinoids behave chemically as if they should fit in group IIIB of the periodic table (where Lu and Lr are found), both groups are separated from the rest of the table and placed together in a block below. Taken together, the lanthanoids and actinoids are called **inner transition elements** because the *f* subshells being filled lie so deep within the remaining electronic structure of their atoms.

Figure 5.11 summarizes the type of subshell in which the distinguishing electron is to be found for atoms of elements in various regions of the periodic table. This summary information makes it relatively simple to use the periodic table to obtain electron configurations, as the following example shows.

EXAMPLE 5.3 Obtain the electron configuration for (a) Nb; (b) Pr.

Solution

a) Nb, element number 41, is found in the fifth period and in a region of the periodic table where a *d* subshell is filling (the second transition series). Moving backward (toward lower atomic numbers) through the periodic table, the nearest noble gas is Kr, and so we use the Kr kernel:

Nb [Kr] ______

The next element after $_{36}$Kr is $_{37}$Rb in which the 5*s* subshell is filling. Moving right one more space, we come to $_{38}$Sr which has a $5s^2$ pair. So far we have

Nb [Kr] ______ $5s^2$

for the Nb configuration. We now move farther right into the *4d subshell* region of the periodic table and count over three spaces (Y, Zr, Nb) to reach Nb. The total electron configuration is thus

$$\text{Nb} \qquad [\text{Kr}]4d^3 5s^2$$

(Note that the principal quantum number of the *d* subshell is *4*—one less than the number of the period.)

b) A similar procedure is followed for Pr, element number 59. Moving backward through the table, the nearest noble gas is Xe, and so we use the Xe kernel. Counting forward again, Cs and Ba correspond to $6s^2$. Then La, Ce, and Pr correspond to three more electrons in the $4f$ subshell. The configuration is thus

$$\text{Pr} \qquad [\text{Xe}]4f^3 6s^2$$

One more point needs to be emphasized about the relationship between electron configuration and the periodic table. *The atoms of elements in the same vertical column of the table have similar electron configurations.* For example, consider the alkaline-earth elements (group IIA). Using our rules for deriving electron configurations (Example 5.3) we have

Element	Electron Configuration	Lewis Diagram
Be	$[\text{He}]2s^2$	Be :
Mg	$[\text{Ne}]3s^2$	Mg :
Ca	$[\text{Ar}]4s^2$	Ca :
Sr	$[\text{Kr}]5s^2$	Sr :
Ba	$[\text{Xe}]6s^2$	Ba :
Ra	$[\text{Rn}]7s^2$	Ra :

Thus the similarities of chemical behavior and valence noted earlier for these elements correlate with the similarities of their outermost electron clouds. Such similarities account for the success of Mendeleev's predictions of the properties of undiscovered elements.

SUMMARY

Ideas about the electronic structures of atoms developed during the first half of the twentieth century. The periodic repetition of chemical properties discovered by Mendeleev led G. N. Lewis to the conclusion that atoms must have a shell structure. This was confirmed by wave mechanics. Only certain specific wave patterns are possible for an electron in an atom, and these electron clouds are arranged in concentric shells.

The energy of each electron in an atom depends on how strongly the electron is attracted by the positive charge on the nucleus and on how much it is repelled by other electrons. Although each electron cannot be assigned a precise trajectory or orbit in an atom, its wave pattern allows us to determine the probability that it will be at a certain location. From this the energy of each electron and the order of filling orbitals can be obtained. Thus we can determine the electron configuration for an atom of any element. Such electron configurations correlate with the periodic table.

Because electrons in inner orbitals screen outer electrons from nuclear charge, the fourth and higher shells begin to fill before *d* (and sometimes *f*) subshells in previous shells are occupied. This overlap in energies of shells explains why Lewis' ideas are less useful for elements in the fourth and subsequent rows of the periodic table. It also accounts for the steady variation in properties of transition metals across the table, and for the nearly identical characteristics of inner transition elements as opposed to the large differences from one group of representative elements to the next.

Although some added complication arises from the wave-mechanical picture, it does confirm Lewis' basic postulate that valence electrons determine chemical properties and influence the bonding of one atom to another. In the next chapter we shall see how rearrangement of valence electrons can hold atoms together, and how different kinds of bonds result in different macroscopic properties.

QUESTIONS AND PROBLEMS

(An asterisk before a number indicates the problem is more difficult.)

5.1 Even though calcium and sulfur have otherwise very different chemical properties, they both have a valence of 2. How can this behavior be explained in terms of Lewis' theory?

5.2 Explain how the shell model of electronic structure proposed by Lewis enables us to explain why the properties of the elements recur periodically.

5.3 A fellow student suggests to you that since the lengths of the periods in the periodic table are 2, 8, 8, 18, 18, and 32, this is also the number of electrons in the first, second, third, fourth, fifth and sixth shells, respectively. Explain why the student is wrong, but not altogether wrong.

5.4 Draw Lewis diagrams showing valence electrons for Rb, C, Al, Kr, Se, Sn, Cl, and Ba.

5.5 Predict the valence of each atom in the previous problem.

5.6 Without drawing Lewis diagrams, predict the normal valence of each of the following elements: Mg, F, Te, B, P, N, Ar, and K.

5.7 In what way were the early Lewis and Bohr theories of electronic structure similar? In which way were they different? Which tried to give the most detailed and accurate model?

5.8 State as clearly as you can what restrictions the uncertainty principle places on our ability to describe the behavior of an electron. Is this true only of electrons?

5.9 De Broglie's equation $\lambda = h/mu$ is often said to be a mathematical expression of the concept of *wave-particle duality*. Explain in what sense this is true.

5.10 In your own words, without referring to the text, explain why Eq. (5.2), $d = n(\lambda/2)$, is valid for a guitar string of length *d*.

5.11 Which of the following values is permitted for the kinetic energy of an electron in a one-dimensional box of length 100 pm?

a 36.0 aJ
b 96.0 aJ
c 23.8 aJ
d 54.0 aJ

5.12 Explain how the shell model for the electronic structure of the atom is in accord with the wave nature of the electron.

5.13 Suggest some experimental method for determining the trajectory of a particle traveling through space. Explain why this method could not be applied to a particle as light as the electron without disturbing its trajectory.

5.14 There is some uncertainty as to whether a baseball pitcher really can pitch a curve ball or whether it is an optical illusion. Can this possibly be a result of Heisenberg's uncertainty principle?

5.15 Why is the question, "How does an electron get from one place to another," impossible to answer?

5.16 Explain what is wrong with the following statement: The particle model of the electron is best for explaining tangible macroscopic events, whereas the wave interpretation is best for rationalizing intangible microscopic events in the atomic world.

5.17 Exactly what information is conveyed by a dot-density diagram like that given in Fig. 5.6?

5.18 List two specific ways in which a real electron in an atom differs from a hypothetical electron in a one-dimensional box.

5.19 Under what circumstances, if any, is it possible to find an electron outside one of the boundary-surface diagrams shown in Fig. 5.8?

5.20 Distinguish between the terms (*a*) orbit and orbital; (*b*) shell and subshell; (*c*) dot-density diagram and boundary-surface diagram.

5.21 Discuss the following statement: The larger the value of the principal quantum number *n*, the larger is the electron.

5.22 How many possible orientations are there in three-dimensional space for *s* orbitals, *p* orbitals, *d* orbitals, and *f* orbitals?

5.23 According to Coulomb's law, how does the force between two oppositely charged particles change in comparison to its original value if (*a*) the charges remain constant, but the distance between them is halved; (*b*) the charge on one particle only is tripled, but the distance between them remains the same as the original distance; (*c*) the charges on both particles are tripled, but the distance remains the same; (*d*) the charges on both particles are doubled from the original, while the distance between them is increased to four times the original.

5.24 Explain why the potential energy of two particles with like charges increases as they move closer together.

5.25 Assuming that Venus and the Earth have the same mass, which has the higher potential energy?

***5.26** If you know some physics, decide whether Venus or Earth has a greater kinetic energy. Which has the greater total energy?

5.27 What would be the effect on the size of a lithium atom if the core electrons were less efficient at screening the nucleus from the 2*s* electron than is actually the case?

5.28 Quantum-mechanical calculations show that the effective nuclear charge experienced by a 1*s* electron in the beryllium atom is 3.685, while that experienced by a 2*s* electron is 1.912. Explain why these two figures are different. Explain also why the one is somewhat less than 4, while the other is somewhat less than 2.

5.29 In a boron atom, calculations show that the effective nuclear charge experienced by a 2*p* electron is 2.421, while that experienced by a 2*s* electron is somewhat larger, namely, 2.576. Explain this difference. Also explain why the effective nuclear charge experienced by a 2*s* electron in a boron atom is larger than that experienced by a 2*s* electron in a beryllium atom.

5.30 On the basis of both the Pauli exclusion principle and Hund's rule, what is wrong with each of the following electron configurations?

a C: $1s^2 2s^2 2p_x^2$	↓↑	↓↓	↓↑		
	$1s$	$2s$	$2p_x$	$2p_y$	$2p_z$
b N: $1s^2 2s^2 2p_x^2 2p_y^1$	↓↑	↓↑	↑↑	↓	
	$1s$	$2s$	$2p_x$	$2p_y$	$2p_z$

5.31 Explain how the operation of Hund's rule produces a lower potential energy for the atom than would otherwise be the case.

5.32 Distinguish between members of each of the following pairs of terms, using both words and drawings:

a A p_x orbital and a p_y orbital.
b A 2*s* orbital and a 3*s* orbital.
c A 3*p* orbital and a 3*d* orbital.

5.33 Discuss the following statement: All the subshells within any shell have lower energies than all the subshells in the next higher shell.

5.34 Identify the atoms with the following electronic structures:

a $1s^22s$
b $1s^22s^22p^4$
c $1s^22s^22p^63s^2$
d $1s^22s^22p^63s^23p^3$
e $1s^22s^22p^63s^23p^63d^24s^2$

5.35 Write out the electron configuration for each of the following atoms in a manner similar to that in the previous problem:

a Si
b Sr
c Se
d Sc
e Kr
f Ra
g Rb
h Pb

5.36 Write out the electron configuration for each of the following atoms, using the appropriate noble-gas kernel for abbreviation. For example, P is [Ne]$3s^23p^3$.

a S
b Co
c Cd
d In
e Ta
f As
g Sm
h Pb

5.37 Differentiate between representative elements and transition elements in terms of the location of their distinguishing electrons.

5.38 Classify the following elements as metals or nonmetals and as representative, transition, or inner transition elements: Li, Ti, Pu, Hg, Se, W, Xe, Fe, Ag, and Ba.

5.39 The chemical properties of an element are determined entirely by the number of electrons in its outermost shell. Discuss the truth of this statement and exceptions there may be to it.

5.40 Explain why the energy of a 4*s* electron is lower than that of a 3*d* electron in an atom of Sc.

5.41 Suggest why many of the transition elements exhibit more than one valence.

5.42 Why is the valence of 2 so common for the first-row transition elements?

5.43 Why are the 14 lanthanoid elements so similar in their chemical properties?

chapter six

CHEMICAL BONDING—ELECTRON PAIRS AND OCTETS

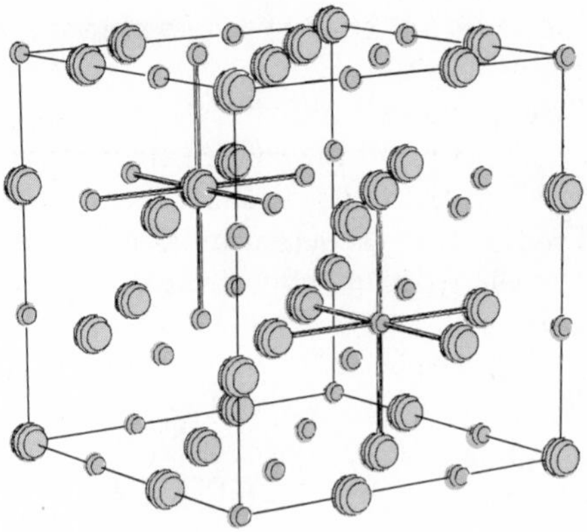

Although we have spent two chapters discussing atomic structure, it is worth realizing that of all the elements only the noble gases are found naturally in a form such that their atoms occur as individuals, widely separated from all other atoms. Under the conditions that prevail on the surface of the earth, *almost all atoms are linked by chemical bonds to other atoms*. Oxygen, for example, is the most common element on earth. It is found in combination with metals in rocks, with hydrogen in water, with carbon and hydrogen in living organisms, or as the diatomic molecule O_2 in the atmosphere, but individual oxygen atoms are quite rare. Most other elements behave in a similar way. Thus, if we want to understand the chemistry of everyday matter, we need to understand the nature of the chemical bonds which hold atoms together.

Theories of chemical bonding invariably involve electrons. When one atom approaches another, the *valence electrons,* found in the outermost regions of the atoms, interact long before the nuclei can come close together. Electrons are the least massive components of an atom, and so they can relocate to produce electrostatic forces which hold atoms together. According to Coulomb's law, such electrostatic or coulombic forces are quite large when charges are separated by distances of a few hundred picometers—the size of an atom. Coulombic forces, then, are quite capable of explaining the strengths of the bonds by which atoms are held together.

An important piece of evidence relating electrons and chemical bonding was noted by G. N. Lewis shortly after the discovery that the atomic number indicated how many electrons were present in each kind of atom. *Most chemical formulas correspond to an even number of electrons summed over all constituent atoms.* Thus H_2O has 2 electrons from 2H and 8 from O for a total of 10, NCl_3 has $7 + (3 \times 17) = 58$ electrons, and so on. This is a bit surprising when you consider that half the elements have odd atomic numbers so that their atoms have an odd number of electrons. Lewis suggested that *when atoms are bonded together, the electrons occur in pairs,* thus accounting for the predominance of even numbers of electrons in chemical formulas.

There are two important ways in which the valence shells of different atoms can interact to produce electron pairs and chemical bonds. When two atoms have the same degree of attraction for their valence electrons, it is possible for them to *share* pairs of electrons in the region between their nuclei. Such shared pairs attract *both* nuclei, holding them together with a **covalent bond.** On the other hand, when two atoms have quite different degrees of attraction for their outermost electrons, one or more electrons may transfer their allegiance from one atom to another, pairing with electrons already present on the second atom. The atom to which electrons are transferred will acquire excess negative charge, becoming a negative *ion,* while the atom which loses electrons will become a positive ion. These oppositely charged ions will be held together by the coulombic forces of attraction between them, forming an **ionic bond.** This chapter will consider the formation of ionic and covalent bonds and the properties of some substances containing each type of bond.

6.1 IONIC BONDING

Since ionic bonding requires that the atoms involved have unequal attraction for their valence electrons, an ionic compound must involve atoms of two different elements. The simplest example is provided by the combination of elements number 1 (H) and number 3 (Li) in lithium hydride, LiH. On a microscopic level the formula LiH contains four electrons. In separate Li and H atoms these electrons are arranged as shown in Fig. 6.1*a*. The H atom has the electron configuration $1s^1$, and Li is $1s^2 2s^1$. When the two atoms are brought close enough together, however, the striking rearrangement of the electron clouds shown in Fig. 6.1*b* takes place. Here the color coding shows clearly that the electron density which was associated with the 2*s* orbital in the individual Li atom has been transferred to a 1*s* orbital surrounding the H atom. As a result, two new microscopic species are formed. The extra electron transforms the H atom into a *negative ion* or **anion,** written H^- and called the *hydride ion.* The two electrons left on the Li atom are not enough to balance the charge of +3 on the Li nucleus, and so removal of an electron produces a *positive ion* or **cation,** written Li^+ and called the *lithium ion.* The electron-transfer process can be summarized in terms of Lewis diagrams as follows:

$$\underset{1s^2 2s^1}{\text{Li}\cdot} + \underset{1s^1}{\cdot\text{H}} \longrightarrow \underset{1s^2}{\text{Li}^+} + \underset{1s^2}{:\text{H}^-}$$

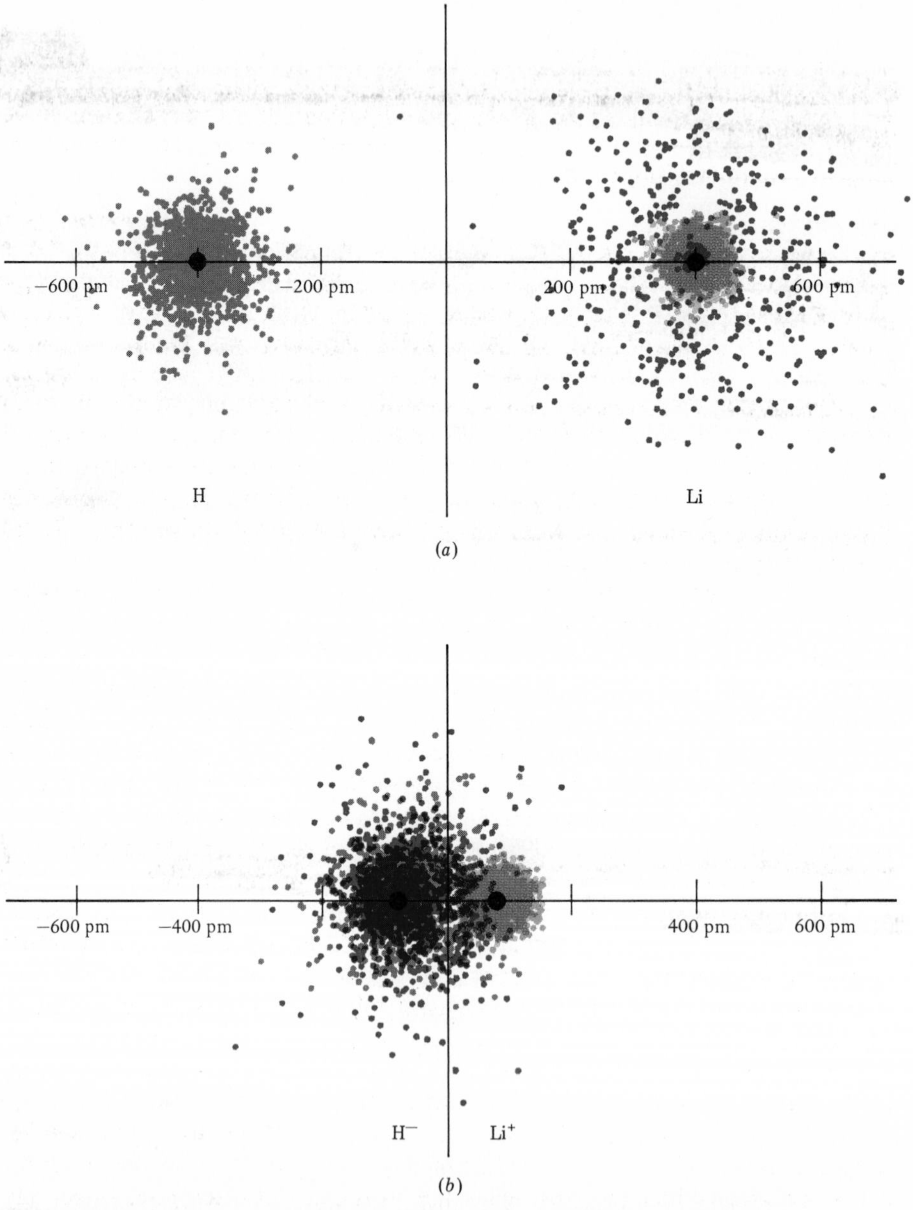

Figure 6.1 Formation of LiH ion pair. (*a*) Dot-density diagrams for separated lithium and hydrogen atoms. (*b*) Dot-density diagram for LiH ion pair. Note that upon formation of the ion pair, electron density has been transferred from the 2*s* orbital of Li to the 1*s* orbital of H. (Computer-generated.) (*Dot-density diagrams copyright © 1975 by W. G. Davies and J. W. Moore.*)

The opposite charges of Li^+ and H^- attract each other strongly, and the ions form an **ion pair** in which the two nuclei are separated by a distance of 160 pm (1.60 Å).

Energy and the Formation of Ions

Formation of an ion pair by transfer of an electron from an Li atom to an H atom results in an overall *lowering of the total energy* of the two nuclei and four electrons involved. How and why this occurs is best seen if we break ion-pair formation into three simpler steps and consider the energy change involved in each. The three steps are

1 Removal of the 2*s* electron from an Li atom to form an Li^+ ion.
2 Addition of that same electron to an H atom to form an H^- ion.
3 The coming together of the two ions to form an ion pair.

The energy required in step 1 to remove an electron completely from an isolated atom is called the **ionization energy.** The ionization energy of lithium is 520 kJ mol^{-1}. In other words, 520 kJ of energy is needed to remove a mole of 2*s* electrons from a mole of isolated lithium atoms in order to form a mole of isolated lithium ions. Alternatively we can say that 520 kJ is needed to **ionize** a mole of lithium atoms.

While energy is needed to accomplish step 1, we find that energy is *released* in step 2 when a hydrogen atom accepts an electron and becomes a hydride ion. The reason for this can be seen in Fig. 6.1. The second electron acquired by the hydrogen atom can pair up with the electron already in the 1*s* orbital without contradicting the Pauli exclusion principle. As a result, the new electron can move in close enough to the hydrogen nucleus to be held fairly firmly, lowering its energy significantly. Although the paired 1*s* electrons repel each other somewhat, this is not enough to offset the attraction of the nucleus for both. Since the energy of the electron is *lowered,* the law of conservation of energy requires that the same quantity of energy must be *released* when a hydrogen atom is transformed into a hydride ion.

The energy released when an electron is acquired by an atom is called the **electron affinity.** The electron affinity of hydrogen is 73 kJ mol^{-1}, indicating that 73 kJ of energy is released when 1 mol of isolated hydrogen atoms each accepts an electron and is converted into a hydride ion.

Since 520 kJ mol^{-1} is *required* to remove an electron from a lithium atom, while 73 kJ mol^{-1} is *released* when the electron is donated to a hydrogen atom, it follows that transfer of an electron from a lithium to a hydrogen atom requires $(520 - 73)$ kJ mol^{-1} = 447 kJ mol^{-1}. At room temperature processes which require such a large quantity of energy are extremely unlikely. Indeed the transfer of the electron would be impossible if it were not for step 3, the close approach of the two ions. When oppositely charged particles move closer to each other, their potential energy decreases and they release energy. The energy released when lithium ions and hydride ions come together to 160 pm under the influence of their mutual attraction is 690 kJ mol^{-1}, more than enough to offset the 447 kJ mol^{-1} needed to transfer the electron. Thus there is a net release of (690 − 447) kJ mol^{-1} = 243 kJ mol^{-1} from the overall process. The transfer of the electron from lithium to hydrogen and the formation of an Li^+H^- ion pair results in an overall lowering of energy.

In Fig. 6.2 the energy change in each step and the overall change are illustrated diagrammatically. As in the case of atomic structure, where electrons occupy orbitals having the lowest allowable energy, a collection of atoms tends to rearrange its constituent electrons so as to minimize its total energy. Formation of a lithium hydride ion pair is energetically "downhill" and therefore favored.

The Ionic Crystal Lattice

Up to this point we have been considering what would happen if a single Li atom and a single H atom were combined. When a large number of atoms of each kind combine, the result is somewhat different. Electrons are again transferred, and ions are formed, but the ions no longer pair off in twos. Instead, under the influence of their mutual attractions and repulsions, they collect together in much larger aggregates, eventually forming a three-dimensional array like that shown in Fig. 6.3. On the macroscopic level a crystal of solid lithium hydride is formed.

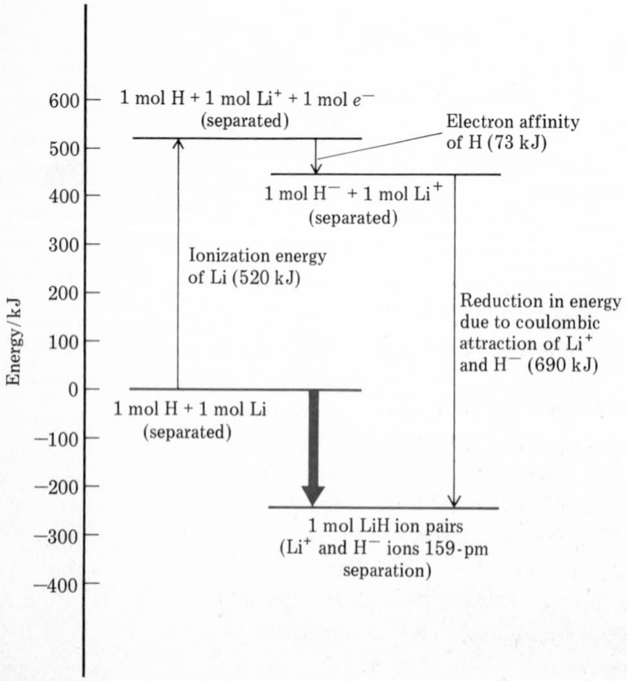

Figure 6.2 Energy changes which occur when 1 mol H atoms and 1 mol Li atoms are transformed into 1 mol LiH ion pairs.

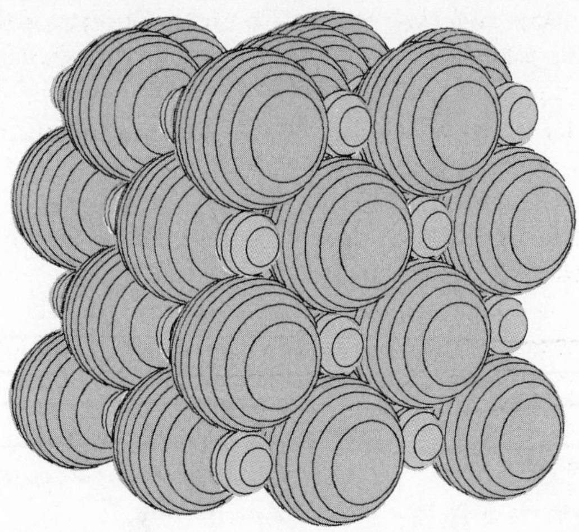

(a)

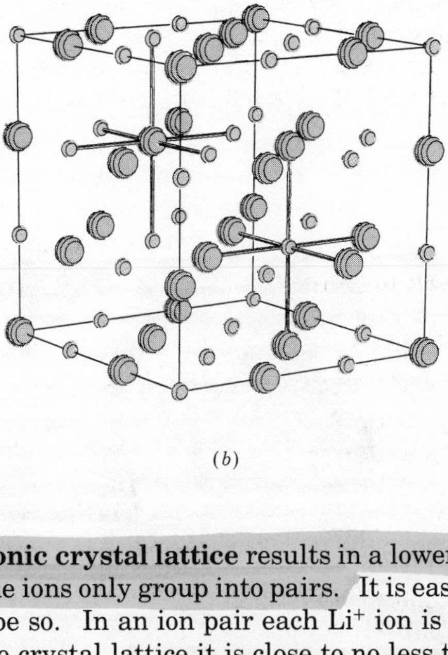

(b)

Figure 6.3 A portion of the ionic crystal lattice of lithium hydride, LiH. (*a*) Lithium ions, Li^+, (color) and hydride ions, H^-, (gray) are shown full size. In a macroscopic crystal the regular array of ions extends indefinitely in all directions. (*b*) "Exploded" view of the lattice, showing that each Li^+ ion (color) is surrounded by six H^- ions (gray), and vice versa. (Computer-generated). (*Copyright © 1976 by W. G. Davies and J. W. Moore.*)

The formation of such an **ionic crystal lattice** results in a lower potential energy than is possible if the ions only group into pairs. It is easy to see from Fig. 6.3 why this should be so. In an ion pair each Li^+ ion is close to only one H^- ion, whereas in the crystal lattice it is close to no less than *six* ions of opposite charge. Conversely each H^- ion is surrounded by six Li^+ ions. In the crystal lattice therefore, more opposite charges are brought closer together than is possible for separate ion pairs and the potential energy is lower by an additional 227 kJ mol^{-1}. The arrangement of the ions in a crystal of LiH corresponds to the lowest possible energy. If there were an alternative geometrical arrangement bringing even more ions of opposite charge even closer together than that shown in Fig. 6.3, the Li^+ ions and H^- ions would certainly adopt it.

Ions and Noble-Gas Electron Configurations

One further aspect of the formation of LiH needs to be explained. If the transfer of one electron from Li to H is energetically favorable, why is the same not true for the transfer of a second electron to produce $Li^{2+}H^{2-}$? Certainly the double charges on Li^{2+} and H^{2-} would attract more strongly than the single charges on Li^+ and H^-, and the doubly charged ions would be held more tightly in the crystal lattice.

The answer to this question can be found by looking back at Fig. 6.1. Removal of a second electron from Li would require much more energy than the removal of the first because this second electron would be a 1*s* electron rather than a 2*s* electron. Not only is this second electron much closer to the nucleus, but it also is very poorly shielded from the nucleus. It is not surprising, therefore, that the **second ionization energy** of Li (the energy required to remove this second electron) is 7297 kJ mol^{-1}, almost *14* times as large as the first ionization energy! Such a colossal energy requirement is enough to insure that only the outermost electron (the *valence* electron) of Li will be removed and that the inner $1s^2$ kernel with its helium-type electron configuration will remain intact.

A similar argument applies to the acceptance of a second electron by the H atom to form the H^{2-} ion. If such an ion were to be formed, the extra electron would have to occupy the 2*s* orbital. Its electron cloud would extend far from the nucleus (even farther than for the 2*s* electron in Li, because the nuclear charge in H^{2-} would only be $+1$, as opposed to $+3$ in Li), and it would be quite high in energy. So much energy would be needed to force a second electron to move around the H nucleus in this way, that only one electron is transferred. The ion formed has the formula H^- and a helium-type $1s^2$ electronic structure.

The simple example of lithium hydride is typical of all ionic compounds which can be formed by combination of two elements. Invariably we find that one of the two elements has a relatively *low ionization energy* and is capable of easily losing one or more electrons. The other element has a relatively *high electron affinity* and is able to accept one or more electrons into its structure. The ions formed by this transfer of electrons almost always have an electronic structure which is the same as that of a *noble gas,* and all electrons are paired in each ion. The resulting compound is *always a solid* in which the ions are arranged in a three-dimensional array or crystal lattice similar to, though not always identical with, that shown in Fig. 6.3. In such a solid the nearest neighbors of each anion are always cations and vice versa, and the solid is held together by the coulombic forces of attraction between the ions of opposite sign.

An everyday example of such an ionic compound is ordinary table salt, sodium chloride, whose formula is NaCl. As we shall see in the next section, sodium is an element with a low ionization energy, and chlorine is an element with a high electron affinity. On the microscopic level crystals of sodium chloride consist of an array of sodium ions, Na^+, and chloride ions, Cl^-, packed together in a lattice like that shown for lithium hydride in Fig. 6.3. The chloride ions are chlorine atoms which have gained an electron and thus have the electronic structure $1s^22s^22p^63s^23p^6$, the same as that of the noble-gas argon. The sodium ions are sodium atoms which have lost an electron, giving them the structure $1s^22s^22p^6$, the same as that of the noble-gas neon. All electrons in both kinds of ions are paired.

6.2 PERIODIC VARIATION OF IONIZATION ENERGY AND ELECTRON AFFINITY

Before we can broaden our discussion of ionic bonding beyond a treatment of lithium hydride, we need to know how ionization energy and electron affinity vary depending on an element's position in the periodic table. We will then have a much clearer idea of what elements are likely to form ionic compounds. At the same time we will find that these properties have some features which are interesting for other reasons.

Ionization Energies

In Fig. 6.4 the ionization energies of the elements are plotted against atomic number. An obvious feature of this graph is that the elements with the highest ionization energies are the noble gases. Since the ionization energy measures the energy which must be *supplied* to remove an electron, these high values mean that it is difficult to remove an electron from an atom of a noble gas. A second obvious feature is that the elements with the

Figure 6.4 Graph of ionization energies of the elements versus atomic number. Rows in the periodic table are indicated at the top of the graph. Representative elements are shaded light gray, transition elements appear in light color, and inner transition elements are in the dark colored areas of the chart. The approximate demarcation between metallic elements (ionization energy below about 800 kJ mol^{-1}) and nonmetallic elements is also shown.

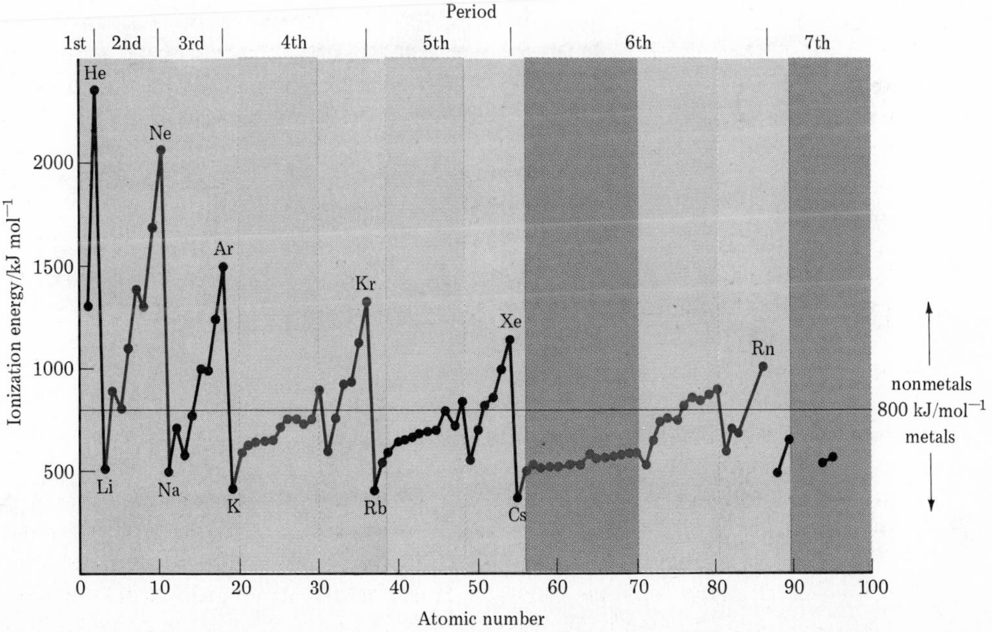

lowest ionization energies are the alkali metals. This means that it is easier to remove electrons from atoms of this group of elements than from any other group. Closer inspection of Fig. 6.4 also reveals the following two general tendencies:

1 As one moves *down a given group in the periodic table*, the *ionization energy decreases*. In group I, for example, the ionization energies decrease in the order Li > Na > K > Rb > Cs. The reason for this is a steady increase in size of the valence electron cloud as the principal quantum number n increases. The 6s valence electron of Cs, for instance, is further from the nucleus and hence easier to remove than the 5s valence electron of Rb.

2 As one moves from *left to right across the periodic table* (from an alkali-metal atom to a noble gas), the *ionization energy increases* on the whole. In such a move the n value of the outermost electrons remains the same, but the nuclear charge increases steadily. This increased nuclear attraction requires that more work be done to remove an electron, and so ionization energy goes up.

Ionization energies can be measured quite accurately for atoms, and the values obtained show some additional features which are less important than the two major trends mentioned above. For example, consider the data in Fig. 6.4 for elements in the second row of the periodic table. Numerical values for the relevant ionization energies are shown in Table 6.1. The general trend of increasing ionization energy across the table is broken at two points. Boron has a smaller value than beryllium, and oxygen has a smaller value than nitrogen. The first break occurs when the first electron is added to a p subshell. As was mentioned several times in the previous chapter, a 2p electron is higher in energy and hence easier to remove than a 2s electron because it is more efficiently shielded from the nuclear charge. Thus the 2p electron in boron is easier to remove than a 2s electron in beryllium. The second exception to the general trend occurs in the case of oxygen, which has one more 2p electron than the half-filled subshell of nitrogen. The last electron in the oxygen atom is forced into an already occupied orbital where it is kept close to another electron. The repulsion between these two electrons makes one of them easier to remove, and so the ionization energy of oxygen is lower than might be expected.

Ionization of Transition and Inner Transition Elements

It can also be seen from Fig. 6.4 that ionization energies increase much more slowly across the transition and inner transition elements than for the representative elements. For example, the ionization energy of the representative-element boron is 800 kJ mol^{-1}. Five elements later we find neon, whose ionization energy is 2080 kJ mol^{-1}, an increase of 160 percent.

TABLE 6.1 Ionization Energies (Black) and Electron Affinities (Color) of the Representative Elements (in kilojoules per mole).*

73 H 1312							He 2372
58 Li 520	−18* Be 899	29* B 801	121* C 1086	−58* N 1402	142 O 1314	331 F 1681	Ne 2080
52 Na 496	−54* Mg 738	48* Al 578	134* Si 786	75* P 1012	200 S 1000	348 Cl 1251	Ar 1520
K 419	Ca 590	Ga 579	Ge 762	65 As 946	207* Se 941	324 Br 1140	Kr 1351
Rb 403	Sr 549	In 558	Sn 708	Sb 834	222* Te 869	296 I 1008	Xe 1170
Cs 376	Ba 503	Tl 589	Pb 715	Bi 703	Po 812	At	Rn 1037

* Electron affinities marked with an asterisk (*) have been obtained from theoretical calculations rather than experimental measurements. The heavy colored line separates metals (ionization energy usually below about 800 kJ mol^{-1}) from nonmetals.

In the fourth period, the transition-element scandium has an ionization energy of 631 kJ mol^{-1}. Five elements later we find iron at 759 kJ mol^{-1}, an increase of only 20 percent. All the lanthanoids have ionization energies from 500 to 600 kJ mol^{-1}, and the actinoids are all between 580 and 680 kJ mol^{-1}.

These similarities among the transition and especially the inner transition elements illustrate statements made in Sec. 5.7. The distinguishing electron for a transition element enters a *d* subshell in the *next-to-outermost* shell, while for an inner transition element it usually enters an *f* subshell in the *third-from-outermost* shell. Thus the distinction between an element and the one preceding it in the periodic table is much smaller than among the representative elements. Furthermore, experimental measurements show that for transition and inner transition elements the electrons lost when ionization occurs are *not* the *last* ones which were added to build up the atomic electron configuration. Instead, *electrons are usually removed first from the subshell having the largest principal quantum number.*

EXAMPLE 6.1 Determine the electron configuration of the Fe^{3+} ion.

Solution Since the charge on the ion is +3, three electrons must have been removed from a neutral iron atom (Fe). The electron configuration of Fe is

$$\text{Fe} \quad 1s^2 2s^2 2p^6 3s^2 3p^6 3d^6 4s^2 \quad \text{or} \quad [\text{Ar}]3d^6 4s^2$$

We now remove electrons successively from subshells having the largest principal quantum number:

Fe^+	$[Ar]3d^6 4s^1$	one 4*s* electron removed
Fe^{2+}	$[Ar]3d^6$	a second 4*s* electron removed
Fe^{3+}	$[Ar]3d^5$	since no electrons are left in the $n = 4$ shell, one 3*d* electron is removed

The behavior described in the previous paragraph and Example 6.1 may be better understood by referring back to Fig. 5.10. Electrons in the subshell having the largest principal quantum number (4*s* in Example 6.1 and Fig. 5.10) are, on the average, farther from the nucleus, and they are first to be removed. The first ionization energy of iron is not much larger than that of scandium because in each case a 4*s* electron is being removed. The iron atom has five more protons in the nucleus, but it also has five more 3*d* electrons which spend most of their time *between* the nucleus and the 4*s* electrons. The screening effect of such 3*d* electrons causes the *effective* nuclear charge to increase very slowly from one transition element to the next. The attraction for 4*s* electrons, and hence ionization energy, also increases very slowly.

Metals

Macroscopic properties such as high thermal and electric conductivity, malleability, and ductility were mentioned in Sec. 4.1 as characteristics of **metals.** In addition, *most metals have low ionization energies,* usually below 800 kJ mol^{-1}. In other words, a metal consists of atoms, each of which has at least one loosely held electron. When such atoms pack close together in a solid metal, the loosely held electrons are relatively free to move from one atom to another. If excess electrons are forced into one end of a metal wire, it is relatively easy for electrons to flow out of the other end. Thus an electric current may be carried through the wire, and the high conductivity of all metals may be understood.

More detailed microscopic interpretations of metallic properties are given in Chap. 22, but for the time being we are primarily interested in the location of metallic elements in the periodic table. Ionization energies are smallest near the bottom and on the left of the periodic table, and so this is where metals are found. Moreover, ionization energies increase slowly from one transition element to the next and hardly at all across the inner transition elements. Therefore *all* transition and inner transition elements are metals. In periodic groups IIIA, IVA, and VA elements near the top of the table have large ionization energies and little metallic character. Ionization energies decrease as one moves downward, however. For example, Al is quite metallic, although the element above it, B, is not. A heavy "stairstep" line on the periodic table inside the front cover of this book separates the nonmetals (above and to the right) from the metals. Elements such as B, Si, Ge, As, Sb, and Te, which are adjacent to the stairstep, have intermediate properties and are called **semimetals.**

Electron Affinities

Electron affinities are more difficult to measure experimentally than are ionization energies, and far fewer values are available. Table 6.1 shows the relationship of the periodic table with those electron affinities that have been measured or estimated from calculations. It is not easy to discern many obvious regularities in this table, especially since some of the electron-affinity values quoted are negative, indicating that energy is sometimes *required* to force an extra electron onto an atom. Nevertheless, it is quite obvious which of the periodic groups correspond to the highest electron affinities. All the halogens have values of about 300 kJ mol^{-1}, while the group VI nonmetals have somewhat lower values, in the region of 200 kJ mol^{-1} or less.

The high electron affinities of the halogens are a result of their having an almost complete outer shell of electrons. The element fluorine, for example, has the structure $1s^2 2s^2 2p^5$, in which one of the 2*p* orbitals contains but one electron. If an extra electron is added to this atom to form a fluoride ion, the electron can pair with the electron in the half-filled 2*p* orbital. The added electron will be shielded from the nucleus by the 1*s* electrons, but the 2*s* and 2*p* electrons are in the same shell and will shield it rather poorly. There will thus be quite a large effective nuclear charge (a rough estimate is +5.5) attracting the added electron. Because of this overall attraction, energy will be *released* when the electron is captured by the fluorine atom. Similar reasoning also explains why oxygen also has a high electron affinity. Here, though, the nuclear charge is smaller, and the attraction for the added electron distinctly less.

6.3 BINARY IONIC COMPOUNDS AND THEIR PROPERTIES

All ionic compounds have numerous properties in common. Consequently, the ability to recognize an ionic compound from its formula will allow you to predict many of its properties. This is often possible in the case of a **binary compound** (one which contains only two elements), because formation of a binary ionic compound places quite severe restrictions on the elements involved. One element must be a metal and must have a very low ionization energy. The other element must be a nonmetal and must have a very high electron affinity.

Even though metals in general have low ionization energies, not all of them are low enough to form binary ionic compounds with a large fraction of the nonmetals. Although it is impossible to draw an exact line of demarcation, a good working rule is that essentially all binary compounds involving metals from periodic group IA, group IIA, group IIIB (Sc, Y, Lu), and the lanthanoids will be ionic. (Hydrogen is not a metal and is therefore an exception to the rule for group IA. Beryllium, whose ionization energy of 899 kJ mol^{-1} is quite high for a metal, also forms many binary compounds which are not ionic. Beryllium is the only exception to the rule from group IIA.) The transition metals to the right of group IIIB in the periodic table form numerous binary compounds which involve covalent bonding, so they cannot be included in our rule. The same is true of the metals in periodic groups IIIA, IVA, and VA.

The number of nonmetals with which a group IA, IIA, or IIIB or lanthanoid metal can combine to form a binary ionic compound is even more limited than the number of appropriate metals. Such nonmetals are found mainly in periodic groups VIIA and VIA. The only other elements which form monatomic anions under normal circumstances are hydrogen (which forms H^- ions) and nitrogen (which forms N^{3-} ions). In addition to combining with metals to form ionic compounds, all of the nonmetals can combine with other nonmetals to form covalent compounds as well. Therefore, presence of a particular nonmetal does not guarantee that a binary compound is ionic. It is necessary, however, for a group VIIA or VIA nonmetal, nitrogen, or hydrogen to be present if a binary compound is to be classified as definitely ionic.

EXAMPLE 6.2 Which of the following compounds can be identified as definitely ionic? Which are definitely not ionic?

a) CuO	**d)** $HgBr_2$	**g)** H_2S
b) CaO	**e)** $BaBr_2$	**h)** InF_3
c) MgH_2	**f)** B_2H_6	**i)** BrCl

Solution According to the guidelines in the previous two paragraphs, only compounds containing metals from groups IA, IIA, and IIIB, or the lanthanoids are definitely ionic, as long as the metal is combined with an appropriate nonmetal. CaO, MgH_2, and $BaBr_2$ fall into this category.

Compounds which do not contain a metallic element, such as B_2H_6, H_2S, and BrCl, cannot possibly be ionic. This leaves CuO, $HgBr_2$, and InF_3 in the category of possibly, but not definitely, ionic.

The Octet Rule

Because binary ionic compounds are confined mainly to group I and II elements on the one hand and group VI and VII elements on the other, we find that they consist mainly of ions having an electronic structure which is the same as that of a noble gas. In calcium fluoride, for example, the calcium atom has lost *two* electrons in order to achieve the electronic structure of argon, and thus has a charge of +2:

$$Ca: \rightarrow Ca^{2+} + 2e^-$$

$$1s^2 2s^2 2p^6 3s^2 3p^6 4s^2 \rightarrow 1s^2 2s^2 2p^6 3s^2 3p^6 + 2e^-$$

By contrast, a fluorine atom needs to acquire but *one* electron in order to achieve a neon structure. The resulting fluoride ion has a charge of -1:

$$:\ddot{\underset{..}{F}}\cdot + e^- \longrightarrow \left[:\ddot{\underset{..}{F}}:\right]^-$$

$$1s^2 2s^2 2p^5 + e^- \longrightarrow 1s^2 2s^2 2p^6$$

The outermost shell of each of these ions has the electron configuration ns^2np^6, where n is 3 for Ca^{2+} and 2 for F^-. Such an ns^2np^6 noble-gas electron configuration is encountered quite often. It is called an **octet** because it contains *eight* electrons.

In a crystal of calcium fluoride the Ca^{2+} and F^- ions are packed together in the lattice shown in Fig. 6.5. Careful study of the diagram shows that each F^- ion is surrounded by *four* Ca^{2+} ions, while each Ca^{2+} ion has *eight* F^- ions as nearest neighbors. Thus there must be twice as many F^- ions as Ca^{2+} ions in the entire crystal lattice. Only a small portion of the lattice is shown in Fig. 6.5, but if it were extended indefinitely in all directions, you could verify the ratio of two F^- for every Ca^{2+}. This ratio makes sense if you consider that two F^- ions (each with a -1 charge) are needed to balance the $+2$ charge of each Ca^{2+} ion, making the net charge on the crystal zero. The formula for calcium fluoride is thus CaF_2.

Newcomers to chemistry often have difficulty in deciding what the formula of an ionic compound will be. A convenient method for doing this is to regard the compound as being formed from its atoms and to use Lewis diagrams. The **octet rule** can then be applied. Each atom must lose or gain electrons in order to achieve an octet. Furthermore, all electrons lost by one kind of atom must be gained by the other.

Figure 6.5 A portion of the ionic crystal lattice of fluorite, calcium fluoride. (*a*) Ca^{2+} ions (color) and F^- ions (gray) are shown full size. (*b*) "Exploded" view shows that each F^- ion is surrounded by *four* Ca^{2+} ions, while each Ca^{2+} ion is surrounded by *eight* F^- ions. The ratio of Ca^{2+} ions to F^- ions is thus 4:8 or 1:2, and the formula is CaF_2. (Computer-generated.) (*Copyright © 1976 by W. G. Davies and J. W. Moore.*)

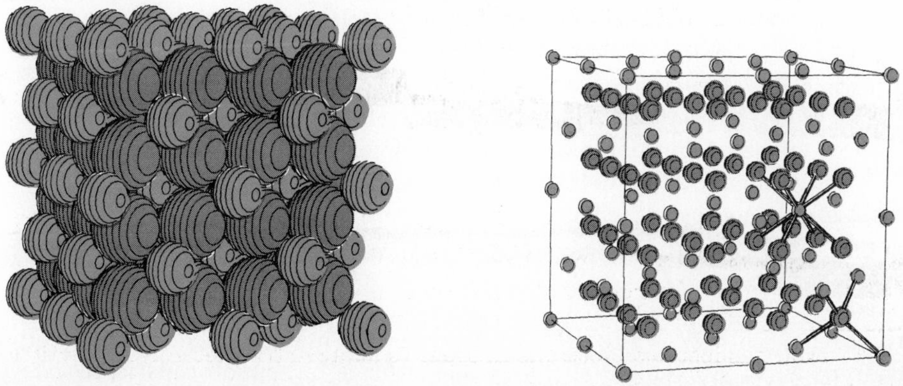

(*a*) (*b*)

EXAMPLE 6.3 Find the formula of the ionic compound formed from O and Al.

Solution We first write down Lewis diagrams for each atom involved:

$$\text{Al}\!:\qquad \text{and} \qquad :\!\ddot{\text{O}}$$

We now see that each O atom needs 2 electrons to make up an octet, while each Al atom has 3 electrons to donate. In order that the same number of electrons would be donated as accepted, we need 2Al atoms ($2 \times 3e^-$ donated) and 3 O atoms ($3 \times 2e^-$ accepted). The whole process is then

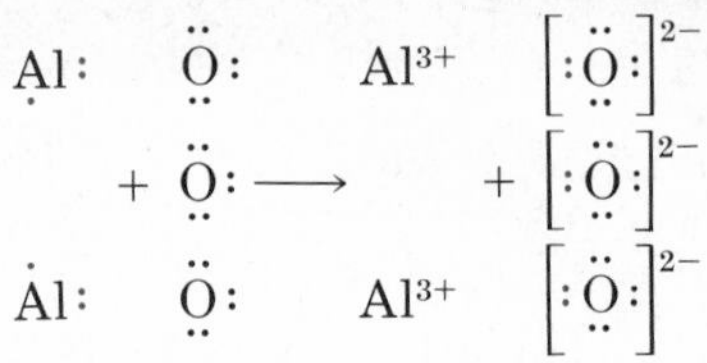

The resultant oxide consists of aluminum ions, Al^{3+}, and oxide ions, O^{2-}, in the ratio of 2:3. The formula is Al_2O_3.

An exception to the octet rule occurs in the case of the three ions having the He $1s^2$ structure, that is, H^-, Li^+, and Be^{2+}. In these cases *two* rather than *eight* electrons are needed in the outermost shell to comply with the rule.

Physical Properties

Binary ionic compounds have many macroscopic properties in common, some of which are easy to explain in terms of the microscopic picture just presented. Because each ion in a crystal lattice, like those shown in Figs. 6.3 and 6.5, is surrounded by many oppositely charged nearest neighbors, each ion is held fairly tightly in its allotted position. At room temperature each ion can vibrate slightly about its mean position, but much more energy must be added before an ion can move fast enough and far enough to begin to escape the attraction of its neighbors. Therefore a fairly high temperature is needed to melt an ionic compound. The melting points of some substances we have considered are: LiH, 680°C; NaCl, 801°C; CaF_2, 1360°C.

Once an ionic solid melts, the individual ions can move around, though most of the nearest neighbors of each anion will still be cations and vice versa. A characteristic feature of such a molten ionic compound is that it *conducts electricity*. If two wires are placed in the melt and are connected to a battery, the positive ions in the melt will gradually move toward the negatively charged wire, while the negative ions will move toward the positively charged wire. Movement of charged ions through the melt constitutes *pas-*

sage of an electrical current just as movement of electrons through a metal wire does. Its occurrence can be detected by an appropriate electrical meter.

Both the melting point of the solid and the electrical conductivity of the liquid provide valuable clues for deciding whether a compound is ionic or not. An interesting example of this is a comparison between aluminum fluoride, AlF_3, and aluminum bromide, $AlBr_3$. We find that AlF_3 has a melting point of 1040°C, and the melt conducts electricity. By contrast $AlBr_3$ melts at 97.5°C, and the melt does not conduct electricity. The obvious inference is that the former compound is ionic, while the latter is not.

The crystals of ionic substances are transparent, hard, brittle, and have characteristic shapes. They also are easily **cleaved.** This means that if a wedge-shaped instrument (such as a knife blade) is properly placed on a crystal and tapped sharply, the crystal will break cleanly in two. The new crystal faces will be nearly perfect planes. Characteristic shapes and cleavages may often be used to distinguish one ionic compound from another.

Hardness of ionic crystals is a consequence of the strong coulombic forces which hold each ion in its allotted position. The shape and cleavage of a crystal are the result of the specific geometric arrangement of the ions in a crystal lattice. For example, since 90° angles occur between layers of ions in the microscopic lattice of NaCl (which is similar to the one shown in Fig. 6.3), 90° angles are also reasonable in the macroscopic crystals of NaCl. (If you examine a few small crystals from a salt shaker, you will see that this is true—some are almost perfect cubes!) When a large force is applied parallel to a layer of ions in the crystal lattice, it can shift that layer with respect to the next. As shown in Fig. 6.6, only a small shift is needed before

Figure 6.6 Microscopic view of cleavage of an ionic crystal. When an external force (large gray arrow) causes one layer of ions to shift slightly with respect to another, positive ions are brought close to other positive ions and negative ions become nearest neighbors of other negative ions. The strong repulsive forces produced by this arrangement of ions (double-headed black arrows) cause the two layers to split apart. (Computer-generated.) (*Copyright © 1976 by W. G. Davies and J. W. Moore.*)

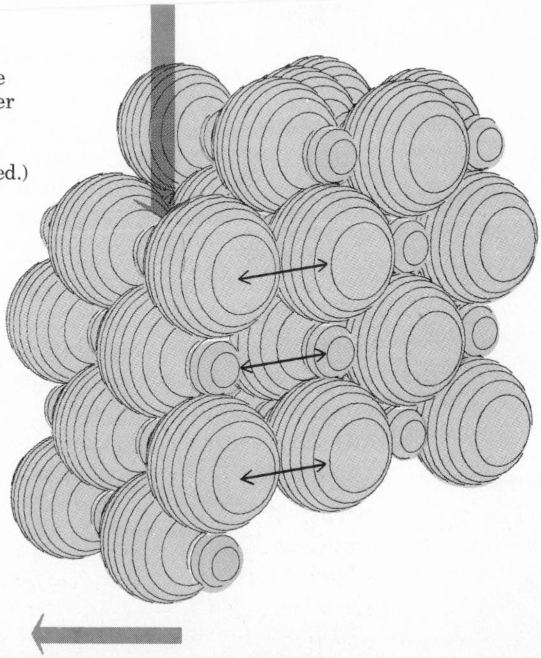

positive ions in one layer are adjacent to positive ions in the other, and negative ions similarly become nearest neighbors. The two portions of the lattice repel one another and fall apart. On a macroscopic level the crystal cleaves, leaving flat faces corresponding to each layer of ions.

Perhaps the most important property of ionic solids is that most of them dissolve in water. Mute testimony to this is provided by the rows and rows of bottles in the average chemistry laboratory which contain aqueous ionic solutions. A characteristic feature of these solutions is that they all *conduct electricity,* showing that the ions retain their charges in solution and are also free to move about in it. When barium chloride, $BaCl_2$, dissolves in water, for example, the Ba^{2+} and Cl^- ions which were held tightly together in the crystal lattice become separated from each other in solution and can move about independently. If two wires are now immersed in this solution and attached to a battery, much the same thing happens as in the case of a molten salt. The Ba^{2+} ions are attracted by the negative charge on the one wire and move toward it, while the Cl^- ions move toward the positive charge on the other wire. This flow of charges in the body of the solution is matched by a flow of electrons in the wire and can be measured by an appropriate meter.

Chemical Properties

The most important chemical characteristic of ionic compounds is that *each ion has its own properties. Such properties are different from those of the atom from which the ion was derived.* In other words, an Na^+ ion is quite different from an Na atom, and a Cl^- ion is unlike an isolated Cl atom or either of the Cl atoms in a Cl_2 molecule. You eat a considerable quantity of Na^+ and Cl^- ions in table salt every day, but Na atoms or Cl_2 molecules would be quite detrimental to your health.

The unique chemical properties of each type of ion are quite evident in aqueous solutions. Most of the reactions of $BaCl_2(aq)$, for example, can be classified as reactions of the $Ba^{2+}(aq)$ ion or the $Cl^-(aq)$ ion. If sulfuric acid, H_2SO_4, is added to a solution of $BaCl_2$, the solution turns milky and very fine crystals of $BaSO_4(s)$ eventually settle out. The reaction can be written as

$$Ba^{2+}(aq) + H_2SO_4(aq) \rightarrow BaSO_4(s) + 2H^+(aq)$$

This reaction is characteristic of the *barium ion.* It will also occur if H_2SO_4 is added to solutions such as $BaI_2(aq)$ or $BaBr_2(aq)$ which contain barium ions but no chloride ions.

By contrast, if a solution of silver nitrate, $AgNO_3$, [which contains silver ions, $Ag^+(aq)$] is added to a $BaCl_2$ solution, a reaction characteristic of the *chloride ion* occurs. A white curdy precipitate of $AgCl(s)$ forms according to the equation

$$Ag^+(aq) + Cl^-(aq) \rightarrow AgCl(s)$$

Other ionic solutions containing chloride ions, such as $LiCl(aq)$, $NaCl(aq)$, or $MgCl_2(aq)$, give an identical reaction.

Many binary ionic solids not only dissolve in water, they also react with it. When the compound contains an anion such as N^{3-}, O^{2-}, or S^{2-}, which has more than one negative charge, the reaction with water produces hydroxide ions, OH^-:

$$O^{2-} + H_2O \rightarrow OH^-(aq) + OH^-(aq) \tag{6.1}$$

$$S^{2-} + H_2O \rightarrow HS^-(aq) + OH^-(aq) \tag{6.2}$$

$$N^{3-} + 3H_2O \rightarrow NH_3(aq) + 3OH^-(aq) \tag{6.3}$$

Thus, when sodium oxide, Na_2O, is added to water, the resulting solution contains sodium ions and hydroxide ions but no oxide ions:

$$Na_2O(s) + H_2O(l) \rightarrow 2Na^+(aq) + 2OH^-(aq) \tag{6.4}$$

The hydride ion also reacts with water to form hydroxide ions. When lithium hydride, LiH, is dissolved in water, for example, the following reaction occurs:

$$LiH(s) + H_2O(l) \rightarrow Li^+(aq) + OH^-(aq) + H_2(g) \tag{6.5}$$

Note that hydrogen gas is evolved in this reaction. Lithium hydride crystals provide a very compact, if somewhat expensive, method for storing hydrogen.

Among the *halide ions* (F^-, Cl^-, Br^-, I^-) only the fluoride ion shows any tendency to react with water, and that only to a limited extent. When sodium fluoride is dissolved in water, for example, faint traces of hydroxide ion can be detected in the solution owing to the reaction

$$F^- + H_2O \rightarrow HF + OH^- \tag{6.6}$$

With sodium chloride, by contrast, no such reaction occurs.

6.4 THE COVALENT BOND

As we have already seen, formation of an ionic bond by complete transfer of an electron from one atom to another is possible only for a fairly restricted set of elements. Covalent bonding, in which neither atom loses complete control over its valence electrons, is much more common. In a **covalent bond** the electrons occupy a region of space *between* the two nuclei and are said to be *shared* by them.

The simplest example of a covalent bond is the bond between the two H atoms in a molecule of H_2. Suppose that two H atoms approach each other until their two 1*s* electron clouds interpenetrate, as shown in Fig. 6.7*a*. In such a situation each of the two electrons cannot continue to move about its own nucleus. The electron indicated in color, for example, will feel the attractive pull of the left-hand nucleus as well as the right-hand nucleus. Since both nuclei have the same charge, the electron is unable to discriminate between them. Accordingly it adopts the new *symmetrical* probability cloud shown in Fig. 6.7*b*. The same argument applies to the electron indicated in gray. Each electron spends the same amount of time in the vicinity of each nucleus.

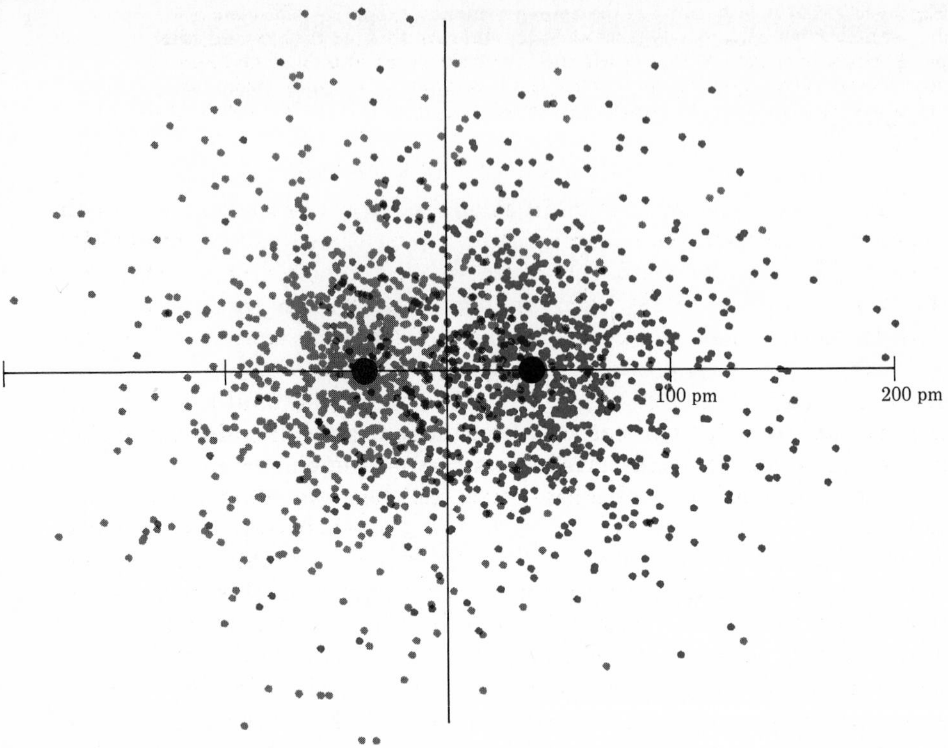

(*a*)

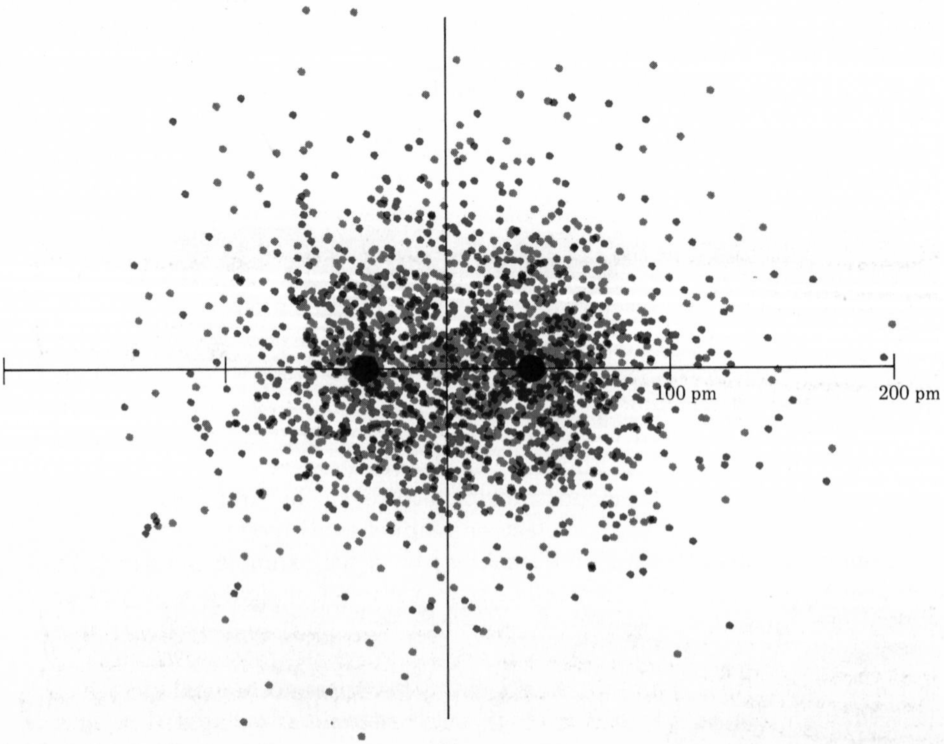

(*b*)

◀ **Figure 6.7** Covalent bonding in the hydrogen molecule, H_2. (*a*) When two hydrogen atoms approach to a distance of 74 pm (0.74 Å), their electron clouds interpenetrate, or overlap. (*b*) Since each nucleus attracts each electron, both electrons occupy a symmetrical molecular orbital. Each electron spends equal time in the vicinity of each nucleus. (Computer-generated.) (*Copyright © 1976 by W. G. Davies and J. W. Moore.*)

An orbital, like that shown in Fig. 6.7*b*, which extends over a whole molecule rather than being restricted to a single atom is called a **molecular orbital.** We can consider it to be the result of a combination or **overlap** of the two 1*s* atomic orbitals. A molecular orbital formed in this way must conform to the Pauli exclusion principle. Only two electrons of opposite spin can occupy each orbital. Since the second (gray) electron from the other H atom is available, it occupies the molecular orbital together with the original (colored) electron. The result is a **shared pair** of electrons moving around both nuclei and holding them together.

The overlap of two 1*s* electron clouds and their spreading over both nuclei when the H_2 molecule forms has the effect of concentrating the electron density *between* the protons. When the molecule forms, the negative charges move *closer* to positive charges than before. There is thus a reduction in potential energy. Since the virial theorem (Sec. 5.4) guarantees that a reduction in the *potential* energy means that the *total* energy (kinetic + potential) is also reduced, we can conclude that the H_2 molecule is *lower in energy* and hence more stable under normal conditions than two separate H atoms. Alternatively we can say that energy is *required* to break apart an H_2 molecule and separate it into two H atoms. This energy (called the **bond energy**) can be measured experimentally. It is found to have a value of 436 kJ mol^{-1} for the H_2 molecule.

6.5 COVALENT MOLECULES AND THE OCTET RULE

The idea that a molecule could be held together by a shared pair of electrons was first suggested by Lewis in 1916. Although Lewis never won the Nobel prize for this or his many other theories, the shared pair of electrons is nevertheless one of the most significant contributions to chemistry of all time. Wave mechanics was still 10 years in the future, and so Lewis was unable to give any mathematical description of exactly how electron sharing was possible. Instead of the detailed picture presented in the previous section, Lewis indicated the formation of a hydrogen molecule from two hydrogen atoms with the aid of his electron-dot diagrams as follows:

$$\mathrm{H}\cdot + \cdot\mathrm{H} \longrightarrow \mathrm{H}{:}\mathrm{H}$$

Lewis also suggested that the tendency to acquire a noble-gas structure is not confined to ionic compounds but occurs among covalent compounds as well. In the hydrogen molecule, for example, each hydrogen atom acquires some control over *two* electrons, thus achieving something resembling the 1*s*^{2} helium structure. (Indeed, if we look at the wave-mechanical picture of the hydrogen molecule in Fig. 6.7*b*, it does resemble an elongated helium atom.) Similarly the formation of a fluorine molecule from its atoms can be represented by

$$:\ddot{\underset{..}{F}}\cdot + \cdot\ddot{\underset{..}{F}}: \longrightarrow :\ddot{\underset{..}{F}}:\ddot{\underset{..}{F}}:$$

Again a pair of electrons is shared, enabling each atom to attain a neon structure with eight electrons (i.e., an octet) in its valence shell. Similar diagrams can be used to describe the other halogen molecules:

$$:\ddot{\underset{..}{Cl}}:\ddot{\underset{..}{Cl}}: \qquad :\ddot{\underset{..}{Br}}:\ddot{\underset{..}{Br}}: \qquad :\ddot{\underset{..}{I}}:\ddot{\underset{..}{I}}:$$

In each case a shared pair of electrons contributes to a noble-gas electron configuration on *both* atoms. Since only the valence electrons are shown in these diagrams, the attainment of a noble-gas structure is easily recognized as the attainment of a full complement of eight electron dots (an *octet*) around each symbol. In other words *covalent* as well as ionic compounds obey the octet rule.

The octet rule is very useful, though by no means infallible, for predicting the formulas of many covalent compounds, and it enables us to explain the usual valence exhibited by many of the representative elements. According to Lewis' theory, hydrogen and the halogens each exhibit a valence of 1 because the atoms of hydrogen and the halogens each contain *one* less electron than a noble-gas atom. In order to attain a noble-gas structure, therefore, they need only to participate in the sharing of *one* pair of electrons. If we identify a shared pair of electrons with a chemical bond, these elements can only form *one* bond.

A similar argument can be extended to oxygen and the group VI elements to explain their valence of 2. Here *two* electrons are needed to complete a noble-gas configuration. By sharing *two* pairs of electrons, i.e., by forming *two* bonds, an octet is attained:

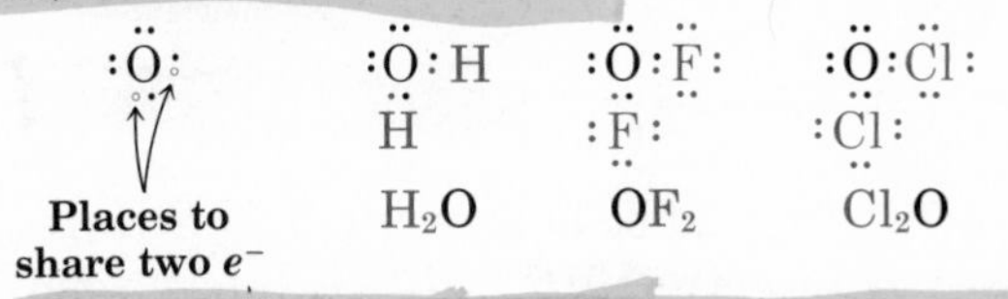

Nitrogen and the group V elements likewise require *three* electrons to complete their octets, and so can participate in *three* shared pairs:

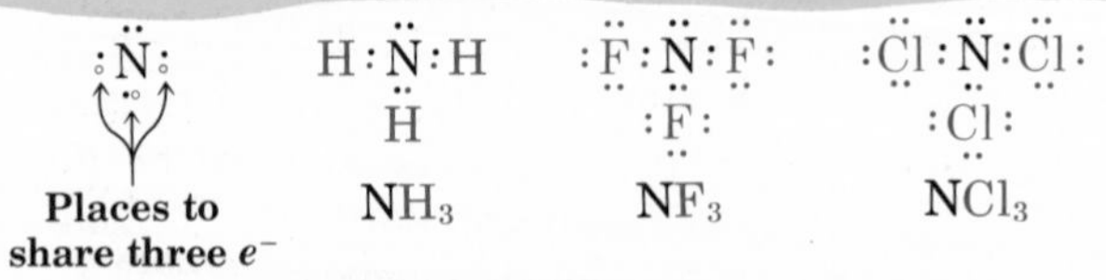

Finally, since carbon and the group IV elements have *four* vacancies in their valence shells, they are able to form *four* bonds:

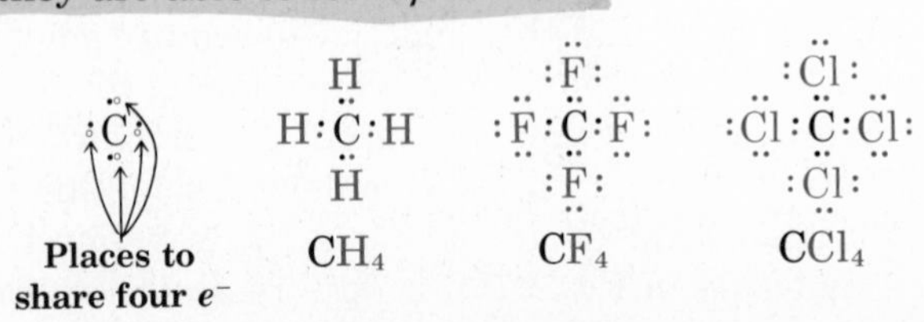

EXAMPLE 6.4 Draw Lewis structures and predict the formulas of compounds containing (a) P and Cl; (b) Se and H.

Solution

a) Draw Lewis diagrams for each atom.

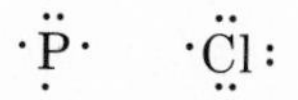

Since the P atom can share three electrons and the Cl atom only one, three Cl atoms will be required, and the formula is

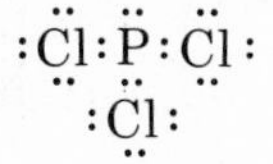 or PCl_3

b) Since Se is in periodic group VI, it lacks two electrons of a noble-gas configuration and thus has a valence of 2. The formula is

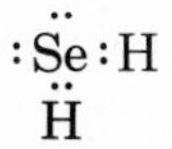 or H_2Se

In drawing Lewis structures, the bonding pairs of electrons are often indicated by a **bond line** connecting the atoms they hold together. Electrons which are not involved in bonding are usually referred to as **lone pairs** or **unshared pairs.** Lone pairs are often omitted from Lewis diagrams, or they may also be indicated by lines. Here are some of the alternative ways in which H_2, F_2, and PCl_3 can be written.

Full Diagram	With Bond Line	With Lone-Pair Lines	Omitting Lone Pairs
H:H	H—H	H—H	H—H
:F̤̈:F̤̈:	:F̤̈—F̤̈:	|F̅—F̅|	F—F
:C̤̈l:P̤̈:C̤̈l: / :C̤̈l:	:C̤̈l—P̈—C̤̈l: / | / :C̤̈l:	|C̅l—P̅—C̅l| / | / |C̲l|	Cl—P—Cl / | / Cl

6.6 WRITING LEWIS STRUCTURES FOR MOLECULES

Because Lewis diagrams are widely used to describe the structures of molecules, it is important that you be able to construct them. The first step in this process is to draw a **skeleton structure** to show which atoms are linked to which. Next, determine how many valence electrons are available by adding the periodic group numbers for all atoms in the molecule. Finally, allocate the available electrons either as shared pairs or as lone pairs so that each atom has an octet (or in the case of H, a $1s^2$ pair).

Deciding on a Skeleton Structure

The skeleton structure of a covalent molecule can often be determined by considering the valences of the constituent atoms. Usually the atom which forms the largest number of bonds is found in the center of the skeleton, where it can connect to the maximum number of other atoms.

EXAMPLE 6.5 Hypochlorous acid has the molecular formula HOCl. Draw a structural formula.

Solution There are several possible ways to link the atoms together

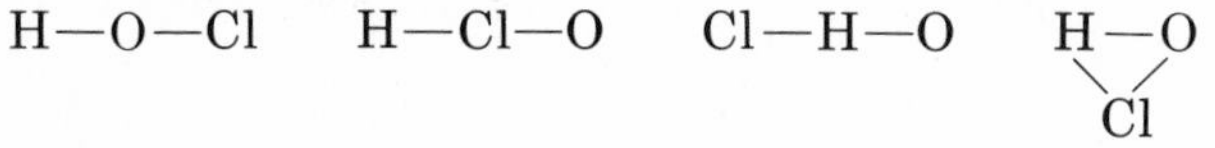

1 2 3 4

The usual valence of H is *1*, and so structures 3 and 4, which have *two* bonds to H, may be eliminated. The usual valence of Cl is also 1, and so structure 2 may also be ruled out. Structure 1 shows H forming one bond, Cl forming one, and O forming two, in agreement with the usual valences, and so it is chosen.

The total number of valence electrons available is 1 from H plus 6 from O plus 7 from Cl, or 14. Filling these into the skeleton we have

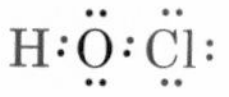

Note that O, which had the largest valence, is in the center of the skeleton.

EXAMPLE 6.6 Draw a structural formula for hydroxylamine, NH_3O.

Solution In this case N has the largest valence (3), followed by O (2) and H (1). Both N and O can form "bridges" between other atoms, but H cannot. Therefore we place N and O in the center of the skeleton to give

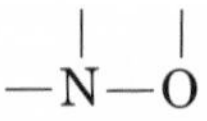

which easily becomes

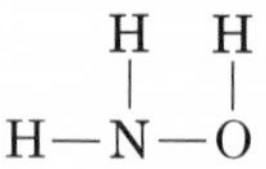

by addition of the three H atoms.

There are a total of 5 + 3 + 6 = 14 valence electrons from N, 3H's and O. These can be placed as follows:

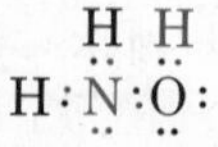

Once the Lewis diagram has been determined, the molecular formula is often rewritten to remind us of what the structural formula is. For example, the molecular formula for hydroxylamine is usually written NH_2OH instead of NH_3O to remind us that two H's are bonded to N and one to O. It is assumed that the person reading the formula will realize that N and O each have one valence electron left to share with each other, connecting —NH_2 with —OH.

In some cases more than one skeleton structure will satisfy the valence of each atom and the octet rule as well. For example, you can verify that the molecular formula C_2H_6O corresponds to both of the following:

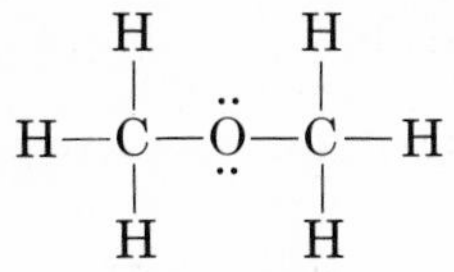

Dimethyl ether
CH_3OCH_3

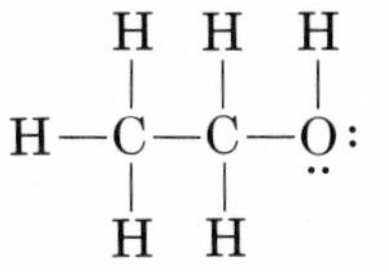

Ethyl alcohol
CH_3CH_2OH

In such a case we can only decide which molecular structure we have by experiment. The properties of ethyl alcohol when diluted with water and imbibed are well known. Dimethyl ether is a gas. Like the diethyl ether used in operating rooms, it is highly explosive and can put you to sleep. Two molecules, such as dimethyl ether and ethyl alcohol, which have the same molecular formula but different structural formulas are said to be **isomers.**

Multiple Bonds

In order to meet the requirements of normal valence, it is sometimes necessary to have *more than one bond,* that is, more than one shared pair of electrons between two atoms. A case in point is formaldehyde, CH_2O. In order to provide carbon with four bonds in this molecule, we must consider carbon as forming *two* bonds to the oxygen as well as one to each of the two hydrogens. At the same time the oxygen atom is also provided with the two bonds its normal valence requires:

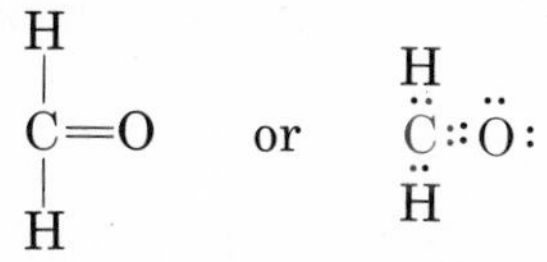

Note that all four of the shared electrons in the carbon-oxygen bond are included both in the octet of carbon and in the octet of oxygen. A bond involving *two* electron pairs is called a **double bond.**

Occasionally the usual valences of the atoms in a molecule do not tell us what the skeleton structure should be. For example, in carbon monoxide, CO, it is hard to see how one carbon atom (usual valence of 4) can be matched with a single oxygen atom (usual valence of 2). In a case like this, where the valences appear to be incompatible, *counting* valence electrons usually leads to a structure which satisfies the octet rule. Carbon has 4 valence electrons and oxygen has 6, for a total of *10*. We want to arrange these 10 electrons in two octets, but two separate groups of 8 electrons would require 16 electrons. Only by *sharing* 16 − 10, or 6, electrons (so that those 6 electrons are part of *each* octet, and, in effect, count twice) can we satisfy the octet rule. This leads to the structure

$$:\mathrm{C}\vdots\vdots\mathrm{O}: \qquad \text{or} \qquad \mathrm{C}{\equiv}\mathrm{O}$$

Here *three* pairs of electrons are shared between two atoms, and we have a **triple bond.**

Double and triple bonds are not merely devices for helping to fit Lewis diagrams into the octet theory. They have an objective existence, and their presence in a molecule often has a profound effect on how it reacts with other molecules. Triple bonds are invariably shorter than double bonds, which in turn are shorter than single bonds. In C≡O, for instance, the carbon-oxygen distance is 113 pm, in $H_2C{=}O$ it is 122 pm, while in both ethyl alcohol and dimethyl ether it is 143 pm. This agrees with the wave-mechanical picture of the chemical bond as being caused by the concentration of electron density between the nuclei. The more pairs of electrons which are shared, the greater this density and the more closely the atoms are pulled together. In line with this, we would also expect multiple bonds to be stronger than single bonds. Indeed, the bond energy of C—O is found experimentally to be 360 kJ mol^{-1}, while that of C═O is 736 kJ mol^{-1}, and that of C≡O is a gigantic 1072 kJ mol^{-1}. The C≡O triple bond in carbon monoxide turns out to be the strongest known covalent bond.

The formation of double and triple bonds is not as widespread among the atoms of the periodic table as one might expect. At least one of the atoms involved in a multiple bond is almost always C, N, or O, and in most cases both atoms are members of this trio. Other elements complete their octets by forming *additional single bonds* rather than multiple bonds.

EXAMPLE 6.7 Draw structural formulas for (a) CO_2 and (b) SiO_2.

Solution

a) Carbon requires four bonds, and each oxygen requires two bonds, and so two C═O double bonds will satisfy the normal valences. The structure is

$$:\ddot{\mathrm{O}}::\mathrm{C}::\ddot{\mathrm{O}}: \qquad \text{or} \qquad \mathrm{O}{=}\mathrm{C}{=}\mathrm{O}$$

b) Silicon also has a normal valence of 4, but it is not an element which readily forms double bonds. Each silicon can form single bonds to four oxygen atoms however,

$$\begin{array}{c} \dot{} \\ :\mathrm{O}: \\ \cdot\ddot{\underset{\cdot\cdot}{\mathrm{O}}}:\ddot{\underset{\cdot\cdot}{\mathrm{Si}}}:\ddot{\underset{\cdot\cdot}{\mathrm{O}}}\cdot \\ :\ddot{\mathrm{O}}: \\ \cdot \end{array}$$

Now the silicon is satisfied, but each oxygen lacks one electron and has only formed one bond. If each of the oxygens link to another silicon, they will be satisfied, but then the added silicon atoms will have unused valences:

The process of adding oxygen or silicon atoms can continue indefinitely, producing a giant lattice of covalently bonded atoms. In this giant molecule each silicon is bonded to four oxygens and each oxygen to two silicons, and so there are twice as many oxygen atoms as silicon. The molecular formula could be written $(SiO_2)_n$, where n is a very large number. A portion of this giant molecule is shown in Fig. 6.8.

The difference in the abilities of carbon and silicon atoms to form double bonds has important consequences in the natural environment. Because C=O double bonds form readily, carbon dioxide consists of individual molecules—there are no "empty spaces" on either the carbon or oxygen atoms where additional electrons may be shared. Hence there is little to hold one carbon dioxide molecule close to another, and at ordinary temperatures the molecules move about independently. On a macroscopic scale this means that carbon dioxide has the properties of a gas. In silicon dioxide, on the other hand, strong covalent bonds link all silicon and oxygen atoms together in a three-dimensional network. At ordinary temperatures the atoms

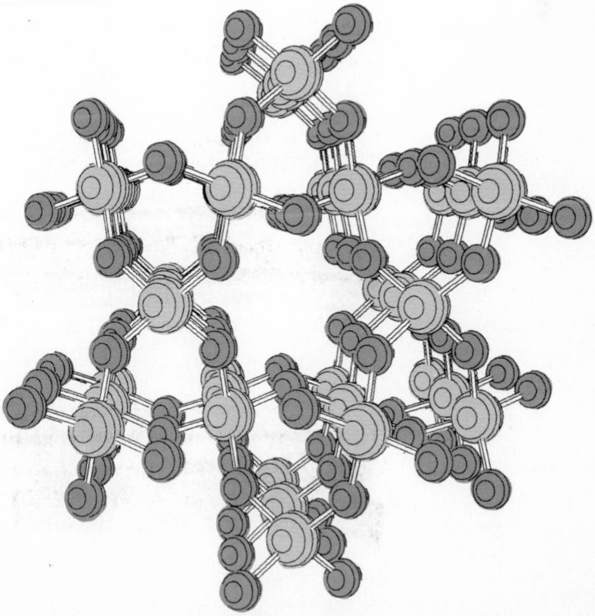

Figure 6.8 A portion of the giant covalent molecule $(SiO_2)_n$. The lattice shown would extend indefinitely in all directions in a macroscopic crystal. Each silicon atom (light color) is covalently bonded to four oxygen atoms (dark color). Each oxygen bonds to two silicons. The ratio of silicon to oxygen is 2:4 or 1:2, in accord with the formula. Computer-generated. (*Copyright © 1976 by W. G. Davies and J. W. Moore.*)

cannot vibrate far from their allotted positions, and silicon dioxide has the macroscopic properties of a solid.

As a gas, carbon dioxide is much freer than silicon dioxide to circulate through the environment. It can be removed from the atmosphere by plants in the photosynthetic process and eventually returned to the air by means of respiration. This is one of the reasons that terrestrial life is based on carbon compounds. If a supply of carbon from atmospheric carbon dioxide were not available, living organisms would be quite different in form and structure from the ones we know on earth.

Science-fiction authors are fond of suggesting, because of the periodic relationship of carbon and silicon, that life on some distant planet might be based on silicon. It is rather hard to imagine, though, the mechanism by which such life forms would obtain silicon from the rocks and soil of their planet's surface. Certainly they would face major difficulties if the combination of silicon with oxygen to form silicon dioxide were to be used as a source of energy. Imagine breathing out a solid instead of the gaseous carbon dioxide which forms when carbon combines with oxygen during respiration in terrestrial organisms! Macroscopic properties which are determined by microscopic structure and bonding are crucial in even such fundamental activities as living and breathing.

An Excess of Bonds

There are a number of cases where the normal valences of the atoms involved do not predict the correct skeleton structure. For example, thionyl chloride, $SOCl_2$, is found *experimentally* to have both chlorines and the oxygen atom bonded to sulfur:

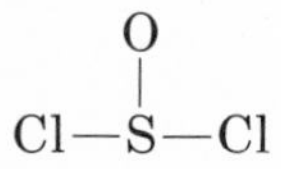

not

This exceeds the usual valence of 2 for sulfur, while oxygen has one less bond than we might have expected. Molecules which deviate in this way from the usual valence rules often contain at least one atom (such as S) from the third row or below in the periodic table. One or more oxygen atoms are usually bonded just to that third-row atom instead of linking a pair of other atoms. Usually the atom which occupies the central position in the skeleton is written *first* in the molecular formula, although sometimes H (which forms only one bond and cannot be the central atom) precedes it. Some examples (with the central atom in italics) are: $\mathit{S}OCl_2$, $\mathit{P}OCl_3$, $H\mathit{Cl}O_4$, $\mathit{S}O_2Br_2$, and N_2O. (In the last case, one of the two nitrogens occupies the central position.)

In such molecules the deviation from the normal valence occurs because at least one electron-pair bond contains two electrons which were originally associated with the *same* atom. Such a bond is called a **coordinate covalent bond** or a **dative bond.** An example is the bond between sulfur and oxygen in $SOCl_2$:

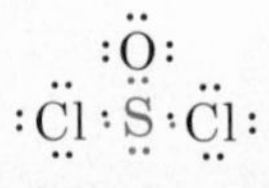

Both electrons in the S—O bond were originally valence electrons of sulfur. Therefore the sharing of this electron pair adds nothing to the valence shell of sulfur, and sulfur can form one more bond than would be predicted by its normal valence. Neither electron in the coordinate covalent bond was originally associated with oxygen, and a single bond (both electrons) is sufficient to provide an octet when added to oxygen's six valence electrons. Hence oxygen forms one less bond than expected.

It should be noted that there is nothing to distinguish a coordinate covalent bond from any other covalent bond once a structural formula has been drawn. A pair of electrons is still a pair of electrons no matter where it came from. The distinction is merely one we make when trying to fit electron pairs into octets around each atom. Structural formulas for the other examples of unusual valence we have mentioned are shown below with coordinate covalent bonds indicated in color:

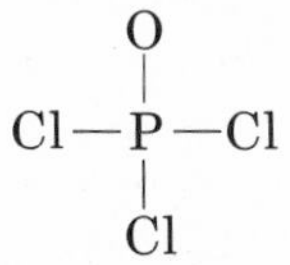

Phosphorus oxychloride, $POCl_3$

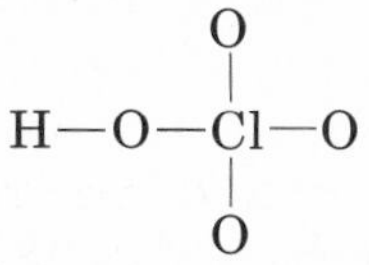

Perchloric acid, $HClO_4$

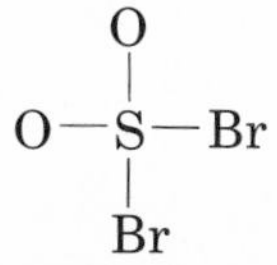

Sulfuryl bromide, SO_2Br_2

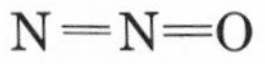

Nitrous oxide, N_2O

It is good practice to draw out the complete Lewis diagram for each of these molecules, differentiating electrons from different nuclei with different symbols such as × and •, and satisfy yourself that they obey the octet rule.

Polyatomic Ions

Our discussion of ionic compounds in Sec. 6.3 was confined to monatomic ions. However, more complex ions, containing several atoms but still having a positive or negative charge, occur quite frequently in chemistry. Well-known examples of such **polyatomic ions** are the sulfate ion (SO_4^{2-}), the hydroxide ion (OH^-), the hydronium ion (H_3O^+), and the ammonium ion (NH_4^+). The atoms in these ions are joined together by covalent electron-pair bonds, and we can draw Lewis structures for the ions just as we can for molecules. The only difference is that the number of electrons in the ion does not exactly balance the sum of the nuclear charges. Either there are too many electrons, in which case we have an anion, or too few, in which case we have a cation.

Consider, for example, the hydroxide ion (OH^-) for which the Lewis structure is

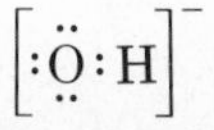

Chemical
Bonding—
Electron
Pairs
and Octets

A neutral molecule containing one O and one H atom would contain only seven electrons, six from O and one from H. The hydroxide ion, though, contains an *octet* of electrons, *one more* than the neutral molecule. The hydroxide ion must thus carry a single negative charge.

In order to draw the Lewis structure for a given ion, we must first determine how many valence electrons are involved. Suppose the structure of H_3O^+ is required. The total number of electrons is obtained by adding the valence electrons for each atom, $6 + 1 + 1 + 1 = 9$ electrons. We must now subtract 1 electron since the species under consideration is not H_3O but H_3O^+. The total number of electrons is thus $9 - 1 = 8$. Since this is an octet of electrons, we can place them all around the O atom. The final structure then follows very easily:

$$\left[\begin{matrix} \text{H} \\ \text{H} : \overset{..}{\underset{..}{\text{O}}} : \text{H} \end{matrix} \right]^+$$

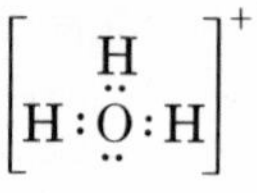

In more complicated cases it is often useful to calculate the number of *shared electron pairs* before drawing a Lewis structure. This is particularly true when the ion in question is an *oxyanion* (i.e., a central atom is surrounded by several O atoms). A well-known oxyanion is the carbonate ion, which has the formula CO_3^{2-}. (Note that the central atom C is written first, as was done earlier for molecules.) The total number of valence electrons available in CO_3^{2-} is

$$4(\text{for C}) + 3 \times 6(\text{for O}) + 2(\text{for the} -2 \text{ charge}) = 24$$

We must distribute these electrons over 4 atoms, giving each an octet, a requirement of $4 \times 8 = 32$ electrons. This means that $32 - 24 = 8$ electrons need to be counted twice for octet purposes; i.e., 8 electrons are shared. The ion thus contains *four* electron-pair bonds. Presumably the C atom is double-bonded to one of the O's and singly bonded to the other two:

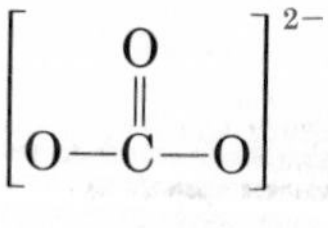

$$\left[\begin{matrix} \text{O} \\ \| \\ \text{O}-\text{C}-\text{O} \end{matrix} \right]^{2-} \quad \text{or} \quad \left[\begin{matrix} {\times}\text{O}{\times} \\ {\times}\overset{\times\times}{\underset{\times\times}{\text{O}}} : \text{C} : \overset{\times}{\underset{\times}{\text{O}}}{\times} \end{matrix} \right]^{2-}$$

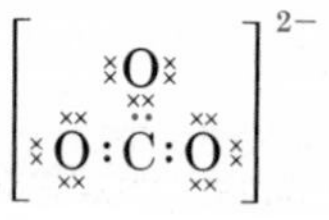

In this diagram the 4C electrons have been represented by dots, the 18 O electrons by × 's, and the 2 extra electrons by colored dots, for purposes of easy reference. Real electrons do not carry labels like this; they are all the same.

There is a serious objection to the Lewis structure just drawn. How do the electrons know which oxygen atom to single out and form a double bond with, since there is otherwise nothing to differentiate the oxygens? The answer is that they do not. To explain the bonding in the CO_3^{2-} ion and some other molecules requires an extension of the Lewis theory. We will pursue this matter further in the next chapter.

We end this section with two examples.

EXAMPLE 6.8 Draw a Lewis structure for the molecule ethylene, C_2H_4.

Solution Since hydrogen atoms are univalent, they must certainly all be bonded to carbon atoms, presumably two to each carbon. Each carbon atom thus has the situation

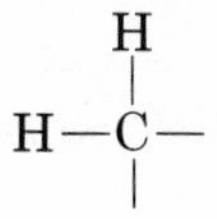

in which two bonds must still be accounted for. By assuming that the two carbon atoms are joined by a double bond, all the valence requirements are satisfied, and we can draw a Lewis structure containing satisfactory octets:

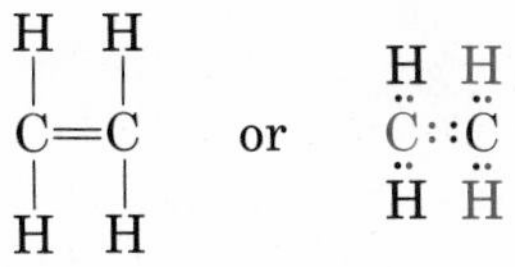

EXAMPLE 6.9 Draw a Lewis structure for the sulfite ion, SO_3^{2-}.

Solution The safest method here is to count electrons. The total number of valence electrons available is

$$6(\text{for S}) + 3 \times 6(\text{for O}) + 2(\text{for the charge}) = 26$$

To make four octets for the four atoms would require 32 electrons, and so the difference, $32 - 26 = 6$, gives the number of shared electrons. There are thus only *three* electron-pair bonds in the ion. The central S atom must be linked by a single bond to each O atom.

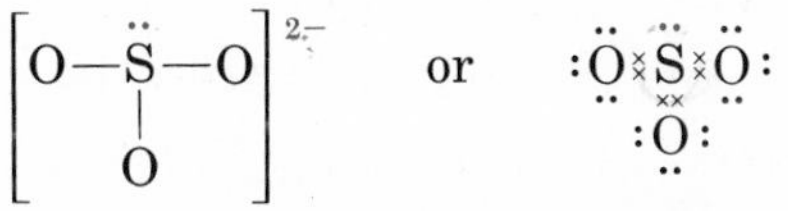

Note that each of the S—O bonds is coordinate covalent.

6.7 IONIC COMPOUNDS CONTAINING POLYATOMIC IONS

When polyatomic ions are included, the number of ionic compounds increases significantly. Indeed, most ionic compounds contain polyatomic ions. Well-known examples are sodium hydroxide (NaOH), calcium car-

TABLE 6.2 Covalently Bonded Polyatomic Ions.

Name	Formula
−3 Charge	
Phosphate	PO_4^{3-}
Arsenate	AsO_4^{3-}
−2 Charge	
Carbonate	CO_3^{2-}
Peroxide	O_2^{2-}
Sulfate	SO_4^{2-}
Sulfite	SO_3^{2-}
Chromate	CrO_4^{2-}
Dichromate	$Cr_2O_7^{2-}$
Hydrogen phosphate	HPO_4^{2-}
−1 Charge	
Hydrogen carbonate (bicarbonate)	HCO_3^-
Superoxide	O_2^-
Hydrogen sulfate	HSO_4^-
Dihydrogen phosphate	$H_2PO_4^-$
Hydroxide	OH^-
Nitrate	NO_3^-
Nitrite	NO_2^-
Acetate	$C_2H_3O_2^-$ or CH_3COO^-
Cyanide	CN^-
Permanganate	MnO_4^-
Perchlorate	ClO_4^-
Chlorate	ClO_3^-
Chlorite	ClO_2^-
Hypochlorite	ClO^-
+1 Charge	
Ammonium	NH_4^+
Hydronium	H_3O^+

bonate ($CaCO_3$), and ammonium nitrate (NH_4NO_3). A list of the more important polyatomic ions is given in Table 6.2. A great many of them are oxyanions.

The properties of compounds containing polyatomic ions are very similar to those of binary ionic compounds. The ions are arranged in a regular lattice and held together by coulombic forces of attraction. The resulting crystalline solids usually have high melting points and all conduct electricity when molten. Most are soluble in water and form conducting solutions in which the ions can move around as independent entities. In general polyatomic ions are colorless, unless, like CrO_4^{2-} or MnO_4^-, they contain a transition-metal atom. The more negatively charged polyatomic ions, like their monatomic counterparts, show a distinct tendency to react with water, producing hydroxide ions; for example,

$$PO_4^{3-} + H_2O \rightarrow HPO_4^{2-} + OH^-$$

It is important to realize that compounds containing polyatomic ions must be *electrically neutral*. In a crystal of calcium sulfate, for instance, there must be equal numbers of Ca^{2+} and SO_4^{2-} ions in order for the charges to balance. The formula is thus $CaSO_4$. In the case of sodium sulfate, by contrast, the Na^+ ion has only a single charge. In this case we need *two* Na^+ ions for each SO_4^{2-} ion in order to achieve electroneutrality. The formula is thus Na_2SO_4.

EXAMPLE 6.10 What is the formula of the ionic compound calcium phosphate?

Solution It is necessary to have the correct ratio of calcium ions, Ca^{2+}, and phosphate ions, PO_4^{3-}, in order to achieve electroneutrality. The required ratio is the *inverse* of the ratio of the charges on ions. Since the charges are in the ratio of 2:3, the numbers must be in the ratio of 3:2. In other words the solid salt must contain three calcium ions for every two phosphate ions:

$$\text{Six positive charges} \begin{cases} Ca^{2+} \\ Ca^{2+} \\ Ca^{2+} \end{cases} \qquad \begin{rcases} PO_4^{3-} \\ PO_4^{3-} \end{rcases} \text{Six negative charges}$$

The formula for calcium phosphate is thus $Ca_3(PO_4)_2$.

6.8 THE SIZES OF ATOMS AND IONS

Atomic Sizes

The sizes of atoms and ions are important in determining the properties of both covalent and ionic compounds. You should already have some appreciation of the factors which govern atomic sizes from the color-coded dot-density diagrams in Plates 4 and 5. By far the largest atom illustrated in these color plates is Li. Because Li has an electron in the $n = 2$ shell, it is larger than H or He whose 1*s* electron clouds are much closer to the nucleus. Li is also larger than Be, B, or C. In the latter atoms, the 2*s* and 2*p* electron clouds are attracted by a greater nuclear charge and hence are held closer to the center of the atom than the 2*s* cloud in Li. Thus two important rules may be applied to the prediction of atomic sizes.

1 As one moves from top to bottom of the periodic table, the principal quantum number *n* increases and electrons occupy orbitals whose electron clouds are successively farther from the nucleus. The atomic radii increase.

2 As one moves from left to right across a horizontal period, the *n* value of the outermost electron clouds remains the same, but the nuclear charge increases steadily. The increased nuclear attraction contracts the electron cloud, and hence the atomic size decreases.

It is difficult to measure the size of an atom very exactly. As the dot-density diagrams show, an atom is not like a billiard ball which has a definite radius. Instead of stopping suddenly, an electron cloud gradually fades out so that one cannot point to a definite radius at which it ends. One way out of this difficulty is to find out how closely atoms are packed together in a crystal lattice. Figure 6.9 illustrates part of a crystal of solid Cl_2 at a very low temperature. The distance *AA′* in this figure has the value of 369 pm. Since this represents the distance between adjacent atoms in *different* Cl_2 molecules, we can take it as the distance at which different Cl atoms just "touch." Half this distance, 184 pm, is called the **van der Waals radius** of Cl. The van der Waals radius gives an approximate idea of how closely atoms in *different* molecules can approach each other. Commonly accepted values of the van der Waals radii for the representative elements are shown in Fig. 6.10. Note how these radii *decrease across* and *increase down* the periodic table.

Also given in Fig. 6.10 are values for the **covalent radius** of each atom. Returning to Fig. 6.9, we see that the distance *AB* between two Cl atoms in the *same* molecule (i.e., the Cl—Cl **bond length**) has a value of 202 pm. The covalent radius is one-half of this bond length, or 101 pm. Covalent radii are approximately additive and enable us to predict rough values for the internuclear distances in a variety of molecules. For example, if we add the covalent radius of C (77 pm) to that of O (66 pm), we obtain an estimate for the length of the C—O bond, namely, 143 pm. This is in exact agreement with the measured value in ethyl alcohol and dimethyl ether discussed in the previous section.

Ionic Sizes

The size of an ion is governed not only by its electronic structure but also by its *charge*. This relationship is evident from Fig. 6.11. Ions in the

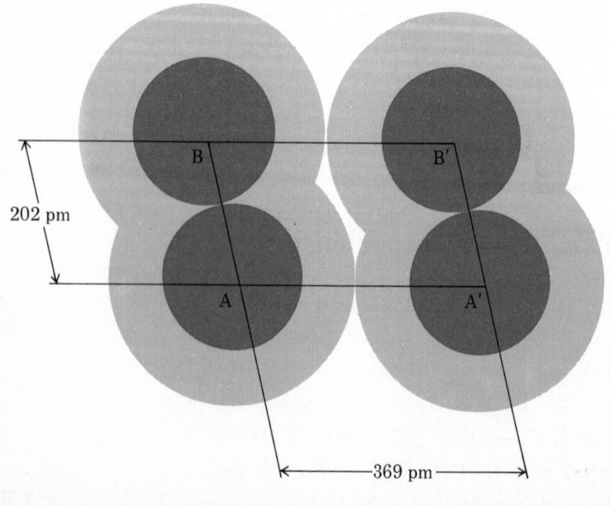

Figure 6.9 The relationship between van der Waals radii and covalent radii for $Cl_2(s)$. In solid chlorine the molecules pack together so that the shortest distance between chlorine nuclei in different molecules (AA′ or BB′) is 369 pm. The van der Waals radius of chlorine is defined as half that distance or 184 pm. The covalent radius of chlorine is half the distance (one-half AB or A′B′) between two chlorine nuclei in the *same* molecule. This is smaller than the van der Waals radius because of the covalent bond in each Cl_2 molecule.

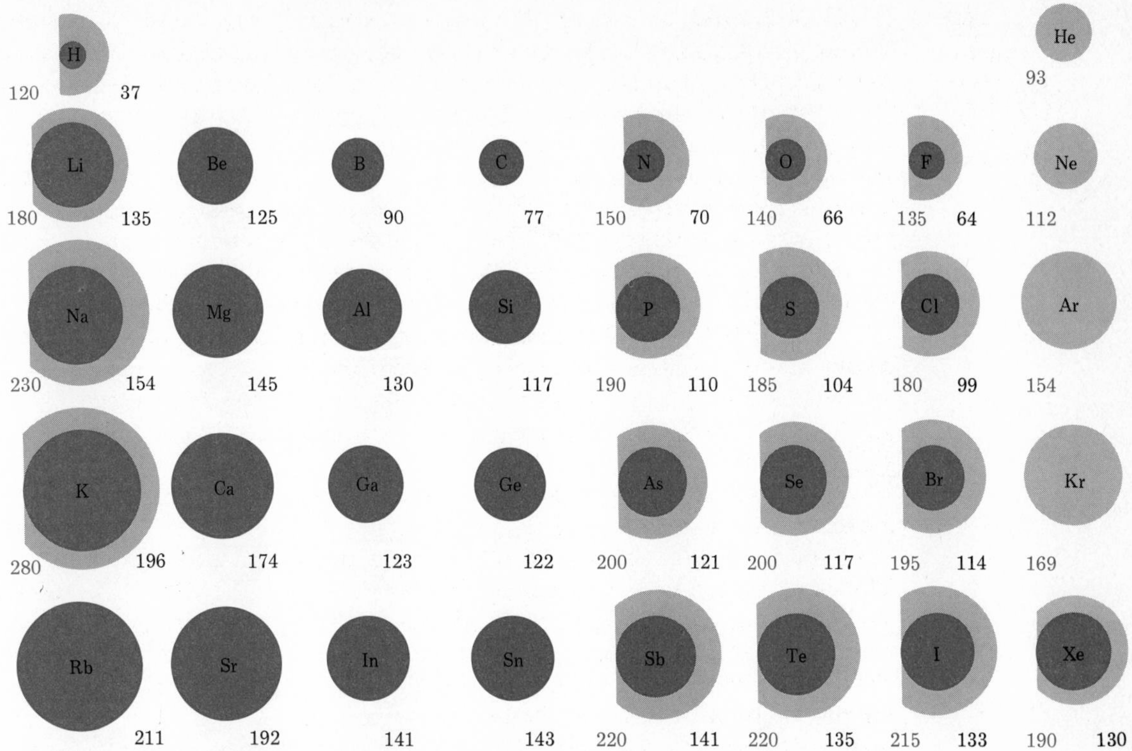

Figure 6.10 Sizes of atoms of the representative elements as a function of their position in the periodic table. Outer (lightly shaded) circles indicate van der Waals radii, while inner (darkly shaded) circles represent covalent radii. Colored numbers are van der Waals radii, and black numbers are covalent radii, both expressed in picometers.

first row of this figure, H^-, Li^+, and Be^{2+}, all have the same $1s^2$ electronic structure as the helium atom. Species which have the same electronic structure but different charges are said to be **isoelectronic.** For any isoelectronic series, such as H^-, He, Li^+, Be^{2+}, in which the nuclear charge increases by 1 each time, we find a progressive *decrease* in size due to the increasingly strong attraction of the nucleus for the electron cloud. Each row in Fig. 6.11 corresponds to an isoelectronic series involving a different noble-gas electron configuration. As we move from the more negative to the more positive ions in each row, there is a steady decrease in size. If we move *down* any of the columns in Fig. 6.11, ionic sizes increase due to the increasing principal quantum number of the outermost electrons. The sizes of singly charged cations, for example, increase in the following order: $Li^+ < Na^+ < K^+ < Rb^+ < Cs^+$.

A further point of interest is the size of an ion relative to the atom from which it was formed. Figure 6.1 at the beginning of this chapter showed that when an Li atom lost its electron and became an Li^+ ion, its size decreased dramatically. The radii given in Fig. 6.10 and 6.11 reveal that this is also true for the other alkali metals. For example, the van der Waals radius of K is 280 pm, while the ionic radius of K^+ is only 133 pm. The large

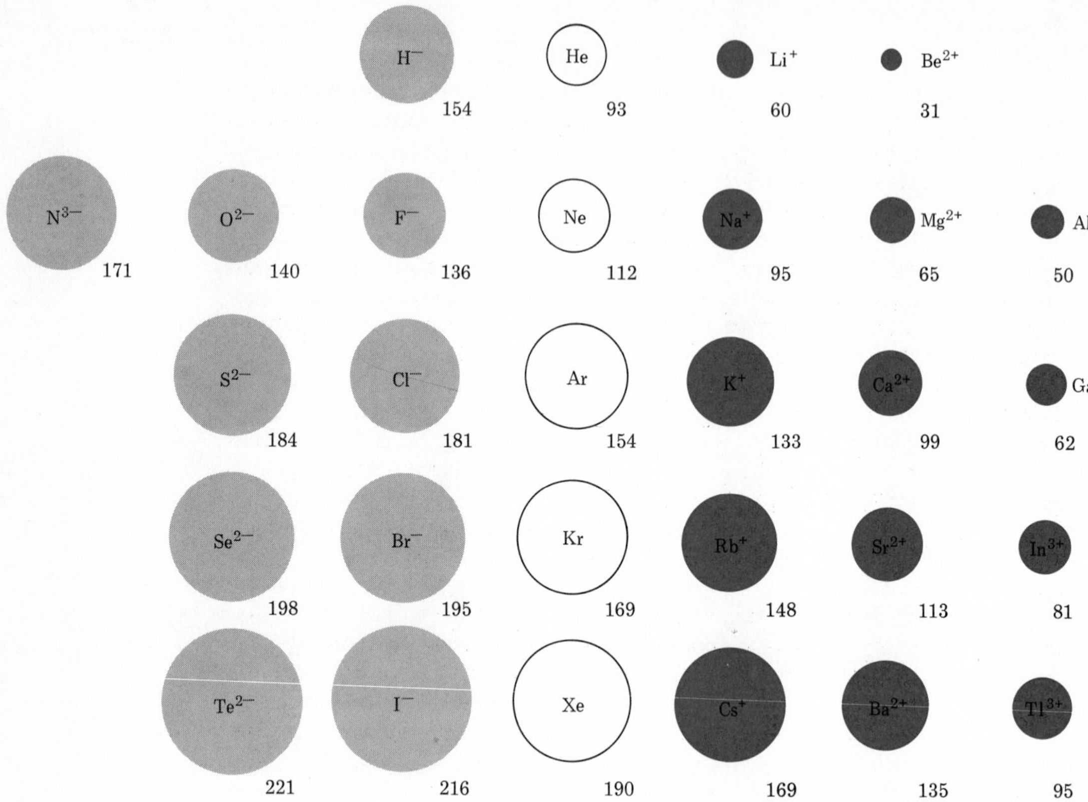

Figure 6.11 Ionic radii (in picometers) as a function of position in the periodic table. Each row consists of ions which are isoelectronic (have the same number of electrons). The van der Waals radius of the noble gas isoelectronic with each row of ions is given for comparison. Note that positive ions (color) are smaller than the corresponding noble gas, while negative ions are somewhat larger. Within an isoelectronic series, ionic radii decrease as nuclear charge increases.

reduction in size is in sharp contrast to what occurs when an atom accepts one or more electrons to attain a noble-gas structure. Since the added electron goes into a subshell which already has occupants, rather than starting on a new subshell, there is very little change in size. This is clearly seen in Fig. 6.1 where formation of an H^- ion from an H atom produces no perceptible increase in size. Comparing Fig. 6.10 and 6.11, we also find that the van der Waals radii of nonmetals are only slightly smaller than the radii of their anions.

The sizes of the ions involved have considerable influence on both the chemical and physical properties of ionic compounds. There is a strong correlation, for example, between ionic size and the melting point of an ionic compound. Among the halides of sodium the melting point decreases in the order of NaF (995°C) > NaCl (808°C) > NaBr (750°C) > NaI (662°C). The larger the anion, the farther it is from the sodium ion, and the weaker the coulombic force of attraction between them. Hence the lower the melting point. When a very small cation combines with a very large anion, the resulting compound is less likely to exhibit the characteristic macroscopic properties of an ionic substance.

SUMMARY

In this chapter we have seen that there are two principal ways in which atoms may be linked to each other. Covalent bonding occurs among elements whose atoms have approximately the same attraction for electrons. Sharing one or more pairs of electrons between two atoms attracts the nuclei together and usually results in an octet around each atom. Covalent bonding often produces individual molecules, like CO_2 or CH_3CH_2OH, which have no net electrical charge and little attraction for one another. Thus covalent substances often have low melting and boiling points and are liquids or gases at room temperature. Occasionally, as in the case of SiO_2, an extended network of covalent bonds is required to satisfy the octet rule. Such giant molecules result in solid compounds with high melting points.

Ionic compounds involve elements which differ considerably in their tendency to attract electrons. Binary ionic compounds are not too common, but the existence of polyatomic ions greatly extends the number of ionic substances. Because oppositely charged ions are held in crystal lattices by strong coulombic forces, ionic compounds are hard, brittle, and have high melting and boiling points. The majority of them dissolve in water, and in solution each ion exhibits its own properties. Like covalent compounds, most ionic substances obey the octet rule.

A number of atomic properties, such as ionization energy, electron affinity, van der Waals and covalent atomic radii, and ionic radius, are important in determining whether certain elements will form covalent or ionic compounds and what properties those compounds will have. Figure 6.12 shows how each of these properties varies according to an atom's position in the periodic table. The figure also includes one atomic property, electronegativity, which will be defined and discussed in the next chapter.

Figure 6.12 Periodic variation of atomic properties.

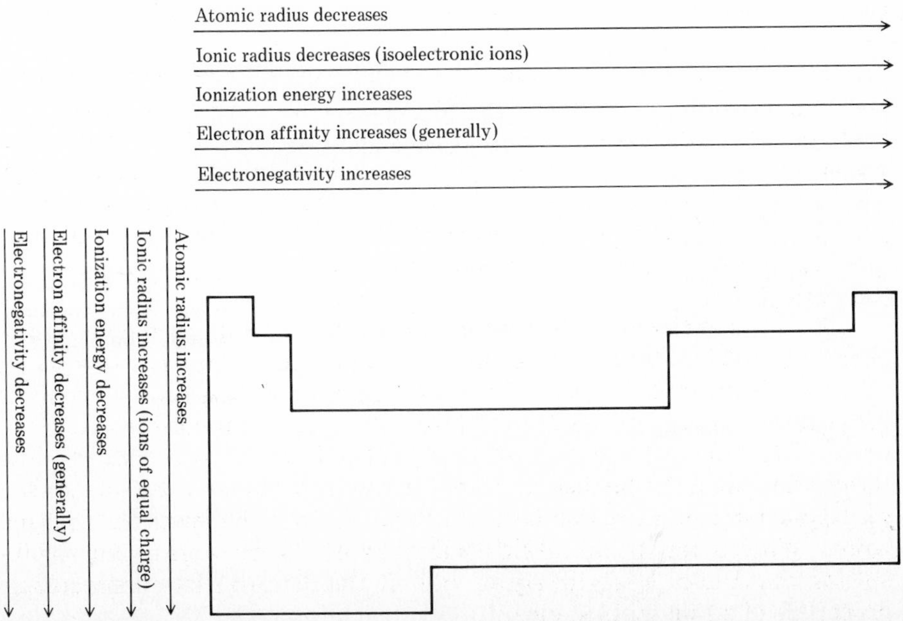

QUESTIONS AND PROBLEMS

6.1 How did Lewis explain the fact that most chemical formulas correspond to an even number of electrons? Choose ten formulas from Table 4.2 and verify that most correspond to an even number of electrons.

6.2 Draw a diagram similar to Fig. 6.2 to illustrate the formation of sodium chloride crystals from sodium atoms and chlorine atoms using the following data:

	ΔH_m/kJ mol^{-1}
$Na(g) \rightarrow Na^+(g) + e^-$	495.4
$Cl(g) + e^- \rightarrow Cl^-(g)$	−348.5
$Na^+(g) + Cl^-(g) \rightarrow NaCl(s)$	−770.3

6.3 What further information would be needed in Prob. 6.2 to find the enthalpy change for the formation of NaCl from solid Na metal and gaseous Cl_2 molecules?

$$Na(s) + \tfrac{1}{2}Cl_2(g) \rightarrow NaCl(s)$$

6.4 Explain carefully why lithium hydride forms a solid ionic crystal lattice rather than gaseous Li^+H^- ion pairs.

6.5 Explain the fallacy in the following statement: When two elements A and B react to form the ionic compound AB, the electron affinity of B must be greater than the ionization energy of A so that energy is *released* by the transfer of an electron.

6.6 Explain in terms of electronic structure why the reaction

$$Cl(g) + e^- \rightarrow Cl^-(g)$$

results in a *release* of energy.

6.7 We normally think of Li^+ ions and H^- ions as attracting each other because of their opposite charges, but if they are closer than 160 pm, they actually *repel* each other. Explain

6.8 An element has the following ionization energies:

First ionization energy	900 kJ mol^{-1}
Second ionization energy	1800 kJ mol^{-1}
Third ionization energy	14 800 kJ mol^{-1}
Fourth ionization energy	21 000 kJ mol^{-1}

To which group in the periodic table does it belong?

6.9 When a sodium atom reacts with a chlorine atom, only one electron is transferred. Basing your argument on energy, explain why two electrons are not transferred.

6.10 Arrange the following elements in order of increasing first ionization energy:

a Ca, Be, Ba *d* K, Ar, Ca
b Ar, S, Cl *e* Kr, Ar, Br
c Ne, Se, Cl *f* N, C, B

6.11 Although ionization energies usually *increase* as one moves across the periodic table, the ionization energy of $_{16}S$ is 1000 kJ mol^{-1}, slightly *less* than that for the preceding element $_{15}P$, which is 1010 kJ mol^{-1}. Explain why this is so.

6.12 The first and second ionization energies of Mg and Al are as follows:

	Mg	Al
First ionization energy/kJ mol^{-1}	740	500
Second ionization energy/kJ mol^{-1}	1500	1800

Why is the first ionization energy of Mg larger than that of Al while the reverse is true for the second ionization energies?

6.13 Explain why the ionization energies of all the first-row transition elements (Sc through Zn) have fairly similar values.

***6.14** Although ionization energies usually decrease down a periodic group, the ionization energy of gallium (579 kJ mol^{-1}) is slightly larger than that of aluminum (578 kJ mol^{-1}) just above it in group IIIA. Explain this discrepancy. See if you can find other examples of this sort of behavior in Table 6.1.

6.15 Which of the following pairs of elements are likely to form ionic compounds? Write appropriate formulas for the compounds you expect to form.

a Chlorine and bromine
b Lithium and tellurium
c Sodium and argon
d Nitrogen and bromine
e Magnesium and fluorine
f Indium and sulfur
g Selenium and bromine

h Barium and iodine
i Barium and nitrogen
j Carbon and oxygen

6.16 Which of the following substances are *not* ionic?
a BaF_2
b Cl_2O
c NH_3
d SO_3
e K_2Se
f H_2

6.17 Which of the following ions have noble-gas electronic structures?
a K^+
b P^{3-}
c Se^{2-}
d Co^{2+}
e Fe^{3+}
f Ca^{2+}
g F^-
h Na^{2+}
i Sc^{3+}
j Ce^{3+}
k Cu^{2+}
l Ti^{4+}

6.18 Which of the following species would not be present in a crystal of $MgCl_2$?
a $MgCl_2$ molecules
b Mg^+ ions
c Cl_2^{2-} ions
d Cl^- ions
e Mg^{2+} ions
f Cl_2^- ions

6.19 Write the electron configuration for
a the Fe^{2+} ion
b the Zn^{2+} ion
c the Cr^{3+} ion

6.20 Write balanced equations for reactions of each of the following anions with water:
a O^{2-}
b S^{2-}
c H^-
d Cl^-
e Br^-

6.21 What characteristics must atoms A and B have if they are able to form a covalent bond A—B with each other?

6.22 Give a brief wave-mechanical description of the bonding in the molecule H_2. In what sense are the two electrons shared?

6.23 Draw Lewis structures for each of the following covalent molecules:
a $AsCl_3$
b H_2Se
c PH_3
d $SnCl_4$
e CH_3Cl
f Cl_2O
g ClF
h BF_3

What is unusual about *h*?

6.24 Draw Lewis structures for each of the following molecules, none of which contain multiple bonds:
a C_2H_6
b C_3H_8
c CH_3NH_2
d N_2H_4
e H_2O_2
f S_2Cl_2

6.25 Draw Lewis structures for the following covalent molecules, each of which contains a multiple bond:
a HCN
b CO_2
c NOCl
d $COCl_2$
e CH_3COCH_3
f N_2F_2
g C_2H_4

6.26 What is a coordinate covalent bond? Is it the same as a dative bond? Once formed, can such a bond be distinguished from an ordinary covalent bond?

6.27 Draw Lewis structures for the following covalent molecules, each of which contains one or more coordinate covalent bonds:
a NO_2Cl
b H_2SO_4
c NH_3BF_3
d $HClO_4$

6.28 Draw Lewis structures for each of the following oxyanions:
a PO_4^{3-}
b ClO_3^-
c ClO_4^-
d SO_4^{2-}
e AsO_3^{3-}
f $P_2O_7^{4-}$

6.29 Which of the following compounds contain only ionic bonds, which contain only covalent bonds, and which contain both ionic and covalent bonds?
a KF
b Na_2CO_3
c HCl
d H_2SO_4
e $K_2Cr_2O_7$
f OF_2
g MgF_2
h $Ba(NO_3)_2$

6.30 Write correct formulas for each of the following ionic compounds:
a Aluminum sulfate
b Magnesium nitrate
c Ammonium sulfite
d Strontium phosphate
e Strontium hydrogen phosphate
f Strontium dihydrogen phosphate
g Potassium sulfite
h Ammonium dichromate
i Calcium arsenate
j Barium peroxide

6.31 Assume that you have an unlabeled bottle containing a white crystalline powder. The powder melts at 310°C. You are told that it is either NH_3, NO_2, or $NaNO_3$. Which do you think it is?

6.32 Draw Lewis structures for each of the following:

a K_2SO_4 *e* $LiBF_4$
b H_3O^+ *f* NO^+
c NH_4Cl *g* C_3H_6
d C_2H_2 *h* N_2F_2

6.33 Differentiate between the van der Waals radius of an atom and its covalent radius. How can these radii be determined?

6.34 Arrange the following species in order of increasing size:

a Sr^{2+}, Ra^{2+}, Ba^{2+}, Ca^{2+} *e* Se, Cl, Ar, Ne
b F^-, Cl^-, Br^-, I^- *f* I^-, Br^-, Kr, Te^{2-}
c Ne, Na^+, F^-, Mg^{2+}
d Cs^+, Ca^{2+}, Sr^{2+}, Rb^+

6.35 Which species in each of the following groups of four is not isoelectronic with the other three?

a O^2, F^-, Ne, Ar *d* H^-, Li^+, Be, He
b K^+, Na^+, Mg^{2+}, Ne *e* Ar, K^+, Ca^{2+}, Ga^{3+}
c Ti^{4+}, Sc^{3+}, V^{3+}, Ca^{2+}

6.36 The C—Br bond length in CBr_4 is 194 pm, while the Br—Br distance in Br_2 is 228.4 pm. Use these values to estimate the length of a C—C single bond. How does your calculated bond length compare with the 154 pm observed for the C—C bond length in solid diamond? What can you conclude from this?

6.37 The Pb—Cl length is found to be 243 pm in $PbCl_4$, while the Cl—Cl distance in Cl_2 has the value 199 pm. From these two distances calculate a value for the covalent radius of Pb and the length of a Pb—Pb bond. Compare this value with the measured Pb—Pb distance of 287 pm in the compound $(CH_3)_3Pb—Pb(CH_3)_3$ and 350 pm in solid Pb metal. The radius of Pb is not given in Fig. 6.10 but is usually taken as 175 pm. Is this really a covalent radius?

6.38 Which of the following solids would you expect to have the highest melting point: MgO, AlN, or NaF? Which would you expect to have the lowest melting point? Explain.

chapter seven

FURTHER ASPECTS OF COVALENT BONDING

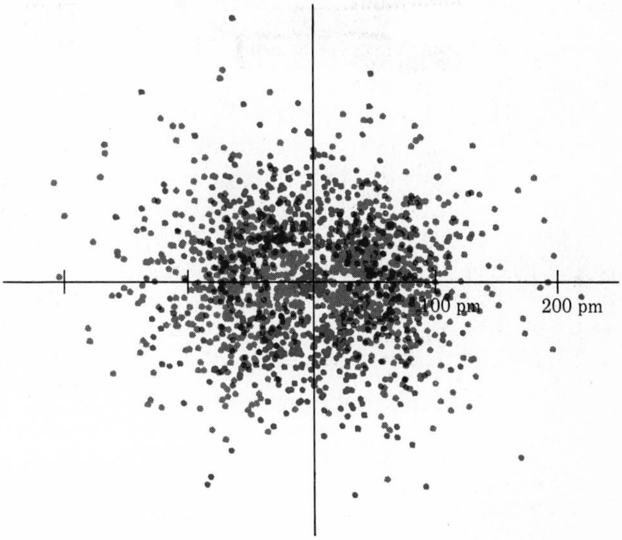

The previous chapter concentrated on the octet rule and Lewis diagrams for simple covalent molecules. There are numerous examples, however, of molecules which are quite stable but contain one or more atoms which do not have a noble-gas electron configuration. Furthermore, structural formulas like those in Chap. 6 only show which atoms are connected to which. They do not tell us how the atoms are arranged in three-dimensional space. In other words a Lewis diagram does not show the shape of a molecule.

In this chapter we will develop a more detailed picture of molecules—including some which do not obey the octet rule. You will learn how both the shapes and bonding of molecules may be described in terms of orbitals. In addition it will become apparent that the distinction between covalent and ionic bonding is not so sharp as it may have seemed in Chap. 6. You will find that many covalent molecules are electrically unbalanced, causing their properties to tend toward those of ion pairs. Rules will be developed so that you can predict which combinations of atoms will exhibit this kind of behavior.

7.1 EXCEPTIONS TO THE OCTET RULE

Considering the tremendous variety in properties of elements and compounds in the periodic system, it is asking a great deal to expect a rule as simple as Lewis' octet theory to be able to predict all formulas or to account for all molecular structures involving covalent bonds. Lewis' theory concentrates on resemblances to noble-gas ns^2np^6 valence octets. Therefore it is most successful in accounting for formulas of compounds of the representative elements, whose distinguishing electrons are also s and p electrons. The octet rule is much less useful in dealing with compounds of the transition elements or inner transition elements, most of which involve some participation of d or f orbitals in bonding.

Even among the representative elements there are some exceptions to the Lewis theory. These fall mainly into two categories: (1) Some stable molecules simply do not have enough electrons to achieve octets around all atoms. This usually occurs in compounds containing Be or B. (2) Elements in the third period and below can accommodate more than an octet of electrons. Although elements such as Si, P, S, Cl, Br, and I obey the octet rule in many cases, under other circumstances they form more bonds than the rule allows.

Good examples of the first type of exception are provided by $BeCl_2$ and BCl_3. Beryllium dichloride, $BeCl_2$, is a covalent rather than an ionic substance. Solid $BeCl_2$ has a relatively complex structure at room temperature, but when it is heated to 750°C, a vapor which consists of separate $BeCl_2$ molecules is obtained. Since Cl atoms do not readily form multiple bonds, we expect the Be atom to be joined to each Cl atom by a single bond. The structure is

$$:\ddot{Cl}—Be—\ddot{Cl}:$$

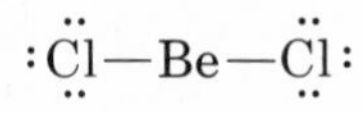

Instead of an octet the valence shell of Be contains only *two* electron pairs. Similar arguments can be applied to boron trichloride, BCl_3, which is a stable gas at room temperature. We are forced to write its structure as

$$
\begin{array}{c}
:\ddot{Cl}: \\
| \\
B \\
\diagup \quad \diagdown \\
\cdot\ddot{Cl} \qquad \ddot{Cl}\cdot
\end{array}
$$

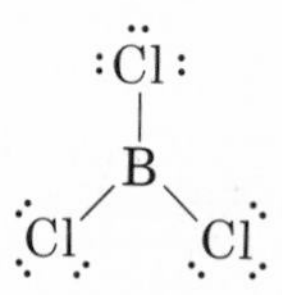

in which the valence shell of boron has only three pairs of electrons.

Molecules such as $BeCl_2$ and BCl_3 are referred to as **electron deficient** because some atoms do not have complete octets. Electron-deficient molecules typically react with species containing lone pairs, acquiring octets by formation of coordinate covalent bonds. Thus $BeCl_2$ reacts with Cl^- ions to form $BeCl_4^-$:

$$
:\ddot{Cl}:Be:\ddot{Cl}: + \begin{array}{c} :\ddot{Cl}:^- \\ \\ :\ddot{Cl}: \end{array} \longrightarrow \left[\begin{array}{c} :\ddot{Cl}: \\ :\ddot{Cl}:Be:\ddot{Cl}: \\ :\ddot{Cl}: \end{array} \right]^{2-}
$$

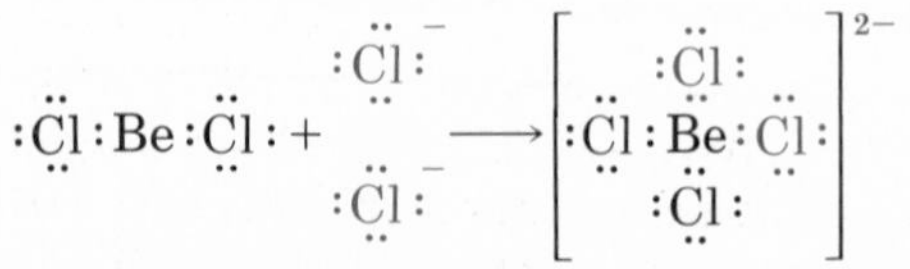

BCl_3 reacts with NH_3 in the following way:

$$
\begin{array}{ccc}
:\ddot{Cl}: & H & \\
| & | & \\
:\ddot{Cl}—B & +:N—H & \longrightarrow \\
| & | & \\
:\ddot{Cl}: & H &
\end{array}
\begin{array}{c}
Cl\ H \\
|\ \ | \\
Cl—B:N—H \\
|\ \ | \\
Cl\ H
\end{array}
$$

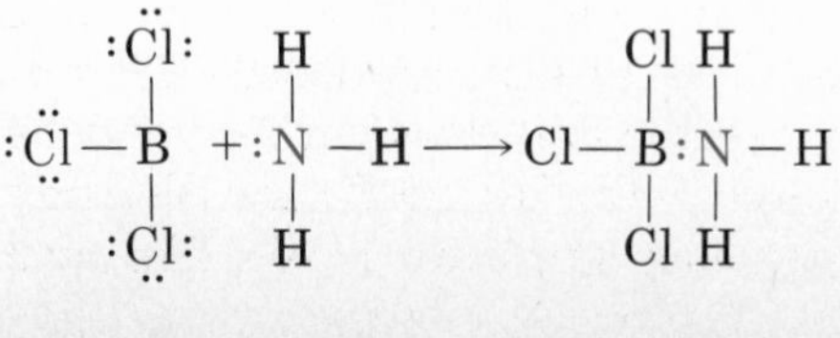

Examples of molecules with more than an octet of electrons are phosphorus pentafluoride (PF_5) and sulfur hexafluoride (SF_6). Phosphorus pentafluoride is a gas at room temperature. It consists of PF_5 molecules in which each fluorine atom is bonded to the phosphorus atom. Since each bond corresponds to a shared pair of electrons, the Lewis structure is

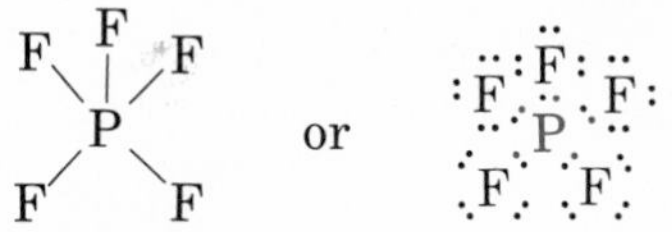

Instead of an octet the phosphorus atom has 10 electrons in its valence shell. Sulfur hexafluoride (also a gas) consists of SF_6 molecules. Its structure is

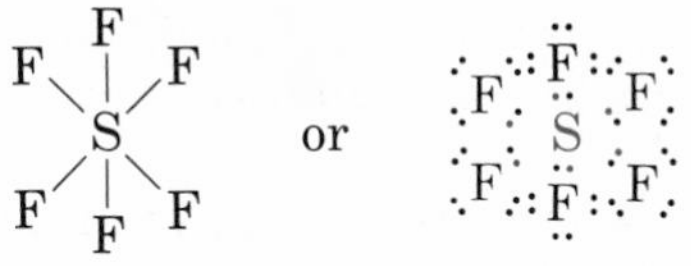

Here the sulfur atom has six electron pairs in its valence shell.

An atom like phosphorus or sulfur which has more than an octet is said to have *expanded its valence shell.* This can only occur when the valence shell has enough orbitals to accommodate the extra electrons. For example, in the case of phosphorus, the valence shell has a principal quantum number $n = 3$. An octet would be $3s^2 3p^6$. However, the $3d$ subshell is also available, and some of the $3d$ orbitals may also be involved in bonding. This permits the extra pair of electrons to occupy the valence ($n = 3$) shell of phosphorus in PF_5.

Expansion of the valence shell is impossible for an atom in the second period because there is no such thing as a $2d$ orbital. The valence ($n = 2$) shell of nitrogen, for example, consists of the $2s$ and $2p$ subshells only. Thus nitrogen can form NF_3 (in which nitrogen has an octet) but not NF_5. Phosphorus, on the other hand, forms both PF_3 and PF_5, the latter involving expansion of the valence shell to include part of the $3d$ subshell.

7.2 THE SHAPES OF MOLECULES

The location in three-dimensional space of the nucleus of each atom in a molecule defines the **molecular shape** or **molecular geometry.** Molecular shapes are important in determining macroscopic properties such as melting and boiling points, and in predicting the ways in which one molecule can react with another. A number of experimental methods are available for finding molecular geometries, but we will not describe them here. Instead we will concentrate on several rules based on Lewis diagrams which will allow you to *predict* molecular shapes.

To provide specific cases which illustrate these rules, "ball-and-stick" models for the four molecules discussed in the previous section are shown in Fig. 7.1. In addition to $BeCl_2$, BCl_3, PF_5, and SF_6, we have included CCl_4, a molecule which does obey the octet rule. The atoms (spheres) in each

(a) Formula	(b) Molecular geometry	(c) Number of electron pairs	(d) Predicted bond angles
$BeCl_2$	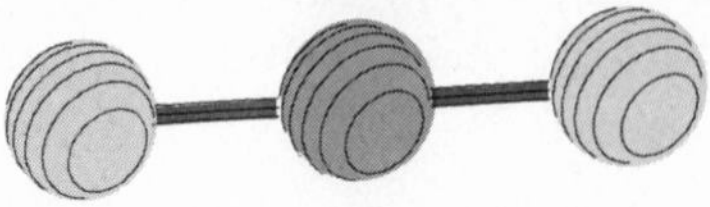	2	180°
BCl_3		3	120°
CCl_4	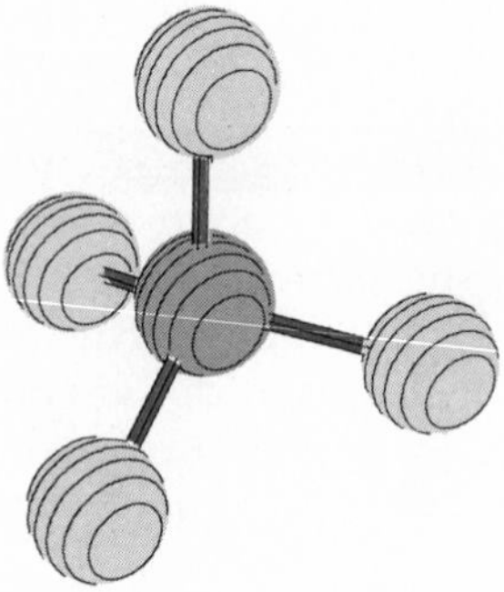	4	109.5°
PF_5	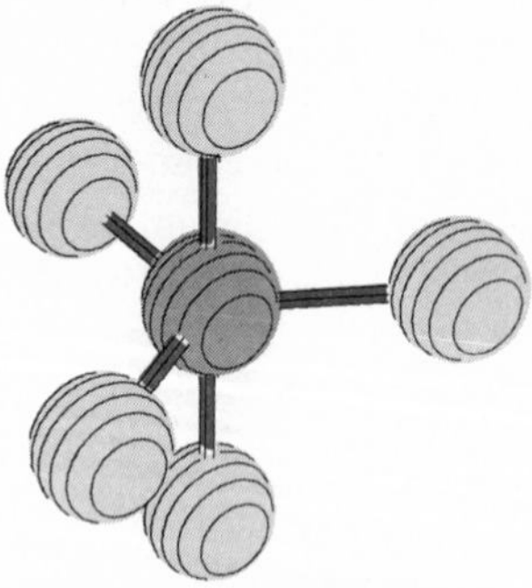	5	120°, 90°
SF_6	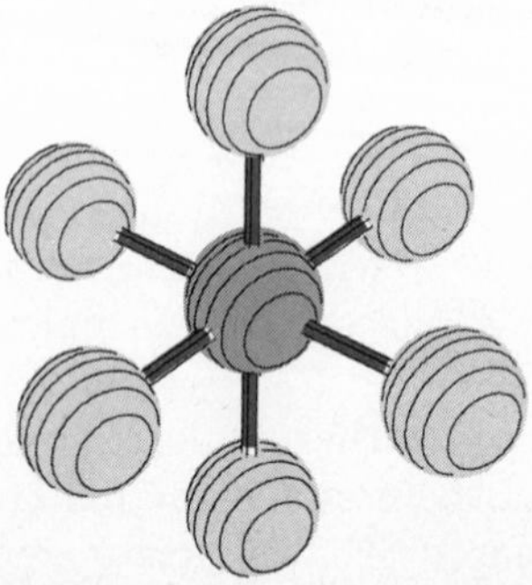	6	90°

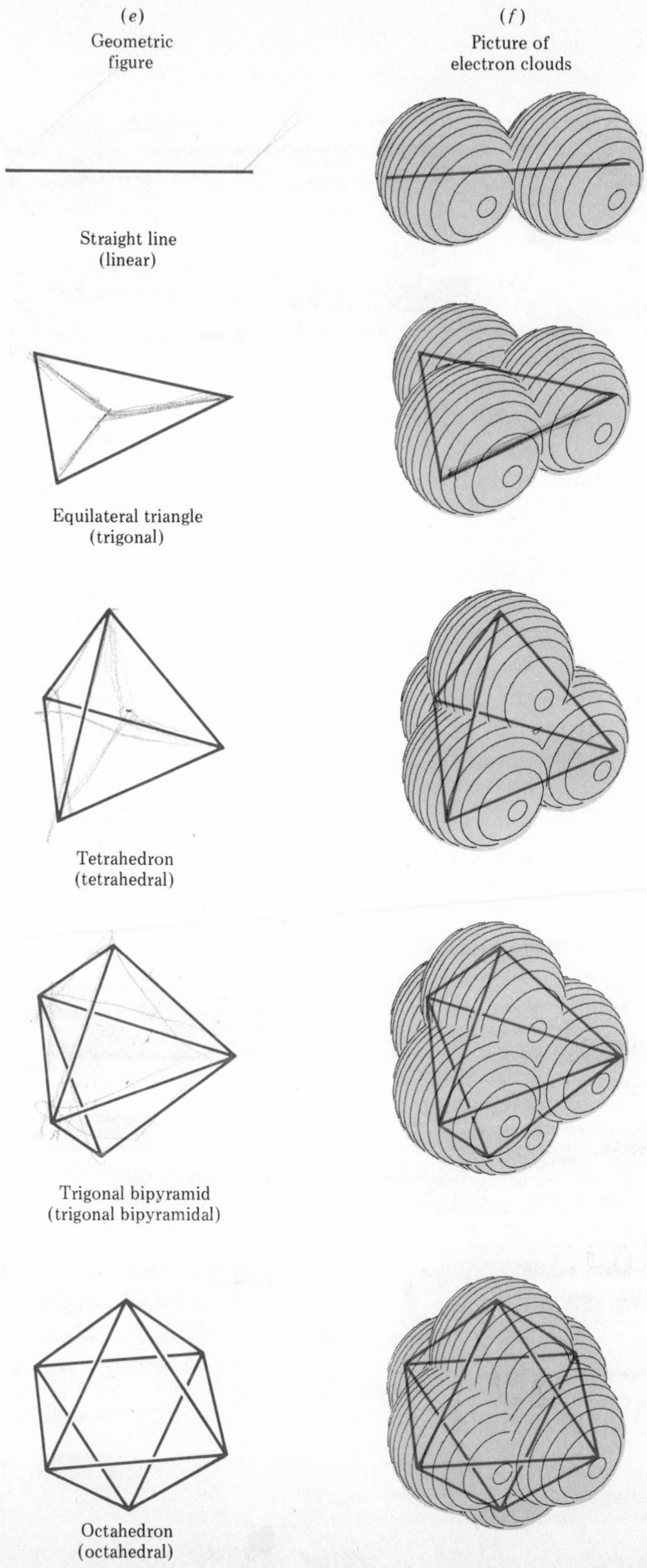

Figure 7.1 The shapes and geometries of molecules which contain no lone pairs. In each case the shape adopted is the one in which the electron-pair bonds are as far apart as possible. Parts (*b*) and (*f*) are computer generated. (*Copyright © 1976 by W. G. Davies and J. W. Moore.*)

ball-and-stick model are held together by bonds (sticks). These *electron-pair bonds determine the positions of the atoms and hence the molecular geometry*.

In each of the molecules shown in Fig. 7.1 the electron-pair bonds are arranged so that they avoid each other in space to the maximum possible extent. This may be understood in terms of the repulsion between electron clouds due to their like charges. During the 1950s the Australian R. S. Nyholm (1917 to 1971) and the Canadian R. J. Gillespie (1924 to) summed up this behavior in terms of the *v*alence-*s*hell–*e*lectron-*p*air *r*epulsion (VSEPR) theory. The VSEPR theory states that, because of their mutual repulsions, *valence electron pairs surrounding an atom stay as far as possible from one another*.

A simple model for demonstrating the behavior of electron pairs under the influence of their mutual repulsion is provided by a set of spherical balloons of equal size. It is a model that you can easily make for yourself. If, say, four balloons are tied together so that they squeeze each other fairly tightly, they inevitably adopt the *tetrahedral* arrangement shown for CCl_4 in col. *f* of Fig. 7.1 Although it is possible to flatten the balloons on a table until they are all in the same plane, they invariably spring back to the tetrahedral configuration as soon as the pressure is removed. A similar behavior is found if two, three, five, or six balloons are tightly tied together, except that in each case a different stable shape is adopted once the balloons are left to themselves. The overall appearance of the balloons is shown in col. *f* of Fig. 7.1 In col. *e* a sketch of the geometrical figure to which these shapes correspond is also drawn.

Since all the shapes described in Fig. 7.1 constantly recur in chemical discussions, it is worth being able to recall them and their names without hesitation. To this end we will discuss the geometry of each of the five molecules.

In **$BeCl_2$** the central Be atom has only two electron pairs in its valence shell. These are arranged on opposite sides of the Be atom in a straight line, and they bond the two Cl atoms to the Be atom. Thus the three nuclei are all in a **straight line,** and the Cl—Be—Cl angle is 180°. A molecule such as $BeCl_2$, whose atoms all lie on the same straight line, is said to be **linear.**

In **BCl_3** the three valence electron pairs, and hence the three Cl nuclei, are arranged in an **equilateral triangle** around the B atom. Each Cl—B—Cl angle is 120°, and all four nuclei (B included) lie in the same plane. The three Cl atoms are said to be **trigonally** arranged around B.

In **CCl_4** the four Cl nuclei are at the four corners of a geometric figure called a **tetrahedron.** A tetrahedron has six equal edges, four equilateral triangular faces, and four identical corners (apices). The C nucleus lies in the exact center of the tetrahedron, equidistant from each corner. All the Cl—C—Cl angles are the same, namely, 109.5°. This important angle is called the **tetrahedral angle.** The four Cl atoms are said to be **tetrahedrally** arranged around the C atom. This tetrahedral arrangement is the most important of the five described in Fig. 7.1.

In **PF_5** the five F nuclei are arranged at the corners of a **trigonal bipyramid.** As drawn in the figure, one F atom lies directly above the P atom and one directly below. The remaining three F atoms are arranged in a triangle around the middle of the P. Some of the F—P—F angles are 90°, while others are 120°.

In $\mathbf{SF_6}$ the six F atoms are arranged at the corners of an **octahedron.** An octahedron has twelve edges, eight equilateral triangular faces, and six identical corners. The name octahedron is derived from the eight faces, but it is usually the *six corners* of this figure which are of interest to chemists. Thus you will have to remember that an octahedral arrangement involves six atoms, not the eight that the name seems to imply.

In SF_6 the six F atoms are **octahedrally** arranged around the S. All the F—S—F angles are 90°. Octahedral arrangements are quite common in chemistry. In crystals of LiH and NaCl, for instance (see Fig. 6.3), six anions are arranged octahedrally around each cation while six cations are arranged octahedrally around each anion.

Molecules with Lone Pairs

The VSEPR theory is also able to explain and predict the shapes of molecules which contain lone pairs. In such a case the lone pairs as well as the bonding pairs are considered to repel and avoid each other. For example, since there are two bonds in the $SnCl_2$ molecule, one might expect it to be linear like $BeCl_2$. If we draw the Lewis diagram, though, we find a *lone pair* as well as two bonding pairs in the valence shell of the Sn atom:

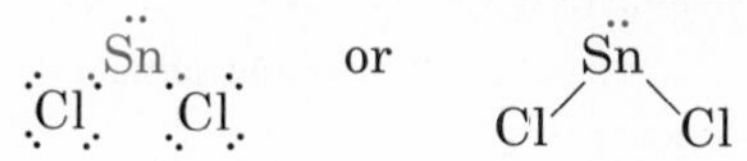

Ideally the three pairs of electrons should arrange themselves trigonally around the Sn atom, giving an angle of 120° between electron pairs, and hence between the two Cl atoms. Experimental measurements on $SnCl_2$ reveal that the molecule is **angular** or V-shaped, as shown, but the Cl—Sn—Cl angle is significantly *less* than the predicted 120°.

This smaller angle occurs because a lone pair of electrons is always "fatter" than a bonding pair. That is, a lone pair is like a bigger balloon which takes up more room and squeezes the bonding pairs closer together. This decreases the angle between bonding pairs in $SnCl_2$, and hence between the bonded Cl atoms, from the expected value of 120°. The "fatness" of a lone pair compared with a bonding pair is shown in Fig. 7.2.

A lone pair also affects the structure of ammonia, NH_3. Since this molecule obeys the octet rule, the N atom is surrounded by *four* electron pairs:

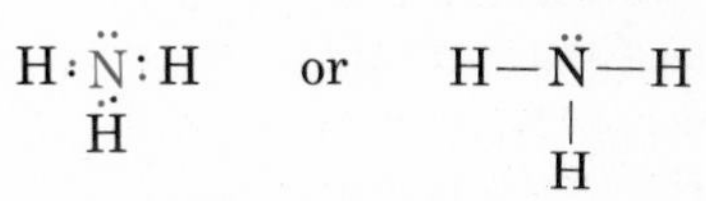

If these pairs were all equivalent, we would expect the angle between them to be the regular tetrahedral angle of 109.5°. Experimentally, the angle is found to be somewhat less, namely, 107°. Again this is because the lone pair is "fatter" than the bonding pairs and able to squeeze them closer together.

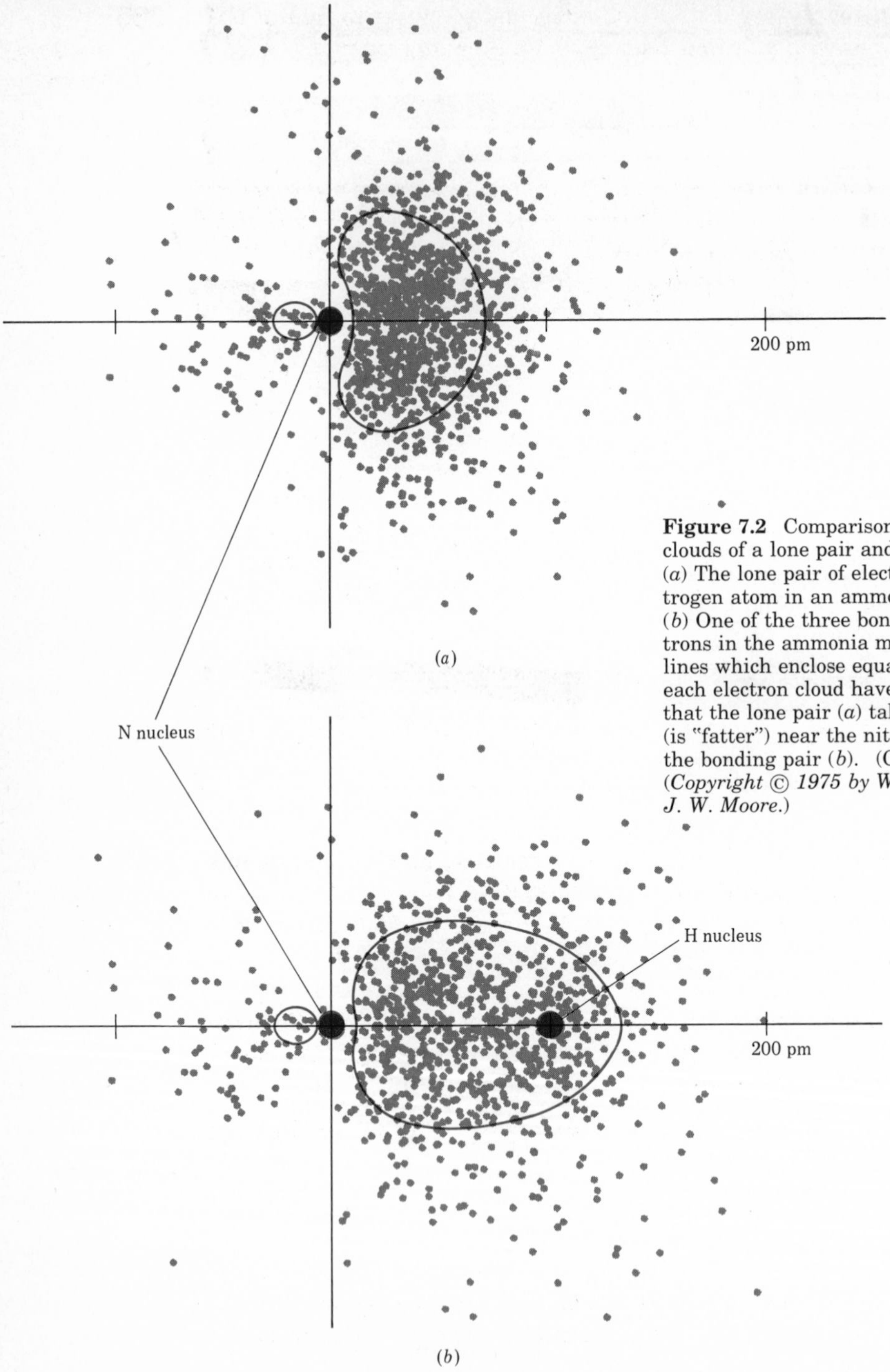

Figure 7.2 Comparison of the electron clouds of a lone pair and a bonding pair. (*a*) The lone pair of electrons on the nitrogen atom in an ammonia molecule, NH_3. (*b*) One of the three bonding pairs of electrons in the ammonia molecule. Boundary lines which enclose equal percentages of each electron cloud have been drawn. Note that the lone pair (*a*) takes up more space (is "fatter") near the nitrogen nucleus than the bonding pair (*b*). (Computer generated.) (*Copyright © 1975 by W. G. Davies and J. W. Moore.*)

The electronic structure of the H_2O molecule is similar to that of NH_3 except that one bonding pair has been replaced by a lone pair:

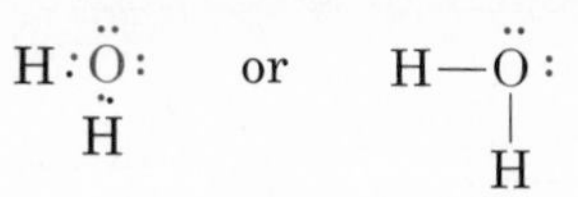

Here there are two "fat" lone pairs, and so the bonding pairs are squeezed even closer together than in NH_3. The H—O—H angle is found to be 104.5°.

The structures of $BeCl_2$, BCl_3, $SnCl_2$, CCl_4, NH_3, and H_2O include all the combinations of lone pairs and bonding pairs and all molecular shapes which are possible for two, three, and four pairs of electrons. These shapes, together with details of their geometries, are summarized in Fig. 7.3. Again, because of their frequent occurrence, it is wise to commit these to memory. Note in particular that the shape of a molecule is described in terms of the geometry of the *nuclei* and not of the electron clouds. For example, the shape of the NH_3 molecule is described as a *trigonal pyramid* since the N nucleus forms the apex of a pyramid, slightly above an equilateral triangle of H nuclei. Although the electron-pair clouds are arranged in an approximate tetrahedron around the N nucleus, it is incorrect to describe the molecular shape as tetrahedral. The atomic *nuclei* are not at the corners of a tetrahedron.

EXAMPLE 7.1 Sketch and describe the geometry of the following molecules: (a) $GaCl_3$, (b) $AsCl_3$, and (c) $AsOCl_3$.

Solution

a) Since the element gallium belongs to group III, it has three valence electrons. The Lewis diagram for $GaCl_3$ is thus

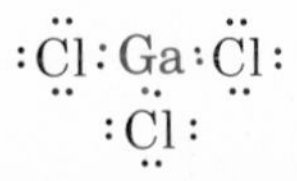

Since there are three bonding pairs and no lone pairs around the Ga atom, we conclude that the three Cl atoms are arranged *trigonally* and that all four atoms are in the same plane.

b) Arsenic belongs to group V and therefore has five valence electrons. The Lewis structure for $AsCl_3$ is thus

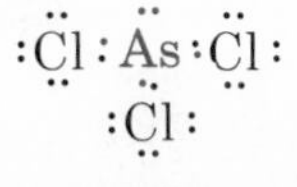

Since a lone pair is present, the shape of this molecule is a trigonal pyramid, with the As nucleus a little above an equilateral triangle of Cl nuclei.

c) The Lewis diagram for $AsOCl_3$ is similar to that of $AsCl_3$:

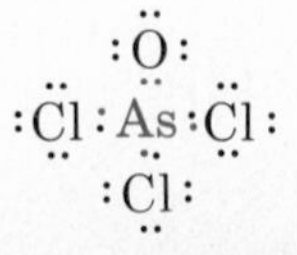

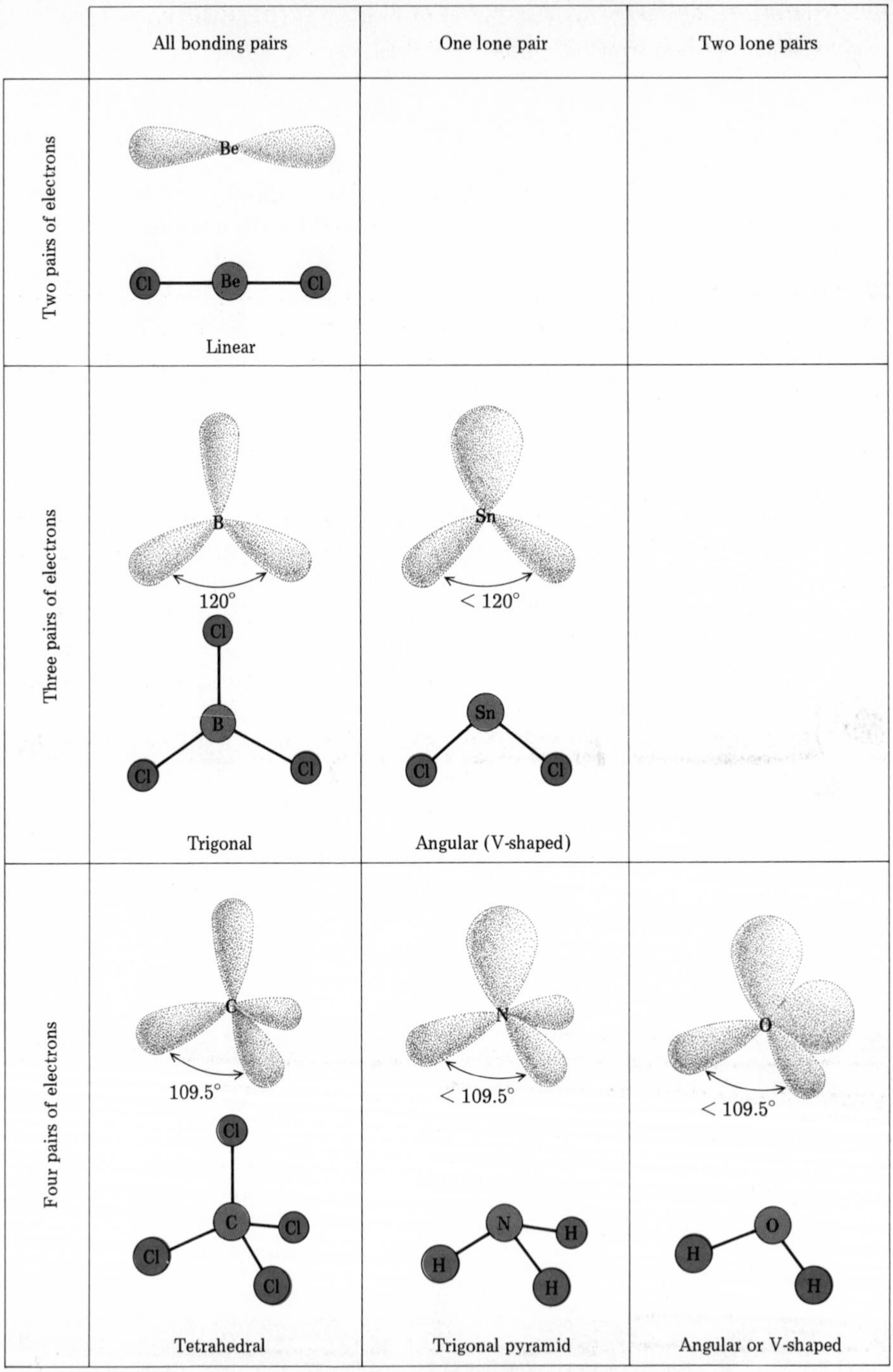

Figure 7.3 The arrangement of electron pairs and the shapes of molecules which contain lone pairs. Bonding pairs are indicated in color and have purposely been made very thin for diagrammatic effect. Lone pairs are indicated in gray. Note that the geometry of these molecules is described in terms of the *nuclei* and not of the electron pairs; i.e., it is described in terms of the *ball-and-stick* diagrams shown in the figure.

Since there are four bonding pairs, the molecule is tetrahedral.
Sketches of each of these molecules are

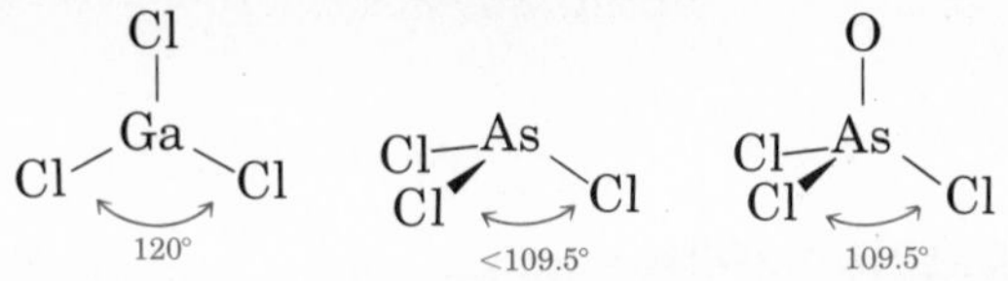

The VSEPR theory can also be applied to molecules which contain five and six pairs of valence electrons, some of which are lone pairs. We have not included such species here because the majority of compounds fall into the categories we have described.

Multiple Bonds and Molecular Shapes

In a double bond, two electron pairs are shared between a pair of atomic nuclei. Despite the fact that the two electron pairs repel each other, they must remain between the nuclei, and so they cannot avoid each other. Therefore, for purposes of predicting molecular geometry, the two electron pairs in a double bond behave as one. They will, however, be somewhat "fatter" than a single electron-pair bond. For the same reason the three electron pairs in a triple bond behave as an "extra-fat" bond.

As an example of the multiple-bond rules, consider hydrogen cyanide, HCN. The Lewis structure is

$$H{-}C{\equiv}N{:}$$

Treating the triple bond as if it were a single "fat" electron pair, we predict a linear molecule with an H—C—N angle of 180°. This is confirmed experimentally. Another example is formaldehyde, CH_2O, whose Lewis structure is

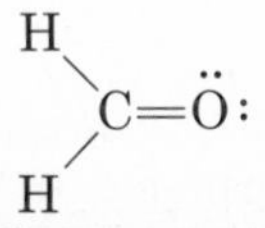

Since no lone pairs are present on C, the two H's and the O should be arranged trigonally, with all four atoms in the same plane. Also, because of the "fatness" of the double bond, squeezing the C—H bond pairs together, we expect the H—C—H angle to be slightly less than 120°. Experimentally it is found to have the value of 117°.

EXAMPLE 7.2 Predict the shape of the two molecules (a) nitrosyl chloride, NOCl, and (b) carbon dioxide, CO_2.

Solution

a) We must first construct a skeleton structure and then a Lewis diagram. Since N has a valence of 3, O a valence of 2, and Cl is monovalent, a probable structure for NOCl is

$$O{=}N{-}Cl$$

Completing the Lewis diagram, we find

$$:\ddot{O}::\ddot{N}:\ddot{\underset{..}{Cl}}:$$

Since N has two bonds and one lone pair, the molecule must be *angular*. The O—N—Cl angle should be about 120°. Since the "fat" lone pair would act to reduce this angle while the "fat" double bond would tend to increase it, it is impossible to predict at this level of argument whether the angle will be slightly larger or smaller than 120°.

b) The Lewis structure of CO_2 was considered in the previous chapter and found to be

$$:\ddot{O}{=}C{=}\ddot{O}:$$

Since C has no lone pairs in its valence shell and each double bond acts as a fat bond pair, we conclude that the two O atoms are separated by 180° and that the molecule is *linear*.

7.3 ORBITALS CONSISTENT WITH MOLECULAR SHAPES

In Chap. 6 we showed that a covalent bond results from an overlap of atomic orbitals—usually one orbital from each of two bonded atoms. Maximum bond strength is achieved when maximum overlap occurs. When we try to integrate this idea of orbital overlap with VSEPR theory, however, a problem arises. For example, the 2*s* and 2*p* atomic orbitals shown in Plate 5 for a C atom are either spherically symmetrical (2*s*) or dumbbell shaped at angles of 90° to each other ($2p_x 2p_y$). The VSEPR theory predicts that the C—Cl bonds in CCl_4 are oriented at angles of 109.5° from one another and that all four C—Cl bonds are equivalent. If each C—Cl bond is formed by overlap of Cl orbitals with C 2*s* and 2*p* orbitals, it is hard to understand how four equivalent bonds can be formed. It is also difficult to see why the angles between bonds are 109.5° rather than the 90° angle between *p* orbitals.

sp Hybrid Orbitals

To answer these questions we first need to consider a simpler case, that of beryllium chloride. As we have already seen in Fig. 7.1, VSEPR theory predicts a linear $BeCl_2$ molecule with a Cl—Be—Cl angle of 180°, in agreement with experiments. However, if we associate two valence electron pairs with a beryllium atom and place them in the lowest energy orbitals, we obtain the configuration $2s^2 2p_x^2$. (Note that since the electrons must be *paired* in order to form bonds, they do not obey Hund's rule.) The electron density distribution of such a configuration is shown in Fig. 7.4*a*. The 2*s* cloud is indicated in red, and the 2*p* cloud in gray.

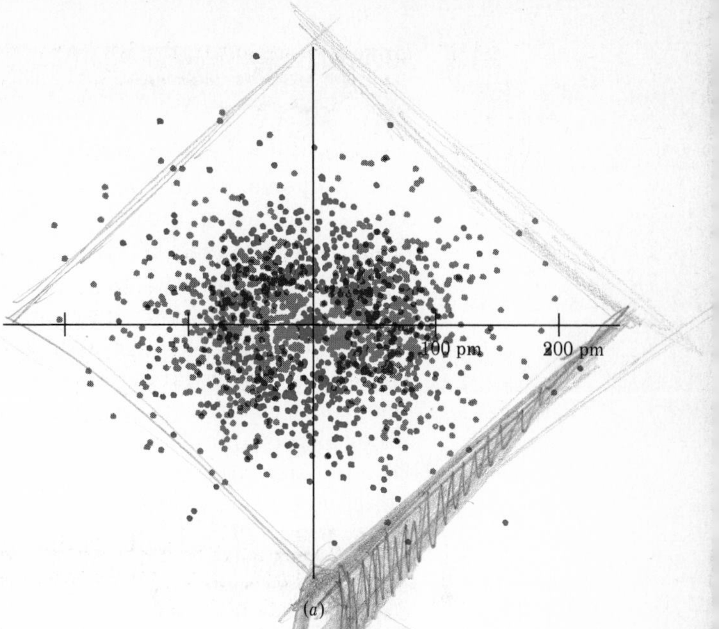

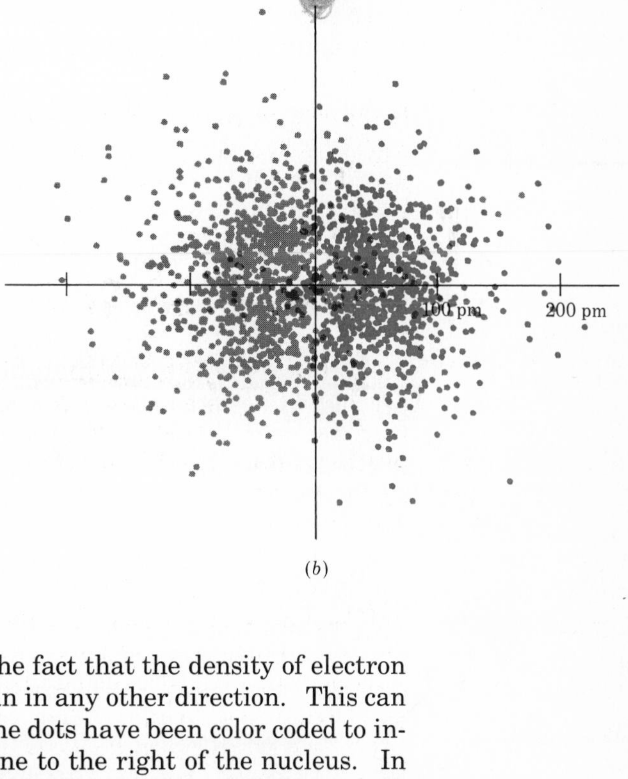

Figure 7.4 Electron-density distribution for the valence electron configuration $2s^2 2p_x^2$. (*a*) Color coded to show *s* (red) and p_x (gray) electron densities. (*b*) Color coded to show left-pointing and right-pointing *sp* hybrid orbital electron densities. (Computer generated.) (*Copyright © 1975 by W. G. Davies and J. W. Moore.*)

An important aspect of Fig. 7.4*a* is the fact that the density of electron probability is greater along the *x* axis than in any other direction. This can be seen more clearly in Fig. 7.4*b* where the dots have been color coded to indicate one electron pair to the left and one to the right of the nucleus. In this case we should certainly predict that chlorine atoms bonded to the beryllium through these electron pairs would lie on a straight line, the *x* axis. Each of the two orbitals whose electron densities are shown in Fig. 7.4*b* is called an ***sp* hybrid.** The word *hybrid* indicates that each orbital is derived from two or more of the atomic orbitals discussed in Chap. 5, and the designation *sp* indicates that a single *s* and a single *p* orbital contributed to each *sp* hybrid.

Careful comparison of the *s* and *p* electron densities with those of the two *sp* hybrid orbitals will reveal another important fact. For every dot in Fig. 7.4*a*, there is a corresponding dot (in the same location) in Fig. 7.4*b*. That is, the overall electron density (due to the four electrons occupying two orbitals) is exactly the same in both cases. We have not created something new with the two *sp* hybrids. Rather, we are looking at the same electron density, but we have color coded it to emphasize its concentration along the *x* axis. In actual fact all electrons are identical—we cannot distinguish one from another experimentally. Labeling one electron cloud as *s* and another as *p* is an aid to our thinking, just as color-coding one *sp*-hybrid electron density red and the other gray is, but it is the *overall electron density* which determines the experimentally observable molecular geometry. In other words, it does not matter *to the molecule* whether we think of the total electron cloud as being formed from *s* and *p* orbitals or from *sp* hybrids, but it does matter *to us*. It is much easier to think of two Be—Cl bonds separated by 180° in terms of *sp* hybrids than in terms of separate *s* and *p* orbitals. By contrast it is much easier to explain the periodic table by using *s* and *p* orbitals rather than hybrids. Since both correspond to the same physical reality, we can use whichever approach suits us best.

When beryllium forms a linear molecule such as beryllium chloride, it is not the *sp* hybrids themselves that form the two bonds but rather an *overlap* between each of these orbitals and some orbital on each other atom. The situation is shown schematically in Fig. 7.5. As a result of each orbital overlap, there is a concentration of electron density between two nuclei. This pulls the nuclei together and forms a covalent bond.

sp^2 Hybrid Orbitals

Somewhat more complex hybrid orbitals are found in BCl_3, where the boron is surrounded by *three* electron pairs in a trigonal arrangement. If we place these three electron pairs in the valence shell of boron, one 2*s* and two 2*p* orbitals ($2p_x$ and $2p_y$, for example) will be filled, giving the electron configuration $2s^22p_x^22p_y^2$. This configuration is illustrated in Plate 6*a* as a dot-density diagram in which each electron pair is represented in a different color. Immediately below this diagram is another (Plate 6*b*) which is dot-for-dot the same as the upper diagram but with a different color-coding. The *total* electron density distribution is the same in both diagrams, but in the bottom diagram the three electron pairs are distributed in a trigonal arrangement. All three electron clouds are identical in shape, and they are

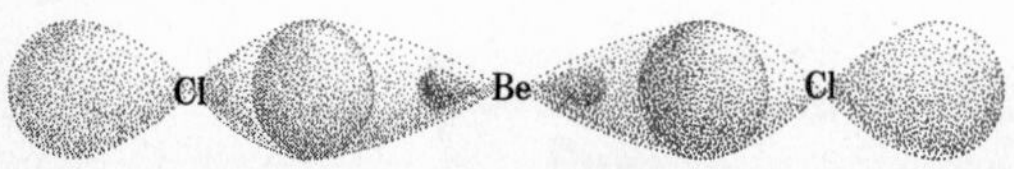

Figure 7.5 Schematic representation of the bonding in $BeCl_2$. Each of the two *sp* hybrids around the Be overlaps with an orbital from a chlorine atom. The result is a concentration of negative charge *between* the beryllium and each of the chlorine nuclei, holding them together.

oriented at angles of 120° with respect to each other. Again we have a set of hybrid orbitals, but this time they are derived from one *s* orbital and *two p* orbitals. Accordingly these orbitals are known as ***sp*² hybrids.**

Because electrons are indistinguishable, and because both the upper and the lower diagrams in Plate 6 correspond to an identical total electron density, we are entitled to use either formulation when it suits our purposes. To explain the trigonal geometry of BCl_3 and similar molecules, the hybrid picture is obviously more suitable than the *s* and *p* picture. The three electron-pair bonds usually formed by boron result from an overlap of each of these three sp^2 hybrids with a suitable orbital in each other atom.

sp^3 Hybrid Orbitals

In addition to the two hybrids just considered, a third combination of *s* and *p* orbitals, called ***sp*³ hybrids,** is possible. As the name suggests, sp^3 hybrids are obtained by combining an *s* orbital with *three p* orbitals (p_x, p_y, and p_z). Suppose we have two *s* electrons, two p_x electrons, two p_y electrons, and two p_z electrons, as shown by the boundary-surface diagrams in Fig. 7.6. When these are all arranged around the nucleus, the total electron cloud is essentially spherical. The boundary-surface diagrams in Fig. 7.6 show a slightly "bumpy" surface, but we must remember that electron clouds are fuzzy and do not stop suddenly at the boundary surface shown. When this fuzziness is taken into account, the four atomic orbitals blend into each other perfectly to form an exactly spherical shape. This blending of *s* and *p* orbitals is much the same as that discussed in the previous cases of *sp* and sp^2 hybrids. It can be clearly seen in two dimensions in Fig. 7.4 or Plate 6*a*.

The total electron probability cloud in Fig. 7.6 can be subdivided in a different way—into four equivalent sp^3 hybrid orbitals, each occupied by two electrons. Each of these four hybrid orbitals has a similar appearance to each of the *sp* and sp^2 hybrids encountered previously, but the sp^3 hybrids are arranged *tetrahedrally* around the nucleus. The four sp^3 orbitals, each occupied by two electrons, also appear bumpy in a boundary-surface diagram, but when the fuzziness of the electron clouds is taken into account, the result is a spherical electron cloud equivalent in every way to an ns^2np^6 configuration. We are thus equally entitled to look at an octet of electrons in terms of four sp^3 hybrids, each doubly occupied, or in terms of one *s* and three *p* orbitals, each doubly occupied. Certainly, nature cannot tell the difference!

According to the VSEPR theory, if an atom has an octet of electrons in its valence shell, these electrons are arranged tetrahedrally in pairs around it. Since sp^3 hybrids also correspond to a tetrahedral geometry, they are the obvious choice for a wave-mechanical description of the octets we find in Lewis structures. Consider, for example, a molecule of methane, CH_4, whose Lewis structure is

```
   H
   ..
H : C : H
   ..
   H
```

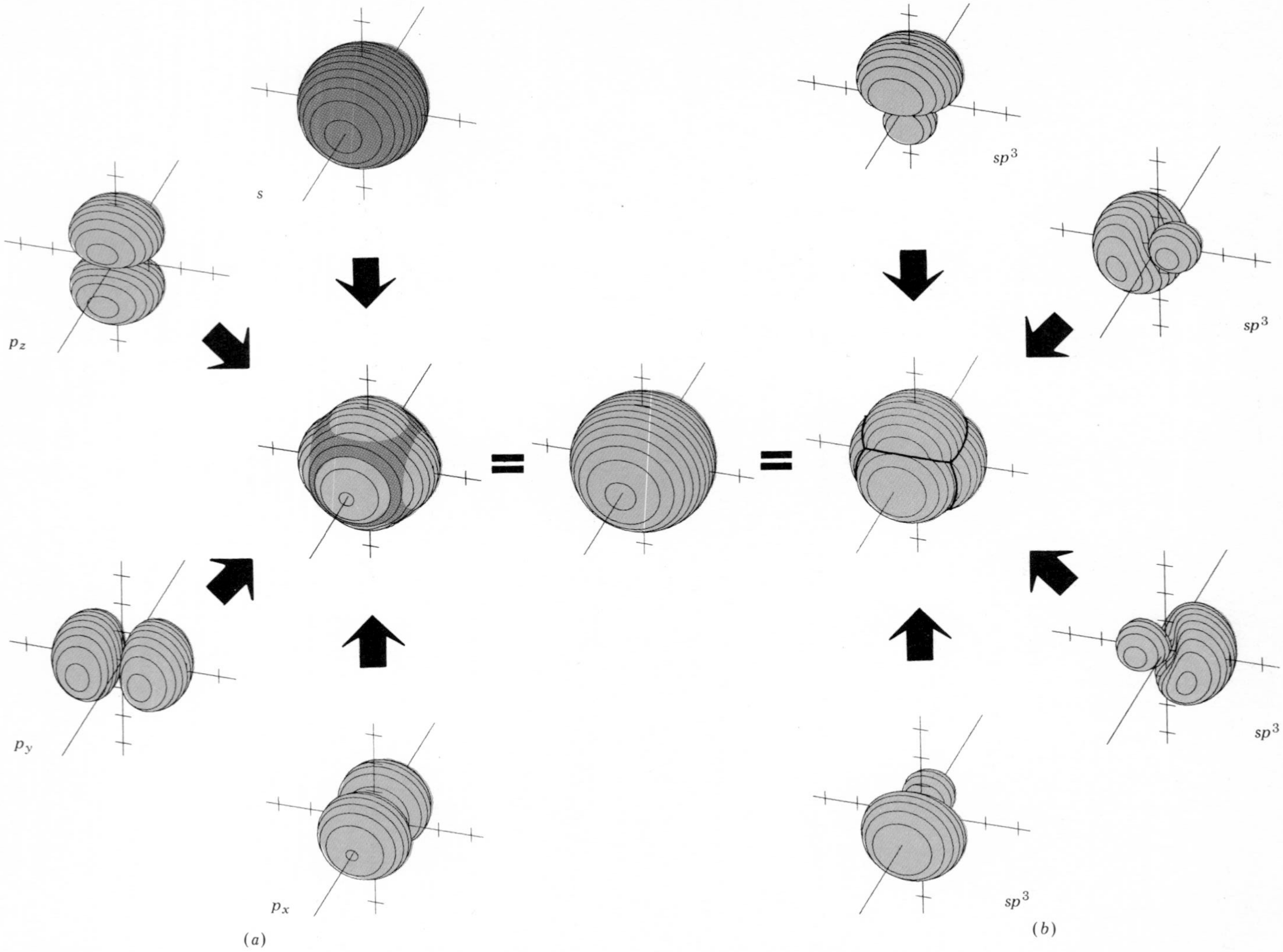

Figure 7.6 An octet may be thought of as *either:* (*a*) an *s* orbital, a p_x orbital, a p_y orbital, and a p_z orbital, each occupied by a pair of electrons; or (*b*) four equivalent sp^3 hybrid orbitals arranged tetrahedrally around the nucleus, each occupied by a pair of electrons. Both formulations are physically equivalent. (Computer-generated.) (*Copyright © 1977 by W. G. Davies and J. W. Moore.*)

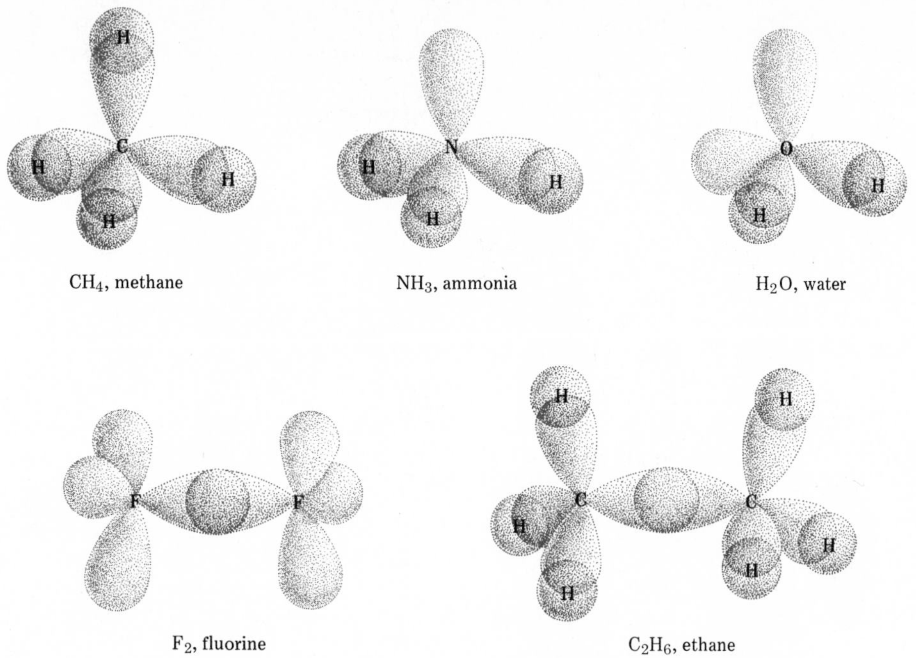

Figure 7.7 The use of sp^3 hybrid orbitals in explaining the bonding in molecules which obey the octet rule. Each octet is represented by four sp^3 hybrids. These hybrids can either overlap with the 1*s* orbital of H or with another sp^3 hybrid as in F—F and $H_3C—CH_3$. Alternatively they can form lone pairs (shown in gray), in which case no overlap occurs.

Here the eight electrons of the octet can be thought of as occupying four orbitals, each formed by the overlap of an sp^3 hybrid centered on the C atom and a 1*s* orbital belonging to an H atom. A schematic diagram of this bonding scheme is shown in Fig. 7.7.

Also shown in this figure are four other simple molecules whose Lewis structures are

$$\mathrm{H}:\overset{\cdot\cdot}{\underset{\cdot\cdot}{\mathrm{N}}}:\mathrm{H} \qquad \mathrm{H}:\overset{\cdot\cdot}{\underset{\cdot\cdot}{\mathrm{O}}}: \qquad :\overset{\cdot\cdot}{\underset{\cdot\cdot}{\mathrm{F}}}:\overset{\cdot\cdot}{\underset{\cdot\cdot}{\mathrm{F}}}: \qquad \mathrm{H}:\overset{\mathrm{H}}{\underset{\mathrm{H}}{\mathrm{C}}}:\overset{\mathrm{H}}{\underset{\mathrm{H}}{\mathrm{C}}}:\mathrm{H}$$

with H below N and H below O.

In each case the octets in the Lewis structure can be translated into sp^3 hybrids. These sp^3 hybrids can either overlap with the 1*s* orbitals of H or with sp^3 hybrids in other atoms. Alternatively they can form lone pairs, in which case no bond is formed and no overlap is necessary.

EXAMPLE 7.3 Suggest which hybrid orbitals should be used to describe the bonding of the central atom in the following molecules: (a) BeF_2, (b) PbF_2, (c) $SbCl_3$, and (d) $InCl_3$.

Solution In making decisions on which hybrids need to be used, we must first decide on the *shape* of the molecule from VSEPR theory. This in turn is best derived from a Lewis structure.

The Lewis structures of these four molecules are

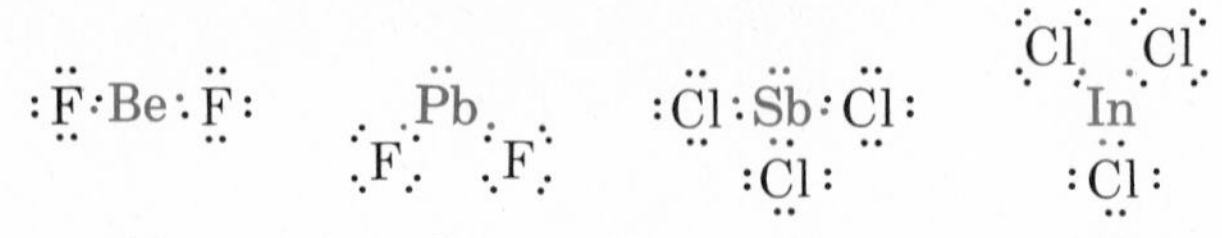

(a) (b) (c) (d)

a) In BeF_2 there are only *two* electron pairs in the valence shell of Be. These must be arranged *linearly* according to VSEPR theory thus necessitating an *sp hybrid.*

b) In PbF_2 there are *three* electron pairs in the valence shell of Pb. According to VSEPR theory these will be arranged *trigonally* and will thus utilize *sp^2 hybrids.*

c) In $SbCl_3$ there is an *octet* of electrons—*four* pairs arranged *tetrahedrally.* Accordingly *sp^3 hybrids* are needed to describe the bonding.

d) The atom of In has only *three* pairs of electrons in its valence shell. As in case *b*, this produces a *trigonal* arrangement of electron pairs and is described by *sp^2 hybridization.*

Other Hybrid Orbitals

Hybrid orbitals can also be used to describe those shapes which occur when there is more than an octet of electrons in an atom's valence shell. The combination of a *d* orbital with an *s* orbital and three *p* orbitals yields a set of five ***dsp^3* hybrids.** These dsp^3 hybrids are directed toward the corners of a trigonal bipyramid. An octahedral arrangement is possible if one more *d* orbital is included. Then ***d^2sp^3* hybrids,** all at 90° to one another, are formed. Note that as soon as we get beyond an octet (whose eight electrons fill all the *s* and *p* orbitals), the inclusion of *d* orbitals is mandatory. This can only occur for atoms in the third row of the periodic table and below. Thus bonding in PCl_5 involves dsp^3 hybrid orbitals which include a P 3*d* orbital. Bonding in SF_6 uses two S 3*d* orbitals in d^2sp^3 hybrids.

7.4 ORBITAL DESCRIPTIONS OF MULTIPLE BONDS

The association of four sp^3 hybrid orbitals with an octet can be applied to multiple bonds as well as single bonds. A simple example is ethene (ethylene), C_2H_4. The Lewis structure for this molecule is

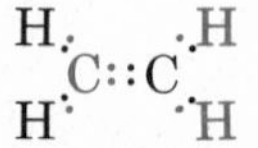

As shown in Fig. 7.8, we can look at the double bond as being formed by *two* overlaps of sp^3 hybrid orbitals, one above the plane of the molecule and one

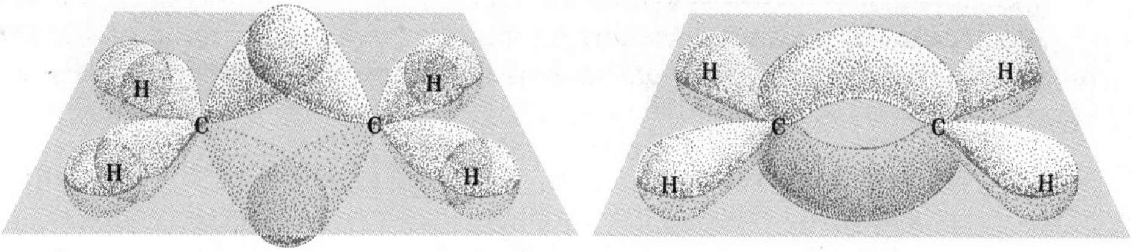

(a) (b)

Figure 7.8 The bent-bond or "banana-bond" model of a double bond. In (a) two sp^3 hybrid orbitals overlap without pointing directly at each other. The resulting bond, shown in (b), is somewhat curved and in this representation resembles two bananas. Note that the tetrahedral arrangement of bonds around each carbon atom requires that all four hydrogens and both carbons lie in the same plane.

below. Since the orbitals which overlap are not pointing directly at each other, each of these bonds is referred to as a **bent bond** or (more frivolously) as a banana bond.

In ball-and-stick models of molecules, a double bond is usually represented by two springs or by curved sticks (shown in Fig. 7.9) joining the two atoms together. In making such a model, it is necessary to bend the springs a fair amount in order to fit them into the appropriate holes in the balls. The ability to bend or stretch is characteristic of all chemical bonds—not just those between doubly bonded atoms. Thus each atom can vibrate about its most stable position. Perhaps ball-and-spring models would be more appropriate than ball-and-stick models in all cases.

The bent-bond picture makes it easy to explain several characteristics of double bonds. As we have noted in the previous chapter, the distance between two atomic nuclei connected by a double bond is shorter than if they were connected by a single bond. In the case of carbon-carbon bonds, for example, the C═C distance is 133 pm, while the C—C distance is 156 pm. This makes sense when we realize that each bent bond extends along a curved path. The distance between the ends of such a path (the C nuclei) is necessarily shorter than the path itself.

Figure 7.9 Ball-and-stick models for the three isomeric forms of $C_2H_2F_2$. Carbon atoms are dark gray, hydrogen atoms are light red, and fluorine atoms are light gray. Note that all three structures are different, although the molecular formula is always $C_2H_2F_2$. Structures (b) and (c) differ only because of the barrier to rotation around the carbon-carbon double bond. (Computer-generated.) *(Copyright © 1977 by W. G. Davies and J. W. Moore.)*

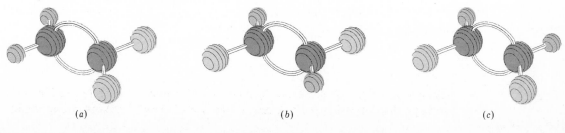

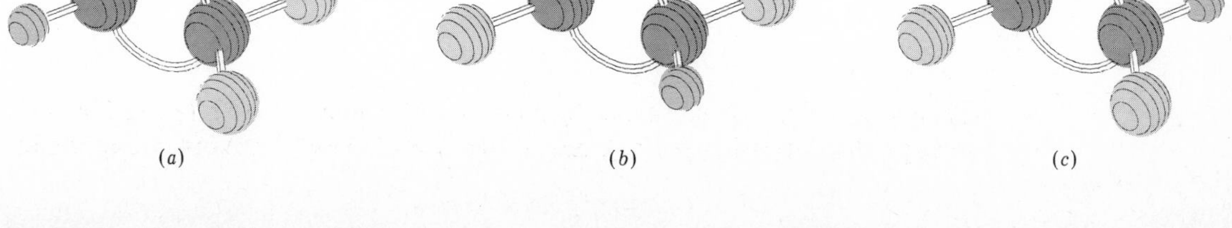

(a) (b) (c)

Another characteristic of double bonds is that they make it difficult to twist one end of a molecule relative to the other. This phenomenon usually is called a **barrier to rotation.** Such a barrier accounts for the fact that it is possible to prepare *three* different compounds with the formula $C_2H_2F_2$. Their structures are shown in Fig. 7.9. Structure (*a*) is unique because both F atoms are attached to the same C atom, but (*b*) and (*c*) differ only by a 180° flip of the right-hand =CHF group. If there were no barrier to rotation around the double bond, structures (*b*) and (*c*) could interconvert very rapidly whenever they collided with other molecules. It would then be impossible to prepare a sample containing only type (*b*) molecules or only type (*c*) molecules.

Since they have the same molecular formula, (*a*), (*b*), and (*c*) are isomers. Structure (*b*) in which the two F atoms are on *opposite* sides of the double bond is called the **trans** isomer, while structure (*c*) in which two like atoms are on the *same* side is called the **cis** isomer. It is easy to explain why there is a barrier to rotation preventing the interconversion of these cis and trans isomers in terms of our bent-bond model. Rotation of one part of the molecule about the line through the C atoms will cause one of the bent-bond electron clouds to twist around the other. Unless one-half of the double bond breaks, it is impossible to twist the molecule through a very large angle.

Sigma and Pi Bonds

An alternative, somewhat more complex, description of the double bond is the **sigma-pi** model shown in Fig. 7.10. In this case only two of the *p* orbitals on each C atom are involved in the formation of hybrids. Consequently sp^2 hybrids are formed, separated by an angle of 120°. Two of these hybrids from each C atom overlap with H 1*s* orbitals, while the third overlaps with an sp^2 hybrid on the other C atom. This overlap directly between the two C atoms is called a **sigma bond,** and is abbreviated by the Greek letter σ. A second carbon-carbon bond is formed by the overlap of the

Figure 7.10 The sigma-pi model of a double bond. Three sp^2 hybrids around each carbon atom are indicated in color. Two of these overlap directly between the carbon atoms to form the σ bond. Two p orbitals, one on each C atom, are shown in gray. These overlap sideways to form a π bond, also shown in gray.

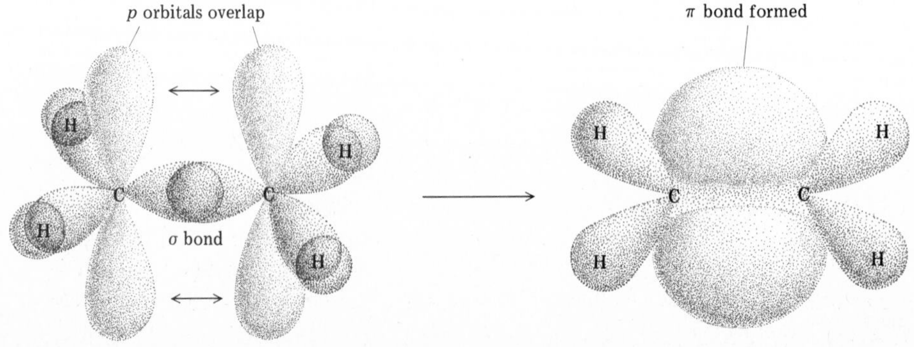

remaining p orbital on one C atom with that on the other. This is called a **pi bond,** Greek letter π. The pi bond (π bond) has two halves—one above the plane of the molecule, and the other below it. Each electron in the pi bond (π bond) spends half its time in the top part and the other half its time in the bottom part. Overall this sigma-pi picture of the double bond is reminiscent of a hot dog in a bun. The sigma bond (σ bond) corresponds to the wiener, while the pi bond corresponds to the bun on either side of it.

Although the sigma-pi picture is more complex than the bent-bond picture of the double bond, it is much used by organic chemists (those chemists interested in carbon compounds). The sigma-pi model is especially helpful in understanding what happens when visible light or other radiation is absorbed by a molecule. We will discuss this subject in some detail in Chap. 21.

In actual fact the difference between the two models of the double bond is more apparent than real. They are related to each other in much the same way as s and p orbitals are related to sp hybrids. Figure 7.11 shows two dot-density diagrams for a carbon-carbon double bond in a plane through both carbon nuclei but at right angles to the plane of the molecule. Figure 7.11a corresponds to a sigma-pi model with the sigma bond (σ bond) in color and the pi bond in gray. Figure 7.11b shows two bent bonds. Careful inspection reveals that both diagrams are dot-for-dot the same. Only the color coding of the dots is different. Thus the bent-bond and sigma-pi models of the double bond are just two different ways of dividing up the same overall electron density.

A similar situation applies to triple bonds, such as that found in a molecule of ethyne (acetylene), H—C≡C—H. As shown in Fig. 7.12a, we can regard this triple bond as being the result of three overlaps of sp^3 hybrids on different carbon atoms forming three bent bonds. Alternatively we can regard it as being composed of one sigma bond and two pi bonds, the sigma bond being due to the overlap of an sp hybrid from each carbon atom. Again both pictures of the bond correspond to the same overall electron density, and hence both are describing the same physical reality. We can use whichever one seems more convenient for the problem under consideration.

7.5 POLARITY OF BONDS: ELECTRONEGATIVITY

In the previous chapter we divided chemical bonds into two classes: covalent bonds, in which electrons are shared between atomic nuclei, and ionic bonds, in which electrons are transferred from one atom to the other. However, a sharp distinction between these two classes cannot be made. Unless both nuclei are the same (as in H_2), an electron pair is never shared *equally* by both nuclei. There is thus some degree of electron transfer as well as electron sharing in most covalent bonds. On the other hand there is never a *complete* transfer of an electron from one nucleus to another. The first nucleus always maintains some slight residual control over the transferred electron.

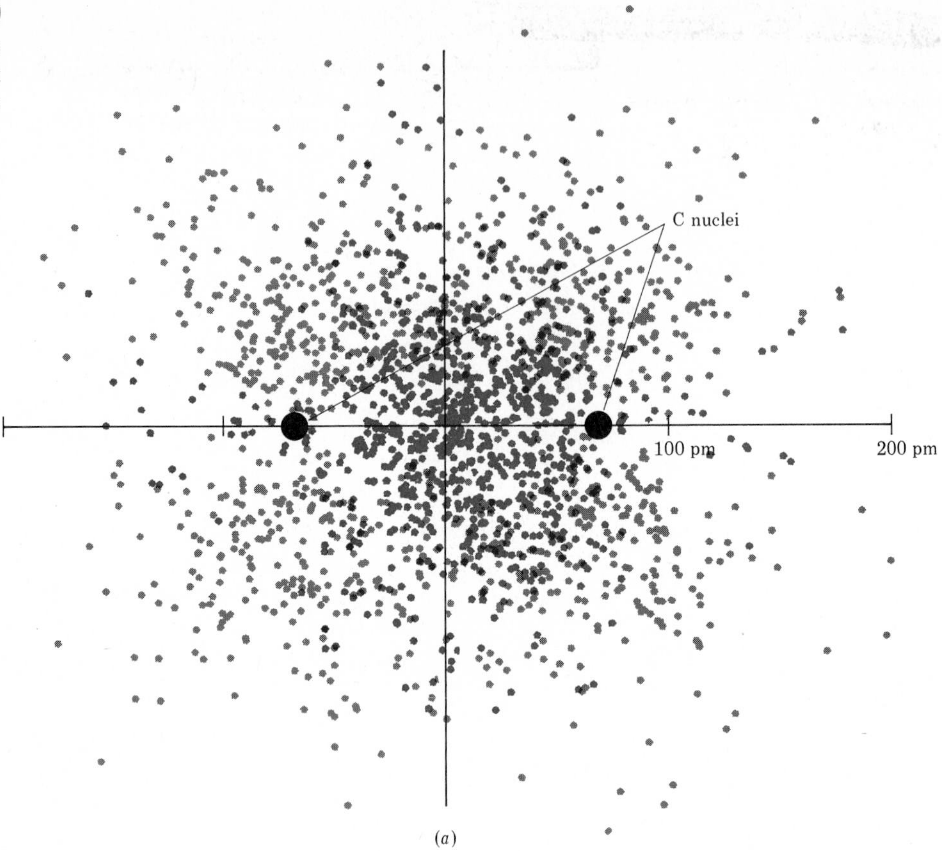

(a)

Figure 7.11 Dot-density diagrams comparing the sigma-pi and bent-bond models of the double bond. (*a*) The sigma-pi model. The σ bond is in color and the π bond is in gray. (*b*) The bent-bond model. One bond lies above the two nuclei and the other below. Since both diagrams are dot-for-dot the same, they are both describing the same physical reality. (Computer generated.) *Copyright © 1975 by W. G. Davies and J. W. Moore.*)

Figure 7.12 Two alternative models for the triple bond in ethyne, H—C≡C—H. (*a*) Three sp^3 hybrids from each carbon atom overlap to form three bent bonds. (*b*) Two *sp* hybrids overlap to form the sigma bond. Two *p* orbitals on one carbon overlap with two on the other to form two pi bonds (one in light gray, the other in dark gray). Though these two models appear to be different, the indistinguishability of electrons makes them exactly equivalent.

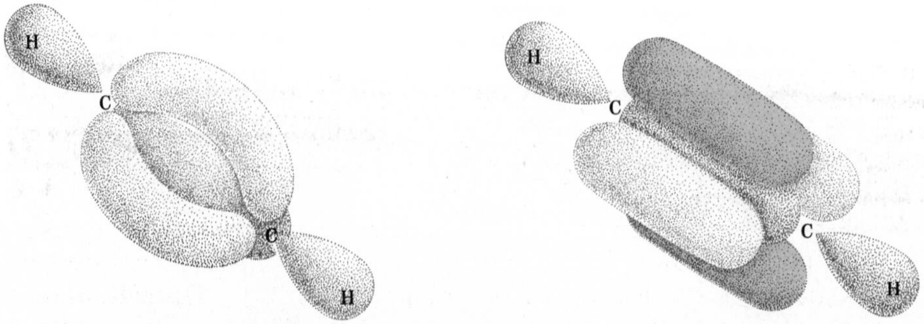

(*a*) (*b*)

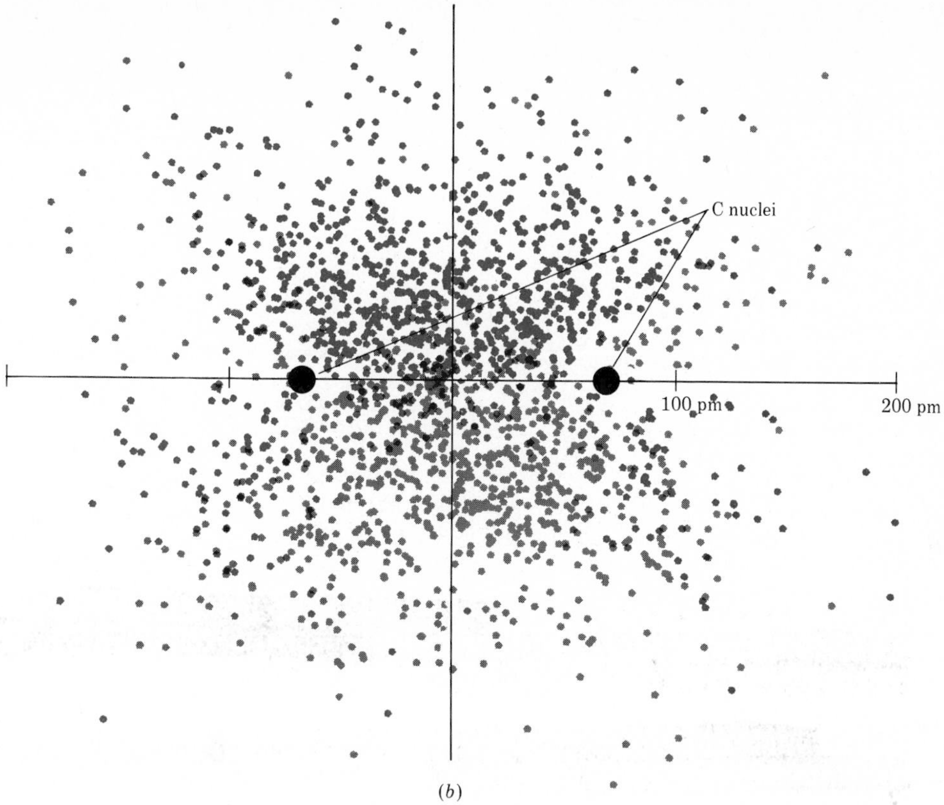

(*b*)

Polarizability

The latter phenomenon can be observed by studying carefully the lithium hydride ion pair discussed in the previous chapter. Figure 7.13 shows a dot-density diagram of the $1s^2$ electron cloud of the hydride ion, H^-, as well as the two nuclei. For the sake of clarity the two electrons around the Li nucleus have been omitted. Were it not for the presence of the Li^+ ion on the right-hand side of the diagram, we could expect a spherical (or, in the two dimensions shown, a circular) distribution of electron density around the H nucleus. As can be seen by comparing the density of dots to the left of the colored circle in Fig. 7.13 with that to the right, the actual distribution is not exactly circular. Instead the electron cloud is distorted by the attraction of the Li^+ ion so that some of the H^- $1s^2$ electron density is pulled into the bonding region between the Li and H nuclei. This contributes **partial covalent character** to the bond.

Distortion of an electron cloud, as described in the previous paragraph, is called **polarization.** The tendency of an electron cloud to be distorted from its normal shape is referred to as its **polarizability.** The polarizability of an ion (or an atom) depends largely on how diffuse or spread out its

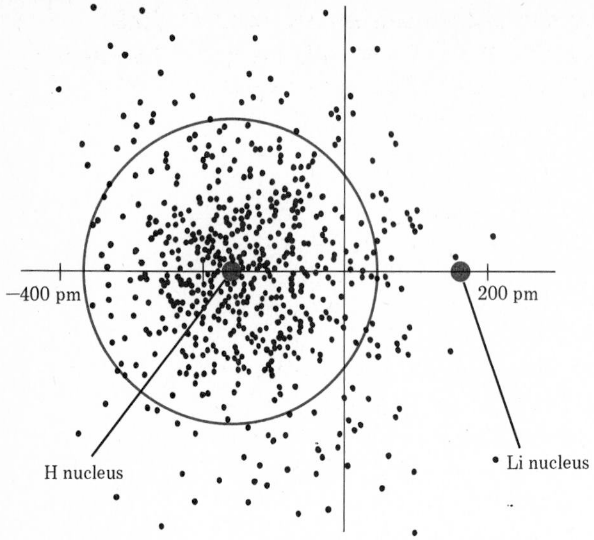

Figure 7.13 Electron dot-density diagram for valence electrons in the LiH ion pair. Note that more dots are found to the right than to the left of the colored circle centered on the H nucleus. This is the result of distortion (polarization) of the electron cloud of H^- toward Li^+. (Computer generated.) (*Copyright © 1975 by W. G. Davies and J. W. Moore.*)

electron cloud is. For example, most positive ions have relatively small radii, and their electrons are held rather tightly by the excess of protons in the nucleus. Thus their polarizabilities are usually small. Only quite large positive ions such as Cs^+ are significantly polarizable. On the other hand, negative ions have excess electrons, large radii, and diffuse electron clouds which can be polarized easily. Thus negative ions, especially large ones, have high polarizabilities. Small, highly charged positive ions can distort them quite extensively.

The slight shift in the electron cloud of H^- shown in Fig. 7.13 can be confirmed experimentally. An ion pair like LiH has a negative end (H^-) and a positive end (Li^+). That is, it has two electrical "poles," like the north and south magnetic poles of a magnet. The ion pair is therefore an electrical **dipole** (literally "two poles"), and a quantity known as its dipole moment may be determined from experimental measurements. The **dipole moment** μ is proportional to the size of the separated electrical charges Q and to the distance r between them:

$$\mu = Qr \tag{7.1}$$

In the LiH ion pair the two nuclei are known to be separated by a distance of 159.5 pm. If the bond were completely ionic, there would be a net charge of -1.6021×10^{-19} C (the electronic charge) centered on the H nucleus and a charge of $+1.6021 \times 10^{-19}$ C centered on the Li nucleus:

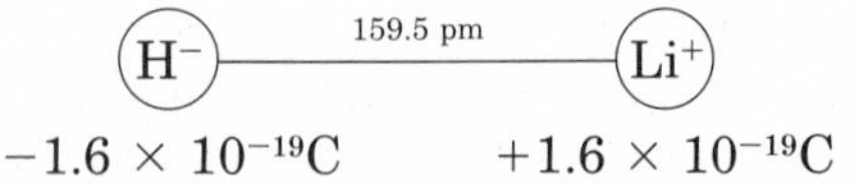

The dipole moment would then be given by

$$\mu = Qr = 1.6021 \times 10^{-19}\text{ C} \times 159.5 \times 10^{-12}\text{ m}$$
$$= 2.555 \times 10^{-29}\text{ C m}$$

The measured value of the dipole moment for the LiH ion pair is only about 77 percent of this value, namely, 1.963×10^{-29} C m. This can only be because the negative charge is *not* centered on the H nucleus but shifted somewhat toward the Li^+ nucleus. This shift brings the opposite charges closer together, and the experimental dipole moment is smaller than would be expected.

If we increase the degree of polarization of an ionic bond, a bond which is more covalent than ionic is eventually obtained. This is illustrated in Fig. 7.14. Three bonds involving hydrogen are shown, and the diagrams are arranged so that the midpoint of each bond lies on the same vertical (dashed) line. We have already discussed the bond between hydrogen and lithium, in which most of the electron density is associated with hydrogen. By comparison, electron density in the bond between hydrogen and carbon is much more evenly distributed—the bond certainly appears to be covalent. The third bond involves fluorine, which is three places farther to the right along the second row of the periodic table than carbon. In the H—F bond, electron density has been distorted away from hydrogen even more. Thus as we move from lithium with a nuclear charge of +3, through carbon with a nuclear charge of +6, to fluorine with a nuclear charge of +9, there is a continual shift in electron density away from hydrogen. The original H^- ion is polarized to the point where much of its electron density has been removed, and it begins to look more like an H^+ ion.

Polar Covalent Bonds

So far we have looked at polarization from the standpoint of ions, but we could consider the same three bonds shown in Fig. 7.14 by assuming that each had been distorted from a purely covalent, electron-pair *sharing* situation. In that case the H—C bond is closest to pure covalency—in it the H has roughly the same electron density it would have in an H_2 molecule. In the H—Li bond, however, the H has almost complete control over both electrons and hence has a negative charge. This situation is often indicated as follows:

$$\overset{\delta^-}{\text{H}}\text{—}\overset{\delta^+}{\text{Li}} \qquad \delta = 0.77$$

The Greek letter δ (delta) is used here to indicate that electron transfer is not complete and that some sharing takes place. The dipole moment of LiH shows that in effect only 77 percent of a full electronic charge has been transferred to H, and so δ = 0.77. If the transfer had been complete, δ would have been 1.0. Because the Li—H bond is only *partially* negative at the one end and *partially* positive at the other, we often say that the bond is **polar** or **polar covalent,** rather than 100 percent ionic.

The H—F bond is also a polar covalent bond, but in this case F rather than H has the partial negative charge, and we can write

$$\overset{\delta^+}{\text{H}}\text{—}\overset{\delta^-}{\text{F}} \qquad \text{where } 0 < \delta < 1$$

Again the value of δ can be found from a dipole-moment measurement, as the following example illustrates.

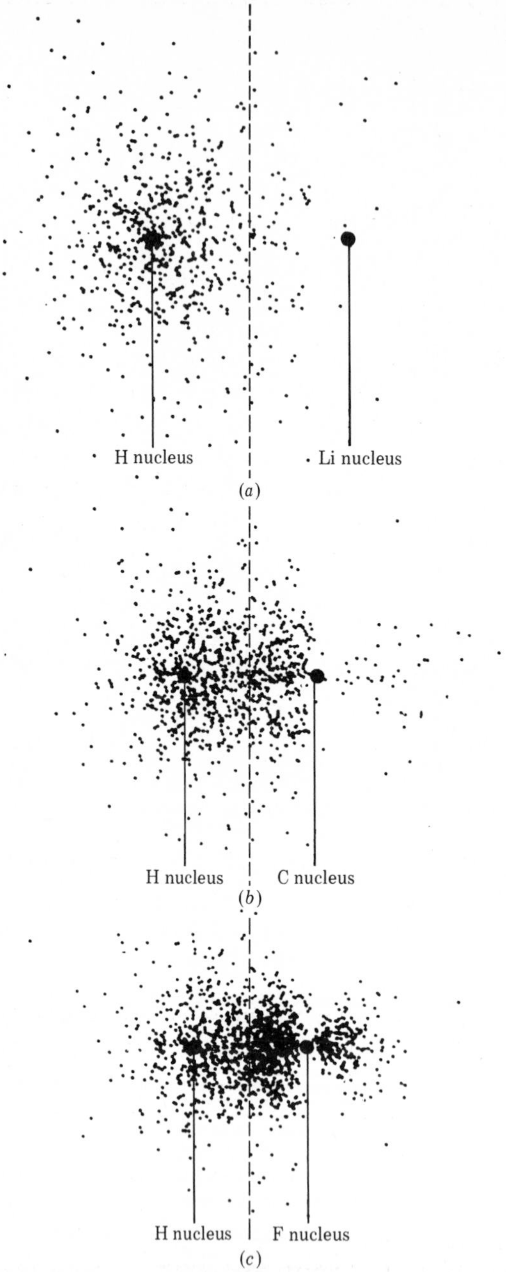

Figure 7.14 Comparison of bonding electron densities in (*a*) LiH bond; (*b*) CH bond; and (*c*) HF bond. Note the considerable shift of electron density away from hydrogen as the electronegativity of its bonding partner increases. (Computer generated.) (*Copyright © 1975 by W. G. Davies and J. W. Moore.*)

EXAMPLE 7.4 The dipole moment of the HF molecule is found to be 6.37×10^{-30} C m, while the H—F distance is 91.68 pm. Find the partial charge on the H and F atoms.

Solution Rearranging Eq. (7.1), we have

$$Q = \frac{\mu}{r}$$

Thus the apparent charge on each end of the molecule is given by

$$Q = \frac{6.37 \times 10^{-30}\ \text{C m}}{91.68 \times 10^{-12}\ \text{m}} = 6.95 \times 10^{-20}\ \text{C}$$

Since the charge on a single electron is 1.6021×10^{-19} C, we have

$$\delta = \frac{6.95 \times 10^{-20}}{1.6021 \times 10^{-19}} = 0.43$$

It is worth noting in the above example that the dipole moment measures the electrical imbalance of the *whole molecule* and not just that of the H—F bonding pair. In the HF molecule there are four valence electron pairs:

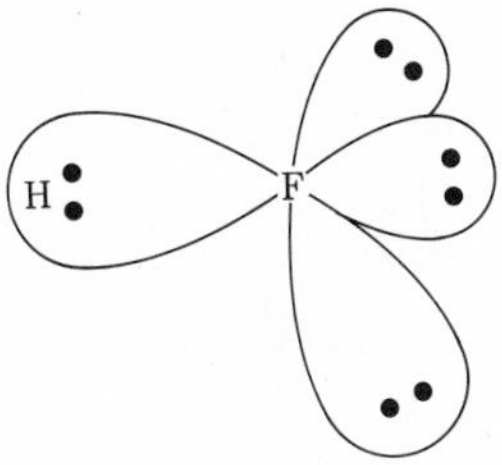

The three lone pairs sticking out on the right of the F atom also contribute to the overall negative charge of F.

Electronegativity

The ability of an atom in a molecule to attract a shared electron pair to itself, forming a polar covalent bond, is called its **electronegativity.** The negative side of a polar covalent bond corresponds to the *more electronegative* element. Furthermore the more polar a bond, the larger the *difference* in electronegativity of the two atoms forming it.

Unfortunately there is no direct way of measuring electronegativity. Dipole-moment measurements tell us about the electrical behavior of *all* electron pairs in the molecule, not just the bonding pair in which we are interested. Also, the polarity of a bond depends on whether the bond is a single, double, or triple bond and on what the other atoms and electron pairs in a molecule are. Therefore the dipole moment cannot tell us quantitatively the difference between the electronegativities of two bonded atoms.

Various attempts have been made over the years to derive a scale of electronegativities for the elements, none of which is entirely satisfactory. Nevertheless most of these attempts agree in large measure in telling us which elements are more electronegative than others. The best-known of these scales was devised by the Nobel prize-winning California chemist

Linus Pauling (1900 to   ) and is shown in Table 7.1. In this scale a value of 4.0 is arbitrarily given to the most electronegative element, fluorine, and the other electronegativities are scaled relative to this value.

As can be seen from this table, elements with electronegativities of 2.5 or more are all nonmetals in the top right-hand corner of the periodic table. These have been color-coded dark red. By contrast, elements with electronegativities of 1.3 or less are all metals on the lower left of the table. These elements have been coded in dark gray. They are often referred to as the most **electropositive** elements, and they are the metals which invariably form binary ionic compounds (Sec. 6.3). Between these two extremes we notice that most of the remaining metals (largely transition metals) have electronegativities between 1.4 and 1.9 (light gray), while most of the remaining nonmetals have electronegativities between 2.0 and 2.4 (light red). Another feature worth noting is the very large differences in electronegativities in the top right-hand corner of the table. Fluorine, with an electronegativity of 4, is by far the most electronegative element. At 3.5 oxygen is a distant second, while chlorine and nitrogen are tied for third place at 3.0.

If the electronegativity values of two atoms are very different, the bond between those atoms is largely ionic. In most of the typical ionic compounds discussed in the previous chapter, the difference is greater than 1.5, although it is dangerous to attach too much significance to this figure since electronegativity is only a semiquantitative concept. As the electronegativity difference becomes smaller, the bond becomes more covalent. An important example of an almost completely covalent bond between two dif-

TABLE 7.1 Relative Electronegativities of the Elements.*

H 2.1																	He —
Li 1.0	Be 1.5											B 2.0	C 2.5	N 3.0	O 3.5	F 4.0	Ne —
Na 0.9	Mg 1.2											Al 1.5	Si 1.8	P 2.1	S 2.5	Cl 3.0	Ar —
K 0.8	Ca 1.0	Sc 1.3	Ti 1.5	V 1.6	Cr 1.6	Mn 1.5	Fe 1.8	Co 1.8	Ni 1.8	Cu 1.9	Zn 1.6	Ga 1.6	Ge 1.8	As 2.0	Se 2.4	Br 2.8	Kr —
Rb 0.8	Sr 1.0	Y 1.2	Zr 1.4	Nb 1.6	Mo 1.8	Tc 1.9	Ru 2.2	Rh 2.2	Pd 2.2	Ag 1.9	Cd 1.7	In 1.7	Sn 1.8	Sb 1.9	Te 2.1	I 2.5	Xe —
Cs 0.7	Ba 0.9	La–Lu 1.1–1.2	Hf 1.3	Ta 1.5	W 1.7	Re 1.9	Os 2.2	Ir 2.2	Pt 2.2	Au 2.4	Hg 1.9	Tl 1.8	Pb 1.8	Bi 1.9	Po 2.0	At 2.2	Rn —

* Data from Linus Pauling, "The Nature of the Chemical Bond," 3d ed., Cornell University Press, Ithaca, N.Y., 1960.

ferent atoms is that between carbon (2.5) and hydrogen (2.1). We will describe the properties of numerous compounds of hydrogen and carbon (hydrocarbons) in the next chapter. These properties indicate that the C—H bond has almost no polar character.

EXAMPLE 7.5 Without consulting Table 7.1, arrange the following bonds in order of decreasing polarity: B—Cl, Ba—Cl, Be—Cl, Br—Cl, Cl—Cl.

Solution We first need to arrange the elements in order of increasing electronegativity. Since the electronegativity increases in going up a column of the periodic table, we have the following relationships:

$$\text{Ba} < \text{Be} \qquad \text{and} \qquad \text{Br} < \text{Cl}$$

Also since the electronegativity increases across the periodic table, we have

$$\text{Be} < \text{B}$$

Since B is a group III element on the borderline between metals and nonmetals, we easily guess that

$$\text{B} < \text{Br}$$

which gives us the complete order

$$\text{Ba} < \text{Be} < \text{B} < \text{Br} < \text{Cl}$$

Among the bonds listed, therefore, the Ba—Cl bond corresponds to the largest difference in electronegativity, i.e., to the most nearly ionic bond. The order of bond polarity is thus

$$\text{Ba—Cl} > \text{Be—Cl} > \text{B—Cl} > \text{Br—Cl} > \text{Cl—Cl}$$

where the final bond, Cl—Cl, is, of course, purely covalent.

Polarity in Polyatomic Molecules

When more than one polar bond is present in the same molecule, the polarity of one bond may cancel that of another. Thus the presence of polar bonds in a polyatomic molecule does not *guarantee* that the molecule as a whole will have a dipole moment. In such a case it is necessary to treat each polar bond mathematically as a *vector* and represent it with an arrow. The *length* of such an arrow shows how large the bond dipole moment is, while the *direction* of the arrow is a line drawn from the positive to the negative end of the bond. Adding the individual bond dipole moments as vectors will give the overall molecular dipole moment.

As an example of this vector addition, consider the BF_3 molecule in Fig. 7.15. The dipole moments of the three B—F bonds are represented by the arrows BF′ (pointing straight left), BF″ (pointing down to the right), and

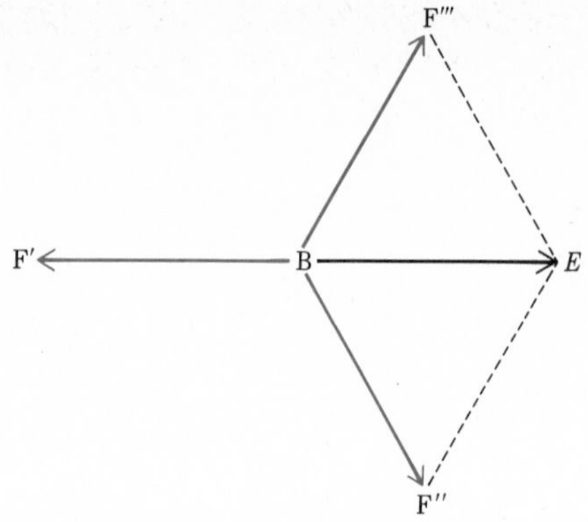

Figure 7.15 Although all three BF bonds in BF_3 are very polar, they are arranged at 120° to each other. The net result is that the three dipole moments (indicated by colored arrows) cancel each other and there is zero dipole moment.

BF‴ (pointing up to the right). The sum of vectors BF″ and BF‴ may be obtained by the parallelogram law—a line from F‴ drawn parallel to BF″ intersects a line from F″ drawn parallel to BF‴ at point *E*. Thus the resultant vector B*E* (the diagonal of the parallelogram BF‴*E*F″) is the sum of BF″ and BF‴. The resultant B*E* is exactly equal in length and exactly opposite in direction to bond dipole BF′. Therefore the net result is zero dipole moment.

Those arrangements of equivalent bonds that give zero dipole moment in this way are shown in Fig. 7.16. In addition to the trigonal arrangement just discussed, there is the obvious case of two equal bonds 180° apart. The other is much less obvious, namely, a tetrahedral arrangement of equal bonds. Any combination of these arrangements will also be nonpolar. The molecule PF_5, for example, is nonpolar since the bonds are arranged in a trigonal bipyramid, as shown in Fig. 7.1. Since three of the five bonds constitute a trigonal arrangement, they will have no resultant dipole moment. The remaining two bonds have equal but opposite dipoles which will likewise cancel.

If we replace any of the bonds shown in Fig. 7.16 with a different bond, or with a lone pair, the vectors will no longer cancel and the molecule will

Figure 7.16 The simplest arrangements of equivalent bonds around a central atom which produce a resultant dipole moment of zero: (*a*) linear; (*b*) trigonal. The two right-hand bonds (black resultant) cancel the left-hand bond. (*c*) Tetrahedral. The three right-hand bonds (black resultant) cancel the left-hand bond.

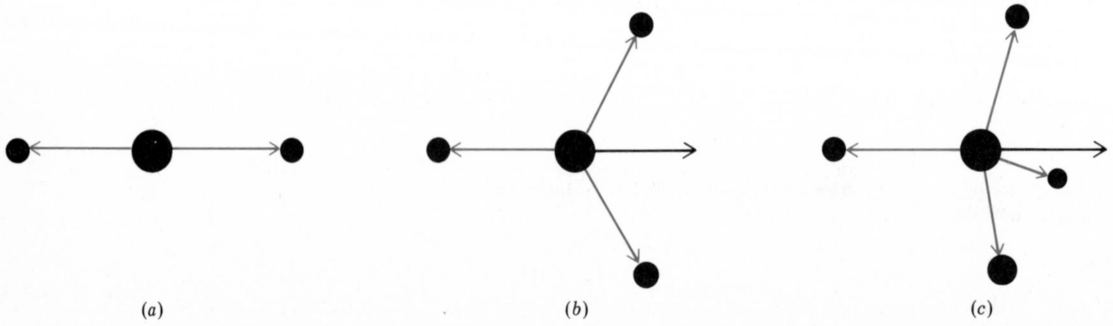

have a resultant dipole moment. Using this rule together with VSEPR theory, you can predict whether a molecule is polar or not. You can also make a rough estimate of how polar it will be.

EXAMPLE 7.6 Which of the following molecules are polar? About how large a dipole moment would you expect for each? (a) CF_4; (b) CHF_3; (c) H_2O; (d) NF_3.

Solution

a) VSEPR theory predicts a tetrahedral geometry for CF_4. Since all four bonds are the same, this molecule corresponds to Fig. 7.16*c*. It has zero dipole moment.

b) For CHF_3, VSEPR theory again predicts a tetrahedral geometry. However, all the bonds are not the same, and so there must be a resultant dipole moment. The C—H bond is essentially nonpolar, but the three C—F bonds are very polar and negative on the F side. Thus the molecule should have quite a large dipole moment:

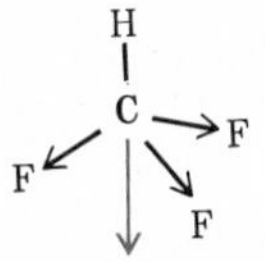

The resultant dipole is shown in color.

c) The O atom in H_2O is surrounded by four electron pairs, two bonded to H atoms and two lone pairs. All four pairs are not equivalent, and so there is a resultant dipole. Since O is much more electronegative than H, the two O—H bonds will produce a partial negative charge on the O. The two lone pairs will only add to this effect.

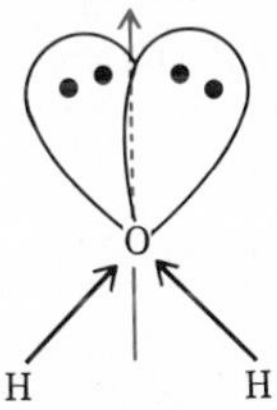

d) In NF_3 the N atom is surrounded by four electron pairs in an approximately tetrahedral arrangement. Since all four pairs are not equivalent, the molecule is polar. The dipole moment, though, is surprisingly small because the lone pair cancels much of the polarity of the N—F bonds:

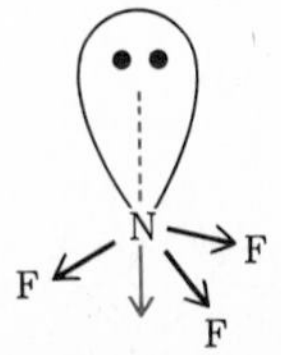

7.6 OXIDATION NUMBERS

A useful concept, closely related to electronegativity, is the **oxidation number** of an element or atom. The oxidation number is a somewhat artificial device—invented by chemists, not by molecules—which enables us to keep track of electrons in complicated reactions. We can obtain oxidation numbers by arbitrarily assigning the electrons of each covalent bond to the more electronegative atom in the bond. When this has been done for all bonds, the charge remaining on each atom is said to be its oxidation number. If two like atoms are joined, each atom is assigned half the bonding electrons.

EXAMPLE 7.7 Determine the oxidation number of each atom in each of the following formulas: (a) Cl_2; (b) CH_4; (c) NaCl; (d) OF_2; (e) H_2O_2.

Solution In each case we begin by drawing a Lewis diagram:

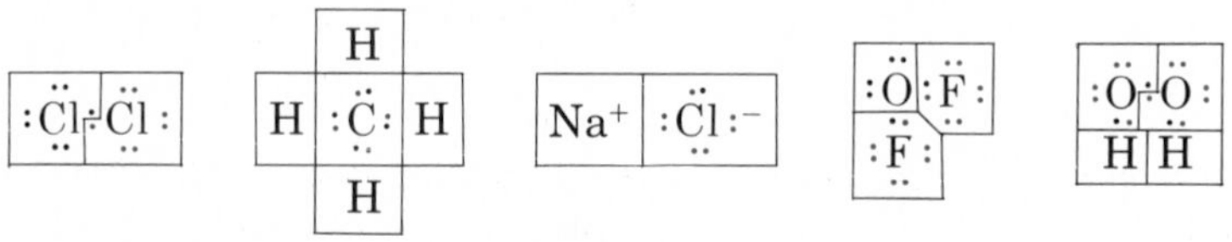

In each Lewis diagram, electrons have been color coded to indicate the atom from which they came originally. The boxes enclose electrons assigned to a given atom by the rules for determining oxidation number.

a) Since the bond in Cl_2 is purely covalent and the electrons are shared equally, one electron from the bond is assigned to each Cl, giving the same number of valence electrons (7) as a neutral Cl atom. Thus neither atom has lost any electrons, and the oxidation number is 0. This is indicated by writing a 0 above the symbol for chlorine in the formula

$$\overset{0}{Cl}_2$$

b) Since C is more electronegative than H, the pair of electrons in each C—H bond is assigned to C. Therefore each H has *lost* the one valence electron it originally had, giving an oxidation number of +1. The C atom has *gained* four electrons, giving it a negative charge and hence an oxidation number of −4:

$$\overset{-4}{C}\overset{+1}{H}_4$$

c) In NaCl each Na atom has lost an electron to form an Na^+ ion, and each Cl atom has gained an electron to form Cl^-. The oxidation numbers therefore correspond to the ionic charges:

$$\overset{+1}{Na}\overset{-1}{Cl}$$

d) Since F is more electronegative than O, the bonding pairs are assigned to F in oxygen difluoride (OF_2). The O is left with four valence electrons, and each F has eight. The oxidation numbers are

$$\overset{+2}{O}\overset{-1}{F_2}$$

e) In hydrogen peroxide (H_2O_2) the O—H bond pairs are assigned to the more electronegative O's, but the O—O bond is purely covalent, and the electron pair is divided equally. This gives each O seven electrons, a gain of 1 over the neutral atom. The oxidation numbers are

$$\overset{+1}{H_2}\overset{-1}{O_2}$$

Although one could always work out Lewis diagrams to obtain oxidation numbers as shown in Example 7.7, it is often easier to use a few simple rules to obtain them. The rules summarize the properties of oxidation numbers illustrated in the example.

1 *The oxidation number of an atom in an uncombined element is 0.* Since atoms of the same element always form pure covalent bonds, they share electrons equally, neither losing nor gaining, e.g., Cl_2 in Example 7.7*a*.

2 *The oxidation number of a monatomic ion equals the charge on that ion,* e.g., Na^+ and Cl^- in Example 7.7*c*.

3 *Some elements have the same oxidation number in nearly all their compounds.*
   a Elements in periodic group IA have oxidation numbers of +1, and elements in periodic group IIA have oxidation numbers of +2, e.g., Na^+ in Example 7.7*c*.
   b The most electronegative element, fluorine, is always assigned both electrons from any bond in which it participates. This gives fluorine an oxidation number of −1 in all its compounds, e.g., OF_2 in Example 7.7*d*.
   c Oxygen usually exhibits an oxidation number of −2, but exceptions occur in peroxides (Example 7.7*e*), superoxides, and when oxygen combines with fluorine (Example 7.7*d*).
   d Hydrogen exhibits an oxidation number of +1 unless it is combined with an element more electropositive than itself, e.g., with lithium, in which case its oxidation number is −1.

4 *The sum of the oxidation numbers of all atoms in a complete formula must be 0;* that is, when an electron is lost by one atom (+1 contribution to oxidation number), the same electron must be gained by another atom (−1 contribution to oxidation number).

5 *If a polyatomic ion is considered by itself, the sum of the oxidation numbers of its constituent atoms must equal the charge on the ion.*

As an illustration of these rules, let us consider a few more examples.

EXAMPLE 7.8 Determine the oxidation number of each element in each of the following formulas: (a) NaClO; (b) ClO_4^-; and (c) MgH_2.

Solution

a) Since Na is a group IA element, its oxidation number is +1 (rule 3*a*). The oxidation number of O is usually −2 (rule 3*c*). Therefore (rule 4), +1 + oxidation number of Cl + (−2) = 0.

$$\text{Oxidation number of Cl} = 2 - 1 = +1$$

Thus we write the formula

$$\overset{+1}{\text{Na}}\overset{+1}{\text{Cl}}\overset{-2}{\text{O}}$$

if oxidation numbers are to be included.

b In this case the oxidation numbers must add to −1, the charge on the polyatomic ion. Since O is usually −2, we have

$$\text{Oxidation number of Cl} + 4(-2) = -1$$

$$\text{Oxidation number of Cl} = -1 + 8 = +7$$

c In MgH_2, H is combined with an element more electropositive than itself, and so its oxidation number is −1. Mg is in group IIA, and so its oxidation number is +2:

$$\overset{+2}{\text{Mg}}\overset{-1}{\text{H}}_2$$

As a check on these assignments, it is wise to make sure that the oxidation numbers sum to 0:

$$+2 + 2(-1) = 0 \qquad \text{OK}$$

Oxidation numbers are mainly used by chemists to identify and handle a type of chemical reaction called a **redox reaction,** or an **oxidation-reduction reaction.** This type of reaction can be recognized because it involves a *change in oxidation number* of at least one element. We will deal with redox reactions in Chap. 11. Oxidation numbers are also used in the names of compounds. The internationally recommended rules of nomenclature involve roman numerals which represent oxidation numbers. For example, the two bromides of mercury, Hg_2Br_2 and $HgBr_2$, are called mercury(I) bromide and mercury(II) bromide, respectively. Here the numeral I refers to an oxidation number of +1 for mercury, and II to an oxidation number of +2.

Oxidation numbers can sometimes also be useful in writing Lewis structures, particularly for oxyanions. In the sulfite ion, SO_3^{2-}, for example, the oxidation number of sulfur is +4, suggesting that only *four* sulfur electrons are involved in the bonding. Since sulfur has six valence electrons, we conclude that two electrons are not involved in the bonding, i.e., that there is a *lone pair*. With this clue, a plausible Lewis structure is much easier to draw:

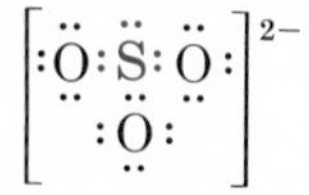

7.7 SOME DIFFICULTIES WITH ELECTRON-PAIR BONDS

Free Radicals

All the molecules we have discussed up to this point have involved *pairs* of electrons. However, there are a few stable molecules which contain an odd number of electrons. In such cases it is impossible to arrange the electrons in pairs, let alone into octets. Three well-known examples of such molecules are nitrogen(II) oxide, nitrogen(IV) oxide, and chlorine dioxide. The most plausible Lewis structures for these molecules are

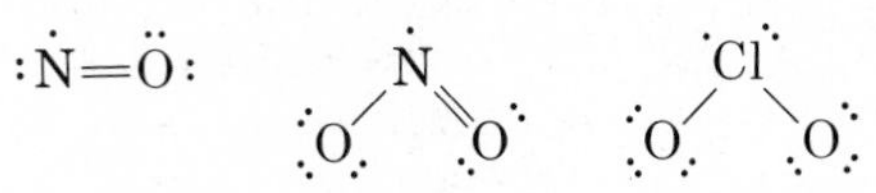

Molecules like this are called **odd-electron molecules** or **free radicals.**

Free radicals are usually more reactive than the average molecule in which all electrons are paired. In particular they tend to combine with other molecules so that their unpaired electron finds a partner of opposite spin. Since most molecules have all electrons paired, such reactions usually produce a new free radical. This is one reason why automobile emissions which cause even small concentrations of NO and NO_2 to be present in the air can be a serious pollution problem. When one of these free radicals reacts with other automobile emissions, the problem does not go away. Instead a different free radical is produced which is just as reactive as the one which was consumed. To make matters worse, when sunlight interacts with NO_2, it produces *two* free radicals for each one destroyed:

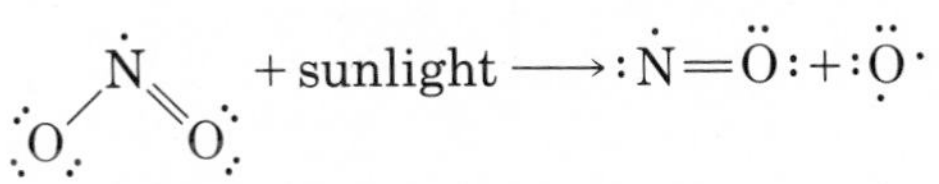

In this way a bad problem is made very much worse.

Resonance

In addition to molecules containing odd numbers of electrons, there is another class of molecules which does not fit easily into the electron-pair theory of the covalent bond that we have developed in the last two chapters. For these molecules it is possible to draw *more than one* Lewis structure which obeys the octet rule but which is unsatisfactory in other ways. A simple example of such a molecule is ozone, an unusual form of oxygen, whose molecular formula is O_3. Like the oxides of nitrogen, ozone is important in a discussion of atmospheric-pollution problems, but for the moment we will confine ourselves to its structure.

We can draw two Lewis diagrams for O_3, both of which obey the octet rule:

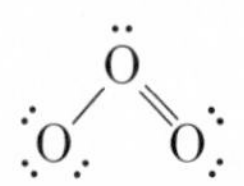

1

and

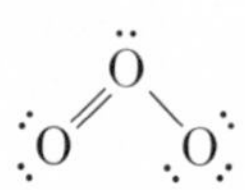

2

Structure 1 suggests that there is an O—O single bond on the left and an O═O double bond on the right side of the molecule. Structure 2 suggests the opposite placement of the double bond. However, it seems unlikely that electrons should be able to distinguish left from right in this way, and experimental evidence confirms this suspicion. *Both* bonds are found to have the *same* length, namely, 128 pm. This is intermediate between the O═O double-bond length of 121 pm in O_2 and the O—O single-bond length of 147 pm in H_2O_2. In other words the structure of O_3 is somehow intermediate in character between the two structures shown.

On a mathematical level we can satisfactorily account for the properties of ozone by regarding its structure as a *hybrid* of the two structures shown above, the term hybrid having exactly the same sense as for *sp* hybrids. We then obtain an electron probability distribution in which both bonds receive equal treatment and are intermediate in character between double and single bonds. Such a structure is called a **resonance hybrid** and is indicated in one of the following ways:

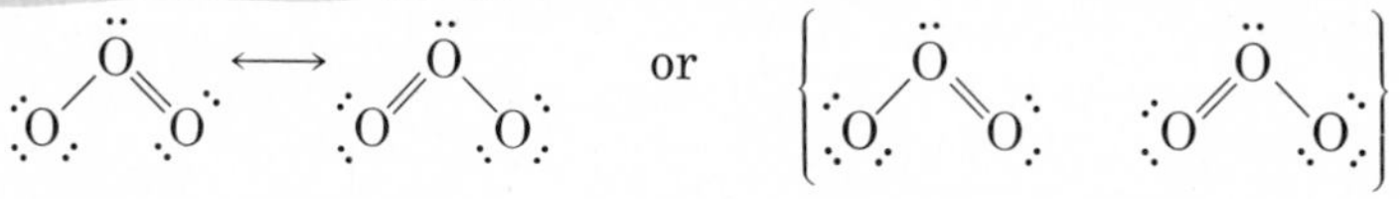

The term resonance and the use of a double-headed arrow, ↔, are both unfortunate since they suggest that the structure is continually oscillating between the two alternatives, so that if only you were fast enough, you could "catch" the double bond on one side or the other. One can no more do this than "catch" an *sp* hybrid orbital instantaneously in the form of an *s* or a *p* orbital.

The most important example of resonance is undoubtedly the compound benzene, C_6H_6, which has the structure

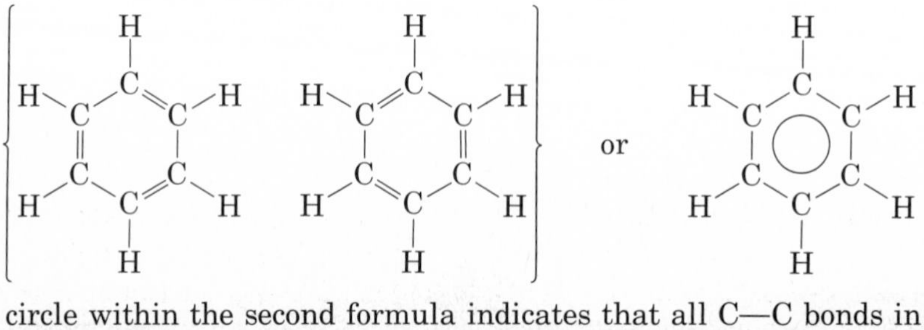

The circle within the second formula indicates that all C—C bonds in the hexagon are equivalent. This hexagonal ring of six carbon atoms is called a **benzene ring.** Each carbon-carbon bond is 139 pm long, intermediate

between the length of a C—C single bond (154 pm) and a C═C double bond (135 pm). Whereas a double bond between two carbon atoms is normally quite reactive, the bonding between the carbons in the benzene ring is difficult to alter. In virtually all its chemical reactions, the ring structure of benzene remains intact.

Even when a molecule exhibits resonance, it is still possible to predict its shape. Any bonds which are intermediate in character can be treated as though they were single bonds, though perhaps a bit fatter. On this basis one would predict that the ozone molecule is angular rather than linear because of the lone pair on the central oxygen atom, with an angle slightly less than 120°. Experimentally the angle is 117°. In the same way each carbon atom in benzene can be expected to be surrounded by three atoms in a plane around it, separated by angles of approximately 120°. Again this agrees with experiment. All the atoms in C_6H_6 lie in the same plane, and all bond angles are 120°.

EXAMPLE 7.9 Write resonance structures to indicate the bonding in the carbonate ion, CO_3^{2-}. Predict the O—C—O angle and the carbon-oxygen bond length.

Solution We must first write a plausible Lewis structure for the ion. Counting valence electrons, we have a total of

$$4(\text{from C}) + 3 \times 6(\text{from O}) + 2(\text{from charge}) = 24 \text{ electrons}$$

There are 4 octets to be filled, making a total of 4 × 8 = 32 electrons. We must thus count 32 − 24 = 8 electrons twice, and so there are 4 shared pairs. Since there are only three oxygen atoms, one must be double bonded to the carbon atom:

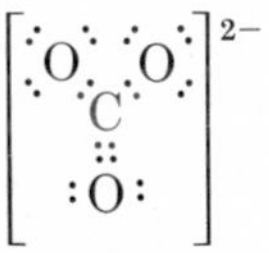

or

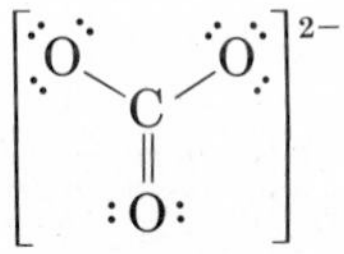

Two other equivalent structures can also be drawn, and so the carbonate ion corresponds to the following resonance hybrid:

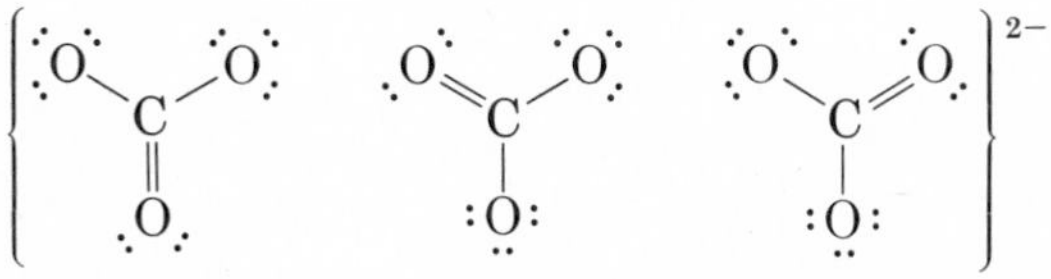

Since the carbon has no lone pairs in its valence shell, the three oxygens should be arranged trigonally around the carbon and all four atoms should lie in a plane. As we saw in the previous chapter, the C—O single-bond length is 143 pm, while the C═O double-bond length is 122 pm. We can expect the carbon-oxygen distance in the carbonate ion to lie somewhere in between these values. Experimentally it is found to be 129 pm.

SUMMARY

In this chapter we have seen that Lewis diagrams may be used to predict the structures for covalently bonded molecules. This applies to some which do not obey the octet rule and to those which contain multiple bonds, as well as to simple molecules. The structure of a molecule is determined by the positions of atomic nuclei in three-dimensional space, but it is repulsions among electron pairs—bonding pairs and lone pairs—which determine molecular geometry.

When dealing with molecular shapes, it is often convenient to think in terms of *sp*, sp^2, sp^3, or other hybrid orbitals. These correspond to the same overall electron density and to the same physical reality as do the *s* and *p* atomic orbitals considered earlier, but hybrid orbitals emphasize the directions in which electron density is concentrated. Hybrid orbitals may also be used to describe multiple bonding, in which case bonds must be bent so that two or three of them can link the same pair of atoms. A second equivalent approach to multiple bonding involves sigma and pi bonds. Electron density in a sigma bond is concentrated directly between the bonded nuclei, while that of the pi bond is divided in two—half on one side and half on the other side of the sigma bond.

No chemical bond can be 100 percent ionic, and, except for those between identical atoms, 100 percent covalent bonds do not exist either. Electron clouds—especially large, diffuse ones—are easily polarized, affecting the electrical balance of atoms, ions, or molecules. Large negative ions are readily polarized by small positive ions, increasing the covalent character of the bond between them. Electron density in covalent bonds shifts toward the more electronegative atom, producing partial charges on each atom and hence a dipole. In a polyatomic molecule, bond dipoles must be added as vectors to obtain a resultant which indicates molecular polarity. In the case of symmetric molecules, the effects of individual bond dipoles cancel and a nonpolar molecule results.

Oxidation numbers are used by chemists to keep track of electrons during the course of a chemical reaction. They may be obtained by arbitrarily assigning valence electrons to the more electronegative of two bonded atoms and calculating the resulting charge as if the bond were 100 percent ionic. Alternatively, some simple rules are available to predict the oxidation number of each atom in a formula. Oxidation numbers are used in the names of compounds and are often helpful in predicting formulas and writing Lewis diagrams.

In addition to deficiency of electrons and expansion of the valence shell, the octet rule is violated by species which have one or more unpaired electrons. Such free radicals are usually quite reactive. A difficulty of another sort occurs in benzene and other molecules for which more than one Lewis diagram can be drawn. Rearranging electrons (but not atomic nuclei) results in several structures which are referred to collectively as a resonance hybrid. Like an *sp* hybrid, a resonance hybrid is a combination of the contributing structures and has properties intermediate between them.

QUESTIONS AND PROBLEMS

7.1 List at least three shortcomings of Lewis diagrams for describing the bonding in simple covalent molecules.

7.2 Which of the following molecules do not obey the octet rule?

a BCl_3	*e* SnF_2
b PCl_3	*f* XeF_2
c CF_4	*g* $POCl_3$
d SF_4	*h* B_2H_6

7.3 Explain why the octet rule cannot be applied with much success to compounds of the transition metals.

7.4 Boron trifluoride (BF_3) reacts with phosphine (PH_3) to produce a compound with the formula BF_3PH_3. Draw the Lewis diagram for a molecule of this compound.

7.5 What species would you expect to result from the reaction between BF_3 and fluoride ions? Draw the Lewis structure for this species.

7.6 While sulfur forms the compounds SF_4 and SF_6, no equivalent compounds of oxygen, OF_4 and OF_6, are known. Suggest why this should be so.

7.7 Explain why silicon tetrachloride, $SiCl_4$, will react with chloride ions to form $SiCl_6^{2-}$, but carbon tetrachloride will not form CCl_6^{2-}.

7.8 In the following compounds:

a $BeCl_2(g)$	*d* $SbCl_5(g)$
b $BCl_3(g)$	*e* $TeF_6(g)$
c $SnI_4(g)$	

which of the following bond angles will occur:

i 90°	*v* 120°
ii 107°	*vi* 123°
iii 109.5°	*vii* 180°
iv 117°	

7.9 For molecules containing nonbonding pairs of electrons (lone pairs) on the central atom, there are really two geometries. First, there is the *molecular geometry* defined by the location of nuclei; and second, there is the *orbital geometry* which describes the arrangement of all the electron pairs bonding and nonbonding around the central atom. Construct a table showing both the molecular geometry and the orbital geometry for the following molecules:

a $BeCl_2$	*d* BF_3
b $SnCl_2$	*e* NF_3
c SCl_2	*f* $SiCl_4$

7.10 It is sometimes the convention when using VSEPR theory to classify molecules according to "AXE" type. The A denotes the central atom, X denotes the noncentral atoms, and E the lone pairs. Thus CH_4 is AX_4E_0. Classify all the molecules in the previous problem according to this convention.

7.11 In each of the following molecules:

a $PbCl_2$	*e* $PbCl_4$
b Cl_2O	*f* $COCl_2$
c PCl_3	*g* $SOCl_2$
d BCl_3	

which of the following angles would you expect to find for the X—A—X bond angle?

i Exactly 90°
ii Somewhat less than 109.5°
iii Exactly 109.5°
iv Somewhat less than 120°
v Exactly 120°
vi Somewhat greater than 120°
vii Somewhat less than 180°
viii Exactly 180°

7.12 In the molecule XeF_4 all atoms are in the same plane and all F—Xe—F angles are 90°. Why is this geometry so different from that of CF_4?

7.13 As we saw in Chap. 2, the $HgBr_2$ molecule is linear. Is it possible to explain this in terms of VSEPR theory?

7.14 Draw ball-and-stick diagrams to describe the geometry of the following species, indicating approximate angles where possible.

a CH_3CH_3 *e* FN=NF
b CH_3NH_2 *f* $CH_3C(=O)—NH_2$
c H_3O^+ *g* BO_3^{3-}
d HONO *h* P_2I_4

7.15 What hybrid orbitals are used in the bonding of the atom indicated in each of the following molecules?

a H—O—O—H	*f* CH_3CH_3
b PbF_2	*g* $BeCl_2$
c PbF_2	*h* BiI_3
d BI_3	*i* H_2S
e SF_6	*j* PF_5

7.16 Figure 7.4*a* shows two electron pairs, one in a 2*s* orbital and the other in a 2*p* orbital. How much energy is needed to transform this situation into the situation shown in Fig. 7.4*b* where each pair occupies an *sp* hybrid orbital?

7.17 What does each individual dot represent in a dot-density diagram like Fig. 7.4?

7.18 Explain how the representation of a covalent bond as the overlap of atomic orbitals is compatible with the idea of a shared pair of electrons.

7.19 Sketch rough diagrams, similar to those shown in Fig. 7.7, of the bonding in the following molecules. Label all hybrid orbitals.

a SnH_4 *b* BCl_3 *c* $H_2N—NH_2$

7.20 Sketch rough diagrams, similar to those shown in Figs. 7.8*a* and 7.10, of the bonding in the following molecules. Sketch two diagrams in each case, one with a bent-bond representation of the multiple bond, and the other with a sigma-pi representation. Label all orbitals and all bond angles.

a $H_2C{=}O$
b N≡N
c O=C=O
d FN=NF
e Cl—N=O
f H—C≡N

7.21 The dipole moment of the HCl molecule is found to be 3.43×10^{-30} C m, while the H—Cl distance is 127.4 pm. Find the partial charge on the H and Cl atoms. Compare this result with that found in Example 7.4 in the text. How does such a comparison show that F is more electronegative than Cl?

7.22 The ion-pair KF is found to have a dipole moment of 28.7×10^{-30} C m and an internuclear distance of 217.2 pm. Use these data to calculate the partial charge on the K and F atoms and hence the percentage ionic character. Before performing the calculation, make a guess as to how close to 100 percent the answer will be. How well does your guess compare with the actual value obtained?

7.23 Why are negative ions more polarizable on the whole than positive ions? How is polarizability related to the partial covalent character of an ionic bond?

7.24 If M is an alkali metal and X is a halogen, which of the M—X bonds would you expect to have most ionic character? Which would have least?

7.25 Define the term *electronegativity*. What is the highest value for electronegativity on Pauling's scale? What is the lowest value? How does electronegativity vary horizontally across a period of the periodic table? How does it vary vertically, within a given group?

7.26 According to the proximity rule, elements which are close together on the periodic table form covalent bonds, while those which are widely separated form ionic bonds. Does the proximity rule agree with the results obtained by using a table of electronegativities?

7.27 Without referring to any table of electronegativity values, arrange the following bonds in order of increasing ionic character:

a H—F, H—C, H—H, H—N
b Li—F, Li—H, Li—Li, Li—S
c K—F, Al—F, C—F, N—F

7.28 Using Table 7.1, arrange the following bonds in order of increasing ionic character:

a C—Br,
b C—H
c C—S
d Si—Br
e Si—H
f S—H
g B—H

7.29 Decide which of the following molecules will have a dipole moment:

a Br_2
b H_2S
c HBr
d BrCl
e CO_2
f BF_3
g AsF_3
h CCl_4

7.30 In the following pairs of molecules, one is always polar while the other is nonpolar. In each case identify which is which and explain why.

a HCl, H_2
b $SnCl_4$, $SnCl_2$
c PF_5, PF_3
d $SnCl_2$, $BeCl_2$
e CF_4, CH_3F
f PF_3, BF_3
g $BeCl_2$, $SeCl_2$
h CH_2Cl_2, CH_4

7.31 What does the term *oxidation number* mean when applied to an element or an atom? How does the oxidation number relate to the actual charge on an ion in an ionic compound? How does it relate to bond polarity in covalent compounds?

7.32 Determine the oxidation number of each element in each of the compounds given below:

a $MgCl_2$
b SO_3
c NH_3
d Al_2O_3
e Cl_2O_7
f PbO_2
g P_4S_{10}
h $NaNO_3$
i $BaSO_4$
j $CaCO_3$

7.33 Determine the oxidation number of each element in each of the compounds or ions given below:

a $KMnO_4$ *f* BaO_2
b $K_2Cr_2O_7$ *g* S_8
c LiH *h* P_4
d H_2O_2 *i* $P_2O_7^{4-}$
e $Na_2S_2O_3$ *j* ICl_2^-

***7.34** What is the oxidation number of sulfur in the tetrathionate ion, $S_4O_6^{2-}$? See if you can draw a Lewis structure for this ion and rationalize the anomalous value of the oxidation number.

7.35 Decide which of the following species are free radicals:

a O_3 *d* CH_3
b NO_2 *e* Cl
c N_2O_4 *f* NO^+

7.36 Explain what is meant by the term *resonance,* using the SO_2 molecule as an example. Why is it impossible to write a single, entirely satisfactory Lewis structure for SO_2? How is this difficulty overcome by invoking resonance?

7.37 Draw all possible resonance structures for the following species:

a SO_3
b NO_3^-
c NNO (The central N is bonded to both other atoms.)
d NO_2Cl (The N is bonded to the three other atoms.)

7.38 Using a sketch as well as words, describe the shape of each of the following species:

a SO_2 *f* SO_3^{-2}
b NO_3^- *g* NO_2^-
c SO_3 *h* NO_2^+
d ClO_3^- *i* N_3^-
e BO_3^{3-} *j* ClO_4^-

7.39 Sketch the geometry of two molecules or ions containing carbon in which the angle between two atoms bonded to carbon is

a Exactly 180°.
b Exactly 120°.
c Somewhat less than 120°.
d Exactly 109.5°.
e Slightly different from 109.5°.

chapter eight

PROPERTIES OF ORGANIC COMPOUNDS AND OTHER COVALENT SUBSTANCES

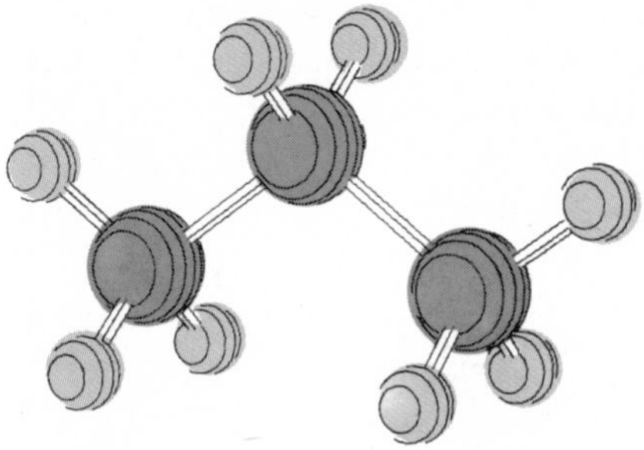

In Chap. 6 we described a number of macroscopic properties of ionic substances, such as electrical conductivity, crystal shape and cleavage, and characteristic chemical behavior of ions. These were understandable in terms of the microscopic picture of individual ions packed into a crystal lattice in a solid ionic compound or able to move past one another in a liquid or solution. The macroscopic properties of covalent and polar covalent substances can likewise be attributed to microscopic structure, and in this chapter we will see how the nature of the molecules in a covalently bonded substance influences its behavior.

The number of covalent substances is far larger than the number of ionic compounds, largely because of the ability of one element, carbon, to form strong bonds with itself. Hydrogen, oxygen, nitrogen, and a number of other elements also bond strongly to carbon, and a tremendous variety of compounds can result. In the early days of chemistry such compounds were obtained from plants or animals rather than being synthesized by chemists, and so they came to be known as **organic compounds.** This distinguished them from the **inorganic compounds** available from nonliving portions of the earth's surface. Today literally millions of carbon compounds can be synthesized in laboratories, and so this historical distinction is no longer valid. Nevertheless, the study of carbon compounds is still referred to as **organic chemistry.** Since organic compounds all involve covalent bonds, we will describe a number of them in this chapter. Many are of considerable commercial importance, and you probably encounter them, perhaps without knowing it, every day.

8.1 COVALENT COMPOUNDS AND INTERMOLECULAR FORCES

The ionic compounds which we discussed in Chap. 6 are almost all solids with melting temperatures above 600°C. By contrast, most substances which contain simple molecules are either gases or liquids at room temperature. They can only be persuaded to solidify at rather low temperatures. The reason for this contrasting behavior is easily explained on the microscopic level. Oppositely charged ions attract each other very strongly and usually require energies of 400 kJ mol^{-1} or more in order to be separated. On the other hand, molecules are electrically neutral and scarcely attract each other at all. The energy needed to separate two simple molecules is usually less than a hundredth of that needed to separate ions.

For example, only 1.23 kJ mol^{-1} is needed to separate two molecules of methane, CH_4. At room temperature virtually all molecules are moving around with energies in excess of this, so that methane is ordinarily a gas. Only if we cool the gas to quite a low temperature can we slow down the molecules to a point where they find it difficult to acquire an energy of 1.23 kJ mol^{-1}. At such a temperature, the molecules will be difficult to separate and the substance will become a liquid or a solid. Experimentally, we find that methane condenses to a liquid at -162°C, and this liquid freezes at -182°C.

Dipole Forces

You may be wondering, why should neutral molecules attract each other at all? If the molecules are polar, the explanation is fairly obvious. When two polar molecules approach each other, they can arrange themselves in such a way that the negative side of one molecule is close to the positive end of the other:

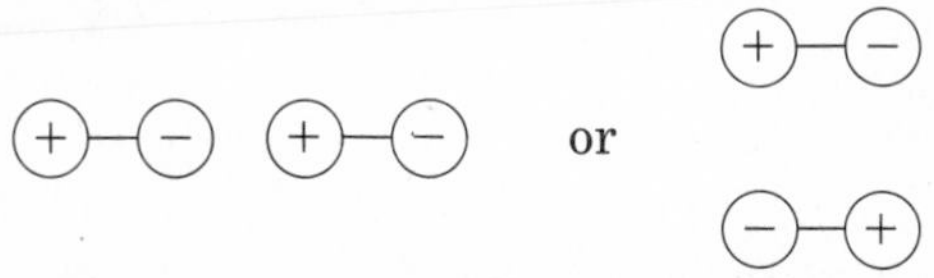

The molecules will then attract each other because the charges which are closest together are opposite in sign. (This behavior is very similar to the attraction between two bar magnets placed end to end or side by side with

TABLE 8.1 Boiling Points of Otherwise Similar Polar and Nonpolar Substances.

Nonpolar Molecules				Polar Molecules			
Molecule	Molar Mass/ g mol^{-1}	Total Number of Electrons	Boiling Point (in °C)	Molecule	Molar Mass/ g mol^{-1}	Total Number of Electrons	Boiling Point (in °C)
N_2	28	14	−196	CO	28	14	−192
SiH_4	32	18	−112	PH_3	34	18	−85
GeH_4	77	36	−90	AsH_3	78	36	−55
Br_2	160	70	59	ICl	162	70	97

the north poles opposite the south poles.) Forces between polar molecules which arise in this way are called **dipole forces.**

The existence of dipole forces explains why polar molecules have higher boiling points and melting points than do nonpolar molecules. In Table 8.1 we compare the boiling points of several pairs of molecules. In each pair, one molecule is polar and the other is nonpolar, but otherwise they are as similar as possible. The polar substance always has the higher boiling point, indicating greater attractive forces between separate molecules, that is, larger **intermolecular forces.**

London Forces

Dipole forces explain how polar molecules can attract each other, but it is a bit harder to account for the forces of attraction which exist between completely *nonpolar* molecules. Even the noble gases, whose atoms do not form chemical bonds with each other, can be condensed to liquids at sufficiently low temperatures. This indicates that the atoms attract each other, though only feebly. An explanation of these attractive forces was first given in 1930 by the Austrian physicist Fritz London (1900 to 1954). According to his theory, when two molecules approach each other very closely, the motion of the electrons in one of the molecules interferes with the motion of the electrons in the other, and the net result is an attractive force.

In order to understand London's ideas better, let us start by considering the hypothetical situation shown in Fig. 8.1. When a dipole approaches a

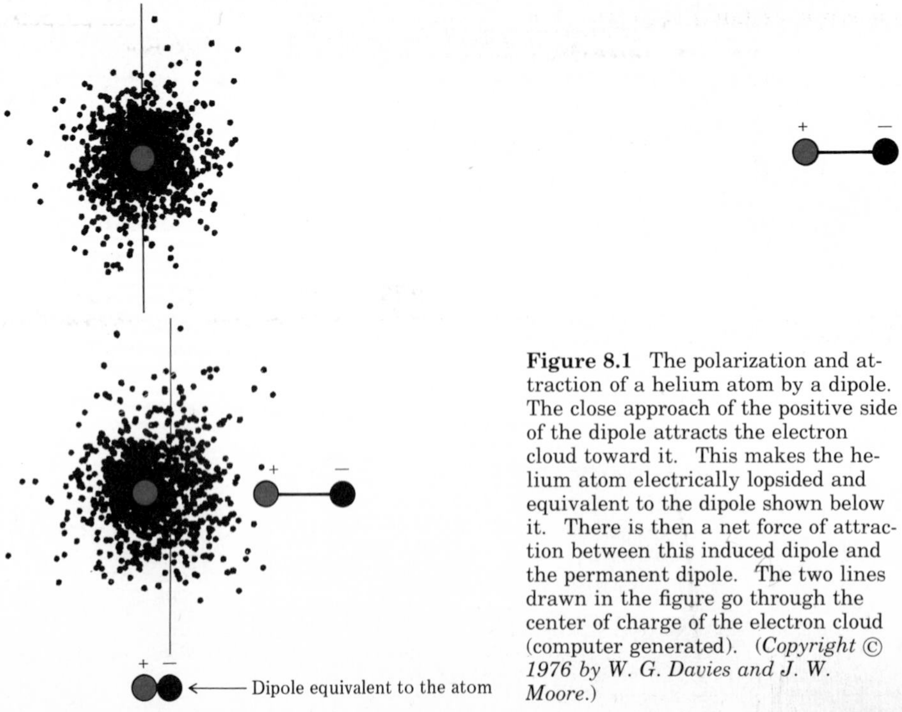

Figure 8.1 The polarization and attraction of a helium atom by a dipole. The close approach of the positive side of the dipole attracts the electron cloud toward it. This makes the helium atom electrically lopsided and equivalent to the dipole shown below it. There is then a net force of attraction between this induced dipole and the permanent dipole. The two lines drawn in the figure go through the center of charge of the electron cloud (computer generated). (*Copyright © 1976 by W. G. Davies and J. W. Moore.*)

helium atom, the electron cloud of the helium atom is attracted toward the positive end of the dipole. The helium atom becomes polarized and behaves electrically as though it were a second dipole, with its negative side pointed toward the positive side of the first dipole. As we have already seen, two dipoles oriented in this fashion *attract* each other.

If instead of a dipole, we now bring up another helium atom, a similar effect occurs. The electrons moving about the nucleus in this second atom will often both be found momentarily on one side of the nucleus or both on the other side. At any given instant, therefore, the approaching helium atom is likely to be slightly polar. It can then behave like the dipole in Fig. 8.1, inducing a dipole in the first atom, and attraction will result. Thus, as the electrons in one atom move around it, they will tend to synchronize to some extent with the motion of the electrons in the other atom. Overall there will be a force of attraction between the two helium atoms.

An argument similar to that just presented can be applied to pairs of atoms of the other noble gases as well. Indeed, it explains why there must be forces of attraction, albeit quite small, between two molecules of *any* kind. Forces caused by the mutual instantaneous polarization of two molecules are called **London forces,** or sometimes **dispersion forces.** When referring to intermolecular forces in general, i.e., to either London or dipole forces or both, the term **van der Waals forces** is generally used. Johannes van der Waals (1837 to 1923) was a Dutch scientist who first realized that neutral molecules must attract each other, even though he was unable to explain these attractions himself.

In general, when we compare substances whose molecules have similar electronic structures, it is always the *larger molecules* which correspond to the *stronger London forces.* This rule is illustrated by the physical properties shown in Table 8.2 for the noble gases and the halogens. Both melting points and boiling points increase in the order He < Ne < Ar < Kr < Xe and $F_2 < Cl_2 < Br_2 < I_2$. This corresponds with the order of increasing van der Waals radius, showing that in each case the larger molecules are more strongly attracted to each other.

TABLE 8.2 Some Physical Properties of Nonpolar Substances.

Substance	van der Waals Radius* /pm	Melting Point (in °C)	Boiling Point (in °C)
He	93	†	−269
Ne	112	−248	−246
Ar	154	−189	−186
Kr	169	−157	−153
Xe	190	−112	−108
F_2	135	−220	−188
Cl_2	180	−101	−34
Br_2	195	−47	58
I_2	215	114	183

* From Fig. 6.10. Note that the halogen molecules are not spherical. Nevertheless the van der Waals radius of the halogen atoms is in proportion to molecular size.

† Only forms a solid at very high pressures.

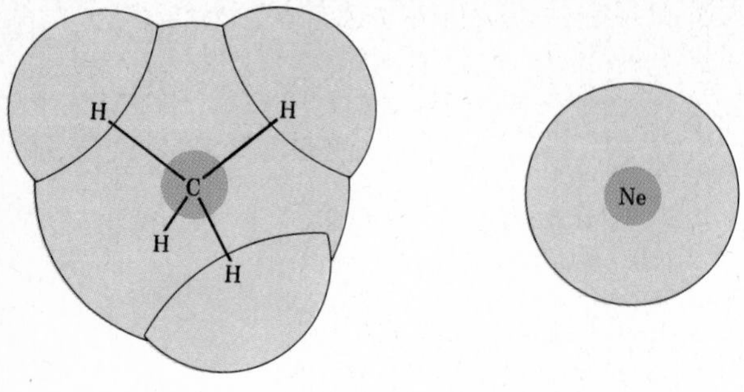

Figure 8.2 The effect of polarizability on London forces. Both CH_4 and Ne have eight electrons in the valence shell. In CH_4 these electrons form a more diffuse, larger, and less tightly held cloud than in Ne. This makes the CH_4 electrons more easily polarizable and the intermolecular forces larger.

This dependence on the size of the molecule is readily explained by London's theory. In larger molecules, the valence electrons are on the whole farther from the nuclei. The electron cloud is more diffuse, less tightly held, and hence *more easily polarizable* than for smaller molecules. The ionization energy for Xe is 1170 kJ mol^{-1}, for example, much less than for Ne (2080 kJ mol^{-1} from Table 6.1). This indicates that the outermost octet in Xe is much less tightly held than in Ne. Thus when two Xe atoms approach each other closely, the motion of the electrons in the one atom can synchronize with the motion in the other more effectively than in the case of two Ne atoms.

The magnitude of London forces is often said to depend on the molar mass of the molecules involved; if we compare molecules of similar electronic structure, the larger molecules are usually the heavier ones. But this is a coincidence rather than a cause-and-effect relationship, and is not always true. If we compare methane, CH_4, ($\mathcal{M}$ = 16 g mol^{-1}) with Ne ($\mathcal{M}$ = 20.2 g mol^{-1}), for example, we find that the *lighter* molecule has the stronger London forces. Both molecules contain 10 electrons, of which 8 are in the valence shell. In an Ne atom, the electrons are tightly held by a single nucleus of charge +10, while in CH_4, this same total positive charge is spread out over one C nucleus of charge +6 and four H nuclei of charge +1. As can be seen from Fig. 8.2, the electrons in CH_4 occupy a much larger electron cloud and are not so tightly constrained as in Ne. They are thus easier to polarize, and the London forces between two CH_4 molecules are expected to be larger than between two Ne molecules. In agreement with this we find a much higher boiling point for CH_4 (−162°C) than for Ne (−246°C).

EXAMPLE 8.1 Decide which substance in each of the following pairs will have the higher boiling point:

a SiH_4, SnH_4 **c** Kr, HBr
b CF_4, CCl_4 **d** C_2H_6, F_2

Solution

a) Both SiH_4 and SnH_4 correspond to the same Lewis diagram. In SnH_4, though, the valence octet is in the $n = 5$ shell, as opposed to the $n = 3$

shell for SiH_4. SnH_4 is the larger molecule and should have the higher boiling point.

b) Again the two molecules have similar Lewis diagrams. Since Cl is larger than F, we conclude that electrons in CCl_4 are more easily polarized, and the boiling point will be higher for this compound.

c) Both Kr and HBr have the same number of electrons. HBr, however, is polar and thus has the higher boiling point.

d) Both molecules have the same total number of electrons, namely, 18, but in C_2H_6 the electron cloud is distributed around eight nuclei rather than two. This larger cloud is more easily polarized so that we can expect stronger London forces. C_2H_6 will thus have the higher boiling point.

Note that it is not always possible to decide which of two substances has the higher boiling point even though their electronic structures are very similar. A case in point is the pair of substances HCl and HI. We can expect HCl to be more polar than HI so that the dipole forces between HCl molecules should be greater than for HI molecules. The London forces, however, will be the other way around since HI is so much larger in size than HCl. It happens that the effect of the London forces is larger—HI is found to have a higher boiling point (−38°C) than HCl (−88°C).

8.2 ORGANIC COMPOUNDS: HYDROCARBONS

Carbon is unique among the elements of the periodic table because of the ability of its atoms to form strong bonds with one another while still having one or more valences left over to link to other atoms. The strength of the carbon-carbon bond permits long chains to form:

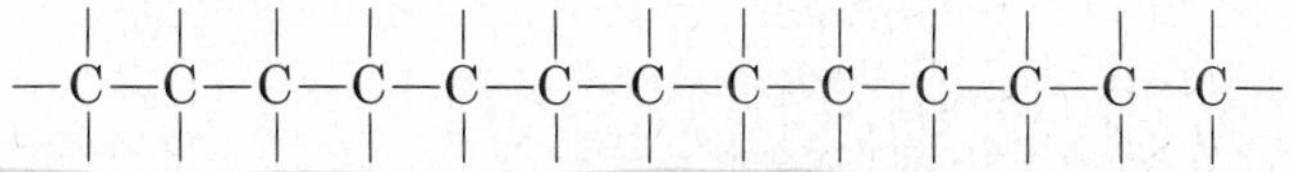

This behavior is referred to as **catenation.** Such a chain contains numerous sites to which other atoms (or more carbon atoms) can bond, leading to a great variety of carbon compounds (organic compounds).

The **hydrocarbons** contain only hydrogen and carbon. They provide the simplest examples of how catenation, combined with carbon's valence of 4, gives rise to a tremendous variety of molecular structures. The hydrocarbons are also extremely important from an economic and geopolitical point of view. The **fossil fuels,** coal, petroleum (or crude oil), and natural gas consist primarily of hydrocarbons and are extremely important in everyday life. In addition, a great many other modern conveniences, such as plastics and synthetic fibers, are derived from hydrocarbon compounds.

Alkanes

Most of the hydrocarbons in petroleum belong to a family of compounds called the **alkanes,** in which all carbon atoms are linked by single bonds.

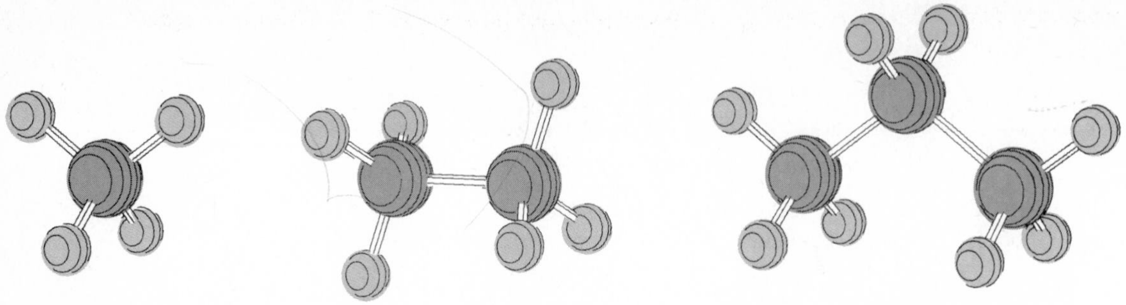

(a) (b) (c)

Figure 8.3 Ball-and-stick models for first three members of the alkane family: (*a*) methane, CH_4; (*b*) ethane, C_2H_6; (*c*) propane, C_3H_8. Carbon atoms are dark gray and hydrogen atoms are light red. (Computer generated.) (*Copyright © 1976 by W. G. Davies and J. W. Moore.*)

Ball-and-stick models of alkane molecules which contain up to three carbon atoms are shown in Fig. 8.3. Methane (Fig. 8.3*a*) has four C—H bonds arranged tetrahedrally around a single carbon atom. Ethane (Fig. 8.3*b*) has a slightly more complicated structure—each carbon is still surrounded tetrahedrally by four bonds, but only three are C—H bonds, while the fourth is a C—C bond. Ethane may be thought of as two methyl groups connected by a single C—C bond. (The **methyl** group, CH_3—, has the same structure as methane except that one hydrogen has been removed.) Thus the formula for ethane is CH_3CH_3, or C_2H_6.

The third member of the alkane family is propane. As can be seen in Fig. 8.3*c*, the three carbon atoms are in a chain. The two methyl groups at the ends of the chain are linked together by a **methylene** group, —CH_2—. The formula is $CH_3CH_2CH_3$, or C_3H_8. Again the tetrahedral arrangement of C—H or C—C bonds around each carbon atom is maintained.

Most of us lack the artistic skill to make good three-dimensional sketches of the molecular structures of alkanes or other organic compounds. Consequently **projection formulas,** which indicate how the atoms are connected but do not show tetrahedral angles of 109.5°, are commonly used. Projection formulas for the first three members of the alkane family are

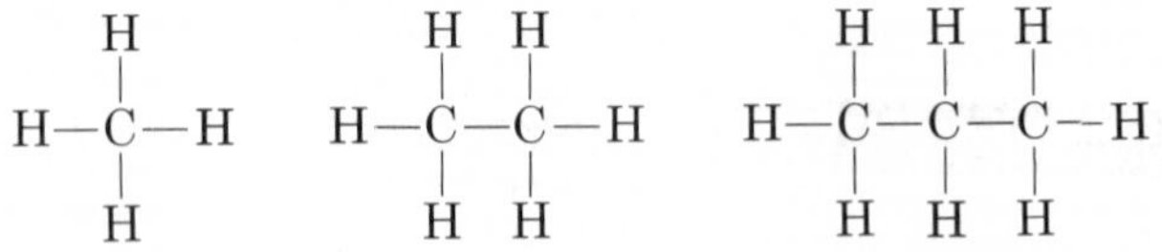

Methane **Ethane** **Propane**

Each of these represents one of the molecular structures in Fig. 8.3. You should compare the projection formulas with the ball-and-stick figures, trying to visualize the flat projections in three dimensions.

Clearly we could go on to chains of four, five, six, or more carbon atoms by adding more methylene groups to the propane molecule. The first 10 compounds whose structures may be derived in this way are listed in Table 8.3. They are called **normal alkanes** or **straight-chain alkanes,** indicating that all contain a single continuous chain of carbon atoms and can be represented by a projection formula whose carbon atoms are in a straight

TABLE 8.3 The First 10 Members of the Family of Straight-Chain or Normal Alkanes.

Name	Molecular Formula	Projection Formula	Condensed Structural Formula	Boiling Point (in °C)
Methane	CH_4	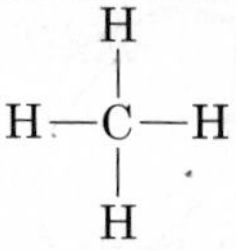	CH_4	−162
Ethane	C_2H_6	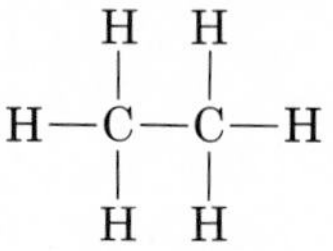	CH_3CH_3	−89
Propane	C_3H_8	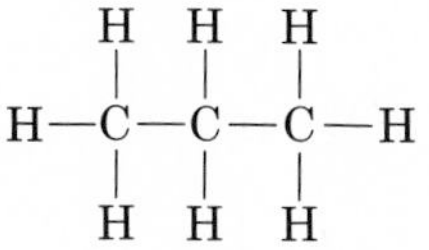	$CH_3CH_2CH_3$	−42
n-Butane*	C_4H_{10}	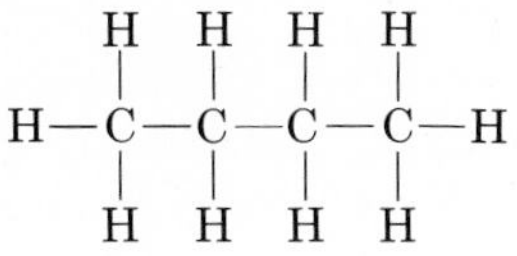	$CH_3CH_2CH_2CH_3$ or $CH_3(CH_2)_2CH_3$	−0.5
n-Pentane*	C_5H_{12}	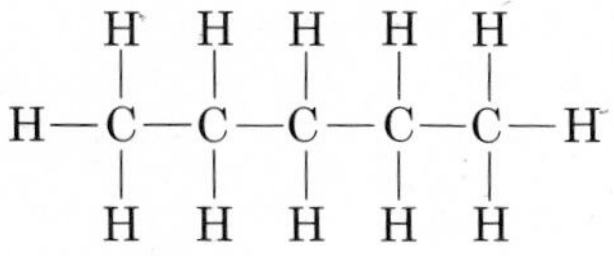	$CH_3CH_2CH_2CH_2CH_3$ or $CH_3(CH_2)_3CH_3$	36
n-Hexane*	C_6H_{14}	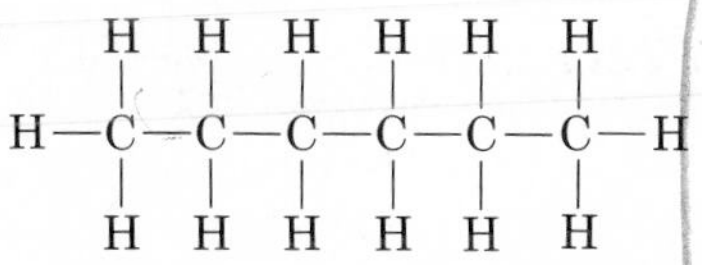	$CH_3CH_2CH_2CH_2CH_2CH_3$ or $CH_3(CH_2)_4CH_3$	69
n-Heptane*	C_7H_{16}	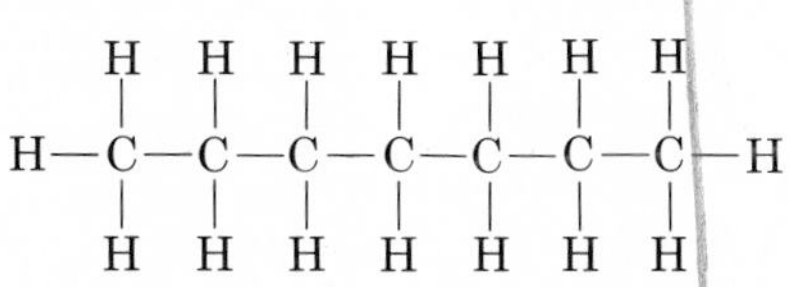	$CH_3(CH_2)_5CH_3$	98
n-Octane*	C_8H_{18}	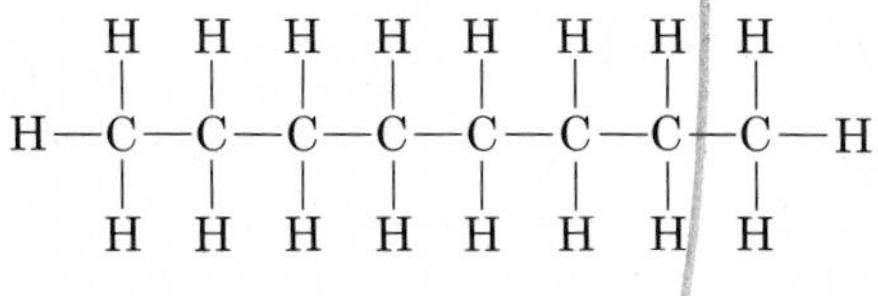	$CH_3(CH_2)_6CH_3$	126
n-Nonane*	C_9H_{20}		$CH_3(CH_2)_7CH_3$	151
n-Decane*	$C_{10}H_{22}$		$CH_3(CH_2)_8CH_3$	174

* The *n* before the name indicates that this is the normal straight-chain isomer.

line. Note that all the projection formulas in Table 8.3 have an initial hydrogen atom followed by a number of CH_2 groups. The chain ends with a second single hydrogen atom. The general formula $H(CH_2)_nH$, or C_nH_{2n+2}, may be written, where n is the number of CH_2 groups, or the number of C atoms. Propane, for example, has $n = 3$. Its formula is C_3H_8, and it is referred to as a C_3 hydrocarbon.

Another important feature of molecular structure is illustrated by the C_4 hydrocarbons. In addition to the straight chain shown for normal butane in Table 8.3, a **branched chain,** in which some carbon atoms are linked to more than two other carbons, is possible. The projection formula for the branched-chain compound isobutane is

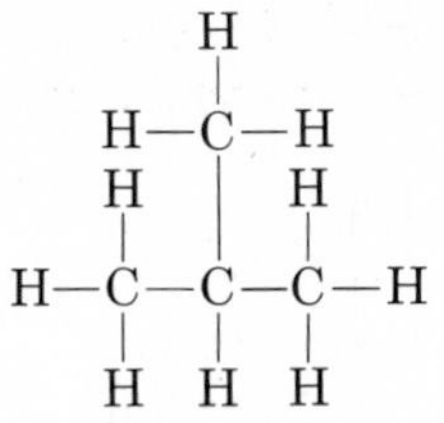

or more compactly $CH_3CH(CH_3)CH_3$

Isobutane

This should be compared with Fig. 8.4, where ball-and-stick models of normal butane and isobutane are both shown.

Unlike normal butane, which has a straight chain of four carbon atoms, the longest chain in isobutane is only three carbon atoms long. The central of these three atoms is bonded to the fourth carbon. Nevertheless, you can verify from the projection formulas or from the ball-and-stick drawings that both normal butane and isobutane have the same molecular formula, C_4H_{10}. The two compounds are isomers (Sec. 6.6), hence the prefix *iso* in the name for one of them.

As the number of carbon atoms in an alkane molecule increases, so do the possibilities for isomerism of this kind. There are three isomeric pentanes, all with the formula C_5H_{12}, five isomeric hexanes, C_6H_{14}, and nine isomeric heptanes, C_7H_{16}. The number of possible isomers of tetracontane,

Figure 8.4 The two isomers of butane, C_4H_{10}. (*a*) *n*-butane; (*b*) isobutane (computer generated). (*Copyright © 1976 by W. G. Davies and J. W. Moore.*)

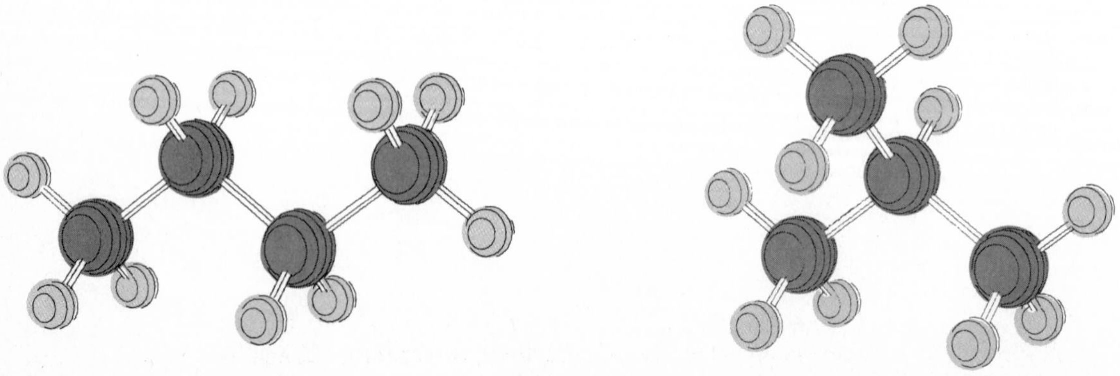

(*a*) (*b*)

alkanes – single bonds

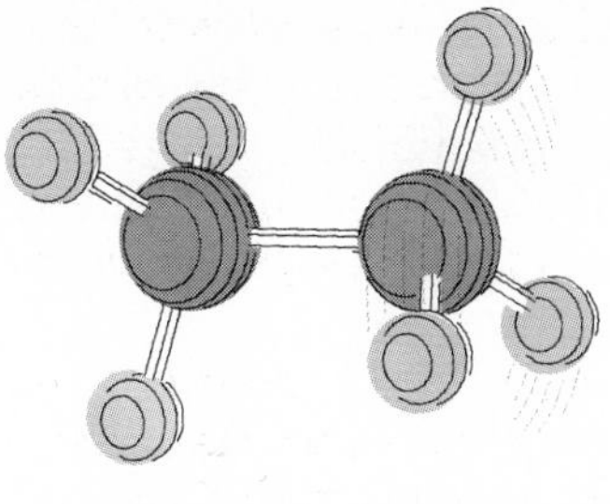

Figure 8.5 In all alkanes, portions of the molecule can flex or rotate about all the single carbon-carbon bonds. Shown in the figure is the rotation of one methyl group, CH_3, with respect to the other in a molecule of ethane, CH_3CH_3 (computer generated). (*Copyright © 1976 by W. G. Davies and J. W. Moore.*)

$C_{40}H_{82}$, is larger than 62 million. Thus an inconceivable variety of different molecular structures is possible for compounds containing only carbon and hydrogen atoms connected by single bonds. In crude oil, the most important source of hydrocarbons in the United States, branched-chain and straight-chain alkanes are about equally common.

Another aspect of the behavior of alkane molecules (and other molecules containing single bonds) is not apparent from ball-and-stick illustrations or from projection formulas. Like small children, molecules cannot stop wriggling, and most alkane structures are not rigid. Groups on either side of a carbon-carbon single bond are able to rotate freely with respect to each other. Rotation of one methyl group with respect to the other in ethane, CH_3CH_3, is shown in Fig. 8.5. Because of this free rotation, and because they often collide with other molecules, alkane molecules are constantly flexing and writhing about their C—C bonds, assuming different shapes (different **conformations**) all the time. Some feeling for the way in which alkane molecules can adopt a variety of conformations is obtained from the following example.

EXAMPLE 8.2 In Fig. 8.6 five ball-and-stick diagrams labeled (*a*) through (*e*) are shown. All five correspond to the formula C_5H_{12} (pentane). Since there are only three isomers of pentane, some of these molecules must correspond to different views or different conformations of the same molecule. Decide which of these diagrams correspond to the same isomer, and which isomer each represents. Draw a projection formula for each isomer.

Solution Molecule (*a*) has five carbon atoms in a single continuous sequence. It corresponds to

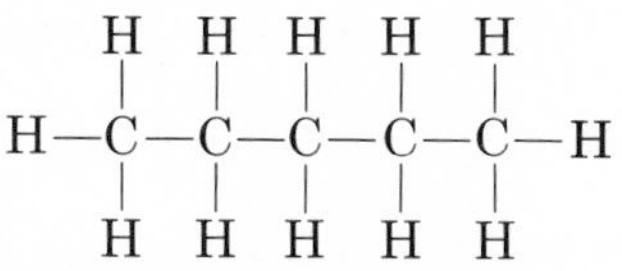

***n*-Pentane**

Careful inspection of (*b*) reveals that again the five carbon atoms form a single chain. No carbon atom is joined to more than two others. Molecule (*b*) is thus also *n*-pentane, but in a different conformation. Molecule (*c*) *does*

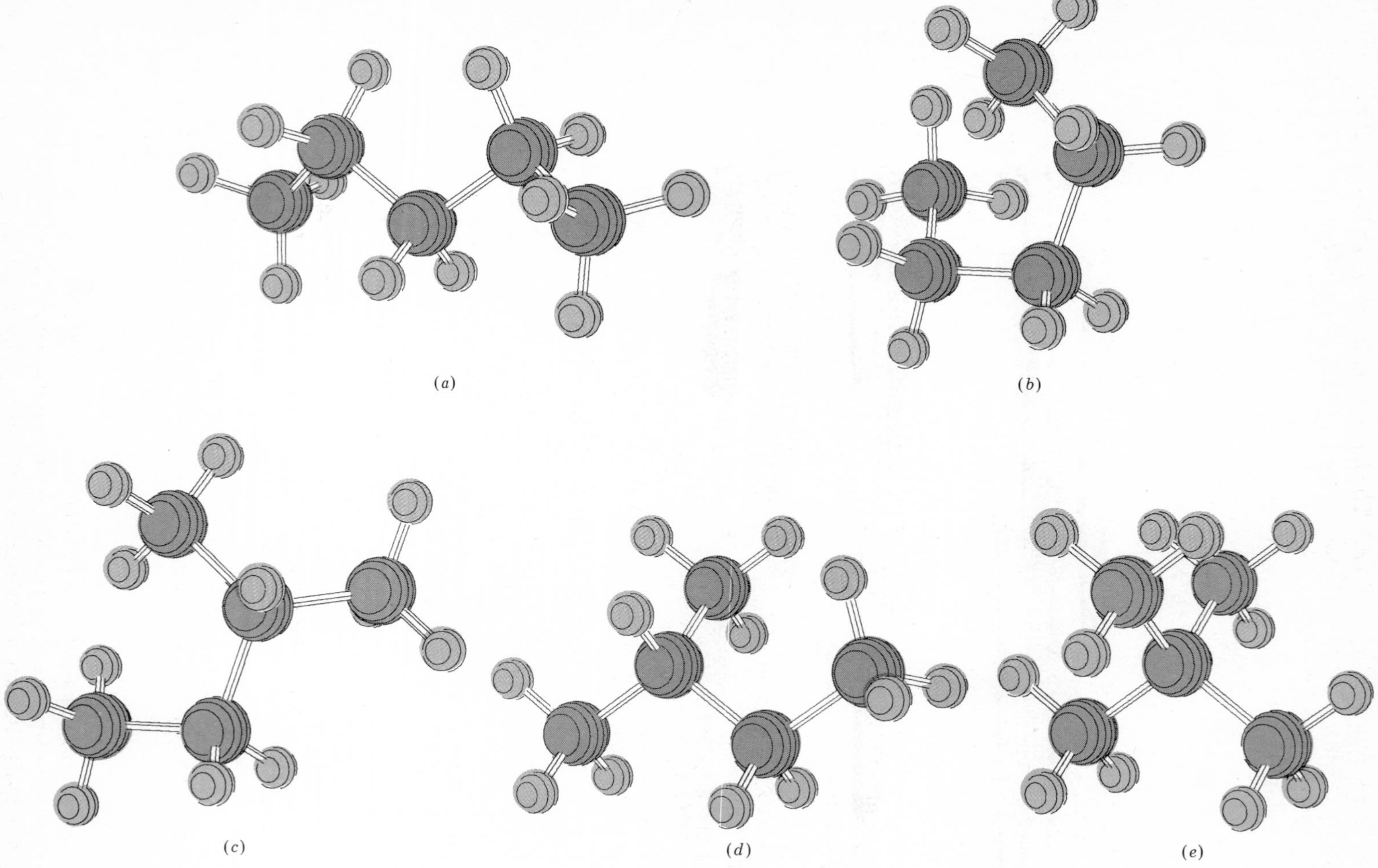

Figure 8.6 As explained in Example 8.2, all these molecules are pentanes with the formula C_5H_{12}. Some are genuine isomers, but others are different conformations of the same isomer. (Computer generated.) (*Copyright © 1976 by W. G. Davies and J. W. Moore.*)

have a carbon atom joined to three others. The longest chain is four carbon atoms long, and an additional carbon atom is attached to the second carbon atom. The projection formula is accordingly

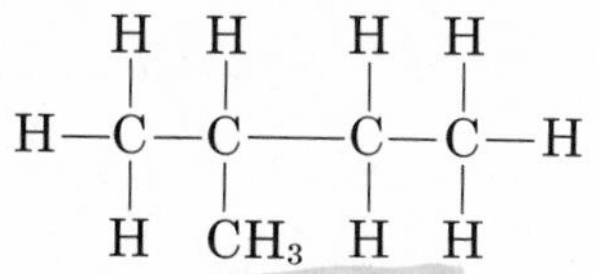

Isopentane

Molecule (*d*) is also isopentane, while molecule (*e*) corresponds to the third isomer, called neopentane:

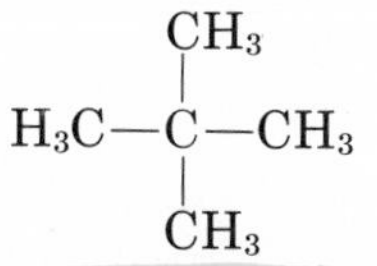

Neopentane

Cycloalkanes

Though petroleum consists mainly of normal and branched alkanes, other classes of compounds also occur. The next best represented are the **cycloalkanes,** which are characterized by a *ring* of carbon atoms. If a hydrocarbon chain is to be made into a ring, a new C—C bond must be formed between carbon atoms at the end of the chain. This requires that two hydrogen atoms be removed to make room for the new bond. Consequently such a ring involves one more C—C bond and two less C—H bonds than the corresponding normal alkane, and the general formula for a cycloalkane is C_nH_{2n}. The most common cycloalkane in petroleum is methylcyclohexane, which has the projection formula

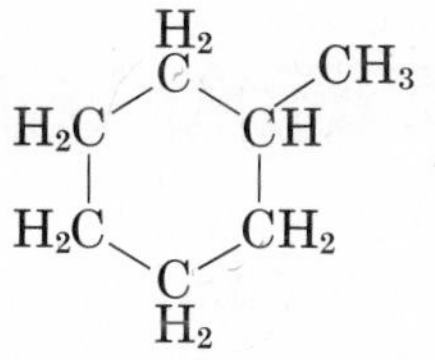

A ball-and-stick model of one of the conformations of methylcyclohexane is shown in Fig. 8.7. Note that the ring of six carbon atoms is puckered to permit tetrahedral arrangement of bonds around the carbon atoms. Also shown in the figure is cyclopentane, C_5H_{10}, which contains a five-membered ring. Virtually all the cycloalkanes in petroleum contain five- or six-membered rings similar to those illustrated.

Properties of Alkanes

Other things being equal, the more carbon atoms, the higher the boiling point of an alkane. As can be seen from Table 8.3, boiling points rise steadily from CH_4 to $C_{10}H_{22}$ as each additional CH_2 group is added to the chain.

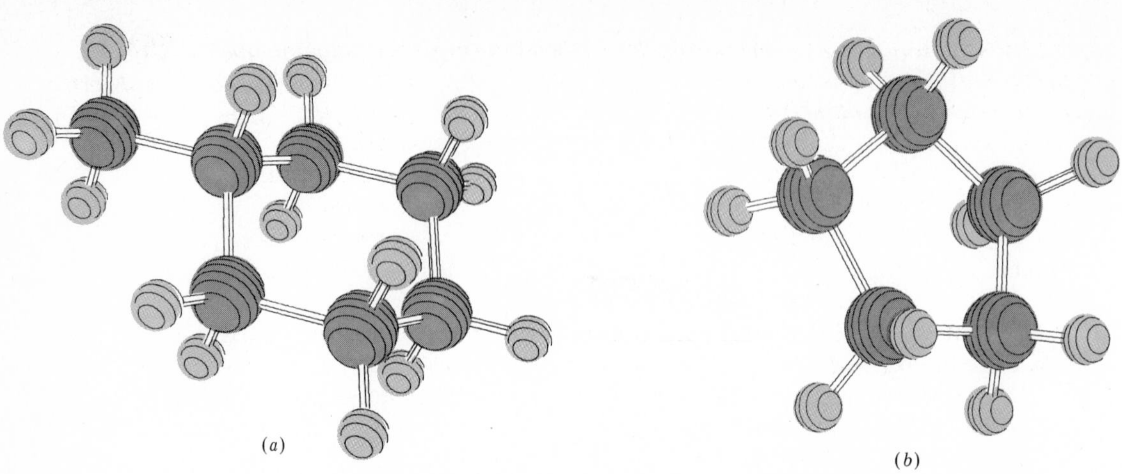

Figure 8.7 Cycloalkanes. (*a*) Methylcyclohexane, C_7H_{14}, the most common cycloalkane found in crude oil. Note that the six-membered ring is puckered to accommodate the tetrahedral bond angles of carbon. (*b*) Cyclopentane, C_5H_{10}, in which the ring of carbon atoms is more nearly planar but is still slightly puckered. (Computer generated.) (*Copyright © 1976 by W. G. Davies and J. W. Moore.*)

The reason for this rise is an increase in the intermolecular forces. As shown in Fig. 8.8, larger molecules can make close contact with each other over a much larger surface area than can smaller molecules. The total force exerted between the two is thus greater. In the same way we can also explain why branched-chain hydrocarbons boil at lower temperatures than straight-chain compounds. The branched molecules are more compact and provide less area over which intermolecular forces can act.

The alkanes are rather unreactive and do not combine readily with other substances. When heated sufficiently, however, they burn in air, a process known as **combustion:**

$$CH_4(g) + 2O_2(g) \rightarrow CO_2(g) + 2H_2O(g) \qquad \Delta H_m = -890.4 \text{ kJ mol}^{-1}$$

Combustion releases large quantities of heat; the most important use of alkanes is as fuels. Natural gas is mainly (approximately 85 percent)

Figure 8.8 Longer-chain alkane molecules can touch each other over a much larger area than short-chain alkanes. The London forces between them are correspondingly larger, and the boiling point is higher. Similar remarks apply, though not so strongly, to a comparison between straight-chain and branch-chain isomers.

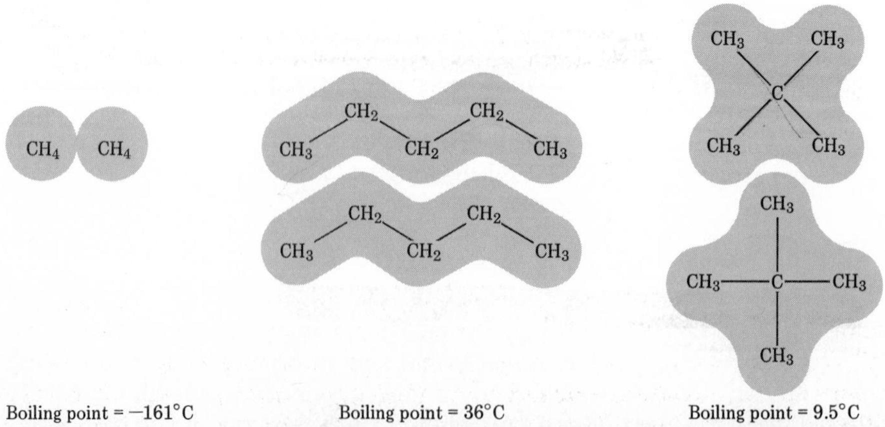

methane. Propane or liquified petroleum gas sold in tanks for portable use is usually a mixture of both propane and butane. Though both are gases at ordinary temperature and pressure, they liquefy under pressure in the tank. Gasoline is a more complex mixture of alkanes having 5 to 12 carbon atoms. Better-quality gasoline usually contains a higher percentage of branch-chain alkanes than the regular grade. Kerosene contains C_{10} to C_{16} alkanes, while heating oil usually involves the C_{12} to C_{18} range. Even longer-chain alkanes are found in lubricating oil, C_{15} to C_{25}, and paraffin wax (used for candles and waxed paper), C_{23} to C_{29}.

The variation of boiling point with chain length in the alkanes provides a simple method for partially separating them from each other in petroleum. When petroleum is heated, the shorter-chain compounds begin to boil off initially. These can be collected by cooling the vapor until it recondenses to a liquid. As boiling proceeds, the temperature rises and longer and longer-chain compounds boil off. Finally only the very long chain compounds are left. Such a process is called **fractional distillation** and will be discussed further in Chap. 10. Fractional distillation plays an important role in **petroleum refining,** which is a series of physical and chemical processes by which the most valuable and useful components are obtained from natural crude oil.

The alkanes are typical examples of nonpolar compounds. The electronegativities of carbon and hydrogen (Table 7.1) are 2.5 and 2.1, respectively, and so the C—H bond has almost no polarity. Consequently, even the most unsymmetrical hydrocarbon molecules have very small dipole moments. We can therefore take the boiling-point behavior of the alkanes to be representative of other nonpolar substances. Simple molecules are gases, while medium-sized molecules are liquids, and only quite large molecules are solids.

Aromatic Hydrocarbons

A third important class of compounds found in petroleum is the **aromatic hydrocarbons.** These are less abundant than the alkanes and cycloalkanes, amounting to only a few percent of the total, but they are quite important commercially. All aromatic hydrocarbons contain a **benzene ring.** You will recall from the previous chapter that benzene, C_6H_6, contains a flat ring of six carbon atoms joined by bonds which are intermediate in character between single and double bonds. The benzene ring is usually indicated by

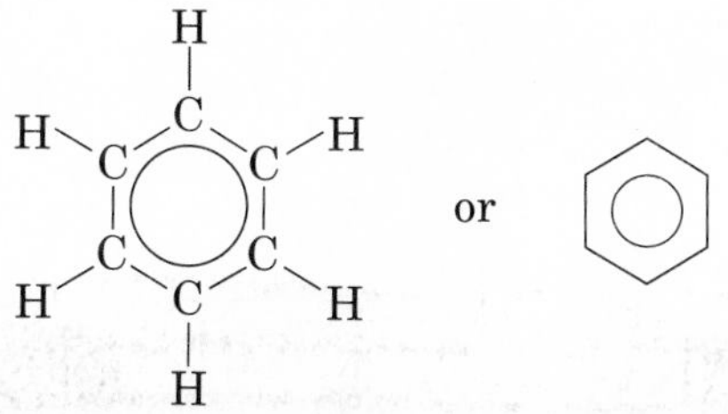

In the latter structure the lines represent C—C bonds, but carbon and hydrogen atoms, as well as C—H bonds, have been omitted. The benzene ring is very stable, surviving unchanged in most chemical reactions. It is very different from the puckered six-membered rings found in cycloalkanes.

Examples of aromatic hydrocarbons found in crude oil are

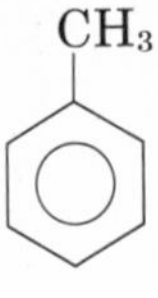

Toluene
C_7H_8

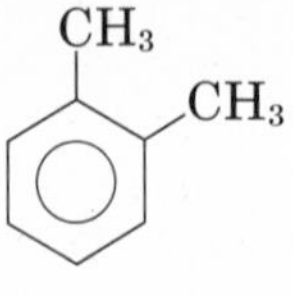

ortho-Xylene
(*o*-Xylene)
C_8H_{10}

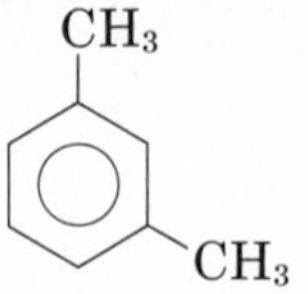

meta-Xylene
(*m*-Xylene)
C_8H_{10}

para-Xylene
(*p*-Xylene)
C_8H_{10}

Note that the three xylenes are also isomers. Compounds containing two benzene rings joined together, such as naphthalene, $C_{10}H_8$, are also found in crude oil, though they are much rarer than benzene-related compounds.

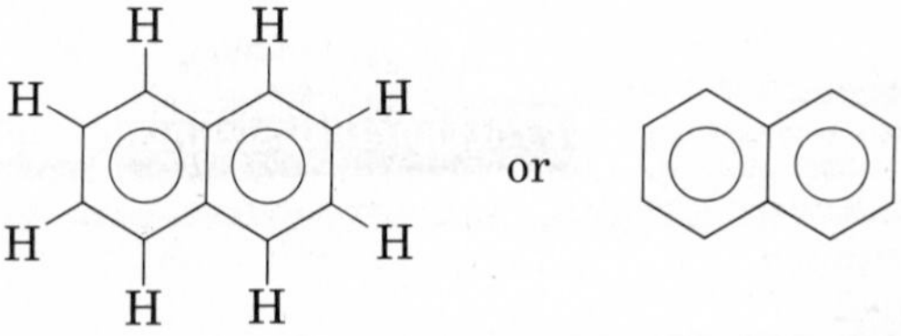

Aromatic hydrocarbons are much more common in coal than in petroleum, though in the United States they are mostly manufactured from the latter. In addition to their use in motor fuel, they may be made into dyes, plastics, explosives, detergents, insecticides, medicines, and many other products. The 3.5×10^9 kg of benzene produced by United States industry in 1975 ranked this compound second among all organic chemicals.

Some aromatic compounds, benzene among them, are toxic. The compound 1,2-benzopyrene was the cause of the first demonstrated case of occupational disease.

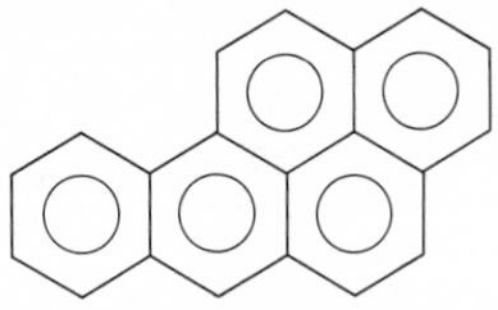

1,2-Benzopyrene

During the eighteenth century chimney sweeps in London were found to have extremely high rates of skin cancer relative to the average person. This was eventually traced to the carcinogenic (cancer-causing) properties of 1,2-benzopyrene in the soot which coated the insides of the chimneys they cleaned. Small quantities of the compound were produced by inefficient combustion of coal in the fireplaces used to heat London houses.

Unsaturated Hydrocarbons

Two important families of hydrocarbons which are not found in petroleum are the **alkenes** (also called olefins) and the **alkynes** (also called acetylenes).[1] Alkene molecules are similar to alkane molecules, except that they

[1] There are two names for many simple organic compounds. Since 1930 the International Union of Pure and Applied Chemistry has developed a systematic method for naming all organic compounds, but many of the earlier names still survive, particularly among industrial chemists. Where appropriate, both names will be given, the older name in parentheses.

contain a carbon-carbon *double* bond (C═C) and two fewer H atoms. They thus have the general formula C_nH_{2n}. Alkyne molecules contain *triple* bonds (C≡C) and have four H atoms less than the corresponding alkane. Their general formula is thus C_nH_{2n-2}. Compounds containing double or triple bonds are often referred to collectively as **unsaturated compounds.** Because of their multiple bonds, alkenes and alkynes are usually more chemically reactive than alkanes and aromatic hydrocarbons.

The presence of a double or triple bond in the molecule opens up many more possibilities for isomerism than is the case for alkanes. There are usually several alternative locations for the multiple bond, and in the case of a double bond there is the possibility of cis-trans isomerism. Thus while there are only *two* alkane molecules possible with four carbon atoms, *four* alkene molecules are possible:

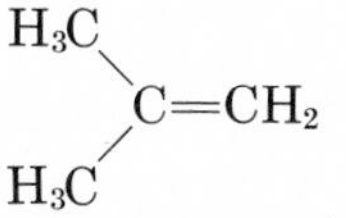

Methylpropene

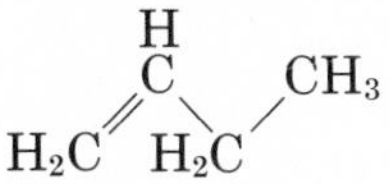

1-Butene

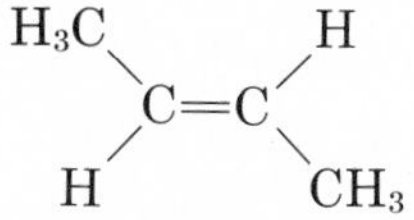

***trans*-2-Butene**

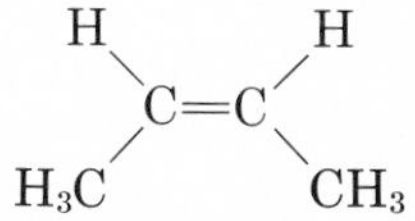

***cis*-2-Butene**

Although alkenes are not present in crude petroleum, they are produced in large quantities in petroleum *refining.* Many of the hydrocarbons in crude oil have very long chains and are solids or thick syrupy liquids at ordinary temperatures because of their relatively large intermolecular forces. The most important petroleum product, gasoline, requires molecules containing from 6 to 12 carbon atoms. These can be obtained by heating the longer-chain compounds. These big molecules writhe around so fast at higher temperatures that they "crack" or break into smaller fragments. Usually a catalyst is added to speed up the reaction, which is called **catalytic cracking.** When cracking occurs, one alkene and one alkane molecule are produced:

$$C_{20}H_{42} \xrightarrow[500°C]{Al_2O_3,\ SiO_2} C_{10}H_{22} + C_{10}H_{20}$$

An alkane **An alkane** **An alkene**

The most important alkenes from an industrial point of view are the two simplest: ethene (ethylene), $H_2C{=}CH_2$, and propene (propylene), $CH_3CH{=}CH_2$. Almost 9×10^9 kg of ethene and 3.5×10^9 kg of propene were produced in the United States in 1975, ranking these chemicals first and third among all organic compounds. Both are used in the manufacture of plastics, a topic we will discuss in Sec. 8.6. They are also raw materials for production of detergents, antifreeze, elastics, and lubricating oils.

Among the alkynes only the two simplest are of any industrial importance. Both ethyne (acetylene), HC≡CH, and propyne (methyl acetylene), $CH_3C{\equiv}CH$, are used in welding and steel cutting where they are burned together with pure oxygen gas in an *oxyacetylene torch.* As supplies of petroleum dwindle, though, acetylene may become more important as a starting material for the manufacture of other chemicals, since it can be made from coal.

8.3 HYDROGEN BONDING: WATER

very strong

It would seem that the London forces and dipole forces discussed in the earlier sections of this chapter should be adequate to account for macroscopic properties of covalently bonded substances. Certainly they can be successfully applied to hydrocarbons and many polar substances. There are some experimental data, however, which cannot be explained by London and dipole forces alone. An example appears in Fig. 8.9, where boiling points are plotted for hydrogen compounds (**hydrides**) of most of the nonmetals.

Hydrides of elements in the fifth period behave as we might predict. SnH_4, which consists of nonpolar molecules, boils at the lowest temperature. SbH_3, H_2Te, and HI, all of which are polar, have somewhat higher boiling points, but all lie within a range of 50°C. Similar behavior occurs among the hydrides of elements in the fourth and third periods. In the second period, however, the polar hydrides NH_3, H_2O, and HF all have boiling points more than 100°C above that of the nonpolar compound CH_4. Clearly these second-row hydrides must have particularly strong intermolecular forces.

In order to see why this happens, let us consider the simplest second-row hydride—HF. Suppose that two HF molecules approach each other, as shown in Fig. 8.10. In each HF molecule the hydrogen nucleus is rather poorly shielded by a thin electron cloud (only two electrons), and much of that electron cloud has been distorted toward the highly electronegative fluorine atom. Consequently the hydrogen nucleus in the HF molecule on

Figure 8.9 The boiling points of the hydrides of the nonmetals plotted against the period in which they occur in the periodic table. Note the anomalously high boiling points of H_2O, HF, and NH_3 in the second period.

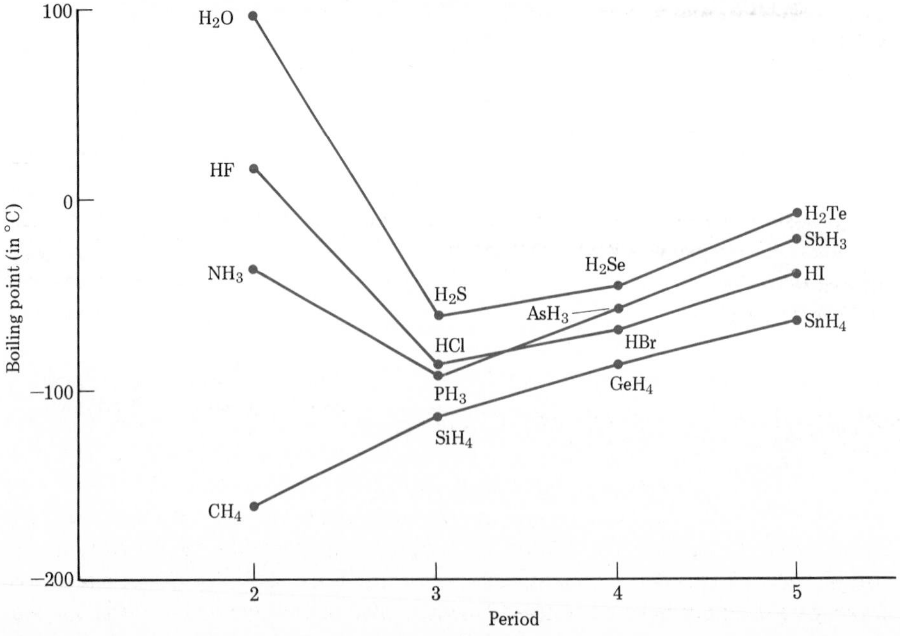

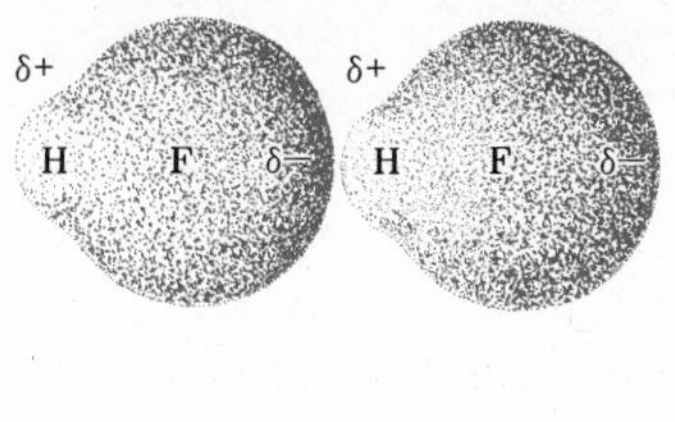

Figure 8.10 Hydrogen bonding between two HF molecules. When two HF dipoles line up in the position of least potential energy, the H nucleus in the one molecule is able to approach the F atom in the other particularly closely owing to the very poor shielding of the two electrons in its valence shell. A particularly strong dipole attraction is set up which is also reinforced by distortion of the electron clouds (not shown in the figure).

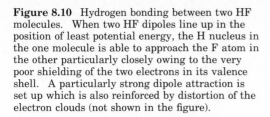

the right can approach much closer to the fluorine side of the left-hand molecule than could any other kind of nucleus. Also, because fluorine is such an electronegative atom, its partial negative charge is particularly large. The attractive force between the electrons of the left-hand fluorine and the right-hand hydrogen nucleus will be unusually strong, and there will be considerable distortion of the electron clouds. Some covalent-bond character develops between fluorine in one molecule and hydrogen in the other.

The close approach of oppositely charged ends of molecular dipoles combines with this small degree of covalent-bond character to produce an abnormally strong intermolecular force called a **hydrogen bond.** In order for hydrogen bonding to occur, there must be a hydrogen atom connected to a small, highly electronegative atom (usually fluorine, oxygen, or nitrogen) in one molecule. The other molecule must have a very electronegative atom (again usually fluorine, oxygen, or nitrogen) which has one or more lone pairs of electrons. Separation of two molecules joined by a hydrogen bond requires 10 to 30 kJ mol^{-1}, roughly 10 times the energy needed to overcome dipole forces. Thus hydrogen bonding can account for the unusually high boiling points of NH_3, H_2O, and HF.

Hydrogen bonding between HF molecules is particularly evident in solid HF, where the atoms are arranged in a zigzag pattern:

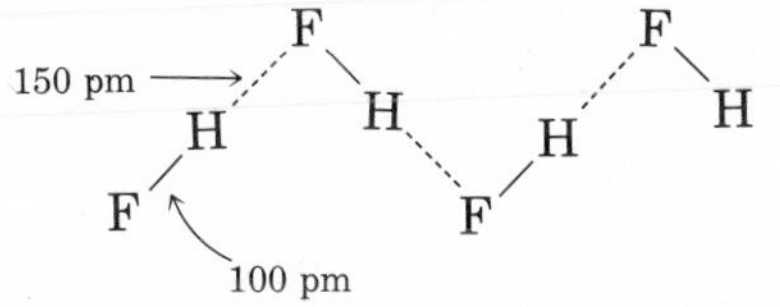

Here the distance between hydrogen and fluorine nuclei in different molecules is only 150 pm. If we add the van der Waals radii for hydrogen and fluorine given in Fig. 6.10, we obtain an expected hydrogen to fluorine distance of (120 + 135) pm = 255 pm, over 100 pm larger than that observed. Obviously the hydrogen and fluorine atoms in adjacent molecules are not just "touching," but must be associated in a much more intimate way.

Ice and Water

The next simplest, and by far the most important, example of hydrogen bonding is that which occurs in H_2O. Again there is clear evidence of hydrogen bonding in the structure of the solid. Figure 8.11 shows two computer-drawn diagrams of the crystal lattice of ice. In the ball-and-stick model we can clearly see that each O atom is surrounded by four H atoms arranged tetrahedrally. Two of these are at a distance of 99 pm and are clearly covalently bonded to the O atom. The other two are at a distance of

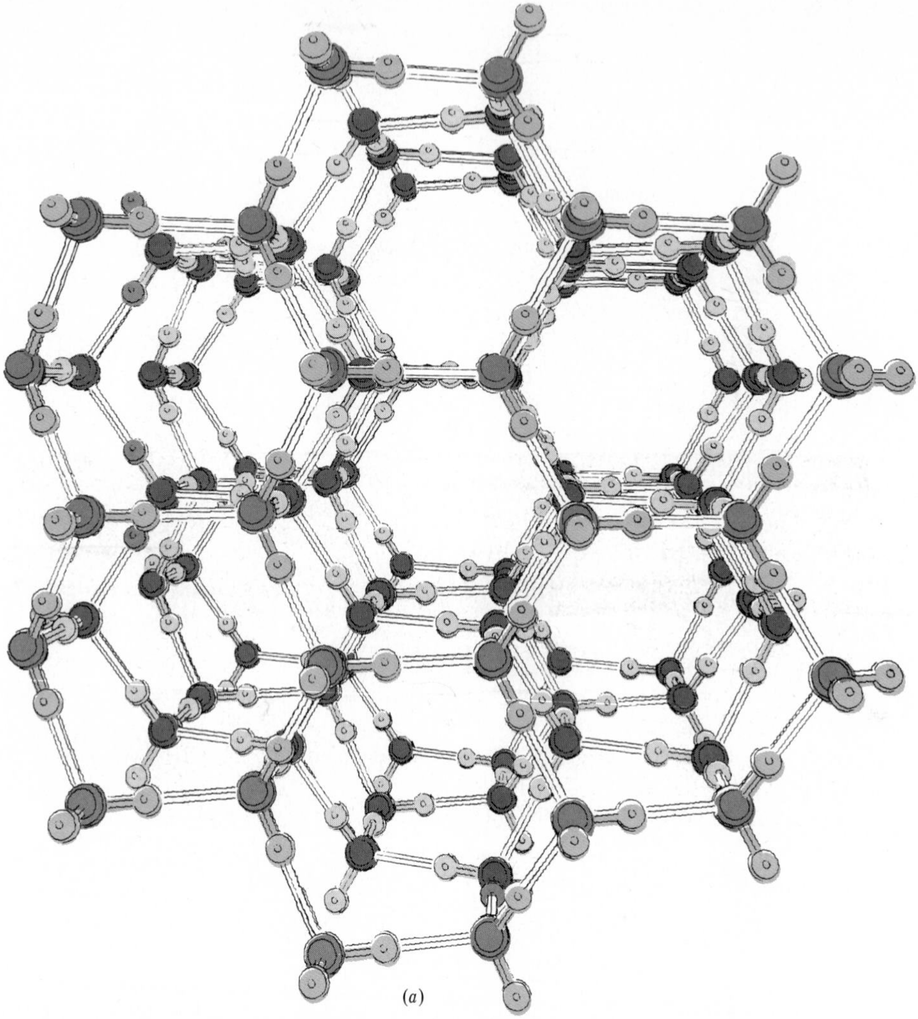

Figure 8.11 Two diagrams of the crystal structure of ice. Oxygen atoms are shown dark and hydrogen atoms light. The hexagonal, layerlike structure has been emphasized by indicating the front layer in color. Note in (*a*) how each O atom is surrounded by four H atoms, two close and two somewhat farther away. The shorter O—H distances correspond to covalent bonds, while the longer ones correspond to hydrogen bonds. The space-filling diagram in (*b*) clearly shows the large hexagonal channels running through the crystal. (Computer generated.) *Copyright © 1976 by W. G. Davies and J. W. Moore.*)

177 pm. They are covalently bonded to other O atoms but are hydrogen bonded to the one in question. The situation is thus:

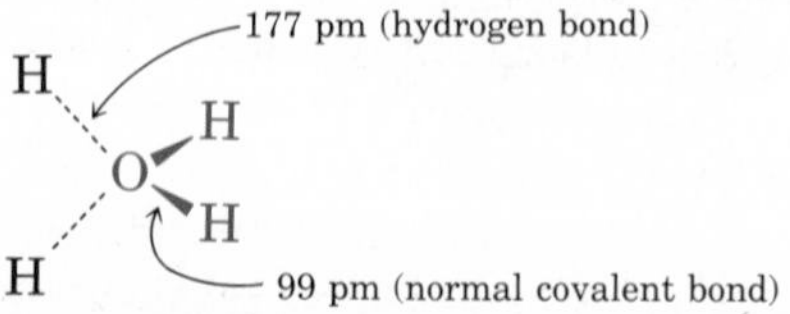

As in the case of HF, the distance between molecules is abnormally short. The sum of the van der Waals radii of H and O is 260 pm, considerably larger than the observed 177 pm.

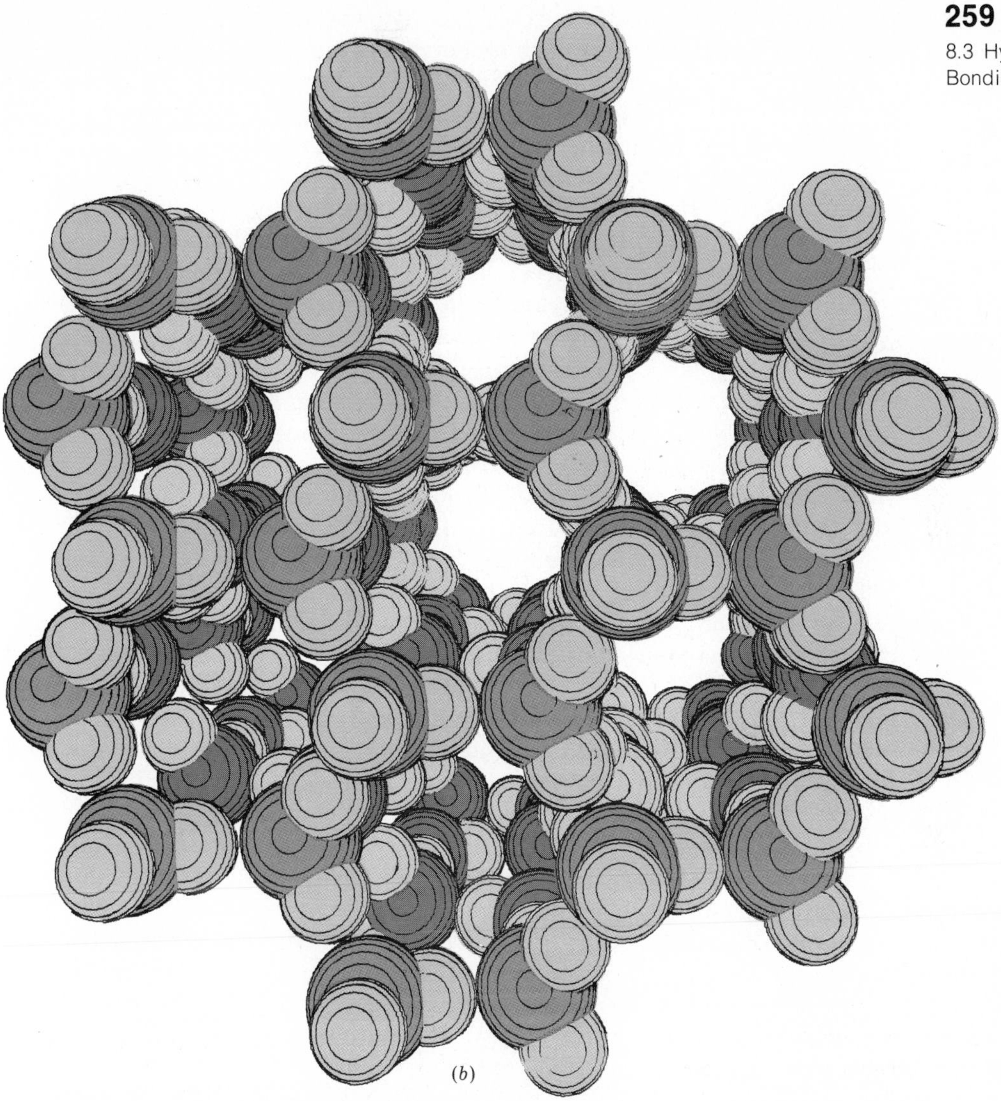

(b)

The tetrahedral orientation of H atoms around O atoms which results from hydrogen-bond formation has a profound effect on the properties of ice and of liquid water. Figure 8.11*b* is a space-filling diagram, in which most of the electron density of each H and O atom is enclosed by a boundary surface. As you can see, hydrogen bonding causes the H_2O molecules to adopt a rather open structure with hexagonal channels running through it. These channels contain an almost perfect vacuum—in them there is a little electron density from the surrounding atoms, but nothing else.

When ice melts, some of the hydrogen bonds are broken and the rigid crystal lattice shown in Fig. 8.11 collapses somewhat. The hexagonal channels become partially filled, and the volume of a given amount of H_2O decreases. This is the reason that ice is less dense than water and will float on it. As the temperature is raised above 0°C, more hydrogen bonds are broken, more empty space becomes occupied, and the volume continues to decrease. By the time 4°C has been reached, increased molecular velocities allow each H_2O molecule to push its neighbors farther away. This counteracts the effect of breaking hydrogen bonds, and the volume of a given amount of H_2O begins to increase with temperature.

Most solids expand when they melt, and the corresponding liquids expand continually with increasing temperature, so the behavior of water is rather unusual. It is also extremely important in the environment. When water freezes in small cracks in a rock, the greater volume of the ice can split the rock into smaller pieces. These eventually become able to support plant life, and so water contributes to the formation of fertile soil. Since water has maximum density at 4°C, water at that temperature sinks to the bottom of a deep lake, providing a relatively uniform environment all year around. If ice sank to the bottom, as most freezing liquids would, the surface of a lake would not be insulated from cold winter air. The remaining water would crystallize much more rapidly than it actually does. In a world where ice was denser than water, fish and other aquatic organisms would have to be able to withstand freezing for long periods.

Hydrogen bonding also contributes to the abnormally large quantities of heat that are required to melt, boil, or raise the temperature of a given quantity of water. Heat energy is required to break hydrogen bonds as well as to make water molecules move faster, and so a given quantity of heat raises the temperature of a gram of water less than for almost any other liquid. Even at 100°C there are still a great many unbroken hydrogen bonds, and almost 4 times as much heat is required to vaporize a mole of water than would be expected if there were no hydrogen bonding. This extra-large energy requirement is the reason that water has a higher boiling point than any of the hydrides in Fig. 8.9.

The fact that it takes a lot of heat to melt, boil, or increase the temperature of water, makes this liquid ideal for transferring heat from one place to another. Water is used by engineers in automobile radiators, hot-water heating systems, and solar-energy collectors. More significantly, circulation (in the bloodstream) and evaporation (from the skin) of water regulate the temperature of the human body. (You are between 55 and 65 percent water if female and between 65 and 75 percent water if male.) Because of this (as well as for many other reasons) water is an important component of living systems. Water's ability to store heat energy is also a major factor affecting world climate. Persons who live near large lakes or oceans experience smaller fluctuations in temperature between winter and summer than those who inhabit places like Siberia, thousands of kilometers from a sizable body of water. Ocean currents, such as the Gulf Stream, convey heat from the tropics to areas which otherwise would be quite cold. It is interesting to ask, for example, whether European civilization could have developed without the aid of warmth transported by the common, but highly unusual, liquid—water.

8.4 ORGANIC COMPOUNDS—SOME ADDITIONAL CLASSES

Although a tremendous number of different hydrocarbons occur as a result of carbon's ability to bond in long chains, an even greater variety of substances is possible when oxygen, nitrogen, and several other elements combine with carbon and hydrogen. The presence of highly electronegative atoms like oxygen or nitrogen in combination with hydrogen permits hydrogen bonds to form between molecules of many of these substances. We have deferred discussion of their properties until now so that you can apply your knowledge of hydrogen bonding to them.

In Sec. 8.2 we mentioned that alkanes were relatively unreactive and that the presence of a double or triple bond made unsaturated molecules more likely to combine chemically. A site which makes an organic molecule more reactive than a simple hydrocarbon chain is called a **functional group.** Many of the important organic functional groups involve oxygen atoms, nitrogen atoms, or both. We will discuss substances containing some of these functional groups in this section.

Alcohols

Molecules of alcohols contain one or more **hydroxyl groups** (OH groups) substituted for hydrogen atoms along the carbon chain. The structure of the simplest alcohol, methanol (methyl alcohol), can be derived from that of methane by putting an OH in place of one of the H's:

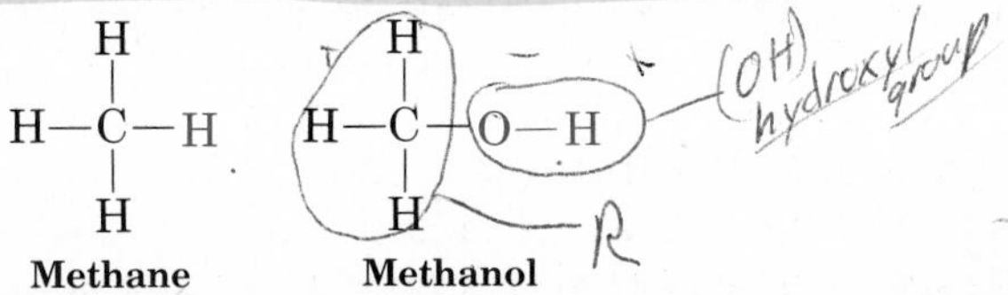

Methane **Methanol**

The name, too, is derived from methane by replacing the final *e* with *ol* (for alcoh*ol*). The general formula for an alcohol may be written as R—OH, where R represents the hydrocarbon (alkane) portion of the molecule and is called an **alkyl group.** In methanol, R is the methyl group CH_3.

Methanol is also called wood alcohol because it can be obtained by heating wood in the absence of air, a process called **destructive distillation.** Methanol vapor given off when the wood is heated can be condensed to a liquid by cooling below its boiling point of 65°C. The effect of polarity and especially hydrogen bonding due to the OH group is evident when this is compared with the temperature of −89°C at which ethane, C_2H_6, boils. Both molecules contain 18 electrons and are nearly the same size, and so London forces should be about the same, but the OH group in one methanol molecule can form strong hydrogen bonds with an OH in another molecule.

Methanol is an important industrial chemical—nearly 2.4×10^9 kg was produced in the United States in 1975. Some was made by destructive distillation, but most was synthesized from hydrogen and carbon monoxide:

$$2H_2(g) + CO(g) \xrightarrow{250 \text{ to } 400°C} CH_3OH(l)$$

This reaction is carried out at pressures several hundred times normal atmospheric pressure, using metal oxides as catalysts. Methanol is mainly used to make other compounds from which plastics are manufactured, but some is consumed as fuel in jet engines and racing cars. Methanol is also a component of nonpermanent antifreeze and automobile windshield-washer solvent.

The second member of the alcohol family is ethanol (ethyl alcohol)—the substance we commonly call *alcohol*. Ethanol is also known as grain alcohol because it is obtained when grain or sugar ferments. **Fermentation** refers to a chemical reaction which is speeded up by enzymes and occurs in the absence of air. (Enzymes, catalysts which occur naturally in yeasts and other living organisms, will be discussed in more detail in Chap. 18.)

Ethanol can also be synthesized by adding H_2O to ethene, obtained during petroleum refining:

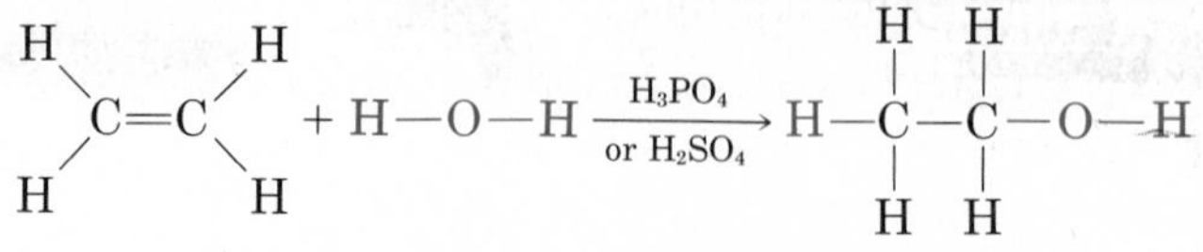

This is a typical example of an **addition reaction.** The H and OH from H_2O are added to the ethene molecule and held there by electrons made available when one-half of the double bond breaks.

Ethanol is used as a solvent, in some special fuels, in antifreeze, and to manufacture a number of other chemicals. You are probably most familiar with it as a component of alcoholic beverages. Ethanol makes up 3 to 6 percent of beer, 12 to 15 percent of most wines, and 40 to 50 percent of distilled liquor. (The "proof" of an alcoholic beverage is just twice the percentage of ethanol.) Alcohol's intoxicating effects are well known, and it is a mild depressant. Prolonged overuse can lead to liver damage. Methanol also produces intoxication but is much more poisonous than ethanol—it can cause blindness and death. Denatured alcohol is ethanol to which methanol or some other poison has been added, making it unfit for human consumption. Most of the ethanol not used in alcoholic beverages is denatured because in that form its sale is taxed at a much lower rate.

Two isomers are possible for alcohols containing three carbon atoms:

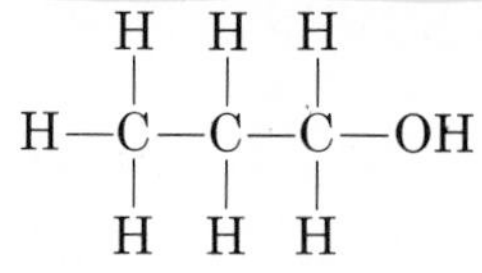

1-Propanol
(Normal propyl alcohol)

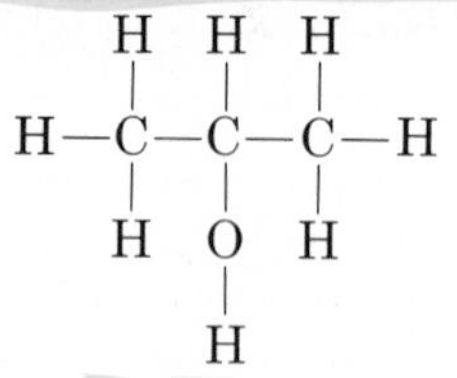

2-Propanol
(Isopropyl alcohol)

The *1* and the *2* in the names of these compounds indicate the position of the OH group along the carbon chain. The propanols are much less important commercially than methanol and ethanol, although 2-propanol is commonly found in rubbing alcohol.

The structures of two more alcohols which are of commercial importance and are familiar to many persons are shown below:

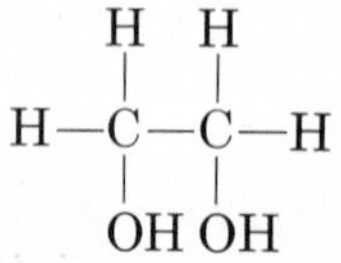

1,2-Ethanediol
(Ethylene glycol, or glycol)

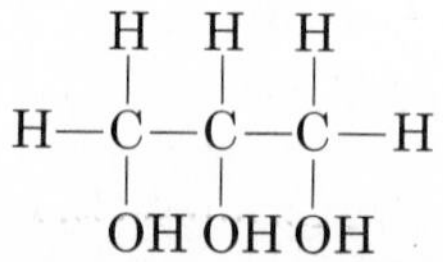

1,2,3-Propanetriol
(Glycerol, or glycerin)

Each ethylene glycol molecule has two hydrogens which can participate in hydrogen bonding, and each glycerol molecule has three. Both substances have rather high boiling points (198°C for ethylene glycol and 290°C for glycerol) and are syrupy, viscous liquids at room temperature. Their resistance to flowing freely is due to the network of hydrogen bonds which links each molecule to several of its fellows, making it more difficult for them to slide past one another.

Ethylene glycol is the principal component of permanent antifreeze for automobiles and is also used to manufacture polyester fibers. More than 1.7×10^9 kg was produced in the United States in 1975. Glycerol is used as a lubricant and in the manufacture of explosives:

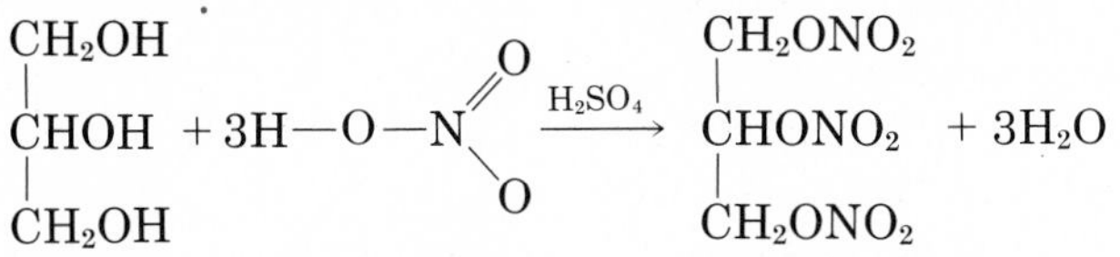

(8.1)

Glycerin **Nitric acid** **Nitroglycerin**

When nitroglycerin is mixed with a solid material such as nitrocellulose (which is made by treating cotton or wood pulp with nitric acid), the product is dynamite.

Ethers

In alcohols one of the two bonds to the oxygen atom involves hydrogen and one involves carbon. When two or more carbon atoms are present, however, isomeric structures in which oxygen is bonded to two different carbons become possible. Such compounds are called ethers. For example, dimethyl ether is isomeric with ethanol, and methyl ethyl ether is isomeric with propanol:

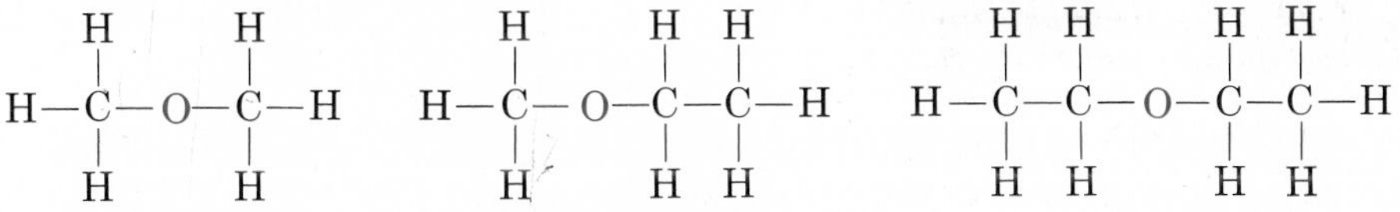

Dimethyl ether **Methyl ethyl ether** **Diethyl ether**

The general formula for an ether is R—O—R′, where R′ signifies that both R groups need not be the same.

EXAMPLE 8.3 Draw projection formulas and name all the isomers which correspond to the molecular formula C_3H_8O.

Solution The formula C_3H_8 would correspond to an alkane. The extra oxygen atom might be added between two carbons, giving an ether, or it might be added between a carbon and a hydrogen, giving an alcohol. The

Properties of
Organic
Compounds
and Other
Covalent
Substances

alcohol molecules might have the hydroxyl group at the end of the three-carbon chain or on the second carbon atom:

$$
\begin{array}{ccc}
\text{H} & \text{H} & \text{H} \\
| & | & | \\
\text{H}-\text{C}-\text{C}-\text{C}-\text{OH} \\
| & | & | \\
\text{H} & \text{H} & \text{H}
\end{array}
\qquad
\begin{array}{ccc}
& \text{H} & \\
& | & \\
\text{H} & \text{O} & \text{H} \\
| & | & | \\
\text{H}-\text{C}-\text{C}-\text{C}-\text{H} \\
| & | & | \\
\text{H} & \text{H} & \text{H}
\end{array}
$$

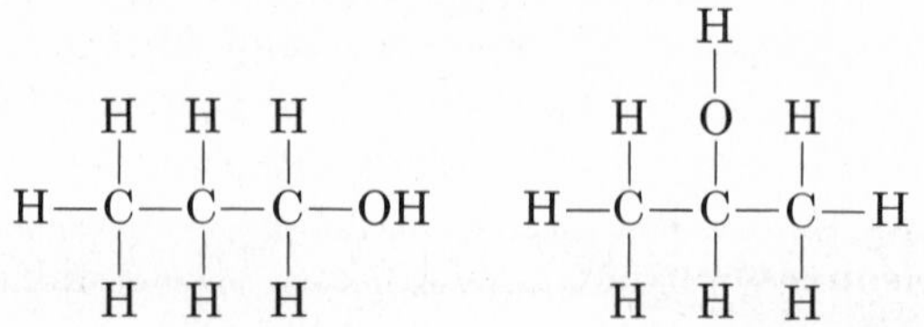

1-Propanol   **2-Propanol**

Only one ether structure is possible—that in which one methyl and one ethyl group are attached to oxygen:

$$
\begin{array}{ccc}
\text{H} & \text{H} & \text{H} \\
| & | & | \\
\text{H}-\text{C}-\text{O}-\text{C}-\text{C}-\text{H} \\
| & | & | \\
\text{H} & \text{H} & \text{H}
\end{array}
$$

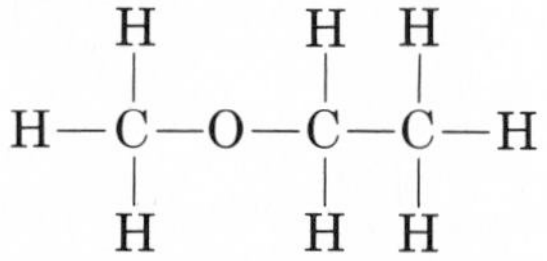

Methyl ethyl ether

In an ether there are no hydrogen atoms connected to a highly electronegative neighbor, and so, unlike alcohols, ether molecules cannot hydrogen bond among themselves. Each C—O bond is polar, but the bonds are at approximately the tetrahedral angle. The polarity of one partially cancels the polarity of the other. Consequently the forces between two ether molecules are not much greater than the London forces between alkane molecules of comparable size. The boiling point of dimethyl ether, for example, is $-23°C$, slightly above that of propane ($-42°C$), but well below that of ethanol ($78.5°C$). All three molecules contain 26 electrons and are about the same size.

The chemical reactivity of ethers is also closer to that of the alkanes than that of the alcohols. Ethers undergo few characteristic reactions other than combustion, and so they are commonly used as solvents. Diethyl ether is also used as an anesthetic, although the flammability of its vapor requires that precautions be taken to prevent fires.

Aldehydes and Ketones

Drinking methanol is harmful, not because of the CH_3OH molecules themselves, but rather because the human body converts these molecules into methanal (formaldehyde) molecules by combination with oxygen:

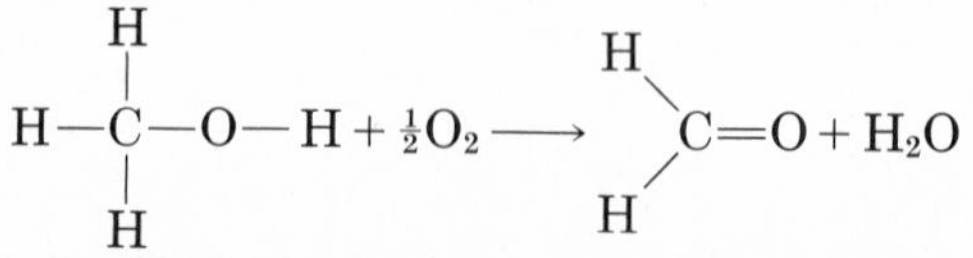

$$
\begin{array}{c}
\text{H} \\
| \\
\text{H}-\text{C}-\text{O}-\text{H} + \tfrac{1}{2}\text{O}_2 \longrightarrow \\
| \\
\text{H}
\end{array}
\quad
\begin{array}{c}
\text{H} \\
\diagdown \\
\quad\text{C}=\text{O} + \text{H}_2\text{O} \\
\diagup \\
\text{H}
\end{array}
\qquad (8.2)
$$

Formaldehyde, H_2CO, is very reactive—in the pure state it can combine explosively with itself, forming much larger molecules. Consequently it is prepared commercially as a water solution, formalin, which contains about 35 to 40 percent H_2CO. Formalin is made by combining methanol with air [Eq. (8.2)] at 550°C over a silver or copper catalyst. It is used as a preservative for biological specimens, in embalming fluids, and as a disinfectant and insecticide—not a very good substance to introduce into your body. The biggest commercial use of formaldehyde is manufacture of Bakelite, melamine, and other plastics.

The C=O functional group found in formaldehyde is called a **carbonyl group.** Two classes of compounds may be distinguished on the basis of the location of the carbonyl group. In **aldehydes** it is at the end of a carbon chain and has at least one hydrogen attached. In **ketones** the carbonyl group is attached to two carbon atoms. Some examples are

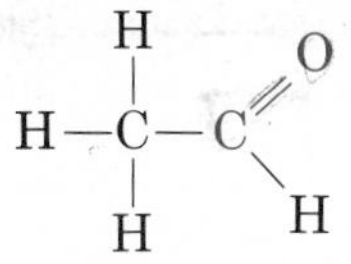

Ethanal
(Acetaldehyde)

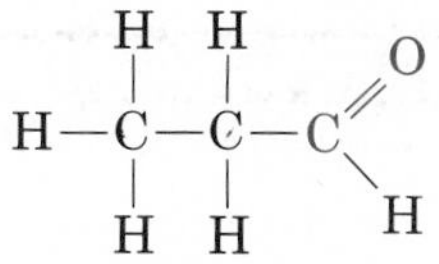

Propanal
(Propionaldehyde)

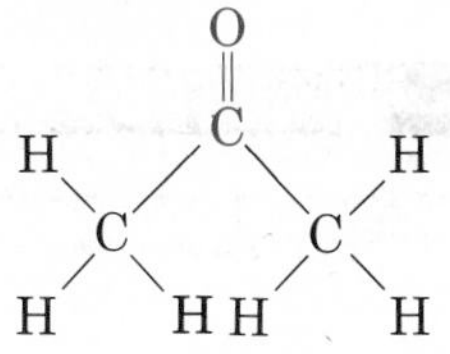

Propanone
(Acetone, or dimethyl ketone)

The endings *al* and *one* signify *al*dehyde and ket*one*, respectively. The general formula for an aldehyde is $R{-}\overset{\displaystyle H}{\overset{|}{C}}{=}O$, while for a ketone it is $R{-}\overset{\displaystyle O}{\overset{\|}{C}}{-}R'$. Note that every ketone is isomeric with at least one aldehyde. Acetone, for example, has the same molecular formula (C_3H_6O) as propanal.

Aldehyde and ketone molecules cannot hydrogen bond among themselves for the same reason that ethers cannot—they do not contain hydrogens attached to highly electronegative atoms. The carbonyl group is rather polar, however, since the difference between the electronegativities of carbon (2.5) and oxygen (3.5) is rather large, and there are usually no other dipoles in an aldehyde or ketone molecule to cancel the effect of C=O. Therefore the boiling points of aldehydes and ketones are intermediate between those of alkanes or ethers on the one hand and alcohols on the other. Acetaldehyde, CH_3CH_2CHO, boils at 20.8°C, midway between propane (−42°C) and ethanol (78.5°C).

Of all the aldehydes and ketones, only formaldehyde and acetone are of major commercial importance. Uses of formaldehyde have already been mentioned. Like other ketones, acetone is mainly useful as a solvent, and you may have used it for this purpose in the laboratory. Acetone and other ketones are somewhat toxic and should not be handled carelessly.

Carboxylic Acids

Serving wine usually involves a rather elaborate ceremony in which the host tastes the wine before pouring it for the guests. One reason for this is the possibility that the wine may have been spoiled by exposure to air.

Certain bacterial enzymes are capable of converting ethanol to ethanoic acid (acetic acid) when oxygen is present:

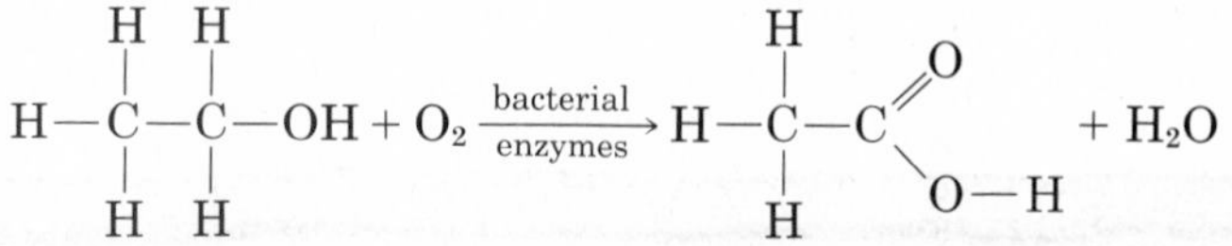

Acetic acid

(8.3)

The same reaction occurs when cider changes into vinegar, which contains 4 to 5 percent acetic acid. Acetic acid gives vinegar its sour taste and pungent odor and can do the same thing to wine.

Acetic acid, CH_3COOH, is an example of the class of compounds called **carboxylic acids,** each of which contains one or more **carboxyl groups,** COOH. The general formula of a carboxylic acid is RCOOH. Some other examples are

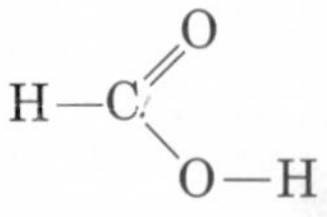

Methanoic acid
(Formic acid)

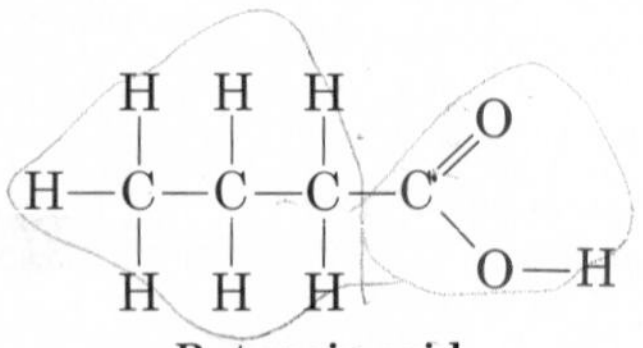

Butanoic acid
(Butyric acid)

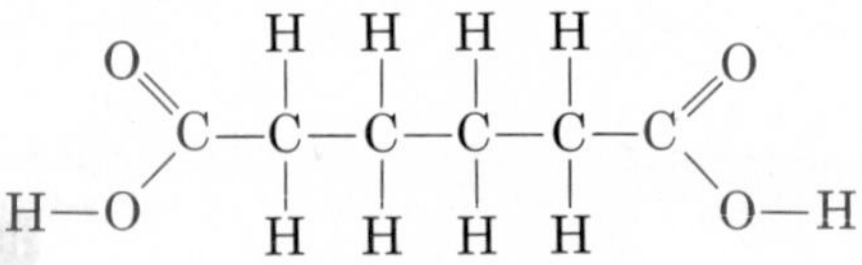

Hexanedioic acid
(Adipic acid)

Formic acid (the name comes from Latin word *formica* meaning "ant") is present in ants and bees and is responsible for the burning pain of their bites and stings. Butyric acid, a component of rancid butter and Limburger cheese, has a vile odor. Adipic acid is an example of a dicarboxylic acid—it has two functional groups—and is used to make nylon. We will describe this process in Sec. 8.6.

Since the carboxyl group contains a highly polar C=O as well as an OH group, hydrogen bonding is extensive among molecules of the carboxylic acids. Pure acetic acid is called *glacial* acetic acid because its melting point of 16.6°C is high enough that it can freeze in a cold laboratory. As you can see from Table 8.4, acetic acid boils at a higher temperature than any other organic substance whose molecules are of comparable size and have but one functional group. It is also quite thick and syrupy because of extensive hydrogen bonding.

Acetic acid is synthesized commercially according to Eq. (8.3), but silver is used as a catalyst instead of enzymes. It is also prepared by reacting air with propane separated from natural gas. The liquid acetaldehyde obtained in this reaction is then combined with oxygen in the presence of manganese(II) acetate to make acetic acid. About half the acetic acid produced in the United States goes into cellulose acetate from which acetate fibers are made.

TABLE 8.4 Boiling Points of Some Organic Compounds Whose Molecules Contain 32 or 34 Electrons.

Name	Projection Formula	Type of Compound	Boiling Point (in °C)
Isobutane	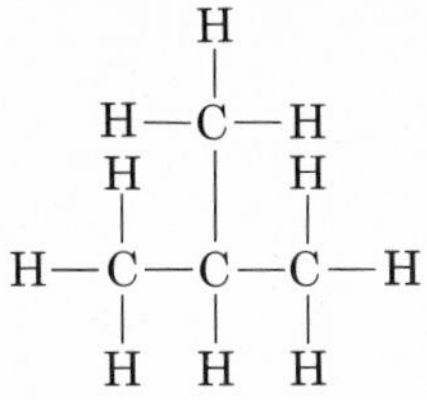	Branched alkane	−10.2
n-Butane	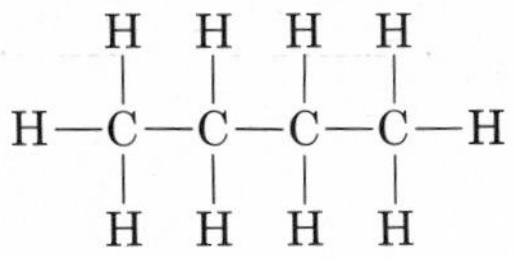	Normal alkane	−0.5
Methyl ethyl ether	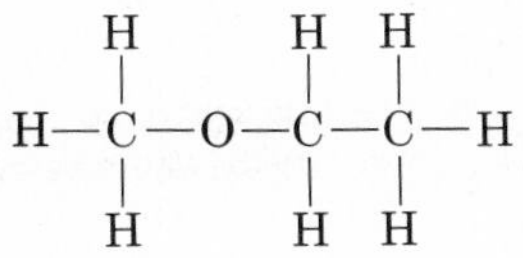	Ether	10.8
Methyl formate	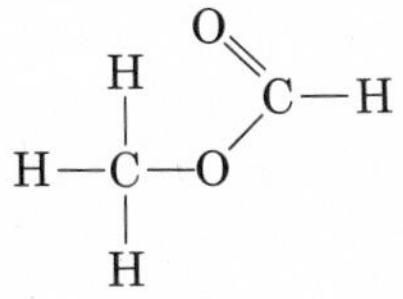	Ester	31.5
Propanal	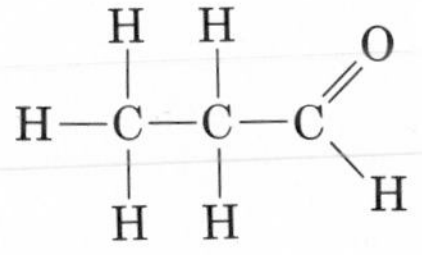	Aldehyde	48.8
Acetone	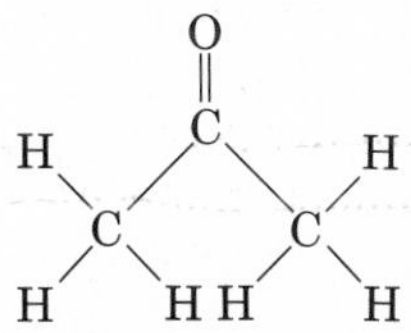	Ketone	56.2
2-Propanol	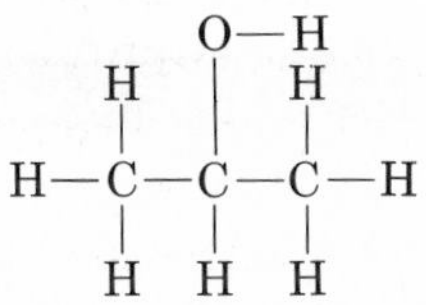	Alcohol	82.4
1-Propanol	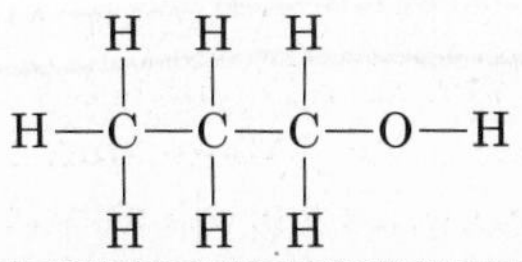	Alcohol	97.4

(*Continued on next page*)

TABLE 8.4 Boiling Points of Some Organic Compounds Whose Molecules Contain 32 or 34 Electrons (*Continued*).

Name	Projection Formula	Type of Compound	Boiling Point (in °C)
Acetic acid	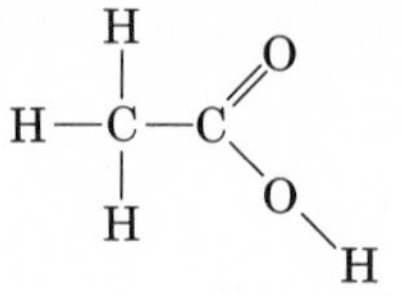	Carboxylic acid	117.9
Ethylene glycol	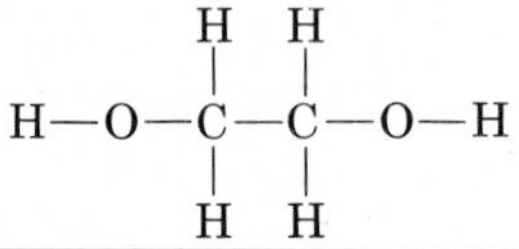	Dialcohol (two OH groups)	198

Esters

Cellulose acetate is a complicated example of another group of organic compounds, **esters,** which can be made by combining alcohols with acids. A simpler case is the reaction of ethanol with acetic acid to give ethyl acetate:

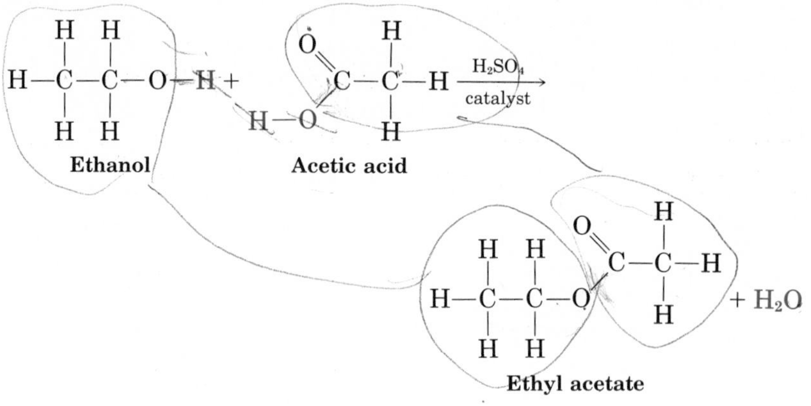

(8.4)

The general formula for an ester can be written as R—O—C(=O)—R′. In the case of ethyl acetate, R is CH_3CH_2 and R′ is CH_3. Equation (8.1), the synthesis of nitroglycerin, was also an example of ester formation, but in that case an inorganic acid, HNO_3, was combined with an alcohol.

Formation of an ester is an example of an important class of reactions called condensations. In a **condensation reaction** a pair of molecules join together, giving off a small, very stable molecule like H_2O or HCl. In both Eqs. (8.1) and (8.4), this small molecule is H_2O. A condensation can often be undone if large numbers of the small molecules are added to the product. In the case of an ester, addition of large quantities of H_2O causes **hydrolysis** (literally, "splitting by means of water"):

$$CH_3CH_2O\overset{\overset{O}{\|}}{C}CH_3 + H_2O \longrightarrow CH_3CH_2OH + HO\overset{\overset{O}{\|}}{C}CH_3$$

This is just the reverse of Eq. (8.4).

Although the ester functional group has a polar carbonyl, it contains no hydrogen atoms suitable for hydrogen bonding. Therefore esters have low boiling points relative to most molecules of similar size (see Table 8.4). In many cases, even though its molecules are almost twice as large as those of the constituent alcohol and acid, an ester is found to have a lower boiling point than either. Ethyl acetate, for example, boils at 77.1°C, lower than ethanol (78.5°C) or acetic acid (117.9°C). By contrast to acids and alcohols which have unpleasant and rather weak odors, respectively, esters usually smell good. The odors of many fruits and flowers are due to esters. Ethyl acetate, for example, is the most important factor in the flavor of pineapples.

Organic Nitrogen Compounds

The tremendous variety of organic compounds which can be derived from carbon, hydrogen, and oxygen is evident from Table 8.4 and the previous discussion in this section. If we include nitrogen as a possible constituent of these molecular structures, many more possibilities arise. Most of the nitrogen-containing compounds are less important commercially, however, and we will only discuss a few of them here.

Amines may be derived from ammonia by replacing one, two, or all three hydrogens with alkyl groups. Some examples are

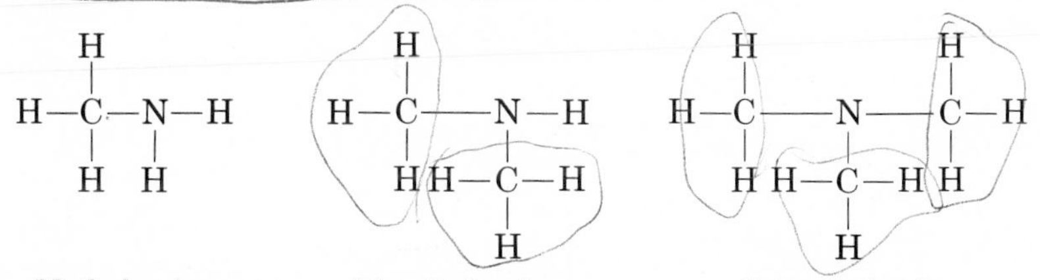

Methylamine (A primary amine) **Dimethylamine (A secondary amine)** **Trimethylamine (A tertiary amine)**

The terms primary (one), secondary (two), and tertiary (three) refer to the number of hydrogens that have been replaced. Both primary and secondary amines are capable of hydrogen bonding with themselves, but tertiary amines have no hydrogens on the electronegative nitrogen atom.

Amines usually have unpleasant odors, smelling "fishy." The three methylamines listed above can all be isolated from herring brine. Amines, as well as ammonia, are produced by decomposition of nitrogen-containing compounds when a living organism dies. The methylamines are obtained commercially by condensation of methanol with ammonia over an aluminum oxide catalyst:

$$CH_3OH + NH_3 \xrightarrow[Al_2O_3]{400^\circ C} CH_3NH_2 + H_2O$$

$$CH_3OH + CH_3NH_2 \xrightarrow[Al_2O_3]{400^\circ C} (CH_3)_2NH + H_2O$$

$$CH_3OH + (CH_3)_2NH \xrightarrow[Al_2O_3]{400^\circ C} (CH_3)_3N + H_2O$$

Dimethylamine is the most important, being used in the preparation of herbicides, in rubber vulcanization, and to synthesize dimethylformamide, an important solvent.

Sometimes an amine group and a carboxyl group are both found in the same molecule. Examples of such **amino acids** are glycine and lysine:

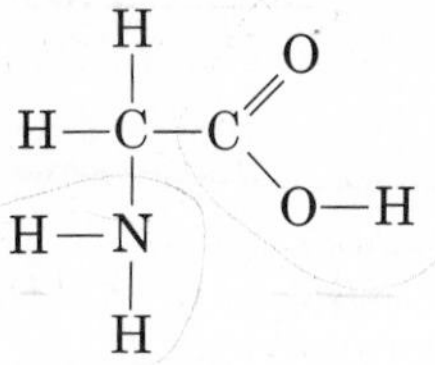

Glycine

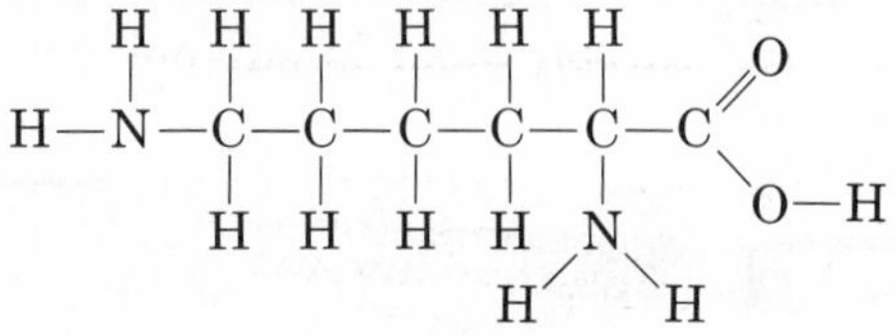

Lysine

Amino acids are the constituents from which proteins are made. Some, like glycine, can be synthesized in the human body, but others cannot. Lysine is an example of an **essential amino acid**—one which must be present in the human diet because it cannot by synthesized within the body. We will discuss amino acids and proteins further in Chaps. 18 and 20.

The intermolecular forces and boiling points of nitrogen-containing organic compounds may be explained according to the same principles used for oxygen-containing substances.

EXAMPLE 8.4 Rationalize the following boiling points: (a) 0°C for $CH_3CH_2CH_2CH_3$; (b) 11°C for $CH_3CH_2OCH_3$; (c) 97°C for $CH_3CH_2CH_2OH$; and (d) 170°C for $NH_2CH_2CH_2OH$.

Solution All four molecules have very similar geometries and the same number of electrons (26 valence electrons plus 8 core electrons), and so their London forces should be about the same. Compound (a) is an alkane and is nonpolar. By contrast compound (b) is an ether and should be slightly polar. This slight polarity results in a slightly higher boiling point. Compound (c) is isomeric with compound (b) but is an alcohol. There is hydrogen bonding between molecules of (c), and its boiling point is much higher. Molecule (d) has both an amino group and a hydroxyl group, each of which can participate in hydrogen bonding. Consequently it has the highest boiling point of all.

8.5 MACROMOLECULAR SUBSTANCES

Up to this point we have confined our discussion to covalently bonded substances consisting of small discrete molecules. The forces between these molecules are quite weak compared with covalent bonds. There are some covalently bonded substances, however, in which no small individual molecules exist. These substances contain extremely large molecules, and so they are called **macromolecular substances.** They are invariably solids at room temperature. Quite often we can regard each crystal of the substance as a *single molecule.*

Diamond and Graphite

The simplest example of a macromolecular solid is *diamond.* Crystals of diamond contain only carbon atoms, and these are linked to each other by covalent bonds in a giant three-dimensional network, as shown in Fig. 8.12. Note how each carbon atom is surrounded tetrahedrally by four bonds.

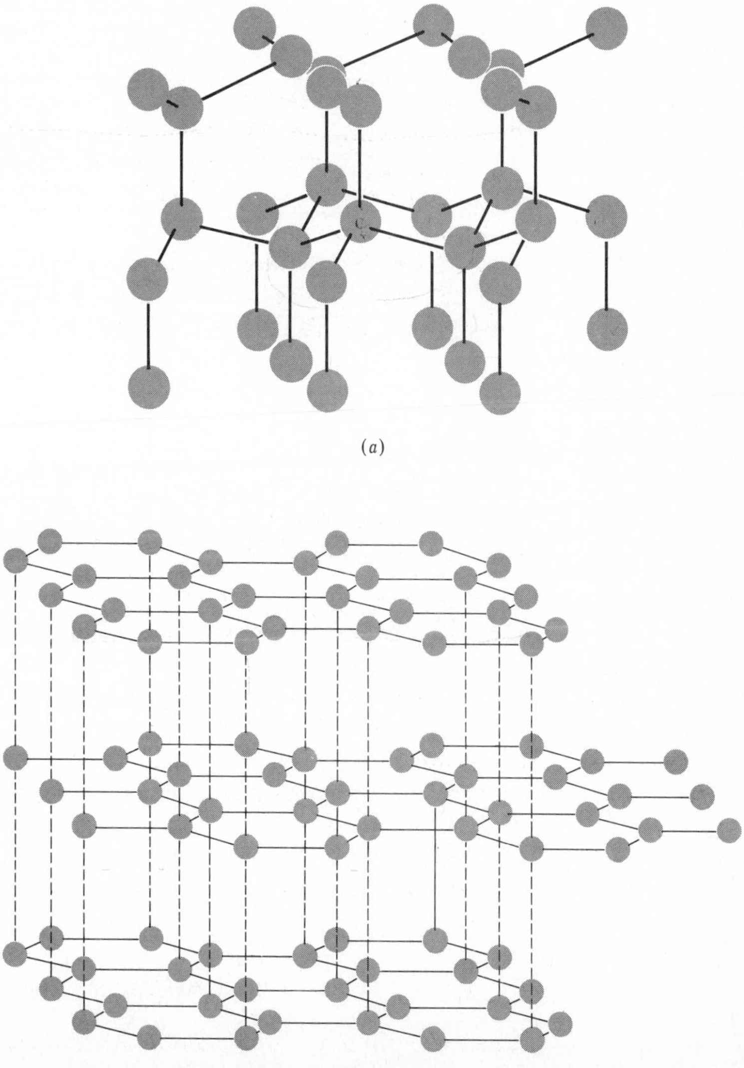

(*a*)

(*b*)

Figure 8.12 The crystal structure of (*a*) diamond and (*b*) graphite.

Such a network of carbon atoms extends throughout the crystal so that the whole diamond is one extremely large covalently bonded entity, i.e., a macromolecule.

Because strong covalent bonds, rather than London forces or dipole forces, hold the carbon atoms together in this crystal, it takes a great deal of energy to separate them. Accordingly, diamond has an extremely high melting point, 3550°C—much higher than any ionic solid. Diamond is also the hardest substance known. Each carbon atom is held firmly in its place from all sides and is thus very difficult to displace or remove.

Carbon also exists in a second, more familiar, crystalline form called *graphite,* whose crystal structure is also shown in Fig. 8.12. You use graphite every time you write with a pencil. (Pencil leads consist of C, not Pb!) The structure of graphite consists of flat layers. In each layer the carbon atoms are arranged in a regular hexagonal array. We can regard each layer as a large number of benzene rings fused together to form a gigantic honeycomb. All carbon-carbon bonds in this honeycomb are equivalent and intermediate in character between a single and a double bond.

While there are strong covalent bonds between the carbon atoms in a given plane, only weak London forces attract the planes together. The various layers can therefore slide past each other quite easily. When a pencil lead rubs across paper, the planes slide past each other and thin plates of crystal are left behind on the paper. These sliding plates also make graphite useful as a lubricant.

When an element can exist in more than one crystalline form, as carbon can in diamond and graphite, each form is said to be an **allotrope.** Other elements, such as sulfur and phosphorus, also form allotropes.

Silicon Dioxide

Silicon dioxide, or silica, (SiO_2) is another important example of a macromolecular solid. Silica can exist in six different crystalline forms. The best known of these is *quartz,* whose crystal structure was shown in Fig. 6.8. Sand consists mainly of small fragments of quartz crystals. Quartz has a very high melting point, though not so high as diamond.

If you refer back to Example 6.7 in Sec. 6.6, you can remind yourself of the reason that SiO_2 is macromolecular. Silicon is reluctant to form multiple bonds, and so discrete O═Si═O molecules, analogous to O═C═O, do not occur. In order to satisfy silicon's valence of 4 and oxygen's valence of 2, each silicon must be surrounded by four oxygens and each oxygen by two silicons. This can be represented schematically by the Lewis diagram

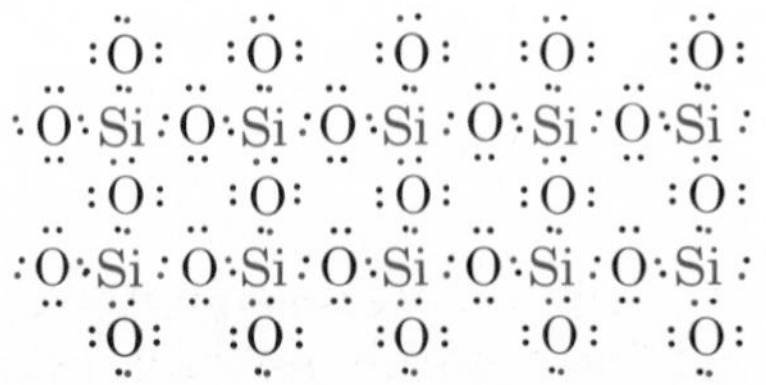

About 12 percent of all the material in the crust of the earth (a layer which extends about 30 to 40 km below the surface) is silica. A group of substances which are closely related to silica, the **feldspars,** are even more important. They constitute about 50 percent of the earth's crust. One of the simplest of the feldspars, and also one of the most abundant, is *orthoclase,* which has the formula $KAlSi_3O_8$. The structure of this mineral is best described in terms of the schematic Lewis diagram

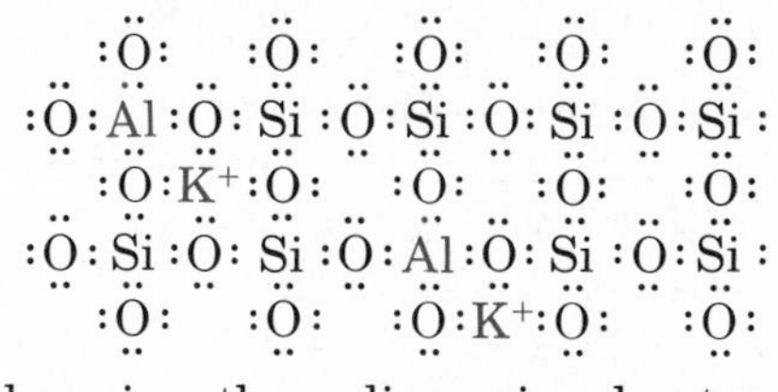

As with silica, there is a three-dimensional network of interconnected tetrahedrons of O atoms. In orthoclase, though, every fourth tetrahedron has an Al rather than an Si atom at its center. Since Al has one less valence electron than Si, an extra electron is needed to preserve the octet structure of the network. This electron can be donated by a K atom, which then becomes a potassium ion, K^+. Thus for every Al atom in the middle of a tetrahedron of O atoms, there is a K^+ ion close by. The three-dimensional framework of Al, Si, and O atoms is thus a giant polyatomic ion with the formula $(AlSi_3O_8^-)_n$. (The number of Al atoms in the crystal is n.) The crystal is electrically neutral because of the presence of $n\,K^+$ ions. The actual network of tetrahedrons into which the K^+ ions fit is somewhat different from that shown for quartz in Fig. 6.8 because the channels in the quartz network are not large enough to accommodate K^+ ions.

The other feldspars are similar to orthoclase in general structure. All consist of a three-dimensional network of interlocked tetrahedrons of O with both Al and Si atoms at their centers. Since these networks are negatively charged, they must also accommodate positive ions. Na^+, Mg^{2+}, and Ca^{2+} ions are found as well as K^+.

In addition to feldspars and silica, a further 20 percent of the earth's crust also consists of silicon compounds. In these minerals the tetrahedrons of O around each Si atom are not as extensively interlinked as in quartz and the feldspars. An example is shown in Fig. 8.13. The flat sheet shown has the formula $(Si_2O_5^{2-})_n$. In minerals such as mica these flat sheets are layered above one another and positive ions occupy the spaces between them. The overall effect is like a giant multilayered sandwich. The sheets in this structure can easily be separated and account for the ease with which crystals of mica cleave into plates a few tenths of a millimeter thick.

8.6 SYNTHETIC MACROMOLECULES: SOME APPLIED ORGANIC CHEMISTRY

Other important classes of substances containing very large molecules are the plastics and artificial fibers, which are such a conspicuous, though not

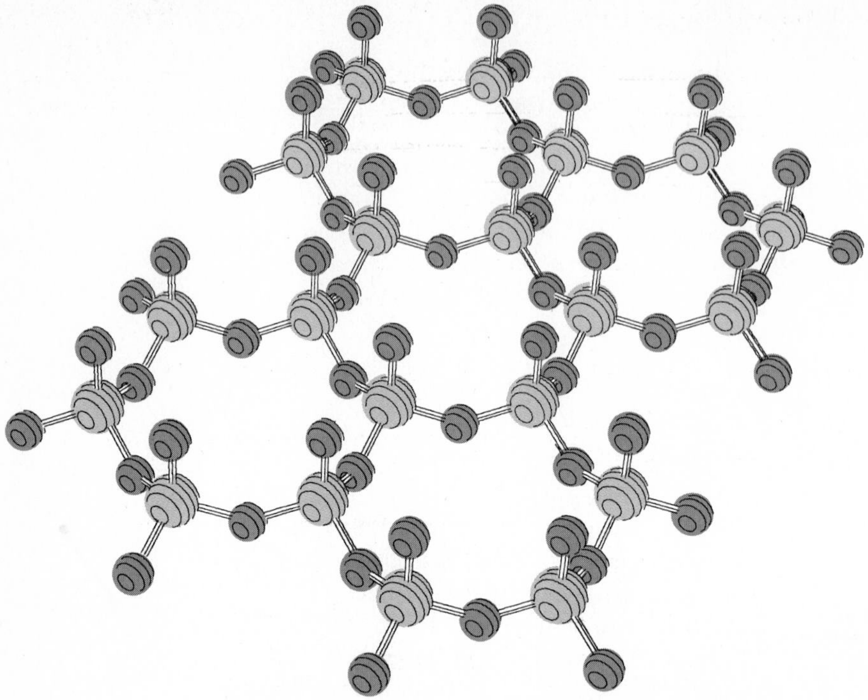

Figure 8.13 Silicate sheet with the formula $Si_2O_5^{2-}$. Sheets like this occur both in micas and in clays. (Computer generated.) (*Copyright © 1977 by W. G. Davies and J. W. Moore.*)

always a positive, feature of modern life. Most of these materials are made in the same basic way. The starting materials or **monomers** are relatively simple molecules—usually carbon compounds derived from petroleum—which can be persuaded to link up with each other in order to form a long chain of repeating units called a **polymer.** If we think of the monomer as a bead, then the polymer corresponds to a string of beads. Polymers are usually classified into two types: **addition polymers** and **condensation polymers,** according to the kind of reaction by which they are made.

Addition Polymers

Addition polymers are usually made from a monomer containing a double bond. We can think of the double bond as "opening out" in order to participate in two new single bonds in the following way:

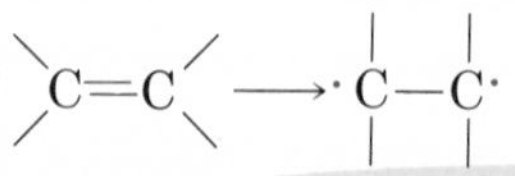

Thus, if ethene is heated at moderate temperature and pressure in the presence of an appropriate catalyst, it polymerizes:

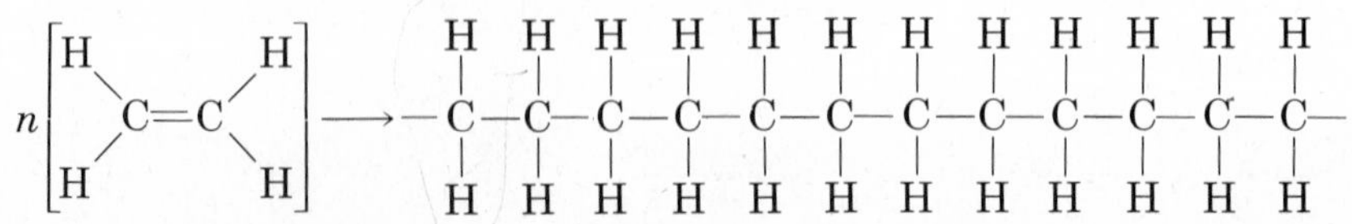

TABLE 8.5 Some Common Addition Polymers.

Monomer	Nonsystematic Name	Polymer	Some Typical Uses
$H_2C{=}CH_2$	Ethylene	Polyethylene	Film for packaging and bags, toys, bottles, coatings
$H_2C{=}CHCH_3$	Propylene	Polypropylene	Milk cartons, rope, outdoor carpeting
$H_2C{=}CH$ 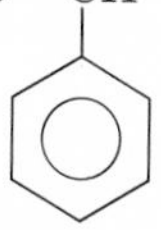	Styrene	Polystyrene	Transparent containers, plastic glasses, refrigerators, styrofoam
$H_2C{=}CHCl$	Vinyl chloride	Polyvinyl chloride, PVC	Pipe and tubing, raincoats, curtains, phonograph records, luggage, floor tiles
$H_2C{=}CHCN$	Acrylonitrile	Polyacrylonitrile (Orlon, Acrilan)	Textiles, rugs
$F_2C{=}CF_2$	Tetrafluoroethylene	Teflon	Nonstick pan coatings, bearings, gaskets

The result is the familiar waxy plastic called polyethylene, which at a molecular level consists of a collection of long-chain alkane molecules, most of which contain tens of thousands of carbon atoms. There is only an occasional short branch chain.

Polyethylene is currently manufactured on a very large scale, larger than any other polymer, and is used for making plastic bags, cheap bottles, toys, etc. Many of its properties are what we would expect from its molecular composition. The fact that it is a mixture of molecules each of slightly different chain length (and hence slightly different melting point) explains why it softens over a range of temperatures rather than having a single melting point. Because the molecules are only held together by London forces, this melting and softening occurs at a rather low temperature. (Some of the cheaper varieties of polyethylene with shorter chains and more branch chains will even soften in boiling water.) The same weak London forces explain why polyethylene is soft and easy to scratch and why it is not very strong mechanically.

Table 8.5 lists some other well-known addition polymers and also some of their uses. You can probably find at least one example of each of them in your home. Except for Teflon, all these polymers derive from a monomer of the form

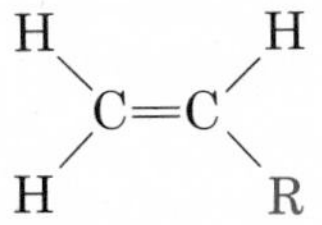

The resulting polymer thus has the general form

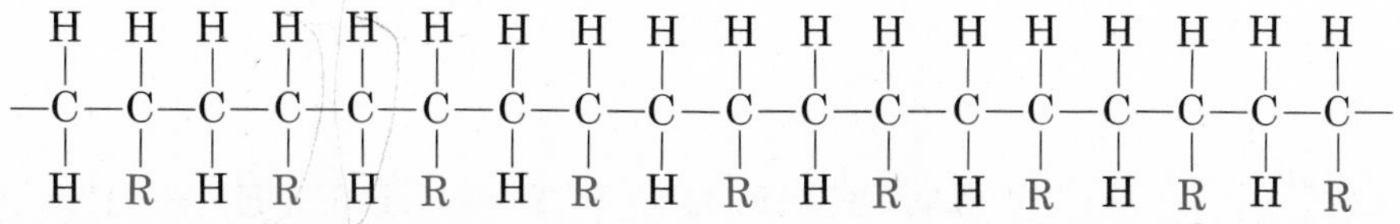

By varying the nature of the R group, the physical properties of the polymer can be controlled rather precisely.

Condensation Polymers

When addition polymers are formed, no by-products result. Formation of a condensation polymer, on the other hand, produces H_2O, HCl, or some other simple molecule which escapes as a gas. A familiar example of a condensation polymer is *nylon,* which is obtained from the reaction of two monomers

$$H_2N{-}CH_2{-}CH_2{-}CH_2{-}CH_2{-}CH_2{-}CH_2{-}NH_2 \qquad \text{and}$$

Hexamethylenediamine

$$HO{-}\overset{\overset{O}{\|}}{C}{-}CH_2{-}CH_2{-}CH_2{-}CH_2{-}\overset{\overset{O}{\|}}{C}{-}OH$$

Adipic acid

These two molecules can link up with each other because each contains a reactive functional group at both ends. They combine as follows:

$$(n+1)H_2N(CH_2)_6NH_2 + (n+1)HO\overset{\overset{O}{\|}}{C}(CH_2)_4\overset{\overset{O}{\|}}{C}OH \longrightarrow$$

$$H_2N(CH_2)_6\overset{\overset{H}{|}}{N}\left[\underset{\underset{O}{\|}}{C}(CH_2)_4\overset{\overset{O}{\|}}{C}{-}\underset{\underset{H}{|}}{N}(CH_2)_6\overset{\overset{H}{|}}{N}\right]_n\underset{\underset{O}{\|}}{C}(CH_2)_4\overset{\overset{O}{\|}}{C}OH + nH_2O \qquad (8.5)$$

Well-known condensation polymers other than nylon are Dacron, Bakelite, melamine, and Mylar.

Nylon makes extremely strong threads and fibers because its long-chain molecules have stronger intermolecular forces than the London forces of polyethylene. Each N—H group in a nylon chain can hydrogen bond to the O of a C═O group in a neighboring chain, as shown in Fig. 8.14. Therefore the chains cannot slide past one another easily. If you pull on both

Figure 8.14 Hydrogen bonds (dotted) between polymer chains in nylon. Since each chain contains thousands of N—H and C═O groups, a great many hydrogen bonds must be broken in order to separate the chains. Note that the diagram includes the ends of two chains. These are held in place because they overlap with and hydrogen bond to adjacent chains.

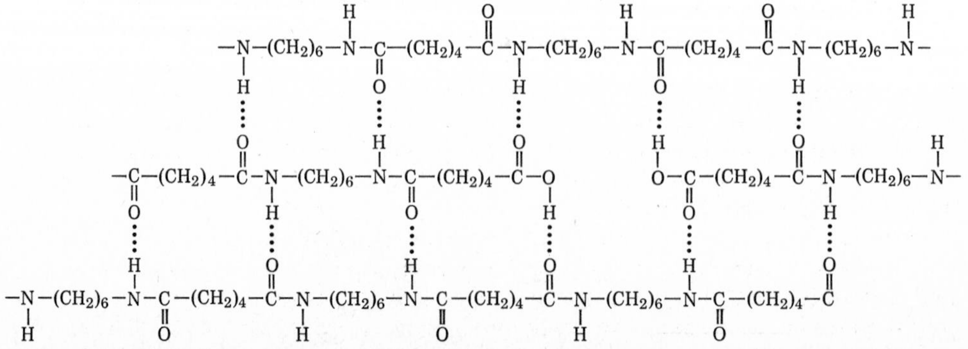

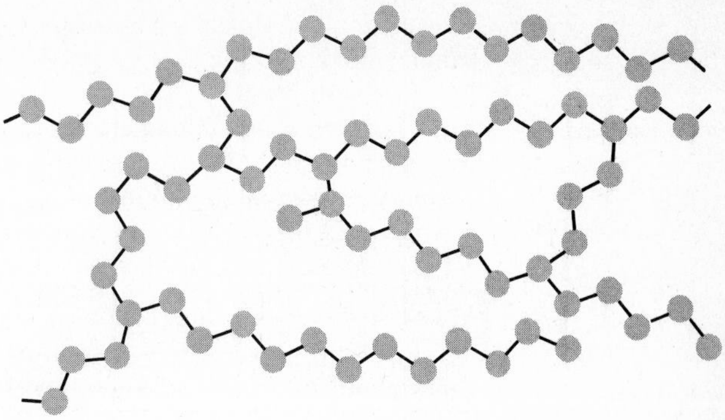

Figure 8.15 A cross-linked polymer. For purposes of clarity, hydrogen atoms and side chains have been omitted, and only the carbon atoms in the chains are shown. Note that the cross links between chains occur at random.

ends of a nylon thread, for example, it will only stretch slightly. After that it will strongly resist breaking because a large number of hydrogen bonds are holding overlapping chains together. The same is not true of a polyethylene thread in which only London forces attract overlapping chains together, and this is one reason that polyethylene is not used to make thread.

Cross-Linking

The formation of covalent bonds which hold portions of several polymer chains together is called **cross-linking.** Extensive cross-linking results in a random three-dimensional network of interconnected chains, as shown in Fig. 8.15. As one might expect, extensive cross-linking produces a substance which has more rigidity, hardness, and a higher melting point than the equivalent polymer without cross-linking. Almost all the hard and rigid plastics we use are cross-linked. These include *Bakelite,* which is used in many electric plugs and sockets, *melamine,* which is used in plastic crockery, and *epoxy resin* glues.

SUMMARY

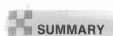

Macroscopic physical properties such as melting and boiling points depend on the strengths of the forces which hold microscopic particles together. In the case of molecules whose atoms are connected by covalent bonds, such intermolecular forces may be of three types. All molecules are attracted together by weak London forces. These depend on instantaneous polarization and increase in strength with the size of the molecular electron cloud. When a molecule contains atoms whose electronegativities differ significantly and the resulting bond dipoles do not cancel each other's effects, dipole forces occur. This results in higher melting and boiling points than for nonpolar substances.

The third type of intermolecular force, the hydrogen bond, occurs when one molecule contains a hydrogen atom connected to a highly electronegative partner. The other molecule must contain an electronegative atom, like fluorine, oxygen, or nitrogen, which has a lone pair. Although each hydrogen bond is weak compared with a covalent bond, large numbers of hydrogen bonds can have very significant effects.

One example of this is in the properties of water. This highly unusual liquid plays a major role in making living systems and the earth's environment behave as they do.

Carbon normally forms four bonds, and carbon-carbon bonds are quite strong, allowing formation of long chains to which side branches and a variety of functional groups may be attached. Hence the number of molecular structures which can be adopted by organic compounds is extremely large. Functional groups containing oxygen atoms, nitrogen atoms, and multiple bonds often determine the chemical and physical properties of carbon compounds. Therefore organic chemistry may be systematized by studying related groups of compounds such as alkanes, aromatic compounds, alkenes, alkynes, alcohols, ethers, aldehydes, ketones, carboxylic acids, esters, amines, and so on. Within each of these categories chemical and physical behaviors are closely related to molecular structures.

Some covalently bonded substances do not consist of individual small molecules. Instead giant macromolecules form. Examples include most of the rocks in the crust of the earth as well as modern plastics. Properties of such substances depend on whether the macromolecules are three dimensional (like diamond), two dimensional (like graphite or sheets of mica) or one dimensional (like polyethylene). In the latter two cases the strengths of forces between adjacent macromolecules have a significant effect on the properties of the substances.

In addition to the covalently bonded substances discussed in this chapter, we have already described metals (Sec. 6.2) and ionic compounds (Secs. 6.1 and 6.3). Each metal atom has one or more loosely held electrons which can easily transfer to nearby metal atoms in a solid. Attractive forces between these metallic valence electrons and the positive nuclei of metal atoms hold the atoms together. Ionic crystal lattices are maintained by attractions among oppositely charged ions. As in the compounds discussed in this chapter, physical and chemical properties of metals and ionic compounds may be more easily understood when their structures and the forces between atomic or ionic structural units are kept in mind. Table 8.6 summarizes the properties of all types of substances in terms of microscopic structures and forces.

QUESTIONS AND PROBLEMS

8.1 What evidence can you give that all neutral molecules exert forces of attraction on each other?

8.2 Explain how (*a*) polar and (*b*) nonpolar molecules can exert forces of attraction on each other.

8.3 Distinguish between London forces, dipole forces, and van der Waals forces.

8.4 Decide which substance in each of the following pairs would be expected to have the higher boiling point. In each case give reasons.

a Ar or Kr	*e* CO_2 or CS_2
b Ar or HCl	*f* SiH_4 or Ar
c KCl or Cl_2	*g* Ne or NH_3
d CO_2 or SiO_2	*h* SCl_2 or $MgCl_2$

8.5 Draw projection formulas for all five isomers of hexane, C_6H_{14}.

8.6 Distinguish between the term *isomer* and the term *conformation*.

8.7 A student writes, "The boiling point of the alkanes always increase with chain length. Because the longer chains are heavier. A heavier molecule needs more energy to work against gravity when it boils." Rewrite these sentences in acceptable standard English and show the fallacy in the argument.

8.8 Write balanced chemical equations for the complete combustion of each of the following hydrocarbons: (*a*) C_8H_{18}; (*b*) C_6H_6.

TABLE 8.6 Physical Properties of Different Types of Substances.

Type of Solid	Structural Unit(s)	Binding Forces	Relative Properties	Examples
Ionic	Positive and negative ions.	Infinite three-dimensional network of balanced coulombic forces among ions. No discrete molecules.	High melting and boiling points; hard and brittle; solubility in water; poor electrical conductivity as solid, but good as liquid.	NaCl, LiH, $CaCl_2$, K_2SO_4, $(NH_4)_3PO_4$
Molecular				
Elemental	Molecules consisting of atoms of the same element.	Nonpolar, covalent, electron-pair bonds within molecules. London forces between molecules.	Low-to-moderate melting and boiling points; soft; poor electrical conductivity in solid and liquid.	Gases: H_2, O_2, N_2, Cl_2 Liquid: Br_2 Solids: I_2, S_8, P_4 (white phosphorus allotrope)
Compound	Molecules consisting of atoms of at least two different elements.	Covalent electron-pair bonds (possibly polar) within molecules. London, dipole, and hydrogen-bond forces possible between molecules.	Low-to-moderate melting and boiling points; soft; poor electrical conductors in solid and liquid.	Gases: CH_4, C_2H_6, CO_2, SO_2, HCl, NH_3 Liquids: C_5H_{12}, C_2H_5OH, H_2O, C_6H_6 Solids: $C_{20}H_{42}$, H_2NCH_2COOH (glycine), $C_{12}H_{22}O_{11}$ (sugar)
Macromolecular				
Three-dimensional	Atoms, perhaps some ions.	Infinite three-dimensional network of directional electron-pair bonds. Sometimes ions found within network, but coulombic forces not important.	Very high melting and boiling points; very hard; poor electrical conductivity.	Elements: C(diamond), Si, Ge Compounds: SiO_2, feldspars
Two-dimensional	Atoms, sometimes ions, two-dimensional macromolecules.	Infinite two-dimensional network of directional electron-pair bonds. Macromolecular layers held together by van der Waals forces, or, if layers are charged, by coulombic forces.	Moderate-to-high melting and boiling points; easily separated into sheets or layers; electrical conductivity varies from one substance to another.	Elements: C(graphite), P (black allotrope) Compound: mica
One-dimensional	Atoms, one-dimensional macromolecules.	Infinite one-dimensional network of directional electron-pair bonds. Macromolecules held together by van der Waals forces.	Moderate melting and boiling points; soft; poor electrical conductivity.	Element: P (red allotrope) Compounds: Polyethylene, nylon.
Metallic	Metal atoms (or positive metal ions surrounded by a loosely held electron cloud).	Coulombic attraction between loosely held electrons and positive metal ions.	Moderate-to-high melting and boiling points; moderate hardness; malleable and ductile; good electrical conductivity in both solid and liquid; good heat conductivity.	Fe, Al, alloys

8.9 Write structural formulas for two cycloalkanes with the formula C_8H_8 and containing a six-membered ring.

8.10 Judging purely from their formulas, decide whether the following molecules are alkanes, cycloalkanes, or aromatic hydrocarbons:

a C_8H_{10}	*d* $C_{10}H_8$
b C_6H_{12}	*e* $C_{10}H_{22}$
c C_6H_{10}	*f* C_8H_{18}

8.11 Draw structural formulas for two alkenes with the formula C_5H_{10} containing a straight chain of five carbon atoms. Are there any other molecules that meet these specifications?

8.12 Write structural formulas for all the alkynes of formula C_4H_6.

8.13 On the basis of formula alone, decide which of the following compounds is an alkane, an alkene, or an alkyne:

a C_3H_8	*d* C_2H_2
b C_6H_{14}	*e* C_9H_{20}
c C_7H_{14}	*f* C_5H_8

8.14 Write a balanced chemical equation illustrating the cracking of $C_{16}H_{34}$ into a C_{10} fragment and a C_6 fragment. Decide which fragment is an alkane and which is an alkene.

8.15 Although cyclohexane, C_6H_{12}, and benzene, C_6H_6, both contain six-membered rings, the two have very different geometries and bond lengths. Explain.

8.16 Account for the following anomalous properties of water: (*a*) Ice is less dense than water. (*b*) A lot of heat is required to raise the temperature of 1 g H_2O by 1°C. (*c*) Water has a higher boiling point than any other nonmetal hydride.

8.17 When the following liquids are boiled, what is the predominant type of intermolecular force which must be overcome in order to separate the molecules in order to form a gas?

a NH_3	*d* H_2S
b C_2H_6	*e* NH_2OH
c CH_3F	*f* $SnCl_2$

8.18 A student writes, "Water has a high boiling point because of the strong bonds between the oxygen and hydrogen atoms." Give a better, less ambiguous, answer.

8.19 Draw structural formulas for at least three compounds with the formula $C_4H_{10}O$, two of which are alcohols and one of which is an ether.

8.20 Decide which of the following compounds is an alcohol, an ether, an aldehyde, a ketone, or a carboxylic acid:

a $\mathrm{C_2H_5-\overset{\overset{O}{\|}}{C}-CH_3}$	*d* $\mathrm{C_2H_5-\overset{\overset{O}{\|}}{C}-C_2H_5}$
b $\mathrm{C_2H_5-O-CH_3}$	*e* $\mathrm{C_2H_5-\overset{\overset{O}{\|}}{C}-OH}$
c $\mathrm{C_2H_5-O-H}$	*f* $\mathrm{C_2H_5-\overset{\overset{O}{\|}}{C}-H}$

8.21 Draw Lewis diagrams for the following three molecules and identify them: (*a*) CH_2O; (*b*) CH_2O_2; (*c*) CH_4O.

8.22 Write the structural formulas for (*a*) the aldehyde and (*b*) the ketone with the formula C_3H_6O.

8.23 Write the structural formulas for (*a*) the ester and (*b*) the carboxylic acid with the formula $C_3H_6O_2$. Which of these two compounds will have the higher boiling point? Give reasons.

8.24 Name at least one important commercial use for each of the following compounds:

a 1,2,3-propanetriol (glycerin)
b ethyl ethanoate (ethyl acetate)
c methanal (formaldehyde)
d methanol (methyl alcohol)
e propene (propylene)
f propane
g propanone (acetone)
h 1,2-ethanediol (glycol)

8.25 Write structural formulas for the following esters: (*a*) *n*-propyl methanoate (*n*-propyl formate); (*b*) methyl butanoate (methyl butyrate). Now write balanced equations describing the hydrolysis of these two esters in acid solution.

8.26 Draw projection formulas for each of the following compounds:

a 1-propanol
b propanal
c propyne
d propanoic acid (propionic acid)
e propanone
f 1-aminopropane

8.27 Indicate which compound in the following pairs has the higher boiling point. Give reasons in each case.

a isobutane, *n*-butane
b 1-propanol, methyl ethyl ether
c 1-propanol, acetic acid
d 1-propanol, acetone
e 1-propanol, ethylene glycol
f *n*-butane, 1-aminopropane

8.28 Which of the following classes of organic compounds exhibit hydrogen bonding in the liquid state:

a ethers	*d* ketones
b alcohols	*e* acids
c aldehydes	*f* amines

8.29 Distinguish between a carbonyl group and a carboxyl group.

8.30 Write an equation for the preparation of the methyl ester of the amino acid glycine.

8.31 Write structural formulas for all the possible amines (primary, secondary, and tertiary) with the formula C_3H_9N.

8.32 Make a perspective sketch of a molecule of propanone (acetone) indicating the approximate value of all bond angles and the approximate lengths of all bonds.

8.33 In graphite the carbon-carbon bonds have a length of 142 pm, while in diamond they have the length of 154 pm. Explain this difference.

8.34 Although the carbon-carbon bonds in graphite are shorter than in diamond, the density of graphite (2.26 g cm^{-3}) is appreciably *less* than that of diamond (3.51 g cm^{-3}). Explain how this can be so.

*8.35 A friend of yours maintains that the silicate sheet shown in Fig. 8.13 has the formula SiO_4. Convince your friend that the formula is actually $(Si_2O_5^{2-})_n$ and that there are five O atoms for each two Si atoms, and a negative charge of 1 for each Si atom in the sheet. (*Hint:* Divide each O atom which is connected to 2 Si atoms in half. Use oxidation numbers to argue the charge.)

8.36 In some silicates tetrahedrons of oxygen are linked in very long chains according to the following scheme:

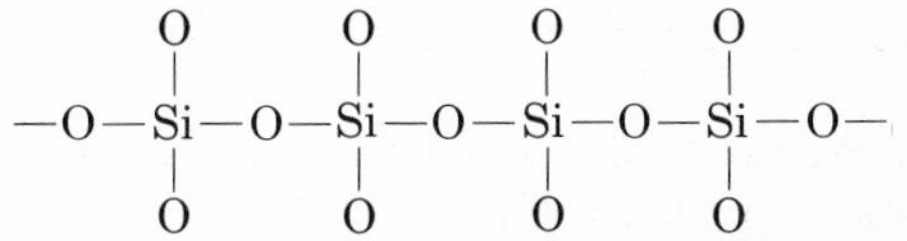

What is the formula (and the charge) for such a chain?

8.37 With the use of Table 8.5, draw the structure of each of the following addition polymers:

a Teflon	*c* polypropylene
b PVC	*d* Orlon

8.38 Explain how and why cross-linking in polymers affects their physical properties.

***8.39** In the ice structure (Fig. 8.11*a*), each water molecule participates in four hydrogen bonds. If we assign half a hydrogen bond to each of the two molecules it holds together, there are two hydrogen bonds per water molecule. (*a*) How many hydrogen bonds are formed per molecule of H—F in the solid state? (*b*) How many hydrogen bonds would you expect per ammonia (NH_3) molecule? (*c*) How do your answers to parts *a* and *b* apply to the data shown in Fig. 8.9?

8.40 Draw structures of monomers which could be reacted to form each of the following condensation polymers.

a

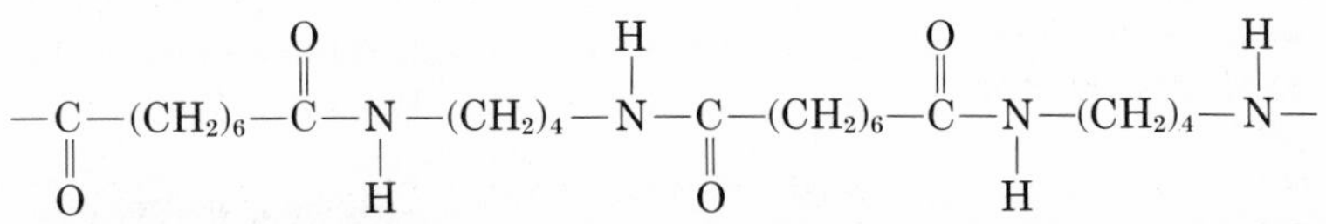

b

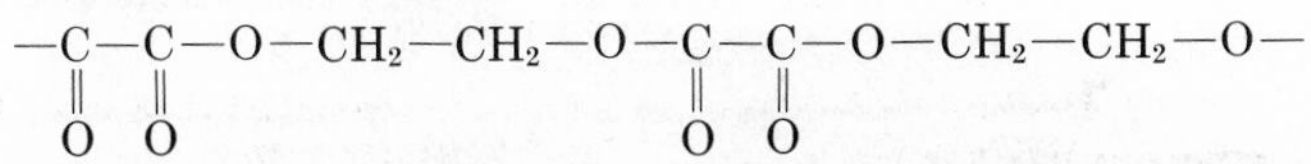

c

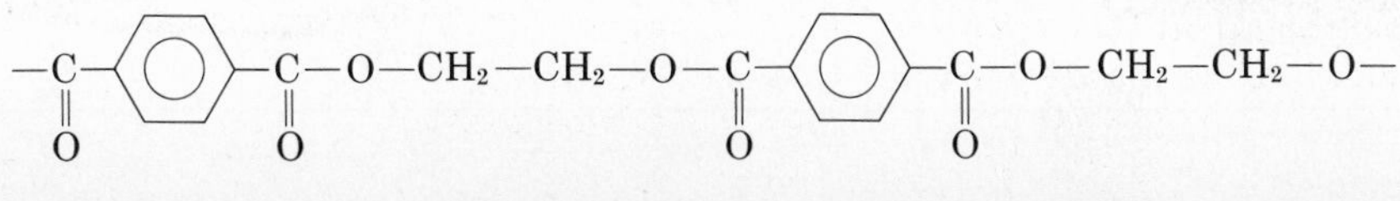

chapter nine

GASES

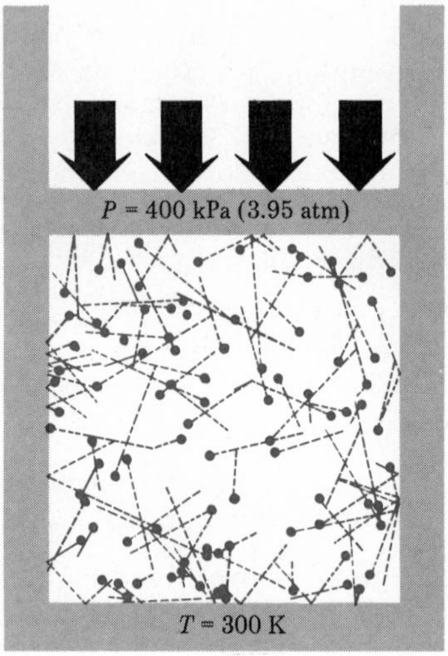

By comparison with solids or liquids, gases are often overlooked or ignored in everyday life. How many times, for example, have you taken an "empty" glass and filled it with water so you could have a drink? If questioned, most people would admit that the glass had been filled with a gas—air—before the water flowed in, but everyday speech has not yet evolved to conform with scientific knowledge. Nevertheless, the air which occupies "empty" glasses, surrounds the surface of the earth to a depth of about 50 km, and fills your lungs every time you breathe is extremely important. If we had to, most of us could survive for weeks without solid food and for days without liquid water. But each of us must have a fresh supply of air every few minutes to go on living.

9.1 PROPERTIES OF GASES

Why does the average person often overlook the presence of gases? Probably because the properties of gases are so unobtrusive. All gases are transparent, and most are colorless. The major exceptions to the second half of this rule are fluorine, F_2, and chlorine, Cl_2, which are pale yellow-green; bromine, Br_2, and nitrogen dioxide, NO_2, which are reddish brown; and iodine, I_2, which is violet.

Another important property of all gases is their mobility. Every gas will disperse to fill all space, unless prevented from doing so by a solid or liquid barrier or a force. (The force of earth's gravity, for example, prevents air from escaping our planet.) Moreover, gases are capable of escaping through small holes (**pores**) in barriers such as plaster of paris or a balloon, even though the human eye sees such materials as continuous and impenetrable. The mobility of gases is also demonstrated by the minimal resistance they present to objects moving through them. You can wave your hand through air much more easily than you can through any liquid.

A third general characteristic of gases is their wide variation in density under various conditions. Densities of solids and liquids change by only a few percent when temperature or pressure is doubled or halved. Similar changes in the conditions of a gas can alter its density by a factor of 2. This occurs because the volume of any gas increases greatly with an increase in temperature or with a reduction in pressure.

Pressure

You are probably familiar with the general idea of pressure from experiences in pumping tires or squeezing balloons. A gas exerts force on any surface that it contacts. *The force per unit surface area* is called the **pressure** and is represented by P. The symbols F and A represent force and area, respectively.

$$\text{Pressure} = \frac{\text{force}}{\text{area}} \qquad P = \frac{F}{A} \tag{9.1}$$

As a simple example of pressure, consider a rectangular block of lead which measures 20.0 cm by 50.0 cm by 100.0 cm (Fig. 9.1). The volume V of the block is 1.00×10^5 cm^3, and since the density ρ of Pb is 11.35 g cm^{-3}, the mass m is

$$m = V\rho = 1.00 \times 10^5 \, \cancel{\text{cm}^3} \times \frac{11.35 \text{ g}}{1 \, \cancel{\text{cm}^3}} = 1.135 \times 10^6 \text{ g} = 1.135 \times 10^3 \text{ kg}$$

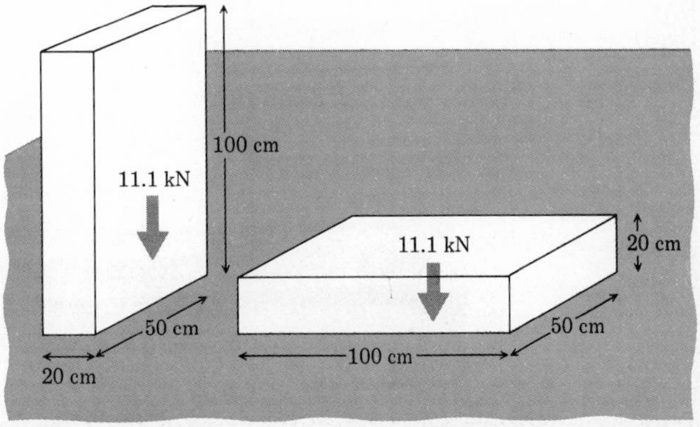

Figure 9.1 The pressure exerted by a block of lead on the floor. When the block stands upright, the weight of the block (11.1 kN) is distributed over an area of 0.1 m^3. If the block is laid flat, this same force is now exerted over an area 5 times larger, namely, 0.5 m^3. The pressure exerted by the block on the floor is 5 times as large in the first case as in the second.

According to the second law of motion, discovered by British physicist Isaac Newton (1643 to 1727), the force on an object is the product of the mass of the object and its acceleration a:

$$F = ma \tag{9.2}$$

At the surface of the earth, the acceleration of gravity is 9.81 m s^{-2}. Substituting into Eq. (9.2), we have

$$F = 1.135 \times 10^3 \text{ kg} \times 9.81 \text{ m s}^{-2} = 11.13 \times 10^3 \text{ kg m s}^{-2}$$

The units kilogram meter per square second are given the name **newton** in the International System and abbreviated N. Thus the force which gravity exerts on the lead block (the **weight** of the block) is 11.13 × 10^3 N. A block that is resting on the floor will always exert a downward force of 11.13 kN.

The *pressure* exerted on the floor depends on the *area* over which this force is exerted. If the block rests on the 20.0 cm by 50.0 cm side (Fig. 9.1*a*), its weight is distributed over an area of 20.0 cm × 50.0 cm = 1000 cm^2. Thus

$$P = \frac{F}{A} = \frac{11.13 \text{ kN}}{1000 \text{ cm}^2} = \frac{11.13 \text{ kN}}{1000 \text{ cm}^2} \times \left(\frac{100 \text{ cm}}{1 \text{ m}}\right)^2$$
$$= \frac{11.13 \text{ kN}}{10^3 \text{ cm}^2} \times \frac{10^4 \text{ cm}^2}{1 \text{ m}^2} = 111.3 \frac{\text{kN}}{\text{m}^2}$$
$$= 111.3 \times 10^3 \text{ N m}^{-2}$$

Thus we see that pressure can be measured in units of newtons (force) per square meter (area). The units newton per square meter are used in the International System to measure pressure, and they are given the name **pascal** (abbreviated Pa). Like the newton, the pascal honors a famous scientist, in this case Blaise Pascal (1623 to 1662), one of the earliest investigators of the pressure of liquids and gases.

If the lead block is laid on its side (Fig. 9.1*b*), the pressure is altered. The area of contact with the floor is now 50.0 cm × 100.0 cm = 5000 cm^2, and so

$$P = \frac{F}{A} = \frac{11.13 \times 10^3 \text{ N}}{5000 \text{ cm}^2} = \frac{11.13 \times 10^3 \text{ N}}{0.500 \text{ m}^2} = 22.26 \times 10^3 \text{ N m}^{-2}$$
$$= 22.26 \text{ kPa}$$

When the block is lying flat, its pressure on the floor (22.26 kPa) is only one-fifth as great as the pressure (111.3 kPa) when it stands on end. This is because the area of contact is 5 times larger.

The air surrounding the earth is pulled toward the surface by gravity in the same way as the lead block we have been discussing. Consequently the air also exerts a pressure on the surface. This is called **atmospheric pressure.**

EXAMPLE 9.1 The total mass of air directly above a 30 cm by 140 cm section of the Atlantic Ocean was 4.34×10^3 kg on July 27, 1977. Calculate the pressure exerted on the surface of the water by the atmosphere.

Solution First calculate the force of gravitational attraction on the air:

$$F = ma = 4.34 \times 10^3 \text{ kg} \times 9.81 \text{ m s}^{-2} = 4.26 \times 10^4 \text{ kg m s}^{-2} = 4.26 \times 10^4 \text{ N}$$

The area is

$$A = 30 \text{ cm} \times 140 \text{ cm} = 4200 \text{ cm}^2 \times \left(\frac{1 \text{ m}}{100 \text{ cm}}\right)^2 = 0.42 \text{ m}^2$$

Thus the pressure is

$$P = \frac{F}{A} = \frac{4.26 \times 10^4 \text{ N}}{0.42 \text{ m}^2} = 1.01 \times 10^5 \text{ Pa} = 101 \text{ kPa}$$

Because winds may add more air or take some away from the vertical column above a given area on the surface, atmospheric pressure will vary above and below the result obtained in Example 9.1. Pressure also decreases as one moves to higher altitudes. The tops of the Himalayas, the highest mountains in the world at about 8000 m (almost 5 miles), are above more than half the atmosphere. The lower pressure at such heights makes breathing very difficult—even the slightest exertion leaves one panting and weak. For this reason jet aircraft, which routinely fly at altitudes of 8 to 10 km, have equipment to maintain air pressure in their cabins artificially.

It is often convenient to express pressure using a unit which is about the same as the average atmospheric pressure at sea level. As we saw in Example 9.1, atmospheric pressure is about 101 kPa, and the **standard atmosphere** (abbreviated atm) is defined as exactly 101.325 kPa. Since this unit is often used, it is useful to remember that

$$1 \text{ atm} = 101.325 \text{ kPa}$$

Measurement of Pressure

The pressure of the atmosphere is exerted in *all directions,* not just downward, at any given altitude. This can easily be demonstrated with a water glass and a flat sheet of cardboard or plastic. Fill the glass to the brim with water and carefully slide the cardboard across the top so that no air is trapped within the glass. While holding the cardboard, turn the glass upside down. Now you can remove your fingers from the cardboard, and atmospheric pressure will hold both cardboard and water up. (Eventually, of course, some air will leak in, there will be an increase in pressure above the cardboard, and the water will spill from the glass. If you try this, be sure a sink is handy.)

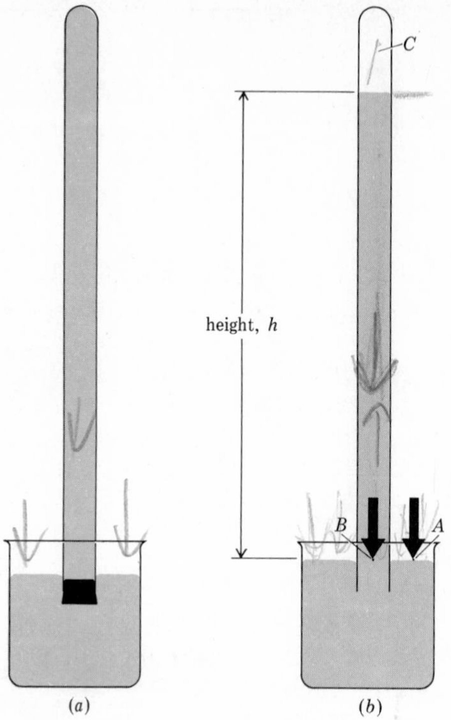

Figure 9.2 A mercury barometer. (*a*) A tube filled with Hg(*l*) is stoppered with a cork and inverted in a beaker of mercury. (*b*) When the cork is removed, mercury flows out of the tube until atmospheric pressure (at point *A*) just balances pressure due to the column of mercury (at point *B*). The region *C* above the mercury in the tube is an almost perfect vacuum with zero pressure. Therefore the pressure of the mercury column, of height *h*, equals atmospheric pressure.

The maximum height of liquid which can be supported by atmospheric pressure provides a measure of that pressure. It turns out that a column of water about 10 m (more than 30 ft) high can be held up by earth's atmosphere. This would be an inconvenient height to measure in the laboratory, and so a much denser liquid, mercury, is used instead. The mercury **barometer,** a device for measuring atmospheric pressure, is shown in Fig. 9.2.

EXAMPLE 9.2 A barometer is constructed as shown in Fig. 9.2. The cross-sectional area of the tube is 1.000 cm², and the height of the mercury column is 760.0 mm. Calculate the atmospheric pressure.

Solution First calculate the volume of mercury above point *B*. Use the density of mercury to obtain the mass, and from this, calculate force and pressure as in Example 9.1.

$$V = 1.000 \text{ cm}^2 \times 760.0 \text{ mm}$$

$$= 1.000 \text{ cm}^2 \times 760.0 \text{ mm} \times \frac{1 \text{ cm}}{10 \text{ mm}} = 76.00 \text{ cm}^3$$

$$m_{\text{Hg}} = 76.00 \text{ cm}^3 \times 13.595 \text{ g cm}^{-3} = 1033.2 \text{ g} = 1.0332 \text{ kg}$$

$$F = ma = 1.0332 \text{ kg} \times 9.807 \text{ m s}^{-2} = 10.133 \text{ kg m s}^{-2} = 10.133 \text{ N}$$

$$P = \frac{F}{A} = \frac{10.133 \text{ N}}{1 \text{ cm}^2} = \frac{10.133 \text{ N}}{1 \text{ cm}^2} \times \left(\frac{100 \text{ cm}}{\text{m}}\right)^2$$

$$= 10.133 \times 10^4 \text{ N m}^{-2} = 101.33 \text{ kPa}$$

The preceding example shows that a mercury column 760.0 mm high and 1.000 cm² in area produces a pressure of 101.33 kPa (1 atm). It can also be shown (see Problem 9.4 at the end of the chapter) that only the height of the mercury column affects its pressure. For a larger cross section there is a greater mass of mercury and therefore a greater force, but this is exerted over a greater area, leaving force per unit area unchanged. For this reason it is convenient to measure pressures of gases in terms of the height of a mercury column that can be supported. That is, we might report the atmospheric pressure in Example 9.2 as 760 mmHg instead of 101.3 kPa or 1.000 atm. It is useful to remember that

$$760 \text{ mmHg} = 1.000 \text{ atm} = 101.3 \text{ kPa}$$

The pressure of a gas in a container is often measured relative to atmospheric pressure using a **manometer.** This is a U-shaped tube containing mercury and connecting the container to the air (Fig. 9.3).

EXAMPLE 9.3 A mercury manometer is used to measure the pressure of a gas in a flask. As shown in Fig. 9.3*b*, the level of mercury is higher in the arm connected to the flask, but the difference in levels is 43 mm. Barometric pressure is 737 mmHg. Calculate the pressure in the container (a) in millimeters of mercury; (b) in kilopascals; and (c) in atmospheres.

Solution

a) $P_{\text{gas}} + P_{\text{Hg}} = P_A \qquad P_{\text{gas}} = P_A - P_{\text{Hg}}$

$$P_{\text{gas}} = 737 \text{ mmHg} - 43 \text{ mmHg} = 694 \text{ mmHg}$$

b) $$P_{\text{gas}} = 694 \text{ mmHg} \times \frac{101.3 \text{ kPa}}{760 \text{ mmHg}} = 92.5 \text{ kPa}$$

c) $$P_{\text{gas}} = 694 \text{ mmHg} \times \frac{1 \text{ atm}}{760 \text{ mmHg}} = 0.913 \text{ atm}$$

Note that essentially the same procedure suffices to convert from millimeters of mercury to either kilopascals or atmospheres. Laboratory measurements are usually made in millimeters of mercury, but further calcula-

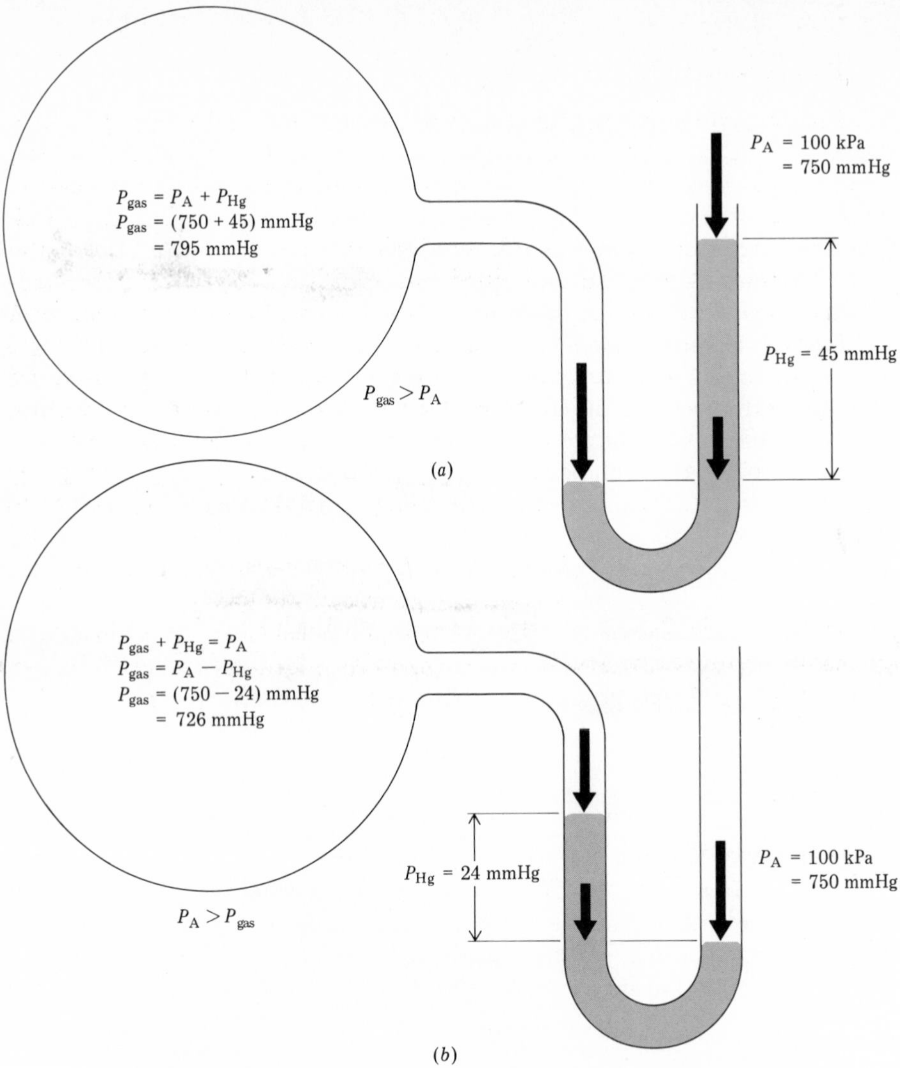

Figure 9.3 Use of a mercury manometer. (*a*) Measuring a pressure greater than atmospheric; (*b*) measuring a pressure less than atmospheric. P_A = atmospheric pressure; P_{Hg} = pressure of mercury column; P_{gas} = pressure of confined gas.

tions almost invariably are more convenient if kilopascals or atmospheres are used. Although the pascal is the accepted SI unit of pressure, it is not yet in general use in the United States. Therefore, one must also be familiar with the atmosphere. The atmosphere is also convenient because 1.000 atm is nearly the same as the atmospheric pressure each of us experiences every day of our lives. This gives a concrete reference with which other pressures can be compared. For these reasons we will usually employ the atmosphere as the unit of pressure for the remainder of this chapter. Nevertheless, there are a number of cases where using the pascal gives a significant insight into gas behavior. In such cases we shall use the newer internationally recognized unit.

Avogadro's Law

For most solids and liquids it is convenient to obtain the amount of substance (and the number of particles, if we want it) from the mass. In Sec. 2.8 numerous such calculations using molar mass were done. In the case of gases, however, accurate measurement of mass is not so simple. Think about how you would weigh a balloon filled with helium, for example. Because it is buoyed up by the air it displaces, such a balloon would force a balance pan *up* instead of down, and a *negative* weight would be obtained. Solids and liquids are also buoyed up, but they have much greater densities than gases. For a given mass of a solid or liquid, the volume is much smaller, much less air is displaced, and the buoyancy effect is negligible.

The mass of a gas can be obtained by weighing a truly empty container (one in which there is a perfect vacuum), and then filling and reweighing the container. But this is a time-consuming, inconvenient, and sometimes dangerous procedure. (Such a container might **implode**—explode inward—due to the difference between atmospheric pressure outside and zero pressure within.)

A more convenient way of obtaining the amount of substance in a gaseous sample is suggested by the data on molar volumes in Table 9.1. Remember that a molar quantity (a quantity divided by the amount of substance) refers to the same number of particles. The data in Table 9.1, then, indicate that for a variety of gases, 6.022×10^{23} molecules occupy almost exactly the same volume. If the temperature is 0°C and the pressure is 1.00 atm (101.3 kPa), this volume is 22.4 liters (22.4 dm^3). (Since the liter, now defined as exactly 1 dm^3, is commonly used as a unit of volume in conjunction with the atmosphere as a unit of pressure, we shall use it that way in this chapter.) That *equal volumes of gases at the same temperature and pressure contain equal numbers of molecules* was first suggested in 1811 by the Italian chemist Amadeo Avogadro (1776 to 1856). Consequently it is called **Avogadro's law** or **Avogadro's hypothesis.**

TABLE 9.1 Molar Volumes of Several Gases at 0°C and 1 atm Pressure.

Substance	Formula	Molar Volume/liter mol^{-1}
Hydrogen	$H_2(g)$	22.43
Neon	$Ne(g)$	22.44
Oxygen	$O_2(g)$	22.39
Nitrogen	$N_2(g)$	22.40
Carbon dioxide	$CO_2(g)$	22.26
Ammonia	$NH_3(g)$	22.09

Avogadro's law has two important messages. First, it says that molar volumes of *all* gases are the same at a given temperature and pressure. Therefore, even if we do not know what gas we are dealing with, we can still find the amount of substance. Second, we expect that if a particular volume corresponds to a certain number of molecules, twice that volume would contain twice as many molecules. In other words, *doubling* the volume corresponds to *doubling* the amount of substance, *halving* the volume corresponds to *halving* the amount, and so on.

In general, if we *multiply* the volume by some factor, say x, then we also *multiply* the amount of substance by that same factor x. Such a relationship is called **direct proportionality** and may be expressed mathematically as

$$V \propto n \tag{9.3}$$

where the symbol $\propto$ means "is proportional to." Any proportion, such as Eq. (9.3), can be changed to an equivalent equation if one side is multiplied by a proportionality constant, such as k_A in Eq. (9.4):

$$V = k_A n \tag{9.4}$$

If we know k_A for a gas, we can determine the amount of substance from Eq. (9.4).

The situation is complicated by the fact that the volume of a gas depends on pressure and temperature, as well as on the amount of substance. That is, k_A will vary as temperature and pressure change. Therefore we need quantitative information about the effects of pressure and temperature on the volume of a gas before we can explore the relationship between amount of substance and volume.

Boyle's Law

You are probably already familiar with the fact that when you squeeze a gas, it will take up less space. In formal terms, *increasing* the pressure on a gas will *decrease* its volume. Studies in 1662 by the English scientist Robert Boyle (1627 to 1691) gave results such as those in Table 9.2. Careful study of such data reveals that if we double the pressure, we halve the volume; if we triple the pressure, the volume is reduced to one-third; and so on. In general, if we *multiply* the pressure by some factor x, then we *divide* the volume by the same factor x. Such a relationship, in which the increase in one quantity produces a proportional decrease in another, is called **inverse proportionality.**

TABLE 9.2 Variation in the Volume of 0.0446 mol $H_2(g)$ with Pressure at 0°C.

Trial	Pressure/kPa	Pressure/atm	Volume/liter
1	152.0	1.50	0.666
2	126.7	1.25	0.800
3	101.3	1.00	1.00
4	76.0	0.750	1.333
5	50.7	0.500	2.00
6	25.3	0.250	4.00
7	10.1	0.100	10.00

The results of Boyle's experiments with gases are summarized in **Boyle's law**—*for a given amount of gas at constant temperature, the volume is inversely proportional to the pressure.* In mathematical terms

$$V \propto \frac{1}{P} \tag{9.5}$$

The reciprocal of P indicates the inverse nature of the proportionality. Using the proportionality constant k_B to convert relationship (9.5) to an equation, we have

$$V = k_B \times \frac{1}{P} = \frac{k_B}{P} \tag{9.6a}$$

Multiplying both sides of Eq. (9.6*a*) by P, we have

$$PV = k_B \tag{9.6b}$$

where k_B represents a constant value for any given temperature and amount (or mass) of gas.

EXAMPLE 9.4 Using the data printed in color in Table 9.2, confirm that Boyle's law is obeyed.

Solution Since the data apply to the same amount of gas at the same temperature, PV should be constant [Eq. (9.6*b*)] if Boyle's law holds.

$$P_1V_1 = 1.50 \text{ atm} \times 0.666 \text{ liter} = 0.999 \text{ atm liter}$$

$$P_4V_4 = 0.750 \text{ atm} \times 1.333 \text{ liters} = 1.000 \text{ atm liter}$$

$$P_6V_6 = 0.250 \text{ atm} \times 4.00 \text{ liters} = 1.00 \text{ atm liter}$$

The first product differs from the last two in the fourth significant digit. Since some data are reported only to three significant figures, PV is constant within the limits of the measurements.

If the units atmosphere liter, in which PV was expressed in Example 9.4, are changed to SI base units, an interesting result arises:

$$\begin{aligned}
1 \text{ atm} \times 1 \text{ liter} &= 101.3 \text{ kPa} \times 1 \text{ dm}^3 \\
&= 101.3 \times 10^3 \text{ Pa} \times 1 \text{ dm}^3 \times \left(\frac{1 \text{ m}}{10 \text{ dm}}\right)^3 \\
&= 101.3 \times \cancel{10^3} \frac{\text{N}}{\cancel{\text{m}^2}} \times 1 \,\cancel{\text{dm}^3} \times \frac{1 \,\cancel{\text{m}^3}}{\cancel{10^3}\,\cancel{\text{dm}^3}} \\
&= 101.3 \text{ N m} \\
&= 101.3 \text{ kg m s}^{-2} \text{ m} \\
&= 101.3 \text{ J}
\end{aligned}$$

In other words, *PV has the same units (joules) as an energy.* While this does not guarantee that *PV is* an energy (recall the warning in Chap. 1 against relying on cancellation of units unless you *know* that a relationship between quantities exists), it does suggest that we should explore the possibility. This will be done in Sec. 9.4. The above argument also shows that the product of the units kilopascals times cubic decimeters is the unit joules. In subsequent discussions you will find it handy to remember that

$$1 \text{ kPa} \times 1 \text{ dm}^3 = 1 \text{ J} \qquad \text{and} \qquad 1 \text{ kPa} = \frac{1 \text{ J}}{1 \text{ dm}^3} \qquad \text{and} \qquad 1 \text{ dm}^3 = \frac{1 \text{ J}}{1 \text{ kPa}}$$

Boyle's law enables us to calculate the pressure or volume of a gas under one set of conditions, provided we know the pressure and volume under a previous set of circumstances.

EXAMPLE 9.5 The volume of a gas is 0.657 liter under a pressure of 729.8 mmHg. What volume would the gas occupy at atmospheric pressure (760 mmHg)? Assume constant temperature and amount of gas.

Solution Two methods of solution will be given.

a) Since *PV* must be constant,

$$P_1V_1 = k_B = P_2V_2$$

Initial conditions: $P_1 = 729.8$ mmHg $\quad V_1 = 0.657$ liter

Final conditions: $P_2 = 760$ mmHg $\quad V_2 = ?$

Solving for V_2, we have

$$V_2 = \frac{P_1V_1}{P_2} = \frac{729.8 \text{ mmHg} \times 0.657 \text{ liter}}{760 \text{ mmHg}} = 0.631 \text{ liter}$$

b) Note that in method *a* the original volume was multiplied by a ratio of pressures (P_1/P_2):

$$V_2 = 0.657 \text{ liter} \times \text{ratio of pressures}$$

Rather than solving algebraically, we can use common sense to decide which of the two possible ratios

$$\frac{729.8 \text{ mmHg}}{760 \text{ mmHg}} \qquad \text{or} \qquad \frac{760 \text{ mmHg}}{729.8 \text{ mmHg}}$$

should be used. The units cancel in either case, and so units are no help. However, if you reread the problem, you will see that we are asked to find the new volume (V_2) produced by an *increase* in pressure. Therefore there must be a *decrease* in volume, and we multiply the original volume by a ratio which is *less than* 1:

$$V_2 = 0.657 \text{ liter} \times \frac{729.8 \text{ mmHg}}{760 \text{ mmHg}} = 0.631 \text{ liter}$$

It is reassuring that both common sense and algebra produce the same answer.

Charles' Law

Having explored the effect of pressure on the volume of a gas, we now turn to the effect of temperature. Again you are probably familiar with the fact that increasing the temperature of a gas will cause the gas to expand. This effect was first studied quantitatively in 1787 by Jacques Charles (1746 to 1823) of France. Typical data from such an experiment are given in Table 9.3. You can see that for 0.0466 mol $H_2(g)$ at constant pressure, a 50°C rise in temperature produces a 0.18-liter increase in volume, whether the temperature increases from 0.0 to 50.0°C or from 100.0 to 150.0°C.

TABLE 9.3 Variation in the Volume of $H_2(g)$ with Temperature.

Temperature (in °C)	Volume/liter
Data for 0.0446 mol $H_2(g)$ at 1 atm (101.3 kPa)	
0.0	1.00
50.0	1.18
100.0	1.37
150.0	1.55
Data for 0.100 mol $H_2(g)$ at 1 atm (101.3 kPa)	
0.0	2.24
50.0	2.65
100.0	3.06
150.0	3.47

When the experimental data of Table 9.3 are graphed, we obtain Fig. 9.4. Notice that the four points corresponding to 0.0446 mol $H_2(g)$ lie on a straight line, as do the points for 0.100 mol $H_2(g)$. If the lines are **extrapolated** (extended beyond the experimental points) to very low temperatures, we find that *both* of them intersect the horizontal axis at −273°C. The behavior of $H_2(g)$ (and of many other gases) at normal temperatures suggests that if we cool a gas sufficiently, its volume will become zero at −273°C.

Of course a real substance would condense to a liquid and freeze to a solid as it was cooled. When the pressure is 1.00 atm (101.3 kPa), $H_2(g)$ liquefies at −253°C and freezes at −259°C, and so all experiments involving $H_2(g)$ would have to be performed above −253°C. If we could find a gas that did not condense, however, it would still be impossible to cool it below −273°C, because at that temperature its volume would be zero. Going to a lower temperature would correspond to a negative volume—something that is very hard to conceive of. Hence −273°C is referred to as the **absolute zero** of temperature—it is impossible to go any lower.

In Fig. 9.4*b* the zero of the temperature axis has been shifted to absolute zero. The temperature scale used in this graph is called **absolute** or **thermodynamic temperature.** It is measured in SI units called **kelvins** (abbreviated K), in honor of the English physicist William Thomson, Lord Kelvin (1824 to 1907). The temperature interval 1 K corresponds to a

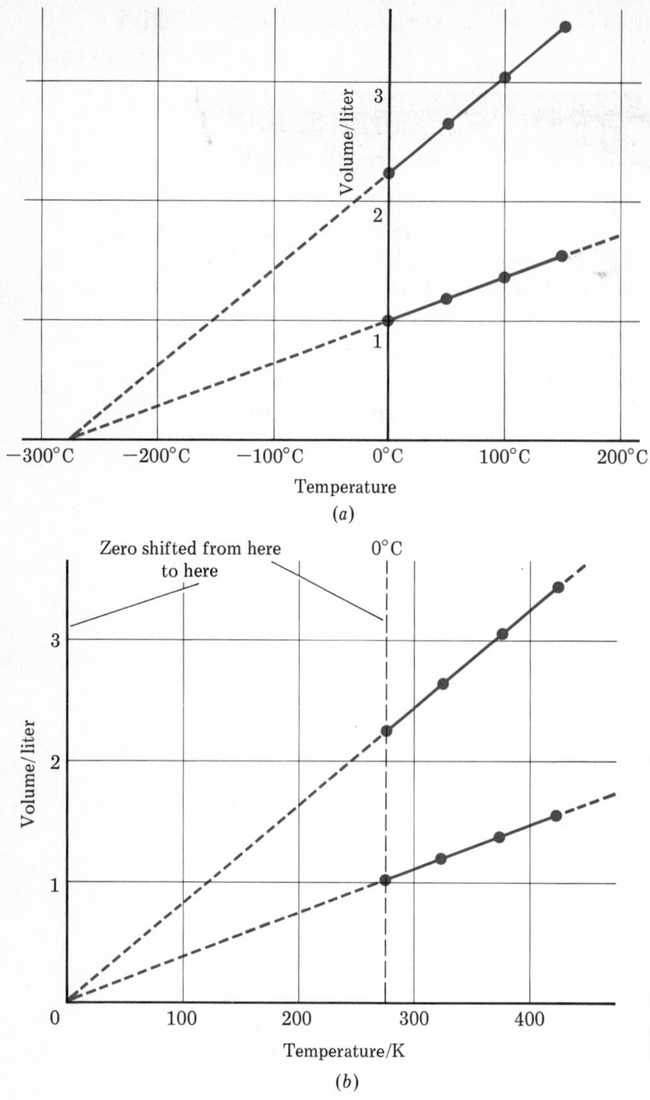

Figure 9.4 Charles' law. (*a*) Volume plotted against Celsius temperature for two samples of H(*g*) at 1.00 atm (101.3 kPa); (*b*) volume plotted against thermodynamic temperature for the same two samples. Note how much simpler graph (*b*) is than graph (*a*).

change of 1°C, but zero on the thermodynamic scale (0 K) is −273.15°C. The freezing point of water at 1.00 atm (101.3 kPa) pressure is thus 273.15 K. The relationship between the Celsius temperature scale and the thermodynamic temperature scale is discussed further in Appendix 1.

By shifting to the absolute temperature scale, we have simplified the graph of gas volume versus temperature. Figure 9.4*b* shows that *the volume of a gas is directly proportional to its thermodynamic temperature, provided that the amount of gas and the pressure remain constant.* This is known as **Charles' law,** and can be expressed mathematically as

$$V \propto T \tag{9.7}$$

where T represents the absolute temperature (usually measured in kelvins).

As in the case of previous gas laws, we can introduce a proportionality constant, in this case, k_C:

$$V = k_C T \qquad \text{or} \qquad \frac{V}{T} = k_C \tag{9.8}$$

EXAMPLE 9.6 A sample of $H_2(g)$ occupies a volume of 69.37 cm^3 at a pressure of exactly 1 atm when immersed in a mixture of ice and water. When the gas (at the same pressure) is immersed in boiling benzene, its volume expands to 89.71 cm^3. What is the boiling point of benzene?

Solution As in the case of Boyle's law (Example 9.5), two methods of solution are possible.

a) Algebraically, we have, from Eq. (9.8),

$$\frac{V_1}{T_1} = k_C = \frac{V_2}{T_2}$$

and substituting into the equation

$$T_2 = \frac{V_2 T_1}{V_1} = \frac{89.71 \text{ cm}^3 \times 273.15 \text{ K}}{69.37 \text{ cm}^3} = 353.2 \text{ K}$$

yields the desired result. (The ice-water mixture must be at 273.15 K, the freezing point of water.)

b) By common sense we argue that since the gas expanded, its temperature must have increased. Thus

$$T_2 = 273.15 \text{ K} \times \text{ratio greater than } 1$$

$$= 273.15 \text{ K} \times \frac{89.71 \text{ cm}^3}{69.37 \text{ cm}^3} = 353.2 \text{ K}$$

Note: In this example we have used the expansion of a gas instead of the expansion of liquid mercury to measure temperature.

Because both temperature and pressure affect the volume of a gas, it is convenient to specify a reference temperature and pressure at which all volumes may be compared. **Standard temperature** and **pressure** (or **STP**) is chosen to be 273.15 K (0°C) and 1 atm (101.325 kPa).

Gay-Lussac's Law

A third gas law may be derived as a corollary to Boyle's and Charles' laws. Suppose we double the thermodynamic temperature of a sample of gas. According to Charles' law, the volume should double. Now, how much pressure would be required at the higher temperature to return the gas to its original volume? According to Boyle's law, we would have to double the pressure to halve the volume. Thus, if the volume of gas is to remain the same, doubling the temperature will require doubling the pressure.

This law was first stated by the Frenchman Joseph Gay-Lussac (1778 to 1850). According to **Gay-Lussac's law,** *for a given amount of gas held at constant volume, the pressure is proportional to the absolute temperature.* Mathematically,

$$P \propto T \quad \text{or} \quad P = k_G T \quad \text{or} \quad \frac{P}{T} = k_G$$

where k_G is the appropriate proportionality constant.

Gay-Lussac's law tells us that it may be dangerous to heat a gas in a closed container. The increased pressure might cause the container to explode.

EXAMPLE 9.7 A container is designed to hold a pressure of 2.5 atm. The volume of the container is 20.0 cm³, and it is filled with air at room temperature (20°C) and normal atmospheric pressure. Would it be safe to throw the container into a fire where temperatures of 600°C would be reached?

Solution Using the common-sense method, we realize that the pressure will increase at the higher temperature, and so

$$P_2 = 1.0 \text{ atm} \times \frac{(273.15 + 600) \not{\text{K}}}{(273.15 + 20) \not{\text{K}}} = 3.0 \text{ atm}$$

This would exceed the safe strength of the container. Note that the volume of the container was not needed to solve the problem.

9.3 GAS VOLUMES AND MOLES

The Ideal Gas Equation

Now that we know quantitatively the effects of pressure and thermodynamic temperature on gas volume, we can return to the relation between volume and amount of substance. Avogadro's law [Eq. (9.3)] tells us that at constant P and T, the volume of a gas is directly proportional to the amount of gas. Boyle's law [Eq. (9.5)] says that volume is inversely proportional to pressure, and Charles' law [Eq. (9.7)] indicates that volume is directly proportional to temperature. These three laws may all be applied at once if we write

$$V \propto n \times \frac{1}{P} \times T \tag{9.9}$$

or, introducing a constant of proportionality R,

$$V = R\,\frac{nT}{P} \tag{9.10}$$

Equation (9.10) applies to *all* gases at low pressures and high temperatures and is a very good approximation under nearly all conditions. The value of R, the **gas constant,** is independent of the kind of gas, the temperature, or the pressure. To calculate R, we rearrange Eq. (9.10):

$$R = \frac{PV}{nT} \tag{9.11}$$

and substitute appropriate values of P, V, n, and T. From Table 9.1 we saw that the molar volumes of several gases at 0°C (273.15 K) and 1 atm (101.3 kPa) were close to 22.4 liters (22.4 dm³). Substituting into Eq. (9.11),

$$R = \frac{1 \text{ atm} \times 22.4 \text{ liters}}{1 \text{ mol} \times 273.15 \text{ K}} = 0.0820 \frac{\text{liter atm}}{\text{mol K}}$$

If we use SI units for pressure and volume, as well as for amount of substance and temperature,

$$R = \frac{101.3 \text{ kPa} \times 22.4 \text{ dm}^3}{1 \text{ mol} \times 273.15 \text{ K}} = 8.31 \frac{\text{kPa dm}^3}{\text{mol K}} = 8.31 \text{ J mol}^{-1} \text{ K}^{-1}$$

Thus the gas constant has units of energy divided by amount of substance and thermodynamic temperature.

Equation (9.10) is usually rearranged by multiplying both sides by P, so that it reads

$$PV = nRT \tag{9.12}$$

This is called the **ideal gas equation** or the **ideal gas law.** With the ideal gas equation we can convert from volume of a gas to amount of substance (provided that P and T are known). This is very useful since the volume, pressure, and temperature of a gas are easier to measure than mass, and because knowledge of the molar mass is unnecessary.

EXAMPLE 9.8 Calculate the amount of gas in a 100-cm³ sample at a temperature of 300 K and a pressure of 750 mmHg.

Solution Either of the two values of the gas constant may be used, so long as the units P, V, and T are adjusted properly. Using R = 0.0820 liter atm mol⁻¹ K⁻¹,

$$V = 100 \cancel{\text{cm}^3} \times \frac{1 \text{ liter}}{1000 \cancel{\text{cm}^3}} = 0.100 \text{ liter}$$

$$P = 750 \cancel{\text{mmHg}} \times \frac{1 \text{ atm}}{760 \cancel{\text{mmHg}}} = 0.987 \text{ atm}$$

$$T = 300 \text{ K}$$

Rearranging Eq. (9.12) and substituting,

$$n = \frac{PV}{RT} = \frac{0.987 \cancel{\text{atm}} \times 0.100 \cancel{\text{liter}}}{0.0820 \cancel{\text{liter}} \cancel{\text{atm}} \text{ mol}^{-1} \cancel{\text{K}^{-1}} \times 300 \cancel{\text{K}}} = 4.01 \times 10^{-3} \text{ mol}$$

Essentially the same calculations are required when SI units are used:

$$V = 100 \text{ cm}^3 \times \left(\frac{1 \text{ dm}}{10 \text{ cm}}\right)^3 = 0.100 \text{ dm}^3$$

$$P = 750 \text{ mmHg} \times \frac{101.3 \text{ kPa}}{760 \text{ mmHg}} = 100.0 \text{ kPa}$$

$$T = 300 \text{ K}$$

$$n = \frac{PV}{RT} = \frac{100.0 \text{ kPa} \times 0.100 \text{ dm}^3}{8.31 \text{ J mol}^{-1} \text{ K}^{-1} \times 300 \text{ K}} = 4.01 \times 10^{-3} \text{ mol}$$

Note: Since 1 kPa dm³ = 1 J, the units cancel as shown. For this reason it is convenient to use cubic decimeters as the unit of volume when kilopascals are used as the unit of pressure in the ideal gas equation.

The ideal gas law enables us to find the amount of substance, provided we can measure V, P, and T. If we can also determine the mass of a gas, it is possible to calculate molar mass (and molecular weight). One way to do this is vaporize a volatile liquid (one which has a low boiling temperature) so that it fills a flask of known volume. When the flask is cooled, the vapor condenses to a liquid and can easily be weighed.

EXAMPLE 9.9 The empirical formula of benzene is CH. When heated to 100°C in a flask whose volume was 247.2 ml, a sample of benzene vaporized and drove all air from the flask. When the benzene was condensed to a liquid, its mass was found to be 0.616 g. The barometric pressure was 742 mmHg. Calculate (a) the molar mass and (b) the molecular formula of benzene.

Solution

a) Molar mass is mass divided by amount of substance. The latter quantity can be obtained from the volume, temperature, and pressure of benzene vapor:

$$V = 247.2 \text{ cm}^3 \times \frac{1 \text{ liter}}{10^3 \text{ cm}^3} = 0.2472 \text{ liter}$$

$$T = (273.15 + 100) \text{ K} = 373 \text{ K}$$

$$P = 742 \text{ mmHg} \times \frac{1.00 \text{ atm}}{760 \text{ mmHg}} = 0.976 \text{ atm}$$

$$n = \frac{PV}{RT} = \frac{0.976 \text{ atm} \times 0.2472 \text{ liter}}{0.0820 \text{ liter atm mol}^{-1} \text{ K}^{-1} \times 373 \text{ K}} = 7.89 \times 10^{-3} \text{ mol}$$

The molar mass is

$$\mathcal{M} = \frac{m}{n} = \frac{0.616 \text{ g}}{7.89 \times 10^{-3} \text{ mol}} = 78.1 \text{ g mol}^{-1}$$

b) The empirical formula CH would imply a molar mass of (12.01 + 1.008) g mol^{-1}, or 13.02 g mol^{-1}. The experimentally determined molar mass is 6 times larger:

$$\frac{78.1\ \text{g mol}^{-1}}{13.02\ \text{g mol}^{-1}} = 6.00$$

and so the molecular formula must be C_6H_6.

The Law of Combining Volumes

In effect, the preceding example used the factor P/RT to convert from volume to amount of gas. The reciprocal of this factor can be used to convert from amount of gas to volume. This is emphasized if we rewrite Eq. (9.10) as

$$V = \frac{RT}{P} n \qquad (9.10)$$

This indicates that when we write a chemical equation involving gases, the coefficients not only tell us what amount of each substance is consumed or produced, they also indicate the relative *volume* of each gas consumed or produced. For example,

$$2H_2(g) \quad + \quad O_2(g) \quad \rightarrow \quad 2H_2O(g) \qquad (9.13)$$

means that for every

2 mol $H_2(g)$ consumed	there will be	1 mol $O_2(g)$ consumed	and	2 mol $H_2O(g)$ produced

It also implies that for every

$\left(2\ \text{mol} \times \frac{RT}{P}\right)$ dm^3 $H_2(g)$ consumed	there will be	$\left(1\ \text{mol} \times \frac{RT}{P}\right)$ dm^3 $O_2(g)$ consumed	and	$\left(2\ \text{mol} \times \frac{RT}{P}\right)$ dm^3 $H_2O(g)$ produced

This is an example of the **law of combining volumes,** which states that *when gases combine at constant temperature and pressure, the volumes involved are always in the ratio of simple whole numbers.* Since the factor RT/P would be the same for all three gases, the volume of $O_2(g)$ consumed must be half the volume of $H_2(g)$ consumed. The volume of $H_2O(g)$ produced would be only two-thirds the total volume [of $H_2(g)$ and $O_2(g)$] consumed, and so at the end of the reaction the total volume must be less than at the beginning.

The law of combining volumes was proposed by Gay-Lussac at about the same time that Dalton published his atomic theory. Shortly thereafter, Avogadro suggested the hypothesis that equal volumes of gases contained equal numbers of molecules. Dalton strongly opposed Avogadro's hypothesis because it required that some molecules contain more than the minimum number of atoms.

For example, according to Dalton, the formula for hydrogen gas should be the simplest possible, e.g., H. Similarly, Dalton proposed the formula O for oxygen gas. His equation for formation of water vapor was

H	+	O	→	HO
1 volume hydrogen		**1 volume oxygen**		**1 volume water vapor**

But experiments showed that twice as great a volume of hydrogen as of oxygen was required for complete reaction. Furthermore, the volume of water vapor produced was twice the volume of oxygen consumed. Avogadro proposed (correctly, as it turned out) that the formulas for hydrogen, oxygen, and water were H_2, O_2, and H_2O, and he explained the volume data in much the same way as we have done for Eq. (9.13).

Dalton, who had originally conceived the idea of atoms and molecules, was unwilling to concede that substances such as hydrogen or water might have formulas more complicated than was absolutely necessary. Partly as a result of Dalton's opposition, it took almost half a century before Avogadro's Italian countryman Stanislao Cannizzaro (1826 to 1910) was able to convince chemists that Avogadro's hypothesis was correct. The blindness of chemists to Avogadro's ideas for so long makes one wonder whether today's Nobel prize winners might not be equally wrong about some other aspect of chemistry. Who knows but that some forgotten Argentinian Avogadro is still waiting for a Cannizzaro to explain his or her ideas to the scientific world.

Chemical Equations and Gas Volumes

Because the amount of gas is related to volume by the ideal gas law, it is possible to calculate the volume of a gaseous substance consumed or produced in a reaction. Molar mass and stoichiometric ratio are employed in the same way as in Sec. 3.1, and the factor RT/P is used to convert from amount of gas to volume.

EXAMPLE 9.10 Oxygen was first prepared by Joseph Priestly by heating red "calx of mercury" [mercury(II) oxide, HgO] according to the equation

$$2HgO(s) \rightarrow 2Hg(l) + O_2(g)$$

What volume (in cubic centimeters) of O_2 can be prepared from 1.00 g HgO? The volume is measured at 20°C and 0.987 atm.

Solution The mass of HgO can be converted to amount of HgO, and this can be converted to amount of O_2 by means of a stoichiometric ratio. Finally, the ideal gas law is used to obtain the volume of O_2. Schematically,

$$m_{HgO} \xrightarrow{\mathcal{M}_{HgO}} n_{HgO} \xrightarrow{S(O_2/HgO)} n_{O_2} \xrightarrow{RT/P} V_{O_2}$$

$$V_{O_2} = 1 \text{ g HgO} \times \frac{1 \text{ mol HgO}}{216.59 \text{ g HgO}} \times \frac{1 \text{ mol } O_2}{2 \text{ mol HgO}}$$
$$\times \frac{0.0820 \text{ liter atm}}{1 \text{ K mol } O_2} \times \frac{293.15 \text{ K}}{0.987 \text{ atm}}$$
$$= 0.0562 \text{ liter} = 56.2 \text{ cm}^3$$

Dalton's Law of Partial Pressures

The ideal gas law can also be rearranged to show that the pressure of a gas is proportional to the amount of gas:

$$P = \frac{RT}{V} n \qquad (9.14)$$

Thus the factor RT/V may be used to interconvert amount of substance and pressure in a container of specified volume and temperature.

Equation (9.14) is also useful in dealing with the situation where two or more gases are confined in the same container (i.e., the same volume). Suppose, for example, that we had 0.010 mol of a gas in a 250-ml container at a temperature of 32°C. The pressure would be

$$P = \frac{RT}{V} n = \frac{0.0820 \text{ liter atm mol}^{-1} \text{ K}^{-1} \times 305 \text{ K}}{0.250 \text{ liter}} \times 0.010 \text{ mol}$$
$$= 1.00 \text{ atm}$$

Now suppose we filled the same container with 0.004 mol $H_2(g)$ at the same temperature. The pressure would be

$$p_{H_2} = \frac{0.0820 \text{ liter atm mol}^{-1} \text{ K}^{-1} \times 305 \text{ K}}{0.250 \text{ liter}} \times 0.004 \text{ mol} = 0.40 \text{ atm}$$

If we put 0.006 mol N_2 in the container,

$$p_{N_2} = \frac{0.0820 \text{ liter atm mol}^{-1} \text{ K}^{-1} \times 305 \text{ K}}{0.250 \text{ liter}} \times 0.006 \text{ mol} = 0.60 \text{ atm}$$

Now suppose we put both the 0.004 mol H_2 and the 0.006 mol N_2 into the same flask together. What would the pressure be? Since the ideal gas law does not depend on *which* gas we have but only on the amount of *any* gas, the pressure of the (0.004 + 0.006) mol, or 0.010 mol, would be exactly what we got in our first calculation. But this is just the sum of the pressure that H_2 would exert if it occupied the container alone plus the pressure of N_2 if it were the only gas present. That is,

$$P_{\text{total}} = p_{H_2} + p_{N_2}$$

We have just worked out an example of **Dalton's law of partial pressures** (named for John Dalton, its discoverer). This law states that *in a mixture of two or more gases, the total pressure is the sum of the partial pressures of all the components*. The **partial pressure** of a gas is the pressure that gas would exert if it occupied the container by itself. Partial pressure is represented by a lowercase letter *p*.

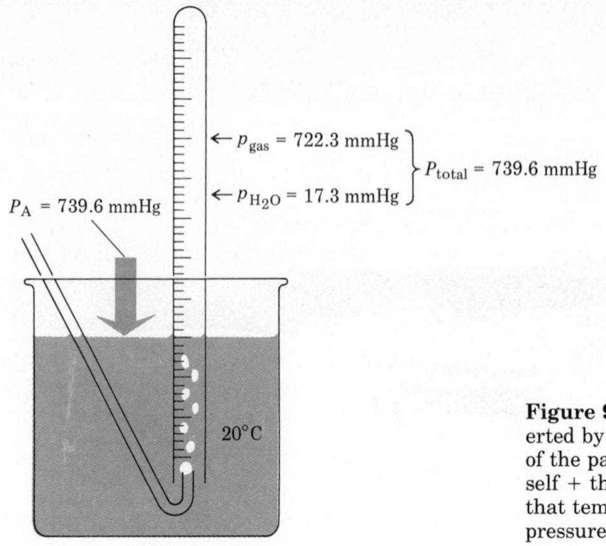

Figure 9.5 The total pressure exerted by a wet gas is equal to the sum of the partial pressure of the gas itself + the vapor pressure of water at that temperature. (At 20°C the vapor pressure of water is 17.3 mmHg.)

Dalton's law of partial pressures is most commonly encountered when a gas is collected by displacement of water, as shown in Fig. 9.5. Because the gas has been bubbled through water, it contains some water molecules and is said to be "wet." The total pressure of this wet gas is the sum of the partial pressure of the gas itself and the partial pressure of the water vapor it contains. The latter partial pressure is called the **vapor pressure** of water. It depends only on the temperature of the experiment and may be obtained from a handbook or from Table 9.4. (Vapor pressure is discussed more fully in Sec. 10.3.)

TABLE 9.4 Vapor Pressure of Water as a Function of Temperature.

Temperature (in °C)	Vapor Pressure/mmHg	Vapor Pressure/kPa
0	4.6	0.61
5	6.5	0.87
10	9.2	1.23
15	12.8	1.71
20	17.5	2.33
25	23.8	3.17
30	31.8	4.24
50	92.5	12.33
70	233.7	31.16
75	289.1	38.63
80	355.1	47.34
85	433.6	57.81
90	525.8	70.10
95	633.9	84.51
100	760.0	101.32

EXAMPLE 9.11 Assume 0.321 g zinc metal is allowed to react with excess hydrochloric acid (an aqueous solution of HCl gas) according to the equation

$$Zn(s) + 2HCl(aq) \rightarrow ZnCl_2(aq) + H_2(g)$$

The resulting hydrogen gas is collected over water at 25°C, while the barometric pressure is 745.4 mmHg. What volume of wet hydrogen will be collected?

Solution From Table 9.4 we find that at 25°C the vapor pressure of water is 23.8 mmHg. Accordingly $p_{H_2} = P_{\text{total}} - p_{H_2O} = 745.4$ mmHg − 23.8 mmHg = 721.6 mmHg. This must be converted to units compatible with R:

$$p_{H_2} = 721.6\ \cancel{\text{mmHg}} \times \frac{1\ \text{atm}}{760\ \cancel{\text{mmHg}}} = 0.949\ \text{atm}$$

The road map for this problem is

$$m_{\text{Zn}} \xrightarrow{\mathcal{M}_{\text{Zn}}} n_{\text{Zn}} \xrightarrow{S(\text{H}_2/\text{Zn})} n_{\text{H}_2} \xrightarrow{RT/P} V_{\text{H}_2}$$

$$\text{Thus } V_{H_2} = 0.321\ \cancel{\text{g Zn}} \times \frac{1\ \cancel{\text{mol Zn}}}{65.38\ \cancel{\text{g Zn}}} \times \frac{1\ \cancel{\text{mol H}_2}}{1\ \cancel{\text{mol Zn}}} \times \frac{0.0820\ \text{liter}\ \cancel{\text{atm}}}{1\ \cancel{\text{K}}\ \cancel{\text{mol H}_2}} \times \frac{298.15\ \cancel{\text{K}}}{0.949\ \cancel{\text{atm}}}$$

$$= 0.126\ \text{liter}$$

9.4 KINETIC THEORY OF GASES

At the beginning of Chap. 2 it was mentioned that many of the properties of solids, liquids, and gases could be accounted for if we assumed that substances are made of atoms or molecules which are constantly in motion. Boyle's law and the other gas laws have now given us much more quantitative information about gases than we had in Chap. 2, and it is worth asking whether with our previous model we can make quantitative predictions in agreement with these laws. In answering this question, we will also gain important insights into the nature of temperature and of heat energy.

Postulates of the Kinetic Theory

The microscopic theory of gas behavior based on molecular motion is called the **kinetic theory of gases.** Its basic postulates are listed in Table 9.5. From them it is possible to derive the following expression for the pressure of a gas in terms of the properties of its molecules:

$$P = \frac{1}{3}\frac{N}{V} m(u^2)_{\text{ave}} \qquad (9.15)$$

where P, V = pressure and volume of the gas
N = number of molecules
m = mass of each molecule
$(u^2)_{ave}$ = *average* (or mean) of the *squares* of all individual molecular velocities

(This mean square velocity must be used because pressure is proportional to the square of molecular velocity, and molecular collisions cause different molecules to have quite different velocities.) Rather than concerning ourselves with the procedure for deriving Eq. (9.15),* let us inspect the equation and see that its general features are much as we would expect. In some ways, the ability to do this with a formula is more useful than the ability to derive it.

First of all, the equation tells us that the pressure of a gas is proportional to the number of molecules divided by the volume. This is shown graphically in Fig. 9.6, where a computer has drawn the same number of gas molecules occupying each of three different volumes. The "tail" on each molecule shows the exact path followed by that molecule in the previous microsecond—the longer the tail, the faster the molecule was going. The average of the squares of the tail lengths is proportional to $(u^2)_{ave}$ and is the same in all three diagrams. It is also assumed that all the molecules have equal masses.

As you can see, reducing the volume of the gas increases the number of collisions per unit area on the walls of the container. Each collision exerts a force on the wall; force per unit area is pressure, and so the number of collisions per unit area is proportional to pressure. Halving the volume doubles the pressure, a prediction which agrees with the experimental facts summarized in Boyle's law.

Equation (9.15) also says that the pressure is proportional to the mass of each gas molecule. Again, this is what we would expect. Heavy molecules give a bigger "push" (the technical term for this is **impulse**) against the wall than do light ones with the same velocity.

Finally, the equation tells us that pressure is proportional to the average of the squares of the molecular velocities. This dependence on the

TABLE 9.5 Postulates of the Kinetic Theory of Gases.

1	The molecules in a gas are small and very far apart. Most of the volume which a gas occupies is empty space.
2	Gas molecules are in constant random motion. Just as many molecules are moving in one direction as in any other.
3	Molecules can collide with each other and with the walls of the container. Collisions with the walls account for the pressure of the gas.
4	When collisions occur, the molecules lose no kinetic energy; that is, the collisions are said to be **perfectly elastic.** The total kinetic energy of all the molecules remains constant unless there is some outside interference with the gas.
5	The molecules exert no attractive or repulsive forces on one another except during the process of collision. Between collisions, they move in straight lines.

* A derivation appears in J. Waser, K. N. Trueblood, and C. M. Knobler, "Chem One," pp. 119–122, McGraw-Hill Book Company, New York, 1976.

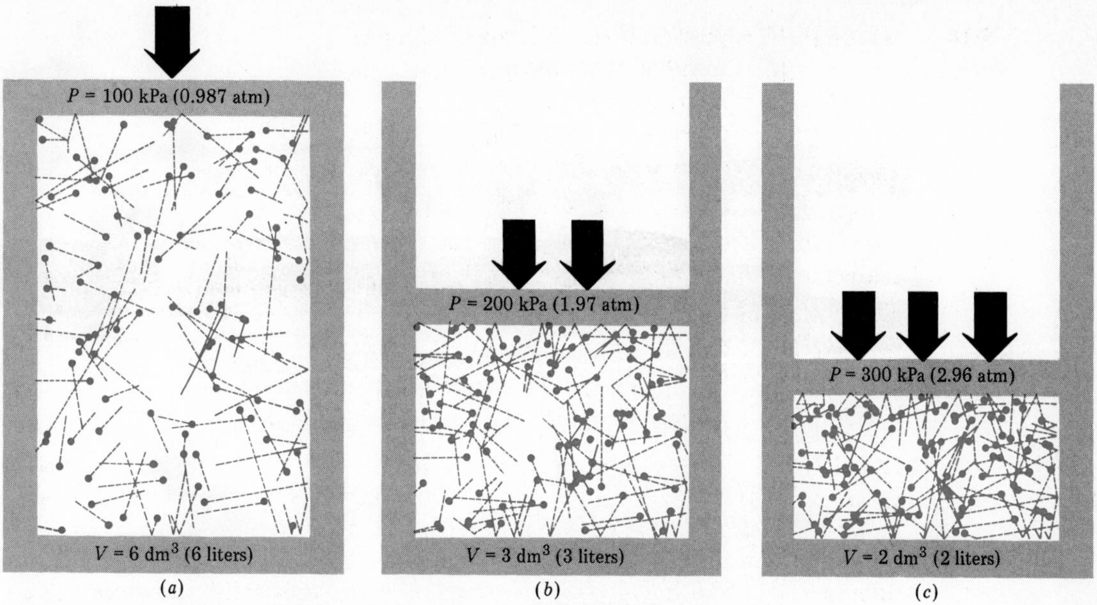

Figure 9.6 Boyle's law. There are equal numbers of molecules in the enclosed portion of each diagram. As the volume of gas decreases, the number of molecules per unit volume increases. The number of molecules striking a given area of the walls per unit time also increases, thus increasing the force on that area. This force per unit area is the pressure, and hence the pressure increases.

square of velocity is reasonable if we realize that doubling the velocity of a molecule has *two* effects. First, the molecule can move farther in a given length of time, doubling the number of collisions with the walls. This would double the pressure. Second, doubling the velocity of a molecule doubles the push or impulse of each collision. This doubles the pressure again. Therefore doubling a molecule's velocity quadruples the pressure, and for a large number of molecules, P is proportional to the mean *square* velocity.

The Total Molecular Kinetic Energy

Equation (9.15) can tell us a lot more than this about gases, however. If both sides are multiplied by V, we have

$$PV = \tfrac{1}{3}Nm(u^2)_{\text{ave}} \tag{9.16}$$

The kinetic energy of an individual molecule (Sec. 3.3) is $\frac{1}{2}mu^2$, and so the average kinetic energy $(E_k)_{\text{ave}}$ of a collection of molecules, all of the same mass m is

$$(E_k)_{\text{ave}} = (\tfrac{1}{2}mu^2)_{\text{ave}} = \tfrac{1}{2}m(u^2)_{\text{ave}}$$

The total kinetic energy E_k is just the number of molecules times this average:

$$E_k = N \times (E_k)_{\text{ave}} = N \times \tfrac{1}{2}m(u^2)_{\text{ave}}$$

or, multiplying both sides by $3/3$ (i.e., by 1),

$$E_k = \tfrac{3}{3} \times \tfrac{1}{2}Nm(u^2)_{\text{ave}} = \tfrac{3}{2} \times \tfrac{1}{3}Nm(u^2)_{\text{ave}}$$

Substituting from Eq. (9.16),

$$E_k = \tfrac{3}{2}PV \tag{9.17}$$

or

$$PV = \tfrac{2}{3}E_k \tag{9.18}$$

The product of the pressure and the volume of a gas is two-thirds the total kinetic energy of the molecules of the gas.

Now we can understand why PV came out in joules in Sec. 9.2—it is indeed an energy. According to postulate 4 of the kinetic theory, gas molecules have constant total kinetic energy. This is reflected on the macroscopic scale by the constancy of PV, or, in other words, by Boyle's law.

The kinetic theory also gives an important insight into what the *temperature* of gas means on a microscopic level. We know from the ideal gas law [Eq. (9.12)] that $PV = nRT$. Substituting this into Eq. (9.18),

$$nRT = \tfrac{2}{3}E_k \tag{9.19}$$

If we divide both sides of Eq. (9.19) by n and multiply by $\tfrac{3}{2}$,

$$\frac{E_k}{n} = \tfrac{3}{2}RT$$

The term E_k/n is the total kinetic energy divided by the amount of substance, that is, the **molar kinetic energy.** Representing molar kinetic energy by E_m, we have

$$E_m = \tfrac{3}{2}RT \tag{9.20}$$

The molar kinetic energy of a gas is proportional to its temperature, and the proportionality constant is $\tfrac{3}{2}$ times the gas constant R.

Molecular Speeds

In Chap. 2 we stated that increasing the temperature increases the speeds at which molecules move. We are now in a position to find just how large that increase is for a gaseous substance. Combining the ideal gas law with Eq. (9.16), we obtain

$$PV = nRT = \tfrac{1}{3}Nm(u^2)_{\text{ave}}$$

or

$$3RT = \frac{Nm}{n}(u^2)_{\text{ave}} \tag{9.21}$$

Since N is the number of molecules and m is the mass of each molecule, Nm is the total mass of gas. Dividing total mass by amount of substance gives molar mass $\mathcal{M}$:

$$\mathcal{M} = \frac{Nm}{n}$$

Substituting in Eq. (9.21), we have

$$3RT = \mathcal{M}(u^2)_{\text{ave}}$$

or

$$(u^2)_{\text{ave}} = \frac{3RT}{\mathcal{M}}$$

so that

$$u_{\text{rms}} = \sqrt{(u^2)_{\text{ave}}} = \sqrt{\frac{3RT}{\mathcal{M}}} \tag{9.22}$$

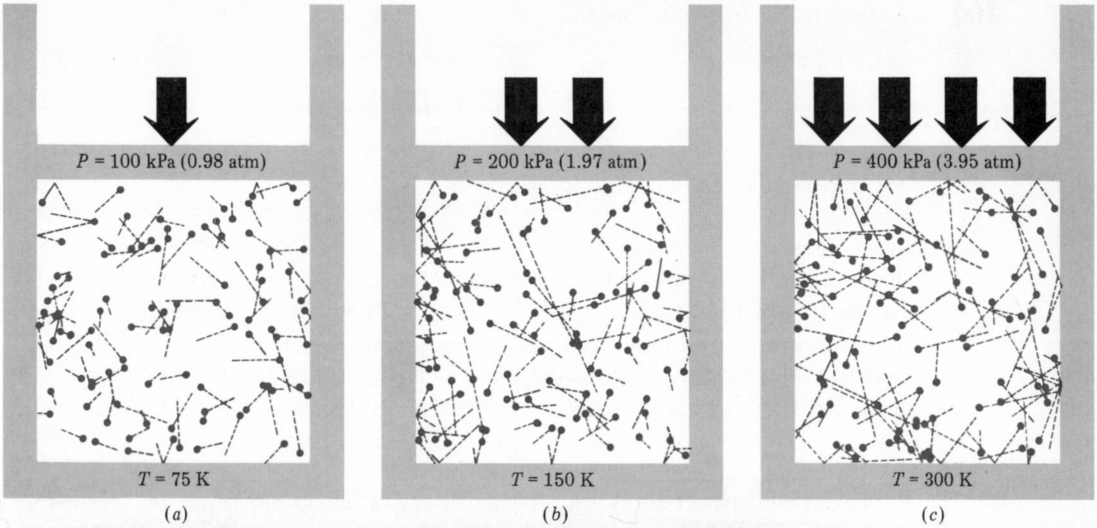

Figure 9.7 A microscopic interpretation of Gay-Lussac's law. As the temperature of a gas is increased, the velocity of the molecules is also increased. More molecules hit the sides of the container, each with a greater impulse, so that the pressure increases.

The quantity u_{rms} is called the **root-mean-square** (rms) **velocity** because it is the square root of the mean square velocity.

The rms velocity is directly proportional to the square root of temperature and inversely proportional to the square root of molar mass. Thus quadrupling the temperature of a given gas doubles the rms velocity of the molecules. Doubling this average velocity doubles the number of collisions between gas molecules and the walls of a container. It also doubles the impulse of each collision. Thus the pressure quadruples. This is indicated graphically in Fig. 9.7. Pressure is thus directly proportional to temperature, as required by Gay-Lussac's law.

The inverse proportionality between root-mean-square velocity and the square root of molar mass means that the heavier a molecule is, the slower it moves. This is illustrated in the next example.

EXAMPLE 9.12 Find the rms velocity for (a) H_2 and (b) O_2 molecules at 27°C.

Solution This problem is much easier to solve if we use SI units. Thus we choose

$$R = 8.314 \text{ J mol}^{-1}\text{ K}^{-1} = 8.314 \text{ kg m}^2\text{s}^{-2}\text{ mol}^{-1}\text{ K}^{-1}$$

a) For H_2

$$u_{rms} = \sqrt{\frac{3RT}{\mathcal{M}}} = \sqrt{\frac{3 \times 8.314 \text{ J } \cancel{\text{mol}^{-1}}\ \cancel{\text{K}^{-1}} \times 300\ \cancel{\text{K}}}{2.016 \text{ g } \cancel{\text{mol}^{-1}}}}$$

$$= \sqrt{3.712 \times 10^3\ \frac{\text{kg m}^2\text{s}^{-2}}{\text{g}}} = \sqrt{3.712 \times 10^3 \times 10^3\ \frac{\cancel{\text{g}}\text{ m}^2\text{s}^{-2}}{\cancel{\text{g}}}}$$

$$u_{rms} = \sqrt{3.712 \times 10^3} \text{ m s}^{-1} = 1.927 \times 10^3 \text{ m s}^{-1}$$

b) For O_2

$$u_{rms} = \sqrt{\frac{3 \times 8.314 \text{ J } \cancel{\text{mol}^{-1}} \, \cancel{\text{K}^{-1}} \times 300 \, \cancel{\text{K}}}{32.00 \text{ g } \cancel{\text{mol}^{-1}}}}$$

$$= 4.836 \times 10^2 \text{ m s}^{-1}$$

The rms velocities 1927 m s^{-1} and 484 m s^{-1} correspond to about 4300 miles per hour and 1080 miles per hour, respectively. The O_2 molecules in air at room temperature move about 50 percent faster than jet planes, and H_2 molecules are nearly 4 times speedier yet. Of course an O_2 molecule would take a lot longer to get from New York to Chicago than a jet would. Gas molecules never go far in a straight line before colliding with other molecules.

Now we can see the microscopic basis for Avogadro's law. Most of the volume in H_2, O_2, or any gas is empty space, and that empty space is the same for a given amount of any gas at the same temperature and pressure. This happens because the total kinetic energy of the molecules is the same for H_2 or O_2 or any other gas. The more energy they have, the more room the molecules can make for themselves by expanding against a constant pressure. This is illustrated in Fig. 9.8, where equal numbers of H_2 and O_2 molecules occupy separate containers at the same temperature and pressure. The volumes are seen to be the same. Because O_2 molecules are 16 times heavier than H_2 molecules, the average speed of H_2 molecules is 4 times faster. H_2 molecules therefore make 4 times as many collisions with the walls. Based on mass, each collision of an H_2 molecule with the wall has one-sixteenth the effect of an O_2 collision, but an H_2 collision has 4 times the effect of an O_2 collision when molecular velocity is considered. The net result is that each H_2 collision is only one-fourth as effective as an O_2 collision.

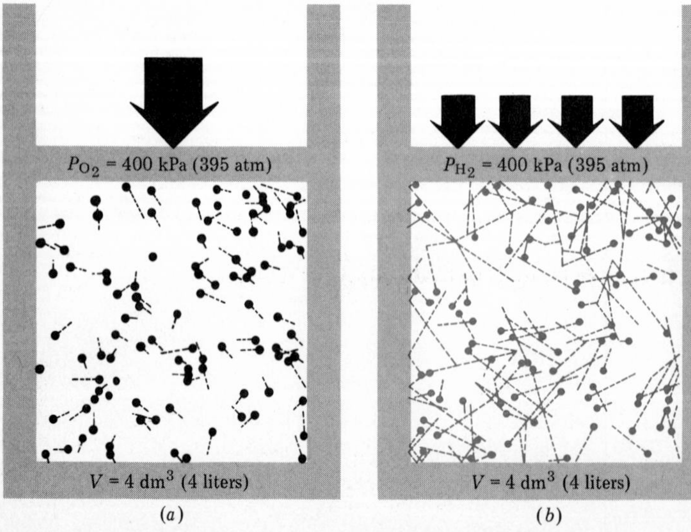

Figure 9.8 Avogadro's law. Equal number of (*a*) O_2 and (*b*) H_2 molecules are shown in separate containers at the same P. The speedier H_2 molecules make 4 times as many collisions with the walls, but each collision by one of the heavier O_2 molecules is 4 times as effective. Therefore both gases push the piston up to the same height and occupy the same volume.

Four times as many collisions, each one-fourth as effective, results in the same pressure. Thus the same number of O_2 molecules as H_2 molecules is required to occupy the same volume at the same temperature and pressure.

Graham's Law of Diffusion

Faster-moving molecules can escape more readily through small holes or pores in containers. Such an escape is called **effusion.** They can also mix more rapidly with other gases by **diffusion.** Such processes are usually carried out at constant temperature, and so the relative rates of diffusion or effusion of two gases A and B depend only on the molar masses $\mathcal{M}_A$ and $\mathcal{M}_B$:

$$(u_A)_{rms} = \sqrt{\frac{3RT}{\mathcal{M}_A}} \qquad (u_B)_{rms} = \sqrt{\frac{3RT}{\mathcal{M}_B}}$$

The rates of effusion or diffusion are proportional to the rms velocities, and so

$$\frac{\text{Rate of diffusion of A}}{\text{Rate of diffusion of B}} = \frac{(u_A)_{rms}}{(u_B)_{rms}} = \frac{\sqrt{\frac{3RT}{\mathcal{M}_A}}}{\sqrt{\frac{3RT}{\mathcal{M}_B}}} = \sqrt{\frac{3RT}{\mathcal{M}_A} \times \frac{\mathcal{M}_B}{3RT}}$$

$$= \sqrt{\frac{\mathcal{M}_B}{\mathcal{M}_A}} \tag{9.23}$$

This result is known as **Graham's law of diffusion** after Thomas Graham (1805 to 1869), a Scottish chemist, who discovered it by observing effusion of gases through a thin plug of plaster of paris.

EXAMPLE 9.13 Calculate the relative rates of effusion of He(g) and O_2(g).

Solution From Eq. (9.23)

$$\frac{\text{Rate of effusion of He}}{\text{Rate of effusion of } O_2} = \sqrt{\frac{\mathcal{M}_{O_2}}{\mathcal{M}_{He}}} = \sqrt{\frac{32.00 \text{ g mol}^{-1}}{4.003 \text{ g mol}^{-1}}} = 2.83$$

In other words we would expect He to escape from a balloon nearly 3 times as fast as O_2.

The Distribution of Molecular Speeds

It should be quite clear from the computer-generated diagrams of Figs. 9.6, 9.7, and 9.8 that molecules in the same sample of gas may have widely varying speeds. Later on we shall find it useful to know just how popular certain speeds are—that is, we might want to know what fraction of the molecules in a sample have speeds between 0 and 400 m s^{-1}, or how many are going 3600 to 4000 m s^{-1}. Such information can be obtained from a graph showing the number of molecules within a certain range of speeds on one axis and molecular speed on the other.

To see what such a graph would look like, let us study Fig. 9.6*c* in some detail. The figure refers to a temperature of 300 K, and we will assume that $H_2(g)$ is involved. Since the tails on the molecules indicate their velocities, we can count how many are traveling between 0 and 400 m s^{-1}, how many between 400 and 800 m s^{-1}, and so on. When these results are plotted as a histogram (bar graph), Fig. 9.9*a* is obtained. You can see from the histogram, for example, that 20 of the 100 H_2 molecules in Fig. 9.6*c* are traveling between 1200 and 1600 m s^{-1}.

Of course any real sample of gas at 300 K and normal atmospheric pressure would contain far more than the 100 molecules in Fig. 9.6*c*. We could narrow the range of speeds within which molecules are counted (i.e., make the bars in the histogram much narrower), and there would still be a very large number of molecules within each bar. When the distribution of molecular speeds is measured for such a large number of molecules, the curve printed in red over the histogram is obtained. This pattern was first deduced theoretically in 1860 by James Clerk Maxwell (1831 to 1879). In 1868 Ludwig Boltzmann (1844 to 1906) used a much better argument to substantiate this curve. It is known as the **Maxwell-Boltzmann distribution** of molecular speeds.

The most important aspect of the Maxwell-Boltzmann curve is that it is

Figure 9.9 The distribution of molecular speeds in a sample of 100 H_2 molecules at (*a*) 300 K and (*b*) 373 K. The colored line on each graph is the theoretical distribution given by the Maxwell-Boltzmann law.

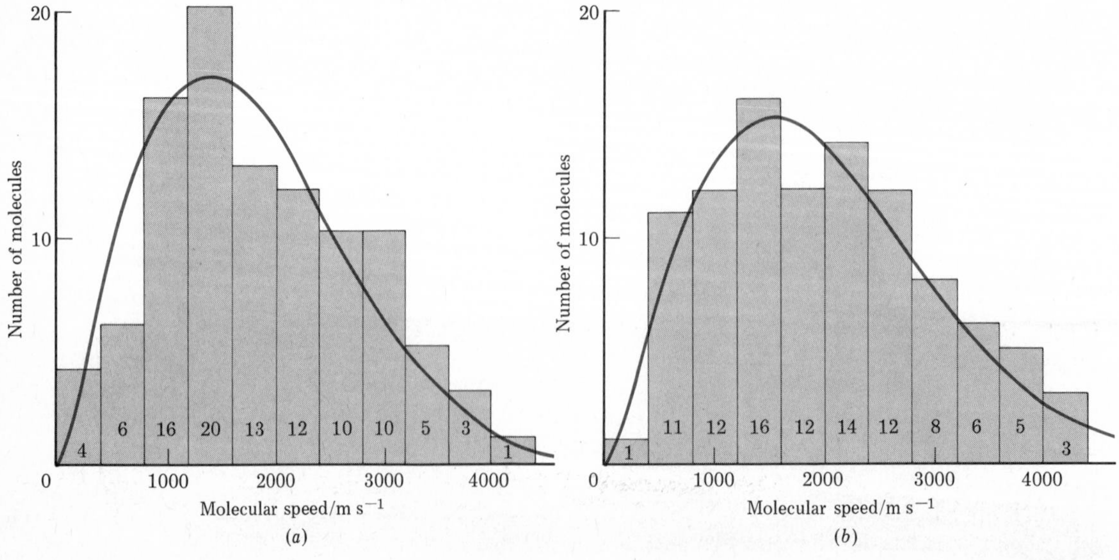

lopsided in favor of higher speeds. The maximum in the colored curve in Fig. 9.9*a* is at 1414 m s^{-1}. This is the most probable speed for an H_2 molecule at 300 K. It is less than the root-mean-square velocity of 1926 m s^{-1} calculated in Example 9.12. More molecules (62 out of 100 used to obtain Fig. 9.9*a*) move faster than the most probable speed than move slower. Moreover, there is a small but very important fraction of molecules with very large velocities. Three have speeds between 3600 and 4000 m s^{-1}, and one is moving faster than 4000 m s^{-1}. This is about 3 times the most probable speed and more than double the average (rms) velocity.

Raising the temperature of a gas raises the average speed of its molecules, shifting the Maxwell-Boltzmann curve toward higher speeds, as shown in Fig. 9.9*b*. Increasing the temperature from 300 to 373 K increases the most probable speed from 1414 to 1577 m s^{-1} and the rms velocity from 1926 to 2157 m s^{-1}, both increases of 11 percent. The number of really fast molecules goes up much more significantly, however. In the range 3600 to 4000 m s^{-1} the number of molecules increases from 3 to 5 (67 percent), and in the range 4000 to 4400 m s^{-1}, the number increases from 1 to 3 (300 percent). If we apply the Maxwell-Boltzmann distribution to 1 mole $H_2(g)$ (instead of the 100 molecules in Fig. 9.9), we find 0.75 million molecules moving faster than 10 000 m s^{-1} at 300 K. At 373 K this number has increased to 465 million—more than 60 000 percent.

There are many things in nature which depend on the average (rms) velocity of molecules. A mercury thermometer is one, and the pressure of a gas is another. Many other things, however, are influenced by the number of very fast molecules rather than by the average velocity. One example is the human finger—in a sample of $H_2(g)$ at 300 K it feels pleasantly warm, but at 373 K it will blister. This important difference in behavior is not caused by an 11-percent increase in the average velocity of the molecules. It is caused by a dramatic increase in the number of very energetic molecules, which occurs when the temperature is raised from 300 to 373 K.

As we shall see in Chap. 18, the rates of chemical reactions respond to the number of really fast molecules instead of the average molecular velocity, just as your finger would. Most chemical reactions can be speeded up tremendously by raising the temperature, and this is why chemists so often boil things in flasks to get reactions to occur.

9.5 DEVIATIONS FROM THE IDEAL GAS LAW

Sufficiently accurate measurement of pressure, temperature, volume, and amount of any gas will reveal that the ideal gas law [Eq. (9.12)] is never obeyed exactly. This is why the molar volumes in Table 9.1 were not all exactly 22.414 liters. A convenient way to detect deviations from the ideal gas law is to calculate PV under various conditions. According to the kinetic theory [Eq. (9.18)], PV is two-thirds the total molecular kinetic energy and should remain constant at a given temperature for a given amount of gas. That it does not is evident from Fig. 9.10, where PV for 1 mol $CH_4(g)$ is plotted versus P. At high pressures, PV is always larger than would be

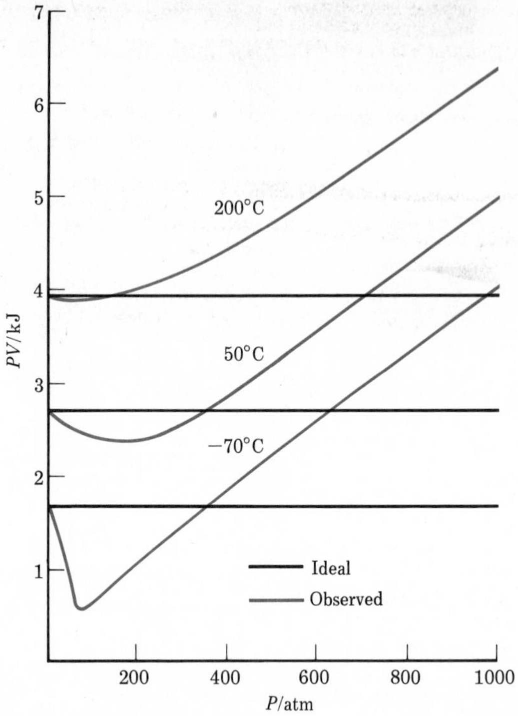

Figure 9.10 Plot of PV versus P for methane (CH_4) gas at three temperatures.

predicted by the ideal gas law. As the temperature decreases, deviations occur at lower pressures, and PV drops below the predicted horizontal line before rising again with pressure.

At high pressures, PV increases above the ideal gas value because the first postulate of the kinetic theory of gases (Table 9.5) is no longer valid. As pressure increases, the molecules are squeezed close to one another, and the volume of the molecules themselves becomes a significant fraction of the volume of the container. This is shown in Fig. 9.11. The space which other molecules are prevented from occupying is called the **excluded volume.** The measured volume of the container, $V_{container}$, is the sum of the volume available to the gas molecules, V_{gas}, and this excluded volume. Since $PV_{container}$ is larger than PV_{gas}, the experimentally measured PV is too large.

Intermolecular forces of the type described in Chap. 8 cause PV to drop below the ideal gas prediction at low temperatures and medium pressures. Consider a gas molecule which is about to hit the wall of the container (Fig. 9.12). Kinetic theory assumes that its neighbors exert no forces on such a

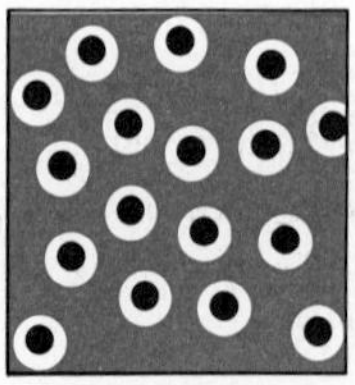

Figure 9.11 The excluded volume. Because the molecules in a gas have a finite volume, they are not free to move about throughout the whole volume of the container. The figure represents the molecules (black) stationary at a given instant of time. If we added a new molecule to the gas, its center would have to be in the volume colored red. The white and black volumes together are called the excluded volume. Note that the excluded volume is larger than the actual volume of the molecules.

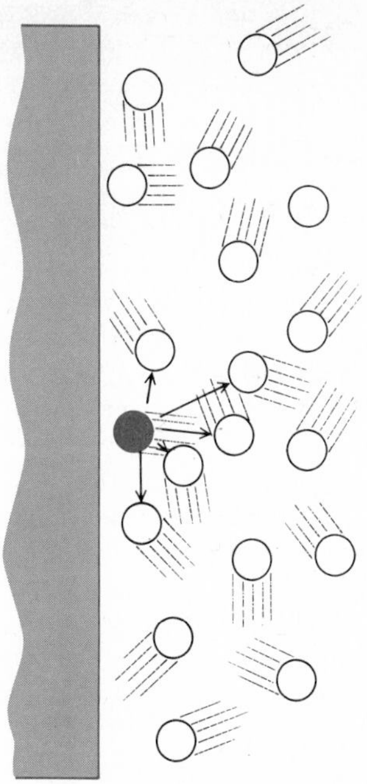

Figure 9.12 The effect of intermolecular forces on the pressure of a gas. A molecule about to hit the wall, such as the one indicated in red, is held back by the attractive forces of its fellow molecules and strikes the wall with less impact than would have been the case if these forces were absent.

molecule except during a collision (postulate 5), but we know that such forces exist. When a molecule is near the wall, the attractions between it and its neighbors are unbalanced, tending to pull it away from the wall. The molecule produces slightly less impact than it would if there were no intermolecular forces. All collisions with the walls are softer, and the pressure is less than would be predicted by the ideal gas law. This effect of intermolecular forces is more pronounced at lower temperatures because under those conditions the kinetic energies of the molecules are smaller. The potential energy of intermolecular attraction is comparable to that kinetic energy and can have a significant effect.

For gases such as hydrogen, oxygen, nitrogen, helium, or neon, deviations from the ideal gas law are less than 0.1 percent at room temperature and atmospheric pressure. Other gases, such as carbon dioxide or ammonia, have stronger intermolecular forces and consequently greater deviation from ideality. Nonideal behavior is quite pronounced for any gas at very high pressures or at temperatures just above the boiling point. Under these conditions molecular volume or intermolecular attractions can have maximum effect.

SUMMARY

The molecules of any substance in the gaseous state are quite far from one another, and so most of the volume of any gas is empty space. Consequently all gases have similar properties. Gases are transparent and extremely mobile, and they have low, but quite variable, densities. The volume of any sample of gas is proportional to the amount of substance, and the proportionality constant is the *same* for all gases, provided temperature and pressure remain constant. The variation introduced by changes in T and P is taken into account by the ideal gas law, which is obeyed to a high degree of accuracy by any gas at sufficiently low pressures and/or sufficiently high temperatures. Deviations from ideal gas behavior occur when molecules are close enough together that their volume is a significant fraction of the volume occupied by the gas, or when intermolecular attractions reduce the impulse of molecular collisions with the walls of a container.

The relationship between gas volume and amount of substance allows determination of the amount of gas from simple measurements of volume, temperature, and pressure. If the mass of a gas can be obtained in the same experiment, molar mass and molecular weight can be calculated. In addition, when gaseous substances participate in chemical reactions, the volumes consumed or produced at the same temperature and pressure are proportional to the coefficients of the balanced chemical equation. These volumes may be calculated using the ideal gas law.

The kinetic theory of gases is capable of predicting quantitatively the properties of gases, based on the assumption that gases consist of rapidly moving, widely separated molecules. An important result of the theory is the proportionality between molar kinetic energy of the gas molecules and the absolute or thermodynamic temperature. The average (root-mean-square) velocities of different types of molecules depend directly on the square root of absolute temperature and inversely on the square root of molar mass. Thus, on average, lighter molecules move faster and diffuse or effuse more readily than heavier ones.

Although the total kinetic energy of a collection of molecules is always the same at a given temperature, the speeds of individual molecules vary considerably as a result of collisions with other molecules. The distribution of molecular speeds is lopsided in favor of high speeds, and a small increase in temperature can greatly increase the fraction of molecules which travels at very high speeds. This can have very significant effects on the rates of chemical reactions.

QUESTIONS AND PROBLEMS

9.1 Calculate the pressure exerted by the block of lead in Fig. 9.1 when it stands on the side measuring 20 cm by 100 cm.

9.2 A steel rod has a mass of 1 lb. Calculate its weight in newtons. If the rod has a cross-sectional area of 1 in^2, calculate the pressure in kilopascals which it exerts when placed upright on a floor. Use this result to calculate how many pounds per square inch are equal to 1 atm pressure.

9.3 Describe the differences between a barometer and a manometer.

9.4 Repeat the calculation in Example 9.2, but assume that the cross-sectional area of the tube is exactly 2 cm^2. What effect does this have on the weight of the mercury column? on the *pressure?* Explain why the pressure exerted by a mercury column is independent of its cross-sectional area.

9.5 Calculate the pressure in kilopascals exerted by a column of water (ρ = 1.000 g cm^{-3}) of height 1 m and cross section 1 cm^2. How high would this column need to be to exert a pressure of 1 atm?

9.6 Calculate the pressure of a gas in the container shown in Fig. 9.3 in millimeters of mercury, kilopascals, and atmospheres in each of the following cases: (*a*) The level of mercury is 5.7 cm higher in the arm of the manometer connected to the flask, and the atmospheric pressure is 775 mmHg. (*b*) The level of mercury is 37 mm lower in the arm of the manometer connected to the flask, and the atmospheric pressure is 0.951 atm.

9.7 According to Boyle's law the pressure of a gas is *inversely proportional* to the volume if the temperature remains constant. Explain in your own words what is meant by the term *inversely proportional.*

9.8 A gas occupies 16.3 cm^3 at a pressure of 720 mmHg. Express the product PV for this gas in joules.

9.9 A gas occupies 175 liters at a pressure of 0.921 atm. What will be its volume if the pressure is increased to 1.326 atm while the temperature remains constant?

9.10 At 300 K, a 2.95-liter sample of CO_2 exerts a pressure of 150 kPa. What volume would this same gas occupy at 1 atm and the same temperature?

9.11 A gas originally at a pressure of 1130 mmHg is compressed from a volume of 675 liters to one of 629 liters without change in temperature. What is the final pressure of the gas?

9.12 The volume of a gas at atmospheric pressure is 11.0 dm^3 at 300 K. What will its volume be at 373 K at the same pressure?

9.13 At 25°C, a gas has a volume of 27.3 cm^3. What will its volume be at 100°C and the same pressure?

9.14 The volume of a gas expands from 28.8 to 36.1 cm^3 when the temperature is raised to 100°C. Assuming constant pressure, what was the original temperature of the gas prior to expansion?

9.15 Find the fallacies in the following argument: Since

$$20°C = 68°F = 293\ K$$

while

$$40°C = 104°F = 313\ K$$

We can conclude that

$$\frac{40°C}{20°C} = \frac{104°F}{68°F} = \frac{313\ K}{293\ K}$$

or

$$2 = 1.529 = 1.068$$

9.16 A 500-cm^3 cylinder of nitrogen gas was originally at a pressure of 0.89 atm at 10°C. What would be the pressure inside the cylinder if it were placed in the sun until the temperature of the gas rose to 40°C?

9.17 What amount of gas is present in these cases: (*a*) 352 cm^3 at 75°C and 680 mmHg pressure; (*b*) 2.63 liters at 23°C and 0.600 atm pressure; (*c*) 3.86 dm^3 at 0°C and 95.3 kPa pressure?

9.18 What volume would each of the samples in the previous problem occupy at STP?

9.19 Calculate the mass of the following gases: (*a*) 653 cm^3 NH_3 at 100°C and 200 mmHg pressure; (*b*) 40.0 liters O_2 at 20°C and 150 atm pressure in an O_2 cylinder; (*c*) 1.82 dm^3 CO_2 at 290 K and 93.7 kPa pressure.

9.20 Calculate the density of the following gases in grams per cubic centimeter or grams per cubic decimeter:

a N_2 at 25°C and 740 mmHg
b O_2 at 20°C and 136 atm pressure

9.21 At what temperature will Ne gas have a density of 1 g dm^{-3} at 1 atm pressure?

9.22 Calculate the molar mass and molecular formula of the following gases, all of which have the empirical formula CH: (*a*) 273 cm^3 at 15°C and 750 mmHg, weighing 1.78 g; (*b*) 1.861 liters at 100°C and 0.912 atm, weighing 1.44 g; (*c*) 0.836 dm^3 at 373 K and 96.5 kPa, weighing 2.03 g.

9.23 A sample of gas occupies 483 cm^2 at 20°C and 775 mmHg pressure. Calculate its volume at

a 80°C and 520 mmHg pressure.
b 0°C and 1.86 atm pressure.
c 400 K and 96 kPa pressure.

9.24 What volume of nitrous oxide, N_2O, (also called laughing gas) can be prepared at 20°C and 737 mmHg pressure by carefully heating 24.0 g ammonium nitrate and allowing it to decompose according to the equation

$$NH_4NO_3(s) \rightarrow N_2O(g) + 2H_2O(g)$$

9.25 Striking a wooden match involves the combustion of P_4S_3 to produce a white smoke of P_4O_{10} and gaseous sulfur dioxide, SO_2. Calculate the volume of SO_2 that can be prepared at 20°C and 772 mmHg pressure from the combustion of 0.157 g P_4S_3.

$$P_4S_3(s) + O_2(g) \rightarrow P_4O_{10}(s) + SO_2(g)$$

9.26 When potassium nitrate reacts with mercury in the presence of sulfuric acid, the following reaction occurs:

$$2KNO_3 + 4H_2SO_4 + 3Hg \rightarrow K_2SO_4 + 3HgSO_4 + 4H_2O + 2NO$$

By using a special apparatus to measure the volume of NO gas produced, the quantity of KNO_3 originally present in the sample can be estimated. If a sample of impure KNO_3 weighing 0.1593 g was used and 34.7 cm^3 of NO was produced at 25°C and a pressure of 741 mmHg, find the percentage purity of the sample.

9.27 A 1-g sample of nitroglycerin is allowed to explode in a casing of volume 2 cm^3 according to the following equation:

$$4C_3H_5(NO_3)_3(l) \rightarrow 12CO_2(g) + 10H_2O(g) + 6N_2(g) + O_2(g)$$

If the gaseous products reach a maximum temperature of 3000 K after the explosion, what is the maximum pressure produced in the casing?

9.28 If 0.3 mol H_2 gas, 0.2 mol N_2, and 0.25 mol O_2 are mixed in a glass container of volume 10.0 dm^3 at 298 K, what will the final pressure be?

9.29 Two glass bulbs, A of volume 500 cm^3 and B of volume 200 cm^3, are connected by a closed stopcock. If A contains N_2 at 50 kPa pressure while B contains O_2 at 100 kPa pressure, what will the pressure be when the stopcock is opened?

9.30 If the temperature in the previous problem was 20°C, calculate the density of the resulting mixture of gases in grams per cubic decimeter.

9.31 Find the amount of the following gas collected "wet" over water: (*a*) 25.6 cm^3 of O_2 collected at 23°C and a total pressure of 760 mmHg; (*b*) 2.91 liters of H_2 collected at 17°C and a total pressure of 1.036 atm; (*c*) 1.831 dm^3 of CO collected at 293 K and a total pressure of 98.6 kPa.

9.32 A 1.225-g sample of potassium chlorate, $KClO_3$, is heated in the presence of an MnO_2 catalyst and completely decomposed according to the following equation:

$$2KClO_3(s) \rightarrow 2KCl(s) + 3O_2(g)$$

What volume of O_2 collected over H_2O at 25°C and a barometric pressure of 752 mmHg will be produced?

9.33 Nitric oxide, NO, can be prepared by the following reaction:

$$3Cu(s) + 8HNO_3(aq) \rightarrow 3Cu(NO_3)_2(aq) + 2NO(g) + 4H_2O(l)$$

What mass of Cu is needed to produce 239 cm^3 of NO gas collected over H_2O at 20°C and a barometric pressure of 629 mmHg?

***9.34** A mole of gas has a volume V, a pressure P, and a temperature T. On the molecular level the rms velocity of the molecules is u, the frequency of collisions is f, and the impulse of each collision is p. How will u, f, and p be altered if the following alterations in the macroscopic properties are made:

a V is reduced to $V/2$, but T remains constant.

b T is raised to $2T$, but V remains constant.

c P, V, and T remain constant, but the mass of the molecules is doubled.

If V is held constant, how will P and T be affected if u is doubled?

9.35 Find the total kinetic energy of the molecules in a gas sample if (*a*) the pressure is 98.1 kPa and the volume is 20.3 dm^3; (*b*) the pressure is 1 atm and the volume is 16 liters.

9.36 Find the rms velocity of

a H_2 molecules at 20°C.

b CO_2 molecules at 20°C.

c CO_2 molecules at 200°C.

Which gas would you expect to conduct heat better, H_2 or CO_2?

***9.37** Many chemical reactions double their speed when the temperature is raised 10°C. Explain why this fact shows that the rate of chemical reactions is not entirely determined by the rate at which the reacting molecules can collide.

9.38 Calculate the relative rates of effusion for H_2 gas and SO_2 gas under similar conditions.

9.39 When an unknown gas at 2.0 atm pressure and 20°C is allowed to diffuse through a porous plug, it is found to do so at the rate of 7.20 mmol s^{-1}. When O_2 at the same pressure and temperature is allowed to effuse through the same plug, the rate is found to be 5.09 mmol s^{-1}. Find the molar mass of the unknown gas.

***9.40** Using the histogram in Fig. 9.9, calculate the mean speed of the 100H_2 molecules at 337 K. Next calculate the mean square velocity of these molecules. Take the square root of this quantity and find the root-mean-square velocity. Compare this value to the theoretical value given by Eq. (9.22). Which is larger, the root-mean-square velocity or the mean speed? (Use of a printing calculator or a computer is recommended for this problem.)

9.41 Explain why you would expect helium to show smaller deviation from the ideal gas behavior than xenon, both at high pressures and at low temperatures.

9.42 Explain why the curve of PV versus P for methane, shown in Fig. 9.10, shows a more pronounced dip at low temperatures than at high temperatures.

chapter ten

SOLIDS, LIQUIDS, AND SOLUTIONS

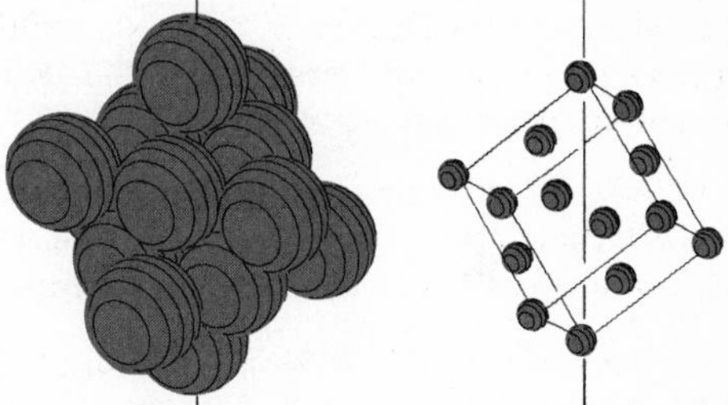

By comparison with gases, solids and liquids have microscopic structures in which the constituent particles are very close together. The volume occupied by a given amount of a solid or liquid is much less than that of the corresponding gas. Consequently solids and liquids collectively are called **condensed phases.** The properties of solids and liquids are much more dependent on intermolecular forces and on atomic, molecular, or ionic sizes and shapes than are the properties of gases.

Despite their greater variation with changes in molecular structure, some properties of condensed phases are quite general. In a solid, for example, microscopic particles are arranged in a regular, repeating crystal lattice. There are only a limited number of different ways such a lattice can form, and so it is worth spending some time to see what they are. Similarly, useful generalizations can be made regarding the properties of liquids and about changes of phase—when a solid melts, a liquid vaporizes, and so on. Many such generalizations will be explored in this chapter.

The liquid phase, where microscopic particles are close together but can still move past one another, provides an ideal medium for chemical reactions. Reactant molecules can move toward one another because they are not held in fixed locations as in a solid, and a great many more collisions between molecules are possible because they are much closer together than in a gas. Such collisions lead to breaking of some bonds and formation of new ones, that is, to chemical reactions. This molecular intimacy without rigidity, combined with ease of handling of liquids in the laboratory, leads chemists to carry out many reactions in the liquid phase. Usually such reactions involve solutions of reactants in liquid solvents. Consequently we shall examine some general properties of solutions in the latter part of this chapter.

10.1 SOLIDS

The most obvious distinguishing feature of a solid is its rigidity. On the microscopic level this corresponds to strong forces between the atoms, ions, or molecules relative to the degree of motion of those particles. The only movements within a solid crystal lattice are relatively restricted vibrations about an average position. Thus we often think and speak of crystalline solids as having atoms, ions, or molecules in fixed positions.

Lattices and Unit Cells

The regular three-dimensional arrangement of atoms or ions in a crystal is usually described in terms of a space lattice and a unit cell. To see what these two terms mean, let us first consider the two-dimensional patterns shown in Fig. 10.1. We can think of each of these three structures as

Figure 10.1 The unit cell for two-dimensional lattices.

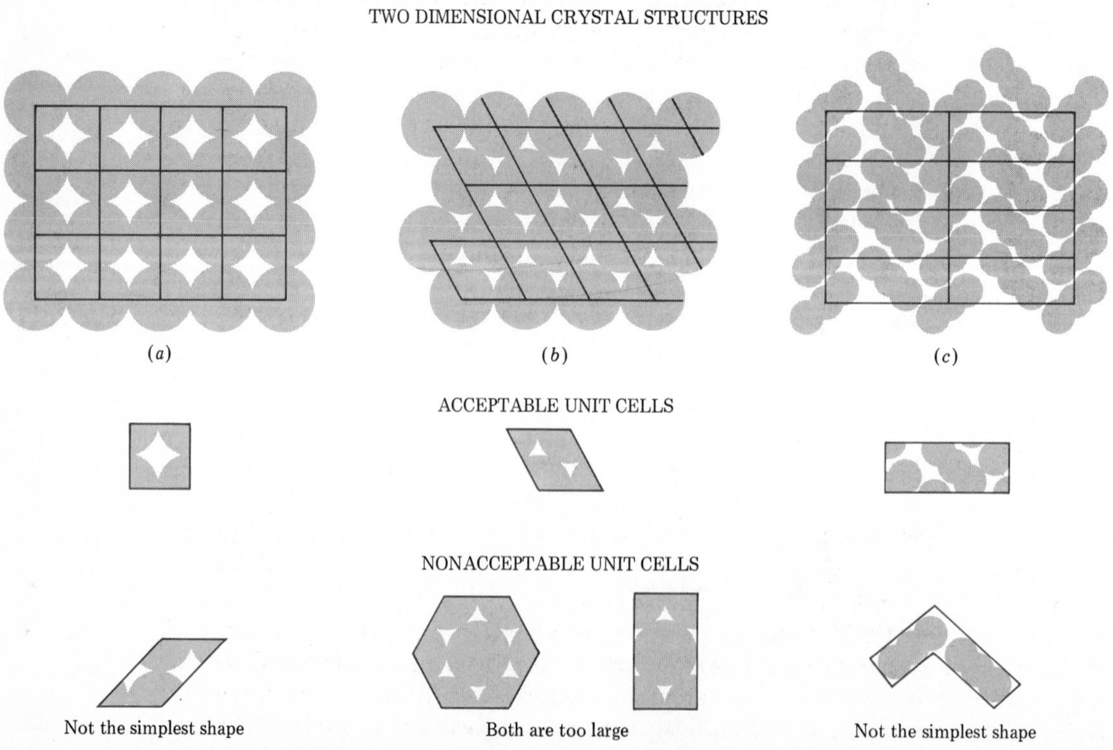

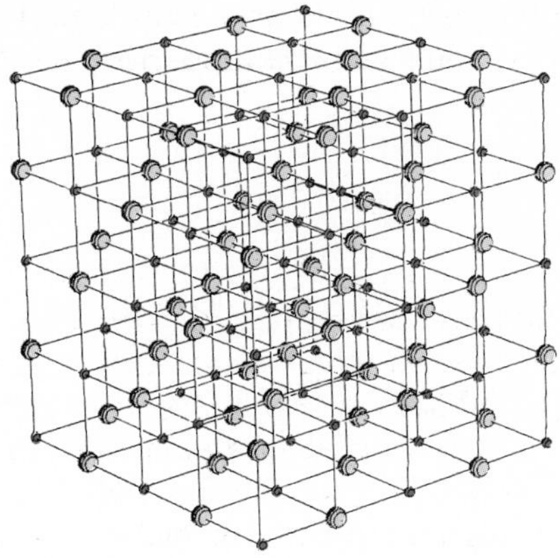
(a)

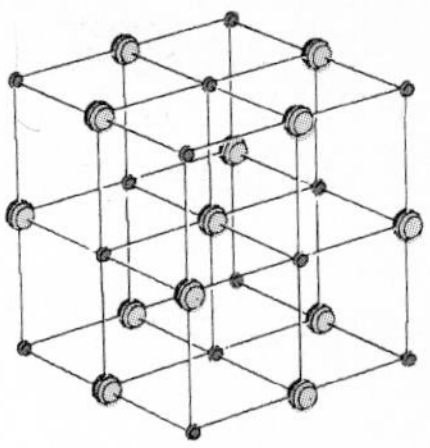
(b)

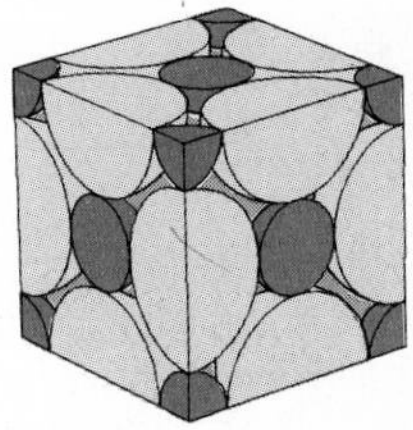
(c)

Figure 10.2 The crystal lattice and unit cell of a crystal of sodium chloride. Sodium ions, Na^+, are shown in color. Chloride ions, Cl^-, are shown in gray. (*a*) Overall crystal lattice; (*b*) unit cell; (*c*) unit cell, showing division of spheres. (Computer-generated.) *(Copyright © 1976 by W. G. Davies and J. W. Moore.)*

a large number of repetitions in two directions of the parallel-sided figure shown immediately below each pattern. This parallel-sided figure is the **unit cell.** It represents the simplest, smallest shape from which the overall structure can be constructed. The pattern of *points* made by the corners of the unit cells when they are packed together is called the **space lattice.** In Fig. 10.1 the lines joining the points of the space lattice are shown in color.

Without some experience, it is quite easy to pick the wrong unit cell for a given structure. Some incorrect choices are shown immediately below the correct choice in the figure. Note in particular that the unit cell for structure *b*, in which each circle is surrounded by six others at the corners of a hexagon, is *not* a hexagon, but a parallelogram of equal sides (a rhombus) with angles of 60 and 120°.

Figure 10.2 illustrates the space lattice and the unit cell for a real three-dimensional crystal structure—that of sodium chloride. This is the same structure that was shown for lithium hydride in Fig. 6.3, except that the sizes of the ions are different. A unit cell for this structure is a cube whose corners are all occupied by sodium ions. Alternatively the unit cell could be chosen with chloride ions at the corners.

The unit cell of sodium chloride contains *four* sodium ions and *four* chloride ions. In arriving at such an answer we must bear in mind that many of the ions are shared by several adjacent cells (Fig. 10.2*c*). Specifically, the sodium ions at the centers of the square faces of the cell are shared by two cells, so that only half of each lies within the unit cell. Since there are six faces to a cube, this makes a total of three sodium ions. In the middle of each edge of the unit cell is a chloride ion which is shared by four adjacent cells and so counts one-quarter. Since there are twelve edges, this makes three chloride ions. At each corner of the cube, a sodium ion is shared by eight other cells. Since there are eight corners, this totals to one more sodium ion. Finally there is a chloride ion in the body of the cube unshared

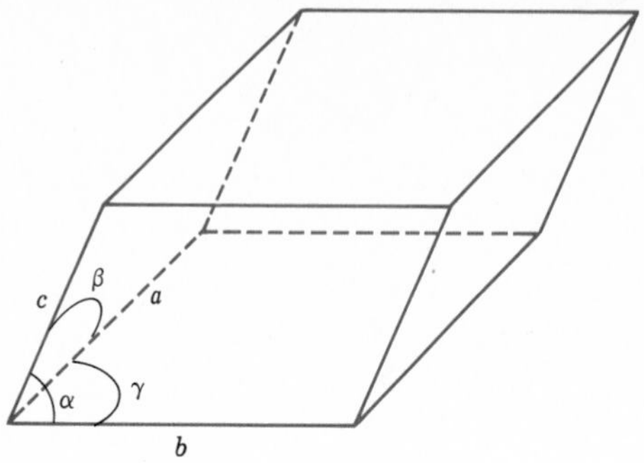

Figure 10.3 A generalized unit cell with sides a, b, and c, and angles α, β, and γ.

by any other cell. The grand total is thus four sodium and four chloride ions.

A general formula can be derived from the arguments just presented for counting N, the number of atoms or ions in a unit cell. It is

$$N = N_{\text{body}} + \frac{N_{\text{face}}}{2} + \frac{N_{\text{edge}}}{4} + \frac{N_{\text{corner}}}{8} \qquad (10.1)$$

Crystal Systems

Unit cells need not be cubes, but they *must* be parallel-sided, three-dimensional figures. A general example is shown in Fig. 10.3. Such a cell can be described in terms of the lengths of three adjacent edges, a, b, and c, and the angles between them, α, β, and γ.

Crystals are usually classified as belonging to one of seven **crystal systems,** depending on the shape of the unit cell. These seven systems are listed in Table 10.1. The simplest is the cubic system, in which all edges of the unit cell are equal and all angles are 90°. The tetragonal and orthorhombic classes also feature rectangular cells, but the edges are not all equal. In the remaining classes some or all of the angles are not 90°. The least symmetrical is the triclinic, in which no edges are equal and no angles are equal to each other or to 90°. Special note should be made of the hexagonal system whose unit cell is shown in Fig. 10.4. It is related to the two-dimensional cell encountered in Fig. 10.1*b* in that two edges of the cell are equal and subtend an angle of 120°. Hexagonal crystals are quite common

TABLE 10.1 The Seven Crystal Systems.

Name	Relationship Between Edges of Unit Cell	Relationship Between Angles of Cell
Cubic	$a = b = c$	$\alpha = \beta = \gamma = 90°$
Tetragonal	$a = b \neq c$	$\alpha = \beta = \gamma = 90°$
Orthorhombic	$a \neq b \neq c$	$\alpha = \beta = \gamma = 90°$
Monoclinic	$a \neq b \neq c$	$\alpha = \beta = 90° \neq \gamma$
Triclinic	$a \neq b \neq c$	$\alpha \neq \beta \neq \gamma \neq 90°$
Rhombohedral	$a = b = c$	$\alpha = \beta = \gamma \neq 90°$
Hexagonal	$a = b \neq c$	$\alpha = \beta = 90°; \gamma = 120°$

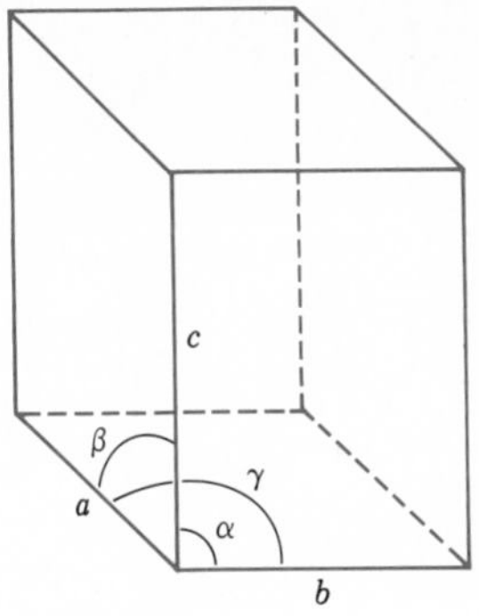

Figure 10.4 The hexagonal unit cell $a = b \neq c$, $\alpha = \beta = 90°$, $\gamma = 120°$.

among simple compounds. We have already encountered three examples, namely, ice (Fig. 8.11), graphite (Fig. 8.12*b*), and quartz (Fig. 6.8).

Closest-Packed Structures

An important class of crystal structures is found in many metals and also in the solidified noble gases where the atoms (which are all the same) are packed together as closely as possible. Most of us are familiar with the process of packing spheres together, either from playing with marbles or BB's as children or from trying to stack oranges or other round fruit into a pyramid. On a level surface we can easily arrange a collection of spheres of the same size into a very compact hexagonal layer in which each sphere is touching six of its fellows (Fig. 10.5). Then we can add a second layer so that each added sphere snuggles into a depression between three spheres in the layer below. Within this second layer each sphere also contacts six neighbors, and the layer is identical to the first one. It appears that we can add layer after layer indefinitely, or until we run out of spheres. Each sphere will be touching *twelve* of its fellows since it is surrounded by six in the same plane and nestles among three in the plane above and three in the plane below. We say that each sphere has a **coordination number** of 12. It is impossible to make any other structure with a larger coordination number, that is, to pack more spheres within a given volume. Accordingly the structure just described is often referred to as a **closest-packed** structure.

Figure 10.5 Closest packing. When a set of spheres is packed together as tightly as possible in a plane, each sphere is surrounded by six others. If a second layer is nested above the first, it is found to have the same arrangement (computer generated). (*Copyright © 1976 by W. G. Davies and J. W. Moore.*)

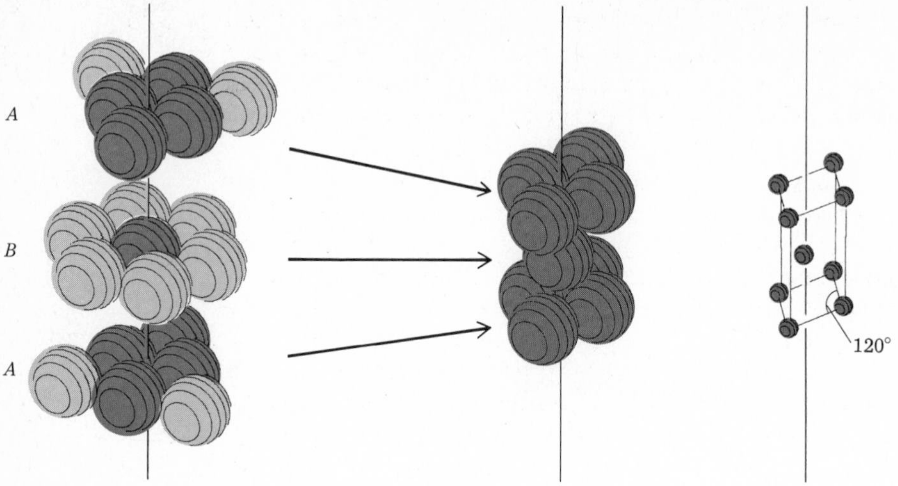

(a)

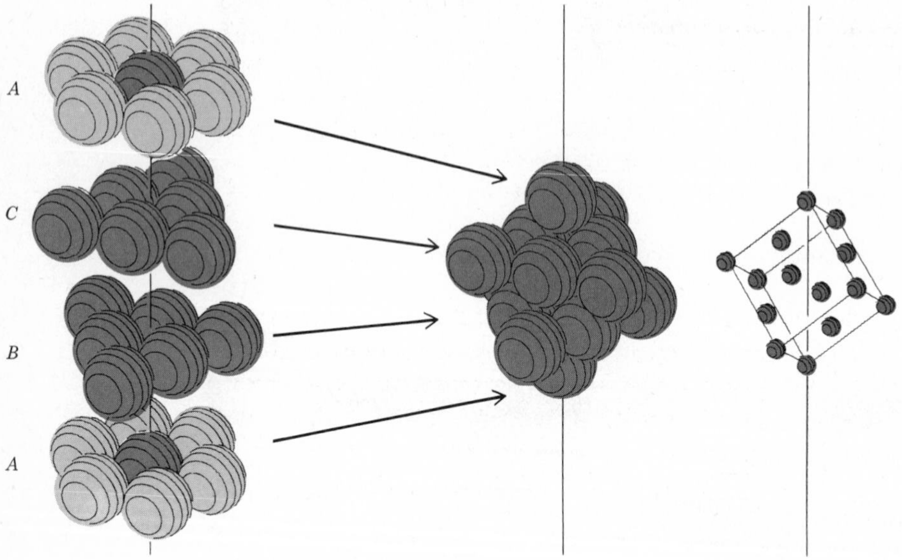

(b)

Figure 10.6 (a) Hexagonal and (b) cubic closest packing. In hexagonal packing every sphere is immediately above a sphere *two* layers below, and in cubic packing every sphere is immediately above a sphere *three* layers below. Spheres corresponding to a unit cell have been highlighted in dark color. (Computer-generated.) (*Copyright © 1976 by W. G. Davies and J. W. Moore.*)

There is a rather subtle complication about our description of closest packing—it actually fits more than one structure. There are two distinct ways of adding a third layer to the two shown in Fig. 10.5. These are illustrated in Fig. 10.6. In part *a* of the figure the first layer of spheres has been labeled *A* and the second *B* to indicate that spheres in the second layer are not directly above those in the first. The third layer is directly above the first, and so it is labeled *A*. If we continue in the fashion shown, adding alternately A, then B, then A layers, we obtain a structure whose unit cell (shown in Fig. 10.6*a*) has two equal sides with an angle of 120° between them. Other angles are 90°, and so the cell belongs to the hexagonal crystal system. Hence this structure is called **hexagonal closest packed** (hcp).

Another way of adding more layers is shown in Fig. 10.6*b*. Spheres in the third layer are not directly above those in the first, even though they snuggle just as tightly into layer B as in the hexagonal case. It is not till we get to the fourth layer that the pattern of layers repeats. Accordingly the third layer is labeled C, and the fourth is labeled A. If we build up a structure according to this $ABCABCABC$ . . . pattern, the unit cell is a cube (Fig. 10.6*b*) whose body diagonal is perpendicular to the layers. Since its unit cell belongs to the cubic system, this structure is **cubic closest packed** (ccp). The ccp structure is also called **face-centered cubic** (fcc) because there is a sphere at the center of each face of the unit cell as well as a sphere at each corner. This can be seen more easily in Fig. 10.7*a* where the ccp unit cell has been turned so its edges line up with the page.

Also shown in Fig. 10.7 is the unit cell of a structure called **body-centered cubic** (bcc). This is similar to the fcc structure except that, instead of spheres in the faces, there is a single sphere in the center of the cube. This central sphere is surrounded by eight neighbors at the corners of the unit cell, giving a coordination number of 8. Hence the bcc structure is not as compact as the closest-packed structures which had a coordination number 12. Nevertheless, some metals are found to have bcc structures.

EXAMPLE 10.1 Count the number of spheres in the unit cell of (a) a face-centered cubic structure, and (b) a body-centered cubic structure.

Solution Referring to Fig. 10.7 and using Eq. (10.1), we find

a) $$N = N_{\text{body}} + \frac{N_{\text{face}}}{2} + \frac{N_{\text{edge}}}{4} + \frac{N_{\text{corner}}}{8}$$

$$= 0 + \frac{6}{2} + 0 + \frac{8}{8} = 4$$

b) $$N = 1 + 0 + 0 + \frac{8}{8} = 2$$

EXAMPLE 10.2 Silicon has the same crystal structure as diamond (Fig. 8.12*a*). Techniques are now available for growing crystals of this element which are virtually flawless. Recently (1976) four scientists at the National Bureau of Standards carried out measurements on some of these perfect crystals and found the side of the unit cell to be 543.106 61 pm long. The unit cell is a cube containing eight Si atoms. The mean density of these crystals was found to be 2.328 993 9 g cm^{-3}, and the relative atomic mass of the Si atoms of which they were composed was found to be 28.085 435. Find the value of the Avogadro constant N_A.

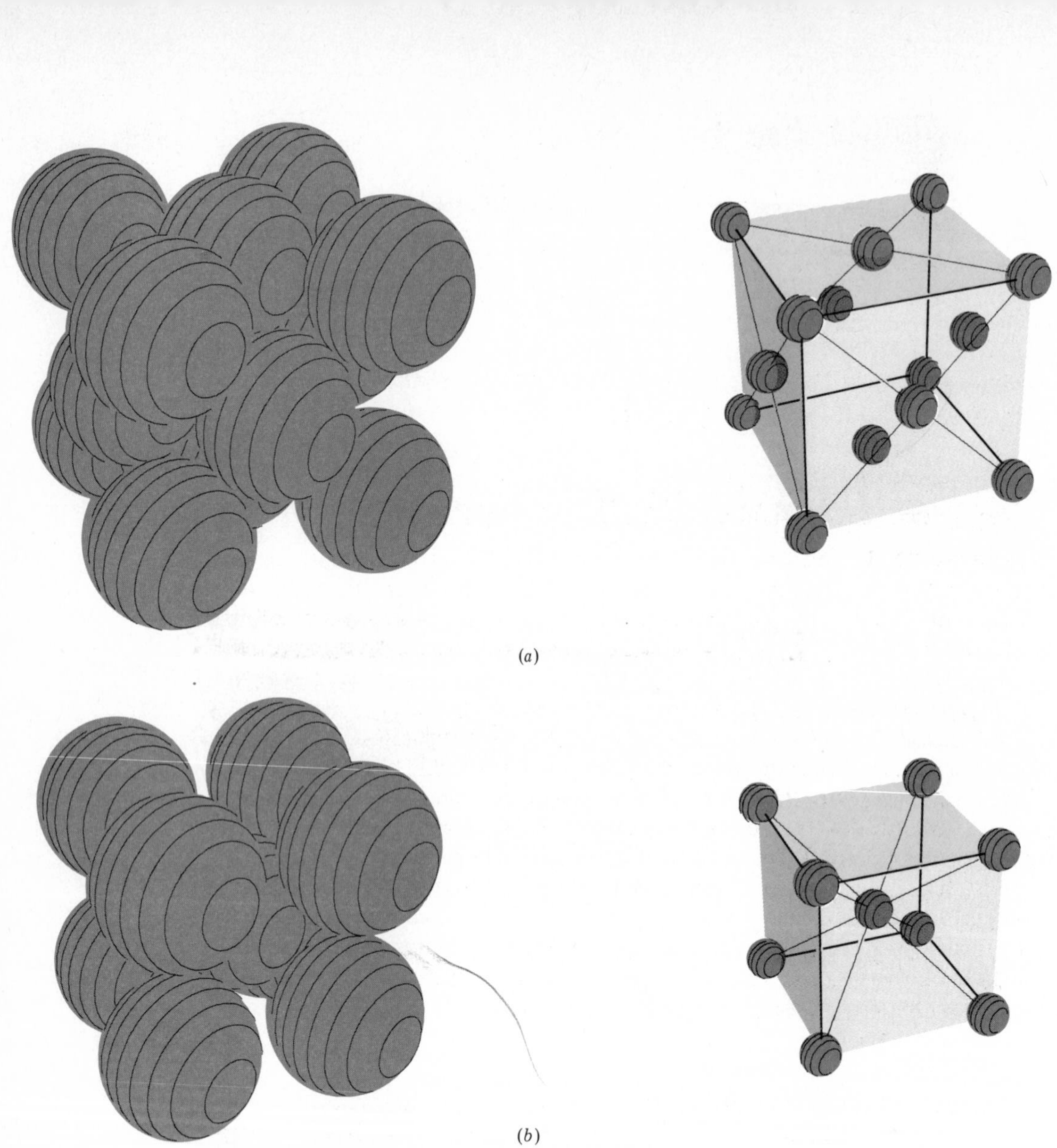

Figure 10.7 Unit cells for (*a*) face-centered cubic and (*b*) body-centered cubic lattices. Note that the face-centered cubic unit cell is the same as that for cubic closest packing in Fig. 10.6*b*. (Computer-generated.) (*Copyright © 1976 by W. G. Davies and J. W. Moore.*)

Solution This problem is very similar to Example 2.7. From the edge length, we can obtain the volume of the cubic unit cell. The density can then be used to calculate the mass of the contents of the unit cell, and molar mass will convert this to amount of substance. We know that the unit cell contains eight atoms, and so we can calculate N_A from Eq. (2.4), which defined the Avogadro constant as the number of particles per unit amount of substance.

$$V_{\text{unit cell}} \xrightarrow{\rho} m_{\text{unit cell}} \xrightarrow{\mathcal{M}} n_{\text{unit cell}}$$

$$n_{\text{unit cell}} = (543.106\ 61 \times 10^{-12}\ \text{m})^3 \times 2.328\ 993\ 9\ \frac{\text{g}}{\text{cm}^3} \times \left(\frac{10^2\ \text{cm}}{\text{m}}\right)^3 \times \frac{1\ \text{mol}}{28.085\ 435\ \text{g}}$$

$$= 1.328\ 441\ 6 \times 10^{-23}\ \text{mol}$$

$$N_A = \frac{N}{n} = \frac{8}{1.328\ 441\ 6 \times 10^{-23}} = 6.022\ 093\ 9 \times 10^{23}\ \text{mol}^{-1}$$

Note: This value is the most accurate determination of the Avogadro constant to date.

10.2 LIQUIDS

When a crystalline solid melts, it loses its rigid form and adopts the shape of its container. At the same time there is usually an increase in volume of a few percent. On the molecular level we can interpret this as a breakdown in the regular structure of the solid. As temperature rises toward the melting point, the molecules vibrate more and more strongly. Above the melting point, these vibrations are so energetic that they overcome the forces holding the molecules in the crystal lattice. The molecules no longer vibrate around an average position but begin to slide past each other. The regular arrangement of the crystal disappears, but the molecules have not escaped each other's attractive influence. The very small volume change which occurs on melting shows that the molecules have moved apart to only a very limited extent and that there can be only a few gaps caused by the less-regular packing. This view is confirmed by the experimental fact that liquids, as opposed to gases, are very difficult to compress. Even at the bottom of the deepest oceans, under pressures of thousands of atmospheres, the density of water is only minutely larger than at the surface.

Viscosity

Because its molecules can slide around each other, a liquid has the ability to flow. The resistance to such flow is called the **viscosity.** Liquids which flow very slowly, like glycerin or honey, have high viscosities. Those like ether or gasoline which flow very readily have low viscosities.

Viscosity is governed by the strength of intermolecular forces and especially by the shapes of the molecules of a liquid. Liquids whose molecules are polar or can form hydrogen bonds are usually more viscous than similar nonpolar substances. Concentrated sulfuric acid, H_2SO_4, is a good example of a liquid which owes its viscosity to hydrogen bonding. Liquids containing long molecules are invariably very viscous. This is because the molecular chains get tangled up in each other like spaghetti—in order for the liquid to flow, the molecules must first unravel. Fuel oil, lubricating grease, and other long-chain alkane molecules are quite viscous for this reason. Glycerol, $CH_2OHCHOHCH_2OH$, is viscous partly because of the length of the chain but also because of the extensive possibilities for hydrogen bonding between the molecules.

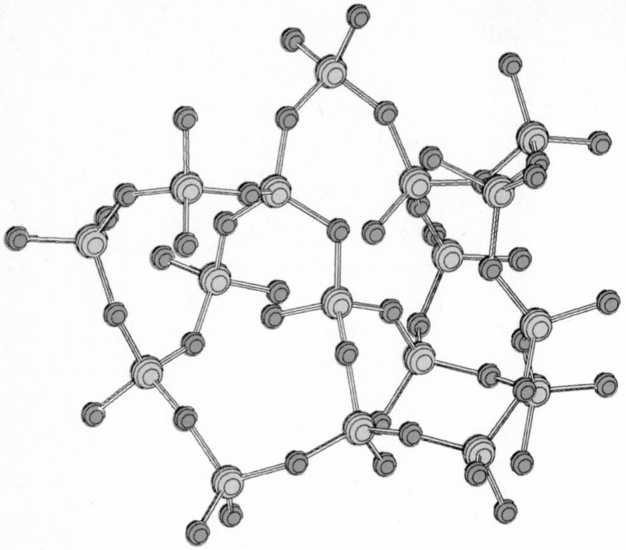

Figure 10.8 A feasible structure for fused silica. Although each Si atom is connected to its regular tetrahedron of O atoms and each O atom to two Si atoms, there is no overall regular arrangement as in a crystal of quartz. This situation is sometimes described as short-range order and long-range disorder and is typical of all glassy materials, including those which are not compounds of Si. (Computer-generated.) (*Copyright © 1976 by W. G. Davies and J. W. Moore.*)

The viscosity of a liquid always decreases as temperature increases. As the molecules acquire more energy, they can escape from their mutual attraction more readily. Long-chain molecules can also wriggle around more freely at a higher temperature and hence disentangle more quickly.

Amorphous Materials: Glasses

Some liquids become extremely viscous as the temperature falls toward their freezing points, often because they consist of macromolecules. An example is quartz, SiO_2, whose macromolecular solid structure was shown in Fig. 6.8. When quartz melts (at 1610°C), a few Si—O bonds break, but most remain intact. The liquid contains large covalently bonded fragments of the original structure and is highly viscous. When the liquid is cooled, the macromolecular fragments cannot readily slide past one another to attain the regular solid structure of quartz. Instead, a collection of interconnected, randomly oriented tetrahedrons of oxygen atoms surrounding silicon atoms is formed, as shown in Fig. 10.8. The material having this structure is known as **fused silica.**

Fused silica is an example of an **amorphous material** or **glass.** It is highly rigid at room temperature, but it does not have the long-range microscopic regularity of a solid crystal lattice. Consequently it cannot be made to cleave along a plane. Instead, like ordinary window glass, it shatters into irregular fragments when struck sharply. (Window glass is primarily silica, but oxides of sodium and calcium are added to lower the melting point.) Since the microscopic structure of a glass is random, like that of a liquid, scientific purists describe glasses as highly viscous liquids, not as solids.

Enthalpy of Fusion and Enthalpy of Vaporization

When heat energy is supplied to a solid at a steady rate by means of an electrical heating coil, we find that the temperature climbs steadily until the melting point is reached and the first signs of liquid formation become evident. Thereafter, even though we are still supplying heat energy to the system, the temperature remains constant as long as both liquid and solid are present. Only when the last vestiges of the solid have disappeared does the temperature start to climb again.

This macroscopic behavior demonstrates quite clearly that energy must be supplied to a solid in order to melt it. On a microscopic level melting involves separating molecules which attract each other. This requires an increase in the potential energy of the molecules, and the necessary energy is supplied by the heating coil.

The heat energy which a solid absorbs when it melts is called the **enthalpy of fusion** or heat of fusion and is usually quoted on a molar basis. (The word *fusion* means the same thing as "melting.") When 1 mol of ice, for example, is melted, we find from experiment that 6.01 kJ are needed. The molar enthalpy of fusion of ice is thus $+6.01 \text{ kJ mol}^{-1}$, and we can write

$$H_2O(s) \rightarrow H_2O(l) \qquad (0°C) \qquad \Delta H_m = 6.01 \text{ kJ mol}^{-1}$$

Selected molar enthalpies of fusion are tabulated in Table 10.2. Solids like ice which have strong intermolecular forces have much higher values than those like CH_4 with weak ones.

When a liquid is boiled, the variation of temperature with the heat energy supplied is similar to that found for melting. When heat is supplied at a steady rate to a liquid at atmospheric pressure, the temperature rises until the boiling point is attained. After this the temperature remains constant until the **enthalpy of vaporization** has been supplied. Once all the liquid has been converted to vapor, the temperature again rises. In the case of water the molar enthalpy of vaporization is $40.67 \text{ kJ mol}^{-1}$. In other words

$$H_2O(l) \rightarrow H_2O(g) \qquad (100°C) \qquad \Delta H_m = 40.67 \text{ kJ mol}^{-1}$$

TABLE 10.2 Molar Enthalpies of Fusion and Vaporization of Selected Substances.

Substance	Formula	ΔH_m (fusion)/ kJ mol^{-1}	Melting Point/K	ΔH_m (vaporization)/ kJ mol^{-1}	Boiling Point/K	$(\Delta H_v/T_b)$/ JK^{-1} mol^{-1}
Neon	Ne	0.33	24	1.80	27	67
Oxygen	O_2	0.44	54	6.82	90.2	76
Methane	CH_4	0.94	90.7	8.18	112	73
Ethane	C_2H_6	2.85	90.0	14.72	184	80
Chlorine	Cl_2	6.40	172.2	20.41	239	85
Carbon tetrachloride	CCl_4	2.67	250.0	30.00	350	86
Water	H_2O	6.01	273.1	40.7	373.1	109
n-Nonane	C_9H_{20}	19.3	353	40.5	491	82
Mercury	Hg	2.30	234	58.6	630	91
Sodium	Na	2.60	371	98	1158	85
Aluminum	Al	10.9	933	284	2600	109

Heat energy is absorbed when a liquid boils because molecules which are held together by mutual attraction in the liquid are jostled free of each other as the gas is formed. Such a separation requires energy. In general the energy needed differs from one liquid to another depending on the magnitude of the intermolecular forces. We can thus expect liquids with strong intermolecular forces to have larger enthalpies of vaporization. The list of enthalpies of vaporization given in Table 10.2 bears this out.

Two other features of Table 10.2 deserve mention. One is the fact that the enthalpy of vaporization of a substance is always higher than its enthalpy of fusion. When a solid melts, the molecules are not separated from each other to nearly the same extent as when a liquid boils. Second, there is a close correlation between the enthalpy of vaporization and the boiling point measured on the thermodynamic scale of temperature. If we divide the one by the other, we find that the result is often in the range of 75 to 90 J K^{-1} mol^{-1}. To a first approximation therefore the *enthalpy of vaporization of a liquid is proportional to the thermodynamic temperature at which the liquid boils.* This interesting result is called **Trouton's rule.** An equivalent rule does not hold for fusion. The energy required to melt a solid and the temperature at which this occurs depend on the structure of the crystal as well as on the magnitude of the intermolecular forces.

Vapor-Liquid Equilibrium

When a liquid such as water or alcohol is exposed to air in an open container, the liquid evaporates. This happens because the distribution of speeds (and hence kinetic energies) among molecules in a liquid is similar to that illustrated for gases in Fig. 9.9. At any given instant a small fraction of the molecules in the liquid phase will be moving quite fast. If one of these is close to the surface and is traveling upward, it can escape the attraction of its fellow molecules entirely and pass into the gas phase. As the higher energy molecules depart, the average energy of the molecules in the liquid decreases and the temperature of the liquid falls. Heat energy will be absorbed from the surroundings, an effect which you can feel if you let water or alcohol evaporate from your skin. Absorption of heat maintains the average molecular speed in the liquid, so that, given enough time, all the liquid can evaporate. The heat absorbed during the entire process corresponds to the enthalpy of vaporization.

If the liquid is placed in a closed, rather than an open, container, we no longer find that it evaporates completely. Once a certain partial pressure of gas has been built up by the evaporation of liquid, no more change occurs, and the amount of liquid remains constant. The partial pressure attained in this way is called the **vapor pressure** of the liquid. It is different for different liquids and increases with temperature for a given liquid. So long as some liquid is present, the vapor pressure is always the same, regardless of the size of the container or the quantity of liquid. For example, we find that any size sample of water held at 25°C will produce a vapor pressure of 23.8 mmHg (3.168 kPa) in any closed container, provided only that all the water does not evaporate.

On the macroscopic level, once the vapor pressure has been attained in

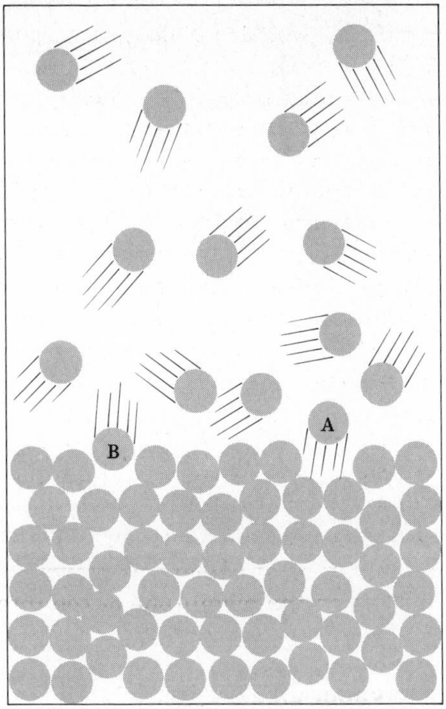

Figure 10.9 Molecular interpretation of vapor pressure. The vapor pressure remains constant because every molecule which escapes (like A) is immediately replaced by another molecule reentering from the vapor (like B). Any given molecule spends some of its time in the vapor and some in the liquid.

a closed container, evaporation appears to stop. On the microscopic level, though, molecules are still escaping from the liquid surface into the vapor above, as shown in Fig. 10.9. The amount of vapor remains the same only because molecules are reentering the liquid just as fast as they are escaping from it. The molecules of the vapor behave like any other gas: They bounce around colliding with each other and the walls of the container. However, one of these "walls" is the surface of the liquid. In most cases a molecule colliding with the liquid surface will enter the body of the liquid, not have enough energy to escape, and be recaptured. When the liquid is first introduced into the container, there are very few molecules of vapor and the rate of recapture will be quite low, but as more and more molecules evaporate, the chances of a recapture will become proportionately larger. Eventually the vapor pressure will be attained, and the rate of recapture will exactly balance the rate of escape. There will then be no net evaporation of liquid or condensation of gas.

Once the vapor-liquid system has attained this state, it will appear on the macroscopic level not to be undergoing any change in its properties. The amount, the volume, the pressure, the temperature, the density, etc., of both liquid and gas will all remain constant with time. When this happens to a system, it is said to be in an **equilibrium state** or to have attained **equilibrium.** In later chapters we will encounter many other quite different examples of equilibrium, but they all have one property in common. The lack of change on the macroscopic level is always the result of two opposing microscopic processes whose rates are equal. The effect of each process is to nullify the effect of the other. Since both microscopic processes are still in motion, such a situation is often referred to as **dynamic equilibrium.**

The magnitude of the vapor pressure of a liquid depends mainly on two factors: the strength of the forces holding the molecules together and the temperature. It is easy to see that if the intermolecular forces are weak, the vapor pressure will be high. Weak intermolecular forces will permit molecules to escape relatively easily from the liquid. The rate at which molecules escape will thus be high. Quite a large concentration of molecules will have to build up in the gas phase before the rate of reentry can balance the escape rate. Consequently the vapor pressure will be large. By contrast, strong intermolecular forces result in a low escape rate, and only a small concentration of molecules in the vapor is needed to balance it. The vapor pressure of a liquid is quite a sensitive indicator of small differences in intermolecular forces. In the case of the alkanes, for example, we find that the vapor pressure of normal pentane at 25°C is 512.3 mmHg (68.3 kPa), while that of normal hexane is 150.0 mmHg (20.0 kPa) and that of normal heptane only 45.7 mmHg (6.1 kPa), despite the fact that the intermolecular forces for the three substances differ by less than 10 percent.

The other major factor governing the magnitude of the vapor pressure of a liquid is temperature. At a low temperature only a minute fraction of the molecules have enough energy to escape from the liquid. As the temperature is raised, this fraction increases very rapidly (recall the discussion of gases in Sec. 9.4) and the vapor pressure increases with it. Moreover, the higher the temperature, the more rapid the rate of increase of the energetic fraction of molecules. The result is a variation of vapor pressure with temperature like that shown for four liquids in Fig. 10.10. Note from this figure how the vapor-pressure increase for a 10°C increase in temperature is larger at higher temperatures.

Boiling Point

When we heat a liquid until it boils, the bubbles that form inside the liquid consist of pure vapor. If the liquid is well stirred while boiling occurs, the vapor in the bubbles will be in equilibrium with the liquid and will have a pressure equal to the vapor pressure at the boiling temperature. However, the pressure inside the bubbles must also be equal to the external pressure above the liquid. If this were not so, the bubbles would either suddenly collapse or suddenly expand. It follows therefore that when a liquid boils, the *vapor pressure of the liquid is equal to the external pressure.*

Normally when we boil a liquid, we do so at atmospheric pressure. If this pressure is the standard pressure of 1 atm (101.3 kPa), then the temperature at which the liquid boils is referred to as its **normal boiling point.** This is the boiling point which is usually quoted in chemical literature. Not everyone lives at sea level, though. Denver, Colorado, for example, is about a mile high, and the average atmospheric pressure there is only 630 mmHg (84 kPa). Liquids attain a vapor pressure of 630 mmHg at a somewhat lower temperature than is required to produce 760 mmHg (1 atm). Consequently liquids in Denver boil some 4 to 5°C lower than the normal boiling point. Since the boiling point is often used to identify a liquid, chemists living at high altitudes must be careful to allow for this difference.

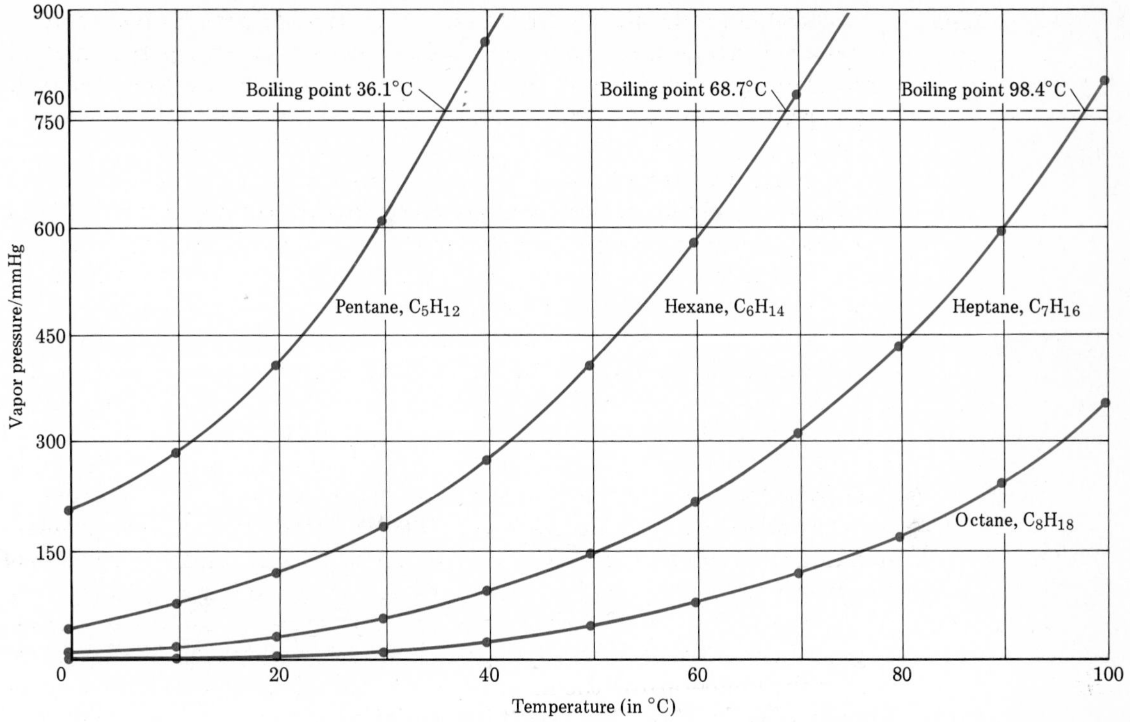

Figure 10.10 The vapor pressure of four liquid alkanes.

The dependence of the boiling point on the external pressure can often be very useful. Chemists often purify liquids by boiling them and collecting the vapor, a process known as **distillation.** Some liquids have such high normal boiling points that they begin to decompose before distillation can be carried out. Such a liquid can often be distilled at reduced pressure. The temperature of boiling is then much lower, and the risk of decomposition considerably less. The reverse procedure is used in a pressure cooker. The pressure inside the sealed cooker builds up until it is larger than atmospheric, and so the water used for cooking boils at a temperature above its normal boiling point. Therefore the cooking proceeds more rapidly.

EXAMPLE 10.3 From Fig. 10.10 estimate the boiling points of the four alkanes when the pressure is reduced to 600 mmHg.

Solution Reading along the 600-mmHg line in the graph, we find that it meets the vapor-pressure curve for pentane at about 29°C. Accordingly this is the boiling point of pentane at 600 mmHg. Similarly we find the boiling point of hexane to be 61°C, and of heptane to be 90°C. The boiling point of octane is above 100°C and cannot be estimated from the graph.

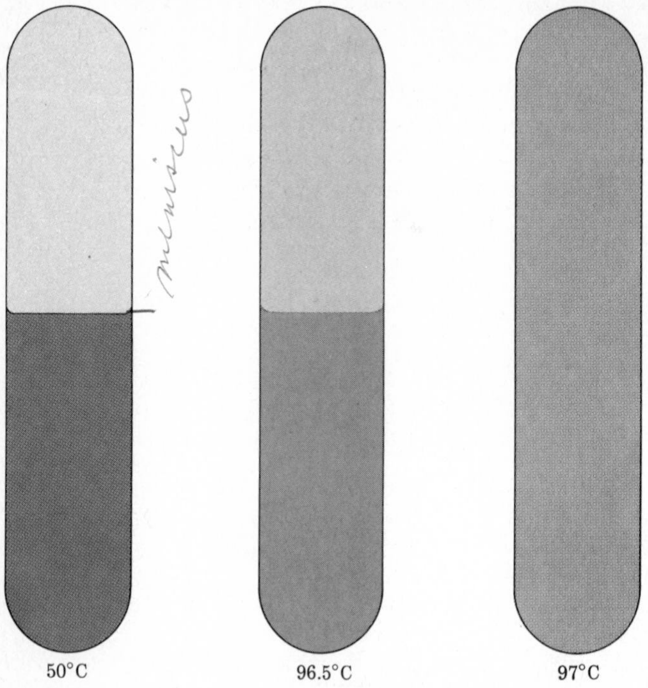

Figure 10.11 The critical point. When a mixture of propane liquid and vapor is heated in a closed tube, the liquid and vapor become more and more alike. At 96.5°C it is almost impossible to tell the difference between them or to distinguish the top surface of the liquid. Finally all distinction disappears at the critical temperature of 97°C. The pressure at this point is 4.24 MPa, about 42 atm.

Critical Temperature and Pressure

Suppose we seal a pure liquid and its vapor in a strong glass tube and heat it to a very high temperature. As we increase the temperature, the vapor pressure will rise. (It is not a good idea to heat a liquid this way unless you are sure the container can withstand the increased pressure.) The rising vapor pressure corresponds to a greater number of molecules in the limited volume of the vapor phase. In other words the vapor becomes considerably denser. Eventually we reach a temperature at which the density of the vapor becomes the *same as that of the liquid*. Since liquids are usually distinguished from gases on the basis of density, at this point both have become identical. The temperature at which this occurs is called the **critical temperature,** and the pressure is called the **critical pressure.**

Figure 10.11 illustrates what happens experimentally in the case of propane. As the temperature nears the critical value of 97°C, liquid and vapor become very similar in appearance and the meniscus between them becomes difficult to distinguish. Finally, at the critical temperature the meniscus disappears completely. Above the critical temperature the sample is quite uniform and it is difficult to know whether to call it a liquid or a gas.

Once a gas is above its critical temperature, it is *impossible* to get it to separate into a liquid layer below and a vapor layer above no matter how great a pressure is applied. Oxygen, for instance, is well above its critical temperature at room temperature. If we increase the pressure on it to a few thousand atmospheres, its density becomes so high that we are forced to classify it as a liquid. Nevertheless as we increase the pressure, there is no point at which drops of liquid suddenly appear in the gas. Instead the oxygen gradually changes from something which is obviously a gas to something which is obviously a liquid. Conversely, if we gradually relax the pressure, there is no point at which the oxygen will start to boil.

Table 10.3 lists the critical temperatures and critical pressures for some well-known gases and liquids. Such data are often quite useful. Many gases are sold commercially in strong steel cylinders at high pressures. The behavior of the gas in such a cylinder depends on whether it is above or below the critical temperature. The critical temperature of propane, for instance, is 97°C, well above room temperature. Thus propane in a high-pressure cylinder consists of a mixture of liquid and vapor, and you can sometimes hear the liquid sloshing about inside.

TABLE 10.3 Critical Temperatures and Pressures of Some Simple Substances.

Name	Critical Temperature/K	Critical Pressure/MPa	Critical Pressure/atm
Hydrogen	33.2	1.30	12.8
Neon	44.5	2.73	26.9
Nitrogen	126.0	3.39	33.5
Carbon dioxide	304.2	7.39	73.0
Propane	370	4.23	41.8
Ammonia	405.5	11.29	111.5
Water	647.1	22.03	217.5

The pressure of the gas in such a cylinder will be the vapor pressure of propane at 20°C, namely, 9.53 atm (965.4 kPa). As long as there is some liquid left in the cylinder, the pressure will remain at 9.53 atm. Only when all the liquid has evaporated will the pressure begin to drop. At that point the cylinder will be virtually empty. A very different behavior is found in the case of a cylinder of oxygen. Since oxygen is above its critical temperature at 20°C, the cylinder will contain a uniform fluid rather than a liquid-vapor mixture. As we use up the oxygen, the pressure will gradually decrease to 1 atm, at which point no more O_2 will escape from the cylinder.

The principles discussed in the preceding paragraph apply to the aerosol sprays most of us encounter every day. Such spray cans contain a small quantity of the active ingredient—hair conditioner, deodorant, shaving cream, and the like—and a large quantity of propellant. The propellant is a substance, such as propane, whose critical temperature is well above room temperature. Therefore it can be liquefied at the high pressure in the spray can. When the valve is opened, the vapor pressure of the liquid propellant causes the active ingredient and the propellant to spray out of the can. As long as liquid propellant remains, the pressure inside the can will be constant (it will be the vapor pressure), and the spray will be reproducible. It should be obvious why such cans always bear a warning against throwing them in a fire—vapor pressure increases more rapidly at higher temperatures (Fig. 10.10), and so heating an enclosed liquid is far more likely to produce an explosion than heating a gas alone. (The latter case was described in Example 9.7.)

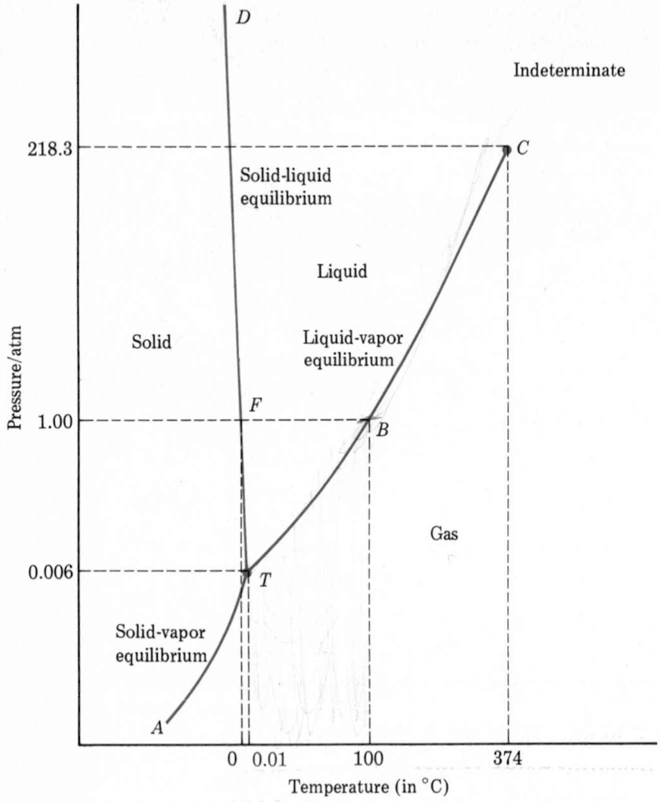

Figure 10.12 Phase diagram of water. The scale has been distorted for purposes of diagrammatic clarity. Point T is the triple point, C the critical point, B the boiling point, and F the freezing point.

10.4 PHASE DIAGRAMS

Solids as well as liquids exhibit vapor pressure. If you hang out a wet cloth in winter, the cloth first freezes as hard as a board, but after sufficient time all the ice evaporates and the cloth becomes soft and dry. Another solid which exhibits evidence of a vapor pressure is para-dichlorobenzene, $C_6H_4Cl_2$, which is used for mothballs. The fact that you can smell this solid from across the room means that some of its molecules must have evaporated into the air and entered your nose. Indeed, given a year or so, moth crystals will evaporate completely. Still another example is dry ice, $CO_2(s)$, whose vapor pressure reaches atmospheric pressure at −78.5°C. Consequently $CO_2(s)$ sublimes, forming vapor without passing through the liquid state. Unlike ordinary ice, it remains dry, making it very convenient as a refrigerant.

Like the vapor pressure of a liquid, the vapor pressure of a solid increases with temperature. This variation is usually presented in combination with other curves in a **phase diagram,** such as that for water in Fig. 10.12. The term *phase* in this context is used to distinguish solid, liquid, and gas, each of which has its own distinctive properties, such as density, viscosity, etc. Each phase is separated from the others by a boundary.

In the phase diagram for water, the variation of the vapor pressure of ice with temperature is shown by the line AT. As might be expected, the vapor pressure of ice is quite small, never rising above 0.006 atm (0.61 kPa). The vapor pressure of liquid water is usually much higher, as is shown by the curve *TC*. The point *C* on this curve corresponds to the critical point, and the vapor pressure is the critical pressure of water, 218.3 atm (22 MPa). The curve stops at this point since any distinction between liquid and gas ceases to exist above the critical temperature.

The other end of this vapor-pressure curve, point *T*, is of particular interest. At the temperature of point *T* (273.16 K or 0.01°C) both ice and water have the same vapor pressure, 0.006 atm. Since the same vapor is in equilibrium with both liquid and solid, it follows that *all three* phases, ice, water, and vapor, are in equilibrium at point *T*. This point is therefore referred to as the **triple point.**

Line *TD* in Fig. 10.12 remains to be discussed. This line describes how the melting point of ice varies with pressure. In other words, it includes temperatures and pressures at which solid and liquid phases are in equilibrium. Note that the line is not exactly vertical. Point *D* corresponds to a slightly lower temperature than point *T*. This means that, as we increase the pressure on ice, its freezing point decreases. Again we can draw on everyday experience for an example of this behavior. A person on a pair of skates standing on ice exerts his or her whole weight on the ice through the thin skate blade. The very high pressure immediately under the blade causes some ice to melt, affording lubrication which enables the skater to glide smoothly over the ice.

It should be emphasized that point *T* is not the normal freezing point of water. When we measure the normal freezing point of water (or any other liquid), we do so at standard atmospheric pressure in a container open to the air (Fig. 10.13*a*). In this container we have not only ice and water in equilibrium with each other but also *air saturated with water vapor*. The total

Figure 10.13 The normal freezing point (*a*) and triple point (*b*) of water.

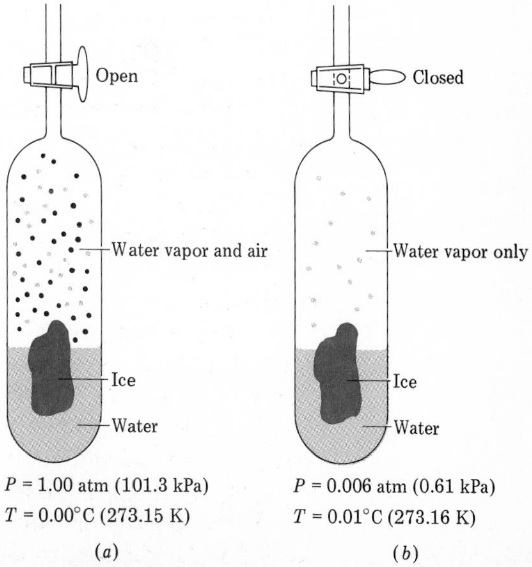

pressure on the contents of this container is 1 atm (101.3 kPa), and its temperature is exactly 273.15 K (0.00°C). As far as liquid and solid are concerned, this corresponds to point *F* in the phase diagram.

If we could now pump all the air out of the container so that only pure water in its three phases was left, we would have the situation shown in Fig. 10.13*b*. Ice, water, and pure water vapor, but no air, now occupy the container. The pressure has dropped from the 1.00 atm of the atmosphere to the vapor pressure (0.006 atm). Because of this decrease in pressure, the melting point of ice increases slightly and the new equilibrium temperature is 0.01°C. This corresponds to the triple point of water, point *T* in the phase diagram.

Because the triple point is more readily attained (one need not be at sea level) and also more reproducible, it is now used as a primary reference temperature for the international thermodynamic scale of temperature. In SI units the temperature of the triple point of water is defined as 273.16 K.

10.5 SOLUTIONS

Up to this point we have discussed only the properties of pure solids and liquids. Of more importance to a chemist, though, are the properties of solutions. Very few chemical reactions involve only pure substances—almost all involve a solution of some sort.

In Sec. 3.5 we defined a solution as a homogeneous mixture of two or more substances, that is, a mixture which appears to be uniform throughout. Under this definition we would refer to sugar or salt dissolved in water as solutions, but we would not apply the term to muddy water or to milk. A close inspection of muddy water reveals that it is not uniform in appearance but consists of small solid particles dispersed in water. We refer to such a mixture as a **suspension.** Under the microscope, milk can also be seen to be nonuniform. It consists of small drops of milk fat dispersed throughout an aqueous phase.

Our definition of a solution in terms of the homogeneity of a mixture is somewhat unsatisfactory since it does not tell us where to draw the line. Field-emission microscopes and electron microscopes have now been developed which can just about "see" a single atom. With such a microscope virtually *all* matter looks nonuniform and hence not homogeneous. If our definition extends to such microscopes, then true solutions do not exist. In practice we draw the line somewhere around the 5 nanometer (nm) mark, even though some molecules (see Fig. 1.5) are larger than this.

On the molecular level a solution corresponds to the random arrangement of one kind of molecule or ion around another. In Fig. 10.14 the illustration on the left corresponds to a solution since each black molecule is randomly surrounded by black and red molecules, and vice versa. The illustration on the right is a suspension. The distribution is not random, and most red molecules have red neighbors, while most black molecules have black neighbors.

Strictly speaking, the term *solution* applies to any homogeneous mixture, but we will concentrate our discussion on those solutions which in-

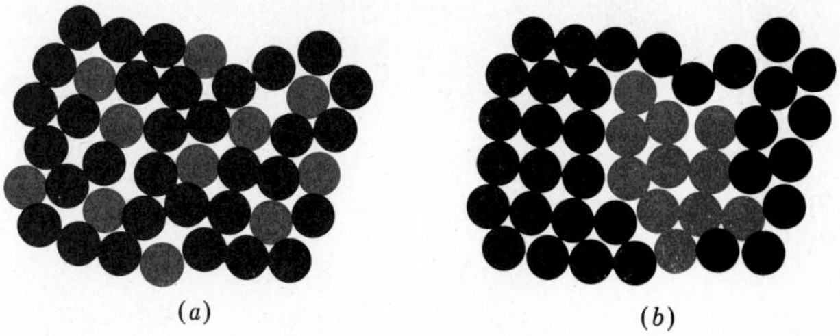

Figure 10.14 Suspensions and solutions viewed on the microscopic level. (*a*) A solution corresponds to a random arrangement of one kind of molecule around the other. (*b*) In a suspension there are small regions of space where only one kind of molecule exists.

volve liquids since these are the most common. It should be realized, though, that other types of solutions also exist. Air is a solution of a large number of gases (oxygen is the most concentrated) in another gas (nitrogen). A 5-cent coin is made from an alloy in which one solid (nickel) is dissolved in another (copper). Solutions of hydrogen gas in solid palladium (Pd) and some other metals are also possible.

As mentioned in the brief discussion of solutions in Chap. 3, it is sometimes difficult to decide which component of a solution is the solute and which is the solvent. Usually the amount of solvent is much larger than that of the solute. If the pure components were initially in separate phases (a gas and a solid, for example), the phase corresponding to the state of the solution is taken to be the solvent. In the case of $H_2(g)$ and $Pd(s)$ mentioned above, for example, Pd would be the solvent because the solution is a solid phase.

Saturated and Supersaturated Solutions

We often find that there is a limit to the quantity of solute which will dissolve in a given quantity of solvent. This is especially true when solids dissolve in liquids. For example, if 36 g KCl crystals is shaken with 100 g H_2O at 25°C, only 35.5 g of the solid dissolves. If we raise the temperature somewhat, all the KCl will dissolve, but on cooling to 25°C again, the extra 0.5 g KCl will precipitate, leaving exactly 35.5 g of the salt dissolved. We describe this phenomenon by saying that at 25°C the **solubility** of KCl in H_2O is 35.5 g KCl per 100 g H_2O. A solution of this composition is also described as a **saturated** solution since it can accommodate no more KCl.

Under some circumstances it is possible to prepare a solution which behaves anomalously and contains more solute than a saturated solution. Such a solution is said to be **supersaturated.** A good example of supersaturation is provided by $Na_2S_2O_3$, sodium thiosulfate, whose solubility at 25°C is 50 g $Na_2S_2O_3$ per 100 g H_2O. If 70 g $Na_2S_2O_3$ crystals is dissolved in 100 g hot H_2O and the solution cooled to room temperature, the extra 20 g $Na_2S_2O_3$ usually does not precipitate. The resulting solution is supersaturated; consequently it is also unstable. It can be "seeded" by adding a crystal of $Na_2S_2O_3$, whereupon the excess salt suddenly crystallizes and heat is given off. After the crystals have settled and the temperature has returned to 25°C, the solution above the crystals is a saturated solution—it contains 50 g $Na_2S_2O_3$.

Miscibility

When a solid dissolves in a liquid, we very seldom find that the liquid has any tendency to dissolve in the solid. In a saturated solution of potassium chloride, for example, essentially no water dissolves in the potassium chloride crystals. With liquids the situation is usually different. If equal quantities of 1-butanol and water are shaken together, the mixture slowly separates into two layers. The bottom layer is a saturated solution of 1-butanol in water—it contains about 8% 1-butanol by weight. The top layer is not pure 1-butanol but a saturated solution of water in 1-butanol. It contains about 32% water by weight. A pair of liquids, like 1-butanol and water, which separates into two layers is said to be partially **miscible**.

By contrast with the solubilities of solids in liquids, a great many liquid pairs are completely miscible. That is, regardless of the proportions in which the two liquids are mixed, each will dissolve completely in the other. There will be no phase boundary as in the case of partially miscible liquids like 1-butanol and water. Ethanol and water provide a good example of two liquids which are completely miscible. If you have a source of pure ethanol, it is possible to mix a drink in any proportions you like—even up to 200 proof—without forming two separate liquid phases.

Measuring the Composition of a Solution

When speaking of solubility or miscibility, or when doing quantitative experiments involving solutions, it is necessary to know the exact composition of a solution. This is invariably given in terms of a *ratio* telling us how much solute is dissolved in a unit quantity of solvent or solution. The ratio can be a ratio of masses, of amounts of substances, or of volumes, or it can be some combination of these. For example, **concentration** was defined in Sec. 3.5 as the *amount of solute per unit volume of solution:*

$$c_{\text{solute}} = \frac{n_{\text{solute}}}{V_{\text{solution}}} \tag{3.15}$$

The two simplest measures of the composition are the **mass fraction** w, which is the *ratio of the mass of solute to the total mass of solution,* and the **mole fraction** x, which is the *ratio of the amount of solute to the total amount of substance in the solution.* If we indicate the solute by A and the solvent by B, the mass fraction and the mole fraction are defined by

$$w_A = \frac{m_A}{m_A + m_B} \quad \text{and} \quad x_A = \frac{n_A}{n_A + n_B}$$

EXAMPLE 10.4 A solution is prepared by dissolving 18.65 g naphthalene, $C_{10}H_8$, in 89.32 g benzene, C_6H_6. Find (a) the mass fraction and (b) the mole fraction of the naphthalene.

Solution

a) The mass fraction is easily calculated from the masses:

$$w_{C_{10}H_8} = \frac{m_{C_{10}H_8}}{m_{C_{10}H_8} + m_{C_6H_6}} = \frac{18.65\ \text{g}}{18.65\ \text{g} + 89.32\ \text{g}} = 0.1727$$

It is sometimes useful to distinguish mass of solute and mass of solution for purposes of calculation. In such a case we can write

$$w_{C_{10}H_8} = 0.1727\ \text{g}\ C_{10}H_8\ \text{per g solution}$$

b) In order to calculate the mole fraction, we must first calculate the amount of each substance. Since

$$\mathcal{M}_{C_{10}H_8} = 128.18\ \text{g mol}^{-1} \qquad \text{and} \qquad \mathcal{M}_{C_6H_6} = 78.11\ \text{g mol}^{-1}$$

we find $$n_{C_{10}H_8} = \frac{18.65\ \text{g}}{128.18\ \text{g mol}^{-1}} = 0.1455\ \text{mol}$$

$$n_{C_6H_6} = \frac{89.32\ \text{g}}{78.11\ \text{g mol}^{-1}} = 1.144\ \text{mol}$$

Thus $$x_{C_{10}H_8} = \frac{n_{C_{10}H_8}}{n_{C_{10}H_8} + n_{C_6H_6}} = \frac{0.1455\ \text{mol}}{0.1455\ \text{mol} + 1.144\ \text{mol}}$$

$$= 0.1128 \text{ or } 0.1128\ \text{mol}\ C_{10}H_8\ \text{per mol solution}$$

The mass fraction is useful because it does not require that we know the exact chemical nature of both solute and solvent. Thus if we dissolve 10 g crude oil in 10 g gasoline, we can calculate $w_{\text{crude oil}} = 0.5$ even though the solute and solvent are both mixtures of alkanes and have no definite molar mass. By contrast, the mole fraction is useful when we want to know the nature of the solution on the microscopic level. In the above example, for instance, we know that for every 100 mol of solution, 11.28 mol is naphthalene. On the molecular level this means that out of each 100 molecules in the solution, 11.28 will, on the average, be naphthalene molecules.

The mass fraction of a solution is often encountered in other disguises. The **weight percentage** (strictly speaking, the *mass* percentage) of a solution is often defined by the formula

$$\text{Weight percentage of } A = \frac{m_A}{m_A + m_B} \times 100\%$$

This definition is really the same as that of the mass fraction because the percent sign means "divided by 100." Thus 100% is merely a synonym for 100/100, that is, the number 1, and we can write

$$w_{C_{10}H_8} = 0.1727 \times 1 = 0.1727 \times 100\% = 17.27\%$$

When the mass fraction is very small, it is often expressed in **parts per million** (ppm) or **parts per billion** (ppb). These symbols can be handled in much the same way as a percentage if you remember how they are related to unity:

$$1 = 100\% = 10^6\ \text{ppm} = 10^9\ \text{ppb}$$

In other words

$1\% = \frac{1}{100} = 10^{-2}$ $\quad$ 1 ppm = 10^{-6} $\quad$ 1 ppb = 10^{-9}

EXAMPLE 10.5 A 1-kg sample of water from Lake Powell, Utah, is found to contain 10 ng mercury. Walleyed pike caught in the lake contain 0.427 ppm mercury. (a) What is the mass fraction of mercury (in ppb) in water from Lake Powell? (b) If you ate 2 lb of walleyed pike caught from the lake, what mass of mercury would you ingest?

Solution

a) The mass fraction of mercury is by definition

$$w_{Hg} = \frac{m_{Hg}}{m_{solution}}$$

Therefore $$w_{Hg} = \frac{10 \text{ ng}}{1 \text{ kg}} = \frac{10 \times 10^{-9} \text{ g}}{1 \times 10^{3} \text{ g}} = 1 \times 10^{-11}$$

In ppb $$w_{Hg} = 1 \times 10^{-11} \times 1 = 1 \times 10^{-11} \times 10^{9} \text{ ppb} = 0.01 \text{ ppb}$$

b) Assuming the mercury to be uniformly distributed throughout the walleyed pike, we have

$$w_{Hg} = 0.427 \text{ ppm} = 0.427 \times 10^{-6} = 4.27 \times 10^{-7}$$

By definition

$$w_{Hg} = \frac{m_{Hg}}{m_{\text{walleyed pike}}}$$

and $$m_{Hg} = w_{Hg} \times m_{\text{walleyed pike}} = 4.27 \times 10^{-7} \times 2.0 \text{ lb}$$

$$= 4.27 \times 10^{-7} \times 2.0 \text{ lb} \times \frac{1 \text{ kg}}{2.2 \text{ lb}} \times \frac{10^{3} \text{ g}}{1 \text{ kg}} = 3.9 \times 10^{-4} \text{ g}$$

$$= 390 \ \mu\text{g}$$

From Example 10.5 you can see that from nearly the same mass of fish, almost 40 000 times as much mercury would be obtained as from the water. Indeed, if you ate 2 lb of walleyed pike every day, you would exceed the minimum dosage (300 μg for a 70-kg human) at which symptoms of mercury poisoning can appear. Fortunately, most of us do not eat fish every day, nor are we as gluttonous as the example suggests. Nevertheless, the much higher mass fraction of mercury in fish than in water shows that very small quantities of mercury in the environment can be magnified many times in living systems. This process of **bioamplification** will be discussed more thoroughly in Sec. 20.3.

Solubility and Molecular Structure

Chemical theory has not reached the point where it can predict exactly how much of one substance will dissolve in another. The best we can do is to indicate in general terms the relationships between solubility and the microscopic structures of solute and solvent.

To begin with, moving particles of any kind tend to become more randomly distributed as time passes. If you put a layer of red marbles in the bottom of a can and cover it with a second layer of white marbles, shaking the can for a short time will produce a nearly random distribution. The same principle applies on the microscopic level. Moving molecules tend to become randomly distributed among one another, unless something holds them back. Thus gases, whose molecules are far apart and exert negligible forces on one another, are all completely miscible with other gases.

In liquid solutions, the molecules are much closer together and the characteristics of different types of molecules are much more important. In particular, if the solute molecules exert large intermolecular forces on each other but do not attract solvent molecules strongly, the solute molecules will tend to group together. This forms a separate phase and leaves the solvent as a second phase. Conversely, if the solvent molecules attract each other strongly but have little affinity for solute molecules, solvent molecules will segregate, and two phases will form.

A classic example of the second situation described in the previous paragraph is the well-known fact that oil and water do not mix—or if they do, they do not stay mixed for long. The reason is that oil consists of alkanes and other nonpolar molecules, while water molecules are polar and can form strong hydrogen bonds with each other. Suppose that alkane or other nonpolar molecules are randomly dispersed among water molecules, as shown in Fig. 10.15*a*. The constant jostling of both kinds of molecules will soon bring two water molecules together. Dipole forces and hydrogen bonding will tend to hold the water molecules together, but there are only weak London forces between water and nonpolar molecules. Before long, clusters of water molecules like those in Fig. 10.15*b* will have formed.

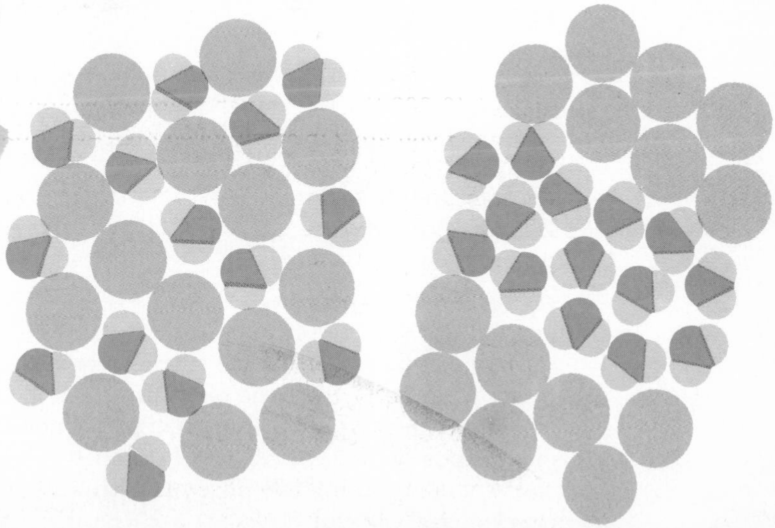

(*a*) (*b*)

Figure 10.15 The insolubility of nonpolar molecules in water. Even if it were possible to mix nonpolar molecules (shown in gray) and water molecules as shown in (*a*), this situation would be unstable. The water molecules would soon congregate together under the influence of their dipole and hydrogen-bonding attractions, attaining the situation of lower energy shown in (*b*).

These clusters will be stable at room temperature because the energy of interaction between the water molecules will be larger than the average energy of molecular motion. Only an occasional molecular collision will be energetic enough to bump two water molecules apart, especially if they are hydrogen bonded. Given enough time, this process of aggregation will continue until the polar molecules are all collected together. If the nonpolar substance is a liquid, this process corresponds on the macroscopic level to the liquids separating from each other and forming two layers.

If instead of mixing substances like oil and water, in which there are quite different kinds of intermolecular attractions, we mix two polar substances or two nonpolar substances, there will be a much smaller tendency for one type of molecule to segregate from the other. Thus two alkanes like *n*-heptane, C_7H_{16}, and *n*-hexane, C_6H_{14}, are completely miscible in all proportions. The C_7H_{16} and C_6H_{14} molecules are so similar (see Table 8.3 for projection formulas) that there are only negligible differences in intermolecular forces. Thus the molecules remain randomly mixed as they jostle among one another.

For a similar reason, methanol, CH_3OH, is completely miscible with water. In this case both molecules are polar and can form hydrogen bonds among themselves, and so there are strong intermolecular attractions within each liquid. However, CH_3OH dipoles can align with H_2O dipoles, and CH_3OH molecules can hydrogen bond to H_2O molecules, and so the attractions among unlike molecules in the solution are similar to those among like molecules in each pure liquid. Again there is little tendency for one type of molecule to become segregated from the other.

All the cases just discussed are examples of the general rule that *like dissolves like*. Two substances whose molecules have very similar structures and consequently similar intermolecular forces will usually be soluble in each other. Two substances whose molecules are quite different will not mix randomly on the microscopic level. In general, polar substances will dissolve other polar substances, while nonpolar materials will dissolve other nonpolar materials. The greater the difference in molecular structure (and hence in intermolecular attractions), the lower the mutual solubility.

EXAMPLE 10.6 Predict which of the following compounds will be most soluble in water:

a) CH_3CH_2OH (**Ethanol**) b) $CH_3CH_2CH_2CH_2CH_2CH_2OH$ (**Hexanol**)

Solution Since ethanol contains an OH group, it can hydrogen bond to water. Although the same is true of hexanol, the OH group is found only at one end of a fairly large molecule. The rest of the molecule can be expected to behave much as though it were a nonpolar alkane. This substance should thus be much less soluble than the first. Experimentally we find that ethanol is completely miscible with water, while only 0.6 g hexanol dissolves in 100 g water.

Our discussion of solubility in terms of microscopic structure concludes with one more point. The solubilities of other substances in solids are usually small. The constituent particles in a solid crystal lattice are packed tightly together in a very specific geometric arrangement. For one particle to replace another in such a structure is very difficult, unless the particles are almost identical. The most common solid solutions are alloys, in which one essentially spherical metal atom replaces another. Thus alloys are easily made by melting two metals and cooling the liquid solution. In many other cases, however, completely miscible liquids separate when a solid phase forms. A good example of this is benzene and naphthalene:

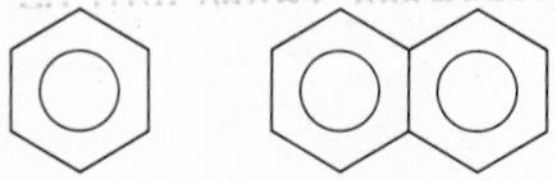

Benzene **Naphthalene**

A naphthalene molecule is almost twice as big as a benzene molecule and cannot fit in the benzene lattice.

Ideal Solutions: Raoult's Law

When two substances whose molecules are very similar form a liquid solution, the vapor pressure of the mixture is very simply related to the vapor pressures of the pure substances. Suppose, for example, we mix 1 mol benzene with 1 mol toluene

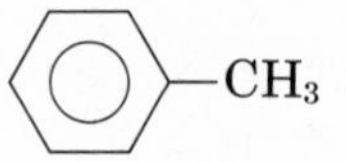

Toluene

as shown in Fig. 10.16. The mole fraction of benzene, x_b, and the mole fraction of toluene, x_t, are both equal to 0.5. At 79.6°C the measured vapor pressure of this mixture is 516 mmHg, slightly less than 517 mmHg, the average of the vapor pressures of pure benzene (744 mmHg) and of pure toluene (290 mmHg) at the same temperature.

It is easy to explain this behavior if we assume that because benzene and toluene molecules are so nearly alike, they behave the same way in solution as they do in the pure liquids. Since there are only half as many benzene molecules in the mixture as in pure benzene, the rate at which benzene molecules escape from the surface of the solution will be half the rate at which they would escape from the pure liquid. In consequence the partial vapor pressure of benzene above the mixture will be one-half the vapor pressure of pure benzene. By a similar argument the partial vapor pressure of the toluene above the solution is also one-half that of pure toluene. Accordingly, we can write

$$p_b = \tfrac{1}{2}P_b^* \qquad \text{and} \qquad p_t \times \tfrac{1}{2}P_t^*$$

where p_b and p_t are the partial pressures of benzene and toluene vapors, respectively, and P_b^* and P_t^* are the vapor pressures of the pure liquids. The total vapor pressure of the solution is

$$P = p_b + p_t = \frac{1}{2}P_b^* + \frac{1}{2}P_t^* = \frac{P_b^* + P_t^*}{2}$$

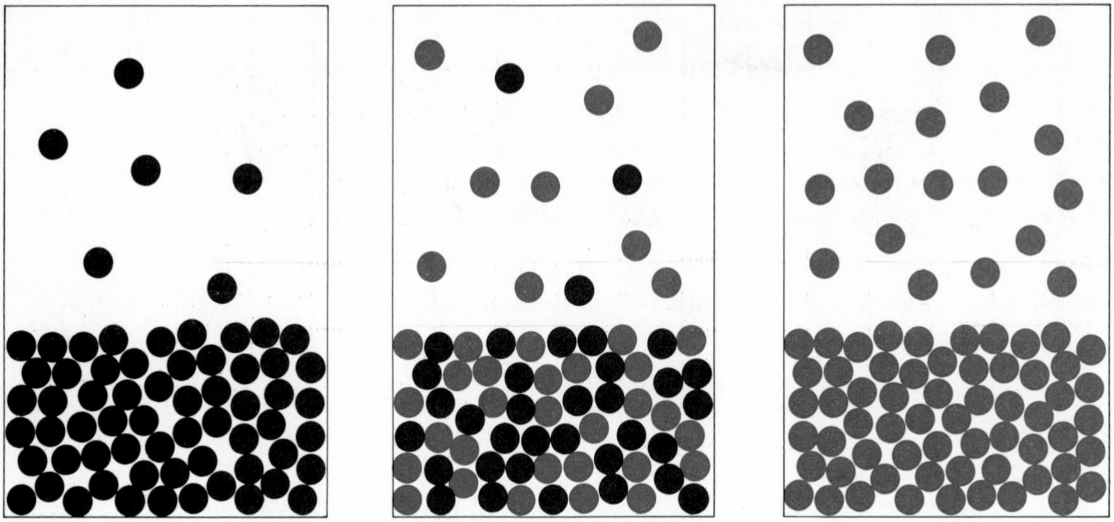

Figure 10.16 Vapor-liquid equilibria for (*a*) pure toluene; (*b*) a mixture of equal amounts of toluene and benzene; and (*c*) pure benzene. In the solution (*b*) only half the molecules are benzene molecules, and so the concentration of benzene molecules in the vapor phase is only half as great as above pure benzene. Note also that although the initial amounts of benzene and toluene in the solution were equal, more benzene than toluene escapes to the gas phase because of benzene's higher vapor pressure.

The vapor pressure of the mixture is equal to the mean of the vapor pressures of the two pure liquids.

We can generalize the above argument to apply to a liquid solution of any composition involving any two substances A and B whose molecules are very similar. The partial vapor pressure of A above the liquid mixture, p_A, will then be the vapor pressure of pure A, P_A^*, multiplied by the fraction of the molecules in the liquid which are of type A, that is, the mole fraction of A, x_A. In equation form

$$p_A = x_A P_A^* \qquad (10.2a)$$

Similarly for component B

$$p_B = x_B P_B^* \qquad (10.2b)$$

Adding these two partial pressures, we obtain the total vapor pressure

$$P = p_A + p_B = x_A P_A^* + x_B P_B^* \qquad (10.3)$$

Liquid solutions which conform to Eqs. (10.2) and (10.3) are said to obey **Raoult's law** and to be **ideal mixtures** or **ideal solutions.**

In addition to its use in predicting the vapor pressure of a solution, Raoult's law may be applied to the solubility of a gas in a liquid. Dividing both sides of Eq. (10.2*a*) by P_A^* gives

$$x_A = \frac{1}{P_A^*} \times p_A = k_A \times p_A \qquad (10.4)$$

Since the vapor pressure of any substance has a specific value at a given temperature, Eq. (10.4) tells us that the mole fraction x_A of a gaseous solute is proportional to the partial pressure p_A of that gas above the solution. For an

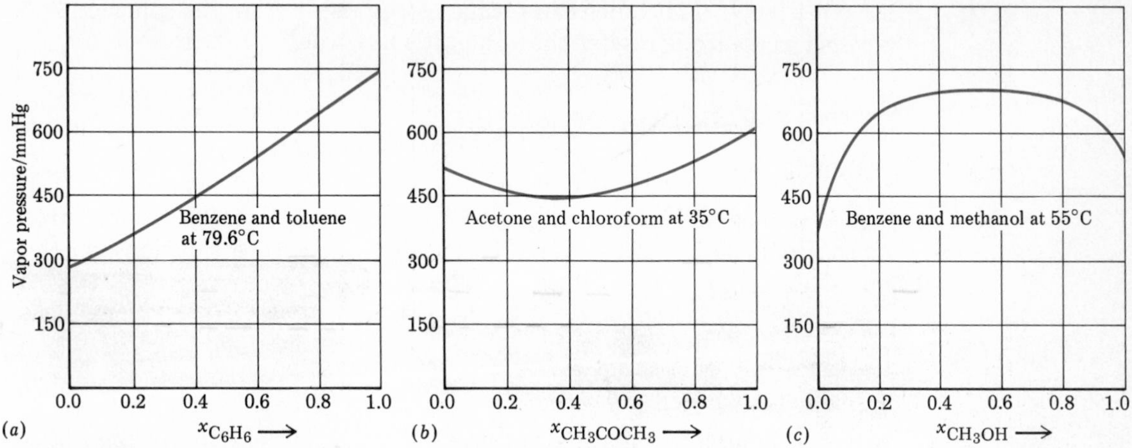

Figure 10.17 Deviations from Raoult's law. (*a*) When Raoult's law is obeyed, a plot of vapor pressure against mole fraction yields a straight line. This is nearly true for the benzene and toluene mixture at 79.6°C. (*b*) A mixture of acetone and chloroform shows negative deviations from Raoult's law at 35°C, indicating that the two different molecules prefer each other's company to their own. (*c*) The opposite behavior is shown at 55°C by a benzene-methanol mixture where the polar and nonpolar molecules prefer the company of their own kind.

ideal solution the proportionality constant k_A is the reciprocal of the vapor pressure of the pure solute at the temperature in question. Since vapor pressure increases as temperature increases, k_A, which is $1/P_A^*$, must decrease. Thus we expect the solubility of a gas in a liquid to increase as the partial pressure of gas above the solution increases, but to decrease as temperature increases. Equation (10.4) is known as **Henry's law.** It also applies to gaseous solutes which do not form ideal solutions, but in such cases the Henry's-law constant k_A does not equal the reciprocal of the vapor pressure.

In actual fact very few liquid mixtures obey Raoult's law exactly. Even for molecules as similar as benzene and toluene, we noted a deviation of 517 mmHg − 516 mmHg, or 1 mmHg at 79.6°C. Much larger deviations occur if the molecules are not very similar. These deviations are of two kinds. As can be seen from Fig. 10.17, a plot of the vapor pressure against the mole fraction of one component yields a straight line for an ideal solution. For nonideal mixtures the actual vapor pressure can be larger than the ideal value (positive deviation from Raoult's law) or smaller (negative deviation). Negative deviations correspond to cases where attractions between unlike molecules are greater than those between like molecules. In the case illustrated, acetone (CH_3COCH_3) and chloroform ($CHCl_3$) can form a weak hydrogen bond:

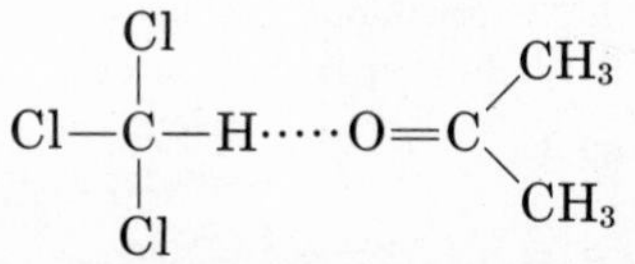

Because of this extra intermolecular attraction, molecules have more difficulty escaping the solution and the vapor pressure is lower. The opposite is true of a mixture of benzene and methanol. When C_6H_6 molecules are randomly distributed among CH_3OH molecules, the latter cannot hydrogen bond effectively. Molecules can escape more readily from the solution, and the vapor pressure is higher than Raoult's law would predict.

10.6 THE SEPARATION OF MIXTURES

Distillation

It is immediately apparent from Fig. 10.16 that Raoult's law requires that the composition of the vapor in equilibrium with a solution and the composition of the solution itself be different. While only 50 percent of the molecules in the liquid phase are benzene molecules, in the vapor phase they represent 80 percent of the molecules. This difference in composition between the two phases is the reason we can separate components of a liquid mixture by distillation.

In the laboratory, distillation is usually carried out in an apparatus similar to that shown in Fig. 10.18. The liquid mixture to be distilled is heated in a round-bottom flask until it boils. The emerging vapor travels into the condenser where it is cooled sufficiently to return to the liquid state. The condensed liquid can then be collected in a suitable container. In practice this container is usually changed several times during the course of the distillation. Each sample collected in this way is called a **fraction,** and the whole process is called **fractional distillation.**

Suppose we use the distillation apparatus shown in the figure to distill an equimolar mixture of benzene and toluene. We would first have to raise the temperature of the mixture to 92.4°C, at which point the total vapor pressure would be 1 atm (101.3 kPa) and the liquid would boil. At 92.4°C the vapor in equilibrium with an equimolar mixture has a mole fraction of

Figure 10.18 A laboratory distillation apparatus for separating liquid mixtures.

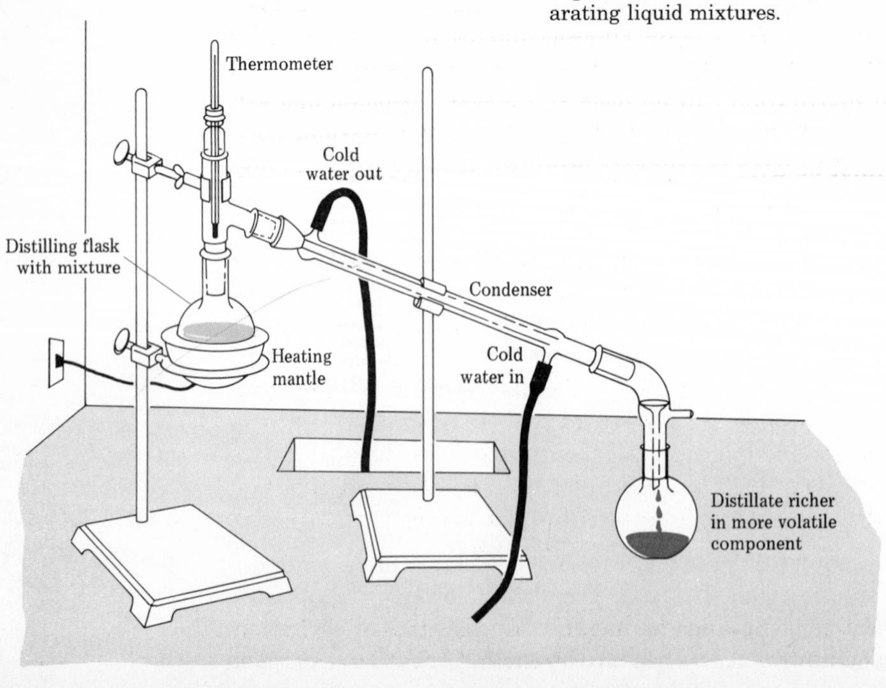

benzene equal to 0.71 (slightly different from that shown in Fig. 10.16). Accordingly, when our mixture is distilled, vapor of this composition will be condensed and the first drops in the collection vessel will contain 71 percent benzene molecules and 29 percent toluene molecules. We will now have a sample which is much richer in benzene than the original. In theory we could go on repeating this process indefinitely. If we redistilled a liquid mixture having $x_b = 0.71$, it would boil at 87°C and yield a condensed vapor whose mole fraction of benzene would be 0.87. A third distillation would yield benzene with $x_b = 0.95$, and a fourth would yield a product with $x_b = 0.99$. If we went on long enough, we could obtain a sample of benzene of any desired purity.

In practice repeated distillation takes up a lot of time, and we would lose part of the sample each time. Fortunately there is a way out of this difficulty. We insert a glass column filled with round beads, or with specially designed packing, between the distillation flask and the condenser. When a vapor reaches the bottom-most glass beads en route to the condenser, part of it will condense. This condensate will be richer in the less volatile component, toluene, and so the vapor which passes on will be richer in benzene than before. As it passes through the column, the vapor will undergo this process several times. In effect the fractionating column allows us to perform several distillations in one operation. A well-designed laboratory column makes it possible to effect several hundred distillations in one pass. By using such a column, it would be possible to distill virtually pure benzene from the equimolar mixture, leaving virtually pure toluene behind.

Fractional distillation is used on a very large scale in the refining of petroleum. Some of the large towers seen in petroleum refineries are gigantic fractionating columns. These columns enable the oil company to separate the petroleum into five or six fractions called **cuts.** The more-volatile cuts can be used directly for gasoline, kerosene, etc. The less-volatile cuts contain the very long-chain alkanes and are usually converted into shorter-chain alkanes by cracking, as described in Sec. 8.2.

While fractional distillation can be used to separate benzene and toluene and most mixtures of alkanes which show only small deviations from ideal behavior, it cannot be used to separate mixtures which exhibit a maximum or a minimum in the vapor-pressure versus mole-fraction curve. Thus any attempt to separate benzene and methanol (see Fig. 10.17) by fractional distillation does not result in methanol distilling over initially but rather results in a mixture with $x_{CH_3OH} = 0.614$. A mixture of this composition is not only more volatile than methanol or benzene but also has a higher vapor pressure than any other mixture, as can be seen from the figure. This is why it distills first. Such a mixture is called a **minimum boiling azeotrope.** Conversely, for a mixture such as chloroform and acetone which exhibits large enough negative deviations from Raoult's law, a **maximum boiling azeotrope** of low volatility distills at the end of a fractional distillation, preventing complete separation of the pure components.

Chromatography

Another useful set of techniques for separating mixtures, called **chromatography,** has been developed in the last 25 years. Perhaps the simplest of these techniques to describe is paper chromatography, an example of which is shown in color in Plate 2. A line of black ink from a felt-tipped pen has been drawn across a strip of special chromatography-grade paper. The edge of the paper parallel to the line is now immersed in an alcohol-water mixture which lies at the bottom of a rectangular glass jar, called a chromatography tank. Due to capillary action, the solvent (called the **mobile phase** because it is moving) creeps up the paper (the **stationary phase**) until it reaches the line of ink. As the solvent front moves past the ink, the initial line separates into a series of bands, each corresponding to a pure component of the ink. Once a good separation has been obtained, the process is stopped by lifting the paper out of the solvent.

Each dye in the ink line can either adhere to the surface of the paper (a process called **adsorption**) or dissolve in the solvent. A dye, such as the red component in Plate 2, which dissolves readily in the solvent and adsorbs poorly on the paper surface will travel along at almost the same speed as the solvent moving up the paper. A dye which is strongly adsorbed and not very soluble in water, like the orange dye in Plate 2, will, by contrast, scarcely move at all. Most dyes will lie somewhere between these two extremes and move at intermediate rates.

All forms of chromatography work on the same general principle as paper chromatography. There is always a stationary phase which does not move and a mobile phase which does. The various components in the mixture being chromatographed separate from each other because they are more strongly held by one phase or the other. Those which have the greatest affinity for the mobile phase move along the fastest.

The most important form of chromatography is **gas chromatography** or vapor-phase chromatography. A long column is packed with a finely divided solid whose surface has been coated with an inert liquid. This liquid forms the stationary phase. The mobile phase is provided by an inert carrier gas, such as He or N_2, which passes continuously through the interstices among the solid particles. A liquid sample can be injected into the gas stream and vaporized just before it enters the tube. As this sample is carried through the tube by the slow stream of gas, those components which are most soluble in the inert liquid are held up, while the less-soluble components move on more rapidly. The components thus emerge one by one from the end of the tube. In this way it is possible to separate and analyze mixtures of liquids which it would be impossible to deal with by distillation or any other technique.

The development of chromatography in the last 25 years is one of the major revolutions in technique in the history of chemistry, comparable to that which followed the development of an accurate balance. Separations which were previously considered impossible are now easily achieved, sometimes with quite simple apparatus. In particular the science of biochemistry, in which complex mixtures are almost always encountered, has been revolutionized by this technique. In the field of environmental chemistry chromatography has helped us separate and detect very low concentrations of contaminants like DDT or PCB (polychlorinated biphenyls). The major drawback to chromatography is that it does not lend itself to large-scale operation. As a result it remains largely a laboratory, rather than an industrial, technique for separating mixtures.

10.7 COLLIGATIVE PROPERTIES OF SOLUTIONS

The **colligative properties** of a solution are those which depend on the number of particles (and hence the amount) of solute dissolved in a given quantity of solvent, irrespective of the chemical nature of those particles. We have already seen from Raoult's law that the vapor pressure of a solution depends on the mole fraction of solute, and now we are in a position to see how this affects several other properties of solutions.

Boiling and Freezing Points

We often encounter solutions in which the solute has such a low vapor pressure as to be negligible. In such cases the vapor above the solution consists only of solvent molecules and the vapor pressure is always *lower* than that of the pure solvent. Consider, for example, the solution obtained by dissolving 0.020 mol sucrose ($C_{12}H_{22}O_{11}$) in 0.980 mol H_2O at 100°C. The sucrose will contribute nothing to the vapor pressure, while we can expect the water vapor, by Raoult's law, to contribute

$$P_{H_2O} = P^*_{H_2O} \times x_{H_2O} = 760 \text{ mmHg} \times 0.980 = 744.8 \text{ mmHg}$$

We would thus expect the vapor pressure to be 744.8 mmHg, in reasonable agreement with the observed value of 743.3 mmHg.

A direct consequence of the lowering of the vapor pressure by a nonvolatile solute is *an increase in the boiling point* of the solution relative to that of the solvent. We can see why this is so by again using the sucrose solution as an example. At 100°C this solution has a vapor pressure which is lower than atmospheric pressure, and therefore it will not boil. In order to increase the vapor pressure from 743.3 to 760 mmHg so that boiling will occur, we need to raise the temperature. Experimentally we find that the temperature must be raised to 100.56°C. We say that the **boiling-point elevation** ΔT_b is 0.56 K.

A second result of the lowering of the vapor pressure is a depression of the freezing point of the solution. Any aqueous solution of a nonvolatile solute, for example, will have a vapor pressure at 0°C which is less than the vapor pressure of ice (0.006 atm or 4.6 mmHg) at this temperature. Accordingly, ice and the aqueous solution will not be in equilibrium. If the temperature is lowered, though, the vapor pressure of the ice decreases more rapidly than that of the solution and a temperature is soon reached when both ice and the solution have the same vapor pressure. Since both phases are now in equilibrium, this lower temperature is also the freezing point of the solution. In the case of the sucrose solution of mole fraction 0.02 described above, we find experimentally that the freezing point is −2.02°C. We say that the **freezing-point depression** ΔT_f is 2.02 K.

The depression of the freezing point of water by a solute explains why the sea does not freeze at 0°C. Because of its high salt content the sea has a freezing point of −2.2°C. If the sea froze at 0°C, larger stretches of ocean would turn into ice and the climate of the earth would be very different. We can now also understand why we add ethylene glycol, CH_2OHCH_2OH, to water in a car radiator in winter. Without any additive the water would freeze at 0°C, and the resulting increase in volume would crack the radiator. Since ethylene glycol is very soluble in water, it can form a solution with a freezing point low enough to prevent freezing even on the coldest winter day.

Both the freezing-point depression and the boiling-point elevation of a solution were once important methods for determining the molar mass of a newly prepared compound. Nowadays a mass spectrometer (Sec. 4.5) is usually used for this purpose, often on an assembly-line basis. Many chemists send samples of newly prepared compounds to a laboratory specializing in these determinations in much the same way as a medical doctor will send a sample of your blood to a laboratory for analysis.

The reason we can use the boiling-point elevation and the freezing-point depression to determine the molar mass of the solute is that both properties are *proportional to the mole fraction and independent of the nature of the solute.* The actual relationship, which we will not derive, is

$$x_A = \frac{\Delta H_m}{RT^2}\,\Delta T \tag{10.5}$$

where x_A is the mole fraction of the solute and ΔT is the boiling-point elevation or freezing-point depression. T indicates either the boiling point or freezing point of the pure solvent, and ΔH_m is the molar enthalpy of vaporization or fusion, whichever is appropriate. This relationship tells us that we can measure the mole fraction of the solute in a solution merely by finding its boiling point or freezing point.

EXAMPLE 10.7 A solution of sucrose in water boils at 100.56°C and freezes at −2.02°C. Calculate the mole fraction of the solution from each temperature.

Solution

a) For boiling we have, from Table 10.2,

$$\Delta H_m = 40.7 \text{ kJ mol}^{-1} \qquad \text{and} \qquad T = 373.15 \text{ K}$$

As in previous examples the units of R should be compatible with the other units appearing in the equation. In this case since ΔH_m is given in units of kJ mol^{-1}, R = 8.314 J K^{-1} mol^{-1} is most appropriate. Since ΔT = 0.56 K, we have

$$x_{\text{sucrose}} = \frac{\Delta H_m}{RT^2}\,\Delta T = \frac{40.7 \times 10^3 \text{ J mol}^{-1} \times 0.56 \text{ K}}{8.314 \text{ J K}^{-1} \text{ mol}^{-1}\,(373.15 \text{ K})^2}$$

$$= 0.0197$$

b) Similarly for freezing we have

$$\Delta H_m = 6.01 \text{ kJ mol}^{-1} \qquad T = 273.15 \text{ K} \qquad \Delta T = 2.02 \text{ K}$$

so that

$$x_{\text{sucrose}} = \frac{6.01 \times 10^3 \text{ J mol}^{-1} \times 2.02 \text{ K}}{8.314 \text{ J K}^{-1} \text{ mol}^{-1} \times (273.15 \text{ K})^2} = 0.0196$$

The two values are in reasonable agreement.

Once we know the mole fraction of the solute, its molar mass is easily calculated from the mass composition of the solution.

EXAMPLE 10.8 33.07 g sucrose is dissolved in 85.27 g H_2O. The resulting solution freezes at $-2.02°C$. Calculate the molar mass of sucrose.

Solution Since the freezing point of the solution is the same as in part *b* of Example 10.7, the mole fraction must be the same. Thus

$$x_{\text{sucrose}} = 0.0196 = \frac{n_{\text{sucrose}}}{n_{\text{sucrose}} + n_{H_2O}}$$

Furthermore

$$n_{H_2O} = \frac{85.27 \text{ g}}{18.02 \text{ g mol}^{-1}} = 4.732 \text{ mol}$$

Thus

$$0.0196 = \frac{n_{\text{sucrose}}}{n_{\text{sucrose}} + 4.732 \text{ mol}}$$

so that

$$4.732 \times 0.0196 \text{ mol} = n_{\text{sucrose}} - 0.0196 n_{\text{sucrose}}$$

$$0.092\ 75 \text{ mol} = 0.9804 n_{\text{sucrose}}$$

or

$$n_{\text{sucrose}} = \frac{0.092\ 75}{0.9804} \text{ mol} = 0.0946 \text{ mol}$$

From which

$$\mathcal{M}_{\text{sucrose}} = \frac{33.07 \text{ g}}{0.0946 \text{ mol}} = 350 \text{ g mol}^{-1}$$

Note: The correct molar mass is 342.3 g mol^{-1}. Neither the freezing point nor the boiling point gives a very accurate value for the molar mass of the solute.

Osmotic Pressure

Suppose we have, as in Fig. 10.19, a solution of sugar in water separated by a membrane from a sample of pure water. The membrane is porous, but the holes are not large enough to allow sucrose molecules to pass through from one side to the other while still being large enough to allow water molecules to pass freely through them. In such a situation, as can be seen in the figure, water molecules will be hitting one side of the membrane more often than the other. As a result water molecules will move more often from left to right through the membrane in the illustration than they

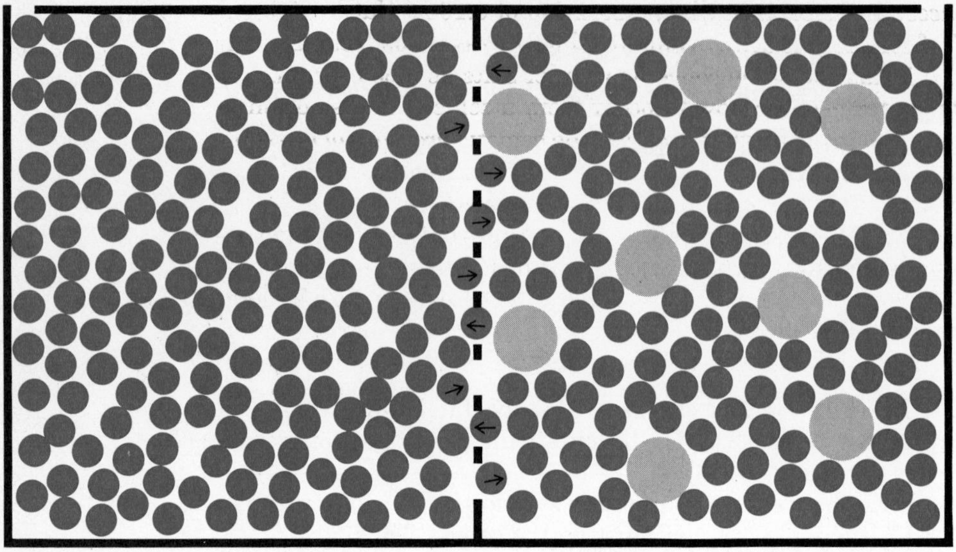

Figure 10.19 Osmotic flow of a solvent through a semipermeable membrane to a solution. The semipermeable membrane is shown acting as a sieve. Many membranes operate in a different way, but the ultimate effect is the same.

will in the reverse direction. There will thus be a net flow of water from the compartment containing pure solvent, through the membrane, into the compartment containing the sucrose. This is another example of the tendency for moving molecules to become more thoroughly mixed together.

The situation just described is a specific example of a more general phenomenon called **osmosis.** In general when two solutions of different concentrations are separated by a membrane which will selectively allow some species through it but not others, then osmosis, or flow of material from the less concentrated to the more concentrated side of the membrane, will occur. A membrane which is selective in the way just described is said to be **semipermeable.** Osmosis is of particular importance in living organisms, since most living tissue is semipermeable in one way or another. In Fig. 10.19 we have depicted a membrane which is selective purely because of the pore size. In biological systems, the semipermeability does not rely solely on this sieve action, but the precise mode of action of these membranes is still not entirely understood.

A simple demonstration of osmosis is provided by the behavior of red blood cells. If these are immersed in water and observed under the microscope, they will be seen to gradually swell and to finally burst. Osmotic flow occurs from the surrounding water into the more concentrated solutions inside the cell. If the blood cells are now immersed in a saturated solution of NaCl, osmosis occurs in the opposite direction, since the solution inside the cell is not as concentrated as that outside. Under the microscope, the blood cells can be seen to shrink and shrivel. In medical practice, any solution which is to be introduced into the blood must take the possibility of osmosis into account. Normal saline, a solution of 0.16 *M* NaCl (0.16 mol NaCl per dm^3 of solution) is always employed for intravenous feeding or injection, because it has the same concentration of salts as blood serum.

The tendency for osmotic flow to occur from a solvent to a solution is usually measured in terms of what is called the osmotic pressure of the solution, symbol Π. This osmotic pressure is not a pressure which the solution itself exerts but is rather the pressure which must be applied to the solution (but not the solvent) from outside in order to just prevent osmosis from occurring. A simple method for measuring the osmotic pressure is shown in Fig. 10.20. The wider end of a funnel-shaped tube is covered with a membrane. The tube is filled with solution and placed in a container of the solvent. The height of the solution above the solvent increases until a maximum value is reached. The osmotic pressure is then the pressure exerted by the column of a solution of height h:

$$\Pi = \rho g h$$

where ρ is the density of the solution and g is the gravitational acceleration.

Experimentally the osmotic pressure is found to obey a law similar in form to the ideal gas law and hence easy to remember:

$$\Pi V = nRT \tag{10.6}$$

where n is the amount of solute in volume V of solution. In practice it is more useful to have Eq. (10.6) in terms of the concentration of solute n/V. Accordingly we rearrange it to read

$$\Pi = \frac{n}{V} RT$$

or

$$\Pi = cRT \tag{10.7}$$

Figure 10.20 A simple apparatus for determining the osmotic pressure of a solution.

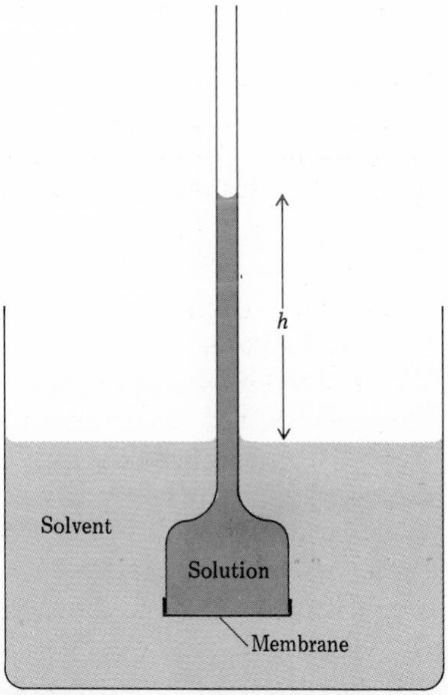

Equation (10.7) is a very useful relationship since it means that we can find the concentration of any solution merely by measuring its osmotic pressure. This, in turn, allows us to find the molar mass of the solute. Suppose we have a solution in which a known mass of solute is dissolved in a known volume of solution. By measuring the osmotic pressure of this solution, we can find the concentration of solute and hence the amount of solute in the total volume. Since we already know the mass of solute, the molar mass follows immediately.

EXAMPLE 10.9 A solution of 20.0 g of polyisobutylene in 1.00 dm^3 of benzene was placed in an osmometer, similar to the one shown in Fig. 10.20, at 25°C. After equilibrium had been obtained, the height h was found to be 24.45 mm of benzene. Find the average molar mass of the polymer. The density of the solution is 0.879 g cm^{-3}.

Solution We must first find the osmotic pressure from the height h with the formula $\Pi = \rho g h$. In doing this, it is most convenient to convert everything to SI base units.

$$\begin{aligned}\Pi = \rho g h &= 0.879\,\frac{\text{g}}{\text{cm}^3} \times 9.807\,\frac{\text{m}}{\text{s}^2} \times 24.45\text{ mm} \\ &= 210.8\,\frac{\cancel{\text{g}}}{\cancel{\text{cm}^3}}\,\frac{\text{m}}{\text{s}^2}\,\cancel{\text{mm}} \times \frac{1\text{ kg}}{1000\,\cancel{\text{g}}} \times \left(\frac{100\,\cancel{\text{cm}}}{1\text{ m}}\right)^3 \times \frac{1\text{ m}}{1000\,\cancel{\text{mm}}} \\ &= 210.8\,\frac{\text{kg m}^2}{\text{m}^3\text{ s}^2} = 210.8\,\frac{\text{kg m}}{\text{s}^2} \times \frac{1}{\text{m}^2} = 210.8\text{ N m}^{-2} = 210.8\text{ Pa}\end{aligned}$$

Knowing Π, we can now calculate the concentration c by rewriting

$$\Pi = cRT$$

as $c = \dfrac{\Pi}{RT} = \dfrac{210.8 \times 10^{-3}\text{ kPa}}{8.3143\text{ J K}^{-1}\text{ mol}^{-1} \times 298.15\text{ K}} = 8.50 \times 10^{-5}\text{ mol dm}^{-3}$

For 1 dm^3 of solution therefore,

$$n_{\text{solute}} = 8.50 \times 10^{-5}\text{ mol}$$

while $$m_{\text{solute}} = 20\text{ g}$$

Thus $$\mathcal{M}_{\text{solute}} = \frac{20\text{ g}}{8.50 \times 10^{-5}\text{ mol}} = 2.35 \times 10^5\text{ g mol}^{-1}$$

Note: An average polyisobutylene molecule thus has a molecular weight of close to a quarter of a million! Such a molecule is made from

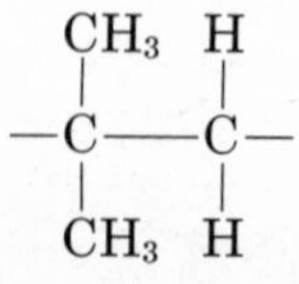

units, each with a molar mass of 56.10 g mol^{-1}. An average polyisobutylene chain is thus $\frac{2.35 \times 10^5 \text{ g mol}^{-1}}{56.10 \text{ g mol}^{-1}} = 4189$ units long. In other words the chain length is over 8000 carbon atoms in this sample of polymer.

SUMMARY

The microscopic structure of a solid may be described in terms of a unit cell—the smallest parallel-sided shape from which the overall structure can be constructed—and a space lattice—the pattern of points made by the corners of many unit cells packed together. The seven crystal systems include all the basic types of unit cells. Closest-packed structures are important because they are ways of squeezing the maximum number of atoms into the minimum volume. There are two types: hexagonal closest packed (hcp) and cubic closest packed (ccp). Most metals adopt hcp, ccp, or body-centered cubic crystal lattices.

Liquids differ from solids in that their constituent particles have sufficient energy to move past one another rather than vibrating about an average location. Some liquids become so viscous just above their melting points that their particles no longer can move. In such a case a glass or amorphous material, in which there is no long-range order, is formed.

Phase transitions from solid to liquid, liquid to gas, or solid to gas all require energy because they involve separation of particles which attract one another. Each phase transition can reach a state of equilibrium under certain conditions of temperature and pressure. Such conditions are summarized in a phase diagram. The liquid-vapor equilibrium differs from the other two in that there is no distinction between liquid and vapor above the critical temperature. Hence the liquid-vapor curve in a phase diagram ends at the critical point.

Liquid solutions are important because of the large number of chemical reactions which occur and are carried out in them. A saturated solution is one in which solute is in equilibrium with solution at a given temperature. If a supersaturated solution is prepared, it is unstable. Adding a seed crystal permits solute molecules to precipitate until the rate of solution equals the rate of precipitation. The solubility of a solute or the composition of any solution which is to be used quantitatively is usually reported in terms of concentration (amount of solute/volume of solution), mass fraction (mass of solute/mass of solution), or mole-fraction (amount of solute/amount of solution).

The solubilities of various substances can be related to the molecular structures of solute and solvent. The most important rule is that *like dissolves like*. When solute and solvent molecules are nearly identical, a solution behaves ideally and obeys Raoult's law. However, nonideal behavior of solutions is much more common than nonideal behavior of gases, because molecules are much closer together in the liquid phase and intermolecular forces are much more important.

The components of liquid solutions may be separated by distillation, provided deviations from ideal behavior are not so large as to produce azeotropes. A second very important method of separation is chromatography. It depends on the relative abilities of different substances to dissolve in or be adsorbed on one phase in preference to another. Thus if intermolecular forces are such that a component of a mixture is strongly held by the mobile phase, while another component is held by the stationary phase, the two can be separated chromatographically.

Colligative properties include lowering of vapor pressure, elevation of boiling point, depression of freezing point, and osmotic pressure. They depend on the number of particles (and hence the amount) of solute dissolved in a given quantity of solvent, irrespective of the nature of the particles. Colligative properties can be used to determine molar masses, although the only one commonly used for that purpose today is osmotic pressure.

QUESTIONS AND PROBLEMS

10.1 Explain why a solid is rigid and a liquid is not.

10.2 When a crystal of sodium chloride is heated from 0 to 100°C, 88 J of heat energy is absorbed by the crystal. What type of energy is this converted to on the microscopic level?

10.3 Draw the unit cell for the following arrays of like atoms extended indefinitely in two dimensions. How many atoms are there in each unit cell?

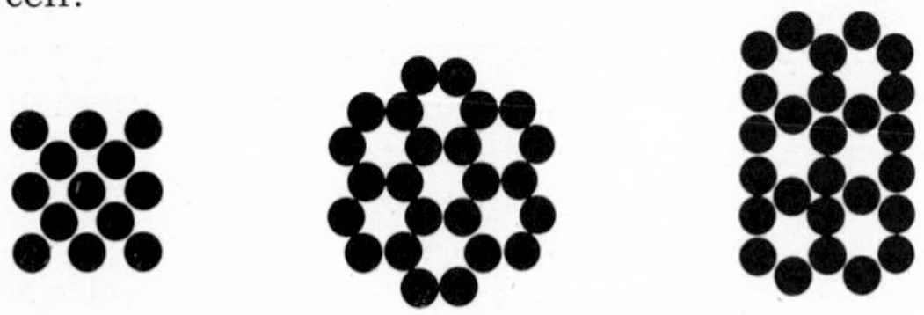

a *b* *c*

10.4 When oranges are stacked in a pyramid in a food store, is this arrangement body-centered cubic, face-centered cubic, or hexagonal closest packed?

10.5 What structure do the chloride ions considered by themselves have in the sodium chloride lattice? (See Fig. 10.2.)

10.6 How many atoms are there in a unit cell of the hexagonal closest-packed structure shown in Fig. 10.6*a*?

10.7 Argon crystallizes with a cubic closest-packed structure of cell length 525.6 pm. Find the van der Waals radius of argon.

10.8 The unit cell of CsCl is a cube of side 411.0 pm. At each corner of the cube is a Cl^- ion, and at the center of the cube is a Cs^+ ion. Using a table of atomic weights, calculate the density of this salt.

10.9 Lithium metal solidifies forming cubic crystals. The edge of the unit cell is found to be 351 pm. The density of lithium is found to be 0.534 g cm^{-3}. Does lithium have a body-centered cubic or a face-centered cubic structure?

10.10 The density of sodium chloride is 2.165 g cm^{-3}, while the length of the unit cell shown in Fig. 10.2*c* is 526.7 pm. Calculate a value for the Avogadro constant from these data.

10.11 Criticize the following statement: "The coordination number of an atom or an ion in a crystal is an indication of its valence."

10.12 How many parameters need to be specified in order to describe the unit cell of a crystal belonging to the triclinic system?

10.13 Explain why most molten plastics are extremely viscous.

10.14 Explain why the viscosity of a liquid decreases with rising temperature.

10.15 Arrange the following liquids in order of increasing viscosity:

a CH_3OH *b* $H_2C{=}O$
c CH_4 *d* CH_2OHCH_2OH

10.16 Match the following substances to the molar heats of vaporization:

Substance	Molar Heat of Vaporization/ kJ mol^{-1}
a $CH_3CH(CH_3)CH_3$	*i* 20.0
b $CH_3CH(CH_3)OH$	*ii* 41.2
c CH_2OHCH_2OH	*iii* 30.3
d Kr	*iv* 49.6
e $CH_3CH_2CH_2CH_3$	*v* 22.2
f $CH_3CH_2CH_2OH$	*vi* 40.0
g CH_3COCH_3	*vii* 9.0

10.17 "The enthalpy of vaporization of $H{-}C{\equiv}C{-}H$ is much larger than that of $N{\equiv}N$. This shows that the carbon-carbon triple bond is much stronger than the nitrogen-nitrogen triple bond." Explain what is wrong with this statement and give a correct explanation in terms of London forces.

10.18 Why is it that the molar enthalpy of vaporization of a substance is always higher than its molar enthalpy of fusion?

***10.19** Use a handbook or other library source to obtain the molar enthalpy of fusion and the molar enthalpy of vaporization for each of the normal alkanes from C_1 (methane, CH_4) to C_{12} (n-dodecane, $C_{12}H_{26}$). Plot ΔH_m (fusion) versus number of carbon atoms and do the same for ΔH_m (vaporization). Note regularities and unusual features of both graphs, and try to explain both in molecular terms.

10.20 A student writes, "The boiling point of CF_4 is bigger than CCl_4 because the C—F bond is stronger than the C—Cl bond." What is wrong with this sentence from a scientific point of view?

10.21 Discuss the statement, "When water boils, it is the potential energy rather than the kinetic energy of the water molecules which changes."

10.22 The molar enthalpy of vaporization of silicon tetrafluoride has the value 151 J K^{-1} mol^{-1}. Estimate the boiling point using Trouton's rule.

10.23 When a liquid is in equilibrium with its vapor, the rate at which molecules escape from the liquid is equal to the rate at which they reenter from the vapor. What would happen to both of these rates if the following changes were made:

a The temperature is raised.

b The surface area of the liquid is doubled.

c Air, as well as vapor, is present above the liquid.

d The intermolecular forces could somehow be doubled.

10.24 When a drop or two of a very volatile liquid is placed on the skin and allowed to evaporate, a cooling sensation is experienced. Give reasons for this effect. Why is this effect so much less noticeable in the case of water?

10.25 When 1 g of liquid water is injected into an evacuated 10 dm^3 flask held at 25°C, the water initially starts to boil but eventually stops. At the same time the pressure inside the flask increases rapidly from zero until it attains a value of 3.17 kPa (23.8 mmHg). Explain this macroscopic behavior in microscopic terms.

10.26 The vapor pressure of n-butane at −20°C is 339 mmHg while that of isobutane [$CH_3CH(CH_3)CH_3$] is 543 mmHg. Explain why isobutane has the higher vapor pressure even though both isomers have the same molar mass.

10.27 Refer to Fig. 10.10 to answer each of the following questions: (*a*) What is the vapor pressure of heptane at 60°C? (*b*) At what temperature will pentane have a vapor pressure of 525 mmHg? (*c*) At what temperature will hexane boil in a high-pressure container maintained at 900 mmHg? (*d*) Which of the liquids featured in the figures has the largest molar heat of vaporization? (*e*) Which of the liquids would you expect to have the lowest critical temperature?

10.28 A student writes, "When there are heavy molecules a substance has a lower vapor pressure. This is because heavy molecules move slower in the liquid. Hence the rate at which they escape is much slower and the vapor pressure lower." What is wrong with this argument from a scientific standpoint?

10.29 Another student writes, "A substance with heavy molecules has a lower vapor pressure because of gravity. More energy is needed to raise the molecules upwards from the liquid to the gas against gravity if they are heavy." How could you disprove this theory of vapor pressure experimentally?

10.30 When water is boiling briskly in a beaker over a bunsen burner, are the bubbles in the liquid air bubbles, pure water vapor, or are they perhaps a mixture of water vapor and air?

10.31 Plot the vapor pressure of water as a function of temperature in the range of 80 to 100°C from the data given in Table 9.4. Use this plot to estimate as accurately as possible the boiling point of water in Denver when the atmospheric pressure is 629 mmHg (83.8 kPa). If you like, plot log P against $1/T$ instead of P against T, since this yields a straight line rather than a curve.

10.32 When water is boiling in a large flask heated at the bottom, one often notices that bubbles produced at the bottom get somewhat larger as they rise to the top. Explain why this is so.

*10.33 Assume 2.904 g of water is introduced into an evacuated flask of 10.00 dm^3 held at 20°C. What mass of water will evaporate? What mass of liquid water will still be left if the temperature is now raised to 50°C? (Use Table 8.4.)

*10.34 To what temperature must the flask in the previous example be heated in order for all the water to evaporate? (Table 8.4 must again be consulted.)

10.35 Sketch a rough curve to show how the pressure inside a cylinder of oxygen at 20°C varies as the oxygen is released. Also sketch a curve showing the behavior of the pressure in a cylinder of ammonia under similar circumstances.

10.36 Is the propellant used in an aerosol spray can above or below its critical temperature? Give reasons for your answer.

***10.37** The triple point of CO_2 is at −56.6°C and 5.11 atm. Solid CO_2 melts at a temperature of −53.2°C and at a pressure of 20 atm, while the vapor pressure of the solid is found to be 0.37 atm at −90°C. The vapor pressure of the liquid is 16.6 atm at −25°C and 63.5 atm at +25°C. Use these data to sketch a rough phase diagram for CO_2. (*a*) From this diagram predict what will happen if solid CO_2 at −90°C is allowed to warm up to 25°C at a pressure of 1 atm. (*b*) What will happen if the solid is warmed up from −90°C at a pressure of 10 atm? (*c*) Solid CO_2 is held at −50°C and 50 atm. What will happen if the pressure is now gradually reduced to 1 atm while the temperature remains constant?

10.38 From your experience with the previous problem, state under what conditions a solid will sublime, i.e., change directly into vapor without first melting.

10.39 Assume 35.31 g glycerol, $CH_2OHCHOHCH_2OH$, is dissolved in 98.31 g water at 20°C. The resulting solution has a density of 1.063 g cm^{-3}. Calculate

a The mole fraction of glycerol.
b The mole fraction of water.
c The mass fraction of glycerol.
d The concentration of glycerol.

10.40 Assume 48.41 g ethanol, C_2H_5OH, was weighed into a volumetric flask and sufficient water was then added to bring the total volume up to the 100-cm^3 mark. The resulting solution was found to have a density of 0.9062 g cm^{-3}. Calculate

a The concentration of ethanol.
b The mole fraction of ethanol.
c The mass fraction of ethanol.

10.41 A 0.1917-*M* solution of $HgCl_2$ has a density of 1.0411 g cm^{-3} at 20°C. Find the mass fraction of $HgCl_2$ of this solution expressed as a percentage. Also find the mass fraction of 1.917×10^{-5} *M* $HgCl_2$ and 1.917×10^{-8} *M* $HgCl_2$ expressed as parts per million and parts per billion, respectively. Assume that these very dilute solutions have the same density as water at 20°C, namely, 0.9982 g cm^{-3}.

10.42 Find the mass fraction of glycerol (expressed as a percent) in a glycerol-water mixture for which the mole fraction of glycerol ($CH_2OHCHOHCH_2OH$) is

a 0.3000 *b* 0.5000 *c* 0.8000

10.43 Find the mole fraction of ethanol, C_2H_5OH, in an ethanol-water mixture for which the mass fraction is

a 30.0% *b* 50.0% *c* 70.0%

10.44 Which substance from each of the following pairs would you expect to be more soluble in heptane, C_7H_{16}?

a CH_3COCH_3 or CH_3COOH
b C_2H_5Cl or C_2H_5OH
c CH_3OCH_3 or CH_3CH_2OH
d $HO(CH_2)_6OH$ or $CH_3(CH_2)_6OH$
e $CH_3CH_2CH_3$ or CH_3OCH_3

10.45 Which substance from each of the following pairs would you expect to be more soluble in water?

a $CH_3(CH_2)_6COOH$ or CH_3COOH
b CH_3Cl or CH_3OH
c CH_3OCH_3 or CH_3CH_2OH
d $HO(CH_2)_6OH$ or $CH_3(CH_2)_6OH$
e $CH_3CH_2CH_3$ or CH_3OCH_3

10.46 As a general rule, partially miscible liquids become more soluble in each other as the temperature is raised. Explain why this is so.

10.47 At 67.5°C the vapor pressure of *n*-hexane is 280.1 mmHg while that of *n*-heptane is 744.5 mmHg. Assuming ideal behavior, calculate the vapor pressure of a solution for which the mole fraction of hexane is 0.3172. Compare this with the measured value of 423.7 mmHg. Does this system show positive or negative deviations from Raoult's law?

***10.48** Calculate the composition of the *vapor* above the hexane-heptane liquid mixture of the previous problem. *Hint:* Assume ideal partial vapor pressures and then calculate the amount of each component in a given volume (say 1 dm^3) of vapor. Do your results show that the vapor is richer in the more volatile component?

10.49 At 55°C the vapor pressure of benzene is 327.1 mmHg while that of methanol is 516.2 mmHg. When 2.033 mol benzene is mixed with 1.000 mol methanol, the vapor pressure of the mixture is found to be 664.2 mmHg. Does this mixture exhibit positive or negative deviations from Raoult's law? Give a molecular explanation for the result.

10.50 When air ($x_{O_2} = 0.2$) at 1 atm pressure is bubbled through water at 20°C, the mass fraction of O_2 is found to be 42 ppm. What would be the mass fraction of O_2 if pure O_2 at a pressure of 5 atm were bubbled through water at 20°C?

***10.51** A mixture of acetone and chloroform for which $x_{CH_3COCH_3}$ is 0.8 is distilled using an efficient fractionating column. Judging from Fig. 10.17*b*, which one of the following alternative behaviors would you expect to observe:

a Virtually pure chloroform distills over until almost pure acetone is left.

b Virtually pure acetone distills over until almost pure chloroform is left.

c An azeotropic mixture distills over until almost pure chloroform is left.

d An azeotropic mixture distills over until almost pure acetone is left.

e Virtually pure acetone distills over until an azeotropic mixture is left.

f Virtually pure chloroform distills over until an azeotropic mixture is left.

10.52 Explain in as few words as possible the principles of fractional distillation.

10.53 Explain what is meant by the terms *stationary phase* and *mobile phase* in the context of chromatography. Identify these phases in the case of both paper chromatography and gas chromatography.

10.54 Calculate the vapor pressure of the solution obtained when 50.0 g of the nonvolatile solid urea, NH_2CONH_2, is dissolved in 1.000 kg H_2O at 30°C.

10.55 Without consulting the text, see if you can explain in your own words why the boiling point of a solution is higher than that of the pure solvent, when the solute is nonvolatile.

10.56 Two aqueous solutions, one of glycerol and the other of sucrose, both boil at 101.32°C. What do these two solutions have in common? At what temperature would these two solutions freeze?

10.57 Find the boiling point and the freezing point of the solution obtained by dissolving 10.0 g sucrose, $C_{12}H_{22}O_{11}$, in 100 g water. $\Delta H_v(H_2O, 100°C) = 40.88$ kJ mol^{-1}; $\Delta H_f(H_2O, 0°C) = 6.01$ kJ mol^{-1}.

10.58 Glycol, CH_2OHCH_2OH, is commonly added to water as an antifreeze for car radiators. What mass of glycol must be added to 20 kg water to produce a solution freezing at −15°C? The molar enthalpy of fusion of water is 6.01 kJ mol^{-1}.

10.59 If 16.8 g sulfur crystals is added to 100.0 g naphthalene, $C_{10}H_8$, the freezing point of the naphthalene is lowered from 79.90 to 75.69°C. If the molar heat of fusion of naphthalene is 19.1 kJ mol^{-1}, calculate the molar mass of sulfur. See if you can explain why the answer is so large.

10.60 If 13.1 g benzoic acid, C_6H_5COOH, is dissolved in 100 g benzene, the boiling point of benzene is raised from 80.10 to 81.46°C. Find the apparent molecular weight of benzoic acid if the molar heat of vaporization of benzene is 30.7 kJ mol^{-1}. Is this the answer you would expect from the formula? How do you explain the difference?

10.61 A solution containing 5.31 g starch per cubic decimeter of solution is found to have an osmotic pressure of 12.7 mmHg (1.69 kPa) at 25°C. Find the molar mass of starch in this solution. (The structure of starch is considered in Chap. 20.)

10.62 Assume 1.21 g polystyrene is dissolved in toluene so as to make a solution of total volume 100 cm^3 at 25°C. When this solution is placed in an osmometer, similar to that shown in Fig. 10.20, the height *h* is found to be 21.32 mm of toluene. The density of toluene is 0.861 g cm^{-3}. Find the molar mass of the polystyrene sample.

chapter eleven

REACTIONS IN AQUEOUS SOLUTIONS

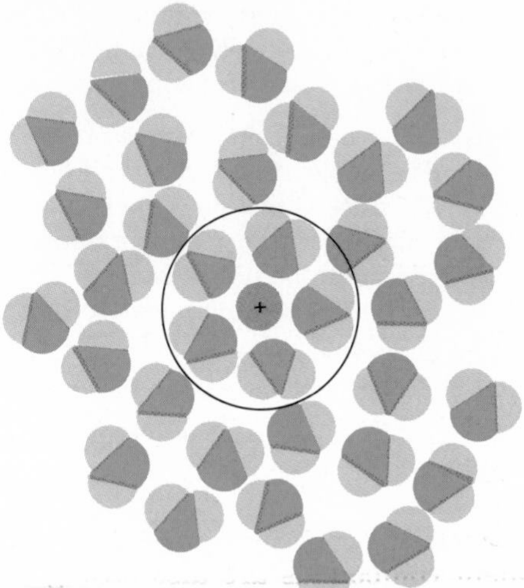

In the previous chapter we emphasized the importance of liquid solutions as a medium for chemical reactions. Water is by far the most important liquid solvent, partly because it is plentiful and partly because of its unique properties (Sec. 8.3). In your body, in other living systems, and in the outside environment a tremendous number of reactions take place in aqueous solutions. Consequently this chapter, as well as significant portions of many subsequent chapters, will be devoted to developing an understanding of reactions which occur in water. Since ionic compounds and polar covalent compounds constitute the main classes which are appreciably soluble in water, reactions in aqueous solutions usually involve these types of substances.

11.1 IONS IN SOLUTION

In Sec. 6.3 we pointed out that when an ionic compound dissolves in water, the positive and negative ions originally present in the crystal lattice persist in solution. Their ability to move nearly independently through the solution permits them to carry positive or negative electrical charges from one place to another. Hence the solution conducts an electrical current.

Electrolytes

Substances whose solutions conduct electricity are called **electrolytes.** All soluble ionic compounds are **strong electrolytes.** They conduct very

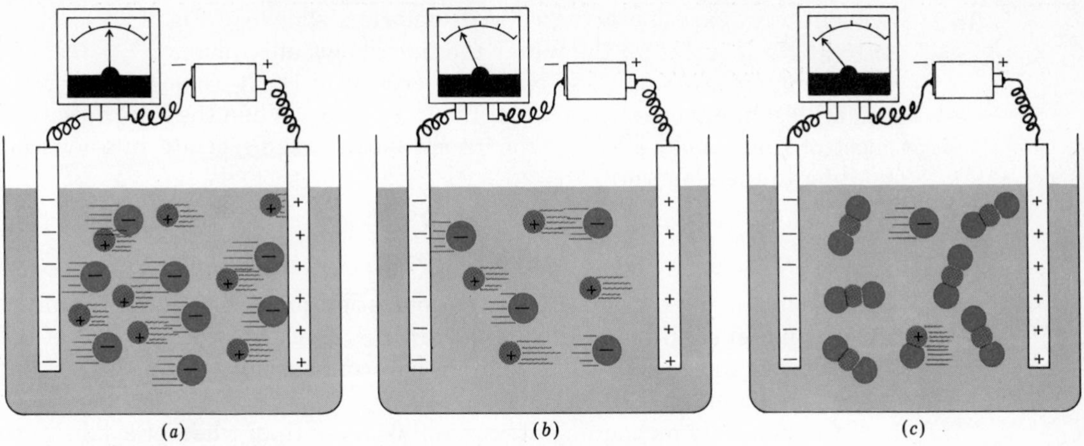

Figure 11.1 The conductivity of electrolyte solutions: (a) 0.1 M NaCl; (b) 0.05 M NaCl; (c) 0.1 M $HgCl_2$. An electrolyte solution conducts electricity because of the movement of ions in the solution. The larger the concentration of ions, the better the solution conducts. Weak electrolytes, such as $HgCl_2$, conduct badly because they produce very few ions when dissolved and exist mainly in the form of molecules.

well because they provide a plentiful supply of ions in solution. Some polar covalent compounds are also strong electrolytes. Common examples are HCl, HBr, HI, and H_2SO_4, all of which react with H_2O to form large concentrations of ions. A solution of HCl, for example, conducts even better than one of NaCl having the same concentration.

The effect of the concentration of ions on the electrical current which can flow through a solution is illustrated in Fig. 11.1. Part *a* of the figure shows what happens when a battery is connected through an electrical meter to two inert metal strips (**electrodes**) dipping in a 0.10-M NaCl solution. Each cubic decimeter of such a solution contains 0.10 mol NaCl (that is, 0.10 mol Na^+ and 0.10 mol Cl^-). An electrical current is carried through the solution both by the Na^+ ions moving toward the negative electrode and by the Cl^- ions which are attracted toward the positive electrode. The quantity of current is indicated by the dial on the meter.

Figure 11.1*b* shows that if we replace the 0.10-M NaCl solution with a 0.05-M NaCl solution, the meter reading falls to about one-half its former value. Halving the concentration of NaCl halves the number of ions between the electrodes, and half as many ions can only carry half as much electrical charge. Therefore the current is half as great. Because it responds in such a direct way to the concentration of ions, conductivity of electrical current is a useful tool in the study of solutions.

Conductivity measurements reveal that most covalent compounds, if they dissolve in water at all, retain their original molecular structures. Neutral molecules cannot carry electrical charges through the solution, and so no current flows. A substance whose aqueous solution conducts no better than water itself is called a **nonelectrolyte.** Some examples are oxygen, O_2, ethanol, C_2H_5OH, and sugar, $C_{12}H_{22}O_{11}$.

Some covalent substances behave as **weak electrolytes**—their solutions allow only a small current flow, but it is greater than that of the pure solvent. An example is mercury(II) chloride, shown in Fig. 11.1*c*. For a 0.10-*M* $HgCl_2$ solution the meter reading shows only about 0.2 percent as much current as for 0.10 *M* NaCl. A crystal of $HgCl_2$ consists of discrete molecules, like those shown for $HgBr_2$ in Plate 3. When the solid dissolves, most of these molecules remain intact, but a few dissociate into ions according to the equation

$$\underset{99.8\%}{HgCl_2} \rightleftharpoons \underset{0.2\%}{HgCl^+} + Cl^- \qquad (11.1)$$

(The double arrows indicate that the ionization proceeds only to a limited extent and an equilibrium state is attained.) Since only 0.2 percent of the $HgCl_2$ forms ions, the 0.10 *M* solution can conduct only about 0.2 percent as much current as 0.10 *M* NaCl.

Conductivity measurements can tell us more than whether a substance is a strong, a weak, or a nonelectrolyte. Consider, for instance, the data in Table 11.1 which shows the electrical current conducted through various aqueous solutions under identical conditions. At the rather low concentration of 0.001 *M*, the strong electrolyte solutions conduct between 2500 and 10 000 times as much current as pure H_2O and about 10 times as much as the weak electrolytes $HC_2H_3O_2$ (acetic acid) and NH_3 (ammonia). Closer examination of the data for strong electrolytes reveals that some compounds which contain H or OH groups [such as HCl or $Ba(OH)_2$] conduct unusually well. We will discuss this anomalous behavior later in this chapter. If these compounds are excluded, we find that 1:1 electrolytes (compounds which consist of equal numbers of +1 ions and −1 ions) usually conduct about half as much current as 2:2 electrolytes, 1:2 electrolytes, or 2:1 electrolytes.

TABLE 11.1 Electrical Current Conducted Through Various 0.001 *M* Aqueous Solutions at 18°C.*

Substance	Current/mA
1:1 Electrolytes (+1 ion: −1 ion)	
NaCl	1.065
NaI	1.069
KCl	1.273
KI	1.282
$AgNO_3$	1.131
HCl	3.77
HNO_3	3.75
NaOH	2.08
KOH	2.34
1:2 Electrolytes (+1 ion: −2 ion)	
Na_2SO_4	2.134
Na_2CO_3	2.24
K_2CO_3	2.660
Pure Water	
H_2O	3.69×10^{-4}

Substance	Current/mA
2:1 Electrolytes (+2 ion: −1 ion)	
$MgCl_2$	2.128
$CaCl_2$	2.239
$SrCl_2$	2.290
$BaCl_2$	2.312
$Ba(OH)_2$	4.14
2:2 Electrolytes (+2 ion: −2 ion)	
$MgSO_4$	2.00
$CaSO_4$	2.086
$CuSO_4$	1.97
$ZnSO_4$	1.97
Weak Electrolytes	
$HC_2H_3O_2$	0.41
NH_3	0.28

* All measurements refer to a cell in which the distance between the electrodes is 1.0 mm and the area of each electrode is 1.0 cm². A potential difference of 1.0 V is applied to produce the tabulated currents.

There is a simple reason for this behavior. Under similar conditions, most ions move through water at comparable speeds. This means that ions like Mg^{2+} or SO_4^{2-}, which are doubly charged, will carry twice as much current through the solution as will singly charged ions like Na^+ or Cl^-. Consequently, a 0.001 *M* solution of a 2:2 electrolyte like $MgSO_4$ will conduct about twice as well as a 0.001 *M* solution of a 1:1 electrolyte like NaCl. A similar argument applies to solutions of 1:2 and 2:1 electrolytes. A solution like 0.001 *M* Na_2SO_4 conducts about twice as well as 0.001 *M* NaCl partly because there are twice as many Na^+ ions available to move when a battery is connected, but also because SO_4^{2-} ions carry twice as much charge as Cl^- ions when moving at the same speed. These differences in conductivity between different types of strong electrolytes can sometimes be very useful in deciding what ions are actually present in a given electrolyte solution as the following example makes clear.

EXAMPLE 11.1 At 18°C a 0.001-*M* aqueous solution of potassium hydrogen carbonate, $KHCO_3$, conducts a current of 1.10 mA in a cell of the same design as that used to obtain the data in Table 11.1. What ions are present in solution?

Solution Referring to Table 6.2 which lists possible polyatomic ions, we can arrive at three possibilities for the ions from which $KHCO_3$ is made:

a) K^+ and H^+ and C^{4+} and three O^{2-}

b) K^+ and H^+ and CO_3^{2-}

c) K^+ and HCO_3^-

Since the current conducted by the solution falls in the range of 1.0 to 1.3 mA characteristic of 1:1 electrolytes, possibility *c* is the only reasonable choice.

A second, slightly more subtle, conclusion can be drawn from the data in Table 11.1. When an electrolyte dissolves, each type of ion makes an independent contribution to the current the solution conducts. This can be seen by comparing NaCl with KCl, and NaI with KI. In each case the compound containing K^+ conducts about 0.2 mA more than the one containing Na^+. If we apply this observation to Na_2CO_3 and K_2CO_3, each of which produces twice as many Na^+ or K^+ ions in solution, we find that the difference in current is also twice as great—about 0.4 mA. Thus conductivity measurements confirm our statement in Sec. 6.3 that each ion exhibits its own characteristic properties in aqueous solutions, independent of the presence of other ions. One such characteristic property is the quantity of electrical current that a given concentration of a certain type of ion can carry.

Precipitation Reactions

The independent behavior of each type of ion in solution was illustrated in Chap. 6 by means of precipitation reactions. **Precipitation** is a process in which a solute separates from a supersaturated solution. In a chemical laboratory it usually refers to a solid crystallizing from a liquid solution, but in weather reports it applies to liquid or solid water separating from supersaturated air.

A typical precipitation reaction occurs when an aqueous solution of barium chloride is mixed with one containing sodium sulfate. The equation

$$BaCl_2(aq) + Na_2SO_4(aq) \rightarrow BaSO_4(s) + 2NaCl(aq) \qquad (11.2a)$$

can be written to describe what happens, and such an equation is useful in making chemical calculations such as those in Chap. 3. However, Eq. (11.2*a*) does not really represent the microscopic particles (that is, the ions) present in the solution. Thus we might write

$$Ba^{2+}(aq) + 2Cl^-(aq) + 2Na^+(aq) + SO_4^{2-}(aq) \rightarrow BaSO_4(s) + 2Na^+(aq) + 2Cl^-(aq) \qquad (11.2b)$$

Equation (11.2*b*) is rather cumbersome and includes so many different ions that it may be confusing. In any case, we are often interested in the independent behavior of ions, not the specific compound from which they came. A precipitate of $BaSO_4(s)$ will form when *any* solution containing $Ba^{2+}(aq)$ is mixed with *any* solution containing $SO_4^{2-}(aq)$ (provided concentrations are not extremely small). This happens independently of the $Cl^-(aq)$ and $Na^+(aq)$ ions in Eq. (11.2*b*). These ions are called **spectator ions** because they do not participate in the reaction. When we want to emphasize the independent behavior of ions, a **net ionic equation** is written, omitting the spectator ions. For precipitation of $BaSO_4$ the net ionic equation is

$$Ba^{2+}(aq) + SO_4^{2-}(aq) \rightarrow BaSO_4(s) \qquad (11.2c)$$

EXAMPLE 11.2 When a solution of $AgNO_3$ is added to a solution of $CaCl_2$, insoluble AgCl precipitates. Write three equations to describe this process.

Solution Both $AgNO_3$ and $CaCl_2$ are soluble ionic compounds, and so they are strong electrolytes. The three equations are

$$2AgNO_3(aq) + CaCl_2(aq) \rightarrow 2AgCl(s) + Ca(NO_3)_2(aq) \qquad (11.3a)$$

$$2Ag^+(aq) + 2NO_3^-(aq) + Ca^{2+}(aq) + 2Cl^-(aq) \rightarrow 2AgCl(s) + Ca^{2+}(aq) + 2NO_3^-(aq) \qquad (11.3b)$$

$$Ag^+(aq) + Cl^-(aq) \rightarrow AgCl(s) \qquad (11.3c)$$

The occurrence or nonoccurrence of precipitates can be used to detect the presence or absence of various species in solution. $BaCl_2$ solution, for instance, is often used as a test for $SO_4^{2-}(aq)$ ion. There are several insoluble salts of Ba, but they all dissolve in dilute acid except for $BaSO_4$. Thus, if

$BaCl_2$ solution is added to an unknown solution which has previously been acidified, the occurrence of a white precipitate is proof of the presence of the SO_4^{2-} ion. $AgNO_3$ solution is often used in a similar way to test for halide ions. If $AgNO_3$ solution is added to an acidified unknown solution, a white precipitate indicates the presence of Cl^- ions, a cream-colored precipitate indicates the presence of Br^- ions, and a yellow precipitate indicates the presence of I^- ions. Further tests can then be made to see whether perhaps a mixture of these ions is present. When $AgNO_3$ is added to tap water, a white precipitate is almost always formed. The Cl^- ions in tap water usually come from the Cl_2 which is added to municipal water supplies to kill microorganisms.

Precipitates are also used for quantitative analysis of solutions, that is, to determine the amount of solute or the mass of solute in a given solution. For this purpose it is often convenient to use the first of the three types of equations described above. Then the rules of stoichiometry developed in Chap. 3 may be applied.

EXAMPLE 11.3 When a solution of 0.1 *M* $AgNO_3$ is added to 50.0 cm^3 of a $CaCl_2$ solution of unknown concentration, 2.073 g AgCl precipitates. Calculate the concentration of the unknown solution.

Solution We know the volume of the unknown solution, and so only the amount of solute is needed to calculate the concentration. This can be found using Eq. (11.3*a*) in Example 11.2. From the equation the stoichiometric ratio $S(CaCl_2/AgCl)$ may be obtained. A road map to the solution of the problem is

$$m_{AgCl} \xrightarrow{\mathcal{M}_{AgCl}} n_{AgCl} \xrightarrow{S(CaCl_2/AgCl)} n_{CaCl_2}$$

$$n_{CaCl_2} = 2.073 \text{ g AgCl} \times \frac{1 \text{ mol AgCl}}{143.32 \text{ g AgCl}} \times \frac{1 \text{ mol CaCl}_2}{2 \text{ mol AgCl}}$$

$$= 7.23 \times 10^{-3} \text{ mol CaCl}_2$$

$$c_{CaCl_2} = \frac{n_{CaCl_2}}{V_{soln}} = \frac{7.23 \times 10^{-3} \text{ mol CaCl}_2}{50.0 \text{ cm}^3} \times \frac{10^3 \text{ cm}^3}{1 \text{ dm}^3}$$

$$= 0.145 \text{ mol dm}^{-3}$$

Thus the concentration of the unknown solution is 0.145 *M*.

Because of the general utility of precipitates in chemistry, it is worth having at least a rough idea of which common classes of compounds can be precipitated from solution and which cannot. Table 11.2 gives a list of rules which enable us to predict the solubility of the most commonly encountered substances. Use of this table is illustrated in the following example.

TABLE 11.2 Solubility Rules.

Soluble in Water	Important Exceptions (insoluble)
All Na^+, K^+, and NH_4^+ salts	
All nitrates and perchlorates	
All acetates	CH_3COOAg
All sulfates	$BaSO_4$, $SrSO_4$, $PbSO_4$
All chlorides, bromides, and iodides	AgX, Hg_2X_2, PbX_2 (X = Cl, Br, or I)
Sparingly Soluble in Water	**Important Exceptions (soluble)**
All carbonates and phosphates	Group IA and NH_4^+ salts
All hydroxides	Group IA, Ba^{2+}, Sr^{2+}
All sulfides	Group IA and IIA
The following electrolytes are of only moderate solubility in water: $CaSO_4$, $Ca(OH)_2$, Ag_2SO_4, $KClO_4$. They will precipitate only if rather concentrated solutions are used.	

EXAMPLE 11.4 Write balanced net ionic equations to describe any reactions which occur when the following solutions are mixed:

a) 0.1 *M* Na_2SO_4 + 0.1 *M* NH_4I

b) 0.1 *M* K_2CO_3 + 0.1 *M* $SrCl_2$

c) 0.1 *M* $FeSO_4$ + 0.1 *M* $Ba(OH)_2$

Solution

a) If any precipitate forms, it will be either a combination of Na^+ ions and I^- ions, namely, NaI, or a combination of ammonium ions, NH_4^+, and sulfate ions, SO_4^{2-}, namely, $(NH_4)_2SO_4$. From Table 11.2 we find that NaI and $(NH_4)_2SO_4$ are both soluble. Thus no precipitation reaction will occur, and there is no equation to write.

b) Possible precipitates are KCl and $SrCO_3$. From Table 11.2 we find that $SrCO_3$ is insoluble. Accordingly we write the net ionic equation as

$$Sr^{2+}(aq) + CO_3^{2-}(aq) \rightarrow SrCO_3(s)$$

omitting the spectator ions K^+ and Cl^-.

c) Possible precipitates are $Fe(OH)_2$ and $BaSO_4$. Both are insoluble. The net ionic equation is thus

$$Fe^{2+}(aq) + SO_4^{2-}(aq) + Ba^{2+}(aq) + 2OH^-(aq) \rightarrow Fe(OH)_2(s) + BaSO_4(s)$$

Hydration of Ions

We have already stated several times that solubility in water is a characteristic property of many ionic compounds, and Table 11.2 provides further confirmation of this fact. We have also presented experimental evidence that ions in solution are nearly independent of one another. This

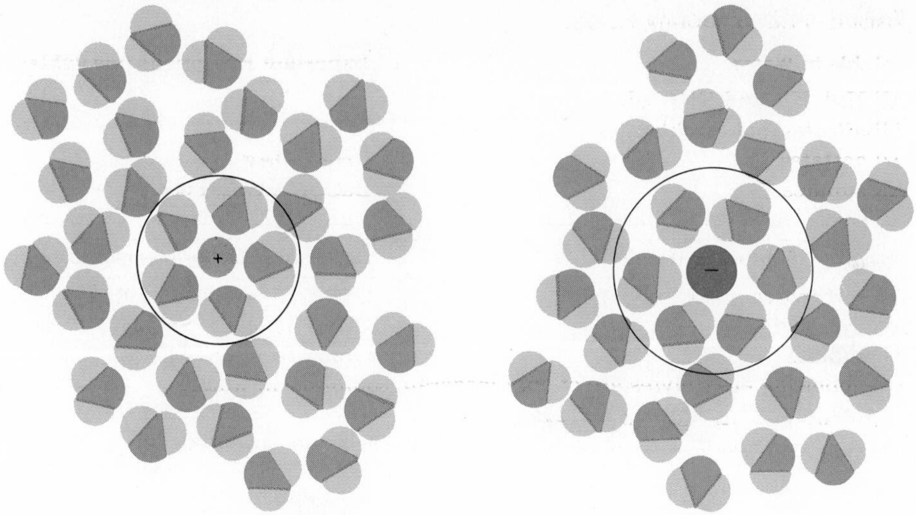

(a) (b)

Figure 11.2 The hydration of (*a*) a positive ion; (*b*) a negative ion. When ions are dissolved in water, they attract and hold several water dipoles around them as shown in the circular area in the center of each part of the diagram.

raises an important question, though, because we have also stated that attractive forces between oppositely charged ions in a crystal lattice are large. The high melting and boiling points of ionic compounds provide confirmation of the expected difficulty of separating oppositely charged ions. How, then, can ionic compounds dissolve at room temperature? Surely far more energy would be required for an ion to escape from the crystal lattice into solution than even the most energetic ions would possess.

The resolution of this apparent paradox lies in the interactions between ions and the molecules of water or other polar solvents. The negative (oxygen) side of a dipolar water molecule attracts and is attracted by any positive ion in solution. Because of this **ion-dipole force,** water molecules cluster around positive ions, as shown in Figure 11.2*a*. Similarly, the positive (hydrogen) ends of water molecules are attracted to negative ions. This process, in which either a positive or a negative ion attracts water molecules to its immediate vicinity, is called **hydration.**

When water molecules move closer to ions under the influence of their mutual attraction, there is a net lowering of the potential energy of the microscopic particles. This counteracts the increase in potential energy which occurs when ions are separated from a crystal lattice against their attractions for other ions. Thus the process of dissolving an ionic solid may be divided into the two hypothetical steps shown in Fig. 11.3. First, the crystalline salt is separated into gaseous ions. The heat energy absorbed when the ions are separated this way is called the **lattice enthalpy** (or sometimes the lattice energy). Next, the separate ions are placed in solution; that is, water molecules are permitted to surround the ions. The enthalpy change for this process is called the **hydration enthalpy.** Since there is a lowering

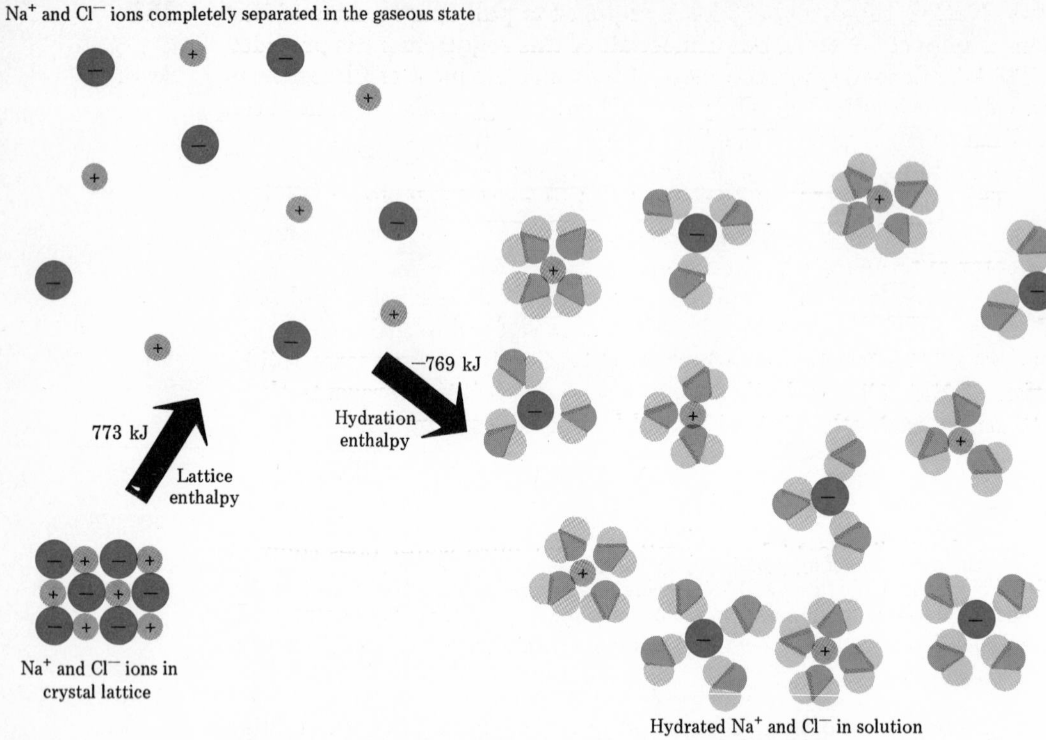

Figure 11.3 Enthalpy changes and solution. There is usually very little energy or enthalpy change when ionic solids like NaCl dissolve in H_2O since the energy needed to separate the ions from each other is not very different from the energy liberated when the ions become hydrated by attracting H_2O dipoles around them.

of the potential energy of the ions and water molecules, heat energy is given off and hydration enthalpies are invariably negative.

The heat energy absorbed when a solute dissolves (at a pressure of 1.00 atm) is called the **enthalpy of solution.** It can be calculated using Hess' law (Sec. 3.3), provided the lattice enthalpy and hydration enthalpy are known.

EXAMPLE 11.5 Using data given in Fig. 11.3, calculate the enthalpy of solution for NaCl(*s*).

Solution According to the figure, the lattice enthalpy is 773 kJ mol^{-1}. The hydration enthalpy is −769 kJ mol^{-1}. Thus we can write the thermochemical equations

$$\begin{array}{rl} NaCl(s) \rightarrow Na^+(g) + Cl^-(g) & \Delta H_\ell = 773 \text{ kJ mol}^{-1} \\ \underline{Na^+(g) + Cl^-(g) \rightarrow Na^+(aq) + Cl^-(aq)} & \Delta H_h = -769 \text{ kJ mol}^{-1} \\ NaCl(s) \rightarrow Na^+(aq) + Cl^-(aq) & \Delta H_s = \Delta H_\ell + \Delta H \end{array}$$

$$\Delta H_s = (773 - 769) \text{ kJ mol}^{-1} = +4 \text{ kJ mol}^{-1}$$

When NaCl(*s*) dissolves, 773 kJ is required to pull apart a mole of Na^+ ions from a mole of Cl^- ions, but almost all of this requirement is provided by the 769 kJ released when the mole of Na^+ and the mole of Cl^- becomes surrounded by water dipoles. Only 4 kJ of heat energy is absorbed from the surroundings when a mole of NaCl(*s*) dissolves. You can verify the small size of this enthalpy change by putting a few grains of salt on your moist tongue. The quantity of heat energy absorbed as the salt dissolves is so small that you will feel no cooling, even though your tongue is quite a sensitive indicator of temperature changes.

Few molecules are both small enough and sufficiently polar to cluster around positive and negative ions in solution as water does. Consequently water is one of the few liquids which readily dissolves many ionic solids. Hydration of Na^+, Cl^-, and other ions in aqueous solution prevents them from attracting each other into a crystal lattice and precipitating.

Hydrogen and Hydroxide Ions

If you refer to Table 11.1, you will see that pure water does conduct some electrical current, albeit much less than even the weak electrolytes listed there. This is because water itself is a very weak electrolyte. It ionizes to hydrogen ions and hydroxide ions to an extremely small extent:

$$H_2O(l) \rightleftharpoons H^+(aq) + OH^-(aq) \qquad (11.4a)$$

Careful measurements show that at 25°C the concentrations of $H^+(aq)$ and $OH^-(aq)$ are each 1.005×10^{-7} mol dm^{-3}. At higher temperatures more $H^+(aq)$ and $OH^-(aq)$ are produced while at lower temperatures less ionization of water occurs. Nevertheless, in pure water the concentration of $H^+(aq)$ always equals the concentration of $OH^-(aq)$. As we shall see in the next section, dissolving acids or bases in water can change the concentrations of both $H^+(aq)$ and $OH^-(aq)$, causing them to differ from one another. The special case of a solution in which these two concentrations remain equal is called a **neutral solution.**

A hydrogen ion, H^+, is a hydrogen atom which has lost its single electron; that is, a hydrogen ion is just a proton. Because a proton is only about one ten-thousandth as big as an average atom or ion, water dipoles can approach very close to a hydrogen ion in solution. Consequently the proton can exert a very strong attractive force on a lone pair of electrons in a water molecule—strong enough to form a coordinate covalent bond:

$$H^+ + :\underset{\textstyle H}{\overset{\cdot\cdot}{\underset{\cdot\cdot}{O}}}:H \longrightarrow H:\underset{\textstyle H}{\overset{\cdot\cdot}{\underset{\cdot\cdot}{O}}}:H^+$$

The H_3O^+ is formed in this way is called a **hydronium ion.** All three of its O—H bonds are exactly the same, and the ion has a pyramidal structure as predicted by VSEPR theory (Fig. 11.4*a*). To emphasize the fact that a proton cannot exist by itself in aqueous solution, Eq. (11.4*a*) is often rewritten as

$$2H_2O(l) \rightleftharpoons H_3O^+(aq) + OH^-(aq) \qquad (11.4b)$$

Like other ions in aqueous solution, both hydronium and hydroxide ions are hydrated. Moreover, hydrogen bonds are involved in attracting water

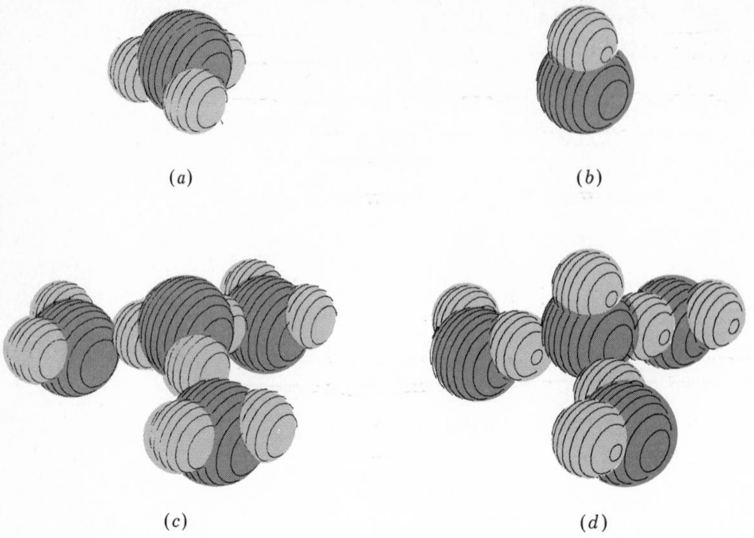

Figure 11.4 Hydrogen and hydroxide ions in aqueous solution. (*a*) Hydronium ion, H_3O^+; (*b*) hydroxide ion, OH^-; (*c*) hydrated hydronium ion, $H_9O_4^+$; (*d*) hydrated hydroxide ion, $H_7O_4^-$. (Computer-generated.) *(Copyright © 1976 by W. G. Davies and J. W. Moore.)*

molecules to hydronium and hydroxide ions. In both cases three water molecules appear to be rather tightly held, giving formulas $H_3O(H_2O)_3^+$ (or $H_9O_4^+$) and $OH(H_2O)_3^-$ (or $H_7O_4^-$). Possible structures for the hydrated hydronium and hydroxide ions are shown in Fig. 11.4.

Hydrogen bonding of hydronium and hydroxide ions to water molecules accounts rather nicely for the unusually large electrical currents observed for some electrolytes containing H and OH (Table 11.1). The case of the hydronium ion is illustrated in Fig. 11.5. When a hydronium ion collides with one end of a hydrogen-bonded chain of water molecules, a different hydronium ion can be released at the other end. Only a slight movement of six protons and a rearrangement of covalent and hydrogen bonds is needed. In effect a hydronium ion can almost instantaneously "jump" the length of several water molecules. It need not elbow its way through a crowd as other ions must. The same is true of aqueous hydroxide ions. Since both ions move faster, they can transfer more electrical charge per unit time, that is, more current.

11.2 ACID-BASE REACTIONS

Early in the history of chemistry it was noted that aqueous solutions of a number of substances behaved very similarly, although the substances themselves did not at first seem to be related. Solutions were classified as **acids** if they had the following characteristics: sour taste; ability to dissolve metals such as Zn, Mg, or Fe; ability to release a gas from solid limestone ($CaCO_3$) or other carbonates; ability to change the color of certain dyes (litmus paper turns red in the presence of acid). Another group of substances called **bases** or **alkalies** can also be distinguished by the properties of their aqueous solutions. These are bitter taste, slippery or soapy feel,

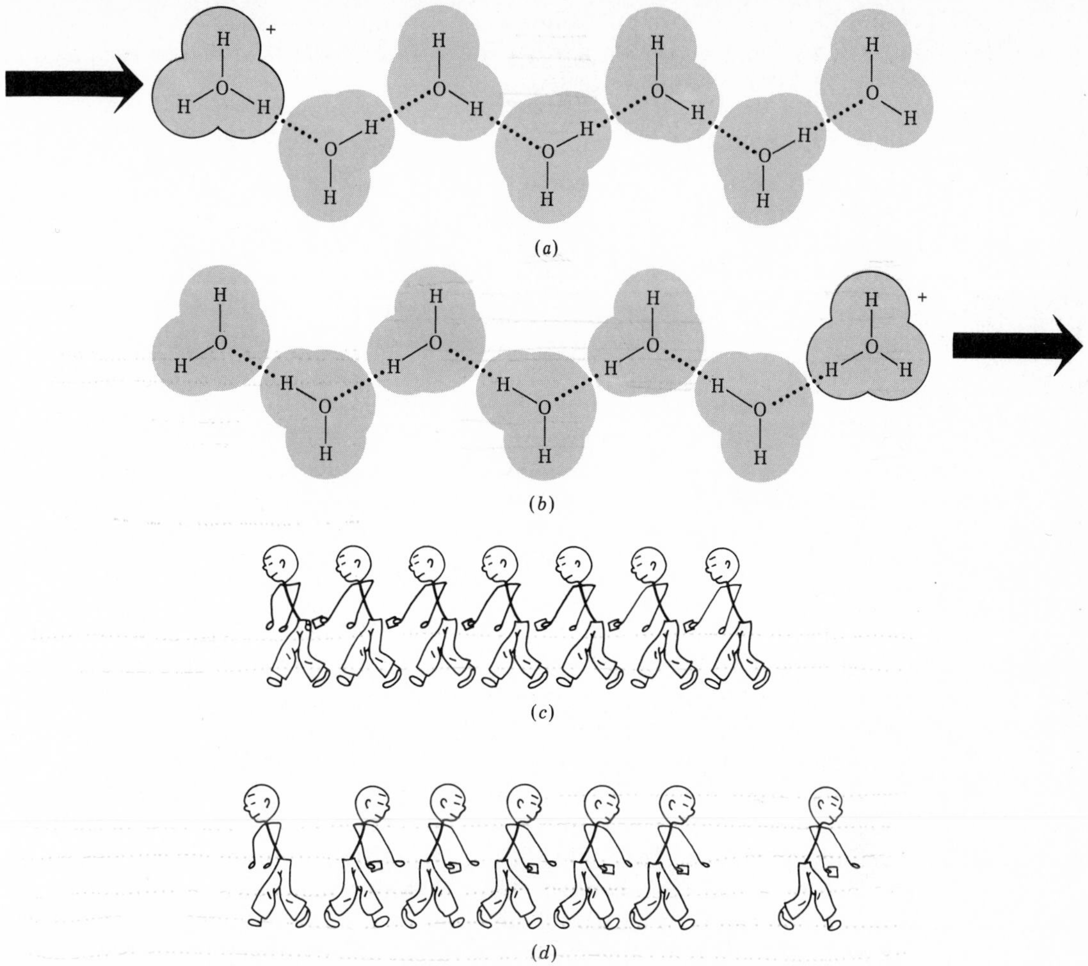

Figure 11.5 Rapid transfer of protons through aqueous solution. (*a*) Hydronium ion hydrogen bonded to a chain of water molecules. Minor rearrangement of the positions of H nuclei and hydrogen bonds results in situation (*b*) where the hydronium ion is at the right end of the chain. (*c*) An analogous situation would be a row of pickpockets in a crowd. The net result is transfer of a wallet from the person at the left of the row to the right (*d*). Only the ends of the chain have gained or lost wallets.

and the ability to change the color of certain dyes (litmus paper turns blue in base). Most important of all, acids and bases appear to be opposites. Any acid can counteract or **neutralize** the properties of a base. Similarly any base can neutralize an acid.

Acids

A typical example of an acid is hydrogen chloride gas, HCl(g). When it dissolves in water, HCl reacts to form hydronium ions and chloride ions:

$$HCl(g) + H_2O(l) \rightarrow H_3O^+(aq) + Cl^-(aq)$$

Thus the concentration of hydronium ions is increased above the value of 1.00×10^{-7} mol dm^{-3} characteristic of pure water. Other acids, such as nitric acid, HNO_3, behave in the same way:

$$HNO_3(l) + H_2O(l) \rightarrow H_3O^+(aq) + NO_3^-(aq)$$

Thus the characteristic properties of solutions of acids are due to the presence of hydronium ions (or hydrogen ions). Whenever the concentration of hydronium ions exceeds 1.00×10^{-7} mol dm^{-3}, an aqueous solution is said to be **acidic.** In 1884 a Swedish chemist, Svante Arrhenius (1859 to 1927), first recognized the importance of hydrogen ions. He defined an acid as any substance which increases the concentration of hydrogen (or hydronium) ions in aqueous solution.

The formation of a hydronium ion involves transfer of a proton from an acid molecule to a water molecule. This process is immediate—there are no free protons in solution which have left an acid molecule but have not yet attached themselves to a water molecule. To put it another way, a proton transfer is like a quarterback hand-off as opposed to a forward pass in football. The proton is always under the control and influence of one molecule or another. In the case of HCl we can indicate the transfer as

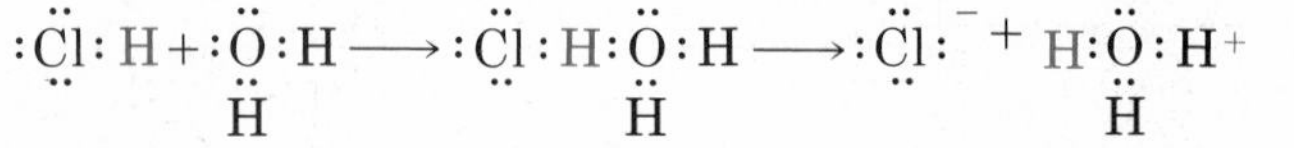

As the HCl molecule collides with an H_2O molecule, a hydrogen bond forms between the H and O atoms: Cl—H---OH_2. When it begins to bounce away from the H_2O molecule, the Cl atom loses control of the proton, leaving it attached to the O atom. The Cl atom retains control over both electrons which were in the H—Cl bond and thus ends up as a Cl^- ion. The H_2O molecule ends up with an extra proton, becoming H_3O^+.

EXAMPLE 11.6 Write balanced equations to describe the proton transfer which occurs when each of the following acids is dissolved in H_2O: (a) $HClO_4$ (perchloric acid) and (b) HBr (hydrogen bromide, or hydrobromic acid).

Solution Although a free proton is never actually produced in solution, it is often convenient to break the proton-transfer process into two hypothetical steps: (1) the loss of a proton by the acid, and (2) the gain of a proton by H_2O.

a) When $HClO_4$ loses a proton, H^+, the valence electron originally associated with the H atom is left behind, producing a negative ion, ClO_4^-. The proton can then be added to a water molecule in the second hypothetical step. Summing the two steps gives the overall proton transfer:

$$\begin{array}{rll} HClO_4 \rightarrow \cancel{H^+} + ClO_4^- & & \text{step 1} \\ \cancel{H^+} + H_2O \rightarrow H_3O^+ & & \text{step 2} \\ \hline HClO_4 + H_2O \rightarrow H_3O^+ + ClO_4^- & & \text{overall} \end{array}$$

b) Proceeding as in part *a*, we have

$$\begin{array}{ll} HBr \rightarrow H^+ + Br^- & \text{step 1} \\ H^+ + H_2O \rightarrow H_3O^+ & \text{step 2} \\ \hline HBr + H_2O \rightarrow H_3O^+ + Br^- & \text{overall} \end{array}$$

With practice, you should be able to write overall proton transfers without having to write steps 1 and 2 every time.

Another point to note about proton transfers is that in any equation involving ions, the sum of the ionic charges on the left side must equal the sum of the ionic charges on the right. For example, the last overall equation in Example 11.6 has HBr and H_2O on the left. Neither is an ion, and so the sum of the ionic charges is zero. On the right we have H_3O^+ and Br^-, which satisfy the rule because $+1 + (-1) = 0$. An equation which does not satisfy this rule of charge balance will involve creation or destruction of one or more electrons and therefore cannot be valid. For example, the equation

$$2HBr \rightarrow 2H^+ + Br_2$$

cannot describe a valid proton transfer because the charges sum to zero on the left but +2 (because of $2H^+$ ions) on the right. Careful examination reveals that there are 16 valence electrons (two octets in 2HBr) on the left but only 14 valence electrons (none in $2H^+$ and 14 in $:\ddot{\underset{..}{Br}}:\ddot{\underset{..}{Br}}:$) on the right. Two electrons have been destroyed—something which does not happen. Therefore the equation must be incorrect.

Because hydronium ions can be formed by transferring protons to water molecules, it is convenient when dealing with aqueous solutions to define an *acid* as a *proton donor.* This definition was first proposed in 1923 by the Danish chemist Johannes Brönsted (1879 to 1947) and the English chemist Thomas Martin Lowry (1874 to 1936). It is called the **Brönsted-Lowry definition** of an acid, and we will use it for the remainder of this book. The Brönsted-Lowry definition has certain advantages over Arrhenius' idea of an acid as a producer of $H_3O^+(aq)$. This is especially true when acid strengths are compared, a subject we shall come to a bit later. Consequently, when we speak of an acid, we will mean a proton donor, unless some qualification, such as Arrhenius acid, is used.

Bases

Bases have characteristics opposite those of acids, and bases can be neutralized by acids. Therefore it is logical, in the Brönsted-Lowry scheme, to define a *base* as a *proton acceptor,* that is, a species which can incorporate an extra proton into its molecular or ionic structure. For example, when barium oxide, BaO, dissolves in water, oxide ions accept protons from water molecules according to the equation

$$BaO + H_2O \rightarrow Ba^{2+} + 2OH^-$$

i.e.,

$$O^{2-} + H_2O \rightarrow OH^- + OH^-$$

The added proton transforms the oxide ion, O^{2-}, into a hydroxide ion. Removal of a proton from a water molecule leaves behind a hydroxide ion also, accounting for the $2OH^-$ on the right side of the equation. Since it can accept protons, barium oxide (or, more specifically, oxide ion) serves as a base.

When a base is added to water, its molecules or ions accept protons from water molecules, producing hydroxide ions. Thus the general properties of solutions of bases are due to the presence of hydroxide ions [$OH^-(aq)$]. Any aqueous solution which contains a concentration of hydroxide ions greater than the 1.00×10^{-7} mol dm^{-3} characteristic of pure water is said to be **basic.**

Unlike the hydronium ion, which forms very few solid compounds, hydroxide ions are often present in solid crystal lattices. Therefore it is possible to raise the hydroxide-ion concentration above 1.00×10^{-7} mol dm^{-3} by dissolving compounds such as NaOH, KOH, or $Ba(OH)_2$. Hydroxide ions can accept protons from water molecules, but of course such a proton transfer has no net effect because the hydroxide ion itself becomes a water molecule:

$$HOH + OH^- \rightarrow HO^- + HOH \qquad (11.5)$$

Nevertheless, the hydroxide ion fits the Brönsted-Lowry definition of a base as a proton acceptor.

EXAMPLE 11.7 Write a balanced equation to describe the proton transfer which occurs when the base sodium hydride, NaH, is added to water.

Solution NaH consists of Na^+ and H^- ions. Since positive ions repel protons, the H^- ion is the only likely base. Again it may be useful to use two hypothetical steps: (1) donation of a proton by an H_2O molecule, and (2) acceptance of a proton by the base. As in Example 11.6, we can then sum the steps

$$H_2O \rightarrow \cancel{H^+} + OH^- \qquad \text{step 1}$$
$$H^- + \cancel{H^+} \rightarrow H_2 \qquad \text{step 2}$$
$$H^- + H_2O \rightarrow H_2 + OH^- \qquad \text{overall}$$

Note that adding a proton to H^- balances the excess electron of that ion, producing a neutral H_2 molecule. Note also that charges balance in all equations.

11.3 ACID AND BASE STRENGTH

Strong Acids and Bases

So far all our examples have involved strong acids and strong bases. **Strong acids,** like HCl or HNO_3, are such good proton donors that none of their own molecules can remain in aqueous solution. All HCl molecules, for example, transfer their protons to H_2O molecules, and so the solution contains only $H_3O^+(aq)$ and $Cl^-(aq)$ ions. Similarly, the ions of **strong bases,** like BaO or NaH, are such good proton acceptors that they cannot remain in aqueous solution. All O^{2-} ions, for example, are converted to OH^- ions by accepting protons from H_2O molecules, and the H_2O molecules are also converted to OH^-. Therefore a solution of BaO contains only $Ba^{2+}(aq)$ and $OH^-(aq)$ ions.

Table 11.3 lists molecules and ions which act as strong acids and bases in aqueous solution. In addition to those which react completely with H_2O to form H_3O^+ and OH^-, any compound which itself contains these ions will serve as a strong acid or base. Note that the strength of an acid refers only to its ability to donate protons to H_2O molecules and the strength of a base to its ability to accept protons from H_2O molecules. The acidity or basicity of a solution, on the other hand, depends on the concentration as well as the strength of the dissolved acid or base.

As a general rule, strong proton donors are molecules in which a hydrogen is attached to a rather electronegative atom, such as oxygen or a halogen. Considerable electron density is shifted away from hydrogen in such a molecule, making it possible for hydrogen ions to depart without taking along any electrons. The strong acids in Table 11.3 fit this rule nicely. They are either hydrogen halides (HCl, HBr, HI) or oxyacids (whose general formula is H_nXO_m). The Lewis structures

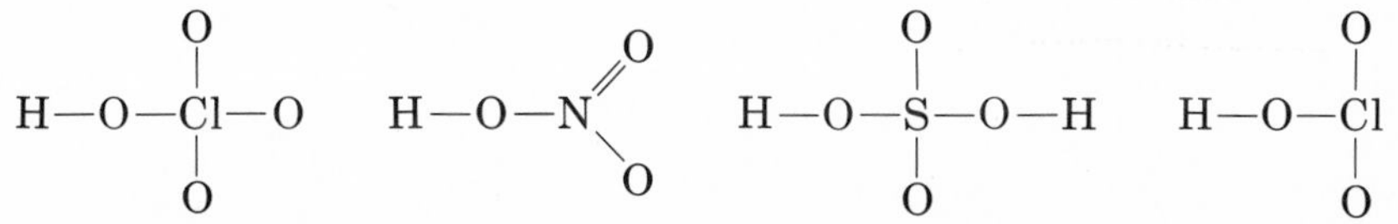

Perchloric acid **Nitric acid** **Sulfuric acid** **Chloric acid**

indicate a proton bonded to oxygen in each of the oxyacids, hence their general name. Note that for a strong oxyacid the number of oxygens is always larger by two or more than the number of hydrogens. That is, in the general formula H_nXO_m, $m \geq n + 2$.

The strength of a base depends on its ability to attract and hold a proton. Therefore bases often have negative charges, and they invariably have at least one lone pair of electrons which can form a coordinate covalent bond to a proton. The strong bases in Table 11.3 might be thought of as being derived from neutral molecules by successive removal of protons. For example, OH^- can be obtained by removing H^+ from H_2O, and O^{2-} can be obtained by removing H^+ from OH^-. When the strong bases are considered this way, it is not surprising that they are good proton acceptors.

TABLE 11.3 Species Which Are Strong Acids and Bases in Aqueous Solution.

Strong Acids	Strong Bases
H_3O^+ (Only a few compounds like H_3OCl and H_3ONO_3 are known to contain hydronium ions.)	OH^- [Only LiOH, NaOH, KOH, RbOH, CsOH, $Ca(OH)_2$, $Sr(OH)_2$, and $Ba(OH)_2$ are sufficiently soluble to produce large concentrations of $OH^-(aq)$.]
HCl, HBr, HI	O^{2-} (Li_2O, Na_2O, K_2O, Rb_2O, Cs_2O, CaO, SrO, and BaO are soluble.)
$HClO_4$	
HNO_3, H_2SO_4, $HClO_3$	H^-, S^{2-}, NH_2^-, N^{3-}, P^{3-}

Weak Acids

Not all molecules which contain hydrogen are capable of donating protons. For example, methane (CH_4) and other hydrocarbons show no acidic properties at all. Carbon is not highly electronegative, and so electron density is fairly evenly shared in a C—H bond, and the hydrogen atom is unlikely to depart without at least one electron. Even when it is bonded to highly electronegative atoms like oxygen or fluorine, a hydrogen atom is not *always* strongly acidic.

EXAMPLE 11.8 Acetic acid has the projection formula

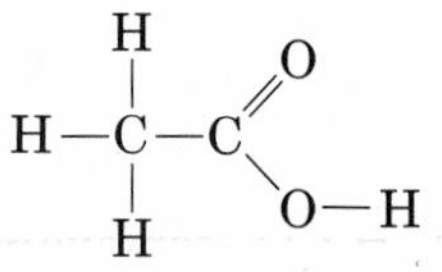

Write an equation for transfer of a proton from acetic acid to water.

Solution By the electronegativity rule given above, only the hydrogen attached to oxygen should be acidic. The equation is

$$CH_3COOH + H_2O \rightleftharpoons H_3O^+ + CH_3COO^- \qquad (11.6)$$

Note: To emphasize that only one hydrogen atom is acidic, the formula for acetic acid is often written $HC_2H_3O_2$. You may also find the formulas HAc or HOAc for acetic acid, where Ac^- or OAc^- represents the acetate ion, CH_3COO^-.

The equation for the proton transfer between acetic acid and water is written with a double arrow because it occurs to only a limited extent. Like $HgCl_2$ [Eq. (11.1)], acetic acid is a weak electrolyte. According to Table 11.1, 0.001 *M* $HC_2H_3O_2$ conducts slightly more than one-tenth as much current as the same concentration of the strong acids HCl or HNO_3. Therefore we can conclude that at a given instant only a little over 10 percent of the acetic acid molecules have donated protons to water molecules according to Eq. (11.6). Nearly 90 percent are in molecular form as CH_3COOH (or $HC_2H_3O_2$) and make no contribution to the current. Because acetic acid is not a strong enough proton donor to be entirely converted to hydronium ions in aqueous solution, it is called a **weak acid**. A given concentration of a weak acid produces fewer hydronium ions per unit volume and therefore less acidity than the same concentration of a strong acid.

There are a large number of weak acids, but fortunately they fall into a few well-defined categories.

Carboxylic acids These compounds have the general formula RCOOH and were described in Sec. 8.4. All react with water in the same way as acetic acid [Eq. (11.6)].

Weak oxyacids These have the same general formula H_nXO_m as strong oxyacids, but the number of hydrogens is equal to or one less than the number of oxygens. For a weak oxyacid, in other words, $m \leqslant n + 1$. Some examples are

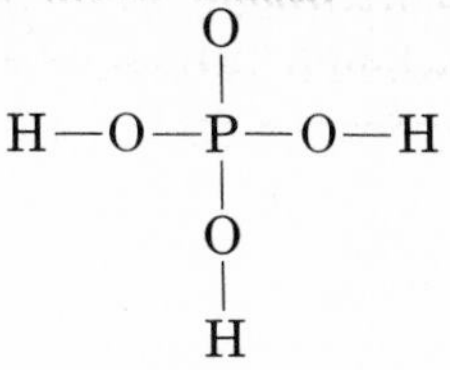

Phosphoric acid

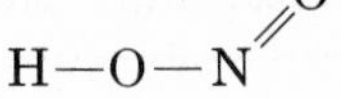

Nitrous acid

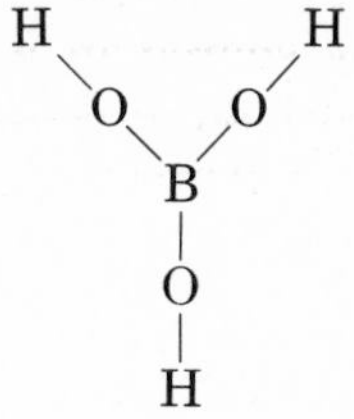

Boric acid

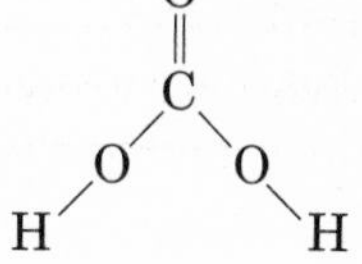

Carbonic acid

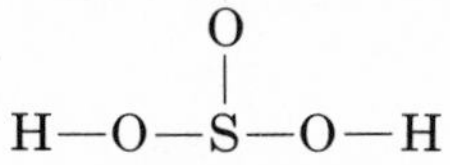

Sulfurous acid

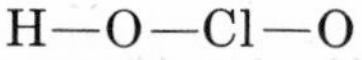

Chlorous acid

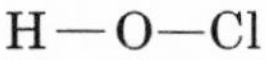

Hypochlorous acid

Some of the weak oxyacids, H_2CO_3 for example, are very unstable and cannot be separated in pure form from aqueous solution.

Other molecules containing acidic hydrogen atoms Hydrogen fluoride (HF) has a very strong bond and does not donate its proton as readily as other hydrogen halides. Other molecules in this category are hydrogen sulfide (H_2S) and hydrogen cyanide (HCN). In the latter case, even though H is bonded to C, the electronegative N atom pulls some electron density away, and the HCN molecule is a very weak proton donor.

Hydrated cations Cations, especially those of charge +3 or more or of the transition metals, are surrounded closely by four to six water molecules in aqueous solution. An example is $Cr(H_2O)_6^{3+}$, shown in Fig. 11.6. The positive charge of the metal ion pulls electron density away from the surrounding water molecules, weakening the hold of the oxygen atoms for the hydrogen atoms. The latter can consequently be more easily donated as protons:

$$Cr(H_2O)_6^{3+} + H_2O \rightleftharpoons Cr(H_2O)_5(OH)^{2+} + H_3O^+$$

Figure 11.6 Space-filling and ball-and-stick diagrams of the hydrated chromium ion, $Cr(H_2O)_6^{3+}$. The Cr atom is linked in an identical way to the six oxygen atoms arranged octahedrally around it. (Computer-generated.) (*Copyright © 1976 by W. G. Davies and J. W. Moore.*)

Ions having acidic protons Certain other ions can donate protons. One example is the ammonium ion, NH_4^+:

$$NH_4^+ + H_2O \rightleftharpoons NH_3 + H_3O^+$$

Anions formed when some acids donate protons can lose yet another H^+. An example of this is the hydrogen sulfate ion formed when sulfuric acid donates a proton:

$$H_2SO_4 + H_2O \rightarrow H_3O^+ + HSO_4^-$$

$$HSO_4^- + H_2O \rightleftharpoons H_3O^+ + SO_4^{2-}$$

Although sulfuric acid is strong, the negative charge on the hydrogen sulfate ion holds the proton tighter, and so the ion is a considerably weaker acid. Acids such as H_2SO_4, H_2S, H_2SO_3, and H_2CO_3 are called **diprotic** because they can donate two protons. Phosphoric acid, H_3PO_4, is **triprotic**—it can donate three protons.

Weak Bases

By analogy with weak acids, **weak bases** are not strong enough proton acceptors to react completely with water. A typical example is ammonia, which reacts only to a limited extent:

$$NH_3 + H_2O \rightleftharpoons NH_4^+ + OH^-$$

As in the case of acetic acid, the current conducted by 0.001 M NH_3 (Table 11.1) is slightly above 10 percent that of 0.001 M NaOH or KOH, indicating that somewhat more than one-tenth of the NH_3 molecules have been converted to NH_4^+ and OH^- ions.

Weak bases fall into two main categories.

Ammonia and amines Amines may be derived from ammonia by replacing one or more hydrogens with alkyl groups (Sec. 8.4). They react with water in the same way as ammonia. Trimethyl amine behaves as follows:

$$(CH_3)_3N + H_2O \rightleftharpoons (CH_3)_3NH^+ + OH^-$$

Anions of weak acids The fact that the molecules of a weak acid are not entirely converted into hydronium ions and anions implies that those anions must have considerable affinity for protons. For example, the anion of acetic acid, acetate ion, can accept protons as follows:

$$C_2H_3O_2^- + H_2O \rightleftharpoons HC_2H_3O_2 + OH^-$$

Thus when a compound like sodium acetate, $NaC_2H_3O_2$, dissolves, some hydroxide ions are produced and the solution becomes slightly basic. (Sodium ions do not react with water at all, and so they have no effect on acidity or basicity.) Other examples of weak bases which are anions of weak acids are CN^-, CO_3^{2-}, and PO_4^{3-}.

Amphiprotic Species

Molecules or ions which can either donate or accept a proton, depending on their circumstances, are called **amphiprotic species.** The most important amphiprotic species is water itself. When an acid donates a proton to water, the water molecule is a proton acceptor, and hence a base. Conversely, when a base reacts with water, a water molecule donates a proton, and hence acts as an acid.

Another important group of amphiprotic species is the amino acids, which we mentioned in Sec. 8.4. Each amino acid molecule contains an acidic carboxyl group and a basic amino group. In fact the amino acids usually exist in *zwitterion* (German for "double ion") form, where the proton has transferred from the carboxyl to the amino group. In the case of glycine, for example, the zwitterion is

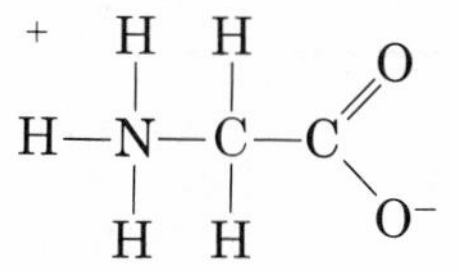

The zwitterion can donate one of the protons from the N, just as an NH_4^+ ion can donate a proton. On the other hand, its COO^- end can accept a proton, just as a CH_3COO^- ion can.

Other common amphiprotic species are HCO_3^-, $H_2PO_4^-$, HPO_4^{2-}, and other anions derived from diprotic or triprotic acids.

EXAMPLE 11.9 Write equations to show the amphiprotic behavior of (a) $H_2PO_4^-$; (b) H_2O.

Solution To make an amphiprotic species behave as an acid requires a fairly good proton acceptor. Conversely, to make it behave as a base requires a proton donor.

a) Acid: $H_2PO_4^- + OH^- \rightarrow HPO_4^- + H_2O$

Base: $H_2PO_4^- + H_3O^+ \rightarrow H_3PO_4 + H_2O$

b) Acid: $H_2O + S^{2-} \rightarrow OH^- + HS^-$

Base: $H_2O + H_2SO_4 \rightarrow H_3O^+ + HSO_4^-$

Conjugate Acid-Base Pairs

In Examples 11.6 and 11.7 we broke proton-transfer processes into two hypothetical steps: (1) donation of a proton by an acid, and (2) acceptance of a proton by a base. (Water served as the base in Example 11.6 and as the acid in Example 11.7.) The hypothetical steps were useful because they made it easy to see what species was left after an acid donated a proton and what species was formed when a base accepted a proton. We shall use hypothetical steps or **half-equations** again in this section, but you should bear in mind that free protons never actually exist in aqueous solution.

Suppose we first consider one of the weak acids mentioned earlier, the ammonium ion. When it donates a proton to *any* other species, we can write the half-equation

$$NH_4^+ \rightarrow H^+ + NH_3$$

But NH_3 is one of the compounds we already know as a weak base. In other words, when it donates a proton, the weak acid NH_4^+ is transformed into a weak base NH_3. Another example, this time starting with a weak base, is provided by fluoride ion:

$$F^- + H^+ \rightarrow HF$$

The proton converts a weak base into a weak acid.

The situation just described for NH_4^+ and NH_3 or for F^- and HF applies to *all* acids and bases. Whenever an acid donates a proton, the acid changes into a base, and whenever a base accepts a proton, an acid is formed. An acid and a base which differ only by the presence or absence of a proton are called a **conjugate acid-base pair.** Thus NH_3 is called the conjugate base of NH_4^+, and NH_4^+ is the conjugate acid of NH_3. Similarly, HF is the conjugate acid of F^-, and F^- the conjugate base of HF.

EXAMPLE 11.10 What is the conjugate acid or the conjugate base of (a) HCl; (b) CH_3NH_2; (c) OH^-; (d) HCO_3^-.

Solution

a) HCl is a strong acid. When it donates a proton, a Cl^- ion is produced, and so Cl^- is the conjugate base.

b) CH_3NH_2 is an amine and therefore a weak base. Adding a proton gives $CH_3NH_3^+$, its conjugate acid.

c) Adding a proton to the strong base OH^- gives H_2O, its conjugate acid.

d) Hydrogen carbonate ion, HCO_3^-, is derived from a diprotic acid and is amphiprotic. Its conjugate acid is H_2CO_3, and its conjugate base is CO_3^{2-}.

The use of conjugate acid-base pairs allows us to make a very simple statement about relative strengths of acids and bases. *The stronger an acid, the weaker its conjugate base,* and, conversely, *the stronger a base, the weaker its conjugate acid.* A strong acid like HCl donates its proton so read-

TABLE 11.4 Some Important Conjugate Acid-Base Pairs Arranged in Order of Increasing Strength of the Base.

ACID STRENGTH (↑)	Acid				Base	BASE STRENGTH (↓)
Strong	HCl	→	H^+	+	Cl^-	Negligible
Strong	H_2SO_4	→	H^+	+	HSO_4^-	Negligible
Strong	HNO_3	→	H^+	+	NO_3^-	Negligible
Strong	H_3O^+	→	H^+	+	H_2O	Very Weak
Medium	HSO_4^-	→	H^+	+	SO_4^{2-}	Very Weak
Medium	H_3PO_4	→	H^+	+	$H_2PO_4^-$	Very Weak
Medium	HF	→	H^+	+	F^-	Very Weak
Weak	CH_3COOH	→	H^+	+	CH_3COO^-	Weak
Weak	H_2CO_3	→	H^+	+	HCO_3^-	Weak
Weak	H_2S	→	H^+	+	HS^-	Weak
Weak	$H_2PO_4^-$	→	H^+	+	HPO_4^{2-}	Weak
Weak	NH_4^+	→	H^+	+	NH_3	Weak
Very Weak	HCO_3^-	→	H^+	+	CO_3^{2-}	Medium
Very Weak	HPO_4^{2-}	→	H^+	+	PO_4^{3-}	Medium
Very Weak	H_2O	→	H^+	+	OH^-	Strong
Negligible	HS^-	→	H^+	+	S^{2-}	Strong
Negligible	H_2	→	H^+	+	H^-	Strong

ily that there is essentially no tendency for the conjugate base Cl^- to reaccept a proton. Consequently, Cl^- is a very weak base. A strong base like the H^- ion accepts a proton and holds it so firmly that there is no tendency for the conjugate acid H_2 to donate a proton. Hence, H_2 is a very weak acid.

Table 11.4 gives a list of some of the more important conjugate acid-base pairs in order of increasing strength of the base. This table enables us to see how readily a given acid will react with a given base. The reactions with most tendency to occur are between the strong acids in the top left-hand corner of the table and the strong bases in the bottom right-hand corner. If a line is drawn from acid to base for such a reaction, it will have a *downhill* slope. By contrast, reactions with little or no tendency to occur (between the weak acids at the bottom left and the weak bases at the top right) correspond to a line from acid to base with an *uphill* slope. When the slope of the line is not far from horizontal, the conjugate pairs are not very different in strength, and the reaction goes only part way to completion. Thus, for example, if the acid HF is compared with the base CH_3COO^-, we expect the reaction to go part way to completion since the line is barely downhill.

EXAMPLE 11.11 Write a balanced equation to describe the reaction which occurs when a solution of potassium hydrogen sulfate, $KHSO_4$, is mixed with a solution of sodium bicarbonate, $NaHCO_3$.

Solution The Na^+ ions and K^+ ions have no acid-base properties and function purely as spectator ions. Therefore any reaction which occurs must be between the hydrogen sulfate ion, HSO_4^-, and the hydrogen carbonate ion, HCO_3^-. Both HSO_4^- and HCO_3^- are amphiprotic, and either could act as an acid or as a base. The reaction between them is thus either

$$HCO_3^- + HSO_4^- \rightarrow CO_3^{2-} + H_2SO_4$$

or

$$HSO_4^- + HCO_3^- \rightarrow SO_4^{2-} + H_2CO_3$$

Table 11.4 tells us immediately that the second reaction is the correct one. A line drawn from HSO_4^- as an acid to HCO_3^- as a base is downhill. The first reaction cannot possibly occur to any extent since HCO_3^- is a very weak acid and HSO_4^- is a base whose strength is negligible.

EXAMPLE 11.12 What reactions will occur when an excess of acetic acid is added to a solution of potassium phosphate, K_3PO_4?

Solution The line joining CH_3COOH to PO_4^{3-} in Table 11.4 is downhill, and so the reaction

$$CH_3COOH + PO_4^{3-} \rightarrow CH_3COO^- + HPO_4^{2-}$$

should occur. There is a further possibility because HPO_4^{2-} is itself a base and might accept a second proton. The line from CH_3COOH to HPO_4^{2-} is also downhill, but just barely, and so the reaction

$$CH_3COOH + HPO_4^{2-} \rightleftharpoons CH_3COO^- + H_2PO_4^-$$

can occur, but it does not go to completion. Hence double arrows are used. Although $H_2PO_4^-$ is a base and might be protonated to yield phosphoric acid, H_3PO_4, a line drawn from CH_3COOH to $H_2PO_4^-$ is *uphill*, and so this does not happen.

11.4 LEWIS ACIDS AND BASES

We stated earlier that many oxyacids are rather unstable and cannot be isolated in pure form. An example is carbonic acid, H_2CO_3, which decomposes to water and carbon dioxide:

$$H_2CO_3(aq) \rightleftharpoons H_2O(l) + CO_2(g)$$

Since it can be made by removing H_2O from H_2CO_3, CO_2 is called the **acid anhydride** of H_2CO_3. (The term anhydride is derived from anhydrous, meaning "not containing water.") Acid anhydrides are usually oxides of nonmetallic elements. Some common examples and their corresponding oxyacids are: SO_2—H_2SO_3; SO_3—H_2SO_4; P_4O_{10}—H_3PO_4; N_2O_5—HNO_3. Any of these anhydrides increases the hydronium-ion concentration when dissolved in water; for example,

$$P_4O_{10}(s) + 6H_2O(l) \rightarrow 4H_3PO_4(aq)$$

$$H_3PO_4(aq) + H_2O(l) \rightleftharpoons H_3O^+(aq) + H_2PO_4^-(aq)$$

In the Arrhenius sense, then, acid anhydrides are acids, but according to the Brönsted-Lowry definition, they are not acids because they contain no hydrogen.

In 1923, at the same time that the Brönsted-Lowry definition was proposed, G. N. Lewis suggested another definition which includes the acid anhydrides and a number of other substances as acids. According to the **Lewis definition,** an acid is any species which can accept a lone pair of electrons, and a base is any species which can donate a lone pair of electrons. An acid-base reaction in the Lewis sense involves formation of a coordinate covalent bond.

The Lewis definition has little effect on the types of molecules we expect to be basic. All the Brönsted-Lowry bases, for example, NH_3, O^{2-}, H^-, contain at least one lone pair. Lewis' idea does expand the number of acids, though. The proton is not the only species which can form a coordinate covalent bond with a lone pair. Cations of the transition metals, which are strongly hydrated, do the same thing:

$$\underset{\textbf{Lewis acid}}{Cr^{3+}} + \underset{\textbf{Lewis base}}{6H_2O} \longrightarrow Cr(H_2O)_6^{3+}$$

$$\underset{\textbf{Lewis acid}}{Cu^{2+}} + \underset{\textbf{Lewis base}}{4NH_3} \longrightarrow Cu(NH_3)_4^{2+}$$

So can electron deficient compounds such as boron trifluoride:

$$\underset{\textbf{Lewis acid}}{BF_3} + \underset{\textbf{Lewis base}}{:NH_3} \longrightarrow F_3B:NH_3$$

EXAMPLE 11.13 Identify the Lewis acids and bases in the following list. Write an equation for the combination of each acid with the Lewis base H_2O. (a) $BeCl_2(g)$; (b) CH_3OH; (c) SO_2; (d) CF_4.

Solution

a) The Lewis diagram

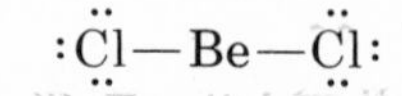

shows that Be is electron deficient. Therefore $BeCl_2(g)$ is a Lewis acid. Because of the lone pairs on the Cl atoms, $BeCl_2$ can also act as a Lewis base, but Cl is rather electronegative and reluctant to donate electrons, so the Lewis base strength of $BeCl_2$ is less than the Lewis acid strength.

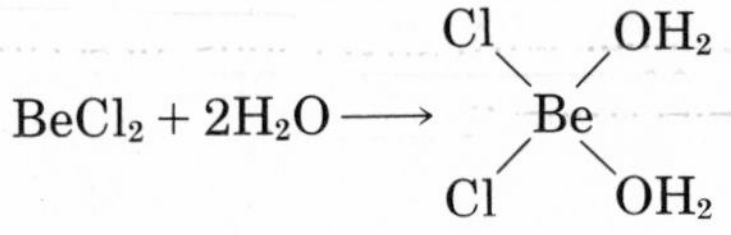

b) There are lone pairs on O in CH_3OH, and so it can serve as a Lewis base.

c) The S atom in SO_2 can accept an extra pair of electrons, and so SO_2 is a Lewis acid. The O atoms have lone pairs but are only weakly basic for the same reason as the Cl atoms in part (a).

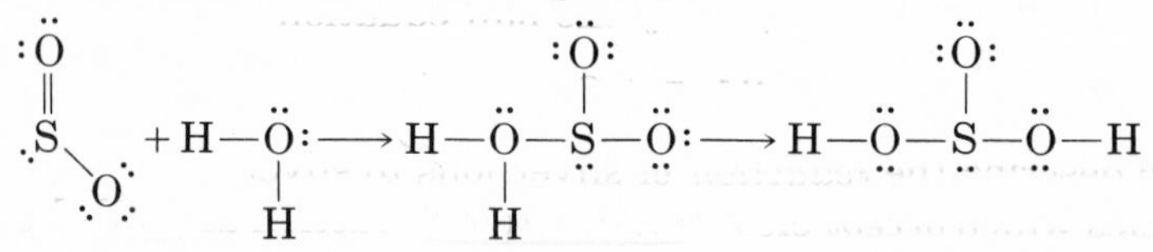

d) Although there are lone pairs on the F atoms, the high electronegativity of F prevents them from being donated to form coordinate covalent bonds. Consequently CF_4 has essentially no Lewis-base character.

Many Lewis acid-base reactions occur in media other than aqueous solution. The Brönsted-Lowry theory accounts for almost all aqueous acid-base chemistry. Therefore the Brönsted-Lowry concept is most often intended when the words acid or base are used. The Lewis definition is useful when discussing transition-metal ions, however, and we shall return to Lewis acids and bases in Chap. 22.

11.5 REDOX REACTIONS

In addition to precipitation and acid-base reactions, a third important class called **oxidation-reduction reactions** is often encountered in aqueous solutions. The terms *red*uction and *ox*idation are usually abbreviated to **redox.** Such a reaction corresponds to the transfer of electrons from one species to another.

A simple redox reaction occurs when copper metal is immersed in a solution of silver nitrate. The solution gradually acquires the blue color characteristic of the hydrated Cu^{2+} ion, while the copper becomes coated with glittering silver crystals. The reaction may be described by the net ionic equation

$$Cu(s) + 2Ag^+(aq) \rightarrow Cu^{2+}(aq) + 2Ag(s) \qquad (11.7)$$

We can regard this equation as being made up from two hypothetical half-equations. In one, each copper atom loses 2 electrons:

$$Cu \rightarrow Cu^{2+} + 2e^- \qquad (11.7a)$$

while in the other, 2 electrons are acquired by 2 silver ions:

$$2e^- + 2Ag^+ \rightarrow 2Ag \qquad (11.7b)$$

If these two half-equations are added, the net result is Eq. (11.7). In other words, the reaction of copper with silver ions, described by Eq. (11.7), corresponds to the *loss* of electrons by the copper metal, as described by half-equation (11.7*a*), and the *gain* of electrons by silver ions, as described by Eq. (11.7*b*).

A species like copper which *donates* electrons in a redox reaction is called a **reducing agent,** or **reductant.** (A mnemonic for remembering this is *r*emember, *e*lectron *d*onor = *red*ucing agent.) When a reducing agent donates electrons to another species, it is said to *reduce* the species to which the electrons are donated. In Eq. (11.7), for example, copper reduces the silver ion to silver. Consequently the half-equation

$$2Ag^+ + 2e^- \rightarrow 2Ag$$

is said to describe the *reduction* of silver ions to silver.

Species which *accept* electrons in a redox reaction are called **oxidizing agents,** or **oxidants.** In Eq. (11.7) the silver ion, Ag^+, is the oxidizing agent. When an oxidizing agent accepts electrons from another species, it is said to *oxidize* that species, and the process of electron removal is called *oxidation*. The half-equation

$$Cu \rightarrow Cu^{2+} + 2e^-$$

thus describes the oxidation of copper to Cu^{2+} ion.

In summary, then, when a redox reaction occurs and electrons are transferred, there is always a reducing agent donating electrons and an oxidizing agent to receive them. The reducing agent, because it loses electrons, is said to be oxidized. The oxidizing agent, because it gains electrons, is said to be reduced.

EXAMPLE 11.14 Write the following reaction in the form of half-equations. Identify each half-equation as an oxidation or a reduction. Also identify the oxidizing agent and the reducing agent in the overall reaction

$$Zn + 2Fe^{3+} \rightarrow Zn^{2+} + 2Fe^{2+}$$

Solution The half-equations are

$$Zn \rightarrow Zn^{2+} + 2e^- \qquad \text{oxidation—loss of electrons}$$

$$2e^- + 2Fe^{3+} \rightarrow 2Fe^{2+} \qquad \text{reduction—gain of electrons}$$

Since zinc metal (Zn) has *donated* electrons, we can identify it as the reducing agent. Conversely, since iron(III) ion (Fe^{3+}) has accepted electrons, we identify it as the oxidizing agent. An alternative method of identification is to note that since zinc has been oxidized, the oxidizing *agent* must have been the other reactant, namely, iron(III). Also, since the iron(III) ion has been reduced, the zinc must be the reducing agent. Observe also that both the oxidizing and reducing agents are the *reactants* and therefore appear on the *left*-hand side of an equation.

A more complex redox reaction occurs when copper dissolves in nitric acid. The acid attacks the metal vigorously, and large quantities of the red-brown gas, nitrogen dioxide (NO_2), are evolved. (NO_2 is poisonous, and so this reaction should be done in a hood.) The solution acquires the blue color characteristic of the hydrated Cu^{2+} ion. The reaction which occurs is

$$Cu + 2NO_3^- + 4H_3O^+ \rightarrow Cu^{2+} + 2NO_2 + 6H_2O \qquad (11.8)$$

Merely by inspecting this net ionic equation, it is difficult to see that a transfer of electrons has occurred. Clearly the copper metal has lost electrons and been oxidized to Cu^{2+}, but where have the donated electrons gone? The matter becomes somewhat clearer if we break up Eq. (11.8) into half-equations. One must be

$$Cu \rightarrow Cu^{2+} + 2e^- \qquad (11.8a)$$

and the other is

$$2e^- + 4H_3O^+ + 2NO_3^- \rightarrow 2NO_2 + 6H_2O \qquad (11.8b)$$

You can verify that these are correct by summing them to obtain Eq. (11.8).

The second half-equation shows that each NO_3^- ion has not only accepted an electron, but it has also accepted two protons. To further complicate matters, a nitrogen-oxygen bond has also been broken, producing a water molecule. With all this reshuffling of nuclei and electrons, it is difficult to say whether the two electrons donated by the copper ended up on an NO_2 molecule or on an H_2O molecule. Nevertheless, it is still meaningful to call this a redox reaction. Clearly, copper atoms have lost electrons, while a combination of hydronium ions and nitrate ions have accepted them. Accordingly, we can refer to the nitrate ion (or nitric acid, HNO_3) as the oxidizing agent in the overall reaction. Half-equation (11.8*b*) is a reduction because electrons are accepted.

Oxidation Numbers

Because redox reactions may involve proton transfers and other bond-breaking and bond-making processes, as well as electron transfers, the equations involved are much more difficult to deal with than those describing acid-base reactions. In order to be able to recognize redox reactions, we need a method for keeping a careful account of all the electrons. This is done by assigning *oxidation numbers* to each atom before and after the reaction. You will recall from Sec. 7.6 that oxidation numbers are the charges each atom would have if all valence electrons in covalent bonds were arbitrarily assigned to the more electronegative atom.

For example, in NO_3^- the nitrogen is assigned an oxidation number of +5 and each oxygen an oxidation number of −2. This arbitrary assignment corresponds to the nitrogen's having lost its original five valence electrons to the electronegative oxygens. In NO_2, on the other hand, the nitrogen has an oxidation number of +4 and may be thought of as having one valence electron for itself, that is, one more electron than it had in NO_3^-. This arbitrarily assigned gain of one electron corresponds to reduction of the nitrogen atom on going from NO_3^- to NO_2. As a general rule, *reduction corresponds to a lowering of the oxidation number* of some atom. *Oxidation* corresponds to *increasing the oxidation number* of some atom.

Applying the oxidation number rules to Eq. (11.8), we have

$$\overset{0}{Cu} + 2\overset{+5-2}{NO_3^-} + 4\overset{+1\ -2}{H_3O^+} \rightarrow \overset{+2}{Cu^{2+}} + 2\overset{+4-2}{NO_2} + 6\overset{+1\ -2}{H_2O}$$

Since the oxidation number of copper increased from 0 to +2, we say that copper was oxidized and lost two negatively charged electrons. This corresponds to half-equation (11.8*a*). The oxidation number of nitrogen went down from 5 to 4, and so the nitrogen (or nitrate ion) was reduced. Each nitrogen gained one electron, so $2e^-$ were needed for the $2NO_3^-$ in half-equation (11.8*b*). The nitrogen was reduced by electrons donated by copper, and so copper was the reducing agent. Copper was oxidized because its electrons were accepted by an oxidizing agent, nitrogen (or nitrate ion).

Although they are useful and necessary for recognizing redox reactions, oxidation numbers are a highly artificial device. The nitrogen atom in NO_3^- does not really have a +5 charge which can be reduced to +4 in NO_2. Instead, there are covalent bonds and electron-pair sharing between nitrogen and oxygen in both species, and nitrogen has certainly not lost its valence electrons entirely to oxygen. Even though this may (and indeed should) make you suspicious of the validity of oxidation numbers, they are undoubtedly a useful tool for spotting electron-transfer processes. So long as they are used for that purpose only, and not taken to mean that atoms in covalent species actually have the large charges oxidation numbers often imply, their use is quite valid.

EXAMPLE 11.15 Identify the redox reactions and the reducing and oxidizing agents from the following:

a) $2MnO_4^- + 5SO_2 + 6H_2O \rightarrow 5SO_4^{2-} + 2Mn^{2+} + 4H_3O^+$

b) $NH_4^+ + PO_4^{3-} \rightarrow NH_3 + HPO_4^{2-}$

c) $HClO + H_2S \rightarrow H_3O^+ + Cl^- + S$

Solution

a) The appropriate oxidation numbers are

$$2\overset{+7-2}{MnO_4^-} + 5\overset{+4-2}{SO_2} + 6\overset{+1\ -2}{H_2O} \rightarrow 5\overset{+6-2}{SO_4^{2-}} + 2\overset{+2}{Mn^{2+}} + 4\overset{+1\ -2}{H_3O^+}$$

The only atoms which change are Mn, from +7 to +2, a reduction, and S, from +4 to +6, an oxidation. The reaction is a redox process. SO_2 has been oxidized by MnO_4^-, and so MnO_4^- is the oxidizing agent. MnO_4^- has been reduced by SO_2, and so SO_2 is the reducing agent.

b) The oxidation numbers

$$\overset{-3+1}{NH_4^+} + \overset{+5-2}{PO_4^{3-}} \rightarrow \overset{-3+1}{NH_3} + \overset{+1+5-2}{HPO_4^{2-}}$$

show that no redox has occurred. This is an acid-base reaction because a proton, but no electrons, has been transferred.

c)

$$\overset{+1+1-2}{HClO} + \overset{+1\ -2}{H_2S} \rightarrow \overset{+1\ -2}{H_3O^+} + \overset{-1}{Cl^-} + \overset{0}{S}$$

H_2S has been oxidized, losing two electrons to form elemental S. Since H_2S donates electrons, it is the reducing agent. HClO accepts these electrons and is reduced to Cl^-. Since it accepts electrons, HClO is the oxidizing agent.

Balancing Redox Equations

Some redox equations may be balanced using the methods developed in Sec. 2.10, but most are rather difficult to handle. Therefore it is useful to have some rules, albeit somewhat arbitrary ones, to help find appropriate coefficients. These rules depend on whether the reaction occurs in acidic or basic solution. In either situation we must make sure that the number of electrons accepted by the oxidizing agent exactly equals the number of electrons donated by the reducing agent.

In acid solution We shall apply the rules to the equation

$$\overset{+5-2}{IO_3^-} + \overset{+4-2}{SO_2} + \overset{+1\ -2}{H_2O} \rightarrow \overset{0}{I_2} + \overset{+6-2}{SO_4^{2-}} + \overset{+1\ -2}{H_3O^+} \qquad (11.9)$$

The changes in oxidation numbers verify that it is a redox equation, and the presence of H_3O^+ indicates that it occurs in acidic solution. The rules are

1 *Write unbalanced half-equations for the oxidation of the reducing agent and for the reduction of the oxidizing agent.*

Oxidation: $SO_2 \rightarrow SO_4^{2-}$

Reduction: $IO_3^- \rightarrow I_2$

2 *Balance the element reduced or oxidized in each half-equation.*

Oxidation: $SO_2 \rightarrow SO_4^{2-}$ S already balanced

Reduction: $2IO_3^- \rightarrow I_2$

3 *Balance oxygen atoms by adding water (solvent) molecules.*

Oxidation: $SO_2 + 2H_2O \rightarrow SO_4^{2-}$

Reduction: $2IO_3^- \rightarrow I_2 + 6H_2O$

4 *Balance hydrogen atoms by adding hydrogen ions (available from the acidic solution).*

Oxidation: $SO_2 + 2H_2O \rightarrow SO_4^{2-} + 4H^+$

Reduction: $12H^+ + 2IO_3^- \rightarrow I_2 + 6H_2O$

5 *Balance electrical charges by adding electrons.*

Oxidation: $SO_2 + 2H_2O \rightarrow SO_4^{2-} + 4H^+ + 2e^-$

(The total charge on the left side was 0, but on the right it was $-2 + 4 = +2$. Therefore $2e^-$ were needed on the right.)

Reduction: $10e^- + 12H^+ + 2IO_3^- \rightarrow I_2 + 6H_2O$

(The total charge on the left was $+12 - 2 = +10$, but on the right it was 0. Therefore $10e^-$ were needed on the left.)

6 *Use oxidation numbers to check that the number of electrons is correct.*
Oxidation: The oxidation number of S increases from +4 to +6, corresponding to a loss of $2e^-$.
Reduction: The oxidation number of I falls from +5 to 0, corresponding to a gain of $5e^-$ for *each* I. Since there are 2 I atoms, $10e^-$ must be added.

7 *Adjust both half-equations so that the number of electrons donated by the reducing agent equals the number of electrons accepted by the oxidizing agent.* Since only 2 electrons are donated in the oxidation half-equation, while 10 are required by the reduction, the oxidation must occur 5 times for each reduction. That is, both sides of the oxidation half-equation must be multiplied by 5:

Oxidation: $5SO_2 + 10H_2O \rightarrow 5SO_4^{2-} + 20H^+ + 10e^-$

Reduction: $10e^- + 12H^+ + 2IO_3^- \rightarrow I_2 + 6H_2O$

8 *Sum the half-equations.* The net equation which results is

$$5SO_2 + 4H_2O + 2IO_3^- \rightarrow 5SO_4^{2-} + 8H^+ + I_2$$

Note that when the half-equations were summed, the number of electrons was the same on both sides, and so no free electrons (which could not exist in aqueous solution) appear in the final result. It also would be more accurate to write H_3O^+ instead of H^+ for the hydronium ion. This can be done by adding $8H_2O$ to both sides of the equation:

$$5SO_2 + 12H_2O + 2IO_3^- \rightarrow 5SO_4^{2-} + 8H_3O^+ + I_2$$

(On the right, the $8H_2O$ molecules are protonated to $8H_3O^+$.) It is also a good idea at this point to check that all atoms, as well as the electrical charges, balance.

In basic solution Potassium permanganate, $KMnO_4$, can be used to oxidize alcohols to carboxylic acids. An example is

$$\overset{+7\,-2}{MnO_4^-} + \overset{-2+1\,-2+1}{CH_3OH} \rightarrow \overset{+4\,-2}{MnO_2} + \overset{+1\,-2}{H_2O} + \overset{+1+2-2-2}{HCOO^-} + \overset{-2+1}{OH^-} \qquad (11.10)$$

Since OH^- is produced, the reaction occurs in basic solution. It clearly involves redox.

1 *Write unbalanced equations for the oxidation of the reducing agent and the reduction of the oxidizing agent (same as for acid solution).*

Oxidation: $CH_3OH \rightarrow HCOO^-$

Reduction: $MnO_4^- \rightarrow MnO_2$

2 *Balance the element reduced or oxidized in each half-equation (same as for acid solution).* (Both C and Mn are already balanced.)

3 *Balance oxygen atoms by adding hydroxide ions (available from the basic solution).*

Oxidation: $CH_3OH + OH^- \rightarrow HCOO^-$

Reduction: $MnO_4^- \rightarrow MnO_2 + 2OH^-$

4 *To the side of each half-equation which lacks hydrogen, add one water molecule for each hydrogen needed. Add an equal number of hydroxide ions to the opposite side.*

Oxidation: $CH_3OH + 5OH^- \rightarrow HCOO^- + 4H_2O$

(Four hydrogens were needed on the right, and so 4 water molecules were added on the right and 4 hydroxide ions on the left.)

Reduction: $MnO_4^- + 2H_2O \rightarrow MnO_2 + 4OH^-$

(Two hydrogens were needed on the left, and so 2 water molecules were added on the left and 2 hydroxide ions were added on the right. Note that the added hydroxide ions are to maintain the balance of oxygen atoms.)

5 *Balance electrical charges by adding electrons (same as for acid solution).*

Oxidation: $CH_3OH + 5OH^- \rightarrow HCOO^- + 4H_2O + 4e^-$

(The total charge on the left was -5, but on the right it was -1, and so $4e^-$ were added on the right.)

Reduction: $MnO_4^- + 2H_2O + 3e^- \rightarrow MnO_2 + 4OH^-$

(The total charge on the left was -1, but on the right it was -4, and so $3e^-$ were added on the left.)

6 *Use oxidation numbers to check that the number of electrons is correct (same as for acid solution).*
Oxidation: C goes from −2 to +2, corresponding to a loss of $4e^-$.
Reduction: Mn goes from +7 to +4, corresponding to a gain of $3e^-$.

7 *Adjust both half-equations so that the number of electrons donated by the reducing agent equals the number of electrons accepted by the oxidizing agent (same as for acid solution).* Multiplying the oxidation half-equation by 3 and the reduction half-equation by 4 adjusts each so it involves $12e^-$.

$$\text{Oxidation:}\quad 3CH_3OH + 15OH^- \rightarrow 3HCOO^- + 12H_2O + 12e^-$$

$$\text{Reduction:}\quad 4MnO_4^- + 8H_2O + 12e^- \rightarrow 4MnO_2 + 16OH^-$$

8 *Sum the half-equations (same as for acid solution).* The net equation which results is

$$3CH_3OH + 4MnO_4^- \rightarrow 3HCOO^- + 4H_2O + 4MnO_2 + OH^-$$

Again, it is worthwhile to check that all atoms and charges balance.

The rules for balancing redox equations involve adding H^+, H_2O, and OH^- to one side or the other of the half-equations. Since these species are present in the solution, they may participate as reactants or products, but usually there is no experiment which can tell whether they do participate. However, the balanced equation derived from our rules does indicate just what role H^+, H_2O, or OH^- play in a given redox process.

11.6 SOME COMMON OXIDIZING AND REDUCING AGENTS

A good reducing agent must be able to donate electrons readily. This means that it must not have very much attraction for electrons. Among the elements, low electronegativity is characteristic of good reducing agents. Molecules and ions which contain relatively electropositive elements which have low oxidation numbers are also good reducing agents. Oxidizing agents, on the other hand, must be able to accept electrons readily. Highly electronegative elements can do this, as can molecules or ions which contain relatively electronegative elements and even some metals which have high oxidation numbers. Bear these general rules in mind as we examine examples of common oxidizing and reducing agents in the following paragraphs.

Oxidizing Agents

Halogens (group VIIA elements) All four elemental halogens, F_2, Cl_2, Br_2, and I_2, are able to accept electrons according to the half-equation

$$X_2 + 2e^- \rightarrow 2X^- \qquad X = F, Cl, Br, I$$

As we might expect from the periodic variation of electronegativity, the oxidizing power of the halogens decreases in the order $F_2 > Cl_2 > Br_2 > I_2$. Fluorine is such a strong oxidizing agent that it can react with water:

$$2F_2 + 6H_2O \rightarrow 4H_3O^+ + 4F^- + O_2$$

Chlorine also reacts with water, but only in the presence of sunlight. Bromine is weaker, and iodine has only mild oxidizing power.

Oxygen Oxygen gas, which constitutes about 20 percent of the earth's atmosphere, is another electronegative element which is a good oxidizing agent. It is very slightly weaker than chlorine, but considerably stronger than bromine. Because the atmosphere contains such a strong oxidant, few substances occur in reduced form at the earth's surface. An oxidized form of silicon, SiO_2, is one of the most plentiful constituents of the crust of the earth. Most metals, too, occur as oxides and must be reduced before they can be obtained in elemental form. When iron rusts, it forms the red-brown oxide $Fe_2O_3 \cdot xH_2O$, which always contains an indeterminate amount of water.

Oxyanions and oxyacids In aqueous solution NO_3^-, IO_3^-, MnO_4^-, $Cr_2O_7^{2-}$, and a number of other oxyanions serve as convenient, strong oxidizing agents. The structure of the last oxyanion mentioned above is shown in Fig. 11.7. The most strongly oxidizing oxyanions often contain an element in its highest possible oxidation state, that is, with an oxidation number equal to the periodic group number. For example, NO_3^- contains nitrogen in a +5 oxidation state, $Cr_2O_7^{2-}$ contains chromium +6, and MnO_4^- has manganese +7. The oxidizing power of the dichromate ion is employed in laboratory cleaning solution, a solution of $Na_2Cr_2O_7$ in concentrated H_2SO_4. This readily oxidizes the organic compounds in grease to carbon dioxide. It is also highly corrosive, eats holes in clothing, and must be handled with care. Dark purple permanganate ion is another very common oxidizing agent. In basic solution [Eq. (11.10)] it is reduced to solid dark brown MnO_2. In acidic solution, however, it forms almost colorless $Mn^{2+}(aq)$, as shown by the equation in Example 11.15*a*.

Reducing Agents

Metals All metals have low ionization energies and are relatively electropositive, and so they can lose electrons fairly easily. Metals on the left of the periodic table exhibit this property to the greatest extent, and some of them, such as Li or Na, can even reduce H_2O:

$$2Li(s) + 2H_2O(l) \rightarrow 2Li^+(aq) + 2OH^-(aq) + H_2(g)$$

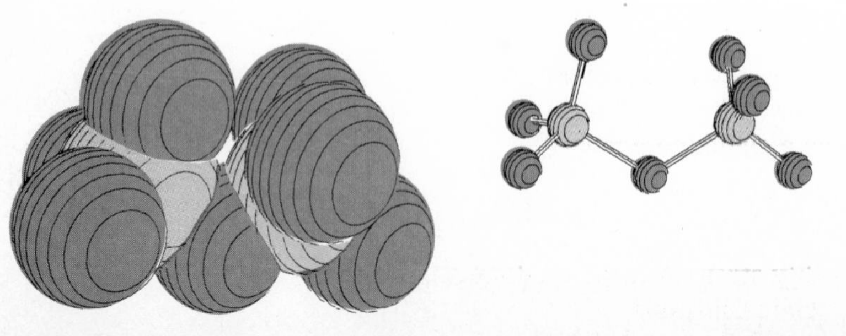

Figure 11.7 Space-filling and ball-and-stick models of dichromate ion, $Cr_2O_7^{2-}$. Chromium atoms are gray and oxygen atoms are dark red. (Computer-generated.) (*Copyright © 1976 by W. G. Davies and J. W. Moore.*)

Other metals, such as Fe or Zn, cannot reduce H_2O but can reduce hydronium ions, and so they dissolve in acid solution:

$$Zn(s) + 2H_3O^+(aq) \rightarrow Zn^{2+}(aq) + 2H_2O(l) + H_2(g)$$

This is one of the characteristic reactions of acids that we mentioned in Sec. 11.2.

There are a few metals that will not dissolve in just any acid but instead require an acid like HNO_3 whose anion is a good oxidizing agent. Cu and Hg are examples:

$$3Hg(s) + 8H_3O^+(aq) + 2NO_3^-(aq) \rightarrow 3Hg^{2+}(aq) + 2NO(g) + 12H_2O$$

Finally, a few metals, such as Au and Pt, are such poor reducing agents that even an oxidizing acid like HNO_3 will not dissolve them. This is the origin of the phrase "the acid test." If a sample of an unknown yellow metal can be dissolved in HNO_3, the metal is not gold. Kings who collected tax payments in gold kept a supply of HNO_3 available to make sure they were not being cheated.

Substances Which Are Both Oxidizing and Reducing Agents

Water We have seen that some oxidizing agents, such as fluorine, can oxidize water to oxygen. There are also some reducing agents, such as lithium, which can reduce water to hydrogen. In terms of redox, water behaves much as it did in acid-base reactions, where we found it to be amphiprotic. In the presence of a strong electron donor (strong reducing agent), water serves as an oxidizing agent. In the presence of a strong electron acceptor (strong oxidizing agent), water serves as a reducing agent. Water is rather weak as an oxidizing or as a reducing agent, however; so there are not many substances which reduce or oxidize it. Thus it makes a good solvent for redox reactions. This also parallels water's acid-base behavior, since it is also a very weak acid and a very weak base.

Hydrogen peroxide, H_2O_2 In this molecule the oxidation number for oxygen is −1. This is halfway between $O_2(0)$ and $H_2O(-2)$, and so hydrogen peroxide can either be reduced or oxidized. When it is reduced, it acts as an oxidizing agent:

$$H_2O_2 + 2H^+ + 2e^- \rightarrow 2H_2O$$

When it is oxidized, it serves as a reducing agent:

$$H_2O_2 \rightarrow O_2 + 2H^+ + 2e^-$$

Hydrogen peroxide is considerably stronger as an oxidizing agent than as a reducing agent, especially in acidic solutions.

11.7 REDOX COUPLES

When a reducing agent donates one or more electrons, its oxidation number goes up, and the resulting species is capable of reaccepting the electrons. That is, the oxidized species is an oxidizing agent. For example, when copper metal dissolves, the copper(II) ion formed can serve as an oxidizing agent:

$$\underset{\textbf{Reducing agent}}{Cu} \longrightarrow \underset{\textbf{Oxidizing agent}}{Cu^{2+}} + 2e^-$$

Similarly, when an oxidizing agent such as silver ion is reduced, the silver metal can donate an electron, serving as a reducing agent:

$$\underset{\textbf{Oxidizing agent}}{Ag^+} + e^- \longrightarrow \underset{\textbf{Reducing agent}}{Ag}$$

This is analogous to what we observed previously in the case of conjugate acids and bases. For every oxidizing agent, there corresponds some reducing agent, and for every reducing agent, there corresponds an oxidizing agent.

An oxidizing and reducing agent which appear on opposite sides of a half-equation constitute a **redox couple.** Redox couples are analogous to conjugate acid-base pairs and behave in much the same way. *The stronger an oxidizing agent, the weaker the corresponding reducing agent,* and *the stronger a reducing agent, the weaker the corresponding oxidizing agent.* Thus the strong oxidizing agent F_2 produces the weak reducing agent F^-. Conversely, the strong reducing agent Li corresponds with the weak oxidizing agent Li^+.

This type of relationship is demonstrated in Table 11.5, where selected redox couples have been arranged in order of increasing strength of the reducing agent. As in the case of Table 11.3 for acids and bases, this table of half-equations can be used to predict which way a redox reaction will proceed. The reactions with the greatest tendency to occur are between strong oxidizing agents from the upper left and strong reducing agents from the lower right. If a line is drawn from oxidizing agent to reducing agent for such a reaction, it will have a *downhill* slope. Reactions with little tendency to occur involve the weak oxidizing agents at the lower left and weak reducing agents at the upper right. In such cases a line from oxidizing to reducing agent has an *uphill* slope. When the slope of the line is not far from horizontal, the oxidizing and reducing agents are similar in strength and their reaction will go only part way to completion.

EXAMPLE 11.16 Predict whether iron(III) ion, Fe^{3+}, will oxidize copper metal. If so, write a balanced equation for the reaction.

Solution A line from the oxidizing agent Fe^{3+} to the reducing agent Cu is downhill, and so the reaction will occur. The balanced equation is the

TABLE 11.5 Selected Redox Couples Arranged in Order of Decreasing Strength of Oxidizing Agent.

← OXIDIZING STRENGTH					← REDUCING STRENGTH
Strong	F_2	$+ 2e^-$	→	$2F^-$	Negligible
	$MnO_4^- + 8H^+$	$+ 5e^-$	→	$Mn^{2+} + 4H_2O$	
	Cl_2	$+ 2e^-$	→	$2Cl^-$	
	$Cr_2O_7^{2-} + 14H^+$	$+ 6e^-$	→	$2Cr^{3+} + 7H_2O$	
	$O_2 + 4H^+$	$+ 4e^-$	→	$2H_2O$	
Medium	$OCl^- + H_2O$	$+ 2e^-$	→	$Cl^- + OH^-$	Weak
	Ag^+	$+ e^-$	→	Ag	
	Fe^{3+}	$+ e^-$	→	Fe^{2+}	
	I_2	$+ 2e^-$	→	$2I^-$	
	Cu^{2+}	$+ 2e^-$	→	Cu	
Weak	Sn^{4+}	$+ 2e^-$	→	Sn^{2+}	Medium
	$2H^+$	$+ 2e^-$	→	H_2	
	Pb^{2+}	$+ 2e^-$	→	Pb	
	Fe^{2+}	$+ 2e^-$	→	Fe	
Negligible	Zn^{2+}	$+ 2e^-$	→	Zn	Strong
	Mg^{2+}	$+ 2e^-$	→	Mg	
	Na^+	$+ e^-$	→	Na	
	Li^+	$+ e^-$	→	Li	

sum of the two half-equations adjusted to equalize the number of electrons transferred.

$$\begin{array}{r} 2Fe^{3+} + \cancel{2e^-} \rightarrow 2Fe^{2+} \\ Cu \rightarrow Cu^{2+} + \cancel{2e^-} \\ \hline 2Fe^{3+} + Cu \rightarrow 2Fe^{2+} + Cu^{2+} \end{array}$$

Note that the half-equation for Cu is reversed from that in Table 11.5 because the reactant Cu is a reducing agent. All half-equations in Table 11.5 have the oxidizing agent on the left.

EXAMPLE 11.17 Use Table 11.15 to find a reagent that will oxidize H_2O (other than F_2 or Cl_2, which were mentioned earlier).

Solution We must find an oxidizing agent from which a line to the reducing agent H_2O has a downhill slope. The only possibilities are MnO_4^- or

$Cr_2O_7^{2-}$. The latter reacts extremely slowly, but aqueous permanganate solutions decompose over a period of weeks. The concentration of MnO_4^- decreases because of the reaction

$$4MnO_4^- + 12H_3O^+ \rightarrow 4Mn^{2+} + 5O_2 + 18H_2O$$

SUMMARY

Reactions in aqueous solutions usually involve ionic or polar covalent compounds. The solubilities of such compounds are enhanced because of their interactions with water molecules, especially the hydration of ions. Measurement of the electrical current conducted by a solution of known concentration enables us to determine whether a solute is an electrolyte, and, if so, of what type. Water itself is an extremely weak electrolyte, producing hydronium ions and hydroxide ions, each at a concentration of 1.00×10^{-7} mol dm^{-3} at 25°C.

There are three important classes of reactions which occur in aqueous solution: precipitation reactions, acid-base reactions, and redox reactions. Whether or not a precipitate will form when solutions of about 0.1 *M* concentration are mixed can be predicted using the solubility rules in Table 11.2. Precipitation reactions are useful for detecting the presence of various ions and for determining the concentrations of solutions.

Acid-base reactions and redox reactions are similar in that something is being transferred from one species to another. Acid-base reactions involve proton transfers, while redox reactions involve electron transfers. Redox reactions are somewhat more complicated, though, because proton transfers and other bond-making and bond-breaking processes occur at the same time as electron transfer. Both proton transfer and electron transfer can be broken into half-equations: one to describe donation of a proton (or electrons), and one to describe acceptance of a proton (or electrons). For acid-base reactions, each half-equation involves a conjugate acid-base pair. For oxidation-reduction reactions, each half-equation involves a redox couple.

Both conjugate acid-base pairs and redox couples may be tabulated in order of acid or base strength (Table 11.4) or according to reducing or oxidizing power (Table 11.5). In each case such tabulations may be used to determine whether a given reaction will go to completion, occur only to a limited extent, or not occur at all. As we study acids, bases, reducing agents, and oxidizing agents in subsequent chapters, you will learn how such tables are derived as well as how they can be made to give quantitative information about the extent to which a given reaction will occur.

QUESTIONS AND PROBLEMS

11.1 What is the difference between a strong electrolyte, a weak electrolyte, and a nonelectrolyte? How can these three classes be distinguished by their conductivities? Give at least two examples of each class.

11.2 Explain why 0.02 *M* HF solution has a conductivity which is only about 15 percent that of 0.02 *M* HCl.

11.3 When a cell consisting of two electrodes each of area 1 cm^2 and 1 mm apart is immersed in a 0.001-*M* solution of perchloric acid, $HClO_4$, and 1 V is applied, a current of 3.70 mA is found to pass at 18°C. By comparing this result with those given in Table 11.1, would you say that $HClO_4$ is a strong or a weak acid?

11.4 When 0.001 *M* $KHCO_3$ is used in the experiment described in the previous problem, a current of 1.31 mA is measured, while with 0.001 *M* $KHSO_4$, a current of 4.82 mA is found. By comparing these values with those given in Table 11.1, suggest what ions are produced when $KHCO_3$ and $KHSO_4$ are dissolved in H_2O and explain the large difference in their conductivities.

11.5 Write balanced net ionic equations to describe the reaction (if any) which occurs when each of the following pairs of solutions are mixed:

a $NaCl(aq) + KBr(aq)$
b $Na_2S(aq) + Pb(NO_3)_2(aq)$
c $AgNO_3(aq) + KBr(aq)$
d $Al_2(SO_4)_3(aq) + Ba(OH)_2(aq)$
e $K_3PO_4(aq) + CuCl_2(aq)$

In each case also list the spectator ions (if any).

11.6 Calculate the amount of K^+ ion in the following solutions:

a 250 cm³ of 0.100 *M* KCl
b 300 cm³ of 0.200 *M* K_2SO_4
c 100 cm³ of 0.150 *M* K_3PO_4

11.7 Find the mass of $PbCl_2$ precipitated when excess $Pb(NO_3)_2$ is added to 400 cm³ of 0.125 *M* HCl.

11.8 What mass of $Mg(OH)_2$ is precipitated when 200 cm³ of 0.1 *M* $MgCl_2$ is mixed with 200 cm³ of 0.1 *M* KOH?

11.9 When excess $BaCl_2$ is added to 200 cm³ of a solution of H_2SO_4, 5.195 g $BaSO_4$ is precipitated. What is the concentration of the H_2SO_4 solution?

11.10 The lattice enthalpy of potassium iodide is −640 kJ mol⁻¹, while the enthalpy of solution is +21 kJ mol⁻¹. Calculate the hydration energy of potassium iodide and explain how you arrived at your answer.

11.11 Explain why the lattice enthalpy of LiF (1021 kJ mol⁻¹) is very much larger than that of CsI (586 kJ mol⁻¹).

11.12 Use the following data to deduce the trend in hydration enthalpies of the sodium, potassium, and rubidium ions:

	Lattice Enthalpy/ kJ mol⁻¹	Enthalpy of Solution/ kJ mol⁻¹
NaCl	−770	+ 5.0
KCl	−703	+17.6
RbCl	−674	+18.8

Give reasons for any trend you observe.

11.13 Discuss the statement, "If ions were not hydrated in solution, all ionic compounds would be insoluble in water."

11.14 Explain why both the hydronium ions and hydroxide ions conduct electricity through water very much better than any other ions.

11.15 Both nitric acid and perchloric acid form crystalline solids containing water of crystallization. The formulas of these solids are usually given as $HNO_3{\cdot}H_2O$ and $HClO_4{\cdot}H_2O$. Suggest an alternative method of formulating these compounds more in accord with modern ideas about acids and bases.

11.16 Write an equation to describe the proton transfer which occurs when each of the following acids is added to water:

a HI   *c* CH_3COOH
b HSO_4^- *d* $HClO_3$

11.17 Write a balanced equation to describe the proton transfer which occurs when each of the following bases is added to water:

a NH_3 *b* O^{2-} *c* PO_4^{3-} *d* F^-

11.18 Write a balanced equation to describe the proton transfer which occurs between acid and base in each of the following cases:

a $HCl + NH_3$ *c* $H_3O^+ + H^-$
b $CH_3COOH + OH^-$ *d* $H_3O^+ + NH_3$

11.19 Identify which reactant in the following proton transfers is an acid and which is a base:

a $HI + H_2O \rightarrow H_3O^+ + I^-$
b $OH^- + NH_4^+ \rightarrow H_2O + NH_3$
c $HCO_3^- + HSO_4^- \rightarrow H_2CO_3 + SO_4^{2-}$
d $HS^- + HF \rightarrow H_2S + F^-$
e $NH_3 + H_2CO_3 \rightarrow NH_4^+ + HCO_3^-$

11.20 Write an appropriate sequence of equations depicting the steps in the reaction of the following polyprotic acids with water:

a H_3PO_4
b H_2CO_3
c $NH_3^+CH_2COOH$ (the glycinium ion, a diprotic acid)

11.21 State what is wrong about the following definitions of a strong acid:

a A strong acid is a very concentrated acid.
b A strong acid has very strong bonds.
c A strong acid has a very strong hold on protons.
d A strong acid is a good proton acceptor.

11.22 Which acid in each of the following pairs is stronger?

a HNO_3 or HNO_2 *c* $HClO_4$ or $HOCl$
b H_2SO_4 or H_2SO_3 *d* HCl or HF

11.23 Classify the following acids as either weak or strong:

a HI *f* H_2S
b H_2CO_3 *g* NH_4^+
c H_3BO_3 *h* HNO_3
d HCN *i* HF
e CH_3COOH *j* H_2SO_4

11.24 Classify the following bases as weak or strong:

a NH_3 *d* OH^-
b CO_3^{2-} *e* acetate ion
c O^{2-}

11.25 Write equations showing the reaction of the following amphiprotic species with *both* H_3O^+ *and* OH^-:

a HCO_3^- *c* HPO_4^{2-}
b $NH_3^+CH_2COO^-$ (glycine)

11.26 Write the formula for the conjugate base of the following acids:

a HBr *d* NH_4^+
b H_2SO_4 *e* H_3O^+
c $H_2PO_4^-$

11.27 Write the formula for the conjugate acid of the following bases:

a NH_3 *d* H^-
b CO_3^{2-} *e* H_2O
c PO_4^{3-}

11.28 Which of the following solutions will be the most acidic, i.e., have the highest hydronium-ion concentration? Which will be the least acidic?

a 0.1 *M* H_2CO_3 *c* 0.1 *M* NH_4Cl
b 0.1 *M* HF *d* 0.1 *M* $KHSO_4$

11.29 Which of the following solutions will be the most basic, i.e., have the highest hydroxide-ion concentration? Which will be the least basic?

a 0.1 *M* NH_3 *c* 0.1 *M* NaF
b 0.1 *M* H_2S *d* 0.1 *M* K_3PO_4

11.30 Using Table 11.4 predict which of the following acid-base reactions will go virtually to completion, which will occur only to a limited extent, and which will scarcely occur at all:

a $H_2O + H_2O$ *e* $NH_4^+ + HPO_4^{2-}$
b $HNO_3 + H_2O$ *f* $CH_3COOH + F^-$
c $H_3PO_4 + H_2O$ *g* $CH_3COOH + OH^-$
d $NH_4^+ + F^-$ *h* $HSO_4^- + H_2PO_4^-$

11.31 Using Table 11.4 predict what reactions will occur when

a H_2SO_4 is added to water.
b Excess CH_3COOH is added to a solution of K_3PO_4.
c $KHSO_4$ is added to a solution of Na_2CO_3.

11.32 Choose which of the following species is a Lewis acid and which is a Lewis base:

a NH_3 *g* $AlCl_3$
b Cu^{2+} *h* H_2O
c $BeCl_2$ *i* BF_3
d O^{2-} *j* Cr^{3+}
e SO_2 *k* CO_2
f SO_3 *l* S^{2-}

11.33 Identify the Lewis acid and the Lewis base in the following reactions:

a $H_2O + SO_3 \rightarrow H_2SO_4$
b $2Cl^- + SnCl_2 \rightarrow SnCl_4^{2-}$
c $H_3BO_3 + OH^- \rightarrow B(OH)_4^-$
d $Ag^+ + 2NH_3 \rightarrow Ag(NH_3)_2^+$
e $SnCl_2 + BF_3 \rightarrow F_3BSnCl_2$

11.34 Which of the reactants is the oxidizing agent and which is the reducing agent in the following reactions:

a $2Ag^+ + Zn \rightarrow Zn^{2+} + 2Ag$
b $2Fe^{3+} + Cu \rightarrow Cu^{2+} + 2Fe^{2+}$
c $Zn + Cl_2 \rightarrow Zn^{2+} + 2Cl^-$
d $CuO + H_2 \rightarrow Cu + H_2O$
e $NO + \frac{1}{2}O_2 \rightarrow NO_2$
f $2Fe^{3+} + Sn^{2+} \rightarrow 2Fe^{2+} + Sn^{4+}$
g $H_2 + Br_2 \rightarrow 2HBr$
h $2Na + 2H_2O \rightarrow 2NaOH + H_2$

11.35 Decide whether the element indicated in color has been (*i*) oxidized, (*ii*) reduced, (*iii*) both, or (*iv*) neither oxidized nor reduced in the following reactions:

a $Mg + 2H^+ \rightarrow Mg^{2+} + H_2$
b $2Ag^+ + Fe \rightarrow 2Ag + Fe^{2+}$
c $Ag^+ + Br^- \rightarrow AgBr$
d $2I^- + Br_2 \rightarrow I_2 + 2Br^-$
e $Cu^{2+} + Cu \rightarrow 2Cu^+$
f $Pb + Cu^{2+} \rightarrow Pb^{2+} + Cu$
g $NH_3 + HSO_4^- \rightarrow NH_4^+ + SO_4^{2-}$
h $4NH_3 + 5O_2 \rightarrow 4NO + 6H_2O$

11.36 Balance the following half-equations:

a $Sn^{2+} \rightarrow Sn^{4+}$ *d* $Hg^{2+} \rightarrow Hg_2^{2+}$
b $Cl_2 \rightarrow Cl^-$ *e* $Hg \rightarrow Hg^{2+}$
c $Hg \rightarrow Hg_2^{2+}$ *f* $H_2S \rightarrow S + H^+$

11.37 Balance the following half-equations:

a $H^+ + NO_3^- \rightarrow NO_2$
b $H^+ + NO_3^- \rightarrow NO$

c $H_3AsO_3 \rightarrow H_3AsO_4 + H^+$
d $Cr^{3+} \rightarrow Cr_2O_7^{2-} + H^+$
e $OH^- + Cl^- \rightarrow ClO_2$
f $CrO_4^{2-} \rightarrow Cr^{3+} + OH^-$
g $H_2O_2 \rightarrow OH^-$
h $H_2O_2 \rightarrow H^+ + O_2$

11.38 Decide which of the following reactions is (*i*) a proton transfer, (*ii*) a redox reaction, (*iii*) neither, or (*iv*) both.

a $Zn + 2H_3O^+ \rightarrow Zn^{2+} + H_2 + 2H_2O$
b $H_2O + S^{2-} \rightarrow 2HS^-$
c $2PbO_2 \rightarrow 2PbO + O_2$
d $N_2O_4 \rightarrow 2NO_2$
e $2H_2O_2 \rightarrow 2H_2O + O_2$
f $2H^- + 2H_2O \rightarrow 2OH^- + H_2$
g $Cl_2 + OH^- \rightarrow HOCl + Cl^-$
h $AgCl + NaBr \rightarrow AgBr + NaCl$
i $NH_4^+ + S^{2-} \rightarrow NH_3 + HS^-$

11.39 Complete and balance the following redox reactions in acid solution:

a $Cr_2O_7^{2-} + SO_3^{2-} \rightarrow Cr^{3+} + SO_4^{2-}$
b $ClO_2^- + I^- \rightarrow Cl^- + I_2$
c $MnO_4^- + H_2O_2 \rightarrow Mn^{2+} + O_2$
d $H_2O_2 + I^- \rightarrow I_2 + H_2O$
e $VO_3^- + Al \rightarrow VO^{2+} + Al^{3+}$
f $Ce^{4+} + As_2O_3 \rightarrow Ce^{3+} + H_3AsO_4$
g $Ce^{4+} + H_2O_2 \rightarrow Ce^{3+} + O_2$
h $BrO_3^- + Br^- \rightarrow Br_2$

11.40 Complete and balance each of the following redox reactions occurring in basic solutions:

a $Zn + NO_3^- \rightarrow ZnO_2^{2-} + NH_3$
b $Br_2 + AsO_2^- \rightarrow AsO_4^{3-} + Br^-$
c $Cr(OH)_3 + OCl^- \rightarrow CrO_4^{2-} + Cl^-$
d $SO_3^{2-} + Cl_2 \rightarrow SO_4^{2-} + Cl^-$
e $Cl_2 \rightarrow Cl^- + ClO_3^-$

11.41 Using Table 11.5 predict which of the following reactions (*i*) will go virtually to completion, (*ii*) will occur to a limited extent only, or (*iii*) will scarcely occur at all:

a $F_2 + 2I^- \rightarrow 2F^- + I_2$
b $2Fe^{3+} + 2I^- \rightarrow 2Fe^{2+} + I_2$
c $Zn^{2+} + 2I^- \rightarrow Zn + I_2$
d $8H^+ + MnO_4^- + 5Fe^{2+} \rightarrow 5Fe^{3+} + Mn^{2+} + 4H_2O$
e $Sn^{4+} + 2Fe^{2+} \rightarrow 2Fe^{3+} + Sn^{2+}$
f $Pb^{2+} + 2Ag \rightarrow 2Ag^+ + Pb$
g $2Fe^{3+} + Fe \rightarrow 3Fe^{2+}$
h $Cu + I_2 \rightarrow Cu^{2+} + 2I^-$

11.42 When 25.00 cm³ of a solution of $Na_2S_2O_3$ was titrated with a 0.051 07-*M* solution of I_2, it required 22.63 cm³ of the latter solution to achieve an endpoint. The unbalanced equation for this reaction is

$$S_2O_3^{2-} + I_2 \rightarrow S_4O_6^{2-} + I^-$$

Balance the equation and find the concentration of the $Na_2S_2O_3$ solution.

11.43 Calculate the concentration of a solution of potassium oxalate, $K_2C_2O_4$, if 35.16 cm³ of it is needed to achieve an endpoint with 46.72 cm³ of 0.0617 *M* $KMnO_4$ solution acidified with H_2SO_4. The unbalanced equation is

$$MnO_4^- + C_2O_4^{2-} \rightarrow Mn^{2+} + CO_2$$

11.44 We find that 30.86 cm³ of 0.2016 *M* $KMnO_4$ is required to achieve an endpoint when a sample of impure iron(II) sulfate pentahydrate, $FeSO_4 \cdot 7H_2O$, weighing 0.8700 g is dissolved in dilute H_2SO_4 and titrated. The unbalanced equation for the titration reaction is

$$MnO_4^- + Fe^{2+} \rightarrow Mn^{2+} + Fe^{3+}$$

Balance the equation and hence calculate the percentage purity of the $FeSO_4 \cdot 7H_2O$.

***11.45** Steel is often "pickled" by treating it with hydrochloric acid to remove iron oxide and other materials from its surface. After being used for some time, the pickling solution loses efficiency and must be discarded. This may cause environmental problems because the solution is strongly acidic and because it contains a high concentration of iron(II) chloride. Suppose you are a chemist analyzing pickling solution in a steel mill. You find that when a 25.0 ml aliquot of the solution is titrated with 0.1967 *M* $Ce(NO_3)_4(aq)$, it takes 43.24 ml to reach the end point. The titration reaction is

$$Fe^{2+}(aq) + Ce^{4+}(aq) \rightarrow Fe^{3+}(aq) + Ce^{3+}(aq)$$

(*a*) What is the concentration of $Fe^{2+}(aq)$ in the pickling solution? (*b*) If your plant disposes of 1500 dm³ of this solution per day into a river which flows at a rate of 200 000 m³ per day, and the solution mixes thoroughly with the river water, will the concentration of $Fe^{2+}(aq)$ exceed the permissible water-quality criterion of 5.3×10^{-6} *M* set by the U. S. Environmental Protection Agency?

chapter twelve

CHEMISTRY OF THE REPRESENTATIVE ELEMENTS

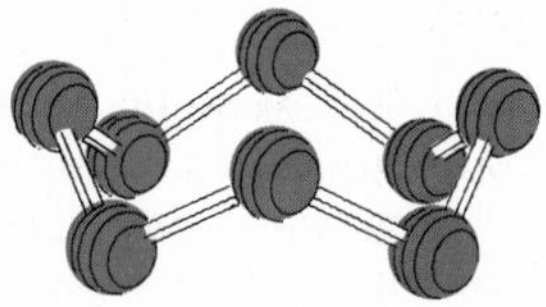

Most of the preceding chapters in this book have been devoted to explaining fundamental concepts and principles such as the atomic theory, electronic structure and chemical bonding, intermolecular forces and their effects on solids, liquids, or gases, and classes of reactions such as redox or acid-base. It is well to remember, though, that all these concepts and principles have been developed and used by chemists in order to better understand, recall, and systematize macroscopic laboratory observations. In other words, although the concepts we have described so far have their own inherent beauty as great ideas, they are primarily important because they reduce the quantity of memorization which is necessary to master descriptive chemistry, allowing people to recall facts that otherwise might be forgotten.

Such a reduction in memory work is only possible if you know how to apply principles to specific elements, their compounds, and the reactions they undergo. This is not as easy as it may seem, but neither is it impossibly difficult. In this chapter we will describe some of the chemistry of the representative elements, showing as we do so how their properties may be rationalized on the basis of concepts and principles. The chapter is organized according to the concept of periodicity, with each section corresponding to one of the eight groups of representative elements. As you read it, you should try to see *why* a certain reaction occurs, as well as what actually happens, and instead of memorizing each specific equation, you should try to organize the chemistry of these elements according to the generalizations you have already learned.

12.1 GROUP IA: ALKALI METALS

The alkali metals, Li, Na, K, Rb, and Cs, have already been described in some detail in Sec. 4.1. They were included there because all are so similar to each other and because they played an important role in Mendeleev's formulation of the periodic table. (You should review Sec. 4.1 now.) All the alkali metals are soft and, except for Cs which is yellow, are silvery-gray in color. All combine swiftly with air to form white oxide, yellow peroxide, or yellow superoxide coatings which protect their surfaces from further oxidation.

Two other elements are found in group IA. Hydrogen, although many of its compounds have formulas similar to the alkali metals, is a nonmetal and is almost unique in its chemical behavior. Therefore it is not usually included in this group. Francium (Fr) is quite radioactive, and only small quantities are available for study; so it too is usually omitted. Its properties, however, appear to be similar to those of Cs and the other alkali metals.

Some general properties of the alkali metals are summarized in Table 12.1. All these metal atoms contain a single *s* electron outside a noble-gas configuration, and so the valence electron is well shielded from nuclear charge and the atomic radii are relatively large. The large volume of each atom results in a low density—small enough that Li, Na, and K float on water as they react with it. (See color Plate 1.) The atoms do not have a strong attraction for the single valence electron, and so it is easily lost (small first-ionization energy) to form a +1 ion. Because they readily donate electrons in this way, all the alkali metals are strong reducing agents. They are quite reactive, even reducing water.

Weak attraction for the valence electron also results in weak metallic bonding, because it is attraction among nuclei and numerous valence electrons that holds metal atoms together. Weak metallic bonding results in low melting points, especially for the larger atoms toward the bottom of the group. Cs, for example, melts just above room temperature. Weak metallic bonding also accounts for the fact that all these metals are rather soft.

That the chemistry of alkali metals is confined to the +1 oxidation state is confirmed by the large second-ionization energies. Removing the first electron from the large, diffuse *s* orbital is easy, but removing a second electron from an octet in an M^+ ion is much too difficult for any oxidizing agent to do.

TABLE 12.1 Properties of the Group IA Alkali Metals.

Element	Symbol	Electron Configuration	Usual Oxidation State	Ionization Energy/ MJ mol^{-1} First	Ionization Energy/ MJ mol^{-1} Second	Radius/pm Atomic	Radius/pm Ionic (M^+)	Density/ g cm^{-3}	Electro-negativity	Melting Point (in °C)
Lithium	Li	[He]$2s^1$	+1	0.526	7.305	122	60	0.534	1.0	179
Sodium	Na	[Ne]$3s^1$	+1	0.502	4.569	157	95	0.97	0.9	98
Potassium	K	[Ar]$4s^1$	+1	0.425	3.058	202	133	0.86	0.8	64
Rubidium	Rb	[Kr]$5s^1$	+1	0.409	2.638	216	148	1.52	0.8	39
Cesium	Cs	[Xe]$6s^1$	+1	0.382	2.430	235	169	1.87	0.7	28

The alkali metals react directly with many elements, as described in Sec. 4.1. Their oxides, hydrides, hydroxides, and sulfides all dissolve in water to give basic solutions, and these compounds are among the strong bases listed in the previous chapter.

The peroxides and superoxides formed when the heavier alkali metals react with O_2 also dissolve to give basic solutions:

$$2Na_2O_2(s) + 2H_2O(l) \rightarrow 4Na^+(aq) + 4OH^-(aq) + O_2(g)$$

$$4KO_2(s) + 2H_2O(l) \rightarrow 4K^+ + 4OH^- + 3O_2(g)$$

Both of the latter equations describe redox as well as acid-base processes, as you can confirm by assigning oxidation numbers. The peroxide and superoxide ions contain O atoms in the unusual (for O) -1 and $-\frac{1}{2}$ oxidation states:

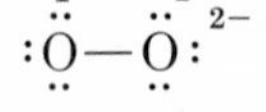

Peroxide ion

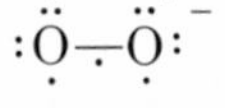

Superoxide ion

Therefore **disproportionation** (simultaneous oxidation and reduction) of O_2^{2-} or O_2^- to the more common oxidation states of 0 (in O_2) and -2 (in OH^-) is possible.

Sodium and potassium are quite abundant, ranking sixth and seventh among all elements in the earth's crust, but the other alkali metals are rare. Sodium and potassium ions are components of numerous silicate crystal lattices (Sec. 8.5), but since most of their compounds are water soluble, they are also important constituents of seawater and underground deposits of brine. Sodium chloride obtained from such brines is the chief commercial source of sodium, while potassium can be obtained from the ores sylvite (KCl) or carnallite ($KCl \cdot MgCl_2 \cdot 6H_2O$).

Both sodium (Na^+) and potassium (K^+) ions are essential to living systems. Na^+ is the main cation in fluids surrounding the cells, while K^+ is most important inside the cells. Na^+ plays a role in muscle contraction, and both K^+ and Na^+ play a role in transmitting nerve impulses. K is more important than Na in plants, and it is one of three elements (K, P, N) which must be supplied in fertilizer to maintain high crop yields. K is especially abundant in trees—wood ashes from kitchen fires (potash) were the major source of this element as recently as a century ago, and they still make good fertilizer for your garden. Wood ashes contain a mixture of potassium oxide and potassium carbonate, the latter formed by combination of K_2O with CO_2 produced when C in the wood combines with O_2:

$$K_2O + CO_2 \rightarrow K_2CO_3$$

Na compounds are obtained commercially from brine or from seawater. When an electrical current is passed through an NaCl solution (a process called **electrolysis**), $Cl_2(g)$, $H_2(g)$, and a concentrated solution of NaOH (caustic soda or lye) are obtained:

$$Na^+(aq) + 2Cl^-(aq) + 2H_2O(l) \xrightarrow{\text{electrolysis}} Cl_2(g) + H_2(g) + Na^+(aq) + 2OH^-(aq)$$

This process will be described in more detail in Chap. 17, but you can see from the equation that the electrical current oxidizes Cl^- to Cl_2 and reduces H_2O to H_2. $NaOH(aq)$ is used as a strong base in numerous industrial processes to make soap, rayon, cellophane, paper, dyes, and many other products. Lye is also used in home drain cleaners. It must be handled with care because it is strongly basic, highly caustic, and can severely burn the skin.

A second important industrial use of brine is the **Solvay process:**

$$CO_2(g) + NH_3(aq) + Na^+(aq) + Cl^-(aq) + H_2O(l) \rightarrow NaHCO_3(s) + NH_4^+(aq) + Cl^-(aq)$$

The Solvay process is an acid-base reaction combined with a precipitation. The acid anhydride, CO_2, reacts with H_2O to produce H_2CO_3. This weak acid donates a proton to NH_3, yielding NH_4^+ and HCO_3^-, and the latter ion precipitates with Na^+. The weakly basic sodium hydrogen carbonate ($NaHCO_3$) produced by the Solvay process can be purified for use as an antacid (bicarbonate of soda), but most of it is converted to sodium carbonate (soda ash) by heating:

$$2NaHCO_3(s) \xrightarrow{\Delta} Na_2CO_3(s) + H_2O(g) + CO_2(g)$$

(The Δ in this equation indicates heating of the reactant.) Sodium carbonate (Na_2CO_3) is used in manufacturing glass and paper, and in some detergents. The carbonate ion is a rather strong base, however, and detergents containing Na_2CO_3 (washing soda) have resulted in severe chemical burns to some small children who, out of curiosity, have eaten them.

12.2 GROUP IIA: ALKALINE EARTHS

Some properties of the alkaline earths have already been described in Sec. 4.1. You should review that section now. The last member of the group, Ra, is radioactive and will not be considered here. All alkaline earths are silvery-gray metals which are ductile and relatively soft. However, Table 12.2 shows that they are much denser than the group IA metals, and their melting points are significantly higher. They are also harder than the alkali metals. This may be attributed to the general valence electron configuration ns^2 for the alkaline earths, which involves two electrons per metal atom in metallic bonding (instead of just one as in an alkali metal).

First and second ionization energies for the alkaline earths (corresponding to removal of the first and second valence electrons) are relatively small, but the disruption of an octet by removal of a third electron is far more difficult (see Table 12.2). Like the alkali metals, the alkaline-earth atoms lose electrons easily, and so they are good reducing agents. Except for Be (which has the largest ionization energy), they react directly with most nonmetallic elements, forming hydrides: MH_2 (M = Mg, Ca, Sr, Ba); halides: MF_2, MCl_2, MBr_2, and MI_2; oxides: MO; and sulfides: MS. In all these compounds the alkaline-earth elements occur as dipositive ions, Mg^{2+}, Ca^{2+}, Sr^{2+}, or Ba^{2+}.

Similar compounds of Be can be formed by roundabout means, but not by direct combination of the elements. Moreover, the Be compounds are

TABLE 12.2 Properties of the Group IIA Alkaline-Earth Metals.

Element	Symbol	Electron Configuration	Usual Oxidation State	Ionization Energy/ $MJ\ mol^{-1}$ First	Second	Third
Beryllium	Be	$[He]2s^2$	+2	0.906	1.763	14.86
Magnesium	Mg	$[Ne]3s^2$	+2	0.744	1.457	7.739
Calcium	Ca	$[Ar]4s^2$	+2	0.596	1.152	4.918
Strontium	Sr	$[Kr]5s^2$	+2	0.556	1.071	4.21
Barium	Ba	$[Xe]6s^2$	+2	0.509	0.972	3.43

more covalent than ionic. The Be^{2+} ion has a very small radius (31 pm) and is therefore capable of distorting (polarizing) the electron cloud of an anion in its vicinity. Therefore all bonds involving Be have considerable covalent character, and the chemistry of Be is significantly different from that of the other members of group IIA.

Other trends among the data in Table 12.2 are what we would expect. Ionization energies and electronegativities decrease from top to bottom of the group, and atomic and ionic radii increase. The radii of +2 alkaline-earth ions are much smaller than the +1 alkali-metal ions of the same period (compare Tables 12.1 and 12.2), because the greater nuclear charge holds the inner shells more tightly. This effect is sufficiently large that an alkaline earth below and to the right of a given alkali metal in the periodic table often has nearly the same ionic radius. Thus Na^+ (95 pm) can fit into exactly the same type of crystal lattice as Ca^{2+} (99 pm), and these two elements are often found in the same minerals. The same is true of K^+ and Ba^{2+}.

Similarity of ionic radii also leads to related properties for Li and Mg. Since these two elements are adjacent along a diagonal line from the upper left to the lower right in the periodic table, their similarity is called a **diagonal relationship.** Diagonal relationships are mainly evident in the second and third periods: Be is similar to Al, and B is like Si in many ways. Farther toward the right-hand side of the table such relationships are less pronounced. The most striking similarity between Li and Mg is their ability to form covalent bonds with elements of average electronegativity, such as C, while forming fairly ionic compounds with more electronegative elements, such as O or F. Two examples of covalent compounds are ethyllithium, CH_3CH_2Li, and diethylmagnesium, $(CH_3CH_2)_2Mg$. Such compounds are likely in the case of Li and Mg but not the alkali or alkaline earths below them, because Li^+ and Mg^{2+} are small enough to be strongly polarizing and thus form bonds with considerable covalent character.

Chemical Reactions and Compounds

As in the case of the alkali metals, the most important and abundant alkaline earths, Mg and Ca, are in the third and fourth periods. Be is rare, although its strength and low density make it useful in certain special

Radius/pm					
Atomic	Ionic (M^{2+})	Density/ g cm^{-3}	Electro-negativity	Melting Point (in °C)
89	31	1.86	1.5	1278
136	65	1.74	1.2	651
174	99	1.54	1.0	839
191	113	2.60	1.0	769
198	135	3.51	0.9	725

alloys. Sr and Ba occur naturally as the relatively insoluble sulfates $SrSO_4$ (strontianite) and $BaSO_4$ (barite), but these two elements are of minor commercial importance.

The most common ores of Mg and Ca are dolomite, $MgCO_3 \cdot CaCO_3$, after which an entire mountain range in Italy is named, and limestone, $CaCO_3$, an important building material. Mg is also recovered from seawater on a wide scale. The oxides of the alkaline earths are commonly obtained by heating the carbonates. For example, lime, CaO, is obtained from limestone as follows:

$$CaCO_3(s) \xrightarrow{\Delta} CaO(s) + CO_2(g)$$

Except for BeO, which is covalently bonded, alkaline-earth oxides contain O^{2-} ions and are strongly basic. When treated with water (a process known as **slaking**), they are converted to hydroxides:

$$CaO(s) + H_2O(l) \rightarrow Ca(OH)_2(s)$$

$Ca(OH)_2$ (slaked lime) is an important strong base for industrial applications, because it is cheaper than NaOH.

MgO has an extremely high melting point (2800°C) because of the close approach and large charges of its constituent Mg^{2+} and O^{2-} ions in the crystal lattice. As a solid it is a good electrical insulator, and so it is used to surround metal-resistance heating wires in electric ranges. MgO is also used to line high-temperature furnaces. When converted to the hydroxide, Mg finds a different use. $Mg(OH)_2$ is quite insoluble in water, and so it does not produce a high enough concentration of hydroxide ions to be caustic. It is basic, however, and gram for gram can neutralize nearly twice the quantity of acid that NaOH can. Consequently a suspension of $Mg(OH)_2$ in water (milk of magnesia) makes an excellent antacid, for those who can stand its taste.

Because the carbonate ion behaves as a Brönsted-Lowry base, carbonate salts dissolve in acidic solutions. In nature, water often becomes acidic because the acidic oxide CO_2 is present in the atmosphere. When CO_2 from the air dissolves in water, it can help dissolve limestone:

$$CO_2(g) + H_2O(l) + CaCO_3(s) \rightleftharpoons Ca^{2+}(aq) + 2HCO_3^-(aq)$$

This reaction often occurs underground as rainwater saturated with CO_2

TABLE 12.3 Properties of the Group IIIA Elements.

Element	Symbol	Electron Configuration	Usual Oxidation State	Ionization Energy/ MJ mol^{-1} First	Second	Third
Boron	B	[He]$2s^22p^1$	+3	0.807	2.433	3.666
Aluminum	Al	[Ne]$3s^23p^1$	+3	0.584	1.823	2.751
Gallium	Ga	[Ar]$4s^23d^{10}4p^1$	+3	0.585	1.985	2.969
Indium	In	[Kr]$5s^24d^{10}5p^1$	+3, +1	0.565	1.827	2.711
Thallium	Tl	[Xe]$6s^24f^{14}5d^{10}6p^1$	+1, +3	0.596	1.977	2.884

seeps through a layer of limestone. Caves from which the limestone has been dissolved are often prevalent in areas where there are large deposits of $CaCO_3$. In addition, the groundwater and well water in such areas becomes hard. **Hard water** contains appreciable concentrations of Ca^{2+}, Mg^{2+}, and certain other metal ions. These form insoluble compounds with soap, causing curdy, scummy precipitates. Hard water can be softened by adding Na_2CO_3, washing soda, which precipitates $CaCO_3$, or by **ion exchange,** a process in which the undesirable Ca^{2+} and Mg^{2+} ions are replaced in solution by Na^+ ions, which do not precipitate soap. Most home water softeners work on the latter principle.

12.3 GROUP IIIA

The elements of this group show considerable variability in properties from top to bottom of the periodic table. B is a semimetal or metalloid, and the element has a covalent network structure in which icosahedrons of boron atoms (Fig. 12.1) are linked together. None of its compounds contain B^{3+} ions—covalent bonding is the rule there as well. Many B compounds, especially those containing O, are similar to those of Si (another diagonal relationship) instead of Al. In elemental Al the atoms are closest packed. Al is definitely metallic, but like Be, to which it is related diagonally, it forms many compounds with extensive covalent character. The other three ele-

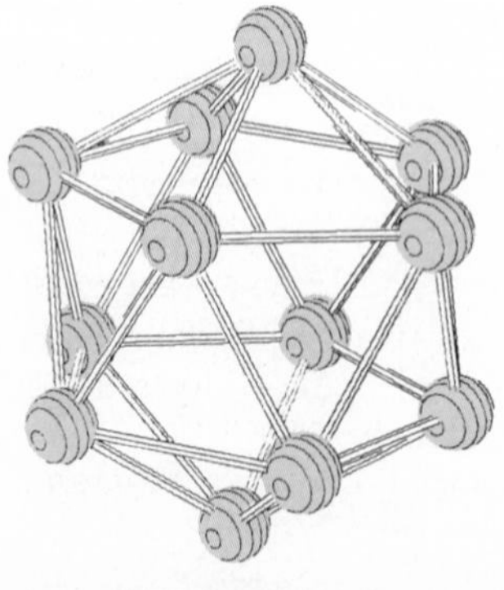

Figure 12.1 Boron icosahedron structural unit. (Computer-generated.) *(Copyright © 1976 by W. G. Davies and J. W. Moore.)*

Radius/pm				
Atomic	Ionic (M^{3+})	Density/ g cm^{-3}	Electro-negativity	Melting Point (in °C)
80	—	2.34	2.0	2300
125	50	2.70	1.5	660
124	62	5.90	1.6	30
142	81	7.31	1.7	157
144	95	11.83	1.8	304

ments in the group, Ga, In, and Tl, are also metals, but their chemistry is affected to some extent by the fact that they follow the transition elements in the periodic system. The presence of a filled *d* subshell, as opposed to only *s* and *p* subshells in B and Al, introduces some new atomic properties.

Some properties of the group IIIA elements are shown in Table 12.3. In this case the trends are not quite what might have been expected on the basis of previous experience with the periodic table. Both ionization energy and electronegativity decrease significantly from B to Al, but as one proceeds on down the group, these atomic properties change very little. Indeed electronegativity increases from Ga to In to Tl. There is also a break in the expected steady increase in atomic radius down the group. Al and Ga have nearly identical radii, and the ionic radii of Al^{3+} and Ga^{3+} differ by only 8 pm. In the case of Tl the most common oxidation state is +1, corresponding to loss of only the $6p$ valence electron, rather than an oxidation state of +3, which would entail loss of the $6s^2$ pair as well.

Many of these anomalies can be understood if we look more closely at the electron configurations of Ga, In, and Tl. In each case a *d* subshell has been filled between the ns^2 pair and the final np^1 valence electron. The $4f$ subshell also intervenes in the building-up process for Tl. Since *d* and *f* electrons are not as efficient as *s* or *p* electrons at screening nuclear charge, the valence electrons of Ga, In, and Tl are more strongly attracted, the atoms are smaller than would be expected, and ionization energies are unusually large.

Another factor affecting the chemistry of the group IIIA elements is the relative sizes of the first, second, and third ionization energies. These are roughly in the ratio 1:3:4.5 for all elements, and the large increase from first to second ionization energy becomes more pronounced toward the bottom of the group. In the case of Tl, whose large radius prevents close approach and strong bonding to other atoms, the energy required to unpair or ionize the $6s^2$ pair of electrons and use them to form bonds is often too large to be compensated by the bond energies of the two additional bonds. Consequently Tl^+ is the more common oxidation state, rather than Tl^{3+}. This reluctance of the $6s^2$ pair of electrons to be used in bonding is called the **inert-pair effect.** It also affects the chemistry of Hg, Pb, and Bi, elements which are adjacent to Tl in the sixth period.

Unlike groups IA and IIA, none of the group IIIA elements react directly with hydrogen to form hydrides. The halides of B, Al, and Ga will react with sodium hydride, however, to form tetrahydro anions:

$$4NaH + BF_3 \rightarrow NaBH_4 + 3NaF$$

The compounds $NaBH_4$, $NaAlH_4$, and $NaGaH_4$ do not contain H^- ions. Instead the hydrogens are covalently bonded to the group IIIA atom:

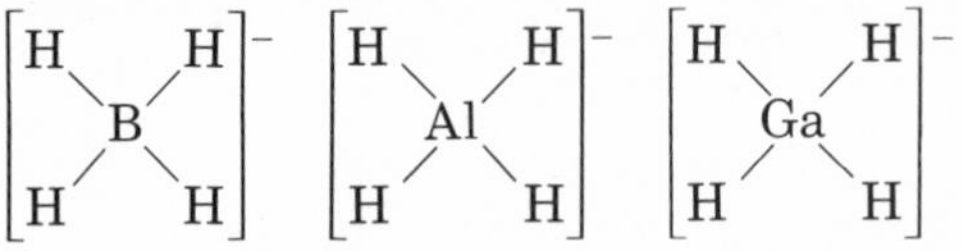

All these anions are tetrahedral, as would be predicted by VSEPR theory (Sec. 7.2).

The tetrahydroaluminate and tetrahydrogallate ions react readily with H_2O, splitting the H_2O molecule so that the H ends up with another H to form H_2, while the OH ends up with the group IIIA atom:

$$AlH_4^-(aq) + 4HOH(l) \rightarrow Al(OH)_3(s) + OH^-(aq) + 4H_2(g)$$

Splitting of an H_2O molecule (hydrolysis) by these compounds is similar to the hydrolysis of esters, discussed in Sec. 8.4. The hydrolysis of AlH_4^- is violently explosive, making it necessary to handle $NaAlH_4$ in a dry environment. The anion in $NaBH_4$, by contrast, involves bonds with greater covalent character and is much less readily hydrolyzed.

Only B forms a wide variety of hydride compounds. The simplest of these is diborane, B_2H_6, a volatile, readily hydrolyzed compound which may decompose explosively. Diborane is made as follows:

$$3NaBH_4 + 4BF_3 \rightarrow 3NaBF_4 + 2B_2H_6$$

Diborane is electron deficient—if you try to draw a Lewis diagram, you will soon find it to be impossible—and for a number of years theoretical chemists were mystified about what held the atoms together.

The current picture of bonding in diborane is shown in Fig. 12.2. Each B is assumed to be surrounded by four sp^3 hybrid orbitals. Four of the eight sp^3 hybrids from the two B's overlap the 1*s* orbitals of four H's, forming four B—H bonds, all of which lie in a plane. The other four sp^3 hybrids overlap pairwise between the B's, forming two banana bonds (like those in ethene, C_2H_4, in Fig. 7.8). Each of the remaining two H nuclei is embedded in the middle of one of the banana bonds. In this way the two electrons in each banana bond hold three nuclei (two B's and one H) together, and the bond is called a **three-center two-electron bond.**

All the group IIIA elements form trihalides, although Tl does so reluctantly, preferring to remain in the +1 oxidation state. The general reaction is

$$2M(s) + 3X_2 \rightarrow 2MX_3 \qquad M = B, Al, Ga, In, Tl \qquad X = F, Cl, Br, I$$

All the boron halides consist of discrete BX_3 molecules, which are electron deficient. VSEPR theory predicts trigonal planar structures, and the boron

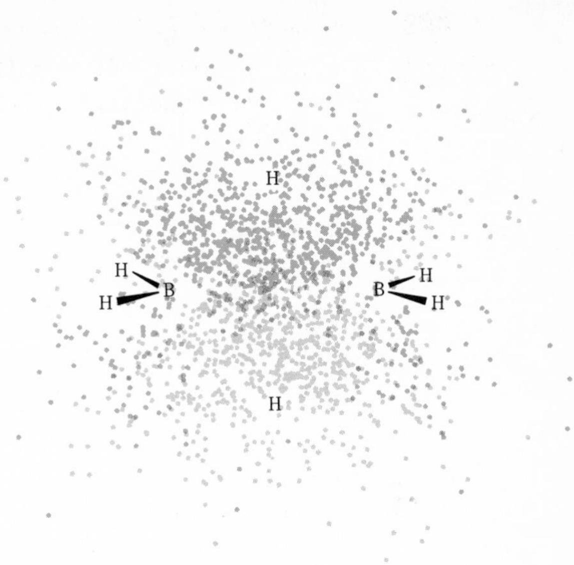

Figure 12.2 Structure and bonding in diborane, B_2H_6. Electron density in the three-center, two-electron bonds is shown in color. (Computer-generated.) *(Copyright © 1977 by W. G. Davies and J. W. Moore.)*

halides are strong Lewis acids because each can accept an electron pair to complete an octet on B. The bromides and iodides of Al, Ga, In, and Tl are primarily covalent, and two MX_3 molecules usually combine to form a **dimer** (dimer means two units):

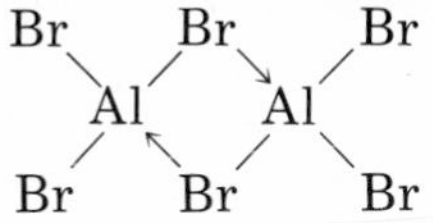

In the dimer structure a lone electron pair from a Br atom in one $AlBr_3$ unit has been donated to the Al of the other $AlBr_3$ unit, and vice versa. This forms two coordinate covalent bonds to hold the dimer together. (These bonds are shown as arrows.)

The fluorides of Al, Ga, In, and Tl are primarily ionic. Only the larger, more easily polarizable Br^- and I^- ions can be distorted enough by the small M^{3+} ions to give mainly covalent bonds. Most of the chlorides are borderline cases. This is reflected in the melting points:

AlF_3, 1291°C $\quad$ $AlCl_3$, 190°C $\quad$ Al_2Br_6, 97.5°C $\quad$ Al_2I_6, 191°C

Solid $AlCl_3$ consists of Al^{3+} and Cl^- ions, but in the liquid and gas phases Al_2Cl_6 dimers predominate.

The oxides and oxyanions of B and Al are the main compounds of commercial importance. Borax, $Na_2B_4O_7{\cdot}10H_2O$, is the principal ore of B, and Al is obtained from bauxite, $Al_2O_3{\cdot}xH_2O$. Bauxite is an example of a **hydrous oxide.** It contains an indeterminate amount of water (xH_2O) in the crystal lattice. Although Al is the most abundant metal in the earth's crust, most of it is found in complicated aluminosilicates (feldspars) of the type described in Sec. 8.5. It is extremely difficult to convert the feldspars to Al metal, and any Si which remains as an impurity greatly degrades the strength and other properties of Al. Consequently the metal is obtained from bauxite, of which there is only a limited supply.

The first step in the recovery of Al is called the **Bayer process.** This depends on the fact that Al_2O_3 is **amphoteric;** that is, it can behave as either an acid or a base. If it behaves as a base, it can react with, and dissolve in, acidic solutions. If it behaves as an acid, it can dissolve in basic solutions. The amphoteric behavior of Al_2O_3 comes about because of the presence of Al^{3+} ions and O^{2-} ions. Al^{3+} has a very small radius and a large ionic charge, and so it strongly attracts H_2O molecules—so strongly that the $Al(H_2O)_6^{3+}$ ion can donate protons. (Other cations which do this were described in Sec. 11.3.) Oxide ion, of course, is a good proton acceptor and a strong base.

The Bayer process makes use of the acidic behavior of Al_2O_3 by dissolving it in strong base:

$$Al_2O_3(s) + 3H_2O(l) + 2OH^-(aq) \rightarrow 2Al(OH)_4^-(aq)$$

The oxides of most metals are strongly basic and do not dissolve, while the oxides of nonmetals such as Si are acidic and more soluble in base than Al_2O_3. Thus a fairly pure sample of aluminum hydroxide, $Al(OH)_3$, can be obtained by filtering off the basic oxides and acidifying the solution slightly. CO_2, an acidic oxide, is used for this purpose:

$$CO_2(g) + Al(OH)_4^-(aq) \rightarrow Al(OH)_3(s) + HCO_3^-(aq)$$

The $Al(OH)_3$ is then heated to drive off H_2O:

$$2Al(OH)_3(s) \xrightarrow{\Delta} Al_2O_3(s) + 3H_2O(g)$$

and Al is obtained by the **Hall-Heroult process:**

$$Al_2O_3(l) \xrightarrow{\text{electrolysis}} 2Al(l) + 3O_2(g)$$

(This process is described in greater detail in Chap. 17.)

The last step in aluminum production requires tremendous quantities of electrical energy. This, as well as the scarcity of its ore, makes aluminum more expensive than iron, the only metal which is more widely used. Nevertheless, aluminum has several advantages over iron. One is its considerably lower density, making possible alloys of comparable strength with considerably lower mass. A second advantage of aluminum is the coating of Al_2O_3 which forms on its surface when it is exposed to air. This coating sticks to

TABLE 12.4 Properties of the Group IVA Elements.

Element	Symbol	Electron Configuration	Usual Oxidation State	Ionization Energy/MJ mol^{-1} First	Second	Third	Fourth
Carbon	C	$[He]2s^22p^2$	+4, +2	1.093	2.359	4.627	6.229
Silicon	Si	$[Ne]3s^23p^2$	+4, +2	0.793	1.583	3.238	4.362
Germanium	Ge	$[Ar]4s^23d^{10}4p^2$	+4, +2	0.768	1.544	3.308	4.407
Tin	Sn	$[Kr]5s^24d^{10}5p^2$	+2, +4	0.715	1.418	2.949	3.937
Lead	Pb	$[Xe]6s^24f^{14}5d^{10}6p^2$	+2, +4	0.722	1.457	3.088	4.089

the surface and is insoluble in neutral solutions, and so it prevents further oxidation. Iron, by contrast, forms a hydrous oxide (rust) which flakes off easily, exposing additional metal to the air. Thus under normal environmental conditions aluminum will last much longer than iron. This is a problem when aluminum beverage cans litter roadsides, but in general it makes aluminum a very favorable candidate for recycling.

12.4 GROUP IVA

Near the middle of the periodic table there is greatest variability of properties among elements of the same group. This is certainly true of group IVA, which contains carbon, a nonmetal, silicon and germanium, both semimetals, and tin and lead, which are definitely metallic. Elemental carbon exists in two allotropic forms, diamond and graphite, whose structures were shown in Chap. 8 (Fig. 8.12). In diamond there is a three-dimensional network of covalent bonds, while graphite consists of two-dimensional layers covalently bonded. Silicon, germanium, and one allotrope of tin (gray tin) also have the diamond structure—each atom is surrounded by four others arranged tetrahedrally. White tin has an unusual structure in which there are four nearest-neighbor atoms at a distance of 302 pm and two others at 318 pm. Only lead has a typical closest-packed metallic structure in which each atom is surrounded by 12 others.

Some properties of the group IVA elements are summarized in Table 12.4. As in the case of group IIIA (and for the same reason), there is a large decrease in ionization energy and electronegativity from carbon to silicon, but little change farther down the group. Note also that ionization energies, especially the third and fourth, are rather large. Formation of true +4 ions is very difficult, and in their +4 oxidation states all group IVA elements form predominantly covalent bonds. The +2 oxidation state, corresponding to use of the np^2, but not the ns^2, electrons for bonding, occurs for all elements. It is most important in the case of tin and especially lead, the latter having an inert pair like that of thallium. In the +4 oxidation state lead is a rather strong oxidizing agent, gaining two electrons ($6s^2$) and being reduced to the +2 state.

Radius/pm				
Covalent	**Ionic (M^{2+})**	**Density/ g cm^{-3}**	**Electro-negativity**	**Melting Point (in °C)**
77	—	3.51	2.5	3550
117	—	2.33	1.8	1410
122	—	5.35	1.8	937
141	122	7.28	1.8	232
146	131	11.34	1.8	327

Carbon's ability to form strong bonds with other carbon atoms and the tremendous variety of organic compounds have already been discussed in Secs. 8.2 and 8.4. (You should review those sections now.) The most important inorganic carbon compounds are carbon monoxide and carbon dioxide. Both are produced by combustion of any fuel containing carbon:

$$C + \tfrac{1}{2}O_2 \rightarrow CO \qquad (12.1)$$

$$CO + \tfrac{1}{2}O_2 \rightarrow CO_2 \qquad (12.2)$$

The triple bond in C≡O is the strongest chemical bond known, and O=C=O contains two double bonds, and so both molecules are quite stable.

Equations (12.1) and (12.2) occur stepwise when a fuel is burned, and the strong C≡O bond makes Eq. (12.2) slow unless the temperature is rather high. If there is insufficient O_2 or if the products of combustion are cooled rapidly, significant quantities of CO can be produced. This is precisely what happens in an automobile engine, and the exhaust contains between 3 and 4% CO unless pollution controls have been installed.

CO is about 200 times better than O_2 at bonding to hemoglobin, the protein which transports O_2 through the bloodstream from the lungs to the tissues. Consequently a small concentration of CO in the air you breathe can inhibit transport of O_2 to the brain, causing drowsiness, loss of consciousness, and death. (After a few minutes of breathing undiluted auto exhaust, more than half your hemoglobin will be incapable of transporting O_2, and you will faint.) CO in automobile exhaust can be used to put animals to sleep. Because CO is colorless and odorless, your senses cannot detect it, and people must constantly be cautioned not to run cars in garages or other enclosed spaces. With the large number of cars and the great number of miles driven today, CO concentrations in many cities have become a problem, and over the past decade pollution-control devices have become mandatory for most automobiles.

Like the organic compounds of carbon, the oxygen compounds of silicon which make up most of the earth's crust have already been described (Sec. 8.5). These substances illustrate a major contrast between the chemistry of carbon and silicon. The latter element does form a few compounds, called silanes, which are analogous to the alkanes, but the Si—Si bonds in silanes are much weaker than Si—O bonds. Consequently the silanes combine readily with oxygen from air, forming Si—O—Si linkages. Unlike the alkanes, which must be ignited with a spark or a match before they will burn, silanes catch fire of their own accord in air:

$$2Si_4H_{10} + 13O_2 \rightarrow 4SiO_2 + 5H_2O$$

Another important group of silicon compounds is the **silicones.** These polymeric substances contain Si—O—Si linkages and may be thought of as derived from silicon dioxide, SiO_2. To make silicones, one must first reduce silicon dioxide to silicon. This can be done using carbon as the reducing agent in a high-temperature furnace:

$$SiO_2(s) + 2C(s) \xrightarrow{3000°C} Si(l) + 2CO(g)$$

The silicon is then reacted with chloromethane:

$$Si(s) + 2CH_3Cl(g) \xrightarrow[\text{Cu catalyst}]{300°C} (CH_3)_2SiCl_2(g)$$

The dichlorodimethylsilane obtained in this reaction polymerizes when treated with water:

$$n(CH_3)_2SiCl_2 + nH_2O \longrightarrow \left(\begin{array}{c} CH_3 \\ | \\ -Si-O \\ | \\ CH_3 \end{array} \right)_n + 2nHCl$$

The silicone polymer consists of a strongly bonded —Si—O—Si—O—Si—O chain, called a **siloxane** chain, with two methyl groups (or other organic groups) on each silicon atom. The strong backbone of a silicone polymer makes it stable to heat and difficult to decompose. Silicone oils make good lubricants and heat-transfer fluids, and rubber made from silicone remains flexible at low temperatures.

Besides the metals themselves, some tin and lead compounds are of commercial importance. Tin(II) fluoride (stannous fluoride), SnF_2, is added to some toothpastes to inhibit dental caries. Tooth decay involves dissolving of dental enamel [mainly $Ca_{10}(PO_4)_6(OH)_2$] in acids synthesized by bacteria in the mouth. Fluoride ions from SnF_2 inhibit decay by transforming tooth surfaces into $Ca_{10}(PO_4)_6F_2$, which is less soluble in acid:

$$Ca_{10}(PO_4)_6(OH)_2 + SnF_2 \rightarrow Ca_{10}(PO_4)_6F_2 + Sn(OH)_2$$

Since F^- is a weaker base than OH^-, the F^- compound has less tendency to react with acids. Note that when tin or lead are in the +2 oxidation state and are combined with a highly electronegative element like fluorine, the compounds formed are rather ionic.

Lead is found in two main commercial applications. One, the lead-acid storage battery, used to start cars and power golf carts, will be discussed in Chap. 17. The other is the lead found in automobile fuel. In the +4 oxidation state lead forms primarily covalent compounds and bonds strongly to carbon. The compound tetraethyllead may be synthesized by reacting chloroethane with a sodium-lead alloy:

$$4NaPb + 4CH_3CH_2Cl \rightarrow (CH_3CH_2)_4Pb + 4Pb + 4NaCl$$

Sodium dissolved in the lead makes the latter more reactive. Tetraethyllead prevents gasoline from igniting too soon or burning unevenly in an automobile engine, circumstances which cause the engine to "knock" or "ping."

12.5 GROUP VA

Although all the elements in this group form compounds in which their oxidation state of +5 equals the group number, their other properties vary considerably. Nitrogen is clearly nonmetallic and consists of diatomic triply-

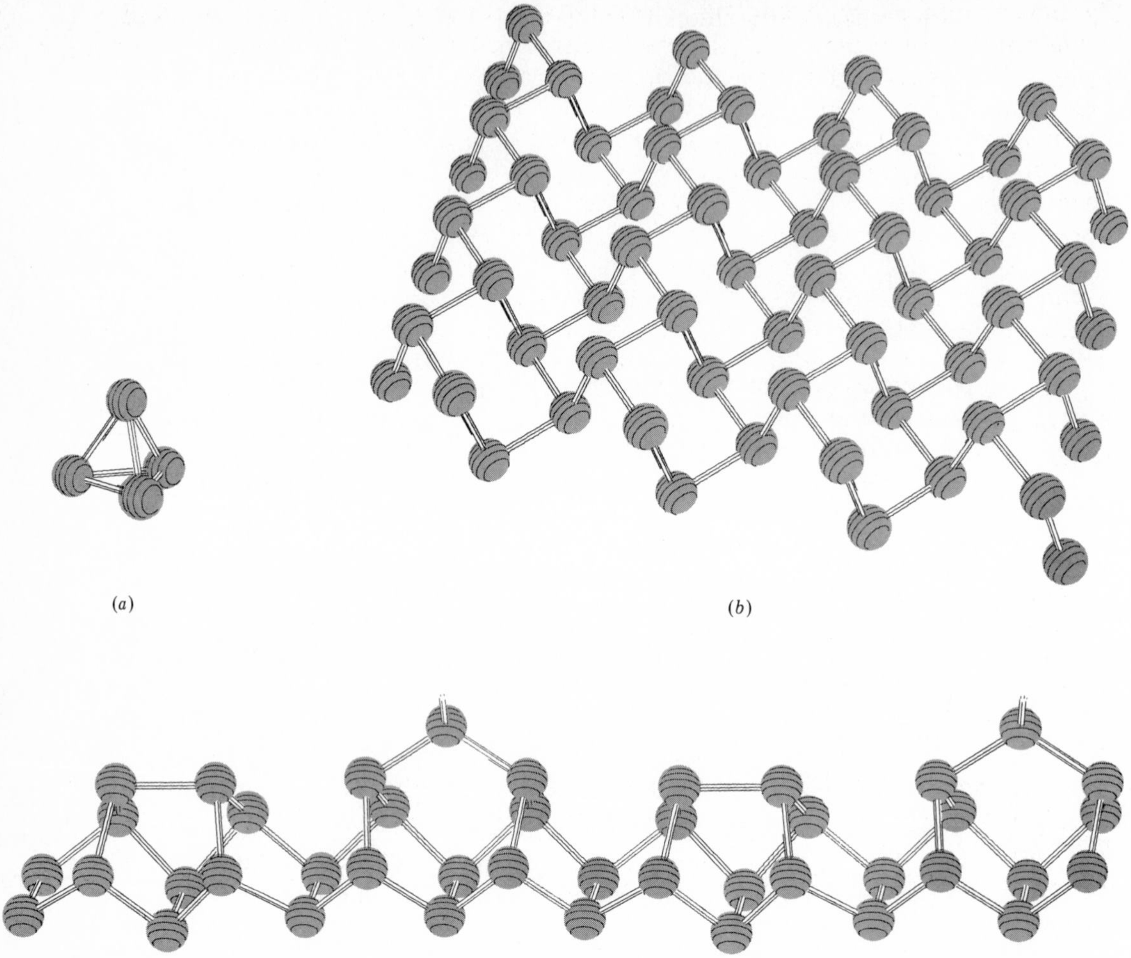

(a) *(b)* *(c)*

Figure 12.3 Allotropic forms of phosphorus. *(a)* P_4 molecule of phosphorus vapor and solid white phosphorus; *(b)* layer structure of black phosphorus; *(c)* chain (polymeric) structure of red phosphorus. Note that in all structures each phosphorus atom forms three bonds. (Computer-generated.) *(Copyright © 1977 by W. G. Davies and J. W. Moore.)*

bonded N_2 molecules. Phosphorus, also a nonmetal, exists as tetrahedral P_4 molecules in the vapor and the white allotropic form of the solid. On standing, white phosphorus slowly changes to the red allotrope, whose structure is shown in Fig. 12.3*c*. The most stable form of the element is black phosphorus, which has a layer structure (Fig. 12.3*b*). Black phosphorus can be made by heating the white form with a mercury catalyst for 8 days at 220 to 370°C.

Arsenic is a semimetal and consists of As_4 molecules in the gas phase. When $As_4(g)$ is condensed to a solid, three allotropes may form. The most stable of these is metallic arsenic, in which each arsenic atom has three nearest neighbors, with three more arsenic atoms somewhat farther away.

Antimony, also a semimetal, has two allotropes, the more stable one being metallic, like arsenic. In the case of bismuth, only the metallic form occurs. Note that for all the group VA elements the $8 - N$ rule is followed. The number of bonds or nearest neighbors for each atom is 8 minus the group number in N≡N, P_4, and even in the metallic forms of As, Sb, and Bi.

Table 12.5 summarizes the atomic properties of the group VA elements. Overall, the trends are what we would expect, based on our experience with previous groups. These elements exhibit a much wider variety of oxidation states, however, especially in the case of nitrogen. This element forms compounds in which it has every possible oxidation number from -3 (the group number minus 8) to $+5$ (the group number). As in previous groups, the oxidation state in which the ns^2 pair of electrons is not used for bonding becomes more prominent toward the bottom of the periodic table. There are a few compounds, $Bi(NO_3)_3$, for example, in which discrete Bi^{3+} ions are present.

Chemical Reactions and Compounds

The most important compounds of the group VA elements are those of nitrogen and phosphorus. Both elements are essential to all living organisms, and both are progressively removed from soil when plants are cultivated and crops harvested. According to Liebig's law of the minimum (Sec. 3.1), an insufficient supply of either element can limit plant growth and reduce crop yields, and so these elements are important components of fertilizer. More recently both elements have been implicated in several kinds of pollution problems. As we discuss the properties of nitrogen and phosphorus compounds, their effects on food production and environmental degradation will also be discussed.

Nitrogen The importance of nitrogen fertilizer was first recognized over a century ago. By the late 1800s the only major ore of nitrogen, Chile saltpeter, $NaNO_3$, was being mined in Chile and shipped to Europe for application to agricultural land, but the supply was obviously limited. Most nitrogen at the earth's surface is in the form of $N_2(g)$, which makes up 78 percent of the atmosphere by volume (or by amount of substance). Therefore chemists began to look for ways of obtaining nitrogen compounds directly from the atmosphere. Any process which does this is called **nitrogen fixation.**

Nitrogen fixation can occur naturally when an electrical discharge (lightning) heats air to a high temperature. The following reaction occurs:

$$N_2(g) + O_2(g) \rightarrow 2NO(g) \qquad (12.3)$$

The nitrogen monoxide (nitric oxide) formed can react further at ordinary temperatures, producing the brown gas, nitrogen dioxide:

$$2NO(g) + O_2(g) \rightarrow 2NO_2(g) \qquad (12.4)$$

The Lewis diagrams for these and other important nitrogen compounds are shown in Fig. 12.4. From the figure you can see that both NO and NO_2 have

TABLE 12.5 Properties of the Group VA Elements.

Element	Symbol	Electron Configuration	Usual Oxidation State	Ionization Energy/MJ mol^{-1}				
				First	Second	Third	Fourth	Fifth
Nitrogen	N	[He]$2s^22p^3$	+5, +3, −3	1.407	2.862	4.585	7.482	9.452
Phosphorus	P	[Ne]$3s^23p^3$	+5, +3, −3	1.018	1.909	2.918	4.963	6.280
Arsenic	As	[Ar]$4s^23d^{10}4p^3$	+5, +3	0.953	1.804	2.742	4.843	6.049
Antimony	Sb	[Kr]$5s^24d^{10}5p^3$	+5, +3	0.840	1.601	2.450	4.271	5.403
Bismuth	Bi	[Xe]$6s^24f^{14}5d^{10}6p^3$	+5, +3	0.710	1.616	2.472	4.380	5.417

an odd number of electrons and violate the octet rule. In such a case it is common for two molecules to combine (dimerize) by pairing their odd electrons. In the case of NO_2, dimerization occurs below room temperature, producing colorless dinitrogen tetroxide:

$$2NO_2(g) \underset{>110°C}{\overset{<10°C}{\rightleftharpoons}} N_2O_4(g)$$

At room temperature, however, the NO_2 and N_2O_4 are in equilibrium, as evidenced by the brown color of the mixture. NO dimerizes only at very low temperatures in the solid state.

The first industrial nitrogen fixation was done by mimicking nature. Reaction (12.3) was carried out in a plant near Niagara Falls, where hydroelectric generation provided inexpensive power to support an electric arc. NO was further oxidized to NO_2, which was dissolved in H_2O to convert it to nitric acid, HNO_3:

$$3NO_2(g) + H_2O(l) \rightarrow 2H^+(aq) + 2NO_3^-(aq) + NO(g) \qquad (12.5)$$

Note that NO_2 is *not* the acid anhydride of HNO_3. This reaction involves disproportionation of NO_2 (which contains N in the +4 oxidation state) to form HNO_3 (N in the +5 state) and NO (N in the +2 state). The NO can be recycled by reoxidizing it to NO_2, and so it was not wasted. The HNO_3 pro-

Figure 12.4 Lewis diagrams for important nitrogen compounds.

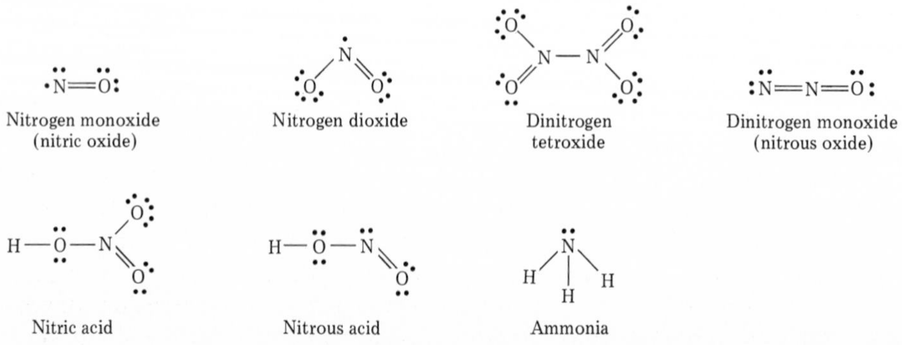

Radius/pm				
Covalent	Ionic (Charge)	Density/ g cm^{-3}	Electro-negativity	Melting Point (in °C)
70	(3−) 171	1.25×10^{-3}	3.0	−210
110	—	1.82	2.1	44
121	—	5.72	2.0	817
141	—	6.69	1.9	631
146	(3+) 108	9.80	1.9	271

duced in Eq. (12.5) was neutralized with NaOH to make a substitute for Chile saltpeter:

$$NaOH(aq) + HNO_3(aq) \rightarrow NaNO_3(aq) + H_2O(l)$$

Fixation of nitrogen by the electric-arc process used a great deal of energy and was rather expensive. Other methods were designed to replace it, and the most successful of these is the **Haber process,** which is the major one used today. Nitrogen is reacted with hydrogen at a high temperature and extremely high pressure over a catalyst consisting of iron and aluminum oxide:

$$N_2(g) + 3H_2(g) \xrightleftharpoons[1000\text{ atm}]{450°\text{C}} 2NH_3(g)$$

The ammonia produced by the Haber process is used directly as a fertilizer. It can be liquefied under pressure and injected through special nozzles about a foot under the soil surface. This prevents loss of gaseous ammonia which would otherwise irritate the nose, throat, and lungs of anyone near a fertilized field. You are probably familiar with the odor of ammonia since it is the most common weak base encountered in the chemical laboratory.

Prior to the recent development of underground injection techniques, most ammonia was converted to ammonium nitrate for fertilizer use:

$$NH_3(aq) + HNO_3(aq) \rightarrow NH_4NO_3(aq) \qquad (12.6)$$

Except for ammonia, ammonium nitrate contains a greater mass fraction of nitrogen than any other compound of comparable cost. Ammonium nitrate manufacture requires that half the ammonia produced in the Haber process be converted to nitric acid. The first step is oxidation of ammonia over a catalyst of platinum metal:

$$4NH_3(g) + 5O_2(g) \xrightarrow[800°\text{C}]{\text{Pt}} 4NO(g) + 6H_2O(g)$$

This is called the **Ostwald process.** It is followed by Eqs. (12.4) and (12.5), yielding nitric acid, which can be combined with ammonia [Eq. (12.6)].

Nitric acid and nitrates have commercial applications other than fertilizer production. Because NO_3^- is a strong oxidizing agent, it reacts vigorously with substances whose elements are in low oxidation states. One example of this is black powder, which consists of charcoal (carbon), sulfur, and potassium nitrate, KNO_3, (saltpeter or nitre). During the American revolution, for example, both armies had numerous persons whose job was to find caves in which the relatively soluble KNO_3 had been deposited as water evaporated. A second example is nitroglycerin (Sec. 8.4) which contains carbon and hydrogen in low oxidation states as well as nitrate. Still another example of an explosive nitrate is NH_4NO_3, which contains nitrogen in its maximum and minimum oxidation state. NH_4NO_3 decomposes as follows:

$$NH_4NO_3(s) \xrightarrow[\text{shock}]{\text{heat or}} N_2O(g) + 2H_2O(g) \qquad \Delta H_m = -37 \text{ kJ mol}^{-1}$$

The reaction is exothermic and produces 3 mol of gaseous products for every mole of solid reactant. This causes a tremendous increase in pressure, and, if the reaction is rapid, an explosion.

The compound dinitrogen monoxide (nitrous oxide or laughing gas), produced by decomposition of NH_4NO_3, is a third important oxide of nitrogen (in addition to NO and NO_2). N_2O is produced during microbial decomposition of organic matter containing nitrogen. Because it is quite unreactive, it is the second most-concentrated nitrogen-containing substance in the atmosphere (after N_2). It is used commercially as an anesthetic, is mildly intoxicating, and is poisonous in large doses.

The other two important oxides of nitrogen, NO and NO_2, play a major role in an air-pollution problem known as **photochemical smog** (or Los Angeles smog). NO is formed by Eq. (12.3) in automobile engines and other high-temperature combustion processes. At normal temperatures NO is oxidized to NO_2 [Eq. (12.4)]. Both these oxides are free radicals (Sec. 7.7) and are rather reactive. Moreover, brown-colored NO_2 absorbs sunlight, and the energy of the absorbed photons breaks a nitrogen-oxygen bond:

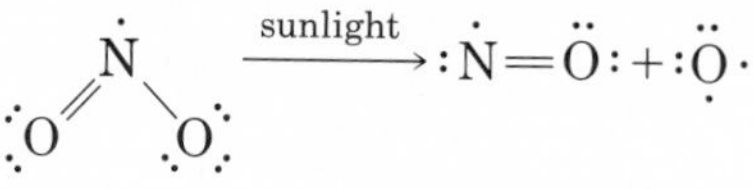

The oxygen atoms produced are highly reactive. They combine with hydrocarbon molecules (from evaporated or unburned gasoline) to form aldehydes, ketones, and a number of other compounds which form an almost foglike cloud and irritate the eyes, throat, and lungs. Photochemical smog is especially bad in cities like Los Angeles and Denver which have lots of sunshine and automobile traffic, but its effects have been observed in every large city in the United States.

Phosphorus As in the case of carbon and silicon, there are major differences between the chemistries of nitrogen and phosphorus. The concentrations of phosphorus compounds in the earth's atmosphere are so small as to be negligible, but phosphorus is more abundant than nitrogen in the solid crust. Here it is found as **phosphate rock,** which is mainly hydroxyapatite, $Ca_{10}(PO_4)_6(OH)_2$, or fluorapatite, $Ca_{10}(PO_4)_6F_2$. (These are the same substances involved in the discussion of dental decay in the previous section.)

Phosphate rock is quite insoluble, and hence its phosphate ions cannot be assimilated by plants. Production of phosphate fertilizer requires treatment of apatite with acid. This protonates the PO_4^{3-} ions, converting them to $H_2PO_4^-$, whose calcium salt is much more soluble:

$$Ca_{10}(PO_4)_6(OH)_2 + 7H_2SO_4 + H_2O \rightarrow 3Ca(H_2PO_4)_2 \cdot H_2O + 7CaSO_4$$

$$Ca_{10}(PO_4)_6(OH)_2 + 14H_3PO_4 \rightarrow 10Ca(H_2PO_4)_2 + 2H_2O$$

The compound $Ca(H_2PO_4)_2 \cdot H_2O$ is known as *superphosphate,* and $Ca(H_2PO_4)_2$ is called *triple superphosphate*.

The phosphoric acid, H_3PO_4, used to make triple superphosphate is also obtained from phosphate rock. The first step is a reduction with carbon (coke) and silicon dioxide in an electric furnace:

$$2Ca_{10}(PO_4)_6(OH)_2 + 18SiO_2 + 30C \rightarrow 3P_4 + 30CO + 2Ca(OH)_2 + 18CaSiO_3$$

The phosphorus obtained this way is then oxidized to phosphorus pentoxide:

$$P_4(s) + 5O_2(g) \rightarrow P_4O_{10}(s)$$

(The name phosphorus pentoxide for P_4O_{10} comes from the empirical formula P_2O_5 of this compound.) Phosphorus pentoxide is the acid anhydride of phosphoric acid:

$$P_4O_{10}(s) + 6H_2O(l) \rightarrow 4H_3PO_4(aq)$$

Although not a very strong acid, phosphoric acid is triprotic. Therefore, 1 mol of this acid can transfer 3 mol of protons to a strong base.

There is another oxide of phosphorus, P_4O_6, which involves the +3 oxidation state, corresponding to use of the $3p^3$, but not the $3s^2$, electrons for bonding. P_4O_6 is the acid anhydride of phosphorous acid, H_3PO_3:

$$P_4O_6(s) + 6H_2O(l) \rightarrow 4H_3PO_3(aq)$$

Phosphorous acid is quite weak, and, contrary to what its formula might suggest, can only donate two protons. This is apparently because its Lewis structure is

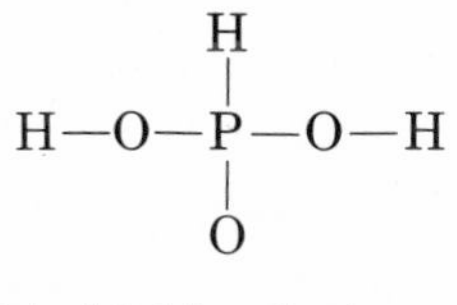

Only the two protons bonded to highly electronegative oxygen atoms are expected to be acidic.

Another major commercial use of phosphates is in laundry detergents. The problem of precipitation of soap by hard-water ions such as Ca^{2+} was mentioned in Sec. 12.2. This can be prevented, and the cleaning power of synthetic detergents can be improved, by adding phosphates. The compound usually used is sodium tripolyphosphate, whose anion is a condensation polymer of hydrogen phosphate and dihydrogen phosphate ions:

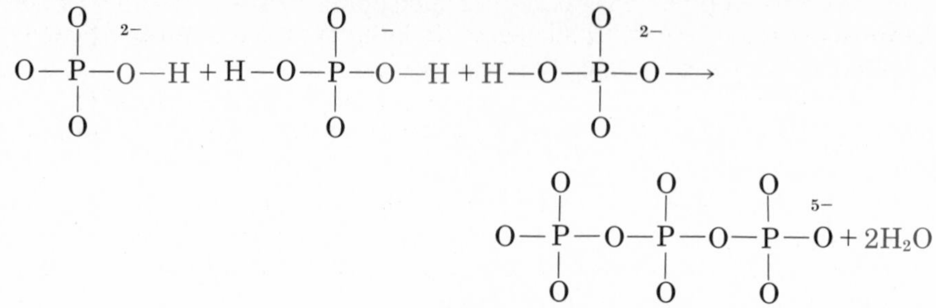

The tripolyphosphate ion has numerous O atoms whose lone pairs of electrons can form coordinate covalent bonds to metal ions like Ca^{2+}:

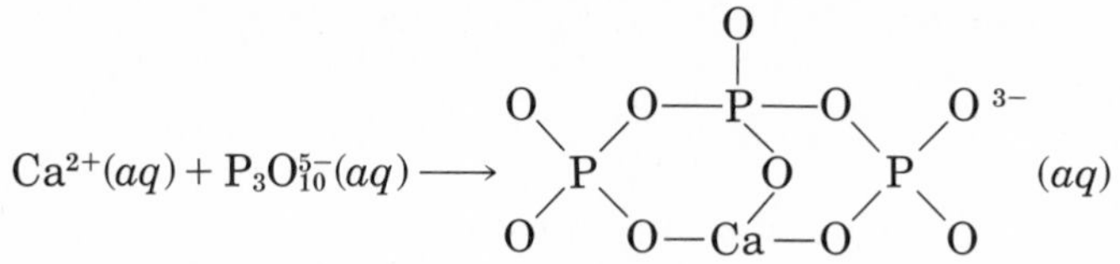

The Ca^{2+} ions are effectively removed from solution (they are said to be **sequestered**) because they are bonded to the tripolyphosphate ion. Consequently $Ca^{2+}(aq)$ is not available to precipitate soap or detergent molecules.

The use of phosphates in detergents is responsible in part for an environmental problem known as **accelerated eutrophication,** or premature aging of bodies of water. Over a period of many thousands of years, a lake or other body of water will slowly accumulate essential nutrient elements such as nitrogen or phosphorus because their compounds dissolve in streams that feed the lake. As the water becomes richer in nutrients, more plants and microorganisms can grow. Some of the organic matter which remains when these organisms die precipitates to the bottom of the lake and is not decomposed. Eventually the lake fills up with debris, becoming a swamp, and finally dry land.

This process of eutrophication can be greatly accelerated by human input of nutrients such as nitrogen or phosphorus fertilizers, or phosphates from detergents. Since reduction in the use of detergent phosphates would appear to have the least negative effects—people's clothes might not look as clean—many have suggested that prohibiting or limiting phosphate content is the way to solve the problem. Many states have passed laws implementing such limitations or bans.

TABLE 12.6 Properties of the Group VIA Elements.

Element	Symbol	Electron Configuration	Usual Oxidation State	Ionization Energy/ MJ mol^{-1} First	Second	Third
Oxygen	O	[He]$2s^22p^4$	−2	1.320	3.395	5.307
Sulfur	S	[Ne]$3s^23p^4$	+6, +4, −2	1.006	2.257	3.367
Selenium	Se	[Ar]$4s^23d^{10}4p^4$	+6, +4, −2	0.947	2.051	2.980
Tellurium	Te	[Kr]$5s^24d^{10}5p^4$	+6, +4, −2	0.876	1.800	2.704

As we approach the right-hand side of the periodic table, similarities among the elements within a group become greater again. This is true of group VIA. Except polonium, which is radioactive and usually omitted from discussion, all members of the group form X^{2-} ions when combined with highly electropositive metals. The tendency to be reduced to the -2 oxidation state decreases significantly from top to bottom of the group, however, and tellurium shows some metallic properties. The group VIA elements are called **chalcogens** because most ores of copper (Greek *chalkos*) are oxides or sulfides, and such ores contain traces of selenium and tellurium. Atomic properties of the chalcogens are summarized in Table 12.6.

At ordinary temperatures and pressures, oxygen is a gas. It exists in either of two allotropic forms: O_2, which makes up 21 percent of the earth's atmosphere, or O_3 (ozone), which slowly decomposes to O_2. O_3 can be prepared by passing an electrical discharge through O_2 or air:

$$3O_2(g) \xrightarrow{\text{electrical discharge}} 2O_3(g)$$

This reaction occurs naturally as a result of lightning bolts. O_3 is also produced by any device which produces electrical sparks. You may have noticed its distinctive odor in the vicinity of an electric motor, for example.

Ozone is formed in the earth's stratosphere (between altitudes of 10 and 50 km) by ultraviolet rays whose wavelengths are shorter than 250 nm:

$$O_2 \xrightarrow{\text{ultraviolet light}} 2O$$

$$O + O_2 \rightarrow O_3 \qquad (12.7)$$

The ozone itself absorbs longer-wavelength ultraviolet radiation (up to 340 nm), preventing these harmful rays from reaching the earth's surface. Otherwise these rays would increase the incidence of human skin cancer and cause other environmental problems. In recent years convincing evidence has been obtained to show that nitrogen oxide emissions from supersonic transport (SST) airplanes (which fly in the stratosphere) can reduce the concentration of ozone. Similar conclusions have been drawn regarding chlorofluoromethanes used as propellants in aerosol hair sprays and deodorants. A 5 percent decrease in stratospheric ozone is expected to produce a 10 per-

Radius/pm				
Covalent	Ionic (X^{2-})	Density/ g cm^{-3}	Electro-negativity	Melting Point (in °C)
66	140	1.43×10^{-3}	3.5	−218
104	184	2.06	2.5	119
117	198	4.82	2.4	217
135	221	6.25	2.1	450

cent increase in skin cancer, and so sizable effects on the stratosphere caused by human activity will have serious consequences.

O_3 is also an important component of photochemical smog. It is produced when O atoms (formed by breaking N—O bonds in NO_2) react with O_2 molecules according to Eq. (12.7). O_3 is a stronger oxidizing agent than O_2. It reacts with unsaturated hydrocarbons (alkenes) in evaporated gasoline to produce aldehydes and ketones which are eye irritants. Rubber is a polymeric material which contains C═C bonds, and so it too reacts with O_3. In the United States it is estimated that O_3 and other strong oxidants from photochemical smog do more than 450 million dollars worth of damage each year to tires and other articles made of rubber.

Sulfur occurs in a variety of allotropic forms. At room temperature the most stable form is rhombic sulfur. This yellow solid consists of S_8 molecules (Fig. 12.5) packed in a crystal lattice which belongs to the orthorhombic system (one of the crystal systems listed in Table 10.1). When heated to 96°C, solid rhombic sulfur changes very slowly into monoclinic sulfur, in which one-third of the S_8 molecules are randomly oriented in the crystal lattice. When either form of sulfur melts, the liquid is at first pale yellow and flows readily, but above 160°C it becomes increasingly viscous. Only near the boiling point of 444.6°C does it thin out again. This unusual change in viscosity with temperature is attributed to opening of the eight-membered ring of S_8 and formation of long chains of sulfur atoms. These intertwine and prevent the liquid from flowing. This explanation is supported by the fact that if the viscous liquid is cooled rapidly by pouring it into water, the amorphous sulfur produced can be shown experimentally to consist of long chains of sulfur atoms.

Both selenium and tellurium have solid structures in which the atoms are bonded in long spiral chains. Both are semiconductors, and the electrical conductivity of selenium depends on the intensity of light falling on the element. This property is utilized in selenium photocells, which are often used in photographic exposure meters. Selenium is also used in recti-

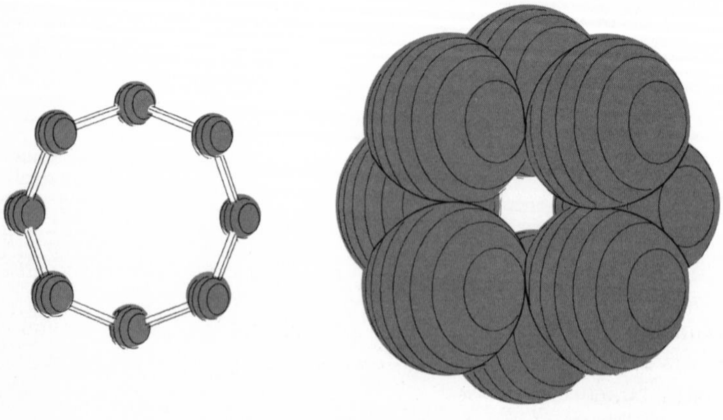

(*a*) (*b*)

Figure 12.5 S_8 molecule: (*a*) ball-and-stick model; (*b*) space-filling model. A side view of the ball-and-stick model appears on the first page of this chapter. (Computer-generated.) (*Copyright © 1976 by W. G. Davies and J. W. Moore.*)

fiers to convert alternating electrical current to direct current. Compounds of selenium and tellurium are of little commercial importance, and they often are toxic. Moreover, many of them have foul odors, are taken up by the body, and are given off in perspiration and on the breath. These properties have inhibited study of tellurium and selenium compounds.

Chemical Reactions and Compounds

Oxygen Since oxygen has the second largest electronegativity among all the elements, it is found in the -2 oxidation state in most compounds. Important oxides have already been discussed in sections dealing with the elements from which they form, and so we will deal only with unusual oxidation states of oxygen here. One of these is the $+2$ state found in OF_2, the most common compound in which oxygen is combined with the more electronegative fluorine. We have already mentioned the $-\frac{1}{2}$ and -1 states observed in alkali-metal superoxides and peroxides, but one important peroxide, hydrogen peroxide (H_2O_2), has not yet been discussed.

H_2O_2 can be prepared by electrolysis of solutions containing sulfate ions. H_2O_2 is a weak acid, and it can serve as an oxidizing agent (oxygen being reduced to the -2 state) or as a reducing agent (oxygen being oxidized to the 0 state). Like the peroxide ion, the H_2O_2 molecule contains an O—O single bond. This bond is rather weak compared with many other single bonds, and this contributes to the reactivity of H_2O_2. The compound decomposes easily, especially if exposed to light or contaminated with traces of transition metals. The decomposition

$$2H_2O_2(l) \rightarrow 2H_2O(l) + O_2(g)$$

can occur explosively in the case of the pure liquid.

Sulfur Although this element is only sixteenth in abundance at the surface of the earth, it is one of the few that has been known and used throughout history. Deposits of elemental sulfur are not uncommon, and, because they were stones that would burn, were originally called brimstone. Burning sulfur produces sulfur dioxide, SO_2:

$$S_8(s) + 8O_2(g) \rightarrow 8SO_2(g)$$

This colorless gas has a choking odor and is more poisonous than carbon monoxide. It is the anhydride of sulfurous acid, a weak diprotic acid:

$$SO_2(g) + H_2O(l) \rightarrow H_2SO_3(aq)$$

SO_2 is also produced when almost any sulfur-containing substance is burned in air. Coal, for example, usually contains from 1 to 4% sulfur, and so burning coal releases SO_2 to the atmosphere. Many metal ores are sulfides, and when they are heated in air, SO_2 is produced. Copper, for example, may be obtained as the element by heating copper(I) sulfide:

$$Cu_2S(s) + O_2(g) \xrightarrow{\Delta} 2Cu(s) + SO_2(g)$$

Since SO_2 is so poisonous, its release to the atmosphere is a major pollution problem. Once in the air, SO_2 is slowly oxidized to sulfur trioxide, SO_3:

$$2SO_2(g) + O_2(g) \rightarrow 2SO_3(g)$$

This compound is the anhydride of sulfuric acid, H_2SO_4:

$$SO_3(g) + H_2O(l) \rightarrow H_2SO_4(aq)$$

Thus if air is polluted with SO_2 and SO_3, a fine mist of dilute droplets of H_2SO_4 can form. All three substances are very irritating to the throat and lungs and are responsible for considerable damage to human health.

The natural mechanism for removal of sulfur oxides from the air is solution in raindrops, followed by precipitation. This makes the rainwater more acidic than it would otherwise be, and acid rain is now common in industrialized areas of the United States and Europe. Acid rain can slowly dissolve limestone and marble, both of which consist of $CaCO_3$:

$$CaCO_3(s) + H_3O^+(aq) \rightarrow Ca^{2+}(aq) + HCO_3^-(aq) + H_2O(l)$$

Thus statues and buildings made of these materials may be damaged.

Despite the fact that a tremendous amount of sulfur is released to the environment by coal combustion and ore smelting, this element is not usually recovered from such processes. Instead it is obtained commercially from large deposits along the U.S. Gulf Coast and from refining of sour petroleum. Sour petroleum contains numerous sulfur compounds, including H_2S, which smells like rotten eggs. The deposits of elemental sulfur in Texas and Louisiana are mined by the **Frasch process.** Water at 170°C is pumped down a pipe to melt the sulfur, and the latter is forced to the surface by compressed air. Most of the H_2S or S_8 obtained from these sources is oxidized to SO_2, passed over a vanadium catalyst to make SO_3, and dissolved in water to make H_2SO_4. In 1975 nearly 28 billion kg of H_2SO_4 was produced in the United States, making H_2SO_4 the most important industrial chemical. About half of it is used in phosphate fertilizer production.

Pure H_2SO_4 is a liquid at room temperature and has a great affinity for H_2O. This is apparently due to the reaction

$$H_2SO_4 + H_2O \rightarrow H_3O^+ + HSO_4^-$$

TABLE 12.7 Properties of the Group VIIA Elements.

Element	Symbol	Electron Configuration	Usual Oxidation State	Ionization Energy/MJ mol^{-1} First	Second	Third
Fluorine	F	$[He]2s^22p^5$	−1	1.687	3.381	6.057
Chlorine	Cl	$[Ne]3s^23p^5$	+7, +5, +3, +1, −1	1.257	2.303	3.828
Bromine	Br	$[Ar]4s^23d^{10}4p^5$	+7, +5, +3, +1, −1	1.146	2.113	3.471
Iodine	I	$[Kr]5s^24d^{10}5p^5$	+7, +5, +3, +1, −1	1.015	1.852	3.184

Formation of H_3O^+ releases energy, and the reaction is exothermic. Concentrated H_2SO_4 is 93% H_2SO_4 and 7% H_2O by mass, corresponding to more than twice as many H_2SO_4 as H_2O molecules. Since many H_2SO_4 molecules still have protons to donate, concentrated H_2SO_4 also has a great affinity for H_2O. It is often used as a drying agent and can be employed in condensation reactions which give off H_2O.

12.7 GROUP VIIA: HALOGENS

Except for astatine, which is radioactive, the properties of the elemental halogens have already been described in Sec. 4.1. A summary of atomic properties is given in Table 12.7. All the elements consist of diatomic molecules. They are strong oxidizing agents and are readily reduced to the X^- ions, and so the halogens form numerous ionic compounds. Fluorine, the most electronegative element, has no positive oxidation states, but the other halogens commonly exhibit +1, +3, +5, and +7 states. Most compounds containing halogens in positive oxidation states are good oxidizing agents, however, reflecting the strong tendency of these elements to gain electrons.

Chemical Reactions and Compounds

Combination of the halogens with alkali, alkaline-earth, and other metals was described in Sec. 4.1. The elements also react directly with hydrogen, yielding the hydrogen halides:

$$H_2 + X_2 \rightarrow 2HX \qquad X = F, Cl, Br, I$$

These compounds are all gases, are water soluble, and, except for HF, are strong acids in aqueous solution. They are conveniently prepared in the laboratory by acidifying the appropriate sodium or other halide:

$$NaCl(s) + H_3O^+(aq) \xrightarrow{\Delta} Na^+(aq) + H_2O(l) + HCl(g) \qquad (12.8)$$

The acid must be nonvolatile so that heating will drive off only the gaseous hydrogen halide. In the case of fluorides and chlorides, H_2SO_4 will do, but bromides and iodides are oxidized to Br_2 or I_2 by hot H_2SO_4, and so H_3PO_4 is used instead.

Radius/pm				
Covalent	Ionic (X^-)	Density/ g cm^{-3}	Electro-negativity	Melting Point (in °C)
64	136	1.73×10^{-3}	4.0	−220
99	181	3.17×10^{-3}	3.0	−101
114	195	3.14	2.8	−7
133	216	4.94	2.5	114

A reaction similar to Eq. (12.8) occurs when phosphate rock containing fluorapatite is treated with H_2SO_4 to make fertilizer:

$$Ca_{10}(PO_4)_6F_2 + 7H_2SO_4 + 3H_2O \rightarrow 3Ca(H_2PO_4)_2 \cdot H_2O + 7CaSO_4 + 2HF$$

The HF produced in this reaction can cause significant air-pollution problems. Fluorides are also emitted to the atmosphere in steelmaking and aluminum production. There is some evidence that fluorides, rather than sulfur dioxide, may have been responsible for human deaths in air-pollution episodes at Donora, Pennsylvania, and the Meuse Valley in Belgium.

The relative oxidizing strengths of the halogens can be illustrated nicely in the laboratory. If, for example, a solution of Cl_2 in H_2O is combined with a solution of NaI, the dark color of I_2 can be observed, showing that the Cl_2 has oxidized the I^-:

$$Cl_2(aq) + 2I^-(aq) \rightarrow 2Cl^-(aq) + I_2(aq)$$

This is especially evident if a nonpolar solvent which is immiscible with H_2O, such as CCl_4, is added. The nonpolar I_2 molecules will concentrate in the CCl_4 layer, and a beautiful violet color can be observed. From such experiments it can be shown that the strongest oxidizing agent is F_2 (at the top of the group). F_2 will react with Cl^-, Br^-, and I^-. The weakest oxidizing agent, I_2, does not react with any of the halide ions.

The extremely high oxidizing power of F_2 makes it the only element which can combine directly with a noble gas. The reactions

$$Xe(g) + F_2(g) \rightarrow XeF_2(s)$$
$$XeF_2(s) + F_2(g) \rightarrow XeF_4(s)$$
$$XeF_4(s) + F_2(g) \rightarrow XeF_6(s)$$

may be used to synthesize the three xenon fluorides, all of which are strong oxidizing agents. When an electrical discharge is passed through a mixture of Kr and F_2 at a low temperature, KrF_2 can be formed. This is the only compound of Kr, and it decomposes slowly at room temperature.

Fluorine is also set apart from the other halogens because of its ability to oxidize water:

$$2F_2 + 6H_2O \rightarrow 4H_3O^+ + 4F^- + O_2$$

Chlorine is also capable of oxidizing water, but it does so very slowly. Instead the reaction

$$Cl_2 + 2H_2O \rightleftharpoons H_3O^+ + Cl^- + HOCl$$

goes partway to completion. Hypochlorous acid, HOCl, is a weak acid. Small concentrations of hypobromous and hypoiodous acids can also be obtained in this way. In basic solution the halogen is completely consumed, producing the hypohalite anion:

$$Cl_2 + 2OH^- \rightarrow Cl^- + H_2O + OCl^-$$

Since hypochlorite, OCl^-, could also be supplied from an ionic compound such as NaOCl, the latter is often used to chlorinate swimming pools.

Larger-scale projects, such as city water supplies, use elemental chlorine, liquefied and stored under pressure. Because it is a strong oxidizing agent, Cl_2 (or OCl^-) can destroy harmful microorganisms, making water safe to drink.

Hypohalite ions disproportionate in aqueous solution:

$$3OCl^- \rightarrow 2Cl^- + ClO_3^-$$

This reaction is rather slow for hypochlorite unless the temperature is above 75°C, but OBr^- and OI^- are consumed immediately at room temperature. Chlorate, ClO_3^-, bromate, BrO_3^-, and iodate, IO_3^-, salts can be precipitated from such solutions. All are good oxidizing agents. Potassium chlorate, $KClO_3$, decomposes, giving O_2 when heated in the presence of a catalyst:

$$2KClO_3 \xrightarrow[\text{MnO}_2\text{ catalyst}]{\Delta} 2KCl + 3O_2$$

This is a standard laboratory reaction for making O_2. It also occurs whenever a match is struck. The O_2 produced helps ignite other components of the match head as well as the match itself. If $KClO_3$ is heated without a catalyst, potassium perchlorate, $KClO_4$, may be formed. Perchlorates oxidize organic matter rapidly and often uncontrollably. They are notorious for exploding unexpectedly and should be handled with great care.

One other interesting group of compounds is the interhalogens, in which one halogen bonds to another. Some interhalogens, such as BrCl, are diatomic, but the larger halogen atoms have room for several smaller ones around them. Thus compounds such as ClF_3, BrF_3 and BrF_5, and IF_3, ICl_3, IF_5, and IF_7 can be synthesized. Note that the largest halogen atom I can accommodate three chlorines and up to seven fluorines around it.

12.8 GROUP VIIIA: NOBLE GASES

The properties of the noble gases are summarized in Table 12.8. The ionization energies of He and Ne are greater than 2000 kJ mol^{-1}, and it is unlikely that these noble gases will ever be induced to form chemical bonds. The same probably applies to Ar. Kr and especially Xe do form compounds (some of which were discussed in the previous section), and Rn might be expected to be even more reactive. Rn is radioactive, however, and study of its chemistry is difficult.

TABLE 12.8 Properties of the Group VIIIA Elements.

Element	Symbol	Electron Configuration	Usual Oxidation State	Ionization Energy/ MJ mol^{-1} First	Second	Radius/ pm Covalent	Density/ 10^{-3} g cm^{-3}	Electro-negativity	Melting Point (in °C)
Helium	He	$1s^2$	0	2.379	5.257	. . .	0.179	. . .	−272
Neon	Ne	$[He]2s^22p^6$	0	2.087	3.959	. . .	0.901	. . .	−249
Argon	Ar	$[Ne]3s^23p^6$	0	1.527	2.672	. . .	1.78	. . .	−190
Krypton	Kr	$[Ar]4s^23d^{10}4p^6$	+2	1.357	2.374	110	3.74	2.6	−157
Xenon	Xe	$[Kr]5s^24d^{10}5p^6$	+8, +6, +4, +2	1.177	2.053	130	5.86	2.4	−112

SUMMARY

This chapter has considered the properties and reactions of the eight groups of representative elements. We have not only presented interesting and important facts about the elements but have also attempted to rationalize those facts in terms of the generalizations and theories of preceding chapters. At this point it is useful to look back and consolidate some of the general trends that have been observed.

First, metals on the far left of the periodic table are good reducing agents, while nonmetals on the far right (excluding noble gases) are strong oxidizing agents. Thus these elements are quite reactive, especially when one from the left combines with one from the right. Hydrogen compounds (hydrides) of the alkali and alkaline-earth metals contain strongly basic H^- ions and produce basic solutions. Toward the middle of the periodic table acid-base properties of hydrogen compounds are harder to predict. Some, like CH_4, are neither acids nor bases, but others, like NH_3, have lone pairs of electrons and can accept protons. Protons can be easily donated and are acidic only when they are bound to halogens or oxygen.

The acidic behavior of oxides also increases from left to right across the periodic table and decreases from top to bottom. The situation is complicated by the fact that the higher the oxidation state of an atom, the more covalent its oxide and the more acidic it will be. Thus SO_3 dissolves in water to give a strong acid, while SO_2 gives a weak one. Taking account of both of these trends, one can fairly well predict which oxides are likely to be basic, which amphoteric, and which acidic.

General rules can also be used to predict which oxidation states will be most common. On the left of the periodic table the group number gives the most common oxidation state. From group IIIA on, the group number minus 2 (for the ns^2 electrons) is also common, especially for elements near the bottom of the table. The group number is a good choice when an element combines with a highly electronegative element, but the group number minus 2 is more common when one element is bonded to another element of intermediate electron-withdrawing power. For example, SF_4 and SF_6 are both stable, but SCl_4 is the most stable sulfur chloride.

From group VA on to the right of the table, the group number minus 8 is an important oxidation state, especially for the first member of a group. Oxidation numbers other than those already mentioned usually differ by increments of 2. For example, chlorine exhibits -1, $+1$, $+3$, $+5$, and $+7$ oxidation states in stable compounds, and sulfur is found in -2, $+2$, $+4$, and $+6$ states.

In conclusion, do not overlook the forest by concentrating too much on individual trees. Look for and try to understand and use the generalizations and correlations that have been developed in this chapter. If you do this, you will retain the facts presented here much more efficiently.

QUESTIONS AND PROBLEMS

12.1 Write the valence electron configuration for each atom below:

a Ga	*d* Ca	*g* Kr
b Ba	*e* S	*h* Sn
c Na	*f* Cl	*i* B

12.2 Describe the trend observed for each of the properties listed below. (Refer to the tables in this chapter, if necessary.) Use your knowledge of atomic electronic structure to explain each trend.

a Ionization energies of group IA elements
b Densities of group IA, IIA, and IIIA elements across the same period
c Radii of M^{3+} ions of group IIIA
d Radii of X^- ions of group VIIA

e Ionization energies of Mg, Al, Si, P, and S
f Electronegativities of group IVA elements
g Ionization energies of Na^+, Mg^{2+}, and Al^{3+}

12.3 Why are the alkali metals strong reducing agents?

12.4 Complete and balance each of the following equations:

a $Na + H_2 \rightarrow$ *d* $Na + O_2 \rightarrow$
b $RbO_2 + H_2O \rightarrow$ *e* $K + O_2 \rightarrow$
c $Li + O_2 \rightarrow$ *f* $LiH + H_2O \rightarrow$

12.5 Complete and balance each of the following equations:

a $Be + H_2 \rightarrow$ *d* $Na + O_2 \rightarrow$
b $SrO + H_2O \rightarrow$ *e* $K + O_2 \rightarrow$
c $MgCO_3 \xrightarrow{\Delta}$ *f* $LiH + H_2O \rightarrow$

12.6 Define the term *disproportionation*. Give balanced equations for two examples of disproportionation.

12.7 True or false? All ionic hydrides dissolve in water to form basic solutions, and all covalent hydrides dissolve in water to form acidic solutions. Give formulas for at least three compounds whose properties either support or refute this statement.

12.8 Addition of sodium to ethanol (C_2H_5OH) evolves a colorless, odorless gas and produces a solution which conducts electrical current. Write a balanced equation for the reaction which occurs.

12.9 Write balanced chemical equations for (*a*) two industrial electrolysis reactions; (*b*) two industrial acid-base reactions.

12.10 What is a diagonal relationship? Give two examples of pairs of elements that are related this way and quote experimental facts which indicate the relationship.

12.11 What makes hard water hard? Why is water obtained from wells in many parts of the United States hard? How can hard water be softened?

12.12 Give formulas for at least two sodium compounds which are so strongly basic that they are caustic to the skin and poisonous if swallowed. Would the analogous compounds of magnesium have the same effect? (*Hint:* Consult the solubility rules in Table 11.2, if necessary.)

*__12.13__ Beryllium chloride, $BeCl_2$, has the following properties:

a It dissolves in H_2O to give a slightly acidic solution.
b Its boiling point is 520°C as compared with 1412°C for $MgCl_2$.
c Molten, it is a poor conductor of electrical current.
d It reacts directly with gaseous NH_3.
e Determination of the molecular weight of gaseous $BeCl_2$ gives a result higher than 80.

Use your knowledge of chemical bonding to account for each of these observations.

12.14 Give formulas for at least three compounds which behave as Lewis acids. Each compound you choose should be from a different periodic group.

12.15 "The first element in any periodic group has significantly different chemical properties from those of the elements below it." Choose four examples from groups IA to VIIA which support this statement.

12.16 Ionization energy and electronegativity decrease significantly from B to Al, but change little as we descend further in group IIIA. What differences in atomic electronic structure can account for this trend?

12.17 Which of the following statements are true? Which are false? Give reasons for your choice in each case.

a B_2O_3 gives a more basic aqueous solution than In_2O_3.
b TlCl is more likely to form than BCl.
c Formation of compounds containing B^{3+} ions is more likely than formation of those containing Al^{3+}.
d The melting point of GaF_3 is higher than that of GaI_3.

12.18 Try to draw a Lewis diagram for diborane, B_2H_6. What is the current description of bonding in this molecule?

12.19 What is meant by the term *amphoteric?* Give two examples of amphoteric oxides and write balanced equations for the reaction of each with $HNO_3(aq)$ and with $NaOH(aq)$.

12.20 Complete and balance each of the following equations:

a $NaH + AlCl_3 \rightarrow$ *d* $SiO_2 + C \xrightarrow{\Delta}$
b $Al_2O_3(s) + OH^-(aq) \rightarrow$ *e* $C_2H_6 + O_2 \rightarrow$
c $B + Cl_2 \rightarrow$

12.21 Complete and balance each of the following equations:

a $Ca_{10}(PO_4)_6(OH)_2 + SnF_2 \rightarrow$
b $Si_4H_{10} + O_2 \rightarrow$
c $In + Br_2 \rightarrow$
d $Ga_2O_3(s) + H_3O^+(aq) \rightarrow$
e $Sn^{2+}(aq) + O_2(g) + H_3O^+(aq) \rightarrow$

12.22 Use VSEPR theory (Sec. 7.2) to predict the shape of each of the following:

a BH_4^- *c* $Pb(CH_3)_4$ *e* CO_2
b BF_3 *d* $SnCl_2$ *f* SiF_5^-

12.23 What is the inert-pair effect? Which elements are affected? Give two examples (each from a different periodic group) of compounds in which there is an inert pair.

12.24 Which of the following compounds is the strongest oxidizing agent: PbO_2, CO_2, or SiO_2? Explain your choice.

12.25 Describe and/or sketch the structure of

a SiO_2 *d* elemental B
b graphite *e* white P
c gray Sn *f* Pb(*s*)

12.26 Write balanced chemical equations for the following industrial reactions:

a Haber process *c* Solvay process
b Bayer process *d* Ostwald process

12.27 Write formulas for two nitrate compounds which are explosive. What molecular components are required in such a compound?

12.28 Predict the most common oxidation states for each of the following elements:

a Sn *c* Tl *e* O *g* Sr
b S *d* P *f* Cl *h* Al

12.29 Complete and balance each of the following equations:

a $P_4O_6 + H_2O \rightarrow$
b $N_2 + O_2 \xrightarrow[\text{temp.}]{\text{high}}$
c $Cl_2(aq) + F^-(aq) \rightarrow$
d $CH_3—S—CH_3 + O_2 \rightarrow$
e $H_2O_2 \rightarrow$
f $Cl_2 + H_2O \rightarrow$

12.30 Complete and balance each of the following equations:

a $Br_2(aq) + I^-(aq) \rightarrow$
b $P_4O_6 + O_2 \rightarrow$
c $Xe + F_2 \rightarrow$
d $SO_3 + H_2O \rightarrow$
e $Cl_2(aq) + OH^-(aq) \rightarrow$
f $NO_2 + H_2O \rightarrow$

12.31 What is photochemical smog? How is it formed? Write formulas for at least three compounds which are found in photochemical smog. What kinds of damage does photochemical smog cause?

12.32 Titration of phosphorous acid requires only 2 mol NaOH per mol H_3PO_3. Why?

*__12.33__ Which of the following compounds would you expect to be very difficult to synthesize? Explain your reasoning in each case.

a PCl_5 *c* $SnCl_5$ *e* SF_4 *g* SBr_6
b NF_5 *d* OF_4 *f* PbI_4 *h* AsF_5

12.34 What is the principal ore of phosphorus? What must be done to it to make it useful as fertilizer? Why do phosphate fertilizer plants sometimes contribute to air pollution? Give balanced equations for the reactions involved.

12.35 Use VSEPR theory (Sec. 7.2) to predict the shapes of each of the following:

a $COCl_2$ *c* PF_5 *e* Cl_2O
b SF_4 *d* SO_3

12.36 Write balanced equations for the synthesis of ozone (*a*) in the laboratory; (*b*) in the stratosphere. Why is the stratospheric ozone layer so important?

12.37 Describe the macroscopic changes which would be observed as elemental sulfur is heated from 0 to 500°C. Interpret these observations in terms of the microscopic structural changes which are occurring.

12.38 A medium-sized electric power–generating plant burns 2.5×10^6 kg of coal per day. Assuming that this coal contains an average of 2% sulfur by weight, calculate

a The mass of sulfur burned each day.
b The mass of sulfur dioxide emitted by the power plant each day.
c The mass of sulfuric acid which would have to be removed from the atmosphere each day as a result of operation of this power plant. (Assume that 75 percent of the emitted sulfur oxides are converted to sulfuric acid.)

12.39 Approximately 300 power plants of the type described in Problem 12.38 were employed to supply United States electric power in 1975. Calculate how much H_2SO_4 was produced by those plants for that entire year. How does your answer compare with the 27.8×10^9 kg H_2SO_4 produced and sold by the United States chemical industry in 1975?

12.40 Write balanced equations for the synthesis of (*a*) HCl; (*b*) HI. Why is the choice of reactions more limited in the case of HI than in the case of HCl?

12.41 What property of chlorine makes it useful as a disinfectant for water supplies and swimming pools? Write an appropriate equation to show that either a solid chlorine compound or compressed $Cl_2(g)$ will produce the same effect in aqueous solution.

12.42 Write formulas and draw the structures for each of the following:

a Hypochlorite ion
b Phosphorous acid
c Chlorate ion
d Aluminum bromide
e Chlorite ion
f Iodine pentafluoride
g Xenon difluoride
h Perchlorate ion

*__12.43__ In what region of the periodic table would you expect to find an element which

a Is a strong reducing agent.
b Forms a hydride which is a strong acid.
c Forms compounds in which the element has a +1 as well as a +3 oxidation state.
d Forms an oxide which is a strong acid.
e Forms a hydride which is a strong base.
f Forms no compounds in which it has a positive oxidation state.
g Is a strong oxidizing agent.
h Forms an oxide which is a strong base.

chapter thirteen

CHEMICAL EQUILIBRIUM

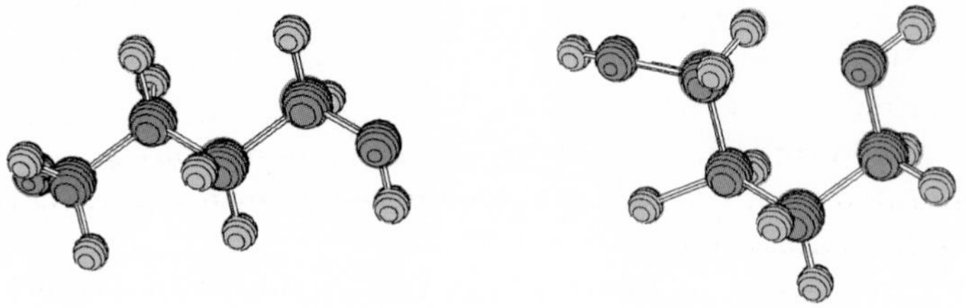

In most of the chemical reactions we have discussed so far, the reactants have been completely transformed into products. Such a reaction is said to go to completion. Not all chemical reactions are like this, though. Quite often the reactants are only partially converted into products. An obvious example of a reaction which does not go to completion is the solution of a weak acid (such as acetic acid) in water. The acid molecules donate their protons to water according to the equation

$$CH_3COOH + H_2O \rightleftharpoons CH_3COO^- + H_3O^+ \qquad (13.1)$$

However, measurements of the conductivity of a solution of acetic acid (Sec. 11.3) indicate that the concentrations of hydrogen ions and acetate ions are much smaller than we would have expected, had the proton-transfer reaction gone to completion. By repeated conductivity measurements, these concentrations may be shown to remain constant over a long period of time.

A system such as we have just described, in which appreciable concentrations of both reactants and products are present and in which concentrations do not change with time, is called an **equilibrium mixture.** The fact that the reaction does not go to completion is usually indicated, as in Eq. (13.1), by a double arrow. In such a mixture the reactants and products are said to be **in equilibrium** with each other.

A simpler and more instructive example of a chemical equilibrium is provided by the interconversion of the cis and trans isomers of difluoroethene:

$$cis\text{-}C_2H_2F_2 \rightleftharpoons trans\text{-}C_2H_2F_2$$

The two molecules involved in this equilibrium were illustrated in Fig. 7.9. The only difference between them is that in the cis isomer the two fluorine atoms are on the *same* side of the molecule, while in the trans isomer they are on *opposite* sides of the molecule. Although their molecules are so similar, these two isomers of difluoroethene are distinct chemical substances. They both condense to liquids at low temperatures, and these liquids have different boiling points. At room temperature both are gases, but they may be separated from each other and analyzed quantitatively by the technique of gas chromatography.

We mentioned in Chap. 7 that the barrier to free rotation about the C=C bond prevents *cis*-$C_2H_2F_2$ from changing rapidly into *trans*-$C_2H_2F_2$. The same applies to the reverse reaction, conversion of the trans isomer to the cis. These reactions do occur very slowly at higher temperatures, but even at 700 K (427°C) several weeks are required before equilibrium is reached and the concentrations of cis and trans species no longer vary with time. To study the reaction conveniently, a catalyst, such as $I_2(g)$, is added, speeding the reaction so that equilibrium is reached in a few minutes. When this is done, we always end up with a mixture which is slightly richer in the cis isomer. Furthermore, at a given temperature, the ratio of concentrations of the two isomers is always the same. For example, at 623 K the ratio

$$\frac{\text{Equilibrium concentration of trans}}{\text{Equilibrium concentration of cis}} = \frac{[trans\text{-}C_2H_2F_2]}{[cis\text{-}C_2H_2F_2]} = 0.50 \quad (13.2)$$

[In the second ratio in Eq. (13.2) square brackets are used to indicate the concentrations of *trans*-$C_2H_2F_2$ and *cis*-$C_2H_2F_2$ once equilibrium has been reached.]

Apart from a change in temperature, nothing will alter this equilibrium ratio from 0.50. Whether we start with the pure cis isomer, the pure trans isomer, or even a mixture of isomers, the same ratio is obtained (see Fig. 13.1). Other variations, such as starting with half the amount of either isomer, changing the volume of the container, or heating the mixture to 1000 K and then cooling it to 623 K, are likewise without effect. Even adding a catalyst has no effect on the equilibrium ratio. If we heat *cis*-$C_2H_2F_2$ and *trans*-$C_2H_2F_2$ to 623 K with iodine added as a catalyst, the only difference is that equilibrium is achieved in a few minutes instead of a few weeks. The final composition is the same as in the uncatalyzed case.

We have described this equilibrium between the cis and trans isomers of difluoroethene because it demonstrates very clearly the four features which are characteristic of *any* chemical situation in which appreciable concentrations of reactants and products are in equilibrium with each other. These four features are

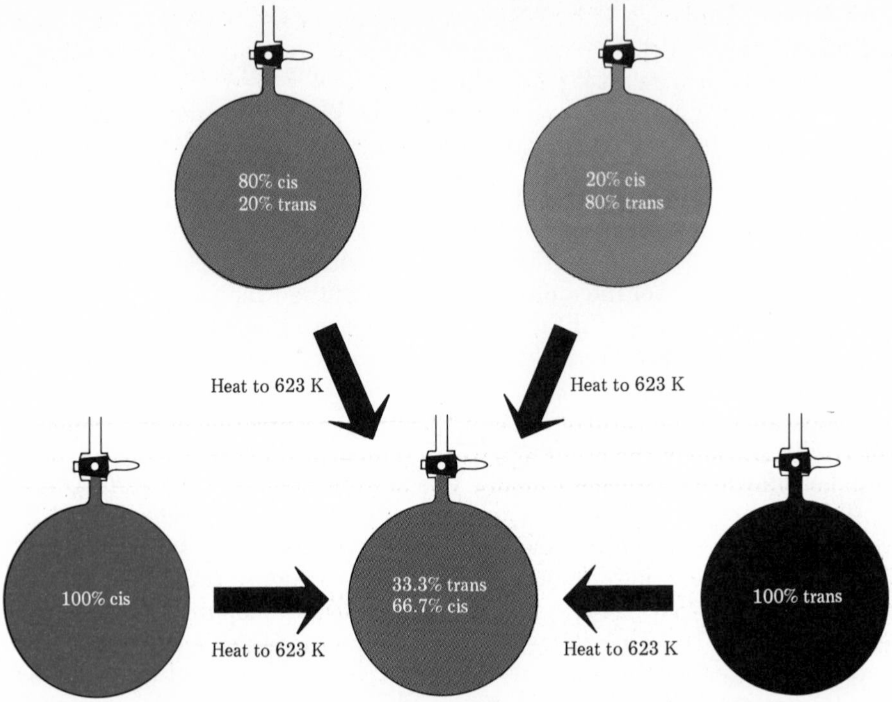

Figure 13.1 Chemical equilibrium. When any mixture of *cis*- and *trans*- difluoroethene is heated to 623 K, it eventually changes into an equilibrium mixture of the isomers with 33.3% of the trans isomer and 66.7% of the cis isomer. That is, the ratio [*trans*-$C_2H_2F_2$]/[*cis*-$C_2H_2F_2$] is always 0.5 at 623 K.

1 Even though the attainment of an equilibrium may be slow, once equilibrium has been achieved, the concentrations of all species participating in the equilibrium remain *constant.*
2 A chemical equilibrium can always be attained by approaching the equilibrium from *more than one* direction. We can begin with pure products or with pure reactants. Alternatively we can approach the equilibrium from either a higher or lower temperature.
3 The final equilibrium concentrations of the reactants are *unaffected* by the presence or absence of a *catalyst.*
4 There is always some mathematical relationship connecting the concentrations of the various species involved in the equilibrium mixture at a constant temperature. In the example just discussed, the concentrations were in a constant ratio. Usually the relationship is more complex, as we will see in the next section.

It is not always easy to tell when a chemical system is in a genuine equilibrium state. We often find mixtures of substances whose compositions do not change with time but which are not really in equilibrium with each other. A mixture of hydrogen and oxygen gas at room temperature is a good example. Although hydrogen and oxygen do react with each other to form water at room temperature, this reaction is so slow that no detectable change is apparent even after a few years. However, if an appropriate catalyst is added, the two gases react explosively and are converted completely to water according to the equation

$$2H_2(g) + O_2(g) \rightarrow 2H_2O(l)$$

13.2 THE EQUILIBRIUM CONSTANT

The constancy of the ratio of the equilibrium concentration of one isomer to the concentration of the other at a given temperature is characteristic of all gaseous equilibria between isomers, i.e., of all reactions of the general type

$$A(g) \rightleftharpoons B(g) \tag{13.3}$$

The constant ratio of concentrations is called the **equilibrium constant** and is given the symbol K_c. For reactions of the type given by Eq. (13.3), the equilibrium constant is thus described by the equation

$$\boxed{K_c = \frac{[B]}{[A]}} \tag{13.4}$$

where, by convention, the concentration of the product B appears in the numerator of the ratio. If, for some reason, we wish to look at this reaction in reverse,

$$B(g) \rightleftharpoons A(g)$$

then the equilibrium constant is denoted as

$$K_c = \frac{[A]}{[B]}$$

i.e., it is the reciprocal of the constant given in Eq. (13.4).

In general the equilibrium constant K_c varies with temperature and also differs from one reaction to another. Examples illustrating this behavior are given in Table 13.1 where the experimentally determined equilibrium constants for various cis-trans isomerization equilibria are recorded at various temperatures.

When we turn our attention to more complex equilibrium reactions, we find that the relationship between the concentrations of the various species is no longer a simple ratio. A good demonstration of this fact is provided by the dissociation of dinitrogen tetroxide, N_2O_4. This compound is a colorless gas, but even at room temperature it dissociates partly into a vivid red-brown gas, NO_2, according to the equation

$$N_2O_4(g) \rightleftharpoons 2NO_2(g) \tag{13.5}$$

TABLE 13.1 The Equilibrium Constant K_c for Some Cis-Trans Interconversions.

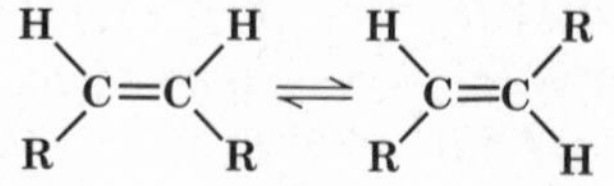

Temperature/K	R = F	R = Cl	R = CH_3
500	0.420	0.608	1.65
600	0.491	0.678	1.47
700	0.549	0.732	1.36

If 1 mol N_2O_4 contained in a flask of volume 1 dm^3 is heated to 407.2 K, exactly one-half of it dissociates into NO_2. If the volume is now increased, the ratio of $[NO_2]$ to $[N_2O_4]$ does not remain constant but increases as more N_2O_4 dissociates. As shown in Table 13.2, if we increase the volume still further, even more dissociation occurs. By the time we have increased the volume to 10 dm^3, the fraction of N_2O_4 molecules dissociated has increased to 0.854 (i.e., to 85.4 percent).

Obviously the situation is now not quite so straightforward as in the previous example. Nevertheless there is a simple relationship between the equilibrium concentrations of the reactant and product in this case too. We find that it is the quantity

$$\frac{[NO_2]^2}{[N_2O_4]}$$

rather than the simple ratio of concentrations, which is now constant. Accordingly we also call this quantity an equilibrium constant and give it the symbol K_c. Thus K_c for Eq. (13.5) is given by the relationship

$$K_c = \frac{[NO_2]^2}{[N_2O_4]}$$

where again by convention the product appears in the numerator.

It is easy to check that K_c actually is a constant quantity with the value 2.00 mol dm^{-3} from the data given in Table 13.2. Thus if we take the result from line **d**, we find that when 1 mol N_2O_4 is placed in a 10 dm^3 flask at 407

TABLE 13.2 The Dissociation of 1 mol N_2O_4 into NO_2 at 407.2 K (134°C) and Various Volumes.

	Amount N_2O_4 Added/mol	Volume of Flask/dm^3	Fraction N_2O_4 Dissociated	Amount of Each Gas at Equilibrium/mol		Equilibrium Concentration/mol dm^{-3}		Equilibrium Constant K_c/mol dm^{-3}
				N_2O_4	NO_2	N_2O_4	NO_2	
a	1	1	0.500	0.500	1.000	0.500	1.000	2.000
b	1	2	0.618	0.382	1.236	0.191	0.618	2.000
c	1	5	0.766	0.234	1.532	0.0468	0.3064	2.006
d	1	10	0.854	0.146	1.708	0.0146	0.1708	1.998

K, 0.854 mol dissociate. Since from Eq. (13.5) each mole which dissociates yields 2 mol N_2O_4, there will be

$$0.854 \text{ mol } N_2O_4 \times \frac{2 \text{ mol } NO_2}{1 \text{ mol } N_2O_4} = 1.708 \text{ mol } NO_2$$

present in the reaction vessel. There will also be (1 − 0.854) mol = 0.146 mol N_2O_4 left undissociated in the flask. Since the total volume is 10 dm³, the equilibrium concentrations are

$$[NO_2] = \frac{1.708 \text{ mol}}{10 \text{ dm}^3} = 0.1708 \text{ mol dm}^{-3}$$

and

$$[N_2O_4] = \frac{0.146 \text{ mol}}{10 \text{ dm}^3} = 0.0146 \text{ mol dm}^{-3}$$

Accordingly

$$K_c = \frac{0.1708 \text{ mol dm}^{-3} \times 0.1708 \text{ mol dm}^{-3}}{0.0146 \text{ mol dm}^{-3}}$$
$$= 2.00 \text{ mol dm}^{-3}$$

In exactly the same way, if we use the data from line **a** in Table 13.2, we find

$$K_c = \frac{[NO_2]^2}{[N_2O_4]} = \frac{(1.00 \text{ mol dm}^{-3})^2}{0.5 \text{ mol dm}^{-3}}$$
$$= 2.00 \text{ mol dm}^{-3}$$

You can check for yourself that lines **b** and **c** also yield the same value for K_c.

EXAMPLE 13.1 When 2 mol N_2O_4 gas is heated to 407 K in a vessel of volume 5 dm³, it is found that 0.656 of the molecules dissociate into NO_2. Show that these data are in agreement with the value for K_c of 2.00 mol dm⁻³ given in the text.

Solution Many equilibrium problems can be solved in a fairly standardized fashion in three stages.

a) Calculate the amount of each substance transformed by the reaction as it comes to equilibrium, i.e., the amount of each reactant *consumed* by the reaction and the amount of each product *produced* by the reaction. Stoichiometric ratios derived from the equation must always be used in these calculations.

In this particular example we note that 0.656 of the original N_2O_4 dissociates. Since 2 mol was used, a total of 0.656 × 2 mol = 1.312 mol N_2O_4 is *consumed*. The amount of NO_2 *produced* is accordingly

$$n_{NO_2} = 1.312 \text{ mol } N_2O_4 \times \frac{2 \text{ mol } NO_2}{1 \text{ mol } N_2O_4}$$
$$= 2.624 \text{ mol } NO_2$$

b) Use the amounts calculated in the first stage to calculate the amount of each substance present at equilibrium. Dividing by the volume, we can obtain the equilibrium concentrations.

Since 1.312 mol N_2O_4 dissociated out of an original 2 mol, we have (2 − 1.312) mol = 0.688 mol N_2O_4 left. The equilibrium concentration of N_2O_4 is thus

$$[N_2O_4] = \frac{0.688 \text{ mol}}{5.00 \text{ dm}^3} = 0.1376 \text{ mol dm}^{-3}$$

Since no NO_2 was originally present, the amount of NO_2 present at equilibrium is the amount produced by the dissociation, namely, 2.624 mol NO_2. Thus

$$[NO_2] = \frac{2.624 \text{ mol}}{5.00 \text{ dm}^3} = 0.525 \text{ mol dm}^{-3}$$

It is usually worthwhile tabulating these calculations, particularly in more complex examples.

Note that a *negative* quantity in the column headed Amount Produced indicates that a given substance (such as N_2O_4 in this example) has been *consumed*. There is less of that substance when equilibrium is reached than was present initially.

Substance	Initial Amount	Amount Produced	Equilibrium Amount	Equilibrium Concentration
N_2O_4	2.00 mol	−1.312 mol	0.688 mol	$\frac{0.688}{5}$ mol dm^{-3}
NO_2	0.00 mol	2.624 mol	2.624 mol	$\frac{2.624}{5}$ mol dm^{-3}

c) In the third stage we insert the equilibrium concentrations in an expression for the equilibrium constant:

$$K_c = \frac{[NO_2]^2}{[N_2O_4]} = \frac{0.525 \cancel{\text{ mol dm}^{-3}} \times 0.525 \text{ mol dm}^{-3}}{0.1376 \cancel{\text{ mol dm}^{-3}}}$$
$$= 2.00 \text{ mol dm}^{-3}$$

The Law of Chemical Equilibrium

The two examples of an equilibrium constant we have so far dealt with, namely,

$$K_c = \frac{[\textit{trans-}C_2H_2F_2]}{[\textit{cis-}C_2H_2F_2]}$$

for the reaction $\quad \textit{cis-}C_2H_2F_2 \rightleftharpoons \textit{trans-}C_2H_2F_2$

and $$K_c = \frac{[NO_2]^2}{[N_2O_4]}$$

for the reaction $\quad N_2O_4 \rightleftharpoons 2NO_2$

are both particular examples of a more general law governing chemical equilibrium in gases. If we write an equation for a gaseous equilibrium in general in the form

$$aA(g) + bB(g) \rightleftharpoons cC(g) + dD(g) \tag{13.6}$$

then the equilibrium constant defined by the equation

$$K_c = \frac{[C]^c[D]^d}{[A]^a[B]^b} \tag{13.7}$$

is found to be a constant quantity depending only on the temperature and the nature of the reaction. This general result is called the **law of chemical equilibrium,** or the **law of mass action.**

EXAMPLE 13.2 Write expressions for the equilibrium constant for the following reactions:

a) $2HI(g) \rightleftharpoons H_2(g) + I_2(g)$

b) $N_2(g) + 3H_2(g) \rightleftharpoons 2NH_3(g)$

c) $O_2(g) + 4HCl(g) \rightleftharpoons 2H_2O(g) + 2Cl_2(g)$

Solution

a) $$K_c = \frac{[H_2][I_2]}{[HI]^2}$$

b) $$K_c = \frac{[NH_3]^2}{[N_2][H_2]^3}$$

c) $$K_c = \frac{[H_2O]^2[Cl_2]^2}{[O_2][HCl]^4}$$

EXAMPLE 13.3 A mixture containing equal concentrations of methane and steam is passed over a nickel catalyst at 1000 K. The emerging gas has the composition [CO] = 0.1027 mol dm^{-3}, [H_2] = 0.3080 mol dm^{-3}, [CH_4] = [H_2O] = 0.8973 mol dm^{-3}. Assuming this mixture is at equilibrium, calculate the equilibrium constant K_c for the reaction

$$CH_4(g) + H_2O(g) \rightleftharpoons CO(g) + 3H_2(g)$$

Solution The equilibrium constant is given by the following equation:

$$K_c = \frac{[CO][H_2]^3}{[CH_4][H_2O]}$$

$$= \frac{0.1027 \text{ mol dm}^{-3} \times (0.3080 \text{ mol dm}^{-3})^3}{0.8973 \text{ mol dm}^{-3} \times 0.8973 \text{ mol dm}^{-3}}$$

$$= 3.727 \times 10^{-3} \text{ mol}^2 \text{ dm}^{-6}$$

Note: The yield of H_2 at this temperature is quite poor. In the commercial production of H_2 from natural gas, the reaction is run at a somewhat higher temperature where the value of K_c is larger.

As the above example shows, the equilibrium constant K_c is not always a dimensionless quantity. In general it has the units $(\text{mol dm}^{-3})^{\Delta n}$, where Δn is the increase in the number of molecules in the equation. In the above case $\Delta n = 2$, since 4 molecules ($3H_2$ and 1CO) have been produced from 2 molecules (CH_4 and H_2O). Only if $\Delta n = 0$, as is the case for the cis-trans isomerization considered above, is the equilibrium constant a dimensionless quantity.

We can also apply the equilibrium law to reactions which involve pure solids and pure liquids as well as gases. We find in such cases that as long as some solid or liquid is present, the actual amount does not affect the position of equilibrium. Accordingly, only the concentrations of gaseous species are included in the expression for the equilibrium constant. For example, the equilibrium constant for the reaction

$$CaCO_3(s) \rightleftharpoons CaO(s) + CO_2(g) \tag{13.8}$$

is given by the expression

$$K_c = [CO_2] \tag{13.9}$$

in which only the concentration of the gas appears. Equation (13.9) suggests that if we heat $CaCO_3$ to a high temperature so that some of it decomposes, the concentration of CO_2 at equilibrium will depend only on the temperature and will not change if the ratio of amount of solid $CaCO_3$ to amount of solid CaO is altered. Experimentally this is what is observed.

EXAMPLE 13.4 Write expressions for the equilibrium constants for the following reactions:

a) $C(s) + H_2O(g) \rightleftharpoons CO(g) + H_2(g)$

b) $C(s) + CO_2(g) \rightleftharpoons 2CO(g)$

c) $Fe_3O_4(s) + H_2(g) \rightleftharpoons 3FeO(s) + H_2O(g)$

Solution Since only gaseous species need be included, we obtain

a) $K_c = \dfrac{[CO][H_2]}{[H_2O]}$

b) $K_c = \dfrac{[CO]^2}{[CO_2]}$

c) $K_c = \dfrac{[H_2O]}{[H_2]}$

The equilibrium law can be shown experimentally to apply to dilute liquid solutions as well as to mixtures of gases, and the equilibrium-constant expression for a solution reaction can be obtained in the same way as for a gas-phase reaction. In solution only the concentrations of species in the liquid phase need be included. In some solution reactions, the solvent may be a reactant or product. Acetic acid, for example, reacts as follows when it dissolves in water:

$$CH_3COOH + H_2O \rightleftharpoons CH_3COO^- + H_3O^+ \qquad (13.10)$$

As long as the solution is dilute, however, the concentration of the solvent is hardly affected by addition of solutes, even if they react with it. (The concentration of pure water may be calculated from the density:

$$c_{H_2O} = \frac{1.0\ \text{g}}{1\ \text{cm}^3} \times \frac{10^3\ \text{cm}^3}{1\ \text{dm}^3} \times \frac{1\ \text{mol}}{18.0\ \text{g}} = 55.5\ \text{mol dm}^{-3}$$

Even if 0.1 mol dm^{-3} of acetic acid were added, the concentration of water would be affected by much less than 1 percent.)

Because the concentration of solvent remains essentially constant, it is usually incorporated into the equilibrium constant. Following the usual rules, Eq. (13.10) would give

$$K_c = \frac{[CH_3COO^-][H_3O^+]}{[CH_3COOH][H_2O]}$$

This can be rearranged to

$$K_a = K_c \times [H_2O] = K_c \times 55.5\ \text{mol dm}^{-3} = \frac{[CH_3COO^-][H_3O^+]}{[CH_3COOH]}$$

Thus the concentration of water is conventionally included in the equilibrium constant K_a for a reaction in aqueous solution. Since it applies to a weak acid, K_a is called an **acid constant.** (The *a* stands for *acid*.) Other equilibrium constants which contain a constant concentration in this way are the **base constant,** K_b, for ionization of a weak base and the **solubility product constant,** K_{sp}, for dissolution of a slightly soluble compound.

EXAMPLE 13.5 Write out expressions for the equilibrium constants for the following ionic equilibria in dilute aqueous solution:

a) $HF(aq) + H_2O \rightleftharpoons F^-(aq) + H_3O^+(aq)$

b) $H_2O + NH_3(aq) \rightleftharpoons OH^-(aq) + NH_4^+(aq)$

c) $H_2O + CO_3^{2-}(aq) \rightleftharpoons HCO_3^-(aq) + OH^-(aq)$

d) $BaSO_4(s) \rightleftharpoons Ba^{2+}(aq) + SO_4^{2-}(aq)$

Solution We leave out the concentration of H_2O in the first three examples and the concentration of solid $BaSO_4$ in the fourth.

a) $K_a = K_c \times [H_2O] = \dfrac{[H_3O^+][F^-]}{[HF]}$

b) $K_b = K_c \times [H_2O] = \dfrac{[NH_4^+][OH^-]}{[NH_3]}$

c) $K_b = K_c \times [H_2O] = \dfrac{[HCO_3^-][OH^-]}{[CO_3^{2-}]}$

d) $K_{sp} = K_c \times [BaSO_4] = [Ba^{2+}][SO_4^{2-}]$

EXAMPLE 13.6 Measurements of the conductivities of acetic acid solutions indicate that the fraction of acetic acid molecules converted to acetate and hydronium ions is

a) 0.0296 at a concentration of 0.020 00 mol dm^{-3}.

b) 0.5385 at a concentration of 2.801×10^{-5} mol dm^{-3}.

Use these data to calculate the equilibrium constant for Eq. (13.10) at each concentration.

Solution Consider first 1 dm^3 of solution *a*. This originally contained 0.02 mol CH_3COOH, of which the fraction 0.0296 has ionized. Thus $(1 - 0.0296) \times 0.02$ mol undissociated CH_3COOH is left, while 0.0296×0.02 mol H_3O^+ and CH_3COO^- have been produced. In tabular form

Substance	Original Amount	Amount Produced	Equilibrium Amount	Equilibrium Concentration
CH_3COOH	0.02 mol	-0.0296×0.02 mol	(0.02 − 0.000 592) mol	0.0194 mol dm^{-3}
H_3O^+	. . . . . .	$+0.0296 \times 0.02$ mol	0.000 592 mol	5.92×10^{-4} mol dm^{-3}
CH_3COO^-	. . . . . .	$+0.0296 \times 0.02$ mol	0.000 592 mol	5.92×10^{-4} mol dm^{-3}

Substituting into the expression for K_a gives

$$K_a = \frac{[CH_3COO^-][H_3O^+]}{[CH_3COOH]} = \frac{(5.92 \times 10^{-4} \text{ mol dm}^{-3})^2}{0.0194 \text{ mol dm}^{-3}} = 1.81 \times 10^{-5} \text{ mol dm}^{-3}$$

A similar calculation on the second solution yields

$$K_a = \frac{(1.5083 \times 10^{-5} \text{ mol dm}^{-3})^2}{1.2926 \times 10^{-5} \text{ mol dm}^{-3}} = 1.760 \times 10^{-5} \text{ mol dm}^{-3}$$

Note: The two values of the equilibrium constant are only in approximate agreement. In more concentrated solutions the agreement is worse. If the concentration is 1 mol dm^{-3}, for instance, K_a has the value 1.41×10^{-5} mol dm^{-3}. This is the reason for our statement that the equilibrium law applies to *dilute* solutions.

The Equilibrium Constant in Terms of Pressure

Some equilibria involve physical instead of chemical processes. One example is the equilibrium between liquid and vapor in a closed container. In Sec. 10.3 we stated that the vapor pressure of a liquid was always the same at a given temperature, regardless of how much liquid was present. This can be seen to be a consequence of the equilibrium law if we recognize that the pressure of a gas is related to its concentration through the ideal gas law. Rearranging $PV = nRT$, we obtain

$$P = \frac{n}{V}RT = cRT \qquad (13.11)$$

since c = amount of substance/volume = n/V. Thus if the vapor pressure is constant at a given temperature, the concentration must be constant also. Equation (13.11) also allows us to relate the equilibrium constant to the vapor pressure. In the case of water, for example, the equilibrium reaction and K_c are given by

$$H_2O(l) \rightleftharpoons H_2O(g) \qquad K_c = [H_2O(g)]$$

Substituting for the concentration of water vapor from Eq. (13.11), we obtain

$$K_c = \frac{P_{H_2O}}{RT}$$

At 25°C, for example, the vapor pressure of water is 17.5 mmHg (2.33 kPa), and so we can calculate

$$K_c = \frac{2.33 \text{ kPa}}{(8.314 \text{ J K}^{-1} \text{ mol}^{-1})(298.15 \text{ K})} = 9.40 \times 10^{-4} \text{ mol dm}^{-3}$$

For some purposes it is actually more useful to express the equilibrium law for gases in terms of partial pressures rather than in terms of concentrations. In the general case

$$aA(g) + bB(g) \rightleftharpoons cC(g) + dD(g)$$

The **pressure-equilibrium constant K_p** is defined by the relationship

$$K_p = \frac{p_C^c p_D^d}{p_A^a p_B^b} \qquad (13.12)$$

where p_A is the partial pressure of component A, p_B of component B, and so on. Since $p_A = [A] \times RT$, $p_B = [B] \times RT$, and so on, we can also write K_p as follows:

$$K_p = \frac{p_C^c p_D^d}{p_A^a p_B^b} = \frac{([C] \times RT)^c([D] \times RT)^d}{([A] \times RT)^a([B] \times RT)^b} = \frac{[C]^c[D]^d}{[A]^a[B]^b} \times \frac{(RT)^c(RT)^d}{(RT)^a(RT)^b}$$
$$= K_c \times (RT)^{(c+d-a-b)} = K_c \times (RT)^{\Delta n}$$

Again Δn is the increase in the number of gaseous molecules represented in the equilibrium equation. If the number of gaseous molecules does not change, $\Delta n = 0$, $K_p = K_c$, and both equilibrium constants are dimensionless quantities.

EXAMPLE 13.7 In what SI units will the equilibrium constant K_c be measured for the following reactions? Also predict for which reactions $K_c = K_p$.

a) $2NOBr(g) \rightleftharpoons 2NO(g) + Br_2(g)$

b) $H_2O(g) + C(s) \rightleftharpoons CO(g) + H_2(g)$

c) $N_2(g) + 3H_2(g) \rightleftharpoons 2NH_3(g)$

d) $H_2(g) + I_2(g) \rightleftharpoons 2HI(g)$

Solution We apply the rule that the units are given by $(\text{mol dm}^{-3})^{\Delta n}$.

a) Since $\Delta n = 1$, units are moles per cubic decimeter.

b) Since $\Delta n = 1$, units are moles per cubic decimeter (the solid is ignored).

c) Here $\Delta n = -2$ since two gas molecules are produced from four. Accordingly the units are $\text{mol}^{-2}\ \text{dm}^6$.

d) Since $\Delta n = 0$, K_c is a pure number. In this case also $K_c = K_p$.

13.3 CALCULATING THE EXTENT OF A REACTION

Once we know the equilibrium constant for a reaction, we can calculate what will happen when any arbitrary mixture of reactants and products is allowed to come to equilibrium. To take a simple case: What would happen if we mixed 1 mol cis isomer with 1 mol trans isomer of difluoroethene in a 10-dm³ flask at 623 K? From what we know of this reaction, it is easy to guess that some of the trans isomer will be converted to cis isomer, since an equilibrium mixture of the isomers always contains more of the cis than of the trans form. If we want to know the exact amount of trans isomer converted in this way, we can use the equilibrium constant to calculate it.

In all calculations of this sort we need to concentrate on the *increase in the amount of one of the products or the reactants.* Since this quantity is unknown, we label it algebraically, calling it *x mol,* where *x* indicates a pure number.[1] In the case under consideration, we would label the amount of cis isomer produced as the system moves to equilibrium as *x* mol. Once this step has been taken, the amount of each of the other products and reactants transformed by the reaction can be deduced from the equation and the appropriate stoichiometric factors. In our current case the equation is

$$cis\text{-}C_2H_2F_2 \rightleftharpoons trans\text{-}C_2H_2F_2$$

[1] If *x* is used to indicate a quantity rather than a number, the units become much more difficult to handle in the algebra which follows.

and it is obvious that if x mol cis isomer have been produced, x mol trans isomer have been *consumed*.

We can now construct a table showing the initial amounts, the amounts transformed by reaction, the amounts present at equilibrium, and finally the equilibrium concentrations of each product and reagent:

Substance	Initial Amount	Amount Produced	Equilibrium Amount	Equilibrium Concentration
cis-$C_2H_2F_2$	1 mol	x mol	$(1 + x)$ mol	$((1 + x)/10)$ mol dm^{-3}
trans-$C_2H_2F_2$	1 mol	$-x$ mol	$(1 - x)$ mol	$((1 - x)/10)$ mol dm^{-3}

Once the final concentration of each species has been obtained in this way, an algebraic equation can be set up linking x to the value of the equilibrium constant:

$$K_c = 0.500 = \frac{[trans\text{-}C_2H_2F_2]}{[cis\text{-}C_2H_2F_2]} = \frac{((1 - x)/10)\ \text{mol dm}^{-3}}{((1 + x)/10)\ \text{mol dm}^{-3}}$$

$$0.500 = \frac{1 - x}{1 + x}$$

At this stage we are left with an algebraic equation to solve for x. Inevitably this equation involves *only numbers*. If any units remain, a mistake must have been made. From

$$\frac{1 - x}{1 + x} = 0.500$$

by cross-multiplying, we have

$$1 - x = 0.500 + 0.500x$$

which rearranges to

$$0.500 = 1.500x$$

so that

$$x = \frac{0.500}{1.500} = 0.333$$

We thus conclude that 0.333 mol trans isomer is converted to the cis form when the original mixture is allowed to equilibrate. The final equilibrium concentrations are as follows:

$$[cis\text{-}C_2H_2F_2] = \frac{1 + x}{10}\ \text{mol dm}^{-3} = \frac{1.333}{10}\ \text{mol dm}^{-3}$$

$$= 0.1333\ \text{mol dm}^{-3}$$

While

$$[trans\text{-}C_2H_2F_2] = \frac{1 - x}{10}\ \text{mol dm}^{-3} = 0.0667\ \text{mol dm}^{-3}$$

We can easily cross check that the ratio of these two concentrations is actually equal to the equilibrium constant, that is, to 0.5.

EXAMPLE 13.8 When colorless hydrogen iodide gas is heated, a beautiful purple color appears, indicating that some iodine gas has been produced and that the compound has decomposed partially into its elements according to the equation

$$2HI(g) \rightleftharpoons H_2(g) + I_2(g)$$

At 745 K (471.8°C), K_c for this reaction has the value 0.0200. Calculate the concentration of I_2 produced when 1.00 mol HI is heated to this temperature in a flask of volume 10.0 dm^3. Also calculate the fraction of the HI which has dissociated.

Solution Let us denote the amount of I_2 produced by the reaction as x mol. The equation then tells us that the amount of H_2 produced will also be x mol, while the amount of HI consumed by the decomposition will be $2x$ mol. The initial amount of HI, 1 mol, will thus be reduced to $(1 - 2x)$ mol at equilibrium. Dividing the above amounts by the volume 10 dm^3, we easily obtain the equilibrium concentrations

Substance	Initial Amount	Amount Produced	Equilibrium Amount	Equilibrium Concentration
I_2	0.00 mol	x mol	x mol	$(x/10)$ mol dm^{-3}
H_2	0.00 mol	x mol	x mol	$(x/10)$ mol dm^{-3}
HI	1.00 mol	$-2x$ mol	$(1 - 2x)$ mol	$((1 - 2x)/10)$ mol dm^{-3}

We can now write an expression for the equilibrium constant

$$K_c = \frac{[H_2][I_2]}{[HI]^2} = \frac{(x/10) \text{ mol dm}^{-3} \times (x/10) \text{ mol dm}^{-3}}{((1 - 2x)/10) \text{ mol dm}^{-3} \times ((1 - 2x)/10) \text{ mol dm}^{-3}}$$

or $$0.0200 = \frac{x^2}{(1 - 2x)^2}$$

which is the required algebraic expression, free of units. This equation is easily solved by taking the square root of both sides.

$$\sqrt{2 \times 10^{-2}} = \sqrt{2} \times 10^{-1} = 0.1414 = \frac{x}{1 - 2x}$$

Thus $$0.1414 - 0.2828x = x$$

or $$1.2828x = 0.1414$$

so that $$x = \frac{0.1414}{1.2828} = 0.110$$

Thus $$[I_2] = \frac{x}{10} \text{ mol dm}^{-3} = 1.10 \times 10^{-2} \text{ mol dm}^{-3}$$

Since $2x$ mol HI dissociated and 1 mol HI was originally present, we concluded that the fraction of HI which dissociated is

$$\frac{2x \text{ mol}}{1 \text{ mol}} = 0.220$$

It is wise at this point to check the answer. We found $x = 0.110$. If this is the correct value, we should also find that

$$\frac{x^2}{(1 - 2x)^2} = 0.0200$$

The value obtained using a calculator is 0.019 888 2 (which rounds to 0.0199 to three significant figures). The difference is due to errors introduced by rounding off during the calculation.

EXAMPLE 13.9 A mixture of 1.00 mol HI gas and 1.00 mol H_2 gas is heated in a 10.0-dm³ flask to 745 K. Calculate the concentration of I_2 produced in the equilibrium mixture and also the fraction of HI which dissociates.

Solution Again we let x mol represent the amount of I_2 produced. Our table then becomes

Substance	Initial Amount	Amount Produced	Equilibrium Amount	Equilibrium Concentration
I_2	0.00 mol	x mol	x mol	$(x/10)$ mol dm^{-3}
H_2	1.00 mol	x mol	$(1 + x)$ mol	$((1 + x)/10)$ mol dm^{-3}
HI	1.00 mol	$-2x$ mol	$(1 - 2x)$ mol	$((1 - 2x)/10)$ mol dm^{-3}

Substituting the final concentrations in an expression for the equilibrium constant, we then have

$$K_c = \frac{[H_2][I_2]}{[HI]^2} = \frac{((1 + x)/10) \text{ mol dm}^{-3} \times (x/10) \text{ mol dm}^{-3}}{((1 - 2x)/10) \text{ mol dm}^{-3} \times ((1 - 2x)/10) \text{ mol dm}^{-3}}$$

or

$$0.0200 = \frac{(1 + x)x}{(1 - 2x)^2}$$

Because of the added H_2, it is no longer possible to take a square root as in the previous example. Instead we need to multiply out and rearrange in order to obtain a quadratic equation of the form $ax^2 + bx + c = 0$. Accordingly we have

$$0.0200(1 - 2x)^2 = (1 + x)x$$

or

$$0.0200(1 - 4x + 4x^2) = x + x^2$$

multiplying both sides by 50, we obtain

$$1 - 4x + 4x^2 = 50x + 50x^2$$

which on rearrangement has the required form

$$46x^2 + 54x - 1 = 0$$

where $a = 46$
$b = 54$
$c = -1$

We can now use the conventional quadratic formula

$$x = \frac{-b \pm \sqrt{b^2 - 4ac}}{2a}$$

$$= \frac{-54 \pm \sqrt{54^2 + 4 \times 46 \times 1}}{2 \times 46}$$

$$= \frac{-54 \pm 55.678}{92} = 0.0182 \qquad \text{or} \qquad -1.192$$

The negative root, $x = -1.192$, implies that 1.192 mol I_2 was consumed. Since no I_2 was present to begin with, this is impossible. We conclude that the positive root, namely $x = 0.0182$, is the correct one. Thus

$$[I_2] = \frac{x}{10} \text{ mol dm}^{-3}$$

$$= 1.82 \times 10^{-3} \text{ mol dm}^{-3}$$

Again the fraction dissociated is given by $2x$ mol/1 mol and is thus equal to 0.0364. To check this solution, we can substitute $x = 0.0182$ in the expression

$$\frac{(1 + x)x}{(1 - 2x)^2}$$

We then obtain the value 0.019 96, which rounds to the correct value of 0.0200.

Note: The inclusion of one of the products (H_2 gas) in the mixture reduces the extent to which the hydrogen iodide dissociates quite appreciably. We will explore this phenomenon more extensively in the next section.

In calculating the extent of a chemical reaction from an equilibrium constant, it is often useful to realize that if the equilibrium constant is very small, the reaction proceeds to only a limited extent, while if it is very large, the reaction goes almost to completion. This point is easiest to see in the case of an equilibrium between two isomers of the type

$$A \rightleftharpoons B$$

If K_c for this reaction is very small, say 10^{-6}, then the ratio $[B]/[A] = 10^{-6}$. There will thus be a million times more molecules of the A isomer than of the B isomer in the equilibrium mixture. For most purposes we can regard the equilibrium mixture as being pure A. Conversely if K_c has a very large value like 10^6, the very opposite is true. In an equilibrium mixture governed by this second constant, there would be a million times more B isomer than A isomer, and for most purposes the equilibrium mixture could be regarded as pure B.

The realization that an equilibrium mixture can contain only small concentrations of some of the reactants or products is often very useful in solving equilibrium problems, as the following example shows.

EXAMPLE 13.10 The equilibrium constant K_c for the dissociation of dinitrogen tetroxide according to the equation

$$N_2O_4(g) \rightleftharpoons 2NO_2(g)$$

changes from a very small to a very large value as the temperature is increased, as shown in the table. Calculate the fraction of N_2O_4 dissociated at

Temperature/K	K_c/mol dm^{-3}
200	1.09×10^{-7}
400	1.505
600	1.675×10^{3}

each temperature if 1.00 mol N_2O_4 is held in a container of volume 4.00 dm^3.

Solution Let the amount of N_2O_4 dissociated at the temperature under consideration be x mol. From the equation, $2x$ mol NO_2 will be produced. In tabular form we then have

Substance	Initial Amount	Amount Produced	Equilibrium Amount	Equilibrium Concentration
N_2O_4	1.00 mol	$-x$ mol	$(1 - x)$ mol	$((1 - x)/4)$ mol dm^{-3}
NO_2	0.00 mol	$+2x$ mol	$2x$ mol	$(2x/4)$ mol dm^{-3}

We thus have

$$K_c = \frac{[NO_2]^2}{[N_2O_4]} = \frac{(2x/4) \text{ mol dm}^{-3} \times (2x/4) \text{ mol dm}^{-3}}{((1 - x)/4) \text{ mol dm}^{-3}}$$

$$= \frac{4x^2}{16} \times \frac{4}{1 - x} \text{ mol dm}^{-3} = \frac{x^2}{1 - x} \text{ mol dm}^{-3}$$

a) At 200 K, $K_c = 1.09 \times 10^{-7}$ mol dm^{-3}, so that

$$\frac{x^2}{1 - x} \text{ mol dm}^{-3} = 1.09 \times 10^{-7} \text{ mol dm}^{-3}$$

or

$$\frac{x^2}{1 - x} = 1.09 \times 10^{-7}$$

Since K_c is so small, we guess that very little N_2O_4 is dissociated at this temperature. This means that x is very small, and it is probably valid to make the approximation

$$1 - x \approx 1$$

(The symbol $\approx$ means approximately equal.) With this approximation our equation becomes

$$x^2 = 1.09 \times 10^{-7} \text{ mol}$$

or

$$x = \sqrt{10.9 \times 10^{-8}} = \sqrt{10.9} \times 10^{-4}$$

$$= 3.30 \times 10^{-4}$$

Our guess about x was thus correct. It is a small number, especially in comparison with 1. The approximation $1 - x \approx 1$ is valid to three decimal-place accuracy since $1 - x = 0.9997$.

Since x mol has dissociated, the fraction of N_2O_4 dissociated is given by x mol/1 mol $= x$. The fraction dissociated is thus 3.30×10^{-4}.

b) At 400 K, $K_c = 1.505$ mol dm^{-3}, so that

$$\frac{x^2}{1 - x} \text{ mol dm}^{-3} = 1.505 \text{ mol dm}^{-3}$$

or

$$\frac{x^2}{1 - x} = 1.505$$

Since K_c is not small, we can no longer use the approximation

$$1 - x \approx 1$$

Indeed, such an assumption leads to a ridiculous conclusion, since if $1 - x \approx 1$, then $x^2 = 1.505$, or $x = 1.227$. This result is impossible since it tells us that more N_2O_4 has dissociated (1.227 mol) than was originally present (1 mol).

Accordingly we return to the orthodox method for solving quadratic equations. Multiplying out our equation, we obtain

$$x^2 = 1.505 - 1.505x$$

or $$x^2 + 1.505x - 1.505 = 0$$

so that

$$x = \frac{-1.505 \pm \sqrt{(1.505)^2 + 4(1.505)}}{2}$$

$$= \frac{-1.505 \pm 2.878}{2}$$

$$= -2.19 \quad \text{or} \quad 0.6865$$

Since a negative result has no physical meaning, we conclude that 0.6865 is the correct answer. (As a cross check we can feed this result back into our original equation.) The fraction of N_2O_4 dissociated at this temperature is accordingly 0.6865.

c) At 600 K, $K_c = 1.675 \times 10^3$ mol dm^{-3}, so that

$$\frac{x^2}{1 - x} = 1.675 \times 10^3$$

Since K_c is fairly large, we guess that almost all the N_2O_4 has dissociated at this temperature and that x is accordingly close to 1. A valid approximation for these circumstances is then $x \approx 1$. With this approximation our equation becomes

$$\frac{1}{1 - x} = 1.675 \times 10^3$$

or

$$1 - x = \frac{1}{1.675 \times 10^3} = 5.97 \times 10^{-4}$$

Thus

$$x = 1 - 5.97 \times 10^{-4} = 0.999\,40$$

and our approximation is a good one. Since the fraction dissociated is x, we conclude that 0.9994 of the original N_2O_4 has dissociated.
Note: We did not use the approximation $x \approx 1$ to say that $1 - x \approx 0$. This is not a productive way to solve the problem because we would end up dividing by zero. Common sense tells us that it will not work. Similarly, in part **a** of this example we used the fact that x was very small to say that $1 - x \approx 1$. Saying that $x^2 \approx 0$ would not make sense because it would lead to the algebraic equation

$$\frac{0}{1 - x} = 1.09 \times 10^{-7}$$

that is, that $0 = 1.09 \times 10^{-7} - (1.09 \times 10^{-7})x$. This would lead to the solution $x = 1$ which is rather far from the original assumption.

Successive Approximations

An approximation is often useful even when it is not a very good one, because we can use the initial inaccurate approximation to calculate a better one. A good example of this occurs in Example 13.9 where it was necessary to solve the equation

$$0.0200 = \frac{(1 + x)x}{(1 - 2x)^2}$$

The conditions of the problem suggest that x may well be much smaller than 1. Accordingly we can approximate

$$1 + x \approx 1 \approx 1 - 2x$$

from which we obtain the approximate result (called the first approximation, x_1):

$$0.0200 \approx \frac{1x_1}{1^2} \qquad \text{or} \qquad x_1 \approx 0.0200$$

Although for some purposes this is a sufficiently accurate result, a much better approximation can be obtained by feeding this one back into the formula. If we write the formula as

$$x = \frac{0.0200\,(1 - 2x)^2}{1 + x}$$

we can now substitute $x_1 = 0.0200$ on the right-hand side, giving the second approximation:

$$x_2 \approx \frac{0.0200\,(1 - 2 \times 0.0200)^2}{1 + 0.0200}$$

$$= \frac{0.0200 \times 0.96^2}{1.0200}$$

$$= 0.0181$$

If we repeat this process, a third approximation is obtained:

$$x_3 \approx 0.0182$$

in exact agreement with the accurate result obtained from the quadratic formula in Example 13.9.

With practice, using this method of successive approximations is much faster than using the quadratic formula. It also has the advantage of being self-checking. A mistake in any of the calculations almost always leads to an obviously worse approximation. In general, if the last approximation for x differs from the next to last by less than 5 percent, it can be assumed to be accurate, and the successive-approximation procedure can be stopped. In the example just given,

$$\frac{x_2 - x_1}{x_1} = \frac{0.0181 - 0.0200}{0.0200} = \frac{-0.0019}{0.0200} \approx -0.10 = -10 \text{ percent}$$

so a third approximation was calculated. This third approximation was almost identical to the second and so was taken as the final result.

13.4 LE CHATELIER'S PRINCIPLE

Often it is useful to predict *qualitatively* (without doing calculations such as those just described) what will happen to a system at equilibrium when conditions such as temperature or volume change or when a reactant or product is added or removed from the reaction mixture. Fortunately a simple rule, **Le Chatelier's principle,** enables us to make such qualitative predictions. This rule states that *if a system is in equilibrium and some factor in the equilibrium conditions is altered, then the system will (if possible) adjust to a new equilibrium state so as to counteract this alteration to some degree.*

The Effect of a Change in Pressure

As an example of the application of Le Chatelier's principle, consider the effect of tripling the pressure on an equilibrium mixture of NO_2 and N_2O_4:

$$N_2O_4(g) \rightleftharpoons 2NO_2(g) \tag{13.13}$$

This could be done using the piston and cylinder shown in Fig. 13.2, in which case tripling the pressure would be expected to reduce the volume of the mixture to one-third its former value. Under the new conditions, however, Le Chatelier's principle tells us that a new equilibrium will be achieved which counteracts the alteration of pressure. That is, the concentrations of N_2O_4 and NO_2 should change in such a way as to lessen the pressure increase. This can happen if some of the NO_2 reacts to form N_2O_4, because two molecules of NO_2 are consumed for every one molecule of N_2O_4 produced. This reduction in the number of gas molecules will reduce the pressure at the new volume. Thus Le Chatelier's principle predicts that the reverse of Eq. (13.13) will occur, producing more N_2O_4 and using up some

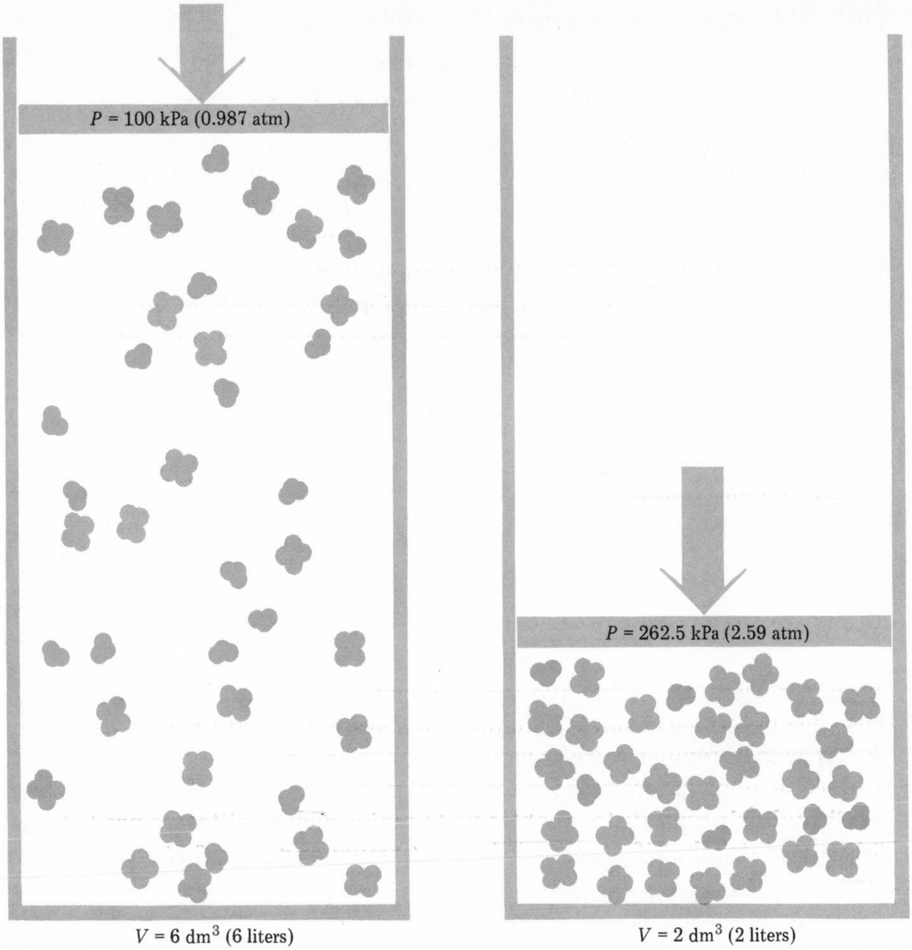

Figure 13.2 Le Chatelier's principle: effect of pressure. If the pressure on a equilibrium mixture of N_2O_4 (gray) and NO_2 (red) molecules is increased, some of the NO_2 molecules combine to form more N_2O_4. In this way the number of molecules is decreased and the increase of pressure counteracted to some extent. If none of the NO_2 molecules had recombined, the final pressure would have been 300 kPa rather than the 262 kPa actually observed.

NO_2. We say that *increasing the pressure on the N_2O_4 and NO_2 causes the equilibrium to shift to the left.* This agrees with the experimental data on this equilibrium, already given in Table 13.2. Note that the value of the equilibrium constant remains the same, even though the equilibrium shifts.

Notice that the effect just described occurs because the gas volume decreased and the concentrations of NO_2 and N_2O_4 both increased. If we had increased the total pressure on the equilibrium system by pumping in an inert gas such as $N_2(g)$, the volume would have remained the same, as would the partial pressures and concentrations of NO_2 and N_2O_4. In such a case no shift in the equilibrium would be expected. Note also the words *if possible* in the statement of Le Chatelier's principle. If the equilibrium reaction had not involved a change in the number of molecules in the gas phase, no shift in the concentrations could have made any difference in the pressure. Thus for the reaction

$$2HI(g) \rightleftharpoons H_2(g) + I_2(g)$$

changing the pressure by changing the size of the container would have no effect. You can check this for yourself by redoing Example 13.8 using several different volumes and the same initial amount of HI.

In general, whenever a gaseous equilibrium involves a change in the number of molecules ($\Delta n \neq 0$), increasing the pressure by reducing the volume will shift the equilibrium in the direction of fewer molecules. This applies even if pure liquids or solids are involved in the reaction. An example is the reaction

$$C(s) + H_2O(g) \rightleftharpoons CO(g) + H_2(g)$$

in which superheated steam is passed over carbon obtained from coal to produce carbon monoxide and hydrogen. Since the volume of solid carbon is negligible compared with the volumes of the gases, we need consider only the latter. Hence $\Delta n = 1$ (two gas molecules on the right for every one on the left), and an increase in pressure should favor the reverse reaction. This reaction is an important industrial process, and for the reason we have just outlined, is carried out at a low pressure.

The Effect of a Change in Temperature

In a chemical equilibrium there is almost always a difference in energy, and hence in enthalpy, between the reactants and the products. The thermochemical equation for dissociation of N_2O_4, for example, is

$$N_2O_4(g) \rightleftharpoons 2NO_2(g) \qquad \Delta H_m = 54.8 \text{ kJ mol}^{-1}$$

Because of this enthalpy difference, any shift in the equilibrium toward further dissociation will result in the absorption of heat energy and a momentary decrease in temperature. Conversely, a shift in the reverse direction will cause a small rise in temperature. If we increase the temperature of a mixture of N_2O_4 and NO_2, the mixture should respond in such a way as to oppose the rise in temperature. This can happen if some N_2O_4 in the mixture dissociates, since the resulting absorption of energy will produce a cooling effect. We would therefore expect that by raising the temperature of the equilibrium mixture, we would shift the equilibrium in favor of dissociation. Indeed we already know from the result of Example 13.10 that raising the temperature from 200 to 600 K changes an equilibrium mixture which is almost pure N_2O_4 into an equilibrium mixture which is almost pure NO_2.

In the general case, if we raise the temperature of any mixture of species which are in chemical equilibrium with each other, Le Chatelier's principle tells us that we will shift the equilibrium in the direction of those species with the higher energy. Thus, if the reaction is *endothermic,* as in the dissociation just discussed, raising the temperature will swing the equilibrium toward the *products*, and the value of the equilibrium constant K_c will increase with temperature. Conversely, if the reaction is *exothermic,* a rise in temperature will favor the *reactants,* and K_c will get smaller as the temperature increases. We can also turn the argument around. If we find a reaction for which K_c increases with temperature, we know immediately that the reaction must be endothermic. Conversely, if K_c decreases as temperature increases, the reaction must be exothermic.

Effect of Adding a Reactant or Product

If we have a system which is already in equilibrium, addition of an extra amount of one of the reactants or one of the products throws the system out of equilibrium. Either the forward or the reverse reaction will then occur in order to restore equilibrium conditions. We can easily tell which of these two possibilities will happen from Le Chatelier's principle. If we add more of one of the *products,* the system will adjust in order to offset the gain in concentration of this component. The *reverse* reaction will occur to a limited extent so that some of the added product can be consumed. Conversely, if one of the *reactants* is added, the system will adjust by allowing the *forward* reaction to occur to some extent. In either case *some of the added component will be consumed.*

Actually, we have already seen this principle in operation in the case of the decomposition of HI at high temperatures:

$$2HI \rightleftharpoons H_2(g) + I_2(g)$$

In Example 13.8 we saw that if 1 mol HI is heated to 745 K in a 10-dm^3 flask, some of the HI will decompose, producing an equilibrium mixture of composition

$$[HI] = 0.0780 \text{ mol dm}^{-3};\ [I_2] = 0.0110 \text{ mol dm}^{-3};\ [H_2] = 0.0110 \text{ mol dm}^{-3} \quad (1)$$

This is a genuine equilibrium mixture since it satisfies the equilibrium law

$$\frac{[H_2][I_2]}{[HI]^2} = \frac{(0.011 \text{ mol dm}^{-3})(0.011 \text{ mol dm}^{-3})}{(0.078 \text{ mol dm}^{-3})^2} = 0.020 = K_c$$

If an extra mole of H_2 is added to this mixture, the concentrations become

$$c_{HI} = 0.0780 \text{ mol dm}^{-3};\ c_{I_2} = 0.0110 \text{ mol dm}^{-3};\ c_{H_2} = 0.1110 \text{ mol dm}^{-3} \quad (2)$$

The system is no longer in equilibrium (hence the lack of square brackets to denote equilibrium concentrations) as we can easily check from the equilibrium law

$$\frac{c_{H_2} \times c_{I_2}}{c_{HI}^2} = \frac{0.011 \text{ mol dm}^{-3} \times 0.111 \text{ mol dm}^{-3}}{(0.078 \text{ mol dm}^{-3})^2} = 0.201 \neq K_c$$

The addition of H_2 has increased the concentration of this component. Accordingly, Le Chatelier's principle predicts that the system will achieve a new equilibrium in such a way as to reduce this concentration. The reverse reaction

$$H_2 + I_2 \rightarrow 2HI$$

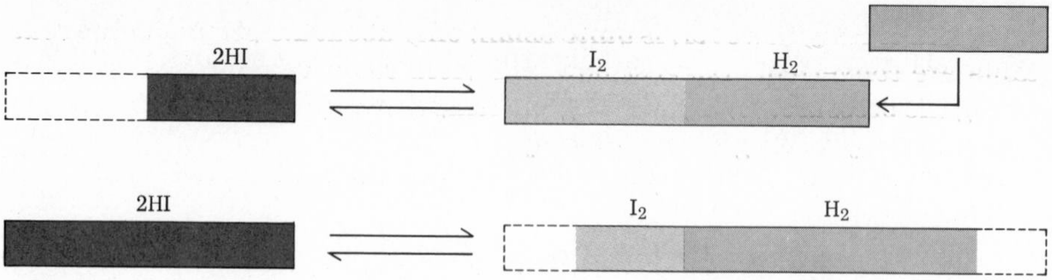

Figure 13.3 Le Chatelier's principle: effect of adding a component. At 745 K, HI is partially decomposed into H_2 and I_2: $2HI \rightleftharpoons H_2 + I_2$. If extra hydrogen (gray) is added to the equilibrium mixture, the system responds in such a way as to reduce the concentration of H_2. Some I_2 reacts with the H_2, and more HI is formed. The equilibrium is shifted to the left. Note, however, that some of the I_2 has been consumed, and its concentration is smaller than before.

occurs to a limited extent. This not only reduces the concentration of H_2 but the concentration of I_2 as well. At the same time the concentration of HI is increased. The system finally ends up with the concentrations calculated in Example 13.9, namely,

3 $[HI] = 0.0963 \text{ mol dm}^{-3}$; $[I_2] = 0.001\ 82 \text{ mol dm}^{-3}$; $[H_2] = 0.1018 \text{ mol dm}^{-}$

This is again an equilibrium situation since it conforms to the equilibrium law

$$\frac{[H_2][I_2]}{[HI]^2} = \frac{0.1018 \text{ mol dm}^{-3} \times 0.001\ 82 \text{ mol dm}^{-3}}{(0.0963 \text{ mol dm}^{-3})^2} = 0.02 = K_c$$

The way in which this system responds to the addition of H_2 is also illustrated schematically in Fig. 13.3. The actual extent of the change is exaggerated in this figure for diagrammatic effect.

Le Chatelier's principle can also be applied to cases where one of the components is *removed*. In such a case the system responds by *producing more of the component removed*. Consider, for example, the ionization of the weak diprotic acid H_2S:

$$H_2S + 2H_2O \rightleftharpoons 2H_3O^+ + S^{2-}$$

Since H_2S is a weak acid, very few S^{2-} ions are produced, but a much larger concentration of S^{2-} ions can be obtained by adding a strong base. The base will consume most of the H_3O^+ ions. As a result, more H_2S will react with H_2O in order to make up the deficiency of H_3O^+, and more S^{2-} ions will also be produced. This trick of removing one of the products in order to increase the concentration of *another product* is often used by chemists, and also by living systems.

EXAMPLE 13.11 When a mixture of 1 mol N_2 and 3 mol H_2 is brought to equilibrium over a catalyst at 773 K (500°C) and 10 atm (1.01 MPa), the mixture reacts to form NH_3 according to the equation

$$N_2(g) + 3H_2(g) \rightleftharpoons 2NH_3(g) \qquad \Delta H_m = -94.3 \text{ kJ}$$

The yield of NH_3, however, is quite small; only about 2.5 percent of the reactants are converted. Suggest how this yield could be improved (a) by altering the pressure; (b) by altering the temperature; (c) by removing a component; (d) by finding a better catalyst.

Solution

a) Increasing the pressure will drive the reaction in the direction of fewer molecules. Since $\Delta n = -2$, the forward reaction will be encouraged, increasing the yield of NH_3.

b) Increasing the temperature will drive the reaction in an endothermic direction, in this case in the reverse direction. In order to increase the yield, therefore, we need to *lower* the temperature.

c) Removing the product NH_3 will shift the reaction to the right. This is usually done by cooling the reaction mixture so that $NH_3(l)$ condenses out. Then more $N_2(g)$ and $H_2(g)$ are added, and the reaction mixture is recycled to a condition of sufficiently high temperature that the rate becomes appreciable.

d) While a better catalyst would speed up the *attainment* of equilibrium, it would not affect the position of equilibrium. It would therefore have no effect on the yield.

Note: As mentioned in Chaps. 3 and 12, NH_3 is an important chemical because of its use in fertilizers. In the design of a Haber-process plant to manufacture ammonia, attempts are made to use as high a pressure and as low a temperature as possible. The pressure is usually of the order of 150 atm (15 MPa), while the temperature is not usually below 750 K. Although a lower temperature would give a higher yield, the reaction would go too slowly to be economical, at least with present-day catalysts. The discoverer of a better catalyst for this reaction would certainly become a millionaire overnight.

13.5 THE MOLECULAR VIEW OF EQUILIBRIUM

Up to this point we have treated chemical equilibrium strictly on the macroscopic level. We have shown that equilibrium constants exist, how they can be determined, and how they can be used, but we have not explained why they are sometimes large and sometimes small. In order to answer this and related questions we must switch our point of view from the macroscopic to the microscopic level.

The cis-trans isomerism of difluoroethene provides a simple example for examination on the molecular level. As we saw earlier, if 1 mol cis isomer of this compound is heated to 623 K, it is gradually transformed into the equilibrium mixture of $\frac{2}{3}$ mol cis isomer and $\frac{1}{3}$ mol trans isomer. From a macroscopic point of view, we might conclude that the reaction stops once equilibrium is attained, but on the microscopic level, such a statement is obviously untrue. As cis molecules move around a container at 623 K, they occasionally suffer a collision with sufficient energy to flip them over to the trans conformation. This conversion process does not suddenly stop when equilibrium has been attained. There are plenty of cis molecules present in the equilibrium mixture, and they continue to collide with their fellow molecules and hence to flip over from the cis to the trans conformation. Thus the forward reaction

$$cis\text{-}C_2H_2F_2 \rightarrow trans\text{-}C_2H_2F_2$$

must *continue to occur* in the equilibrium mixture.

If this were the only reaction taking place in the flask, the eventual result would be that all the cis isomer would be converted to the trans isomer and no equilibrium mixture would result. This does not happen because the *reverse reaction*

$$trans\text{-}C_2H_2F_2 \rightarrow cis\text{-}C_2H_2F_2$$

is also occurring at the same time. The trans molecules in the equilibrium mixture are also continually being bombarded by their fellow molecules. Many of these collisions have sufficient energy to flip the trans molecule back into the cis form. Thus cis molecules are not only being consumed but also being produced in the flask. The concentration of the cis molecules remains constant because the rate at which they are being consumed is exactly balanced by the rate at which they are being produced. In other words, for every cis molecule which flips to the trans conformation in one part of the container, there will be on the average a molecule of trans isomer flipping in the reverse direction in another part of the container. This constant reshuffle of molecules between the cis and the trans forms can then continue indefinitely without any net change in the concentration of either species.

This molecular interpretation of equilibrium is not confined to the particular example of cis-trans isomerism or even just to chemical reactions. In Sec. 10.3 we described vapor-liquid equilibria in which the condensation of vapor just balanced the evaporation of liquid. Any *state of equilibrium corresponds to a situation in which the rate at which the forward reaction is occurring is exactly equal to the rate at which the reverse reaction is occurring.* Since these rates are equal, there is no net change in the concentration of any of the reactants or products with time. Because an equilibrium corresponds to a balance between the forward and the reverse reaction in this way, we use double arrows ($\rightleftharpoons$) in the equation.

As another example of chemical equilibrium, let us take a 0.001 *M* solution of acetic acid at 25°C. As we saw in Chap. 11, slightly more than 10 percent of the acetic acid molecules in this solution have ionized according to the equation

$$CH_3COOH + H_2O \rightleftharpoons CH_3COO^- + H_3O^+$$

Although the concentrations of all species in this solution remain constant with time, this does not mean that acetic acid molecules have stopped transferring protons to water molecules. The concentrations remain constant because the *reverse* reaction (transfer of a proton from a hydronium ion to an acetate ion) is occurring at the same rate as the forward reaction. For every hydronium ion produced by a proton transfer somewhere in the solution, another hydronium ion is losing its proton somewhere else. The net result is that the concentration of the hydronium ion remains constant, and with it the concentrations of the other species involved.

The microscopic view of equilibrium as a constant reshuffle of chemical species is often given a special name and referred to as a **dynamic equilibrium.** It is important to realize that once a dynamic equilibrium has been set up, a particular atom will sometimes turn up as part of one of the product molecules and sometimes as part of one of the reactant molecules. In the ionization equilibrium of acetic acid just considered, suppose for the sake of argument that we could identify a particular oxygen atom in some way.[1] If we could now follow the history of this atom, we would find that it would sometimes form part of an acetic acid molecule, CH_3CO^*OH, and sometimes part of an acetate ion, $CH_3CO^*O^-$. Since the acid is about 10 percent ionized, we would find on the average that our labeled oxygen would spend only 10 percent of its time as part of an acetate ion and the remaining 90 percent as part of an acetic acid molecule.

Figure 13.4 gives a computer simulation of this dynamic view of equilibrium for the cis-trans isomerization equilibrium of difluoroethene at 623 K. Suppose that it were possible to color one of the molecules in the equilibrium mixture and to photograph it at regular intervals thereafter under a high-power microscope. The figure shows the kind of results we could expect. In each "photograph" the molecule appears at a different orientation—between photos it has been bounced around quite a bit by collisions with other molecules. A careful inspection reveals a more important fact. Three of the nine photographs show the screened molecule in the *trans* conformation, while in the other six the molecule has the *cis* conformation. If we went on taking a large number of photographs of this molecule, we would find on the average that this ratio would be preserved. On the average one-third of the photographs would show the trans conformation, while the other two-thirds would show the cis form. The situation is much as though we had a conventional six-sided cubic die[2] with the word *trans* printed on two sides and the word *cis* printed on the other four. If we roll this die a sufficient number of times, it will come up cis twice as often as trans.

We can now begin to see why it is that the equilibrium constant for this reaction is 0.5 at 623 K. Not only the screened molecule, but every other molecule in the equilibrium mixture as well, spends one-third of its time in the trans conformation and two-thirds in the cis conformation. At any given time, therefore, we will find a mixture in which one-third of the molecules are trans while the other two-thirds are cis. The ratio of trans to cis will be 1:2, or 0.5, as found experimentally.

Such an explanation is still not the whole story. We still need to explain *why* the dice are loaded against the trans conformation of a molecule of difluoroethene. Why is it that this isomer is less likely to occur in the constant reshuffle between the two conformations caused by the continual

[1] In the laboratory the heavy isotope of oxygen, $^{18}_{8}O$, could be used.

[2] When we roll dice, each of them singly is called a die.

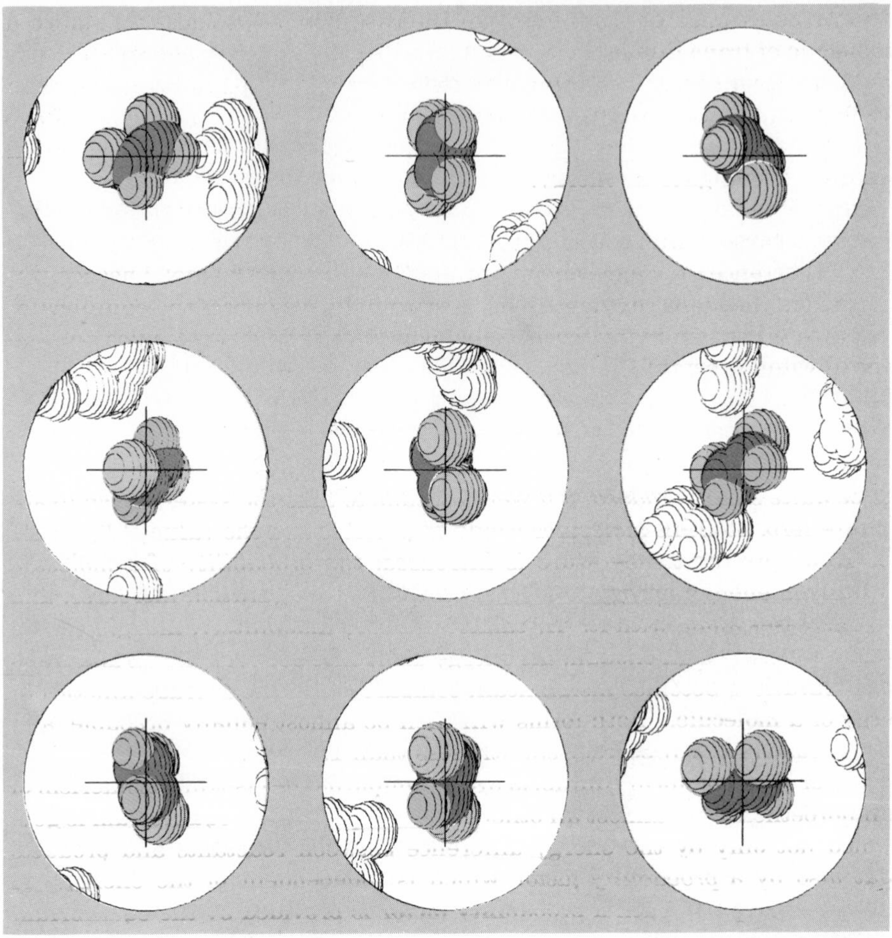

Figure 13.4 The equilibrium between the cis and trans isomers of difluoroethene at 623 K. If we could somehow photograph a given molecule in a equilibrium mixture at regular intervals, we could find it sometimes in the trans conformation but about twice as often in the cis conformation. Note that the molecule undergoes very many collisions between photographs. (Computer-generated.) (*Copyright © 1976 by W. G. Davies and J. W. Moore.*)

random collisions of molecules with each other in the equilibrium mixture? The reason is that the trans isomer is slightly *higher in energy* than the cis form. The enthalpy change ΔH_m for the reaction

$$cis\text{-}C_2H_2F_2 \rightarrow trans\text{-}C_2H_2F_2$$

has the small, but nevertheless positive, value of $+3.88$ kJ mol^{-1}.

In any collection of molecules, energy is constantly being transferred from one molecule to another as they collide. In this continual interchange of energy, inevitably some molecules acquire more kinetic or potential energy than their fellow molecules. However, as we saw in Sec. 9.4, when considering the distribution of kinetic energy, the probability of a molecule's acquiring a given energy depends on the magnitude of that energy. The higher the energy, the less likely it is to occur. In the constant reshuffle of energies, molecules which are even slightly higher in energy than their fellows occur slightly less often. In the case under consideration, since a molecule of *trans*-difluoroethene is 3.88 kJ mol^{-1} *higher* in energy than the cis form, it will occur *less* often than the cis form in an equilibrium mixture in which each molecule is constantly flipping back and forth between the two forms. In other words, the die is loaded against the trans configuration because it is higher in energy.

We can now also understand why the position of equilibrium for this reaction varies with temperature. At a very low temperature it will be a rare occurrence for a molecule to acquire the extra 3.88 kJ mol^{-1} needed if it is to have the trans configuration. Accordingly, we expect the equilibrium mixture to contain many fewer trans molecules than cis molecules and the equilibrium constant

$$K_c = \frac{[trans\text{-}C_2H_2F_2]}{[cis\text{-}C_2H_2F_2]}$$

to be quite small. Indeed, if it were possible to take the system down to absolute zero, no trans molecules would be possible and the value of K_c would be zero. As the temperature is increased, the probability of a molecule acquiring enough energy to assume the trans configuration increases, and K_c increases along with it. K_c cannot increase indefinitely, though. If the temperature is high enough, the energy difference between the cis and trans configurations becomes insignificant compared with the average kinetic energy of a molecule. Both forms will then be almost equally probable, and the value of K_c will be very slightly less than 1.

Very few chemical equilibria are as simple as the cis-trans isomerism of difluoroethene. In almost all other cases the position of equilibrium is governed not only by the energy difference between reactants and products but also by a *probability factor* which is independent of the energy. A simple example of such a probability factor is provided by the equilibrium between the ring and chain forms of 1,4-butanediol. The projection formula for this alcohol is

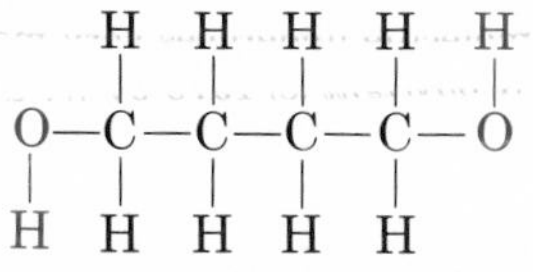

When a dilute solution (about 0.04 *M*) of this compound is prepared in carbon tetrachloride, the two ends of the molecule hydrogen-bond together to form a ring containing seven atoms, and an equilibrium between the ring form and the chain form is set up:

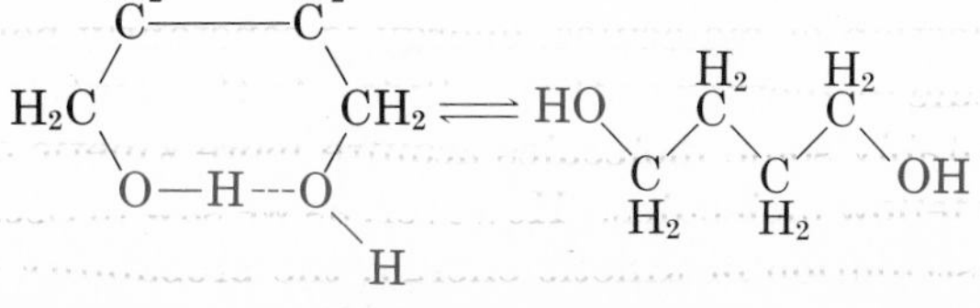

(13.14)

Ring ⇌ chain

At 70°C (343 K) the equilibrium constant for this reaction is found by experiment to be

$$K_c = \frac{[\text{chain}]}{[\text{ring}]} = 1.22$$

so that there are about 22 percent more chain molecules than ring molecules in the equilibrium mixture at this temperature. This is contrary to what one would expect if energy were the only factor involved in determining the position of equilibrium. The chain form of the molecule is obviously higher in energy than the ring form, since energy must be supplied in order to break the hydrogen bond. Indeed ΔH_m for Eq. (13.14) is found to be +13.5 kJ mol^{-1}. On the basis of our previous argument, if the chain form is higher in energy, it should be less probable than the ring form and K_c should be less than 1. The experimental fact that K_c is greater than 1 means that there is some other factor making the chain form more probable than would be the case if energy alone were responsible.

In order to see what this other factor is, let us suppose that we can somehow investigate our system at a temperature which is so high that the energy difference between the ring and the chain form is insignificant. (In practice, the carbon tetrachloride solvent would boil off long before this temperature was obtained.) At such a temperature, if the energy difference between the two forms were the only factor determining their relative probability, we would expect the ring and chain forms to be equally probable. It is easy to see, though, why this would not actually be the case.

The chain form of 1,4-butanediol is a very flexible molecule because of the free rotation which is possible around the three carbon-carbon bonds and the two carbon-oxygen bonds. As it collides with its fellow molecules, it can adopt a very large number of different conformations. Figure 13.5 illustrates some of these conformations, selected at random by a computer. Again we can take these as representing a series of photographs of the same molecule taken at regular intervals, say every 5 min. For the sake of clarity, though, we have kept the position of two of the carbon atoms the same for each photograph. In actual fact the molecule would be rotating as well as flexing, and the orientation of the bond joining these two carbon atoms would be continually changing. We should also remember that the molecule has undergone very many collisions between successive photographs.

The important thing to notice about this random selection of nine conformations shown in Fig. 13.5 is that in *none* of them are the two oxygen atoms close enough to allow the formation of a hydrogen bond. At this high temperature, because of the large number of ways in which a butanediol molecule can arrange itself in space, it will only occasionally adopt a ring conformation like that shown in Fig. 13.6 in which the two oxygen atoms are close enough to hydrogen bond to each other. In the equilibrium mixture, therefore, there will be many more chain molecules than ring molecules and K_c will be much larger than 1.

This situation begins to change as the temperature is reduced. With decreasing temperature, a smaller and smaller fraction of the molecules will have enough energy to break open a hydrogen bond. In consequence the

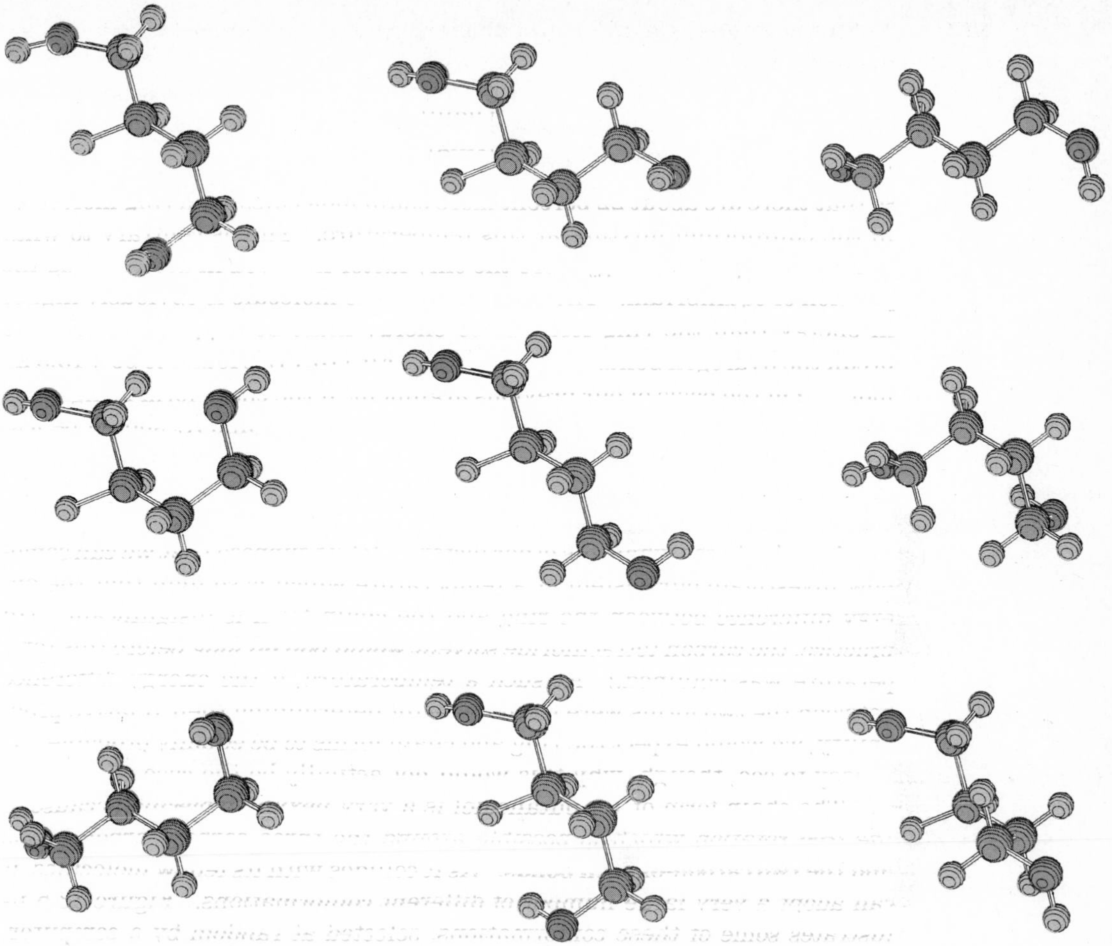

Figure 13.5 The chain form of butanediol. Because various segments of the molecule are free to rotate about carbon-carbon and carbon-oxygen bonds, the molecule can adopt a very large number of conformations. A random selection of nine of these is shown. Note that in none of these are the two oxygen atoms close enough to hydrogen bond together and form a ring. (Computer-generated.) *(Copyright © 1976 by W. G. Davies and J. W. Moore.)*

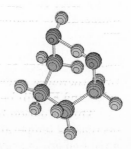

Figure 13.6 The ring form of butanediol. The two oxygen atoms at the top of the ring are hydrogen bonded together. From a strictly geometrical point of view this is a rather improbable conformation for the molecule. It is only because of the *lowering in energy* caused by the formation of a hydrogen bond that this ring form turns out to be slightly more probable in the equilibrium mixture than the chain form at low temperatures. (Computer-generated.) *(Copyright © 1976 by W. G. Davies and J. W. Moore.)*

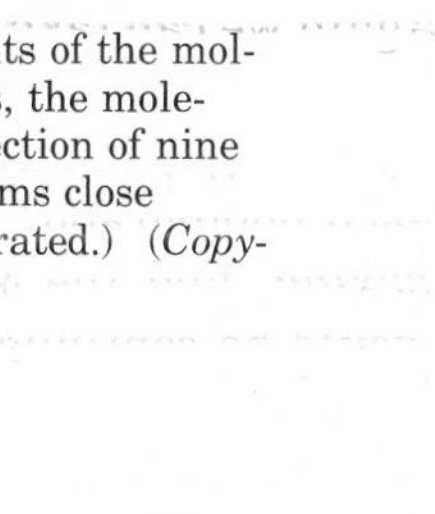

ring form will become more probable. Once having formed, a ring will be able to survive quite a few collisions with other molecules before it is broken open again. Indeed, if we could reduce the temperature of our system close to absolute zero without the solvent freezing, then we would find that virtually all the butanediol molecules were in the ring form. Once formed, a ring would almost never experience a collision which was sufficiently energetic to break it open. At such a low temperature the value of K_c would be very close to zero.

The two equilibria we have just described have both been very simple examples, especially since they each involved only one reactant and one product. Nevertheless the principles we have discovered in discussing them apply to all chemical equilibria. We can always interpret the equilibrium constant of a chemical reaction as the product of two factors, each of which has a meaning on the microscopic level:

$$K = \text{energy factor} \times \text{probability factor} \qquad (13.15)$$

The **energy factor** takes account of the fact that either the products or the reactants are higher in energy and hence less probable in the constant reshuffling of energy that goes on between molecules. The **probability factor** takes account of the fact that even if there were no energy difference between products and reactants, one or the other would still be more probable. This greater probability arises because of the larger number of ways in which the products (or reactants) can be realized on the molecular level.

In general the energy factor has the greatest effect on the value of the equilibrium constant at low temperatures. The lower the temperature, the very much lower the probability of a high-energy species occurring by chance in the constant energy reshuffle. At higher temperatures the effect of the energy factor becomes less pronounced and the probability factor becomes more important. If the temperature can be raised high enough, the probability factor eventually predominates.

SUMMARY

Many reactions do not go to completion but instead reach a state of dynamic equilibrium in which the concentrations of all participating substances remain constant. The equilibrium state can be reached in several ways, beginning with reactants or products or from a higher or lower temperature or with or without a catalyst. In all cases, however, the equilibrium concentrations at a given temperature will be related by the mathematical equilibrium-constant expression K_c. According to the law of chemical equilibrium, the equilibrium-constant expression contains the equilibrium concentration of each product raised to a power equal to the coefficient of that product in the chemical equation. This is divided by the equilibrium concentration of each reactant raised to its appropriate power. Once the value of K_c has been measured at a given temperature, we can use it to calculate the equilibrium concentrations of reactants and products. The equilibrium constant may also be expressed in terms of partial pressures in the case of reactions involving gases.

In addition to its use in quantitative calculations, the equilibrium law permits qualitative predictions. A very large equilibrium constant corresponds to a reaction which goes nearly to completion, while a small equilibrium constant suggests that almost no reaction takes place. For a specific equilibrium reaction, Le Chatelier's principle can be used to predict the effect of a change in conditions of temperature, pressure, volume, or concentration of the species involved. A system always adjusts to a new equilibrium so as to counteract such a change of conditions to some degree.

On a molecular level there are two factors which affect the position of an equilibrium reaction. The first of these is energy. The lower the energy of a molecule, the more likely the occurrence of that molecule, and therefore the greater its concentration will be in an equilibrium mixture. The second factor has to do with the number of structural arrangements possible for a given species. The greater the number of ways of arranging the atoms of a given molecule in three-dimensional space, the greater the probability that that particular molecule will exist. Both this probability factor and the energy factor mentioned earlier affect the size of any equilibrium constant. At low temperatures the energy factor predominates, while at high temperatures the probability factor is most important. In Chap. 16 we will discuss these two factors further and learn how they can be measured on the macroscopic level.

QUESTIONS AND PROBLEMS

13.1 What are the four features characteristic of chemical equilibrium listed in the text?

13.2 A chemist heats a mixture of two isomers, A and B, to 500 K and finds that the mixture is 60% A and 40% B. After heating the mixture to 900 K, the chemist finds the composition is now 70% A and 30% B. On cooling to 500 K, the mixture resumes its 60:40 composition. Is this a genuine equilibrium mixture, or should more experiments be done? If so, what type of experiments?

13.3 A chemist is investigating the equilibrium between two isomers, C and D. On heating pure C to 500 K, the chemist obtains a mixture which is 80% C and 20% D. Exactly the same mixture is obtained when pure D is heated to 500 K. Is this a genuine equilibrium mixture, or should more experiments be done?

13.4 If the equilibrium constant K_c for the reaction

$$cis\text{-}C_2H_2Cl_2 \rightleftharpoons trans\text{-}C_2H_2Cl_2$$

has the value 0.608 at 500 K, what is the value of K_c for the reaction

$$trans\text{-}C_2H_2Cl_2 \rightleftharpoons cis\text{-}C_2H_2Cl_2$$

at the same temperature?

13.5 Using the data given on lines *b* and *c* in Table 13.2, verify that K_c for the equilibrium

$$N_2O_4(g) \rightleftharpoons 2NO_2(g)$$

has the value 2.00 mol dm^{-3} at 407 K.

13.6 Using the results of the previous problem, find the value of K_c for the equilibrium

$$2NO_2(g) \rightleftharpoons N_2O_4(g)$$

at 407 K. Also calculate the value of K_c for

$$\tfrac{1}{2}N_2O_4(g) \rightleftharpoons NO_2(g)$$

at the same temperature.

13.7 Write the equilibrium-constant expression for each of the following reactions:

a $4NH_3(g) + 5O_2(g) \rightleftharpoons 4NO(g) + 6H_2O(g)$

b $2C_2H_6(g) + 7O_2(g) \rightleftharpoons 4CO_2(g) + 6H_2O(g)$

c $2N_2O(g) \rightleftharpoons 2N_2(g) + O_2(g)$

d $4NO(g) + F_2(g) \rightleftharpoons 2NO_2F(g) + N_2(g)$

e $2CO(g) + O_2(g) \rightleftharpoons 2CO_2(g)$

f $3O_2(g) \rightleftharpoons 2O_3(g)$

What units (mol dm^{-3}, mol^2 dm^{-6}, etc.) would be used to describe each of these equilibrium constants?

13.8 An equilibrium mixture contains 3.00 mol CO, 2.00 mol Cl_2, and 9.00 mol $COCl_2$ in a 50-dm^3 reaction flask at 800 K. Calculate the value of the equilibrium constant K_c for the reaction

$$CO(g) + Cl_2(g) \rightleftharpoons COCl_2(g)$$

at this temperature.

13.9 At 667 K, HI is found to be 11.4 percent dissociated into its elements:

$$2HI(g) \rightleftharpoons H_2(g) + I_2(g)$$

If 1.00 mol HI is placed in a 1.00-dm^3 container and heated to 667 K, calculate (*a*) the equilibrium concentration of all three substances; (*b*) the value of K_c for this equilibrium at this temperature.

13.10 A sample of nitrosyl bromide is heated to 100°C in a 10.00-dm^3 container in order to decompose it partially according to the equation

$$2NOBr(g) \rightleftharpoons 2NO(g) + Br_2(g)$$

The container is found to contain 6.44 g NOBr, 3.15 g NO, and 8.38 g Br_2 at equilibrium. (*a*) Find the value of K_c at this temperature. (*b*) Find the total pressure exerted by the mixture of gases.

13.11 Exactly 5.0 mol ammonia, NH_3, was placed in a 2.0-dm^3 flask which was then heated to 473 K. When equilibrium was established, 0.2 mol nitrogen had been formed according to the decomposition reaction $2NH_3(g) \rightleftharpoons N_2(g) + 3H_2(g)$. (*a*) Calculate the value of the equilibrium constant K_c for this reaction at 473 K. (*b*) Calculate the total pressure exerted by the mixture of gases inside the 2.0-dm^3 flask at this temperature.

*13.12 A sample of pure SO_3 weighing 0.8312 g was placed in a flask of volume 1.00 dm^3 and heated to 1100 K in order to decompose it partially:

$$2SO_3(g) \rightleftharpoons 2SO_2(g) + O_2(g)$$

If a pressure of 131.1 kPa (1.295 atm) was developed, find the value of K_c for this reaction at this temperature.

13.13 Write the equilibrium-constant expression for each of the following heterogeneous systems:

a $CaSO_4 \cdot 5H_2O(s) \rightleftharpoons CaSO_4 \cdot 3H_2O(s) + 2H_2O(g)$

b $SiF_4(g) + 2H_2O(g) \rightleftharpoons SiO_2(s) + 4HF(g)$

c $LaCl_3(s) + H_2O(g) \rightleftharpoons LaClO(s) + 2HCl(g)$

d $N_2O_4(g) + O_3(g) \rightleftharpoons N_2O_5(s) + O_2(g)$

e $C(s) + 2N_2O(g) \rightleftharpoons CO_2(g) + 2N_2(g)$

f $H_2O(l) \rightleftharpoons H_2O(g)$

13.14 The vapor pressure of water at 80°C is 47.34 kPa (0.467 atm). Find the value of K_c for the reaction

$$H_2O(l) \rightleftharpoons H_2O(g)$$

at this temperature.

13.15 The value of K_c for the reaction

$$N_2(g) + 3H_2(g) \rightleftharpoons 2NH_3(g)$$

is 2.00 $mol^{-2}\ dm^6$ at 400°C. Find the value of K_p for this reaction at this temperature (*a*) using kilopascals; (*b*) using atmospheres as units.

13.16 The equilibrium constant K_c for the cis-trans isomerization of gaseous 2-butene has the

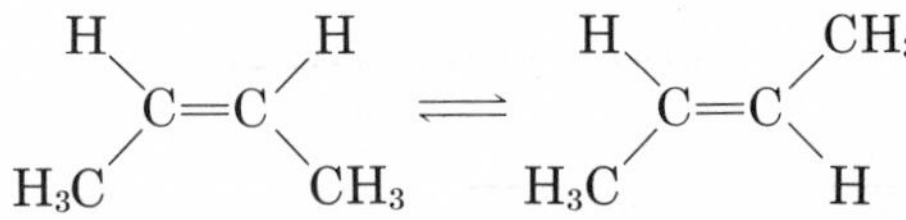

value 1.50 at 580 K. Calculate the amount of trans isomer produced when 1 mol *cis*-2-butene is heated to 580 K in the presence of a catalyst in a flask of volume 1.00 dm^3. What would the answer be if the flask had a volume of 10.0 dm^3?

13.17 (*a*) The equilibrium constant K_c for the reaction

$$CO(g) + H_2O(g) \rightleftharpoons CO_2(g) + H_2(g)$$

has the value 4.00 at 500 K. If a mixture of 1 mol CO and 1 mol H_2O is allowed to come to equilibrium in a flask of volume 1.000 dm^3 at 500 K, calculate the final concentrations of all four species: CO, H_2O, CO_2, and H_2. (*b*) What would be the equilibrium concentrations if a further mole of both CO and H_2O were added to the flask?

13.18 (*a*) At 503 K the equilibrium constant K_c for the dissociation of N_2O_4,

$$N_2O_4(g) \rightleftharpoons 2NO_2(g)$$

has the value 40.0 mol dm^{-3}. Calculate the fraction of N_2O_4 left undissociated when 1 mol of this gas is heated to 500 K in a 10-dm^3 container. (*Hint:* Assume the fraction to be small.) (*b*) If the volume is now reduced to 2 dm^3, what will be the new fraction of N_2O_4 which is undissociated? (*Hint:* Assume the fraction is small and use successive approximations.)

13.19 The equilibrium constant K_c for the reaction

$$N_2(g) + 3H_2(g) \rightleftharpoons 2NH_3(g)$$

has the value 5.97×10^{-2} mol^{-2} dm^6 at 500°C. If 1 mol N_2 gas and 1 mol H_2 gas are heated to 500°C in a flask of volume 10.00 dm^3 together with a catalyst, calculate the percentage of N_2 converted to NH_3. (*Hint:* Assume that this percentage is small.)

***13.20** What percentage of N_2 would be converted to NH_3 in the previous problem if the volume of the flask were 1.00 dm^3?

13.21 (*a*) When 1.00 mol I_2 and 3.00 mol H_2 are allowed to come to equilibrium at 745 K in a flask of volume 10.00 dm^3, what amount of HI will be produced? The equilibrium constant K_c for the reaction

$$H_2(g) + I_2(g) \rightleftharpoons 2HI(g)$$

has the value 50.0 at this temperature. (*b*) What amount of HI would have been produced if the volume of the flask had been 5.00 dm^3? (*c*) What will be the total amount of HI present at equilibrium if a further 3.00 mol H_2 is added?

13.22 The equilibrium constant K_c has a value of 3.30 mol dm^{-3} at 760 K for the decomposition of phosphorous pentachloride, $PCl_5(g) \rightleftharpoons PCl_3(g) + Cl_2(g)$. (*a*) Calculate the equilibrium concentrations of all three species arising from the decomposition of 0.75 mol PCl_5 in a 5.00-dm^3 vessel. (*b*) Calculate the equilibrium concentrations of all three species resulting from an initial mixture of 0.75 mol PCl_5 and 0.75 mol PCl_3 in a 5.00-dm^3 vessel.

13.23 (*a*) 1 mol CO_2 is heated to 1000 K with excess solid graphite in a container of volume 40 dm^3. What will be the composition of the gaseous equilibrium mixture if K_c for the reaction

$$C(graphite) + CO_2(g) \rightleftharpoons 2CO(g)$$

has the value 2.11×10^{-2} mol dm^{-3} at 1000 K? (*b*) The volume of the flask is now changed and a new equilibrium established. The amount of CO_2 and the amount of CO are found to be identical. What is the new volume of the flask?

13.24 Using Le Chatelier's principle as a guide, predict whether the equilibria listed below will be shifted to the left or the right when the following changes are inflicted on them: (*i*) the temperature is increased; (*ii*) the pressure is decreased; (*iii*) more of the substance indicated in color is added.

a $C(s) + H_2O(g) \rightleftharpoons CO(g) + H_2(g)$, $\Delta H_m(298\text{ K}) = +131.3$ kJ mol^{-1}
b $3Fe(s) + 4H_2O(g) \rightleftharpoons Fe_3O_4(s) + 4H_2(g)$, $\Delta H_m(298\text{ K}) = -149.9$ kJ mol^{-1}
c $C(s) + CO_2(g) \rightleftharpoons 2CO(g)$, $\Delta H_m(298\text{ K}) = +172.5$ kJ mol^{-1}
d $N_2O_4(g) \rightleftharpoons 2NO_2(g)$, $\Delta H_m(298\text{ K}) = +54.8$ kJ mol^{-1}
e $N_2(g) + O_2(g) \rightleftharpoons 2NO(g)$, $\Delta H_m(298\text{ K}) = +180.0$ kJ mol^{-1}
f $CH_4(g) + 2O_2(g) \rightleftharpoons CO_2(g) + 2H_2O(g)$, $\Delta H_m(298\text{ K}) = -802.3$ kJ mol^{-1}
g $CaCO_3(s) \rightleftharpoons CaO(s) + CO_2(g)$ $\Delta H_m(298\text{ K}) = +177.9$ kJ mol^{-1}

13.25 When pressure is applied to ice at 0°C, the ice melts. Explain how this is an example of Le Chatelier's principle.

13.26 The solution of lithium chloride in water is an exothermic process. Would you expect lithium chloride to be more or less soluble as the temperature is increased?

13.27 From the data given in Table 13.1, predict whether the conversion from the cis to the trans isomer of 2-butene is exothermic or endothermic.

13.28 Explain why the addition of hydroxide ions to any solution containing hydrogen sulfide will increase the concentration of the sulfide ions.

13.29 The following amounts of HI, H_2, and I_2 are introduced into a 10.00-dm^3 reaction flask and heated to 745 K.

	n_{HI}/mol	n_{H_2}/mol	n_{I_2}/mol
Case *a*	1	0.1	0.1
Case *b*	10	1	1
Case *c*	10	10	1
Case *d*	5.62	0.381	1.75

The equilibrium constant for the reaction

$$2HI(g) \rightleftharpoons H_2(g) + I_2(g)$$

has the value 0.0200 at 745 K. In which cases will the concentration of HI increase as equilibrium is attained and in which cases will the concentration of HI decrease?

13.30 Discuss the statement, "No true chemical equilibrium can exist unless reactant molecules are constantly changing into product molecules and vice versa."

13.31 A small sample of *cis*-dichloroethene in which one carbon atom is the radioactive isotope ^{14}C is added to an equilibrium mixture of the cis and trans isomers at a certain temperature. Eventually 40 percent of the radioactive molecules are found to be in the trans configuration at any given time. What is the value of K_c for the cis $\rightleftharpoons$ trans equilibrium? What would have occurred if a small sample of radioactive trans isomer had been added instead of the cis isomer?

13.32 A small amount of D_2O is added to a warm saturated solution of benzoic acid, C_6H_5COOH. When the solution is allowed to cool, the crystals of benzoic acid which separate are found to contain C_6H_5COOD. Explain how this is possible. By contrast, if water containing the heavy isotope of oxygen ^{18}O is added to the benzoic acid solution, *none* of the benzoic acid crystals obtained from the solution contain any ^{18}O. (Recall that D is the heavy isotope of H, ^2_1H.)

13.33 In a 0.002-*M* solution of acetic acid any acetate species spends 9 percent of its time as an acetate ion, CH_3COO^-, and the remaining 91 percent of its time as acetic acid, CH_3COOH. What is the value of K_a for the reaction

$$CH_3COOH + H_2O \rightleftharpoons CH_3COO^- + H_3O^+$$

13.34 Samples of N_2O_4 can be prepared in which both N atoms consist of the heavy isotope ^{15}N. We can write these molecules as $O_2N\text{—}NO_2$ to distinguish them from the ordinary form of N_2O_4, written $O_2N\text{—}NO_2$. When a small sample of $O_2N\text{—}NO_2$ is introduced into ordinary N_2O_4, most of it is converted immediately into molecules of $O_2N\text{—}NO_2$, containing one heavy and one light N atom. What does this behavior tell us about the microscopic behavior of the equilibrium

$$N_2O_4(g) \rightleftharpoons 2NO_2(g)$$

***13.35** Using the symbolism of the previous question, we can write the equilibrium

$$O_2N\text{—}NO_2 + O_2N\text{—}NO_2 \rightleftharpoons 2O_2N\text{—}NO_2$$

Assuming that ^{15}N and ^{14}N atoms behave identically, what is the value of K_c for this equilibrium? (*Hint:* Take a molecule at random from the mixture and consider probabilities.)

13.36 By varying the temperature of a cis-trans equilibrium mixture of difluoroethene, it is possible to obtain a mixture containing 90 percent of the cis form. It is impossible, however, to obtain an equilibrium mixture containing 90 percent of the trans form. Explain why this is so in terms of molecular probabilities.

13.37 In the case of the ring and the chain forms of butanediol, it is possible by varying the temperature to obtain an equilibrium mixture which is 90 percent ring form. A temperature can also be found for which the equilibrium mixture is 90 percent chain form. Explain why this behavior is so different from that encountered in the previous problem.

chapter fourteen

IONIC EQUILIBRIA IN AQUEOUS SOLUTIONS

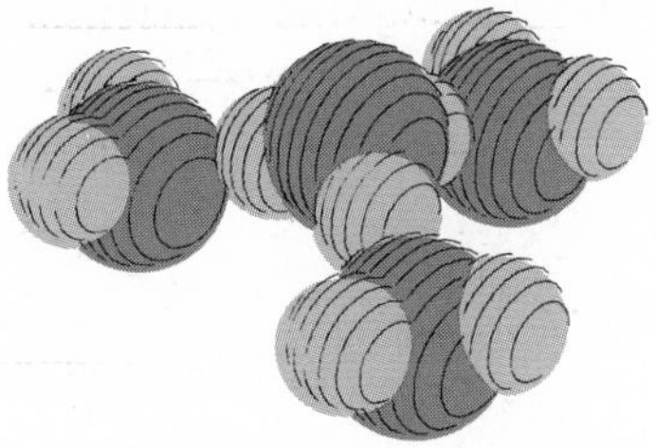

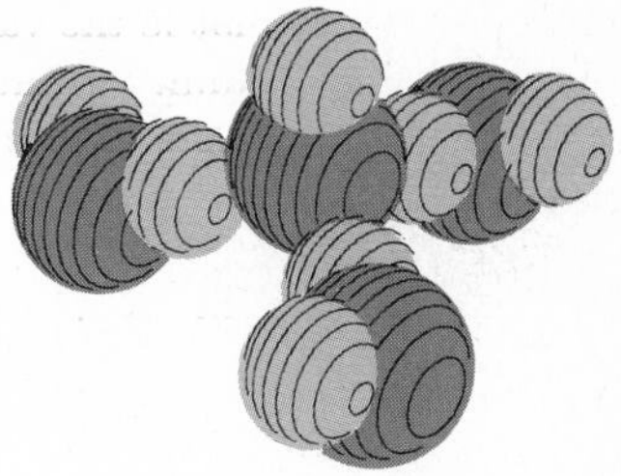

We have already noted the importance of reactions in aqueous solutions in the chemical laboratory, in the natural environment, and in the human body. Many reactions in aqueous solutions involve weak acids or bases or slightly soluble substances, and in such cases one or more equilibria are achieved in solution. Furthermore, the equilibrium state is usually reached almost instantaneously, and so we can use the equilibrium law to calculate the concentrations and amounts of substance of different species in solution. Such information enables us to understand, predict, and control what will happen in solution, and it has numerous practical applications. This chapter will illustrate how equilibrium constants may be used to obtain information about reactions in solution, and in many cases the results of equilibrium calculations will be applied to practical problems.

14.1 IONIZATION OF WATER

In Chap. 11 we saw that water can act as a very weak acid and a very weak base, donating protons to itself to a limited extent:

$$2H_2O(l) \rightleftharpoons H_3O^+(aq) + OH^-(aq)$$

Applying the equilibrium law to this reaction, we obtain

$$K_c = \frac{[H_3O^+][OH^-]}{[H_2O]^2}$$

However, as we have already seen in Sec. 13.2, the concentration of water has a constant value of 55.5 mol dm^{-3}, and so its square can be multiplied by K_c to give a new constant K_w, called the **ion-product constant** of water:

$$K_w = K_c(55.5 \text{ mol dm}^{-3})^2 = [H_3O^+][OH^-] \qquad (14.1)$$

Measurements of the electrical conductivity of carefully purified water indicate that at 25°C $[H_3O^+] = [OH^-] = 1.00 \times 10^{-7}$ mol dm^{-3}, so that

$$K_w = (1.00 \times 10^{-7} \text{ mol dm}^{-3})(1.00 \times 10^{-7} \text{ mol dm}^{-3})$$
$$= 1.00 \times 10^{-14} \text{ mol}^2 \text{ dm}^{-6}$$

(Since the equilibrium law is not obeyed exactly, even in dilute solutions, results of most equilibrium calculations are rounded to three significant figures. Hence the value of $K_w = 1.00 \times 10^{-14}$ mol^2 dm^{-6} is sufficiently accurate for all such calculations.)

The equilibrium constant K_w applies not only to pure water but to *any* aqueous solution at 25°C. Thus, for example, if we add 1.00 mol of the strong acid HNO_3 to H_2O to make a total volume of 1 dm^3, essentially all the HNO_3 molecules donate their protons to H_2O:

$$HNO_3 + H_2O \rightarrow NO_3^- + H_3O^+$$

and a solution in which $[H_3O^+] = 1.00$ mol dm^{-3} is obtained. Although this solution is very acidic, there are still hydroxide ions present. We can calculate their concentration by rearranging Eq. (14.1):

$$[OH^-] = \frac{K_w}{[H_3O^+]} = \frac{1.00 \times 10^{-14} \text{ mol}^2 \text{ dm}^{-6}}{1.00 \text{ mol dm}^{-3}} = 1.00 \times 10^{-14} \text{ mol dm}^{-3}$$

The addition of the HNO_3 to H_2O not only increases the hydronium-ion concentration but also reduces the hydroxide-ion concentration from an initially minute 10^{-7} mol dm^{-3} to an even more minute 10^{-14} mol dm^{-3}.

EXAMPLE 14.1 Calculate the hydronium-ion concentration in a solution of 0.306 *M* $Ba(OH)_2$.

Solution Since 1 mol $Ba(OH)_2$ produces 2 mol OH^- in solution, we have

$$[OH^-] = 2 \times 0.306 \text{ mol dm}^{-3} = 0.612 \text{ mol dm}^{-3}$$

Then

$$[H_3O^+] = \frac{K_w}{[OH^-]} = \frac{1.00 \times 10^{-14} \text{ mol}^2 \text{ dm}^{-6}}{0.612 \text{ mol dm}^{-3}} = 1.63 \times 10^{-14} \text{ mol dm}^{-3}$$

Note that since strong acids like HNO_3 are completely converted to H_3O^+ in aqueous solution, it is a simple matter to determine $[H_3O^+]$, and from it, $[OH^-]$. Similarly, when a strong base dissolves in H_2O, it is entirely converted to OH^-, so that $[OH^-]$, and from it $[H_3O^+]$, are easily obtained.

pH and pOH

The calculations we have just done show that the concentrations of hydronium and hydroxide ions in aqueous solution can vary from about 1 mol dm^{-3} down to about 1×10^{-14} mol dm^{-3}, and perhaps over an even wider range. The numbers used to express $[H_3O^+]$ and $[OH^-]$ in the units mole per cubic decimeter will often include large negative powers of 10. Consequently it is convenient to define the following:

$$\text{pH} = -\log \frac{[H_3O^+]}{1 \text{ mol dm}^{-3}} \qquad \text{pOH} = -\log \frac{[OH^-]}{1 \text{ mol dm}^{-3}}$$

Note carefully what these equations tell us to do. To obtain pH, for example, we divide $[H_3O^+]$ by the units mole per cubic decimeter. This gives a pure number, and so we can take its logarithm. (It does not make sense to take the logarithm of a unit, such as mole per cubic decimeter.) The minus sign insures that we will obtain a positive result most of the time.

The logarithm of a number is the power to which 10 must be raised to give the number itself.[1] Therefore the definitions of pH and pOH mean that we can deal with powers of 10 rather than numerical values. Since the numbers needed to express $[H_3O^+]$ and $[OH^-]$ are usually between 1 and 10^{-14}, pH and pOH values are usually between 0 and 14.

EXAMPLE 14.2 Calculate the pH and the pOH of each of the following aqueous solutions: (a) 1.00 *M* HNO_3; (b) 0.306 *M* $Ba(OH)_2$.

Solution

a) Our previous discussion showed that for this solution $[H_3O^+] = 1.00$ mol dm^{-3} and $[OH^-] = 1.00 \times 10^{-14}$ mol dm^{-3}. Applying the definitions of pH and pOH, we have

$$\text{pH} = -\log \frac{1.00 \cancel{\text{mol dm}^{-3}}}{1 \cancel{\text{mol dm}^{-3}}} = -\log(10^0) = -(0) = 0.00$$

$$\text{pOH} = -\log \frac{1.00 \times 10^{-14} \cancel{\text{mol dm}^{-3}}}{1 \cancel{\text{mol dm}^{-3}}} = -\log(10^{-14}) = -(-14) = 14.00$$

b) In Example 14.1 we found for this solution $[H_3O^+] = 1.63 \times 10^{-14}$ mol dm^{-3} and $[OH^-] = 6.12 \times 10^{-1}$ mol dm^{-3}. Thus

$$\begin{aligned} \text{pH} &= -\log(1.63 \times 10^{-14}) = -[\log(1.63) + \log(10^{-14})] \\ &= -[0.212 + (-14)] = -(-13.788) = 13.788 \\ \text{pOH} &= -\log(6.12 \times 10^{-1}) = -[\log(6.12) + \log(10^{-1})] \\ &= -[0.787 + (-1)] = -(-0.213) = 0.213 \end{aligned}$$

[1] Logarithms are discussed and a table of logarithms appears in Appendix 2.

Note that to obtain the logarithm of a number which is not an exact power of 10, we first express the number in scientific notation. The logarithm of the power of 10 is just the exponent; that is, $\log(10^{-14}) = -14$, and the logarithm of the number multiplied times the power of 10 can be obtained from a table of logarithms (Appendix 2). If you have a calculator with a log key, all these steps are not necessary—simply enter the number (after dividing out the units) and hit the log key. Do not forget to change the sign, though, since it is minus the logarithm that you want.

In the laboratory it is convenient to measure the pH of a solution using a pH meter. Such a device works on a different principle from the conductivity measurements we have already mentioned, and an accurate explanation of how it works is beyond the scope of the present discussion. Suffice it to say that unless great care and special instruments are used, pH is usually measured to an accuracy of ± 0.01. Therefore pH values are usually rounded to the second decimal place; i.e., the results of Example 14.2*b* would commonly be rounded to pH = 13.79 and pOH = 0.21.

Because pH measurements are so easily made, it is essential that you be able to convert from pH to $[H_3O^+]$. This is the reverse of finding pH from $[H_3O^+]$. Consequently it involves antilogs instead of logs. From the definition

$$\text{pH} = -\log \frac{[H_3O^+]}{\text{mol dm}^{-3}}$$

we have

$$-\text{pH} = \log \frac{[H_3O^+]}{\text{mol dm}^{-3}}$$

Taking the antilog of both sides, we have

$$\text{antilog}(-\text{pH}) = \text{antilog}\left\{\log \frac{[H_3O^+]}{\text{mol dm}^{-3}}\right\}$$

so that

$$\text{antilog}(-\text{pH}) = \frac{[H_3O^+]}{\text{mol dm}^{-3}}$$

remembering that antilog $x = 10^x$, we can write this expression as

$$10^{-\text{pH}} = \frac{[H_3O^+]}{\text{mol dm}^{-3}}$$

or

$$[H_3O^+] = 10^{-\text{pH}} \text{ mol dm}^{-3} \qquad (14.2a)$$

An alternative method of writing this equation is

$$[H_3O^+] = \frac{1}{10^{\text{pH}}} \text{ mol dm}^{-3} \qquad (14.2b)$$

EXAMPLE 14.3 The pH of a solution is found to be 3.40. Find the hydronium-ion concentration of the solution.

Solution If you have a calculator which has an antilog or 10^x button, the problem is very simple. You enter -3.40 and hit the button. The number thus obtained, $3.9822 - 04$ is the number of moles of hydronium ion per cubic decimeter. This follows from Eq. (14.2*a*):

$$[H_3O^+] = 10^{-pH} \text{ mol dm}^{-3} = 10^{-3.4} \text{ mol dm}^{-3}$$
$$= 3.98 \times 10^{-4} \text{ mol dm}^{-3}$$

The same result is almost as easy to find using Eq. (14.2*b*) and a table of logarithms.

$$10^{pH} = \text{antilog(pH)} = \text{antilog } 3.40$$
$$= \text{antilog } 3 \times \text{antilog } 0.40$$
$$= 10^3 \times 2.51$$

Thus

$$10^{-pH} = \frac{1}{10^{pH}} = \frac{1}{2.51 \times 10^3} = 3.98 \times 10^{-4}$$

in other words,

$$[H_3O^+] = 3.98 \times 10^{-4} \text{ mol dm}^{-3}$$

There is a very simple relationship between the pH and the pOH of an aqueous solution at 25°C. We know that at this temperature

$$[H_3O^+][OH^-] = K_w = 10^{-14} \text{ mol}^2 \text{ dm}^{-6}$$

Dividing both sides by $\text{mol}^2 \text{ dm}^{-6}$, we obtain

$$\frac{[H_3O^+]}{\text{mol dm}^{-3}} \times \frac{[OH^-]}{\text{mol dm}^{-3}} = 10^{-14}$$

Taking logs and multiplying both sides by -1, we then have

$$-\log \frac{[H_3O^+]}{\text{mol dm}^{-3}} - \log \frac{[OH^-]}{\text{mol dm}^{-3}} = -\log(10^{-14})$$

or

$$\text{pH} + \text{pOH} = 14.00 \qquad (14.3)$$

This simple relationship is often useful in finding the pH of solutions containing bases, as the following example shows.

EXAMPLE 14.4 If 3.53 g of pure NaOH is dissolved in 10 dm³ of H_2O, find the pH of the resulting solution.

Solution We first calculate the concentration of the NaOH.

$$n_{NaOH} = 3.53 \text{ g} \times \frac{1 \text{ mol}^{-1}}{40.0 \text{ g}} = 0.088\,25 \text{ mol}$$

so that

$$c_{NaOH} = \frac{n_{NaOH}}{V} = \frac{0.088\,25 \text{ mol}}{10 \text{ dm}^3}$$
$$= 8.82 \times 10^{-3} \text{ mol dm}^{-3}$$

Since NaOH is a strong base, each mole of NaOH dissolved produces 1 mol OH^- ions, so that

$$[OH^-] = 8.82 \times 10^{-3} \text{ mol dm}^{-3}$$

Thus
$$\text{pOH} = -\log(8.82 \times 10^{-3})$$
$$= -(0.95 - 3.00) = +2.05$$

From which pH = 14.00 − pOH = 11.95

Practical Aspects of pH

While the ability to calculate the pH of a solution from the hydronium-ion concentration and vice versa is useful, it is not the only thing we need to understand about pH. If someone gives you a solution whose pH is 14.74, it is true that the hydronium-ion concentration must be 1.82×10^{-15} mol dm^{-3}, but it is perhaps more important to know that the solution is corrosively basic and should be handled with respect. In general, then, we need not only to be able to calculate a pH but also to have some realization of what kind of solutions have what kind of pH. Table 14.1 is designed to meet this need.

In pure water at 25°C the hydronium-ion concentration is close to 1.00×10^{-7} mol dm^{-3}, so that the pH is 7. In consequence any solution, not only pure water, which has a *pH of 7* is described as being *neutral*. An *acidic* solution, as we know, is one in which the hydronium-ion concentration is greater than that of pure water, i.e., *greater than* 10^{-7} mol dm^{-3}. In pH terms this translates into a pH which is *less than 7* (because the pH is a negative logarithm). Small pH values are thus characteristic of acidic solutions; the smaller the pH, the more acidic the solution.

By contrast, a *basic* solution is one in which the hydroxide-ion concentration is greater than 10^{-7} mol dm^{-3}. In such a solution (see Table 14.1) the hydronium-ion concentration is *less than* 10^{-7} mol dm^{-3}, so that the pH

TABLE 14.1 The pH Scale.

	pH	$[H_3O^+]$	$[OH^-]$	pOH	
Battery acid	0	1	10^{-14}	14	
	1	10^{-1}	10^{-13}	13	**Strongly acidic**
Stomach acid	2	10^{-2}	10^{-12}	12	
Lemon juice	3	10^{-3}	10^{-11}	11	**Weakly acidic**
Soda water	4	10^{-4}	10^{-10}	10	
Black coffee	5	10^{-5}	10^{-9}	9	**Barely acidic**
	6	10^{-6}	10^{-8}	8	
Pure water	7	10^{-7}	10^{-7}	7	**Neutral**
Seawater	8	10^{-8}	10^{-6}	6	**Barely basic**
Baking soda	9	10^{-9}	10^{-5}	5	
Toilet soap	10	10^{-10}	10^{-4}	4	**Mildly basic**
Laundry water	11	10^{-11}	10^{-3}	3	
	12	10^{-12}	10^{-2}	2	
Household ammonia	13	10^{-13}	10^{-1}	1	**Very basic**
Drain cleaner	14	10^{-14}	1	0	

of a basic solution is *greater than 7*. Large pH values are thus characteristic of basic solutions. The larger the pH, the more basic the solution.

We have some direct experience of pH through our sense of taste, which responds to the concentration of hydronium ions. Most of us are able to detect a sour, tart, acidic taste in a solution with a pH of between 4 and 5. Black coffee with a pH of 5 does not taste acidic to most people, whereas carbonated water (soda water) with a pH of 4 does. A somewhat more acidic solution with a pH of 3 or even 2 tastes pleasantly acidic, especially if sweetened with sugar. Most fruit juices and most soft drinks have a pH in this range. If the pH of the solution is lower than 2, the taste is too tart for us to tolerate it for long. Occasionally during a stomach upset we experience the taste of a solution with a pH of 1.4, since this is the normal pH of stomach acid. Any pH below this value is not only unpleasant to the taste but acidic enough to attack the skin. Note that a sufficiently concentrated solution of strong acid can actually have a *negative* pH. Sulfuric acid in a car battery has a concentration of about 5 mol H_2SO_4 dm^{-3} and a pH of about -0.7.

Since we do not often intentionally swallow basic solutions, we cannot rely on the sense of taste to guide us through pH values which are greater than 7. Examples of very weakly basic solutions are sodium bicarbonate (baking soda, $NaHCO_3$), which has a pH close to 8, and a solution of soap, which usually has a pH slightly greater than 9. Solutions with a pH of 10 or 11 (which we can describe as mildly basic) are to be found in the weekly wash. Most laundry powders contain a weak base such as a phosphate or a carbonate, which raises the pH to this value. Solutions with a pH of 11 feel noticeably "soapy" to the touch and also cause a characteristic wrinkling of the skin. If the pH is greater than 12 or 13, the solution attacks skin rapidly enough to be dangerous. Very basic solutions are more corrosive to skin than very acidic solutions. A solution of drain cleaner (mainly NaOH with scent and coloring matter added) has a pH of between 14 and 15 and should be handled only with the utmost caution.

14.2 THE pH OF SOLUTIONS OF WEAK ACIDS AND WEAK BASES

Weak Acids

When any weak acid, which we will denote by the general formula HA, is dissolved in water, the reaction

$$HA + H_2O \rightleftharpoons H_3O^+ + A^- \qquad (14.4)$$

proceeds to only a limited extent, and we must allow for this in calculating the hydronium-ion concentration and hence the pH of such a solution. In general the pH of a solution of a weak acid depends on only two factors, the concentration of the acid, c_a, and the magnitude of an equilibrium constant K_a, called the acid constant, which measures the strength of the acid. The acid constant is defined by the relationship:

$$K_c \times 55.5 \text{ mol dm}^{-3} = K_a = \frac{[H_3O^+][A^-]}{[HA]} \qquad (14.5)$$

(The acid constant is K_c multiplied by the constant concentration of water, as already defined in Sec. 13.2. It is also called the ionization constant or the dissociation constant of the acid.)

In Example 13.6 we showed how measurements of the conductivity of acetic acid solutions could be used to find the acid constant for acetic acid, $K_a(CH_3COOH)$. You will recall from that example that K_a was only approximately a constant, varying from a value of 1.81×10^{-5} mol dm^{-3} in very dilute solutions to a value of 1.41×10^{-5} mol dm^{-3} in a 1 *M* solution. A similar variation is found for other weak acids, so that most of the calculations we do using K_a are only approximate. Only two or possibly three significant figures should be retained.

Table 14.2 gives the K_a values for a few selected acids arranged in order of their strength. It is at once apparent from this table that the larger the K_a value, the stronger the acid. The strongest acids, like HCl and H_2SO_4, have K_a values which are too large to measure, while another strong acid, HNO_3, has a K_a value close to 20 mol dm^{-3}. Typical weak acids such as HF and CH_3COOH have acid constants with a value of 10^{-4} or 10^{-5} mol dm^{-3}. Acids like the ammonium ion, NH_4^+, and hydrogen cyanide, HCN, for which K_a is less than 10^{-9} mol dm^{-3}, are very weakly acidic.

Before we can go on to discuss how the hydronium-ion concentration and the pH of a solution of a weak acid depend on the concentration of the acid, we need to clarify a point of terminology. In order to do this let us take as an example a 0.0010 *M* solution of acetic acid. As we saw in Sec. 11.3, conductivity measurements show that only about 10 percent of the acid molecules have donated protons to water at any given time. We thus have a situation which can be summarized schematically in the following way:

$$\underset{\textbf{0.0009 mol dm}^{-3}}{\overset{90\%}{CH_3COOH}} + H_2O \rightleftharpoons \underset{\textbf{0.0001 mol dm}^{-3}}{\overset{10\%}{CH_3COO^-}} + \underset{\textbf{0.0001 mol dm}^{-3}}{H_3O^+} \qquad (14.6)$$

In such a solution there is some ambiguity as to what we mean by the phrase *concentration of acetic acid*. Do we mean 0.0010 mol dm^{-3}, or do we mean 90 percent of this value, namely, 0.0009 mol dm^{-3}? In order to resolve this difficulty, we will use the term **stoichiometric concentration** of acid and the symbol c_a to indicate the quantity 0.0010 mol dm^{-3}, that is, to indicate the total amount of acetic acid originally added per unit volume of solution. On the other hand we will use the term **equilibrium concentration**

TABLE 14.2 The Acid Constants for Some Acids at 25°C.

Name	Proton Transfer	K_a/mol dm^{-3}
Hydrochloric acid	$HCl + H_2O \rightarrow H_3O^+ + Cl^-$	Large
Sulfuric acid	$H_2SO_4 + H_2O \rightarrow H_3O^+ + HSO_4^-$	Large
Nitric acid	$HNO_3 + H_2O \rightarrow H_3O^+ + NO_3^-$	About 20
Hydrogen sulfate ion	$HSO_4^- + H_2O \rightleftharpoons H_3O^+ + SO_4^{2-}$	1.2×10^{-2}
Hydrofluoric acid	$HF + H_2O \rightleftharpoons H_3O^+ + F^-$	6.7×10^{-4}
Acetic acid	$CH_3COOH + H_2O \rightleftharpoons H_3O^+ + CH_3COO^-$	1.8×10^{-5}
Aluminum ion	$Al(H_2O)_6^{3+} + H_2O \rightleftharpoons H_3O^+ + Al(H_2O)_5OH^{2+}$	7.2×10^{-6}
Carbonic acid	$H_2CO_3 + H_2O \rightleftharpoons H_3O^+ + HCO_3^-$	4.3×10^{-7}
Hypochlorous acid	$HOCl + H_2O \rightleftharpoons H_3O^+ + OCl^-$	3.1×10^{-8}
Hydrogen cyanide	$HCN + H_2O \rightleftharpoons H_3O^+ + CN^-$	4.9×10^{-10}
Ammonium ion	$NH_4^+ + H_2O \rightleftharpoons H_3O^+ + NH_3$	5.6×10^{-10}

and the symbol $[CH_3COOH]$ to indicate the quantity 0.0009 mol dm^{-3}, that is, the final concentration of this species in the equilibrium mixture.

Let us now consider the general problem of finding $[H_3O^+]$ in a solution of a weak acid HA whose acid constant is K_a and whose stoichiometric concentration is c_a. According to the equation for the equilibrium,

$$HA + H_2O \rightleftharpoons H_3O^+ + A^- \qquad (14.4)$$

for every mole of H_3O^+ produced, there must also be a mole of A^- produced. At the same time a mole of HA and a mole of H_2O must be consumed. Since the volume which all these species occupy is the same, any increase in $[H_3O^+]$ must be accompanied by an equal increase in $[A^-]$ and an equal decrease in [HA]. Consequently we can draw up the following table (in which equilibrium concentrations of all species have been expressed in terms of $[H_3O^+]$:

Species	Initial Concentration	Change in Concentration	Equilibrium Concentration
H_3O^+	10^{-7} mol dm^{-3}	$[H_3O^+]$*	$[H_3O^+]$
A^-	0	$[H_3O^+]$*	$[H_3O^+]$
HA	c_a	$-[H_3O^+]$*	$c_a - [H_3O^+]$

* The hydronium-ion concentration actually increases from 10^{-7} mol dm^{-3} to the equilibrium concentration, and so the change in each of the concentrations is $\pm([H_3O^+] - 10^{-7}$ mol $dm^{-3})$. However, the concentration of hydronium ions produced by the weak acid is usually so much larger than 10^{-7} mol dm^{-3} that the latter quantity can be ignored. In the case of 0.0010 *M* acetic acid, for example, $[H_3O^+] \approx 1 \times 10^{-4}$ mol dm^{-3}. Subtracting gives:

$$\begin{array}{r} 0.000\,100\,0 \text{ mol dm}^{-3} \\ -0.000\,000\,1 \text{ mol dm}^{-3} \\ \hline 0.000\,099\,9 \text{ mol dm}^{-3} \end{array}$$

which is very close to 1×10^{-4} mol dm^{-3}.

We can now substitute the equilibrium concentrations into the expression

$$K_a = \frac{[H_3O^+][A^-]}{[HA]} = \frac{[H_3O^+]^2}{c_a - [H_3O^+]} \qquad (14.7)$$

This could be solved for $[H_3O^+]$ by means of the quadratic formula, but in most cases a quicker approximate method is available. Since the acid is weak, only a small fraction of the HA molecules will have donated protons to form H_3O^+ ions. Therefore $[H_3O^+]$ is only a small fraction of c_a and can be ignored when we calculate $c_a - [H_3O^+]$. That is,

$$c_a - [H_3O^+] \approx c_a \qquad (14.8)$$

Equation (14.7) then becomes

$$K_a \approx \frac{[H_3O^+]^2}{c_a}$$

which rearranges to

$$[H_3O^+]^2 \approx K_a c_a$$

Taking the square root of both sides gives an important approximate formula:

$$[H_3O^+] \approx \sqrt{K_a c_a} \tag{14.9}$$

EXAMPLE 14.5 Use Eq. (14.9) to calculate the pH of a 0.0200-*M* solution of acetic acid. Compare this with the pH obtained using the $[H_3O^+]$ of 5.92×10^{-4} mol dm^{-3} derived from accurate conductivity measurements.

Solution From Table 14.2, $K_a = 1.8 \times 10^{-5}$ mol dm^{-3}. Since $c_a = 2.00 \times 10^{-2}$ mol dm^{-3}, we have

$$\begin{aligned}
[H_3O^+] &= \sqrt{K_a c_a} \\
&= \sqrt{(1.8 \times 10^{-5}\ \text{mol dm}^{-3})(2.00 \times 10^{-2}\ \text{mol dm}^{-3})} \\
&= \sqrt{3.6 \times 10^{-7}\ \text{mol}^2\ \text{dm}^{-6}} = \sqrt{36 \times 10^{-8}}\ \text{mol dm}^{-3} \\
[H_3O^+] &= 6.0 \times 10^{-4}\ \text{mol dm}^{-3} \\
\text{pH} &= -\log(6.0 \times 10^{-4}) = -(0.78 - 4) = 3.22
\end{aligned}$$

Using the accurate $[H_3O^+]$ from conductivity measurements,

$$\text{pH} = -\log(5.92 \times 10^{-4}) = -(0.772 - 4) = 3.228$$

Note that the approximate equation gives an $[H_3O^+]$ which differs by 1 in the second significant digit from the accurate value. The calculated pH differs by 1 in the second place to the right of the decimal—roughly the same as the accuracy of simple pH measurements.

In a few cases, if the acid is quite strong or the solution very dilute, it turns out that Eq. (14.9) is too gross an approximation. A convenient rule of thumb for determining when this is the case is to take the ratio $[H_3O^+]/c_a$. If this is larger than 5 percent or so, we need to make a second approximation, and then the rules for successive approximations developed in the previous chapter can be applied. Equation (14.7) can be converted to a convenient form for successive approximations by multiplying both sides by $c_a - [H_3O^+]$:

$$[H_3O^+]^2 = K_a(c_a - [H_3O^+])$$

or

$$[H_3O^+] = \sqrt{K_a(c_a - [H_3O^+])} \tag{14.10}$$

To get a second approximation for $[H_3O^+]$, we can feed the first approximation into the right side of this equation. The exact procedure is detailed in the following example.

EXAMPLE 14.6 Find the pH of 0.0200 *M* HF using Table 14.2.

Solution Using Eq. (14.9) we find

$$[H_3O^+] = \sqrt{K_a c_a}$$
$$= \sqrt{6.7 \times 10^{-4}\ \text{mol dm}^{-3} \times 0.0200\ \text{mol dm}^{-3}}$$
$$= 3.66 \times 10^{-3}\ \text{mol dm}^{-3}$$

Checking we find

$$\frac{[H_3O^+]}{c_a} = \frac{0.003\ 66}{0.0200} = 0.18, \text{ that is, 18 percent}$$

A second approximation is thus necessary. We feed our first approximation of $[H_3O^+] = 3.66 \times 10^{-3}$ mol dm^{-3} into Eq. (14.10):

$$[H_3O^+] = \sqrt{K_a(c_a - [H_3O^+])}$$
$$= \sqrt{6.7 \times 10^{-4}\ \text{mol dm}^{-3}(0.02 - 0.003\ 66)\text{mol dm}^{-3}}$$
$$= 3.31 \times 10^{-3}\ \text{mol dm}^{-3}$$

Taking a third approximation we find

$$[H_3O^+] = \sqrt{6.7 \times 10^{-4}\ \text{mol dm}^{-3}(0.02 - 0.003\ 31)\text{mol dm}^{-3}}$$
$$= 3.34 \times 10^{-3}\ \text{mol dm}^{-3}$$

Since this differs from the second approximation by less than 5 percent, we take it as the final result. The pH is

$$\text{pH} = -\log(3.34 \times 10^{-3}) = 2.48$$

Cross check: Since HF is a stronger acid than acetic acid, we expect this solution to have a lower pH than 0.02 *M* CH_3COOH, and indeed it does.

Weak Bases

The pH of a solution of a weak base can be calculated in a way which is very similar to that used for a weak acid. Instead of an acid constant K_a, a base constant K_b must be used. If a weak base B accepts protons from water according to the equation

$$B + H_2O \rightleftharpoons BH^+ + OH^- \tag{14.11}$$

then the base constant is defined by the expression

$$K_b = \frac{[BH^+][OH^-]}{[B]} \tag{14.12}$$

A list of K_b values for selected bases arranged in order of strength is given in Table 14.3.

TABLE 14.3 The Base Constants for Some Bases at 25°C.

Name	Proton Transfer	K_b/mol dm^{-3}
Hydride ion	$H^- + H_2O \rightarrow H_2 + OH^-$	Large
Amide ion	$NH_2^- + H_2O \rightarrow NH_3 + OH^-$	Large
Phosphate ion	$PO_4^{3-} + H_2O \rightleftharpoons HPO_4^{2-} + OH^-$	5.9×10^{-3}
Carbonate ion	$CO_3^{2-} + H_2O \rightleftharpoons HCO_3^- + OH^-$	2.1×10^{-4}
Ammonia	$NH_3 + H_2O \rightleftharpoons NH_4^+ + OH^-$	1.8×10^{-5}
Hydrazine	$N_2H_4 + H_2O \rightleftharpoons N_2H_5^+ + OH^-$	9.5×10^{-7}
Aniline	$C_6H_5NH_2 + H_2O \rightleftharpoons C_6H_5NH_3^+ + OH^-$	4.2×10^{-10}

To find the pH we follow the same general procedure as in the case of a weak acid. If the stoichiometric concentration of the base is indicated by c_b, the result is entirely analogous to Eq. (14.7); namely,

$$K_b = \frac{[OH^-]^2}{c_b - [OH^-]} \tag{14.13}$$

Under most circumstances we can make the approximation

$$c_b - [OH^-] \approx c_b$$

in which case Eq. (14.13) reduces to the approximation

$$[OH^-] \approx \sqrt{K_b c_b} \tag{14.14}$$

which is identical to the expression obtained in the acid case [approximation shown in Eq. (14.9)] except that OH^- replaces H_3O^+ and b replaces a. Once we have found the hydroxide-ion concentration from this approximation, we can then easily find the pOH, and from it the pH.

EXAMPLE 14.7 Using the value for K_b listed in Table 14.3, find the pH of 0.100 M NH_3.

Solution It is not a bad idea to guess an approximate pH before embarking on the calculation. Since we have a dilute solution of a weak base, we expect the solution to be only mildly basic. A pH of 13 or 14 would be too basic, while a pH of 8 or 9 is too close to neutral. A pH of 10 or 11 seems reasonable. Using Eq. (14.14) we have

$$\begin{aligned}[OH^-] &= \sqrt{K_b c_b} \\ &= \sqrt{1.8 \times 10^{-5} \text{ mol dm}^{-3} \times 0.100 \text{ mol dm}^{-3}} \\ &= \sqrt{1.8 \times 10^{-6} \text{ mol}^2 \text{ dm}^{-6}} = 1.34 \times 10^{-3} \text{ mol dm}^{-3}\end{aligned}$$

Checking the accuracy of the approximation, we find

$$\frac{[OH^-]}{c_b} = \frac{1.34 \times 10^{-3}}{0.1} \approx 1 \text{ percent}$$

The approximation is valid, and we thus proceed to find the pOH.

$$\mathrm{pOH} = -\log \frac{[\mathrm{OH^-}]}{\mathrm{mol\ dm^{-3}}} = -\log(1.34 \times 10^{-3}) = 2.87$$

From which

$$\mathrm{pH} = 14.00 - \mathrm{pOH} = 14.00 - 2.87 = 11.13$$

This calculated value checks well with our initial guess.

Occasionally we will find that the approximation

$$c_b - [\mathrm{OH^-}] \approx c_b$$

is not valid, in which case we must use a series of successive approximations similar to that outlined above for acids. The appropriate formula can be derived from Eq. (14.13) and reads

$$[\mathrm{OH^-}] = \sqrt{K_b(c_b - [\mathrm{OH^-}])} \qquad (14.15)$$

Polyprotic Acids and Bases

In the case of polyprotic acids and bases we can write down an equilibrium constant for each proton lost or gained. These constants are subscripted 1, 2, etc., to distinguish them. For sulfurous acid, a diprotic acid, we can, for example, write

$$H_2SO_3 + H_2O \rightleftharpoons H_3O^+ + HSO_3^-$$

$$K_{a1} = \frac{[H_3O^+][HSO_3^-]}{[H_2SO_3]} = 1.7 \times 10^{-2}\ \mathrm{mol\ dm^{-3}}$$

and

$$HSO_3^- + H_2O \rightleftharpoons H_3O^+ + SO_3^{2-}$$

$$K_{a2} = \frac{[H_3O^+][SO_3^{2-}]}{[HSO_3^-]} = 5.6 \times 10^{-8}\ \mathrm{mol\ dm^{-3}}$$

The carbonate ion, CO_3^{2-}, is an example of a diprotic base for which the appropriate base constants are

$$CO_3^{2-} + H_2O \rightleftharpoons HCO_3^- + OH^-$$

$$K_{b1} = \frac{[HCO_3^-][OH^-]}{[CO_3^{2-}]} = 2.1 \times 10^{-4}\ \mathrm{mol\ dm^{-3}}$$

$$HCO_3^- + H_2O \rightleftharpoons H_2CO_3 + OH^-$$

$$K_{b2} = \frac{[H_2CO_3][OH^-]}{[CO_3^-]} = 2.4 \times 10^{-8}\ \mathrm{mol\ dm^{-3}}$$

A general treatment of the pH of solutions of polyprotic species is beyond the scope of this text, but it is worth noting that in many cases we can treat polyprotic species as monoprotic. In the case of H_2SO_3, for example, K_{a1} is very much larger than K_{a2}, indicating that H_2SO_3 is a very much stronger acid than HSO_3^-. This means that when H_2SO_3 is dissolved in water, we can treat it as a monoprotic acid and ignore the possible loss of a second proton. Solutions of salts containing the carbonate ion, such as Na_2CO_3 or K_2CO_3, can be treated similarly.

EXAMPLE 14.8 Find the pH of a 0.100-*M* solution of sodium carbonate, Na_2CO_3. Use the base constant $K_{b1} = 2.10 \times 10^{-4}$ mol dm^{-3}.

Solution We ignore the acceptance of a second proton and treat the carbonate ion as a monoprotic base. We then have

$$[OH^-] = \sqrt{K_b c_b} = \sqrt{2.10 \times 10^{-4} \text{ mol dm}^{-3} \times 0.100 \text{ mol dm}^{-3}}$$
$$= 4.58 \times 10^{-3} \text{ mol dm}^{-3}$$

Checking, we find that

$$\frac{[OH^-]}{c_b} = \frac{4.58 \times 10^{-3}}{0.1} = 4.6 \text{ percent}$$

so that our approximation is only just valid.

We now find

$$pOH = -\log(4.58 \times 10^{-3}) = 2.34$$

while
$$pH = 14.00 - 2.34 = 11.66$$

Since the carbonate ion is a somewhat stronger base than NH_3, we expect a 0.1-*M* solution to be somewhat more basic, as actually found.

A glance at Tables 14.2 and 14.3 reveals that most acid and base constants involve numbers having negative powers of 10. As in the case of $[H_3O^+]$ and $[OH^-]$, then, it is convenient to define

$$pK_a = -\log \frac{K_a}{\text{mol dm}^{-3}} \qquad \text{and} \qquad pK_b = -\log \frac{K_b}{\text{mol dm}^{-3}}$$

Using these definitions, the larger K_a or K_b is (i.e., the stronger an acid or base, respectively), the smaller pK_a or pK_b will be. For a strong acid like HNO_3, $K_a = 20$ mol dm^{-3} and

$$pK_a = -\log 20 = -(1.30) = -1.3$$

Thus for very strong acids or bases p*K* values can even be negative.

14.3 CONJUGATE ACID-BASE PAIRS

One of the more useful aspects of the Brönsted-Lowry definition of acids and bases in helping us deal with the pH of solutions is the concept of the conjugate acid-base pair. We argued qualitatively in Chap. 11 that the strength of an acid and its conjugate base are inversely related. The stronger one is, the weaker the other will be. This relationship can be expressed quantitatively in terms of a very simple mathematical equation involving the appropriate acid and base constants.

Suppose in the general case we have a weak acid HA whose conjugate base is A^-. If either or both of these species are dissolved in H_2O, we will have *both* the following equilibria set up simultaneously:

$$HA + H_2O \rightleftharpoons H_3O^+ + A^- \qquad \text{in which HA acts as acid}$$

and $$A^- + H_2O \rightleftharpoons HA + OH^- \qquad \text{in which } A^- \text{ acts as base}$$

To the first of these equilibria we can apply the equilibrium constant $K_a(\text{HA})$:

$$K_a(\text{HA}) = \frac{[H_3O^+][A^-]}{[HA]}$$

while to the second we can apply the equilibrium constant $K_b(A^-)$:

$$K_b(A^-) = \frac{[HA][OH^-]}{[A^-]}$$

Multiplying these two constants together, we obtain a simple relationship between them.

$$K_a(\text{HA}) \times K_b(A^-) = \frac{[H_3O^+][A^-]}{[HA]} \times \frac{[HA][OH^-]}{[A^-]}$$

$$= [H_3O^+][OH^-]$$

so that $$K_a(\text{HA}) \times K_b(A^-) = K_w \tag{14.16}$$

If we divide both sides of this equation by the units $\text{mol}^2\ \text{dm}^{-6}$ and take negative logarithms of both sides, we obtain

$$-\log \frac{K_a(\text{HA})}{\text{mol dm}^{-3}} - \log \frac{K_b(A^-)}{\text{mol dm}^{-3}} = -\log \frac{10^{-14}\ \text{mol}^2\ \text{dm}^{-6}}{\text{mol}^2\ \text{dm}^{-6}}$$

$$pK_a(\text{HA}) + pK_b(A^-) = 14 \tag{14.17}$$

Thus the product of the acid constant for a weak acid and the base constant for the conjugate base must be K_w, and the sum of pK_a and pK_b for a conjugate acid-base pair is 14.

Equation (14.16) or (14.17) enables us to calculate the base constant of a conjugate base from the acid constant of the acid, and vice versa. Given the acid constant for a weak acid like HOCl, for instance, we are able to calculate not only the pH of HOCl solutions but also the pH of solutions of salts like NaOCl or KOCl which are, in effect, solutions of the conjugate base of HOCl, namely, the hypochlorite ion, OCl^-.

EXAMPLE 14.9 Find the pH of (a) 0.1 *M* HOCl (hypochlorous acid) and (b) 0.1 *M* NaOCl (sodium hypochlorite) from the value for K_a given in Table 14.2.

Solution

a) For 0.1 *M* HOCl, we find in the usual way that

$$[H_3O^+] = \sqrt{K_a c_a} = \sqrt{3.1 \times 10^{-8} \times 0.1 \text{ mol}^2 \text{ dm}^{-6}}$$
$$= 5.57 \times 10^{-5} \text{ mol dm}^{-3}$$

so that $\quad \text{pH} = 4.25$

b) For 0.1 *M* NaOCl, we must first calculate K_b:

$$K_b(\text{OCl}^-) = \frac{K_w}{K_a(\text{HOCl})} = \frac{1.00 \times 10^{-14} \text{ mol}^2 \text{ dm}^{-6}}{3.1 \times 10^{-8} \text{ mol dm}^{-3}}$$
$$= 3.22 \times 10^{-7} \text{ mol dm}^{-3}$$

Thus $\quad [OH^-] = \sqrt{K_b c_b} = \sqrt{3.22 \times 10^{-7} \times 0.1 \text{ mol}^2 \text{ dm}^{-3}}$
$$= 1.79 \times 10^{-4} \text{ mol dm}^{-3}$$

From which

$$\text{pOH} = 3.75$$

and $\quad \text{pH} = 14.00 - \text{pOH} = 10.25$

In this, as in all pH problems, it is worth checking that the answers obtained are not wildly unreasonable. A pH of 4 for a weak acid is reasonable, though a little high, but then HOCl is among the weaker acids in Table 14.2. A pH of 10 corresponds to a mildly basic solution—reasonable enough for a weak base like OCl^-.

Not only can we use Eq. (14.16) to find the value of K_b for the base conjugate to a given acid, we can also employ it in the reverse sense to find the value of K_a for the acid conjugate to a given base, as the following example shows.

EXAMPLE 14.10 Find the pH of 0.05 *M* NH_4Cl (ammonium chloride), using the value $K_b(NH_3) = 1.8 \times 10^{-5}$ mol dm^{-3}.

Solution We regard this solution as a solution of the weak acid NH_4^+ and start by finding K_a for this species:

$$K_a(NH_4^+) = \frac{K_w}{K_b(NH_3)} = \frac{1.00 \times 10^{-14} \text{ mol}^2 \text{ dm}^{-6}}{1.8 \times 10^{-5} \text{ mol dm}^{-3}}$$
$$= 5.56 \times 10^{-10} \text{ mol dm}^{-3}$$

We can now evaluate the hydronium-ion concentration with the usual approximation:

$$[H_3O^+] = \sqrt{K_a c_a}$$
$$= \sqrt{5.56 \times 10^{-10} \text{ mol dm}^{-3} \times 0.05 \text{ mol dm}^{-3}}$$
$$= 5.27 \times 10^{-6} \text{ mol dm}^{-3}$$

whence $$pH = -\log(5.27 \times 10^{-6}) = 5.28$$

Note: The ammonium ion is a very weak acid (see Table 14.2). A solution of NH_4^+ ions will thus not produce a very acidic solution. A pH of 5 is about the same pH as that of black coffee, not very acidic.

Before the Brönsted-Lowry definition of acids and bases and the idea of conjugate acid-base pairs became generally accepted, the interpretation of acid-base behavior revolved very much around the equation

$$\text{Acid} + \text{base} \rightarrow \text{salt} + \text{water}$$

In consequence the idea prevailed that when an acid reacted with a base, the resultant salt should be neither acidic or basic, but neutral. In order to explain why a solution of sodium acetate was basic or a solution of ammonium chloride was acidic, a special term called **hydrolysis** had to be invoked. Thus, for instance, sodium acetate was said to be hydrolyzed because the acetate ion reacted with water according to the reaction

$$CH_3COO^- + H_2O \rightleftharpoons CH_3COOH + OH^-$$

From the Brönsted-Lowry point of view there is, of course, nothing special about such a hydrolysis. It is a regular proton transfer. Nevertheless you should be aware of the existence of the term hydrolysis since it is still often used in this context.

Because the Brönsted-Lowry definition is so successful at explaining why some salt solutions are acidic and some basic, one must beware of making the mistake of assuming that no salt solutions are neutral. Many are. A good example is 0.10 *M* $NaNO_3$. This solution is neutral because neither the Na^+ ion nor the NO_3^- ion shows any appreciable acidic or basic properties. Since NO_3^- is the conjugate base of HNO_3, we might expect it to produce a basic solution, but NO_3^- is such a weak base that it is almost impossible to detect such an effect. Just how weak a base NO_3^- is can be demonstrated using the value of $K_a(HNO_3) = 20 \text{ mol dm}^{-3}$ obtained from Table 14.3.

$$K_b(NO_3^-) = \frac{K_w}{K_a(HNO_3)} = \frac{1.00 \times 10^{-14} \text{ mol}^2 \text{ dm}^{-6}}{20 \text{ mol dm}^{-3}}$$
$$= 5.0 \times 10^{-16} \text{ mol dm}^{-3}$$

If we now apply the conventional formula [Eq. (14.14)] to calculate $[OH^-]$ in 0.10 *M* $NaNO_3$, we obtain

$$[OH^-] = \sqrt{K_b c_b} = \sqrt{5.0 \times 10^{-16} \times 1.0 \times 10^{-1}} \text{ mol dm}^{-3}$$
$$= 7.1 \times 10^{-9} \text{ mol dm}^3$$

But this is less than one-tenth the concentration of OH^- ion which would have been present in pure H_2O, with no added $NaNO_3$. Essentially all the OH^- ions are produced by H_2O, and the pH turns out to be only slightly above 7.00. (Note also that the derivation of Eq. (14.14) assumed that the

TABLE 14.4 The Acid-Base Properties of Some Common Ions.

	Cations	Anions
Acidic	Cr^{3+}, Fe^{3+}, Al^{3+} Hg^{2+}, Be^{2+} NH_4^+, H_3O^+	HSO_4^-
Neutral	Mg^{2+}, Ca^{2+}, Sr^{2+}, Ba^{2+} Li^+, Na^+, K^+ Ag^+	NO_3^-, ClO_4^- Cl^-, Br^-, I^- SO_4^{2-} (very weakly basic)
Basic	None	PO_4^{3-}, CO_3^{2-}, SO_3^{2-} F^-, CN^-, OH^-, S^{2-} CH_3COO^-, HCO_3^-

$[OH^-]$ produced by H_2O was negligible. To get an accurate result in this case requires a completely different equation.)

In general all salts in which group I and group II cations are combined with anions which are the conjugate bases of strong acids yield neutral solutions when dissolved in water. Examples are CaI_2, $LiNO_3$, KCl, $Mg(ClO_4)_2$. There is only one exception to this rule. The hydrated beryllium ion, $Be(H_2O)_4^{2+}$, is a weak acid ($K_a = 3.2 \times 10^{-7}$ mol dm^{-3}) so that solutions of beryllium salts are acidic.

Table 14.4 lists the acid-base properties of some of the more frequently encountered ions and provides a quick reference for deciding whether a given salt will be acidic, basic, or neutral in solution. Note that the table tells us nothing about the strength of any acid or base. If we need to know more about the pH, other than whether it is above, below, or equal to 7, we need information about the actual value of the acid or base constant. The table also lists the SO_4^{2-} ion as neutral, though classifying it as very feebly basic would be more accurate.

EXAMPLE 14.11 Classify the following solutions as acidic, basic, or neutral: (a) 1 *M* KBr; (b) 1 *M* calcium acetate; (c) 1 *M* MgF_2; (d) 1 *M* $Al(NO_3)_3$; (e) 1 *M* $KHSO_4$; (f) 1 *M* NH_4I.

Solution

a) Both cation and anion are neutral: neutral.

b) Cation is neutral but anion basic: basic.

c) Cation is neutral but anion basic: basic.

d) Cation is acidic and anion neutral: acidic.

e) Cation is neutral but anion acidic: acidic.

f) Cation is acidic but anion neutral: acidic.

EXAMPLE 14.12 Without actually doing any calculations, match the following solutions and pH values, using Tables 14.2, 14.3, and 14.4.

Solution	pH
1 *M* NH_4NO_3	8.0
1 *M* KCN	11.7
1 *M* $Ca(NO_3)_2$	9.4
1 *M* $MgSO_4$	7.0
1 *M* CH_3COONa (sodium acetate)	1.0
1 *M* $KHSO_4$	4.6

Solution The pH of 7.0 is easiest to pick. Only one of the salt solutions given has both a neutral anion and a neutral cation. This is $Ca(NO_3)_2$. In the case of Mg_2SO_4 the Mg^{2+} ion is neutral but the SO_4^{2-} ion is very feebly basic; this would agree with a pH of 8.0, only slightly basic. The SO_4^{2-} ion is such a feeble base because its conjugate acid, HSO_4^-, is quite a strong acid, certainly the most acidic of all the ions featured. Accordingly we expect 1 *M* $KHSO_4$ to correspond to the lowest pH, namely, 1.0. The only other acidic solution is 4.6, and this must correspond to 1 *M* NH_4NO_3 since NH_4^+ is the only other acidic ion present. Among basic ions the cyanide ion, CN^-, is the strongest. The most basic pH, 11.7, thus corresponds to 1 *M* KCN. Only one solution is left: 1 *M* CH_3COONa. This should be feebly basic and so matches the remaining pH of 9.4 rather well.

14.4 BUFFER SOLUTIONS

So far in discussing pH we have dealt only with solutions obtained by adding a single acid, such as acetic acid, or a single base, such as the acetate ion, to water. We must now turn to a consideration of solutions to which both an acid and a base have been added. The simplest case of such a solution occurs when the acid and base are conjugate to each other and also present in comparable amounts. Solutions of this special kind are called **buffer solutions** because, as we shall shortly see, it is difficult to change their pH even when an appreciable amount of strong acid or strong base is added.

As a typical example of a buffer solution, let us consider the solution obtained when 3.00 mol acetic acid ($HC_2H_3O_2$) and 2.00 mol sodium acetate ($NaC_2H_3O_2$) are added to sufficient water to produce a solution of total volume 1 dm^3. The stoichiometric concentration of acetic acid, namely, c_a, is then 3.00 mol dm^{-3}, while the stoichiometric concentration of sodium acetate, c_b, is 2.00 mol dm^{-3}. As a result of mixing the two components, some of the acetic acid, say x mol dm^{-3}, is converted to acetate ion and hydronium ion. We can now draw up a table in order to find the equilibrium concentrations in the usual way.

Species	Initial Concentration/mol dm^{-3}	Change in Concentration/ mol dm^{-3}	Equilibrium Concentration/mol dm^{-3}
H_3O^+	10^{-7} (negligible)	x	x
CH_3COO^-	2.00	x	$2.00 + x$
CH_3COOH	3.00	$-x$	$3.00 - x$

We can now substitute concentrations in the equilibrium expression

$$K_a = \frac{[CH_3COO^-][H_3O^+]}{[CH_3COOH]}$$

from which we obtain

$$1.8 \times 10^{-5} \text{ mol dm}^{-3} = \frac{(2.00 + x)x}{3.00 - x} \text{ mol dm}^{-3} \qquad (14.18)$$

In order to solve this equation, we make the approximation that x is negligibly small compared with both 2.00 and 3.00, that is, that only a minute fraction of acetic acid has been converted to acetate ion. We then have

$$\frac{2.00x}{3.00} = 1.8 \times 10^{-5}$$

or

$$x = \frac{3.00}{2.00} \times 1.8 \times 10^{-5} = 2.7 \times 10^{-5}$$

Obviously, our approximation is a very good one. Since x is only 0.001 percent of 2.00 or 3.00, there really is no point in obtaining a second approximation by feeding x back into Eq. (14.18). We can thus safely conclude that

$$[H_3O^+] = 2.7 \times 10^{-5} \text{ mol dm}^{-3} \qquad \text{and} \qquad \text{pH} = 4.57$$

The example we have just considered demonstrates two obvious features:

1 When the acid and its conjugate base are mixed, very little of the acid is converted to base, or vice versa. (x was small compared with 2.00 and 3.00.)
2 In a buffer mixture, the hydronium-ion concentration and the hydroxide-ion concentration are small compared with the concentrations of acid and conjugate base. ($[H_3O^+] = 2.7 \times 10^{-5}$ mol dm^{-3}; $[OH^-] = 3.7 \times 10^{-10}$ mol dm^{-3} as compared with $[CH_3COO^-] = 2.00$ mol dm^{-3} and $[CH_3COOH] = 3.00$ mol dm^{-3})

By assuming that these features are common to all buffer solutions, we make it very easy to handle them from a mathematical standpoint.

Let us now consider the general problem of finding the pH of a buffer solution which is a mixture of a weak acid HA, of stoichiometric concentration c_a, and its conjugate base A^-, of stoichiometric concentration c_b. We can rearrange the expression for K_a of the weak acid [Eq. (14.5)] as follows:

$$[H_3O^+] = K_a \times \frac{[HA]}{[A^-]} \qquad (14.19)$$

Taking negative logarithms of both sides, we obtain

$$-\log[H_3O^+] = -\log K_a - \log \frac{[HA]}{[A^-]}$$

or

$$pH = pK_a + \log \frac{[A^-]}{[HA]} \tag{14.20}$$

Equation (14.20) is called the **Henderson-Hasselbalch equation** and is often used by chemists and biologists to calculate the pH of a buffer.

As we saw in the case of the acetic acid–sodium acetate buffer described earlier, the equilibrium concentrations of HA and A^- are usually almost identical to the stoichiometric concentrations. That is,

$$[HA] \approx c_a \qquad \text{and} \qquad [A^-] \approx c_b$$

We can substitute these values into Eqs. (14.19) and (14.20) to obtain the very useful approximations

$$[H_3O^+] \approx K_a \times \frac{c_a}{c_b} \tag{14.21}$$

and

$$pH \approx pK_a + \log \frac{c_b}{c_a} \tag{14.22}$$

EXAMPLE 14.13 Find the pH of the solution obtained when 1.00 mol NH_3 and 0.40 mol NH_4Cl are mixed to give 1 dm³ of solution. $K_b(NH_3) = 1.8 \times 10^{-5}$ mol dm⁻³.

Solution In order to use Eq. (14.21), we need first to have the value of $K_a(NH_4^+)$.

$$K_a(NH_4^+) = \frac{K_w}{K_b(NH_3)} = \frac{1.00 \times 10^{-14} \text{ mol}^2 \text{ dm}^{-6}}{1.8 \times 10^{-5} \text{ mol dm}^{-3}}$$
$$= 5.59 \times 10^{-10} \text{ mol dm}^{-3}$$

We also have $c_a = 0.40$ mol dm^{-3} and $c_b = 1.00$ mol dm^{-3}. Thus

$$[H_3O^+] = K_a \times \frac{c_a}{c_b}$$
$$= 5.56 \times 10^{-10} \text{ mol dm}^{-3} \times \frac{0.4 \text{ mol dm}^{-3}}{1.0 \text{ mol dm}^{-3}}$$
$$= 2.22 \times 10^{-10} \text{ mol dm}^{-3}$$

from which

$$pH = 9.65$$

To see why a mixture of an acid and its conjugate base is resistant to a change in pH, let us go back to our first example: a mixture of acetic acid (3 mol dm^{-3}) and sodium acetate (2 mol dm^{-3}). What would happen if we now added 0.50 mol sodium hydroxide to 1 dm^3 of this mixture? The added hydroxide ion will attack both the acids present, namely, the hydronium ion and acetic acid. Since the hydronium-ion concentration is so small, very little hydroxide ion will be consumed by reaction with the hydronium ion. Most will be consumed by reaction with acetic acid. Further, since the hydroxide ion is such a strong base, the reaction

$$CH_3COOH + OH^- \rightarrow CH_3COO^- + H_2O$$

will go virtually to completion, and 0.50 mol acetic acid will be consumed. The same amount of acetate ion will be produced. In tabular form:

Species	Initial Concentration/mol dm^{-3}	Change in Concentration/mol dm^{-3}	Equilibrium Concentration/mol dm^{-3}
H_3O^+	2.7×10^{-5}	Small	$\approx 2.7 \times 10^{-5}$
CH_3COO^-	2.00	0.50	$2.50 + 2.7 \times 10^{-5} \approx 2.50$
CH_3COOH	3.00	−0.50	$2.50 - 2.7 \times 10^{-5} \approx 2.50$

Substituting the equilibrium concentrations of base (acetate ion) and conjugate acid (acetic acid) into the Henderson-Hasselbalch equation [Eq. (14.20)], we have

$$\begin{aligned} pH &= pK_a + \log \frac{[A^-]}{[HA]} \\ &= -\log(1.8 \times 10^{-5}) + \log\left(\frac{2.50 \text{ mol dm}^{-3}}{2.50 \text{ mol dm}^{-3}}\right) \\ &= -(0.26 - 5) + \log(1) = 4.74 + 0 = 4.74 \end{aligned}$$

The addition of 0.5 mol sodium hydroxide to the buffer mixture has thus succeeded in raising its pH from 4.57 to only 4.74. If the same 0.5 mol had been added to a cubic decimeter of pure water, the pH would have jumped all the way from 7.00 up to 13.7! The buffer is extremely effective at resisting a change in pH because the added hydroxide ion attacks the *weak acid* (in very high concentration) rather than the hydronium ion (in very low concentration). The major effect of the addition of the hydroxide ion is thus to change the ratio of acid to conjugate base, i.e., to change the value of

$$\frac{[CH_3COOH]}{[CH_3COO^-]}$$

As long as the amount of weak acid is much larger than the amount of base added, this ratio is not altered by very much. Since the hydronium-ion concentration is governed by

$$[H_3O^+] = K_a \frac{[CH_3COOH]}{[CH_3COO^-]}$$

the hydronium-ion concentration and pH are also altered to only a small extent.

The ability of a buffer solution to resist large changes in pH has a great many chemical applications, but perhaps the most obvious examples of buffer action are to be found in living matter. If the pH of human blood, for instance, gets outside the range 7.2 to 7.6, the results are usually fatal. The pH of blood is controlled by the buffering action of several conjugate acid-base pairs. The most important of these is undoubtedly the H_2CO_3/HCO_3^- pair, but side chains of the amino acid histidine in the hemoglobin molecule also play a part. (Hemoglobin, a protein, is the red substance in the blood. It is responsible for carrying oxygen away from the lungs.) Most enzymes (biological catalysts) can only function inside a rather limited pH range and must therefore operate in a buffered environment. The enzymes which start the process of digestion in the mouth at a pH of around 7 become inoperative in the stomach at a pH of 1.4. The stomach enzymes in turn cannot function in the slightly basic environment of the intestines.

14.5 ACID-BASE TITRATIONS

Skip

Indicators

In Sec. 3.5 we mentioned that in most titrations it is necessary to add an indicator which produces a sudden color change at the equivalence point. A typical indicator for acid-base titrations is phenolphthalein, $HC_{20}H_{13}O_4$. Phenolphthalein is a weak acid ($K_a = 3 \times 10^{-10}$ mol dm^{-3}) which is colorless. Its conjugate base, $C_{20}H_{13}O_4^-$, has a strong pinkish-red color. In the interest of brevity, we will write the phenolphthalein molecule as HIn and its pink conjugate base as In^-. In aqueous solution, then the equilibrium

$$HIn + H_2O \rightleftharpoons In^- + H_3O^+ \qquad (14.23)$$

will be set up whenever phenolphthalein is present.

According to Le Chatelier's principle, equilibrium Eq. (14.23) will be shifted to the left if H_3O^+ is added. Thus in a strongly acidic solution we expect nearly all the pink In^- to be consumed, and only colorless HIn will remain. On the other hand, if the solution is made strongly basic, the equilibrium will shift to the right because OH^- ions will react with HIn molecules, converting them to In^-. Thus the phenolphthalein solution will become pink.

Clearly there must be some intermediate situation where half the phenolphthalein is in the acid form and half in the colored conjugate-base form. That is, at some pH

$$[HIn] = [In^-]$$

This intermediate pH can be calculated by applying the Henderson-Hasselbalch equation to the indicator equilibrium:

$$pH = pK_a + \log \frac{[In^-]}{[HIn]}$$

Thus at the point where half the indicator is conjugate acid and half conjugate base,

$$pH = pK_a + \log 1 = pK_a$$

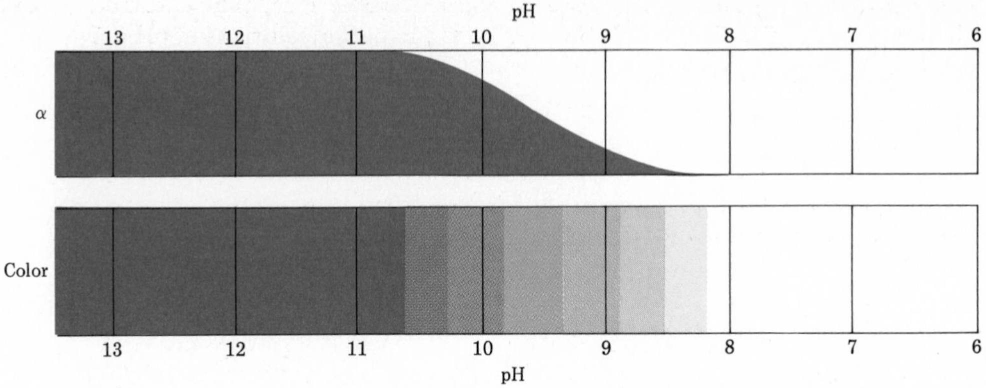

Figure 14.1 The variation of the color of the indicator phenolphthalein with pH. In the upper section of the figure, α, the fraction of the indicator species present as the red base, is plotted against pH. The lower section shows the effect of this variation on the overall color of the indicator. Note that the color change is confined to a fairly narrow pH range.

For phenolphthalein, we have

$$\text{pH} = \text{p}K_a = -\log(3 \times 10^{-10}) = 9.5$$

so we expect phenolphthalein to change color in the vicinity of pH = 9.5.

The way in which both the color of phenolphthalein and the fraction present as the conjugate base varies with the pH is shown in detail in Fig. 14.1. The change of color occurs over quite a limited range of pH—roughly p$K_a \pm 1$. In other words the color of phenolphthalein changes perceptibly between about pH 8.3 and 10.5. Other indicators behave in essentially the same way, but for many of them both the acid and the conjugate base are colored. Their pK_a's also differ from phenolphthalein, as shown in Table 14.5. The indicators listed have been selected so that their pK_a values are approximately two units apart. Consequently, they offer a series of color changes spanning the whole pH range.

TABLE 14.5 Properties of Selected Indicators.

Name	pK_a	Effective pH Range	Color: Acid form	Color: Base form
Thymol blue	1.6	1.2–2.8	Red	Yellow
Methyl orange	4.2	3.1–4.4	Red	Orange
Methyl red	5.0	4.2–6.2	Red	Yellow
Bromothymol blue	7.1	6.0–7.8	Yellow	Blue
Phenolphthalein	9.5	8.3–10.0	Colorless	Red
Alizarin yellow	11.0	10.1–12.1	Yellow	Red

Indicators are often used to make measurements of pH which are precise to about 0.2 or 0.3 units. Suppose, for example, we add two drops of bromothymol blue to a sample of tap water and obtain a green-blue solution. Since bromothymol blue is green at a pH of 6 and blue at a pH of 8, we conclude that the pH is between these two limits. A more precise result could be obtained by comparing the color in the tap water with that obtained when two drops of indicator solution are added to buffer solutions of pH 6.5 and 7.5.

EXAMPLE 14.14 What indicator, from those listed in Table 14.5, would you use to determine the approximate pH of the following solutions:

a) 0.1 *M* CH_3COONa (sodium acetate)

b) 0.1 *M* CH_3COOH (acetic acid)

c) A buffer mixture of sodium acetate and acetic acid

d) 0.1 *M* NH_4Cl (ammonium chloride)

Solution

a) A solution of sodium acetate will be mildly basic with a pH of 9 or 10. Phenolphthalein would probably be best.

b) A solution of acetic acid, unless very dilute, has a pH in the vicinity of 3. Both thymol blue and methyl orange should be tried.

c) Since K_a for acetic acid is 1.8×10^{-5} mol dm^{-3}, we can expect this buffer to have a pH not far from 5. Methyl red would be a good indicator to try.

d) Since NH_4^+ is a very weak acid, this solution will be only faintly acidic with a pH of 5 or 6. Again methyl red would be a good indicator to try, though bromothymol blue is also a possibility.

If a careful choice of both colors and pK_a is made, it is possible to mix several indicators and obtain a **universal indicator** which changes color continuously over a very wide pH range. With such a mixture it is possible to find the approximate pH of any solution within this range. So-called pH paper is impregnated with one or several indicators. When a strip of this paper is immersed in a solution, its pH can be judged from the resulting color.

Titration Curves

When an acid is titrated with a base, there is typically a sudden change in the pH of the solution at the endpoint. If a few drops of indicator solution have been added, this sharp increase in pH causes an abrupt change in color. The actual magnitude of this jump and the pH range which it covers depend on the strength of both the acid and the base involved, and so the choice of indicator can vary from one titration to another. To learn how to choose an appropriate indicator, we need to study in some detail the variation of pH during a titration.

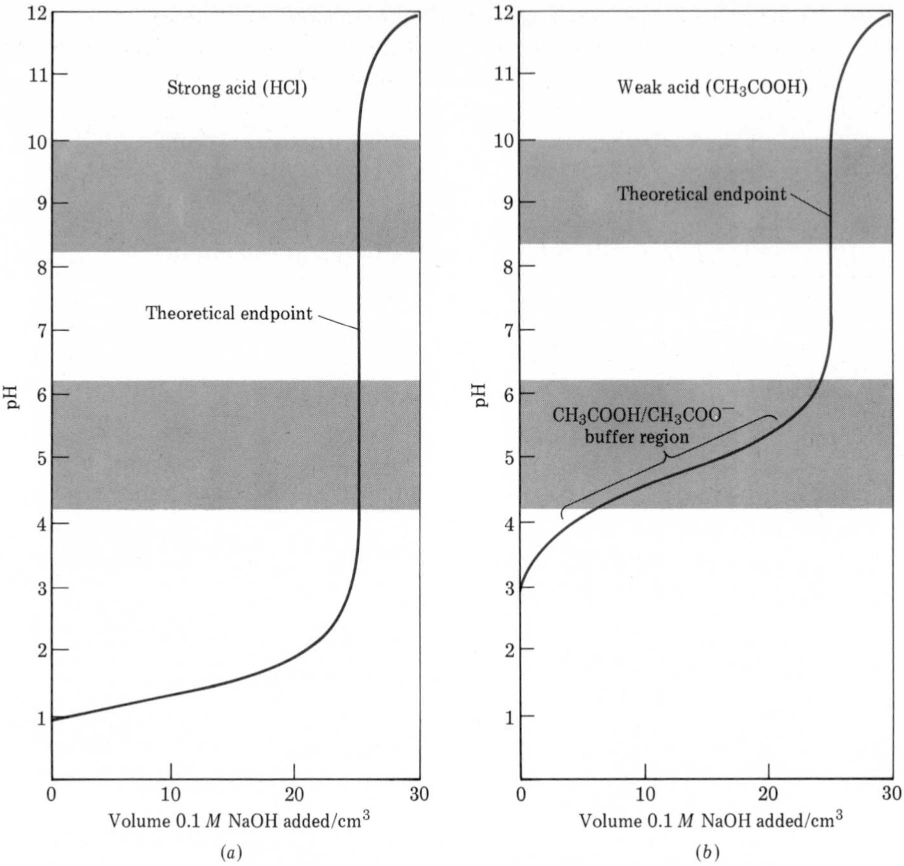

Figure 14.2 Variation of pH during titration of strong and weak acids with strong base. (*a*) 25.00 cm^3 of 0.10 M HCl titrated with 0.10 M NaOH. (*b*) 25.00 cm^3 of 0.10 M acetic acid titrated with 0.10 M NaOH. The pH ranges over which the indicators phenolphthalein (pH 8.3 to 10.0) and methyl red (pH 4.2 to 6.2) change color are indicated in gray.

First we shall consider titration of a strong acid such as HCl with a strong base such as NaOH. Suppose we place 25.00 cm^3 of 0.10 M HCl solution in a flask and add 0.10 M NaOH from a buret. The pH of the solution in the flask varies with added NaOH, as shown in Fig. 14.2*a*. The pH changes quite slowly at the start of the titration, and almost all the increase in pH takes place in the immediate vicinity of the endpoint.

The pH change during this titration is caused by the proton-transfer reaction

$$H_3O^+ + OH^- \rightarrow H_2O + H_2O \tag{14.24}$$

which occurs as hydroxide ions are added from the buret. Though hydronium ions are being consumed by hydroxide ions in the early stages of the titration, the hydronium-ion concentration remains in the vicinity of 10^{-1} or 10^{-2} mol dm^{-3}. As a result, the pH remains in the range 1 to 2. As an example of this behavior let us consider the situation halfway to the endpoint, i.e., when exactly 12.50 cm^3 of 0.10 *M* NaOH have been added to 25.00 cm^3 of 0.10 *M* HCl in the flask. The amount of hydronium ion has been reduced at this point from an original 2.5 mmol to half this value, 1.25 mmol. At the same time the volume of solution has increased from 25 cm^3 to (25 + 12.50) cm^3 = 37.50 cm^3. Therefore, the hydronium-ion concentration is 1.25 mmol/37.50 cm^3 = 0.0333 mol dm^{-3}, and the resultant pH is 1.48. Though the titration is half completed, this is not very different from the initial pH of 1.00.

The pH of the solution in the flask will only change drastically when we reach that point in the titration when only a minute fraction of the hydronium ions remain unconsumed, i.e., as we approach the endpoint. Only then will we have reduced the hydronium-ion concentration by several powers of 10, and consequently increased the pH by several units. When 24.95 cm^3 of base have been added, we are only 0.05 cm^3 (approximately one drop) short of the endpoint. At this point 24.95 cm^3 × 0.10 mmol cm^{-3} = 2.495 mmol hydroxide ions have been added. These will have consumed 2.495 mmol hydronium ions, leaving (2.5 − 2.495) mmol = 0.005 mmol hydronium ions in a volume of 49.95 cm^3. The hydronium-ion concentration will now be

$$[H_3O^+] = \frac{0.005 \text{ mmol}}{49.95 \text{ cm}^3} = 1.00 \times 10^{-4} \text{ mol dm}^{-3}$$

giving a pH of 4.00. Because almost all the hydronium ions have been consumed, only a small fraction (one five-hundredth) remains and the volume of solution has nearly doubled. This reduces the hydronium-ion concentration by a factor of 10^{-3}, and the pH increases by three units from its original value of 1.00.

When exactly 25.00 cm^3 of base have been added, we have reached the theoretical endpoint,[1] and the flask will contain 2.5 mmol of both sodium and chloride ions in 50 cm^3 of solution; i.e., the solution is 0.05 *M* NaCl. Furthermore its pH will be *exactly 7.00,* since neither the sodium ion nor the chloride ion exhibits any appreciable acid-base properties.

Immediately after this endpoint the addition of further NaOH to the flask results in a sudden increase in the concentration of hydroxide ions, since there are now virtually no hydronium ions left to consume them. Thus even one drop (0.05 cm^3) of base added to the endpoint solution adds 0.005 mmol hydroxide ions and produces a hydroxide-ion concentration of 0.005 mmol/50.05 cm^3 = 1.00 × 10^{-4} mol dm^{-3}. The resultant pOH = 4.00, and the pH = 10.00. The addition of just two drops of base results in a pH jump from 4.00 to 7.00 to 10.00.

Titration of a strong base with a strong acid can be handled in essentially the same way as the strong acid–strong base situation we have just described.

[1] The theoretical endpoint is also called the equivalence point.

EXAMPLE 14.15 Suppose that 25.00 cm³ of 0.010 *M* KOH is titrated with 0.01 *M* HNO_3. Find the pH of the solution in the flask (a) before any HNO_3 is added; (b) halfway to the endpoint; (c) one drop (0.05 cm³) before the endpoint; (d) one drop after the endpoint.

Solution

a) Since KOH is a strong base, $[OH^-] = 0.010 = 10^{-2}$. Thus

$$pOH = 2 \qquad \text{and} \qquad pH = 14 - 2 = 12$$

b) Halfway to the endpoint means that half the OH^- has been consumed. The original amount of OH^- was

$$n_{OH-} = V_{NaOH} \times c_{NaOH} = 25.00 \text{ cm}^3 \times \frac{0.010 \text{ mmol}}{1 \text{ cm}^3} = 0.25 \text{ mmol}$$

so the amount remaining is 0.125 mmol OH^-. It must have required 0.125 mmol HNO_3 to consume the other 0.125 mmol OH^-, and so the volume of HNO_3 added is

$$V_{HNO_3} = \frac{n_{HNO_3}}{c_{HNO_3}} = \frac{0.125 \text{ mmol}}{0.010 \text{ mmol cm}^{-3}} = 12.50 \text{ cm}^3$$

The total volume of solution is thus (12.50 + 25.00) cm³ = 37.50 cm³, and

$$[OH^-] = \frac{0.125 \text{ mmol}}{37.50 \text{ cm}^3}$$

$$= 3.33 \times 10^{-3} \text{ mmol cm}^{-3}$$

$$= 3.33 \times 10^{-3} \text{ mol dm}^{-3}$$

$$pOH = -\log(3.33 \times 10^{-3}) = 2.48$$

and

$$pH = 14 - 2.48 = 11.52$$

c) When 24.95 cm³ of HNO_3 solution has been added, the amount of H_3O^+ added is

$$n_{H_3O^+} = 24.95 \text{ cm}^3 \times \frac{0.010 \text{ mmol}}{1 \text{ cm}^3} = 0.2495 \text{ mmol}$$

This would consume an equal amount of OH^-, and so the amount of OH^- remaining is (0.2500 − 0.2495) mmol = 0.0005 mmol. Thus

$$[OH^-] = \frac{0.0005 \text{ mmol}}{(25.00 + 24.95)\text{cm}^3} = 1.0 \times 10^{-5} \text{ mmol cm}^{-3}$$

$$= 1.0 \times 10^{-5} \text{ mol dm}^{-3}$$

$$pOH = 5.0 \qquad \text{and} \qquad pH = 9.0$$

d) An excess of 0.05 cm³ of acid will add

$$n_{H_3O^+} = 0.05 \text{ cm}^3 \times \frac{0.010 \text{ mmol}}{1 \text{ cm}^3} = 5 \times 10^{-4} \text{ mmol}$$

of H_3O^+ to a neutral solution whose volume is (25.00 + 25.05) cm^3 = 50.05 cm^3. Thus

$$[H_3O^+] = \frac{5 \times 10^{-4}\ \text{mmol}}{50.05\ \text{cm}^3} = 1 \times 10^{-5}\ \text{mol dm}^{-3}$$

$$\text{pH} = 5$$

Note: In this case, because the solutions were one-tenth as concentrated as in the titration of HCl with NaOH worked out in the text, the jump in pH (from 9 to 5) at the endpoint is smaller.

Because there is such an abrupt jump in pH at the endpoint of any titration involving a strong acid and a strong base, a wide choice of indicators is suitable for detecting the pH change. For the titration of 25 cm^3 of 0.10 *M* HCl with 0.10 *M* NaOH, as we have seen, the pH changes from 4.00 a drop before the endpoint to 10.00 a drop after, and so any indicator which changes color within this range will enable us to estimate the endpoint to within one drop. Three of the indicators listed in Table 14.5 fall into this category: methyl red (4.2 to 6.2), bromothymol blue (6.0 to 7.8), and phenolphthalein (8.3 to 10.0). Any of these three could serve as an indicator for this titration. The range of pH over which two of these indicators change color is shown in gray in Fig. 14.2*a*. In both cases these gray areas coincide with the vertical portion of the pH curve at the endpoint very nicely.

A wide choice of indicators like this is not possible for titrations involving weak acids or bases. When 25.00 cm^3 of 0.10 *M* CH_3COOH is titrated with 0.10 *M* NaOH, for instance, there is a very much smaller change in pH at the endpoint, as shown in Fig. 14.2*b*, and the choice of indicators is correspondingly narrowed. The behavior of the pH in this case is very different from that of the titration of HCl with NaOH, because the acid-base reaction is different. When CH_3COOH is titrated with NaOH, the OH^- ions consume *CH_3COOH molecules* according to the equation

$$OH^- + CH_3COOH \rightarrow H_2O + CH_3COO^- \qquad (14.25)$$

As a result, the solution in the titration flask soon becomes a buffer mixture with appreciable concentrations of the CH_3COO^- ion as well as its conjugate acid. The $[H_3O^+]$ and the pH are then controlled by the ratio of acid to conjugate base [Eqs. (14.19) and (14.20)]. When we are halfway to the endpoint, for example, $[CH_3COOH]$ will be essentially the same as $[CH_3COO^-]$, and

$$[H_3O^+] = K_a \times \frac{[CH_3COOH]}{[CH_3COO^-]} \approx K_a = 1.8 \times 10^{-5}\ \text{mol dm}^{-3}$$

while the pH will be given by the Henderson-Hasselbalch equation as

$$\text{pH} = \text{p}K_a + \log \frac{[CH_3COO^-]}{[CH_3COOH]} \approx \text{p}K_a = 4.74$$

Comparing this to the pH of 1.78 calculated above for the halfway stage in the titration of HCl, we find a difference of roughly three pH units. The effect of the buffering action of the CH_3COOH/CH_3COO^- conjugate pair is thus to keep the pH some three units higher than before and hence to cut the jump in pH at the endpoint by approximately this amount.

Exactly at the endpoint we no longer have a buffer mixture but a 0.05-*M* solution of sodium acetate. This solution is slightly basic, and its pH of 8.72 can be calculated from Eq. (14.14). Beyond this endpoint, the story is much the same as in the strong-acid case. Addition of even a drop (0.05 cm^3) of excess base raises the OH^- concentration to 10^{-4} mol dm^{-3} and the pH to 10.

Of the three indicators which could be used in the titration of HCl, only one is useful for acetic acid. This is phenolphthalein, which changes color in the pH range 8.3 to 10.0. This range lies within the pH jump from 7 to 10 which occurs ± 0.05 cm^3 (one drop) from the endpoint. Both the other two indicators would give a color change *before* the true endpoint. As shown in Fig. 14.2*b*, the color of methyl red would start changing after only about 4 cm^3 of base had been added!

The titration of a weak base with a strong acid also involves a buffer solution and consequently requires a more careful choice of indicator.

EXAMPLE 14.16 For the titration of 25.00 cm^3 of 0.010 *M* NH_3 with 0.010 *M* HCl, calculate the pH (a) before any acid is added; (b) after 12.50 cm^3 of HCl has been added; (c) at the equivalence point; (d) after 30.00 cm^3 of HCl has been added.

Solution

a) Before any acid is added, we have a solution of a weak base, and Eq. (14.14) applies:

$$[OH^-] = \sqrt{K_b c_b}$$
$$= \sqrt{1.8 \times 10^{-5} \text{ mol dm}^{-3} \times 1.0 \times 10^{-2} \text{ mol dm}^{-3}}$$
$$= 4.24 \times 10^{-4} \text{ mol dm}^{-3}$$
$$\text{pOH} = -\log(4.24 \times 10^{-4}) = 3.37 \qquad \text{pH} = 14 - 3.37 = 10.63$$

b) Since the concentrations of base and acid are equal, 12.50 cm^3 of HCl is enough to consume half the NH_3, converting it to the conjugate acid NH_4^+. The Henderson-Hasselbalch equation may be used to obtain the pH of this buffer system, provided we use $pK_a(NH_4^+)$. Rearranging Eq. (14.17),

$$pK_a(NH_4^+) = 14 - pK_b(NH_3) = 14 + \log(1.8 \times 10^{-5})$$
$$= 14 - 4.74 = 9.25$$

Since $[NH_4^+] \approx [NH_3]$,

$$\text{pH} = pK_a + \log\frac{[A^-]}{[HA]} = pK_a + \log\frac{[NH_3]}{[NH_4^+]} \approx pK_a = 9.25$$

c) At the equivalence point *all* the NH_3 has been converted to NH_4^+, and so we have a solution of NH_4Cl. The volume of solution has doubled because 25.00 cm^3 of HCl was added, and so $c_{NH_4^+}$ must be half the original c_{NH_3}; that is, $c_{NH_4^+} = 0.005$ mol dm^{-3}. $K_a(NH_4^+)$ is obtained from K_b (NH_3) using Eq. (14.16):

$$K_a(NH_4^+) = \frac{K_w}{K_b(NH_3)} = \frac{10^{-14}\ \text{mol}^2\ \text{dm}^{-6}}{1.8 \times 10^{-5}\ \text{mol dm}^{-3}}$$

$$= 5.56 \times 10^{-10}\ \text{mol dm}^{-3}$$

Thus $$[H_3O^+] = \sqrt{K_a c_a}$$

$$= \sqrt{5.56 \times 10^{-10}\ \text{mol dm}^{-3} \times 5.0 \times 10^{-3}\ \text{mol dm}^{-3}}$$

$$= 1.67 \times 10^{-6}\ \text{mol dm}^{-3}$$

$$\text{pH} = -\log(1.67 \times 10^{-6}) = 5.78$$

d) By adding an excess 5.0 cm³ of HCl, we have added a strong acid to a solution which previously contained the very weak acid NH_4^+. To a first approximation, then, we consider only H_3O^+ produced by HCl:

$$[H_3O^+] = \frac{5.0\ \cancel{\text{cm}^3} \times 0.010\ \text{mol dm}^{-3}}{55.00\ \cancel{\text{cm}^3}} = 9.1 \times 10^{-4}\ \text{mol dm}^{-3}$$

$$\text{pH} = -\log(9.1 \times 10^{-4}) = 3.04$$

Figure 14.3 Variation of pH during titration of strong and weak bases with strong acid: (*a*) 25.00 cm³ of 0.010 *M* KOH titrated with 0.010 *M* HNO_3; (*b*) 25.00 cm³ of 0.010 *M* NH_3 titrated with 0.010 *M* HCl. The pH ranges over which the indicators phenolphthalein (pH 8.3 to 10.0) and methyl red (pH 4.2 to 6.2) change color are indicated in gray.

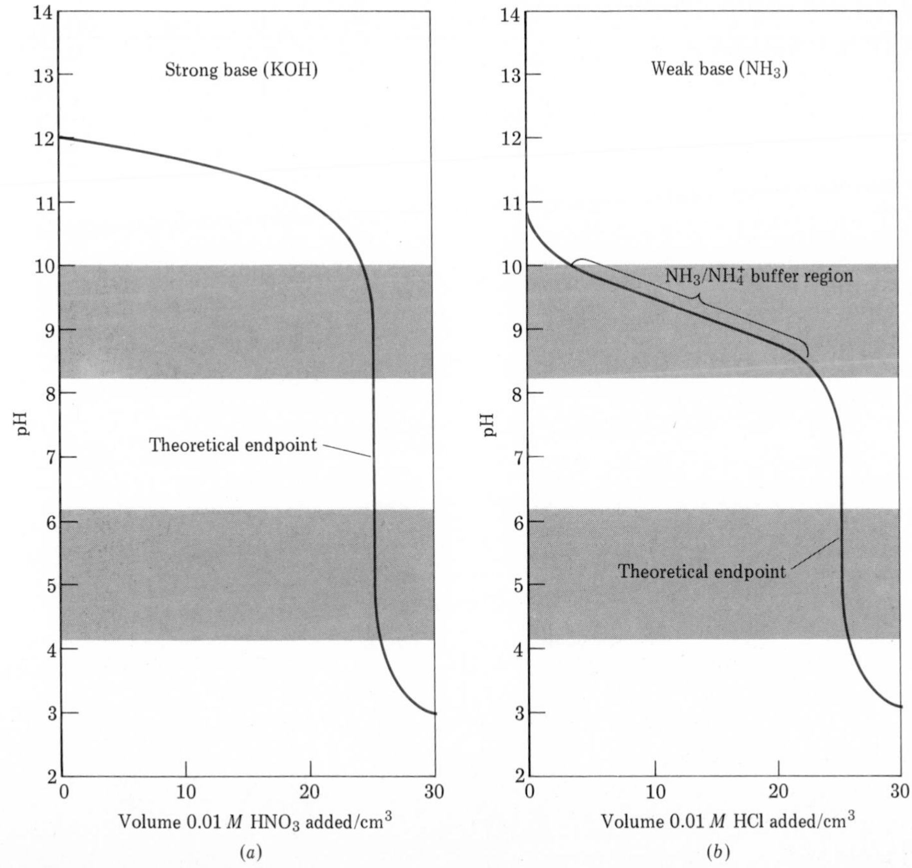

The pH variation during titrations of strong and weak bases with strong acid are shown in Fig. 14.3. In the case of the titration of 0.010 M NH_3 with 0.010 M HCl, methyl red, but not phenolphthalein, would be a suitable indicator. In general the best indicator for a given titration is the one whose pK_a most nearly matches the pH calculated at the theoretical endpoint.

14.6 THE SOLUBILITY PRODUCT

In Sec. 11.1 we saw that there are some salts which dissolve in water to only a very limited extent. For example, if $BaSO_4$ crystals are shaken with water, so little dissolves that it is impossible to see that anything has happened. Nevertheless, the few $Ba^{2+}(aq)$ and $SO_4^{2-}(aq)$ ions that do go into solution increase the conductivity of the water, allowing us to measure their concentration. We find that at 25°C

$$[Ba^{2+}] = 0.97 \times 10^{-5} \text{ mol dm}^{-3} = [SO_4^{2-}] \tag{14.26}$$

so that we would describe the solubility of $BaSO_4$ as 0.97×10^{-5} mol dm^{-3} at this temperature. The solid salt and its ions are in dynamic equilibrium, and so we can write the equation

$$BaSO_4(s) \rightleftharpoons Ba^{2+}(aq) + SO_4^{2-}(aq) \tag{14.27}$$

As in other dynamic equilibria we have discussed, a particular Ba^{2+} ion will sometimes find itself part of a $BaSO_4$ crystal and at other times find itself hydrated and in solution.

Since the concentration of $BaSO_4$ has a constant value, it can be incorporated into K_c for Eq. (14.27). This gives a special equilibrium constant called the **solubility product** K_{sp}:

$$K_{sp} = K_c[BaSO_4] = [Ba^{2+}][SO_4^{2-}] \tag{14.28}$$

For $BaSO_4$, K_{sp} is easily calculated from the solubility by substituting Eq. (14.26) into (14.28):

$$K_{sp} = (0.97 \times 10^{-5} \text{ mol dm}^{-3})(0.97 \times 10^{-5} \text{ mol dm}^{-3}) = 0.94 \times 10^{-10} \text{ mol}^2 \text{ dm}^{-6}$$

TABLE 14.6 The Solubility Product K_{sp} for Some Sparingly Soluble Salts at 25°C.

Compound	K_{sp}	Compound	K_{sp}
BaF_2	2.4×10^{-5} mol^3 dm^{-9}	PbS	7×10^{-29} mol^2 dm^{-6}
$BaSO_4$	1.0×10^{-10} mol^2 dm^{-6}	$Mg(OH)_2$	8.9×10^{-12} mol^3 dm^{-9}
$CaCO_3$	4.7×10^{-9} mol^2 dm^{-6}	HgS	1.6×10^{-54} mol^2 dm^{-6}
CaF_2	3.9×10^{-11} mol^3 dm^{-9}	NiS	3×10^{-21} mol^2 dm^{-6}
$CaSO_4$	2.4×10^{-5} mol^2 dm^{-6}	$AgC_2H_3O_2$*	2.3×10^{-3} mol^2 dm^{-6}
$Fe(OH)_2$	1.8×10^{-15} mol^3 dm^{-9}	AgCl	1.7×10^{-10} mol^2 dm^{-6}
$Fe(OH)_3$	6×10^{-38} mol^4 dm^{-12}	AgBr	5.0×10^{-13} mol^2 dm^{-6}
FeS	4×10^{-19} mol^2 dm^{-6}	AgI	8.5×10^{-17} mol^2 dm^{-6}
$PbCl_2$	1.7×10^{-5} mol^3 dm^{-9}	Ag_2CrO_4	1.9×10^{-12} mol^3 dm^{-9}
$PbSO_4$	1.3×10^{-8} mol^2 dm^{-6}	ZnS	2.5×10^{-22} mol^2 dm^{-6}

* Silver acetate

In the general case of an ionic compound whose formula is A_xB_y, the equilibrium can be written

$$A_xB_y(s) \rightleftharpoons xA^{m+}(aq) + yB^{n+}(aq) \qquad (14.29)$$

The solubility product is then

$$K_{sp} = [A^{m+}]^x[B^{n+}]^y \qquad (14.30)$$

Solubility products for some of the more common sparingly soluble compounds are given in Table 14.6.

EXAMPLE 14.17 When crystals of $PbCl_2$ are shaken with water at 25°C, it is found that 1.62×10^{-2} mol $PbCl_2$ dissolves per cubic decimeter of solution. Find the value of K_{sp} at this temperature.

Solution We first write out the equation for the equilibrium:

$$PbCl_2(s) \rightleftharpoons Pb^{2+}(aq) + 2Cl^-(aq) \qquad (14.31)$$

so that

$$K_{sp}(PbCl_2) = [Pb^{2+}][Cl^-]^2 \qquad (14.32)$$

Since 1.62×10^{-2} mol $PbCl_2$ dissolves per cubic decimeter, we have

$$[Pb^{2+}] = 1.62 \times 10^{-2} \text{ mol dm}^{-3}$$

while $$[Cl^-] = 2 \times 1.62 \times 10^{-2} \text{ mol dm}^{-3}$$

since *2* mol Cl^- ions are produced for each mol $PbCl_2$ which dissolves. Thus

$$K_{sp} = (1.62 \times 10^{-2} \text{ mol dm}^{-3})(2 \times 1.62 \times 10^{-2} \text{ mol dm}^{-3})^2$$
$$= 1.70 \times 10^{-5} \text{ mol}^3 \text{ dm}^{-9}$$

EXAMPLE 14.18 The solubility product of silver chromate, Ag_2CrO_4, is 1.0×10^{-12} mol³ dm⁻⁹. Find the solubility of this salt.

Solution Again we start by writing the equation

$$Ag_2CrO_4(s) \rightleftharpoons 2Ag^+(aq) + CrO_4^{2-}(aq)$$

from which

$$K_{sp} = [Ag^+]^2[CrO_4^{2-}] = 1.0 \times 10^{-12} \text{ mol}^3 \text{ dm}^{-9}$$

Let the solubility be x mol dm^{-3}. Then

$$[CrO_4^{2-}] = x \text{ mol dm}^{-3} \qquad \text{and} \qquad [Ag^+] = 2x \text{ mol dm}^{-3}$$

Thus $$K_{sp} = (2x \text{ mol dm}^{-3})^2 x \text{ mol dm}^{-3} = (2x)^2 x \text{ mol}^3 \text{ dm}^{-9}$$
$$= 1.0 \times 10^{-12} \text{ mol}^3 \text{ dm}^{-9}$$

or $\quad 4x^3 = 1.0 \times 10^{-12}$

and $\quad x^3 = \dfrac{1.0}{4} \times 10^{-12} = 2.5 \times 10^{-13} = 250 \times 10^{-15}$

so that $\quad x = \sqrt[3]{250} \times \sqrt[3]{10^{-15}} = 6.30 \times 10^{-5}$

Thus the solubility is 6.30×10^{-5} mol dm^{-3}.
Note: If your calculator does not have a $\sqrt[y]{x}$ or a x^y button, a cube root can be found from the relationship $\sqrt[3]{x} = x^{1/3} = \text{antilog}\,(\frac{1}{3}\log x)$. If your calculator cannot find a log, it is not difficult to find the cube root of 250 by trial and error. Since $6^3 = 216$ and $7^3 = 343$, the cube root is between 6 and 7, probably closer to 6. Continuing in this fashion we quickly find that x is very close to 6.3.

The Common-Ion Effect

Suppose we have a saturated solution of lead chloride in equilibrium with the solid salt:

$$PbCl_2(s) \rightleftharpoons Pb^{2+}(aq) + 2Cl^-(aq)$$

If we increase the chloride-ion concentration, Le Chatelier's principle predicts that the equilibrium will shift to the *left*. More lead chloride will precipitate, and the *concentration of lead ions will decrease*. A decrease in concentration obtained in this way is often referred to as the **common-ion effect.**

The solubility product can be used to calculate how much the lead-ion concentration is decreased by the common-ion effect. Suppose we mix 10 cm^3 of a saturated solution of lead chloride with 10 cm^3 of concentrated hydrochloric acid (12 *M* HCl). Because of the twofold dilution, the chloride-ion concentration in the mixture will be 6 mol dm^{-3}. Feeding this value into Eq. (14.32), we then have the result

$$K_{sp} = [Pb^{2+}][Cl^-]^2$$

or
$$1.70 \times 10^{-5} \text{ mol}^3\text{ dm}^{-9} = [Pb^{2+}](6 \text{ mol dm}^{-3})^2$$

so that
$$[Pb^{2+}] = \frac{1.70 \times 10^{-15} \text{ mol}^3\text{ dm}^{-9}}{36 \text{ mol}^2\text{ dm}^{-6}}$$
$$= 4.72 \times 10^{-7} \text{ mol dm}^{-3}$$

We have thus lowered the lead-ion concentration from an initial value of 1.62×10^{-2} mol dm^{-3} (see Example 14.17) to a final value of 4.72×10^{-7} mol dm^{-3}, a decrease of about a factor of 30 000! As a result, we have at our disposal a very sensitive test for lead ions. If we mix equal volumes of 12 *M* HCl and a test solution, and no precipitate occurs, we can be certain that the lead-ion concentration in the test solution is below $2 \times 4.72 \times 10^{-7}$ mol dm^{-3}.

Because it tells us about the conditions under which equilibrium is attained, the solubility product can also tell us about those cases in which equilibrium is *not* attained. If extremely dilute solutions of $Pb(NO_3)_2$ and KCl are mixed, for instance, it may be that the concentrations of lead ions and chloride ions in the resultant mixture are both too low for a precipitate to form. In such a case we would find that the product Q, called the **ion product** and defined by

$$Q = (c_{Pb^{2+}})(c_{Cl^-})^2 \qquad (14.33)$$

has a value which is *less* than the solubility product 1.70×10^{-5} mol³ dm⁻⁹. In order for equilibrium between the ions and a precipitate to be established, either the lead-ion concentration or the chloride-ion concentration or both must be increased until the value of Q is exactly equal to the value of the solubility product. The opposite situation, in which Q is larger than K_{sp}, corresponds to concentrations which are too large for the solution to be at equilibrium. When this is the case, precipitation occurs, lowering the concentration of both the lead and chloride ions, until Q is exactly equal to the solubility product.

To determine in the general case whether a precipitate will form, we set up an ion-product expression Q which has the same form as the solubility product, except that the *stoichiometric* concentrations rather than the *equilibrium* concentrations are used. Then if

$$Q > K_{sp} \qquad \text{precipitation occurs}$$

while if

$$Q < K_{sp} \qquad \text{no precipitation occurs}$$

EXAMPLE 14.19 Decide whether $CaSO_4$ will precipitate or not when (a) 100 cm³ of 0.02 *M* $CaCl_2$ and 100 cm³ of 0.02 *M* Na_2SO_4 are mixed, and also when (b) 100 cm³ of 0.002 *M* $CaCl_2$ and 100 cm³ of 0.002 *M* Na_2SO_4 are mixed. $K_{sp}(CaSO_4) = 2.4 \times 10^{-5}$ mol² dm⁻⁶.

Solution

a) After mixing, the concentration of each species is halved. We thus have

$$c_{Ca^{2+}} = 0.01 \text{ mol dm}^{-3} = c_{SO_4^{2-}}$$

so that the ion-product Q is given by

$$Q = c_{Ca^{2+}} \times c_{SO_4^{2-}} = 0.01 \text{ mol dm}^{-3} \times 0.01 \text{ mol dm}^{-3}$$

or

$$Q = 10^{-4} \text{ mol}^2 \text{ dm}^{-6}$$

Since Q is larger than $K_{sp}(2.4 \times 10^{-5} \text{ mol}^2 \text{ dm}^{-6})$, precipitation will occur.

b) In the second case

$$c_{Ca^{2+}} = 0.001 \text{ mol dm}^{-3} = c_{SO_4^{2-}}$$

and

$$Q = c_{Ca^{2+}} \times c_{SO_4^{2-}} = 1 \times 10^{-6} \text{ mol}^2 \text{ dm}^{-6}$$

Since Q is now less than K_{sp}, no precipitation will occur.

EXAMPLE 14.20 Calculate the mass of $CaSO_4$ precipitated when 100 cm³ of 0.0200 *M* $CaCl_2$ and 100 cm³ of 0.0200 *M* Na_2SO_4 are mixed together.

Solution We have already seen in part *a* of the previous example that precipitation does actually occur. In order to find how much is precipitated, we must concentrate on the amount of each species. Since 100 cm³ of 0.02 *M* $CaCl_2$ was used, we have

$$n_{Ca^{2+}} = 0.0200 \frac{\text{mmol}}{\text{cm}^3} \times 100 \text{ cm}^3 = 2.00 \text{ mmol}$$

similarly $$n_{SO_4^{2-}} = 0.0200 \frac{\text{mmol}}{\text{cm}^3} \times 100 \text{ cm}^3 = 2.00 \text{ mmol}$$

If we now indicate the amount of $CaSO_4$ precipitated as x mmoles, we can set up a table in the usual way:

Species	Initial Amount	Amount Reacted	Equilibrium Amount	Equilibrium Concentration
$Ca^{2+}(aq)$	2.00 mmol	$-x$ mmol	$(2 - x)$ mmol	$\frac{2-x}{200}$ mmol cm^{-3}
$SO_4^{2-}(aq)$	2.00 mmol	$-x$ mmol	$(2 - x)$ mmol	$\frac{2-x}{200}$ mmol cm^{-3}

Thus

$$K_{sp} = [Ca^{2+}][SO_4^{2-}]$$

or $$2.4 \times 10^{-5} \text{ mol}^2 \text{ dm}^{-6} = \left(\frac{2-x}{200} \text{ mol dm}^{-3}\right)\left(\frac{2-x}{200} \text{ mol dm}^{-3}\right)$$

Rearranging,

$$200^2 \times 2.4 \times 10^{-5} = 0.96 = (2 - x)^2$$

or $$2 - x = \sqrt{0.96} = 0.980$$

so that $$x = 2 - 0.980 = 1.020$$

Since 1.020 mmol $CaSO_4$ is precipitated, the mass precipitated is given by

$$m_{CaSO_4} = 1.020 \text{ mmol} \times 136.12 \frac{\text{mg}}{\text{mmol}}$$

$$= 138.8 \text{ mg} = 0.139 \text{ g}$$

Note: Because the solutions are so dilute and because $CaSO_4$ has a fairly large solubility product, only about half (1.02 mmol out of a total of 2.00 mmol) the Ca^{2+} ions are precipitated. If we wished to determine the concentration of Ca^{2+} ions in tap water or river water, where it is quite low, it would be foolish to try to precipitate the Ca as $CaSO_4$. Another method would have to be found.

The Solubilities of Salts of Weak Acids

In many chemical operations it is an advantage not only to be able to form a precipitate but to be able to redissolve it. Fortunately, there is a wide class of sparingly soluble salts which can almost always be redissolved by adding acid. These are precipitates in which the anion is basic; i.e., they are the salts of weak acids. An example of such a precipitate is calcium carbonate, whose solubility equilibrium is

$$CaCO_3(s) \rightleftharpoons Ca^{2+}(aq) + CO_3^{2-}(aq) \tag{14.34}$$

If acid is now added to this solution, some of the carbonate ions become protonated and transformed into HCO_3^- ions. As a result, the concentration of the carbonate ion is reduced. In accord with Le Chatelier's principle, the system will respond to this reduction by trying to produce more carbonate ions. Some solid $CaCO_3$ will dissolve, and the equilibrium will be shifted to the right. If enough acid is added, the carbonate-ion concentration in the solution can be reduced so as to make the ion product ($Q = c_{Ca^{2+}} \times c_{CO_3^{2-}}$) smaller than the solubility product K_{sp} so that the precipitate dissolves.

A similar behavior is shown by other precipitates involving basic anions. Virtually all the carbonates, sulfides, hydroxides, and phosphates which are sparingly soluble in water can be dissolved in acid. Thus, for instance, we can dissolve precipitates like ZnS, $Mg(OH)_2$, and $Ca_2(PO_4)_3$ because all the following equilibria

$$ZnS(s) \rightleftharpoons Zn^{2+}(aq) + S^{2-}(aq)$$
$$Mg(OH)_2(s) \rightleftharpoons Mg^{2+}(aq) + 2OH^-(aq)$$
$$Ca_3(PO_4)_2(s) \rightleftharpoons 3Ca^{2+}(aq) + 2PO_4^{3-}(aq)$$

can be shifted to the right by attacking the basic species S^{2-}, OH^-, and PO_4^{3-} with hydronium ions. Very occasionally we find an exception to this rule. Mercury(II) sulfide, HgS, is notorious for being insoluble. The solubility product for the equilibrium

$$HgS(s) \rightleftharpoons Hg^{2+}(aq) + S^{2-}(aq)$$

is so minute that not even concentrated acid will reduce the sulfide ion sufficiently to make Q smaller than K_{sp}.

Occasionally the shift in a solubility-product equilibrium caused by a decrease in pH may be undesirable. One example of this was mentioned in Sec. 12.6. Acid rainfall can occur when oxides of sulfur and other acidic air pollutants are removed from the atmosphere. In some parts of the United States pH values as low as 4.0 have been observed. These acid solutions dissolve marble and limestone ($CaCO_3$), causing considerable property damage. This is especially true in Europe, where some statues and other works of art have been almost completely destroyed over the last half century.

SUMMARY

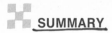

Acid-base reactions in aqueous solutions are intimately related to water's ability to act as both a weak acid and a weak base, producing H_3O^+ and OH^- by proton transfer. In any aqueous solution at 25°C $[H_3O^+][OH^-] = K_w = 1.00 \times 10^{-14}$ mol^2 dm^{-6}, and concentrations of H_3O^+ and OH^- can vary from roughly 10^0 to 10^{-14} mol dm^{-3}. This makes it convenient to define pH $= -\log([H_3O^+]/\text{mol dm}^{-3})$ and pOH $= -\log([OH^-]/\text{mol dm}^{-3})$. Since molecules of a strong acid transfer their protons to water molecules completely, $[H_3O^+]$ (and hence pH) can be obtained directly from the stoichiometric concentration of the solution. Similarly $[OH^-]$ and pOH may be obtained from the stoichiometric concentration of a strong base. In the case of weak acids and weak bases, proton-transfer reactions proceed to only a limited extent and a dynamic equilibrium is set up. In such cases an acid constant K_a or a base constant K_b as well as the stoichiometric concentration of weak acid or base are required to calculate $[H_3O^+]$, $[OH^-]$, pH, or pOH. K_a and K_b for a conjugate acid-base pair are related, and their product is always K_w.

Often it is necessary or desirable to restrict the pH of an aqueous solution to a narrow range. This can be accomplished by means of a buffer solution—one which contains a conjugate weak acid–weak base pair. If a small amount of strong base is added to a buffer, the OH^- ions are consumed by the conjugate weak acid, so they have little influence on pH. Similarly, a small amount of strong acid can be consumed by the conjugate weak base in a buffer. To a good approximation the $[H_3O^+]$ in a buffer solution depends only on K_a for the weak acid and the stoichiometric concentrations of the weak acid and weak base.

Indicators for acid-base titrations are conjugate acid-base pairs, each member of which is a different color. An indicator changes from the color of the conjugate acid to the color of the conjugate base as pH increases from approximately $pK_{In} - 1$ to $pK_{In} + 1$. For titrations involving only strong acids and strong bases, several indicators are usually capable of signaling the endpoint because there is a large jump in pH within ± 0.05 cm^3 of the exact stoichiometric volume of titrant. In the case of titrations which involve a weak acid or a weak base, a buffer solution is involved and the jump in pH is smaller. Consequently greater care is required in selection of an appropriate indicator.

A dynamic equilibrium is set up when a solid compound is in contact with a saturated solution. In the case of an ionic solid, the equilibrium constant for such a process is called the solubility product. K_{sp} can be determined by measurement of the solubility of a compound, and it is useful in predicting whether the compound will precipitate when ionic solutions are mixed. The common-ion effect, in which an increase in the concentration of one ion decreases the concentration of the other ion of an insoluble compound, can be interpreted quantitatively using solubility products. It is also true that removal of one ion of an insoluble compound from solution will increase the concentration of the other ion, and hence the solubility. It is for this reason that salts of weak acids often dissolve in acidic solutions—protonation of the anion effectively reduces its concentration to the point where the solubility product is not exceeded.

QUESTIONS AND PROBLEMS

14.1 Classify each of the following as a strong acid, weak acid, strong base, weak base, amphiprotic substance, or neither acid nor base:

a HCl   *b* $C_2H_3O_2^-$ *c* NH_4^+ *d* CH_3CH_3 *e* H_2O *f* $HC_2H_3O_2$ *g* NH_3 *h* Na_2O *i* $Ba(OH)_2$ *j* H_2SO_4 *k* HSO_4^- *l* CO_3^{2-}

14.2 Calculate $[H_3O^+]$ and $[OH^-]$ in each of the following aqueous solutions:

a 0.10 *M* HNO_3 *b* 0.034 *M* KOH   *c* 0.10 *M* $Ba(OH)_2$ *d* 0.034 *M* Na_2O

14.3 Calculate the pH and pOH of each of the solutions in Problem 14.2.

14.4 Complete the table below for each aqueous solution listed.

Solution	$[H_3O^+]$/mol dm^{-3}	$[OH^-]$/mol dm^{-3}	pH	pOH
1	1.67×10^{-3}	____	____	____
2	____	____	____	4.83
3	____	____	5.69	____
4	____	7.63×10^{-9}	____	____
5	8.13×10^{-5}	____	____	____
6	____	____	7.01	____

14.5 Calculate the pH and the pOH of aqueous solutions prepared in each of the following ways: (*a*) 42.35 cm^3 of 0.100 *M* NaOH is diluted to a final volume of 250.0 cm^3. (*b*) 7.81 dm^3 of HCl(*g*) at 25°C and 1 atm is dissolved in pure H_2O and the solution made up to 50.0 cm^3 in a volumetric flask. (*c*) 0.4375 g of pure, dry K_2O is dissolved in pure H_2O and made up to 250.0 cm^3.

14.6 Give an example of a solution which might be encountered in everyday life and whose pH would be within ±1 of each of the following:

a −0.5   *b* 2.0   *c* 4.5   *d* 7.5   *e* 10.0   *f* 13.0

14.7 Write the chemical equation for proton transfer to water and the equilibrium-constant expression for K_a for each of the following acids. In the case of a polyprotic acid, write a separate equation and equilibrium constant for loss of each acidic proton.

a $HC_2H_3O_2$ *b* HCN   *c* H_2CO_3 *d* H_3PO_4 *e* NH_4^+ *f* H_3PO_3

14.8 Write the chemical equation for proton transfer from water and the equilibrium-constant expression for K_b for each of the following bases. In the case of a base which can accept two or more protons, write a separate equation and equilibrium constant for each.

a SO_3^{2-} *b* NH_3 *c* PO_4^{3-} *d* $C_2H_3O_2^-$ *e* F^- *f* CH_3NH_2

14.9 Using data from Tables 14.2 and 14.3, calculate $[H_3O^+]$ and $[OH^-]$ in each of the following aqueous solutions at 25°C.

a 0.100 *M* HCN
b 0.432 *M* NH_4Cl
c 0.057 *M* NH_3
d 0.087 *M* Al^{3+}(*aq*)

14.10 Calculate the pH and pOH of each solution in Problem 14.9.

14.11 Using data from Tables 14.2 and 14.3, calculate $[H_3O^+]$ and $[OH^-]$ in each of the following aqueous solutions at 25°C:

a 0.021 *M* NaOCl
b 0.213 *M* N_2H_5Cl
c 0.513 *M* KF
d 0.065 *M* $Ca(C_2H_3O_2)_2$

14.12 Calculate the pH and pOH of each solution in Problem 14.11.

14.13 For each solution listed below, calculate $[H_3O^+]$ and pH:

a 0.050 *M* *n*-C_3H_7COOH; $pK_a = 4.81$.
b 0.200 *M* C_6H_5COOH; $pK_a = 4.19$.
c 0.175 *M* $Ca(HCOO)_2$; $pK_a(HCOOH) = 3.75$.

14.14 Calculate pK_b and K_b for the anion obtained by removing a proton from each of the acids whose pK_a's are given in Problem 14.13.

14.15 Calculate the pH of each of the following solutions at 25°C. Use Tables 14.2 and 14.3 to obtain necessary acid or base constants.

a 0.015 *M* HF
b 0.0050 *M* NaCN
c 0.115 *M* $NaHSO_4$
d 0.037 *M* $CH_2ClCOOH$; $pK_a(CH_2ClCOOH) = 2.85$

14.16 Choose the response which makes each of the following statements correct. In each case give a reason for your choice. (*a*) In acid solution $[H_3O^+]$ is (greater, less) than that of pure H_2O.

(*b*) In acid solution at 25°C the pH is (greater, less) than 7.00. (*c*) The (larger, smaller) the K_a value for a weak acid, the stronger the acid. (*d*) The (larger, smaller) the K_a value for a weak acid, the smaller the K_b for the conjugate base. (*e*) The (larger, smaller) the pK_a of a weak acid, the stronger the acid. (*f*) The (larger, smaller) the pK_a of a weak acid, the stronger the conjugate base.

14.17 For each of the acid solutions below, a pH value measured at 25°C is given. Calculate K_a for each acid. Check your results against the values in Table 14.2.

a 0.173 *M* $HC_2H_3O_2$; pH = 2.75.
b 0.075 *M* HF; pH = 2.17.
c 0.207 *M* $Al(NO_3)_3$; pH = 2.91.

14.18 For each of the basic solutions below, a pH value measured at 25°C is given. Calculate pK_b for each base.

a 0.113 *M* NH_3; pH = 11.15.
b 0.074 *M* N_2H_4; pH = 10.42.
c 0.175 *M* NaCN; pH = 11.28.
d 0.437 *M* $NaC_2H_3O_2$; pH = 9.19.

14.19 Use data from Tables 14.2 and 14.3 to calculate the pH of each of the following aqueous solutions:

a 0.100 *M* H_2CO_3
b 0.100 *M* Na_2CO_3
c 0.075 *M* K_3PO_4

14.20 For phosphoric acid, H_3PO_4, (*a*) write balanced chemical equations for successive donation of each of the three protons to water molecules; (*b*) identify conjugate acid-base pairs in each equation; (*c*) based on Coulomb's law (Sec. 5.4), which of the three conjugate bases, $H_2PO_4^-$, HPO_4^{2-}, or PO_4^{3-}, would you expect to have the greatest attraction for protons? Which should have the least? (*d*) based on your answer to part *c*, which of the phosphorus-containing conjugate acids should be strongest? weakest? (*e*) Do your answers to parts *c* and *d* agree with the statements about polyprotic acids and bases made at the end of Sec. 14.2?

14.21 For phosphoric acid, $K_{a_1} = 7.5 \times 10^{-3}$ mol dm^{-3} and $K_{a_2} = 6.2 \times 10^{-8}$ mol dm^{-3} at 25°C. (*a*) Calculate $[H_3O^+]$ and pH for a 0.100-*M* aqueous solution of H_3PO_4. Assume that only a single proton is lost from each molecule. (*b*) Calculate $[H_3O^+]$ and pH for a 0.100-*M* aqueous solution of NaH_2PO_4. Assume that only a single proton is lost from each $H_2PO_4^-$ ion. (*c*) Do your results in part *b* confirm the validity of the assumption made in part *a*? Explain.

***14.22** Use Eq. (14.9) to derive a relationship between the pK_a of a weak acid and the pH of a solution of that acid whose stoichiometric concentration is c_A. Derive a similar equation relating pH of a solution of a weak base to pK_b and c_B.

14.23 By multiplying the appropriate equilibrium-constant expressions, verify that (*a*) for HF and F^-, $K_a \times K_b = K_w$; (*b*) for H_2SO_4, $K_{a1} \times K_{a2} = K_a$, where K_a refers to the donations of *both* protons to water:

$$H_2SO_4 + 2H_2O \rightleftharpoons 2H_3O^+ + SO_4^{2-}$$

14.24 What is meant by hydrolysis of a salt? How is hydrolysis explained in terms of the Brönsted-Lowry acid-base definition? What kinds of ions must be present in a salt for hydrolysis to occur?

14.25 Without doing any calculations, assign each of the following 1.0-*M* aqueous solutions to one of the pH ranges listed:

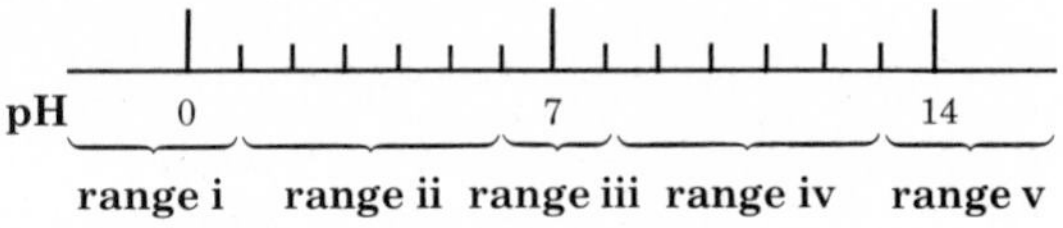

a HNO_3
b NH_4Cl
c NaF
d $Mg(C_2H_3O_2)_2$
e BaO
f $KHSO_4$
g $NaHCO_3$
h $SrCl_2$

14.26 Calculate the pH of each solution in Problem 14.25 to verify your predictions.

14.27 Distinguish clearly between the stoichiometric concentration of an acid, c_A, and the equilibrium concentration, [HA]. How is the difference between these quantities affected as the strength of the acid changes? Give an example of an acid from Table 14.2 for which $c_A \approx$ [HA] in a 0.1-*M* aqueous solution. Give another example for which c_A would be quite different from [HA].

14.28 What components must be present in a buffer solution? From the list of solutions below, choose those which are not buffers and explain why they would not maintain approximately constant pH when small quantities of strong acid or base are added.

a 0.1 mol $HC_2H_3O_2$ and 0.1 mol $NaC_2H_3O_2$ in 1 dm^3 solution.

b 0.05 mol NH_4Cl and 0.08 mol NH_3 in 1 dm^3 solution.

c 0.15 mol NH_4Cl and 0.10 mol NaCl in 1 dm^3 solution.

d 0.1 mol NaOH and 0.1 mol HNO_3 in 1 dm^3 solution.

14.29 Calculate the pH for each of the buffer solutions in Problem 14.28.

14.30 What mass (in grams) of $NaC_2H_3O_2$ must be dissolved in 500 cm^3 of 0.1 *M* $HC_2H_3O_2$ to produce a buffer solution whose pH = 5? (Assume the change in volume of the solution is negligibly small.)

14.31 A buffer solution is prepared by dissolving 2.62 g NH_4Cl in 250 cm^3 of 0.1 *M* NH_3. (*a*) Calculate the pH of this buffer. (*b*) Calculate the pH of the same solution after 0.04 g NaOH(*s*) has been added.

14.32 Calculate the change in pH produced by the addition of 500 cm^3 of 0.02 *M* NaOH to 500 cm^3 of each of the following buffer solutions:

a 0.1 *M* $HC_2H_3O_2$ and 0.1 *M* $NaC_2H_3O_2$.

b 1.0 *M* $HC_2H_3O_2$ and 1.0 *M* $NaC_2H_3O_2$.

c 0.1 *M* NaH_2PO_4 and 0.05 *M* Na_2HPO_4; $pK_{a_2}(H_2PO_4^-) = 7.21$.

14.33 Construct a graph of the pH of 1.00 dm^3 of 0.1 *M* $HC_2H_3O_2$ solution as a function of the amount of $NaC_2H_3O_2(s)$ added, from 0.0 mol $NaC_2H_3O_2(s)$ up to 0.20 mol $NaC_2H_3O_2(s)$ in increments of 0.02 mol. (Assume negligible change in volume as the solid is added.) Why does the graph have the shape you have obtained?

14.34 What volume (in cubic centimeters) of 0.1012 *M* NaOH is required to titrate each of the following solutions to the equivalence point? (Assume that phenolphthalein is used as an indicator.)

a 25.0 cm^3 of 0.0973 *M* HCl.

b 15.0 cm^3 of 0.237 *M* $HC_2H_3O_2$.

c 50.0 cm^3 of 0.0437 *M* H_2SO_4.

14.35 Use Table 14.5 to select an indicator to measure the approximate pH of each of the following solutions:

a 0.01 *M* HCN

b 0.10 *M* NaOH

c 0.50 *M* NaCN

d 0.10 *M* $NaHCO_3$

e 0.001 *M* HCl

f 0.1 *M* NH_3 + 0.1 *M* NH_4Cl

14.36 Plot pH versus volume of base added for each of the following titrations. Plot from 10 to 15 points, including the initial point, the equivalence point, and several points within 0.5 cm^3 of the equivalence point.

a 25.0 cm^3 of 0.100 *M* HCl with 0.100 *M* NaOH.

b 25.0 cm^3 of 0.113 *M* $HC_2H_3O_2$ with 0.879 *M* NaOH.

c 25.0 cm^3 of 0.100 *M* HNO_3 with 0.100 *M* NH_3.

14.37 Plot pH versus volume of acid added for each of the following titrations:

a 25.0 cm^3 of 0.100 *M* KOH with 0.100 *M* HBr.

b 25.0 cm^3 of 0.344 *M* NH_3 with 0.241 *M* HNO_3.

c 25.0 cm^3 of 0.100 *M* KOH with 0.100 *M* $HC_2H_3O_2$.

14.38 Use Table 14.5 to select an appropriate indicator for the endpoint in each of the titrations in Problems 14.36 and 14.37.

14.39 Magnesium hydroxide, $Mg(OH)_2$, is only slightly soluble in water. Suggest two different methods by which the concentrations of Mg^{2+} and OH^- ions in equilibrium with $Mg(OH)_2(s)$ could be determined. Solid $Mg(OH)_2$ absorbs water from the air very readily, and so weighing would not be a good way to determine how much went into solution.

14.40 Write an equation for dissolution and the expression for the solubility-product constant K_{sp} for

a $AgC_2H_3O_2$	*d* $Mg(OH)_2$
b $PbSO_4$	*e* $Fe(OH)_3$
c Ag_2CrO_4	*f* BaF_2

14.41 Using K_{sp} values from Table 14.6, calculate the solubility of each of the compounds in Problem 14.40.

14.42 The concentration of AsO_4^{3-} ions in an aqueous solution in equilibrium with each solid compound is given in the list below. For each compound write an equation for the dissolution reaction and calculate K_{sp}.

a $FeAsO_4(s)$;
$[AsO_4^{3-}] = 7.55 \times 10^{-11}$ mol dm^{-3}.

b $Ag_3AsO_4(s)$;
$[AsO_4^{3-}] = 1.40 \times 10^{-6}$ mol dm^{-3}.

c $Pb_3(AsO_4)_2(s)$;
$[AsO_4^{3-}] = 6.57 \times 10^{-2}$ mol dm^{-3}.

14.43 For each compound listed below, calculate the amount of substance that would dissolve in 500 cm^3 of the indicated solution.

a $BaSO_4(s)$ in 0.1 *M* Na_2SO_4.

b $PbCl_2(s)$ in 0.5 *M* HCl.

c Ag_2CrO_4 in 0.2 *M* K_2CrO_4.

d $Fe(OH)_3$ in 0.1 *M* $Ba(OH)_2$.

14.44 Determine whether precipitation will occur when each of the following pairs of solutions is mixed. (Assume that the total volume of the mixed solution is the sum of the volumes of the separate solutions.)

a 100 cm^3 of 0.01 *M* $Ba(NO_3)_2$ and 100 cm^3 of 0.01 *M* KF.

b 100 cm^3 of 0.01 *M* $AgNO_3$ and 100 cm^3 of 0.01 *M* HCl.

c 50 cm^3 of 0.01 *M* $Pb(NO_3)_2$ and 150 cm^3 of 0.10 *M* HCl.

d 100 cm^3 of 0.01 *M* $Fe(NO_3)_3$ and 100 cm^3 of 0.10 *M* NaOH.

*__14.45__ Determine whether precipitation will occur when each of the following sets of solutions is mixed. What substance will precipitate? What mass of that substance will be found as precipitate?

a 50 cm^3 of 0.10 *M* $AgNO_3$ + 50 cm^3 of 0.10 *M* $NaC_2H_3O_2$ + 100 cm^3 of 0.05 *M* HCl.

b 100 cm^3 of 0.04 *M* $MgCl_2$ + 100 cm^3 of 0.08 *M* $NaC_2H_3O_2$.

c 50 cm^3 of 0.02 *M* $Ba(NO_3)_2$ + 50 cm^3 of 0.20 *M* $NaHSO_4$.

d 50 cm^3 of 0.08 *M* $Ba(NO_3)_2$ + 50 cm^3 of 0.08 *M* HF.

14.46 What is the maximum pH which can be obtained by dissolving $Mg(OH)_2$ in pure water? How does this compare with the maximum pH which can be attained with NaOH? (The solubility of NaOH is about 1.20 g cm^{-3} at 25°C.) How is this difference related to the use of milk of magnesia as an antacid?

14.47 Apply Le Chatelier's principle to each of the following situations: (*a*) the change in solubility of PbS when excess Na_2S is added to a solution; (*b*) the change in solubility of PbS when excess HCl is added to a solution.

14.48 A solution containing Fe^{2+}, Pb^{2+}, Hg^{2+}, Ni^{2+}, and Zn^{2+}, each at a concentration of 0.1 *M*, is saturated with $H_2S(g)$, and then base is slowly added to convert the H_2S to HS^- and S^{2-} ions. In what order will the above metal ions precipitate as sulfides?

14.49 Use data from Tables 14.2 and 14.3 to determine which base, NH_3 or $C_2H_3O_2^-$, is stronger. Apply this knowledge to the glycinate ion, $NH_2CH_2COO^-$, which is formed when the amino acid glycine is placed in strongly basic solution. Write equations for the successive proton transfers to glycinate ion which occur as such a solution is acidified. Be sure to specify *where* each proton is located in the molecular structure.

chapter fifteen

THERMODYNAMICS: ATOMS, MOLECULES, AND ENERGY

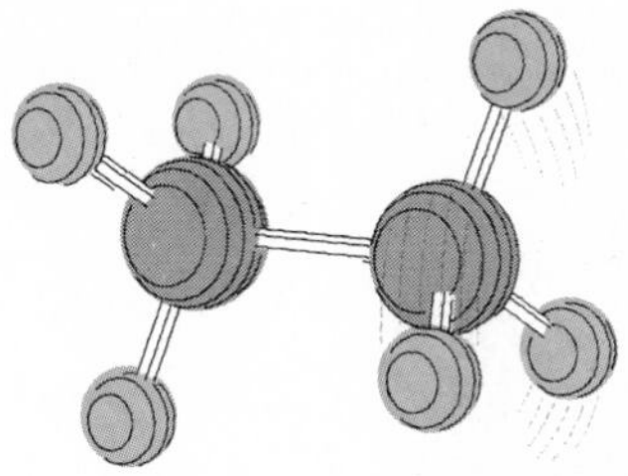

In Chap. 3 we indicated that heat is a form of energy and showed how the quantity of heat energy absorbed or released by a chemical change could be related to the corresponding chemical equation. We also stated the law of conservation of energy, and arguments in subsequent chapters have often been based on the idea that energy can neither be created nor destroyed. The law of conservation of energy is the first of three important laws involving energy and matter, which were discovered over a century ago. These laws were originally based on the movement or transfer (dynamics) of heat (thermo), and the law of conservation of energy is therefore referred to as the **first law of thermodynamics.** In this chapter we will discuss many of the important applications of the first law to chemistry. Chapter 16 will define and apply the second and third laws.

In Chap. 3 we assigned the symbol ΔH and the name enthalpy change to the quantity of heat absorbed by a chemical or physical change under conditions of constant pressure. You may have wondered at that time just how heat energy could be absorbed or given off when atoms and molecules changed position and structure during a chemical reaction, but we had not yet developed theories of chemical bonding, molecular structure, intermolecular forces, and molecular motion to the point where a satisfactory explanation could be given. Now we are in a position to investigate what can happen to molecules when matter absorbs or releases heat. One result of this study will be a clearer understanding of enthalpy. At the same time we will begin to appreciate what molecular factors contribute to making a reaction exothermic or endothermic. This will give us a solid basis for discussing several aspects of what is probably the most important problem facing our technological society today—the energy crisis.

15.1 HEAT CAPACITIES

When we supply heat energy from a bunsen burner or an electrical heating coil to a sample of matter, a rise in temperature usually occurs. Provided that no chemical changes or phase changes take place, the rise in temperature is proportional to the quantity of heat energy supplied. If q is the quantity of heat supplied and the temperature rises from T_1 to T_2, then

$$q = C(T_2 - T_1) \tag{15.1}$$

where the constant of proportionality C is called the **heat capacity** of the sample.

The most convenient way to supply a known quantity of heat energy to a sample is to use an electrical coil. The heat supplied is the product of the applied potential V, the current I flowing through the coil, and the time t during which the current flows:

$$q = VIt \tag{15.2}$$

If the SI units volt for applied potential, ampere for current, and second for time are used, the energy is obtained in joules. This is because the volt is defined as one joule per ampere per second:

$$1 \text{ volt} \times 1 \text{ ampere} \times 1 \text{ second} = 1 \frac{\text{J}}{\cancel{\text{A}}\,\cancel{\text{s}}} \times 1\,\cancel{\text{A}} \times 1\,\cancel{\text{s}} = 1 \text{ J}$$

EXAMPLE 15.1 An electrical heating coil, 230 cm^3 of water, and a thermometer are all placed in a polystyrene coffee cup. A potential difference of 6.23 V is applied to the coil, producing a current of 0.482 A which is allowed to pass for 483 s. If the temperature rises by 1.53 K, find the heat capacity of the contents of the coffee cup. Assume that the polystyrene cup is such a good insulator that no heat energy is lost from it.

Solution The heat energy supplied by the heating coil is given by

$$q = VIt$$
$$= 6.23 \text{ V} \times 0.482 \text{ A} \times 483 \text{ s} = 1450 \text{ V A s} = 1450 \text{ J}$$

However,
$$q = C(T_2 - T_1)$$

or
$$1450 \text{ J} = C \times 1.53 \text{ K}$$

so that
$$C = \frac{1450 \text{ J}}{1.53 \text{ K}} = 948 \text{ J K}^{-1}$$

Note: The heat capacity found applies to the complete contents of the cup—water, coil, and thermometer taken together, not just the water.

If our sample of matter is homogeneous, such as a pure substance or a solution, the quantity of heat needed to raise its temperature is proportional to the mass as well as to the rise in temperature. That is,

$$q = cm(T_2 - T_1) \tag{15.3}$$

The new proportionality constant c is the heat capacity per unit mass. It is called the **specific heat capacity** (or sometimes the specific heat), where the word *specific* means "per unit mass."

An important specific heat capacity is that of water. At 15°C it is 4.184 J K^{-1} g^{-1}, and at other temperatures it varies from 4.178 to 4.218 J K^{-1} g^{-1}. You may recall from Sec. 3.3 that an older, non-SI energy unit, the calorie, was defined as the heat energy required to raise the temperature of 1 g H_2O from 14.5 to 15.5°C. Thus at 15°C the specific heat capacity of water is 1.00 cal K^{-1} g^{-1}. This value is accurate to three significant figures between about 4 and 90°C.

If the sample of matter we are heating is a pure substance, then the quantity of heat needed to raise its temperature is proportional to the amount of substance. The heat capacity per unit amount of substance is called the **molar heat capacity,** symbol C_m. Thus the quantity of heat needed to raise the temperature of an amount of substance n from T_1 to T_2 is given by

$$q = C_m n(T_2 - T_1) \tag{15.4}$$

The molar heat capacity is usually given a subscript to indicate whether the substance has been heated at constant pressure (C_p) or in a closed container at constant volume (C_V).

EXAMPLE 15.2 0.854 mol neon gas is heated in a closed container by means of an electrical heating coil. 5.26 V was applied to the coil causing a current of 0.336 A to pass for 30.0 s. The temperature of the gas was found to rise by 4.98 K. Find the molar heat capacity of the neon gas, assuming no heat losses.

Solution The heat supplied by the heating coil is given by

$$\begin{aligned} q &= VIt \\ &= 5.26 \text{ V} \times 0.336 \text{ A} \times 30.0 \text{ s} = 53.0 \text{ V A s} = 53.0 \text{ J} \end{aligned}$$

Rearranging Eq. (15.4), we then have

$$\begin{aligned} C_m &= \frac{q}{n(T_2 - T_1)} \\ &= \frac{53.0 \text{ J}}{0.854 \text{ mol} \times 4.98 \text{ K}} = 12.47 \text{ J K}^{-1} \text{ mol}^{-1} \end{aligned}$$

However, since the process occurs at constant volume, we should write

$$C_V = 12.47 \text{ J K}^{-1} \text{ mol}^{-1}$$

Heat Capacity and Microscopic Changes

Now let us turn our attention from the macroscopic to the microscopic level. According to the first law of thermodynamics, the heat energy absorbed as we raise the temperature of a substance cannot be destroyed. But where does it go? In the case of a monatomic gas, like the neon in the previous example, this question is easy to answer. All the energy absorbed is converted into the kinetic energy of the neon molecules (atoms).

Recall from Sec. 9.4 that the kinetic energy of the molecules in a sample of gas is given by the expression

$$E_k = \tfrac{3}{2}nRT \tag{9.19}$$

Thus if the temperature of a sample of neon gas is raised from T_1 to T_2, the kinetic energy of the molecules increases from $\frac{3}{2}nRT_1$ to $\frac{3}{2}nRT_2$, a total change of

$$\tfrac{3}{2}nR(T_2 - T_1) = (\tfrac{3}{2}R)n(T_2 - T_1) \tag{15.5}$$

Inserting the value of R in appropriate units, we obtain

$$\frac{3}{2}\left(8.314 \frac{\text{J}}{\text{K mol}}\right) n(T_2 - T_1) = \left(12.47 \frac{\text{J}}{\text{K mol}}\right) n(T_2 - T_1)$$

This is the same quantity that is obtained by substituting the experimental value of C_V for neon (calculated in Example 15.2) into Eq. (15.4). In other words the quantity of heat found *experimentally* exactly matches the increase in kinetic energy of the molecules required by the kinetic *theory* of gases.

Table 15.1 lists the C_V values not only for neon but for some other gases as well. We immediately notice that only the noble gases and other monatomic gases such as Hg and Na have molar heat capacities equal to $\frac{3}{2}R$, or 12.47 J K^{-1} mol^{-1}. All other gases have higher molar heat capacities than this. Moreover, as the table shows, the more complex the molecule, the higher the molar heat capacity of the gas. There is a simple reason for this

TABLE 15.1 Molar Heat Capacities at Constant Volume C_V for Various Gases. (Values at 298 K Unless Otherwise Stated.)

Gas	C_V/J K^{-1} mol^{-1}	Gas	C_V/J K^{-1} mol^{-1}
Monatomic Gases		Triatomic Gases	
Ne	12.47	CO_2	28.81
Ar	12.47	N_2O	30.50
Hg	12.47 (700 K)	SO_2	31.56
Na	12.47 (1200 K)		
Diatomic Gases		Alkanes	
N_2	20.81	CH_4	27.42
O_2	21.06	C_2H_6	44.32
Cl_2	25.62	C_3H_8	65.20
		C_4H_{10}	89.94

behavior. A molecule which has two or more atoms is not only capable of moving from one place to another (**translational motion**), it can also *rotate* about itself, and it can change its shape by *vibrating*. When we heat a mole of Cl_2 molecules, for example, we not only need to supply them with enough energy to make them move around faster (increase their translational kinetic energy), we must also supply an additional quantity of energy to make them rotate and vibrate more strongly than before. For a mole of more complex molecules like *n*-butane, even more energy is required since the molecule is capable of changing its shape in all sorts of ways. In the butane molecule there are three C—C bonds around which segments of the molecule can rotate freely. All the bonds can bend or stretch, and the whole molecule can rotate as well. Such a molecule is constantly flexing and writhing at room temperature. As we raise the temperature, this kind of movement occurs more rapidly and extra energy must be absorbed in order to make this possible.

When we heat solids and liquids, the situation is somewhat different than for gases. The rapid increase of vapor pressure with temperature makes it virtually impossible to heat a solid or liquid in a closed container, and so heat capacities are always measured at constant pressure rather than at constant volume. Some C_p values for selected simple liquids and solids at the melting point are shown in Table 15.2. In general the heat capacities of solids and liquids are higher than those of gases. This is because of the intermolecular forces operating in solids and liquids. When we heat solids and liquids, we need to supply them with potential energy as well as kinetic energy. Among the solids, the heat capacities of the metals are easiest to explain since the solid consists of individual atoms. Each atom can only vibrate in three dimensions. According to a theory first suggested by Einstein, this vibrational energy has the value $3RT$, while the heat capacity is given by $3R = 24.9 \text{ J K}^{-1} \text{ mol}^{-1}$. As can be seen from the table, most

TABLE 15.2 Molar Heat Capacities at Constant Pressure C_p for Various Solids and Liquids at the Melting Point.

Substance	C_p(solid)/J K^{-1} mol^{-1}	C_p(liquid)/J K^{-1} mol^{-1}
	Monatomic Substances	
Hg	27.28	27.98
Pb	29.40	30.33
Na	28.20	31.51
	Diatomic Substances	
Br_2	53.8	75.7
I_2	54.5	80.7
HCl	50.5	62.2
HI	47.5	68.6
	Polyatomic Substances	
H_2O	37.9	76.0
NH_3	49.0	77.0
Benzene	129.0	131.8
n-Heptane	146.6	203.1

monatomic solids have C_p values slightly larger than this. This is because solids expand slightly on heating. The atoms get farther apart and thus increase in potential as well as vibrational energy.

Solids which contain molecules rather than atoms have much higher heat capacities than $3R$. In addition to the vibration of the whole molecule about its site in the crystal lattice, the individual atoms can also vibrate with respect to each other. Occasionally molecules can rotate in the crystal, but usually rotation is only possible when the solid melts. As can be seen from the values for molecular liquids in Table 15.2, this sudden ability to rotate causes a sharp increase in the heat capacity. For monatomic substances, where there is no motion corresponding to the rotation of atoms around each other, the heat capacity of the liquid is only very slightly higher than that of the solid.

15.2 INTERNAL ENERGY

When matter absorbs or releases heat energy, we cannot always explain this energy change on the microscopic level in terms of a speeding up or a slowing down of molecular motion as we have just done for heat capacities. This is particularly true when the heat change accompanies a chemical change. Here we must consider changes in the kinetic and potential energies of electrons in the atoms and molecules involved, that is, changes in the **electronic energy.**

As a simple example of a chemical change, let us consider an exothermic reaction involving only one kind of atom, the decomposition of ozone, O_3:

$$2O_3(g) \rightarrow 3O_2(g) \qquad 25°C \tag{15.6}$$

O_3 is a gas which occurs in very low but very important concentrations in the upper atmosphere. It can be produced in somewhat higher concentrations in the laboratory by using an electric spark discharge and can then be concentrated and purified. The result is a blue gas which is dangerously unstable and liable to explode without warning. Let us now suppose that we have a pure sample of 2 mol $O_3(g)$ in a closed container at 25°C and are able to measure the quantity of heat evolved when it subsequently explodes to form O_2 gas according to Eq. (15.5) (see Fig. 15.1). It is found that 287.9 kJ is released. This energy heats up the surroundings.

Where do these 287.9 kJ come from? Certainly not entirely from the translational kinetic energy of the molecules. To begin with we had 2 mol O_3 at 25°C and at the end 3 mol O_2 at 25°C. Since the translational kinetic energy of any gas is $\frac{3}{2}nRT$, this corresponds to an increase of $\frac{3}{2}$(3 mol − 2 mol)$RT = \frac{3}{2} \times 1 \text{ mol} \times 8.314 \text{ J K}^{-1} \text{ mol}^{-1} \times 298 \text{ K} = 3.72$ kJ. After the reaction the translational energy of the molecules is higher, and so heat should have been absorbed, not given off. In any case 3.73 kJ is only 1.3 percent of the total heat change. The changes in rotational and vibrational energy are even smaller, accounting for a decrease in energy of the substance in the container by only 0.88 kJ.

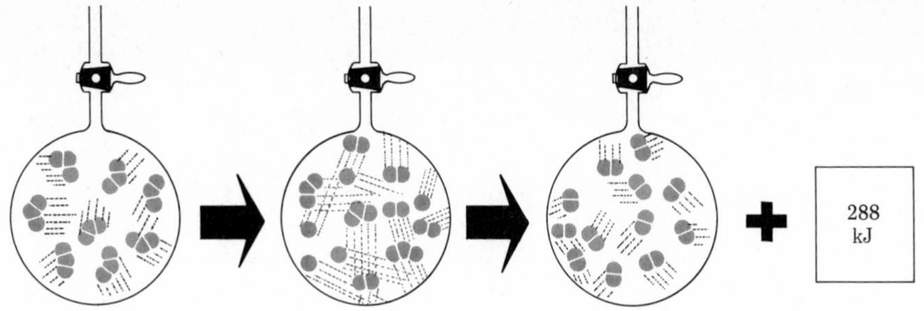

Figure 15.1 The reaction $2O_3(g) \rightarrow 3O_2(g)$ (25°C). When the atoms in the O_3 molecules rearrange to form O_2 molecules, the result is a lowering in the electronic energy and a consequent release of 288 kJ of heat energy to the surroundings.

From the standpoint of energy, the most important thing that happens as 2 mol O_3 is converted into 3 mol O_2 is rearrangement of the valence electrons so that they are closer to positively charged O nuclei. We have already seen (Sec. 5.4) that the closer electrons are to nuclei, the lower their total energy. Thus three O_2 molecules have less electronic energy than the two O_3 molecules from which they were formed. The remaining energy first appears as kinetic energy of the O_2 molecules. Immediately after the reaction the O_2 is at a very high temperature. Eventually this energy finds its way to the surroundings, and the O_2 cools to room temperature.

A detailed summary of the various energy changes which occur when O_3 reacts to form O_2 is given in Table 15.3. The important message of this table is that 99 percent of the energy change is attributable to the change in electronic energy. This is a typical figure for gaseous reactions. What makes a gaseous reaction exothermic or endothermic is the *change in the bonding*. Changes in the energies of molecular motion can usually be neglected by comparison.

TABLE 15.3 Detailed Balance Sheet of the Energy Changes Occurring in the Reaction

$2O_3(g) \rightarrow 3O_2(g)$ 25°C, constant volume

Type of Energy	Initial Value/kJ	Final Value/kJ	Change* in Energy/kJ
Electronic	x†	$x - 290.70$	−290.70
Translational	7.43	11.15	+3.72
Rotational and vibrational	8.32	7.44	−0.88
Total	$x + 15.75$	$x - 272.11$	−287.86

* Final value − initial value.

† There is no experimental means of determining the initial or final value of the electronic energy—only the change in electronic energy can be measured. Highly accurate calculations of electronic energy from wave-mechanical theory require complicated mathematics and a great deal of computer power. Therefore we have represented the initial electronic energy by x.

The sum of all the different kinds of energy which the molecules of a substance can possess is called the **internal energy** and given the symbol U. (The symbol E is also widely used.) In a gas we can regard the internal energy as the sum of the electronic, translational, rotational, and vibrational energies. In the case of liquids and solids the molecules are closer together, and we must include the potential energy due to their interactions with each other. In addition, the motion of one molecule now affects its neighbors, and we can no longer subdivide the energy into neat categories as in the case of a gas.

Thermodynamic Terms and Conventions

Although we have defined the internal energy in molecular terms, it is important to realize that we can deal with this property purely in macroscopic terms as well. Thus we can talk about the internal energy of a substance without even knowing its chemical formula. More importantly, we can usually measure the change in the internal energy which accompanies a physical or chemical change without having the slightest idea of what is happening on the molecular level. In order to show how this is done, we first need to introduce the following conventions and terms:

1 *System.* The term system is used to describe any sample of matter in which we are particularly interested. Thus, in the conversion of O_3 to O_2 described in Fig. 15.1, the system would denote the *contents of the flask* before, after, or during the chemical change which occurs. Initially the system would thus consist of 2 mol O_3 and finally of 3 mol O_2.

2 *Initial and final states.* In thermodynamics our principal concern is with the initial state, before any changes begin, and the final state, when no more changes occur. By convention we refer to the initial state with the subscript 1 and to the final state with the subscript 2. Referring again to the reaction illustrated in Fig. 15.1, if the volume of the flask is 24.47 dm^3, the initial pressure P_1 will be 202.7 kPa (2.00 atm) and the final pressure P_2 will be 304.0 kPa (3.00 atm). Likewise we could refer to the initial value of the internal energy as U_1 and the final value as U_2.

3 *The delta convention.* When a chemical or physical change occurs, many of the properties of the system change. We conventionally refer to a *change* in a property with the symbol Δ (delta). In the O_3–O_2 reaction, for example, the pressure P increases by 101.3 kPa, and we can indicate this fact by writing

$$\Delta P = 101.3 \text{ kPa}$$

More formally, if a property X changes from an initial value X_1 to a final value X_2, then ΔX is defined as

$$\Delta X = X_2 - X_1$$

Thus, in the above case

$$\Delta P = P_2 - P_1$$
$$= 304.0 \text{ kPa} - 202.7 \text{ kPa}$$
$$= 101.3 \text{ kPa}$$

We can also describe changes in internal energy using the delta convention. It is apparent from Table 15.3 that the conversion of 2 mol O_3 to 3 mol O_2 results in a *decrease* of 287.9 kJ in the internal energy. In the delta convention we express this as a *negative* quantity:

$$\Delta U = U_2 - U_1 = -287.9 \text{ kJ}$$

since the initial value of U is larger than the final value.

4 *Heat energy*. Our final convention refers to the heat energy q absorbed by the system. If the heat change occurs at constant volume, the symbol q_V is used, if at constant pressure, q_p. By convention we refer to the heat energy *absorbed* by the system as being *positive* and the heat energy *released* by the system as being *negative*. In the O_3 to O_2 conversion described in Fig. 15.1, for example, we would write

$$q_V = -287.9 \text{ kJ}$$

since the system *releases* 287.9 kJ of heat energy to the surroundings.

Now that we are familiar with these conventions, we can use them to state an important general principle. When a chemical or physical change occurs *at constant volume,* the heat energy absorbed by the system is equal to the increase in its internal energy, or in mathematical language

$$q_V = \Delta U \qquad (15.7)$$

This general behavior of matter is an immediate consequence of the law of conservation of energy. If an endothermic reaction occurs at constant volume and heat energy is absorbed by the reaction system, then this heat energy cannot disappear: It all goes toward increasing U. Alternatively, if the internal energy decreases, as in Fig. 15.1, no energy is lost: It *all* appears as heat energy and is transferred to the surroundings.

Equation (15.7) tells us how to detect and measure changes in the internal energy of a system. If we carry out any process in a closed container (so the volume remains constant), the quantity of heat absorbed by the system equals the increase in internal energy. A convenient device for making such measurements is a bomb calorimeter (Fig. 15.2), which contains a steel-walled vessel (bomb) with a screw-on gas-tight lid. In the bomb can be placed a weighed sample of a combustible substance together with $O_2(g)$ at about 3 MPa (30 atm) pressure. When the substance is ignited by momentarily passing electrical current through a heating wire, the heat energy released by its combustion raises the temperature of water surrounding the bomb. Measurement of the change in temperature of the water permits calculation of q_V (and thus ΔU), provided the heat capacity of the calorimeter is known. The heat capacity can be determined as in Example 15.1 or by igniting a substance for which ΔU is already known.

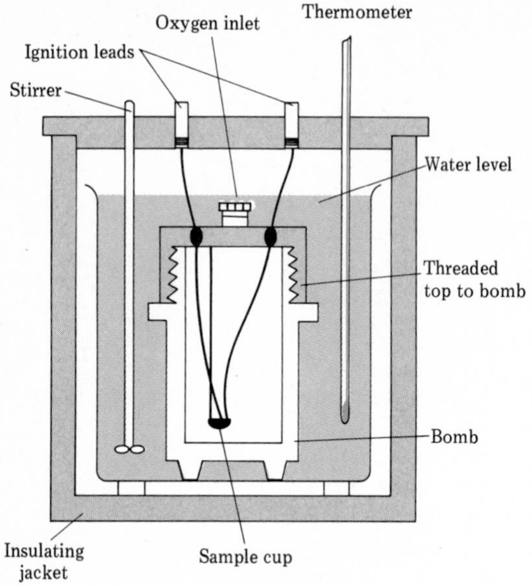

Figure 15.2 A bomb calorimeter. *(By permission from W. G. Davies, "Introduction to Chemical Thermodynamics," p. 161, W. B. Saunders Company, Philadelphia, 1972.)*

EXAMPLE 15.3 When 0.7943 g of glucose, $C_6H_{12}O_6$, is ignited in a bomb calorimeter, the temperature rise is found to be 1.841 K. The heat capacity of the calorimeter is 6.746 kJ K^{-1}. Find ΔU_m for the reaction

$$C_6H_{12}O_6(s) + 6O_2(g) \rightarrow 6CO_2(g) + 6H_2O(l)$$

under the prevailing conditions.

Solution The heat energy absorbed by the calorimeter in increasing its temperature by 1.841 K is given by

$$q = C\,\Delta T = 6.746\,\frac{\text{kJ}}{\text{K}} \times 1.841\text{ K} = 12.42\text{ kJ}$$

Since this heat energy was *released* by the reaction system, we must regard it as negative. Accordingly,

$$q_V = -12.42\text{ kJ} = \Delta U$$

We need now only to calculate the change in internal energy per mole, that is, ΔU_m. Now

$$n_{\text{glucose}} = 0.7943\text{ g} \times \frac{1\text{ mol}}{180.16\text{ g}} = 4.409 \times 10^{-3}\text{ mol}$$

Thus $$\Delta U_m = \frac{-12.42\text{ kJ}}{4.409 \times 10^{-3}\text{ mol}} = -2817\text{ kJ mol}^{-1}$$

In chemistry we are interested not only in those changes occurring in a closed container at constant volume but also in those occurring in an open container at constant (i.e., atmospheric) pressure. When a change occurs at constant pressure, there is another energy factor we must consider in addition to the heat absorbed and the change in internal energy. This is the *expansion work* w_{exp} which the system does as its volume expands against the external pressure.

In order to understand the nature and magnitude of this expansion work, let us consider the simple example illustrated in Fig. 15.3. Here a sample of oxygen gas is heated at a constant pressure P from an initial temperature T_1 to a final temperature T_2 by means of an electrical heating coil. The gas is confined in a cylinder by a piston, and the pressure is maintained by placing a weight of the correct magnitude on top of the piston. The whole apparatus is maintained in a vacuum so that there is no atmospheric pressure to consider in addition to the effect of the weight and piston.

When the heating coil in this apparatus is switched on, the temperature rises and the gas expands in compliance with Charles' law, lifting the piston and weight in the process. Energy must be supplied from the heating coil not only to increase the energy of the oxygen molecules but also to *lift the weight*. In other words,

$$q_p = \Delta U + w_{exp} \tag{15.8}$$

where q_p = heat absorbed
ΔU = change in internal energy
w_{exp} = expansion work, i.e., work done in lifting the weight

To calculate the magnitude of this expansion work, we begin with the definition of pressure:

$$P = \frac{\text{force}}{\text{area}} = \frac{F}{A}$$

or

$$F = PA$$

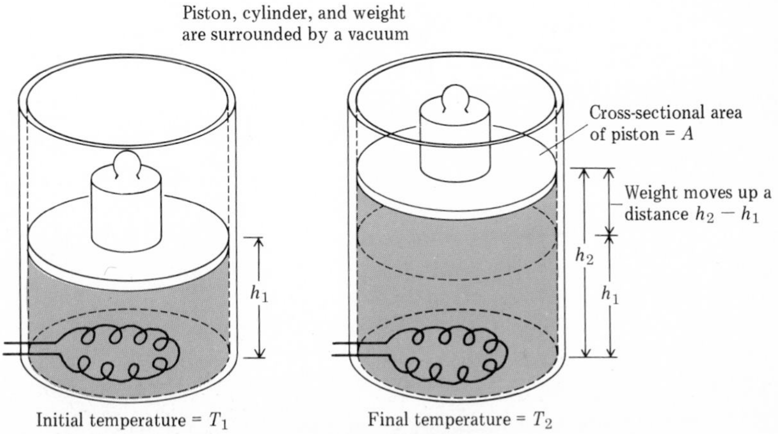

Figure 15.3 When a sample of gas is heated at constant pressure, energy must be supplied to expand the gas as well as to increase its internal energy.

where F is the total force exerted by the weight and piston on the oxygen gas and A is the area of the piston. As the gas sample is heated, the volume increases from an initial value V_1 to a final value V_2, while the piston and its weight move from a height h_1 to a height h_2. The work done is thus given by the expression

$$\begin{aligned} w_{\text{exp}} &= \text{force exerted} \times \text{distance moved} \\ &= (P \times A) \times (h_2 - h_1) \\ &= P \times (A \times h_2 - A \times h_1) \end{aligned}$$

However, since the volume of a cylinder is the area of the base times the height,

$$A \times h_1 = V_1 \qquad \text{initial volume of gas}$$

and

$$A \times h_2 = V_2 \qquad \text{final volume}$$

Thus

$$w_{\text{exp}} = P \times (V_2 - V_1)$$

or

$$w_{\text{exp}} = P\,\Delta V \tag{15.9}$$

Inserting this value for w_{exp} into Eq. (15.8), we obtain the final result

$$q_p = \Delta U + P\,\Delta V \tag{15.10}$$

It is important to realize that the expansion work $P\,\Delta V$ does not depend on our sample being in the apparatus of Fig. 15.3 in which there is an obvious gain in the potential energy of a weight. If instead of a weight we allowed the atmosphere to exert a pressure P on the gas, the result would still be $P\,\Delta V$. In this second case, instead of lifting a visible weight, the expanding gas would push back the atmosphere and hence be lifting invisible air molecules. The work done, and hence the *increase in the potential energy of the atmosphere,* would still be $P\,\Delta V$.

This simple example of an expanding gas helps us to see what is involved in the general case of a chemical or physical change occurring at constant pressure. In any such case the heat energy absorbed by the system, q_p, will always exactly account for the increase in internal energy ΔU and the expansion work $P\,\Delta V$. In other words the relationship

$$q_p = \Delta U + P\,\Delta V \tag{15.10}$$

is valid in the general case.

EXAMPLE 15.4 When 1 mol liquid H_2O is boiled at 100°C and 101.3 kPa (1.000 atm) pressure, its volume expands from 19.8 cm^3 in the liquid state to 30.16 dm^3 in the gaseous state. The heat energy absorbed by the vaporization process is found experimentally to be 40.67 kJ. Calculate the increase in internal energy ΔU of H_2O.

Solution We must first calculate the expansion work $P\,\Delta V$.

$$w_{\text{exp}} = P(V_2 - V_1)$$
$$= 101.3\ \text{kPa} \times (30.16\ \text{dm}^3 - 0.0198\ \text{dm}^3) = 3053\ \text{kPa dm}^3$$

Recalling from Sec. 9.2 that 1 kPa × 1 dm^3 = 1 J, we have

$$w_{\text{exp}} = 3053\ \text{J} = 3.053\ \text{kJ}$$

Also, since

$$q_p = \Delta U + P\,\Delta V$$

we have

$$\Delta U = q_p - P\,\Delta V$$
$$= 40.67\ \text{kJ} - 3.05\ \text{kJ} = 37.62\ \text{kJ}$$

Note: As we saw in Chaps. 8 and 10, the vaporization of a liquid is always an endothermic process. Since the molecules attract each other, energy must be supplied to separate them as vaporization occurs. However, not all the energy supplied when a liquid boils goes to increasing the potential energy of the molecules. A significant proportion is needed to increase the potential energy of the air as well.

It is far easier to carry out most chemical reactions in a container open to the atmosphere than in a closed system like a bomb calorimeter. However, when a reaction is carried out at constant atmospheric pressure, it is necessary to measure P, V_1, and V_2 (as in Example 15.4) as well as q_p in order to calculate the change in internal energy from the equation $\Delta U = q_p - P\,\Delta V$. These extra measurements are a nuisance, and they can be avoided by introducing a new quantity which is related to the internal energy but also includes the potential energy of the atmosphere. This quantity has already been used in Chap. 3. It is called the **enthalpy,** symbol H, and is defined by

$$H = U + PV \tag{15.11}$$

From Eq. (15.11) we can see that the enthalpy is always larger than the internal energy by a quantity PV. This extra energy is added to take account of the fact that whenever a body of volume V is introduced into the atmosphere, the potential energy of the atmosphere is increased by PV [by the same argument used to derive Eq. (15.9)]. The enthalpy is thus of the form

Enthalpy = internal energy + potential energy of atmosphere (15.12)

When a system undergoes a chemical or physical change at constant pressure, the change in enthalpy is given by

Change in enthalpy = change in internal energy
+ change in potential energy of atmosphere (15.13)

The change in enthalpy thus includes *both* the energy changes for which heat energy must be supplied from the surroundings. In more formal language

$$\Delta H = \Delta U + \Delta(PV) = \Delta U + (P_2V_2 - P_1V_1) \qquad (15.14)$$

If we consider conditions of constant pressure, $P_2 = P_1 = P$, and

$$\Delta H = \Delta U + (PV_2 - PV_1) = \Delta U + P(V_2 - V_1) = \Delta U + P\,\Delta V$$

but from Eq. (15.10)

$$q_p = \Delta U + P\,\Delta V$$

Thus

$$q_p = \Delta H \qquad (15.15)$$

In other words, *the heat energy absorbed by a system in any change at constant pressure is equal to the increase in its enthalpy.* The change in enthalpy ΔH can be obtained from a single experimental measurement: the heat energy absorbed at constant pressure.

Measuring the Enthalpy Change

In an elementary laboratory, enthalpy changes are often measured in a "coffee-cup calorimeter" such as that shown in Fig. 15.4. Suppose the reaction to be measured is between two solutions. One of these solutions is introduced into the coffee cup, and the temperature of both solutions is measured. The second solution is now introduced, the mixture stirred, and the rise in temperature recorded. Since the cup is made of polystyrene foam, a very good insulator, very little heat energy escapes. In order to find the quantity of heat energy evolved, we need to know not only the rise in temperature but also the heat capacity of the calorimeter. This can be found in a separate experiment. After the final solution has cooled to room temperature, it can be reheated to the higher temperature by means of an electrical coil. The heat supplied in this second experiment can be calculated in the usual way from the applied potential, the current flowing, and the time.

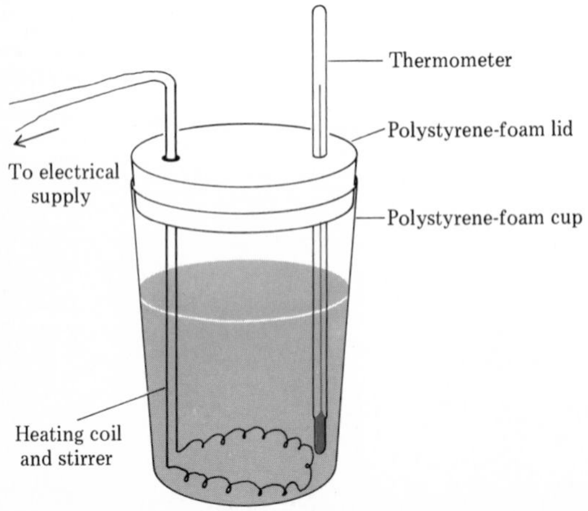

Figure 15.4 A coffee-cup calorimeter.

EXAMPLE 15.5 250.0 cm^3 of 1.000 *M* HCl at 20.38°C and 250.0 cm^3 of 1.000 *M* NaOH also at 20.38°C are mixed in a coffee-cup calorimeter similar to that shown in Fig. 15.4. The temperature of the mixture rises to 27.80°C. The mixture is then cooled to 20.45°C after which 24.06 V is applied to the heating coil, allowing 2.13 A to flow for 300.0 s. The temperature rises to 28.23°C. Find ΔH_m for the reaction

$$H_3O^+(aq) + OH^-(aq) \rightarrow 2H_2O(l)$$

Solution We first calculate the heat capacity of the contents of the calorimeter. The heat supplied by the coil q_2 is given by

$$q_2 = 24.06 \text{ V} \times 2.13 \text{ A} \times 300.0 \text{ s} = 15\ 374 \text{ J}$$

so that
$$C = \frac{q_2}{\Delta T} = \frac{15\ 374 \text{ J}}{(28.23 - 20.45)\text{K}} = 1976 \text{ J K}^{-1}$$

Knowing the heat capacity, we can now calculate q_1, the heat change in the first part of the experiment:

$$q_1 = -C\,\Delta T = -1976\,\frac{\text{J}}{\text{K}}\,(27.80 - 20.38)\text{K}$$
$$= -14.66 \text{ kJ}$$

where the sign is *negative* since the system has released energy. Because the change occurs at *constant pressure,* we can equate q_1 to the enthalpy change ΔH. However, the question asks for the enthalpy change per *mole* of acid (that is, ΔH_m), while only 250 cm^3 × 1 mol dm^{-3} = 0.250 mol acid was used. Accordingly,

$$\Delta H_m = \frac{-14.66 \text{ kJ}}{0.250 \text{ mol}} = 58.7 \text{ kJ mol}^{-1}$$

Even though most chemical reactions are more conveniently studied at constant pressure than at constant volume, there are some reactions for which the opposite is true. When one of the reactants is a gas, for instance, the reaction is much easier to carry out in a closed container. If a reaction is measured at constant volume, of course, we find ΔU rather than ΔH. However, almost all thermochemical data are recorded in terms of the enthalpy rather than the internal energy. It is therefore often necessary to convert a measured ΔU value to a ΔH value. This is done as shown in the following example:

EXAMPLE 15.6 Use the data in Table 15.3 to find ΔH for the reaction

$$2O_3(g) \rightarrow 3O_2(g) \qquad 25°\text{C, 1 atm}$$

Solution ΔH for a constant-pressure change is given by

$$\Delta H = \Delta U + P\,\Delta V$$

Since ΔU is known from Table 15.3, we need only calculate $P\,\Delta V$. The initial volume of the system V_1 is that of 2 mol O_3 at 25°C.

$$V_1 = \frac{nRT}{P} = 2 \text{ mol } \frac{RT}{P}$$

Similarly, the final volume is that of 3 mol O_2 at 25°C:

$$V_2 = 3 \text{ mol } \frac{RT}{P}$$

Thus $$\Delta V = V_2 - V_1 = \frac{RT}{P}(3 - 2)\text{mol}$$

and $$P\,\Delta V = \cancel{P}\left[\frac{RT}{\cancel{P}}(3 - 2)\text{mol}\right] = RT(3 - 2)\text{mol}$$

$$= 8.3143 \frac{\text{J}}{\text{K mol}} \times 298.15 \text{ K} \times 1 \text{ mol}$$

$$= 2479 \text{ J} = 2.479 \text{ kJ}$$

Thus $$\Delta H = \Delta U + P\,\Delta V$$

$$= -287.9 \text{ kJ} + 2.5 \text{ kJ} = -285.4 \text{ kJ}$$

Perhaps because values of ΔH are usually easier to measure than values of ΔU, chemists have chosen to concentrate exclusively on ΔH rather than on ΔU as a way of recording thermochemical data. Though enthalpies of formation like those given in Table 3.1 are easy to find, equivalent tables of internal energies are nonexistent. In many ways this insistence on ΔH rather than on ΔU is a pity. In particular, it suggests that somehow the enthalpy H has more fundamental significance on the molecular level than the internal energy U. It is important to realize that this is *not* the case. It is the internal energy which has a simple molecular interpretation, namely, the total energy of all the molecules in the system. By contrast the enthalpy includes not only the total energy of the molecules in the system but the potential energy of the atmosphere *outside the system* as well. We use the enthalpy so often because of its convenience rather than because of its molecular significance.

A further point worth making about the enthalpy is that the difference between ΔH and ΔU is not often of great chemical importance. This is particularly true of reactions which involve only gases, such as the decomposition of ozone. In a gaseous reaction the main factor determining both ΔU and ΔH is the *change in electronic energy*. Changes in molecular energy and also the expansion work $P\,\Delta V$ are usually small compared with this change. In the decomposition of ozone, for example, the change in electronic energy is -290.7 kJ per 2 mol ozone. The value of ΔU is -287.9 kJ, while that of ΔH is -285.4 kJ. The three quantities are all within a few percent of each other. For many purposes, differences of this order of magnitude are immaterial. When this is the case, we can equate both ΔU and ΔH to the change in electronic energy.

State Functions

Both enthalpy and the internal energy are often described as **state functions.** This means that they depend only on the *state* of the system, i.e., on its pressure, temperature, composition, and amount of substance, but not on its previous history. Thus any solution of NaCl at 25°C and 101.3 kPa (1.000 atm) which contains a mixture of 1 mol NaCl and 50 mol H_2O has the same internal energy and the same enthalpy as any other solution with the same specifications. It does not matter whether the solution was prepared by simply dissolving NaCl(*s*) in H_2O, by reacting NaOH(*aq*) with HCl(*aq*), or by some more exotic method.

The fact that the internal energy and the enthalpy are both state functions has an important corollary. It means that when a system undergoes any change whatever, then the alteration in its enthalpy (or its internal energy) depends only on the initial state of the system and its final state. The initial value of the enthalpy will be H_1, and the final value will be H_2. No matter what pathway we employ to get to state 2, we will always end up with the value H_2 for the enthalpy. The enthalpy change $\Delta H = H_2 - H_1$ will thus be independent of the path used to travel from state 1 to state 2. This corollary is of course the basis of Hess' law (Sec. 3.3). The change in enthalpy for a given chemical process is the same whether we produce that change in one or in several steps.

Standard Pressure

You will often find enthalpy changes indicated by $\Delta H^{\ominus}$ (or ΔH^0) and called *standard enthalpy changes*. The superscript is added to indicate that the enthalpy change has occurred at the *standard pressure of 101.3 kPa (1.000 atm)*. Unless very high pressures are involved, ΔH changes very little with a change in pressure, and so we have ignored superscripts up to this point. However, two other properties of matter called the entropy, symbol S, and the free energy, symbol G, will be discussed in Chap. 16. These are quite sensitive to pressure, and the inclusion of this superscript is important. For reasons of consistency therefore we will indicate standard enthalpy changes as $\Delta H^{\ominus}$ from now on.

15.4 BOND ENTHALPIES

The heat changes which accompany a chemical reaction are caused largely by changes in the electronic energy of the molecules. If we restrict our attention to gases, and hence to fairly simple molecules, we can go quite a long way toward predicting whether a reaction will be exothermic by considering the bonds which are broken and made in the course of the reaction. In order to do this we must first become familiar with the idea of a bond enthalpy.

In Sec. 6.4 we pointed out that when a chemical bond forms, negative charges move closer to positive charges than before, and so there is a lowering of the energy of the molecule relative to the atoms from which it was made. This means that energy is required to break a molecule into its constituent atoms. The **bond enthalpy** D_{X-Y} of a diatomic molecule X—Y is the enthalpy change for the (usually hypothetical) process:

$$XY(g) \rightarrow X(g) + Y(g) \qquad \Delta H^{\ominus}(298\ K) = D_{X-Y} \qquad (15.16)$$

We have already used the term **bond energy** to describe this quantity, though strictly speaking the bond energy is a measure of ΔU rather than ΔH. As we have already seen, ΔU and ΔH are nearly equal, and so either term may be used.

As an example, let us consider the bond enthalpy for carbon monoxide. It is possible to establish the thermochemical equation

$$CO(g) \rightarrow C(g) + O(g) \qquad \Delta H^{\ominus}(298\ K) = 1076.4\ kJ\ mol^{-1} \quad (15.17)$$

Accordingly we can write

$$D_{C\equiv O} = 1076.4\ kJ\ mol^{-1}$$

even though the process to which Eq. (15.17) corresponds is hypothetical: Neither carbon nor oxygen exists as a monatomic gas at 298 K.

For triatomic and polyatomic molecules, the bond enthalpy is usually defined as a mean. In the case of water, for instance, we have

$$H_2O(g) \rightarrow 2H(g) + O(g) \qquad \Delta H^{\ominus} = 927.2\ kJ\ mol^{-1}$$

Since it requires 927.2 kJ to break open *two* O—H bonds, we take *half* this value as the mean bond enthalpy and write

$$D_{O-H} = 463.6\ kJ\ mol^{-1}$$

In methanol, CH_3OH, however, a value of 427 kJ mol^{-1} for the O—H bond enthalpy fits the experimental data better. In other words the *strength* of the O—H bond varies somewhat from compound to compound. Because of this fact, we must expect to obtain only approximate results, accurate only to about ± 50 kJ mol^{-1}, from the use of bond enthalpies. Bond enthalpies for both single and multiple bonds are given in Table 15.4.

As an example of how a table of bond enthalpies can be used to predict the ΔH value for a reaction, let us take the simple case

$$H_2(g) + F_2(g) \rightarrow 2HF(g) \qquad 298\ K,\ 1\ atm \qquad (15.18)$$

We can regard this reaction as occurring in two stages (Fig. 15.5). In the first stage all the reactant molecules are broken up into atoms:

$$H_2(g) + F_2(g) \rightarrow 2H(g) + 2F(g) \qquad 298\ K,\ 1\ atm \qquad (15.18a)$$

For this stage

$$\Delta H_I = D_{H-H} + D_{F-F}$$

since 1 mol H_2 and 1 mol F_2 have been dissociated.

TABLE 15.4 Mean Bond Enthalpies/kJ mol^{-1}.

Single Bonds							
C—H	413	N—H	391	O—H	463	F—F	159
C—C	348	N—N	163	O—O	146		
C—N	293	N—O	201	O—F	190	Cl—F	253
C—O	358	N—F	272	O—Cl	203	Cl—Cl	242
C—F	485	N—Cl	200	O—I	234		
C—Cl	328	N—Br	243			Br—F	237
C—Br	276			S—H	339	Br—Cl	218
C—I	240	H—H	436	S—F	327	Br—Br	193
C—S	259	H—F	567	S—Cl	253		
		H—Cl	431	S—Br	218	I—Cl	208
Si—H	323	H—Br	366	S—S	266	I—Br	175
Si—Si	226	H—I	299			I—I	151
Si—C	301						
Si—O	368						
Multiple Bonds							
C=C	614	N=N	418	O=O	498		
C≡C	839	N≡N	945				
C=N	615			S=O	323		
C≡N	891			S=S	418		
C=O	804*						
C≡O	1076						

* The value for CO_2.

Figure 15.5 Bond-breaking–bond-making diagram for the reaction $H_2 + F_2 \rightarrow 2HF$. When H_2 reacts with F_2, a strong H—H bond and a weak F—F bond are broken, while two extra-strong H—F bonds are made. The reaction is exothermic since more energy is released by the formation of the H—F bonds than is required to break the H—H and H—F bonds.

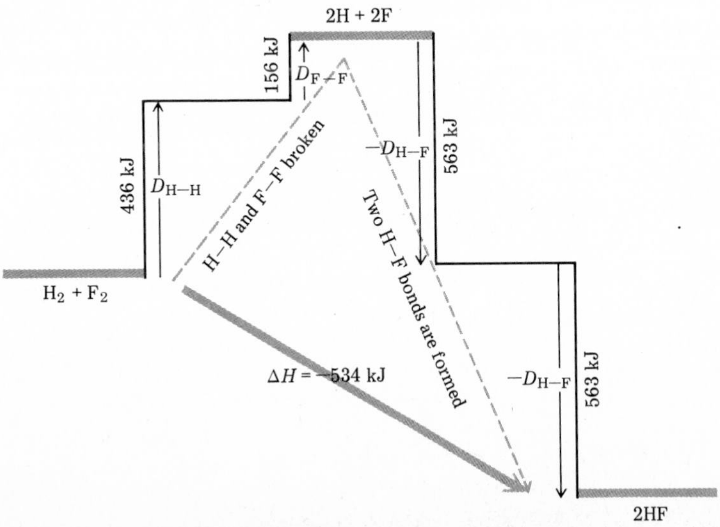

In the second stage the H and F atoms are reconstituted to form HF molecules:

$$2H(g) + 2F(g) \rightarrow 2HF(g) \qquad 298\text{ K, 1 atm} \qquad (15.18b)$$

For which

$$\Delta H_{II} = -2D_{H-F}$$

where a negative sign is necessary since this stage corresponds to the *reverse* of dissociation.

Since Eq. (15.18) corresponds to the sum of Eqs. (15.18*a*) and (15.18*b*), Hess' law allows us to add ΔH values:

$$\begin{aligned}\Delta H^{\ominus}_{\text{reaction}} &= \Delta H_I + \Delta H_{II} \\ &= D_{H-H} + D_{F-F} - 2D_{H-F} \\ &= (436 + 159 - 2 \times 567)\text{ kJ mol}^{-1} \\ &= -539\text{ kJ mol}^{-1}\end{aligned}$$

We can work this same trick of subdividing a reaction into a bond-breaking stage followed by a bond-making stage for the general case of any gaseous reaction. In the first stage all the bonds joining the atoms in the reactant molecules are broken and a set of gaseous atoms results. For this stage

$$\Delta H_I = \Sigma D(\text{bonds broken})$$

The enthalpy change is the sum of the bond enthalpies for all bonds broken. In the second stage these gaseous atoms are reconstituted into the product molecules. For this second stage therefore

$$\Delta H_{II} = -\Sigma D(\text{bonds formed})$$

where the negative sign is necessary because the *reverse* of bond breaking is occurring in this stage. The total enthalpy change for the reaction at standard pressure is thus

$$\Delta H^{\ominus} = \Delta H_I + \Delta H_{II}$$

or

$$\Delta H^{\ominus} = \Sigma D(\text{bonds broken}) - \Sigma D(\text{bonds formed}) \qquad (15.19)$$

The use of this equation is illustrated in the next example.

EXAMPLE 15.7 Using Table 15.4 calculate the value of $\Delta H^{\ominus}(298\text{ K})$ for the reaction

$$CH_4(g) + 2O_2(g) \rightarrow CO_2(g) + 2H_2O(g)$$

Solution It is best to sketch the molecules and their bonds in order to make sure that none are missed.

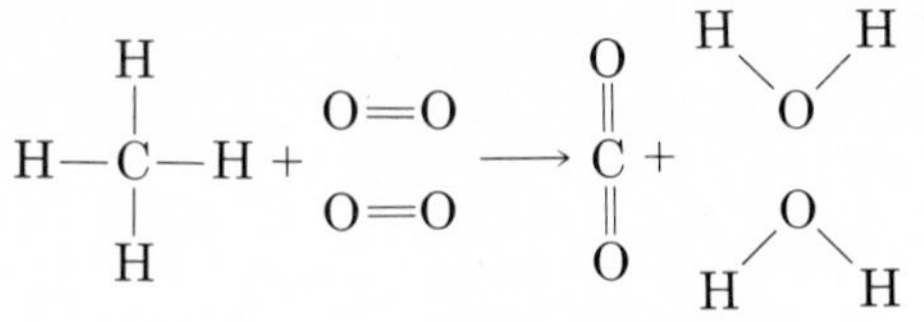

$$
\begin{aligned}
\text{Thus}\qquad \Delta H^{\ominus} &= \Sigma D(\text{bonds broken}) - \Sigma D(\text{bonds formed})\\
&= 4D_{\mathrm{C-H}} + 2D_{\mathrm{O=O}} - 2D_{\mathrm{C=O}} - 4D_{\mathrm{O-H}}\\
&= (4 \times 413 + 2 \times 498 - 2 \times 804 - 4 \times 463)\ \mathrm{kJ\ mol^{-1}}\\
&= -812\ \mathrm{kJ\ mol^{-1}}
\end{aligned}
$$

The experimental value for this enthalpy change can be calculated from standard enthalpies of formation. It is -802.4 kJ mol^{-1}. The discrepancy is due to the unavoidable use of *mean* bond enthalpies in the calculation.

Bond Enthalpies and Exothermic or Endothermic Reactions

We are now in a position to appreciate the general principle which determines whether a gaseous reaction will be exothermic or endothermic. If less energy is needed to break up the reactant molecules into their constituent atoms than is released when these atoms are reconstituted into product molecules, then the reaction will be exothermic. Usually an exothermic reaction corresponds to the breaking of weak bonds (with small bond enthalpies) and the making of strong bonds (with large bond enthalpies). Both the examples we have considered are of this type. In the hydrogen-fluorine reaction (Fig. 15.5) one quite strong bond ($D_{\mathrm{H-H}}$ = 436 kJ mol^{-1}) and one very weak bond ($D_{\mathrm{F-F}}$ = 159 kJ mol^{-1}) are broken, while two very strong H—F bonds are made. (Note that with a bond enthalpy of 567 kJ mol^{-1}, H—F is the strongest of all single bonds.) The combustion of methane discussed in Example 15.7 is another example of the formation of stronger bonds at the expense of weaker ones. The bond enthalpy of the O—H bond is not much different in magnitude from those of the C—H and O═O bonds which it replaces: All lie between 400 and 500 kJ mol^{-1}. The determining factor making this reaction exothermic is the exceedingly large bond enthalpy of the C═O bond which at 804 kJ mol^{-1} is almost twice as great as for the other bonds involved in the reaction. Not only this reaction but *virtually all* reactions in which CO_2 with its two very strong C═O bonds is produced are exothermic.

In other cases the number of bonds broken or formed can be important. A nice example of this is the highly exothermic [$\Delta H^{\ominus}$(298 K) = -483.7 kJ mol^{-1}] reaction between hydrogen and oxygen to form water:

$$2H_2(g) + O_2(g) \rightarrow 2H_2O(g)$$

All three types of bonds involved have comparable bond enthalpies:

$$D_{\mathrm{H-H}} = 436\ \mathrm{kJ\ mol^{-1}} \qquad D_{\mathrm{O=O}} = 498\ \mathrm{kJ\ mol^{-1}} \qquad D_{\mathrm{O-H}} = 463\ \mathrm{kJ\ mol^{-1}}$$

but the reason the reaction is exothermic becomes obvious if we rewrite it to make the bonds visible:

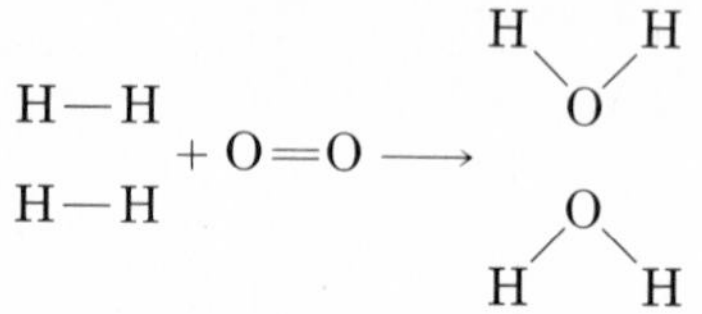

While three bonds must be broken (two H—H and one O═O bond), a total of four bonds are made (four O—H bonds). Since all the bonds are similar in strength, making more bonds than are broken means the release of energy. In mathematical terms

$$\begin{aligned}\Delta H^{\ominus} &= 2D_{\text{H-H}} + D_{\text{O=O}} - 4D_{\text{O-H}} \\ &= 2 \times \frac{436 \text{ kJ}}{\text{mol}} + 1 \times \frac{498 \text{ kJ}}{\text{mol}} - 4 \times \frac{463 \text{ kJ}}{\text{mol}} \\ &= -482 \text{ kJ mol}^{-1}\end{aligned}$$

In summary, there are two factors which determine whether a gaseous reaction will be exothermic or not: (1) the relative *strengths* of the bonds as measured by the bond enthalpies, and (2) the relative *number* of bonds broken and formed. An exothermic reaction corresponds to the formation of *more* bonds, or *stronger* bonds, or both.

Since the strength of chemical bonds is a factor in determining whether a reaction will release energy or not, it is obviously important to know which kinds of bonds will be strong and which weak, and we can make some empirical generalizations about the magnitudes of bond enthalpies. The first and most obvious of these is that triple bonds are stronger than double bonds which in turn are stronger than single bonds. As can be seen from Table 15.4, triple bonds have bond enthalpies in the range of 800 to 1000 kJ mol^{-1}. Double bonds range between 400 and 800 kJ mol^{-1}, and single bonds are in the range of 150 to 500 kJ mol^{-1}.

A second generalization is that the strengths of bonds usually increase with polarity. The bond enthalpies of the hydrogen halides, for instance, increase in the order HI $<$ HBr $<$ HCl $<$ HF, and a similar order can be noted for bonds between carbon and the halogens. There are exceptions to this rule, though. One would expect the C—N bond to be intermediate in strength between the C—C bond and the C—O bond. As the table shows, it is actually weaker than either.

A third factor affecting the strength of bonds is the size of the atoms. For the most part smaller atoms form stronger bonds. The smallest atom, hydrogen, forms four of the five strongest single bonds in Table 15.4. This is not entirely a matter of size since hydrogen is also the most electropositive element featured. Difference in electronegativity, however, does not explain why the H—H bond enthalpy is so large. If we look at the halogens (VIIA elements), we find that the bond enthalpies of I—I, Br—Br, and Cl—Cl increase as expected with decreasing size. The F—F bond is an exception, though: $D_{\text{F-F}}$ (159 kJ mol^{-1}) is significantly *smaller* than $D_{\text{Cl-Cl}}$ (242 kJ mol^{-1}). Other notable exceptions to this rule are the N—N and O—O bonds. The occurrence of these weak single bonds is of considerable importance to the chemistry of compounds which contain them.

The value of a particular bond enthalpy in a given molecule can sometimes be very informative about the nature of the bonding in the molecule. This is especially true of molecules in which resonance is a possibility, as the following example shows.

EXAMPLE 15.8 When benzene is burned in oxygen according to the equation

$$C_6H_6(g) + \tfrac{15}{2}\,O_2(g) \rightarrow 6CO_2(g) + 3H_2O(g)$$

calorimetric measurements give the value of $\Delta H^\ominus(298\text{ K})$ as -3169 kJ mol^{-1} benzene. Use this information together with Table 15.4 to calculate the mean bond enthalpy for the carbon-carbon bond in benzene.

Solution Indicating the required bond enthalpy by the symbol D_{CC} and carefully counting the bonds broken and bonds formed, we have

$$\Delta H^\ominus = 6D_{\text{C—H}} + 6D_{\text{CC}} + \tfrac{15}{2}D_{\text{O=O}} - 12D_{\text{C=O}} - 6D_{\text{O—H}}$$

or $$-3169\,\frac{\text{kJ}}{\text{mol}} = 6\left(413\,\frac{\text{kJ}}{\text{mol}}\right) + 6D_{\text{CC}} + \frac{15}{2}\left(498\,\frac{\text{kJ}}{\text{mol}}\right) - 12\left(804\,\frac{\text{kJ}}{\text{mol}}\right)$$

$$- 6\left(463\,\frac{\text{kJ}}{\text{mol}}\right) = -6213\,\frac{\text{kJ}}{\text{mol}} + 6D_{\text{CC}}$$

Thus $$6D_{\text{CC}} = (-3169 + 6213)\,\frac{\text{kJ}}{\text{mol}} = 3044\,\frac{\text{kJ}}{\text{mol}}$$

or $$D_{\text{CC}} = 507\text{ kJ mol}^{-1}$$

As expected for a resonance structure the bond-enthalpy value is intermediate between that given in Table 15.4 for a C—C single bond (348 kJ mol^{-1}) and that given for a C═C double bond (614 kJ mol^{-1}).

15.5 FOSSIL FUELS AND THE ENERGY CRISIS

A chemical fuel is any substance which will react exothermically with atmospheric oxygen, is available at reasonable cost and quantity, and produces environmentally acceptable reaction products. During the past century the most important sources of heat energy in the United States and other industrialized countries have been the **fossil fuels:** coal, petroleum, and natural gas. In 1971, for example, the United States obtained 43.5 percent of its energy from oil, 34.7 percent from natural gas, and 19.7 percent from coal. Only 1.5 percent was obtained from hydroelectric power and 0.6 percent from nuclear power. Other industrialized countries also obtain 95 percent or more of their energy from fossil fuels.

Coal, petroleum, and natural gas consist primarily of carbon and hydrogen (Sec. 8.2), and so it is not hard to see why they make excellent fuels. When they burn in air, the principal products are water and carbon dioxide, compounds which contain the strongest double bond ($D_{C=O}$ = 804 kJ mol^{-1}) and the third-strongest single bond (D_{O-H} = 463 kJ mol^{-1}) in Table 15.4. Thus more energy is liberated by bond formation than is needed to break the weaker C—C and C—H bonds in the fuel.

EXAMPLE 15.9 Use the bond enthalpies given in Table 15.4 to estimate the enthalpy change when 1 mol heptane, C_7H_{16}, is burned completely in oxygen:

$$C_7H_{16}(g) + 11O_2 \rightarrow 7CO_2 + 8H_2O$$

Solution Remembering that the projection formula for heptane is

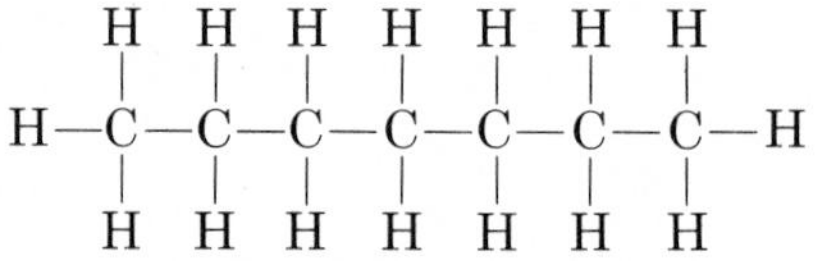

we can make up the following list of bonds broken and formed:

Bonds Broken		Bonds Formed	
6 C—C	2 088 kJ mol^{-1}	14 C═O	−11 256 kJ mol^{-1}
16 C—H	6 608 kJ mol^{-1}	16 H—O	− 7 408 kJ mol^{-1}
11 O═O	5 478 kJ mol^{-1}		
Total	14 174 kJ mol^{-1}	Total	−18 664 kJ mol^{-1}

Thus

$$\Delta H = 14\ 174 \text{ kJ mol}^{-1} - 18\ 664 \text{ kJ mol}^{-1}$$
$$= -4490 \text{ kJ mol}^{-1}$$

Apart from the hydrocarbon compounds in fossil fuels, there are few substances which fulfill the criteria for a good fuel. One example is hydrogen gas:

$$2H_2(g) + O_2(g) \rightarrow 2H_2O(l) \qquad \Delta H^\ominus(298 \text{ K}) = -571.7 \text{ kJ mol}^{-1}$$

Hydrogen does not occur as the element at the surface of the earth, however, so it must be manufactured. Right now much of it is made as a by-product of petroleum refining, and so hydrogen will certainly not be an immediate panacea for our current petroleum shortage. Eventually, though, it may be possible to generate hydrogen economically by electrolysis of water with current provided by nuclear power plants, and so this fuel does merit consideration.

Photosynthesis

The adjective *fossil* describes the fossil fuels very aptly, because all of them are derived from the remains of plants or animals which lived on earth millions of years ago. Coal, for example, began as plant matter in prehistoric swamps, where it was able to decompose in the absence of air. Present-day peat bogs are examples of this first stage in coal formation, and in countries such as Ireland dried peat is an important fuel. Over long periods of time, at high temperatures and pressures under the earth's surface, peat can be transformed into lignite, a brown, soft form of coal. Continued action of geological forces converts lignite into bituminous, or soft coal, and eventually into anthracite, or hard coal. When burned, these latter two types of coal release considerably more heat per unit mass than do lignite or peat.

A crucial point to realize about fossil fuels is that the energy we release by burning them came originally from the sun. The plants from which the fuels were derived grew as a result of **photosynthesis,** the combination of carbon dioxide and water under the influence of sunlight to form organic compounds whose empirical formula is approximately CH_2O:

$$CO_2(g) + H_2O(l) \rightarrow [CH_2O](s) + O_2(g) \qquad \Delta H \approx 469 \text{ kJ mol}^{-1} \qquad (15.20)$$

Since a number of different substances are formed by photosynthesis, the empirical formula $[CH_2O]$ and the ΔH are only approximate.

Photosynthesis is endothermic, and the necessary energy is supplied by the absorption of solar radiant energy. This energy can be released by carrying out the reverse of Eq. (15.20), a process which is exothermic. When we burn paper, wood, or dried leaves, the heat given off is really a stored form of sunlight. Plants and animals obtain the energy they need to grow or move about from the oxidation of substances produced by photosynthesis. This oxidation process is called **respiration.**

After millions of years of geological change, the fossil fuels are significantly different in chemical structure from newly photosynthesized plant or animal material. The changes which occur can be approximated by the equation for formation of methane (natural gas):

$$2[CH_2O] \rightarrow CH_4(g) + CO_2(g) \qquad \Delta H \approx -47 \text{ kJ mol}^{-1}$$

This reaction is only slightly exothermic, and so very little of the energy captured from sunlight is lost. However, about half the carbon and all the oxygen are lost as carbon dioxide gas, and so a fossil fuel like methane can release more heat per carbon atom (and per gram) than can wood or other organic materials. This is why anthracite and bituminous coals are better fuels than the peat from which they are formed. The enthalpy changes which occur during photosynthesis, respiration, and formation and combustion of fossil fuels are summarized in Fig. 15.6.

The Energy Crisis

Although a great deal of plant material is produced by photosynthesis every year, most of it is converted back to CO_2 and H_2O by respiration or combustion processes in a very short time. Only about 0.04 percent manages to escape reoxidation so that it can eventually be transformed into fossil fuel. Thus the rate at which fossil fuel deposits are being replenished

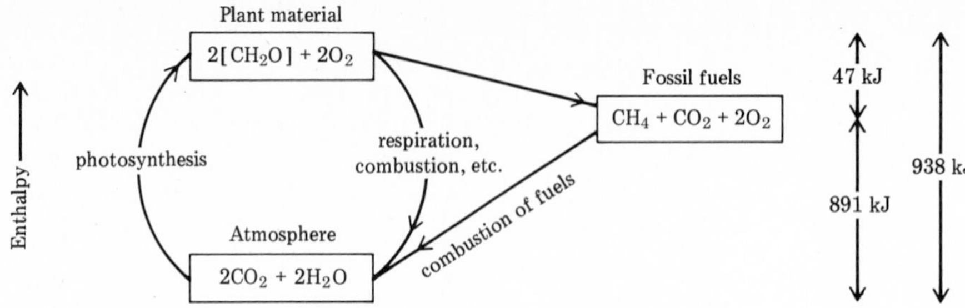

Figure 15.6 Diagram to show enthalpy relationships of the atmosphere, plant materials, and the fossil fuels.

naturally is so slow that these materials must be classed as **nonrenewable resources**—our current reserves have been accumulated over millions of years, and if we continue to use them at our present rate, we will eventually exhaust the supply. Current estimates indicate that the world's supply of coal ought to last for a century at least, but a recent report of the National Academy of Sciences[1] predicts that world supplies of petroleum and natural gas will be essentially depleted by the first quarter of the next century, if world trends of consumption continue.

For the United States the oil and gas situation is even more critical. This is shown graphically for petroleum production in the 48 contiguous United States in Fig. 15.7. The smooth red curve is a prediction first made in 1956 by geologist M. King Hubbert (1904 to). Using data on the number and depth of new wells needed to find an additional barrel of oil, Hubbert predicted that production would peak in 1970 and then gradually tail off as oil became more difficult and expensive to find. In 1956 Hubbert was accused of crying wolf, but as you can see from the dark line in Fig. 15.7, actual oil production has followed his prediction rather closely, except for minor fluctuations. Moreover, even an error as large as 25 percent in Hubbert's estimate of the total quantity of oil available will only change the date of the peak in the red curve by about 5 years. Thus even the Alaska pipeline or possible discoveries off the Atlantic coast can postpone the inevitable decline in United States oil production by a decade at most.

Although domestic production peaked in 1970, the consumption of gasoline and other petroleum products has continued to climb in the United States, and imports from oil-rich nations have grown with them. The situation with respect to natural gas is similar—United States production peaked in 1973—but this fuel is harder to transport, and imports have not grown as rapidly. Consequently shortages are beginning to be apparent. Diminishing domestic supplies of these fuels, on which we currently depend for about three-quarters of our energy, are a very serious problem, and one that the United States must face up to. Some of the options that are available to us in the immediate future are: (1) to continue to import oil and later

[1] "Mineral Resources and the Environment," Committee on Mineral Resources and the Environment, National Academy of Sciences, p. 8, Washington, D.C., 1975.

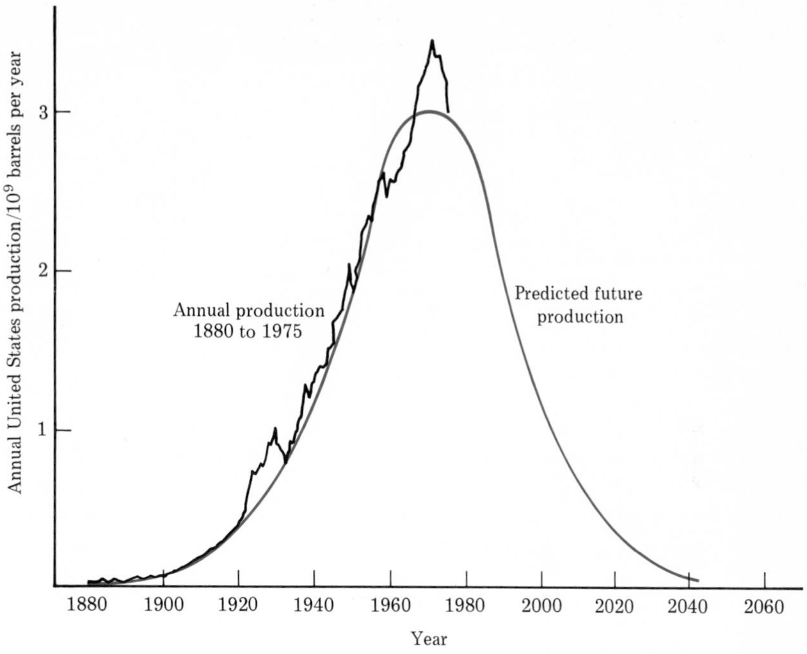

Figure 15.7 Oil production in the United States, (48 contiguous states only).

to import gas; (2) to conserve energy, particularly that derived from oil and gas; (3) to use coal to a much greater extent; and (4) to exploit fossil fuels such as oil shale or tar sands, heretofore regarded as too expensive. Over the longer term we must obviously find an alternative to fossil fuels, such as nuclear power (Chap. 19) or solar energy, or abandon industrial technology altogether.

Of the short-term options from the list above, conservation may well have the quickest and greatest impact. Sweden and Switzerland both have slightly greater gross national products per capita than the United States, but Sweden uses less than one-half and Switzerland less than one-third as much energy per person. Clearly there is considerable latitude for energy savings.

Greater use of coal will also be important because the United States has large reserves of this fuel. Coal is already well established in large stationary applications such as electric power plants. However, a solid like coal would not work for the smaller engines currently used in automobiles or aircraft because it would be quite difficult to handle. This problem can be bypassed by transforming coal into a liquid or gaseous hydrocarbon fuel. The chemical processes to do this are called **coal liquefaction** and **coal gasification,** respectively, and the technology for carrying them out is already fairly well developed. Gasification of coal corresponds to its conversion to methane, the necessary hydrogen being supplied from water in the form of steam. In the first step in this process, **volatilization,** the coal is heated at a high temperature and the volatile products, chiefly methane, are collected:

$$4CH(s) \rightarrow CH_4(g) + 3C(s) \qquad 500 \text{ to } 700°C \qquad (15.21)$$

(Since coal contains approximately equal numbers of C and H atoms, we have used its approximate empirical formula, CH.) The solid carbon is then reacted with steam in a three-stage process. First the **water-gas reaction** is carried out:

$$2C(s) + 2H_2O(g) \rightarrow 2CO(g) + 2H_2(g) \qquad 1300°C \qquad (15.22a)$$

The gas resulting from this stage is not rich enough in hydrogen; thus more steam is added, and the gases are passed over an iron oxide catalyst to produce more hydrogen at the expense of turning carbon monoxide into carbon dioxide:

$$H_2O(g) + CO(g) \xrightarrow[\text{catalyst}]{\text{iron oxide}} CO_2(g) + H_2(g) \qquad 400°C \qquad (15.22b)$$

This is called the **water-gas shift reaction.** Just enough steam is added at this stage to react with exactly half the carbon monoxide produced in Eq. (15.22*a*). The net result of the two reactions is then

$$2C(s) + 3H_2O(g) \rightarrow CO(g) + 3H_2(g) + CO_2(g) \qquad (15.22)$$

We now have a mixture of gases in which there are 3 mol H_2 for every 1 mol CO. This is exactly the right proportion for the third and final stage called **methanation:**

$$CO(g) + 3H_2(g) \xrightarrow[\text{catalyst}]{\text{Ni}} CH_4(g) + H_2O(g) \qquad (15.23)$$

The resulting methane is contaminated with steam and also with the carbon dioxide produced by Eq. (15.22*b*). These two impurities are easily removed, and methane of a quality only a little inferior to natural gas results. The final product is called (with a sublime disrespect for the meaning of words) **synthetic natural gas,** or SNG.

The liquefaction of coal follows a path similar to gasification, except that a different catalyst is used in the last stage so that long-chain hydrocarbons are produced, for example,

$$8CO + 17H_2 \rightarrow C_8H_{18} + 8H_2O$$

In practice a mixture of hydrocarbons of varying chain length is obtained, and this mixture must be further processed before gasoline is obtained.

The liquefaction and gasification of coal sound like very attractive propositions when described on paper, but in practice there are several difficulties. In the first place these fuels are more expensive than those they replace. As the price of imported oil increases, though, this difficulty is likely to disappear. A more important problem is the effect that large-scale gasification and liquefaction of coal will have on the environment. The gasification of coal is estimated to be about 65 percent efficient. The remaining 35 percent of the energy will be lost as heat energy and must go somewhere. Usually this "somewhere" is the nearest river or stream. Many gasification proposals involve coal from arid states such as North Dakota and Wyoming which will find it difficult to supply the required quantity of water. More drastic than this, though, will be the effect of strip mining. In order to replace oil and gas completely with coal-derived fuel, we would eventually have to mine coal at more than *5 times* the present rate, and it is doubtful whether this would be possible without strip mining large portions of Appalachia and the northern plains states. Whether it would be worth creating this devastation in order to maintain our present standard of living and to avoid an economic dependence on the oil-producing nations is, of course, not a question which chemists, as chemists, can answer, but it is one which you, as a citizen, may have to face.

SUMMARY

The first law of thermodynamics (the law of conservation of energy) states that when heat energy is supplied to a substance, that energy cannot disappear—it must still be present in the atoms or molecules of the substance. Some of the added energy makes the atoms or molecules move faster. This is called translational energy. In the case of molecules, which can rotate and vibrate, some of the added energy increases the rotational and vibrational energies of the molecules. Finally, any atom or molecule will have a certain electronic energy which depends on how close its electron clouds are to positively charged nuclei.

The total of translational, rotational, vibrational, and electronic energies is the internal energy U of an atom or molecule. When chemical reactions occur, the internal energy of the products is usually different from that of the reactants, and the difference appears as heat energy in the surroundings. If the reaction is carried out in a closed container (bomb calorimeter, for example), the increase in internal energy of the atoms and molecules is exactly equal to the heat energy absorbed from the surroundings. If the internal energy decreases, the energy of the surroundings must increase; i.e., heat energy is given off.

When a chemical reaction occurs at constant pressure, as in a coffee-cup calorimeter, there is a change in potential energy of the atmosphere (given by $P\,\Delta V$) as well as a change in heat energy of the surroundings. Because the heat energy absorbed can be measured more easily than $P\,\Delta V$, it is convenient to define the enthalpy as the internal energy plus the increased potential energy of the atmosphere. Thus the enthalpy increase equals the heat absorbed at constant pressure.

Enthalpy changes for a variety of reactions may be calculated from standard enthalpies of formation, as described in Chap. 3. They may also be estimated by summing the bond enthalpies of all bonds broken and subtracting the bond enthalpies of all bonds formed. Because the dissociation enthalpy for the same type of bond varies from one molecule to another, the second method is not as accurate as the first. However, it has the advantage that enthalpy changes for reactions of a particular compound can be estimated even if the compound has not yet been synthesized.

The enthalpy change for a reaction depends on the relative strengths of the bonds broken and formed and on the relative number of bonds broken and formed. A good fuel is a substance which can combine with oxygen from the air, forming more bonds and/or stronger bonds than were originally present. The fossil fuels, coal, petroleum, and natural gas consist mainly of carbon and hydrogen. When they burn in air, strong O—H and C═O bonds are formed in the resulting H_2O and CO_2 molecules. The supply of fossil fuels is limited, and they constitute a nonrenewable resource. Coal supplies ought to last another century or two, but petroleum and natural-gas supplies will be essentially depleted in half a century or less. During the next few decades it will be possible to gasify or liquefy coal to extend our supply of gaseous and liquid fuels. Conservation of these fuels can also make a major contribution toward continuing their use. Eventually, however, it will be necessary to develop nuclear or solar energy or some unknown source of energy if we are to continue our current energy-intensive way of life.

QUESTIONS AND PROBLEMS

15.1 Define and give the correct symbol of each of the following quantities:

a Heat capacity
b Specific heat capacity
c Molar heat capacity at constant volume
d Molar heat capacity at constant pressure

15.2 Explain in microscopic terms why the quantity of heat required to raise the temperature of a pure, monatomic ideal gas is proportional to the amount of substance. What happens to the molecules as the gas is heated?

15.3 Would your answer to Problem 15.2 be different if the gas were diatomic, like O_2? Would you expect the monatomic or the diatomic gas to have a larger C_V? Explain your choice in molecular terms.

15.4 A potential difference of 7.25 V is applied to an electrical heating coil which is immersed in 0.150 dm^3 of H_2O in a polystyrene cup. As a result, a current of 1.00 A passes for 5.0 min, producing a temperature rise of 2.25 K. Assuming no heat loss through the cup, calculate the heat capacity of the contents of the cup.

15.5 How long must a voltage of 1.75 V be applied to an electrical heating coil to produce 3.50 kJ of heat energy, assuming a current of 0.500 A?

15.6 A sample of N_2 gas is heated in a closed container using an electrical heating coil (10.00 V causing 0.575 A to flow for 2.00 min). The temperature of the gas rises by 9.65 K. Using data from Table 15.1 and assuming no heat losses, calculate the amount of N_2 in the container.

15.7 Which is of greater importance in determining how large C_V is for a gas: number of atoms in each gas molecule or molar mass of the gas? Use data from Table 15.1 to support your answer.

15.8 State the first law of thermodynamics in your own words. Write two mathematical equations that were used in this chapter and are based on the first law.

15.9 State the original definition of the calorie. Why is this unit very convenient for experiments in which water is heated? What is the definition of the calorie in terms of SI energy units?

15.10 Describe what happens to the molecules, atoms, and/or subatomic particles in $H_2O(g)$ when each of the following occurs:

a Increase in translational energy
b Increase in the electronic energy
c Decrease in rotational energy
d Increase in vibrational energy

15.11 Which of the following gases would have the largest C_V? Which would have the smallest?

a $Xe(g)$
b $CF_3CF_3(g)$
c $S_2Cl_2(g)$

15.12 For which of the following would you expect C_p (liquid) to be significantly larger than C_p (solid)? Explain your reasoning.

a mercury
b *n*-heptane
c water
d krypton
e aluminum bromide

15.13 Which is most important in determining ΔU for a reaction in the gas phase: translational, rotational, vibrational, or electronic energy? Cite data in support of your answer.

15.14 Criticize the following statement: The change in internal energy during a chemical reaction can be determined only if we have complete, quantitative, and accurate knowledge of structure and bonding for all reactant and product molecules.

15.15 Describe or define and give an example of the use of each of the following thermodynamic conventions or terms:

a System
b Initial and final states
c Delta convention
d State function
e Hess' law
f Standard pressure

15.16 For each process described below, calculate the change in the indicated quantity:

a Temperature increases from 5 to 35°C; ΔT = ______.

b 0.257 mol $CaCO_3(s)$ decomposes to $CaO(s)$ and $CO_2(g)$ at 1000 K and 1 atm; ΔV = ______. (Assume that the volumes of the solids are negligible and that $CO_2(g)$ behaves ideally.)

c 36.63 kJ of heat energy is released in a calorimeter whose heat capacity is 5.85 kJ K^{-1}; ΔT = ______.

d 1 mol of an ideal monatomic gas is heated from 25 to 75°C, at constant volume; ΔU = ______.

e 1 mol of an ideal monatomic gas is heated from 25 to 75°C, at constant pressure; ΔU = ______.

15.17 When 0.6576 g of phenol, C_6H_5OH, undergoes combustion in a bomb calorimeter (heat capacity = 7.150 kJ K^{-1}), a temperature rise of 2.912 K is observed. Calculate the change in internal energy, ΔU_m, for the reaction

$$C_6H_5OH(s) + 7O_2(g) \rightarrow 6CO_2(g) + 3H_2O(l)$$

15.18 When 0.0275 mol benzoic acid is ignited in a bomb calorimeter, the temperature rises from 24.75 to 26.43°C. Ignition of 7.32 g methanol, CH_3OH, in the same calorimeter increases the temperature from 24.47 to 25.76°C. Calculate ΔU_m for combustion of methanol:

$$CH_3OH(l) + \tfrac{3}{2}O_2(g) \rightarrow CO_2(g) + 2H_2O(g)$$

ΔU_m for combustion of benzoic acid is -3225 kJ mol^{-1}.

15.19 Using the ideal gas law, show that Eq. (15.14) can be written as

$$\Delta H = \Delta U + \Delta n(RT)$$

where Δn is the change in amount of gaseous substance when a reaction occurs.

15.20 Use the equation derived in Problem 15.19 to calculate ΔH for the reactions in Problems 15.17 and 15.18 at 25°C.

15.21 Name and define the physical quantity conventionally represented by each symbol:

a U *b* H *c* q_V *d* w_{exp} *e* q_p

15.22 When a sample of $Zn(s)$ is reacted with $HCl(aq)$, 750 cm^3 $H_2(g)$ is released. Calculate the work done by the $H_2(g)$ as it pushes back the atmosphere. Express your answer in (*a*) joules; (*b*) liter atmospheres; (*c*) calories.

15.23 When 2.178 g NH_4Cl was dissolved in 250 cm^3 H_2O in a coffee-cup calorimeter (Fig. 15.4), the temperature decreased from 23.70 to 22.47°C. When 6.00 V was applied to the heating coil, a current of 0.987 A flowed for 240 s, raising the temperature from 22.47 to 25.35°C. Calculate the heat capacity of the calorimeter and ΔH_m for the reaction

$$NH_4Cl(s) \rightarrow NH_4^+(aq) + Cl^-(aq)$$

15.24 Using standard enthalpies of formation from Table 3.1, calculate $\Delta H^\ominus$ for each of the following

a $CH_4(g) + 2O_2(g) \rightarrow CO_2(g) + 2H_2O(g)$
b $H_2(g) + Br_2(g) \rightarrow 2HBr(g)$
c $H_2S(g) + \tfrac{3}{2}O_2(g) \rightarrow SO_2(g) + H_2O(g)$
d $C_2H_2(g) + H_2(g) \rightarrow C_2H_4(g)$
e $CO(g) + H_2O(g) \rightarrow CO_2(g) + H_2(g)$

15.25 Calculate $\Delta U^\ominus$ for each reaction in Problem 15.24 at 25°C. What is the percent difference, $[(\Delta H^\ominus - \Delta U^\ominus)/\Delta H^\ominus] \times 100\%$, in each case?

15.26 Use bond enthalpies from Table 15.4 to estimate $\Delta H^\ominus$ for each reaction in Problem 15.24. (In part *c* assume both S=O bonds in SO_2 are double bonds.) What is the percent difference between the experimental $\Delta H^\ominus$ (calculated from $\Delta H_f^\ominus$) and your estimated value in each case?

15.27 Why does the $\Delta H^\ominus$ estimated using bond enthalpies differ from the experimental value calculated using $\Delta H_f^\ominus$ and Hess' law?

15.28 Distinguish clearly between the definitions of the terms *bond enthalpy* and *bond energy*. In Sec. 15.4 it was stated that for practical purposes the distinction between these terms is negligible. Do your results from Problems 15.25 and 15.26 confirm or deny that statement?

15.29 Without looking up bond enthalpies or standard enthalpies of formation, predict which of the following are exothermic reactions. Give reasons for your predictions.

a $CH_4(g) + 2F_2(g) \rightarrow CH_2F_2(g) + 2HF(g)$
b $2CO_2(g) + 4H_2O(g) \rightarrow 2CH_3OH(g) + 3O_2(g)$
c $N_2H_2(g) + 2H_2(g) \rightarrow 2NH_3(g)$

15.30 Confirm your predictions in Problem 15.29 by using bond enthalpies to estimate $\Delta H^\ominus$ for each reaction.

15.31 Without looking up bond enthalpies, predict which bond is stronger in each of the pairs below. Then use Table 15.4 to check your predictions.

a C—F or C—I *d* C≡O or N≡N
b C≡C or C≡N *e* O—O or S—S
c C═N or C—N

15.32 For the reaction

$$O_3(g) \rightarrow 3O(g) \qquad \Delta H^\ominus = 604.3 \text{ kJ mol}^{-1}$$

calculate the mean bond enthalpy for the oxygen-oxygen bond in ozone. Compare your result with values from Table 15.4. Explain the lack of agreement.

15.33 Write balanced equations for photosynthesis and for respiration. Which process is endothermic?

15.34 Which of the following would probably do least to alleviate the United States' energy crisis in the immediate future? In each case assume that your action saves fuel equivalent to 1 GJ of energy per year.

a Driving your car less.
b Reducing your use of electricity.
c Lowering the thermostat in a house with gas heat.

15.35 Name and write balanced equations for each of the four steps in coal gasification.

***15.36** For each process listed below, state whether heat energy is absorbed or given off and whether there is an increase or decrease in the temperature of the system. In each case try to account for the change in heat energy and/or temperature on the microscopic level. Try to state in your own words the difference between heat energy and temperature.

a 0.1 mol $N_2(g)$ is warmed from −10° to −9°C.

b 0.1 mol $N_2(l)$ is vaporized at its normal boiling point.

c 0.1 mol $N_2(g)$ expands from a volume of 2.44 liters and a pressure of 1.00 atm to a volume of 2.63 liters and a pressure of 0.90 atm in a perfectly insulated container.

***15.37** Evaluate each of the potential fuels listed below with respect to (*a*) enthalpy change per unit amount of substance burned in air; (*b*) enthalpy change per unit mass when burned in air; (*c*) enthalpy change per unit volume when burned in air. Based on the data given, which fuel would be most suitable for use in an airplane? Give reasons for your choice.

Name	Formula	$\Delta H_f^\ominus$/kJ mol^{-1}	Density†/g cm^{-3}
Anthracene‡	$C_{14}H_{10}(s)$	155.7	1.28
Benzene	$C_6H_6(l)$	−49.1	0.88
Glucose§	$C_6H_{12}O_6(s)$	−2987	1.56
Hydrogen	$H_2(g)$	0	8.2×10^{-5}
Methane	$CH_4(g)$	−74.8	6.5×10^{-4}
Methanol	$CH_3OH(l)$	−252.0	1.32
n-Octane	$C_8H_{18}(l)$	−283.4	1.40

† Densities given at 25°C and 1 atm.
‡ Anthracene is a component of coal.
§ Glucose is the principal product of photosynthesis.

***15.38** Foods consist primarily of fat, carbohydrate, protein, and water. The first three types of substance are oxidized by O_2 during respiration, and the quantities of heat obtained are: fat, 37 kJ g^{-1}; carbohydrate, 17 kJ g^{-1}; and protein, 17 kJ g^{-1}. Assuming the composition indicated, calculate how much heat would be given off when each portion of food listed below is metabolized. Express your answers in both joules and kilocalories (nutritional calories).

Portion	% Fat	% Carbohydrate	% Protein
1 pint (1.07 lb) milk	4.1	4.9	3.3
1 tablespoon (14 g) butter	78.6	Trace	Trace
3 oz hamburger	20.0	0	24.7
1 cup (180 g) dandelion greens	0.6	8.9	2.8
10 (20 g) potato chips	35.0	50.0	5.0

chapter sixteen

ENTROPY AND SPONTANEOUS REACTIONS

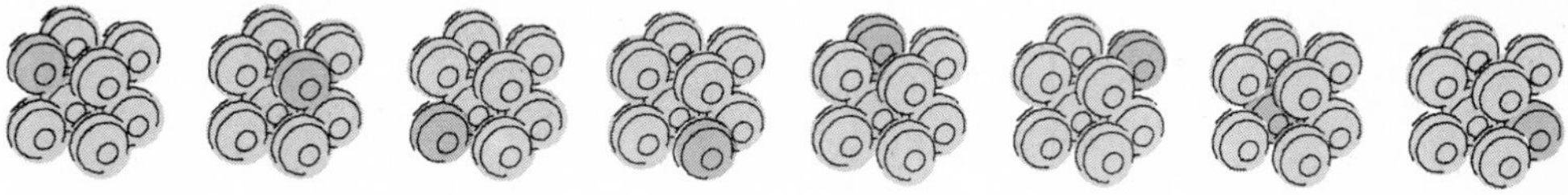

The experiences you have had in the chemical laboratory have probably already taught you that there is an uphill character to some reactions and a downhill character to others. A simple example is the combination of mercury with bromine which we considered in detail in Sec. 2.4 and which was shown in Plate 3:

$$Hg(l) + Br_2(l) \rightarrow HgBr_2(s)$$

This reaction occurs of its own accord, much as a ball rolls downhill. On the other hand, decomposition of $HgBr_2$ to the elements is an uphill process, and we must "push" this reaction to force it to occur. One way to do this is to dissolve $HgBr_2$ in water and pass an electrical current through the solution. Mercury will appear at one electrode and bromine at the other.

This chapter is designed to show you *why* some chemical reactions are downhill while others are uphill. In particular we will indicate what kinds of atomic and molecular processes will occur of their own accord. We will also develop criteria based on macroscopic measurements which will permit you to predict which reactions will occur and which must be pushed. This should help to sharpen the intuition you have already developed about what is likely to happen when you mix certain chemicals together.

16.1 SPONTANEOUS PROCESSES AND MOLECULAR PROBABILITY

Processes like formation of $HgBr_2$, which have a downhill character, are called **spontaneous processes.** When we attempt to reverse a spontaneous process, as in decomposition of $HgBr_2$, an uphill battle must invariably be fought and we are dealing with a **nonspontaneous process.** Figure 16.1 shows three examples of spontaneous processes. In each case our everyday experience tells us that the reverse of the process shown does not occur of its own accord. A gas in two connected flasks will not suddenly collect in one, a metal bar will not suddenly get hot at one end and cold at the other, nor will a book suddenly jump from the floor to a table, unless there is some outside intervention.

Atoms, Molecules, and Probability

On a microscopic level we can easily explain why some processes occur of their own accord while others do not. *A spontaneous process corresponds to rearrangement of atoms and molecules from a less-probable situation to a*

Figure 16.1 Three examples of a spontaneous process. (*a*) A gas expands to fill a vacuum. (*b*) Heat flows from a hotter region to a colder region until the temperature is uniform. (*c*) A book falls to the floor.

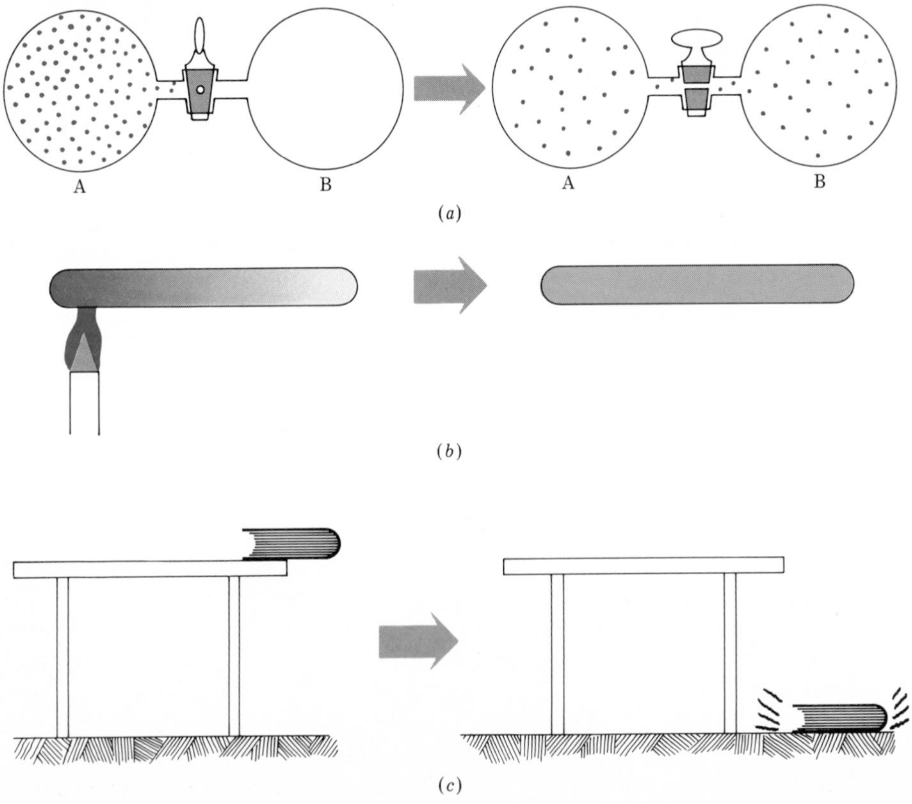

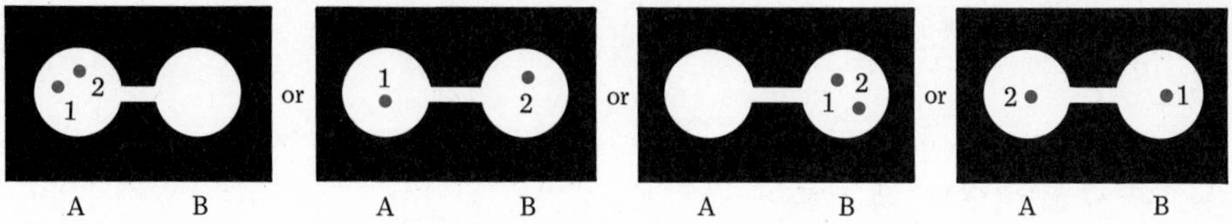

Figure 16.2 When two molecules are placed in two containers of equal volume, four different arrangements are possible.

more-probable one. A nonspontaneous process, by contrast, corresponds to movement from a probable situation to an improbable one.

An example of what probability has to do with spontaneity is provided by expansion of a gas into a vacuum. Let us calculate the probability that the process shown in Fig. 16.1*a* will reverse itself, that is, the probability that the gas molecules will all collect again in flask *A*. If we choose a particular molecule and label it number 1, we find that it is sometimes in flask *A* and sometimes in flask *B*. Since the molecule's motion is random and the two flasks contain the same volume, the molecule should spend half its time in each container. The probability of finding molecule 1 in container *A* is therefore 1/2.

Next let us consider the probability that two molecules, labeled 1 and 2, are both in flask *A*. Figure 16.2 shows the four possible ways these two molecules can be arranged in the two flasks. All four are equally likely, but only one has both molecules in flask *A*. Thus there is one chance in four that molecules 1 and 2 are both in flask *A*. This probability of 1/4 equals $1/2 \times 1/2$; i.e., it is the product of the probability that molecule 1 was in flask *A* times the probability that molecule 2 was in flask *A*. By a similar argument we can show that the probability that three given molecules are all in flask *A* is $1/2 \times 1/2 \times 1/2 = (1/2)^3$, and, in general, that the probability of all N gas molecules being in flask *A* at once is $(1/2)^N$.

If we had 1 mol gas in the flasks, there would be 6.022×10^{23} molecules. The probability p that all of them would be in flask *A* at the same time would be

$$p = \left(\frac{1}{2}\right)^{6.022 \times 10^{23}} = \frac{1}{2^{6.022 \times 10^{23}}} = \frac{1}{10^{1.813 \times 10^{23}}}$$

This unimaginably small number could be written as 0.000 000 . . . , where there would be 1.813×10^{23} zeros and then a 1. It would take over a thousand million million years to write that many zeros! Because there are so many molecules in a mole of gas (or any other macroscopic quantity), the probability that the spontaneous expansion will reverse itself is inconceivably small. The reversal is so improbable as to be *impossible* in any practical sense.

Similar remarks apply to the probability of reversing the other two spontaneous processes in Fig. 16.1. As shown in Fig. 16.3*a*, some of the atoms in a bar of metal at uniform temperature will be vibrating more than

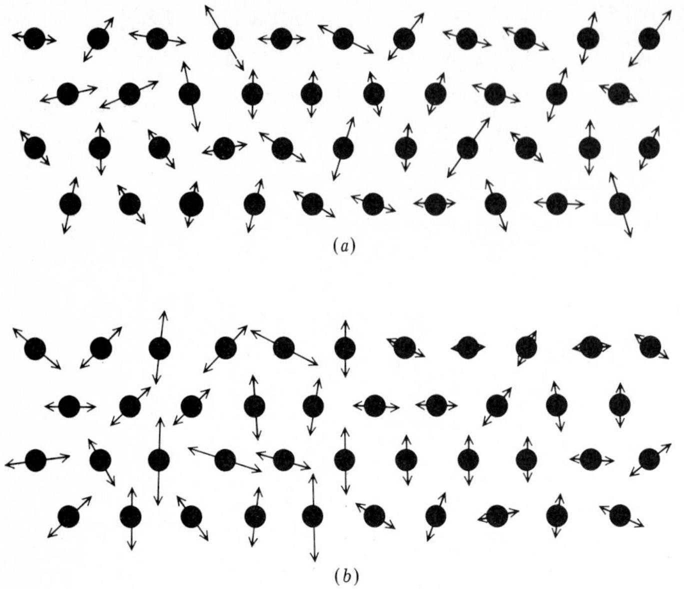

Figure 16.3 On the molecular level the reversal of a spontaneous heat flow is an extremely improbable event. (*a*) When a body is at uniform temperature, we expect some atoms to have more energy than others. These high-energy atoms will be distributed more or less evenly throughout the volume of the bar. (*b*) It is conceivable that by some set of freak collisions all the more-energetic atoms will be found on the left-hand side of the body, making that side warmer. Such an event is so improbable that we can regard it as, in effect, impossible.

others, but the unusually energetic atoms will be distributed fairly evenly throughout the bar. We will not suddenly find all the energetic metal atoms on the left end of the bar and all the weakly vibrating ones on the right, as in Fig. 16.3*b*. The possibility exists that a freak series of collisions between vibrating atoms might produce a high concentration of energetic atoms on the left, but such an occurrence is inconceivably improbable.

When a falling book hits the floor, its kinetic energy is converted to heat energy. The floor warms up slightly, and the molecules there start vibrating a little more energetically. For such a process to reverse itself spontaneously, all the floor molecules under the book would suddenly have to become more energetic and vibrate in unison in a vertical direction, flinging the book into the air. As in the previous two examples, this would require a freak series of molecular collisions which is so improbable as never to occur in the entire lifetime of the universe.

The three simple cases we have described show how spontaneous and nonspontaneous processes can be considered from a microscopic and statistical viewpoint. In any real sample of matter there are a great many molecules jostling each other about, exchanging energy, and sometimes exchanging atoms. This constant jostle is like shuffling a gigantic deck of cards. Because the numbers involved are so large, the laws of probability are inexorable. Some probabilities are large enough to be virtual certainties, while others are small enough to be unthinkable. Invariably the reversal of a spontaneous process turns out to involve movement from an almost certain situation to one which is unimaginably improbable. Conversely, a spontaneous process occurs when a sample of matter finds itself momentarily in a highly improbable situation. As fast as possible, it will adjust on the molecular level until maximum probability is attained.

Rates of Spontaneous Processes

The phrase *as fast as possible* points up a major difficulty in dealing with spontaneous processes. Some of them occur quite rapidly, but others are so slow as to be imperceptible. A rapid spontaneous process occurs when 2 mol H_2O is mixed with 2 mol "heavy water," D_2O, made from the isotope deuterium, ^2_1H, or D. The two species start to transfer protons and deuterons (D^+ ions) as soon as they are stirred together, and we rapidly obtain a mixture consisting of 2 mol H—O—D and 1 mol each of H—O—H and D—O—D. Assuming that deuterium atoms behave the same chemically as ordinary hydrogen atoms, this is what the laws of probability would predict. There are four equally likely possibilities for a randomly selected water molecule:

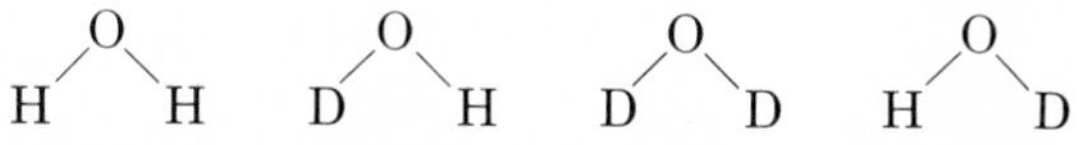

H_2O HDO D_2O HDO

Two of the four possibilities have the molecular formula HDO, and so the probability of finding an HDO molecule in our mixture is 1/2. Half the molecules (2 mol) will be HDO. Similarly 1/4 of the 4 mol water will be H_2O and 1/4 will be D_2O.

The shift from the improbable situation of 2 mol H_2O + 2 mol D_2O to the more probable 2 mol HDO + 1 mol H_2O + 1 mol D_2O occurs rapidly because of the ease with which protons and deuterons can transfer from one water molecule to another. When such a shuffling process is slow, however, the situation is quite different. For example, we would expect that mixing 2 mol H_2 with 2 mol D_2 would produce 2 mol HD and a mole each of H_2 and D_2. At room temperature, though, nothing happens, even over a period of days, because there is no easy way for H or D atoms to swap partners. Reshuffling requires breaking an H—H or a D—D bond, and this takes some 400 kJ mol^{-1}. The molecules are stuck in a situation of low probability because there is no pathway by which they can attain higher probability. If such a pathway is provided, by raising the temperature or adding a catalyst, the molecules start exchanging H and D and move toward the most probable situation.

The moral of this story is that saying a reaction is spontaneous is not the same as saying it *will* occur if the reactants are mixed. Rather, it means the reaction *can* occur but may be so slow that nothing seems to happen. In the case of a slow spontaneous reaction it is worthwhile to look for a catalyst, but if we know the reaction is nonspontaneous, there is no point in even mixing the reactants, let alone searching for a catalyst. A nonspontaneous reaction cannot occur of itself without outside intervention.

16.2 THERMODYNAMIC PROBABILITY AND ENTROPY

Thermodynamic Probability W

The previous section has shown that if we want to predict whether a chemical change is spontaneous or not, we must find some general way of determining whether the final state is more probable than the initial. This can be done using a number W, called the **thermodynamic probability.** W is defined as the number of alternative microscopic arrangements which correspond to the same macroscopic state. The significance of this definition becomes more apparent once we have considered a few examples.

Figure 16.4*a* illustrates a crystal consisting of only eight atoms at the absolute zero of temperature. Suppose that the temperature is raised slightly by supplying just enough energy to set one of the atoms in the crystal vibrating. There are eight possible ways of doing this, since we could supply the energy to any one of the eight atoms. All eight possibili-

Figure 16.4 The thermodynamic probability W of a crystal containing eight atoms at three different temperatures. (*a*) At 0 K there is only one way in which the crystal can be arranged, so that $W = 1$. (*b*) If enough energy is added to start just one of the atoms vibrating (color), there are eight different equally likely arrangements possible, and $W = 8$. (*c*) If the energy is doubled, two different atoms can vibrate simultaneously (light color) or a single atom can have all the energy (dark color). The number of equally likely arrangements is much larger than before; $W = 36$. (Computer-generated.) (*Copyright © 1976 by W. G. Davies and J. W. Moore.*)

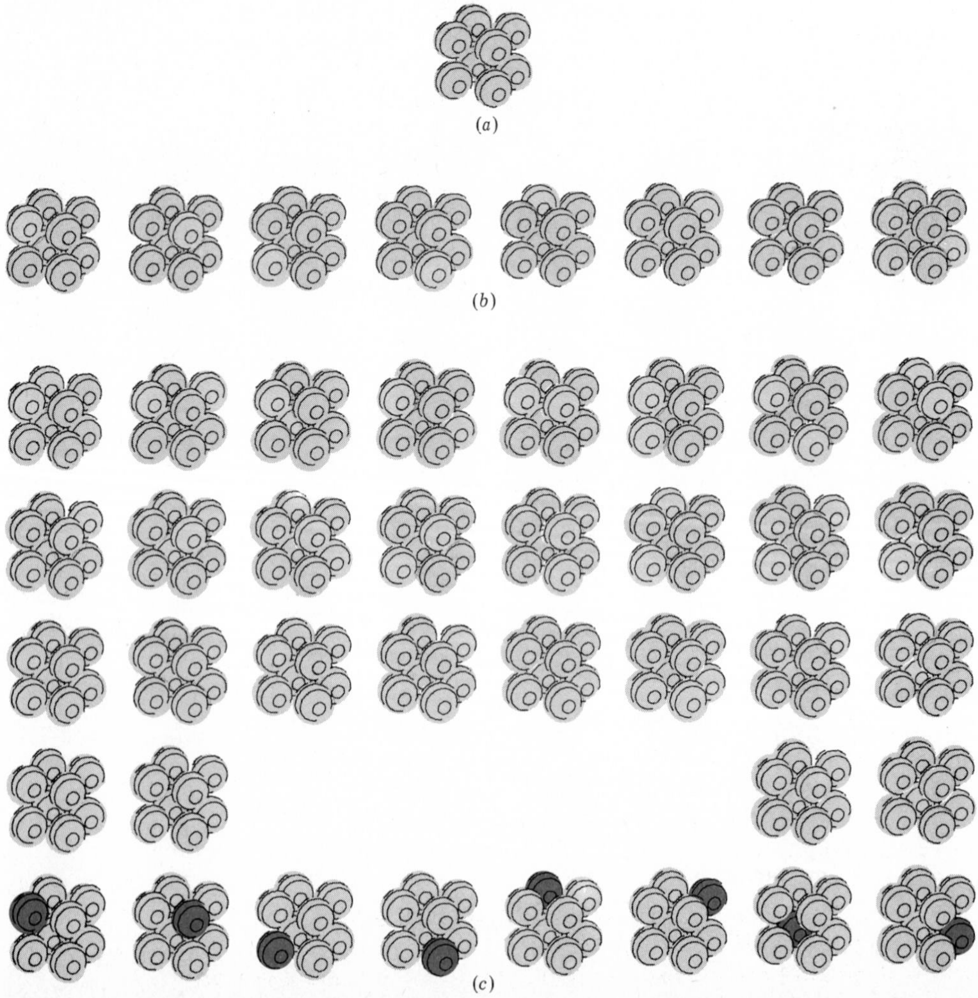

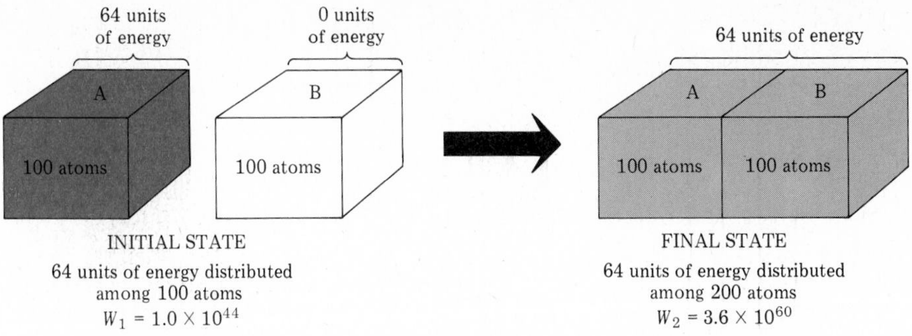

Figure 16.5 Heat flow and thermodynamic probability. When two crystals, one containing 64 units of vibrational energy and the other (at 0 K) containing none are brought into contact, the 64 units of energy will distribute themselves over the two crystals since there are many more ways of distributing 64 units among 200 atoms than there are of distributing 64 units over only 100 atoms.

ties are shown in Fig. 16.4*b*. Since all eight possibilities correspond to the crystal having the same temperature, we say that $W = 8$ for the crystal at this temperature. Also, we must realize that the crystal will not stay perpetually in any of these eight arrangements. Energy will constantly be transferred from one atom to the other, so that all the eight arrangements are *equally probable*.

Let us now supply a second quantity of energy exactly equal to the first, so that there is just enough to start two molecules vibrating. There are 36 different ways in which this energy can be assigned to the eight atoms (Fig. 16.4*c*). We say that $W = 36$ for the crystal at this second temperature. Because energy continually exchanges from one atom to another, there is an equal probability of finding the crystal in any of the 36 possible arrangements.

A third example of W is our eight-atom crystal at the absolute zero of temperature. Since there is no energy to be exchanged from atom to atom, only one arrangement is possible, and $W = 1$. This is true not only for this hypothetical crystal, but also presumably for a real crystal containing a large number of atoms, perfectly arranged, at absolute zero.

The thermodynamic probability W enables us to decide how much more probable certain situations are than others. Consider the flow of heat from crystal A to crystal B, as shown in Fig. 16.5. We shall assume that each crystal contains 100 atoms. Initially crystal B is at absolute zero. Crystal A is at a higher temperature and contains 64 units of energy—enough to set 64 of the atoms vibrating. If the two crystals are brought together, the molecules of A lose energy while those of B gain energy until the 64 units of energy are evenly distributed between both crystals.

In the initial state the 64 units of energy are distributed among 100 atoms. Calculations show that there are 1.0×10^{44} alternative ways of making this distribution. Thus W_1, the initial thermodynamic probability, is 1.0×10^{44}. The 100 atoms of crystal A continually exchange energy among themselves and transfer from one of these 1.0×10^{44} arrangements to another in rapid succession. At any instant there is an equal probability of finding the crystal in any of the 1.0×10^{44} arrangements.

When the two crystals are brought into contact, the energy can distribute itself over twice as many atoms. The number of possible arrangements rises enormously, and W_2, the thermodynamic probability for this new situation, is 3.6×10^{60}. In the constant reshuffle of energy among the 200 atoms, each of these 3.6×10^{60} arrangements will occur with equal probability. However, only 1.0×10^{44} of them correspond to all the energy being in crystal A. Therefore the probability of the heat flow reversing itself and all the energy returning to crystal A is

$$\frac{W_1}{W_2} = \frac{1.0 \times 10^{44}}{3.6 \times 10^{60}} = 2.8 \times 10^{-17}$$

In other words the ratio of W_1 to W_2 gives us the relative probability of finding the system in its initial rather than its final state.

This example shows how we can use W as a general criterion for deciding whether a reaction is spontaneous or not. Movement from a less probable to a more probable molecular situation corresponds to movement from a state in which W is smaller to a state where W is larger. In other words *W increases for a spontaneous change.* If we can find some way of calculating or measuring the initial and final values of W, the problem of deciding in advance whether a reaction will be spontaneous or not is solved. If W_2 is greater than W_1, then the reaction will occur of its own accord.

Although there is nothing wrong in principle with this approach to spontaneous processes, in practice it turns out to be very cumbersome. For real samples of matter (as opposed to 200 atoms in the example of Fig. 16.5) the values of W are on the order of $10^{10^{24}}$—so large that they are difficult to manipulate. The logarithm of W, however, is only on the order of 10^{24}, since $\log 10^x = x$. This is more manageable, and chemists and physicists use a quantity called the **entropy** which is proportional to the logarithm of W. This way of handling the extremely large thermodynamic probabilities encountered in real systems was first suggested in 1877 by the Austrian physicist Ludwig Boltzmann (1844 to 1906). The equation

$$S = k \ln W \tag{16.1}$$

is now engraved on Boltzmann's tomb. The proportionality constant k is called, appropriately enough, the **Boltzmann constant.** It corresponds to the gas constant R divided by the Avogadro constant N_A:

$$k = \frac{R}{N_A} \tag{16.2}$$

and we can regard it as the gas constant *per molecule* rather than per mole. In SI units, the Boltzmann constant k has the value 1.3805×10^{-23} J K^{-1}. The symbol ln in Eq. (16.1) indicates a *natural logarithm,* i.e., a logarithm taken to the base e. Since base 10 logarithms and base e logarithms are related by the formula

$$\ln x = 2.303 \log x$$

it is easy to convert from one to the other. Equation (16.1), expressed in base 10 logarithms, thus becomes

$$S = 2.303k \log W \tag{16.1a}$$

EXAMPLE 16.1 The thermodynamic probability W for 1 mol propane gas at 500 K and 101.3 kPa has the value $10^{10^{25}}$. Calculate the entropy of the gas under these conditions.

Solution Since

$$W = 10^{10^{25}}$$

$$\log W = 10^{25}$$

Thus

$$S = 2.303k \log W = 1.3805 \times 10^{-23}\ \text{J K}^{-1} \times 2.303 \times 10^{25}$$
$$= 318\ \text{J K}^{-1}$$

Note: The quantity 318 J K^{-1} is obviously much easier to handle than $10^{10^{25}}$. Note also that the dimensions of entropy are energy/temperature.

One of the properties of logarithms is that if we increase a number, we also increase the value of its logarithm. It follows therefore that if the thermodynamic probability W of a system increases, its entropy S must increase too. Further, since W always increases in a spontaneous change, it follows that S must also increase in such a change.

The statement that the *entropy increases when a spontaneous change occurs* is called the **second law of thermodynamics.** (The first law is the law of conservation of energy.) The second law, as it is usually called, is one of the most fundamental and most widely used of scientific laws. In this book we shall only be able to explore some of its chemical implications, but it is of importance also in the fields of physics, engineering, astronomy, and biology. Almost all environmental problems involve the second law. Whenever pollution increases, for instance, we can be sure that the entropy is increasing along with it.

The second law is often stated in terms of an entropy difference ΔS. If the entropy increases from an initial value of S_1 to a final value of S_2 as the result of a spontaneous change, then

$$\Delta S = S_2 - S_1 \tag{16.3}$$

Since S_2 is larger than S_1, we can write

$$\Delta S > 0 \tag{16.4}$$

Equation (16.4) tells us that for any spontaneous process, ΔS is greater than zero.

As an example of this relationship and of the possibility of calculating an entropy change, let us find ΔS for the case of 1 mol of gas expanding into a vacuum, as shown in Fig. 16.1*a*. We have already argued for this process that the final state is $10^{1.813 \times 10^{23}}$ times more probable than the initial state. This can only be because there are $10^{1.813 \times 10^{23}}$ times more ways of achieving the final state than the initial state. In other words

$$\frac{W_2}{W_1} = 10^{1.813 \times 10^{23}}$$

Taking logs, we have

$$\log \frac{W_2}{W_1} = 1.813 \times 10^{23}$$

Thus

$$\Delta S = S_2 - S_1 = 2.303 \times k \times \log W_2 - 2.303 \times k \times \log W_1$$

$$= 2.303 \times k \times \log \frac{W_2}{W_1}$$

$$= 2.303 \times 1.3805 \times 10^{-23} \text{ J K}^{-1} \times 1.813 \times 10^{23}$$

$$\Delta S = 5.76 \text{ J K}^{-1}$$

As entropy changes go, this increase in entropy is quite small. Nevertheless, it corresponds to a gargantuan change in probabilities.

16.3 GETTING ACQUAINTED WITH ENTROPY

You can experience directly the mass, volume, or temperature of a substance, but you cannot experience its entropy. Consequently you may have the feeling that entropy is somehow less real than other properties of matter. In point of fact, as we hope to show in this section, it is quite easy to predict whether the entropy under one set of circumstances will be larger than under another set of circumstances, and also to explain why. With a little practice in making such predictions in simple cases you will acquire an intuitive feel for entropy and it will lose its air of mystery.

The entropy of a substance depends on two things: first, the *state* of a substance—its temperature, pressure, and amount; and second, how the substance is *structured* at the molecular level. We will discuss these two kinds of factors in turn.

Dependence of *S* on *T*, *P*, and *n*

Temperature As we saw in the last section, there should be only one way of arranging the energy in a perfect crystal at 0 K. If $W = 1$, then $S = k \ln W = 0$; so that the *entropy should be zero at the absolute zero of temperature.* This rule, known as the **third law of thermodynamics,** is obeyed by all solids unless some randomness of arrangement is accidentally "frozen" into the crystal. As energy is fed into the crystal with increasing temperature, we find that more and more alternative ways of dividing the energy between the atoms become possible. W increases, and so does S. Without exception the entropy of any pure substance *always increases with temperature.*

Volume and Pressure We argued earlier that when a gas doubles its volume, the number of ways in which the gas molecules can distribute themselves in space is enormously increased and the entropy increases by 5.76 J K^{-1}. More generally the entropy of a gas always *increases with* *increasing volume* and *decreases with increasing pressure.* In the case of solids and liquids the volume changes very little with the pressure and so the entropy also changes very little.

Amount of Substance One of the main reasons why the entropy is such a convenient quantity to use is that its magnitude is *proportional to the amount of substance.* Thus the entropy of 2 mol of a given substance is twice as large as the entropy of 1 mol. Properties which behave in this way are said to be **extensive properties.** The mass, the volume, and the enthalpy are also extensive properties, but the temperature, pressure, and thermodynamic probability are not.

Standard Molar Entropies

Because the entropy of a substance depends on the amount of substance, the pressure, and the temperature, it is convenient to describe the entropy of a substance in terms of $S_m^\ominus$, its **standard molar entropy,** i.e., as the entropy of 1 mol of substance at the standard pressure of 1 atm (101.3 kPa) and a given temperature. Values of the standard molar entropies of various substances at 298 K (25°C) are given in Table 16.1. A table like this can be used in much the same way as a table of standard enthalpies of formation in order to find the entropy change $\Delta S_m^\ominus$ for a reaction occurring at standard pressure and at 298 K.

Dependence of *S* on Molecular Structure

Since they all refer to the same temperature and pressure, and also to 1 mol of substance, any differences in the entropy values listed in Table 16.1 must be due to *differences in the molecular structure* of the various substances listed. There are two aspects of the molecular structure of a substance which affect the value of its entropy: (1) The degree to which the movement of the atoms and molecules in the structure is restricted—the less restricted this movement, the greater the entropy. (2) The mass of the atoms and molecules which are moving—the greater the mass, the larger the entropy. We will consider each of these factors in turn.

Restrictions on the Movement of Atoms and Molecules We have already encountered two examples of how the removal of a restriction on the motion of molecules allows an increase in the number of alternative ways in which the molecules can arrange themselves in space. The first of these is the case of the ring and chain forms of 1,4-butanediol discussed in Chap. 13 and illustrated in Figs. 13.5 and 13.6. When the two ends of this molecule hydrogen bond to form a ring, the result is a fairly rigid structure. However, if the restricting influence of the hydrogen bond is removed, the more flexible chain form of the molecule is capable of many more alternative configurations. As a result, we find that *W* for a mole of molecules in the chain form is very much larger than *W* for a mole of molecules in the ring form. The chain form turns out to have a molar entropy which is 41 J K^{-1} mol^{-1} higher than that for the ring form.

TABLE 16.1 The Standard Molar Entropies of Selected Substances at 298.15 K (25°C).

Solids

Substance	$S^{\ominus}_{m}$/ J K^{-1} mol^{-1}
C	2.4
Si	18.8
Ge	31.1
Sn	44.1
Pb	64.8
Li	28.0
Na	51.2
K	64.2
Rb	69.5
Cs	82.8
NaF	51.5
MgO	26.8
AlN	20.2
NaCl	72.8
KCl	82.8
Mg	32.6
Ag	42.7
I_2	116.1
MgH_2	31.1
AgN_3	99.2

Liquids

Substance	$S^{\ominus}_{m}$/ J K^{-1} mol^{-1}
Hg	76.0
Br_2	152.2
H_2O	69.9
H_2O_2	109.6
CH_3OH	126.8
C_2H_5OH	160.7
C_6H_6	124.5
BCl_3	206.3

Monatomic Gases

Substance	$S^{\ominus}_{m}$/ J K^{-1} mol^{-1}
He	126.0
Ne	146.2
Ar	154.7
Kr	164.0
Xe	169.6

Diatomic Gases

Substance	$S^{\ominus}_{m}$/ J K^{-1} mol^{-1}
H_2	130.6
D_2	144.9
HCl	186.8
HBr	198.6
HI	206.5
N_2	191.5
O_2	205.0
F_2	202.7
Cl_2	223.0
Br_2	245.3
I_2	260.6
CO	197.6

Triatomic Gases

Substance	$S^{\ominus}_{m}$/ J K^{-1} mol^{-1}
H_2O	188.7
NO_2	239.9
H_2S	205.7
CO_2	213.6
SO_2	248.1
N_2O	219.7
O_3	238.9

Polyatomic Gases

Substance	$S^{\ominus}_{m}$/ J K^{-1} mol^{-1}
CH_4	187.9
C_2H_6	229.5
C_3H_8	269.9
C_4H_{10}	310.1
C_5H_{12}	348.9
C_2H_4	219.5
N_2O_4	304.2
B_2H_6	232.0
BF_3	254.0
NH_3	192.2

A second example of the effect of relaxing a restriction on molecular motion is the expansion of a gas into a vacuum. When the molecules have more freedom of movement, many more alternative arrangements are possible, and hence W is larger. As we have already calculated, doubling the volume of a gas increases its molar entropy by 5.76 J K^{-1} mol^{-1}.

Mass of the Atoms and Molecules In Sec. 5.2 we showed that when quantum mechanics is applied, the energies of a moving particle are confined to certain specific values and not those in between. The allowed energies, or **energy levels,** for a particle in a box were given by Eq. (5.4):

$$E_k = n^2 \frac{h^2}{8md^2} \qquad (5.4)$$

From this equation we can see that the larger the mass m of a particle, the smaller the value of its energy E_k for a given value of the quantum number n. Furthermore, the larger the value of m, the smaller the *difference* between E_{k_2} (for $n = 2$) and E_{k_1} (for $n = 1$). No matter what kind of energy is involved—rotational, vibrational, or translational—this is always found to be true. *The greater the mass of a particle, the closer together its energy levels.*

The effect of closeness of energy levels on the entropy is shown in Fig. 16.6. The more closely spaced the levels, the more different ways the same quantity of energy can be distributed among them. This applies in general for any number of particles and any quantity of energy. Therefore, *the heavier the molecules of a substance, the larger its molar entropy.* This effect is quite obvious among the noble gases in Table 16.1—their molar entropies increase steadily with molar (and hence molecular) mass.

Some Trends in Entropy Values

A close inspection of the entropy values in Table 16.1 reveals several trends which can be explained in terms of the two factors just described.

Solids, Liquids, and Gases Perhaps the most obvious feature of Table 16.1 is a general increase in the molar entropy as we move from solids to liq-

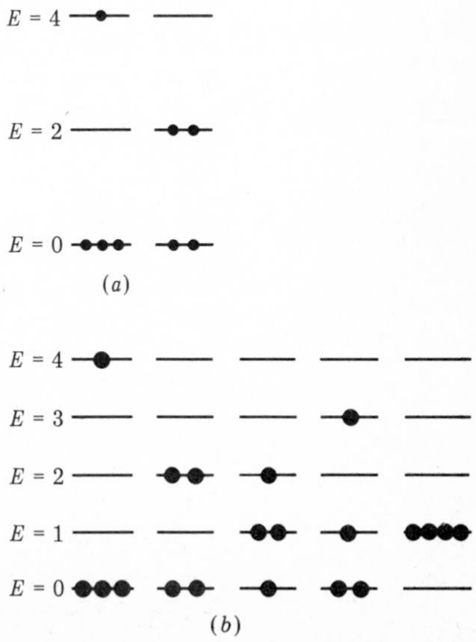

Figure 16.6 Effect of molecular mass on entropy. (*a*) The energy levels available to light molecules are widely spaced. If the spacing is two units, and if four units of energy are supplied to four molecules, only two arrangements are possible: either one molecule has all the energy, or two molecules have two units each. (*b*) For heavier molecules the energy levels are closer together. If the spacing is one unit, and four units of energy are supplied, three more ways of distributing the energy are possible. Therefore the entropy is greater.

uids to gases. In a solid, the molecules are only capable of restricted vibrations around a fixed point, but when a solid melts, the molecules, though still hampered by their mutual attraction, are much freer to move around. Thus when a solid melts, the molar entropy of the substance increases. When a liquid vaporizes, the restrictions on the molecules' ability to move around are relaxed almost completely and a further and larger increase in the entropy occurs. When 1 mol of ice melts, for example, its entropy increases by 22 J K^{-1}, while on boiling the entropy increase is 110 J K^{-1}.

Molecular Complexity A second clear trend in Table 16.1 is the higher molar entropy of substances with more complex molecules. To some extent this is due to the mass since on the whole more complex molecules are heavier than simpler ones. However, we still find an increase of entropy with complexity when we compare molecules of very similar masses:

Substance	$Ar(g)$	$F_2(g)$	$CO_2(g)$	$C_3H_8(g)$
$S_m^\ominus$/J K^{-1} mol^{-1}	155	202.7	213.6	269.9
Molar mass/g mol^{-1}	40	38	44	44
Number of atoms	1	2	3	11

The more atoms there are in a molecule, the more ways the molecule can change its shape by vibrating. In consequence there are more ways in which the energy can be distributed among the molecules.

Strength of Bonding Another trend in entropy, most noticeable in the case of solids, is the decrease in the entropy as the forces between the atoms, molecules, or ions increases. A good example is the three solid compounds

Substance	$NaF(s)$	$MgO(s)$	$AlN(s)$
$S_m^\ominus$(298 K)/J K^{-1} mol^{-1}	51.5	26.8	20.2
Molar mass/g mol^{-1}	42.0	40.3	41.0

which are isoelectronic with sodium fluoride. Since there is very little difference in the molar masses, the entropy decrease can only be attributed to the increase in the coulombic attraction between the ions as we move from the singly charged ions Na^+ and F^-, through the doubly charged ions Mg^{2+} and O^{2-}, to the triply charged ions Al^{3+} and N^{3-}. (While it is true that there is a fair degree of covalent character to the bonding in AlN, the effect of this will be to *increase* the strength of the bonding.)

EXAMPLE 16.2 From each of the following pairs of compounds choose the one with the higher standard molar entropy at 25°C. Give brief reasons for your choice.

a) $HBr(g)$, $HCl(g)$ **c)** $ND_3(g)$, $Ne(g)$ **e)** $C_2H_6(g)$, $C_2H_4(g)$

b) $Cs(s)$, $Cs(l)$ **d)** $KCl(s)$, $CaS(s)$

Solution

a) HBr and HCl are very similar except for their mass. HBr will have a higher entropy because of its greater mass.

b) At the same temperature, the liquid form of a substance always has a higher entropy than the solid.

c) ND_3(D = deuterium) and Ne have almost identical molar masses (20 g mol^{-1}). However, since ND_3 is more complex, it can vibrate and rotate while Ne cannot. ND_3 will have the higher entropy.

d) KCl and CaS are isoelectronic. Because both anion and cation are doubly charged in CaS, the ions are more tightly held to each other and can vibrate less readily. Thus KCl must have the higher entropy.

e) On all counts C_2H_6 must have a higher entropy than C_2H_4. C_2H_6 is heavier and more complex than C_2H_4. In addition there is free rotation about the C—C bond in C_2H_6 but hindered rotation about the C═C bond in C_2H_4.

Entropy Changes in Gaseous Reactions

Knowing what factors affect the magnitude of the entropy often enables us to predict whether the entropy of the products will be greater or less than that of the reactants in a given chemical reaction. This is particularly true for gaseous reactions. In a dissociation reaction like

$$N_2O_4(g) \rightarrow NO_2 + NO_2 \qquad \Delta S_m^\ominus(298\ \text{K}) = +176\ \text{J K}^{-1}\ \text{mol}^{-1}$$

for instance, it is easy to see that ΔS must be positive. The two halves of the N_2O_4 molecule are forced to move around together before dissociation, but they can move around independently as NO_2 molecules once dissociation has occurred.

A similar argument applies to reactions like

$$2O_3 \rightarrow 3O_2 \qquad \Delta S_m^\ominus(298\ \text{K}) = +137\ \text{J K}^{-1}\ \text{mol}^{-1}$$

In the form of O_3, O atoms are constrained to move around in groups of three, but in the form of O_2, only two atoms need move around together, a lesser restriction. Accordingly we expect ΔS to be positive for this reaction.

A further extension of this argument leads us to the general conclusion that in any reaction involving gases if the amount of substance in the gaseous phase increases, ΔS will be positive, while if it decreases, so will ΔS. For example, in the reaction

$$2CO(g) + O_2(g) \rightarrow 2CO_2(g)$$

The amount of gas decreases from 3 to 2 mol (i.e., $\Delta n = -1$ mol). The entropy change should thus be negative for this reaction. From Table 16.1 we can readily find that $\Delta S_m^\ominus(298\ \text{K})$ has the value -173 J K^{-1} mol^{-1}.

Entropy, Randomness, and Disorder

A very useful, though somewhat rough, description of the entropy of a substance is as a measure of the *randomness or disorder* of the atoms and molecules which constitute that substance. In these terms the second law of thermodynamics is seen as a tendency for the disorder of the universe to increase. This way of looking at entropy is entirely compatible with the approach presented above. A situation which we intuitively recognize as being orderly is also one which can only be achieved in a limited number of ways. By contrast, situations which we recognize as disordered, random, or chaotic, can be achieved in a whole variety of ways. In other words, W, and hence S, is small for an ordered situation but large for a disordered situation.

There are limits to the lengths one can take this order-disorder approach to entropy, though. It does not lend itself to a quantitative treatment, and it is also difficult to explain some things like the effect of mass in these terms. There is nothing in our intuition about order, for example, which suggests that 1 mol Xe gas is more disordered than 1 mol He gas, even though its entropy is in fact larger.

Measuring the Entropy

Up to this point in the chapter we have often quoted values of the entropy without giving any indication of how such values may be obtained. Alas, there is no convenient black box labeled *entropy meter* into which we can put a substance and read off its entropy value on a dial. Determining the entropy turns out to be both difficult and laborious. In the case of a simple gas, if we know enough about its molecular structure and enough quantum mechanics, we can actually *calculate* its entropy. For most substances, though, we are forced to derive the entropy from a series of calorimetric measurements, most of them at very low temperatures.

This method for determining the entropy centers around a very simple relationship between q, the heat energy absorbed by a body, the temperature T at which this absorption takes place, and ΔS, the resultant increase in entropy:

$$\Delta S = \frac{q}{T} \qquad (16.5)$$

It is possible to derive this relationship from our original definition of entropy, namely, $S = k \ln W$, but the proof is beyond the level of this text.

It is easy to see how Eq. (16.5) can be used to measure the entropy. We start with our substance as close to the absolute zero of temperature as is technically feasible and heat it in many stages, measuring the heat absorbed at each stage, until we arrive at the desired temperature, say 298 K. The initial value of the entropy is zero, and we can calculate the entropy increase for each stage by means of Eq. (16.5), and so the sum of all these increases is the entropy value for 298 K. In the case of simple gases, values of entropy measured in this way agree very well with those calculated from a knowledge of molecular structure.

Equation (16.5) was discovered long before the statistical nature of entropy was realized. Scientists and engineers began to appreciate the importance of the quantity q/T very early in the nineteenth century because of its connection with the efficiency of steam engines. These arguments were developed by both Lord Kelvin in England and Rudolf Clausius (1822 to 1888) in Germany. It was Clausius who first formulated the second law in terms of the entropy S, but Clausius had only a vague idea that entropy was in any way connected with molecules or probability. The statistical nature of entropy was first suggested by Boltzmann in 1877 and then developed into an elegant system in 1902 by Josiah Willard Gibbs (1839 to 1903), one of the real giants among American scientists.

An important feature of Eq. (16.5) is the *inverse* relationship between the entropy increase and the temperature. A given quantity of heat energy produces a very *large* change of entropy when absorbed at a very *low* temperature but only a *small* change when absorbed at a *high* temperature.

EXAMPLE 16.3 Calculate the increase in entropy which a substance undergoes when it absorbs 1 kJ of heat energy at the following temperatures: (a) 3 K; (b) 300 K; (c) 3000 K.

Solution

a) At 3 K we have

$$\Delta S = \frac{1000 \text{ J}}{3 \text{ K}} = 333.3 \text{ J K}^{-1}$$

b) At 300 K, similarly,

$$\Delta S = \frac{1000 \text{ J}}{300 \text{ K}} = 3.33 \text{ J K}^{-1}$$

c) At 3000 K

$$\Delta S = \frac{1000 \text{ J}}{3000 \text{ K}} = 0.33 \text{ J K}^{-1}$$

An amusing analogy to this behavior can be drawn from everyday life. If a 10-year-old boy is allowed to play in his bedroom for half an hour, the increase in disorder is scarcely noticeable because the room is already disordered (i.e., at a higher "temperature"). By contrast, if the boy is let loose for half an hour in a neat and tidy living room (i.e., at a lower "temperature"), the effect is much more dramatic.

16.4 INCLUDING THE SURROUNDINGS

In order to determine whether a reaction is spontaneous or not, it is not sufficient just to determine ΔS_m, the entropy difference between products and reactants. As an example, let us take the reaction

$$2\text{Mg}(s) + \text{O}_2(g) \rightarrow 2\text{MgO}(s) \qquad 1 \text{ atm, } 298 \text{ K} \tag{16.6}$$

Since this reaction occurs at the standard pressure and at 298 K, we can find ΔS from the standard molar entropies given in Table 16.1

$$\begin{aligned}\Delta S &= \Delta S_m^\ominus(298 \text{ K}) = 2S_m^\ominus(\text{MgO}) - 2S_m^\ominus(\text{Mg}) - S_m^\ominus(\text{O}_2) \\ &= (2 \times 26.8 - 2 \times 32.6 - 205.0) \text{ J K}^{-1} \text{ mol}^{-1} \\ &= -216.6 \text{ J K}^{-1} \text{ mol}^{-1}\end{aligned}$$

This result would suggest that the reaction is not spontaneous, but in fact it is. Once ignited, a ribbon of magnesium metal burns freely in air to form solid magnesium oxide in the form of a white powder. The reaction is plainly spontaneous even though ΔS is negative. Why is this not a contradiction of the second law?

The answer is that we have failed to realize that the entropy change which the magnesium and oxygen atoms undergo as a result of the reaction is not the only entropy change which occurs. The oxidation of magnesium is a highly exothermic reaction, and the heat which is evolved flows into the surroundings, *increasing their entropy as well.* There are thus *two* entropy changes which we must take into account in deciding whether a reaction will be spontaneous or not: (1) the change in entropy of the *system* actually undergoing the chemical change, which we will indicate with the symbol ΔS_{sys}; and (2) the change in entropy of the *surroundings,* ΔS_{surr}, which occurs as the surroundings absorb the heat energy liberated by an exothermic reaction or supply the heat energy absorbed by an endothermic reaction.

Of these two changes, the first is readily obtained from tables of entropy values like Table 16.1. Thus, for the oxidation of magnesium, according to Eq. (16.6), ΔS_{sys} has the value already found, namely, -216.6 J K^{-1} mol^{-1}. The second entropy change, ΔS_{surr}, can also be derived from tables, as we shall now show.

When a chemical reaction occurs at atmospheric pressure and its surroundings are maintained at a constant temperature T, then the surroundings will absorb a quantity of heat, q_{surr} equal to the heat energy given off by the reaction.

$$q_{surr} = -\Delta H \tag{16.7}$$

(The negative sign before ΔH is needed because q_{surr} is positive if the surroundings *absorb* heat energy, but ΔH is negative if the system gives off heat energy for them to absorb.) If we now feed Eq. (16.7) into Eq. (16.5), we obtain an expression for the entropy change of the surroundings in terms of ΔH:

$$\Delta S_{surr} = \frac{q_{surr}}{T} = \frac{-\Delta H}{T} \tag{16.8}$$

Using this equation it is now possible to find the value of ΔS_{surr} from tables of standard enthalpies of formation. In the case of the oxidation of magnesium, for example, we easily find from Table 3.1 that the enthalpy change for

$$2\text{Mg}(s) + \text{O}_2(g) \rightarrow 2\text{MgO}(s) \qquad 1 \text{ atm, } 298 \text{ K}$$

is given by

$$\begin{aligned}\Delta H &= \Delta H_m^{\ominus}(298 \text{ K}) = 2 \times \Delta H_f^{\ominus}(\text{MgO}) = 2 \times -601.8 \text{ kJ mol}^{-1} \\ &= -1204 \text{ kJ mol}^{-1}\end{aligned}$$

Substituting this result into Eq. (16.8), we then find

$$\begin{aligned}\Delta S_{\text{surr}} &= \frac{-\Delta H}{T} = \frac{1204 \times 10^3 \text{ J mol}^{-1}}{298 \text{ K}} \\ &= 4040 \text{ J K}^{-1} \text{ mol}^{-1}\end{aligned}$$

If a reaction is spontaneous, then it is the *total* entropy change ΔS_{tot}, given by the *sum of* ΔS_{surr} and ΔS_{sys}, which must be positive in order to conform to the second law. In the oxidation of magnesium, for example, we find that the total entropy change is given by

$$\begin{aligned}\Delta S_{\text{tot}} &= \Delta S_{\text{surr}} + \Delta S_{\text{sys}} = (4040 - 216.6) \text{ J K}^{-1} \text{ mol}^{-1} \\ &= 3823 \text{ J K}^{-1} \text{ mol}^{-1}\end{aligned}$$

This is a positive quantity because the entropy increase in the surroundings is more than enough to offset the decrease in the system itself, and the second law is satisfied. In the general case the total entropy change is given by

$$\Delta S_{\text{tot}} = \Delta S_{\text{surr}} + \Delta S_{\text{sys}} = \frac{-\Delta H}{T} + \Delta S_{\text{sys}}$$

The second law requires that this sum must be positive; i.e.,

$$\frac{-\Delta H}{T} + \Delta S_{\text{sys}} > 0 \tag{16.9}$$

This simple inequality gives us what we have been looking for since the beginning of the chapter: a simple criterion for determining whether a reaction is spontaneous or not. Since both ΔH and ΔS can be obtained from tables, and T is presumably known, we are now able to predict in advance whether a reaction will be uphill or downhill.

EXAMPLE 16.4 Using Table 3.1 and Table 16.1 find $\Delta H_m^{\ominus}(298 \text{ K})$ and $\Delta S_m^{\ominus}(298 \text{ K})$ for the reaction

$$\text{N}_2(g) + 3\text{H}_2(g) \rightarrow 2\text{NH}_3(g) \qquad 1 \text{ atm}$$

Predict whether the reaction will be spontaneous or not at a temperature of (a) 298 K, and (b) 1000 K.

Solution We find from Table 3.1 that

$$\begin{aligned}\Delta H_m^\ominus(298\text{ K}) &= \Sigma\Delta H_f^\ominus(\text{products}) - \Sigma\Delta H_f^\ominus(\text{reactants})\\ &= 2\Delta H_f^\ominus(NH_3) - \Delta H_f^\ominus(N_2) - 3\Delta H_f^\ominus(H_2)\\ &= (-2\times 46.1 - 0.0 - 0.0)\text{ kJ mol}^{-1}\\ &= -92.2\text{ kJ mol}^{-1}\end{aligned}$$

and from Table 16.1

$$\begin{aligned}\Delta S_m^\ominus(298\text{ K}) &= \Sigma S_m^\ominus(\text{products}) - \Sigma S_m^\ominus(\text{reactants})\\ &= 2S_m^\ominus(NH_3) - S_m^\ominus(N_2) - 3S_m^\ominus(H_2)\\ &= (2\times 192.2 - 191.5 - 3\times 130.6)\text{ J K}^{-1}\text{ mol}^{-1}\\ &= -198.9\text{ J K}^{-1}\text{ mol}^{-1}\end{aligned}$$

a) At 298 K the total entropy change per mol N_2 is given by

$$\begin{aligned}\Delta S_{\text{tot}} &= \frac{-\Delta H_m^\ominus}{T} + \Delta S_{\text{sys}}^\ominus\\ &= \frac{92.2\times 10^3\text{ J mol}^{-1}}{298\text{ K}} - 198.9\text{ J K}^{-1}\text{ mol}^{-1}\\ &= (309.4 - 198.9)\text{ J K}^{-1}\text{ mol}^{-1}\\ &= 110.5\text{ J K}^{-1}\text{ mol}^{-1}\end{aligned}$$

Since the total entropy change is positive, the reaction is spontaneous.

b) Since tables are available only for 298 K, we must make the approximate assumption that neither ΔH nor ΔS varies greatly with temperature. Accordingly we assume

$$\Delta H_m^\ominus(1000\text{ K}) = \Delta H_m^\ominus(298\text{ K}) = -92.2\text{ kJ mol}^{-1}$$

and
$$\Delta S_m^\ominus(1000\text{ K}) = \Delta S_m^\ominus(298\text{ K}) = -198.9\text{ J K}^{-1}\text{ mol}^{-1}$$

Thus for 1 mol N_2 reacted,

$$\begin{aligned}\Delta S_{\text{tot}} &= \frac{-\Delta H_m^\ominus}{T} + \Delta S_{\text{sys}}^\ominus\\ &= \frac{92.2\times 10^3\text{ J mol}^{-1}}{1000\text{ K}} - 198.9\text{ J K}^{-1}\text{ mol}^{-1}\\ &= (92.2 - 198.9)\text{ J K}^{-1}\text{ mol}^{-1}\\ &= -106.7\text{ J K}^{-1}\text{ mol}^{-1}\end{aligned}$$

Since the total entropy change is negative at this high temperature, we conclude that N_2 and H_2 will not react to form NH_3, but that rather NH_3 will decompose into its elements.

Apart from enabling us to predict the direction of a chemical reaction from tables of thermodynamic data, the inequality [Eq. (16.9)] shows that three factors determine whether a reaction is spontaneous or not: the enthalpy change ΔH, the entropy change ΔS, and the temperature T. Let us examine each of these to see what effect they have and why.

The enthalpy change, ΔH As we well know, if ΔH is negative, heat will be released by the reaction and the entropy of the surroundings will be increased, while if ΔH is positive, the surroundings will decrease in entropy. At room temperature we usually find that this change in the entropy of the surroundings as measured by $-\Delta H/T$ is the major factor in determining the direction of a reaction since ΔS is almost always small by comparison. This explains why *at room temperature most spontaneous reactions are exothermic*. On the molecular level, as we saw in Chap. 15, an exothermic reaction corresponds to a movement from a situation of weaker bonding to a situation of stronger bonding. The formation of more and/or stronger bonds is thus a big factor in tending to make a reaction spontaneous.

The entropy change, ΔS_{sys} If the system itself increases in entropy as a result of the reaction (i.e., if ΔS_{sys} is positive), this will obviously contribute toward making the total entropy change positive and the reaction spontaneous. As we saw in the previous section of this chapter, reactions for which ΔS is positive correspond to the relaxation of some of the constraints on the motion of the atoms and molecules in the system. In particular, dissociation reactions and reactions in which the amount of substance in the gas phase increases correspond to reactions for which ΔS is positive.

The temperature, T Because it alters the magnitude of $-\Delta/T$ relative to ΔS, the temperature regulates the relative importance of the enthalpy change and the entropy change in determining whether a reaction will be spontaneous or not. As we lower the temperature, the effect of a reaction on the entropy of its surroundings becomes more and more pronounced because of the operation of the "boy in the living room" effect. As we approach absolute zero, the value of $-\Delta H/T$ begins to be an infinitely large positive or negative quantity, and ΔS becomes insignificant by comparison. At a very low temperature, therefore, whether the reaction is spontaneous or not will depend on the sign of ΔH, i.e., on whether the reaction is exothermic or endothermic. By contrast, as we raise the temperature to very high values, the "boy in the bedroom" effect takes over and the reaction affects the entropy of its surroundings to an increasingly smaller extent until finally it is only the value of ΔS_{sys} which determines the behavior of the reaction. In short, *whether a reaction is spontaneous or not is controlled by the sign of ΔH at very low temperatures and by the sign of ΔS at very high temperatures.*

Since ΔH can be positive or negative and so can ΔS, there are four possible combinations of these two factors, each of which exhibits a different behavior at high and low temperatures. All four cases are listed and described in Table 16.2, and they are also illustrated by simple examples in Fig. 16.7. In this figure the surroundings are indicated by a shaded border around the reaction system. If the reaction is exothermic, the border changes from gray to pink, indicating that the surroundings have absorbed heat energy and thus increased in entropy. When the border changes from

TABLE 16.2 Effect of Temperature on the Spontaneity of a Reaction.

Case	Sign of ΔH	Sign of ΔS	Behavior
1	−	+	Spontaneous at all temperatures
2	−	−	Spontaneous only at low temperatures
3	+	+	Spontaneous only at high temperatures
4	+	−	Never spontaneous

pink to gray, this indicates an endothermic reaction and a decrease in the entropy of the surroundings. In each case the change in entropy of the system should be obvious from an increase or decrease in the freedom of movement of the molecules and/or atoms.

Figure 16.7 The four different thermodynamic types of chemical reactions.

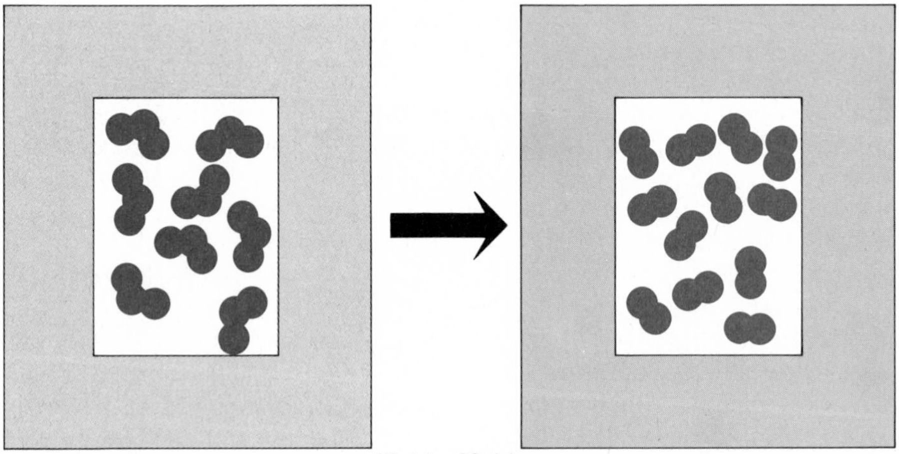

$2O_3(g) \rightarrow 3O_2(g)$

Case I. *ΔH negative and ΔS positive.* These reactions produce an increase in entropy both in the system itself and in the surroundings, and are spontaneous at all temperatures.

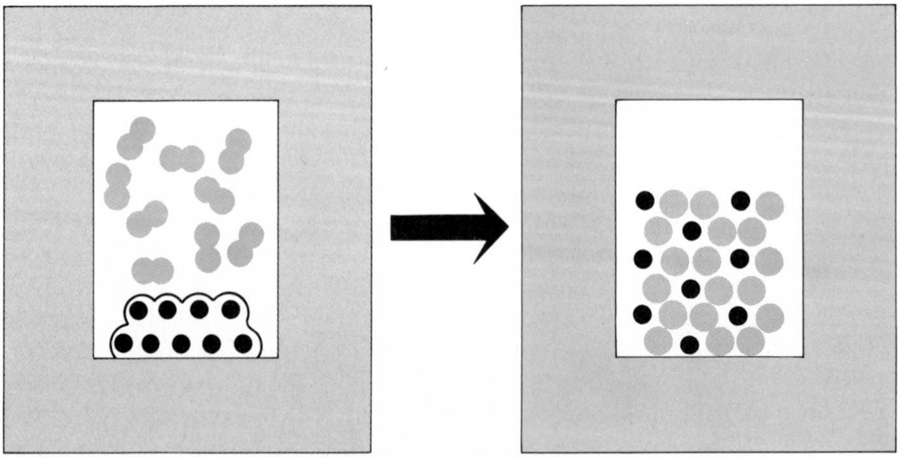

$Mg(s) + H_2(g) \rightarrow MgH_2(s)$

Case II. *ΔH negative and ΔS negative.* At low temperatures reactions in this class are spontaneous because of the entropy increase they cause in the surroundings. At high temperatures the decrease in entropy of the system becomes the controlling factor, and the reaction ceases to be spontaneous.

Case 1: Reaction is exothermic, and ΔS_{sys} is positive The example illustrated in Fig. 16.7(I) is the decomposition of ozone to oxygen.

$$2O_3(g) \rightarrow 3O_2(g) \qquad 1\ \text{atm}$$

for which $\Delta H_m^\ominus(298\ \text{K}) = -285\ \text{kJ mol}^{-1}$ and $\Delta S_m^\ominus(298\ \text{K}) = +137\ \text{J K}^{-1}\ \text{mol}^{-1}$. A reaction of this type is always spontaneous because the entropy of both the surroundings and the system are increased by its occurrence.

Case 2: Reaction is exothermic, and ΔS_{sys} is negative The example illustrated in Fig. 16.7(II) is the reaction of magnesium metal with hydrogen gas to form magnesium hydride:

$$Mg(s) + H_2(g) \rightarrow MgH_2(g) \qquad 1\ \text{atm}$$

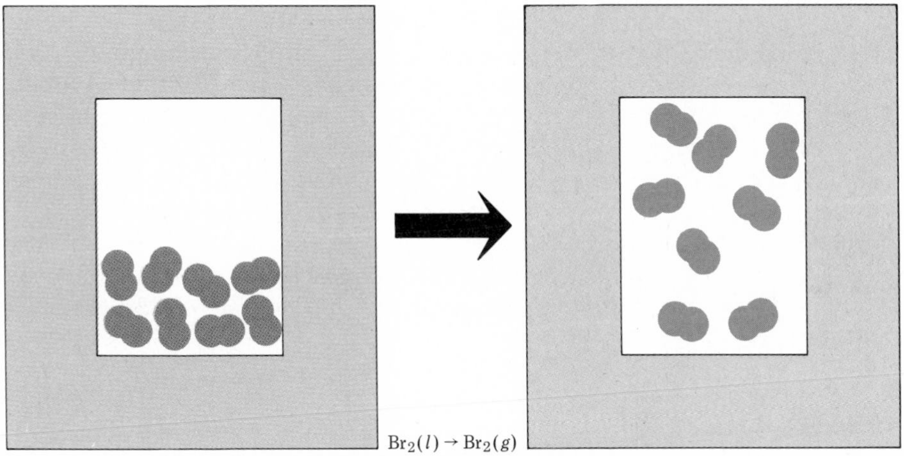

$Br_2(l) \rightarrow Br_2(g)$

Case III. *ΔH positive and ΔS positive.* At low temperatures reactions in this class are nonspontaneous, because their occurrence decreases the entropy of the surroundings. As the temperature is raised, the entropy increase of the system becomes more important and the reaction becomes spontaneous.

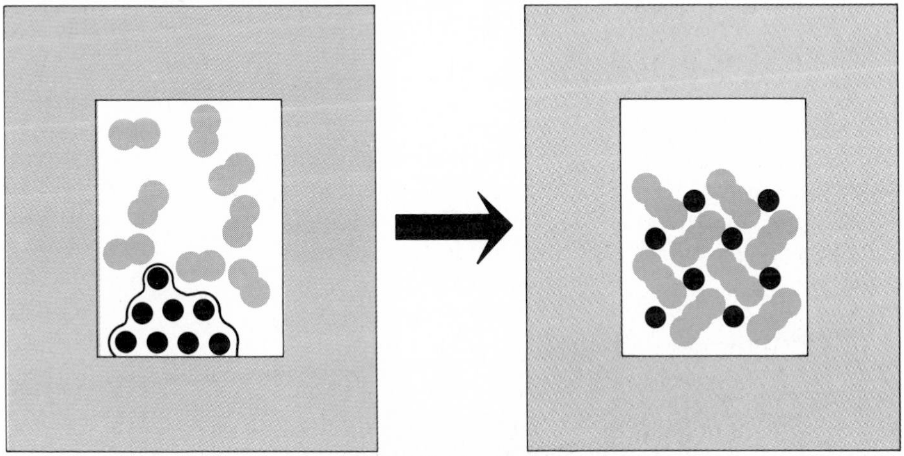

$2Ag(s) + 3N_2(g) \rightarrow 2AgN_3(s)$

Case IV. *ΔH positive and ΔS negative.* The entropy of both surroundings and system is decreased by this type of reaction which is thus nonspontaneous at all temperatures.

for which $\Delta H_m^\ominus = -76.1$ kJ mol^{-1} and $\Delta S_m^\ominus = -132.1$ J K^{-1} mol^{-1}. Reactions of this type can be either spontaneous or nonspontaneous depending on the temperature. At low temperatures when the effect on the surroundings is most important, the exothermic nature of the reaction makes it spontaneous. At high temperatures the effect of ΔS_{sys} predominates. Since ΔS_{sys} is negative (free H_2 molecules becoming fixed H^- ions), the reaction must become nonspontaneous at high temperatures. Experimentally, solid MgH_2 will not form from its elements above 560 K, and any formed at a lower temperature will decompose.

Case 3: Reaction is endothermic, and ΔS_{sys} is positive The example illustrated in Fig. 16.7(III) is the vaporization of liquid bromine:

$$Br_2(l) \rightarrow Br_2(g) \qquad 1 \text{ atm}$$

for which $\Delta H_m^\ominus(298\text{ K}) = +31.0$ kJ mol^{-1} and $\Delta S_m^\ominus(298\text{ K}) = 93.1$ J K^{-1} mol^{-1}. This example is usually classified as a physical rather than as a chemical change, but such distinctions are not important in thermodynamics. As in the previous case this reaction can be spontaneous or nonspontaneous at different temperatures. At low temperatures, bromine will not boil because the entropy increase occurring in the bromine as it turns to vapor is not enough to offset the decrease in entropy which the surroundings experience in supplying the heat energy which is needed for the change in state. At higher temperatures the entropy effect on the surroundings becomes less pronounced, and the positive value of ΔS_{sys} makes the reaction spontaneous. At 101.3 kPa (1 atm) pressure, bromine will not boil below 331 K (58°C), but above this temperature it will.

Case 4: Reaction is endothermic, and ΔS_{sys} is negative The example illustrated in Fig. 16.7(IV) is the reaction between silver and nitrogen to form silver azide, AgN_3:

$$2Ag(s) + 3N_2(g) \rightarrow 2AgN_3(s) \qquad 1 \text{ atm}$$

for which $\Delta H_m^\ominus(298\text{ K}) = +620.6$ kJ mol^{-1} and $\Delta S_m^\ominus(298\text{ K}) = -461.5$ J K^{-1} mol^{-1}.

Reactions of this type can never be spontaneous. If this reaction were to occur, it would reduce the entropy of both the system and the surroundings in contradiction of the second law. Since the forward reaction is nonspontaneous, we expect the reverse reaction to be spontaneous. This prediction is borne out experimentally. When silver azide is struck by a hammer, it decomposes explosively into its elements!

EXAMPLE 16.5 Classify the following reactions as one of the four possible types (cases) just described. Hence suggest whether the reaction will be spontaneous at (i) a very low temperature, and (ii) at a very high temperature.

	$\Delta H_m^\ominus$(298 K)/ kJ mol^{-1}	$\Delta S_m^\ominus$(298 K)/ J K^{-1} mol^{-1}
a) $N_2(g) + 3F_2(g) \rightarrow 2NF_3(g)$	−249	−277.8
b) $N_2(g) + 3Cl_2(g) \rightarrow 2NCl_3(g)$	+460	−275
c) $N_2F_4(g) \rightarrow 2NF_2(g)$	+93.3	+198.3
d) $C_3H_8(g) + 5O_2(g) \rightarrow 3CO_2(g) + 4H_2O(g)$	−2044.7	+101.3

Solution

a) This reaction is exothermic, and ΔS is negative. It belongs to type 2 and will be spontaneous at low temperatures but nonspontaneous at high temperatures.

b) Since this reaction is endothermic and ΔS is negative, it belongs to type 4. It cannot be spontaneous at any temperature.

c) Since this reaction is endothermic but ΔS is positive, it belongs to type 3. It will be spontaneous at high temperatures and nonspontaneous at low temperatures. (All dissociation reactions belong to this class.)

d) This reaction belongs to type 1 and is spontaneous at all temperatures.

16.5 THE FREE ENERGY

In the previous section, we were careful to differentiate between the entropy change occurring in the reaction system ΔS_{sys}, on the one hand, and the entropy change occurring in the surroundings, ΔS_{surr}, given by $-\Delta H/T$, on the other. By doing this we were able to get a real insight into what controls the direction of a reaction and why. In terms of calculations, though, it is a nuisance having to look up both entropy and enthalpy data in order to determine the direction of a reaction. For reasons of convenience, therefore, chemists usually combine the entropy and the enthalpy into a new function called the **Gibbs free energy,** or more simply the free energy, which is given the symbol G. If free-energy tables are available, they are all that is needed to predict the direction of a reaction at the temperature for which the tables apply.

In order to introduce free energy, let us start with the inequality

$$\frac{-\Delta H}{T} + \Delta S_{sys} > 0 \qquad (16.9)$$

This inequality must be true if a reaction occurring at constant pressure in surroundings at constant temperature is to be spontaneous. It is convenient to multiply this inequality by T; it then becomes

$$-\Delta H + T\,\Delta S > 0$$

(From now on we will abandon the subscript sys.) If $-\Delta H + T\,\Delta S$ is greater than zero, it follows that multiplying it by -1 produces a quantity which is *less than* zero, that is,

$$\Delta H - T\,\Delta S < 0 \qquad (16.10)$$

This latest inequality can be expressed very neatly in terms of the free energy G, which is defined by the equation

$$G = H - TS \tag{16.11}$$

When a chemical reaction occurs at constant temperature, the free energy will change from an initial value of G, given by

$$G_1 = H_1 - TS_1$$

to a final value

$$G_2 = H_2 - TS_2$$

The change in free energy ΔG will thus be

$$\Delta G = G_2 - G_1 = H_2 - H_1 - T(S_2 - S_1)$$

or

$$\Delta G = \Delta H - T\,\Delta S \tag{16.12}$$

Feeding this result back into inequality (16.10) gives the result

$$\Delta G = \Delta H - T\,\Delta S < 0$$

or

$$\Delta G < 0 \tag{16.13}$$

This very important and useful result tells us that when a spontaneous chemical reaction occurs (at constant temperature and pressure), *the free-energy change is negative.* In other words a *spontaneous change corresponds to a decrease in the free energy of the system.*

TABLE 16.3 Some Standard Free Energies of Formation at 298.15 K (25°C).

Compound	ΔG_f°/kJ mol^{-1}	Compound	ΔG_f°/kJ mol^{-1}
$AgCl(s)$	−109.7	$H_2O(g)$	−228.6
$AgN_3(s)$	378.5	$H_2O(l)$	−237.2
$Ag_2O(s)$	−10.8	$H_2O_2(l)$	−114.0
$Al_2O_3(s)$	−1576.4	$H_2S(g)$	−33.0
$Br_2(l)$	0.0	$HgO(s)$	−58.5
$Br_2(g)$	3.1	$I_2(s)$	0.0
$CaO(s)$	−604.2	$I_2(g)$	19.4
$CaCO_3(s)$	−1128.8	$KCl(s)$	−408.3
C—graphite	0.0	$KBr(s)$	−393.1
C—diamond	2.9	$MgO(s)$	−569.6
$CH_4(g)$	−50.8	$MgH_2(s)$	−76.1
$C_2H_2(g)$	209.2	$NH_3(g)$	−16.7
$C_2H_4(g)$	68.2	$NO(g)$	86.7
$C_2H_6(g)$	−32.9	$NO_2(g)$	51.8
$C_6H_6(l)$	124.5	$N_2O_4(g)$	98.3
$CO(g)$	−137.3	$NF_3(g)$	−124.7
$CO_2(g)$	−394.4	$NaCl(s)$	−384.0
$CuO(s)$	−127.2	$NaBr(s)$	−347.6
$Fe_2O_3(s)$	−741.0	$O_3(g)$	163.4
$HBr(g)$	−53.2	$SO_2(g)$	−300.4
$HCl(g)$	−95.3	$SO_3(g)$	−370.4
$HI(g)$	1.3	$ZnO(s)$	−318.2

If we have available the necessary free-energy data in the form of tables, it is now quite easy to determine whether a reaction is spontaneous or not. We merely calculate ΔG for the reaction using the tables. If ΔG turns out to be positive, the reaction is nonspontaneous, but if it turns out to be negative, then by virtue of Eq. (16.13) we can conclude that it is spontaneous.

Data on free energy are usually presented in the form of a table of values of **standard free energies of formation.** The standard free energy of formation of a substance is defined as the free-energy change which results when 1 mol of substance is prepared from its elements at the standard pressure of 1 atm and a given temperature, usually 298 K. It is given the symbol $\Delta G_f^\ominus$. A table of values of $\Delta G_f^\ominus(298 \text{ K})$ for a limited number of substances is given in Table 16.3.

Table 16.3 is used in exactly the same way as a table of standard enthalpies of formation, like Table 3.1. This type of table enables us to find ΔG values for any reaction occurring at 298 K and 1 atm pressure, provided only that all the substances involved in the reaction appear in the table. The two following examples illustrate such usage.

EXAMPLE 16.6 Determine whether the following reaction is spontaneous or not:

$$4NH_3(g) + 5O_2(g) \rightarrow 4NO(g) + 6H_2O(l) \qquad 1 \text{ atm, } 298 \text{ K}$$

Solution Following exactly the same rules used for standard enthalpies of formation, we have

$$\begin{aligned}\Delta G_m^\ominus &= \Sigma \Delta G_f^\ominus(\text{products}) - \Sigma \Delta G_f^\ominus(\text{reactants}) \\ &= 4\Delta G_f^\ominus(NO) + 6\Delta G_f^\ominus(H_2O) - 4\Delta G_f^\ominus(NH_3) - 5\Delta G_f^\ominus(O_2)\end{aligned}$$

Inserting values from Table 16.3, we then find

$$\begin{aligned}\Delta G_m^\ominus &= [4 \times 86.7 + 6 \times (-237.2) - 4 \times (-16.7) - 5 \times 0.0] \text{ kJ mol}^{-1} \\ &= -1010 \text{ kJ mol}^{-1}\end{aligned}$$

Since $\Delta G_m^\ominus$ is very negative, we conclude that this reaction is spontaneous.

The reaction of NH_3 with O_2 is very slow, so that when NH_3 is released into the air, no noticeable reaction occurs. In the presence of a catalyst, though, NH_3 burns with a yellowish flame in O_2. This reaction is very important industrially, since the NO produced from it can be reacted further with O_2 and H_2O to form HNO_3:

$$2NO + 1\tfrac{1}{2}O_2 + H_2O \rightarrow 2HNO_3$$

Nitric acid, HNO_3, is used mainly in the manufacture of nitrate fertilizers but also in the manufacture of explosives.

EXAMPLE 16.7 Determine whether the following reaction is spontaneous or not:

$$2NO(g) + 2CO(g) \rightarrow 2CO_2(g) + N_2(g) \qquad 1 \text{ atm, } 298 \text{ K}$$

Solution Following previous procedure we have

$$\Delta G_m^\ominus = (-2 \times 394.4 + 0.0 - 2 \times 86.7 + 2 \times 137.3) \text{ kJ mol}^{-1}$$
$$= -687.6 \text{ kJ mol}^{-1}$$

The reaction is thus spontaneous.

This example is an excellent illustration of how useful thermodynamics can be. Since both NO and CO are air pollutants produced by the internal-combustion engine, this reaction provides a possible way of eliminating both of them in one reaction, killing two birds with one stone. Unfortunately the reaction is very slow, and no one has yet succeeded in finding a catalyst to speed it up sufficiently. Nevertheless, our calculation shows that it is worthwhile looking for a catalyst. If $\Delta G_m^\ominus$ had turned out to be $+695$ kJ mol^{-1}, the reaction would be nonspontaneous and there would be no point at all in such a search.

We quite often encounter situations in which we need to know the value of $\Delta G_m^\ominus$ for a reaction at a temperature other than 298 K. Although extensive thermodynamic tables covering a large range of temperatures are available, we can also obtain approximate values for ΔG from the relationship

$$\Delta G_m^\ominus = \Delta H_m^\ominus - T\,\Delta S_m^\ominus \qquad (16.12)$$

If we assume, as we did previously, that neither $\Delta H_m^\ominus$ nor $\Delta S_m^\ominus$ varies much as the temperature changes from 298 K to the temperature in question, we can then use the values of $\Delta H_m^\ominus$ (298 K) obtained from Table 3.1 and $\Delta S_m^\ominus$ (298 K) obtained from Table 16.1 to calculate $\Delta G_m^\ominus$ for the temperature in question.

EXAMPLE 16.8 Using the enthalpy values in Table 3.1 and the entropy values in Table 16.1, calculate $\Delta H_m^\ominus(298 \text{ K})$ and $\Delta S_m^\ominus(298 \text{ K})$ for the reaction

$$CH_4(g) + H_2O(g) \rightarrow 3H_2(g) + CO(g) \qquad 1 \text{ atm}$$

Calculate an approximate value for $\Delta G_m^\ominus$ for this reaction at 600 and 1200 K and determine whether the reaction is spontaneous at either temperature.

Solution From the tables we find

$$\Delta H_m^\ominus(298 \text{ K}) = 3\Delta H_f^\ominus(H_2) + \Delta H_f^\ominus(CO) - \Delta H_f^\ominus(CH_4) - \Delta H_f^\ominus(H_2O)$$
$$= (3 \times 0.0 - 110.5 + 74.8 + 241.8) \text{ kJ mol}^{-1}$$
$$= +206.1 \text{ kJ mol}^{-1}$$

and similarly

$$\Delta S_m^\ominus(298\text{ K}) = (3 \times 130.6 + 197.6 - 187.9 - 188.7)\text{ J K}^{-1}\text{ mol}^{-1}$$
$$= +212.8\text{ J K}^{-1}\text{ mol}^{-1}$$

At 600 K we estimate

$$\Delta G_m^\ominus = \Delta H^\ominus(298) - T\,\Delta S^\ominus(298\text{ K})$$
$$= 206.1\text{ kJ mol}^{-1} - 600 \times 212.8\text{ J mol}^{-1}$$
$$= (206.1 - 127.7)\text{ kJ mol}^{-1}$$
$$= +78.4\text{ kJ mol}^{-1}$$

Since ΔG is positive, the reaction is not spontaneous at this temperature.
At 1200 K by contrast

$$\Delta G_m^\ominus = 206.1\text{ kJ mol}^{-1} - 1200 \times 212.8\text{ J mol}^{-1}$$
$$= (206.1 - 255.4)\text{ kJ mol}^{-1}$$
$$= -49.3\text{ kJ mol}^{-1}$$

At this higher temperature, therefore, the reaction is spontaneous.
Note:

1 From more extensive tables we find that accurate values of the free-energy change are $\Delta G_m^\ominus(600\text{ K}) = +72.6\text{ kJ mol}^{-1}$ and $\Delta G_m^\ominus(1200\text{ K}) = -77.7\text{ kJ mol}^{-1}$. Our approximate value at 1200 K is thus about 50 percent in error. Nevertheless it predicts the right *sign* for ΔG, a result which is adequate for most purposes.

2 The fact that this reaction can be run in either direction is of considerable industrial importance. Currently the forward reaction is the major method for producing hydrogen gas. As our results show, this can only be done at a high temperature. In the future, natural gas is bound to be scarcer and more costly, and we can expect the reverse reaction to be employed for the *production* of methane from carbon monoxide, which in turn has been produced from coal, as described in Chap. 15.

Maximum Useful Work

The Gibbs free energy has another very useful property. When a spontaneous chemical reaction occurs, *the decrease in free energy, $-\Delta G$, corresponds to the maximum possible quantity of useful work, w_{max}, which can be obtained.* Symbolically,

$$-\Delta G = w_{\text{max}}$$

For a reaction which is not spontaneous ΔG is positive and w_{max} is negative. This means that work must be done on the system (through some outside intervention) to force the nonspontaneous reaction to occur. The minimum work that must be done is given by ΔG.

As an example of the utility of this interpretation of ΔG, consider the recovery of Al from Al_2O_3 ore:

$$Al_2O_3 \rightarrow 2Al + \tfrac{3}{2}O_2 \qquad \Delta G_m^\ominus(298\ \text{K}) = 1576.4\ \text{kJ mol}^{-1}$$

The positive ΔG tells us that at least 1576.4 kJ of work must be done on 1 mol Al_2O_3 to effect this change. In a modern aluminum manufacturing plant this work is supplied electrically, and the electricity is often provided by burning coal. Assuming coal to be mainly carbon, we can write

$$C(s) + O_2(g) \rightarrow CO_2(g) \qquad \Delta G_m^\ominus(298\ \text{K}) = -394.4\ \text{kJ mol}^{-1}$$

Thus 1 mol C can do almost exactly one-quarter the work required to decompose 1 mol Al_2O_3, and we must burn *at least* 4 mol C to process each 1 mol Al_2O_3 ore. (In practice the aluminum smelting process is only 17 percent efficient, so it is necessary to burn nearly 6 times the theoretical 4 mol C.)

In the context we have just described, *free* energy is energy that is *available to do useful work,* not energy that we can get for nothing. When a spontaneous process occurs and there is a free energy decrease, it is the *availability of useful energy* which decreases. According to the first law of thermodynamics, energy *cannot* be consumed in any process, but according to the second law, free (or available) energy is *always* consumed in a spontaneous process.

When we talk about consuming energy resources by burning fossil fuels, it is the availability of energy that is used up. The energy originally stored in a fuel is converted to heat energy and dispersed to the surroundings. Once this has happened its usefulness is lost. There is no way of abstracting this energy from the surroundings and using it to lift a weight or do other useful work, because that would correspond to the reversal of a spontaneous process. The second law thus adds a very important qualification to the first law. While the first law tells us that we cannot destroy energy, the second law tells us that we cannot recycle it either.

16.6 EQUILIBRIUM CONSTANTS REVISITED

In the course of this chapter we have tended to treat chemical reactions as though they had no other alternatives than either going to completion or not occurring at all. In point of fact, it is only when both reactants and products are pure solids or pure liquids that we find this all-or-nothing-at-all type of behavior. Reactions which involve gases or solutions are governed by an equilibrium constant, and as we saw in Chap. 13, this means that there is always some reactant and some product in the equilibrium mixture. Consequently such reactions must always occur to some extent, no matter how minute, and can never go quite to completion.

Figure 16.8 illustrates how the free energy G varies as the reaction proceeds in the two cases. If only pure solids and liquids are involved, then a plot of G against the extent of the reaction is a straight line, as shown in part *a* of the figure for the reaction

$$Hg(l) + HgBr_2(s) \rightarrow Hg_2Br_2(s) \qquad 1\ \text{atm, } 298\ \text{K}$$

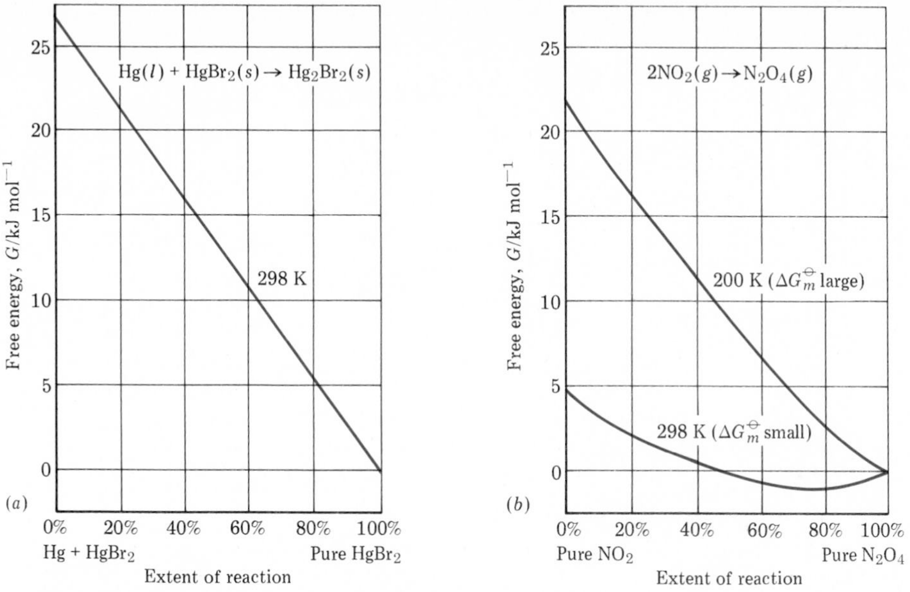

Figure 16.8 The variation of G with the extent of a reaction is a straight-line plot only when no gases or solutions are involved as in a. When gases or solutions are involved, the plot exhibits a minimum as in b.

In such a case, if $\Delta G_m^\ominus$, the free-energy difference between reactants and products, is *negative,* then the reaction will attain the lowest value of G possible by going to completion.

When gases and solutions are involved in the reaction, a plot of G against the extent of the reaction is no longer a straight line but exhibits a "sag," as shown in Fig. 16.8b for the reaction

$$2NO_2(g) \rightarrow N_2O_4(g) \qquad 1\text{ atm} \tag{16.14}$$

at two different temperatures. In such a case, even though $\Delta G_m^\ominus$ is negative, the reaction will not go to completion but will end up at the lowest point of the curve. If the value of $\Delta G_m^\ominus$ is quite small, below about 10 or 20 kJ mol^{-1}, this results in an eventual equilibrium mixture containing an appreciable proportion of both reactants and products. This is the case for Eq. (16.14) at 298 K, when the reaction attains only 81 percent completion at equilibrium. A more usual situation is that shown for this same reaction at 200 K. Because the free-energy difference is larger at this temperature, the "sag" in the curve is much smaller, and as a result the minimum value of G lies extremely close to 100 percent completion (actually 99.9 percent). As a rough rule of thumb, therefore, we can say that if $\Delta G_m^\ominus$ is negative and numerically greater than 20 kJ mol^{-1}, the reaction will go virtually to completion; if $\Delta G_m^\ominus$ is positive and larger than 20 kJ mol^{-1}, the reaction will scarcely occur at all. Between the limits $\Delta G_m^\ominus = \pm 20$ kJ mol^{-1}, we can expect measurable quantities of both reactant and product to be present at equilibrium.

The above discussion suggests that there must be some relationship between the free-energy change $\Delta G_m^\ominus$ and the equilibrium constant. The derivation of this relationship is too complex to produce here, so that only the result will be given. For gases the relationship has the form

$$\Delta G_m^\ominus = -RT \ln K^\ominus \tag{16.15a}$$

or, if base 10 logarithms are used,

$$\Delta G_m^\ominus = -2.303RT \log K^\ominus \tag{16.15b}$$

where $K^\ominus$ (K standard) is called the **standard equilibrium constant.** $K^\ominus$ is closely related to K_p and differs from it only by being a dimensionless number. Recall from Sec. 13.2 that K_p is defined for the general chemical reaction

$$a\text{A} + b\text{B} + \cdots \rightleftharpoons c\text{C} + d\text{D} + \cdots$$

by the equation

$$K_p = \frac{p_\text{C}^c \times p_\text{D}^d \times \cdots}{p_\text{A}^a \times p_\text{B}^b \times \cdots} \tag{13.12}$$

where p_A, p_B, etc., are partial pressures. The definition of $K^\ominus$ is identical except that it involves pressure *ratios* (i.e., pure numbers) rather than pressures:

$$K^\ominus = \frac{\left(\dfrac{p_\text{C}}{p^\ominus}\right)^c \times \left(\dfrac{p_\text{D}}{p^\ominus}\right)^d \times \cdots}{\left(\dfrac{p_\text{A}}{p^\ominus}\right)^a \times \left(\dfrac{p_\text{B}}{p^\ominus}\right)^b \times \cdots} \tag{16.16}$$

where $p^\ominus$ is a standard pressure—almost always 1 atm (101.325 kPa). Thus if the partial pressures p_a, p_b, etc., are expressed in atmospheres, $K^\ominus$ involves the same number as K_p.

EXAMPLE 16.9 The equilibrium constant K_p for the reaction

$$2NO_2 \rightleftharpoons N_2O_4$$

is 0.0694 kPa at 298 K. Use this value to find $\Delta G_m^\ominus$(298 K) for the reaction.

Solution We must first express K_p in atmospheres:

$$K_p = 0.0694 \text{ kPa}^{-1} = \frac{0.0694}{\text{kPa}} \times \frac{101.3 \text{ kPa}}{1 \text{ atm}} = 7.03 \text{ atm}^{-1}$$

Thus

$$K^\ominus = 7.03 \qquad \text{a pure number}$$

From Eq. (16.15*a*) we now have

$$\Delta G_m^\ominus = -2.303RT \log K^\ominus$$

$$= \frac{-8.3143 \text{ J}}{\text{K mol}} \times 298 \text{ K} \times 2.303 \times \log 7.03 = -4833 \text{ J mol}^{-1}$$

$$= -4.833 \text{ kJ mol}^{-1}$$

Note: You may have a calculator which has a ln x key as well as a log x key. If so, you can use Eq. (16.15a) directly:

$$\Delta G_m^\ominus = -RT \ln K^\ominus = -RT \times \ln 7.03 = -RT \times 1.950$$
$$= -4.833 \text{ kJ mol}^{-1}$$

In Sec. 13.5 we argued that the value of an equilibrium constant is the product of two factors:

$$K = \text{energy factor} \times \text{probability factor} \qquad (13.15)$$

The energy factor takes account of the fact that a higher-energy species is less likely to occur in an equilibrium mixture than a lower-energy species, especially at low temperatures. The probability factor reflects the fact that if there are a larger number of ways in which a molecule can arrange itself in space, that molecule is more likely to occur than one for which a smaller number of spatial arrangements is possible.

We are now in a position to make quantitative the qualitative argument presented at the end of Chap. 13. Combining

$$\Delta G_m^\ominus = \Delta H_m^\ominus - T\,\Delta S_m^\ominus \qquad (16.12)$$

with

$$\Delta G_m^\ominus = -RT \ln K^\ominus \qquad (16.15a)$$

we find

$$-RT \ln K^\ominus = \Delta H_m^\ominus - T\,\Delta S_m^\ominus$$

giving us

$$\ln K^\ominus = \frac{-\Delta H_m^\ominus}{RT} + \frac{\Delta S_m^\ominus}{R} \qquad (16.17a)$$

or, in base 10 logarithms

$$\log K^\ominus = \frac{-\Delta H_m^\ominus}{2.303RT} + \frac{\Delta S_m^\ominus}{2.303R} \qquad (16.17b)$$

If we now take the logarithm of each side of Eq. (13.15), we have

$$\log K = \log(\text{energy factor}) + \log(\text{probability factor})$$

This equation has the same form as Eq. (16.17b), and if we confine ourselves to the standard equilibrium constant $K^\ominus$, we can say that

$$\log(\text{energy factor}) = \frac{-\Delta H_m^\ominus}{2.303RT} \qquad (16.18a)$$

and

$$\log(\text{probability factor}) = \frac{\Delta S_m^\ominus}{2.303R} \qquad (16.18b)$$

Although we did not formally derive Eq. (16.15a), on which these results are based, we can examine the last two equalities [Eqs. (16.18a) and (16.18b)] to see that they agree with our qualitative expectations. If $\Delta H_m^\ominus$ is negative, we know that the products of a reaction have a lower enthalpy (and hence lower energy) than the reactants. Thus we would predict that products would be favored by the energy factor, and this is what Eq. (16.18a) says—the more negative $\Delta H_m^\ominus$, the more positive the logarithm of the energy factor and the larger the standard equilibrium constant. Also, since T appears in the denominator of the right-hand side of Eq. (16.18a), the smaller the value of T, the larger (and hence more important) the energy factor for a given value of $\Delta H_m^\ominus$. We have also seen in Sec. 16.2 that an increase in entropy of a system corresponds to an increase in thermodynamic probability. This is precisely what Eq. (16.18*b*) says. The larger the value of $\Delta S_m^\ominus$, the larger the probability factor and hence the larger the standard equilibrium constant. Thus our qualitative description of the two factors affecting the equilibrium constant has been refined to the point where macroscopic quantities, $\Delta H_m^\ominus$ and $\Delta S_m^\ominus$, can be related to what is happening on the microscopic level.

In addition to the feature we have just mentioned, Eq. (16.17*b*) is useful in another way. If we can measure or estimate $\Delta H_m^\ominus$ and $\Delta S_m^\ominus$ at temperatures other than the usual 298.15 K, we can obtain $K^\ominus$ and calculate the extent of reaction as shown in the next example.

EXAMPLE 16.10 Find the concentration of NO in equilibrium with air at 1000 K and 1 atm pressure.

$$N_2(g) + O_2(g) \rightarrow 2NO(g) \qquad \Delta H_m^\ominus = 181 \text{ kJ mol}^{-1}$$
$$\Delta S_m^\ominus = 25.6 \text{ J K}^{-1} \text{ mol}^{-1}$$

Solution We first find $K^\ominus$:

$$\log K^\ominus = \frac{-\Delta H_m^\ominus}{2.303RT} + \frac{\Delta S_m^\ominus}{2.303R}$$
$$= -9.45 + 1.34 = -8.11$$

Thus $$K^\ominus = 10^{-8.11} = 7.71 \times 10^{-9}$$

However, since $\Delta n = 0$, $K^\ominus = K_p = 7.71 \times 10^{-9}$

Assuming the air to be 80% N_2 and 20% O_2, we can write

$$p_{N_2} = 0.8 \text{ atm} \qquad \text{and} \qquad p_{O_2} = 0.2 \text{ atm}$$

Thus $$K_p = \frac{p_{NO_2}^2}{(p_{N_2})(p_{O_2})} = \frac{p_{NO_2}^2}{0.8 \text{ atm} \times 0.2 \text{ atm}} = 7.71 \times 10^{-9}$$

Giving $$p_{NO_2}^2 = 7.71 \times 10^{-9} \times 0.16 \text{ atm}^2 = 1.23 \times 10^{-9} \text{ atm}^2$$
$$p_{NO_2} = \sqrt{1.23 \times 10^{-9} \text{ atm}^2} = 3.5 \times 10^{-5} \text{ atm}$$
$$= 3.5 \times 10^{-3} \text{ kPa}$$

But $$c_{NO_2} = \frac{n_{NO_2}}{V} = \frac{p_{NO_2}}{RT}$$

Thus $$c_{NO_2} = \frac{3.5 \times 10^{-3} \text{ kPa}}{8.3143 \text{ J K}^{-1} \text{ mol}^{-1}} \times \frac{1}{1000 \text{ K}}$$
$$= 4.2 \times 10^{-7} \text{ mol dm}^{-3}$$

Almost all combustion processes heat N_2 and O_2 to high temperatures, producing small concentrations of NO. Under most circumstances this is not of great consequence, but in the presence of sunlight and partially burned gasoline, NO_2 can initiate a form of air pollution called photochemical smog. The presence of minute concentrations of NO in the upper atmosphere from high-flying supersonic jet airplanes could also conceivably deplete the ozone layer there.

SUMMARY

A chemical reaction which occurs of its own accord, without outside intervention, is said to be spontaneous. Such a process corresponds to a rearrangement of atoms and molecules from a less probable to a more probable situation, as measured by the thermodynamic probability W of the reactants and products. W is defined as the number of alternative microscopic arrangements which correspond to the same macroscopic state. Because most macroscopic samples of matter contain 10^{15} particles or more, very large values of W are encountered. Therefore it is convenient to use the entropy S, which is proportional to the logarithm of W, as a measure of spontaneity. According to the second law of thermodynamics, when a spontaneous process occurs, there must be an increase in total entropy.

The entropy of a perfectly ordered crystal at absolute zero is zero according to the third law of thermodynamics. Increasing the temperature, volume, or amount of substance increases the entropy. For a given amount of substance, the heavier and more complex the molecules, the greater the entropy, while weaker forces between atoms or molecules result in lower entropy. In general the greater the randomness or disorder of the atoms or molecules the greater the entropy of a substance.

To determine whether a chemical reaction is spontaneous we must calculate the change in total entropy of the chemical system *and* its surroundings. This can be done directly using $\Delta S^{\ominus}_{sys}$ and $-\Delta H^{\ominus}/T$ or indirectly using the change in Gibbs free energy $\Delta G^{\ominus}$. For a spontaneous reaction occurring at constant temperature and pressure, $\Delta G^{\ominus}$ must be negative. Since $\Delta G^{\ominus} = \Delta H^{\ominus} - T\,\Delta S^{\ominus}$, a negative enthalpy change is the most important factor governing spontaneity of a reaction at low temperatures. At high values of T, $-T\,\Delta S^{\ominus}$ becomes large and an increase in entropy of the system is essential for a spontaneous process.

The change in Gibbs free energy at standard pressure can be calculated from tables of $\Delta G_f^{\ominus}$. Once obtained it can be used to determine the standard equilibrium constant $K^{\ominus}$. For a reaction in the gas phase $K^{\ominus}$ is numerically the same as K_p, provided the latter is expressed in atmospheres. Knowledge of $\Delta G^{\ominus}$ thus allows calculation of the equilibrium partial pressures of reactants and products.

QUESTIONS AND PROBLEMS

16.1 For each of the spontaneous processes shown in Fig. 16.1, account for all the energy before and after the change. Would reversal of any of these processes violate the first law of thermodynamics? Does the first law tell us which is the spontaneous direction for any of the processes?

16.2 Suppose you flip three coins at once. What is the probability that all three will come up heads? What is the general formula for the probability that when N coins are flipped at once, all N will come up heads?

16.3 Extend the example illustrated in Fig. 16.2 to the case of two gas molecules in a series of three interconnected containers A, B, and C. What is the probability that both molecules will be in container A simultaneously? For the case of N molecules, what is the probability that all N will be in container B at once?

16.4 Criticize the following statement: The reversal of any of the spontaneous processes in Fig. 16.1 is absolutely impossible. Give quantitative data to support your criticism, if you can.

16.5 Deduce the equilibrium constant K_c for each of the following reactions:

a $C_2H_2 + C_2D_2 \rightleftharpoons 2C_2HD$

b $NH_3 + ND_3 \rightleftharpoons NH_2D + NHD_2$

16.6 Criticize the following statement, giving data or examples which support your criticism: Molecules always rearrange rapidly from a situation of high probability to one of low probability.

16.7 For a hypothetical crystal consisting of four atoms, one at each corner of a tetrahedron, calculate the thermodynamic probability in each of the following situations: (*a*) The crystal is at 0 K. (*b*) Just enough energy is added to set one of the four atoms vibrating. (*c*) Twice as much energy as in part *b* has been added. (*d*) Three times as much energy as in part *b* has been added.

16.8 Calculate the entropy of the crystal in each of the four situations in Problem 16.7.

16.9 The thermodynamic probability *W* for 1 mol $CO_2(g)$ at 325 K and 1.00 atm is $10^{(8.355 \times 10^{24})}$. Calculate the standard entropy $S_m^\ominus$ of $CO_2(g)$ at 325 K. Why are the thermodynamic probability and the entropy so much larger in this case than in Problems 16.7 and 16.8?

16.10 The entropy of 1 mol of an unknown gas at 400 K and 1.00 atm is 275 J K^{-1}. Calculate the thermodynamic probability.

16.11 State the first, second, and third laws of thermodynamics. Write at least two mathematical equations that are based on the second law and explain how these equations are related to the second law.

16.12 Table 16.1 lists standard molar entropies $S_m^\ominus$, not standard entropies of formation $\Delta S_f^\ominus$. Why is this possible for entropy but not for internal energy, enthalpy, or Gibbs free energy?

16.13 From each pair of substances listed below, select the one having the larger standard molar entropy at 25°C. Give reasons for your choice.

a $Ga(s)$ or $Ga(l)$
b $AsH_3(g)$ or $Kr(g)$
c $NaF(s)$ or $MgO(s)$
d $H_2O(g)$ or $H_2S(g)$
e $CH_3OH(l)$ or $C_2H_5OH(l)$
f *n*-butane or cyclobutane

16.14 Without consulting a table of standard molar entropies predict whether $\Delta S_{sys}^\ominus$ will be positive or negative for each of the following reactions:

a $2CO(g) + O_2(g) \rightarrow 2CO_2(g)$
b $2H_2(g) + O_2(g) \rightarrow 2H_2O(l)$
c $2O_3(g) \rightarrow 3O_2(g)$
d $2NH_3(g) \rightarrow N_2(g) + 3H_2(g)$
e $2Na(s) + Cl_2(g) \rightarrow 2NaCl(s)$
f $H_2(g) + I_2(s) \rightarrow 2HI(g)$

16.15 Using values of $S_m^\ominus$ from Table 16.1, calculate $\Delta S_{sys}^\ominus$ for each reaction in Problem 16.14, thereby checking your predictions.

16.16 When calculating $\Delta S_m^\ominus$ from $S_m^\ominus$ values, as in Problem 16.15, it is necessary to look up all substances, including elements in their most stable state, such as $O_2(g)$, $H_2(g)$, and $N_2(g)$. When calculating $\Delta H_m^\ominus$ from $\Delta H_f^\ominus$ values, however, elements in their most stable state can be ignored. Why is the situation different for $S_m^\ominus$ values?

16.17 What are the advantages and limitations of using the terms *randomness* and *disorder* as synonyms for entropy? Give two examples of pairs of substances whose randomness would appear to be the same but which have different standard molar entropies.

16.18 Calculate ΔS for each of the following substances when the quantity of heat energy indicated is absorbed at the temperature specified:

a $H_2(g)$, 0.775 kJ, 295 K
b $KCl(s)$, 5.00 kJ, 500 K
c $N_2(g)$, 2.45 kJ, 1000 K
d $NaCl(s)$, 5.00 kJ, 500 K
e $N_2O(g)$, 0.30 kJ, 300 K

Assume that you have enough of each substance so that its temperature remains constant as the heat energy is absorbed.

16.19 Calculate $\Delta S_{sys}^\ominus$ at 25°C for the reaction

$$C_2H_4(g) + H_2O(g) \rightarrow C_2H_5OH(l)$$

Can you tell from the result of this calculation whether this reaction is spontaneous? If you cannot tell, what additional information do you need? Obtain that information and decide whether the reaction is spontaneous.

16.20 Use data from Table 3.1 to calculate $\Delta H^\ominus$ for each reaction in Problem 16.14. Combine these $\Delta H^\ominus$ values with your results from Problem 16.15 to decide which of these reactions are spontaneous at 25°C and 1 atm.

16.21 Assuming that all substances remain in the phases indicated, use your data from Problem 16.20 to predict which reactions would be spontaneous at 2000 K and 1 atm.

16.22 Using the reaction

$$2H_2(g) + O_2(g) \rightarrow 2H_2O(g \text{ or } l)$$

as an example, explain why it is dangerous to assume for reactions involving solids or liquids that $\Delta S^\ominus$ and $\Delta H^\ominus$ do not change appreciably with increasing temperature.

16.23 Use data from Table 16.3 to calculate $\Delta G_m^\ominus$ for each of the following reactions at 25°C. Which are spontaneous?

a $2AgN_3(s) \rightarrow 2Ag(s) + 3N_2(g)$
b $C_2H_2(g) + H_2(g) \rightarrow C_2H_4(g)$
c $2SO_3(g) \rightarrow 2SO_2(g) + O_2(g)$
d $4NH_3(g) + 5O_2(g) \rightarrow 4NO(g) + 6H_2O(g)$

*16.24 Evaluate $\Delta H_m^\ominus$ for each reaction in Problem 16.23. Use your results to calculate standard entropies at 25°C for

a $AgN_3(s)$ *c* $SO_3(g)$
b $C_2H_2(g)$ *d* $NO(g)$

16.25 Estimate $\Delta G_m^\ominus$ at 2000 K for each reaction in Problem 16.23.

16.26 Explain in your own words why the sign of $\Delta H_m^\ominus$ determines whether a reaction is spontaneous at a low temperature, while the sign of $\Delta S_m^\ominus$ is most important at a high temperature.

16.27 Without consulting tables of $\Delta H_f^\ominus$, $S_m^\ominus$, or $\Delta G_f^\ominus$ values, predict which of the following reactions will be (i) always spontaneous; (ii) spontaneous at low temperatures, but nonspontaneous at high; (iii) nonspontaneous at low temperatures, but spontaneous at high; (iv) never spontaneous.

a $2NO_2(g) \rightarrow N_2O_4(g)$
b $C_5H_{12}(g) + 8O_2(g) \rightarrow 5CO_2(g) + 6H_2O(g)$
c $P_4(g) + 10F_2(g) \rightarrow 4PF_5(g)$
d $2NaF(s) \rightarrow 2Na(s) + F_2(g)$
e $2H_2O(l) + O_2(g) \rightarrow 2H_2O_2(l)$

Hint: Use the qualitative rules described in Sec. 15.4 to predict the sign of $\Delta H_m^\ominus$.

16.28 Explain clearly the meaning of the word *free* in Gibbs free energy.

16.29 Many metals are recovered by reducing their oxide ores with carbon. For example,

$$Fe_2O_3(s) + 3C(s) \rightarrow 2Fe(s) + 3CO(g)$$

Given the data below and ignoring possible effects of phase changes, which metals could be recovered in this way (*a*) at 800 K; (*b*) at 1500 K?

Metal	$S_m^\ominus$(298 K)/ J K^{-1} mol^{-1}	Ore	$S_m^\ominus$(298 K)/ J K^{-1} mol^{-1}	$\Delta H_f^\ominus$(298 K)/ kJ mol^{-1}
Fe(*s*)	27.2	$Fe_2O_3(s)$	90.0	−822.2
Ag(*s*)	42.7	$Ag_2O(s)$	121.7	− 30.6
Cr(*s*)	23.8	$Cr_2O_3(s)$	81.2	−1128.4
Al(*s*)	28.3	$Al_2O_3(s)$	51.0	−1669.8

16.30 Criticize the following statement: Provided it occurs at an appreciable rate, any chemical reaction for which $\Delta G_m < 0$ will proceed until all reactants have been converted to products.

16.31 Reword the statement in Problem 16.30 so that it is always true.

16.32 Use data from Table 16.3 to obtain the equilibrium constant K_p for each of the following reactions at 298 K.

a $2HCl(g) \rightleftharpoons H_2(g) + Cl_2(g)$
b $N_2(g) + O_2(g) \rightleftharpoons 2NO(g)$
c $CH_4(g) + 2O_2(g) \rightleftharpoons CO_2(g) + 2H_2O(g)$
d $2NO_2(g) \rightleftharpoons N_2O_4(g)$

*16.33 The Haber process for the synthesis of ammonia involves the reaction

$$N_2(g) + 3H_2(g) \rightleftharpoons 2NH_3(g)$$

Using data from Tables 3.1 and 16.1, estimate the amount of $NH_3(g)$ that would be produced from 1 mol $N_2(g)$ and 3 mol $H_2(g)$ once equilibrium is reached at 450°C and a total pressure of 1000 atm.

16.34 Quite often a graph of log $K^\ominus$ versus $1/T$ is a straight line. Use Eq. (16.17*b*) to show how $\Delta H_m^\ominus$ and $\Delta S_m^\ominus$ can be determined from such a graph. Does the fact that such a graph is straight tell you anything about the dependence of $\Delta H_m^\ominus$ and $\Delta S_m^\ominus$ on temperature?

*16.35 Assuming that $\Delta H_m^\ominus$ and $\Delta S_m^\ominus$ do not vary with temperature, use Eq. (16.17*b*) to derive a formula relating $K_1^\ominus$ at temperature T_1 to $K_2^\ominus$ at temperature T_2.

*16.36 For each reaction below, estimate $K^\ominus$ at the temperature indicated.

a $2H_2(g) + O_2(g) \rightleftharpoons 2H_2O(g)$ at 800 K
b $2SO_2(g) + O_2(g) \rightleftharpoons 2SO_3(g)$ at 500 K
c $2HF(g) \rightleftharpoons H_2(g) + F_2(g)$ at 2000 K

16.37 For each of the reactions below, an equilibrium constant at 298 K is given. Calculate $\Delta G_m^\ominus$ for each reaction.

a $Br_2(g) + H_2(g) \rightleftharpoons 2HBr(g)$; $K_p = 4.4 \times 10^{18}$
b $H_2O(l) \rightleftharpoons H_2O(g)$; $K_p = 3.17$ kPa
c $N_2(g) + 3H_2(g) \rightleftharpoons 2NH_3(g)$; $K_c = 5 \times 10^8$ mol^{-2} dm^6

16.38 Which of the following reactions are capable of being harnessed to do useful work at 298 K and 1 atm? Which require that work be done to make them occur?

a $2C_6H_6(l) + 15O_2(g) \rightarrow 12CO_2(g) + 6H_2O(g)$

b $2NF_3(g) \rightarrow N_2(g) + 3F_2(g)$

c $Al_2O_3(s) \rightarrow 2Al(s) + 3O_2(g)$

d $2CO(g) + O_2(g) \rightarrow 2CO_2(g)$

16.39 For each of the reactions in Problem 16.38 that require work to be done, calculate the minimum mass of graphite that would have to be oxidized to $CO_2(g)$ to provide the necessary work.

16.40 Why does the United States government have a major program of energy conservation? Does not the first law of thermodynamics *require* that energy be conserved?

chapter seventeen

ELECTROCHEMICAL CELLS

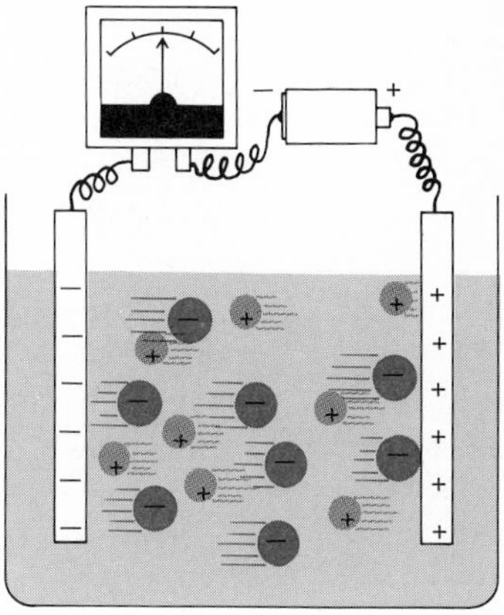

When an electrical current flows through matter, permanent chemical changes often occur. In some cases electrical energy supplied from an outside source can cause a chemical reaction to take place. Such a process is called **electrolysis,** and the system to which electricity is supplied is called an **electrolytic cell.** A typical example of electrolysis is the laboratory preparation of $H_2(g)$ and $O_2(g)$ by passing electrical current through water. Electrolysis is quite important in chemical industry, being involved in manufacture of aluminum, chlorine, copper, and numerous other substances.

It is also possible to produce a flow of electricity as a result of a spontaneous chemical reaction. A chemical system which can cause a current to flow in this way is called a **galvanic cell** or a **voltaic cell.** An example of a galvanic cell with which you are almost certainly familiar is a flashlight battery. Since an electrical current is a flow of electrons or other charged particles, it should come as no surprise that both electrolytic and galvanic cells involve redox reactions. Our discussion in this chapter will expand upon what was learned about redox reactions in Chap. 11. In particular we will be more quantitative about the strengths of reducing and oxidizing agents, and will see how to predict whether such reactions can occur spontaneously.

17.1 ELECTROLYSIS

A typical electrolytic cell can be made as shown in Fig. 17.1. Two electrical conductors (**electrodes**) are immersed in the liquid to be electrolyzed. These electrodes are often made of an inert material such as stainless steel, platinum, or graphite. The liquid to be electrolyzed must be able to conduct electricity, and so it is usually an aqueous solution of an electrolyte or a molten ionic compound. The electrodes are connected by wires to a battery or other source of direct current. This current source may be thought of as an "electron pump" which takes in electrons from one electrode and forces them out into the other electrode. The electrode from which electrons are removed becomes positively charged, while the electrode to which they are supplied has an excess of electrons and a negative charge.

The negatively charged electrode will attract positive ions (cations) toward it from the solution. It can donate some of its excess electrons to such cations or to other species in the liquid being electrolyzed. Hence this electrode is in effect a reducing agent. In any electrochemical cell (electrolytic or galvanic) *the electrode at which reduction occurs* is called the **cathode.**

The positive electrode, on the other hand, will attract negative ions (anions) toward itself. This electrode can accept electrons from those negative ions or other species in the solution and hence behaves as an oxidizing agent. In any electrochemical cell the **anode** is *the electrode at which oxidation occurs.* An easy way to remember which electrode is which is that anode and oxidation begin with vowels while cathode and reduction begin with consonants.

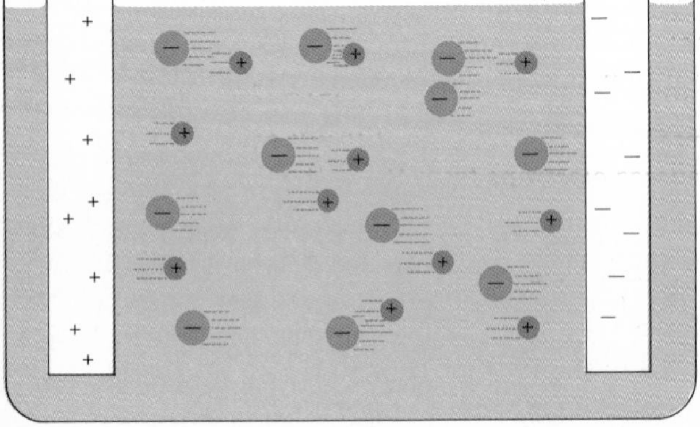

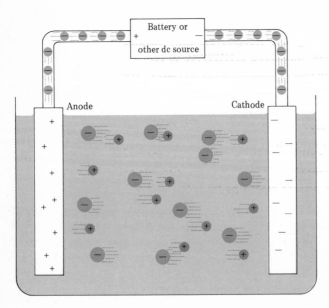

Figure 17.1 An electrolytic cell. The battery pumps electrons away from the anode (making it positive) and into the cathode (making it negative). The positive anode attracts anions toward it, while the negative cathode attracts cations toward it. Electrical current is carried by electrons in the wire and electrodes, but it is carried by anions and cations moving in opposite directions in the cell itself. Since the anode can accept electrons, oxidation occurs at that electrode. The cathode is an electron donor and can cause reduction to occur.

As an example of how electrolysis can cause a chemical reaction to occur, suppose we pass a direct electrical current through 1 *M* HCl. The H_3O^+ ions in this solution will be attracted to the cathode, and the Cl^- ions will migrate toward the anode. At the cathode, H_3O^+ will be reduced to H_2 gas according to the half-equation

$$2H^+ + 2e^- \rightarrow H_2 \tag{17.1a}$$

(As was done in Chap. 11, we shall write H^+ instead of H_3O^+ in half-equations to save time.) At the anode, electrons will be accepted from Cl^- ions, oxidizing them to Cl_2:

$$2Cl^- \rightarrow Cl_2 + 2e^- \tag{17.1b}$$

During electrolysis $H_2(g)$ and $Cl_2(g)$ bubble from the cathode and anode, respectively. The overall equation for the electrolysis is the sum of Eqs. (17.1*a*) and (17.1*b*):

$$\begin{aligned} & 2H^+(aq) + 2Cl^-(aq) \rightarrow H_2(g) + Cl_2(g) \\ \text{or} \quad & 2H_3O^+(aq) + 2Cl^-(aq) \rightarrow H_2(g) + Cl_2(g) + 2H_2O(l) \end{aligned} \tag{17.1}$$

The net reaction [Eq. (17.1)] is the *reverse* of the spontaneous combination of $H_2(g)$ with $Cl_2(g)$ to form $HCl(aq)$. Such a result is true of electrolysis in general: *electrical current supplied from outside the system causes a non-spontaneous chemical reaction to occur.*

Although electrolysis always reverses a spontaneous redox reaction, the result of a given electrolysis may not always be the reaction we want. In an aqueous solution, for example, there are always a great many water molecules in the vicinity of both the anode and cathode. These water molecules can donate electrons to the anode or accept electrons from the cathode just as anions or cations can. Consequently the electrolysis may oxidize and/or reduce water instead of causing the dissolved electrolyte to react.

An example of this problem is electrolysis of lithium fluoride, LiF. We might expect reduction of Li^+ at the cathode and oxidation of F^- at the anode, according to the half-equations

$$Li^+(aq) + e^- \rightarrow Li(s) \tag{17.2a}$$

$$2F^-(aq) \rightarrow F_2(g) + 2e^- \tag{17.2b}$$

However, Li^+ is a very poor electron acceptor, and so it is very difficult to force Eq. (17.2*a*) to occur. Consequently, excess electrons from the cathode are accepted by water molecules instead:

$$2H_2O(l) + 2e^- \rightarrow 2OH^-(aq) + H_2(g) \tag{17.3a}$$

A similar situation arises at the anode. F^- ions are extremely weak reducing agents—much weaker than H_2O molecules—so the half-equation is

$$2H_2O(l) \rightarrow O_2(g) + 4H^+(aq) + 4e^- \tag{17.3b}$$

The overall equation can be obtained by multiplying Eq. (17.3*a*) by 2, adding it to Eq. (17.3*b*), and combining H^+ with OH^- to form H_2O:

$$2H_2O(l) \rightarrow 2H_2(g) + O_2(g) \tag{17.3}$$

Thus this electrolysis reverses the spontaneous combination of H_2 and O_2 to form H_2O.

In Secs. 11.6 and 11.7 we mentioned several oxidizing agents, such as F_2, which are strong enough to oxidize H_2O. At the same time we described reducing agents which are strong enough to reduce H_2O, such as the alkali metals and the heavier alkaline earths. As a general rule such substances cannot be produced by electrolysis of aqueous solutions because H_2O is oxidized or reduced instead. Substances which undergo spontaneous redox reaction with H_2O are usually produced by electrolysis of molten salts or in some other solvent. There are some exceptions to this rule, however, because some electrode reactions are slower than others. Using Table 11.5, for example, we would predict that H_2O is a better reducing agent than Cl^-. Hence we would expect O_2, not Cl_2, to be produced by electrolysis of 1 *M* HCl, in contradiction of Eq. (17.1). It turns out that O_2 is produced more *slowly* than Cl_2, and the latter bubbles out of solution before the H_2O can be oxidized. For this reason Table 11.5 cannot always be used to predict what will happen in an electrolysis.

17.2 COMMERCIAL APPLICATIONS OF ELECTROLYSIS

Electrolysis is of major industrial importance. Of the 10 chemicals whose annual production exceeds 1 million tons, two, sodium hydroxide and chlorine, are produced using this technique. A number of metals, such as aluminum and magnesium, are also produced by electrolysis. In 1975 more than 6.5 percent of all electrical energy produced in the United States was used for electrolysis.

Electrolysis of Brine

Three important chemicals, NaOH, Cl_2, and H_2, can be obtained by electrolyzing an aqueous NaCl solution (brine). This forms the basis of the **chlor-alkali industry.** The diaphragm cell (also called a Hooker cell) in which the electrolysis is carried out is shown schematically in Fig. 17.2. At the cathode, water is reduced:

$$2H_2O + 2e^- \rightarrow H_2 + 2OH^- \tag{17.4a}$$

Chlorine is produced at the anode:

$$2Cl^- \rightarrow Cl_2 + 2e^- \tag{17.4b}$$

Thus the overall reaction is

$$2H_2O(l) + 2Cl^-(aq) \rightarrow H_2(g) + Cl_2(g) + 2OH^-(aq) \tag{17.4}$$

Since the $H_2(g)$ and $Cl_2(g)$ might recombine explosively should they come in

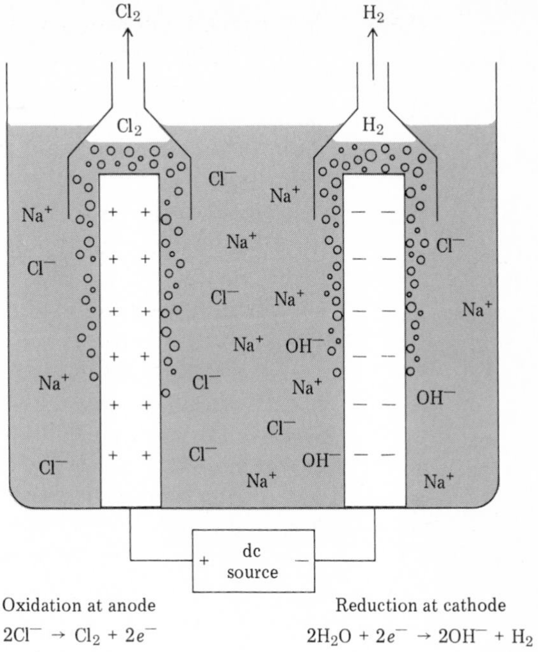

Figure 17.2 Diaphragm or Hooker cell for electrolysis of brine (schematic). Since chloride ions are removed and hydroxide ions produced by the electrolysis, the electrolyte gradually changes from a solution of sodium chloride to a solution of sodium hydroxide.

contact, the cathode must be entirely surrounded by a porous diaphragm of asbestos. Hence the name of this type of cell.

Both the $H_2(g)$ and $Cl_2(g)$ produced in Eq. (17.4) are dried, purified, and compressed into cylinders. Fresh brine is continually pumped into the cell, and the solution which is forced out contains about 10% NaOH together with a good deal of NaCl. [Remember that the spectator ions, $Na^+(aq)$, are not included in a net ionic equation such as Eq. (17.4).] H_2O is allowed to evaporate from this solution until the concentration of the solution reaches 50% NaOH, by which time most of the NaCl has crystallized out and can be recycled to the electrolysis cell. The NaOH is sold as a 50% solution or further dried to give crystals whose approximate formula is $NaOH{\cdot}H_2O$.

The considerable effort required to concentrate the NaOH solution obtained from diaphragm cells can be avoided by using mercury cells. The cathode in such a cell is mercury, and the cathode reaction is

$$Na^+(aq) + e^- + x\text{Hg}(l) \rightarrow \text{NaHg}_x(l)$$

The sodium metal produced in this reaction dissolves in the liquid mercury, producing an **amalgam.** The liquid amalgam is then transferred to another part of the cell and reacted with water:

$$2\text{NaHg}_x(l) + 2H_2O(l) \rightarrow 2Na^+(aq) + 2OH^-(aq) + H_2(g) + x\text{Hg}(l)$$

The 50% sodium hydroxide solution produced by this reaction contains no sodium chloride and can be sold directly, without being concentrated further. Up until 1970, however, chlor-alkali plants using mercury cells did not have adequate controls to prevent losses of mercury to the environment. About 100 to 200 g mercury was lost for each 1000 kg chlorine produced—apparently a small quantity until one realizes that 2 500 000 kg chlorine was produced by mercury cells *every day* during 1960 in the United States. Thus every 2 to 4 days 1000 kg mercury entered the environment, and by 1970 sizable quantities were being found in fish. Since 1970 adequate controls have been installed on mercury cells and most new chlor-alkali plants use diaphragm cells, but the very large quantities of mercury introduced into rivers and lakes prior to 1970 are expected to remain for a century or more.

Aluminum Production

Some of the difficulties involved in obtaining Al(*s*) from the minerals of the earth's crust were mentioned in Sec. 12.3. Al is easily oxidized, and so its ore, Al_2O_3, is difficult to reduce. In fact water is reduced rather than $Al^{3+}(aq)$, and so electrolysis must be carried out in a molten salt. Even this is difficult because the melting point of Al_2O_3 is above 2000°C—a temperature which is very difficult to maintain.

The first successful method for reducing Al_2O_3 is the one still used today. It was developed in the United States in 1886 by Charles Hall (1863 to 1914), who was then 23 years old and fresh out of Oberlin College. Hall realized that if Al_2O_3 were dissolved in another molten salt, the melting point of the mixture would be lower than for either pure substance. The substance Hall used was cryolite, Na_3AlF_6, in which the Al_2O_3 can be dissolved at just over 1000°C. The electrolytic cell used for the Hall process (Fig. 17.3) consists of a steel box lined with graphite. This contains the molten Na_3AlF_6 and Al_2O_3 and also serves as the cathode. The anode is a large cylinder of carbon. Passage of electrical current maintains the high temperature of the cell and causes the following half-equations to occur:

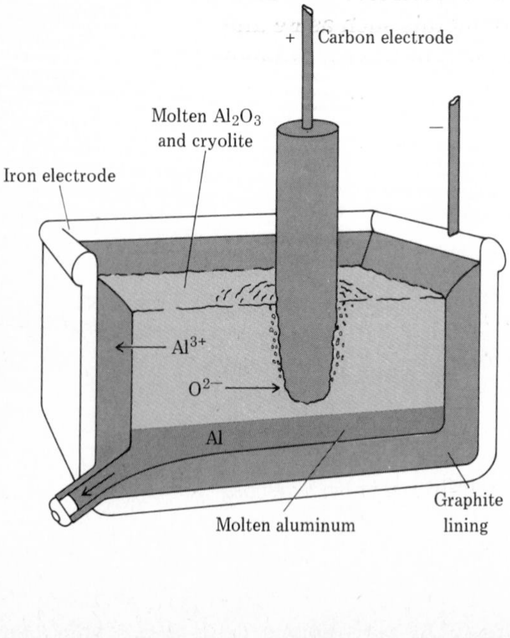

Figure 17.3 The Hall process.

$$Al^{3+} + 3e^- \rightarrow Al(l) \qquad (17.5a)$$

$$2O^{2-} + C(s) \rightarrow CO_2(g) + 4e^- \qquad (17.5b)$$

Since the carbon anode is consumed by the oxidation half-equation, it must be replaced periodically.

Aluminum production requires vast quantities of electrical energy, both to maintain the high temperature and to cause half-equations (17.5*a*) and (17.5*b*) to occur. Currently about 5 percent of the total electrical energy produced in the United States goes into the Hall process. Much of this energy comes from combustion of fossil fuels and hence consumes a valuable, nonrenewable resource. Since Al is protected from oxidation back to Al_2O_3 by a surface coating of oxide, it is a prime candidate for recycling, as well as for applications such as house siding, where it is expected to remain for a long time. Throwing away aluminum beverage cans, on the other hand, is a tremendous waste of energy.

Several other easily oxidized metals are currently produced by electrolysis, but not in such large quantities as Al. Mg is obtained by electrolyzing molten $MgCl_2$ which is derived from seawater, and Na and Ca are produced together from a molten mixture of NaCl and $CaCl_2$.

Refining of Copper

Unrefined or "blister" copper is about 99 percent pure when obtained from the ore, but it is desirable to increase this to 99.95 percent if the copper is to be used in electrical wiring. Even small concentrations of impurities noticeably lower copper's electrical conductivity. Such a high degree of purity can be obtained by electrolytic refining in a cell similar to that shown in Fig. 17.4.

In such a cell a thin sheet of high-purity Cu serves as the cathode, and the anode is the impure Cu which is to be refined. The electrolyte is a solution of copper(II) sulfate. Some of the impurities are metals such as Fe and Zn which are more easily oxidized than Cu. When current passes through the cell, these impurities go into solution from the anode, along with Cu:

$$Cu(s) \rightarrow Cu^{2+}(aq) + 2e^-$$

$$Fe(s) \rightarrow Fe^{2+}(aq) + 2e^-$$

$$Zn(s) \rightarrow Zn^{2+}(aq) + 2e^-$$

These ions all migrate toward the cathode, but $Cu^{2+}(aq)$ is more readily reduced than $Fe^{2+}(aq)$ or $Zn^{2+}(aq)$, and so it is the only one that plates out. The impurity ions remain in solution. Other impurities, such as Ag, Au, and Pt, are less easily oxidized than Cu. These remain in metallic form and fall to the bottom of the cell, forming "anode sludge" from which they can later be recovered. The great value of Ag, Au, and Pt helps to offset the cost of refining.

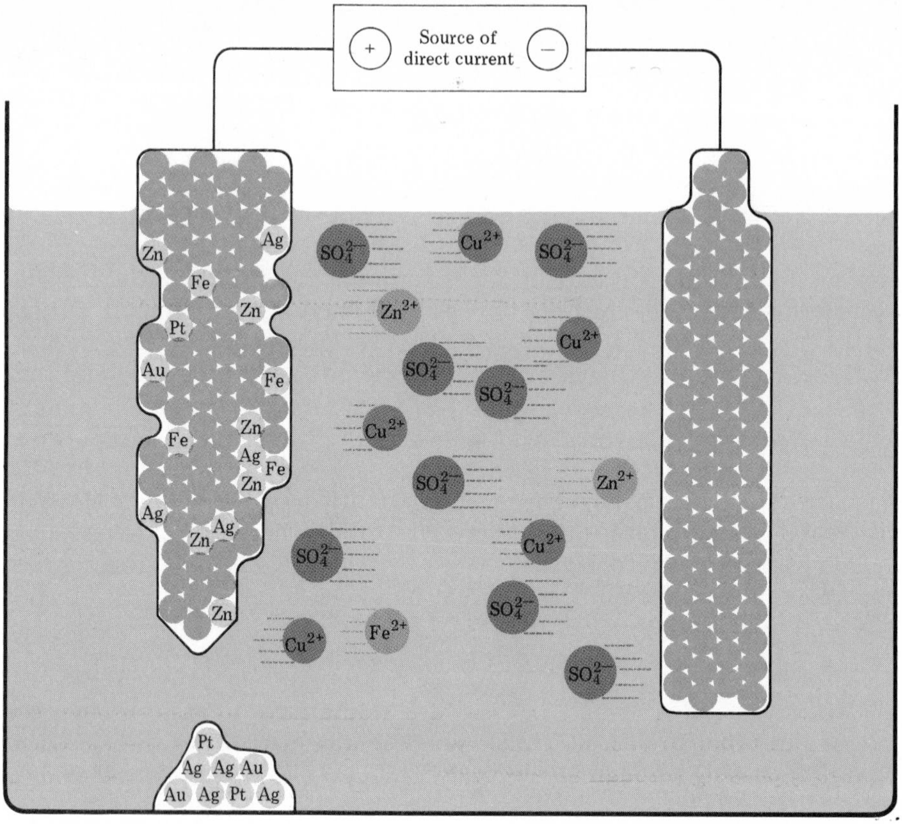

Figure 17.4 The electrolytic purification of copper. See text for a description of the reactions occurring.

Electroplating

Another important industrial application of electrolysis is the plating of one metal on top of another. A typical example is the bumper of a car. This is made from steel and then plated with a thin layer of chromium to make it resistant to rusting and scratching. Many other metal objects, such as pins, screws, watchbands, and doorknobs, are made of one metal with another plated on the surface.

An electroplating cell works in much the same way as the cell used to purify copper. The object to be plated is used as the cathode, and the electrolyte contains some ionic compound of the metal to be plated. As current flows, this compound is reduced to the metal and deposits on the surface of the cathode. In chromium plating, for instance, the electrolyte is usually a solution of potassium dichromate, $K_2Cr_2O_7$, in fairly concentrated sulfuric acid. In this very acidic solution $Cr_2O_7^{2-}$ ions are completely protonated, and so the reduction half-equation is

$$H_2Cr_2O_7(aq) + 12H^+(aq) + 12e^- \rightarrow 2Cr(s) + 7H_2O(l) \qquad (17.6)$$

Other metals which are often electroplated are silver, nickel, tin, and zinc.

In the case of silver the electrolyte must contain the polyatomic ion $Ag(CN)_2^-$ rather than Ag^+. Otherwise the solid silver will be deposited as jagged crystals instead of a shiny uniform layer.

17.3 QUANTITATIVE ASPECTS OF ELECTROLYSIS

As we mentioned in Sec. 4.3, Michael Faraday discovered in 1833 that there is always a simple relationship between the amount of substance produced or consumed at an electrode during electrolysis and the quantity of electrical charge Q which passes through the cell. For example, the half-equation

$$Ag^+ + e^- \rightarrow Ag \qquad (17.7)$$

tells us that when 1 mol Ag^+ is plated out as 1 mol Ag, 1 mol e^- must be supplied from the cathode. Since the negative charge on a single electron is known to be 1.6022×10^{-19} C, we can multiply by the Avogadro constant to obtain the charge per mole of electrons. This quantity is called the **Faraday constant,** symbol F:

$$F = 1.6022 \times 10^{-19}\ \text{C} \times 6.0221 \times 10^{23}\ \text{mol}^{-1} = 9.649 \times 10^4\ \text{C mol}^{-1}$$

Thus in the case of Eq. (17.7), 96 490 C would have to pass through the cathode in order to deposit 1 mol Ag. For any electrolysis the electrical charge Q passing through an electrode is related to the amount of electrons n_{e^-} by

$$F = \frac{Q}{n_{e^-}}$$

Thus F serves as a conversion factor between n_{e^-} and Q.

EXAMPLE 17.1 Calculate the quantity of electrical charge needed to plate 1.386 mol Cr from an acidic solution of $K_2Cr_2O_7$ according to half-equation (17.6).

Solution According to Eq. (17.6), 12 mol e^- is required to plate 2 mol Cr, giving us a stoichiometric ratio $S(e^-/\text{Cr})$. Then the Faraday constant can be used to find the quantity of charge. In road-map form

$$n_{\text{Cr}} \xrightarrow{S(e^-/\text{Cr})} n_{e^-} \xrightarrow{\quad F \quad} Q$$

$$Q = 1.386\ \text{mol Cr} \times \frac{12\ \text{mol}\ e^-}{2\ \text{mol Cr}} \times \frac{9.649 \times 10^4\ \text{C}}{1\ \text{mol}\ e^-} = 8.024 \times 10^5\ \text{C}$$

Often the electrical current rather than the quantity of electrical charge is measured in an electrolysis experiment. Since a **coulomb** is defined as the quantity of charge which passes a fixed point in an electrical circuit when a current of one ampere flows for one second, the charge in coulombs can be calculated by multiplying the measured current (in amperes) by the time (in seconds) during which it flows:

$$Q = It \tag{17.8}$$

In this equation I represents current and t represents time. If you remember that

$$1 \text{ coulomb} = 1 \text{ ampere} \times 1 \text{ second} \qquad 1 \text{ C} = 1 \text{ A s}$$

you can adjust the time units to obtain the correct result.

EXAMPLE 17.2 Hydrogen peroxide, H_2O_2, can be manufactured by electrolysis of cold concentrated sulfuric acid. The reaction at the anode is

$$2H_2SO_4 \rightarrow H_2S_2O_8 + 2H^+ + 2e^- \tag{17.9a}$$

When the resultant peroxydisulfuric acid, $H_2S_2O_8$, is boiled at reduced pressure, it decomposes:

$$2H_2O + H_2S_2O_8 \rightarrow 2H_2SO_4 + H_2O_2 \tag{17.9b}$$

Calculate the mass of hydrogen peroxide produced if a current of 0.893 A flows for 1 h.

Solution The product of current and time gives us the quantity of electricity, Q. Knowing this we easily calculate the amount of electrons, n_{e^-}. From half-equation (17.9a) we can then find the amount of peroxydisulfuric acid. Equation (17.9b) then leads to $n_{H_2O_2}$ and finally to $m_{H_2O_2}$. The road map to describe this logic is as follows:

$$I \xrightarrow{t} Q \xrightarrow{F} n_{e^-} \xrightarrow{S_e} n_{H_2S_2O_8} \xrightarrow{S} n_{H_2O_2} \xrightarrow{\mathcal{M}} m_{H_2O_2}$$

so that

$$m_{H_2O_2} = 0.893 \text{ A} \times 3600 \text{ s} \times \frac{1 \text{ mol } e^-}{96\ 490 \text{ C}} \times \frac{1 \text{ mol } H_2S_2O_8}{2 \text{ mol } e^-} \times \frac{1 \text{ mol } H_2O_2}{1 \text{ mol } H_2S_2O_8} \times \frac{34.01 \text{ g } H_2O_2}{1 \text{ mol } H_2O_2}$$

$$= 0.5666 \frac{\text{A s}}{\text{C}} \times \text{g } H_2O_2$$

$$= 0.5666 \text{ g } H_2O_2$$

17.4 GALVANIC CELLS

In an electrolytic cell electrical energy is consumed and an otherwise spontaneous redox reaction is reversed. A galvanic cell, on the other hand, produces electrical energy as a result of a spontaneous redox process. The electron transfer characteristic of such a process is made to occur in two separate half-cells. Electrons released during an oxidation half-equation must flow through a wire or other external circuit before they can be accepted in a reduction half-equation. Consequently an electrical current is made to flow.

A typical galvanic cell, the **Daniell cell,** was used to power telegraphs 100 years ago. This cell is based on the spontaneous redox reaction

$$Zn(s) + Cu^{2+}(aq) \rightarrow Zn^{2+}(aq) + Cu(s) \qquad (17.10)$$

(You can verify that this reaction is spontaneous by dipping a piece of zinc metal in a copper sulfate solution. In a short time the surface of the zinc will become plated with red-brown copper metal.) The half-equations

$$Zn(s) \rightarrow Zn^{2+}(aq) + 2e^- \qquad (17.10a)$$

$$Cu^{2+}(aq) + 2e^- \rightarrow Cu(s) \qquad (17.10b)$$

indicate that for each mole of zinc which is oxidized and goes into solution as zinc ions, 2 mol electrons are transferred to copper ions, converting them to copper atoms.

To produce electrical current we must prevent the Zn(*s*) from contacting the $Cu^{2+}(aq)$ ions and transferring the electrons directly. This is done in the Daniell cell by pouring a concentrated copper sulfate solution into the bottom of a glass jar and then carefully pouring a layer of less concentrated zinc sulfate solution above it. Because it contains less solute per unit volume, the zinc sulfate solution is less dense. It floats on the copper sulfate and does not mix with it. Therefore a copper electrode placed in the bottom of the jar contacts only $Cu^{2+}(aq)$ ions, and a zinc electrode suspended in the zinc sulfate solution contacts only $Zn^{2+}(aq)$ ions.

In the laboratory it is more convenient to set up a cell based on the Zn + Cu^{2+} reaction, as shown in Fig. 17.5. Two electrodes or half-cells are separated by a **salt bridge.** This contains an electrolyte, KCl, so that current can flow from one half-cell to the other, but the contents of the two half-cells cannot mix. The left-hand electrode in Fig. 17.5 is a Zn rod dipping in a solution of $ZnSO_4$. Thus both components of the Zn^{2+}/Zn redox couple are present, and the metal electrode can conduct electrons produced by Eq. (17.10*a*) to the wire in the external circuit. Since *oxidation* of Zn to Zn^{2+} occurs at the left-hand electrode, this electrode is the *anode*.

The right-hand electrode is a strip of Cu dipping in a solution of $CuSO_4$. Here both components of the Cu^{2+}/Cu redox couple are present, and Eq. (17.10*b*) can occur. Electrons supplied by the external circuit are conducted through Cu to the electrode surface, where they combine with Cu^{2+} ions to produce more Cu. Since *reduction* occurs at this right-hand electrode, this electrode is the *cathode*. The net effect of the two half-cells is that electrons are forced into the external circuit at the anode and withdrawn from it at the cathode. This will cause current to flow, or, if current is prevented from flowing by a device such as the voltmeter in Fig. 17.5, it will cause an electrical potential difference (voltage) to build up.

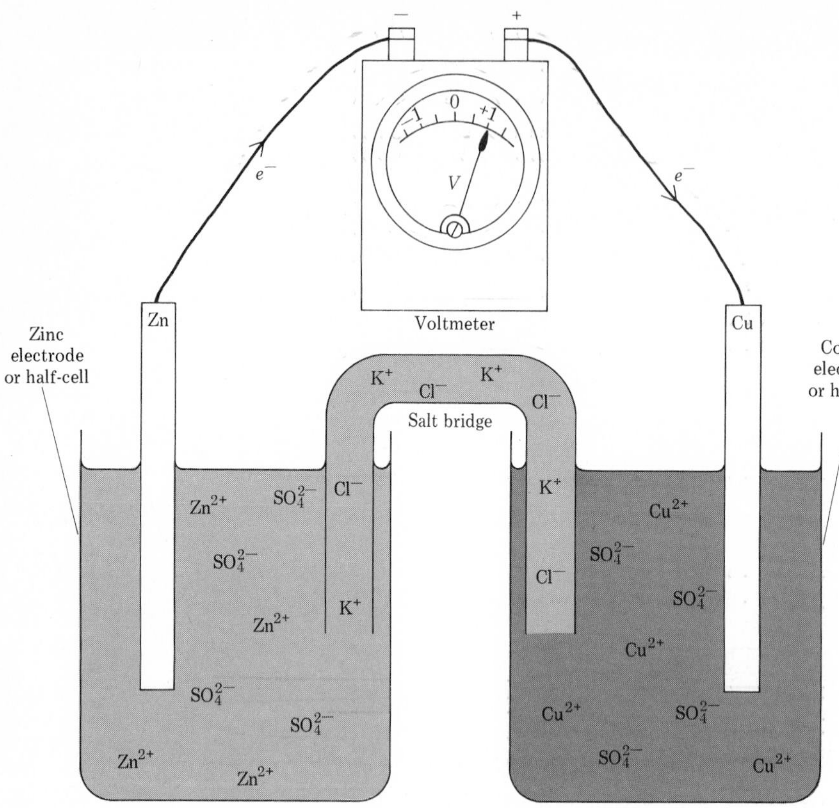

Figure 17.5 A simple galvanic cell.

The components of the redox couples at the electrodes in a galvanic cell need not always be a solid and a species in solution. This is evident from Fig. 17.6. In this case the spontaneous redox reaction

$$2Fe^{2+}(aq) + Cl_2(g) \rightarrow 2Fe^{3+}(aq) + 2Cl^-(aq) \qquad (17.11)$$

is involved. The oxidation half-equation at the anode is

$$Fe^{2+}(aq) \rightarrow Fe^{3+}(aq) + e^- \qquad (17.11a)$$

Thus at the right-hand electrode in Fig. 17.6 both components of the redox couple are in aqueous solution. Reaction (17.11*a*) occurs at the surface of the platinum wire, which conducts the released electrons to the external circuit.

The left-hand electrode in Fig. 17.6 is a **gas electrode.** It consists of a platinum strip dipping in a solution which contains chloride ions. The electrode is surrounded by a glass tube through which chlorine gas can be pumped. At this electrode the reaction is a reduction:

$$Cl_2(g) + 2e^- \rightarrow 2Cl^-(aq) \qquad (17.11b)$$

Therefore the left-hand electrode is the cathode. Since electrons are forced into the external circuit at the anode and withdrawn at the cathode, electrons flow from right to left in this cell.

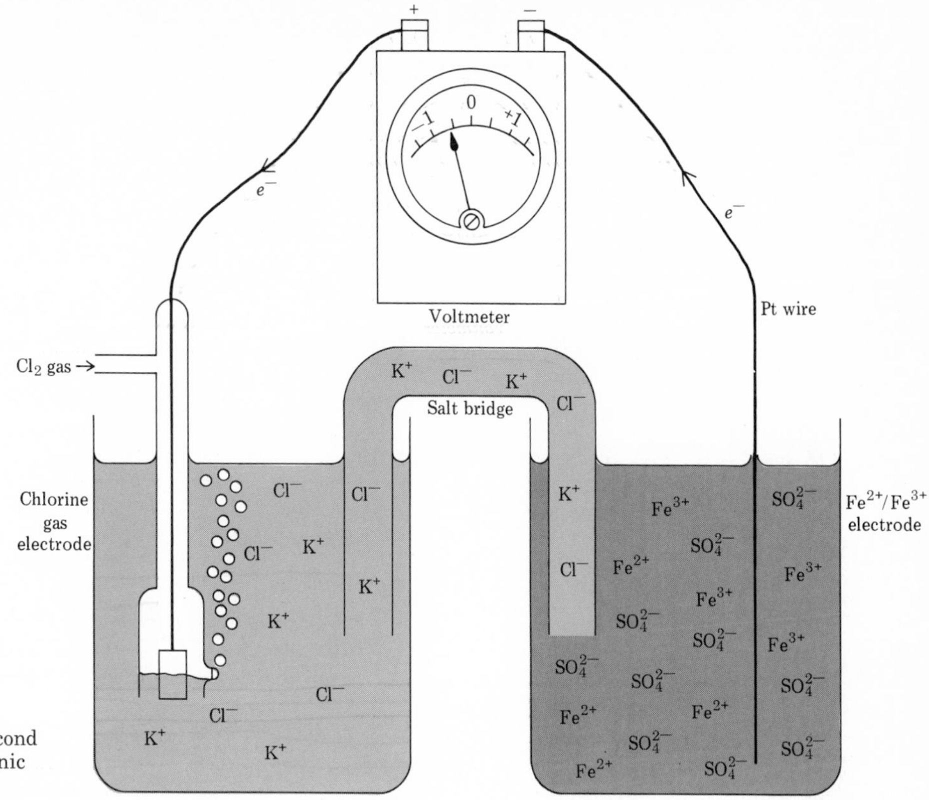

Figure 17.6 A second example of a galvanic cell.

Cell Notation and Conventions

Rather than drawing a complete diagram like those in Fig. 17.5 and 17.6, it is convenient to specify a galvanic cell in shorthand form. The two cells we have just described would be written as

$$Zn\,|\,Zn^{2+}(1\ M)\,\|\,Cu^{2+}(1\ M)\,|\,Cu \qquad (17.12)$$

and $$Pt,\ Cl_2(g)\,|\,Cl^-(1\ M)\,\|\,Fe^{2+}(1\ M),\ Fe^{3+}(1\ M)\,|\,Pt$$

The components of the cell are written in order, starting with the left-hand electrode and moving across the salt bridge to the right. A single vertical line indicates a phase boundary, such as that between the solid Zn electrode and $Zn^{2+}(aq)$, or between $Cl_2(g)$ and $Cl^-(aq)$. The double vertical line represents a salt bridge. Spectator ions, like $SO_4^{2-}(aq)$ in the Zn–Cu cell, are usually omitted.

By convention, the electrode written to the *left* of the salt bridge in this cell notation is always taken to be the *anode*, and the associated half-equation is always written as an *oxidation*. The right-hand electrode is therefore always the *cathode*, and the half-equation is always written as a *reduction*. This is easy to remember, because reading from left to right gives anode and cathode in alphabetical order. The **cell reaction** corresponding to a given shorthand description is obtained by summing the half-equations after multiplying by any factors necessary to equalize the number of electrons lost at the anode with the number gained at the cathode.

EXAMPLE 17.3 Write the half-equations and cell reactions for each of the following cells:

a) $Ag|Ag^+||H^+|H_2$, Pt

b) $Pt|Cr_2O_7^{2-}, Cr^{3+}, H^+||Br^-|Br_2(l)$, Pt

Solution

a) Since the half-equation at the left electrode is taken by convention to be oxidation, we have

$$Ag \rightarrow Ag^+ + e^-$$

At the right-hand electrode, then, we must have a reduction:

$$2H^+ + 2e^- \rightarrow H_2$$

Multiplying the first half-equation by 2 and summing gives the cell reaction

$$2Ag + 2H^+ \rightarrow 2Ag^+ + H_2$$

b) Following the same procedure as in part *a*, we obtain

$2Cr^{3+} + 7H_2O \rightarrow Cr_2O_7^{2-} + 14H^+ + 6e^-$	Left electrode
$(Br_2(l) + 2e^- \rightarrow 2Br^-) \times 3$	Right electrode
$2Cr^{3+} + 7H_2O + 3Br_2(l) \rightarrow Cr_2O_7^{2-} + 14H^+ + 6Br^-$	Cell reaction

Note: The procedures developed in Sec. 11.5 for balancing redox equations were used in arriving at the above results.

EXAMPLE 17.4 Describe in shorthand notation a galvanic cell for which the cell reaction is

$$Cu(s) + 2Fe^{3+}(aq) \rightarrow Cu^{2+}(aq) + 2Fe^{2+}(aq)$$

Solution First divide the cell reaction into half-equations:

Oxidation: $Cu \rightarrow Cu^{2+} + 2e^-$

Reduction: $Fe^{3+} + e^- \rightarrow Fe^{2+}$

Then write the oxidation as the left-hand electrode and the reduction on the right:

$$Cu|Cu^{2+}||Fe^{2+}, Fe^{3+}|Pt$$

(Since both Fe^{2+} and Fe^{3+} are in solution, a Pt electrode is used.)

The conventions we have developed can be used to decide whether the cell reaction is actually spontaneous. If it is, an oxidation will release electrons to the external circuit at the left-hand electrode. If a voltmeter is placed in the circuit, these electrons will make its left-hand terminal negative. Since the right-hand electrode corresponds to reduction, electrons will be withdrawn from the right-hand terminal of the voltmeter. This is shown for the Zn–Cu cell in Fig. 17.5. You can readily confirm that the spontaneous cell reaction [Eq. (17.10)] corresponds to the shorthand cell notation of Eq. (17.12).

For the cell shown in Fig. 17.6, the shorthand notation is

$$\text{Pt, } Cl_2(g)\,|\,Cl^-(1\ M)\,\|\,Fe^{2+}(1\ M),\ Fe^{3+}(1\ M)\,|\,\text{Pt}$$

According to the conventions we have just developed, this corresponds to the cell reaction

$$2Cl^-(aq) + 2Fe^{3+}(aq) \rightarrow Cl_2(g) + 2Fe^{2+}(aq)$$

But this is the *reverse* of the spontaneous cell reaction we described before [Eq. (17.11)]. Since the cell reaction is nonspontaneous, electrons will *not* be forced into the external circuit at the left-hand electrode, and they will not be withdrawn at the right. In fact the *reverse* will actually occur. Thus if a voltmeter is connected to this cell, its right-hand terminal will become more negative and its left-hand terminal will become more positive. This was shown in Fig. 17.6.

In general, if a galvanic cell is connected to a voltmeter, the electrode connected to the negative terminal of the meter must be the anode. If our shorthand cell notation shows that electrode on the left, then the corresponding cell reaction must be spontaneous. Electrons will be released by the oxidation half-equation on the left and accepted by the reduction on the right. If, on the other hand, the voltmeter shows that the right-hand electrode is releasing electrons, then we must have written our shorthand notation *backwards*. This means that the reverse of the cell reaction obtained by our rules must actually be occurring, and it is that reverse reaction which is spontaneous. Thus by simply observing which electrode in the cell releases electrons and which accepts them, that is, by finding which electrode is negative and which positive, we can determine whether the cell reaction is spontaneous.

17.5 ELECTROMOTIVE FORCE OF GALVANIC CELLS

Using a voltmeter to measure the **electrical potential difference** (commonly called voltage) between two electrodes provides a quantitative indication of just how spontaneous a redox reaction is. The potential difference is measured in **volts** (V), an SI unit which corresponds to one joule per ampere-second ($1\ V = 1\ J\ A^{-1}\ s^{-1}$). The voltage indicates the tendency for current to flow in the external circuit, that is, it shows how strongly the anode reaction can push electrons into the circuit and how strongly the cathode reaction can pull them out. The potential difference is greatest when a large electrical resistance in the external circuit prevents any current from flowing. The maximum potential difference which can be measured for a given cell is called the **electromotive force,** abbreviated emf and represented by the symbol E.

By convention, when a cell is written in shorthand notation, its *emf is given a positive value if the cell reaction is spontaneous.* That is, if the electrode on the left forces electrons into the external circuit and the electrode on the right withdraws them, then the dial on the voltmeter gives the cell emf. On the other hand, if the half-cell on the right side of the shorthand cell notation is releasing electrons, making the right-hand terminal of the voltmeter negative, the cell emf is *minus* the reading of the meter. This corresponds to a nonspontaneous cell reaction, written in the conventional way.

EXAMPLE 17.5 When the galvanic cell shown in Fig. 17.6 is connected to a voltmeter, the reading is 0.59 V. The shorthand notation for this cell is

$$\mathrm{Pt, Cl_2}(g)\,|\,\mathrm{Cl^-}(1\ M)\,\|\,\mathrm{Fe^{2+}}(1\ M),\ \mathrm{Fe^{3+}}(1\ M)\,|\,\mathrm{Pt}$$

What is the value of the cell emf?

Solution We have already seen that this cell as written corresponds to a nonspontaneous reaction. Therefore the emf must be negative and $E = -0.59$ V.

EXAMPLE 17.6 If the voltmeter in Fig. 17.5 reads 1.10 V, what is the emf for the cell

$$\mathrm{Cu}\,|\,\mathrm{Cu^{2+}}(1\ M)\,\|\,\mathrm{Zn^{2+}}(1\ M)\,|\,\mathrm{Zn}$$

Solution In this case the shorthand notation corresponds to the reverse of Eq. (17.10); that is, it refers to the nonspontaneous cell reaction

$$\mathrm{Cu + Zn^{2+} \rightarrow Cu^{2+} + Zn}$$

Consequently the emf for this cell must be negative and $E = -1.10$ V.

Example 17.6 shows that if the cell notation is written in reverse, the cell emf changes sign, since for the spontaneous reaction shown in Eq. (17.10) the emf would have been +1.10 V.

Experimentally measured cell emf's are found to depend on the concentrations of species in solution and on the pressures of gases involved in the cell reaction. Consequently it is necessary to specify concentrations and pressures when reporting an emf, and we shall only consider cells in which all concentrations are 1 mol dm^{-3} and all pressures are 1 atm (101.3 kPa).

The emf of such a cell is said to be its **standard electromotive force** and is given the symbol $E^\ominus$.

The electromotive forces of galvanic cells are found to be additive. That is, if we measure the emf's of the two cells

$$\text{Zn}|\text{Zn}^{2+}(1\ M)\|\text{H}^+(1\ M)|\text{H}_2(1\text{ atm}),\text{ Pt} \qquad E^\ominus = 0.76\text{ V} \qquad (17.13)$$

$$\text{Pt, H}_2(1\text{ atm})|H^+(1\ M)\|\text{Cu}^{2+}(1\ M)|\text{Cu} \qquad E^\ominus = 0.34\text{ V} \qquad (17.14)$$

the sum of the $E^\ominus$ values corresponds to the measured emf for a third cell with which we are already familiar:

$$\text{Zn}|\text{Zn}^{2+}(1\ M)\|\text{Cu}^{2+}(1\ M)|\text{Cu} \qquad E^\ominus = 1.10\text{ V} \qquad (17.12)$$

Whenever the right-hand electrode of one cell is identical to the left-hand electrode of another, we can add the emf's in this way, canceling the electrode which appears twice. This additivity makes it possible to store a large amount of emf data in a small table. By convention such data are tabulated as **standard electrode potentials.** These refer to the emf of a cell whose left-hand electrode is the hydrogen-gas electrode and whose right-hand electrode is the electrode whose emf is being sought. Table 17.1 contains a number of useful standard electrode potentials.

As an example of the use of the table, the entry corresponding to the electrode $Cu^{2+}(1\ M)|Cu$ is +0.34 V. Thus when this electrode is written

TABLE 17.1 Standard Electrode Potentials at 298.15 K.

Electrode	Half-Equation	$E^\ominus$/V
$F^-\|F_2(g)$, Pt	$F_2(g) + 2e^- \rightarrow 2F^-(aq)$	+2.87
H^+, $H_2O_2\|$Pt	$H_2O_2(aq) + 2H^+(aq) + 2e^- \rightarrow 2H_2O$	+1.77
H^+, MnO_4^-, $Mn^{2+}\|$Pt	$MnO_4^-(aq) + 8H^+(aq) + 5e^- \rightarrow Mn^{2+}(aq) + 4H_2O$	+1.52
$Cl^-\|Cl_2(g)$, Pt	$Cl_2(g) + 2e^- \rightarrow 2Cl^-(aq)$	+1.36
$Cr_2O_7^{2-}$, $Cr^{3+}\|$Pt	$Cr_2O_7^{2-}(aq) + 14H^+(aq) + 6e^- \rightarrow 2Cr^{3+}(aq) + 7H_2O$	+1.33
H^+, $H_2O\|O_2(g)$, Pt	$O_2(g) + 4H^+(aq) + 4e^- \rightarrow 2H_2O$	+1.19
$Br^-\|Br_2(l)$, Pt	$Br_2(l) + 2e^- \rightarrow 2Br^-$	+1.07
OCl^-, $Cl^-\|$Pt	$OCl^-(aq) + H_2O + 2e^- \rightarrow Cl^-(aq) + 2OH^-(aq)$	+0.89
$Hg^{2+}\|Hg$	$Hg^{2+} + 2e^- \rightarrow Hg$	+0.85
$Ag^+\|Ag$	$Ag^+ + e^- \rightarrow Ag$	+0.80
Fe^{3+}, $Fe^{2+}\|$Pt	$Fe^{3+} + e^- \rightarrow Fe^{2+}$	+0.77
$I^-\|I_2(s)$, Pt	$I_2(s) + 2e^- \rightarrow 2I^-$	+0.54
$Cu^{2+}\|Cu$	$Cu^{2+}(aq) + 2e^- \rightarrow Cu(s)$	+0.34
$Cl^-\|AgCl$, Ag	$AgCl(s) + e^- \rightarrow Ag(s) + Cl^-(aq)$	+0.22
Sn^{4+}, $Sn^{2+}\|$Pt	$Sn^{4+}(aq) + 2e^- \rightarrow Sn^{2+}$	+0.15
$H^+\|H_2(g)$, Pt	$H^+(aq) + e^- \rightarrow \frac{1}{2}H_2(g)$	0.00
$Pb^{2+}\|Pb$	$Pb^{2+}(aq) + 2e^- \rightarrow Pb(s)$	−0.13
$Sn^{2+}\|Sn$	$Sn^{2+}(aq) + 2e^- \rightarrow Sn(s)$	−0.14
$Fe^{2+}\|Fe$	$Fe^{2+}(aq) + 2e^- \rightarrow Fe(s)$	−0.44
$Zn^{2+}\|Zn$	$Zn^{2+}(aq) + 2e^- \rightarrow Zn(s)$	−0.76
$Al^{3+}\|Al$	$Al^{3+}(aq) + 3e^- \rightarrow Al(s)$	−1.66
$Mg^{2+}\|Mg$	$Mg^{2+}(aq) + 2e^- \rightarrow Mg(s)$	−2.36
$Na^+\|Na$	$Na^+(aq) + e^- \rightarrow Na(s)$	−2.71
$Li^+\|Li$	$Li^+(aq) + e^- \rightarrow Li(s)$	−3.05

to the right of Pt, H_2(1 atm)|H^+(1 *M*), as in Eq. (17.14) above, the $E^\ominus$ is +0.34 V. For the Zn^{2+}|Zn redox couple, we find $E^\ominus = -0.76$ V in Table 17.1. This means that for the cell

$$\text{Pt, H}_2(1\text{ atm})|\text{H}^+(1\ M)\|\text{Zn}^{2+}(1\ M)|\text{Zn} \qquad E^\ominus = -0.76\text{ V}$$

Since Eq. (17.13) shows this cell in reverse, we change the sign of $E^\ominus$, obtaining +0.76 V. Thus we can combine standard electrode potentials from Table 17.1 to obtain emf's for cells like Eq. (17.12) so long as both electrodes are given in the table.

EXAMPLE 17.7 Find the standard emf for the cell

$$\text{Hg}(l)|\text{Hg}^{2+}(1\ M)\|\text{Br}^-(1\ M)|\text{Br}_2(l)\text{, Pt}$$

Solution From Table 17.1 we have

$$\text{Pt, H}_2(1\text{ atm})|\text{H}^+(1\ M)\|\text{Hg}^{2+}(1\ M)|\text{Hg}(l) \qquad E^\ominus = +0.85\text{ V}$$

Since we want Hg(l)|Hg^{2+}(1 *M*) to be the left-hand electrode, this must be reversed and the sign of $E^\ominus$ must be changed:

$$\text{Hg}(l)|\text{Hg}^{2+}(1\ M)\|\text{H}^+(1\ M)|\text{H}_2(1\text{ atm})\text{, Pt} \qquad E^\ominus = -0.85\text{ V} \qquad (17.15)$$

For the other electrode Table 17.1 gives

$$\text{Pt, H}_2(1\text{ atm})|\text{H}^+(1\ M)\|\text{Br}^-(1\ M)|\text{Br}_2(l)\text{, Pt} \qquad E^\ominus = +1.07\text{ V} \qquad (17.16)$$

Adding the cells of Eqs. (17.15) and (17.16), we obtain

$$\text{Hg}(l)|\text{Hg}^{2+}(1\ M)\|\text{Br}^-(1\ M)|\text{Br}_2(l)\text{, Pt} \qquad E^\ominus = (1.07 - 0.85)\text{ V} = +0.22\text{ V}$$

The positive value of the standard emf obtained in Example 17.7 indicates that the corresponding cell reaction is spontaneous:

$$\text{Hg}(l) + \text{Br}_2(l) \rightarrow \text{Hg}^{2+}(1\ M) + 2\text{Br}^-(1\ M)$$

In other words, bromine is a strong enough oxidizing agent to convert mercury metal to mercury(II) ions in aqueous solution, assuming the concentrations of mercury(II) and bromide ions to be 1 mol dm^{-3}. This corresponds to the observations made in color Plate 3, where liquid mercury combined with liquid bromine to form mercury(II) bromide. Thus the standard electrode potentials in Table 17.1 can be used to predict whether a particular reaction will take place, just as Table 11.5 was used in our earlier discussion of redox reactions. The advantage of Table 17.1 is that it gives quantitative as well as qualitative information. It not only tells us that Br_2(l) is a stronger oxidizing agent than Hg^{2+}(1 *M*) [because Br_2(l) is above Hg^{2+}(1 *M*)], but it also tells us how much stronger, in terms of the cell emf of +0.22 V.

Very few of the cells obtained by combining the electrodes in Table 17.1 are suitable for everyday use as a source of electrical energy. The chief reason for this is that most of them can only deliver a very small current per unit area of electrode and need to be made very large before they become useful. A less serious difficulty is that they all involve solutions, many of them poisonous or corrosive which must be stored in a robust splash-proof container. Neither of these two difficulties is encountered with the common flashlight battery, known as the **Leclanché cell** or **dry cell.**

Despite the name dry cell, this battery does contain an electrolyte solution but only in the form of a thick paste. A saturated solution of ammonium chloride also containing zinc chloride is mixed with ammonium chloride crystals, as well as some inert filler like diatomaceous earth. As shown in Fig. 17.7, the center of the dry cell is occupied by a graphite rod which functions purely as a conductor in much the same way as a platinum wire in laboratory cells described in the previous section. This rod is surrounded by a powdered mixture of graphite and manganese dioxide, MnO_2. This powder in turn is surrounded by the NH_4Cl–$ZnCl_2$ paste, and the whole is encased in a thin zinc cylinder which acts as the second electrode and also as the container for the cell. In some cases a further cylinder of steel is wrapped around the zinc for added mechanical strength.

In shorthand notation the dry cell corresponds to

$$\text{Zn}|\text{Zn}^{2+}, \text{NH}_4^+|\text{MnO}_2(s), \text{C}$$

while the electrode half-equations can be represented as

$$\text{Zn} \rightarrow \text{Zn}^{2+}(aq) + 2e^-$$

and $$2\text{MnO}_2(s) + 2\text{NH}_4^+ + 2e^- \rightarrow 2\text{MnO(OH)}(s) + 2\text{NH}_3(aq)$$

Figure 17.7 The Leclanché or dry cell.

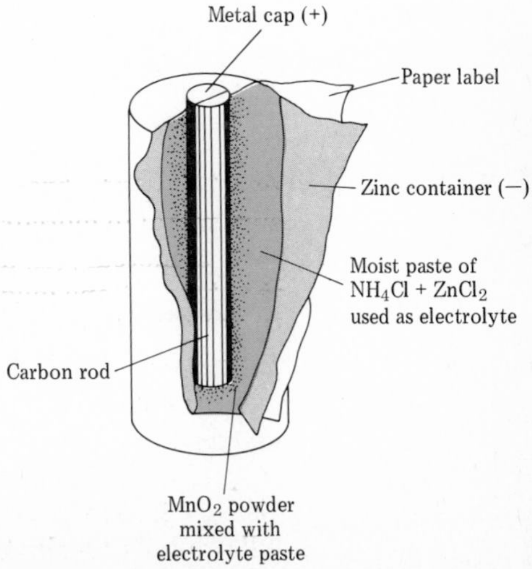

in which the manganese is reduced from Mn(IV) in MnO_2 to Mn(III) in MnO(OH). Note that this cell contains no salt bridge or other device for separating the two electrodes. This is possible because both the oxidizing agent (MnO_2) and the reducing agent (Zn) are solids and so cannot diffuse toward each other and react. Nevertheless dry cells have a limited life. The zinc electrode is eventually eaten away by the slightly acidic ammonium chloride solution, which can also ruin a flashlight if old batteries are left inside. A recent improvement to the dry cell uses sodium hydroxide in place of ammonium chloride. Such an alkaline battery delivers a larger quantity of electrical energy but is also more expensive because of the necessity for a stronger casing to prevent leakage. Both types of batteries deliver a potential difference of about 1.5 V.

Another commonly used modern battery, found in electric and electronic watches, hearing aids, and light meters is the **mercury cell.** This cell has the form

$$\text{Zn(Hg), ZnO}(s)\,|\,\text{OH}^-(8\ M)\,|\,\text{HgO}(s),\ \text{Hg}(l)$$

with the following electrode reactions:

$$\text{Zn} + 2\text{OH}^-(aq) \rightarrow \text{ZnO}(s) + \text{H}_2\text{O} + 2e^-$$

and

$$\text{HgO}(s) + \text{H}_2\text{O} + 2e^- \rightarrow \text{Hg}(l) + 2\text{OH}^-$$

Again the use of a solid reducing agent and a solid oxidizing agent obviates the necessity of separating the contents of the two electrodes. A mercury cell delivers a potential difference of about 1.34 V.

Storage Batteries

A great disadvantage of the two cells we have described is that they must be discarded once the cell reaction has gone to completion. Since they contain valuable metals like mercury and manganese whose supplies will soon become more limited, we can expect that batteries like this will become progressively more expensive and the impetus will grow to recycle the materials they contain, perhaps through a system of returnable deposits. A much more satisfactory solution to this problem, though, is the increased use of batteries which are rechargeable. Such batteries are called **storage batteries,** and they have the property that once the cell reaction has gone to completion, it can easily be reversed by electrolysis.

The most familiar example of a cell of this type is the **lead storage battery** used in automobiles, illustrated in Fig. 17.8. The shorthand description of this cell is

$$\text{Pb, PbSO}_4(s)\,|\,\text{H}_2\text{SO}_4(5\ M)\,|\,\text{PbSO}_4(s),\ \text{PbO}_2(s),\ \text{Pb}$$

and the electrode reactions are

$$\text{Pb}(s) + \text{SO}_4^{2-}(aq) \rightarrow \text{PbSO}_4(s) + 2e^- \qquad (17.17a)$$

and

$$2e^- + 4\text{H}^+ + \text{SO}_4^{2-} + \text{PbO}_2 \rightarrow \text{PbSO}_4(s) + 2\text{H}_2\text{O} \qquad (17.17b)$$

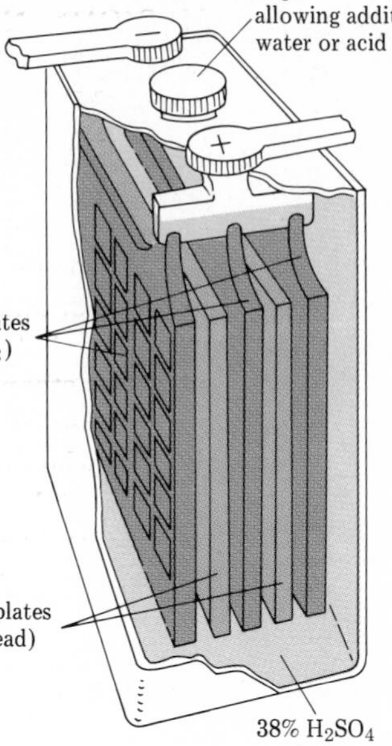

Figure 17.8 The lead storage battery.

giving an overall cell reaction

$$Pb(s) + PbO_2(s) + 4H^+ + 2SO_4^{2-} \rightarrow 2PbSO_4 + 2H_2O \qquad (17.17)$$

Both electrodes are similarly constructed of a flat grid of lead. For the left-hand electrode, the interstices are filled with finely divided spongy Pb of high surface area, while for the right-hand electrode chocolate-brown lead dioxide powder, PbO_2, is used.

As the cell discharges, both electrodes become plated with white $PbSO_4$. Some storage batteries are made with transparent polystyrene cases. When this is the case, it is very apparent when the battery has been discharged since both electrodes begin to acquire a similar white appearance. The discharge of the battery also results in the consumption of H_2SO_4 [see Eq. (17.17)]. As the H_2SO_4 becomes more dilute, its density decreases, and so by measuring the density of the acid, we can decide whether the battery needs recharging or not.

Once a lead storage battery has been discharged, it can be recharged by electrolyzing it with a source of direct current. This results in the cell reaction reversing itself. The $PbSO_4$ disappears from both electrodes, and the concentration of H_2SO_4 increases. Even when the cell becomes fully charged, little damage is done by continued electrolysis. H_2 gas is evolved at one electrode and O_2 at the other. Lead storage cells do not last indefinitely, however. Some of the $PbSO_4$ formed at the electrodes becomes dislodged and falls to the bottom of the container where it is no longer available to take part in the recharging electrolysis. If lead storage batteries are allowed to discharge completely, this loss of $PbSO_4$ is particularly liable to occur. Batteries which are not mistreated in this way inevitably last longer. In a car battery three or six lead cells are connected in series. Since each produces 2.0 V when fully charged, the resultant potential difference is 6 or 12 V.

A second everyday example of a storage battery is the nickel-cadmium battery now commonly used in electronic calculators. These cells have the following construction:

$$Cd(s),\ CdO(s)\,|\,KOH(1\ M)\,|\,Ni(OH)_3(s),\ Ni(OH)_2(s),\ Ni$$

with the following electrode half-equations:

$$Cd(s) + 2OH^- \rightarrow CdO + 2H_2O + 2e^-$$

$$2e^- + 2Ni(OH)_3 \rightarrow 2Ni(OH)_2 + 2OH^-$$

This cell produces a potential difference of 1.25 V and can be used in place of the dry cell for many purposes. A simple contribution anyone can make toward conserving the world's supply of zinc is to invest in a simple battery charger and use nickel-cadmium storage batteries to replace dry cells.

Fuel Cells

A type of galvanic cell which promises to become increasingly important in the future is the **fuel cell.** By contrast to a conventional cell, where only limited quantities of oxidizing agent and reducing agent are available, a continuous supply of both is provided to a fuel cell, and the reaction product is continually removed. A somewhat oversimplified diagram of a fuel cell in which the cell reaction is the production of water from hydrogen and oxygen is shown in Fig. 17.9. Hydrogen enters the cell through a porous

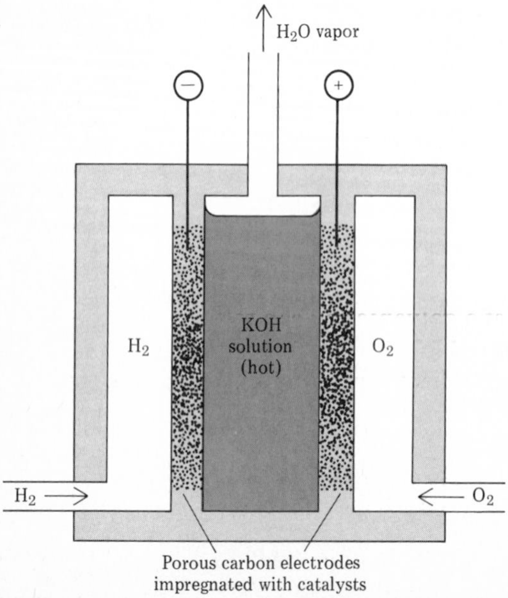

Figure 17.9 A hydrogen-oxygen fuel cell.

carbon electrode which also contains a platinum catalyst. Oxygen is supplied to a similar electrode except that the catalyst is silver. The electrolyte is usually a warm solution of potassium hydroxide, and the two electrode reactions can be written as

$$H_2(g) + 2OH^-(aq) \rightarrow 2H_2O(l) + 2e^- \qquad (17.18a)$$

and

$$\tfrac{1}{2}O_2(g) + H_2O + 2e^- \rightarrow 2OH^-(aq) \qquad (17.18b)$$

giving the overall result

$$H_2(g) + \tfrac{1}{2}O_2(g) \rightarrow H_2O \qquad (17.18)$$

Unless it is removed, water produced by the reaction will gradually dilute the potassium hydroxide, rendering the cell inoperative. Hence the electrolyte is kept warm enough that water evaporates just as fast as it is produced by the cell reaction. A fuel cell like this will continue to operate and produce electrical energy as long as a supply of hydrogen and oxygen are available.

Fuel cells have an important advantage over all other devices which burn fuel in order to obtain useful energy: their efficiency. While an internal-combustion engine is only about 25 percent efficient and a steam engine about 35 percent efficient, the H_2–O_2 cell just described can already operate at an efficiency of 45 percent. The theoretically highest possible efficiency of such a cell, set by the second law of thermodynamics, is 83 percent. Because of this high efficiency many possible uses and developments for fuel cells have been proposed. One of these scenarios for the future envisions large nuclear power plants floating on the sea producing hydrogen gas by the decomposition of water rather than producing electrical power. This hydrogen gas could then be piped to individual homes where it could either be burned for heat or converted to electricity with the aid of a fuel cell. A second scenario involves automobiles powered by cells fueled by conventional gasoline or perhaps hydrogen. These automobiles would run virtually noiselessly without any pollution problems and deliver twice as many kilometers per liter of fuel as a conventional vehicle. Alas for such scenarios, many technological problems still intervene, but further development of fuel cells is certainly one approach to our current energy problems that should be thoroughly investigated.

17.7 GALVANIC CELLS AND FREE ENERGY

We have already seen in this chapter that the emf of a galvanic cell can tell us whether the cell reaction is spontaneous. In Sec. 16.5 we showed that the free-energy change ΔG of a chemical process also indicates whether that process is spontaneous. It is quite reasonable, then, to expect some relationship between ΔG and E, and indeed one exists.

In Sec. 16.5 we stated that the free-energy change corresponds to the maximum quantity of useful work which can be obtained when a chemical reaction occurs. In other words,

$$\Delta G = -w_{max}$$

where the minus sign is necessary because the free energy decreases as the chemical system does useful work on its surroundings. If we are referring to a redox reaction, that work can be obtained in electrical form by means of an appropriate galvanic cell. It can be measured readily, because when a quantity of charge Q moves through a potential difference ΔV, the work done is given by

$$w = Q\,\Delta V$$

Thus if one coulomb passes through a potential difference of one volt, the work done is

$$w = 1\text{ C} \times 1\text{ V} = 1\text{ A s} \times 1\text{ J A}^{-1}\text{ s}^{-1} = 1\text{ J}$$

Now suppose we construct a Zn–Cu cell of the type described earlier:

$$Zn\,|\,Zn^{2+}(1\ M)\,\|\,Cu^{2+}(1\ M)\,|\,Cu$$

and suppose we make the cell large enough that the concentrations of Cu^{2+} and Zn^{2+} will not change significantly even though 1 mol Zn is oxidized to 1 mol Zn^{2+} according to the cell reaction

$$Zn(s) + Cu^{2+}(aq) \rightarrow Zn^{2+}(aq) + Cu(s) \qquad (17.10)$$

If this cell is discharged through a large enough resistance, the potential difference will have its maximum value, namely, the cell emf, $E^\ominus$; so if we know how much charge is transferred, we can calculate the electrical work done. For the oxidation of 1 mol Zn [that is, for the occurrence of 1 mol of reaction (17.10)], there must be 2 mol e^- transferred according to the half-equation

$$Zn(s) \rightarrow Zn^{2+}(aq) + 2e^-$$

Therefore the quantity of electrical charge transferred per mole of reaction is

$$Q_m = 2 \times F = 2 \times 9.649 \times 10^4\text{ C mol}^{-1} = 1.930 \times 10^5\text{ C mol}^{-1}$$

The maximum useful work per mole of reaction which the cell can perform while discharging is thus

$$\begin{aligned} w_{\text{max}} &= Q_m E^\ominus = 2FE^\ominus = 2 \times 9.649 \times 10^4\text{ C mol}^{-1} \times 1.10\text{ V} \\ &= 212\text{ kJ mol}^{-1} \end{aligned}$$

The standard molar free energy change for the cell reaction is thus

$$\Delta G_m^\ominus = -w_{\text{max}} = -2FE^\ominus = -212\text{ kJ mol}^{-1}$$

A similar argument can be applied to any cell in which the reactants and products are all at their standard concentrations or pressures. If the standard emf of such a cell is $E^\ominus$, while $\Delta G_m^\ominus$ is the standard molar free energy change for the cell reaction, then these two quantities are related by the equation

$$\Delta G_m^\ominus = -zFE^\ominus \qquad (17.19)$$

where z (a dimensionless number) corresponds to the number of moles of electrons transferred per mole of cell reaction.

A similar relationship holds even when reactants and products are not at standard concentrations and pressure:

$$\Delta G_m = -zFE$$

This connection between cell emf and free-energy change provides a means of measuring ΔG_m directly, rather than by determining ΔH_m and ΔS_m and then combining them.

EXAMPLE 17.8 The emf of the cell

$$\text{Pt, } H_2(1 \text{ atm}) | H^+(1\ M) | O_2(1 \text{ atm}), \text{ Pt}$$

is 1.229 V at 298.15 K. Calculate $\Delta G_f^\ominus$ for liquid water at this temperature

Solution The half-equations for the cell are

$$2H_2(g) \rightarrow 4H^+(aq) + 4e^-$$
$$4e^- + 4H^+(aq) + O_2(g) \rightarrow 2H_2O(l)$$

so that the cell reaction is

$$2H_2(g) + O_2(g) \rightarrow 2H_2O(l) \qquad 1 \text{ atm, } 298.15 \text{ K}$$

Since there are 4 mol e^- transferred per mol cell reaction, $z = 4$ and

$$\Delta G_m = -zFE = -4 \times \frac{9.649 \times 10^4 \text{ C}}{1 \text{ mol}} \times 1.229 \text{ V}$$
$$= -474.3 \text{ kJ mol}^{-1}$$

The reaction produces liquid water at standard pressure and the desired temperature, but 1 mol reaction produces 1 mol $2H_2O(l)$, that is, 2 mol H_2O. Therefore

$$\Delta G_f^\ominus[H_2O(l), 298 \text{ K}] = \tfrac{1}{2}\Delta G_m = -237.2 \text{ kJ mol}^{-1}$$

SUMMARY

In an electrolytic cell electric energy supplied from an outside source causes a nonspontaneous reaction to occur. A galvanic (or voltaic) cell, on the other hand, harnesses a spontaneous reaction to produce electric current. In either kind of cell the electrode at which oxidation occurs is called the anode and the electrode at which reduction occurs is the cathode.

Electrolytic cells have numerous commercial applications. Chlorine, sodium hydroxide, hydrogen, aluminum, magnesium, sodium, calcium, and high-purity copper are some of the more important chemicals produced by electrolysis. Electroplating of metals such as chromium, silver, nickel, zinc, and tin is also quite important. In any electrolysis reaction the amount of substance consumed or produced can be related to the electric charge which passes through the cell by means of the Faraday constant F, which equals 9.649×10^4 C mol^{-1}.

A galvanic cell may be represented by an abbreviated notation such as

$$Zn|Zn^{2+}(1\ M)\|Ag^{+}(1\ M)|Ag$$

When a cell is written this way, it is always assumed that the left-hand electrode is an anode and an oxidation half-equation occurs there. The right-hand electrode must then be taken as the cathode and a reduction half-equation is assumed to occur there. The cell reaction is the sum of these two half-equations. If it is spontaneous, our assumptions about anode on the left and cathode on the right were correct. Electrons will be forced into an external circuit on the left and the cell emf is taken to be positive. If the cell reaction written according to the above convention turns out to be nonspontaneous, then its reverse will be spontaneous. Our assumptions about which electrode is the anode and which the cathode must also be reversed, and the cell emf is given a negative sign.

Because cell emf values indicate whether a process is spontaneous, they are quite useful. They are additive and are conventionally reported as standard electrode potentials. These refer to the emf of a cell with a hydrogen-gas electrode on the left and the electrode whose potential is reported on the right. The standard electrode potential is directly related to the standard free energy change for a reaction, thus allowing direct determination of $\Delta G^{\ominus}$.

Many galvanic cells are of commercial importance. These include dry cells, mercury cells, rechargeable Ni-Cd batteries, and lead storage cells. Fuel cells, in which a continuous supply of both oxidizing and reducing agent is supplied, may eventually become important because of their high efficiencies.

QUESTIONS AND PROBLEMS

17.1 Define each of the following terms in your own words:

a Galvanic cell
b Cathode
c Oxidation
d Electrolytic cell
e Voltaic cell
f Cathode
g Reduction

17.2 In an electrolytic cell, anions are attracted toward the electrode called the ______, where they undergo ______. Cations are attracted to the ______ and undergo ______.

17.3 By tracing the flow of current from one terminal of the battery to the other, show that an electrolytic cell such as the one in Fig. 17.1 constitutes a complete electrical circuit.

17.4 Calcium reacts spontaneously with chlorine to form calcium chloride, $CaCl_2$. (*a*) Sketch an electrolytic cell in which molten $CaCl_2$ could be decomposed. (*b*) Label the electrodes as anode or cathode and write a balanced half-equation for the reaction at each electrode. (*c*) What is being reduced in the cell? What is being oxidized?

17.5 Suppose the electrolysis of CaF_2 is carried out on a 0.1 *M* solution. (*a*) What reaction would occur at the anode? (*b*) What reaction would occur at the cathode? (*c*) Write a balanced equation for the overall reaction during electrolysis of 0.1 *M* CaF_2.

17.6 Write a balanced chemical equation for the overall reaction during electrolysis of each of the following:

a 0.1 *M* $H_2SO_4(aq)$
b 0.1 *M* $CuSO_4(aq)$
c $KF(l)$
d 0.1 *M* $FeBr_2(aq)$
e $CaS(l)$
f 0.1 *M* $KF(aq)$

Hint: Since SO_4^{2-} contains S in its highest oxidation state, SO_4^{2-} is very difficult to oxidize.

17.7 Arrange the species listed below in order of decreasing strength as oxidizing agents. (You may wish to refer to either Table 11.5 or Table 17.1.)

$Cl_2(g)$, $Fe^{3+}(aq)$, $Zn^{2+}(aq)$, $I_2(s)$, $Pb(s)$, $AgCl(s)$

17.8 Arrange the species listed below in order of decreasing strength as reducing agents:

$Cr^{3+}(aq)$, $Cu(s)$, $Mg(s)$, $H_2(g)$, $Br^-(aq)$, $Na(s)$

17.9 For electrolysis of brine in a diaphragm cell, (*a*) write a net ionic equation for the overall reaction; (*b*) explain why the anode reaction involves Cl^- rather than H_2O or OH^-; and (*c*) explain why Na metal is not formed at the cathode.

17.10 For electrolysis of brine in a mercury cell, (*a*) write the anode reaction; (*b*) write the cathode reaction; (*c*) describe at least one advantage and one disadvantage of the mercury cell when compared with the diaphragm cell.

17.11 In a mercury cell the cathode is liquid mercury, while in a diaphragm cell it consists of steel mesh. How can this difference account for the difference in cathode reactions in the two types of cells? [*Hint:* Compare the reducing strength of pure $Na(s)$ with that of Na dissolved in Hg. Then use the rule developed in Sec. 11.7 that the stronger a reducing agent, the weaker its conjugate oxidizing agent.]

17.12 Two compounds of aluminum are involved in the Hall process. What are their names and formulas? Which is actually the source of the aluminum? What is the function of the other compound?

17.13 Write balanced half-equations and an overall equation for the Hall process.

17.14 Write names and formulas for at least two aluminum compounds which have lower melting points than Al_2O_3. If these were available at prices comparable to Al_2O_3, would you consider investing in a company which planned to extract aluminum from them? Why or why not?

17.15 Write half-equations, label the anode and cathode reactions, and write the overall equation for electrolytic refining of copper.

17.16 Would it be feasible to refine Zn, separating it from Pb impurities, by a method similar to electrolytic refining of Cu? Explain how it could be done or why it could not be done.

17.17 Use Table 11.5 to explain why $Ag(s)$ does not dissolve during electrolytic purification of $Cu(s)$. Why does $Fe(s)$ dissolve but not plate out on the cathode?

17.18 You obtain the same information about which metals will or will not dissolve from Table 17.1 as you did from Table 11.5 in the previous problem. From Table 17.1 choose one metal which would dissolve but not plate out and one which would not dissolve during electrolytic purification of Cu. Neither metal should be one that is listed in Table 11.5.

17.19 Suppose you have a sample of Fe which contains Al, Sn, and Cu impurities. After electrolytic purification, which metals would be in the anode sludge? Which metal ions would be in solution?

17.20 Predict the major products which will be obtained when each of the following aqueous solutions is electrolyzed. Each solution contains both solutes at 1 *M* concentration.

a $CuSO_4$ and HBr *c* KCl and HBr
b $ZnSO_4$ and $CuSO_4$ *d* SO_3 and $SnSO_4$

17.21 Sketch a cell for electroplating tin onto an iron surface from a solution of tin(II) ion, $Sn^{2+}(aq)$. What problem might arise if the $Sn^{2+}(aq)$ solution were open to the atmosphere?

17.22 Calculate the electric charge (in coulombs) needed to plate each amount of substance listed below onto a cathode, assuming the indicated redox couple.

a 2.75 mol Sn; Sn^{2+}/Sn
b 0.475 mol Mn; MnO_4^-/Mn
c 9.67 mol V; VO^{2+}/V
d 1.50×10^{-3} mol S; $S_2O_3^{2-}/S$

17.23 Assuming a current of 50.0 A, how long would it take to electroplate each amount of substance in Problem 17.18?

17.24 Criticize the following statement: To electroplate an object with chromium, the object is made the anode in a concentrated solution of $Cr^{3+}(aq)$.

17.25 What current would be required to plate 5.00 g of each of the metals listed below in 1.00 h, assuming the indicated redox couple.

a Ni; Ni^{2+}/Ni *c* Zn; Zn^{2+}/Zn
b Ag; Ag^+/Ag *d* Au; Au^{3+}/Au

17.26 What volume of $O_2(g)$, measured at 747 mmHg and 23.0°C, would be liberated by electrolysis of aqueous Na_2SO_4 when a current of 0.500 A flows for 40.0 min?

17.27 What will be the concentration of Ni^{2+} after 250 cm^3 of 0.100 *M* $NiSO_4$ has been electrolyzed by a current of 0.200 A passing for 30.0 min?

17.28 Three electrolytic cells are connected in series, so that the same current flows through all of them. In cell *A* 0.0257 g Cu plates out from a $CuSO_4$ solution. Cell *B* contains $AgNO_3(aq)$, and cell *C* contains $Cr(NO_3)_3(aq)$. What mass of Ag(*s*) will plate on the cathode in cell *B*? What amount of Cr(*s*) will plate out in cell *C*?

17.29 During operation of a Hall process aluminum cell, 500.0 kg of Al is produced. What will be the change in mass of the carbon anode in the cell?

17.30 Criticize the following statement: In order for a chemical reaction to be used in a galvanic cell, there must be an increase in the entropy of the chemical system.

17.31 Why is a salt bridge usually needed in a galvanic cell? In which direction would each type of ion move (on average) in the salt bridge shown in Fig. 17.5?

17.32 For each spontaneous reaction given below, write the half-equations that would occur at the anode and the cathode if the reaction were used in a voltaic cell:

a $2Fe^{3+}(aq) + Mg(s) \rightarrow 2Fe^{2+}(aq) + Mg^{2+}(aq)$

b $2AgCl(s) + Zn(s) \rightarrow 2Ag(s) + 2Cl^-(aq) + Zn^{2+}(aq)$

c $Cr_2O_7^{2-}(aq) + 14H^+(aq) + 6I^-(aq) \rightarrow 2Cr^{3+}(aq) + 3I_2(s) + 7H_2O$

d $OCl^-(aq) + H_2O(l) + Sn^{2+}(aq) \rightarrow Cl^-(aq) + 2OH^-(aq) + Sn^{4+}(aq)$

e $Cl_2(g) + Sn(s) \rightarrow Sn^{2+}(aq) + 2Cl^-(aq)$

17.33 Draw a diagram of a galvanic cell in which each of the reactions in Problem 17.32 could be used to produce an electrical current. Label the anode and the cathode and show which direction electrons would flow in an external circuit.

17.34 Criticize the following statement: A salt bridge is necessary in most galvanic cells because the cell reaction will not occur unless it is separated into two half-reactions in separate half-cells.

17.35 Shorthand notation and the standard emf are given below for several galvanic cells. In each case draw a picture of the cell as it might be set up in the laboratory. Label the anode and cathode and show the direction of movement of electrons in the external circuit. Write the anode and cathode half-equations and the overall cell reaction.

a $Pt|NO, NO_3^-, H^+\|AuCl_4^-, Cl^-|Au$; $E^\ominus = +2.04$ V

b $Pt|Sn^{2+}, Sn^{4+}\|Ag^+, Ag^{2+}|Pt$; $E^\ominus = +1.773$ V

c $Pt|V^{3+}, VO^{2+}, H^+\|Au^+|Au$; $E^\ominus = +1.39$ V

d $Al|Al^{3+}\|Zn(CN)_4^{2-}, CN^-|Zn$; $E^\ominus = +0.40$ V

17.36 For each of the cells in Problem 17.35 identify the chemical species which functions as a reducing agent and the species which functions as an oxidizing agent.

17.37 Write the shorthand cell notation corresponding to each of the spontaneous reactions in Problem 17.32.

17.38 Use Table 17.1 to calculate the standard emf for each cell whose shorthand notation is written below. Write the corresponding cell reaction and state whether it is spontaneous or nonspontaneous.

a $Pt, O_2(g)|H^+(1\ M), H_2O(l)\|I^-(1\ M)|I_2(s), Pt$

b $Pt|Sn^{2+}(1\ M), Sn^{4+}(1\ M)\|Pb^{2+}(1\ M)|Pb$

c $Zn|Zn^{2+}(1\ M)\|Fe^{3+}(1\ M), Fe^{2+}(1\ M)|Pt$

d $Pt|Cr^{3+}(1\ M), Cr_2O_7^{2-}(1\ M)\|Cu^{2+}(1\ M)|Cu$

e $Pt, Br_2(l)|Br^-(1\ M)\|Cl^-(1\ M)|Cl_2(g), Pt$

17.39 Write a balanced net ionic equation for the reaction which *actually occurs* when each of the cells in Problem 17.38 is set up in the laboratory and current is allowed to flow.

17.40 Predict whether each of the reactions listed below will occur spontaneously when all reagents are at standard concentrations and pressures. In each case calculate $\Delta G_m^\ominus$ to confirm your prediction.

a $2Ag^+(aq) + Pb(s) \rightarrow 2Ag(s) + Pb^{2+}(aq)$

b $2AgCl(s) + 2Br^-(aq) \rightarrow 2Ag(s) + 2Cl^-(aq) + Br_2(l)$

c $Sn(s) + Hg^{2+}(aq) \rightarrow Sn^{2+}(aq) + Hg(l)$

d $2H^+(aq) + Mg(s) \rightarrow H_2(g) + Mg^{2+}(aq)$

e $4Mn^{2+}(aq) + 6H_2O + 5O_2(g) \rightarrow 4MnO_4^-(aq) + 12H^+(aq)$

17.41 For each type of commercial cell listed below, write the shorthand cell notation, write the cell reaction, and describe the construction of the cell.

a Leclanché cell (dry cell)
b Lead storage cell
c Ni–Cd battery
d Mercury cell

17.42 Describe two ways in which the state of charge of a lead storage cell in a clear plastic case could be checked. Why do such cells eventually wear out and refuse to hold a charge?

***17.43** Suppose a fuel cell is constructed which uses the reaction

$$CH_4(g) + 2O_2(g) \rightarrow CO_2(g) + 2H_2O(l)$$

If such a cell is supplied with sufficient oxygen, what is the maximum electrical work it could do for every 1000 MCF of natural gas supplied? (1 MCF = 1000 ft^3 at 15.5°C and 1 atm.)

***17.44** Calculate the theoretical minimum electrical potential difference needed to cause electrolysis of each of the following solutions:

a $CuSO_4$(1 *M*) in a solution 1 *M* in H_3O^+.
b $ZnCl_2$(1 *M*) in a solution 1 *M* in H_3O^+.
c A solution of H_2SO_4 that is 1 *M* in H_3O^+.

Assume that Pt electrodes are used in each case.

chapter eighteen

CHEMICAL KINETICS

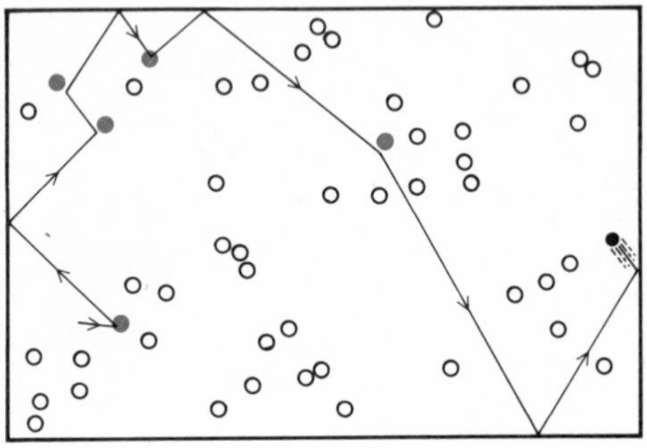

Chemical kinetics is concerned with the rates of chemical reactions, that is, whether reactions proceed quickly or slowly. As we have already mentioned, some spontaneous reactions are extremely slow. An example is the Haber-process synthesis of ammonia:

$$N_2 + 3H_2 \rightarrow 2NH_3 \qquad \Delta G_m^\ominus(298\ \text{K}) = -33.27\ \text{kJ mol}^{-1}$$

Even though a negative $\Delta G_m^\ominus$ predicts that this reaction can occur at room temperature, it is of little value unless chemists can find some way to speed it up. On the other hand we often want to slow down undesirable reactions, such as spoilage of food or decomposition of wood. Hence it is quite useful to know how factors such as temperature, concentrations of reactants and products, and catalysts will affect the rates of reactions. Moreover, studying these factors gives valuable information about the sequence of microscopic events by which a reaction occurs. Knowledge of when and where bonds are formed and broken as well as how molecular structures change during a reaction can be very useful in helping us to devise ways to speed up or slow down that reaction.

18.1 EXPERIMENTAL MEASUREMENT OF RATES

The Rate of Reaction

A chemist who investigates the rate of a reaction is often faced with a situation similar to that illustrated in Fig. 18.1. A solution of a dye is

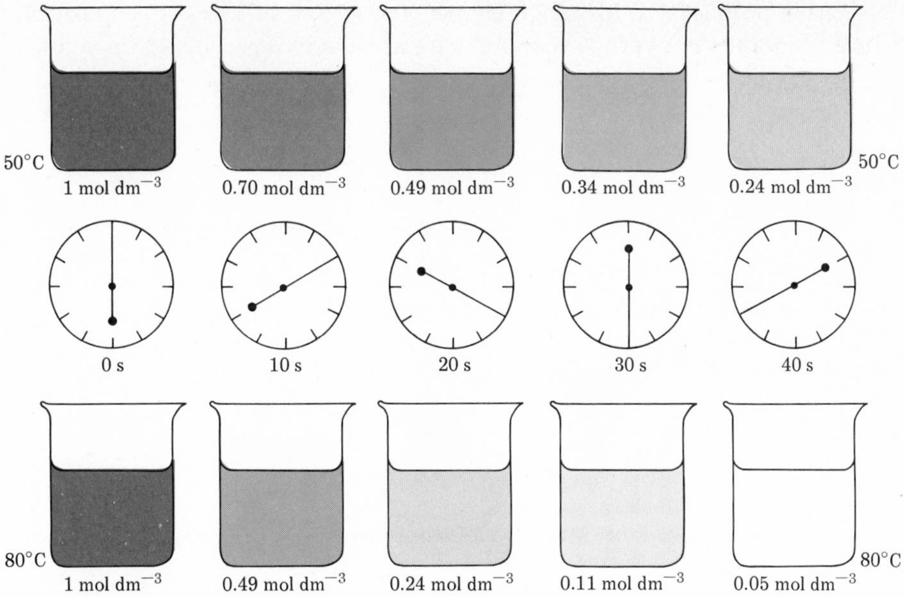

Figure 18.1 The decomposition of a colored dye at two different temperatures. The color of the solution fades gradually because the concentration of the dye decreases with time. The more gradual fading of the color at 50°C indicates that the dye decomposes more slowly at this temperature. How slow or fast a reaction occurs is measured by the rate at which the concentration changes with time.

shown decomposing into colorless products at two different temperatures. As the dye decomposes, its concentration decreases and the color of the solution becomes fainter. This happens in a shorter time at 80°C than at 50°C, indicating that the dye decomposes faster at the higher temperature. By convention, then, the rate of the reaction is described in terms of the *change in concentration per unit time* for a reactant or product. The faster that concentration changes, the faster the reaction is going.

As an example of the use of this definition of reaction rate, consider the first 10 s during decomposition of the dye at 50°C. According to Fig. 18.1 the concentration of dye drops to 0.70 mol dm^{-3} from an initial value of 1.0 mol dm^{-3}. If we represent the change in the concentration of the reactant dye by Δc_R, then

$$\Delta c_R = c_2 - c_1 = (0.70 - 1.00) \text{ mol dm}^{-3} = -0.30 \text{ mol dm}^{-3}$$

In other words the concentration of dye has *decreased* by 0.30 mol dm^{-3}. We can calculate an average rate of reaction by dividing this concentration decrease by the time interval ($\Delta t = t_2 - t_1 = 10 \text{ s} - 0 \text{ s} = 10 \text{ s}$) during which it occurred:

$$\text{Average rate} = \frac{0.30 \text{ mol dm}^{-3}}{10 \text{ s}} = 0.03 \text{ mol dm}^{-3} \text{ s}^{-1}$$

Clearly, no matter what reactant we observe, its concentration will decrease with time and Δc_R will be negative. On the other hand, the concentration of a reaction product will always increase, and so Δc_P will always be positive. Since we want the average reaction rate always to be positive, we define it as

$$\text{Rate} = \frac{-\Delta c_R}{\Delta t} = \frac{\Delta c_P}{\Delta t}$$

EXAMPLE 18.1 Calculate the average rate of decomposition of the dye in Fig. 18.1 at 80°C during the time interval (a) 0 to 10 s; (b) 10 to 20 s; (c) 20 to 30 s; (d) 0 to 30 s.

Solution

a) In the interval 0 to 10 s,

$$\Delta c_R = 0.49 \text{ mol dm}^{-3} - 1.00 \text{ mol dm}^{-3} = -0.51 \text{ mol dm}^{-3}$$

$$\Delta t = 10 \text{ s} - 0 \text{ s} = 10 \text{ s}$$

$$\text{Rate} = \frac{-\Delta c_R}{\Delta t} = \frac{-(-0.51 \text{ mol dm}^{-3})}{10 \text{ s}} = 0.051 \text{ mol dm}^{-3} \text{ s}^{-1}$$

b) $$\text{Rate} = \frac{-\Delta c_R}{\Delta t} = \frac{-(0.24 - 0.49) \text{ mol dm}^{-3}}{(20 - 10) \text{ s}} = 0.025 \text{ mol dm}^{-3} \text{ s}^{-1}$$

c) $$\text{Rate} = \frac{-\Delta c_R}{\Delta t} = \frac{-(0.11 - 0.24) \text{ mol dm}^{-3}}{(30 - 20) \text{ s}} = 0.013 \text{ mol dm}^{-3} \text{ s}^{-1}$$

d) $$\text{Rate} = \frac{-\Delta c_R}{\Delta t} = \frac{-(0.11 - 1.00) \text{ mol dm}^{-3}}{(30 - 0) \text{ s}} = 0.030 \text{ mol dm}^{-3} \text{ s}^{-1}$$

This example illustrates two important points about reaction rates. The first is that the rate of a reaction usually *decreases* with time, reaching a value of zero when the reaction is complete. This is usually because the rate depends in some way on the concentrations of one or more reactants, and as those concentrations decrease, the rate also decreases. The second point is a corollary to the first. Because the reaction rate changes with time, the rate we measure depends on the time interval used. For example, the average rate over the period 0 to 30 s was 0.030 mol dm^{-3} s^{-1}, but the average rate over the middle 10 s of that period was 0.025 mol dm^{-3} s^{-1}. The different average for the 30-s interval reflects the fact that the rate dropped from 0.051 to 0.013 mol dm^{-3} s^{-1} during that period.

To measure a reaction rate as accurately as possible at a particular time, we need to make the time interval Δt as small as we can so that there is the least possible change in rate over the interval. If you are familiar

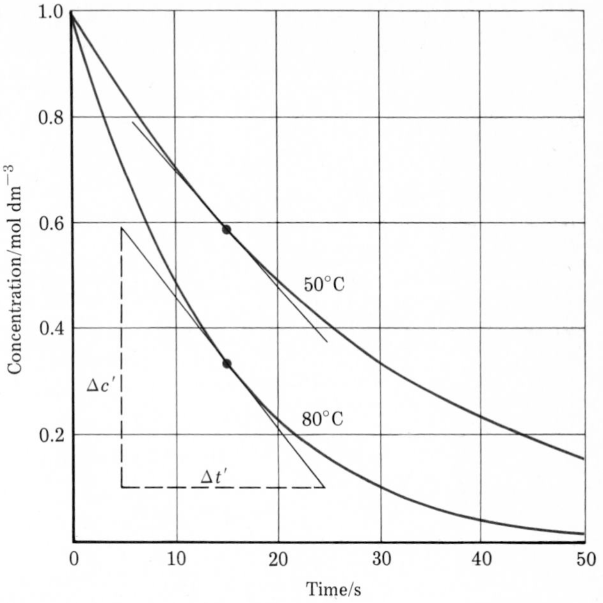

Figure 18.2 The concentration of the dye illustrated in Fig. 18.1 is plotted against time. The rate of the reaction at any time equals the slope of the tangent to the curve corresponding to that time. In the figure the tangents at t = 15 s have been drawn. The more downhill the slope of the tangent, the faster the reaction.

with the branch of mathematics known as the differential calculus, you will recognize that as Δt becomes smaller, $\Delta c/\Delta t$ approaches a limit known as the *derivative*. Thus the rate of reaction can be defined exactly as

$$\text{Rate} = \frac{-dc_R}{dt} = \frac{dc_P}{dt}$$

The derivative dc_R/dt also gives the slope of the tangent to a graph of c_R versus t. Such a graph is shown in Fig. 18.2, with a tangent line drawn at t = 15 s. The slope of this line is -0.0245 mol dm^{-3} s^{-1}, giving

$$\text{Rate} = \frac{-dc_R}{dt} = -\text{slope of tangent} = 0.0245 \text{ mol dm}^{-3}\text{ s}^{-1}$$

This exact value is quite close to the average rate of 0.025 mol dm^{-3} s^{-1} calculated for the interval 10 to 20 s, but it is farther from the 0.030 mol dm^{-3} s^{-1} calculated over the 0 to 30-s interval. As expected, the larger the time interval, the less accurate the calculated rate.

The Rate Equation

As we mentioned earlier, the rate of a reaction usually depends on the concentration of one or more of the reactants. A good example is the decomposition of the dye that we have already seen in Figs. 18.1 and 18.2. In this case it can be shown that the rate is directly proportional to the concentration of dye, c_D. That is,

$$\boxed{\text{Rate} = k_1 c_D} \qquad (18.1)$$

The constant of proportionality k_1 is called the **rate constant.** It is independent of the concentration of dye, but increases as temperature increases. This accounts for the greater rate of reaction at 80°C than at 50°C.

EXAMPLE 18.2 In Example 18.1 average rates were calculated for three 10-s time intervals for the decomposition of a dye at 80°C. Using the graph in Fig. 18.2 to obtain concentrations of the dye at various times, show that Eq. (18.1) is valid for this dye.

Solution Our previous discussion has indicated that the average reaction rate over a small time interval is very close to the actual rate at the midpoint of that interval. Thus we obtain the following table:

Average Rate/ mol dm^{-3} s^{-1}	Time Interval	Time from Start to Midpoint of Interval/s	Concentration of Dye/mol dm^{-3}
0.051	0–10 s	5	0.70
0.025	10–20 s	15	0.34
0.013	20–30 s	25	0.18

If Eq. (18.1) is valid, k_1 should not depend on the concentration of dye. Rearranging Eq. (18.1) to calculate the rate constant, we find that at 5 s,

$$k_1 = \frac{\text{rate}}{c_D} = \frac{0.051 \cancel{\text{mol dm}^{-3}}\ \text{s}^{-1}}{0.70 \cancel{\text{mol dm}^{-3}}} = 0.073\ \text{s}^{-1}$$

At 15 s,

$$k_1 = \frac{\text{rate}}{c_D} = \frac{0.025 \cancel{\text{mol dm}^{-3}}\ \text{s}^{-1}}{0.34 \cancel{\text{mol dm}^{-3}}} = 0.074\ \text{s}^{-1}$$

At 25 s,

$$k_1 = \frac{\text{rate}}{c_D} = \frac{0.013 \cancel{\text{mol dm}^{-3}}\ \text{s}^{-1}}{0.18 \cancel{\text{mol dm}^{-3}}} = 0.072\ \text{s}^{-1}$$

Thus the rate constant is constant to two significant digits. This is within the accuracy of the measurements.

The rate of a reaction may depend on the concentration of one or more products as well as reactants. In some cases it may even be influenced by the concentration of a substance which is neither a reactant nor a product in the overall stoichiometric equation for the reaction. An example of this latter situation is provided by the conversion of *cis*-2-butene to *trans*-2-butene:

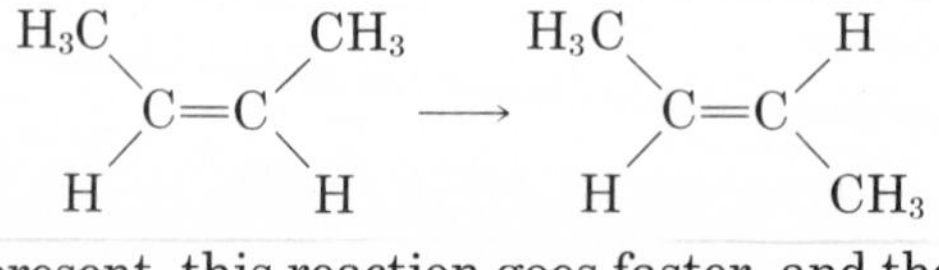

(18.2)

If some iodine is present, this reaction goes faster, and the rate law is found to be

$$\text{Rate} = k_2(c_{cis\text{-}2\text{-butene}})(c_{I_2})^{1/2} \tag{18.3}$$

Although iodine does not appear in Eq. (18.2), its concentration does affect the reaction rate. Consequently iodine is called a **catalyst** for the reaction. Later on in this chapter we will see how this catalytic effect of iodine actually works.

It should be clear from the examples we have just given that we cannot tell from a stoichiometric equation like Eq. (18.2) which reactants, products, or catalysts will affect the rate of a given reaction. Such information must be obtained from experiments in which the concentrations of various species are altered, and the effects of those alterations on the rate of reaction are observed. The results of such experiments are usually summarized in a **rate equation** or **rate law** for a given reaction. Equations (18.1) and (18.3), for example, are the rate laws for decomposition of a dye and for cis-trans isomerization of 2-butene (in the presence of the catalyst, I_2).

In general a rate equation has the form

$$\text{Rate} = k(c_A)^a(c_B)^b \ldots \tag{18.4}$$

It is necessary to determine experimentally which substances (A, B, etc.) affect the rate and to what powers (a, b, etc.) their concentrations must be raised. Then the rate constant k can be calculated. The exponents a, b, etc., are usually positive integers, but they may sometimes be fractions [as in Eq. (18.3)] or negative numbers. We say that the **order** of the reaction with respect to component A is a, while the order with respect to B is b. The overall order of the reaction is $a + b$.

EXAMPLE 18.3 For each reaction and experimentally measured rate equation listed below determine the order with respect to each reactant and the overall order:

a) $14H_3O^+ + 2HCrO_4^- + 6I^- \rightarrow 2Cr^{3+} + 3I_2 + 22H_2O$

$$\text{Rate} = k(c_{HCrO_4^-})(c_{I^-})^2(c_{H_3O^+})^2$$

b) $2I^- + H_2O_2 + 2H_3O^+ \rightarrow I_2 + 4H_2O$

$$\text{Rate} = k(c_{H_2O_2})(c_{I^-}) \qquad 3 \leqslant \text{pH} \leqslant 5$$

Solution

a) The reaction is second order in H_3O^+ ion, second order in I^- ion, and first order in $HCrO_4^-$. The overall order is $2 + 2 + 1 = 5$.

b) The rate law is first order in I^- and first order in H_2O_2, and so the overall order is 2. The concentration of H_3O^+ does not appear at all. When this happens the reaction is said to be zero order in H_3O^+, because $(c_{H_3O^+})^0 = 1$, and a factor of 1 in the rate law has no effect.

How can a rate law be obtained from experimental measurements? One way has already been illustrated in Example 18.2. There we guessed a rate law and then used it to calculate the rate constant k. Since k remained the same at various points during the reaction, we concluded that the rate equation was correct. Another way of obtaining a rate equation is illustrated in the next example.

EXAMPLE 18.4 The rate of the reaction

$$2NO + O_2 \rightarrow 2NO_2$$

was measured at 25°C, with various initial concentrations of NO and O_2. The following results were obtained for the initial rate of the reaction:

Experiment Number	Initial Concentrations		Initial Rate/ mol dm^{-3} s^{-1}
	c_{NO}/mol dm^{-3}	c_{O_2}/mol dm^{-3}	
1	0.020	0.010	0.028
2	0.020	0.020	0.057
3	0.020	0.040	0.114
4	0.040	0.020	0.227
5	0.010	0.020	0.014

Find the order of this reaction with respect to each reactant, the overall order, and the rate constant.

Solution In experiments 1, 2, and 3 the concentration of NO remains constant while the concentration of O_2 increases from 0.010 to 0.020 to 0.040 mol dm^{-3}. Each doubling of c_{O_2} also doubles the rate (from 0.028 to 0.057 to 0.114 mol dm^{-3} s^{-1}). Therefore the rate is proportional to c_{O_2}.

In experiments 2, 4, and 5 the concentration of O_2 remains constant. Comparing experiments 2 and 4, we find that doubling c_{NO} increases the rate from 0.057 to 0.227 mol dm^{-3} s^{-1}. This is a factor of 4, or 2^2. The same factor is observed comparing experiment 5 to experiment 2. Thus we conclude that the rate is proportional to the *square* of c_{NO}, and the rate equation must be

$$\text{Rate} = k(c_{O_2})(c_{NO})^2$$

The reaction is third-order overall—first order in O_2 and second order in NO.

The rate constant can be calculated by rearranging the rate equation and substituting the concentrations and rate from any one of the five experiments. From experiment 3

$$k = \frac{\text{rate}}{(c_{O_2})(c_{NO})^2} = \frac{0.114 \text{ mol dm}^{-3} \text{ s}^{-1}}{(0.040 \text{ mol dm}^{-3})(0.020 \text{ mol dm}^{-3})^2}$$

$$= \frac{0.114 \text{ mol dm}^{-3} \text{ s}^{-1}}{1.6 \times 10^{-5} \text{ mol}^3 \text{ dm}^{-9}} = 7.1 \times 10^3 \text{ mol}^{-2} \text{ dm}^6 \text{ s}^{-1}$$

A better value of the rate constant could be obtained by calculating for all five experiments and averaging the results.

The reaction described in the previous example is of considerable importance in the air above cities. As we mentioned in Sec. 12.5, automobile engines emit NO which is then oxidized to brown NO_2, an important precursor of photochemical air pollution. As we have just seen, this oxidation is second order in NO, and so when heavy traffic increases the concentration of NO by a factor of 10, the rate of production of NO_2 goes up by the considerably larger factor of 10^2, or 100.

18.2 MICROSCOPIC VIEW OF CHEMICAL REACTIONS

Now that we know something about how reaction rates are defined, measured, and related to the concentrations of substances which participate in a reaction, we would like to be able to interpret these macroscopic observations in terms of some microscopic model. On the microscopic level a chemical reaction involves transformation of reactant atoms, ions, and/or molecules into product atoms, ions, and/or molecules. This requires that some bonds be broken, other bonds be formed, and some nuclei be moved to new locations. There are a limited number of categories into which such microscopic transformations can be classified, and each of these can be related to a macroscopic rate law. Therefore studies of reaction rates provide some insight into what the atoms and molecules of reactants and products are doing as a reaction occurs.

Unimolecular Processes

A reaction is said to be **unimolecular** if, on the microscopic level, rearrangement of the structure of a single molecule produces the appropriate product molecules. An example of a unimolecular process is conversion of *cis*-2-butene to *trans*-2-butene (in the absence of any catalyst):

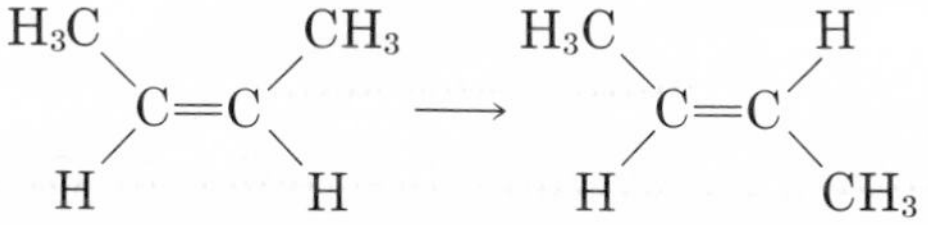

All that is required for this reaction to occur is a twist or rotation around the double bond, interchanging the methyl group with the hydrogen atom on the right-hand side. Only one *cis*-2-butene molecule need be involved as a reactant in this process.

Rotating part of a molecule about a double bond is not easy, however, because it involves a distortion of the electron clouds forming the double bond. This barrier to rotation was described in Sec. 7.4. A considerable increase in energy is required to twist one end of *cis*-2-butene around the other. This is shown in Fig. 18.3, where the energy has been plotted as a function of the angle of rotation. The maximum energy is reached when one end of the molecule has been rotated by 90° with respect to the other. This conformation is 262 kJ mol^{-1} higher than the energy of the original planar molecule. From this maximum it is downhill energetically on either side; so if the molecule has twisted this far, it should keep on twisting, eventually becoming *trans*-2-butene when the angle of rotation reaches 180°. *Trans*-2-butene is slightly lower in energy than *cis*-2-butene, as indicated by the enthalpy change of -4 kJ mol^{-1} for the overall reaction.

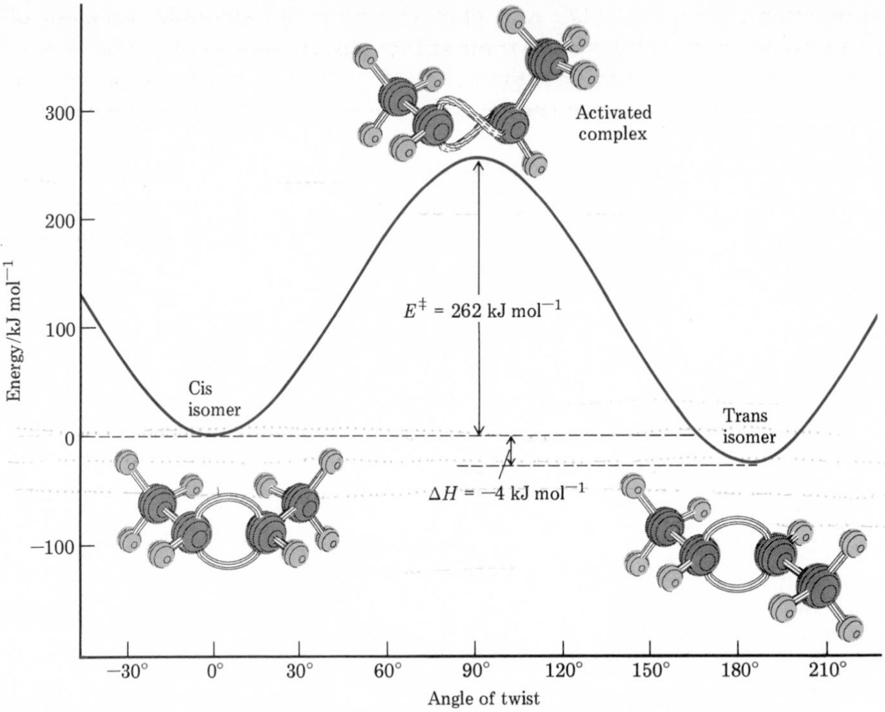

Figure 18.3 Energy profile for the conversion of *cis*- to *trans*-2-butene. As one-half the molecule is twisted relative to the other half against the restraining influence of the double bond, the potential energy increases to a maximum of 262 kJ mol^{-1}. A molecule of *cis*-2-butene must have at least this quantity of energy before it can change to the trans isomer. (Computer-generated.) (*Copyright © 1976 by W. G. Davies and J. W. Moore.*)

Figure 18.3 shows that the barrier to rotation around a double bond is an energy barrier. At least 262 kJ mol^{-1} must be supplied to transform *cis*-2-butene into *trans*-2-butene by a rotation such as we have described. The minimum quantity of energy required to surmount an energy barrier during a chemical reaction is called the **activation energy,** and the molecular species at the top of the barrier is called the **activated complex** or the **transition state.** Quantities associated with this activated complex are usually denoted by a double dagger (‡). For example, the activation energy is given the symbol $E^{\ddagger}$.

In the sample of gaseous *cis*-2-butene at room temperature, only a tiny fraction of molecules have enough energy to surmount the activation-energy barrier. (Recall from Sec. 9.4 and Fig. 9.9 that only a very small fraction of all gas molecules are traveling at very high speeds and hence have large kinetic energies. The same applies to the energy a molecule has because it is vibrating or rotating.) Not only do few molecules have enough energy to overcome the activation-energy barrier, but fewer still have that energy concentrated so that it can cause the atomic movements needed for the reaction to occur. In the case of *cis*-2-butene, for example, very few of the high-energy molecules have their energy distributed so that most of it is causing a twist around the double bond. Thus over a given period of time only a very small fraction of the *cis*-2-butene molecules will be converted to *trans*-2-butene.

Now suppose that we double the concentration of a sample of *cis*-2-butene. This means that there will be twice as many molecules in each cubic decimeter. At a given temperature the fraction of the molecules which can react during a given time interval will be the same, but with twice as many molecules there will be twice as many conversions to *trans*-2-butene. Therefore in a period of 1 s the change in the amount of substance per unit volume will be twice as great, and this means that the reaction rate is twice as great.

What we have just said applies to *any* unimolecular process. The reaction rate must always be directly proportional to the concentration of the reacting species. That is, for a general unimolecular process, A → products, the rate equation must be first order in A:

$$\text{Rate} = kc_A$$

Bimolecular Processes

A second type of microscopic process which can result in a chemical reaction involves collision of two particles. Such a process is called a **bimolecular** process. A typical example of a bimolecular process is the reaction between nitrogen dioxide and carbon monoxide:

$$NO_2 + CO \rightarrow NO + CO_2 \qquad (18.5)$$

Here an O atom is transferred from NO_2 to CO when the two molecules collide.

Several factors affect the rate of a bimolecular reaction. The first of these is the frequency of collisions between the two reactant molecules. Suppose we have a single molecule of type A (shown in black in Fig. 18.4*a*) moving around in a gas which otherwise consists entirely of molecules of kind B (indicated in white). If we double the concentration of B molecules (Fig. 18.4*b*), the number of collisions during the same time period doubles, because there are now twice as many B molecules to get in the way. Similarly, if we put twice as many A molecules into the original container, each of them collides with B molecules the same number of times, again giving twice as many A-B collisions.

The argument in the previous paragraph applies to *any* bimolecular process. The reaction rate must always be directly proportional to the concentration of each of the two reacting species. Thus for the general bimolecular process, A + B → products, the rate equation must be first order in A and first order in B:

$$\text{Rate} = k(c_A)(c_B)$$

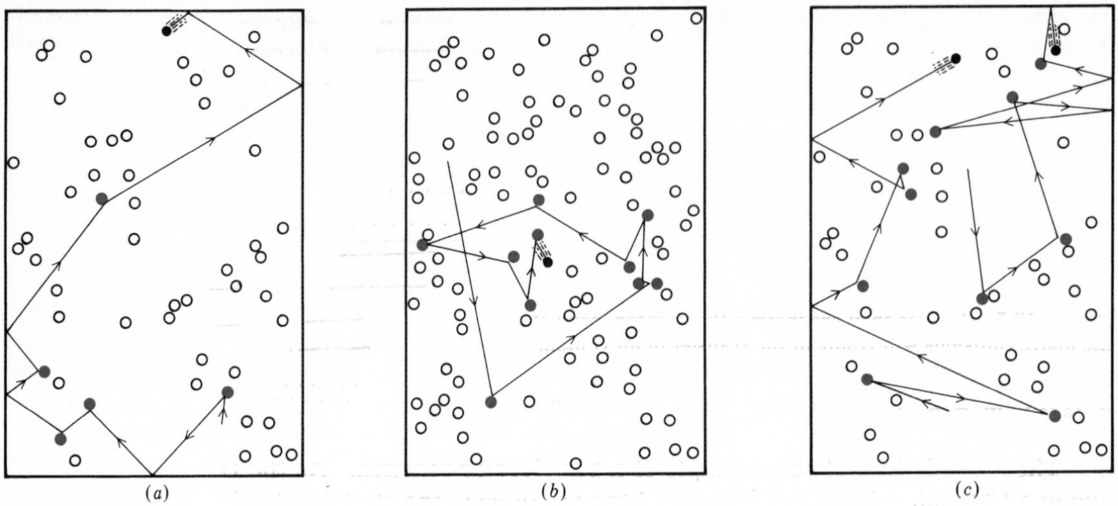

Figure 18.4 The effect of concentration on the frequency of collisions. (*a*) A single black molecule moving among 50 white molecules collides with 5 of them in 1 s. Each white molecule that has been struck by a moving black molecule is shown in color. (*b*) If the number of white molecules is doubled, the frequency of collisions rises to 10 s^{-1} (*c*) Two black molecules moving among 50 white ones produce 10 collisions per second. The number of collisions can thus be seen to be proportional to *both* the concentration of white molecules *and* the concentration of black molecules.

Collision of two molecules is a necessary but not a sufficient condition for a bimolecular process to occur. Returning to the reaction of NO_2 with CO, in which an O atom is transferred from the N atom to the C atom, we can see that the orientation of the two molecules as they collide is important. This introduces what is called a **steric factor.** None of the collisions depicted in Fig. 18.5 would result in reaction, for example, because none of them involve close contact between the C atom in CO and one of the O's in NO_2. For the reaction of CO with NO_2, the steric factor is estimated to be about one-sixth, meaning that only one collision in six involves an appropriate orientation. For more complex molecules, the steric factor is often much smaller. In the reaction

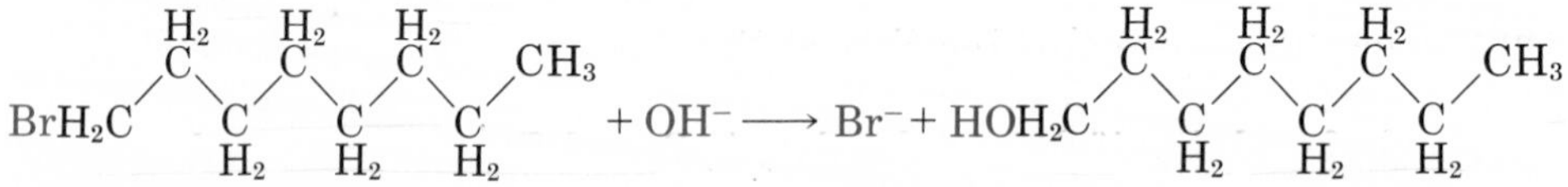

the fraction of favorable collisions is extremely small, because OH^- ions almost always hit the hydrocarbon chain at a point too far from the Br atom to cause a reaction.

The existence of a steric factor implies that there is a fairly well-defined pathway along which the reactant molecules must travel in order to produce an activated complex and then the reaction products. This pathway is called the **reaction coordinate,** and it applies to unimolecular as well as bimolecular reactions. In the case of *cis*-2-butene, for example, the reaction coordinate is the angle of rotation about the double bond, and proceeding along that pathway requires an increase in energy.

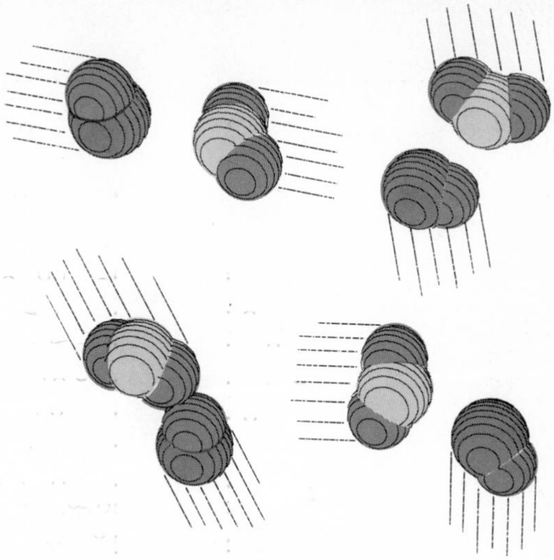

Figure 18.5 Because they are incorrectly oriented, none of the above collisions will result in the transfer of an O atom from an NO_2 molecule to a CO molecule, even if the collision is sufficiently energetic. (Computer-generated.) (*Copyright © 1976 by W. G. Davies and J. W. Moore.*)

The same is true of our example of a bimolecular reaction. As the C end of a CO molecule approaches one of the O's in NO_2, the electron clouds surrounding the molecules begin to repel each other and the energy of the system rises. This is shown in Fig. 18.6. Only if the total energy of the two molecules exceeds the 116 kJ mol^{-1} activation energy can they squeeze close enough together for a C—O bond to start to form. As this occurs the N—O bond begins to break, so that the activated complex has the structure

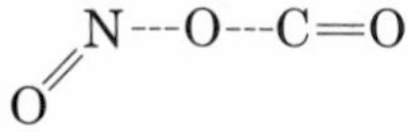

The dashed lines indicate bonds which are just beginning to form or are in the process of breaking. As the reaction occurs, the N—O bond lengthens and the O—C distance continually gets shorter. Consequently the difference between the N—O bond length and the C—O bond length $(r_{N-O} - r_{C-O})$ becomes larger and larger during the reaction. This difference, then, makes a convenient reaction coordinate.

Termolecular Processes

A **termolecular process** is one which involves simultaneous collision of three microscopic particles. In the gas phase, termolecular processes occur much less often than bimolecular processes, because the probability of three molecules all coming together at the same time is less than a thousandth the probability that two molecules will collide. Consequently it is commonly found that if three molecules, A, B, and C, must combine during the course of a reaction, they will do so stepwise. That is, B and C might first combine in a bimolecular process to form BC, which would then combine with A in a second bimolecular process. This can usually happen many times over before a successful termolecular collision would occur.

18.3 REACTION MECHANISMS

To summarize what was said in the previous section, there are two main types of microscopic processes by which chemical reactions can occur. A uni-

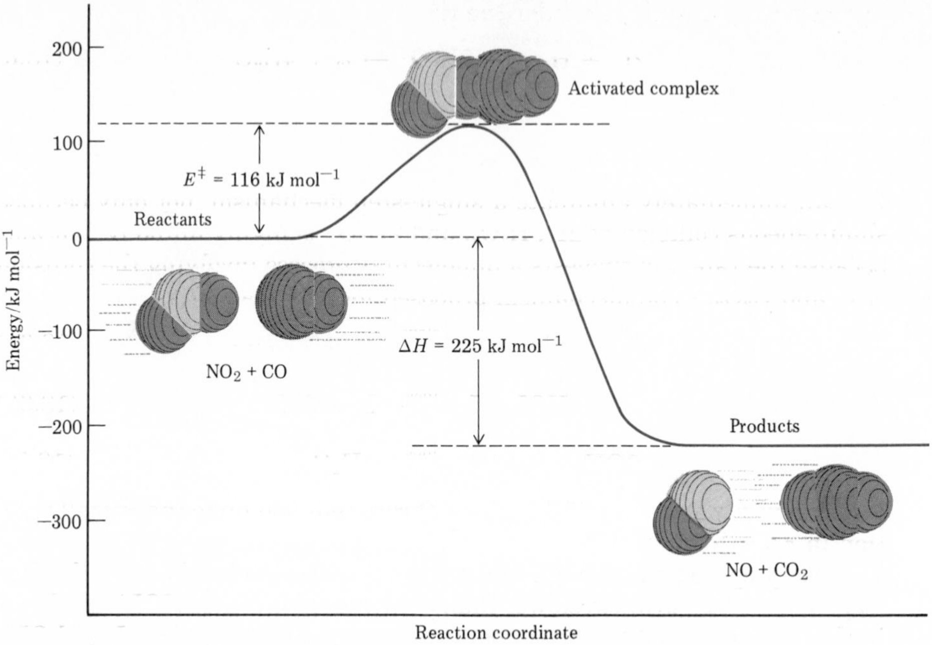

Figure 18.6 Energy profile for the reaction between NO_2 and CO. As the two molecules approach closely, energy is needed to squeeze them together against the mutual repulsion of their outer electrons until the C atom and an O atom are close enough to form a bond. If the two colliding molecules do not have at least 116 kJ mol^{-1} between them, the O atom cannot be forced close enough to the C atom to transfer its allegiance from the N atom. The progress of a reaction like this is usually measured in terms of a convenient *reaction coordinate,* which in this case could be the N—O distance minus the C—O distance. (Computer-generated.) (*Copyright © 1976 by W. G. Davies and J. W. Moore.*)

molecular process involves a single molecule as a reactant, and its rate law is first order in that reactant. A bimolecular process involves collision of two molecules. Its rate law is first order in each of the colliding species and therefore second order overall. Based on this we might expect all rate laws to be first order or second order, but this conclusion does not agree with several of the experimental rate laws described earlier. (Example 18.3 had one rate law which was *fifth* order overall!)

The reason for this discrepancy is that we have not considered the possibility that an overall reaction may be the sum of several unimolecular and/or bimolecular steps. The sequence of steps by which a reaction occurs is called the **mechanism** of the reaction. Each unimolecular or bimolecular step in that mechanism is called an **elementary process.** The term elementary is used to indicate that such steps cannot be broken down into yet simpler processes. In most mechanisms some species which are produced in the earlier steps serve as reactants in later elementary processes. Such species are called **intermediates.**

The mechanism proposed for a given reaction must be able to account for the overall stoichiometric equation, for the rate law, and for any other facts which are known. As an example of how a mechanism can be devised to meet these criteria, consider the reaction

$$2I^- + H_2O_2 + 2H_3O^+ \rightarrow I_2 + 4H_2O \qquad (18.6)$$

When the pH of the solution is between 3 and 5, the rate law is

$$\text{Rate} = k(c_{I^-})(c_{H_2O_2})$$

We can immediately eliminate a single-step mechanism, not only because simultaneous collision of $2I^-$, H_2O_2, and $2H_3O^+$ is highly unlikely, but also because the rate law suggests a bimolecular process involving the collision of I^- and H_2O_2. The mechanism proposed for this reaction is

$$\text{HOOH} + I^- \xrightarrow{\text{slow}} \text{HOI} + OH^- \qquad (18.6a)$$

$$\text{HOI} + I^- \xrightarrow{\text{fast}} I_2 + OH^- \qquad (18.6b)$$

$$2OH^- + 2H_3O^+ \xrightarrow{\text{fast}} 4H_2O \qquad (18.6c)$$

You can verify for yourself that these three steps add up to the overall reaction in Eq. (18.6).

The proposed mechanism can account for the rate law because the first step [Eq. (18.6*a*)] is much slower than the latter two. Once HOI is produced in that first step, it is transformed almost instantaneously into I_2 and OH^- by the second step. Similarly the OH^- produced by the first and second steps reacts immediately with H_3O^+ to form H_2O. Therefore the rate of the overall reaction is limited by the rate of the first step, and the rate law must be second order since that first step is bimolecular.

What we have just said applies to any reaction mechanism. The rate of reaction is limited by the rate of the slowest step. This elementary process is called the **rate-limiting step,** and the rate law gives us information about the activated complex in that step. All the species whose concentrations appear in the rate law must be part of the activated complex, and the amount of each species must be given by the corresponding exponent in the rate law.

It must be emphasized that any reaction mechanism is a *theory* about what is happening on the microscopic level and, as such, cannot be proved to be true. Thus we can say that a proposed mechanism accounts for all the known experimental facts relating to a reaction, but this does not mean it is the *only* mechanism which can account for those facts. A case in point is the reaction

$$H_2(g) + I_2(g) \rightarrow 2HI(g)$$

for which the rate law is

$$\text{Rate} = k(c_{H_2})(c_{I_2}) \qquad (18.7)$$

This was first established experimentally in 1894, and for over 70 years chemists thought that the reaction occurred via a bimolecular collision of an H_2 molecule with an I_2 molecule. This agrees with the rate law since the activated complex would have the formula H_2I_2, containing $1H_2$ and $1I_2$.

In 1967, however, it was shown that this reaction was speeded up considerably by yellow light from a powerful lamp. Such light is capable of dissociating I_2 molecules into atoms:

$$I_2 \xrightarrow{\text{light}} I + I$$

The fact that the light speeded up the reaction suggested that I atoms might be involved as intermediates in the mechanism, and the currently accepted mechanism is

$$I_2 \rightleftharpoons 2I \qquad \text{fast} \tag{18.8a}$$

$$I + H_2 \rightleftharpoons H_2I \qquad \text{fast} \tag{18.8b}$$

$$H_2I + I \rightarrow 2HI \qquad \text{slow} \tag{18.8c}$$

In this case the rate-limiting step is the last one in the mechanism. It is preceded by two rapidly established equilibria.

The rate law for the bimolecular step (18.8*c*) would be

$$\text{Rate} = k'(c_{H_2I})(c_I) \tag{18.9}$$

but since neither H_2I nor I are reactants in the overall reaction, we do not know their concentrations. These concentrations can be obtained, however, by applying the equilibrium law to Eqs. (18.8*a*) and (18.8*b*):

$$K_{(18.8a)} = \frac{c_I^2}{c_{I_2}} \qquad K_{(18.8b)} = \frac{c_{H_2I}}{(c_I)(c_{H_2})}$$

Rearranging these equations we obtain

$$c_I^2 = K_{(18.8a)}(c_{I_2}) \qquad c_{H_2I} = K_{(18.8b)}(c_I)(c_{H_2})$$

These may be substituted into Eq. (18.9):

$$\begin{aligned} \text{Rate} &= k'K_{(18.8b)}(c_I)(c_{H_2})(c_I) = k'K_{(18.8b)}(c_{H_2})(c_I)^2 \\ &= k'K_{(18.8b)}(c_{H_2})K_{(18.8a)}(c_{I_2}) \\ &= k'K_{(18.8b)}K_{(18.8a)}(c_{H_2})(c_{I_2}) \end{aligned} \tag{18.10}$$

The rate constant k in the rate law [Eq. (18.7)] can be identified with the product of constants $k'K_{(18.8b)}K_{(18.8a)}$, and so Eqs. (18.10) and (18.7) are the same. This confirms our previous statement that the rate law tells us what species participate in the activated complex during the rate-limiting step. It also shows how more than one mechanism can lead to an activated complex having the same composition.

EXAMPLE 18.5 The reaction between nitric oxide and oxygen:

$$2NO + O_2 \rightarrow 2NO_2$$

is found experimentally to obey the rate law

$$\text{Rate} = k(c_{NO})^2(c_{O_2})$$

Decide which of the following mechanisms is compatible with this rate law:

a) $NO + NO \rightleftharpoons N_2O_2$ fast

$N_2O_2 + O_2 \rightarrow 2NO_2$ slow

b) $NO + NO \rightarrow NO_2 + N$ slow

$$N + O_2 \rightarrow NO_2 \qquad \text{fast}$$

c) $NO + O_2 \rightleftharpoons NO_3$ fast

$NO_3 + NO \rightarrow 2NO_2$ slow

d) $NO + O_2 \rightarrow NO_2 + O$ slow

$NO + O \rightarrow NO_2$ fast

Solution The slow step in mechanism *a* involves 2N atoms and 4 O atoms, i.e., an activated complex with the formula $N_2O_4^{\ddagger}$. The same is true of mechanism *c*. Both are thus compatible with the rate law which also involves 2N and 4 O atoms. The other two mechanisms are not compatible with the measured rate law.

18.4 INCREASING THE RATE OF A REACTION

Aside from changing the concentrations of reactants or products which appear in the rate law, there are two main techniques by which reaction rates can be increased. As a rough rule of thumb, raising the temperature by an increment of 10 K usually increases the rate by a factor between 2 and 4. The effect of a catalyst can be even more striking. Certain highly efficient catalysts known as enzymes are capable of speeding up reactions to hundreds of thousands of times their normal rates. Both the effect of temperature and the effect of a catalyst are related to activation energy, and we shall consider each in turn.

The Effect of Temperature

The fact that a small increase in temperature can double the rate of a reaction is primarily dependent on an observation we made at the end of Chap. 9. A seemingly minor temperature rise causes a major increase in the number of molecules whose speeds (and hence molecular energies) are far above average. This happens because the distribution of molecular speeds (Fig. 9.9) is lopsided and tails off very slowly at the high-speed end. Since only those molecules whose energy exceeds the activation energy $E^{\ddagger}$ can react, a significant increase in the fraction of molecules having high energy causes a significant increase in rate.

The fraction of molecules which have enough kinetic energy to react depends on the activation energy $E^{\ddagger}$, the temperature T, and the gas constant R in the following way:

$$\text{Fraction of molecules having enough energy to react} = 10^{-E^{\ddagger}/2.303RT} \qquad (18.11)$$

Although a derivation of this result is beyond the scope of our current discussion, we can look at this equation to see that it agrees with common sense. The larger the $E^{\ddagger}$, the more negative the power of 10, and the smaller the fraction of molecules which can react. Thus if two similar reactions are studied under the same conditions of temperature and concentration, the one with the larger activation energy is expected to be slower. As a corollary to this, if we can find some way to reduce the activation energy of a particular reaction, then we can get that reaction to go faster.

Equation (18.11) also shows why changing the temperature has such a large effect on reaction rates. Increasing T decreases the negative power of 10 in the equation. This increases the fraction of molecules which can react, as shown by the following example:

EXAMPLE 18.6 The reaction

$$CO + NO_2 \rightarrow CO_2 + NO$$

has an activation energy of 116 kJ mol^{-1}. Calculate the fraction of molecules whose kinetic energies are large enough that they can collide with sufficient energy to react (a) at 298 K, and (b) at 600 K.

Solution In both cases we use Eq. (18.11).

a) At 298 K, the power of 10 will be

$$\frac{-E^{\ddagger}}{2.303RT} = \frac{-116 \text{ kJ mol}^{-1}}{2.303 \times 8.314 \text{ J K}^{-1} \text{ mol}^{-1} \times 298 \text{ K}}$$

$$= \frac{-116 \times 10^3 \text{ J}}{2.303 \times 8.314 \text{ J} \times 298} = -20.3$$

Then the fraction of molecules is

$$\text{Fraction} = 10^{-20.3} = 5.0 \times 10^{-21}$$

b) At 600 K,

$$\text{Fraction} = 10^{-E^{\ddagger}/2.303RT} = 10^{-10.1} = 8.0 \times 10^{-11}$$

Note: Since $10^x = e^{2.303x}$, where e is the base of the natural logarithm system, the calculations above could be done in an alternative way. Using part b as an example we have

$$\text{Fraction} = 10^{-E^{\ddagger}/2.303RT} = e^{-E^{\ddagger}/RT} = e^{-23.25} = 8.0 \times 10^{-11}$$

If you have a calculator with an e^x key, this latter method may be more convenient to use.

In this case approximately doubling the temperature causes the fraction of molecular collisions which can result in reaction to increase by a factor of

$$\frac{8.0 \times 10^{-11}}{5.0 \times 10^{-21}} = 1.6 \times 10^{10}$$

that is, by about 16 billion. This means that the reaction is expected to be about 16 billion times faster at the higher temperature. This corresponds to slightly more than doubling the rate for each of the 30 intervals of 10 K between 298 and 600 K.

Equation (18.11) is also useful if we want to measure the activation energy of a reaction experimentally because it shows how the rate constant should vary with temperature. We have already seen that the rate of a unimolecular process should be proportional to the concentration of reactant. If only a fraction of those reactant molecules have enough energy to overcome the activation barrier, then the rate should also be proportional to that fraction. Thus we can write

$$\text{Rate} = k(c_A) = k'10^{-E^\ddagger/2.303RT}(c_A)$$

From this equation it is clear that

$$k = k'10^{-E^\ddagger/2.303RT}$$

If we divide both sides by the units per second so that we can take logs, we obtain

$$\log \frac{k}{1\ \text{s}^{-1}} = \log \frac{k'}{1\ \text{s}^{-1}} - E^\ddagger/2.303RT$$

or

$$\log \frac{k}{1\ \text{s}^{-1}} = \left(\frac{-E^\ddagger}{2.303R}\right)\left(\frac{1}{T}\right) + \log \frac{k'}{1\ \text{s}^{-1}}$$

$$y \quad = \quad m \quad \cdot \quad x \quad + \quad b$$

This is the standard form for the equation of a straight line whose slope m is $-E^\ddagger/2.303R$ and whose intercept b is $\log\ (k'/\text{s}^{-1})$. Thus if we plot $\log\ (k/\text{s}^{-1})$ versus $1/T$ and a straight line is obtained, we can determine $E^\ddagger$.

A similar situation applies to the second-order rate equation associated with a bimolecular process, where

$$\text{Rate} = k(c_A)(c_B) = k'10^{-E^\ddagger/2.303RT}(c_A)(c_B)$$

In this case the rate constant has units cubic decimeter per mole per second, and so we must divide by these before taking logs. The equation that is obtained is

$$\log \frac{k}{\text{dm}^3\ \text{mol}^{-1}\ \text{s}^{-1}} = \left(\frac{-E^\ddagger}{2.303R}\right)\left(\frac{1}{T}\right) + \log \frac{k'}{\text{dm}^3\ \text{mol}^{-1}\ \text{s}^{-1}}$$

Again $E^\ddagger$ can be obtained from the slope, provided a straight-line plot is obtained. An example of such a plot for the reaction

$$2HI \rightarrow H_2 + I_2$$

is shown in Fig. 18.7. From its slope we can determine that $E^\ddagger = 186$ kJ mol^{-1}. This means of determining activation energy was first suggested in 1889 by Svante Arrhenius and is therefore called an **Arrhenius plot.**

It should be pointed out that the equations just derived for Arrhenius plots apply strictly to unimolecular and bimolecular elementary processes. If a reaction mechanism involves several steps, there is no guarantee that the same elementary process will be rate limiting at several widely different temperatures. The observed rate constant may also be a product of several constants, as in the $H_2 + I_2$ reaction [Eq. (18.10)]. If either a different rate-limiting step or the product of constants has a different temperature de-

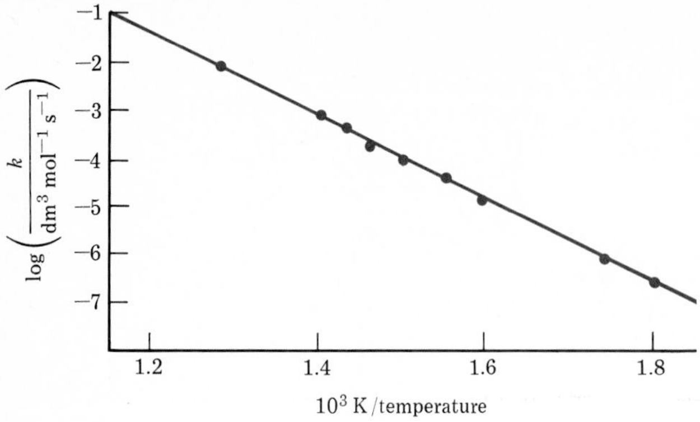

Figure 18.7 Arrhenius plot for the reaction $2HI \rightarrow H_2 + I_2$. From the slope of the graph an activation energy of 186 kJ mol^{-1} may be calculated.

pendence, then an Arrhenius plot may not be linear. This also makes it dangerous to use such a plot to predict reaction rates at temperatures quite different from the conditions under which experimental measurements were made.

Catalysis

Altering the mechanism of a reaction so that the activation energy is lower is the second major way to speed up the reaction. A catalyst can do this by participating in the activated complex for the rate-limiting step, even though the catalyst itself is neither a reactant nor a product in the overall stoichiometric equation.

A good example of catalysis is provided by the effect of I_2 on the rate of isomerization of *cis*-2-butene. You will recall from the discussion earlier in this chapter that the rate law for the catalyzed reaction involves the concentration of I_2 raised to the one-half power. This implies that half a molecule of I_2 (that is, an I atom) is involved in the activated complex, which probably has the structure

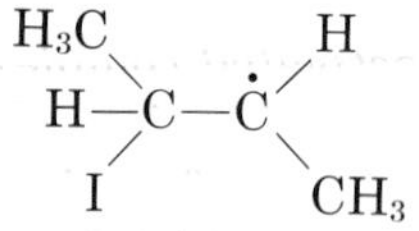

Since there is no double bond between the two central C atoms, one end of the activated complex can readily twist around the other.

The currently accepted mechanism for this catalyzed reaction involves three steps:

$$\tfrac{1}{2}I_2 \rightleftharpoons I$$

$$I + cis\text{-}C_4H_8 \rightarrow C_4H_8I^{\ddagger} \rightarrow trans\text{-}C_4H_8 + I$$

$$I \rightleftharpoons \tfrac{1}{2}I_2$$

The first and last steps have coefficients of one-half associated with I_2 because each I_2 molecule that dissociates produces two I atoms, only one of which is needed to help a given *cis*-2-butene molecule to react. Note also

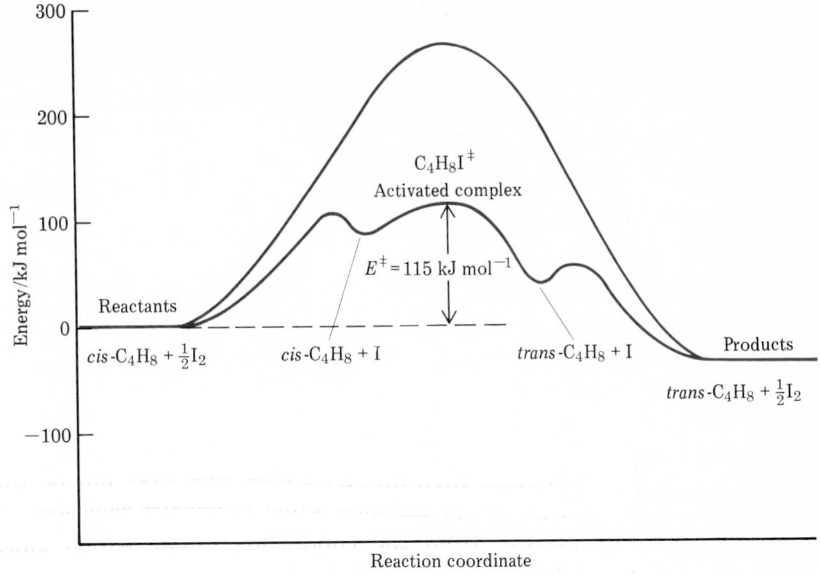

Figure 18.8 Energy profile for the cis-trans conversion of 2-butene catalyzed by iodine. Iodine atoms must first be produced from I_2 molecules after which the addition of an I atom to a butene molecule produces the activated complex. Since the activated complex contains no double bond, one part of the molecule can now swivel easily around the other, producing the trans isomer after the loss of the I atom. Since this pathway requires a much lower activation energy, it allows the reaction to occur much more rapidly.

that for every I_2 molecule which dissociates in the first step of the mechanism, an I_2 molecule is eventually regenerated by the last step. As a result, the concentration of I_2 after the reaction remains exactly the same as before.

If we consider the energetics of each step in the proposed mechanism, we find that much less than the 262 kJ mol^{-1} activation energy of the uncatalyzed reaction is required. A complete energy profile for the catalyzed process is compared with that of the uncatalyzed one in Fig. 18.8. The I—I bond enthalpy is 151 kJ mol^{-1} (from Table 15.4), and so 75.5 kJ mol^{-1} will be required for formation of the I atom in the first step. An additional increase in energy occurs as the I atom collides with *cis*-2-butene and bonds with it. Then about 12 kJ mol^{-1} is required for twisting around the C—C single bond in the activated complex. All told 115 kJ mol^{-1} is required to go from the initial molecules to the activated complex. When rotation to a trans structure is complete, the I atom dissociates from *trans*-2-butene and eventually reacts with another I atom to form I_2. These last two processes involve an overall decrease in energy which is nearly the same as the increase required to achieve the activated complex.

The reduction in activation energy illustrated in Fig. 18.8 is the chief factor in speeding up the catalyzed reaction.

EXAMPLE 18.7 Based on activation energies given in the text, how many times larger would the rate constant be for the catalyzed rather than for the uncatalyzed isomerization of *cis*-2-butene at a temperature of 500 K?

Solution The fraction of molecules which have sufficient energy to achieve the activated complex can be calculated using Eq. (18.11):

$$\text{Fraction (catalyzed)} = 10^{-115\,\text{kJ mol}^{-1}/(2.303 \times 8.314\,\text{J K}^{-1}\,\text{mol}^{-1} \times 500\,\text{K})}$$
$$= 10^{-12.0} = 1 \times 10^{-12}$$
$$\text{Fraction (uncatalyzed)} = 10^{-262\,\text{kJ mol}^{-1}/(2.303 \times 8.314\,\text{J K}^{-1}\,\text{mol}^{-1} \times 500\,\text{K})}$$
$$= 10^{-27.4} = 4 = 10^{-28}$$

Since the respective rate constants should be proportional to these fractions,

$$\frac{k\ \text{(catalyzed)}}{k\ \text{(uncatalyzed)}} = \frac{1 \times 10^{-12}}{4 \times 10^{-28}} = 2.5 \times 10^{15}$$

Thus at unit concentrations of *cis*-2-butene and iodine, the catalyzed reaction will be more than 10^{15} times faster.

Another example of catalysis is involved in the reaction between H_2O_2 and the I^- ion. In discussing this reaction earlier we noted that it is first order in H_2O_2 and first order in I^- ion between pH = 3 and pH = 5. At lower pH values, however, a different mechanism takes over, and the rate law becomes

$$\text{Rate} = k(c_{H_2O_2})(c_{I^-})(c_{H^+})$$

This indicates that the activated complex contains an additional proton when compared with the uncatalyzed case. This proton apparently adds to the H_2O_2 molecule, forming $H_3O_2^+$. Because of the positive charge, less energy increase occurs as an I^- ion approaches $H_3O_2^+$ and the reaction

$$H_3O_2^+ + I^- \rightarrow H_2O + HOI$$

has a lower activation energy than

$$H_2O_2 + I^- \rightarrow OH^- + HOI$$

The latter is the rate-limiting step in the mechanism at higher pH [Eqs. (18.6*a*), (18.6*b*), and (18.6*c*)], and so protonating H_2O_2 results in a faster overall reaction. Measurements over a range of temperatures show that the activation energy is lowered from 56 to 43.6 kJ mol^{-1} by this change of mechanism.

The peroxide-iodide reaction is one example of a great many *acid-catalyzed* reactions. In most of these the hydrogen ion concentration appears in the rate law, indicating that the activated complex contains an extra proton. The positive charge of this proton allows negative ions to combine more readily with the protonated species. The proton may also shift electron density toward itself and away from some other site in the protonated molecule, allowing a negative species to bond to that site more readily. The net result is a lower activation energy and a more rapid reaction.

The preceding examples are illustrative of the three main features of catalysis:

1 The catalyst allows the reaction to proceed via an *alternative mechanism*.
2 The catalyst is directly involved in this mechanism, but for every step in which a catalyst molecule is a reactant, there is another step where the catalyst appears as a product. Thus there is *no net consumption of the catalyst.*
3 The catalyzed mechanism results in a faster reaction, usually because the overall *activation energy is lowered.*

18.5 SOME IMPORTANT TYPES OF CATALYSTS

Heterogeneous Catalysis

Our previous discussion has concentrated on catalysts which are in the same phase as the reaction being catalyzed. This kind of catalysis is called **homogeneous catalysis.** Many important industrial processes rely on **heterogeneous catalysis,** in which the catalyst is in a different phase. Usually the catalyst is a solid and the reactants are gases, and so the rate-limiting step occurs at the solid surface. Thus heterogeneous catalysis is also referred to as **surface catalysis.**

The detailed mechanisms of most heterogeneous reactions are not yet understood, but certain sites on the catalyst surface appear to be able to weaken or break bonds in reactant molecules. These are called **active sites.** One example of heterogeneous catalysis is hydrogenation of an unsaturated organic compound such as ethene (C_2H_4) by metal catalysts such as Pt or Ni:

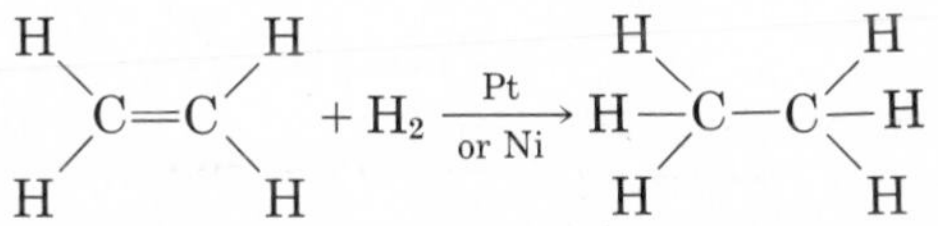

The currently accepted mechanism for this reaction involves weak bonding of both H_2 and C_2H_4 to atoms on the metal surface. This is called **adsorption.** The H_2 molecules dissociate to individual H atoms, each of which is weakly bonded to a Pt atom:

$$H_2 + 2\text{Pt(surface)} \rightarrow 2\text{H---Pt(surface)}$$

These adsorbed H atoms can move across the metal surface, and eventually they combine with a C_2H_4 molecule, completing the reaction. Because adsorption and dissociation of H_2 on a Pt surface is exothermic ($\Delta H_m^\ominus = -160$ kJ mol^{-1}), it can provide H atoms for further reaction without a large activation energy. By contrast, dissociation of gaseous H_2 molecules without a metal surface would require the full bond enthalpy ($\Delta H_m^\ominus = +436$ kJ mol^{-1}). Clearly the metal surface makes a major contribution in lowering the activation energy.

Heterogeneous catalysts are used extensively in the petroleum industry. One example is the combination of SiO_2 and Al_2O_3 used to speed up cracking of long-chain hydrocarbons into the smaller molecules needed for gasoline. Another is the Pt catalyst used to reform hydrocarbon chains into aromatic ring structures. This improves the octane rating of gasoline, making it more suitable for use in automobile engines. Other industries also make effective use of catalysts. SO_2, obtained by burning sulfur (or even from burning coal), can be oxidized to SO_3 over vanadium pentoxide, V_2O_5. This is an important step in manufacturing H_2SO_4, and it may eventually be used to prevent air pollution from coal-fired power plants as well. Another important heterogeneous catalyst is used in the Haber process for synthesis of NH_3 from N_2 and H_2. As with most industrial catalysts, its exact composition is a trade secret, but it is mainly Fe with small amounts of Al_2O_3 and K_2O added.

The surface catalyst you are most likely to be familiar with is found in the exhaust systems of many automobiles constructed since 1975. Such a catalytic converter contains from 1 to 3 g Pt in a fine layer on the surface of a honeycomblike structure or small beads made of Al_2O_3. The catalyst speeds up oxidation of unburned hydrocarbons and CO which would otherwise be emitted from the exhaust as air pollutants. It apparently does this by adsorbing and weakening the bond in the O_2 molecule. Individual O atoms are then more readily transferred to CO or hydrocarbon molecules, producing CO_2 and H_2O. This action of the catalytic surface can be inhibited or poisoned if lead atoms (from tetraethyllead in leaded gasoline) react with the surface. Hence the prohibition of use of leaded fuel in cars equipped with catalytic converters.

Enzymes

The most efficient and amazing catalysts of all are the biological molecules known as enzymes. One of these, nitrogenase, is found in algae, bacteria, and microorganisms associated with leguminous plants such as soybeans. Nitrogenase can fix nitrogen from the atmosphere at room temperature (instead of at 500°C as in the Haber process). It does so efficiently enough that only a few kilograms of enzyme are necessary to fix about 44×10^{12} g of nitrogen worldwide each year. Other enzymes carry out other reactions with comparable efficiencies.

All enzymes are large protein molecules whose molar masses exceed 20 000 g mol^{-1}. They are condensation polymers of amino acids, and a typical section of a long protein chain may be represented as

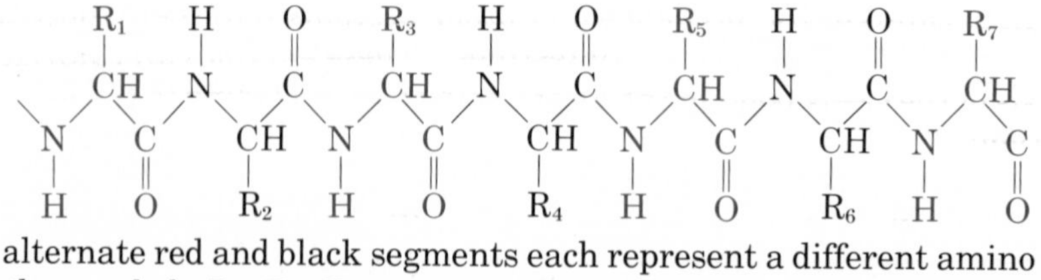

The alternate red and black segments each represent a different amino acid, and the symbols R_1, R_2, R_3, etc., stand for different organic groups called **side chains.** In the amino acid glycine, for instance, the side chain is a hydrogen atom, while for glutamic acid it is the group $—CH_2—CH_2—COOH$. A typical enzyme molecule contains as many as 20 different kinds of amino acids, while the chain may be as long as 200 or 300 amino acid units. (More details on the amino acids, their side chains, and the structures of proteins are given in Sec. 20.5.)

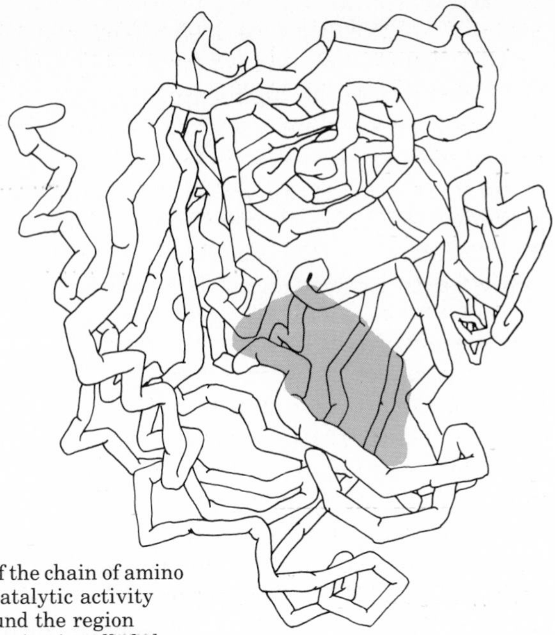

Figure 18.9 The complex folding of the chain of amino acids in the enzyme trypsin. The catalytic activity of the molecule centers largely around the region shaded in color. Accordingly this region is called the active site.

The polymer chain of amino acid units in an enzyme is usually wrapped around itself in a complex fashion to form an overall shape which is approximately spherical. The conformation of the protein chain in the enzyme *trypsin* has been determined experimentally and is shown in Fig. 18.9. The chain appears quite random, almost as though it were a ball of tangled string or yarn, but there are certain molecular features in the region shown in color in the diagram which are essential to its operation as a catalyst. Accordingly, this region of the molecular surface is called the active site.

An important feature of the catalytic behavior of enzymes is their *specificity*. Most enzymes are able to catalyze only one particular chemical reaction or at most only a few closely related reactions. Some interesting examples of specificity involve the enzymes which, as part of the process of digestion, catalyze the breakdown of other proteins into their constituent amino acids in the stomach and intestines. These enzymes all catalyze hydrolysis of the C—N bond in the amino acid chain according to the following equation:

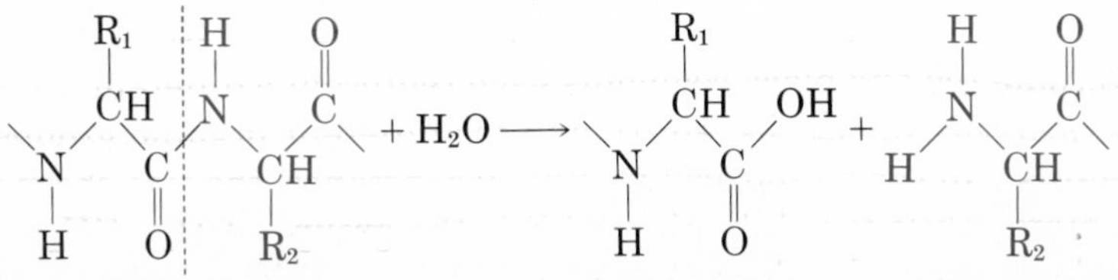

When all the C—N bonds in a protein are broken, it is transformed entirely into amino acids. Amino acids can penetrate the walls of the intestine and be utilized by the body to build up other protein chains.

Hydrolysis of this type of C—N bond is also catalyzed by acids and bases. Thus, for instance, a standard laboratory method for breaking up a protein into its constituent amino acids is to heat it overnight at 100°C in 1 *M* HCl. Despite the higher temperature, this acid-catalyzed hydrolysis is much slower than one catalyzed by an enzyme, and all the C—N bonds are broken, no matter what amino acids are involved. When an enzyme is used, by contrast, only certain C—N bonds are hydrolyzed. The enzyme carboxypeptidase will only catalyze the hydrolysis of a C—N bond at the *end* of an amino acid chain. On the other hand, trypsin will only operate if one of the amino acids involved in the bond is lysine or arginine.

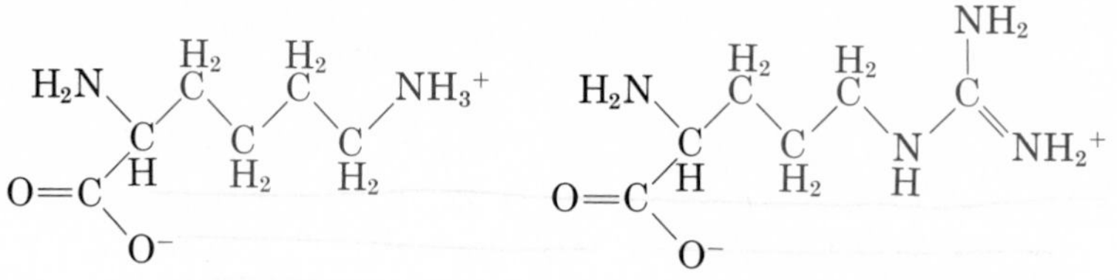

Lysine **Arginine**

Enzymes usually behave very simply from the point of view of chemical kinetics. The rate law is usually found to be first order with respect to the enzyme concentration and also first order with respect to the **substrate** (the substance on which the enzyme acts). In other words

$$\text{Rate} = k(c_{\text{enzyme}})(c_{\text{substrate}})$$

Thus the activated complex consists of a combination of one enzyme molecule and one substrate molecule. Since the enzyme molecule is usually much larger than the substrate molecule, we can think of the substrate mol-

Figure 18.10 A schematic description of the operation of the enzyme trypsin in hydrolyzing a protein chain. (*a*) The protein chain approaches the active site, and an arginine side chain is attracted toward the pocket. (*b*) The arginine side chain lodges in the pocket, while a conveniently situated acidic proton bonds to the N atom of the C—N bond which is about to be hydrolyzed. (*c*) The proton catalyzes the breaking of the C—N bond, and one fragment of the C—N chain moves away. An H_2O molecule can now enter and bond to the positively charged C atom. (*d*) The H_2O molecule has added an OH group to the end of the chain and replaced the lost proton on the enzyme. The arginine side chain is released from the pocket, and the enzyme is ready for the next customer.

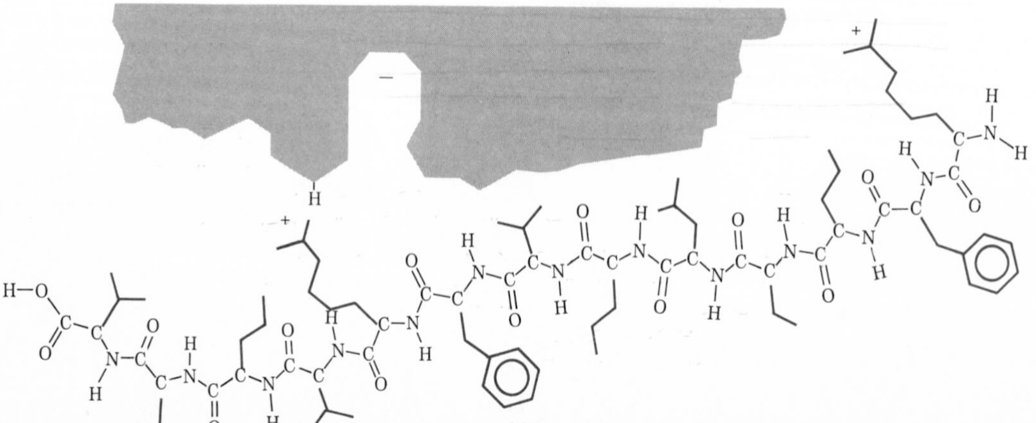

(*a*)

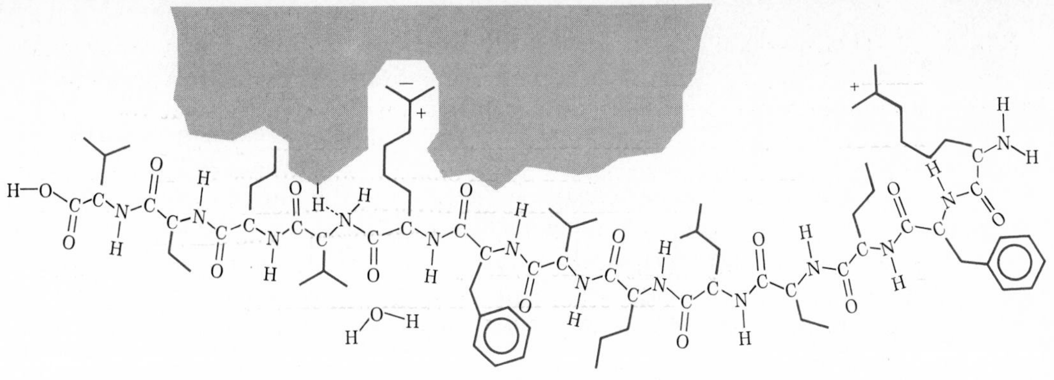

(*b*)

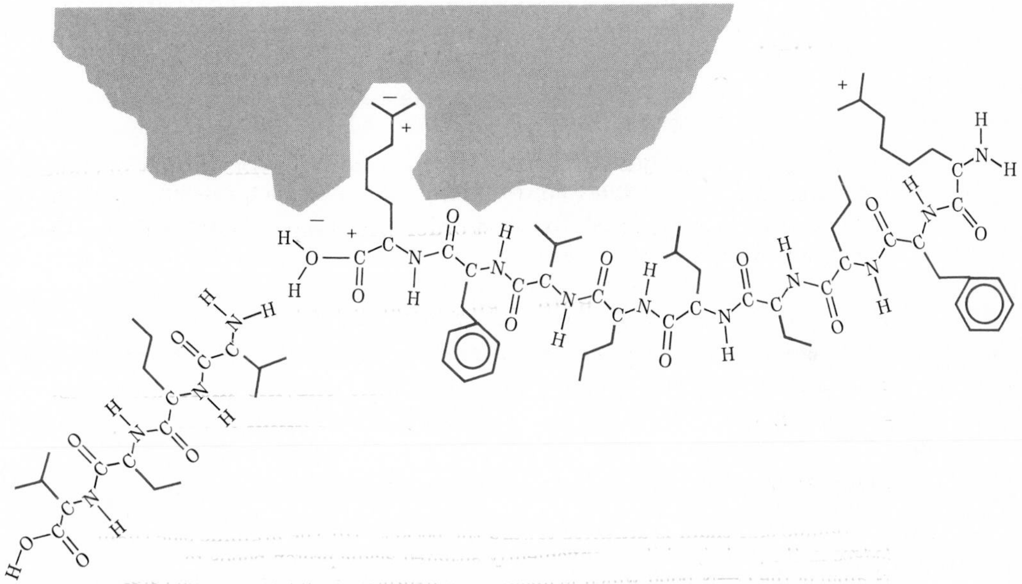

(*c*)

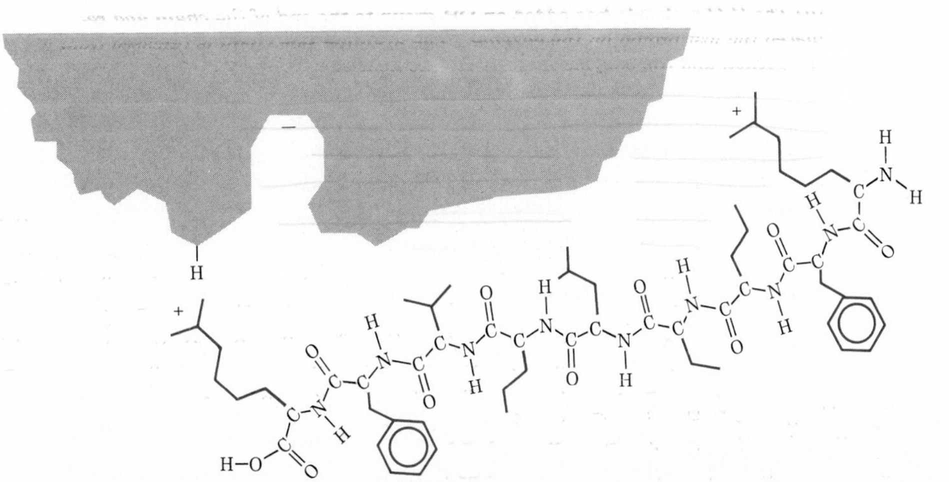

(*d*)

ecule as being held down on the *surface* of the enzyme and enzyme catalysis as being related to heterogeneous catalysis.

In recent years, largely because developments in experimental technique have provided accurate data on the three-dimensional structures of enzymes, biochemists have begun to understand the finer details of how an enzyme is able to operate so efficiently as a catalyst. Among the best understood is the enzyme trypsin, which, as we have just mentioned, catalyzes the hydrolysis of C—N bonds attached only to the amino acids lysine and arginine. A somewhat oversimplified and schematic account of the operation of this enzyme is given in Fig. 18.10.

The active site on the surface of the enzyme, shown in gray, contains an opening or pocket which is just large enough to accommodate the long side chains of lysine or arginine, and which also contains a negative ionic charge deep inside it. Side chains belonging to other amino acids are the wrong size or shape or are not positively charged, and so they cannot be held in this pocket, but an arginine or lysine side chain can be. Once the side chain is held by the pocket, there is an acidic hydrogen atom on the enzyme in *exactly the right position* to protonate the nitrogen atom in a C—N bond in the protein chain. The enzyme thus functions as an acid catalyst for the hydrolysis of the C—N bond. It is, of course, much more efficient than a nonenzymic catalyst such as HCl. Partly this is because the proton is in *exactly the right place* to donate once the side chain has entered the pocket. Partly, also, it is because there is considerable lowering in the activation energy due to the ionic attraction between the enzyme and substrate.

SUMMARY

Chemical kinetics is concerned with the rates of chemical reactions, the dependence of those rates on temperature, concentration, and catalysts, and the microscopic mechanisms by which reactions occur. The rate of a reaction is defined in terms of the change in concentration of a reactant or product per unit time, and it usually decreases as the reaction progresses. The reaction rate ordinarily is proportional to the concentrations of reactants and/or products, each raised to a power called the order with respect to that reactant or product. When the concentration of a species which is not a reactant or product in the overall reaction affects the rate, that species is called a catalyst. An equation expressing the dependence of reaction rate on concentrations is called a rate equation or rate law.

On a microscopic level, a reaction usually involves unimolecular processes, in which a single molecule changes structure, or bimolecular processes, in which two molecules collide. Collision of three or more molecules simultaneously is much less probable. In both uni- and bimolecular elementary processes an activation energy barrier must be surmounted before product molecules can be produced. The species at the top of a graph of energy versus reaction coordinate is called the activated complex or transition state. Because of this energy barrier only a small fraction of the molecules are energetic enough to reach the transition state, but that fraction increases rapidly as temperature increases, and so reaction rates are strongly dependent on temperature. The temperature dependence of the rate constant can be used to obtain the activation energy by means of an Arrhenius plot.

Most reactions occur in two or more steps. Such a sequence of elementary processes is called a reaction mechanism, and the overall rate is determined by the slowest, or rate-limiting, step. The experimental rate law tells us the composition of the activated complex for the rate-limiting step, but often several mechanisms are possible which agree with the rate law. Other evidence must then be used to decide among these mechanisms.

A catalyst speeds up a reaction by changing the mechanism so that the activation energy is lowered. Many heterogeneous catalysts are of great industrial importance, but the most efficient catalysts known are the enzymes in living organisms. An enzyme operates by adsorbing a substrate molecule at an active site whose structure is exactly right to stretch bonds which are to be broken or to hold atoms in position while new bonds form. This almost ideal structure of the active site makes enzymes highly specific and extremely efficient catalysts.

QUESTIONS AND PROBLEMS

18.1 In your own words describe the scope of chemical kinetics. Is this field concerned solely with the study of rates of reactions?

18.2 How is the rate of a chemical reaction defined? How can it be measured experimentally?

18.3 Using data given below for the reaction

$$N_2O_5 \rightarrow 2NO_2 + \tfrac{1}{2}O_2$$

calculate the average rate of reaction during each of the following intervals:

a 0 to 0.5 h
b 0.5 to 1.0 h
c 1.0 to 2.0 h
d 2.0 to 3.0 h
e 3.0 to 4.0 h
f 4.0 to 5.0 h

Time/h	$c_{N_2O_5}$/mol dm^{-3}
0	0.849
0.50	0.733
1.00	0.633
2.00	0.472
3.00	0.353
4.00	0.265
5.00	0.197

18.4 Using all your calculated rates from Problem 18.3, show that the reaction obeys the rate law:

$$\text{Rate} = -\frac{\Delta c_{N_2O_5}}{\Delta t} = k c_{N_2O_5}$$

Evaluate the rate constant k as an average of the values obtained for the six intervals.

18.5 Using data from Problem 18.3, calculate the average rate over the interval 0 to 5.0 h. Compare your result with the average rates over the intervals 1.0 to 4.0 h and 2.0 to 3.0 h, all of which have the same midpoint (2.5 h from the start).

18.6 Using the rate law and the rate constant you calculated in Problem 18.4, calculate the reaction rate exactly 2.5 h from the start. Do your results from this and the previous problem agree with the statement in the text that the smaller the time interval, the more accurate the average rate?

18.7 For each of the rate laws below, what is the order of the reaction with respect to the hypothetical substances X, Y, and Z? What is the overall order?

a Rate = $k c_X c_Y c_Z$
b Rate = $k(c_X)^2(c_Y)^{1/2}c_Z$
c Rate = $k(c_X)^{1.5}(c_Y)^{-1}$
d Rate = $k c_X/(c_Y)^2$

18.8 The hypothetical reaction

$$2A + 2B \rightarrow C + 3D$$

was studied by measuring the initial rate of appearance of C and yielded the following data:

c_A/mol dm^{-3}	c_B/mol dm^{-3}	Initial Rate/ mol dm^{-3} s^{-1}
6.0×10^{-3}	1.0×10^{-3}	0.012
6.0×10^{-3}	2.0×10^{-3}	0.024
2.0×10^{-3}	1.5×10^{-3}	0.0020
4.0×10^{-3}	1.5×10^{-3}	0.0080

a What is the order of the reaction with respect to substance A?
b What is the order with respect to B?
c What is the overall order?
d Write the rate equation.
e Calculate the rate constant.

18.9 For the reaction

$$2NO(g) + 2H_2(g) \rightarrow N_2(g) + 2H_2O(g)$$

the following data were obtained at 1100 K:

c_{NO}/mol dm^{-3}	c_{H_2}/mol dm^{-3}	Initial Rate/ mol dm^{-3} s^{-1}
5.00×10^{-3}	2.50×10^{-3}	3.0×10^{-5}
15.0×10^{-3}	2.50×10^{-3}	9.0×10^{-5}
15.0×10^{-3}	10.0×10^{-3}	3.6×10^{-4}

a What is the order with respect to NO? with respect to H_2? the overall order?

b Write the rate law.

c Calculate the rate constant.

d Calculate the initial rate of reaction when $c_{NO} = c_{H_2} = 8.0 \times 10^{-3}$ mol dm^{-3}.

18.10 In your own words define each of the following terms. Give an example of each term.

a Unimolecular process

b Activated complex

c Steric factor

d Bimolecular process

e Transition state

f Catalyst

g Reaction coordinate

h Activation energy

i Termolecular process

18.11 Explain why the rate law must be second order for a reaction whose mechanism is a single bimolecular step.

18.12 Draw an energy-versus-reaction coordinate diagram (similar to Fig. 18.3) for each of the reactions whose activation energy and enthalpy change are given below:

a $\Delta H_m^{\ominus} = -145$ kJ mol^{-1}; $E^{\ddagger} = 75$ kJ mol^{-1}

b $\Delta H_m^{\ominus} = -70$ kJ mol^{-1}; $E^{\ddagger} = 65$ kJ mol^{-1}

c $\Delta H_m^{\ominus} = +70$ kJ mol^{-1}; $E^{\ddagger} = 85$ kJ mol^{-1}

d $\Delta H_m^{\ominus} = -150$ kJ mol^{-1}; $E^{\ddagger} = 15$ kJ mol^{-1}

18.13 Which of the reactions in Problem 18.12 would be expected to occur fastest? Which would be expected to occur slowest? (Assume equal temperatures, equal concentrations, and the same rate law for all reactions.)

18.14 For which of the reactions in Problem 18.12 would the *reverse* reaction be fastest? For which would it be slowest?

18.15 Criticize the following statement: For the reaction

$$2NO(g) + O_2(g) \rightarrow 2NO_2(g)$$

the most likely mechanism is simultaneous collision of 2NO molecules and 1O_2 molecule.

18.16 Define and give an example of each of the following:

a Mechanism of reaction

b Intermediate

c Elementary process

d Rate-limiting step

18.17 For the reaction

$$2NO(g) + Cl_2(g) \rightarrow 2NOCl(g)$$

the currently accepted mechanism is

$$NO + Cl_2 \rightleftharpoons NOCl_2 \qquad \text{fast}$$

$$NOCl_2 + NO \rightarrow 2NOCl \qquad \text{slow}$$

What is the rate law for this reaction? (Be sure to express it in terms of concentrations of reactants or products of the overall reaction, not in terms of intermediates.) Suggest another mechanism which would agree with the same rate law.

18.18 For the reaction

$$2N_2O_5(g) \rightarrow 4NO_2(g) + O_2(g)$$

the currently accepted mechanism is

$$N_2O_5 \rightleftharpoons NO_2 + NO_3 \qquad \text{fast}$$

$$NO_2 + NO_3 \rightarrow NO_2 + O_2 + NO \qquad \text{slow}$$

$$NO + NO_3 \rightarrow 2NO_2 \qquad \text{fast}$$

What is the rate law for this reaction?

18.19 For the following reaction mechanism:

$$\overset{\overset{\textstyle O}{\|}}{CH_3C}-O-CH_3 + H_3O^+ \rightleftharpoons$$

$$\overset{\overset{\textstyle H}{\diagup}}{\underset{\textstyle CH_3C}{O^+}}-O-CH_3 + H_2O \qquad \text{fast}$$

$$\overset{\overset{\textstyle H}{\diagup}}{\underset{\textstyle CH_3C}{O^+}}-O-CH_3 + H_2O \longrightarrow$$

$$\overset{\overset{\textstyle O}{\|}}{CH_3C}-OH + HOCH_3 + H_3O^+ \qquad \text{slow}$$

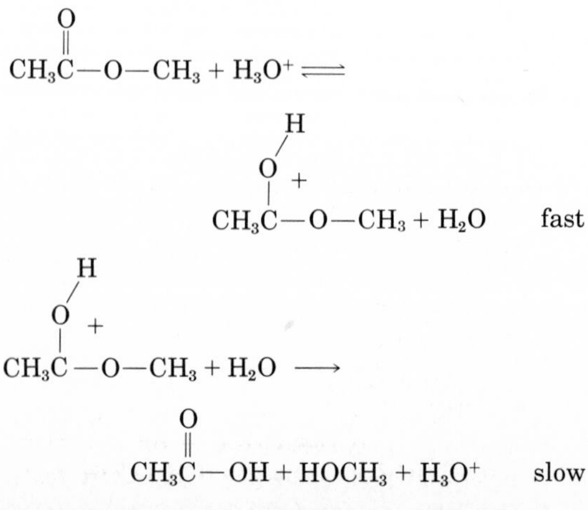

a Write the overall equation for the reaction.

b Write the rate law for the reaction.

c What kind of catalysis is involved in this reaction?

18.20 Suggest a possible structure for the activated complex for the reaction

$$2NO(g) + O_2(g) \rightarrow 2NO_2(g)$$

The rate equation is

$$\text{Rate} = k(c_{NO})^2(c_{O_2})$$

18.21 Which of the following mechanisms are compatible with the rate law given? (More than one mechanism may fit the rate law.)

$$\text{Rate} = k(c_{Cl_2})^{3/2}(c_{CO})$$

a	$\frac{1}{2}Cl_2 \rightleftharpoons Cl$	fast
	$Cl + Cl_2 \rightarrow Cl_3$	fast
	$Cl_3 + CO \rightarrow COCl_2 + Cl$	slow
	$Cl \rightleftharpoons \frac{1}{2}Cl_2$	fast
b	$Cl_2 + CO \rightarrow CCl_2 + O$	slow
	$O + Cl_2 \rightarrow Cl_2O$	fast
	$Cl_2O + CCl_2 \rightarrow COCl_2 + Cl_2$	fast
c	$\frac{1}{2}Cl_2 \rightleftharpoons Cl$	fast
	$Cl + CO \rightleftharpoons COCl$	fast
	$COCl + Cl_2 \rightarrow COCl_2 + Cl$	slow
	$Cl \rightleftharpoons \frac{1}{2}Cl_2$	fast
d	$Cl_2 + CO \rightleftharpoons COCl + Cl$	fast
	$COCl + Cl_2 \rightarrow COCl_2 + Cl$	slow
	$Cl + Cl \rightarrow Cl_2$	fast

18.22 For the reaction

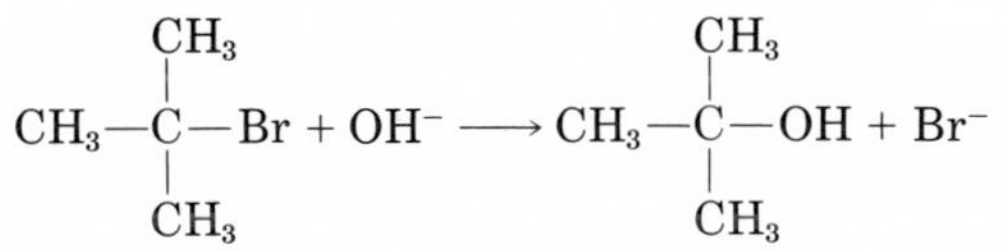

the rate law is

$$\text{Rate} = k(c_{(CH_3)_3CBr})$$

Which of the following mechanisms are compatible with the rate law?

a $(CH_3)_3CBr \rightarrow (CH_3)_3C^+ + Br^-$ slow
$(CH_3)_3C^+ + OH^- \rightarrow (CH_3)_3COH$ fast

b $(CH_3)_3CBr + OH^- \rightarrow (CH_3)_3COH + Br^-$

c $(CH_3)_3CBr + OH^- \rightarrow$
$(CH_3)_2(CH_2)CBr^- + H_2O$ fast
$(CH_3)_2(CH_2)CBr^- \rightarrow$
$(CH_3)_2(CH_2)C + Br^-$ slow
$(CH_3)_2(CH_2)C + H_2O \rightarrow (CH_3)_3COH$ fast

18.23 Assume that each pair of the following reactions occurs via a single bimolecular step. For which member of each pair would you expect the steric factor to be more important?

a $Cl + O_3 \rightarrow ClO + O_2$ or
$NO + O_3 \rightarrow NO_2 + O_2$

b $H_2C{=}CH_2 + H_2 \rightarrow H_3C{-}CH_3$ or
$(CH_3)_2C{=}CH_2 + HBr \rightarrow (CH_3)_2CBr{-}CH_3$

18.24 The activation energy $E^\ddagger$ is 139.7 kJ mol^{-1} for the reaction

$$HI + CH_3I \rightarrow CH_4 + I_2$$

Calculate the fraction of the molecules whose collisions would be energetic enough to react at (*a*) 100°C; (*b*) 200°C; (*c*) 500°C; (*d*) 1000°C.

18.25 Make an Arrhenius plot and calculate the activation energy for the reaction

$$2NOCl \rightarrow 2NO + Cl_2$$

Temperature/K	Rate Constant/dm^3 mol^{-1} s^{-1}
400	6.95×10^{-4}
450	1.98×10^{-2}
500	2.92×10^{-1}
550	2.60
600	16.3

18.26 What are the three main features of catalysis?

18.27 Define and describe an example of each of the following:

a Arrhenius plot
b Substrate
c Heterogeneous catalysis
d Side chain
e Enzyme
f Homogeneous catalysis
g Adsorption

18.28 Name at least three important heterogeneous catalysts and describe the processes in which they are used. Write a balanced chemical equation for a reaction catalyzed by each of your three choices.

18.29 Using data from Fig. 18.2 at 50°C, calculate the time required for the concentration of the dye to fall from

a 1.0 to 0.5 mol dm^{-3}
b 0.8 to 0.4 mol dm^{-3}
c 0.6 to 0.3 mol dm^{-3}
d 0.4 to 0.2 mol dm^{-3}

Repeat your calculations for decomposition of the dye at 80°C. What conclusions can you draw regarding the time required for the concentration of the dye to fall to one-half its former value?

18.30 Results similar to those in Problem 18.29 are obtained for all first-order reactions. The constant time interval required for the concentration of the reactant to fall to one-half its former value is called the half-life, $t_{1/2}$. If the initial concentration of the reactant in a first-order reaction is 0.64 mol and the half-life is 30 s, (*a*) calculate the concentration of the reactant 60 s after initiation of the reaction. (*b*) How long would it take for the concentration of the reactant to drop to one-eighth its initial value? (*c*) How long would it take for the concentration of the reactant to drop to 0.04 mol dm^{-3}?

***18.31** (Before attempting this problem you should do Problems 18.29 and 18.30.) Let x represent the number of half-lives that have elapsed during the course of a first-order reaction. That is, the elapsed time t is x times the half-life $t_{1/2}$:

$$t = xt_{1/2}$$

(*a*) Show that at time t the concentration c of reactant is related to the initial concentration c_0 by

$$c = c_0(\tfrac{1}{2})^x$$

(*b*) Use your result in part *a* to show that

$$\log\left(\frac{c}{c_0}\right) = \frac{t}{t_{1/2}}(-\log 2) = \frac{t}{t_{1/2}}(-0.301)$$

(*c*) Use your result in part *b* to show that for any first-order reaction a plot of log c versus t will be a straight line and that half-life can be obtained from the slope of the line.

***18.32** Given the following data, determine the half-life and the initial concentration of the reactant for the reaction

trans-CHClCHCl → *cis*-CHClCHCl

c_{trans}/mol dm^{-3}	Time/s
9.23×10^{-4}	30
8.51×10^{-4}	60
7.86×10^{-4}	90
7.25×10^{-4}	120
6.19×10^{-4}	180
3.82×10^{-4}	360

***18.33** The reaction

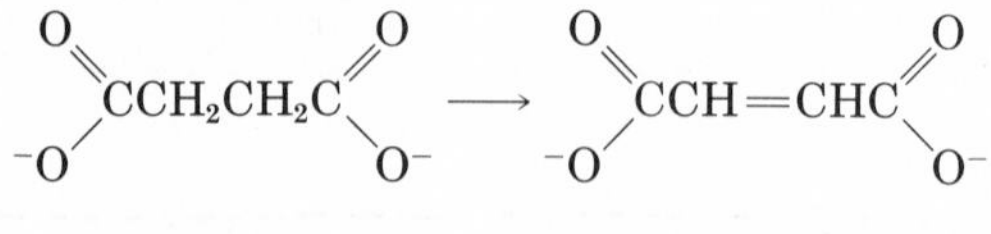

is catalyzed by the enzyme succinate dehydrogenase. When malonate ions or oxalate ions are added to the reaction mixture, the rate decreases significantly. Try to account for this observation in terms of the description of enzyme catalysis given in the text. The structures of malonate and oxalate ions are

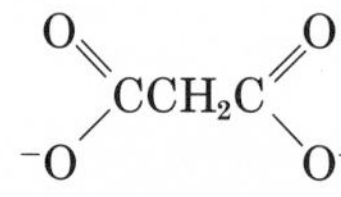

Malonate ion

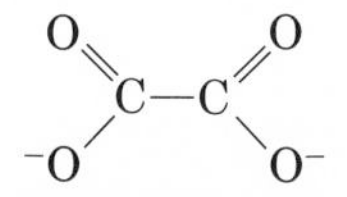

Oxalate ion

***18.34** Suppose that you have a container whose volume is 1.000 dm^3 and which contains 0.300 mol H_2 and 0.100 mol N_2. These substances react according to the equation

$$N_2(g) + 3H_2(g) \rightarrow 2NH_3(g)$$

(*a*) After 0.001 mol N_2 has been consumed, what is the change in concentration Δc of N_2, H_2, and NH_3? (*b*) If this change in concentration required 5.00 min, what is the rate of disappearance of N_2? of H_2? What is the rate of formation of NH_3? (*c*) Show that for a general chemical reaction whose equation is

$$x\text{X} + y\text{Y} \rightarrow z\text{Z}$$

the rate can be defined unambiguously as

$$\text{Rate} = -\frac{1}{x}\frac{dc_\text{X}}{dt} = -\frac{1}{y}\frac{dc_\text{Y}}{dt} = \frac{1}{z}\frac{dc_\text{Z}}{dt}$$

Is this definition better than the one given in Sec. 18.1?

***18.35** For the reaction

$$4HBr(g) + O_2(g) \rightarrow 2Br_2(g) + 2H_2O(g)$$

which of the following statements are true? (*a*) The rate of disappearance of O_2 is one-fourth the rate of disappearance of HBr. (*b*) The rate of appearance of Br_2 is one-half the rate of disappearance of O_2. (*c*) The rate of formation of H_2O is one-half the rate of consumption of HBr. (*d*) The rates of formation of Br_2 and H_2O are equal.

chapter nineteen

NUCLEAR CHEMISTRY

$$
{}^{235}_{92}\mathrm{U} + {}^{1}_{0}n \longrightarrow
\begin{cases}
{}^{140}_{56}\mathrm{Ba} + {}^{93}_{36}\mathrm{Kr} + 3\,{}^{1}_{0}n \\
{}^{144}_{55}\mathrm{Cs} + {}^{90}_{37}\mathrm{Rb} + 2\,{}^{1}_{0}n \\
{}^{144}_{54}\mathrm{Xe} + {}^{90}_{38}\mathrm{Sr} + 2\,{}^{1}_{0}n \\
{}^{146}_{57}\mathrm{La} + {}^{87}_{35}\mathrm{Br} + 3\,{}^{1}_{0}n \\
{}^{160}_{62}\mathrm{Sm} + {}^{72}_{30}\mathrm{Zn} + 4\,{}^{1}_{0}n
\end{cases}
$$

In Chap. 4 we described the experimental basis for the idea that each atom has a small, very massive nucleus which contains protons and neutrons. Surrounding the nucleus are one or more electrons which occupy most of the volume of the atom but make only a small contribution to its mass. Electrons (especially valence electrons) are the only subatomic particles which are involved in ordinary chemical changes, and we have spent considerable time describing the rearrangements they undergo when atoms and molecules combine. However, another category of reactions is possible in which the structures of atomic nuclei change. In such **nuclear reactions** electronic structure is incidental—we are primarily interested in how the protons and neutrons are arranged before and after the reaction. Nuclear reactions are involved in transmutation of one element into another and in natural radioactivity, both of which were described briefly in Chap. 4. In this chapter we will consider nuclear reactions in more detail, exploring their applications to nuclear energy, to the study of reaction mechanisms, to qualitative and quantitative analysis, and to estimation of the ages of objects as different as the Dead Sea scrolls and rocks from the moon.

19.1 NATURALLY OCCURRING NUCLEAR REACTIONS

Radioactivity

When the discovery of natural radioactivity was described in Chap. 4, we mentioned the properties of α particles, β particles, and γ rays, but little attention was paid to the atoms which were left behind when one of these forms of radiation was emitted. Now we can consider the subject of radioactivity in more detail.

α emission An α particle corresponds to a helium nucleus. It consists of two protons and two neutrons, and so it has a mass number of 4 and an atomic number (nuclear charge) of 2. From a chemical point of view we would write it as ${}^4_2He^{2+}$, indicating its lack of two electrons with the superscript 2+. In writing a nuclear reaction, though, it is unnecessary to specify the charge, because the presence or absence of electrons around the nucleus is usually unimportant. For these purposes an α particle is indicated as

$${}^4_2He \qquad \text{or} \qquad {}^4_2\alpha$$

In certain nuclei α particles are produced by combination of two protons and two neutrons which are then emitted.

An example of naturally occurring emission of an α particle is the disintegration of one of the isotopes of uranium, ${}^{238}_{92}U$. The equation for this process is

$${}^{238}_{92}U \rightarrow {}^{234}_{90}Th + {}^4_2He \qquad (19.1)$$

Note that if we sum the mass numbers on each side of a **nuclear equation,** such as Eq. (19.1), the total is the same. That is, 238 on the left equals 234 + 4 on the right. Similarly, the atomic numbers (subscripts) must also balance (92 on the left and 90 + 2 on the right). This is a general rule which must be followed in writing any nuclear reaction. The total number of nucleons (i.e., protons and neutrons) remains unchanged, and electrical charge is neither created nor destroyed in the process.

When a nucleus emits an α particle, its atomic number is reduced by 2 and it becomes the nucleus of an element two places earlier in the periodic table. That one element could transmute into another in this fashion was first demonstrated by Rutherford and Soddy in 1902. It caused a tremendous stir in the scientific circles of the day since it quite clearly contradicted Dalton's hypothesis that atoms are immutable. It gave Rutherford, who was then working at McGill University in Canada, an international reputation.

The type of nucleus that will spontaneously emit an α particle is fairly restricted. The mass number is usually greater than 209 and the atomic number greater than 82. In addition the nucleus must have a lower ratio of neutrons to protons than normal. The emission of an α particle raises the neutron/proton ratio as illustrated by the nuclear equation

$${}^{210}_{84}Po \rightarrow {}^{206}_{82}Pb + {}^4_2He \qquad (19.2)$$

The nucleus of ${}^{210}_{84}Po$ contains 210 − 84 = 126 neutrons and 84 protons, giving a ratio of 126:84 = 1.500. This is increased to 124:82 = 1.512 by the α-particle emission.

β emission A β particle is an electron which has been emitted from an atomic nucleus. It has a very small mass and is therefore assigned a mass number of 0. The β particle has a negative electrical charge, and so its nuclear charge is taken to be -1. Thus it is given the symbol

$$_{-1}^{0}e \quad \text{or} \quad _{-1}^{0}\beta$$

in a nuclear equation. Two examples of unstable nuclei which emit β particles are

$$^{234}_{90}\text{Th} \rightarrow {}^{234}_{91}\text{Pa} + {}_{-1}^{0}e \tag{19.3}$$

and

$$^{14}_{6}\text{C} \rightarrow {}^{14}_{7}\text{N} + {}_{-1}^{0}e \tag{19.4}$$

Note that both of these equations accord with the conservation of mass number and atomic number, showing again that both the total number of nucleons and the total electrical charge remain unchanged.

We can consider a β particle emitted from a nucleus to result from the transformation of a neutron into a proton and an electron according to the reaction

$$^{1}_{0}n \rightarrow {}^{1}_{1}p + {}_{-1}^{0}e \tag{19.5}$$

(Indeed, free neutrons unattached to any nucleus soon decay in this way.) Thus when a nucleus emits a β particle, the nuclear charge rises by 1 while the mass number is unaltered. Therefore the disintegration of a nucleus by β decay produces the nucleus of an element one place *further along* in the periodic table than the original element.

β decay is a very common form of radioactive disintegration and, unlike α decay, is found among both heavy and light nuclei. Nuclei which disintegrate in this way usually have a neutron/proton ratio which is higher than normal. When a β particle is lost, a neutron is replaced by a proton and the neutron/proton ratio decreases. For example, in the decay process

$$^{210}_{82}\text{Pb} \rightarrow {}^{210}_{83}\text{Bi} + {}_{-1}^{0}e \tag{19.6}$$

the neutron/proton ratio changes from 1.561 to 1.530.

γ radiation γ rays correspond to electromagnetic radiation similar to light waves or radiowaves except that they have an extremely short wavelength—about a picometer. Because of the wave-particle duality we can also think of them as particles or photons having the same velocity as light and an extremely high energy. Since they have zero charge and are not nucleons, they are denoted in nuclear equations by the symbol $^{0}_{0}\gamma$ or, more simply, γ.

Virtually all nuclear reactions are accompanied by the emission of γ rays. This is because the occurrence of a nuclear transformation usually leaves the resultant nucleus in an unstable high-energy state. The nucleus then loses energy in the form of a γ-ray photon as it adopts a lower-energy more stable form. Usually these two processes succeed each other so rapidly that they cannot be distinguished. Thus when ^{238}U decays by α emission, it also emits a γ ray. This is actually a two-stage process. In the first stage a high-energy (or excited) form of ^{234}Th is produced:

$$^{238}_{92}\text{U} \rightarrow {}^{234}_{90}\text{Th}^{*} + {}^{4}_{2}\text{He} \tag{19.7}$$

This excited nucleus then emits a γ ray:

$$^{234}_{90}Th^* \rightarrow ^{234}_{90}Th + ^{0}_{0}\gamma \qquad (19.8)$$

Usually when a nuclear reaction is written, the γ ray is omitted. Thus Eqs. (19.7) and (19.8) are usually combined to give

$$^{238}_{92}U \rightarrow ^{234}_{90}Th + ^{4}_{2}He \qquad (19.1)$$

Radiation and Human Health

Alpha particles, beta particles, gamma rays, and some other types of radiation such as x-rays are injurious to humans and other living organisms. A single particle or photon usually has sufficient energy to break one or more chemical bonds or to ionize a molecule in living tissue. The free radicals or ions produced in this way are highly reactive chemically. They can disrupt cell membranes, reduce the effectiveness of enzymes, or even damage genes and chromosomes. The greatest harm of this type is caused by the heavier, more highly charged alpha particles, which produce considerable disruption when they collide with molecules in living tissue. Beta particles are less harmful because of their lesser charge and mass, and gamma or x-rays have the smallest effect.

When dealing with radioactive materials, it is necessary for humans to shield their bodies from harmful radiation. In the case of alpha particles even a single sheet of paper serves to absorb a large fraction. Heavier shielding—on the order of 1 mm of aluminum—is needed to stop the lighter beta particles, and the uncharged gamma rays or x-rays require 5 cm or more of a dense metal such as lead to stop them. Thus although alpha particles will do more damage once inside the human body, one's skin is a fairly good shield against them. If breathed in or swallowed, however, alpha emitters are highly poisonous. Outside the body beta particles are more dangerous than alpha particles, and gamma rays or x-rays, which can penetrate all the way through one's tissues to the internal organs, are most harmful of all.

Radioactive Series

Naturally occurring uranium contains more than 99% $^{238}_{92}U$, an isotope which decays to $^{234}_{90}Th$ by α emission, as shown in Eq. (19.1). The product of this reaction is also radioactive, however, and undergoes β decay, as already shown in Eq. (19.3). The $^{234}_{91}Pa$ produced in this second reaction also emits a β particle:

$$^{234}_{91}Pa \rightarrow ^{234}_{92}U + ^{0}_{-1}e \qquad (19.9)$$

These three reactions [Eqs. (19.1), (19.3), and (19.9)] are only the first of 14 steps. After emission of eight α particles and six β particles, the isotope $^{206}_{82}Pb$ is produced. It has a stable nucleus which does not disintegrate further. The complete process may be written as follows:

$$\begin{array}{c} {}^{238}_{92}U \xrightarrow{\alpha} {}^{234}_{90}Th \xrightarrow{\beta} {}^{234}_{91}Pa \xrightarrow{\beta} {}^{234}_{92}U \xrightarrow{\alpha} {}^{230}_{90}Th \xrightarrow{\alpha} {}^{226}_{88}Ra \xrightarrow{\alpha} {}^{222}_{86}Rn \\ \downarrow \alpha \\ {}^{206}_{82}Pb \xleftarrow{\alpha} {}^{210}_{84}Po \xleftarrow{\beta} {}^{210}_{83}Bi \xleftarrow{\beta} {}^{210}_{82}Pb \xleftarrow{\alpha} {}^{214}_{84}Po \xleftarrow{\beta} {}^{214}_{83}Bi \xleftarrow{\beta} {}^{214}_{82}Pb \xleftarrow{\alpha} {}^{218}_{84}Po \end{array} \qquad (19.10a)$$

While the net reaction is

$${}^{238}_{92}U \rightarrow {}^{206}_{82}Pb + 8{}^{4}_{2}He + 6{}^{0}_{-1}e \qquad (19.10b)$$

Such a series of successive nuclear reactions is called a **radioactive series.**

Two other radioactive series similar to the one just described occur in nature. One of these starts with the isotope ${}^{232}_{90}Th$ and involves 10 successive stages, while the other starts with ${}^{235}_{92}U$ and involves 11 stages. Each of the three series produces a different stable isotope of lead.

EXAMPLE 19.1 The first four stages in the uranium–actinium series involve the emission of an α particle from a ${}^{235}_{92}U$ nucleus, followed successively by the emission of a β particle, a second α particle, and then a second β particle. Write out equations to describe all four nuclear reactions.

Solution The emission of an α particle lowers the atomic number by 2 (from 92 to 90). Since element 90 is thorium, we have

$${}^{235}_{92}U \rightarrow {}^{231}_{90}Th + {}^{4}_{2}He$$

The emission of a β particle now increases the atomic number by 1 to give an isotope of element 91, protactinium:

$${}^{231}_{90}Th \rightarrow {}^{231}_{91}Pa + {}^{0}_{-1}e$$

The next two stages follow similarly:

$${}^{231}_{91}Pa \rightarrow {}^{227}_{89}Ac + {}^{4}_{2}He \qquad \text{and} \qquad {}^{227}_{89}Ac \rightarrow {}^{227}_{90}Th + {}^{0}_{-1}e$$

EXAMPLE 19.2 In the thorium series, ${}^{232}_{90}Th$ loses a total of six α particles and four β particles in a 10-stage process. What isotope is finally produced in this series?

Solution The loss of six α particles and four β particles:

$$6{}^{4}_{2}He + 4{}^{0}_{-1}e$$

involves the total loss of 24 nucleons and $6 \times 2 - 4 = 8$ positive charges from the ${}^{232}_{90}Th$ nucleus. The eventual result will be an isotope of mass number $232 - 24 = 208$ and a nuclear charge of $90 - 8 = 82$. Since element 82 is Pb, we can write

$${}^{232}_{90}Th \rightarrow {}^{208}_{82}Pb + 6{}^{4}_{2}He + 4{}^{0}_{-1}e$$

In 1919 Rutherford performed the first artificial nuclear reaction. He was able to demonstrate that when α particles are introduced into a closed sample of N_2 gas, an occasional collision led to the formation of an isotope of O and the release of a proton:

$$^{4}_{2}He + ^{14}_{7}N \rightarrow ^{17}_{8}O + ^{1}_{1}H \qquad (19.11)$$

Since then many thousands of nuclear reactions have been studied, most of them produced by the bombardment of stable forms of matter with a beam of nucleons or light nuclei as projectiles. Particles which have been used for this purpose include protons, neutrons, deuterons ($^{2}_{1}H$), α particles, and B, C, N, and O nuclei.

Bombardment with Positive Ions

When the bombarding particle is positively charged, which is usually the case, it must have a very high kinetic energy to overcome the coulombic repulsion of the nucleus being bombarded. This is particularly necessary if the nucleus has a high nuclear charge. To give these charged particles the necessary energy, an accelerator or "atom smasher" such as a **cyclotron** must be used. The cyclotron was developed mainly by E. O. Lawrence (1901 to 1958) at the University of California. A schematic diagram of a cyclotron is shown in Fig. 19.1. Two hollow D-shaped plates (dees) are enclosed in an evacuated chamber between the poles of a powerful electromagnet. The two dees are connected to a source of high-frequency alternating current, so that when one is positive, the other is negative. Ions are introduced at the center and are accelerated because of their alternate attraction to the left- and right-hand dees. Since the magnetic field would make ions traveling at constant speed move in a circle, the net result is that they follow a spiral path as they accelerate until they finally emerge at the outer edge of one of the dees.

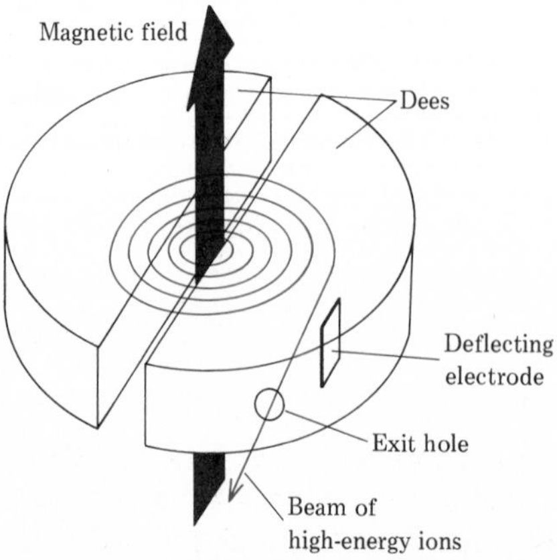

Figure 19.1 A cyclotron. The spiral path of the ions is shown in color.

Some examples of the kinds of nuclear reactions which are possible with the aid of an accelerator are as follows:

$$^{24}_{12}\text{Mg} + ^{1}_{1}\text{H} \rightarrow ^{21}_{11}\text{Na} + ^{4}_{2}\text{He} \quad (19.12)$$

$$^{7}_{3}\text{Li} + ^{2}_{1}\text{H} \rightarrow ^{8}_{4}\text{Be} + ^{1}_{0}n \quad (19.13)$$

$$^{106}_{46}\text{Pd} + ^{4}_{2}\text{He} \rightarrow ^{109}_{47}\text{Ag} + ^{1}_{1}\text{H} \quad (19.14)$$

A particularly interesting type of nuclear reaction performed in an accelerator is the production of the transuranium elements. These elements have atomic numbers greater than that of uranium (92) and are too unstable to persist for long in nature. The heaviest of them can be prepared by bombarding nuclei which are already heavy with some of the lighter nuclei:

$$^{238}_{92}\text{U} + ^{12}_{6}\text{C} \rightarrow ^{246}_{98}\text{Cf} + 4^{1}_{0}n \quad (19.15)$$

$$^{238}_{92}\text{U} + ^{14}_{7}\text{N} \rightarrow ^{247}_{99}\text{Es} + 5^{1}_{0}n \quad (19.16)$$

$$^{252}_{98}\text{Cf} + ^{10}_{5}\text{B} \rightarrow ^{257}_{103}\text{Lr} + 5^{1}_{0}n \quad (19.17)$$

Neutron Bombardment

Since a neutron has no charge, it is not repelled by the nucleus it is bombarding. Because of this, neutrons do not need to be accelerated to high energies before they can undergo a nuclear reaction. Nuclear reactions involving neutrons are thus easier and cheaper to perform than those requiring positively charged particles.

Though neutron-bombardment reactions are often carried out in a nuclear reactor (which will be described later), they can also be very conveniently performed in a small laboratory using a *neutron source.* Usually a neutron source consists of an α emitter such as $^{222}_{86}\text{Rn}$ mixed with Be, an element whose nuclei produce neutrons when bombarded by α particles:

$$^{9}_{4}\text{Be} + ^{4}_{2}\text{He} \rightarrow ^{12}_{6}\text{C} + ^{1}_{0}n \quad (19.18)$$

This reaction was originally used in 1932 by Sir James Chadwick (1891 to) to demonstrate the existence of the neutron. (Previous to this it was believed that electrons were present in the nucleus together with protons.) The neutrons produced by Eq. (19.18) have a very high energy and are called **fast neutrons.** For many purposes the neutrons are more useful if they are first slowed down or **moderated** by passing them through paraffin wax or some other substance containing light nuclei in which they can dissipate most of their energy by collision. The **slow neutrons** produced by a moderator are then able to participate in a larger number of **neutron-capture reactions** of which the following two are typical:

$$^{34}_{16}\text{S} + ^{1}_{0}n \rightarrow ^{35}_{16}\text{S} + \gamma \quad (19.19)$$

$$^{200}_{80}\text{Hg} + ^{1}_{0}n \rightarrow ^{201}_{80}\text{Hg} + \gamma \quad (19.20)$$

In such a reaction a different isotope (often an unstable isotope) of the element being bombarded is produced, with the emission of a γ ray. Radioactive isotopes of virtually all the elements can be produced in this way.

An important neutron-capture reaction is that undergone by the most common isotope of uranium, namely, ^{238}U:

$$^{238}_{92}U + ^{1}_{0}n \rightarrow ^{239}_{92}U + \gamma \qquad (19.21)$$

The uranium-239 produced in this way decays by β emission to produce the first and most important of the transuranium elements, namely, neptunium:

$$^{239}_{92}U \rightarrow ^{239}_{93}Np + ^{0}_{-1}e \qquad (19.22)$$

When nuclei are bombarded by fast neutrons, a secondary particle is emitted—usually a proton or an α particle:

$$^{11}_{5}B + ^{1}_{0}n \rightarrow ^{11}_{4}Be + ^{1}_{1}H \qquad (19.23)$$

$$^{27}_{13}Al + ^{1}_{0}n \rightarrow ^{24}_{11}Na + ^{4}_{2}He \qquad (19.24)$$

Further Modes of Decay

Isotopes produced by nuclear reactions which do not occur in nature (**artificial isotopes**) are invariably unstable and radioactive. They exhibit two kinds of decay not found among naturally occurring radioactive elements. The first is **positron emission** (also called β^+ emission) in which a fundamental particle we have not previously discussed is ejected from the nucleus. The **positron** is identical with the electron except that it has a positive rather than a negative charge. Its symbol is $^{0}_{+1}e$. An example of positron emission is

$$^{11}_{6}C \rightarrow ^{11}_{5}B + ^{0}_{+1}e \qquad (19.25)$$

Positron emission is common among isotopes having a low neutron-to-proton ratio.

The second new method of decay is called **electron capture.** The nucleus absorbs one of the electrons from its own innermost core. An example is the following reaction:

$$^{0}_{-1}e + ^{7}_{4}Be \xrightarrow{ec} ^{7}_{3}Li \qquad (19.26)$$

Again this results in an increased neutron/proton ratio.

19.3 NUCLEAR STABILITY

Why is it that certain combinations of nucleons are stable in a nucleus while others are not? A complete answer to this question cannot yet be given, largely because the exact nature of the forces holding the nucleons together is still only partially understood. We can, however, point to several factors which affect nuclear stability. The most obvious is the neutron/proton ratio. As we have already seen, if this is too high or too low, it makes for an unstable nucleus.

If we plot the number of neutrons against the number of protons for all known stable (i.e., nonradioactive) nuclei, we obtain the result shown in

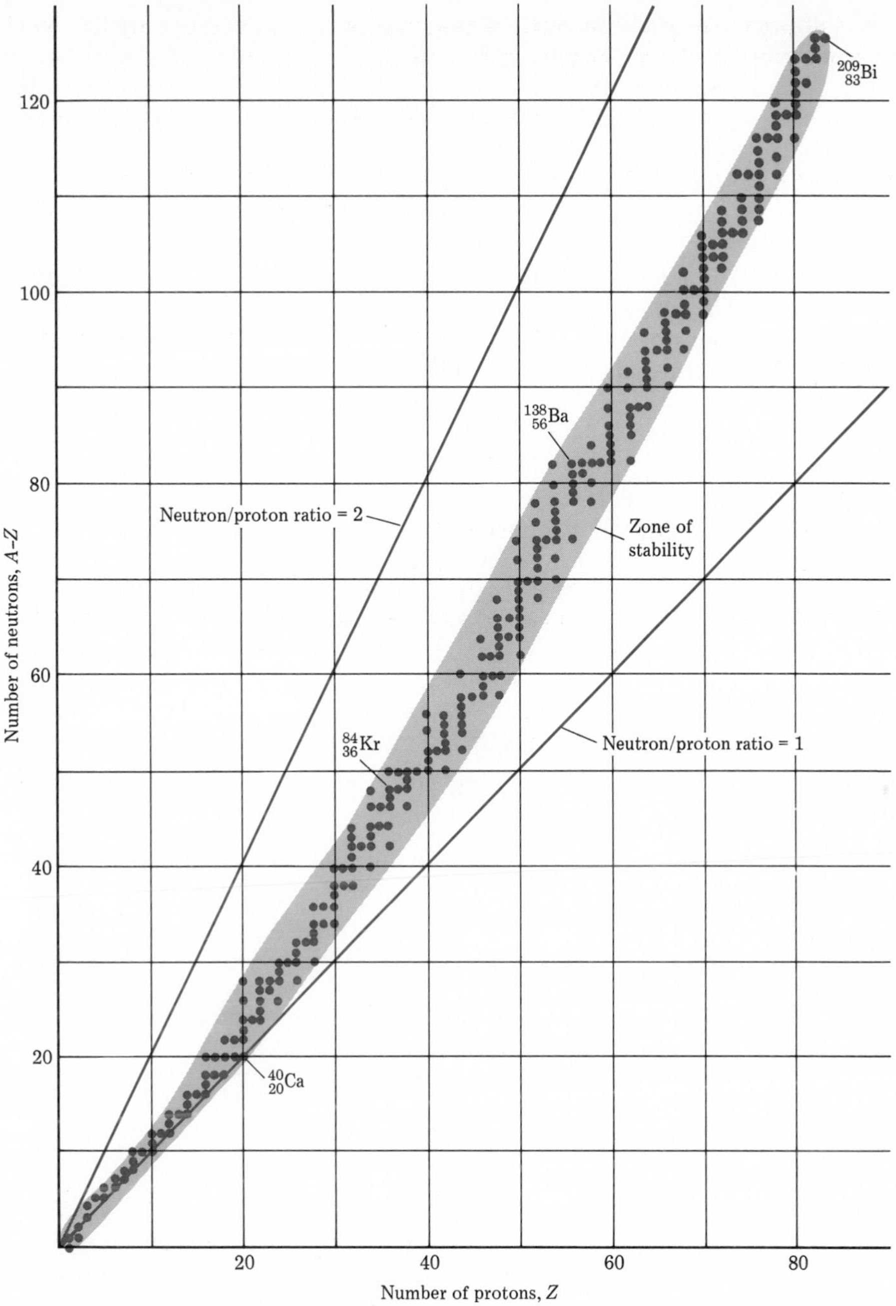

Figure 19.2 The number of neutrons plotted against the number of protons for all the stable nuclei. Note how the neutron/proton ratio increases for the heavier elements.

Fig. 19.2. All the stable nuclei lie within a definite area called the **zone of stability.** For low atomic numbers most stable nuclei have a neutron/proton ratio which is very close to 1. As the atomic number increases, the zone of stability corresponds to a gradually increasing neutron/proton ratio. In the case of the heaviest stable isotope, $^{209}_{83}Bi$, for instance, the neutron/proton ratio is 1.518. If an unstable isotope lies to the left of the zone of stability in Fig. 19.2, it is neutron rich and decays by β emission. If it lies to the right of the zone, it is proton rich and decays by positron emission or electron capture.

Another factor affecting the stability of a nucleus is whether the number of protons and neutrons is even or odd. Among the 354 known stable isotopes, 157 (almost half) have an even number of protons and an even number of neutrons. Only five have an odd number of both kinds of nucleons. In the universe as a whole (with the exception of hydrogen) we find that the even-numbered elements are almost always much more abundant than the odd-numbered elements close to them in the periodic table.

Finally there is a particular stability associated with nuclei in which either the number of protons or the number of neutrons is equal to one of the so-called "magic" numbers 2, 8, 20, 28, 50, 82, and 126. These numbers correspond to the filling of shells in the structure of the nucleus. These shells are similar in principle but different in detail to those found in electronic structure. Of particular stability, and also of high abundance in the universe, are nuclei in which both the number of protons and the number of neutrons correspond to magic numbers. Examples are $^{4}_{2}He$, $^{16}_{8}O$, $^{40}_{20}Ca$, and $^{208}_{82}Pb$.

EXAMPLE 19.3 Find which element has the largest number of isotopes, using Fig. 19.2. Likewise find which is the number of neutrons which occurs most frequently. What do you notice about the numbers of protons and neutrons in each case?

Solution Tin has 10 isotopes, and its atomic number 50 is a magic number. A total of 7 stable isotopes have 82 neutrons in the nucleus, more than for any other number of neutrons. Again the number is a magic number.

19.4 THE RATE OF RADIOACTIVE DECAY

So far we have labeled all isotopes which exhibit radioactivity as *unstable,* but radioactive isotopes vary considerably in their degree of instability. Some decay so quickly that it is difficult to detect that they are there at all before they have changed into something else. Others have hardly decayed at all since the earth was formed.

The process of radioactive decay is governed by the uncertainty principle, so that we can never say exactly when a particular nucleus is going to disintegrate and emit a particle. We can, however, give the probability that a nucleus will disintegrate in a given time interval. For a large number of nuclei we can predict what fraction will disintegrate during that interval. This fraction will be independent of the amount of isotope but will vary from isotope to isotope depending on its stability. We can also look at the matter from the opposite point of view, i.e., in terms of how long it will take a given fraction of isotope to dissociate. Conventionally the tendency for the nuclei of an isotope to decay is measured by its **half-life,** symbol $t_{1/2}$. This is the time required for exactly half the nuclei to disintegrate. This quantity, too, varies from isotope to isotope but is *independent of the amount of isotope.*

Figure 19.3 shows how a 1-amol (attomole) sample of $^{128}_{53}$I, which has a half-life of 25.0 min, decays with time. In the first 25 min, half the nuclei disintegrate, leaving behind 0.5 amol. In the second 25 min, the remainder is reduced by one-half again, i.e., to 0.25 amol. After a third 25-min period, the remainder is $(\frac{1}{2})^3$ amol, after a fourth it is $(\frac{1}{2})^4$ amol, and so on. If x intervals of 25.0 min are allowed to pass, the remaining amount of ^{128}I will be $(\frac{1}{2})^x$ amol.

This example enables us to see what will happen in the general case. Suppose the initial amount of an isotope of half-life $t_{1/2}$ is n_0 and the isotope decays to an amount n in time t, we can measure the time in terms of the number of $t_{1/2}$ intervals which have elapsed by defining a variable x such that

$$x = \frac{t}{t_{1/2}} \tag{19.27}$$

Figure 19.3 Radioactive decay of ^{128}I. In the course of each 25-min period, the amount of the isotope decreases by one-half.

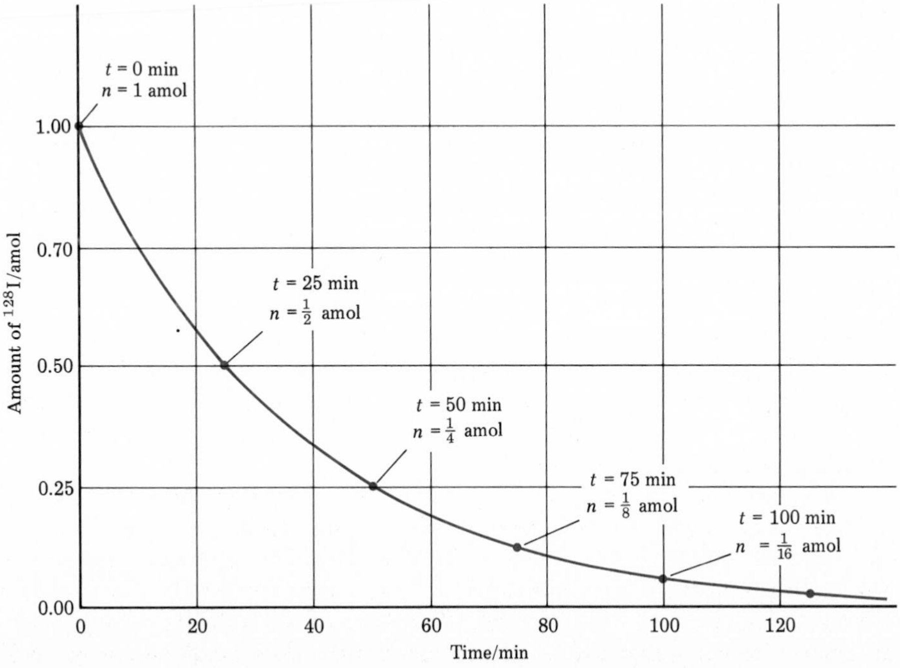

Thus after time t our sample will have been reduced to a fraction $(\frac{1}{2})^x$ of the original amount. In other words

$$\frac{n}{n_0} = \left(\frac{1}{2}\right)^x \tag{19.28}$$

Taking logs we then have

$$\log \frac{n}{n_0} = \log \left(\frac{1}{2}\right)^x = x \log \frac{1}{2} = -0.3010x$$

Substituting from Eq. (19.27), we thus obtain

$$\log \frac{n}{n_0} = \frac{-0.3010}{t_{1/2}} t \tag{19.29}$$

EXAMPLE 19.4 What amount of ^{128}I will be left when 3.65 amol of this isotope is allowed to decay for 15.0 min. The half-life of ^{128}I is 25.0 min.

Solution Substituting in Eq. (19.29), we have

$$\log \frac{n}{n_0} = \frac{-0.3010t}{t_{1/2}} = \frac{-0.3010 \times 15.0 \text{ min}}{25.0 \text{ min}}$$
$$= -0.1806$$

Thus

$$\frac{n}{n_0} = \text{antilog}(-0.1806) = 10^{-0.1806}$$
$$= 0.6598$$

or

$$n = 0.6598n_0 = 0.6598 \times 3.65 \text{ amol} = 2.41 \text{ amol}$$

Equation (19.28) describes how the amount of a radioactive isotope decreases with time, but similar formulas can also be written for the mass m and also for the rate of disintegration r. This is because both the mass and the rate are proportional to the amount of isotope. Thus the rate at which an isotope decays is given by

$$\log \frac{r}{r_0} = \frac{-0.3010}{t_{1/2}} t \tag{19.30}$$

where r_0 is the initial rate at time zero.

The decrease over time of the rate of decay of a radioactive isotope can be used to establish the ages of various objects and thus is important in fields such as archaeology, paleontology, and geology. The best known of these dating techniques involves the isotope $^{14}_{6}C$, a β emitter with a half-life of 5770 years. There is one atom of $^{14}_{6}C$ for every 7.49×10^{11}C atoms in the CO_2 of the atmosphere and in all living things. The proportion does not change with time because as fast as ^{14}C nuclei are destroyed by radioactive decay, they become replenished by the action of cosmic-ray neutrons on N atoms in the upper atmosphere:

$$^{14}_{7}N + ^{1}_{0}n \rightarrow ^{14}_{6}C + ^{1}_{1}H \qquad (19.31)$$

The carbon produced in this way soon becomes part of a CO_2 molecule and enters the carbon cycle. Thus any sample of carbon derived from a living plant or animal or from the atmosphere has the same rate of decay—15.3 disintegrations per minute per gram of carbon.

Once a plant or an animal dies, it is removed from the carbon cycle and the rate of radioactive decay begins to decrease. By measuring the disintegration rate one can estimate how long it has been since the sample was removed from the carbon cycle.

EXAMPLE 19.5 A sample of carbon derived from one of the Dead Sea scrolls was found to be decaying at the rate of 12.1 disintegrations per minute per gram of carbon. Estimate the age of the Dead Sea scrolls.

Solution Since the original rate of decay of the material from which the scrolls were made was 15.3 disintegrations per minute per gram of carbon, we have $r_0 = 15.3\ \text{min}^{-1}\ \text{g}^{-1}$, while $r = 12.1\ \text{min}^{-1}\ \text{g}^{-1}$. Substituting into Eq. (19.30), we then have

$$\log \frac{r}{r_0} = \frac{-0.3010}{t_{1/2}}\, t$$

$$\log \frac{12.1}{15.3} = \frac{-0.3010}{5770 \text{ years}}\, t$$

$$-0.102 = -5.22 \times 10^{-5} \text{ years}^{-1}\, t$$

$$t = 1950 \text{ years}$$

There are several dating techniques which can be used to determine the age of rocks. The simplest of these is perhaps the determination of the ratio of the amount of ^{238}U to the amount of ^{206}Pb in a given rock. As we have already seen, ^{238}U decays to ^{206}Pb in a series of 14 steps for which the net process is

$$^{238}_{92}U \rightarrow ^{206}_{82}Pb + 8\,^{4}_{2}He + 6\,^{0}_{-1}e \qquad (19.10b)$$

The overall rate of this process is governed by its slowest step which has a half-life of 4.5×10^9 years. The assumption is made that all the ^{206}Pb in the rock derives from the ^{238}U and that none was present when the rock was initially formed. If this assumption is correct, the ratio of ^{238}U to ^{206}Pb will decrease with time. By measuring this ratio we can estimate how long ago the rock was formed.

EXAMPLE 19.6 Analysis of a rock revealed that it contained 0.753 μg of $^{238}_{92}U$ and 0.241 μg of $^{206}_{82}Pb$. Calculate the age of the rock.

Solution

$$\text{Amount of } ^{238}\text{U} = n_{\text{U}} = \frac{0.753\ \mu\text{g}}{238\ \text{g mol}^{-1}} = 3.16 \times 10^{-3}\ \mu\text{mol}$$

$$\text{Amount of } ^{206}\text{Pb} = n_{\text{Pb}} = \frac{0.241\ \mu\text{g}}{206\ \text{g mol}^{-1}} = 1.17 \times 10^{-3}\ \mu\text{mol}$$

Since each mol of ^{206}Pb was originally a mole of ^{238}U, the original amount of ^{238}U, n_0, is given by

$$n_0 = (3.16 + 1.17) \times 10^{-3}\ \mu\text{mol} = 4.33 \times 10^{-3}\ \mu\text{mol}$$

Substituting into Eq. (19.29), we have

$$\log \frac{n}{n_0} = \frac{-0.3010}{t_{1/2}}\, t$$

$$\log \frac{3.16 \times 10^{-3}\ \mu\text{mol}}{4.33 \times 10^{-3}\ \mu\text{mol}} = \frac{-0.3010}{4.5 \times 10^{9}\ \text{years}}\, t$$

$$-0.137 = -6.7 \times 10^{-11}\ \text{years}^{-1}\, t$$

or

$$t = 2.0 \times 10^{9}\ \text{years}$$

The majority of the age measurements made on earth rocks, and in recent years on moon rocks, yield values in the range of 1 to 4.5 billion years. On this basis the ages of both the earth and the moon seem to be similar, and the theory that the moon was a fragment of a previously formed earth becomes difficult to support.

19.5 DETECTION AND MEASUREMENT OF RADIATION

Because radiation is harmful to humans and other organisms, it is very important that we be able to detect it and measure how much is present. Such measurements are complicated by two factors. First, we cannot see, hear, smell, taste, or touch radiation, and so special instruments are required to measure it. Second, as we have already mentioned, different types of radiation are more dangerous than others, and corrections must be made for the relative harm done by α particles as opposed to, say, γ rays.

Instruments for Radiation Detection

Perhaps the most common instrument for measuring radiation levels is the **Geiger-Müller counter** (the same Geiger who worked with Rutherford to discover the atomic nucleus.) A schematic diagram of a Geiger-Müller counter is shown in Fig. 19.4. A metal tube containing Ar gas is sealed at

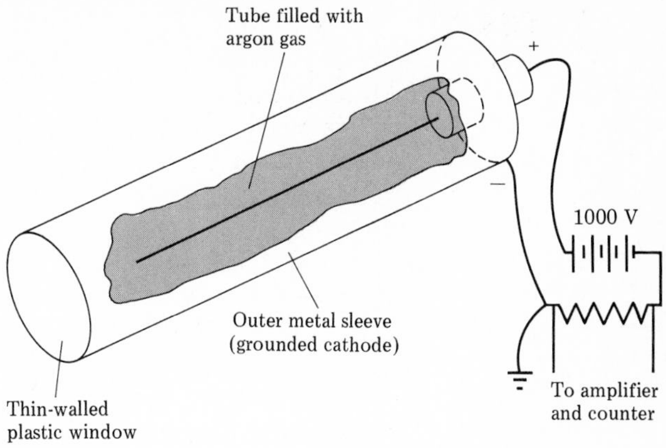

Figure 19.4 A Geiger-Müller tube.

one end with a thin glass or plastic window and contains a central wire well insulated from it. A potential difference of about 1000 V is applied between the central wire and the tube. Any incoming α, β, or γ ray will ionize some of the Ar atoms. These Ar^+ ions are quickly accelerated to a high velocity by the large potential difference, high enough for them in turn to start ionizing further Ar atoms. Thus, for every ray that enters the tube, a large number of ions is formed and a pulse of electrical current is produced. This pulse is amplified and allowed to drive a digital electronic counter which operates on a principle similar to that of a digital watch. The number of particles passing through the tube in a given time can thus be found. Alternatively, the tube can be made to operate a meter indicating the *rate* at which radiation is passing into the Geiger-Müller tube.

Another type of detector, much used for γ rays, is the **scintillation counter.** When a γ ray penetrates a special crystal or solution, it produces a momentary flash of light (called a scintillation) which is detected by a photoelectric cell. Again the output can be amplified and fed into a counter or a meter. A third kind of detector is used to monitor how much exposure laboratory workers have been subjected to in the course of their work. This is simply a strip of photographic film. The degree to which this film is darkened is a measure of the total quantity of radiation to which the worker has been subjected.

Units of Radiation Dose

A variety of units have been designed to measure how much radiation has been absorbed by a given sample of human or animal tissue. The simplest to understand is the **radiation-absorbed dose,** abbreviated **rad.** This corresponds to absorption of 10^{-5} J of energy per gram of tissue. A more useful unit is the **rem** (**roentgen-equivalent man**), which is the same as the rad except that it is corrected for the relative harmfulness of each type of radiation. For example, an α particle having a kinetic energy of 1.6×10^{-22} J can produce about 10 times as many ions as a γ ray of equal energy. Consequently 1 rad of α radiation would be corrected to 10 rem, while 1 rad of γ radiation would correspond to 1 rem.

Once radiation detectors were developed, it was discovered that there is nowhere that one can be entirely free of radiation. That is, there is a natural **background radiation** impinging on all of us every day of our lives. This comes from natural radioactive isotopes in our surroundings and from cosmic radiation which enters the earth's atmosphere from outer space. The average United States citizen receives just over 0.1 rem per year from natural background, although this varies from place to place. In Colorado, for example, background radiation is much higher because of the altitude (less atmosphere to block cosmic rays) and because of naturally occurring deposits of uranium.

Current estimates indicate that the actual radiation dose received by the average person is about 80 percent higher than natural background. The major portion of this increase is due to medical uses—a chest x-ray, for example, contributes about 0.2 rem. Other contributions are made by radioactive fallout from nuclear bombs (about 4 percent of background), and miscellaneous sources such as TV sets (about 2 percent).

There is evidence that the effects of small doses of radiation are cumulative, at least to some degree, and that there is no lower limit to the dose which can cause some damage. Thus even background radiation may be harmful to some extent, but it is hard to determine just how harmful because we have no way of turning it off to see how much difference it makes. In the absence of more accurate information it would seem prudent for each individual and for a society as a whole to minimize unnecessary radiation exposures.

19.6 USES OF ARTIFICIAL ISOTOPES IN CHEMISTRY

Tracers

A very large number of isotopes which do not occur naturally can now be made fairly readily by neutron capture using an atomic reactor or a laboratory neutron source. Many of these artificial isotopes have proved very useful in chemistry since they provide a way of identifying atoms from a particular source, a technique known as **labeling** or **tracer study.** This technique is particularly easy to use if the isotope is radioactive. Thus, for example, if a small quantity of the radioactive isotope ^{131}I is added in the form of iodide ion to a saturated solution of lead iodide, one soon finds that the solid lead iodide in contact with the solution, as well as the solution, become radioactive. This clearly demonstrates that the solution equilibrium

$$PbI_2 \rightleftharpoons Pb^{2+} + 2I^- \qquad (19.32)$$

is a dynamic process involving the constant interchange of iodide ions between the solution and the solid.

Tracer studies are also possible with isotopes which are not radioactive. The isotope ^{18}O is often used in this way, since no convenient radioactive isotope of oxygen is available. In naturally occurring oxygen the isotope ^{18}O is only 0.2 percent of the total. If extra ^{18}O is added, its presence can be detected by mass spectrometry. An interesting and important example of the use of ^{18}O is in the study of photosynthesis. If the water in this reaction is enriched with ^{18}O, then the isotope is found in the oxygen produced:

$$6CO_2(g) + 6H_2O(l) \rightarrow C_6H_{12}O_6(s) + 6O_2(g) \qquad (19.33)$$

By contrast, if the carbon dioxide is enriched with ^{18}O, none of this enrichment appears in the oxygen produced. Another example of the use of ^{18}O comes from inorganic chemistry. It is the reaction between the sulfite ion and the chlorate ion in aqueous solution:

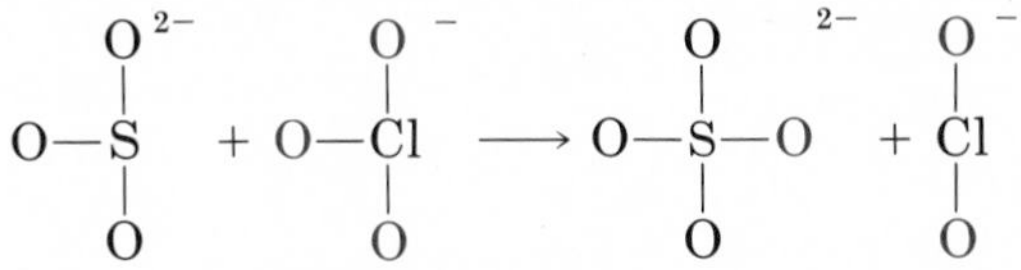

(19.34*a*)

By labeling the oxygens in the chlorate ion, it is found that all the ^{18}O lost from the one species is gained by the other and none of it is transferred to the solvent water. The mechanism of this reaction must thus be a direct transfer of oxygen and is quite unrelated to the two half-equations we conventionally use when balancing the redox equation:

$$\begin{aligned} SO_3^{2-} + H_2O &\rightarrow SO_4^{2-} + 2H^+ + 2e^- \\ 2H^+ + 2e^- + ClO_3^- &\rightarrow ClO_2^- + H_2O \end{aligned} \qquad (19.34b)$$

Neutron Activation Analysis

An important use of radioisotopes in detecting small amounts of certain elements in a sample is neutron activation analysis. The sample being analyzed is irradiated by a neutron source. Nuclei of the element being analyzed capture neutrons, and an unstable nucleus is formed which emits a γ ray. Since the wavelength of this γ ray is characteristic of the element, it can be distinguished from other elements in the sample. This method of analysis has the advantage of being nondestructive. The sample being analyzed is scarcely altered by being irradiated. Neutron activation is also among the most sensitive of analytical techniques. As little as a picogram (10^{-12} g) of arsenic, for example, can be detected. This is about 10 000 times more sensitive than Marsh's test—so often used by the fabled detective Sherlock Holmes. Neutron activation analysis is used by many modern detectives to find evidence of air and water pollution as well as the types of crimes with which Holmes was involved.

19.7 MASS-ENERGY RELATIONSHIPS

In a nucleus the protons and the neutrons are held very tightly by forces whose nature is still imperfectly understood. When the nucleons are very close to each other, these forces are strong enough to counteract the coulombic repulsion of the protons, but they fall off very rapidly with distance and are essentially undetectable outside the nucleus.

Because the energies involved in binding the nucleons together are very large, they give rise to an effect which makes it possible to measure them. According to Einstein's special theory of relativity, when the energy of a body increases, so does its mass, and vice versa. If the change in energy is indicated by ΔE and the change in mass by Δm, these two quantities are related by the equation

$$\Delta E = \Delta mc^2 \tag{19.35}$$

where c is the velocity of light (2.998×10^8 m s^{-1}).

In ordinary chemical reactions this change in mass with energy is so small as to be undetectable, but in nuclear reactions we invariably find that products and reactants have different masses. As a simple example, let us take the dissociation of a deuteron into a proton and a neutron:

$$^2_1\text{D} \rightarrow {}^1_1p + {}^1_0n$$

The molar mass of a deuteron is found experimentally to be 2.013 55 g mol^{-1} (see Table 19.1), but if we add the molar masses of a neutron and a proton,

TABLE 19.1 The Molar Masses of Some Selected Nuclei (Electrons Are *Not* Included in These Masses).

Nucleus	$\mathcal{M}$/g mol^{-1}	Nucleus	$\mathcal{M}$/g mol^{-1}
^1_0n (neutron)	1.008 67	$^{56}_{26}$Fe	55.920 66
^1_1p (proton)	1.007 28	$^{59}_{27}$Co	58.918 37
$^2_1$D (deuteron)	2.013 55	$^{84}_{36}$Kr	83.8917
$^3_1$T (tritium)	3.015 50	$^{120}_{50}$Sn	119.8747
$^4_2$He	4.001 50	$^{138}_{56}$Ba	137.8743
$^7_3$Li	7.014 36	$^{194}_{78}$Pt	193.9200
$^{12}_6$C	11.996 71	$^{209}_{83}$Bi	208.9348
$^{15}_8$O	14.998 68	$^{235}_{92}$U	234.9934
$^{16}_8$O	15.990 52	$^{239}_{94}$Pu	239.0006
$^{17}_8$O	16.994 74		

we obtain a somewhat higher value, namely, (1.007 28 + 1.008 67) g mol^{-1} = 2.015 95 g mol^{-1}. The change in mass using the usual delta convention is thus (2.015 95 − 2.013 55) g mol^{-1} = 0.002 40 g mol^{-1}. From Eq. (19.35) we then have

$$\begin{aligned}\Delta E &= \Delta mc^2\\ &= 0.002\,40\ \frac{\text{g}}{\text{mol}} \times \frac{1\ \text{kg}}{10^3\ \text{g}} \times (2.998 \times 10^8\ \text{m s}^{-1})^2\\ &= 2.16 \times 10^{11}\ \text{kg m}^2\ \text{s}^{-2}\ \text{mol}^{-1}\\ &= 216 \times 10^9\ \text{J mol}^{-1} = 216\ \text{GJ mol}^{-1}\end{aligned}$$

Since expansion work or even electronic energies are negligible compared to this change in nuclear energy, we can equate the change in nuclear energy either to the change in internal energy or the enthalpy; that is,

$$\Delta E = \Delta U_m = \Delta H_m = 216\ \text{GJ mol}^{-1}$$

The energy needed to separate a nucleus into its constituent nucleons is called its **binding energy.** The binding energy of the $^2_1$D nucleus is thus

216 GJ mol^{-1}. Notice how very much larger this is than the bond energy of an average molecule, which is about 200 or 300 kJ mol^{-1}. Since a gigajoule is 1 million kJ, the energies involved in holding the nucleons together in a nucleus are something like a *million times larger* than those holding the atoms together in a molecule.

Since the number of nucleons in a nucleus is quite variable, it is useful to calculate the average energy of each nucleon by dividing the total binding energy by the number of nucleons, A. This gives the **binding energy per nucleon.** In the case of the nucleus $^{56}_{26}Fe$, for instance, we can easily find from Table 19.1 that Δm for the process

$$^{56}_{26}\text{Fe} \rightarrow 26^{1}_{1}p + 30^{1}_{0}n$$

has the value 0.528 72 g mol^{-1}, giving a value for ΔH_m from Eq. (19.35) of 4750 GJ mol^{-1}. Since $A = 56$ for this nucleus, the binding energy per nucleon has the value

$$\frac{\Delta H_m}{A} = \frac{4750}{56} \text{ GJ mol}^{-1} = 848 \text{ GJ mol}^{-1}$$

The binding energy of a nucleus tells us not only how much energy must be expended in pulling the nuclei apart but also how much energy is released when the nucleus is formed from protons and neutrons. In the case of the $^{56}_{26}Fe$, for instance, we have

$$26^{1}_{1}p + 30^{1}_{0}n \rightarrow {}^{56}_{26}\text{Fe} \qquad \Delta H_m = -4.75 \times 10^3 \text{ GJ mol}^{-1}$$

that is, the energy of formation of $^{56}_{26}Fe$ is equal to the negative of the binding energy. In Fig. 19.5 the energy of formation on a per nucleon basis has been plotted against the mass number for the most stable isotope of each element. The zero energy axis in this plot corresponds to the energy of completely separated protons and neutrons, while the points on the graph correspond to the average energy of a nucleon in the nucleus in question. Obviously, the lower the energy, the more stable the nucleus.

As we can see from Fig. 19.5, the most stable nuclei are those of mass number close to 60, the nucleus with the lowest energy being the $^{56}_{26}Fe$ nucleus. As the mass number rises above 60, the nuclei become slightly higher in energy, i.e., less stable. Decreasing the mass number below 60 also brings us into a region of high-energy nuclei. With the exception of the $^{4}_{2}He$ nucleus, the nuclei of highest energy belong to the very lightest elements.

Figure 19.5 shows us that in principle there are two ways in which we can obtain energy from the nuclei of the elements. The first of these is by the splitting up or **fission** of a very heavy nucleus into two lighter nuclei. In such a case each nucleon will move from a situation of higher to lower energy and energy will be released. Even more energy will be released by the **fusion** of two very light nuclei, each containing only a few nucleons, into a single heavier nucleus. Though fine in principle, neither of these methods of obtaining energy is easy to achieve in practice in a controlled way with due respect to the environment.

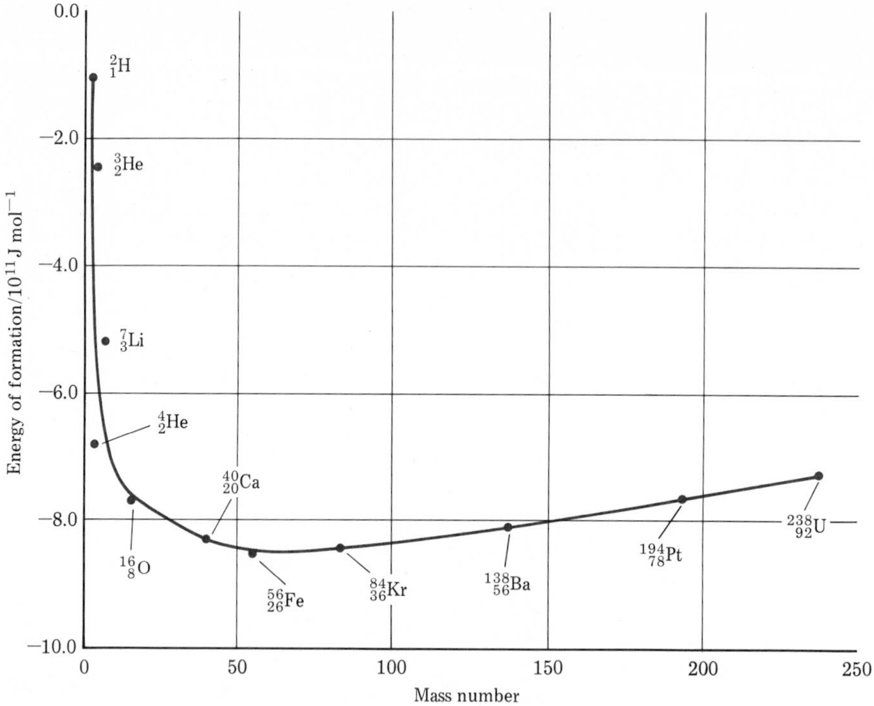

Figure 19.5 Energy of formation per nucleon (from protons and neutrons) as a function of mass number.

19.8 NUCLEAR FISSION

The first time that nuclear fission was achieved in the laboratory was by the Italian physicist Enrico Fermi (1901 to 1954) in 1934. Fermi was among the first to use the neutron in nuclear reactions, following its discovery by Chadwick in 1932. He hoped, by bombarding uranium with slow neutrons, to be able to prepare the first transuranium element. Instead he obtained a product which seemed to be a group II element which he identified incorrectly as radium. It remained for the experienced German radiochemist Otto Hahn to correct Fermi's mistake. (In the meantime Fermi had been awarded the Nobel Prize.) Somewhat reluctantly, Hahn published a paper early in 1939 showing that the element produced by the bombardment of uranium was not radium at all but the very much lighter group II element *barium,* 36 places earlier in the periodic table! It then became clear that instead of knocking a small chip off the uranium nucleus as had been expected, the bombarding neutron had broken the nucleus into two large fragments, one of which was barium. We now know that the initial step in this process is the formation of an unstable isotope of uranium which then fissions in a variety of ways, some of which are shown below:

$$ {}^{235}_{92}U + {}^{1}_{0}n \longrightarrow \left[{}^{236}_{92}U\right] \longrightarrow \begin{cases} {}^{140}_{56}Ba + {}^{93}_{36}Kr + 3\,{}^{1}_{0}n \\ {}^{144}_{55}Cs + {}^{90}_{37}Rb + 2\,{}^{1}_{0}n \\ {}^{144}_{54}Xe + {}^{90}_{38}Sr + 2\,{}^{1}_{0}n \\ {}^{146}_{57}La + {}^{87}_{35}Br + 3\,{}^{1}_{0}n \\ {}^{160}_{62}Sm + {}^{72}_{30}Zn + 4\,{}^{1}_{0}n \end{cases} \tag{19.36} $$

EXAMPLE 19.7 Using Fig. 19.5, make a rough estimate of the energy released by the fission of 1 g of uranium-235 according to the equation

$$^{235}_{92}\text{U} + ^{1}_{0}n \rightarrow ^{140}_{56}\text{Ba} + ^{93}_{36}\text{Kr} + 3^{1}_{0}n$$

Solution From Fig. 19.5 we can make the following estimates of the energies of formation per nucleon for the four species involved:

$$\Delta H_f(^{140}\text{Ba}) = -810 \text{ GJ mol}^{-1} \qquad \Delta H_f(^{235}\text{U}) = -730 \text{ GJ mol}^{-1}$$

$$\Delta H_f(^{93}\text{Kr}) = -840 \text{ GJ mol}^{-1} \qquad \Delta H_f(^{1}_{0}n) = 0$$

Using these quantities in the same way as enthalpies of formation for chemical reactions, we obtain

$$\Delta H_m = [140(-810) + 93(-840) - 235(-730)] \text{ GJ mol}^{-1}$$
$$= -20\ 000 \text{ GJ mol}^{-1}$$

The enthalpy change per gram is then given by

$$\Delta H = -20\ 000 \frac{\text{GJ}}{\text{mol}} \times \frac{1 \text{ mol}}{235 \text{ g}} = -85 \text{ GJ g}^{-1}$$

Note: This is about the same quantity of heat energy as that produced by burning 3 *tons* of bituminous coal!

Calculations similar to that just performed soon persuaded scientists in 1939 that the fission of uranium was highly exothermic and could possibly be used in a super bomb. Adolph Hitler had been in power in Germany for 6 years, and Europe was teetering on the brink of World War II. The possibility that Nazi Germany might develop such a bomb and use it did not seem remote, especially to those scientists who had recently fled Nazi and Fascist Europe and come to the United States. Albert Einstein, himself one of these refugees, was persuaded to write a letter to President Franklin Roosevelt in August 1939 in which the alarming possibilities were outlined. Roosevelt heeded Einstein's advice and established the so-called Manhattan Project, a super-secret research effort to develop an atomic bomb if at all possible. After 5 years of intense effort and the expenditure of more money than had ever been spent on a military-scientific project before, the first bomb was tested in New Mexico in July 1945. Shortly thereafter two atom bombs were dropped on the Japanese cities of Hiroshima and Nagasaki, and World War II ended almost immediately.

Some Features of Nuclear Fission

A crucial feature of the fission of uranium without which an atom bomb is impossible is that fission *produces more neutrons than it consumes.* As can be seen from Eqs. (19.36), for every neutron captured by a $^{235}_{92}\text{U}$ nucleus, between two and four neutrons are produced. Suppose now that we have a very large sample of the pure isotope $^{235}_{92}\text{U}$ and a stray neutron enters this sample. As soon as it hits a ^{235}U nucleus, fission will take place and about three neutrons will be produced. These in turn will fission three more ^{235}U nuclei, producing a total of nine neutrons. A third repetition will produce 27 neutrons, a fourth 81, and so on. This process (which is called a chain reaction) escalates very rapidly. Within a few microseconds a very large number of nuclei fission, with the release of a tremendous amount of energy, and an atomic explosion results.

There are two reasons why a normal sample of uranium metal does not spontaneously explode in this way. In the first place natural uranium consists mainly of the isotope $^{238}_{92}U$ while the fissionable isotope $^{235}_{92}U$ comprises only 0.7 percent of the total. Most of the neutrons produced in a given fission process are captured by $^{238}_{92}U$ nuclei without any further production of neutrons. The escalation of the fission process thus becomes impossible. However, even a sample of pure ^{235}U will not always explode spontaneously. If it is sufficiently small, many of the neutrons will escape into the surroundings without causing further fission. The sample must exceed a **critical mass** before an explosion results. In an atomic bomb several pieces of fissionable material, all of which are below the critical mass, are held sufficiently far apart for no chain reaction to occur. When these are suddenly brought together, an atomic explosion results immediately.

A great deal of the 5 years of the Manhattan Project was spent in separating the 0.7 percent of ^{235}U from the more abundant ^{238}U. This was done by preparing the gaseous compound UF_6 and allowing it to effuse through a porous screen. (Recall from Sec. 9.4 that the rate of effusion is inversely proportional to the square root of molar mass.) Each effusion resulted in a gas which was slightly richer in the lighter isotope. Repeating this process eventually produced a compound rich enough in ^{235}U for the purposes of bomb manufacture.

Only the first bomb dropped on Japan used uranium. The second bomb used the artificial element *plutonium,* produced by the neutron bombardment of ^{238}U [Eqs. (19.21) and (19.22)]:

$$^{238}_{92}U + ^{1}_{0}n \rightarrow ^{239}_{94}Pu + 2\,^{0}_{-1}e$$

Fission of $^{239}_{94}Pu$ occurs in much the same way as for $^{235}_{92}U$, giving a variety of products; for example,

$$^{239}_{94}Pu + ^{1}_{0}n \rightarrow ^{90}_{38}Sr + ^{147}_{56}Ba + 3\,^{1}_{0}n \qquad (19.37)$$

Again this is a highly exothermic reaction yielding about the same energy per mole (20 000 GJ mol^{-1}) as ^{235}U.

Nuclear Power Plants

Even before the atomic bomb had been produced, scientists and engineers had begun to think about the possibility of using the energy released by the fission process for the production of electrical energy. Shortly after World War II confident predictions were made that human beings would soon depend almost entirely on atomic energy for electricity. Alas, we are now 30 years into the future from then and no such miracle has occurred. In the United States only 4 percent of the electrical energy is currently produced by this method. The proportion is a little higher in some other countries, notably Great Britain, but nowhere is nuclear power even on the verge of replacing the fossil fuels. The unfortunate truth is that producing power from atomic fission has turned out to be much more expensive than was previously expected. Even in these days of high prices for the fossil fuels it is still only barely competitive.

A schematic diagram of a typical nuclear reactor is given in Fig. 19.6. The uranium is present in the form of pellets of the oxide U_3O_8 enclosed in long steel tubes about 2 cm in diameter. The uranium is mainly $^{238}_{92}U$ slightly enriched with the fissionable $^{235}_{92}U$. The rate of fission can be regulated by inserting or withdrawing control rods made of cadmium, which is a very efficient neutron absorber. In addition a moderator such as graphite or water must be present to slow down the neutrons, since slow neutrons are more efficient at causing fission than fast ones.

The energy released by the fisson of the uranium is carried off by a coolant, usually superheated steam at about 320°C. This steam cannot be used directly since it becomes slightly radioactive. Instead it is passed through a heat exchanger so as to produce further steam which can then be used to power a conventional steam turbine. The whole system is enclosed in a strong containment vessel (not shown in the figure). This vessel prevents the spread of radioactivity in case of a serious accident.

Figure 19.6 Schematic diagram of a nuclear power reactor.

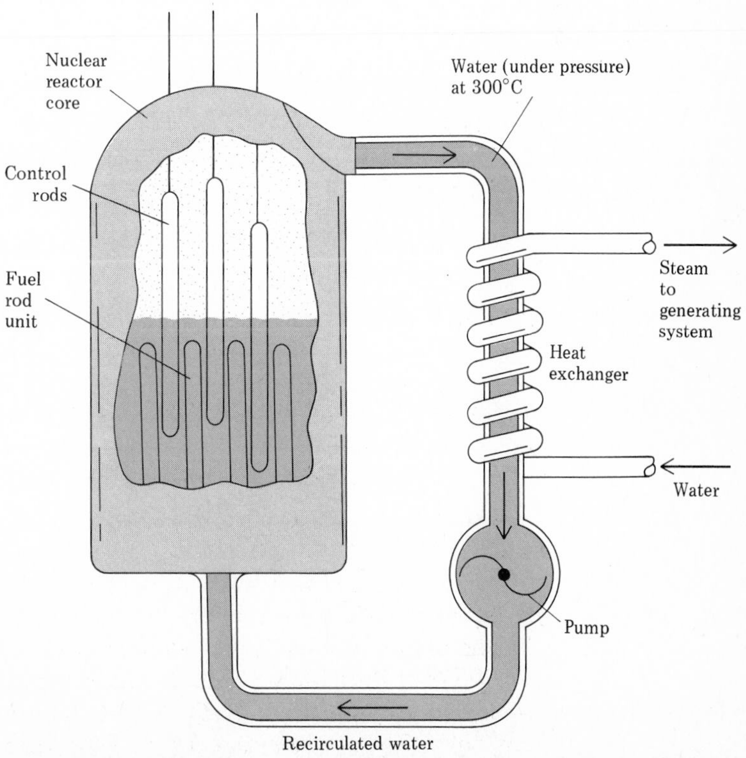

Nuclear power plants have two advantages over conventional power plants using fossil fuels. First, for a given energy output they consume much less fuel. Second, they produce far smaller quantities of toxic effluents. Fossil-fueled plants produce sulfur dioxide, oxides of nitrogen, and smoke particles, all of which are injurious to health.

Despite the much lower cost for fuel, nuclear power plants are very expensive to build. This is largely because of their chief disadvantage, the extremely dangerous nature of the radioactive products of nuclear fission. Fission products consist of a great many neutron-rich, unstable nuclei, ranging in atomic number from 25 to 60. Particularly dangerous are the long-lived isotopes $^{90}_{38}Sr$, $^{137}_{55}Cs$, and the shorter-lived $^{131}_{53}I$, all of which can be incorporated into the human body. Extreme precautions must be taken against accidental release of even traces of these materials into the environment. It should be realized in this connection that the possibility of a nuclear reactor running out of control and becoming a Hiroshima-type bomb is zero. The fuel used in nuclear reactors is not rich enough in $^{235}_{92}U$ for this to occur, and the worst possible accident is complete meltdown of the reactor core. Such an accident could be very serious because a great many highly radioactive isotopes would be scattered by the wind. Fortunately such an accident is highly improbable.

Even if fission products are handled successfully during normal operation of a nuclear plant, there still remains the difficulty of their eventual disposal. Although many of the unstable nuclei produced by fission are short-lived, some, like $^{90}_{38}Sr$ (25 years) and $^{137}_{55}Cs$ (30 years), have quite long half-lives. Accordingly these wastes must be stored for many hundreds of years before enough nuclei decompose to reduce their radioactivity to a safe level. At the present time, most of these wastes are stored as solutions in underground tanks near Richland, Washington. If the number of nuclear plants increases appreciably, the disposal of these wastes will become very difficult. Originally it was planned to store these wastes in solid form in underground salt deposits, but no entirely satisfactory site has yet been found.

Breeder Reactors

Because $^{235}_{92}U$ is only 0.7 percent of naturally occurring uranium, its supply is fairly limited and could well only last for about 50 years of full-scale use. The other 99 percent of the uranium can also be utilized if it is first converted into plutonium by neutron bombardment:

$$^{238}_{92}U + ^{1}_{0}n \rightarrow ^{239}_{94}Pu + 2\,^{0}_{-1}e$$

As we have already seen, $^{239}_{94}Pu$ is also fissionable, and so it could be used in a nuclear reactor as well as $^{235}_{92}U$.

The production of plutonium can be carried out in a **breeder reactor** which not only produces energy like other reactors but is designed to allow some of the fast neutrons to bombard the $^{238}_{92}U$, producing plutonium at the same time. More fuel is then produced than is consumed. Small breeder reactors are now functioning in Britain, France, and the Soviet Union.

Breeder reactors present additional safety hazards to those already outlined. They operate at higher temperatures and use very reactive liquid metals such as sodium in their cooling systems, and so the possibility of a serious accident is higher. In addition the large quantities of plutonium which would be produced in a breeder economy would have to be carefully safeguarded. Plutonium is an α emitter and is very dangerous if taken internally. Its half-life is 24 000 years, and so it will remain in the environment for a long time if dispersed. Moreover, $^{239}_{94}Pu$ can be separated chemically (not by the much more expensive gaseous diffusion used to concentrate $^{235}_{92}U$) from fission products and used to make bombs. Such a material will obviously be attractive to terrorist groups, as well as to countries which are not currently capable of producing their own atomic weapons.

19.9 NUCLEAR FUSION

In addition to fission, a second possible method for obtaining energy from nuclear reactions lies in the fusing together of two light nuclei to form a heavier nucleus. As we have already seen in discussing Fig. 19.5, such a process results in nucleons which are more firmly bonded to each other and hence lower in potential energy. This is particularly true if $^{4}_{2}He$ is formed, because this nucleus is very stable. Such a reaction occurs between the nuclei of the two heavy isotopes of hydrogen, deuterium and tritium:

$$^{2}_{1}D + {}^{3}_{1}T \rightarrow {}^{4}_{2}He + {}^{1}_{0}n \tag{19.38}$$

For this reaction $\Delta m = -0.018\ 88\ g\ mol^{-1}$ so that $\Delta H_m = -1700\ GJ\ mol^{-1}$.

Although very large quantities of energy are released by a reaction like Eq. (19.38), such a reaction is very difficult to achieve in practice. This is because of the very high activation energy, about 30 GJ mol^{-1}, which must be overcome to bring the nuclei close enough to fuse together. This barrier is created by coulombic repulsion between the positively charged nuclei. The only place where scientists have succeeded in producing fusion reactions on a large scale is in a **hydrogen bomb.** Here the necessary activation energy is achieved by exploding a fission bomb to heat the reactants to a temperature of about 10^8 K. Attempts to carry out fusion in a more controlled way have met with only limited success. At the very high temperatures required, all molecules dissociate and most atoms ionize. A new state of matter called a **plasma** is formed. It is neither solid, liquid, or gas and behaves much like the universal solvent of the alchemists by converting any solid material which it contacts into vapor.

Two techniques for producing a controlled fusion reaction are currently being explored. The first is to restrict the plasma by means of a strong magnetic field rather than the walls of a container. This has met with some success but has not yet been able to contain a plasma long enough for usable energy to be obtained. The second technique involves the sudden compression and heating of pellets of deuterium and tritium by means of a sharply focused laser beam. Again, only a limited success has been obtained.

Though these attempts at a controlled fusion reaction have so far been only partially successful, they are nevertheless worth pursuing. Because of the much readier availability of lighter isotopes necessary for fusion as opposed to the much rarer heavier isotopes required for fission, controlled nuclear fusion would offer the human race an essentially limitless supply of energy. There would still be some environmental difficulties with the production of isotopes such as tritium, but these would be nowhere near the seriousness of the problem caused by the production of the witches brew of radioactive isotopes in a fission reactor. It must be confessed, though, that at the present rate of progress, the prospect of limitless clean energy from fusion seems unlikely in the next decade or two.

SUMMARY

Nuclear reactions involve rearrangements of the protons and neutrons within atomic nuclei. During naturally occurring nuclear reactions α particles, β particles, and γ rays are emitted, often in a radioactive series of successive reactions. Nuclear reactions may also be induced by bombarding nuclei with positive ions or neutrons. Artificial isotopes produced in this way may decay by positron emission or electron capture as well as by α, β, or γ emission. Stability of nuclei depends on the neutron/proton ratio (usually between 1 and 1.6) and magic numbers of protons and neutrons.

Radioactive decay obeys a first-order rate law, and its rate is often reported in terms of half-life, the time necessary for half the radioactive nuclei to decompose. Known half-lives of isotopes such as ${}^{14}_{6}C$ and ${}^{238}_{92}U$ may be used to establish the ages of objects containing these elements, provided accurate measurements can be made of the quantity of radiation emitted. Geiger-Müller counters or scintillation counters are often used for such measurements. Other important applications of radioactive isotopes include tracer studies, where a particular type of atom can be labeled and followed throughout a reaction, and neutron activation analysis, which can determine extremely low concentrations of many elements.

The relative stability of a nucleus is given by the energy of formation per nuclear particle. This may be determined from the difference between the molar mass of the nucleus and the sum of the molar masses of its constituent protons and neutrons. Both fission, breaking apart of a heavy nucleus, and fusion, combining of two light nuclei, can result in release of energy. Fission usually involves ${}^{235}_{92}U$ or ${}^{239}_{94}Pu$, and these isotopes have been used in nuclear explosives and nuclear power plants. Fission products are highly radioactive. Because of the considerable damage done to living tissue by the ability of α, β, and γ radiation to break bonds and form ions, emission of radioactive materials must be carefully controlled and fission power plants are quite expensive to construct. Although it promises much larger quantities of free energy and fewer harmful by-products than fission, nuclear fusion has not yet been shown to be feasible for use in power plants. So far its only application has been in hydrogen bombs.

QUESTIONS AND PROBLEMS

19.1 How are nuclear reactions different from ordinary chemical reactions?

19.2 During medieval times alchemists expended a great deal of effort trying to turn inexpensive metals such as Pb into expensive ones such as Au. Beginning with Lavoisier, the father of modern chemistry, this idea of transmutation of the elements was ridiculed. In light of modern knowledge, how ridiculous does the alchemists' dream appear to be?

19.3 For the α particle, β particle, and γ ray (*a*) write the symbol used to represent each in nuclear equations; (*b*) contrast the abilities of these forms of radiation to penetrate matter; (*c*) discuss the relative effect of each form of radiation on human health.

19.4 Complete the following nuclear equations. (Assume that a single nucleus is formed in each case.)

a $^{194}_{76}\text{Os} \rightarrow$ ______ $+ \ ^{0}_{-1}e$
b $^{238}_{94}\text{Pu} \rightarrow$ ______ $+ \ ^{4}_{2}\text{He}$
c $^{156}_{66}\text{Dy} \rightarrow$ ______ $+ \ ^{4}_{2}\text{He}$
d $^{186}_{75}\text{Re} \rightarrow$ ______ $+ \ ^{0}_{0}\gamma$
e $^{0}_{-1}e + \ ^{249}_{99}\text{Es} \xrightarrow{ec}$ ______
f $^{88}_{41}\text{Nb} \rightarrow$ ______ $+ \ ^{0}_{+1}e$

19.5 The isotope $^{237}_{93}\text{Np}$ disintegrates via a radioactive series which includes 11 steps. In order, these steps involve $\alpha, \beta, \alpha, \alpha, \beta, \alpha, \alpha, \alpha, \beta, \alpha$, and β emissions. Write the entire series of nuclear equations and identify the final isotope formed.

19.6 The natural radioactive series which begins with $^{235}_{92}\text{U}$ involves the following emissions of particles: $\alpha, \beta, \alpha, \beta, \alpha, \alpha, \alpha, \alpha, \beta, \alpha$, and β. Write the entire series of nuclear equations and identify the final isotope formed.

19.7 What is a positron? What symbol is used to represent a positron in a nuclear equation?

19.8 The isotopes listed below decay by loss of the indicated particle. Write a balanced nuclear equation for each reaction.

a $^{144}_{60}\text{Nd}$, α emission
b $^{49}_{21}\text{Sc}$, β emission
c $^{178}_{74}\text{W}$, α emission
d $^{8}_{5}\text{B}$, positron emission
e $^{93}_{40}\text{Zr}$, β emission
f $^{123}_{55}\text{Cs}$, positron emission
g $^{138}_{54}\text{Xe}$, β emission
h $^{190}_{78}\text{Pt}$, α emission
i $^{84}_{39}\text{Y}$, positron emission
j $^{179}_{71}\text{Lu}$, β emission

19.9 Calculate the neutron/proton ratio for each isotope in Problem 19.8. What type of emission is favored by a large neutron/proton ratio? What type is most likely when the neutron/proton ratio is smaller than usual?

19.10 What happens during the nuclear reaction known as electron capture? Write a balanced equation showing an example of electron capture.

19.11 If atomic nuclei are assumed to be spherical, nuclear radii are found to be approximately proportional to the cube root of the mass number:

$$r_{\text{nucleus}} = (1.3 \times 10^{-13} \text{ cm})A^{1/3}$$

(*a*) Verify that an average density of 1.8×10^{14} g cm^{-3} is reasonable for several different nuclei. (*b*) What would be the mass of an object 1 ft in diameter which has the same density as an atomic nucleus?

19.12 Describe the principle of operation of a cyclotron and sketch the main features of such a device.

19.13 Suggest a reaction which might be carried out in a cyclotron in an attempt to produce each of the isotopes below:

a $^{254}_{102}\text{No}$ *b* $^{256}_{101}\text{Md}$ *c* $^{253}_{100}\text{Fm}$

19.14 Name and describe two types of apparatus in which nuclei can be bombarded by neutrons. Why is the cyclotron *not* one of them?

19.15 Distinguish between fast and slow neutrons. How are slow neutrons obtained?

19.16 Define and discuss each of the terms listed below:

a Band of stability *c* Rem
b Half-life *d* Magic number

19.17 For each pair of isotopes listed below, predict which is less stable. Explain each choice.

a $^{173}_{78}\text{Pt}$ or $^{194}_{78}\text{Pt}$ *c* $^{16}_{6}\text{C}$ or $^{16}_{8}\text{O}$
b $^{20}_{12}\text{Mg}$ or $^{26}_{12}\text{Mg}$ *d* $^{48}_{20}\text{Ca}$ or $^{48}_{21}\text{Sc}$

19.18 For each pair of elements listed below, predict which has the larger number of stable isotopes. Explain each choice.

a Pb or Bi *b* F or O *c* Ni or Cu

19.19 You are given 100.0 g of each isotope whose half-life is listed below. How much will remain at the end of the indicated time?

a $^{189}_{75}Re$; $t_{1/2}$ = 24 h; 5 days
b $^{153}_{68}Er$; $t_{1/2}$ = 36 s; 2.4 h
c $^{221}_{89}Ac$; $t_{1/2}$ = 52 ms; 1.664 s

19.20 A sample of an unknown radioactive isotope initially measured 4250 disintegrations per minute. Exactly 30 min later the disintegration rate had dropped to 3365 disintegrations per minute. (*a*) What is the half-life of the isotope? (*b*) How long would it take for the disintegration rate to drop to 50 min^{-1}?

19.21 The half-life of $^{131}_{53}I$ is 8.05 days. If exactly 10.0 mg I_2 is injected into an individual's thyroid gland, how much will remain after 30 days? (Assume that no I escapes from the thyroid gland.)

19.22 How much time must elapse for 80 percent of each of the following samples of $^{211}_{82}Pb$ ($t_{1/2}$ = 36.1 min) to disintegrate?

a 1 mol $^{211}_{82}Pb$ *b* 1 g $^{211}_{82}Pb$
c 1 g $^{211}_{82}PbCl_2$

19.23 A pure sample of a radioisotope has an activity of 1555 disintegrations per minute at 1:00 P.M. At 2:00 P.M. on the same day its activity has fallen to 1069 min^{-1}. What is the half-life of the isotope?

19.24 Strontium-90, $^{90}_{38}Sr$, is present in radioactive fallout produced by explosions of nuclear weapons in the atmosphere. The half-life of $^{90}_{38}Sr$ is 29 years. What fraction of the $^{90}_{38}Sr$ released by the Hiroshima and Nagasaki bombs in August 1945 will remain in the earth's biosphere in August 1985?

19.25 Explain the principle of radiocarbon-14 dating. If the activity of a sample of carbon freshly obtained from living matter is 15.3 disintegrations per gram per minute and the minimum activity which can be detected accurately above normal background radiation is 10.0 disintegrations per minute, use the half-life for $^{14}_{6}C$ given in the text to estimate the age of the oldest sample which could be determined reliably by this method if the sample contains 1.0 oz of carbon.

19.26 Calculate the approximate age of each of the following items collected during an archaeological expedition: (*a*) A piece of cloth whose measured decay rate for $^{14}_{6}C$ is 10.1 disintegrations per minute per gram. (*b*) A piece of parchment whose $^{14}_{6}C$ decay rate is 0.690 that of a living plant. (*c*) A rock which contains exactly 1 g $^{238}_{92}U$ for each g $^{206}_{82}Pb$.

19.27 What is meant by the term background radiation? Where could you go to escape it? How does it complicate evaluation of health effects of low levels of radiation?

19.28 Describe a Geiger counter and explain how it works.

19.29 Explain how isotopic tracers might be used for each of the following purposes:

a To measure the solubility of the sparingly soluble salt $BaSO_4$.

b To measure the solubility of water in a nonpolar organic solvent.

c To study the mechanism of the reaction

$$3ClO^- \rightarrow ClO_3^- + 2Cl^-$$

d To show that *dynamic* equilibrium is achieved in the reaction

$$CH_3COOH + H_2O \rightleftharpoons H_3O^+ + CH_3COO^-$$

19.30 Use data from Table 19.1 to calculate the binding energy and the binding energy per nucleon for each of the following

a $^{235}_{92}U$ *d* $^{239}_{94}Pu$
b $^{7}_{3}Li$ *e* $^{120}_{50}Sn$
c $^{17}_{8}O$ *f* $^{194}_{78}Pt$

19.31 Calculate ΔH_m for the fusion reaction

$$^{2}_{1}H + ^{2}_{1}H \rightarrow ^{3}_{2}He + ^{1}_{0}n$$

(The molar mass of $^{3}_{2}He$ is 3.016 03 g mol^{-1}.) Compare the heat energy released by this reaction per gram of hydrogen with that released per gram of hydrogen in the chemical reaction

$$H_2(g) + \tfrac{1}{2}O_2(g) \rightarrow H_2O(g)$$

Obtain $\Delta H_f^\ominus$ [$H_2O(g)$] from Table 3.1.

19.32 Define each of the following terms:

a Binding energy
b Fission
c Plasma
d Chain reaction
e Fusion
f Critical mass

19.33 What are some advantages and some disadvantages of nuclear power plants compared with coal-burning plants?

19.34 What are the most important isotopes employed in nuclear explosives? Describe clearly the differences between an atomic bomb and a hydrogen bomb.

19.35 Name two techniques being developed for use in generating power from controlled nuclear fusion. What major problem must be overcome before fusion will be technologically feasible?

19.36 Describe clearly how a breeder reactor can produce electrical power and at the same time produce more nuclear fuel than it burns.

19.37 Using techniques developed in Chap. 9, calculate the relative rates of effusion of $^{238}_{92}UF_6$ and $^{235}_{92}UF_6$.

chapter twenty

MOLECULES IN LIVING SYSTEMS

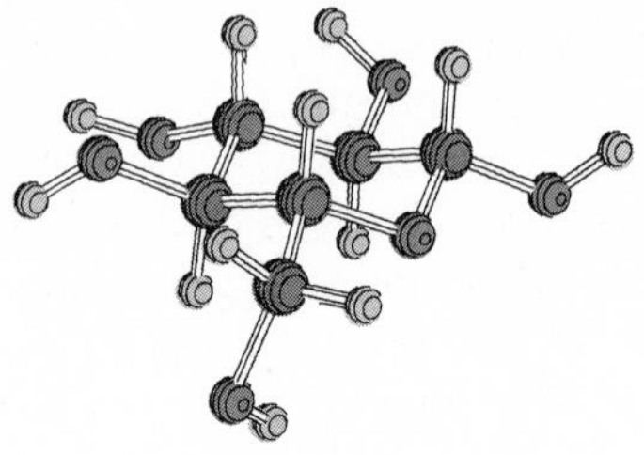

Most of us have little difficulty distinguishing living organisms from inanimate matter. The former are capable of reproducing nearly exact copies of themselves; they can appropriate both matter and energy from their surroundings, moving, growing, and repairing damage caused by external factors; and groups of them evolve and adapt in response to long-term environmental changes. On a macroscopic scale the differences are sufficiently striking that early philosophers and scientists postulated the existence of a vital force without which living organisms would be inanimate. It was thought that organic compounds could only be manufactured in living organisms, and chemistry was divided into the subfields of inorganic and organic on this basis.

This subdivision persists today, but the definition of *organic* has changed in response to the discovery of numerous ways to make organic compounds from inorganic starting materials. As we saw in Chap. 8, *organic chemistry* now means "chemistry of compounds containing carbon." No restriction is placed on the origin of the compounds, and hundreds of thousands of organic compounds which are foreign to all living systems have been produced in laboratories around the world. Indeed, concern about the effects of some of these synthetic substances on the environment has led to yet another definition of organic. The general public now takes it to mean "free of substances produced as a result of human activities."

Just as the division between organic and inorganic chemistry has become more arbitrary with the advance of knowledge, the distinction between life and nonlife has also blurred. Living organisms are made up of atoms and molecules which follow the same chemical principles that we have already discussed in this book. Yet there is a difference—these atoms, molecules, and even groups of molecules are organized to a much greater degree than in any of the cases we have discussed so far. Above a certain level of complexity a collection of chemicals begins to exhibit most of the behavior patterns that we associate with life. A virus, for example, may consist of fewer than 100 associated large molecules. It is the structures of these molecules and the ways in which they are associated that determine a virus' behavior and make it appear to be on the threshold of life.

This chapter is devoted to a consideration of the chemical elements found in living systems, and how these elements combine to form molecules and collections of molecules which carry out the biological functions and behaviors that we associate with life. Our treatment must of necessity be brief, but even if it were not, the complexity of biochemical systems would insure that it would be incomplete. Much of the future of chemistry, both inorganic and organic, will involve the extension to complex biological systems of principles and facts gleaned from studies of the much simpler chemical behavior we have described in previous chapters.

20.1 THE ELEMENTS OF LIFE

Hardly a year goes by without the appearance in the media of the old clichè that the average human body is worth about \$3.57 (or perhaps more, allowing for inflation). Such figures are based on the elemental composition of the human body reported in Table 20.1. Considering just the market

TABLE 20.1 Elemental Composition of the Human Body.

Element	Weight %	Atom %	Period
H	10.2	63.5	I
O	66.0	25.6	II
C	17.5	9.1	II
N	2.4	1.06	II
Subtotal	96.1	99.3	
Ca	1.6	0.25	IV
P	0.9	0.18	III
Na	0.3	0.07	III
K	0.4	0.06	IV
Cl	0.3	0.05	III
S	0.2	0.04	III
Mg	0.05	0.01	III
Subtotal	99.85	99.92	
Fe	0.005	0.0006	IV
Zn	0.002	0.0002	IV
Cu	0.0004	0.000 06	IV
Mn	0.000 05	0.000 006	IV

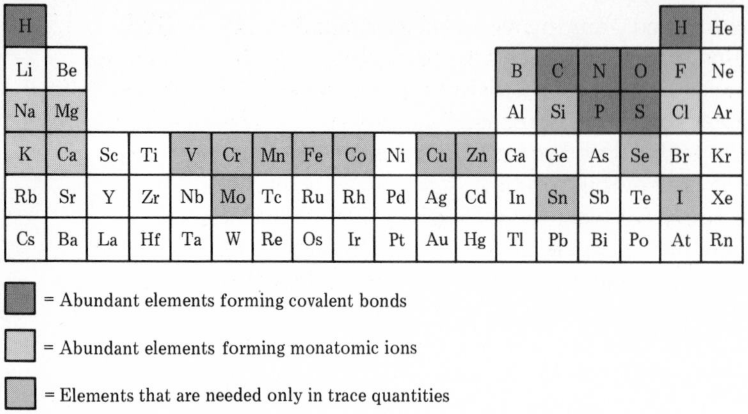

Figure 20.1 The biological periodic table.

value of the elements, one obtains a ridiculously low price. What has been ignored, of course, is how the atoms of those elements are put together. The raw materials for a fine watch are not worth much either—what we pay for is mainly the skill and intelligence with which they are combined.

Before considering *how* the elements from which we all are made have been combined, however, it is worth thinking a little about *why* those particular elements in Table 20.1 are involved. This is especially so because *the same elements in very similar ratios are found in nearly all living systems.* As shown in the biological periodic table (Fig. 20.1), a great many other elements are simply not involved in the chemistry of life—at least not to the extent that the necessity of their presence can be demonstrated experimentally.

More than 99 percent of the atoms in the human body, or any other organism for that matter, are H, O, N, and C. This does not appear to be mere happenstance. The indications are that life as we know it could not be based on any other four elements. These elements have the lightest and smallest atoms which can form one, two, three, and four covalent bonds, respectively. Because of their small radii these atoms can approach very closely and bond very strongly to each other. The most important of these four elements is undoubtedly carbon. As we have already seen in Chap. 8, carbon atoms have a special capacity for binding with each other to form long chains and rings. This capacity allows carbon to form a very large number of stable compounds whose molecular structures are different but nevertheless closely related.

No other element has this capacity. Perhaps the closest contender is silicon. Although a number of compounds containing Si—Si bonds such as

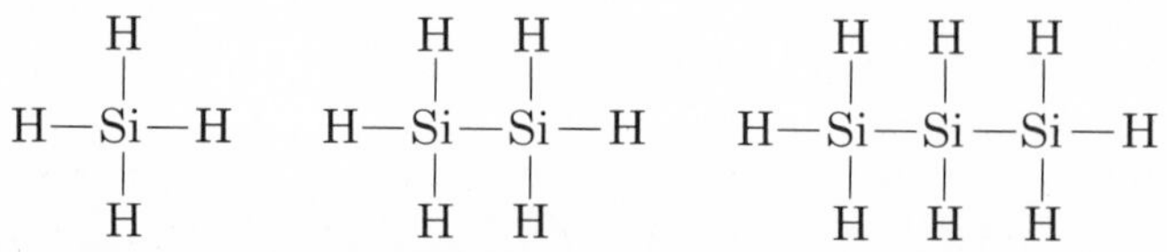

Silane **Disilane** **Trisilane**

are known, none of these compounds is very stable. All are readily converted to compounds containing Si—O bonds. The bond enthalpy of the Si—O bond (368 kJ mol^{-1}) is much larger than that of the Si—Si bond (176 kJ mol^{-1}) and also larger than that of the Si—H bond (318 kJ mol^{-1}). The production of Si—O bonds is thus exothermic and thermodynamically favorable. Another factor is the rapid *rate* at which these chemical reactions can occur. Silicon is capable of expanding its valence shell through the use of *d* orbitals to allow more than four bonds. This enables it to form an activated complex in which both the bond being made and the bond being broken feature, but which requires very little activation energy. Equivalent reactions involving carbon require very much higher activation energies and usually proceed so slowly as to be imperceptible at room temperature, even when thermodynamically permitted.

Carbon is not the only element with unique properties in biological molecules. Hydrogen is also special and plays two important roles. You will recall from Chap. 8 that alkanes are chemically unreactive. We can attribute this to the large C—H bond enthalpy of 413 kJ mol^{-1}. Only fluorine forms a stronger bond than this with carbon. It is thus quite difficult to replace a hydrogen atom attached to a carbon atom with a more stable alternative. When an organic compound reacts chemically, it is almost always a *functional group* which undergoes a change. The presence of C—H bonds renders a large proportion of the carbon chain unreactive and restricts reaction almost entirely to those sites which include an atom other than carbon or hydrogen.

In biological molecules these other atoms are almost invariably oxygen and nitrogen, and this is no accident either. Oxygen and nitrogen are the two most electronegative elements which have a valence greater than 1. They are able to form bonds both to carbon on the one hand and to hydrogen on the other. In groups like —C—O—H and —C—N—H hydrogen atoms (with the remaining bonds of C and N drawn vertically) fulfill their second important role in biological molecules—*forming hydrogen bonds* between different molecules or between different parts of the same molecule.

This ability of different functional groups on a carbon chain to hydrogen bond with each other is a particularly important aspect of biological molecules. You will recall from our discussion in Chap. 13 (see Fig. 13.5) that a molecule containing a chain of carbon atoms is capable of a very large number of conformations due to free rotation around the single bonds. Since many biological molecules contain very long chains, they are capable of adopting an almost infinite number of shapes. In practice only a few of these shapes are useful, and the molecules can be "frozen" into such a useful conformation through hydrogen-bonding between various segments of the chain. A very good example of this is the enzyme trypsin, which we discussed in Chap. 18 (see Fig. 18.9). If the amino acid chain in this molecule were not held in the particular conformation shown in the figure by means of hydrogen bonds between adjacent parts of the chain, the various segments making up the active site would no longer be grouped together and the molecule would be unable to function as a catalyst.

20.2 THE BUILDING BLOCKS OF BIOCHEMISTRY

It has been estimated that even a unicellular organism may contain as many as 5000 different substances, and the human body probably has well over 5 000 000. Only a few of these are exactly the same in both species, and so the total number of different compounds in the living portion of the earth (the **biosphere**) is approximately (10^5 compounds/species) $\times$ 10^6 species = 10^{11} compounds. If it were possible for chemists to synthesize one of these every second, 24 hours a day, 7 days a week, about 3000 years would be required to make all of them—obviously a hopeless task, even if we knew the composition and structure of each one.

How then can we make sense out of the chemistry of living systems? Fortunately nearly all the substances found in living cells are *polymeric*—they are built up by different combinations of a limited number of relatively small molecules. For example, the basic structures of all proteins in all organisms consist of covalently linked chains containing 100 or more amino acid residues. Only 20 different amino acids are found in proteins, but the number of ways of arranging 100 of these in a chain taking any of the amino acids at random for each place in the chain is $20^{100} \cong 10^{130}$, allowing an almost infinite variety of structures. Just as an understanding of the properties of atoms and their bonding characteristics was a significant aid in predicting the chemistry of molecules, a knowledge of the properties of simple molecular building blocks gives us a starting point for the study of biochemistry.

The essential building blocks of all terrestrial organisms are listed in Table 20.2. In addition, polymers formed from these substances have been indicated. Each of the building blocks and their polymeric forms has at least one major role to play in the chemistry of life. Most are quite versatile, serving several functions. Water, for example, provides a solvent medium for all chemical reactions, maintains rigidity of cell walls, and serves as a thermoregulator. By virtue of its large polarity it is a good sol-

TABLE 20.2 Classification of Substances in Living Organisms.

Building-Block Substances	Polymers Formed
Water	*
Calcium phosphates and carbonates; other dissolved salts	Crystalline salts (bones, teeth, shells, etc.);† polyphosphates
Glycerol; organic and inorganic acids; other organic compounds	Lipids‡
Sugars	Carbohydrates
Amino acids	Proteins (including enzymes)
Phosphate; sugars; organic nitrogen bases	Nucleic acids
Vitamins; trace elements	§

* Though held together by weaker hydrogen-bonding interactions, ice and the clusters of water molecules present in the liquid phase might be taken as polymers.

† An ionic solid can be thought of as a three-dimensional polymer built up by repetition of the unit cell.

‡ Lipids usually involve combination of no more than four molecular building blocks. In a strict sense, they are not polymers.

§ Some trace elements are incorporated into protein structures.

vent for ionic substances and therefore provides a means of transporting inorganic nutrients such as NH_4^+, NO_3^-, CO_3^{2-}, PO_4^{3-}, and monatomic ions throughout higher organisms. Its ability to dissolve a wide variety of substances also makes it useful for disposing of wastes. Many of the human body's defense mechanisms against external toxic substances involve conversion into water-soluble forms and elimination via urine.

A similar lengthy list of biological functions could be made for most of the substances in Table 20.2. Subsequent sections of this chapter will be devoted to discussion of the structures, properties, and functions of the building blocks and polymers which constitute lipids and fats, carbohydrates, proteins and enzymes, and nucleic acids.

20.3 FATS AND LIPIDS

The term **lipid** applies to any water-insoluble substance which can be extracted from cells by organic solvents such as chloroform, ether, or benzene. Two major categories may be identified. *Nonpolar lipids* have molecular structures which contain no electrically charged sites, few polar groups, and large amounts of carbon and hydrogen. They are similar to hydrocarbons in being almost completely insoluble in water, and so they are said to be **hydrophobic** (from the Greek, meaning water-hater). On the other hand, *polar lipids* consist of molecules which have polar groups (such as —OH) or electrically charged sites at one end, and hydrocarbon chains at the other. Since polar or charged groups can hydrogen bond to or electrostatically attract water molecules, one end of a polar lipid molecule is said to be **hydrophilic** (water-loving). Some typical structures of both types of lipids are shown in Fig. 20.2.

Figure 20.2 Structures of some typical lipids: (*a*) nonpolar; (*b*) polar. (Hydrophilic portions indicated in color. Only carbon-carbon bonds, but not the carbon and hydrogen atoms, are shown in long carbon chains. Therefore those chains appear as zigzag lines.)

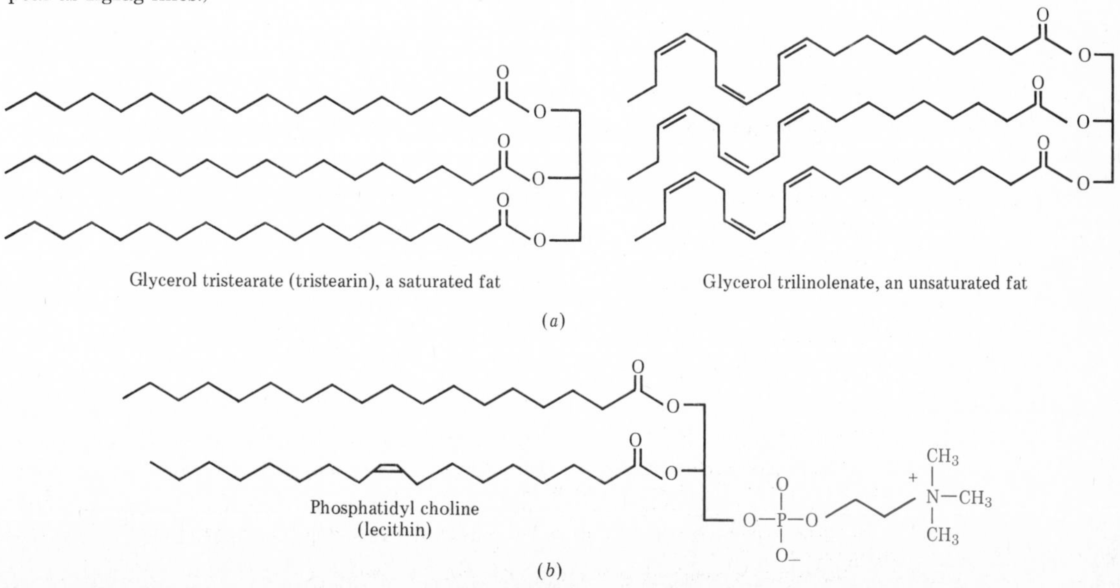

Nonpolar Lipids

A good example of a nonpolar lipid is the neutral fat glycerol tristearate. This most-common form of animal fat serves as a storehouse for energy and as insulation against heat loss. On a molecular level it is constructed from three molecules of stearic acid and one of glycerol:

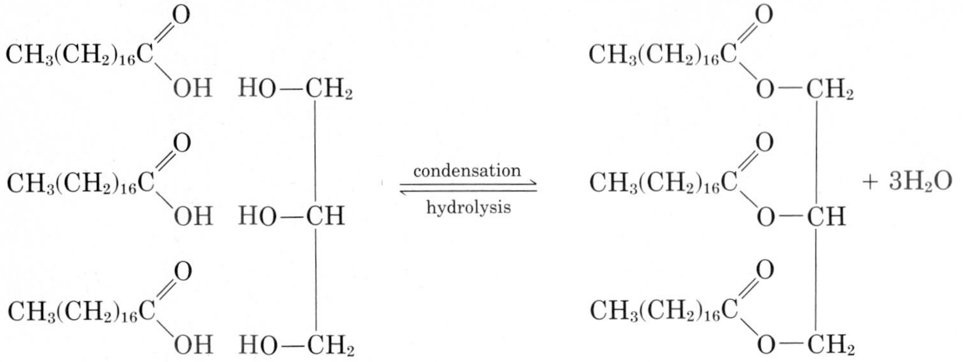

(20.1)

Stearic acid **Glycerol** **Glycerol tristearate**

A great many nonpolar lipids can be made by combining different long-chain acids with glycerol. Because these acids were originally derived from fats, they are collectively referred to as **fatty acids.**

Notice that for each stearic or other fatty acid molecule which combines with one of the —OH groups of glycerol, a molecule of water is given off, and so the reaction is a condensation. It turns out that a great many important biological molecules are put together by condensation reactions during which water is given off. The reverse of Eq. (20.1), in which water reacts with a large molecule and splits it into smaller pieces, is called *hydrolysis.* By carrying out hydrolysis living organisms can break down molecules manufactured by other species. The simple building blocks obtained this way can then be recombined by condensation reactions to form structures appropriate to their new host.

By contrast with the glycerol tristearate found in animals, vegetable fats contain numerous double bonds in their long hydrocarbon chains. This polyunsaturation introduces "kinks" in the hydrocarbon chains because of the barrier to rotation and the 120° angles associated with the double bonds. Consequently it is more difficult to align the chains side by side (see Fig. 20.2), and the unsaturated fats do not pack together as easily in a crystal lattice. Most unsaturated fats (like corn oil) are liquids at ordinary temperatures, while saturated fats (like butter) are solids. Vegetable oils can be converted by hydrogenation to compounds which are solids. This process involves adding H_2 catalytically to the double bonds:

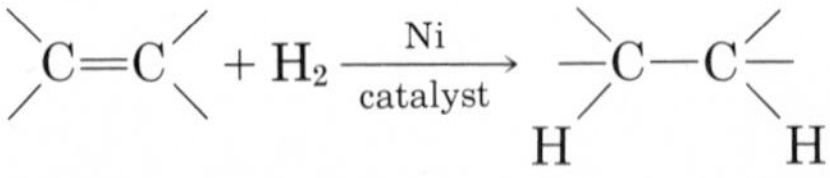

Hydrolysis of fats [the reverse of Eq. (20.1)] is important in the manufacture of soaps. It can be speeded up by the addition of a strong base like NaOH or KOH, in which case the reaction is called **saponification.** Since saponification requires that the pH of the reaction mixture be high, the fatty acid that is produced will dissociate to its anion. When glycerol tristearate is saponified with NaOH, for example, sodium stearate, a relatively water-soluble substance and a common soap, is formed.

The ability of soaps to clean grease and oil from soiled surfaces is a result of the dual hydrophobic-hydrophilic structures of their molecules. The stearate ion, for example, consists of a long nonpolar hydrocarbon chain

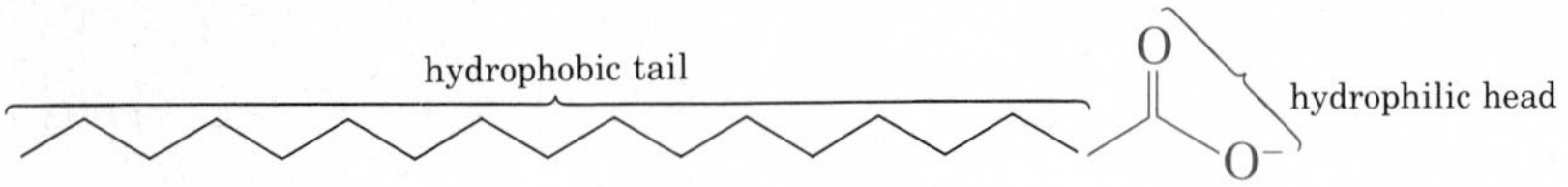

Stearate ion

with a highly polar —COO^- group at one end. The hydrophobic hydrocarbon chain tries to avoid contact with aqueous media, while the anionic group readily accommodates the dipole attractions and hydrogen bonds of water molecules.

The two main ways that the hydrophobic portions of stearate ions can avoid water are to cluster together on the surface or to dissolve in a small quantity of oil or grease (see Fig. 20.3). In the latter case the hydrophilic

Figure 20.3 Behavior of soap molecules in aqueous solution. Micelles containing grease molecules are solubilized by the polar ends of soap molecules.

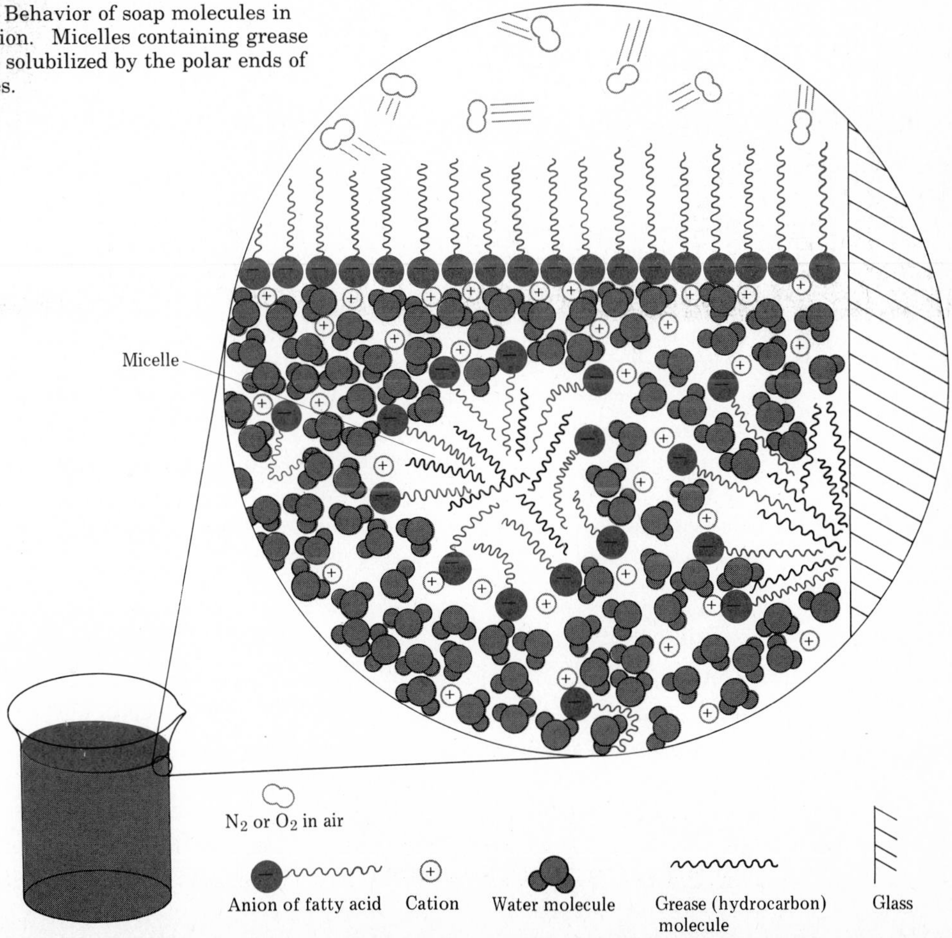

heads of the soap molecules contact the water outside the grease, forming a structure known as a **micelle.** Since the outsides of the micelles are negatively charged, they repel one another and prevent the grease droplets from recombining. The grease is therefore suspended (emulsified) in the water and can be washed away easily.

Natural soaps, such as sodium stearate, were originally made in the home by heating animal fat with wood ashes, which contained potash, K_2CO_3. Large quantities are still produced industrially, but to a considerable extent soaps have been replaced by detergents. This is a consequence of the undesirable behavior of soaps in hard water. Calcium, magnesium, and other hard-water cations form insoluble compounds when combined with the anions of fatty acids. This produces scummy precipitates and prevents the soap molecules from emulsifying grease unless a large excess is used.

Detergents such as alkylbenzenesulfonates (ABS) and linear alkylbenzenesulfonates (LAS) have structures very similar to sodium stearate except that the charged group in their hydrophilic heads is $—SO_3^-$ attached to a benzene ring. The ABS detergents also have methyl (CH_3) groups branching off their hydrocarbon chains. Such molecules do not precipitate

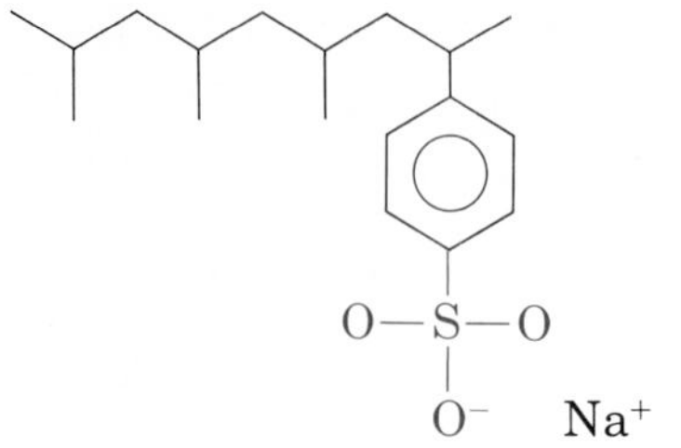

ABS detergent

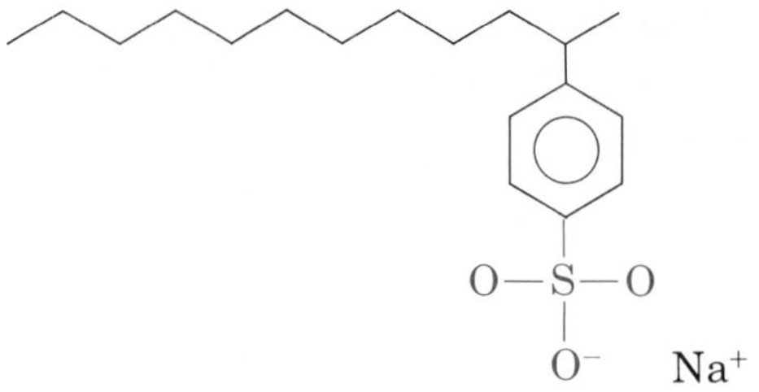

LAS detergent

with hard-water cations and therefore are more suitable for machine-washing of clothes. The LAS detergents replaced ABS during the mid-nineteen-sixties when it was discovered that the latter were not biodegradable. They were causing rivers and even tap water to become covered with detergent suds and foam. Apparently the enzymes in microorganisms which had evolved to break down the unbranched hydrocarbon chains in natural fats and fatty acids were incapable of digesting the branched chains of ABS molecules. LAS detergents, though manufactured by humans, mimic the structures of naturally occurring molecules and are biodegradable.

Polar Lipids

As was true of most nonpolar lipids, the structures of polar lipids are based on condensation of fatty acids with glycerol. The main difference is that only two of the three OH groups on glycerol are involved. The third (as shown by the structures in Fig. 20.2*b*) is combined with a highly polar molecule. In one sense the polar lipids are like the anions of fatty acids, only more so. They contain two hydrophobic hydrocarbon tails and a head which may have several electrically charged sites. As in the case of soap and detergent molecules, the tails of polar lipids tend to avoid water and

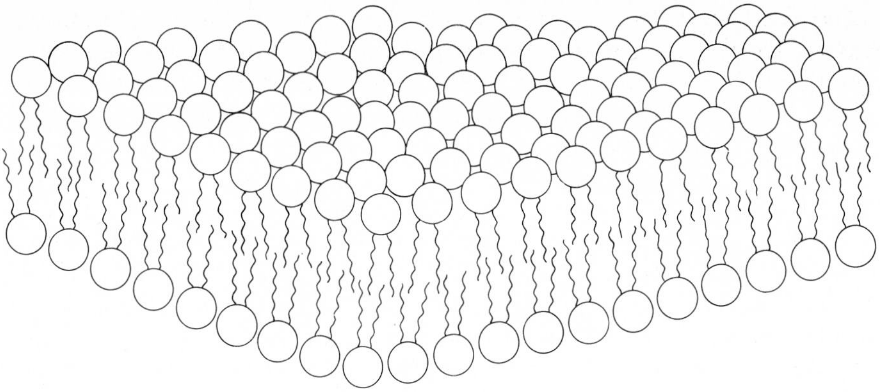

Figure 20.4 Lipid-bilayer model for membranes.

other polar substances, but the heads are quite compatible with such environments.

The polar lipids are most commonly found as components of cell walls and other membranes. Nearly all hypotheses regarding membrane structure take as a fundamental component a lipid bilayer (Fig. 20.4). Bilayers made in the laboratory have many properties in common with membranes. Ions such as Na^+, K^+, and Cl^- cannot pass through them, but water molecules can. The hydrocarbon core of such a bilayer should have large electrical resistance, as does a membrane. Certain carrier molecules can transport K^+ and other ions across a bilayer, apparently by wrapping a hydrophobic cloak around them to disguise their charges.

Bioamplification

One important consequence of the hydrophobic character of lipids is the concentration of nonpolar substances along ecological food chains, a process known as **bioamplification.** As an example, consider organisms in an aqueous environment such as a river or lake. Any substance which is more soluble in living tissues than in the surrounding water will tend to concentrate in even the simplest plants and animals. These plants and animals are often the food supply for more complex life forms—a food supply which contains a greater concentration of the substance in question. As we proceed up the food chain to larger predatory animals, the concentrations of some substances can be increased by factors of 10 000 or more.

What kinds of substances are likely to undergo bioamplification? Clearly those which are more soluble in living systems. Since the surroundings are dilute aqueous solutions while organisms contain both aqueous and lipid phases, nonpolar substances which dissolve in lipids are most likely to be concentrated. This is a problem with DDT, polychlorinated biphenyls (PCB's), and other long-lived synthetic organic compounds.

It also applies to metal ions which can combine with organic groups to form uncharged molecules. A good example of such an **organometallic** compound involves mercury. Certain aquatic microorganisms can convert relatively inert mercury metal into chloromethylmercury(II) (commonly known as methylmercuric chloride):

$$Hg \xrightarrow[\text{microorganisms}]{\text{aquatic}} CH_3\text{—}Hg\text{—}Cl$$

Since the charge of Hg^{2+} is neutralized by CH_3^- and Cl^-, this organometallic compound is quite soluble in lipid tissues. Concentration of mercury along aquatic food chains has led to several episodes where hundreds of Japanese people whose diet consisted mainly of fish became ill and many died. It also accounts for the relatively high level of mercury observed in some species of fish taken from the Great Lakes of the United States and Canada.

Humans and other organisms do have mechanisms for eliminating toxic and otherwise undesirable organic substances. A wide variety of enzymes in the human liver can convert hydrophobic molecules to more polar forms which are less soluble in lipid tissues and more readily excreted. However, there is often a problem with synthetic substances which have been synthesized by chemists and whose structures are quite different from any the liver enzymes are equipped to handle. As in the case of the ABS detergents, the reactions which decompose such substances may be relatively slow. If they are not converted into structures which can be eliminated easily, even small quantities may remain in the body for long periods, sometimes causing chronic illness. It is for this reason that toxic effects of synthetic organic or organometallic chemicals should be thoroughly tested before large quantities are released to the environment.

20.4 CARBOHYDRATES

Carbohydrates are sugars and sugar derivatives whose formulas can be written in the general form: $C_x(H_2O)_y$. (The subscripts x and y are whole numbers.) Some typical carbohydrates are sucrose (ordinary cane sugar), $C_{12}H_{22}O_{11}$; glucose (dextrose), $C_6H_{12}O_6$; fructose (fruit sugar), $C_6H_{12}O_6$; and ribose, $C_5H_{10}O_5$. Since the atom ratio H/O is 2/1 in each formula, these compounds were originally thought to be hydrates of carbon, hence their general name.

Glucose, by far the most-common simple sugar, is a primary source of energy for both animals and plants. Because they contain free glucose molecules, dextrose tablets or foods such as grapes and honey can provide a noticeable "lift" for persons tired by physical exertion. The individual glucose molecules pass rapidly into the bloodstream when such foods are eaten. Glucose is also the monomer from which the polymers cellulose and starch are built up. The structural material of plants, from the woody parts of trees to the cell walls of most algae, is cellulose. Plants store energy in starch, providing a source of glucose for all but the simplest organisms. The energy in starchy foods is not as rapidly available, however, since the polymeric structure must be broken down before glucose is released. As a consequence of the ubiquity of starch and especially cellulose, carbohydrates are by far the most plentiful organic compounds in the biosphere.

Simple Sugars

The structure of some simple sugars is shown in Fig. 20.5. They all contain five or six carbon atoms, and each carbon atom, except for one, is attached to a hydroxyl group (OH). The remaining carbon is double-bonded to oxygen, forming a carbonyl group (C═O). Sugars are thus all aldehydes and ketones and are usually referred to as *aldoses* or *ketoses*. A molecule like glucose is called an alde*hex*ose since it contains *six* carbon atoms. The difference between different sugars can be very subtle. Note, for instance, that mannose and glucose differ only in the geometrical arrangement of the OH group on one carbon atom. Subtle as this difference is, molecules in a living cell can tell the difference.

In Fig. 20.5 we have indicated the structure of the sugars in linear or chain form, but sugars usually occur in one of several ring or cyclic structures. We will only consider two of these, the α and β form of glucose. Because of the flexibility of the carbon chain, the linear form of glucose can easily adopt a conformation in which carbon atom 1 lies adjacent to the oxygen atom on carbon atom 5. When this happens, a proton can be transferred and a carbon-oxygen bond formed:

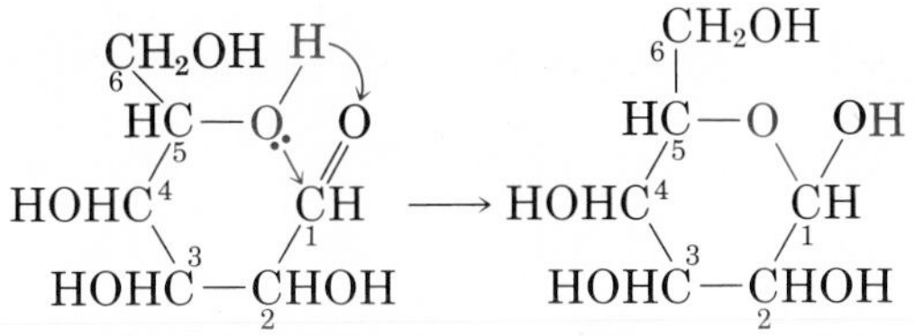

A careful consideration of the geometry of this structure reveals that not one but two cyclic structures are possible. These are called α- and β-glucose and are shown in Fig. 20.6. In the β form the C—O bond on carbon atom 1 (shown in dark color) is *parallel* to the C—O bond on carbon atom 4, while in the α form these two bonds are at an angle of $180° - 109.5° = 70.5°$. This geometric difference may seem relatively trivial, but it turns out to be important when glucose molecules are used as building blocks to form larger entities.

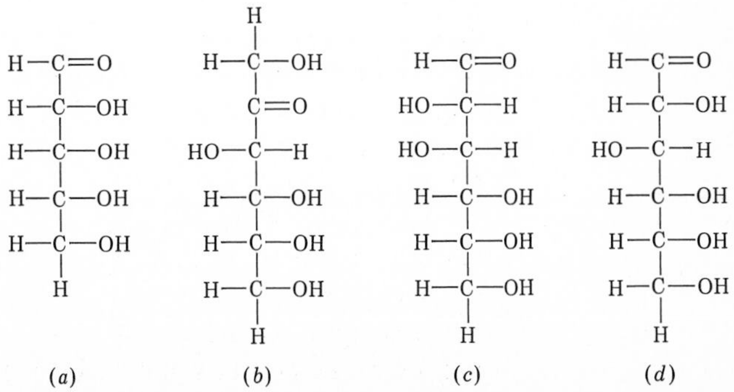

Figure 20.5 The linear form of some important monosaccharides: (*a*) ribose; (*b*) fructose; (*c*) mannose; (*d*) glucose.

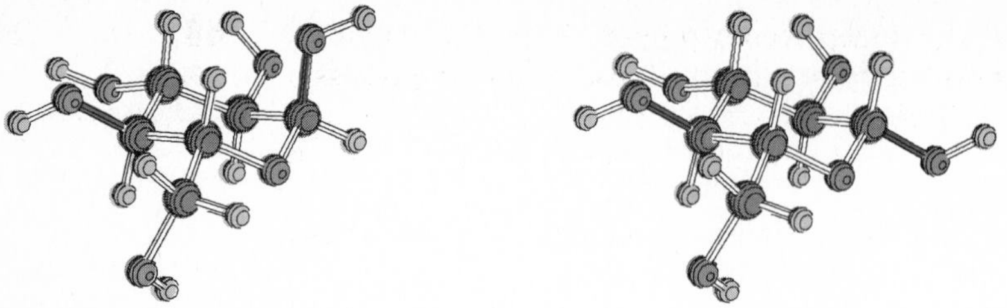

(a) (b)

Figure 20.6 (*a*) α-glucose; (*b*) β-glucose. Note how the C—O bonds protruding from the left- and right-hand sides of the molecule (indicated in color) are parallel to each other in the β form but not in the α form. (Computer-generated.) (*Copyright © 1976 by W. G. Davies and J. W. Moore.*)

Disaccharides

The sugar molecules listed in Fig. 20.5 are usually referred to as **monosaccharides.** This distinguishes them from the **disaccharides** which are made up by condensing two sugar units. A familiar example of a disaccharide is ordinary cane sugar, sucrose, which may be obtained by condensing a molecule of α-glucose with one of the cyclic forms of fructose called β-fructose. The structure of sucrose is shown in Fig. 20.7. Other, less familiar, examples of disaccharides are lactose, which occurs in milk, and maltose. In order to digest a disaccharide like sucrose or lactose, the human body must have an enzyme which can catalyze hydrolysis of the linkage between the two monosaccharide units. Many Asians, Africans, and American Indians are incapable of synthesizing lactase, the enzyme that speeds hydrolysis of lactose. If such persons drink milk, the undigested lactose makes them sick.

Polysaccharides

As the name suggests, polysaccharides are substances built up by the condensation of a very large number of monosaccharide units. Cellulose, for example, is a polymer of β-glucose, containing upwards of 3000 glucose units in a chain. Starch is largely a polymer of α-glucose. These two substances are a classic example of how a minor difference in the monomer can

Figure 20.7 The formation of sucrose from glucose and fructose.

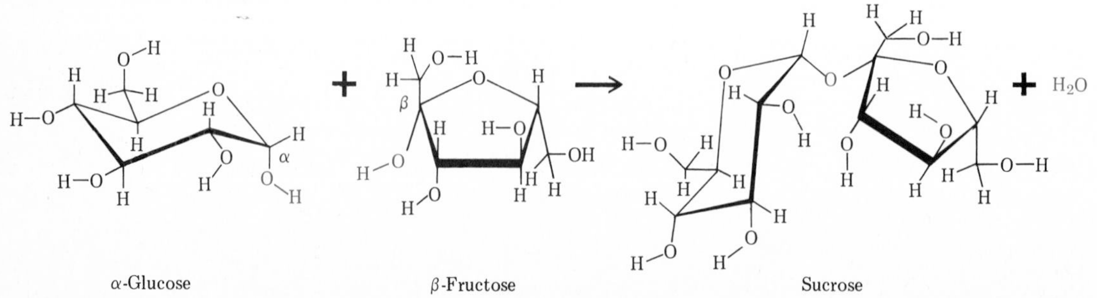

lead to major differences in the macroscopic properties of the polymer. Good-quality cotton and paper are almost pure cellulose, and they give us a good idea of its properties. Cellulose forms strong but flexible fibers and does not dissolve in water. By contrast, starch has no mechanical strength at all, and some forms are water soluble.

Part of the molecular structure of cellulose is shown in Fig. 20.8. Note

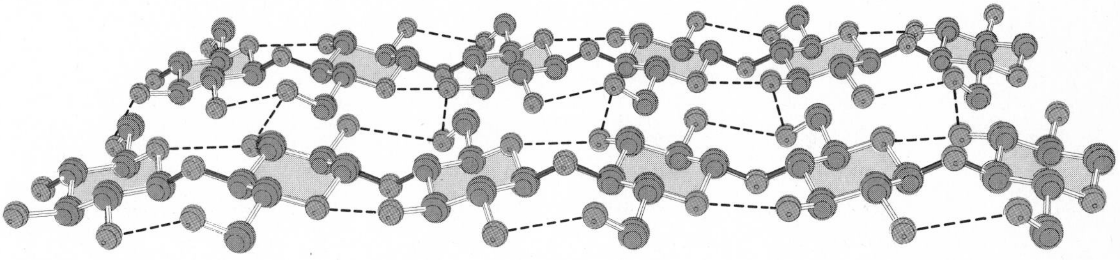

(*a*)

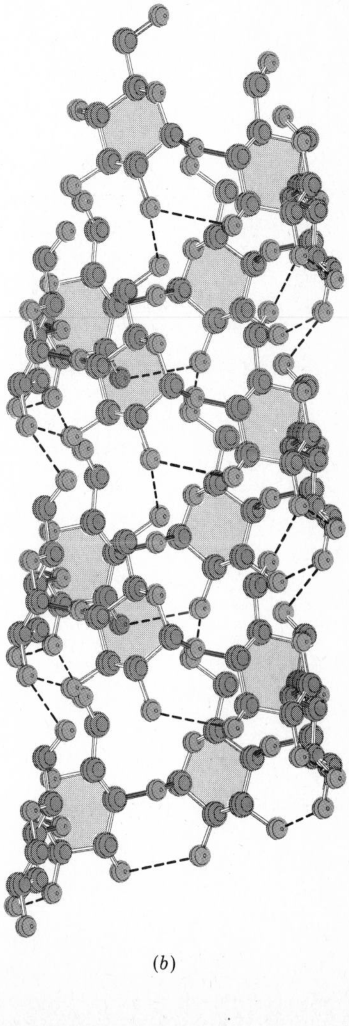

(*b*)

Figure 20.8 Structures of (*a*) cellulose and (*b*) amylose (starch). The C—O bonds linking glucose units are shown in color and hydrogen bonds are shown as single black lines. Hydrogen atoms have been omitted to simplify the structure. Note how the β-glucose units in cellulose form a linear chain because the C—O bonds on either side of the glucose unit are parallel. In amylose, where α-glucose is the monomer, the chain is forced to curve. Note also the numerous hydrogen bonds between the two cellulose chains shown and the hydrogen bonds connecting successive turns of the spiral chain of amylose. When iodine contacts starch, I_2 molecules can fit lengthwise within the spiral of each amylose molecule. The starch-iodine complex which results is responsible for the dark blue color observed when iodine contacts starch. (Computer-generated.) (*Copyright © 1977 by W. G. Davies and J. W. Moore.*)

how the β-glucose units can condense in a straight chain because the C—O bonds on opposite sides of the ring (shown in color) are *parallel* to each other. Each of the chains shown in the figure is hydrogen-bonded (dashed lines) to other chains lying parallel to it and fitting snugly against it. This hydrogen bonding involves all the —$\ddot{\text{O}}$—H groups in the structure. As a consequence none are left exposed to hydrogen bond to solvent molecules except on the surface of the fiber. This explains why cellulose is insoluble in water.

Naturally occurring starch is usually a mixture of two different substances called amylose and amylopectin. Of these two, amylose has the simpler molecular structure. This is also shown in Fig. 20.8. Because amylose is a polymer of α-glucose, the C—O bonds at each end of the monomer are no longer parallel, and this prevents them from forming a straight chain. Instead, a spiral structure like that shown in the figure is produced. This spiral is stabilized to some extent by the occurrence of hydrogen bonding between each glucose ring and the rings immediately above it and below it in the spiral. Each spiral contains between 100 and 1000 glucose rings. The other form of starch, amylopectin, is also a polymer of α-glucose but contains many branch chains. Since these branches occur randomly in the molecule, amylopectin has no overall regular structure.

Cellulose and starch are different not only in overall structure and macroscopic properties. From a biochemical point of view they behave so differently that it is difficult to believe that they are both polymers of the same monosaccharide. Enzymes which are capable of hydrolyzing starch will not touch cellulose, and vice versa. From a plant's point of view this is just as well since cellulose makes up structural material while starch serves as a storehouse for energy. If there were not a sharp biochemical distinction between the two, the need for a bit more energy by the plant might result in destruction of cell walls or other necessary structural components.

Bacteria, protozoa, termites, some cockroaches, and ruminant mammals (cattle, sheep, etc.) are capable of digesting cellulose. Most organisms, including humans, are not. If our digestive enzymes could hydrolyze cellulose, humans would have available a much larger food supply. Quite literally we would be able to eat sawdust! It is possible to hydrolyze cellulose in the laboratory either with strong acid or with cellulase, the enzyme used by bacteria. So far, however, such processes produce more expensive (and less tasty!) food than we already have available.

20.5 PROTEINS

Earlier in the chapter the ability of most biological molecules to serve a variety of functions was pointed out. This versatility is nowhere better exemplified than by the proteins. Perhaps because of their many functions, proteins are the most abundant organic molecules in living cells, constituting more than 50 percent of the mass once water is removed. It is estimated that the human body contains well over a million different kinds of protein, and even a single-cell organism probably contains thousands. Each of these is a polymer of amino acids which has a highly specific composition, a unique molecular weight (usually in the range from 6000 to 1 000 000), and its own sequence of different amino acids along the polymer chain.

Proteins may be divided into two major classes on the basis of their behavior when reacted with water. The products obtained upon hydrolysis of **simple proteins** are all amino acids. In the case of **conjugated proteins** other organic and/or inorganic substances are obtained. The non-amino acid portions of conjugated proteins may consist of metals, lipids, sugars, phosphate, or other types of molecules. These components are referred to as **prosthetic groups.**

Proteins may also be subdivided on the basis of their molecular shape or conformation. In the **fibrous proteins** long polymer chains are arranged parallel or nearly parallel to one another to give long fibers or sheets. This arrangement results in physically tough materials which do not dissolve in water. The fibrous proteins are fundamental components of structural tissues such as tendons, bone, hair, horn, leather, claws, and feathers.

By contrast, polymer chains of the **globular proteins** fold back on themselves to produce compact, nearly spherical shapes. Most globular proteins are water soluble and hence are relatively mobile within a cell. Some examples are enzymes, antibodies, hormones, toxins, and substances such as hemoglobin whose function is to transport simple molecules or even electrons from one place to another. The enzyme trypsin, described in Chap. 18 and illustrated in Fig. 18.9, is a typical globular protein.

Polypeptide Chains

The backbone of any protein molecule is a polypeptide chain obtained by the condensation of a large number of amino acids with the elimination of water. You will recall from Chap. 8 that the amino acids are bifunctional organic compounds containing an acid group, —COOH, and an amine group, —NH_2. The amine group is attached to the carbon atom adjacent to the —COOH (the **α carbon atom**). The three simplest amino acids are

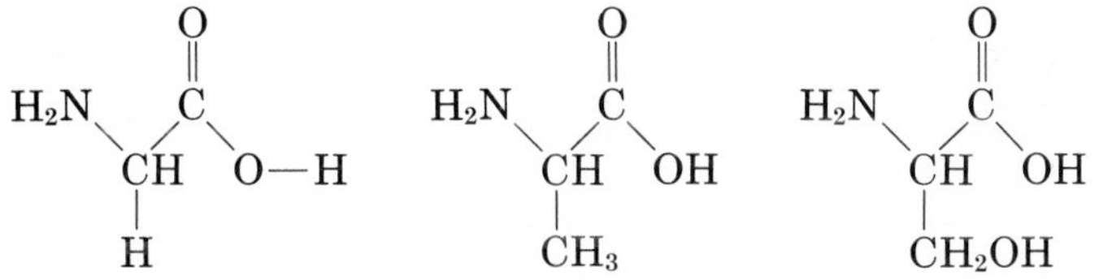

Glycine **Alanine** **Serine**

In practice, though, these acids are usually in the form of their zwitterions, and we should write them

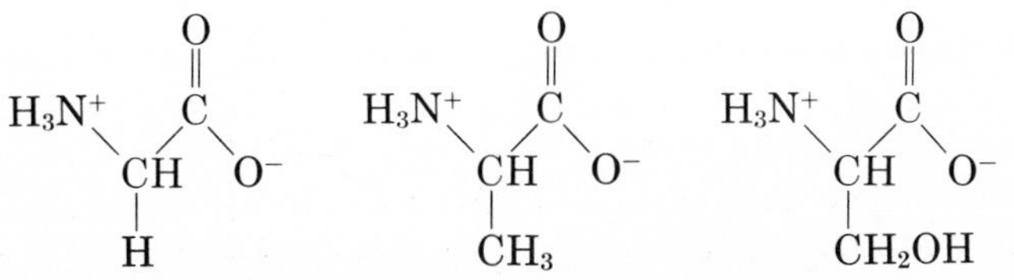

If these three amino acids are now condensed, water is eliminated and a simple polypeptide containing three amino acids is obtained:

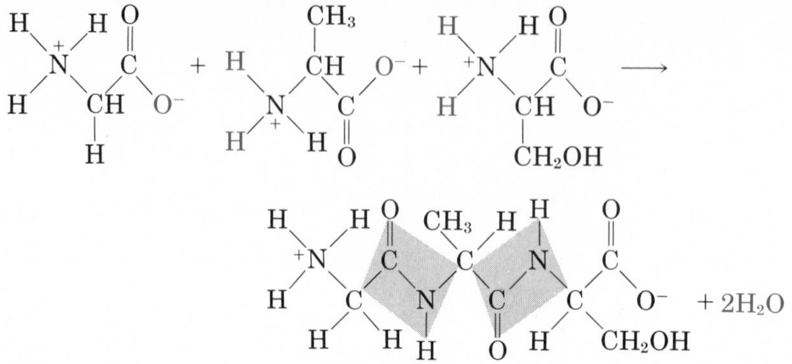

(20.2)

The two CO—NH bonds produced by this reaction are called **peptide bonds.** An important feature of such a peptide bond is that it is *planar.* This is because of the existence of two resonance structures

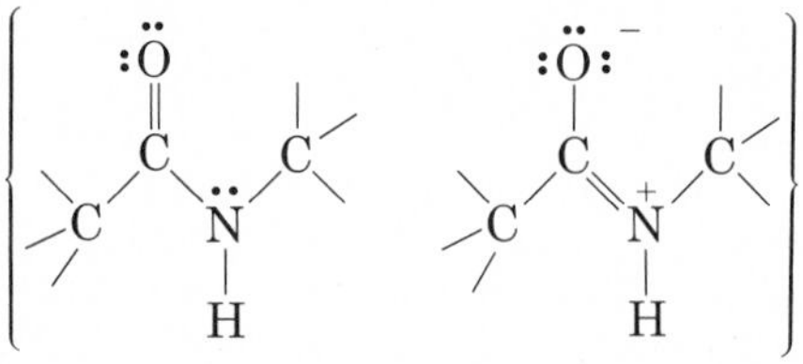

I II

In the second of these structures, there is a double bond between the carbon and the nitrogen atoms but no lone pairs around either atom. This insures that all the bonds involving the C and N atoms must lie in the same plane.

The planarity of the peptide bond places considerable constraint on the conformations available to a polypeptide chain. Rotation can only occur around the single bonds involving the α carbon atoms. That is, the gray shaded areas in the polypeptide chain of Eq. (20.2) can twist relative to one another, but they must remain planar. This limitation greatly simplifies analysis of the conformations available.

Another important aspect of the peptide bond is the opportunity it provides for hydrogen bonding. The oxygen on the >C=O carbonyl group can bond to the hydrogen on an H—N< group further along the chain:

>C=Ö:···H—N<

Such a bond is somewhat stronger than a normal hydrogen bond because of the partially negative character of the oxygen atom and the partially positive character of the nitrogen atom conferred by resonance structure II.

The Amino Acids

Altogether there are 20 amino acids which commonly occur in all organisms. These include the three simple ones mentioned above. Under most circumstances amino acids exist as zwitterions and have the general formula

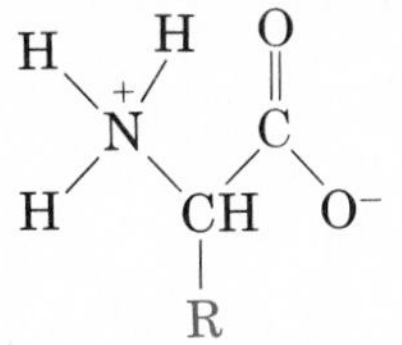

R represents a group called a **side chain** which varies from one amino acid to another. Structures of the 20 different R groups are given in Fig. 20.9, where they are shown as part of a polypeptide chain.

Although each amino acid side chain has its own individual properties, it is useful to divide them into several categories. In Fig. 20.9 this has been done on the basis of how strongly hydrophilic or hydrophobic they are. On the extreme left of the figure are the six most hydrophilic side chains. All are polar, and four are actually ionic at a pH of 7. Next in the figure are the six most hydrophobic side chains. These are all large and nonpolar and contain no highly electronegative atoms like nitrogen and oxygen. The remaining eight side chains either contain small nonpolar groups or groups of fairly limited polarity, and are therefore not very strongly hydrophilic or very strongly hydrophobic.

Whether a side chain is hydrophilic or hydrophobic has considerable influence on the conformation of the polypeptide chain. Like the hydrocarbon tails of fatty acid molecules, hydrophobic amino acid side chains tend to avoid water and cluster together with other nonpolar groups. In a globular protein like trypsin (Fig. 18.9), for example, most hydrophobic residues are found among twisted chains deep within the molecule. Hydrophilic side chains, by contrast, tend to occupy positions on the outside surface, where they contact the surrounding water molecules. Hydrogen bonds and dipole forces attract these water molecules and help solubilize the globular protein. Were all the nonpolar R groups exposed to the aqueous medium, the protein would be much less soluble.

Primary Protein Structure

Proteins which occur in nature differ from each other primarily because their side chains are different. Partly this is a matter of composition. In wool, for example, 11 percent of the side chains are cysteine, while no cysteine occurs in silk at all. To a much larger extent though, the differences between different proteins is a matter of the *sequence* in which the different side chains occur. This is especially true of globular proteins like enzymes. The sequence of side chains along the backbone of peptide bonds in a polypeptide is said to constitute its **primary structure.**

The 20 different amino acids permit construction of a tremendous variety of primary structures. Consider, for example, how many tripeptides similar to that shown in Eq. (20.2) can be constructed from the 20 amino acids. In the example shown, the first amino acid in the chain is glycine, but it might just as well be proline or any other of the 20 amino acids. Thus

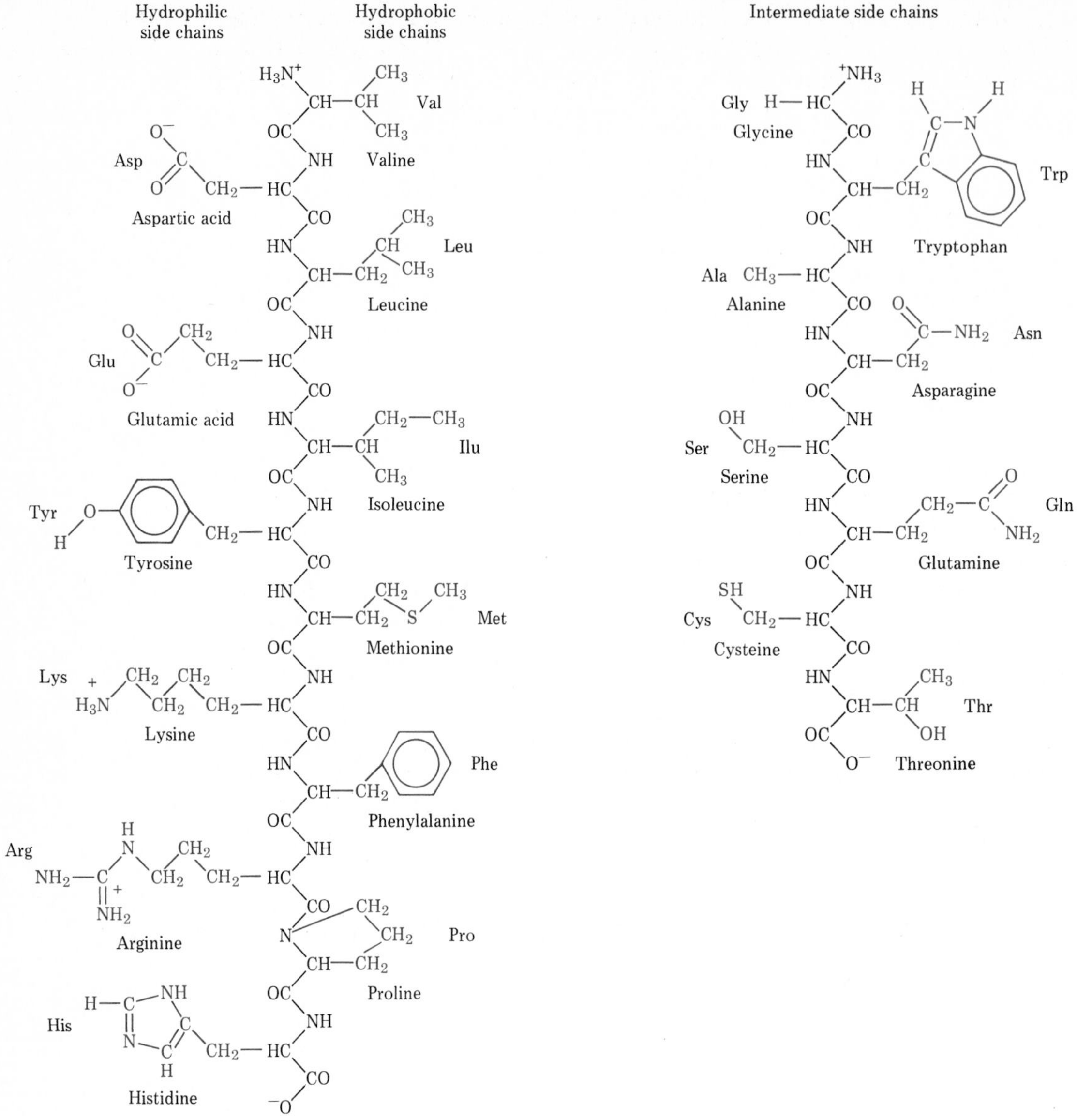

Figure 20.9 Structures of the 20 common amino acid side chains in proteins. Acidic and basic groups are shown with the degree of protonation they would have at pH = 7.

there are 20 possibilities for the first place in the chain. Similarly there are 20 possibilities for the second place in the chain, making a total of $20 \times 20 = 400$ possible combinations. For each of these 400 structures we can again choose from among 20 amino acids for the third place in the chain, giving a grand total of $400 \times 20 = 20^3 = 8000$ possible structures for the tripeptide.

A general formula for the number of primary structures for a polypeptide containing n amino acid units is 20^n—a very large number indeed when you consider that most proteins contain at least 50 amino acid residues. $[20^{50} = (2 \times 10)^{50} = 2^{50} \times 10^{50} \cong 10^{15} \times 10^{50} = 10^{65}]$ Primary structure is conventionally specified by writing three-letter abbreviations for each amino acid, starting with the $—NH_3^+$ end of the polymer. For example, the structure

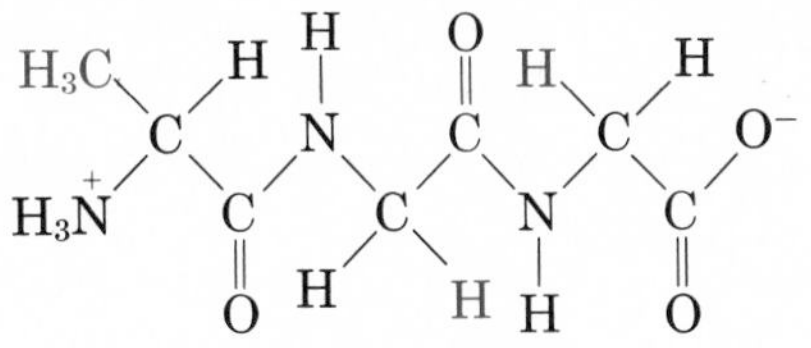

would be specified as

Ala-Gly-Gly

(Note that Ala-Gly-Gly is not the same as Gly-Gly-Ala. In the latter case glycine rather than alanine has the free $—NH_3^+$ group. Because its ends are different, there is a directional character in the polypeptide chain.)

Determination of the primary structure of a protein is a difficult and complicated problem. It also is a rather important one—the sequence of amino acids governs the three-dimensional shape and ultimately the biological function of the protein. Consequently much effort has gone into methods by which primary structure can be elucidated.

Insulin was the first protein whose amino acid sequence was determined. This pioneering work, completed in 1953 after some 10 years of effort, earned a Nobel Prize for British biochemist Frederick Sanger (1918 to). He found the primary structure to be

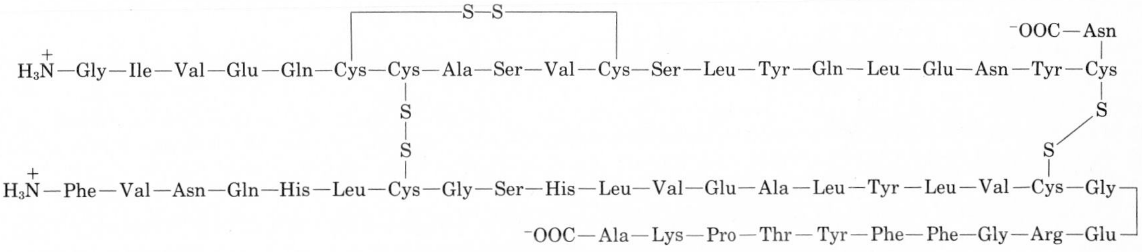

Note how there are two chains in this structure, one with 21 side chains and the other with 30. These two chains are linked in two places with disulfide (—S—S—) bridges, each connecting two cysteine residues in different chains.

In order to determine the primary structure of a protein, a known mass of pure sample is first boiled in acid or base until it is completely hydrolyzed to individual amino acids. The amino acid mixture is then separated chromatographically and the exact amount of each acid determined. In this way, for instance, one can find that for every 3 mol serine in the insulin molecule, there are 6 mol leucine. The next step is to break down the protein into smaller fragments. Disulfide bridges are broken by oxidation after which the protein is selectively hydrolyzed by an enzyme such as trypsin or chymotrypsin. In a favorable case one will then have several fragments each containing 10 or 20 amino acid residues. These can then be separated and analyzed individually.

The sequence of amino acids in one of these polypeptide fragments is usually determined using phenylisothiocyanate, $C_6H_5N{=}C{=}S$, which selectively attacks the $—NH_3^+$ end of the polypeptide chain. Addition of acid then splits off the terminal amino acid, and it can be identified. Since the rest of the polypeptide chain is left intact, this process can now be repeated, and the second amino acid in the sequence can be attacked, removed, and identified. By snipping off amino acids one at a time in this way, one eventually finds the complete sequence for the fragment. This whole process can be automated and hence speeded up considerably. Despite these aids the determination of the exact sequence of amino acids in a protein is still a very lengthy procedure.

As methods for determining primary structure have become more advanced, a great many proteins have been sequenced, and some interesting comparisons can be made. A particularly intriguing example is that of cytochrome *c*, an electron carrier which is found in all organisms that use oxygen for respiration. When samples of cytochrome *c* from different organisms are compared, it is found that the amino acid sequence is usually different in each case. Moreover, the more widely separated two species are in their macroscopic features, the greater the degree of difference in their protein sequences. When horse cytochrome *c* is compared with that of yeast, 45 out of 104 residues are different. Only two substitutions are found between chicken and duck, and cytochrome *c* is identical in the pig, cow, and sheep. The magnitudes of these changes coincide quite well with biological taxonomy based on macroscopically observable differences. Cytochrome *c* can be used to trace biological evolution from unicellular organisms to today's diverse species, and even to estimate times at which branching occurred in the family tree of life.

Secondary Protein Structure

One might expect a long-chain protein molecule to be rather floppy, adopting a variety of molecular shapes and changing rapidly from one conformation to another. In practice this seldom happens. Instead the protein chain stays more or less in the same conformation all the time. It is held in this shape by the cooperative effect of a large number of hydrogen bonds between different segments of the chain.

A particularly important conformation of the polypeptide chain is the spiral structure shown in Fig. 20.10. This is called an **α helix.** Many fibrous proteins like hair, skin, and nails consist almost entirely of α helices. In globular proteins too, although the overall structure is more complex, short lengths of the chain often have this configuration. In an α helix the polypeptide chain is twisted into a right-hand spiral—the chain turns around clockwise as one moves along it. The spiral is held together by hydrogen bonds from the amido (>NH) group of one peptide bond to the car-

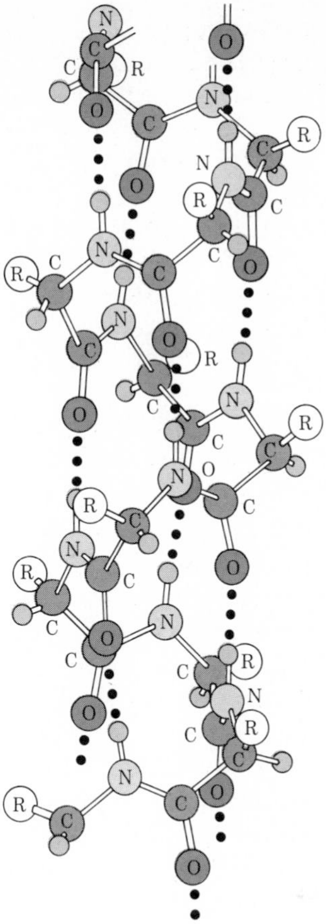

Figure 20.10 Ball-and-stick model of the α helix. Hydrogen bonds are shown as dotted bonds. Note that R groups extend almost perpendicular from the axis.

bonyl group ($>$C═O) of a peptide bond three residues farther along the chain. Two factors contribute toward making this a particularly stable structure. One is the involvement of *all* the $>$N—H and $>$C═O groups in the chain in the hydrogen bonding. Spirals with slightly more or slightly less twist do not permit this. The second factor is the way in which the side chains project outward from an α helix. Bulky side chains therefore do not interfere with the hydrogen bonding, enabling a fairly rigid cylinder to be formed.

A second regular arrangement of the polypeptide chain is the **pleated sheet** or β-keratin structure found in silk and shown in Fig. 20.11. As in the α helix, this structure allows all the amido and carbonyl groups to participate in hydrogen bonds. Unlike the α helix, though, the side chains are squeezed rather close together in a pleated-sheet arrangement. In consequence very bulky side chains make the structure unstable. This explains why silk is composed almost entirely of glycine, alanine, and serine, the three amino acids with the smallest side chains. Most other proteins contain a much more haphazard collection of amino acid residues.

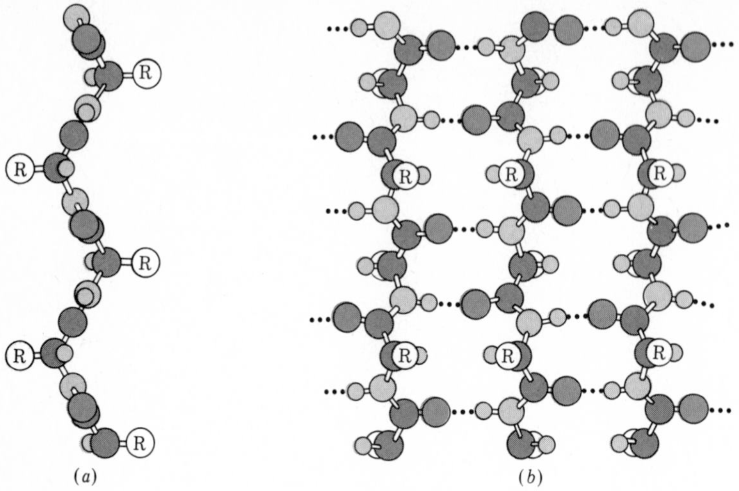

Figure 20.11 The β keratin, pleated-sheet structure. (*a*) Edge view; rotating by 90° gives (*b*) the top view. (Hydrogen bonds are shown as dotted bonds.)

Higher-Order Structure

Several of the amino acid side chains are difficult to fit into either the α helix or β-sheet types of structure. An obvious example is proline, in which the R group is a ring and includes nitrogen bonded to the α carbon. When proline is involved in a peptide bond, all hydrogens are gone from the nitrogen, leaving no site for hydrogen bonding.

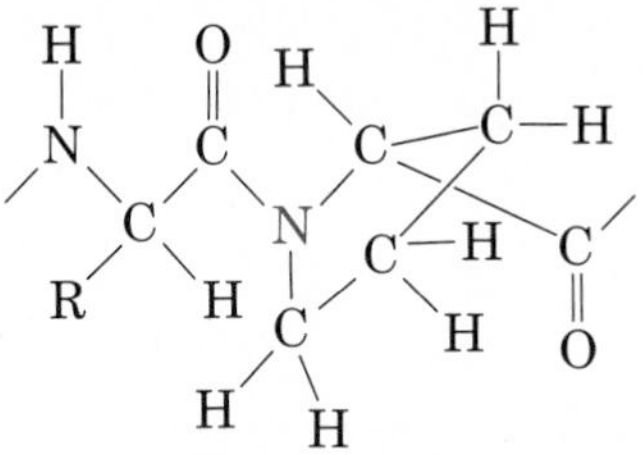

Another problem involves side chains having the same charge, such as those of lysine and arginine or glutamic acid and aspartic acid, which repel one another considerably. When they occur close together, these groups may also destabilize an α helix or β sheet structure.

The presence of several groups of the type just mentioned allows a protein chain to bend sharply instead of laying flat or curling regularly into a spiral. Such sharp bends connect sections of α helix or β sheet structures in the globular proteins. They allow the polypeptide chain to curl back upon itself, folding the protein into a very compact, nearly spherical shape. The nature of this folding is referred to as **tertiary structure,** since it involves a third organizational level above the primary amino acid sequence and the secondary α helix or β sheet.

An excellent example of tertiary structure in a globular protein is provided by the three-dimensional view of sperm-whale myoglobin in Fig. 20.12. The molecular backbone consists of eight relatively straight segments of α helix. The longest of these contain 23 amino acids, the shortest

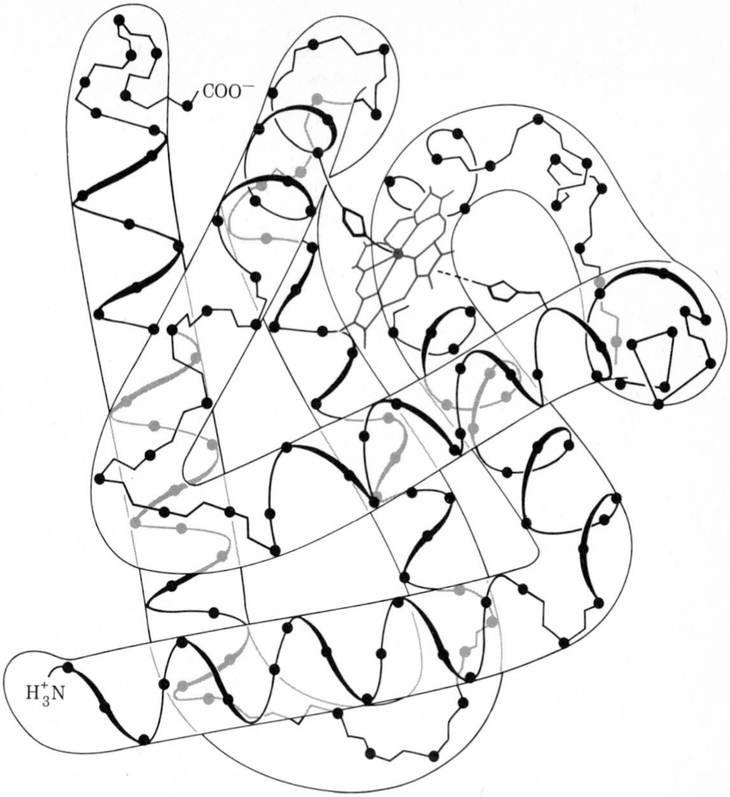

Figure 20.12 Structure of myoglobin. The dots represent α carbon atoms along the polypeptide chain. Side groups are not shown except for two histidine residues which interact with the heme group. (The latter is shown in color.)

but 7. Fewer than 50 of myoglobin's 153 amino acids are found in the bends between the helixes, but all the proline residues fall into this category, as do many of the others which destabilize the α helix or β sheet. (Although only the α helix is prominent in the structure of myoglobin, other globular proteins are found to have regions consisting of β sheets.)

Several other important generalizations may be made about the structure of myoglobin. Even in regions where the chain twists into an α helix, nearly all the nonpolar R groups point toward the interior of the molecule. Here they crowd so closely together that only four water molecules can squeeze their way in. The outside surface of the protein, however, contains all the polar R groups. These interact strongly with the many water molecules which normally surround myoglobin in muscle tissue.

Although myoglobins isolated from a variety of mammals differ slightly in their primary structure, they all seem to have nearly the same overall molecular shape. Apparently some of the amino acids are much more important than others in determining the bend points and other crucial features of tertiary structure. Substitutions at less-important positions do not cause great variations in the ability of the protein to carry out its biological function.

One final point. Figure 20.12 shows clearly that myoglobin contains a prosthetic group. In this case it is the heme group, which contains an iron atom surrounded by four nitrogens in a flat ring structure known as a porphyrin. The porphyrin ring is not covalently bonded to the protein chain, but rather fits snugly into a "pocket" surrounded by several segments of α helix. A nitrogen atom in a histidine side chain on one helix does form a coordinate covalent bond with iron in the heme group, but apart from this the prosthetic group is positioned solely by the way the protein chain is folded.

The iron in the heme group marks the active site at which the oxygen-storage function of myoglobin is accomplished. Iron, like other first-row transition-metal ions, ordinarily forms six bonds directed toward the corners of an octahedron (Sec. 22.3). In myoglobin only five bonds to iron are found. The sixth position (opposite the histidine nitrogen) can be occupied by an oxygen molecule, providing a convenient storage site. If the concentration of oxygen near the protein falls, this binding is reversed, releasing oxygen to replenish the supply.

If an iron ion can bond to an oxygen molecule, you may wonder why the complicated porphyrin and protein structures are necessary. It is to *prevent the oxidation* of the iron in the heme group from the iron(II) to the iron(III) oxidation state. Because the oxygen molecule needs to gain *two* electrons to be oxidized while iron(II) can only supply *one,* no electron transfer occurs. If the iron(II) atom in heme were not surrounded by the protein chain, a water molecule would be able to take part as an intermediate in the electron-transfer mechanism and reduction of the oxygen could occur. As matters stand, though, the oxygen molecule cannot be reduced until it is released by the myoglobin. Thus the combined effects of the tertiary structure of the protein, the prosthetic group, and a specialized active site allow the myoglobin to fulfill its biological function of storing oxygen molecules until needed, without allowing them to be reduced during storage.

In some proteins there is yet a fourth level of organization, labeled **quaternary structure.** This may be illustrated by hemoglobin, whose function as an oxygen carrier in the bloodstream is well known. Hemoglobin consists essentially of four myoglobin molecules packed together in a single unit. Four separate polypeptide chains are each folded as in myoglobin and then nested together. The way these four subunits fit together constitutes the quaternary structure.

Like the other type of structures, quaternary structure contributes to the function of a protein. In the case of hemoglobin, an oxygen molecule attached to one subunit causes slight shifts in tertiary and quaternary structure which make it easier for other oxygen molecules to bond to the other subunits. Consequently hemoglobin in the lungs can be loaded with its full complement of four oxygen molecules rather easily, a factor which increases its efficiency in carrying oxygen to body tissues. The converse is also true—loss of one oxygen molecule causes slight structural rearrangements which allow the remaining three to depart more readily.

Enzymes

The enzymes are the most extensive and highly specialized class of proteins. The structure and mode of action of trypsin, a typical enzyme, were described in Sec. 18.5. Most enzyme-catalyzed processes occur from 10^8 to 10^{11} times faster than the uncatalyzed reactions. Specificity is so great in some cases that only one particular molecule can serve as the enzyme's substrate. Even closely related structures are unable to fit the active site, and so their reactions cannot be speeded up.

Enzymes are of crucial importance to living organisms because they catalyze nearly every important reaction in cell metabolism. In cases where it is necessary to carry out a nonspontaneous reaction, enzymes are capable of coupling a reaction having a negative free-energy change to the desired one. When a living system reproduces, grows, or repairs damage caused by the external environment, enzymes catalyze the necessary reactions. Some enzymes are even stimulated or inhibited by the presence of smaller molecules. This permits regulation of the concentrations of certain substances because the latter can turn off enzymes which initiate their synthesis. On a molecular level enzymes are the means by which the mechanism of a cell is kept running.

20.6 NUCLEIC ACIDS

Given the tremendously important role of proteins in the functioning of any organism, the question arises of how cells synthesize appropriate polypeptides. Although more than 1000 enzymes are currently known, these constitute a very small fraction of the possible combinations of the 20 amino acids in chains of 100 or more. Clearly a great many of these combinations are worthless, in the sense that they are not adapted to any biological function. It would be pointless for any organism to construct them. Indeed, if they did occur, it would be useful for a living system to hydrolyze them to their individual amino acids and use these to build up proteins which did carry out necessary functions. Therefore it is reasonable to expect a living cell to contain some kind of "blueprint" which specifies the structures of those proteins which are essential for the cell's normal functioning. Without such a guide an incredible amount of effort would be wasted in synthesizing unusable polypeptides just to get small amounts of those that worked.

The blueprint just described is contained in the molecular architecture of deoxyribonucleic acid (DNA). The structure of DNA also provides an obvious mechanism by which the information necessary to specify protein structure can be reproduced and passed from a parent cell to its progeny. The complicated task of building protein structures from the DNA blueprint involves several types of ribonucleic acids (RNA's) whose structures are closely related to that of DNA. Hence the storage, reproduction, and application of information about protein structure depends on **nucleic acids.**

Nucleic Acid Structure

Nucleic acids were first isolated from cell nuclei (hence the name) in 1870. Since then they have been found in other portions of cells as well, es-

Figure 20.13 The structure of the monosaccharides (*a*) ribose and (*b*) deoxyribose found in nucleic acids.

pecially the ribosomes, which are the sites of protein synthesis. Most nucleic acids are extremely long-chain polymers—some forms of DNA have molecular weights greater than 10^9.

Nucleic acids are made up from three distinct structural units. These are

1 *A five-carbon sugar.* Only two sugars are involved. These are ribose (used in RNA) and deoxyribose (used in DNA). Their structures are shown in Fig. 20.13. Note that deoxyribose, as its name implies, has one oxygen *less* than ribose in the 2 position.
2 *A nitrogenous base.* There are five of these bases. All are shown in Fig. 20.14. Three of them, adenine, guanine, and cytosine, are common to both DNA and RNA. Thymine occurs only in DNA, and uracil only in RNA.
3 *Phosphoric acid.* H_3PO_4 provides the unit that holds the various segments of the nucleic acid chain to each other.

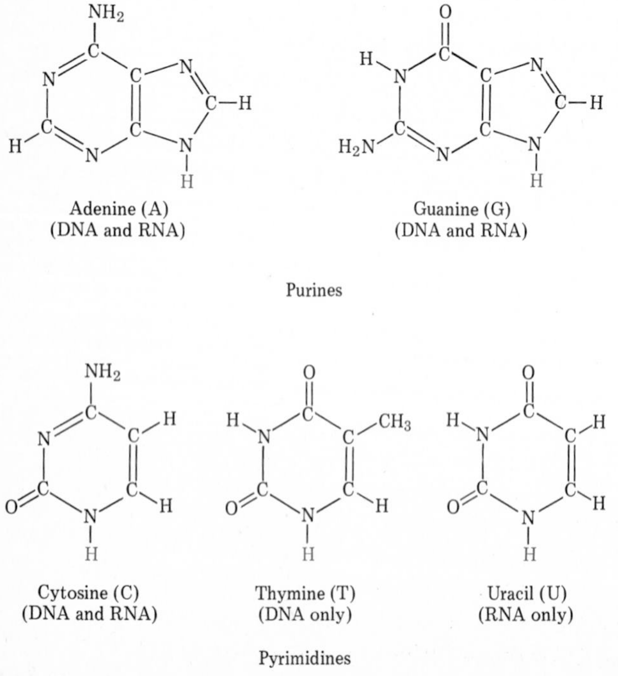

Figure 20.14 Structures of the principal nitrogenous bases obtained by hydrolysis of nucleic acids. The hydrogen lost when the base is condensed with a sugar is shown in color. (Thymine occurs only in DNA, while the very similar uracil occurs in its place in RNA. Adenine, guanine, and cytosine occur in both DNA and RNA.)

The combination of a sugar and a nitrogenous base is called a **nucleoside.** A typical example of a nucleoside is adenosine, derived by the condensation of ribose and adenine:

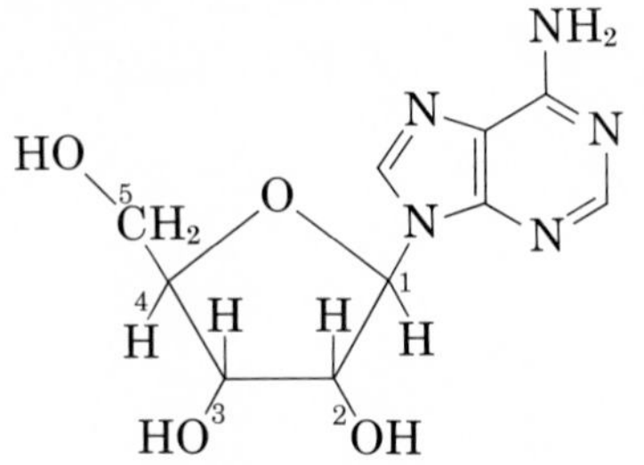

Nucleosides in turn can be condensed with phosphoric acid by means of the —OH group at carbon atom 5. The result is a structure called a **nucleotide.** Thus when adenosine condenses with phosphoric acid, the nucleotide formed is adenosine-5-monophosphoric acid, usually indicated by the acronym AMP:

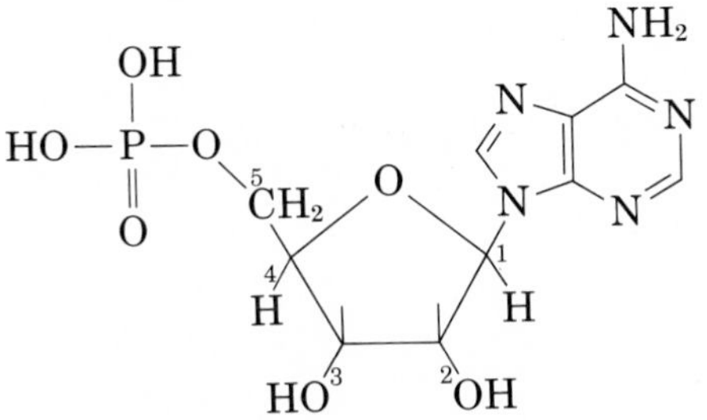

Finally the presence of —OH groups on both the phosphoric acid and sugar portions allows one nucleotide to link with another. This results in the polymeric structure shown in Fig. 20.15. Note that directionality is introduced along the chain because a sugar connected to one phosphate through carbon atom 5 always joins to the next phosphate via carbon atom 3.

Information Storage

How can DNA and RNA molecules act as blueprints for the manufacture of proteins? The exact details were unraveled in the early 1960s mainly by Marshall Nirenberg (1927 to) at the National Institutes of Health and H. G. Khorana (1922 to) at the University of Wisconsin, work which earned them the Nobel prize in 1968. They showed that each amino acid in a protein is determined by a specific **codon** of three nitrogenous bases in the DNA or RNA chain. The details of this **genetic code** are given in Table 20.3. As an example of how this code works, let us take the section of RNA shown in Fig. 20.15. This has the sequence UCAUGG. This is part of the instructions for building a polypeptide chain containing the amino acid serine (UCA) followed by the amino acid tryptophan (UGG).

Since each codon corresponds to three places in the nucleic acid chain and since there are four kinds of nitrogenous bases to fill each place, there

TABLE 20.3 The Genetic Code for RNA.

	Second Base in Codon							
First Base in Codon	U		C		A		G	
U	UUU	Phe	UCU	Ser	UAU	Tyr	UGU	Cys
	UUC		UCC		UAC		UGC	
	UUA	Leu	UCA		UAA	TERM	UGA	TERM
	UUG		UCG		UAG	TERM	UGG	Trp
C	CUU	Leu	CCU	Pro	CAU	His	CGU	Arg
	CUC		CCC		CAC		CGC	
	CUA		CCA		CAA	Gln	CGA	
	CUG		CCG		CAG		CGG	
A	AUU	Ile	ACU	Thr	AAU	Asn	AGU	Ser
	AUC		ACC		AAC		AGC	
	AUA		ACA		AAA	Lys	AGA	Arg
	AUG	Met	ACG		AAG		AGG	
G	GUU	Val	GCU	Ala	GAU	Asp	GGU	Gly
	GUC		GCC		GAC		GGC	
	GUA		GCA		GAA	Glu	GGA	
	GUG		GCG		GAG		GGG	

Notes: (*a*) A termination codon is indicated by TERM. (*b*) The code for DNA is identical except that tyrosine (T) replaces uracil (U).

are a total of $4^3 = 64$ different possible codons. Since there are only 20 amino acids, the genetic code is *degenerate*—several different codons correspond to the same amino acid. This degeneracy acts as a safeguard against errors in reading the code. Thus UCU, UCC, UCA, and UCG all correspond to serine. If a mistake is made in reading the third base in this triplet, no harm is done since serine is still produced.

An important feature of the genetic code is the existence of three **termination codons.** These correspond to an instruction for ending a polypeptide chain. How they work is best illustrated by an example.

EXAMPLE 20.1 Decode the RNA fragment

Head A G G C A G A U U U U U C C G G U U G A A C G A tail

Solution We notice that this fragment contains a codon for terminating a peptide chain, namely, UGA. Marking off triplets in both directions from this codon, we have

AG|GCA|GAU|UUU|UCC|GGU|UGA|ACG|A

The amino acid sequence is thus

Ala-Asp-Phe-Ser-Gly STOP Thr-

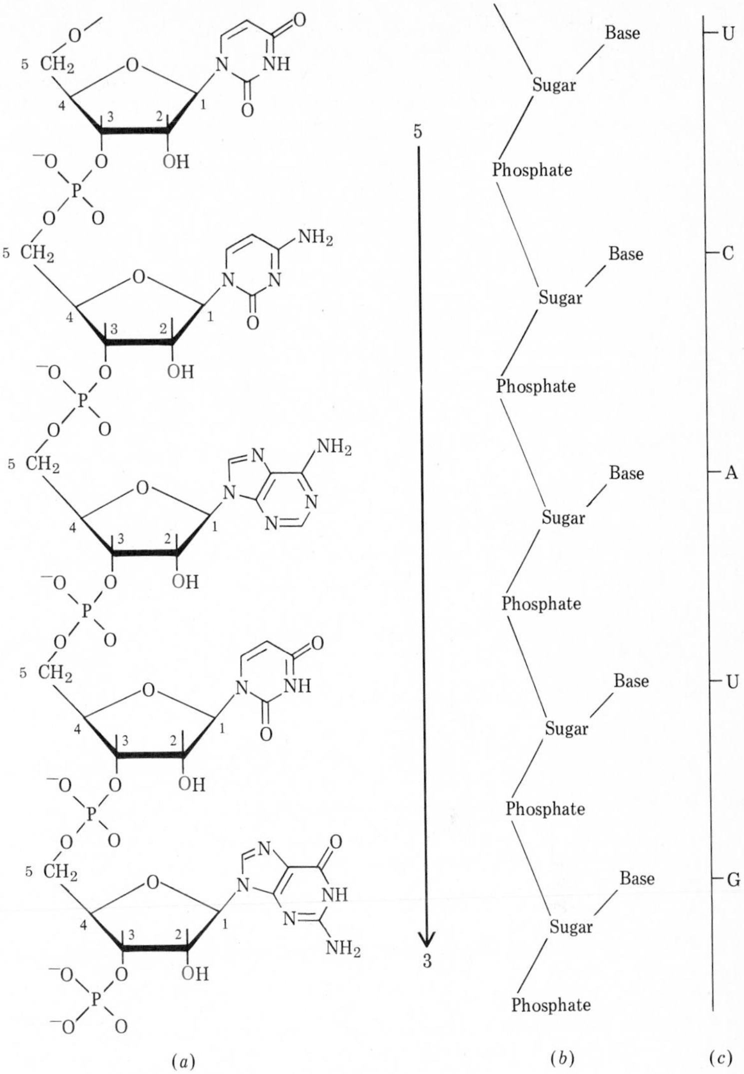

Figure 20.15 (*a*) A portion of a ribonucleic acid chain. (In DNA the oxygens marked in color would not be present and thymine would replace uracil at the two positions where the latter appears along the chain.) (*b*) Locations of sugar, phosphate, and nitrogenous bases. (*c*) Representation of each base as a letter of the alphabet.

Note: We cannot blindly assume that the first nitrogenous base in the chain is the first base in a codon triplet. If we did this, we would end up with an entirely different set of blueprints for a polypeptide chain:

AGG | CAG | AUU | UUU | CCG | GUU | GAA | CGA
Arg- Gln- Ile- Phe- Pro- Val- Glu- Arg

Very recently research has uncovered some facts which suggest that some features of the genetic code are incompletely understood. Sanger and his coworkers at Cambridge University in England have determined the complete sequence of 5375 bases in the DNA molecule of a simple virus, bacteriophage ØX174. They can match this sequence codon for codon for all those proteins which the virus can produce for which the amino acid sequence is known. There are some puzzling features in these results though. A few short segments of DNA chain appear to have no coding function at all. More surprising is the discovery that some segments of the DNA chain correspond to the coding for more than one protein at the same time. The situation is as though both methods of reading the nucleic acid chain in Example 20.1 were equally valid. How this can be the case for some portions of the DNA chain but not for others still remains a mystery.

The Double Helix

There is more to the structure of DNA than just the primary sequence of nitrogenous bases. Secondary structure also plays a crucial biochemical role. Each DNA molecule consists of *two* nucleotide chains wrapped around each other in a double helix and held together by hydrogen bonds. This hydrogen bonding involves only the nitrogenous bases. Each of the purine bases can hydrogen bond with *one and only one* of the pyrimidine bases. Thus adenine can hydrogen bond with thymine and guanine with cytosine, as shown in Fig. 20.16. Note that in both cases there is an exact match of hydrogen atoms on the one base with nitrogen or oxygen atoms on the other. Note also that the distance from sugar linkage to sugar linkage across each of the base pairs in Fig. 20.16 is almost exactly the same. This explains why only these two combinations occur in DNA. Other combinations (i.e., adenine-cytosine) are not nearly so favorable energetically.

The overall geometry of the two nucleotide chains in the DNA molecule is in the form of the double helix shown in Fig. 20.17. Each helix corresponds to a nucleotide chain, and the two chains are joined throughout their length by adenine-thymine or guanine-cytosine pairs. These base pairs are stacked one above the other with their planes perpendicular to the axis of the two spirals. This places the hydrophobic base pairs inside the structure and allows the hydrophilic sugar and phosphate groups to contact water on the exterior. The whole helix will just fit inside a cylinder 2200 pm in diam-

Figure 20.16 Hydrogen-bonded base pairs of DNA. Note the nearly equal separations between points where the bases connect to sugars in the DNA backbone. A pair of purines would have much larger and a pair of pyrimidines much smaller separation, making it difficult for such pairs to fit between the two strands.

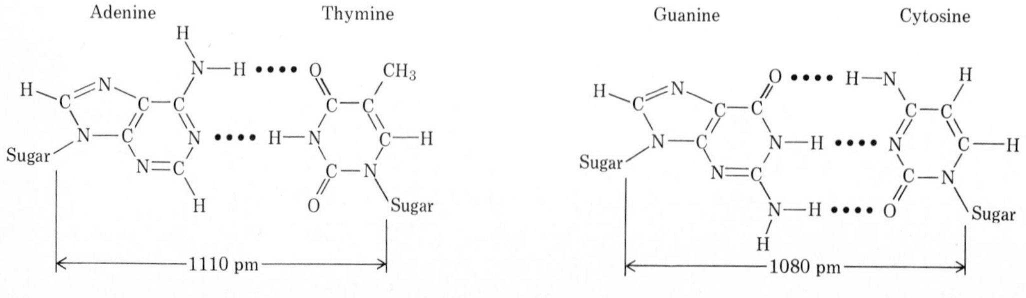

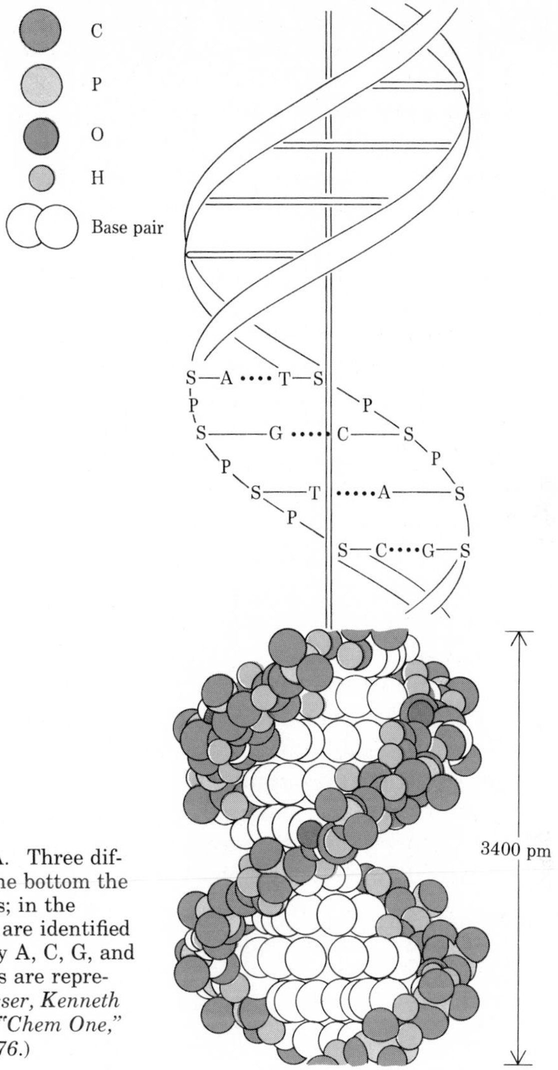

Figure 20.17 The double helix of DNA. Three different representations are shown. At the bottom the atoms are shown as space-filling spheres; in the middle the sugar and phosphate groups are identified by S and P and the nitrogenous bases by A, C, G, and T. In the topmost section the base pairs are represented by crossbars. (*Source: Jürg Wasser, Kenneth N. Trueblood, and Charles M. Knobler, "Chem One," p. 735, McGraw-Hill Book Company, 1976.*)

eter. The spacing between base pairs is 340 pm, and there are exactly 10 pairs in one full turn of the helix.

The two nucleotide chains in the double helix are said to be **complementary** to each other. Because of the exact pairing of the bases we can always tell the sequences of bases in the one chain from that in the other. Thus if the first six bases in one chain are AGATCC, we know that the first six bases in the other will be TCTAGG. Both chains are therefore alternative representations of the same information. If one or two bases become misplaced in either strand, this can be recognized because of mismatching with the complementary strand. Repair enzymes can then correct the sequence of bases along the incorrect strand.

This double-helix model for DNA was first suggested in 1953 by James

D. Watson (1928 to) and Francis Crick (1916 to). It was an important milestone in the history of science, since it marked the birth of a new field, molecular biology, in which the characteristics of living organisms could at last begin to be explained in terms of the structure of their molecules. In 1962 Crick and Watson shared the Nobel Prize with M. F. H. Wilkins, whose x-ray crystallographic data had helped them to formulate their model. A fascinating account of this discovery, which does not always put the author in a favorable light, can be found in Watson's book "The Double Helix."

The Double Helix

According to the *central dogma of molecular genetics,* DNA is the genetically active component of the chromosomes of a cell. That is, DNA in the cell nucleus contains all the information necessary to control synthesis of the proteins, enzymes, and other molecules which are needed as that cell grows, carries on metabolism, and eventually reproduces. It has been known for some time that when a cell divides, its chromosomes also divide. If DNA is to provide information to both daughter cells, it too must somehow be able to divide into duplicate copies. This process is called **replication.**

Given the complementary double strands of DNA, it is relatively easy to imagine ways in which replication can occur. The currently accepted model involves continuous unwinding of the double helix, separating one base pair after another. At the same time new nucleotides condense together to form strands complementary to each of the separated old ones. This is shown schematically at the bottom of Fig. 20.18. This process is controlled by a remarkable enzyme, DNA polymerase, which can add about 100 nucleotides per minute to form a second strand complementary to an existing one.

A number of advantages of the double-stranded structure held together by hydrogen bonds is evident in the process of replication. Complementary base pairing insures that the two new DNA molecules will be the same as the original. The large number of hydrogen bonds, each of which is relatively weak, makes complete separation of the two strands unlikely, but one hydrogen bond, or even a few, can be broken rather easily. The proper enzyme can therefore separate the two strands in much the same way that a zipper operates. Like the teeth of a zipper, hydrogen bonds provide great strength when all work together, but the proper tool can separate them one at a time.

Although DNA contains the necessary instructions for synthesizing all the proteins necessary for the functioning of a cell, it does not take part directly in the synthesis itself. Sections of the DNA chain are first copied into a type of RNA called **messenger RNA,** mRNA. This process is known as **transcription.** The mRNA molecules differ from DNA in that the base uracil (U) replaces thymine (T). They are also considerably smaller than DNA, containing the blueprints for only a few proteins at most. As the

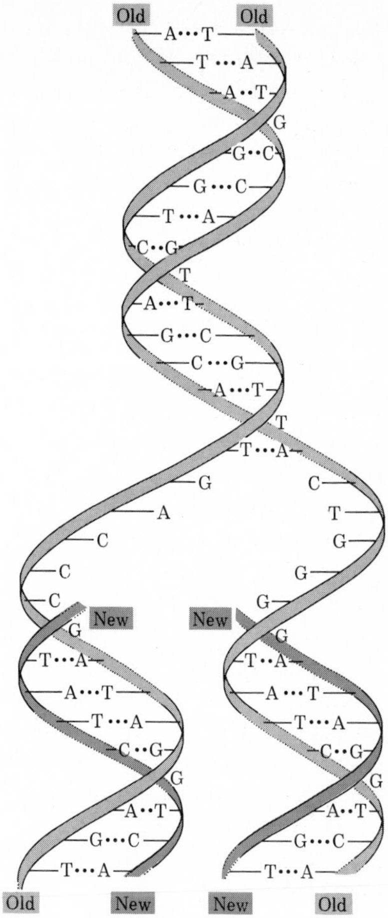

Figure 20.18 The replication of DNA. Replication occurs by means of partial unwinding of the two strands accompanied by synthesis of a new strand complementary to each of the originals.

name implies, mRNA molecules are used to transport their coded instructions from the nucleus of the cell, where the DNA is situated, to the **ribosomes,** where the process of protein synthesis actually takes place.

When an mRNA molecule reaches a ribosome, a process called **translation** takes place in which still smaller segments of RNA called **transfer RNA,** tRNA, are produced. Each tRNA molecule corresponds to one codon in the genetic code and hence to the production of one amino acid. It is these tRNA molecules, lined up in the correct sequence on the mRNA chain, which finally link amino acids together in polypeptide chain.

Mutation and Molecular Disease

If the central dogma of molecular genetics is correct and the DNA in the cell nucleus contains all the information necessary for proper cell functioning, it follows that any errors in replication, transcription, or translation will result in some malfunction of the cell. Errors during transcription and translation will usually affect only one or a few protein molecules. Consequently they are not as serious as errors during replication or direct modification of DNA's sequence of bases. The latter will be transmitted from the damaged cell to the progeny by subsequent replications and thus are longer lasting. In multicellular organisms this is most deleterious in the sex or germinal cells which combine to produce a new organism.

A change in the base sequence of DNA is called a **mutation.** Any change it produces in an organism's macroscopic characteristics will be tested by the process of natural selection. Occasionally such a change is favorable: The organism will be better adapted to its environment, and the mutation will become a legacy to subsequent generations. More often a mutation results in a detrimental characteristic, and its host organism dies. Nevertheless, without some small rate of mutation the characteristics of human and other populations would remain fixed. With no means of adaptation to changed environmental conditions such populations would eventually die off.

Knowledge of the function of DNA in the flow of genetic information provides some clues as to what substances or conditions may contribute to the occurrence of mutations. For example, nitrous acid can react directly with DNA, converting cytosine to uracil:

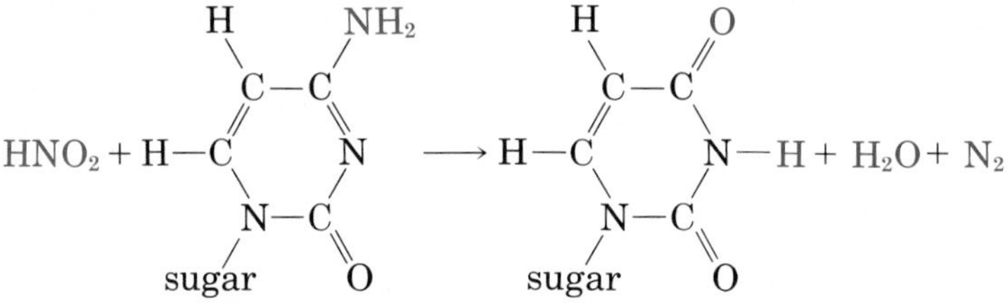

Cytosine **Uracil**

This obviously alters the message of the genetic code. Aromatic hydrocarbons such as β-naphthylamine

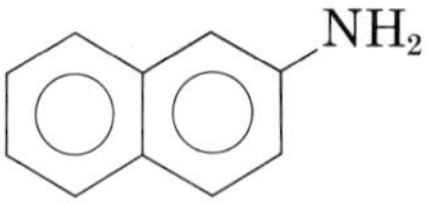

β- Naphthylamine

have flat structures which can squeeze between adjacent base pairs in DNA. This causes the double helix to unwind slightly, and may cause errors during replication or transcription.

Some molecules, as well as high-energy radiation, can dissociate the covalent bonds holding the DNA backbone together, breaking both strands of the double helix. Such breaks can be repaired by appropriate enzymes, but sometimes the strands are put back together the wrong way or an entire section of the molecule can be lost. Since too high a level of mutation will result in numerous deaths of maladapted individuals, dispersal into the environment of agents such as those described above should be avoided.

In some cases what was once a useful mutation may become detrimental and even cause what is known as molecular disease. A good example of this is sickle-cell anemia, a disease in which the red blood cells assume a crescent shape and tend to precipitate to form clots when deoxygenated in the small blood vessels. This impairs circulation and results in premature death of those afflicted. It has been determined that the sickle cells are the result of a mutation in the gene responsible for synthesis of hemoglobin. The change is quite small—only one position out of a chain of 146 amino acids is affected. However, that one mistake involves substitution of glutamic acid (negative, hydrophilic) for valine (hydrophobic). The negative charge on the glutamic acid side chain is enough to disrupt the tertiary and quaternary structure of hemoglobin and to impair its oxygen-carrying function. Thus the cause of the disease is a very small error in the molecular structure of DNA and hence protein.

The sickle cell trait is apparently the result of a chance mutation many thousands of years ago. Since that time it has been transferred from one generation to another by DNA replication. One might have expected this disease to be eliminated by natural selection, but it turns out that the abnormal blood cells are also resistant to a malarial parasite. Even their impaired oxygen-carrying ability was preferable to an invasion by malaria. Consequently the gene which produces sickle cells was somewhat favorable in tropical regions where malaria abounded. Sickle-cell anemia is especially prevalent among black persons, most of whose ancestors lived in the tropics until a few generations ago. Since human medicine can now control malaria, the presence of the DNA base sequence which produces abnormal hemoglobin is no longer of value.

SUMMARY

More than 99 percent of the atoms in most living organisms are H, O, N, or C. These are the smallest atoms which can form one, two, three, and four covalent bonds, respectively, and they are especially suited to make up the more than 10^{11} different, but often related, kinds of molecules estimated to exist in the biosphere. Each such molecule has a specific function to perform in a specific organism, and its molecular structure is very important in determining how that function is carried out.

Many biological molecules are formed by condensation polymerization from small building-block molecules. Reversal of such condensations (hydrolysis) breaks large molecules down into the building blocks again, allowing them to be used by another organism. Examples of condensation reactions are the formation of lipids from fatty acids and glycerol, the formation of cellulose and starch from glucose, the formation of proteins from amino acids, and the formation of nucleic acids from ribose or deoxyribose, phosphate, and nitrogenous bases.

Lipids may be divided into two categories: nonpolar (hydrophobic) and polar (hydrophilic). Both types consist of long hydrocarbon chains, but polar lipids have electrically charged or hydrogen-bonding groups at one end. Lipid bilayers, in which the hydrophilic ends of polar lipids contact an aqueous phase while the hydrophobic tails intertwine, are important components of cell walls and other membranes. Nonpolar substances often dissolve in the hydrophobic portions of lipid tissue and may be concentrated along ecological food chains.

Carbohydrates provide a storehouse for solar energy absorbed during photosynthesis. Simple sugars usually contain five or six C atoms in a ring and a large number of OH groups. Polysaccharides, such as cellulose and starch, are condensation polymers of simple sugars. Their structures and chemical reactivities are very dependent on the exact structure of the simple sugar from which they are made.

Fibrous proteins, in which polypeptide chains are arranged parallel to one another, are fundamental components of structural tissues such as tendons, hair, etc. Globular proteins, on the other hand, have compact, nearly spherical structures in which the polypeptide chain folds back on itself. Enzymes, antibodies, hormones, and hemoglobin are examples of globular proteins. Proteins are made by polymerization of 20 different amino acids. The order of amino acid side chains along the polymer backbone constitutes primary protein structure. Secondary structure involves hydrogen bonding to form α-helix or pleated-sheet structures. The intricate folding of the polypeptide chain in a globular protein is referred to as tertiary structure. Some proteins (hemoglobin, for example) have quaternary structure—several polypeptide chains are nested together.

Nucleic acids, such as DNA and RNA, constitute a blueprint and a mechanism for synthesizing useful proteins. Codons, each consisting of a sequence of three nitrogenous bases along a nucleic acid polymer chain, indicate which amino acid goes where. When cells divide, the double-stranded DNA molecule can also divide, and complementary base pairing insures that each new cell will contain identical DNA. During protein synthesis, information is transcribed from DNA to mRNA and then translated from mRNA to tRNA. The tRNA molecules contain sites appropriate to hold specific amino acids in position while the latter are linked into the polypeptide chain of a protein. In this way the primary protein structure is determined, and secondary and higher order structures then follow directly.

QUESTIONS AND PROBLEMS

20.1 In your own words distinguish between organic and inorganic chemistry.

20.2 The human skeleton is composed primarily of hydroxyapatite, $Ca_{10}(PO_4)_6(OH)_2$. Using data from Table 20.1, calculate what fraction of the total mass of the average human body is bone. If possible, check your result by looking up the total mass and the skeletal mass of the average human in the library.

20.3 List three characteristics of the elements H, O, N, and C which appear to be important in explaining their high abundance in living organisms.

20.4 Why does carbon, of all the elements, form such a large number of different but often closely related compounds?

20.5 Compare the relative strengths of

a A C—C bond and an Si—Si bond.
b A C—H bond and an Si—H bond.
c A C—O bond and an Si—O bond.
d A C—C bond and a C—O bond.
e An Si—Si bond and an Si—O bond.

20.6 Which would you expect to react more rapidly with water, CCl_4 or $SiCl_4$? Give reasons for your choice.

20.7 The C—H bond enthalpy is the second largest of all single bonds involving C (see Table 15.4). Why is this important in both organic and biochemistry?

20.8 Name at least one consequence of hydrogen bonding for each situation listed below:

a Properties of ice
b Properties of liquid water
c Structure of DNA
d Structure of proteins

20.9 List three important biochemical functions of water.

20.10 Criticize the following statement: The tremendously large number of different protein structures requires that there be a similarly large number of different amino acids to make those structures.

20.11 Define in your own words the terms *condensation* and *hydrolysis*. Why is each type of reaction important in biochemistry?

20.12 Draw the structural formulas for each of the following reactants as well as for the triglyceride (fat) product:

$$3C_{15}H_{31}COOH + HOCH_2CHOHCH_2OH \rightarrow$$

20.13 Predict the products arising from the hydrolysis of the compound glycerol trioleate, which is shown below:

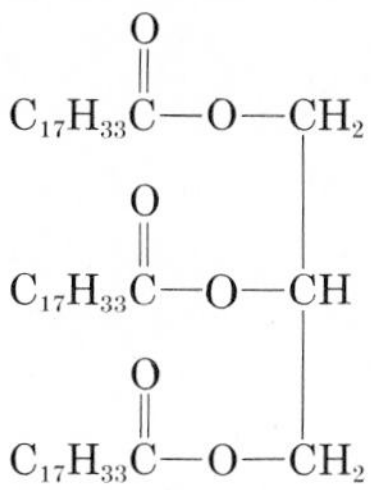

20.14 Complete each of the following reactions and include structural formulas for all the reactants and products:

a $C_9H_{19}COOH + HOCH_2CHOHCH_2OH \rightarrow$

b
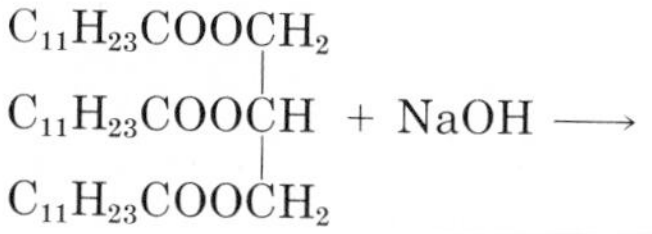

c
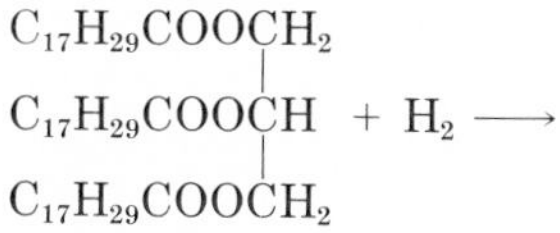

20.15 Explain in your own words why polymeric substances are so important in biochemistry.

20.16 Criticize the following statement: It is impossible for a single molecule to be both hydrophobic and hydrophilic because the two words describe opposite behaviors. Give one example of each of two different types of compounds which contradicts the statement.

20.17 How do soaps differ from detergents? Give one example of the chemical structure of a soap and one of a detergent.

20.18 Complete and balance the following equation and comment on its significance:

$$C_{17}H_{35}COOH(aq) + Ca^{2+}(aq) \rightarrow$$

20.19 Define, describe, or explain the meaning of each of the following terms:
a Bioamplification
b LAS detergent
c Saponification
d Micelle
e Carbohydrate

20.20 Distinguish between each of the following pairs of terms which relate to sugars:
a Monosaccharides and disaccharides
b Aldoses and ketoses
c Aldohexose and aldopentose
d Linear (chain) form and cyclic structure

20.21 Condensation of two molecules of α-glucose gives the disaccharide α-maltose. Using Fig. 20.7 as a guide, draw the structure of α-maltose.

20.22 Criticize the following statement: All polysaccharides hydrolyze rapidly and easily to monosaccharides. If possible, give an example of a compound which contradicts this statement.

20.23 Distinguish between
a Fibrous and globular proteins
b Conjugated and simple proteins
c Prosthetic groups and side chains
d Primary and secondary protein structure
e Tertiary and quaternary protein structure

20.24 Draw the zwitterion form of each of the following amino acids:
a Valine

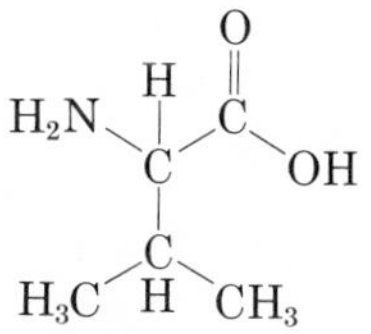

b Leucine

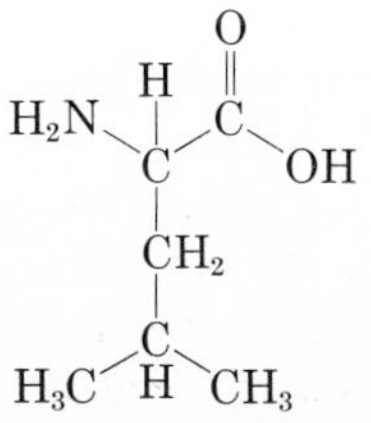

c Threonine

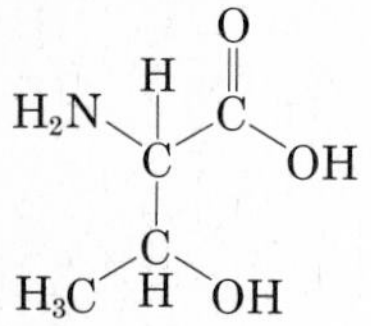

20.25 Suppose you had at least three molecules of each amino acid whose structure is given in Problem 20.24. How many different tripeptides might be formed by condensation of different combinations of the three monomers? Draw the structure of one tripeptide which contains all three different monomers.

20.26 Explain in detail why (*a*) a peptide bond is planar; (*b*) hydrogen bonds formed at a peptide bond are unusually strong.

20.27 Some amino acid side chains are usually found deep within the folds of a globular protein, while others are usually found at the outer surface. (*a*) Give two examples of side chains which would be expected to lie deep within a protein structure. (*b*) Give two examples of side chains which would be expected to be found at the surface. (*c*) Give two examples of side chains which might be found in either type of location. (*d*) Explain your choice of examples in parts *a*, *b*, and *c*.

20.28 Draw the complete structural representation of each of the following amino acid sequences:

a Pro-Ser-Ala
b Tyr-Leu-Met
c Val-Asn-Thr
d Glu-Gly-Trp

20.29 Using data from Table 17.1, calculate $\Delta G_m^\ominus$ for the reaction

$$O_2(g) + 4H^+(aq) + 4Fe^{2+}(aq) \rightarrow 4Fe^{3+}(aq) + 2H_2O(l)$$

(*a*) Is this reaction spontaneous? (*b*) When $O_2(g)$ bonds to Fe^{2+} in the heme group of hemoglobin or myoglobin, no oxidation to Fe^{3+} occurs. Why?

20.30 What is an enzyme? Describe in as much detail as you can how a typical enzyme works.

20.31 Draw the structures of the following nucleosides:

a Ribose-guanine
b Deoxyribose-uracil
c Deoxyribose-thymine
d Ribose-cytosine
e Deoxyribose-adenine

20.32 Draw the structures of the five nucleotides which result from the condensation of the preceding five nucleosides (see Problem 20.31*a* to *e*) with H_3PO_4.

20.33 The genetic code is said to be degenerate. Explain in your own words what this means.

20.34 Decode each of the following RNA fragments:

a Head-GUAGUUGUUGAAAU-tail
b Head-CCUAGGGGGUUGCACUAGCGG-tail
c Head-GAGAGUCUCAUAUGUGG-tail

20.35 Can any of the RNA fragments in Problem 20.34 be read in more than one way? If so, indicate other ways in which they could be read.

20.36 Criticize the following statement: The double helix consists of two nucleosides held together by double bonds.

20.37 For each sequence of nitrogenous bases along a single DNA chain listed below, draw the complementary bases which would be found along the second strand.

a ATTGG
b CCTGA
c ATGCA
d TCGGA

20.38 Distinguish clearly the structures and functions of DNA, mRNA, and tRNA.

20.39 Would a disease such as sickle-cell anemia be likely to result from each of the substitutions listed below? Explain why or why not. (*a*) Replacement of valine by lysine in the amino acid chain of a protein. (*b*) Replacement of glycine by alanine in the amino acid chain of a protein. (*c*) A mistake in DNA which produces mRNA having the sequence CUU instead of AUU.

***20.40** Recently there has been considerable controversy regarding *recombinant DNA* research. (See, for example, *Chemical and Engineering News*, May 30, 1977.) This type of research involves splicing a segment of DNA from one organism, say a virus, onto DNA from another. The latter organism is often *E. coli*, a bacterium found in human intestines. Without reading further, try to suggest some benefits which recombinant DNA research might provide. Also try to predict some types of problems which might be created if this type of research is pursued. Then go to the library and read several articles on recombinant DNA to see what benefits and costs have been predicted by prominent scientists.

chapter twenty-one

SPECTRA AND STRUCTURE OF ATOMS AND MOLECULES

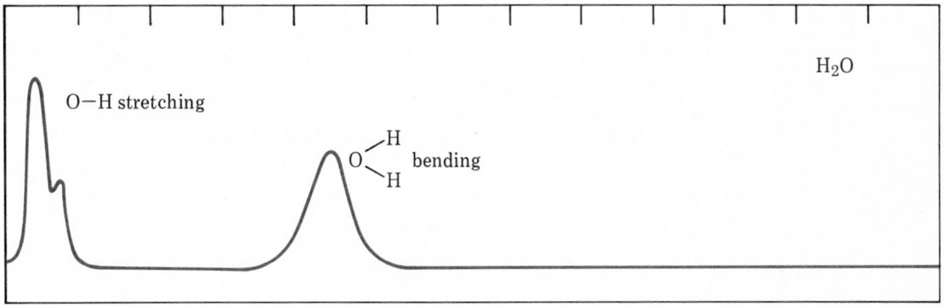

In this chapter we are going to study the way in which matter can both absorb energy and emit it in the form of electromagnetic radiation such as light. The pattern in which matter absorbs or emits radiation is called its **spectrum.** In the past, and still to this day, studies of the spectrum of a substance have furnished important clues to the structure of matter. At the same time, the spectrum of a substance is often a very useful way of characterizing and hence identifying and analyzing that substance.

21.1 THE NATURE OF ELECTROMAGNETIC RADIATION

Visible light, gamma rays, x-rays, ultraviolet ("black") light, infrared radiation, microwaves, and radio waves are all related in that many of their properties can be explained by a wave theory. They may be thought of as periodically varying electric and magnetic fields (electromagnetic waves). Figure 21.1 indicates the relationship of these fluctuating electric and magnetic fields. It also illustrates the maximum **amplitude** A_0 and the **wavelength** λ associated with the wave. The intensity of a wave is associated with the square of its amplitude.

Electromagnetic waves travel through a vacuum at the speed of light, $c = 2.9979 \times 10^8\ \mathrm{m\ s^{-1}}$. The entire wave shown in Fig. 21.1 may be thought of as moving from left to right. Thus at position P, where the electric field

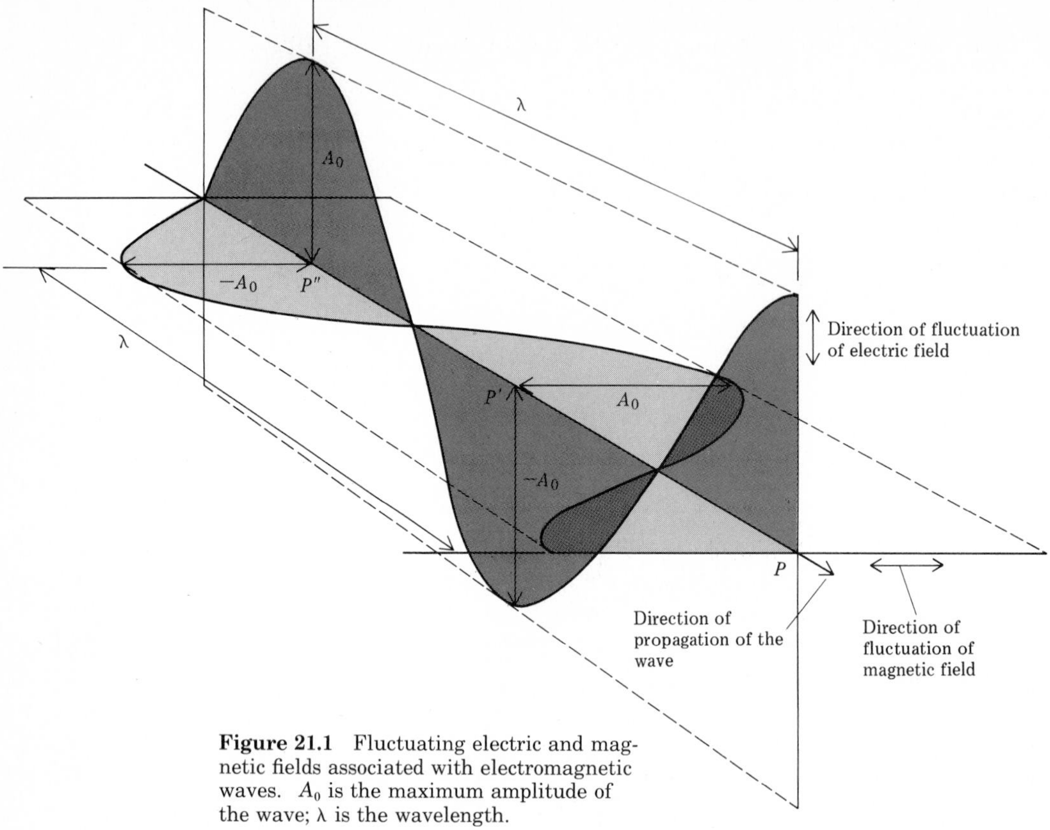

Figure 21.1 Fluctuating electric and magnetic fields associated with electromagnetic waves. A_0 is the maximum amplitude of the wave; λ is the wavelength.

had maximum amplitude at the instant the figure was drawn there is a progressive decrease in amplitude with time. The amplitude reaches its smallest (most negative) value when the wave has moved a distance equal to that separating points P' and P. Eventually the amplitude increases to its maximum value again, corresponding to movement of a distance P'' to P, or one wavelength λ. A moving wave can be characterized by the **frequency** ν at which points of maximum amplitude pass a fixed position. The speed of the wave (distance traveled per unit time) must be the product of the wavelength (distance between maxima) and the frequency (number of maxima passing per unit time):

$$c = \lambda\nu \tag{21.1}$$

Since the speed of electromagnetic radiation in a vacuum is always the same, radiation may be characterized by specifying either λ or ν. The other quantity can always be calculated from Eq. (21.1).

EXAMPLE 21.1 Specify the frequency, wavelength, and speed in a vacuum of each of the types of electromagnetic radiation listed below:

a) Blue-green light; $\lambda = 500$ nm.

b) Heat rays emitted from hot asphalt pavement; $\nu = 1.5 \times 10^{14}\ s^{-1}$.

c) A gamma ray emitted from $^{131}_{53}I$; $\lambda = 3.402$ pm.

d) An FM radio transmission; $\nu = 91.5$ MHz.

Solution In each case we make use of the relationship $c = \lambda\nu = 2.998 \times 10^8\ m\ s^{-1}$.

a) $$\nu = \frac{c}{\lambda} = \frac{2.998 \times 10^8\ \text{m s}^{-1}}{500 \times 10^{-9}\ \text{m}} = 6.00 \times 10^{14}\ \text{s}^{-1}$$

b) $$\lambda = \frac{c}{\nu} = \frac{2.998 \times 10^8\ \text{m s}^{-1}}{1.5 \times 10^{14}\ \text{s}^{-1}} = 2.0 \times 10^{-6}\ \text{m} = 2.0\ \mu\text{m}$$

c) $$\nu = \frac{c}{\lambda} = \frac{2.998 \times 10^8\ \text{m s}^{-1}}{3.402 \times 10^{-12}\ \text{m}} = 8.812 \times 10^{19}\ \text{s}^{-1}$$

d) The unit hertz Hz is 1 s^{-1}, therefore

$$\lambda = \frac{2.998 \times 10^8\ \text{m s}^{-1}}{91.5 \times 10^6\ \text{Hz}} \times \frac{1\ \text{Hz}}{1\ \text{s}^{-1}} = 3.28\ \text{m}$$

The results obtained in Example 21.1 indicate that the frequency and wavelength of electromagnetic radiation can vary over a wide range.

The experiments which did most to convince scientists that light could be described by a wave model are concerned with **interference.** In 1802 Thomas Young (1773 to 1829), an English physicist, allowed light of a

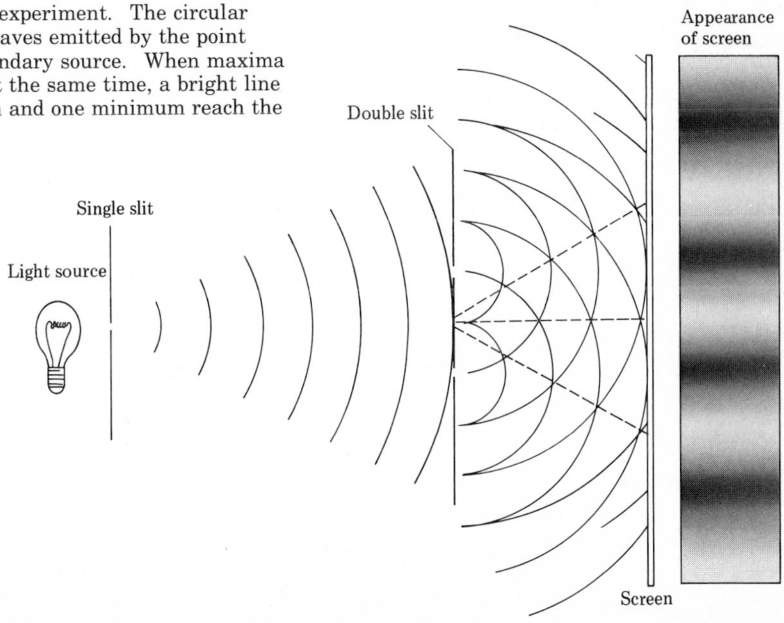

Figure 21.2 Young's double-slit experiment. The circular lines represent maxima in light waves emitted by the point source. Each slit serves as a secondary source. When maxima from both slits reach the screen at the same time, a bright line is produced. When one maximum and one minimum reach the same point, a dark line appears.

single wavelength to pass through a pair of parallel slits very close to each other and then onto a screen. Young observed the interference pattern of alternating dark and bright strips, shown in Fig. 21.2. Instead of two strips of light, three appeared on the screen, the most prominent being in the center.

The appearance of these bright and dark strips on the screen is easy to explain if light is regarded as a wave. The bright areas are the result of **constructive interference,** while the dark ones result from **destructive interference.** Constructive interference occurs when the crests of two waves reach the same point at the same time. The amplitudes of the two waves add together, giving a resultant larger than either. In the case of destructive interference a maximum in one wave and a minimum in the other reach the same point at the same time. Thus one cancels the effect of the other, and the resultant wave is smaller. This is illustrated in Fig. 21.3. When destructive interference occurs between two waves which have the *same* amplitude, the resultant wave has zero amplitude (and zero intensity). Hence the dark strips observed in the double-slit experiment.

Although the behavior of light and other forms of electromagnetic radiation can usually be interpreted in terms of wave motion, this is not always so. When radiation is absorbed or emitted by matter, it is usually more convenient to regard it as a stream of particles called **photons.** Thus electromagnetic radiation has the same kind of wave-particle duality we encoun-

Figure 21.3 Interference of waves: (*a*) constructive interference; (*b*) destructive interference.

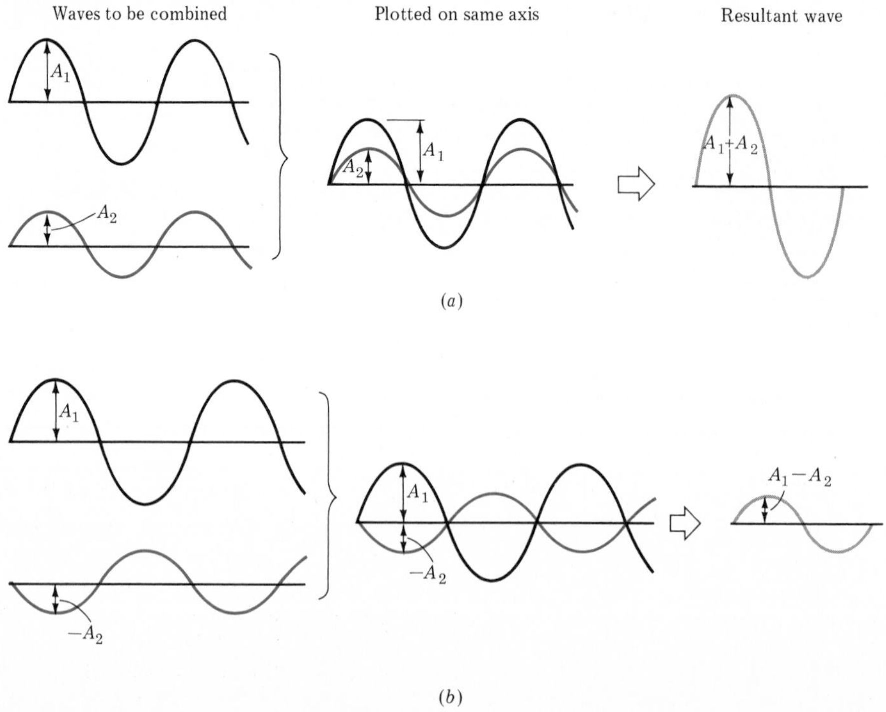

tered in the case of the electron. However, photons have some properties which are very different from those of electrons and other particles. Although photons have mass and energy, and although we can count them, they can travel *only at the speed of light*. We cannot slow down a photon or stop it without changing it into something else.

The wave-particle duality of photons and electromagnetic radiation is enshrined in an equation first proposed by the German physicist Max Planck (1858 to 1947). The energy of a photon E and the frequency of the electromagnetic radiation associated with it are related in the following way:

$$E = h\nu \tag{21.2}$$

where h is a universal constant of nature called **Planck's constant** with the value 6.6262×10^{-34} J s. The application of Eq. (21.2) is best shown by an example.

EXAMPLE 21.2 Calculate the energy of photons associated with each kind of electromagnetic radiation mentioned in Example 21.1. Compare each result with the mean bond enthalpy for a C—C single bond (348 kJ mol^{-1}).

Solution In each case use the formula $E = h\nu$. If wavelengths are given instead of frequencies, the formula $E = hc/\lambda$ may be obtained by combining Eqs. (21.1) and (21.2).

a) $$E = \frac{hc}{\lambda} = \frac{6.626 \times 10^{-34}\ \text{J}\,\cancel{\text{s}} \times 2.998 \times 10^{8}\ \cancel{\text{m}}\,\cancel{\text{s}^{-1}}}{500 \times 10^{-9}\ \cancel{\text{m}}}$$
$$= 3.97 \times 10^{-19}\ \text{J} = 0.397\ \text{aJ}$$

b) $$E = h\nu = 6.626 \times 10^{-34}\ \text{J}\,\cancel{\text{s}} \times 1.5 \times 10^{14}\ \cancel{\text{s}^{-1}}$$
$$= 9.9 \times 10^{-20}\ \text{J} = 0.099\ \text{aJ}$$

c) $$E = \frac{6.626 \times 10^{-34}\ \text{J}\,\cancel{\text{s}} \times 2.998 \times 10^{8}\ \cancel{\text{m}}\,\cancel{\text{s}^{-1}}}{3.402 \times 10^{-12}\ \cancel{\text{m}}}$$
$$= 5.839 \times 10^{-14}\ \text{J} = 58\,390\ \text{aJ}$$

d) $$E = 6.626 \times 10^{-34}\ \text{J}\,\cancel{\text{s}} \times 91.5 \times 10^{6}\ \cancel{\text{s}^{-1}}$$
$$= 6.06 \times 10^{-26}\ \text{J} = 6.06 \times 10^{-8}\ \text{aJ}$$

Since the bond enthalpy quoted refers to 1 mol C—C bonds, we must divide by the Avogadro constant to obtain a quantity which is appropriate to compare with the energy of a single quantum of radiation:

$$\text{Enthalpy to dissociate one C—C bond} = \frac{348\ \text{kJ}\ \cancel{\text{mol}^{-1}}}{6.022 \times 10^{23}\ \cancel{\text{mol}^{-1}}}$$
$$= 5.78 \times 10^{-19}\ \text{J} = 0.578\ \text{aJ}$$

Clearly the energies of visible photons are comparable with the energies of chemical bonds. Infrared and radio waves have far less energy per photon. Gamma-ray photons have enough energy to break open about 100 000 chemical bonds. As a consequence chemical changes often occur when gamma rays or other high-energy photons are absorbed by matter. Such changes are usually detrimental to living systems, and materials such as lead are used to shield humans from sources of high-energy radiation.

The entire spectrum of electromagnetic radiation may be characterized in terms of wavelength, frequency, or energy per photon, as shown in Fig. 21.4. The example calculations we have done so far and the figure both indicate the broad range covered by λ, ν, and E. Electromagnetic radiation which can be detected by the human retina is but a small slice out of the total available spectrum.

Figure 21.4 The electromagnetic spectrum. The visible region has been expanded in the lower part of the figure so that the wavelengths corresponding to different colors can be distinguished.

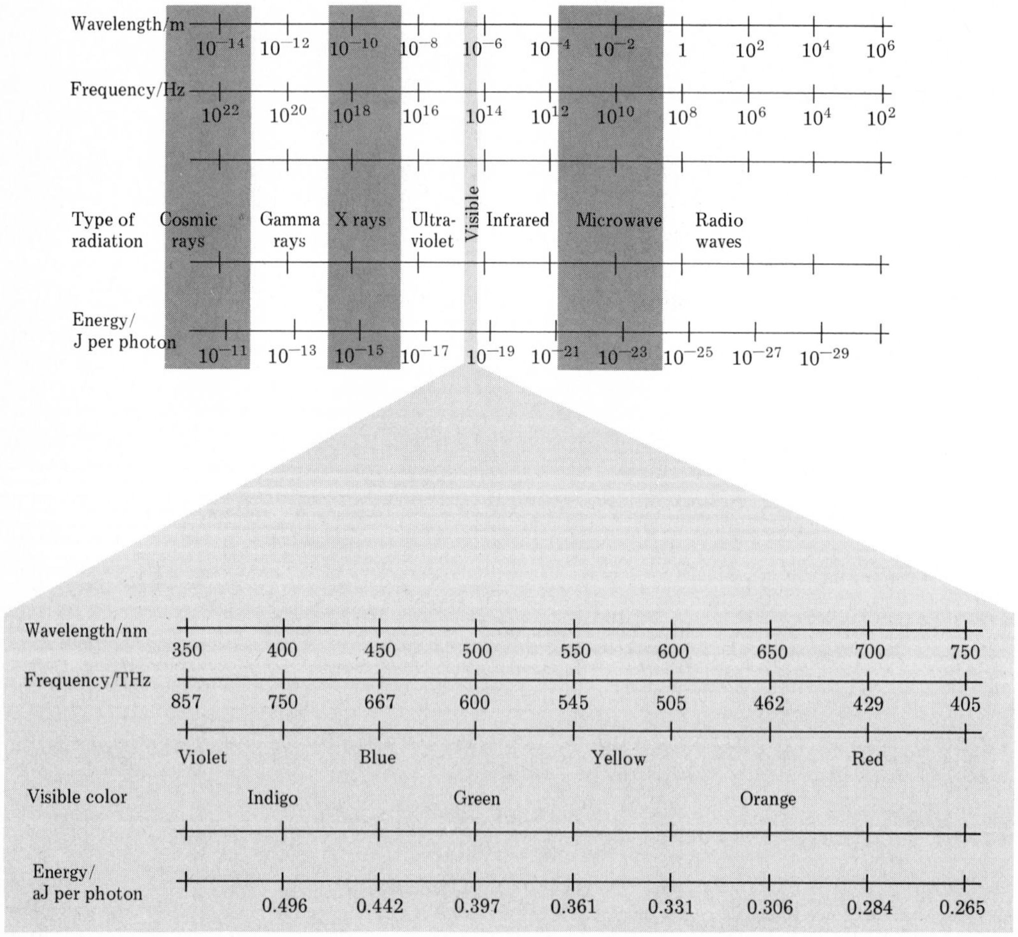

When any element is heated to a sufficiently high temperature, all bonds between atoms are dissociated and the element turns into a monatomic gas. At these or still higher temperatures the single atoms begin to radiate visible and ultraviolet light. The actual pattern of light emitted varies from element to element and is called its **emission spectrum.** Such a spectrum is commonly obtained by passing an electric arc through a powdered sample of the solid element or by applying a voltage to a **discharge tube** containing a gas at low pressure. Neon signs are everyday examples of discharge tubes. Neon produces a bright red glow in the tube, and other gases produce different colors.

If the emission spectrum of an element is passed through a prism and allowed to strike a photographic film, the intensity of emission as a function of wavelength can be measured. The equipment for making such measurements is called a **spectroscope** and is shown schematically in Fig. 21.5. Typical emission spectra obtained in this way are shown in Fig. 21.6 where the intensity is plotted against wavelength. Note from this figure how the light emitted is confined to a few very specific wavelengths. At all other wavelengths there is no emission at all. Spectra of this kind are usually referred to as **line spectra.** The wavelengths of the lines in a line spectrum are unique to each element and are often used, especially in metallurgy, both to identify an element and to measure the amount present. These wavelengths can often be measured to an accuracy as great as one part in a billion (1 in 10^9). Because the spectrum of an element is readily reproducible and can be measured so accurately, it is often used to determine

Figure 21.5 Schematic diagram of a spectroscope. Light emitted by an element in a discharge tube or other source is passed through slits to produce a beam. When the beam strikes a prism, different wavelengths are bent by different amounts, allowing them to strike the film at different locations. The darkness of the exposed film depends on the intensity of the light striking it.

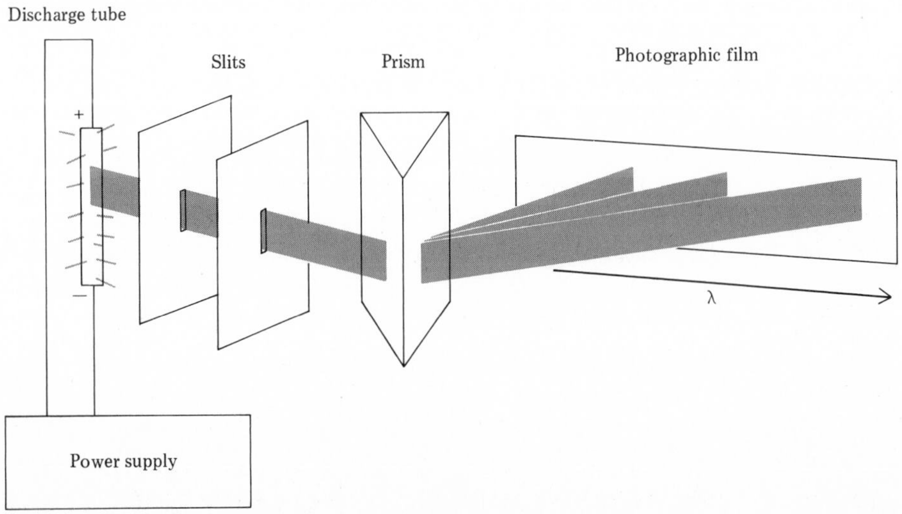

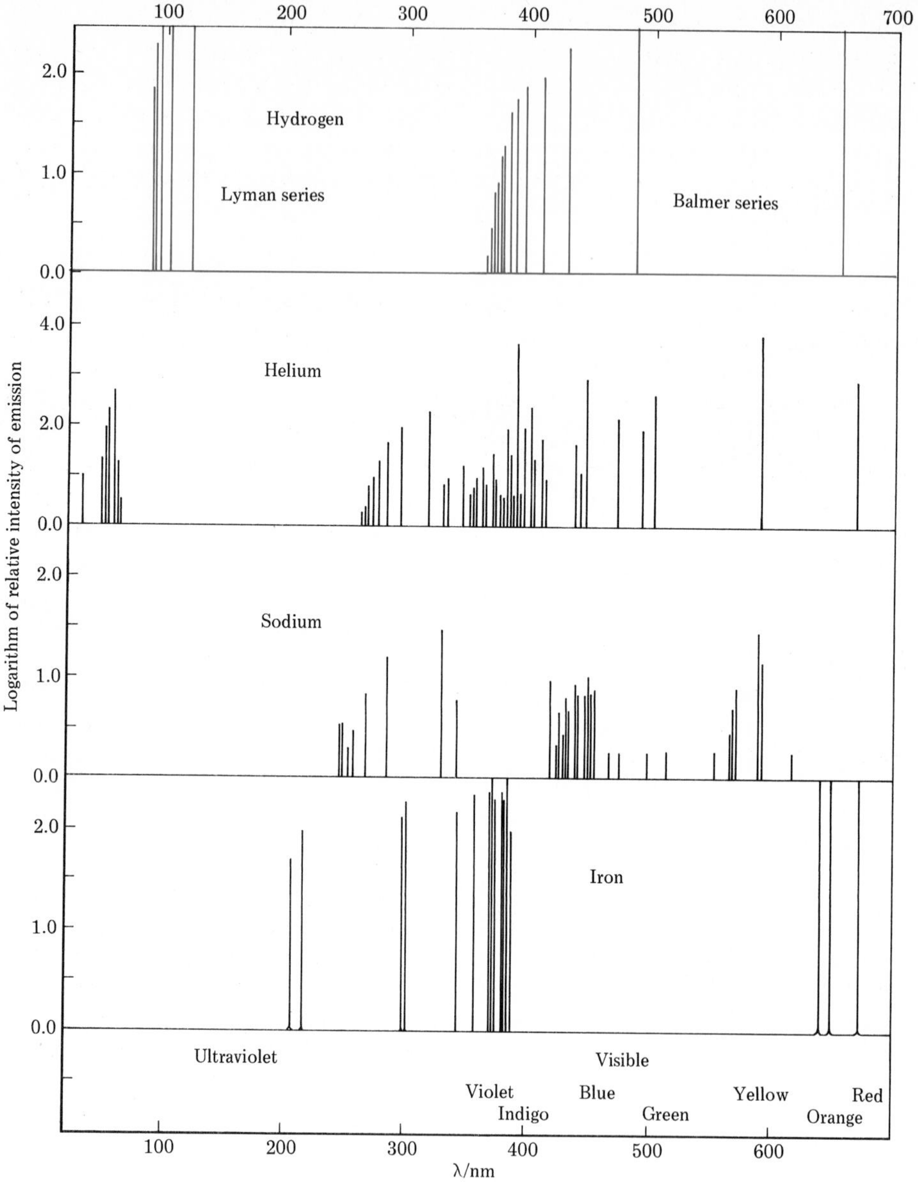

Figure 21.6 The ultraviolet and visible emission spectra of gaseous hydrogen, helium, sodium, and iron. For comparison the colors of visible light are given at the bottom of the figure. All spectral lines indicated in color can be derived from energy levels given in Fig. 21.8.

lengths. For example, the meter was once defined as the distance between two marks on a platinum bar kept at Sevrés in France, but now it is taken as 1 650 763.73 wavelengths of a particular line in the spectrum of $_{86}$Kr.

At first glance the wavelengths of the lines in an emission spectrum, like those shown in Fig. 21.6, appear to have no pattern to them, but upon closer study regularities can be distinguished. For example, you can see that lines in the hydrogen spectrum occur very close together in the region just above 365 nm and then are spaced farther and farther apart as λ increases. In 1885 a Swiss high-school mathematics teacher, J. J. Balmer (1825 to 1898), discovered a formula which accounted for this regularity:

$$\lambda = 364.6 \text{ nm} \frac{n^2}{n^2 - 4} \qquad n = 3, 4, 5, \ldots$$

When the number n is any positive integer greater than 2, the formula predicts a line in the hydrogen spectrum. For $n = 3$, the Balmer formula gives $\lambda = 656.3$ nm; when $n = 4$, $\lambda = 486.1$ nm; and so on. All spectral lines predicted by the Balmer formula are said to belong to the **Balmer series.**

Other similar series of lines are found in the ultraviolet (Lyman series) and infrared (Paschen, Brackett, Pfund series) regions of the spectrum. In each case the wavelengths of the lines can be predicted by an equation similar in form to Balmer's. These equations were combined by Rydberg into the general form

$$\frac{1}{\lambda} = R_\infty \left(\frac{1}{n_1^2} - \frac{1}{n_2^2} \right) \qquad n_2 > n_1 \tag{21.3}$$

where R_∞, called the **Rydberg constant,** has the value $1.097\,094 \times 10^7 \text{ m}^{-1}$. Thus a single equation containing only one constant and two integer parameters is able to predict virtually all the lines in the spectrum of atomic hydrogen.

Bohr Theory of the Atom

In a classic paper published in 1913, the young Niels Bohr, then working with Rutherford in Manchester, England, proceeded to show how Rydberg's formula could be explained in terms of a very simple model of the hydrogen atom. The model was based on the nuclear view of atomic structure which had just been proposed by Rutherford. Bohr's model is shown in Fig. 21.7. An electron of charge $-e$ and mass m moves around a heavy nucleus

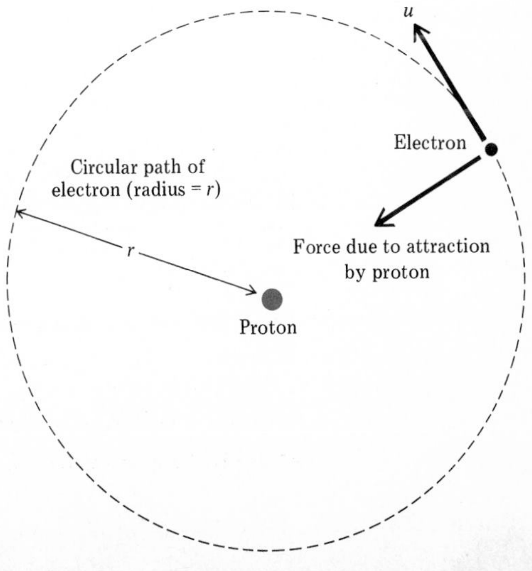

Figure 21.7 Bohr theory of the atom. The force of attraction keeps bending the path of the electron toward the nucleus and away from the straight-line motion. The net result is a circular path.

of charge $+e$. Ordinarily the electron would move in a straight line, but the attraction of the nucleus bends its path so that it moves with a constant velocity u in a perfect circle of radius r around the nucleus. The situation and the mathematics are very similar to that of a planet arcing round the sun. The major difference is that instead of the force of gravity there is an electrostatic force of attraction F between the proton and the electron described by Coulomb's law:

$$F = k\frac{e^2}{r^2} \tag{21.4}$$

where k has the value 8.9876×10^9 J m C^{-2}.

Expressions for both the kinetic and potential energies of the electron can be derived using Eq. (21.4) and the principles of elementary physics. Such a derivation can be found in most introductory physics texts. The two expressions are

$$E_k = \frac{1}{2}mu^2 = \frac{1}{2}k\frac{e^2}{r} \tag{21.5}$$

and

$$E_p = -k\frac{e^2}{r} \tag{21.6}$$

If these are added together, we obtain a simple formula for the total energy of the electron:

$$E = E_k + E_p = -\frac{1}{2}k\frac{e^2}{r} \tag{21.7}$$

If we now insert the known values of e and k, we have the result

$$E = -\frac{1}{2} \times \frac{8.9876 \times 10^9 \text{ J m C}^{-2} \times (1.6022 \times 10^{-19}\text{ C})^2}{r}$$

$$= -\frac{1.1536 \times 10^{-28}\text{ J m}}{r} \tag{21.8}$$

From Eq. (21.8) we see that the total energy of the electron is very negative for an orbit with a small radius but increases as the orbit gets larger.

In addition to suggesting the planetary model just described, Bohr also made two further postulates which enabled him to explain the spectrum of hydrogen. The first of these was the suggestion that an electron of high energy circling the nucleus at a large radius can lose some of that energy and assume an orbit of lower energy closer to the nucleus. The energy lost by the electron is emitted as a photon of light of frequency ν given by Planck's formula

$$\Delta E = h\nu \tag{21.9}$$

where ΔE is the energy lost by the electron.

Bohr's second postulate was that only certain orbits are possible to the electron in a hydrogen atom. This enabled him to explain why it is that only light of a few particular frequencies can be emitted by the hydrogen atom. Since only a limited number of orbits are allowed, when an electron shifts from an outer to an inner orbit, the photon which emerges cannot have just any frequency but only that frequency corresponding to the energy difference between two allowed orbits.

EXAMPLE 21.3 According to Bohr's theory two of the allowed orbits in the hydrogen atom have radii of 52.918 and 211.67 pm. Calculate the energy, the frequency, and the wavelength of the photon emitted when the electron moves from the outer to the inner of these two orbits.

Solution Labeling the outer orbit 2 and the inner orbit 1, we first calculate the energy of each orbit from Eq. (21.8):

$$E_2 = -\frac{1.1536 \times 10^{-28}\ \text{J m}}{211.67 \times 10^{-12}\ \text{m}} = -0.545\ 00\ \text{aJ}$$

$$E_1 = -\frac{1.1536 \times 10^{-28}\ \text{J m}}{52.918 \times 10^{-12}\ \text{m}} = -2.1780\ \text{aJ}$$

Thus $\Delta E = -0.545\ 00\ \text{aJ} - (-2.1780\ \text{aJ}) = 1.6330\ \text{aJ}$
Using Eq. 21.9, we now have

$$\nu = \frac{\Delta E}{h} = \frac{1.6330 \times 10^{-18}\ \not{\text{J}}}{6.6262 \times 10^{-34}\ \not{\text{J}}\ \text{s}} = 2.4645 \times 10^{15}\ \text{s}^{-1}$$

$$= 2.4645\ \text{PHz}$$

Finally $\lambda = \frac{c}{\nu} = 1.2164 \times 10^{-7}\ \text{m} = 121.64\ \text{nm}$

In order to predict the right frequencies for the lines in the hydrogen spectrum, Bohr found that he had to assume that the quantity mur (called the *angular momentum* by physicists) needed to be a multiple of $h/2\pi$. In other words the condition restricting the orbits to only certain radii and certain energies was found to be

$$mur = \frac{nh}{2\pi} \qquad (21.10)$$

where n could have the value 1, 2, 3, etc.

By manipulating both Eqs. (21.10) and (21.5) (see Problem 21.44), it is possible to show that this restriction on the angular momentum restricts the radii of orbits to those given by the expression

$$r = \frac{n^2h^2}{4\pi^2mke^2} \qquad n = 1, 2, 3, \ldots \qquad (21.11)$$

If the known values of h, m, k, and e are inserted, this formula reduces to the convenient form

$$r = n^2 \times 52.918\ \text{pm} \qquad n = 1, 2, 3, \ldots \qquad (21.12)$$

Bohr's postulate thus restricts the electron to orbits for which the radius is 52.9 pm, $2^2 \times 52.9$ pm, $3^2 \times 52.9$ pm, and so on.

If we substitute Eq. (21.11) into Eq. (21.7), we arrive at a general expression for the energy in terms of n:

$$E = -\frac{1}{2} k \frac{e^2}{r} = -\frac{1}{2} ke^2 \times \frac{4\pi^2 mke^2}{n^2h^2} = -\frac{2\pi^2 k^2 e^4 m}{n^2h^2} \qquad (21.13)$$

Again substituting in the known values for all the constants, we obtain

$$E = -\frac{2.1800 \text{ aJ}}{n^2} \qquad (21.14)$$

The integer n thus determines how far the electron is from the nucleus and how much energy it has, just as the principal quantum number n did in Chap 5. It turns out that both n's are the same and that the energy levels of the hydrogen atom predicted by the Bohr theory are the same as those predicted by wave mechanics. The relationship between them is shown in Fig. 21.8.

EXAMPLE 21.4 Using. Eq. (21.13) or (21.14) find the ionization energy of the hydrogen atom.

Solution The ionization energy of the hydrogen atom corresponds to the energy difference between the electron in its innermost orbit ($n = 1$) and the electron when completely separated from the proton. For the completely separated electron $r = \infty$ (infinity) and so does n. Thus

$$E_1 = -\frac{2.1800 \text{ aJ}}{1^2} = -2.1800 \text{ aJ}$$

$$E_\infty = -\frac{2.1800 \text{ aJ}}{\infty^2} = 0.0000 \text{ aJ}$$

The energy difference is thus

$$\Delta E = E_\infty - E_1 = 2.1800 \text{ aJ}$$

which is the ionization energy per atom. On a molar basis the ionization energy is the Avogadro constant times this quantity; namely,

$$2.1800 \times 10^{-18} \text{ J} \times 6.0221 \times 10^{23} \text{ mol}^{-1} = 1312.8 \text{ kJ mol}^{-1}$$

Note: In an atom, the electronic configuration of lowest energy is called the **ground state** while other configurations are called **excited states.**

We can now derive Rydberg's experimental formula from Bohr's theory. Suppose an electron moves from an outer orbit for which the quantum number is n_2 to an inner orbit of quantum number n_1. The energy lost by the electron and emitted as a photon is then given by

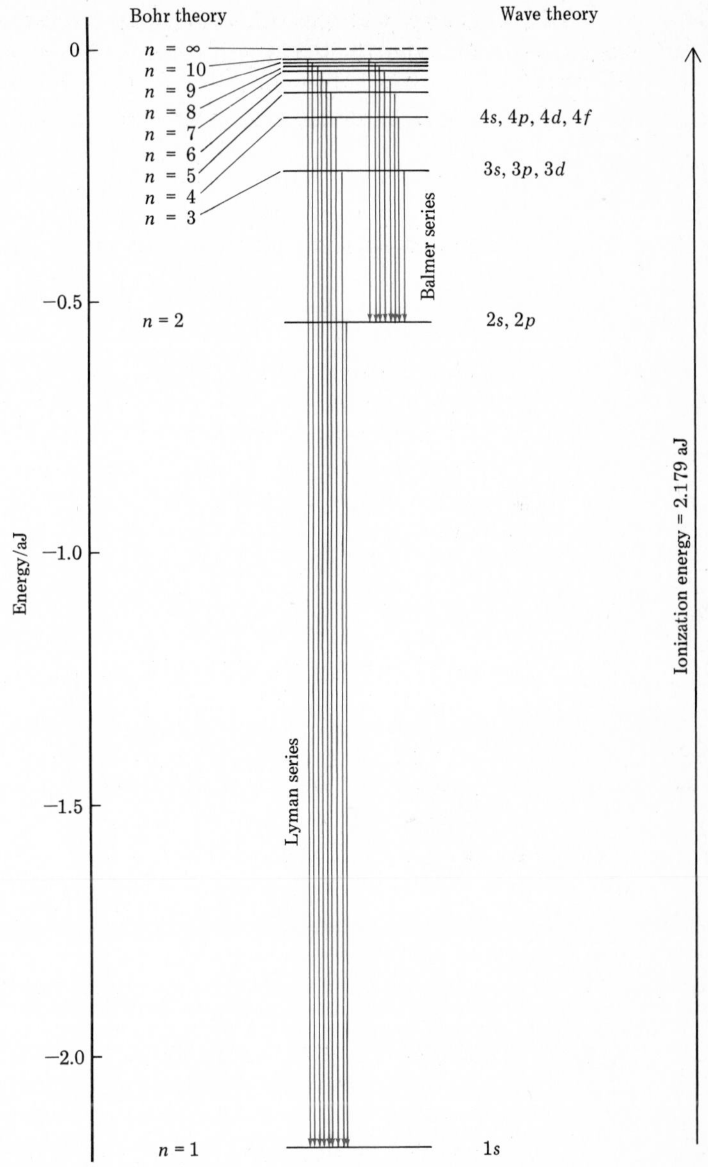

Figure 21.8 Energy-level diagram for H. Transitions producing lines in the Balmer and Lyman series are indicated, and the ionization energy is shown.

$$\Delta E = E_2 - E_1 = -2.1800 \text{ aJ} \left(\frac{1}{n_2^2} - \frac{1}{n_1^2}\right)$$

$$= 2.1800 \text{ aJ} \left(\frac{1}{n_1^2} - \frac{1}{n_2^2}\right) \qquad (21.15)$$

However, $$\Delta E = h\nu = \frac{hc}{\lambda} \qquad (21.16)$$

where λ is the wavelength of the photon. Combining Eqs. (21.16) and (21.15), we obtain

$$\frac{1}{\lambda} = \frac{2.1800 \text{ aJ}}{hc}\left(\frac{1}{n_1^2} - \frac{1}{n_2^2}\right)$$

or

$$\frac{1}{\lambda} = 1.0974 \times 10^7 \text{ m}^{-1}\left(\frac{1}{n_1^2} - \frac{1}{n_2^2}\right) \tag{21.17}$$

This expression is of exactly the same form as that found experimentally by Rydberg [Eq. (21.3)] with a value for R_∞ of $1.0974 \times 10^7 \text{ m}^{-1}$, very close to the experimental value of $1.097\,094 \times 10^7 \text{ m}^{-1}$. Even better agreement can be obtained if allowance is made for the fact that the nucleus is not stationary but that the electron and nucleus revolve around a common center of gravity.

EXAMPLE 21.5 Using Eq. (21.17) calculate the wavelength of the light emitted when the electron in a hydrogen atom drops from the $n = 3$ to the $n = 1$ orbit. In what region of the spectrum does this spectral line lie? To what series does it belong?

Solution From Eq. (21.17) we find

$$\frac{1}{\lambda} = 1.0974 \times 10^7 \text{ m}^{-1} \times \left(1 - \frac{1}{9}\right) = 9.7547 \times 10^6 \text{ m}^{-1}$$

giving $\lambda = 102.51$ nm

As can be seen from Fig. 21.6, this is the second line in the Lyman series and lies in the far ultraviolet. The experimentally determined wavelength is 102.573 nm.

Bohr's success with the hydrogen atom soon led to attempts both by him and by others to extend the same model to other atoms. On the qualitative level these attempts met with some success, and a general picture of electrons occupying orbits in successive levels and sub-levels, similar to that shown in Fig. 5.2, began to emerge. On the quantitative level, however, all attempts to calculate accurate values for the energies of the electrons in their quantized orbitals were dismal failures. It was not until Schrödinger's introduction of wave mechanics in 1926 that these difficulties could be resolved. Suddenly, it seemed, everything fell into place. Since then virtually every line in the spectrum of every element has been accounted for theoretically. As a result, we now have a very exact, though mathematically rather complex, picture of the behavior of electrons in both the ground state and in excited states of atoms. In particular, the study of atomic spectra has allowed us to determine the ionization energies of all the elements very accurately.

Details of the spectra of polyelectronic atoms are complex, and so we will consider only one example: sodium. Excited states of sodium may be obtained by increasing the energy of the atom so that the 3*s* valence electron occupies the 3*p*, 3*d*, 4*s*, 4*p*, 4*d*, 4*f*, or some other orbital. By contrast with the hydrogen atom, however, a sodium atom has other electrons which shield the valence electron from nuclear charge, and this shielding is different for each different orbital shape (*s*, *p*, *d*, *f*, etc.). Consequently the energy of an excited sodium atom whose electron configuration is $1s^22s^22p^64s^1$ is *not the same* as that of an excited sodium atom whose configuration is $1s^22s^22p^64p^1$. Different shielding of the outermost (4*s* or 4*p*) electron results in a different energy. Because of this, four formulas are needed to describe the energy of the sodium atom—one for each of the orbital shapes available to the outermost electron:

$$E_{ns} = -\frac{2.1800 \text{ aJ}}{(n - a_s)^2} \qquad E_{np} = -\frac{2.1800 \text{ aJ}}{(n - a_p)^2}$$

$$E_{nd} = -\frac{2.1800 \text{ aJ}}{(n - a_d)^2} \qquad E_{nf} = -\frac{2.1800 \text{ aJ}}{(n - a_f)^2} \tag{21.18}$$

In all these equations n represents the principal quantum number. It must be 3 or greater since the electron is in the 3*s* orbital to begin with. The different shielding requires a different correction for each type of orbital: $a_s = 1.36$; $a_p = 0.87$; $a_d = 0.012$; and $a_f = 0.001$.

Because there are four different sets of energy levels, the number of transitions between levels (and hence the number of lines in the spectrum) is larger for sodium than for hydrogen. Early spectroscopists were able to distinguish four different types of lines, which they labeled the *sharp, principal, diffuse,* and *fundamental* series. It is from the abbreviation of these terms that we have obtained the modern symbols *s*, *p*, *d*, and *f*.

As most readers will know, when almost any sodium compound is held in a bunsen burner, it imparts a brilliant yellow color to the flame. This yellow color corresponds to the most prominent line in the sodium spectrum. Its wavelength is 589 nm. On the atomic scale this line is caused by the sodium atom moving from an excited state (in which the valence electron is in a 3*p* orbital) to the ground state (in which the electron is in a 3*s* orbital). Using Eq. (21.18), we can obtain approximate values for the two energies involved in the transition:

$$E_{3p} = -\frac{2.1800 \text{ aJ}}{(3 - 0.87)^2} = -0.4805 \text{ aJ}$$

and

$$E_{3s} = -\frac{2.1800 \text{ aJ}}{(3 - 1.36)^2} = -0.8105 \text{ aJ}$$

Thus

$$\Delta E = 0.3300 \text{ aJ}$$

giving

$$\lambda = \frac{hc}{\Delta E} = 602 \text{ nm}$$

This agrees approximately with the experimental result.

Another feature of the sodium spectrum deserves mention. Careful observation reveals that the yellow color of sodium is actually due to two closely spaced lines (a **doublet**). One has a wavelength of 588.995 nm, and the other is at 589.592 nm. When the electron is in a 3*p* orbital, its spin can be aligned in two ways with respect to the axis of the orbital. The small difference in energy between these two orientations results in two slightly different wavelengths.

21.3 THE SPECTRA OF MOLECULES: INFRARED

When we turn from the spectra of atoms to those of molecules, we find that the region of most interest to chemists is no longer the visible and ultraviolet but rather the infrared. As its name implies, the infrared extends beyond the red end of the visible spectrum, from the limit of visibility at roughly 0.8 μm (800 nm) up to about 100 μm where the microwave region begins. Another difference from the spectra discussed in the previous section is that infrared spectra are all *absorption spectra* rather than emission spectra. Infrared light is passed through the sample, and the intensity of light emerging is measured electronically. The energies of infrared photons are very much less than those of visible and ultraviolet photons. A photon of wavelength 10 μm has an energy of only 0.02 aJ (about 12 kJ mol^{-1})—not even enough to break a hydrogen bond, let alone a normal covalent bond. It does have enough energy to make a molecule *vibrate* more strongly, however, and since vibrational energy is quantized, this can only happen at certain discrete frequencies and not at others.

Figure 21.9 shows the infrared spectra of two triatomic molecules, H_2O and CO_2, and also that of a more complex molecule, C_2H_5OH (ethanol). Each of the peaks in these spectra corresponds to a strong absorption of infrared radiation on the macroscopic level and a sudden increase in the amplitude with which the molecule vibrates on the microscopic level. Since a polyatomic molecule can vibrate in a variety of ways, there are several peaks for each molecule. The more complex the molecule, the larger the number of peaks. Note also that not all the vibrations correspond to the stretching and unstretching of bonds. A vibration in a polyatomic molecule is defined as any periodic motion which changes the shape or size of the molecule. In this sense bending and twisting motions also count as vibrations.

A useful feature of the vibrations which occur in polyatomic molecules is that many bonds and some small groups of atoms vibrate in much the same way no matter what molecule they are in. In Fig. 21.9, for example, stretching of the O—H bond gives a peak between 2 and 3 μm for both H_2O and C_2H_5OH. Because of this it is possible to identify many of the functional groups in an organic molecule merely by looking at its infrared spectrum. Table 21.1 shows characteristic wavelengths by which some common functional groups can be identified. On the other hand, each molecule is a unique combination of chemical bonds and functional groups. Quite minor differences in molecular structure can result in noticeable differences in the infrared spectrum. Thus these spectra can be used in the same way the po-

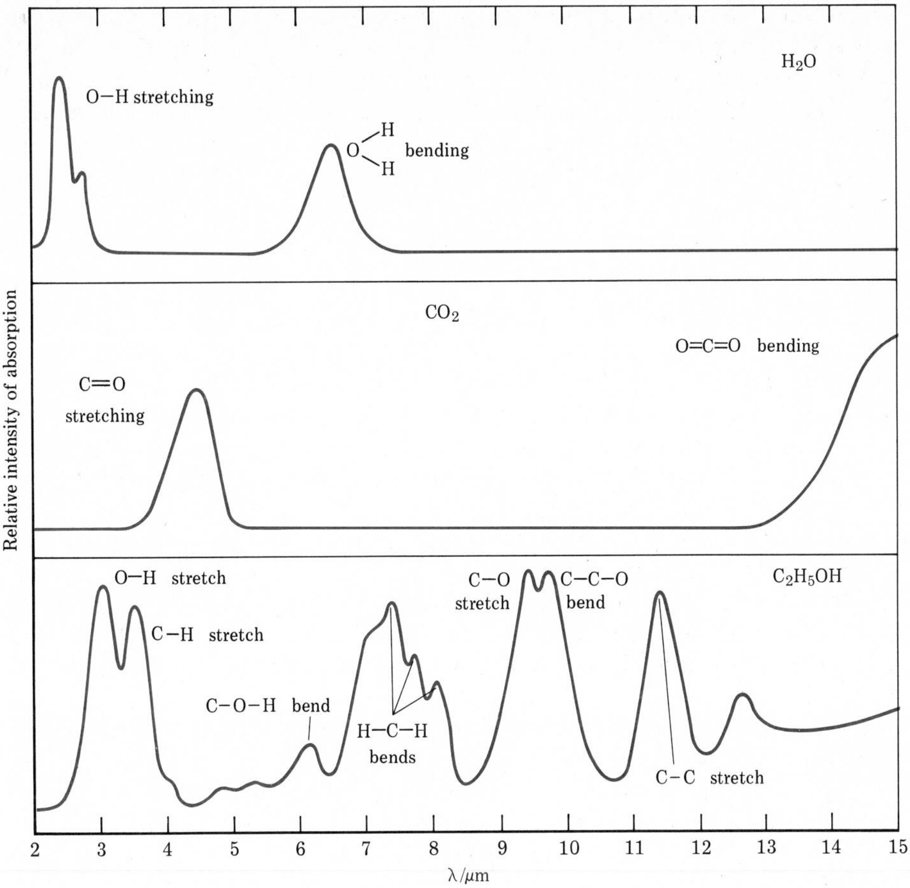

Figure 21.9 Infrared spectra of water, carbon dioxide, and ethanol.

TABLE 21.1 Characteristic Wavelengths of Infrared Absorption for Some Chemical Bonds.

Bond	λ/μm	Bond	λ/μm
—C—H (C with two further single bonds)	3.4–3.5	—C—C— (each C with two further single bonds)	6.6–16.7
=C—H	3.2–3.3	>C=C<	5.9–6.2
		—C≡C—	4.4–4.8
≡C—H	3.0		
—O—H	2.5–2.8	—C—O— (C with two further single bonds)	7.7–10.0
—O—H*	2.9–3.1		
—N—H (N with one further single bond)	2.8–3.0	>C=O	5.7–5.9
		—C≡N	4.3–5.0

* Hydrogen bonded.

lice use fingerprints. When an unknown compound is prepared, one of the first things that is usually measured is its infrared spectrum. If this spectrum should happen to match that of a previously prepared compound, the unknown compound can be readily identified. If not, it may still be possible to identify some of the functional groups that are present.

In order for a molecular vibration to interact with electromagnetic radiation, the dipole moment of the molecule must change as the vibration takes place. The larger this change in dipole moment, the more strongly the substance absorbs the incident radiation. Thus very polar bonds like O—H and C═O usually produce very prominent peaks in an infrared spectrum. Conversely some vibrations do not feature in the infrared at all. In particular, diatomic molecules like N_2 and O_2, in which both atoms are identical, have zero dipole moment at any stage in a vibration. They produce no absorption in the infrared.

Since N_2 and O_2 are the chief constituents of the air, it is just as well that they do not absorb infrared radiation. The atmosphere would become intolerably hot if they did! As it is, only the minor constituents of the atmosphere, CO_2 and H_2O, absorb in the infrared. Nevertheless this absorption still plays an important role in maintaining the surface of the earth at its current temperature.

The earth absorbs energy from the sun by day, and radiates this energy away at night. The inflow and outflow must balance on average, otherwise the earth would heat up or cool down. Most of the sun's radiation is in the visible region of the spectrum, but the radiation which escapes from the much cooler earth is mainly in the infrared, centered around 10 to 12 μm.

As you can see from Fig. 21.9, water absorbs infrared radiation between 2 to 3 μm and 6 to 7.5 μm. Water also absorbs strongly above 18 μm. Thus much of the outgoing infrared radiation is absorbed by water vapor in the earth's atmosphere and prevented from escaping. You may have noticed that after a really humid summer day the temperature does not fall very fast at night. Excess water vapor in the atmosphere prevents radiation from escaping the earth's surface. On the other hand, in a desert area the low humidity allows rapid heat loss. Although rocks may become hot enough to fry an egg by day, temperatures often drop to freezing overnight.

While local concentrations of water vapor may vary from time to time, the total quantity in the earth's atmosphere is buffered by the vast areas of ocean and remains nearly constant. Thus the average absorption of outgoing radiation by water seldom changes. The quantity of CO_2 in the atmosphere is not so well regulated, however, and it appears that human activities are causing it to increase. [Using the data provided in Example 3.4, you can calculate that about 9.4 Pg (9.4×10^{15} g) of CO_2 results from the combustion of fossil fuels in the world each year.] Even in a relatively nonindustrial area such as Hawaii, there has been a steady increase in CO_2 concentration for many years.

Referring to Fig. 21.9 again, we can clearly see that infrared absorption by CO_2 occurs in just those parts of the spectrum that were not blocked by H_2O absorption. Thus increasing the concentration of CO_2 should decrease earth's radiation to outer space and might increase the average surface temperature. On a global scale this is called the **greenhouse effect**—the CO_2 and H_2O act like the glass in a greenhouse, allowing visible light to pass in but blocking the loss of infrared. Climatologists have predicted that during the hundred years human beings have been using fossil fuels, the greenhouse effect should have raised surface temperatures by 0.5 to 1.0 K. Until 1950 that prediction appeared to have been borne out, but measured temperatures have since fallen back to about the 1900 level. Attempts to explain this drop on the basis of additional particulate matter in the atmosphere have met with varying degrees of success and failure. All that can be said for certain is that we know far less about the atmosphere and world climate than we would like.

21.4 THE VISIBLE AND ULTRAVIOLET SPECTRA OF MOLECULES: MOLECULAR ORBITALS

When molecules absorb or emit radiation in the ultraviolet and visible regions of the spectrum, this almost always corresponds to the transition of an electron from a low-energy to a high-energy orbital, or vice versa. One might expect the spectra of molecules to be like the atomic line spectra shown in Fig. 21.6, but in fact molecular spectra are very different. Consider, for example, the absorption spectrum of the rather beautiful purple-violet gas I_2. This molecule strongly absorbs photons whose wavelengths are between 440 and 600 nm, and much of the orange, yellow, and green components of white light are removed. The light which passes through a sample of I_2 is mainly blue and red. When analyzed with an average-quality spectroscope, this light gives the spectrum shown in Fig. 21.10*a*. Instead of the few discrete lines typical of atoms, we now have a broad, apparently continuous, absorption band. This is typical of molecules.

Why is there this difference between atomic and molecular spectra? An answer begins to appear if we use a somewhat more expensive spectroscope. Figure 21.10*b* shows a tracing of the I_2 spectrum made with such an instrument. What originally appeared to be a continuous band is now shown to consist of a very large number of very narrow, closely spaced lines. Thus the broad absorption band of I_2 is actually made up of discrete lines.

The reason molecules give rise to such an enormous number of lines is that molecules can vibrate and rotate in a very large number of ways while atoms cannot. Furthermore both rotational levels and vibrational motion are quantized. When a molecule absorbs a photon of light and an electron is excited to a higher orbital, the molecule will not be stationary either before or after the absorption of the photon. The process of absorption is thus

$$\begin{array}{rcl} \text{Ground-state molecule} & & \text{excited-state molecule} \\ \text{in one of many} & \rightarrow & \text{in another of many} \\ \text{rotational-vibrational states} & & \text{rotational-vibrational states} \end{array}$$

Because of the large number of energy possibilities both before and after the

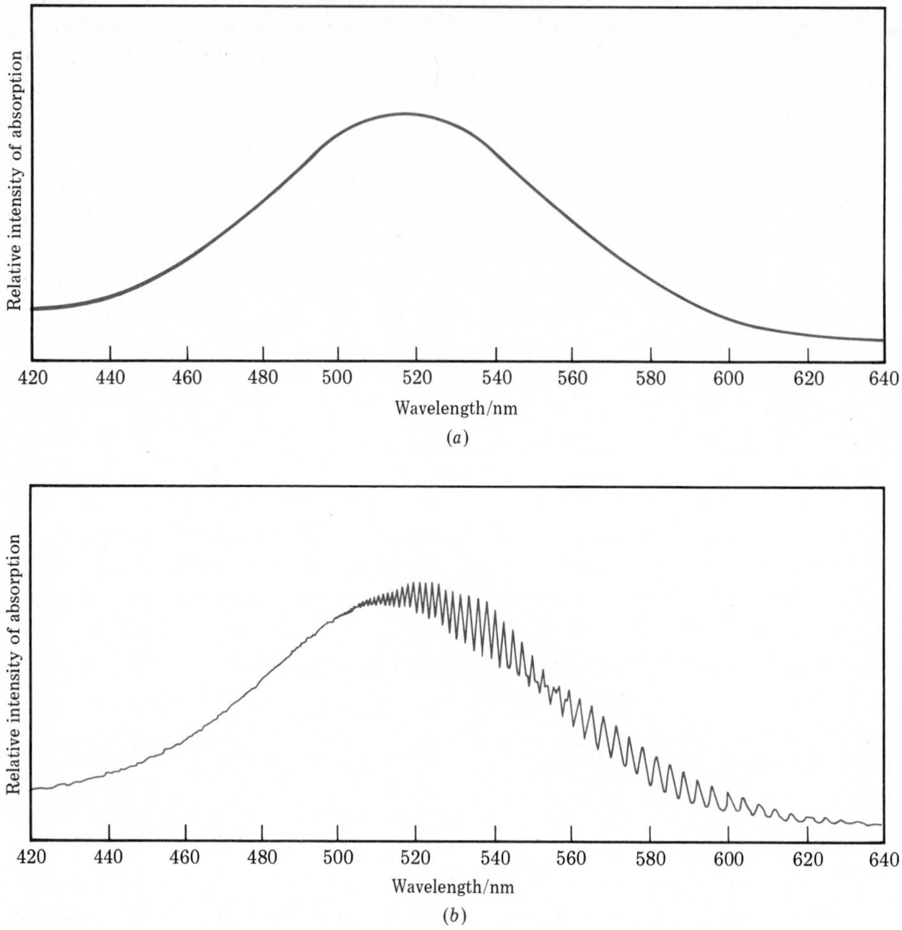

Figure 21.10 Absorption band of $I_2(g)$ in the visible region. (*a*) Result obtained with low-resolution spectroscope; (*b*) result obtained with high-resolution spectroscope.

transition, a very large number of lines of slightly different wavelengths is obtained. A careful analysis of these lines yields much valuable information about the way in which the molecule rotates and vibrates. In particular, very accurate values of bond enthalpies and bond lengths can be obtained from a study of the fine structure of an absorption band like that shown in Fig. 21.10*b*.

Molecular Orbitals

In order to explain both the ground state and the excited state involved in an absorption band in the ultraviolet and visible spectra of molecules, it is necessary to look at the electronic structure of molecules in somewhat different terms from the description given in Chaps. 6 and 7. In those chapters we treated electrons either as bonding pairs located between two nuclei, or as lone pairs associated with a single nucleus. Such a model of electronic structure is known as the **valence-bond** model. It is of very little use in explaining molecular spectra because photons are absorbed by the whole molecule, not an individual atom or bond. Thus we need to look upon electrons in a molecule as occupying orbitals which belong to the *molecule as a whole.* Such orbitals are called **molecular orbitals,** and this way of looking at molecules is referred to as **molecular-orbital** (abbreviated MO) **theory.**

The term molecular orbital was mentioned in Sec. 6.4 when we described formation of a covalent bond in an H_2 molecule as a result of overlap of two 1*s* atomic orbitals—one from each H atom. At that time, however, we did not point out that there are *two* ways in which the 1*s* electron wave of one H atom can combine with the 1*s* electron wave of another. One of these involves *constructive interference* between the two waves and is referred to as **positive overlap.** This results in a bigger electron wave (and hence more electron density) between the two atomic nuclei. This attracts the positively charged nuclei together, forming a bond as described in Sec. 6.4. A molecular orbital formed as a result of positive overlap is called a **bonding MO.**

It is also possible to combine two electron waves so that *destructive interference* of the waves occurs between the atomic nuclei. This situation is referred to as **negative overlap,** and it decreases the probability of finding an electron between the nuclei. In the case of two H atoms this results in a planar node of zero electron density halfway between the nuclei. Without a buildup of negative charge between them, the nuclei repel each other and no chemical bond is possible. A molecular orbital formed as a result of negative overlap is called an **antibonding MO.**

If one or more electrons occupy an antibonding MO, the repulsion of the nuclei increases the energy of the molecule, and so such an orbital is higher in energy than a bonding MO. This is shown in Fig. 21.11. Electron dot-density diagrams for the 1*s* electron in each of two separate H atoms are shown on the left and right sides of the figure. The horizontal lines show the energy each of these electrons would have. A dot-density diagram for a single electron occupying the bonding MO formed by positive overlap of the two 1*s* orbitals is shown in the center of the diagram. This is labeled σ_{1s}. Above it is a dot-density diagram for a single electron occupying the antibonding MO, σ_{1s}^*. (In general, antibonding MO's are distinguished from bonding MO's by adding an * to the label.) The energies of the molecular orbitals are indicated by the horizontal lines in the center of the diagram. The Greek letter σ in the labels for these orbitals refers to the fact that their positive or negative overlap occurs directly between the two atomic nuclei.

Like an atomic orbital, each molecular orbital can accommodate two electrons. Thus the lowest energy arrangement for H_2 would place both electrons in the σ_{1s} MO with paired spins. This molecular electron configuration is written $(\sigma_{1s})^2$, and it corresponds to a covalent electron-pair bond holding the two H atoms together. If a sample of H_2 is irradiated with ultraviolet light, however, an absorption band is observed between 110 and 170 nm. The energy of such an absorbed photon is enough to raise one electron to the antibonding MO, producing an excited state whose electron configuration is $(\sigma_{1s})^1(\sigma_{1s}^*)^1$. In this excited state the effects of the bonding and the

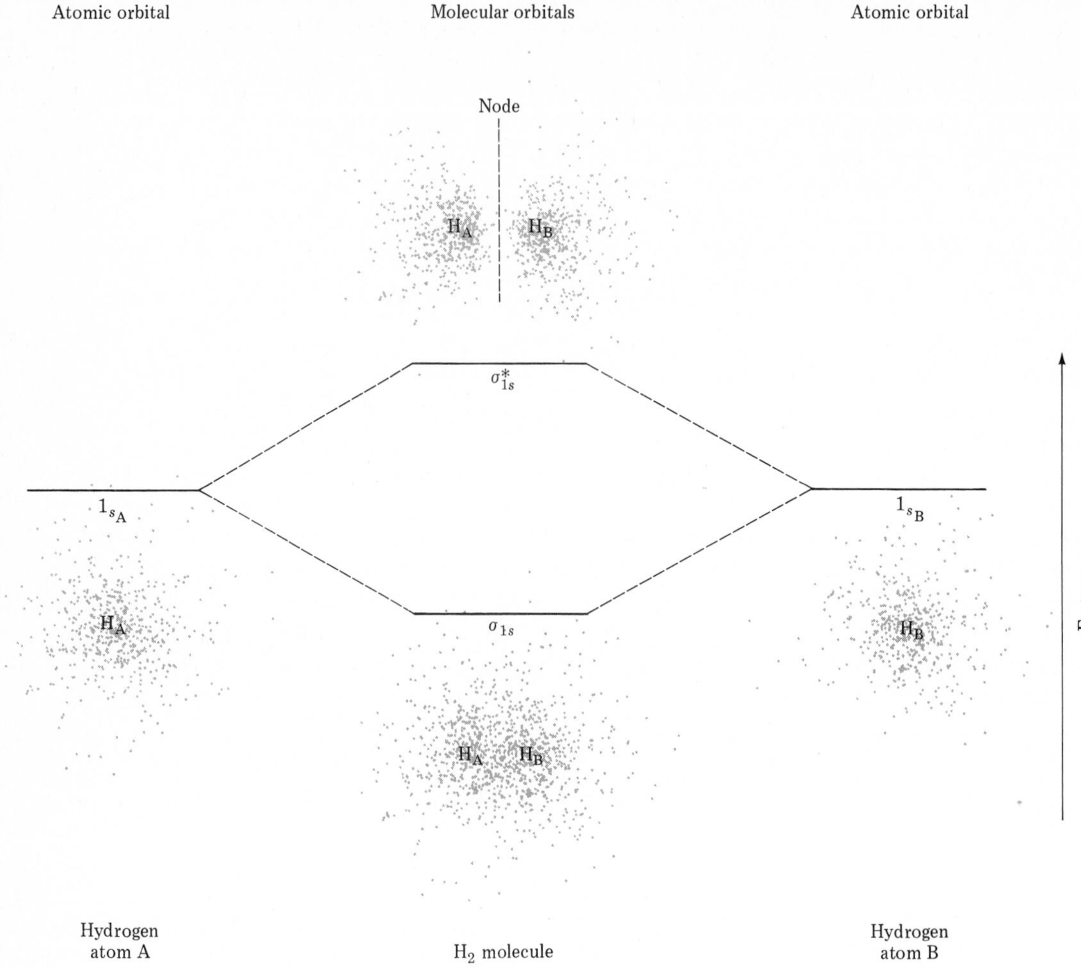

Figure 21.11 Energy levels and electron dot-density diagrams for atomic and molecular orbitals associated with hydrogenic 1*s* wave functions. (Computer-generated.) (*Copyright © 1976 by W. G. Davies and J. W. Moore.*)

antibonding orbitals exactly cancel each other; there is no overall bond between the two H atoms, and the H_2 molecule dissociates. When the absorption of a photon results in the dissociation of a molecule like this, the phenomenon is called **photodissociation.** It occurs quite frequently when UV radiation strikes simple molecules.

The molecular-orbital model we have just described can also be used to explain why a molecule of He_2 cannot form. If a molecule of He_2 were able to exist, the four electrons would doubly occupy both the bonding and the antibonding orbitals, giving the electron configuration $(\sigma_{1s})^2(\sigma_{1s}^*)^2$. However, the antibonding electrons would cancel the effect of the bonding electrons, and there would be no resultant buildup of electronic charge between the nuclei and hence no bond. Interestingly enough, an extension of this argument predicts that if He_2 loses an electron to become the He_2^+ ion, a bond is possible. He_2^+ would have the structure $(\sigma_{1s})^2(\sigma_{1s}^*)^1$, and the single electron in the antibonding orbital would only cancel half the effect of the two electrons in the bonding orbital. This would leave the ion with a "half-bond" joining the two nuclei. The spectrum of He shows bands corresponding to He_2^+, and from them it can be determined that He_2^+ has a bond enthalpy of 322 kJ mol^{-1}.

The molecular-orbital model can easily be extended to other diatomic molecules in which both atoms are identical **(homonuclear diatomic molecules).** Three general rules are followed. First, only the core orbitals and the valence orbitals of the atoms need be considered. Second, only atomic orbitals whose energies are similar can combine to form molecular orbitals. Third, the number of molecular orbitals obtained is always the same as the number of atomic orbitals from which they were derived.

Applying these rules to diatomic molecules which consist of atoms from the second row of the periodic table, such as N_2, O_2, and F_2, we need to consider the $1s$, $2s$, $2p_x$, $2p_y$, and $2p_z$ atomic orbitals. Since the $1s$ orbital of each atom differs in energy from the $2s$, we can overlap the two $1s$ orbitals separately from the $2s$. This gives a σ_{1s} and a σ^*_{1s} MO, as in the case of H_2. Similarly, the $2s$ orbitals can be combined to give σ_{2s} and σ^*_{2s} before we concern ourselves with the higher energy $2p$ orbitals. There are three $2p$ orbitals on each atom, and so we expect a total of six molecular orbitals to be derived from them. The shapes of these six molecular orbitals are shown by the boundary-surface diagrams in Fig. 21.12. Two of them are formed by

Figure 21.12 The formation of molecular orbitals from $2p$ orbitals. In each case the orbitals interact in both a constructive (+) and a destructive (−) sense to produce a bonding and an antibonding orbital. The $2p_x$ and $2p_y$ orbitals overlap side by side and produce two π orbitals of equal energy and two π^* orbitals, also of equal energy. The end-to-end overlap of the $2p_z$ orbitals produces σ and σ^* orbitals. (The positions of atomic nuclei are indicated by the intersection of axis lines in each drawing.) (Computer-generated.) *(Copyright © 1976 by W. G. Davies and J. W. Moore.)*

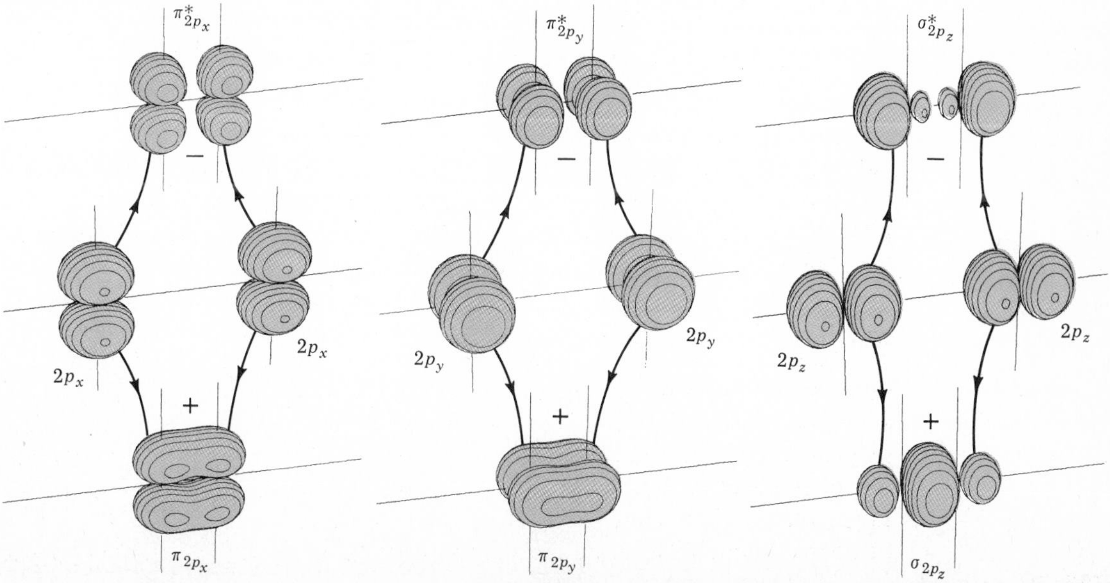

positive and negative overlap of $2p_z$ orbitals directly between the atomic nuclei. Consequently they are labeled σ_{2p_z} and $\sigma^*_{2p_z}$. Two more molecular orbitals are formed by sideways overlap of $2p_x$ atomic orbitals. These are labeled π_{2p_x} and $\pi^*_{2p_x}$, because the molecular orbitals have two parts—one above and one below a nodal plane containing the nuclei. The atomic $2p_y$ orbitals also overlap sideways to form π_{2p_y} and $\pi^*_{2p_y}$ molecular orbitals. These are identical to π_{2p_x} and $\pi^*_{2p_x}$, except for a 90° rotation around the line connecting the nuclei. Consequently π_{2p_x} and π_{2p_y} have the same energy, as do $\pi^*_{2p_x}$ and $\pi^*_{2p_y}$.

The electron configuration for any homonuclear diatomic molecule containing fewer than 20 electrons can be built up by filling electrons into the molecular orbitals we have just derived, starting with the orbital of lowest energy. The relative energies of the molecular orbitals at the time they are being filled are shown in Fig. 21.13. Like the energies of atomic orbitals given in Fig. 5.9, these relative molecular-orbital energies vary somewhat from one diatomic molecule to another. In particular the σ_{2p} orbital is often lower than π_{2p}. Nevertheless Fig. 21.13 gives the correct order of filling the orbitals, and we can use it to determine molecular electron configurations.

EXAMPLE 21.6 Find the electronic configuration of the oxygen molecule, O_2.

Solution Starting with the lowest lying orbitals (σ_{1s} and σ^*_{1s}) we add an appropriate number of electrons to successively higher orbitals in accord with the Pauli principle and Hund's rule. O_2 has 16 electrons, the first 14 of which are easily accommodated in the following way:

$$(\sigma_{1s})^2(\sigma^*_{1s})^2(\sigma_{2s})^2(\sigma^*_{2s})^2(\pi_{2p})^4(\sigma_{2p})^2$$

The remaining two electrons must now be added to the π^*_{2p} orbitals. Since both these orbitals are of equal energy, one electron must be placed in each orbital and the spins must be parallel. The total electronic structure is thus

$$(\sigma_{1s})^2(\sigma^*_{1s})^2(\sigma_{2s})^2(\sigma^*_{2s})^2(\pi_{2p})^4(\sigma_{2p})^2(\pi^*_{2p_x})^1(\pi^*_{2p_y})^1$$

As the previous problem shows, the molecular-orbital model predicts that O_2 has two unpaired electrons. Substances whose atoms, molecules, or ions contain unpaired electrons are weakly attracted into a magnetic field, a property known as **paramagnetism.** (In a few special cases, such as iron, a much stronger magnetic attraction called **ferromagnetism** is also observed.) Most substances have all their electrons paired. Such materials are weakly repelled by a magnetic field, a property known a **diamagnetism.** Hence measurement of magnetic properties can tell us whether all electrons are paired or not. O_2, for example, is found to be paramagnetic, an observation which agrees with the electron configuration predicted in Example 21.6. Before the advent of MO theory, however, the paramagnetism

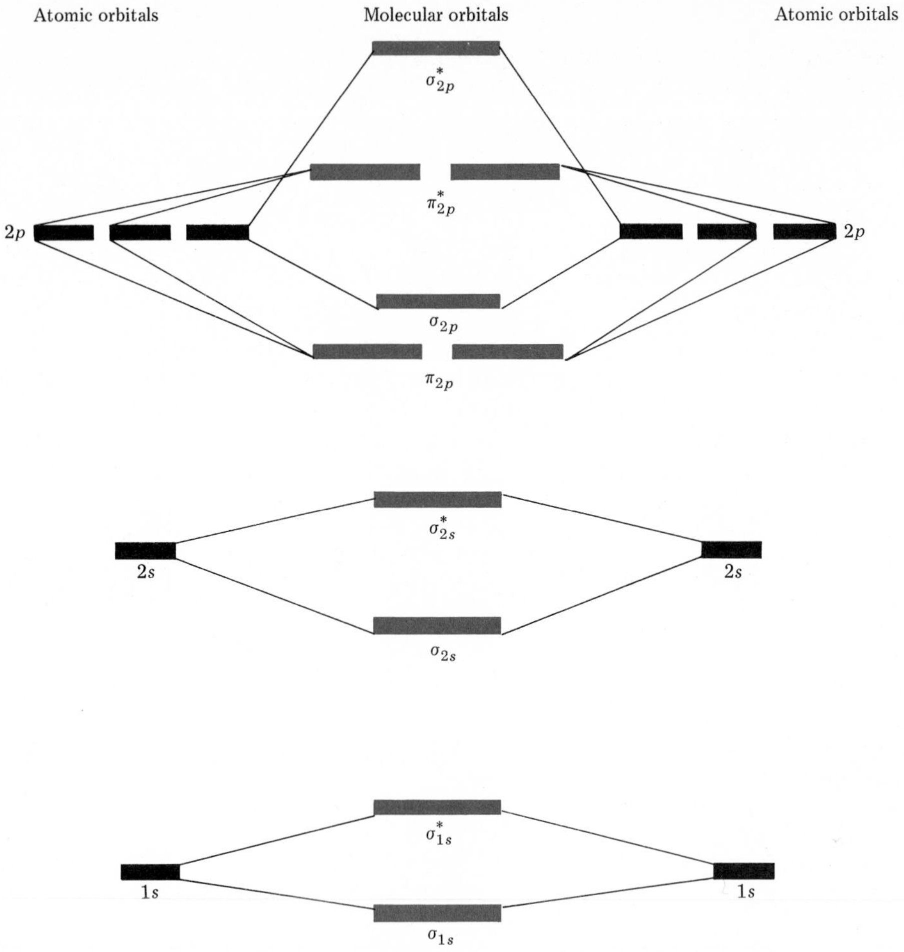

Figure 21.13 Relative energies of molecular orbitals for homonuclear diatomic molecules at the time they are being filled. The light lines connect molecular orbitals with the atomic orbitals from which they were derived.

was a mystery, since the Lewis diagram predicted that all electrons should be paired.

$$:\ddot{O}=\ddot{O}:$$

The molecular-orbital model also allows us to estimate the strengths of bonds in diatomic molecules. We simply count each electron in a bonding orbital as contributing half a bond, while each electron in an antibonding orbital takes away half a bond. Thus if there are B electrons in bonding orbitals and A electrons in antibonding orbitals, the net **bond order** is given by

$$\text{Bond order} = \frac{B - A}{2}$$

The larger the bond order, the more strongly the atoms are held together.

EXAMPLE 21.7 Calculate the bond order for the molecule N_2.

Solution There are 14 electrons, and so the electron configuration is

$$(\sigma_{1s})^2(\sigma_{1s}^*)^2(\sigma_{2s})^2(\sigma_{2s}^*)^2(\pi_{2p})^4(\sigma_{2p})^2$$

There are a total of 2 + 2 + 4 + 2 = 10 electrons in bonding MO's and only 4 in the antibonding orbitals σ_{1s}^* and σ_{2s}^*. Thus

$$\text{Bond order} = \frac{10 - 4}{2} = 3$$

The bond orders derived from the molecular-orbital model for stable molecules agree exactly with those predicted by Lewis' theory. Not only do we find a triple bond for N_2, but we also find a double bond for O_2 and a single bond for F_2. The results of such bond-order calculations are summarized in Table 21.2. Some of the molecules in the table, such as C_2 and B_2, are only stable at high temperatures or only exist transitorily in discharge tubes, and so you are probably not familiar with them. Nevertheless, their spectra can be studied. Also included in the table are values for the bond enthalpies and bond lengths of the various species obtained from their spectra. Note the excellent qualitative agreement with the MO theory. The higher the bond order predicted by the theory, the larger the bond enthalpy and the shorter the bond length.

TABLE 21.2 Bond Orders, Bond Enthalpies, and Bond Lengths of Some Diatomic Molecules.

Molecule*	Bond Order	Bond Enthalpy/kJ mol^{-1}	Bond Length/pm
H_2^+	$\frac{1}{2}$	256	106
H_2	1	432	74
He_2^+	$\frac{1}{2}$	322	108
He_2	0	Molecule not detected	
Li_2	1	110	267
Be_2	0	Molecule not detected	
B_2	1	274	159
C_2	2	603	124
N_2	3	942	110
O_2	2	494	121
F_2	1	139	142
Ne_2	0	Molecule not detected	

* Some molecule-ions such as H_2^+ are included. Except for the number of electrons involved, the MO theory is applied to them in exactly the same way as to molecules.

Delocalized Electrons

One of the most useful aspects of molecular-orbital theory only becomes apparent when we consider molecules containing three or more atoms. According to molecular-orbital theory, electrons occupy orbitals which are **delocalized.** That is, the orbitals spread over the entire molecule. An electron is not confined only to the vicinity of one or two atomic nuclei as in the lone pairs and bond pairs of valence-bond theory. As we saw in Sec. 7.7, there are some molecules, such as O_3 and C_6H_6, which cannot be described adequately by a single Lewis diagram. In such cases valence-bond theory resorts to resonance hybrids, such as

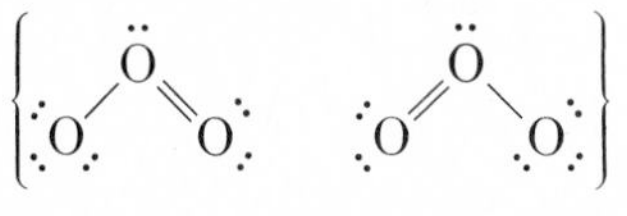

for ozone. The same molecules can be handled by MO theory without the need for several contributing structures, because electrons can occupy orbitals belonging to the molecule as a whole.

As an example of this we shall apply MO theory to the pi orbitals in ozone. The sigma bonds in this molecule may be attributed to overlap of sp^2 hybrid orbitals on each of the three oxygen atoms. You will recall that sp^2 hybrids are directed toward the corners of an equilateral triangle, in reasonable agreement with the 117° angle in ozone. For the sigma bonds and lone pairs, then, we have the Lewis diagram

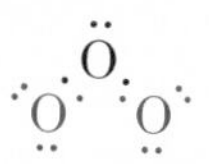

This involves only 14 of the 18 valence electrons of the three oxygens, and so 4 electrons are left for pi bonding. Furthermore, each oxygen has a *p* orbital perpendicular to the plane of its sp^2 hybrids (that is, perpendicular to the plane of the three oxygen atoms), which has not yet been used for bonding.

These three remaining *p* orbitals can overlap sideways to form the three molecular orbitals shown in Fig. 21.14. One of these concentrates electron density between the central oxygen and each of the other two (Fig. 21.14*a*), and is therefore a bonding MO. The second (Fig. 21.14*b*) looks very similar to the original atomic *p* orbitals on the end oxygen atoms. An electron in this orbital neither strengthens nor weakens the attractions between the atoms, and so the MO is said to be **nonbonding.** Finally (Fig. 21.14*c*) there is an antibonding MO which has nodes between each pair of oxygen atoms. Since there are only four electrons, only the two lower-energy MO's (the bonding and nonbonding) will be occupied. The two electrons in the

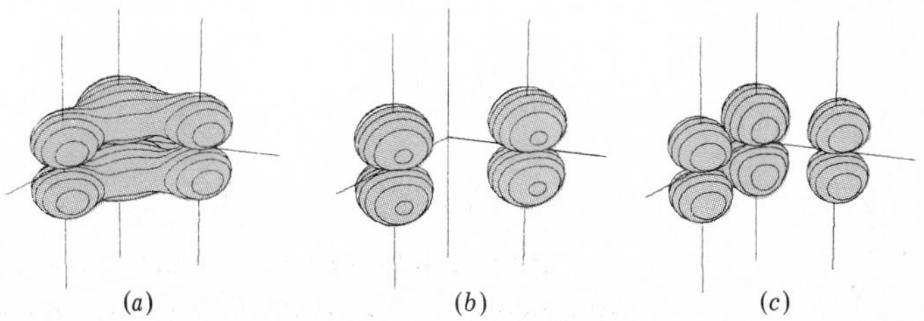

Figure 21.14 The pi molecular orbitals in ozone, O_3. (*a*) bonding; (*b*) nonbonding; (*c*) antibonding. Only (*a*) and (*b*) are occupied in the O_3 molecule, but an electron is excited to orbital (*c*) when ultraviolet light is absorbed. (Positions of atomic nuclei are indicated by the intersections of axis lines.) (Computer-generated.) (*Copyright © 1976 by W. G. Davies and J. W. Moore.*)

bonding MO provide a bond order of 1, but this is spread over both O—O bonds, and so it contributes a bond order of one-half between each pair of oxygens. Similarly the two electrons in the nonbonding MO correspond to a lone pair, half on the left oxygen and half on the right. Including the sigma bonding framework, this gives $1\frac{1}{2}$ bonds between each pair of oxygens and $2\frac{1}{2}$ lone pairs on each end oxygen. This is exactly the average of the two resonance structures already given.

The MO treatment can also be used to interpret the spectrum of ozone. As we mentioned in Sec. 12.6, ozone in the earth's stratosphere absorbs much solar ultraviolet radiation which would otherwise cause damage to the biosphere. This absorption is due to a band centered around 255 nm which corresponds to excitation of an electron to the unfilled antibonding pi orbital shown in Fig. 21.14*c*.

Another important molecule to which MO theory can be applied usefully is benzene. You will recall from Sec. 7.7 that benzene can be represented by the resonance hybrid

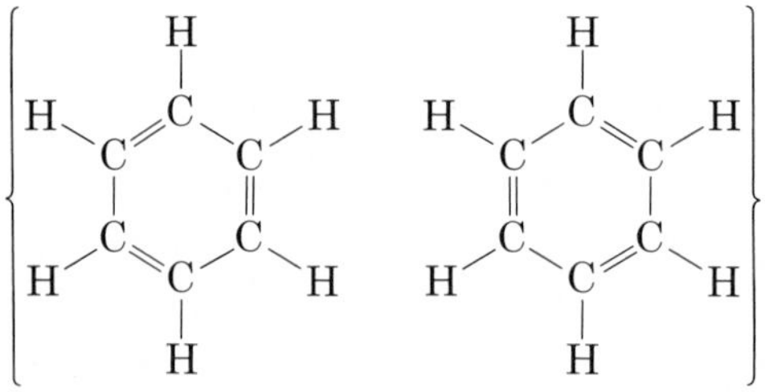

This indicates that all C—C bonds are equivalent and intermediate between a single and a double bond.

As in the case of ozone, we can treat the sigma bonds of benzene in valence-bond terms, dealing only with pi bonding by the molecular-orbital method. The sigma bonding framework is

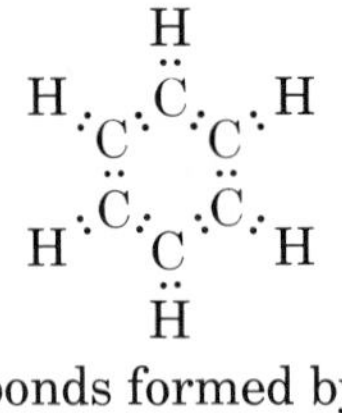

This involves 24 electrons in bonds formed by overlap of sp^2 hybrid orbitals on the carbon atoms with $1s$ orbitals on each hydrogen or with other sp^2 hybrids on other carbons. This gives a planar framework and leaves six p orbitals (one on each carbon) which are perpendicular to the molecular plane and can overlap sideways to form six pi molecular orbitals. Only the three lowest-energy MO's are occupied, and their electron clouds are shown in Fig. 21.15. All these MO's are bonding, while the three not shown are antibonding. When the six valence electrons not used for sigma bonding occupy the pi-bonding MO's, they are evenly distributed around the ring of carbon atoms. This contributes a bond order of one-half between each pair of carbons, or a total bond order of $1\frac{1}{2}$ when the sigma bond is included. Thus the C—C bond is intermediate between a single and double bond, in accord with the resonance hybrid.

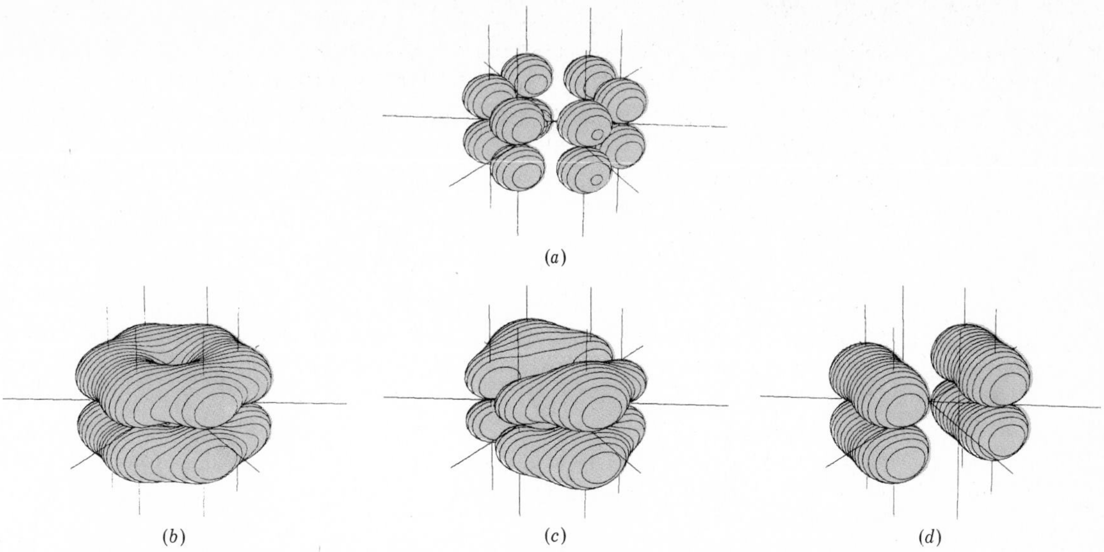

Figure 21.15 The pi molecular orbitals in benzene. When the six *p* orbitals shown in (*a*) are combined, they can form a total of six molecular orbitals. Only three of these are occupied, and they are shown in (*b*), (*c*), and (*d*). The net effect of the three filled orbitals (*b*), (*c*), and (*d*) is a double hexagonal ring similar to (*b*) but slightly larger. This contains six electrons, giving three bonds evenly distributed between all six carbon atoms. (Computer-generated.) *(Copyright © 1976 by W. G. Davies and J. W. Moore.)*

Conjugated Systems

In some molecules the delocalization of electron pairs can be very much more extensive than in ozone and benzene. This is particularly true of carbon compounds containing **conjugated chains,** i.e., long chains of alternating single and double bonds. An example is provided by vitamin A_2 which has the structure

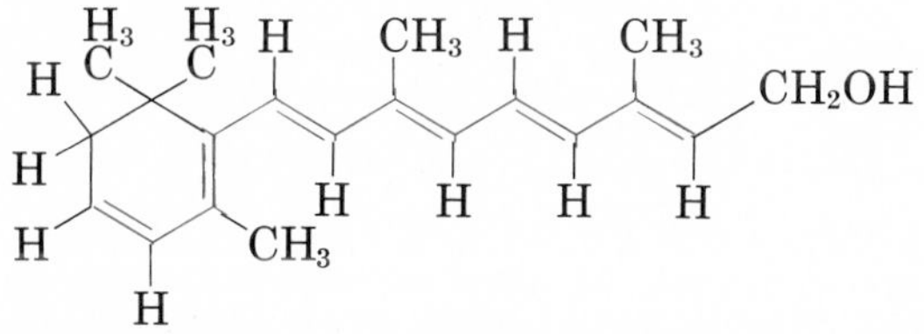

Both the physical and chemical properties of this molecule indicate that all the bonds in the conjugated chain (shown in color) are intermediate in character between single and double bonds and that the pi electrons are free to move over the whole length of the conjugated chain.

A very simple, though approximate, method of handling delocalized electrons mathematically is to treat them as though they were particles in a one-dimensional box of the same length as the chain. In the case of vitamin A_2 the distance from the carbon atom at one end of the conjugated chain to that on the other is about 1210 pm. If we feed this value and the mass of the electron (9.110×10^{-31} kg) into the particle-in-the-box formula [Eq. (5.4)], we can obtain approximate values of the energy levels occupied by the delocalized electrons:

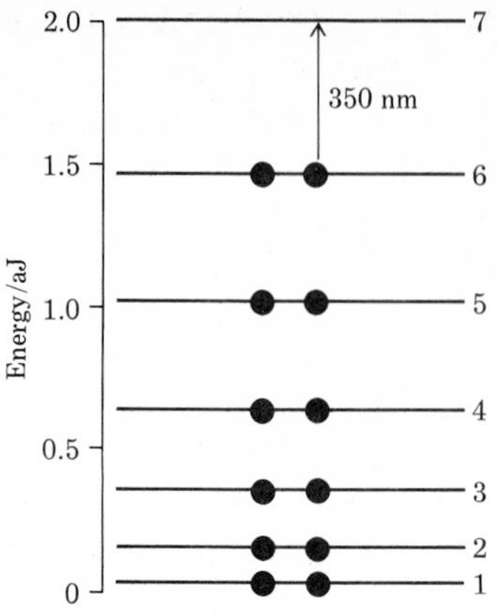

Figure 21.16 The energy levels of the delocalized pi electrons in vitamin A_2 calculated from the particle in a one-dimensional box model. The arrow shows the electronic transition responsible for the absorption band at 350 nm.

$$E_k = \frac{1}{2m}\left(\frac{nh}{2d}\right)^2 \tag{5.4}$$

$$= \frac{1}{2 \times 9.110 \times 10^{-31}\ \text{kg}}\left(\frac{n \times 6.626 \times 10^{-34}\ \text{J s}}{2 \times 1210 \times 10^{-12}\ \text{m}}\right)^2$$

$$E_k = n^2(4.11 \times 10^{-20}\ \text{J}) \tag{21.19}$$

The first seven energy levels derived from this formula are shown in Fig. 21.16. Note that since there are 12 pi electrons (one for each carbon atom on the conjugate chain), these will occupy the six lowest levels in accord with the Pauli principle.

This model allows us to calculate the wavelength of the main absorption band in the spectrum of vitamin A_2. If an electron from the highest occupied level ($n = 6$) is excited to the lowest unoccupied level ($n = 7$), the energy required can be calculated from Eq. (21.19):

$$\Delta E = E_7 - E_6 = (7^2 - 6^2) \times 4.11 \times 10^{-20}\ \text{J}$$

$$= 5.34 \times 10^{-19}\ \text{J}$$

Thus

$$\lambda = \frac{hc}{\Delta E} = \frac{6.626 \times 10^{-34}\ \text{J s} \times 2.998 \times 10^{8}\ \text{m s}^{-1}}{5.34 \times 10^{-19}\ \text{J}}$$

$$= 372\ \text{nm}$$

This is in reasonable agreement with the observed value of 350 nm, considering the approximate nature of the model.

As a general rule, the longer a conjugated chain, the longer the wavelength at which it absorbs. Ethene (C_2H_4), for example, has only one double bond and absorbs at 170 nm. Hexatriene (C_6H_8) has three alternating double bonds and absorbs at 265 nm, while vitamin A_2, with six double bonds, absorbs at 350 nm. It is not difficult to explain this effect in terms of the particle-in-a-box model. According to Eq. (5.4) the energy of a level varies inversely with the square of the length of the box. Thus the longer the conjugated chain, the closer the energy levels will be to each other, and the less energy a photon need have to excite an electron. Naturally the lower the energy of a photon, the longer its wavelength will be.

A similar effect is found for molecules containing several benzene rings. Since these correspond to Lewis formulas of alternating double and single bonds, they can also be regarded as conjugated systems containing delocalized electrons. Experimentally we find that the more benzene rings a molecule contains, the longer the wavelengths at which it absorbs:

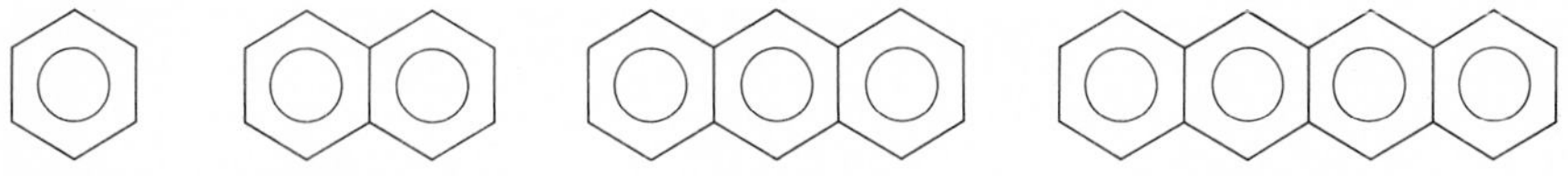

Benzene 260 nm — Naphthalene 280 nm — Anthracene 375 nm — Naphthacene 450 nm

Thus increasing the extent of electron delocalization increases the wavelength at which a molecule will absorb light, whether the electron is delocalized over rings or chains.

This behavior of delocalized electrons is important in the preparation of compounds which strongly absorb visible light, i.e., in the preparation of dyes. Very few compounds which are held together by sigma bonds alone are colored. The electrons are so tightly held that a very energetic photon is needed to excite them. In order for an organic molecule to absorb in the visible region of the spectrum, it must usually contain *very delocalized* pi electrons. Thus most dyes and most colored compounds occurring in living organisms turn out to be large molecules with extensive systems of conjugated double bonds.

SUMMARY

Many of the properties of electromagnetic radiation can be explained if light is thought of as periodically varying electric and magnetic fields (electromagnetic waves). Such waves can be characterized by their frequency ν or their wavelength λ, and their speed of propagation is always $\lambda\nu = c = 2.998 \times 10^8 \text{ m s}^{-1}$. Some properties of light are more easily explained in terms of particles called photons. The energy of a photon is given by $E = h\nu$, where $h = 6.626 \times 10^{-34}$ J s and is called Planck's constant.

When any element is heated to a high temperature or excited in a discharge tube, it gives a line spectrum. Niels Bohr was able to predict the wavelengths of the lines in the spectrum of hydrogen by means of a theory which assigned the single electron to specific energy levels and hence to orbits of specific radius. Absorption of an appropriate quantity of energy can raise the hydrogen atom from a lower to a higher energy level, while emission of electromagnetic radiation corresponds to a change from a higher to a lower energy level. Although Bohr's theory is quantitatively accurate only for hydrogen, his idea of energy levels is useful for all other atoms and even for molecules.

In the case of molecules, energy levels arise because of different speeds and kinds of molecular vibrations and rotations as well as because electrons are moved farther from or closer to positively charged nuclei. In organic compounds some groups of atoms vibrate at much the same frequency no matter what molecule they are in. The energy levels of such vibrations usually differ by roughly the energies of infrared photons, and many organic functional groups can be identified by the characteristic frequencies at which they absorb infrared radiation. When molecules absorb visible or ultraviolet light, band spectra occur. Some of the energy of each absorbed photon goes to excite an electron, but varying amounts also increase vibrational and rotational energies. Thus photons are absorbed over a broad range of frequencies and wavelengths.

The most convenient theory by which the electronic energies of molecules can be predicted is the molecular-orbital theory. It assumes that electrons in a molecule occupy orbitals which are not confined to a single atom but rather extend over the entire molecule. Bonding molecular orbitals involve constructive interference between two electron waves, while antibonding molecular orbitals involve destructive interference. An electron occupying an antibonding MO is higher in energy than it would be if the atoms were not bonded together, and so antibonding electrons cancel the effect of bonding electrons. This explains why molecules such as He_2 or Ne_2 do not form.

Molecular-orbital theory is especially useful in dealing with molecules for which resonance structures must be drawn. Because molecular orbitals can be delocalized over several atoms, there is no need for several resonance structures in the case of molecules like O_3 and C_6H_6. The greater the extent of electron delocalization, the smaller the separation between molecular energy levels and the longer the wavelength at which absorption of ultraviolet or visible light can occur. Thus compounds containing long chains of alternating single and double bonds or having several benzene rings connected together often absorb visible light and are colored.

QUESTIONS AND PROBLEMS

21.1 Compare the terms *electromagnetic radiation* and *light*. Do they both mean exactly the same thing?

21.2 Criticize the following statement: Infrared radiation and x-rays have quite different wavelengths, and so they must travel at different speeds.

21.3 Express the following true statement as a mathematical equation: The frequency of electromagnetic radiation is inversely proportional to the wavelength, and the proportionality constant is the speed of light.

21.4 Fill in the missing values in the following table which refers to electromagnetic radiation in a vacuum:

Radiation Sample	λ	ν	c
1	75.0 pm	______	2.998×10^8 m s^{-1}
2	______	4.0×10^4 Hz	______
3	15.7 m	______	______
4	______	6.5×10^{10} s^{-1}	______
5	______	______	2.998×10^8 m s^{-1}

If it is not possible to calculate one or more of the values, explain why not.

21.5 Indicate the region of the electromagnetic spectrum in which each of the following photons would fall. In the case of visible light, indicate the color.

a $\lambda = 450$ nm
b $\nu = 3 \times 10^{10}$ s^{-1}
c $E = 0.35$ aJ
d $\nu = 8.0 \times 10^{19}$ Hz
e $\lambda = 50$ μm
f $\nu = 4.5 \times 10^{14}$ Hz
g $E = 5.0 \times 10^{-27}$ J
h $\lambda = 5.75 \times 10^{-7}$ m

21.6 Criticize the following statement: The amplitude of a wave represents the distance that whatever is fluctuating increases or decreases from the average value, and so the amplitude must be proportional to wavelength.

21.7 The frequency of electromagnetic radiation can be expressed in reciprocal seconds (per second). This means that the frequency is some number per second. What is it that is counted per second?

21.8 Suppose that two waves have identical wavelengths and are exactly in phase (have nodes at the same positions, as shown in Fig. 21.3). If one wave has 5 times the amplitude of the other, (*a*) what is the amplitude of the resultant wave when the two waves interfere constructively (Express your result in terms of the amplitude of the smaller initial wave.); (*b*) what is the amplitude of the resultant wave when the two waves interfere destructively?

21.9 Repeat Problem 21.8 but assume that you have three waves whose amplitudes are in the ratio 1:2:3. (*a*) What is the resultant amplitude if all three waves interfere constructively? (*b*) What is the resultant amplitude if the first two waves interfere constructively but the third interferes destructively with them?

21.10 Calculate the energy for each of the first four photons listed in Problem 21.4.

***21.11** If it were at rest, a photon would have zero mass and zero kinetic energy. However, photons always move at the speed of light. Calculate the mass of a photon of blue visible light; $\lambda = 450$ nm. (*Hint:* See Sec. 19.7.)

21.12 Criticize the following statement: Photons whose wavelengths are shorter than 50 μm are relatively harmless to humans because they contain insufficient energy to disturb chemical bonds. (There are at least three things wrong with this statement.)

21.13 Express the following true statement as a mathematical equation: The energy of a photon is inversely proportional to its wavelength, and the proportionality constant is the product of the speed of light and Planck's constant.

21.14 Distinguish between an emission spectrum and an absorption spectrum. Which is more commonly used to study atoms? Which is more commonly used to study molecules?

21.15 Explain as clearly as you can, and in your own words, why line spectra are obtained for atoms while broad-band spectra are obtained for molecules in the visible region.

21.16 Use both the Balmer formula and the Rydberg equation to calculate the wavelengths of the spectral emission lines produced when a hydrogen atom drops to the $n = 2$ energy level from each of the following levels: (*a*) $n = 4$; (*b*) $n = 5$; (*c*) $n = 6$; (*d*) $n = 7$. How good is the agreement between the two formulas? (Express this as a percent difference.)

21.17 Calculate the change in energy of a hydrogen atom and the frequency and wavelength of the radiation emitted or absorbed for each of the following transitions. In each case state whether radiation is emitted or absorbed.

a From $n = 5$ to $n = 2$
b From $n = 3$ to $n = 1$
c From $n = 3$ to $n = 7$
d From $n = 4$ to $n = 1$
e From $n = 2$ to $n = 6$
f From $n = 7$ to $n = 9$

21.18 Calculate the radius of the circular orbit predicted by the Bohr theory for each of the first three energy levels of the hydrogen atom. Compare your results with the electron dot-density diagrams and boundary-surface diagrams shown in Figs. 5.7 and 5.8. Roughly, how much electron density seems to be outside Bohr's radius and how much inside?

21.19 Define each of the following terms:

a Balmer series
b Excited state
c Greenhouse effect
d Ground state

21.20 Calculate the ionization energy of a hydrogen atom which is in each of the following excited states: (*a*) $n = 2$; (*b*) $n = 3$; (*c*) $n = 4$; (*d*) $n = 10$.

21.21 Suppose the ionization of a hydrogen atom occurred stepwise, each step corresponding to an increase in the principal quantum number by 1. That is, the first step would be from $n = 1$ to $n = 2$, the second from $n = 2$ to $n = 3$, and so on. How many steps would be required so that the total increase in energy of the hydrogen atom would be 90 percent of the ionization energy? how many to reach 99 percent?

***21.22** What conclusions can you draw from your results in Problem 21.21? Can you apply these conclusions to the hydrogen emission spectrum in Fig. 21.6? What wavelength corresponds with a photon whose energy is exactly the total ionization energy of hydrogen? Locate that wavelength on the hydrogen spectrum in Fig. 21.6.

21.23 What is the historical origin of the labels *s*, *p*, *d*, and *f* used to distinguish subshells when giving electron configurations?

21.24 Calculate the change in energy of a sodium atom and the wavelength of the photon absorbed or emitted when the valence electron undergoes each of the following transitions: (*a*) 3*d* to 3*p*; (*b*) 4*f* to 3*d*; (*c*) 7*s* to 3*p*; (*d*) 7*p* to 3*s*.

21.25 Criticize the following statement: Since infrared spectra are absorption spectra, excited states of molecules are not involved when infrared spectra are measured.

21.26 What similarities would you expect between the infrared spectra of C_2H_5OH and CH_3OH? What differences?

21.27 Criticize the following statement: Careful analysis of the infrared spectrum of a pure substance invariably permits identification of each and every element present as well as determination of the correct empirical and molecular formula.

21.28 Which of the following bonds would be expected to produce prominent absorption peaks in an infrared spectrum? Give reasons for your answer in each case. (*a*) C≡N; (*b*) C—C; (*c*) F—F; (*d*) N—F.

21.29 Explain clearly the importance of atmospheric carbon dioxide and water vapor in controlling the temperature of the surface of the earth. Based on information given in the text (and on outside reading, if possible), what is your opinion as to the seriousness of global increases in CO_2 produced by combustion of fossil fuels?

21.30 Why are visible and ultraviolet spectra of molecules often referred to as *electronic spectra?*

21.31 Define each of the following terms:

a Valence-bond model
b Positive overlap
c Bonding MO
d Bond order
e Band spectrum
f Photodissociation
g Homonuclear diatomic molecule
h Electron delocalization

21.32 Write the molecular electron configuration for each of the following (possibly hypothetical) homonuclear diatomic molecules or ions: (*a*) O_2^{2-}; (*b*) B_2; (*c*) B_2^{2+}; (*d*) C_2; (*e*) Be_2^{2-}; (*f*) F_2^+; (*g*) O_2^-; (*h*) N_2; (*i*) Li_2^{2+}; (*j*) C_2^{2-}.

21.33 Give the bond order and comment on the probable existence of each of the molecules or ions from Problem 21.32.

21.34 What correlation exists between bond orders derived from MO theory and single, double, or triple bonds found in Lewis diagrams? Choose three examples from Problem 21.32, draw their Lewis diagrams, and show the correlation explicitly. Your examples should include a single bond, a double bond, and a triple bond.

21.35 Distinguish between paramagnetism and diamagnetism. Is either the same as the magnetism ordinarily encountered in a bar magnet?

21.36 For those molecules from Problem 21.32 which you predicted would exist, predict which are diamagnetic and which are paramagnetic.

21.37 Most of the benzene rings shown in this textbook have been drawn as a hexagon with a circle inside. In terms of MO theory, why is such a diagram appropriate? What do the hexagon and the circle represent?

21.38 What is a *conjugated chain?* Explain the relationship between the length of a conjugated chain and the wavelength at which maximum absorption of radiation by a molecule occurs.

21.39 The molecular-orbital picture of benzene shows a "double doughnut" of electron density above and below the plane of the six C and six H atoms. (*a*) How many electrons occupy this double doughnut? (*b*) How many molecular orbitals are occupied by those electrons? (*c*) Are all these MO's identical in shape? in energy?

***21.40** In your own words distinguish as clearly as possible between the valence-bond theory and the molecular-orbital theory of chemical bonding. In what ways are both theories similar?

*21.41 A piece of blue paper looks blue because it reflects only blue light (350 nm $< \lambda <$ 500 nm) to your eyes. Suppose you viewed such an object in light obtained by heating NaCl in a bunsen-burner flame. What color would it appear to be?

***21.42** Although water is an excellent solvent for many substances, it is almost never used to dissolve an unknown solid so that an infrared spectrum can be obtained. Why?

*21.43 A mixture of $H_2(g)$ and $Cl_2(g)$ can be stored safely in the dark, but when exposed to high-intensity visible and ultraviolet light, these two substances react explosively. Why?

***21.44** It is stated in the text that Eq. (21.11) can be derived from Eqs. (21.10) and (21.5). Carry out this derivation by using Eq. (21.10) to obtain an expression for u^2. Then substitute this expression into the middle term of Eq. (21.5) and rearrange to obtain Eq. (21.11).

***21.45** It is relatively easy to answer Question 21.41 by doing an experiment. Obtain your instructor's permission and do such an experiment in the laboratory.

chapter twenty-two

METALS

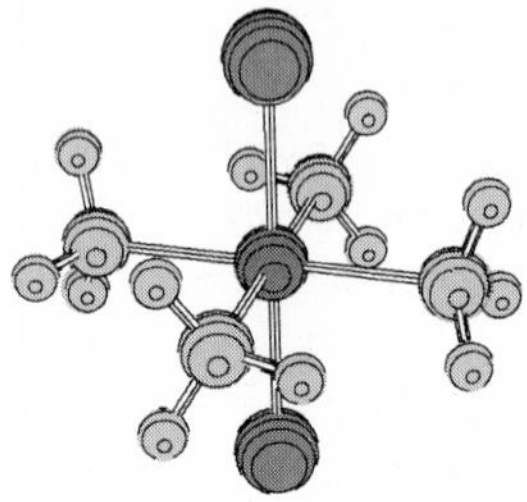

Approximately three-quarters of the known elements display the macroscopic properties characteristic of metals. They conduct both heat and electricity very well; they have shiny surfaces; they are capable of being shaped by hammering (malleable) and also of being drawn into wires (ductile). Chemical properties of the metals include a tendency to lose electrons and form positive ions, and the ability of their oxides to function as bases. The extent of these characteristics varies from one metal to another. Several borderline cases such as B, Si, Ge, As, Sb, and Te are difficult to classify as metals or nonmetals. These elements are usually referred to as the **metalloids** or **semimetals.** As you will recall from Chap. 6 one can draw a zigzag line across the periodic table from B to At which separates the metals from the nonmetals and semimetals. This line is clearly indicated in the periodic table inside the front cover of the book.

In this chapter we will be concerned mainly with the *transition metals*. Metals which are representative elements have already been covered in Chap. 12, while a discussion of the lanthanoid and actinoid metals is beyond the scope of an introductory text. Since transition metals contain *d* electrons in their valence shell, their chemistry is somewhat different from that of the representative elements. In particular they form a family of compounds called **complex compounds** or **coordination compounds** which are very different from those we have encountered up to this point.

22.1 METALLIC BONDING

Most metals have very compact crystal structures involving either the body-centered cubic, face-centered cubic, or hexagonal closest-packed lattices described in Sec. 10.1. Thus every atom in a metal is usually surrounded by 8 or 12 equivalent nearest neighbors. How can each atom be bonded to so many of its fellow atoms? Although there are plenty of electropositive atoms to donate electrons, there are no electronegative atoms to receive them, and so ionic bonding seems unlikely. Ordinary covalent bonding can also be ruled out. Each covalent bond would require one electron from each atom, and no metal has 12 valence electrons.

A valuable clue to the nature of bonding in metals is provided by their ability to conduct electricity. Electrons can be fed into one end of a metal wire and removed from the other end without causing any obvious change in the physical and chemical properties of the metal. To account for this freedom of movement modern theories of metallic bonding assume that the valence electrons are *completely delocalized;* that is, they occupy molecular orbitals belonging to the metallic crystal as a whole. These delocalized electrons are often referred to as an **electron gas** or an **electron sea.** Positive metal ions produced by the loss of these valence electrons can then be thought of as "floating" in this three-dimensional sea. Each ion is held in place by the attraction of the negatively charged electron sea and the repulsion of its fellow positive ions.

In order to see how MO theory can be applied to metals, let us first consider the simplest case, lithium. If two lithium atoms are brought together, the 1*s* core electrons remain essentially unchanged since there is virtually no overlap between them. The 2*s* orbitals, by contrast, overlap extensively and produce both a bonding and an antibonding orbital. Only the bonding orbital will actually be occupied by the two electrons, as shown in Fig. 22.1. Somewhat higher than these two orbitals are a group of six unoccupied orbitals produced by the overlap of six 2*p* atomic orbitals (three on each atom).

Suppose now we add a third atom to the two already considered so that we form a triangular molecule of formula Li_3. As shown in the figure, the overlap of three 2*s* orbitals produces a lower group of three orbitals, while the overlap of three times three 2*p* orbitals produces a higher group of nine orbitals. Again the total number of molecular orbitals is equal to the number of atomic orbitals from which they are derived.

Continuing to add lithium atoms in this fashion, we soon attain a cluster of 25 lithium atoms. The energy-level situation for a cluster this size is a lower group of 25 MO's, all deriving from 2*s* atomic orbitals, and a higher group of 75 MO's, all deriving from 2*p* atomic orbitals. Note how closely spaced these energy levels have become. This is in line with the tendency noted in the previous chapter for the energy levels to get closer the greater the degree of delocalization.

Finally, if we add enough lithium atoms to our cluster to make a visible, weighable sample of lithium, say 10^{20} atoms, the energy spacing between the molecular orbitals becomes so small it is impossible to indicate in the figure or even to measure. In effect an electron jumping among these levels

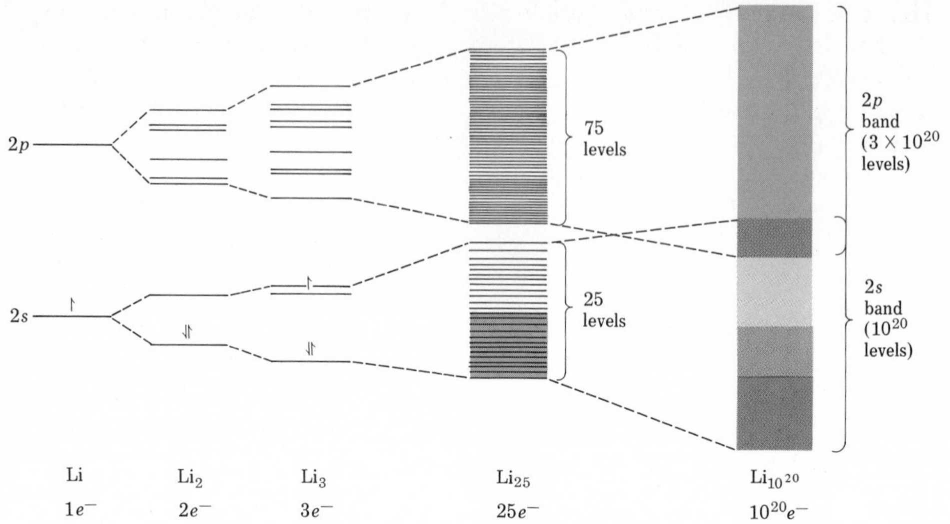

Figure 22.1 Molecular-orbital energies corresponding to delocalization of valence electrons over increasing numbers of Li atoms. A 1-mg sample of Li would contain nearly 10^{20} atoms. The corresponding orbital energies are so closely spaced that they constitute essentially continuous bands.

can have any energy within a broad band from the lowest to highest. In consequence this view of electronic structure in solids is often referred to as the **band theory of solids.**

It should also be clear from Fig. 22.1 that all the available molecular orbitals need not be completely filled with electrons. In the case of lithium, for example, a sample containing 10^{20} atoms would have 10^{20} valence electrons. Since each atom would have a single 2*s* orbital as well as three 2*p* orbitals, there would be 1×10^{20} MO's in the 2*s* band and 3×10^{20} MO's in the 2*p* band. If all electrons were paired, only the 0.5×10^{20} MO's of lowest energy in the 2*s* band would be required to hold them. Note that there is a nice correspondence between the half-filled 2*s* band of the macroscopic sample and the half-filled 2*s* orbital of an individual Li atom.

According to the band theory, it is this partial filling which accounts for the high electrical and thermal conductance of metals. If an electric field is applied to a metallic conductor, some electrons can be forced into one end, occupying slightly higher energy levels than those already there. As a consequence of delocalization this increased electronic energy is available throughout the metal. It therefore can result in an almost instantaneous flow of electrons from the other end of the conductor.

A similar argument applies to the transfer of thermal energy. Heating a small region in a solid amounts to increasing the energy of motion of atomic nuclei and electrons in that region. Since the nuclei occupy specific lattice positions, conduction of heat requires that energy be transferred among nearest neighbors. Thus when the edge of a solid is heated, atoms in that region vibrate more extensively about their average lattice positions.

They also induce their neighbors to vibrate, eventually transferring heat to the interior of the sample. This process can be speeded up enormously if some of the added energy raises electrons to higher energy MO's within an incompletely filled band. Electron delocalization permits rapid transfer of this energy to other atomic nuclei, some of which may be quite far from the original source.

When an energy band is completely filled with electrons, the mechanism just described for electrical and thermal conduction can no longer operate. In such a case we obtain a solid which is a very poor conductor of electricity, or an **insulator.** At first glance we might expect Be, Mg, and other alkaline earths to be insulators like this. Since atoms of these elements all contain filled 2*s* subshells, we would anticipate a filled 2*s* band in the solid for all of them. That this is not the case is due to the relatively small energy difference between the 2*s* and 2*p* levels in these atoms. As you can see from Fig. 22.1, this small separation results in an *overlap* between the 2*s* and 2*p* bands. Thus electrons can move easily from the one band to the other and provide a mechanism for conduction.

Figure 22.2 shows four different possibilities for band structure in a solid. For a solid to be a conductor, a band must be either partially filled or must overlap a higher unfilled band. When there is a very large energy gap between bands and the lower band is filled, we have an insulator. If the gap is quite small, we get an intermediate situation and the solid is a **semiconductor.** All the semimetals found along the stairstep diagonal in the periodic table, notably germanium, have a band structure of this type.

In a semiconductor we find that collisions among atoms and electrons in the crystal are occasionally energetic enough to excite an electron into the top band. As a result there are always a small number of electrons in this band and an equal number of **holes** (orbitals from which electrons have

Figure 22.2 Band structures of conductors, semiconductors, and insulators. Note that conductors may have a partially filled band or a filled band which overlaps an empty one. In semiconductors and insulators, band separation becomes progressively larger (MO's containing electrons are color-coded.)

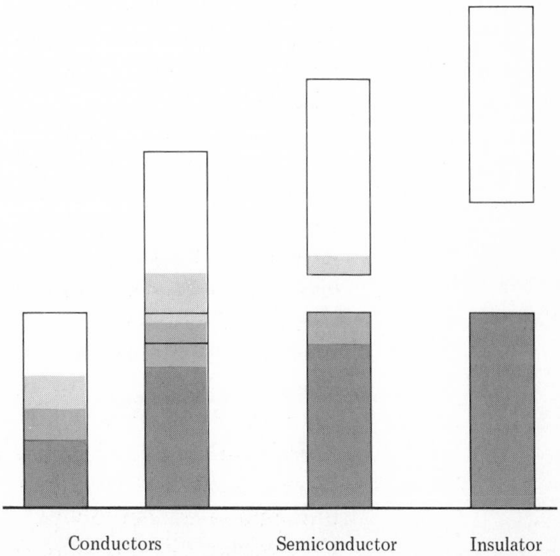

been removed) in the lower-energy band. The excited electrons can carry electrical current because many different energy levels are available to them. So can the holes—other electrons from the nearly filled band can move up or down into them, a process which decreases or increases the energy of the hole.

In a metal, electrical conductivity decreases as the temperature is raised because the nuclei vibrate farther from their rest positions and therefore get in the way of moving valence electrons more often. Exactly the opposite behavior is found for semiconductors. With increasing temperature, more and more electrons are excited to the higher-energy **conduction band** so that more current can be carried. Excitation of electrons to the conduction band can also be accomplished by a photon, a phenomenon known as **photoconduction.** Selenium metal is often used in this way as a photocell in light meters and "electric eyes."

The electron-sea model of metals not only explains their electrical properties but their malleability and ductility as well. When one layer of ions in an electron sea moves along one space with respect to the layer below it, a process we can represent by

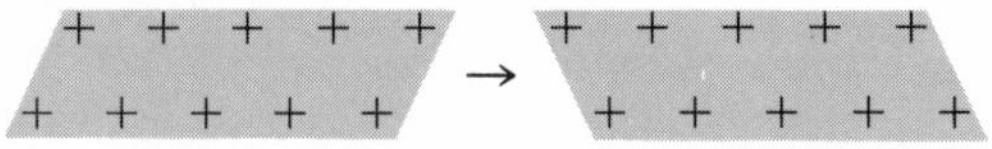

the final situation is much the same as the initial. Thus if we hit a metal with a hammer, the crystals do not shatter but merely change their shape. This is very different from the behavior of ionic crystals that was illustrated in Fig. 6.6.

The electron-sea model also enables us to explain, at least partially, why the metallic bond is noticeably stronger for some metals than others. While the alkali metals and some of the alkaline-earth metals can be cut with a knife, metals like tungsten are hard enough to scratch the knife itself. A good indication of how the strength of the metallic bonding varies with position in the periodic table is given by the melting point. As can be seen from Fig. 22.3, if the melting point of the metals is plotted against the group number for the three long periods, there is a sharp increase from group IA to group VB or VIB, after which there is a leveling off. Finally the melting point again drops to quite low values. A similar behavior is found for other properties such as boiling point, enthalpy of fusion, density, and hardness.

The initial increases in the strength of metallic bonding as we move from group IA to VIB can be explained in terms of the number of valence electrons the metal is capable of contributing to the electron sea. The more electrons an atom loses, the larger will be the charge of the positive ion embedded in the electron sea and the greater will be the electron probability density of the electron sea itself. Thus the more electrons which are lost, the more tightly the ions will be held together. Chromium with six valence electrons is thus much harder than sodium with one.

This trend cannot continue indefinitely, however. The more electrons that are removed from an atom, the more energy it takes to remove the next electron. Eventually we find that more energy is needed to remove an elec-

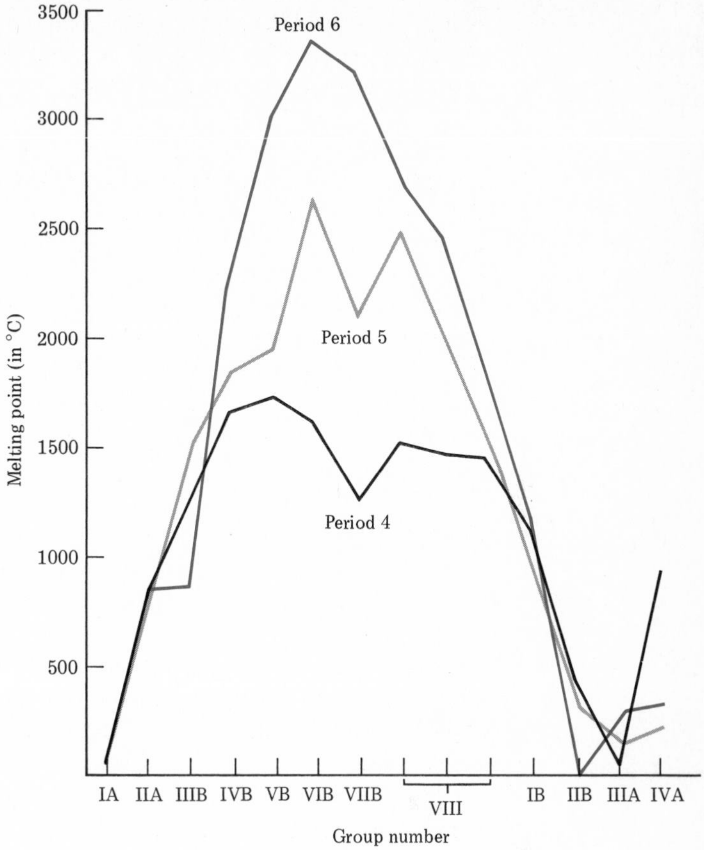

Figure 22.3 Melting points of the metals of the fourth, fifth, and sixth periods.

tron from a metal nucleus than is liberated by placing it in the electron sea. The strength of the bonding thus begins to level off and eventually to drop.

It should be pointed out that metallic bonding strength is not solely dependent on the number of valence electrons (or the periodic group number) of an element. Other factors such as atomic radius and type of crystal lattice are also important. Nevertheless it is useful to remember that melting points and other properties related to metallic bond strength reach their maximum at about the middle of each transition series.

22.2 METALLURGY

Most metals are chiefly useful in elemental form, but they usually occur in compounds on the earth's surface. An **ore** is a naturally occurring material from which one or more useful elements or compounds may be obtained at a cost that is economically feasible. As can be seen from Table 22.1, most metal ores are oxides, carbonates, or sulfides. A few of the least electropositive metals occur as the element.

TABLE 22.1 Occurrence of Metals.

Type of Ore	Examples
Native metals	Cu, Ag, Au, As, Sb, Bi, Pd, Pt
Oxides	Al_2O_3, Fe_2O_3, Fe_3O_4, SnO_2, MnO_2, TiO_2, $FeO{\cdot}Cr_2O_3$, $FeO{\cdot}WO_3$, Cu_2O, ZnO
Carbonates	$CaCO_3$, $CaCO_3{\cdot}MgCO_3$, $MgCO_3$, $FeCO_3$, $PbCO_3$, $BaCO_3$, $SrCO_3$, $ZnCO_3$, $MnCO_3$, $CuCO_3{\cdot}Cu(OH)_2$
Sulfides	Ag_2S, Cu_2S, CuS, PbS, ZnS, HgS, FeS·CuS, FeS_2, Sb_2S_3, Bi_2S, MoS_2, NiS, CdS
Halides	NaCl, KCl, AgCl, $KCl{\cdot}MgCl_2{\cdot}6H_2O$, NaCl and $MgCl_2$ in seawater
Sulfates	$BaSO_4$, $SrSO_4$, $PbSO_4$, $CaSO_4{\cdot}2H_2O$, $CuSO_4{\cdot}2Cu(OH)_2$
Silicates	$Be_3Al_2Si_6O_{18}$, $ZrSiO_4$, $Sc_2Si_2O_7$
Phosphates	$LaPO_4$, $LiF{\cdot}AlPO_4$

Whether a mineral can usefully be regarded as an ore or not depends on such factors as how *concentrated* it is, its exact *location,* and whether there is a suitable *process* for extracting the metal. As the more accessible higher-grade ores become exhausted, less accessible and lower-grade ores are becoming increasingly utilized with a consequent shift in the meaning of the word ore. It is conceivable that many silicates could become sources of metals, notably aluminum. Currently, though, silicates are expensive to decompose chemically, and silicates form ores only for relatively expensive metals like zirconium and beryllium.

The processing of ores may be divided into three steps. Often concentration or other **beneficiation** is required to remove worthless material (**gangue**) or to convert the mineral into an appropriate form for subsequent processing. The second and most-important step is **reduction** of the metal from a positive oxidation state. This may involve elevated temperatures, chemical reducing agents, electrolysis, or some combination of these treatments. Usually a third step, **refining,** is required to achieve the purity (or precise mixed composition in the case of an alloy) desired in the final product.

Beneficiation

Beneficiation may involve physical or chemical processes. Often, as in the case of panning for gold, the desired ore or metal is denser than the gangue. The latter can be suspended in a stream of water and flushed away. The iron ore magnetite, Fe_3O_4, is ferrimagnetic. It can be separated from abundant deposits of taconite by grinding to a fine slurry in water. Passing this suspension over powerful electromagnets removes the Fe_3O_4, more than doubling the concentration of Fe. Both the physical beneficiation processes just described produce large quantities of tailings—water suspensions of the gangue. Usually the silicate and other particles are trapped in a settling basin, but in one case large quantities of taconite tailings have been dumped into Lake Superior for several decades.

Flotation This is a partly physical and partly chemical beneficiation used to concentrate ores of copper, lead, and zinc. These metals have considerable affinity for sulfur. (Table 22.1 lists sulfide ores for each of them.) The ore is ground and suspended with water in large tanks. Flotation agents such as the xanthates are added and a stream of air bubbles passed

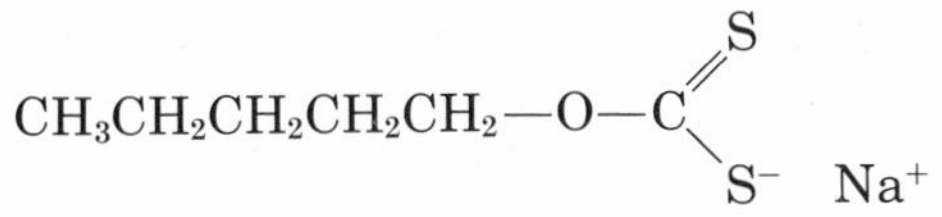

A xanthate

up through the tank. The sulfur-containing end of the xanthate molecule is attracted to the desired metal, while the hydrocarbon end tends to avoid water (like the hydrocarbon end of a soap molecule, Sec. 20.3). The metal-containing particles are swept to the top of the tank in a froth much akin to soap bubbles. The concentrated ore may then be skimmed from the water surface. By means of flotation, ores containing as little as 0.3% copper can be concentrated to 20 to 30% copper, making recovery of the metal economically feasible.

Chemical Processes Many ores can be chemically separated from their gangue by means of acid or base. Thus copper ore is often *leached* with acid which dissolves copper oxides and carbonates and leaves behind most silicates:

$$CuCO_3 + H_2SO_4(aq) \rightarrow CuSO_4(aq) + H_2O + CO_2 \qquad (22.1)$$

In the *Bayer process* bauxite is dissolved in 30% NaOH:

$$Al_2O_3(s) + 2OH^-(aq) + 3H_2O(l) \rightarrow 2Al(OH)_4^-(aq) \qquad (22.2)$$

and then reprecipitated by adding acid. Impurities such as Fe_2O_3 and SiO_2 are eliminated in this way.

Many sulfide and carbonate ores are *roasted* in air in order to convert them into oxides, a form more suitable for further processing:

$$2ZnS(s) + 3O_2(g) \rightarrow 2ZnO(s) + SO_2(g) \qquad (22.3)$$

$$PbCO_3(s) \rightarrow PbO(s) + CO_2(g) \qquad (22.4)$$

The production of SO_2 in Eq. (22.3) is a notorious source of air pollution.

Reduction of Metals

The ease with which a metal may be obtained from its ore varies considerably from one metal to another. Since the majority of ores are oxides or can be converted to oxides by roasting, the free-energy change accompanying the decomposition of the oxide forms a convenient measure of how readily a metal may be obtained from its ore. Values of the free energy change per mol O_2 produced are given in Table 22.2 for a representative sample of metals at 298 and 2000 K. A high positive value of $\Delta G_m^\ominus$ in this table indicates a very stable oxide from which it is difficult to remove the oxygen and obtain the metal, while a negative value of $\Delta G_m^\ominus$ indicates an oxide which will spontaneously decompose into its elements. Note how the

TABLE 22.2 Free-Energy Changes for the Decomposition of Various Oxides at 298 and 2000 K.

Reaction	$\Delta G_m^\ominus$(298 K)/kJ mol^{-1}	$\Delta G_m^\ominus$(2000 K)/kJ mol^{-1}
$\frac{2}{3}Al_2O_3 \rightarrow \frac{4}{3}Al + O_2$	+1054	+691
$2MgO \rightarrow 2Mg + O_2$	+1138	+643
$\frac{2}{3}Fe_2O_3 \rightarrow \frac{4}{3}Fe + O_2$	+ 744	+314
$SnO_2 \rightarrow Sn + O_2$	+ 520	+ 43
$2HgO \rightarrow 2Hg + O_2$	+ 118	−381
$2Ag_2O \rightarrow 4Ag + O_2$	+ 22	−331
	Formation of CO_2	
$C(s) + O_2(g) \rightarrow CO_2(g)$	− 394	−396

value of $\Delta G_m^\ominus$ decreases with temperature in each case. This is because a gas (oxygen) is produced by the decomposition, and ΔS is accordingly positive.

The two metals in Table 22.2 which are easiest to obtain from their oxide ores are Hg and Ag. Since the $\Delta G_m^\ominus$ value for the decomposition of these oxides becomes negative when the temperature is raised, simple heating will cause them to break up into O_2 and the metal. The next easiest metals to obtain are Sn and Fe. These can be reduced by coke, an impure form of C obtained by heating coal. Coke is the cheapest readily obtainable reducing agent which can be used in metallurgy. When C is oxidized to CO_2, the free-energy change is close to -395 kJ mol^{-1} over a wide range of temperatures. This fall in free energy is not quite enough to offset the free-energy rise when Fe_2O_3 and SnO_2 are decomposed at 298 K, but is more than enough if the temperature is 2000 K. Thus, for example, if Fe_2O_3 is reduced by C at 2000 K, we have, from Hess' law,

$$\begin{array}{ll} \frac{2}{3}Fe_2O_3(s) \rightarrow \frac{4}{3}Fe(l) + O_2(g) & \Delta G_m^\ominus = +314 \text{ kJ mol}^{-1} \\ C(s) + O_2(g) \rightarrow CO_2(g) & \Delta G_m^\ominus = -396 \text{ kJ mol}^{-1} \\ \hline \frac{2}{3}Fe_2O_3 + C(s) \rightarrow \frac{4}{3}Fe(l) + CO_2 & \Delta G_m^\ominus = -82 \text{ kJ mol}^{-1} \end{array}$$

Thus $\Delta G_m^\ominus$ for the reduction is negative, and the reaction is spontaneous.

The two metals in Table 22.2 which are most difficult to obtain from their ores are Mg and Al. Since they cannot be reduced by C or any other readily available cheap reducing agent, they must be reduced electrolytically. The electrolytic reduction of bauxite to yield Al (the Hall process) has already been described in Chap. 17.

Reduction of Iron

Since iron is the most important metal in our industrial civilization, its reduction from iron ore in a blast furnace (Fig. 22.4) deserves a detailed description. The oxides present in most iron ores are Fe_2O_3 and Fe_3O_4. These oxides are reduced stepwise: first to FeO and then to Fe. Ore, coke, and limestone are charged to the furnace through an airlock-type pair of valves at the top. Near the bottom a blast of air, preheated to 900 to 1000 K, enters through blowpipes called tuyères. Oxygen from the air blast

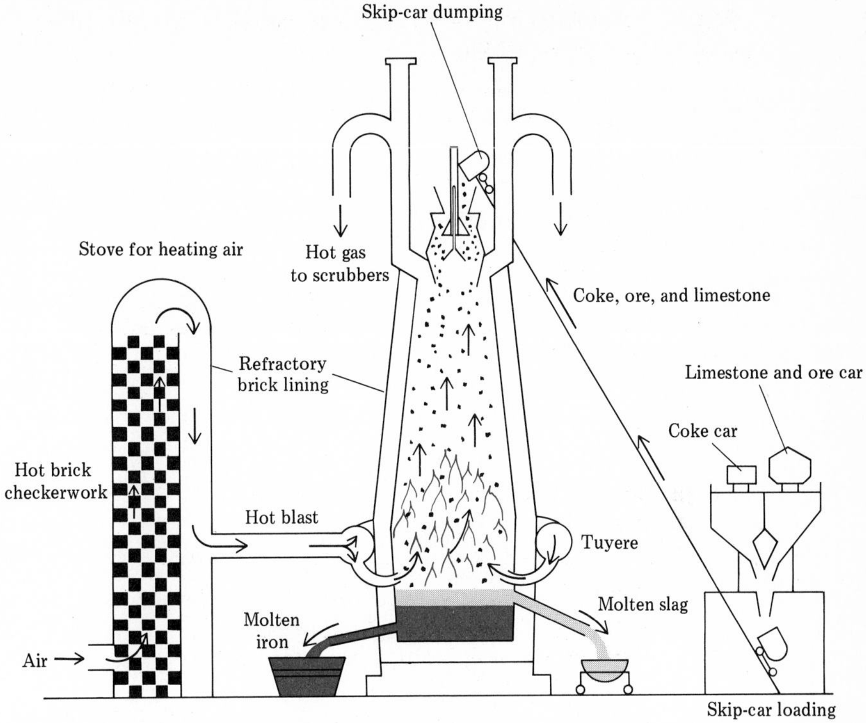

Figure 22.4 Schematic diagram of a blast furnace for smelting iron.

reacts with carbon in the coke to form carbon monoxide and carbon dioxide, releasing considerable heat. The blast carries these gases up through the ore, coke, and limestone, and they exit from the top of the furnace.

By the time the ore works its way to the lower part of the furnace, most of the Fe_2O_3 has already been reduced to FeO. In this region, temperatures reach 1500 to 2000 K, high enough to melt FeO and bring it into close contact with the coke. Most of the FeO is reduced by direct reaction with carbon, the latter being oxidized to carbon monoxide:

$$2C(s) + 2FeO(l) \rightarrow 2Fe(l) + 2CO$$

$$\Delta G_m^\ominus\ (2000\ \text{K}) = -280\ \text{kJ mol}^{-1} \qquad (22.5)$$

The molten iron produced by this reaction drips to the bottom of the furnace where it is collected and occasionally tapped off.

Higher in the furnace, temperatures fall below the melting points of the iron oxides. Because there is little contact between solid chunks of ore and of coke, direct reduction by solid carbon is rather slow. Gaseous carbon monoxide contacts all parts of the ore, however, and reacts much more rapidly:

$$CO(g) + Fe_2O_3(s) \rightarrow CO_2(g) + 2FeO(s) \qquad (22.6)$$

$$CO(g) + FeO(s) \rightarrow CO_2(g) + Fe(s) \qquad (22.7)$$

Thus much of the "carbon reduction" in ironmaking is actually carried out by carbon monoxide.

The gangue in iron ore consists mainly of silicates and silica, SiO_2. These impurities are removed in **slag.** Limestone added with coke and ore is calcined (decomposed to the oxide) by the high temperatures of the blast furnace:

$$CaCO_3(s) \xrightarrow{1100\text{ K}} CaO(s) + CO_2(g)$$

Lime (CaO) serves as a flux, reducing the melting points (mp) of silica (SiO_2) and silicates:

$$\underset{\text{mp = 2853 K}}{CaO(s)} + \underset{\text{mp = 1986 K}}{SiO_2(s)} \rightarrow \underset{\text{mp = 1813 K}}{CaSiO_3(l)}$$

The liquid silicates flow rapidly down through the hottest part of the furnace. This helps to prevent reduction of silica to silicon, hence yielding purer iron. The slag is less dense than molten iron and immiscible with it. Therefore the slag floats on the surface of the iron and can easily be tapped off.

Although most blast-furnace iron now goes directly to a steelmaking furnace in molten form, much of it used to be run into molds where it hardened into small ingots called *pigs* because of their shape. Consequently blast-furnace iron is still referred to as pig iron. A single large blast furnace may produce more than 10^6 kg iron per day. For each kilogram of iron, 2 kg iron ore, 1 kg coke, 0.3 kg limestone, 4 kg air, 63 kg water, and 19 MJ of fossil-fuel energy are required. The furnace produces 0.6 kg slag and 5.7 kg flue gas per kg iron. Nearly 5 percent of the iron ore is lost in the form of small particles suspended in the flue gas unless, as in the furnace shown in Fig. 22.4, air-pollution controls are installed. The latter trap FeO particles for recycling to the furnace and also make the flue gas (which contains about 12% CO and 1% H_2) suitable as a fuel for preheating air fed to the tuyères. Thus control of blast-furnace air pollution (a major contributor to the one-time "smoky city" reputations of Pittsburgh, Pennsylvania and Gary, Indiana) also conserves ore supplies and energy resources.

Refining of Metals

Once a metal is reduced, it is still not necessarily pure enough for all uses to which it might be put. An obvious example is the brittleness and low tensile strength of pig iron, characteristics which make it suitable for casting, but little else. These adverse properties are due to the presence of impurities, a typical analysis of blast-furnace iron showing about 4% C, 2% P, 2.5% Si, 2.5% Mn, and 0.1% S by weight. Further refining to remove these impurities (especially carbon) produces **steel,** a much stronger and consequently more useful material.

Steelmaking involves oxidation of the impurities in basic oxygen, open hearth, or electric furnaces. Some oxidation products (CO, CO_2, and SO_2) are volatile and easily separated. The others end up, along with some iron oxides, in a slag which floats on the surface of the molten steel.

The **basic oxygen furnace** is rapidly replacing other steelmaking methods. It accounted for 65 percent of United States production in 1974, up from only 2 percent 15 years earlier. Pure oxygen is directed onto the surface of molten pig iron in a large crucible. Some of the iron is oxidized to Fe_3O_4 and Fe_2O_3, forming an oxidizing slag. The impurities, namely, C, P, Si, Mn, and S, are all oxidized at the same time. Since all these reactions are spontaneous and exothermic, they provide enough heat so that up to 25 percent solid scrap iron may be melted in the crucible without cooling it to the point where solid iron would remain. Oxidation of one batch of pig iron and scrap normally takes slightly more than half an hour.

In the **open-hearth furnace** the same chemical reactions occur, but oxidation is due to air at the surface of the molten metal. This requires much longer—up to 12 hours—and natural gas or other fuel must be burned to keep the metal liquid. Thus the open hearth wastes roughly 3 times as much free energy as does the basic oxygen furnace. The use of fossil fuel does make it possible to recycle as much as 50 percent scrap iron, however, and the longer melting time allows somewhat greater control over the composition of a batch of steel.

Computers are now used to interpret spectroscopic analyses (Sec. 21.2) of steel in basic oxygen furnaces, indicating in a few minutes what metals must be added to obtain the desired composition. This has largely eliminated the last advantage of the open hearth and speeded up changeovers to basic oxygen. It has also decreased recycling of iron because the latter furnace cannot handle as much scrap. Much recycling of iron is now done in **electric-arc furnaces** which can melt a charge of 100 percent scrap.

In addition to the chemical oxidations used in steelmaking, electrolytic oxidation and reduction is quite important in refining metals. The electrolytic refining of copper has already been described in Sec. 17.2, and so no examples need be given here.

Corrosion

An important aspect of the use of some metals, particularly of iron, is the possibility of corrosion. It is estimated that about one-seventh of all iron production goes to replace the metal lost to corrosion. Rust is apparently a hydrated form of iron(III) oxide. The formula is approximately $Fe_2O_3 \cdot \frac{3}{2}H_2O$, although the exact amount of water is variable. (Note that this is about halfway between iron(III) hydroxide, $Fe(OH)_3$ or $\frac{1}{2}[Fe_2O_3 \cdot 3H_2O]$, and anhydrous Fe_2O_3).

Rusting requires both air and water, and it is usually speeded up by acids, strains in the iron, contact with less-active metals, and the presence of rust itself. In addition, observation of a rusted object, such as an iron nail from an old wooden building, shows that rust will deposit in one location (near the head of the nail) while the greatest loss of metallic iron will occur elsewhere (near the point). These facts suggest that the mechanism of rusting involves a galvanic cell like those discussed in Sec. 17.4. The half-equations involved are

$$2Fe(s) \rightarrow 2Fe^{2+}(aq) + 4e^- \qquad (22.8a)$$

$$4e^- + 4H^+(aq) + O_2(g) \rightarrow 2H_2O \qquad (22.8b)$$

Once $Fe^{2+}(aq)$ is formed, it can migrate freely through the aqueous solution to another location on the metal surface. At that point the iron can precipitate:

$$4Fe^{2+}(aq) + O_2(g) + 7H_2O(l) \rightarrow 2Fe_2O_3 \cdot \tfrac{3}{2}H_2O(s) + 8H^+(aq) \qquad (22.9)$$

Hydrogen ions liberated by this reaction are then partially consumed by Eq. (22.8*b*). The electrons required for half-equation (22.8*b*) are supplied from Eq. (22.8*a*) via metallic conduction through the iron or by ionic conduction if the aqueous solution contains a significant concentration of ions. Thus iron rusts faster in contact with salt water than in fresh.

The mechanism proposed in the preceding paragraph implies that some regions of the iron surface become *cathodic,* i.e., that reduction of oxygen to water occurs there. Other locations are *anodic;* oxidation of Fe to Fe^{2+} occurs. The chief way in which such regions may be set up depends on restriction of oxygen supply, because oxygen is required for the cathodic reaction shown in Eq. (22.8*b*). In the case of the iron nail, for example, rust forms near the head because more oxygen is available. Most of the loss of metal takes place deep in the wood, however, near the point of the nail. At this location Eq. (22.8*a*) but not (22.8*b*) can occur. A similar situation occurs when a drop of moisture adheres to an iron surface (Fig. 22.5). Pitting occurs near the center of the drop, while hydrated iron(III) oxide deposits near the edge.

A second way in which anodic and cathodic regions may be set up involves the presence of a second metal which has a greater attraction for electrons (is less easily oxidized) than iron. Such a metal can drain off electrons left behind in the iron when Fe^{2+} dissolves. This excess of electrons makes the less-active metal an ideal site for Eq. (22.8*b*), and so a cell is set up at the intersection of the metals. Rust may actually coat the surface of the less-active metal while pits form in the iron.

The most important technique for rust prevention is simply to exclude

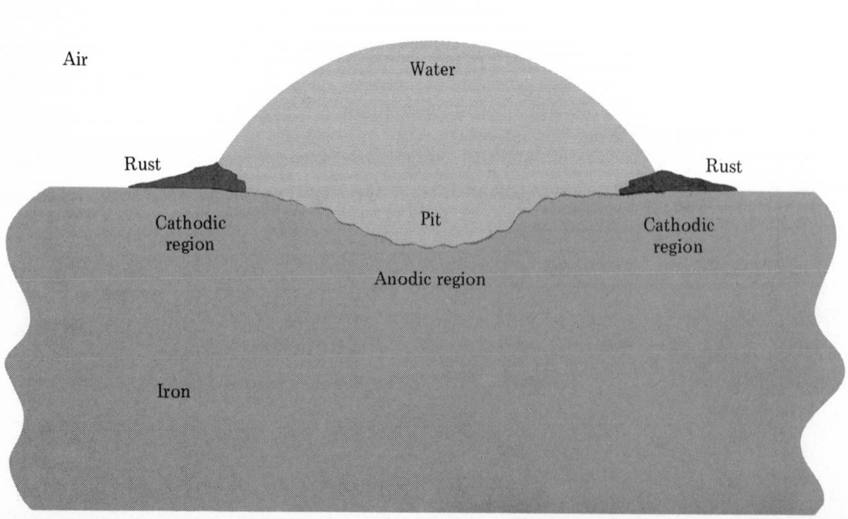

Figure 22.5 Corrosion of iron under a drop of water. Reaction (22.8*a*) occurs in the anodic region, Eq. (22.8*b*) in the cathodic region.

water and oxygen by means of a protective coating. This is the principle behind oiling, greasing, painting, or metal plating of iron. The coating must be complete, however, or rusting may be accelerated by exclusion of oxygen from part of the surface. This is especially true when iron is coated with a less-active metal such as tin. Even a pinhole in the coating on a tin can will rust very quickly, since the tin becomes cathodic due to its larger electrode potential and to the oxygen exclusion from the iron beneath.

A second technique involves bringing the iron object in contact with a more active metal. This is called *cathodic protection* because the more active metal donates electrons to the iron, strongly inhibiting Eq. (22.8*a*). Both cathodic protection and a surface coating are provided by *galvanizing,* a process in which zinc is plated onto steel electrolytically or by dipping in the molten metal. Like many other metals, zinc is self-protective—it reacts with oxygen and carbon dioxide from air to form an adherent impervious coating of zinc hydroxycarbonate, $Zn_2(OH)_2CO_3$. Should there be a scratch in the zinc plate, the iron still cannot rust because zinc will be preferentially oxidized. The hydroxycarbonate formed will then cover the opening, preventing further contact of oxygen with the iron or zinc.

A third technique applies to situations (such as an automobile radiator) where aqueous solutions are in contact with the iron. *Corrosion inhibitors* include chromate salts and organic compounds such as tributylamine, $(C_4H_9)_3N$. Chromates apparently form an impervious coating of $FeCrO_4(s)$ as soon as any iron is oxidized to iron(II). Tributylamine, a derivative of ammonia, reacts with organic acids formed by decomposition of antifreeze at the high temperatures of an automobile engine. The tributylammonium salts produced are insoluble and coat the inside of the cooling system. Thus tributylamine neutralizes acid which would accelerate corrosion and provides a surface coating as well.

22.3 COORDINATION COMPOUNDS

A characteristic feature of the transition metals is their ability to form a group of compounds called **coordination compounds, complex compounds,** or sometimes simply **complexes.** A typical coordination compound is the intensely blue solid substance $Cu(NH_3)_4SO_4$ which can be crystallized from solutions of $CuSO_4$ to which a very large excess of concentrated NH_3 has been added. These crystals contain two polyatomic ions, one of which is the sulfate ion, SO_4^{2-}, and the other of which is the **complex ion** $Cu(NH_3)_4^{2+}$ which is responsible for the blue color.

We can regard a complex ion such as $Cu(NH_3)_4^{2+}$ as being the result of the interaction of $:NH_3$ acting as a Lewis base with the Cu^{2+} ion acting as a Lewis acid. Each NH_3 molecule can be considered as donating a pair of electrons to a central Cu^{2+}, thus forming four coordinate covalent bonds to it:

$$\left[\begin{array}{c} NH_3 \\ \ddot{} \\ H_3N:\underset{\cdot\cdot}{Cu}:NH_3 \\ NH_3 \end{array}\right]^{2+}$$

Most coordination compounds contain a complex ion similar to $Cu(NH_3)_4^{2+}$. This ion can be either positively charged like $Cr(H_2O)_6^{3+}$, or it can be negatively charged like $CoCl_6^{3-}$. Neutral complexes like $Pt(NH_3)_2Cl_2$ are also known. All these species contain a *central metal ion* attached by coordinate covalent bonds to several **ligands.** These ligands are invariably Lewis bases. Some typical examples of ligands are H_2O, NH_3, Cl^-, OH^-, CN^-, Br^-, and SCN^-. The number of ligands attached to the central metal ion is said to be its **coordination number** and is usually 2, 4, or 6. The group of ligands bonded to the metal taken collectively is said to constitute the metal's **coordination sphere.**

When writing the formula of a coordination compound containing complex ions, square brackets are usually used to enclose the coordination sphere. Examples are

$[Cu(NH_3)_4]SO_4$ $\quad$ $[Cr(H_2O)_6]Cl_3$ $\quad$ $[Pt(NH_3)_2Cl_2]$

$[Cu(NH_3)_4](NO_3)_2$ $\quad$ $K_3[Fe(CN)_6]$ $\quad$ $[Pt(NH_3)_4][PtCl_4]$

When such compounds are dissolved in H_2O, each of the ions present in the solid becomes an independent species with its own chemical and physical properties. Thus, when 1 mol $[Cr(H_2O)_6]Cl_3$ crystal is dissolved in H_2O, the solution contains 1 mol $Cr(H_2O)_6^{3+}$ ion which can be recognized by its characteristic grayish-violet color and 3 mol Cl^- which can be detected by the precipitate of AgCl which forms when $AgNO_3$ is added to the solution.

An even better example of how the various ions in a coordination compound can behave independently when dissolved in water is provided by the set of Pt(II) complexes shown in Table 22.3. The first of these compounds contains the complex ion $[Pt(NH_3)_4]^{2+}$, and in each subsequent compound

TABLE 22.3 Observations on Complex Compounds Containing $PtCl_2$, NH_3, and KCl.

Compound	Molar Conductivity/ $A\ V^{-1}\ dm^2\ mol^{-1}$	Moles AgCl Precipitated per Mole Compound	Electrode to which Pt Migrates During Electrolysis
1. $[Pt(NH_3)_4]Cl_2$	3.0	2 immediately	Cathode
2. $[Pt(NH_3)_3Cl]Cl$	1.2	1 immediately; 1 after several hours	Cathode
3. $Pt(NH_3)_2Cl_2$	~0	2 after several hours	Does not migrate
4. $K[Pt(NH_3)Cl_3]$	1.1	3 after several hours	Anode
5. $K_2[PtCl_4]$	2.8	4 after several hours	Anode

one of the NH_3 ligands in the coordination sphere of the Pt is replaced by a Cl^- ligand. As a result each compound contains a Pt complex of different composition and also of different charge, and when dissolved in H_2O, it shows just the conductivity and other properties we would expect from the given formula. When 1 mol $[Pt(NH_3)_3Cl]Cl$ is dissolved in H_2O, it furnishes 1 mol $Pt(NH_3)_3Cl^+$ ions and 1 mol Cl^- ions. The strongest evidence for this is the molar conductivity of the salt (1.2 A V^{-1} dm^2 mol^{-1}), which is very similar to that of other electrolytes like NaCl (1.3 A V^{-1} dm^2 mol^{-1}) which also yield a +1 ion and a −1 ion in solution, but very different from that of electrolytes like $MgCl_2$ (2.5 A V^{-1} dm^2 mol^{-1}) which yield one +2 ion and two −1 ions in solution. The conductivity behavior also suggests that the Pt atom is part of a cation, since the Pt moves toward the cathode during electrolysis. The addition of $AgNO_3$ to the solution serves to confirm this picture. One mol AgCl is precipitated immediately, showing 1 mol free Cl^- ions. After a few hours a further mole of AgCl is precipitated, the Cl^- this time originating from the coordination sphere of the Pt atom due to the slow reaction

$$[Pt(NH_3)_3Cl]^+(aq) + Ag^+(aq) + H_2O \rightarrow [Pt(NH_3)_3H_2O]^{2+}(aq) + AgCl(s)$$

It is worth noting that in all these compounds, Pt has an oxidation number of +2. Thus the combination of Pt with one NH_3 ligand and three Cl^- ligands yields an overall charge of 2(for Pt) − 3(for Cl) + 0(for NH_3) = −1. The ion is thus the anion $[PtNH_3Cl_3]^-$ found in compound 4.

EXAMPLE 22.1 What is the oxidation state of Pt in the compound $Ca[Pt(NH_3)Cl_5]_2$?

Solution Since there are two complex ions for each Ca^{2+} ion, the charge on each must be −1. Adding the charge on each ligand, we obtain −5(for Cl^-) + 0(for NH_3) = −5. If the oxidation number of Pt is x, then x − 5 must equal the total charge on the complex ion:

$$x - 5 = -1$$

or

$$x = +4$$

The compound in question is thus a Pt(IV) complex.

Geometry of Complexes

The geometry of a complex is governed almost entirely by the coordination number. We will consider only the most common coordination numbers, namely, 2, 4, and 6.

Coordination number = 2 Complexes with two ligands are invariably *linear*. The best-known examples of such compounds are Ag(I) and Au(I) complexes such as

$$[N{\equiv}C{-}Au{-}C{\equiv}N]^-$$

and

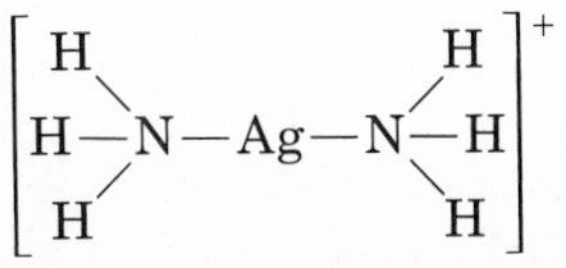

Both of these complexes are important. The $Au(CN)_2^-$ complex is used to extract minute gold particles from the rock in which they occur. The crushed ore is treated with KCN solution and air is blown through it:

$$4Au(s) + 8CN^-(aq) + O_2(g) + 2H_2O(l) \rightarrow 4[Au(CN)_2]^-(aq) + 4OH^-(aq)$$

The resultant complex is water soluble. The silver complex is also water soluble and affords a method for dissolving AgCl, which is otherwise very insoluble.

$$AgCl(s) + 2NH_3(aq) \rightarrow [Ag(NH_3)_2]^+(aq) + Cl^-(aq)$$

This reaction is often used in the laboratory to be sure a precipitate is $AgCl(s)$.

Coordination number = 4 Two geometries are possible for this coordination number. Some complexes, like the $[Pt(NH_3)_4]^{2+}$ ion shown in Fig. 22.6, are *square planar,* while others, like $Cd(NH_3)_4^{2-}$, are *tetrahedral.* Most of the four-coordinated complexes of Zn(II), Cd(II), and Hg(II) are tetrahedral, while the square planar arrangement is preferred by Pd(II), Pt(II), and Cu(II) complexes.

Because the square planar geometry is less symmetrical than the tetrahedral geometry, it offers more possibilities for isomerism. A well-known example of such isomerism is given by the two square planar complexes

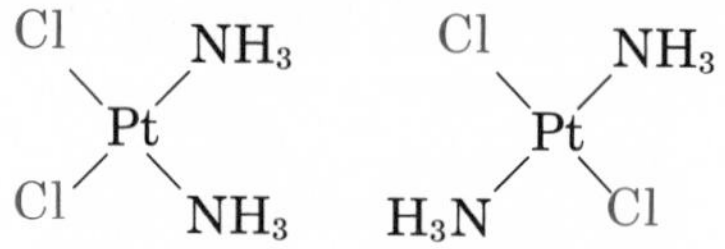

Cis isomer **Trans isomer**

These two isomers are called **geometrical isomers.** That isomer in which two identical ligands are *next* to each other is called the cis isomer, while that in which they are on opposite sides is called the trans isomer. Though these two isomers have some properties which are similar, no properties are identical and some are very different. For example, the cis isomer of the above complex is currently under investigation as an anticancer drug. The trans form, by contrast, shows no similar biological activity.

It is worth noting that cis-trans isomerism is not possible in the case of tetrahedral complexes. As you can quickly verify by examining any three-dimensional tetrahedral shape, any given corner of a tetrahedron is

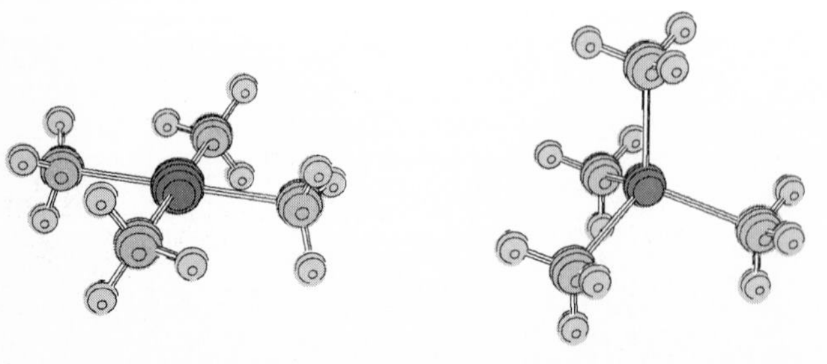

(*a*) (*b*)

Figure 22.6 Structure of (*a*) $Pt(NH_3)_4^{2+}$ (square planar) and (*b*) $Zn(NH_3)_4^{2+}$ (tetrahedral). Metal ions are dark gray, nitrogen atoms are light gray, and hydrogen atoms are light color. Geometrical isomers can exist for square planar complexes but not for tetrahedral complexes. (Computer-generated.) (*Copyright © 1976 by W. G. Davies and J. W. Moore.*)

adjacent to the other three. Since all the corners are cis to each other, none are trans.

Coordination number = 6 When there are six ligands, the geometry of the complex is almost always *octahedral*, like the geometry of SF_6, shown earlier in Fig. 7.1, or of $[Cr(H_2O)_6]^{3+}$, shown in Fig. 11.6. All ligands are equidistant from the central atom, and all ligand-metal-ligand angles are 90°. An octahedral complex may also be thought of as being derived from a square planar structure by adding a fifth ligand above and a sixth below on a line through the central metal ion and perpendicular to the plane.

The octahedral structure also gives rise to geometrical isomerism. For example, two different compounds, one violet and one green, have the formula $[Co(NH_3)_4Cl_2]Cl$. The violet complex turns out to have the cis structure and the green one trans, as shown in Fig. 22.7.

Chelating Agents

Although we have confined our discussion so far to simple ligands such as Cl^-, NH_3, or H_2O, much larger and more complicated molecules can also donate electron pairs to metal ions. An important and interesting example of this is the **chelating agents**—ligands which are able to form two or more coordinate covalent bonds with a metal ion. One of the most common of these is 1,2-diaminoethane (usually called *ethylenediamine* and abbreviated *en*.)

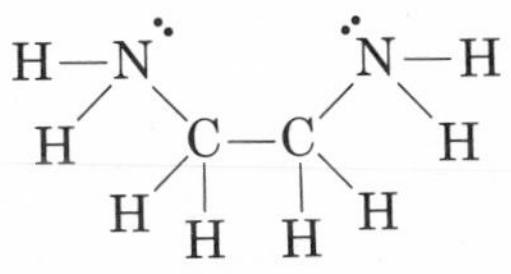

Ethylenediamine; en

When both nitrogens coordinate to a metal ion, a stable five-member ring is formed. The word *chelating*, derived from the Greek *chele*, "claw," describes the pincerlike way in which such a ligand can grab a metal ion.

A chelating agent which forms several bonds to a metal without unduly straining its own structure is usually able to replace a similar simpler ligand. For example, although both form coordinate covalent bonds via

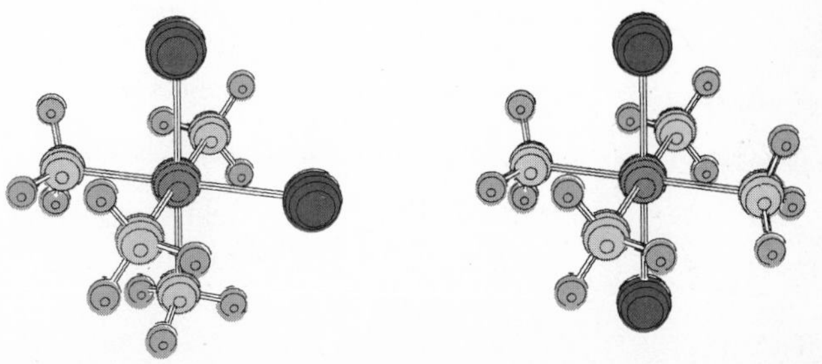

(*a*) (*b*)

Figure 22.7 (*a*) Cis isomer and (*b*) trans isomer of $[Co(NH_3)_4Cl_2]^+$. On the macroscopic level the cis form is violet while the trans form is green. (Computer-generated.) (*Copyright © 1976 by W. G. Davies and J. W. Moore.*)

$\geqslant$N: groups, ethylenediamine can readily replace ammonia from most complexes:

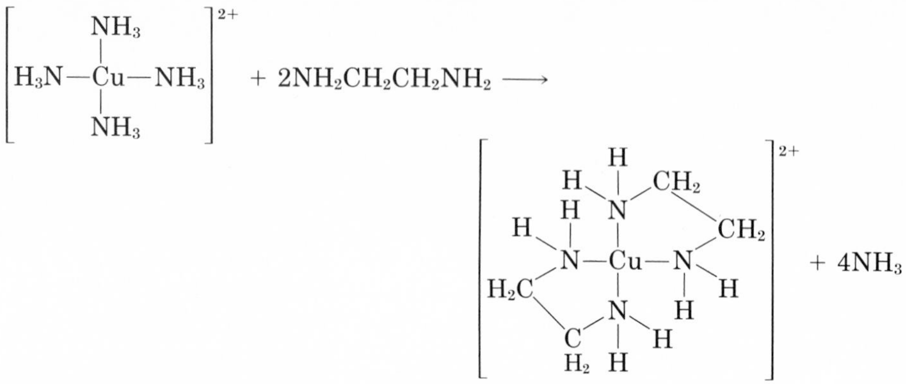

For metals which display a coordination number of 6, an especially potent ligand is *e*thylene*d*iamine*t*etra*a*cetate ion (abbreviated EDTA):

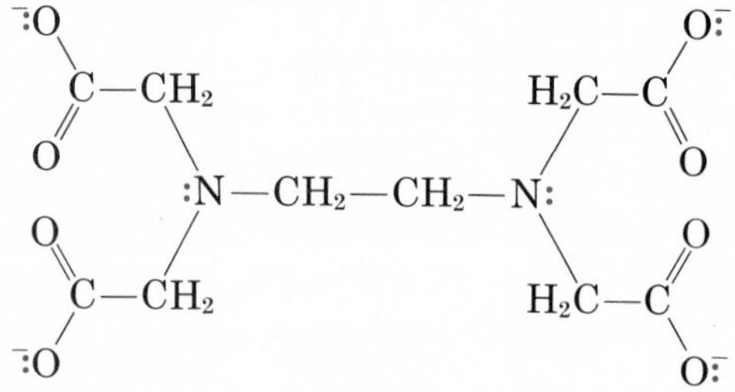

EDTA

All six electron pairs marked in color are capable of coordinating to a metal ion, in which case the EDTA ion wraps completely around the metal and is very difficult to dislodge. EDTA is used to treat lead and mercury poisoning because of its ability to chelate these metals and aid their removal from the body.

Chelate complexes are often important in living systems. The coordination of iron in proteins such as myoglobin or hemoglobin involves four nitrogens of the heme group and one from a histidine side chain (Sec. 20.5). Since iron normally has a coordination number of 6, this leaves one open site, to which oxygen can bond. The presence of carbon monoxide, a stronger ligand than oxygen, causes displacement of oxygen from hemoglobin. This prevents transport of oxygen from the lungs to the brain, causing drowsiness, loss of consciousness, and even death upon long exposure to carbon monoxide. Consequently operating an automobile in a closed garage, a cookstove in a tent, or burning any fossil fuel incompletely in an enclosed space may be hazardous to one's health.

Another important application of chelates is transport of metal ions across membranes. You will recall from Sec. 20.3 that the interior of a biological membrane contains the nonpolar, hydrophobic tails of lipid molecules. This makes it quite difficult for ionic species such as K^+ and Na^+ to travel from one side of a membrane to the other. In this way certain ions may be kept in a particular location until some carrier molecule helps them through the membrane.

One such carrier is the antibiotic *nonactin,* a medium-sized organic molecule with the formula

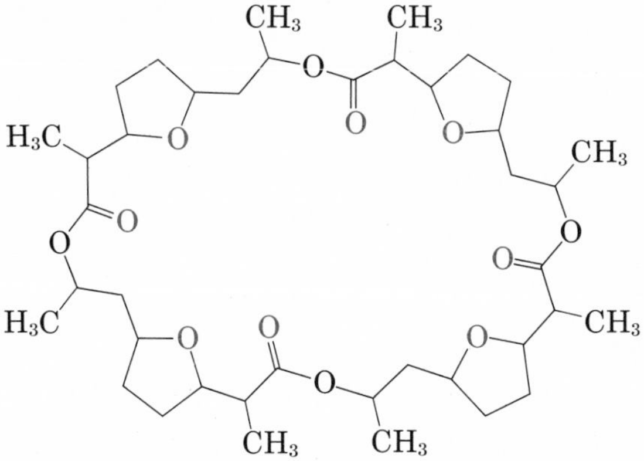

This molecule is able to transport K^+ ions but not Na^+ ions. Apparently the Na^+ ion is too small to fit in among the eight coordinating O's, while the K^+ ion can (see Fig. 22.8). Other than these O's, most of the nonactin molecule is a hydrocarbon chain. Therefore once K^+ is chelated, the outer part of the complex is quite hydrophobic. It can easily pass through the interior of a membrane, releasing K^+ on the other side. The toxic effect of nonactin and several related antibiotics is apparently the result of their ability to transport alkali-metal ions to regions of a cell where they should not be. Consequently the cell wastes energy pumping K^+ and other ions out again.

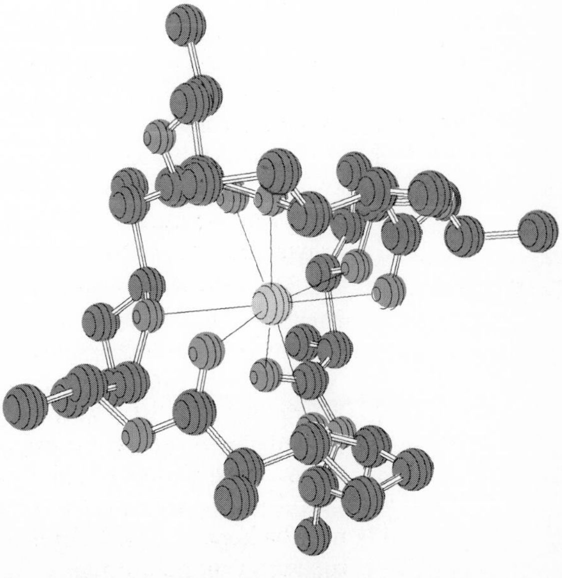

Figure 22.8 The K^+ complex of nonactin (K^+ is light gray). Coordinate covalent bonds are indicated as thin solid lines. Hydrogen atoms have been omitted for clarity. (Computer-generated.) *(Copyright © 1976 by W. G. Davies and J. W. Moore.)*

22.4 TRANSITION-METAL IONS IN AQUEOUS SOLUTION

We often write transition-metal ions in aqueous solution with symbols such as Cr^{3+}, Cu^{2+}, and Fe^{3+} as though they were monatomic, but this is far from being the case. These ions are actually hydrated in solution and can be regarded as complex ions. Thus, for example, the grayish-violet color of many chromium(III) salts when dissolved in H_2O is due to the species $[Cr(H_2O)_6]^{3+}$ rather than to a bare Cr^{3+} ion. The same color is evident in many crystalline solids such as $[Cr(H_2O)_6]Cl_3$ which are known to contain the Cr^{3+} ion surrounded octahedrally by six H_2O molecules. In much the same way the blue color of many solutions of copper(II) salts can be attributed to the species $[Cu(H_2O)_4]^{2+}$ and the pale violet color of some solutions of iron(III) salts to the $[Fe(H_2O)_6]^{3+}$ ion.

Not all salts of transition-metal ions yield the hydrated ion when dissolved in H_2O. Thus when $CuCl_2$ is dissolved in H_2O, a beautiful green color due mainly to the complex $[CuCl_2(H_2O)_2]$ is produced. This is obviously different from the sky-blue color of $[Cu(H_2O)_4]^{2+}$ which is obtained when copper(II) sulfate or copper(II) nitrate are dissolved. This is because the Cl^- ion is a stronger Lewis base with respect to the Cu^{2+} ion than is H_2O. Thus, if there is a competition between H_2O and Cl^- to bond as a ligand to Cu^{2+}, the Cl^- ion will usually win out over the H_2O.

The superior strength of the Cl^- as a Lewis base is easily demonstrated by adding Cl^- ions to a sky-blue solution of copper(II) sulfate. A green color immediately appears due to the formation of chloro complexes:

$$[Cu(H_2O)_4]^{2+} + Cl^- \rightleftharpoons [Cu(H_2O)_3Cl]^+ + H_2O$$

$$[Cu(H_2O)_3Cl]^+ + Cl^- \rightleftharpoons \underset{\textbf{Green}}{[Cu(H_2O)_2Cl_2]} + H_2O$$

If a large excess of Cl^- ion is added, the solution changes color again from green to yellow. This is because of even further displacement of H_2O ligands by Cl^- ligands:

$$[Cu(H_2O)_2Cl_2] + Cl^- \rightleftharpoons [Cu(H_2O)Cl_3]^- + H_2O$$

$$[Cu(H_2O)Cl_3]^- + Cl^- \rightleftharpoons \underset{\textbf{Yellow}}{[CuCl_4]^{2-}} + H_2O$$

Under favorable circumstances yellow crystals of salts like $Cs_2[CuCl_4]$, containing the complex ion $CuCl_4^{2-}$, can be obtained from these solutions.

Because they might very possibly form complexes with it, one must be careful about what ions are added to a solution containing hydrated transition-metal ions. Not only the chloride ions, but the other halide ions are liable to complex, and the same is true of species like NH_3 and CN^-. These ligands differ quite a lot in their affinity for a particular metal ion, but the rules governing this situation are not simple. One finds, for instance, that although NH_3 will complex very readily with Cu^{2+}, it has little or no affinity for Fe^{3+}. In other words, a ligand which is a strong Lewis base with respect to one metal ion is not necessarily a strong base with respect to another. There are some ions, however, which almost always function as very weak Lewis bases. The perchlorate ion, ClO_4^-, in particular, forms almost no complexes. The nitrate ion, NO_3^-, and sulfate ion, SO_4^{2-}, only occasionally form complexes.

The addition of ligands to a solution in order to form a highly colored complex is often used to detect the presence or absence of a given metal in solution. The deep blue color of $[Cu(NH_3)_4]^{2+}$ produced when excess NH_3 is added to solution of Cu(II) salts is a case in point. Other well-known color reactions are the blood-red complex formed between Fe(III) ions and the thiocyanate ion, SCN^-, as well as the pink-red complex of Ni(II) with dimethylglyoxime.

While most of the reactions we have been describing are very fast and occur just as quickly as the solutions are mixed, this is not always the case. With certain types of complexes, ligand substitution is quite a slow process. For example, if Cl^- ions are added to a solution containing $[Cr(H_2O)_6]^{3+}$ ions, it is a few days before the grayish-violet color of the original ion is replaced by the green color of the chloro complexes $[Cr(H_2O_5)Cl]^{2+}$ and $[Cr(H_2O)_4Cl_2]^+$. Alternatively the solution may be heated, in which case the green color will usually appear within 10 min. The reaction

$$[Cr(H_2O)_6]^{3+} + Cl^- \rightarrow [Cr(H_2O)_5Cl]^{2+} + H_2O$$

is thus a slow reaction with a high activation energy. Ligand substitution reactions of other Cr(III) complexes behave similarly. In consequence Cr(III) complexes are said to be **inert,** as opposed to a complex like $Fe(H_2O)_6^{3+}$ which swaps ligands very quickly and is said to be **labile.** Other examples of inert complexes are those of Co(III), Pt(IV), and Pt(II). Almost all the compounds which were used to establish the nature and the geometry of coordination compounds were inert rather than labile. There is very little point in trying to prepare cis and trans isomers of a labile complex, for example, because either will quickly react to form an equilibrium mixture of the cis and trans forms.

A final complication in dealing with aqueous solutions of transition-metal complexes is their acid-base behavior. As was mentioned in Chap. 11, hydrated metal ions like $[Cr(H_2O)_6]^{3+}$ are capable of donating protons to water and acting as weak acids. Most hydrated ions with a charge of +3, like Al^{3+} and Fe^{3+}, behave similarly and are about as strong as acetic acid. The hydrated Hg(II) ion is also noticeably acidic in this way. Perhaps the most obvious of these cationic acids is the hydrated Fe(III) ion. When most Fe(III) salts are dissolved in water, the color of the solution is yellow or brown, though the $Fe(H_2O)_6^{3+}$ ion itself is pale violet. The yellow color is due to the conjugate base produced by the loss of a proton. The equilibrium involved is

$$\underset{\textbf{Pale violet}}{[Fe(H_2O)_6]^{3+}} + H_2O \rightleftharpoons \underset{\textbf{Brown}}{[Fe(H_2O)_5OH]^{2+}} + H_3O^+$$

If solutions of Fe(III) salts are acidified with perchloric acid or nitric acid, the brown base is protonated and the yellow color disappears from the solution entirely.

SUMMARY

Nearly three-quarters of the known elements display metallic properties. These properties include good thermal and electrical conductivity, shiny surface appearance, malleability, ductility, and low-ionization energies. These properties can be understood in terms of metallic bonding in which valence electrons are delocalized over an entire metallic crystal. Positive metal ions formed by loss of valence electrons are held together by an electron sea. The strength of metallic bonding varies roughly as the number of electrons available in this sea.

A great many metals and alloys are of commercial importance, but most metals occur naturally in oxide, carbonate, or sulfide ores. Such ores must be concentrated (beneficiated) before they can be reduced to the metal, and usually the raw metal must be purified (refined) in a third step. An excellent example of these processes involves iron which can be readily beneficiated because its ore is ferromagnetic. Iron ore is then reduced in a blast furnace and purified in a steelmaking furnace. Since ore-reduction is a nonspontaneous process, its reverse, oxidation or corrosion of a metal, is often a problem. This is especially true in the case of iron because the oxide coating which forms on the metal surface does not protect the remaining metal from atmospheric oxidation.

Transition metals are often found in coordination compounds (complexes). Usually several ligands which can serve as Lewis bases are bonded to a metal ion which serves as a Lewis acid. The number of ligands is called the coordination number. For a coordination number of 2 the complex is usually linear. Both square planar and tetrahedral structures occur for coordination number 4, and coordination number 6 usually involves an octahedral structure. Square planar and octahedral structures give rise to cis-trans isomerism.

Some ligands, called chelating agents, can coordinate-covalent bond to metal ions at more than one site. Chelate complexes are often important in biological systems because they can disguise the charge of a metal ion, stabilizing the ion in a hydrophobic environment.

In aqueous solution transition-metal ions are usually octahedrally coordinated by water molecules, but often other ligands which are stronger Lewis bases replace water. Such reactions often produce color changes, and they are usually rapid. A few metal ions, such as Cr(III), Co(III), Pt(IV), and Pt(II), undergo ligand substitution rather slowly and are said to be inert. Metal ions whose reactions are rapid are said to be labile.

QUESTIONS AND PROBLEMS

22.1 What are the characteristic chemical and physical properties of metals?

22.2 Differentiate among the electron configurations of metals in each of the following categories: (*a*) representative elements; (*b*) transition elements; (*c*) inner transition elements.

22.3 Explain why neither ionic nor covalent (valence-bond) models are adequate to account for bonding in metals.

22.4 What are the meaning and significance of the term *electron gas,* or *electron sea?* How does this concept account for the high electric conductivity of metals?

22.5 Refer to the electron dot-density diagrams for Li, Be, and V in color Plates 4 and 5 and Fig. 5.10. Based on these figures, does it seem reasonable to you that the outermost electrons of a metal atom can be shared with a large number of other atoms (delocalized) in a metallic crystal? Explain why or why not.

22.6 How many MO's and how many electrons would be present in the 2*s* band of a perfect crystal of Li which weighed 0.694 g? in the 2*p* band?

22.7 Make a diagram (similar to Fig. 22.1) to show occupancy of electronic energy bands for each of the following metals: (*a*) Na; (*b*) Mg; (*c*) Al; (*d*) Sc; (*e*) Pr.

***22.8** In some cases it is possible to determine the energy required to ionize an electron from a particular orbital. In the case of oxygen, for example, the ionization energy of 1314 kJ mol^{-1} corresponds to removal of a 2*p* electron. The 2*s* electrons are much lower in energy, and so the ionization energy of 3124 kJ mol^{-1} for removal of a 2*s* electron is much larger. For each group of three elements listed below, indicate which would have the largest and which would have the smallest difference between the ionization energies for *ns* and *np* electrons. Give a reason for each choice.

a Be F B; 2*s* versus 2*p*
b As Sc Br; 4*s* versus 4*p*
c Xe Sb Rb; 5*s* versus 5*p*

22.9 Criticize the following statement: Since an electron can have energies within a broad band and is not restricted to a single energy, the band theory of solids contradicts the wave-mechanical theory of the atom.

22.10 Account for the high electric and thermal conductivities of metals in terms of the band theory of solids.

22.11 Differentiate among conductors, semiconductors, and insulators in terms of their electronic band structure. How does the electric conductivity of each class of substances vary with temperature?

22.12 Explain in your own words why the melting points of the metals in the fourth period increase rapidly, level off, and then decrease as one moves from left to right across the periodic table.

***22.13** Use a handbook to obtain the melting points and densities of the lanthanoids. Use your data to defend or attack each of the following statements. Whenever possible give examples from your data which agree with or contradict the statement. (*a*) The increase in melting points per *f* electron added among the lanthanoids is considerably smaller than the increase per *d* electron among transition elements up to group VI (see Fig. 22.3). (*b*) The *f* orbitals appear to be very important in metallic bonding for lanthanoids. (*c*) Densities of the lanthanoids decrease with increasing metallic bond strength. (*d*) When an *f* subshell is half-filled or completely filled, the *f* electrons make a smaller contribution to metallic bonding.

22.14 What characteristics distinguish an ore from any other mineral?

22.15 Criticize the following statement: Silicate minerals are the most common solid substances on the face of the earth, and so they are used as ores for a wide variety of metals.

22.16 What three types of processes are usually necessary to obtain a metal from its ore?

22.17 Define or describe each of the following:

a Gangue
b Flotation
c Leaching
d Bayer process
e Roasting of ore
f Slag

22.18 Apply LeChatelier's principle to Eq. (22.2) to explain why the reaction can be forced to the left when $CO_2(g)$ is bubbled through $Al(OH)_4^-(aq)$.

22.19 Which of the metal oxides in Table 22.2 can be reduced by carbon [assuming $CO_2(g)$ is produced] at 298 K? Which can be reduced by carbon at 2000 K?

22.20 Using the equation

$$\Delta G_m^\ominus = \Delta H_m^\ominus - T\Delta S_m^\ominus$$

and assuming that $\Delta H_m^\ominus$ and $\Delta S_m^\ominus$ do not vary with temperature, calculate $\Delta H_m^\ominus$ and $\Delta S_m^\ominus$ for each reaction in Table 22.2.

22.21 Use your results from Problem 22.20 to calculate the minimum temperature at which each of the metals in Table 22.2 could be obtained simply by heating its oxide.

22.22 Repeat the calculation in Problem 22.21, but assume that the ore is to be reduced by heating it with carbon.

22.23 The reaction

$$2C(s) + O_2(g) \rightarrow 2CO(g)$$

has $\Delta H_m^\ominus(298\ K) = -221.0$ kJ mol^{-1} and $\Delta S_m^\ominus(298\ K) = 185.4$ J K^{-1} mol^{-1}. Assuming no variation of $\Delta H_m^{\ominus}$ or $\Delta S_m^{\ominus}$ with temperature, calculate the temperature at which this reaction becomes more useful for the purpose of ore reduction than oxidation of C(*s*) to $CO_2(g)$.

***22.24** The preceding four problems have all assumed that $\Delta H_m^{\ominus}$ and $\Delta S_m^{\ominus}$ do not vary with temperature. Under what circumstances is this assumption likely to be significantly wrong? For the reaction

$$2MgO(s) \rightarrow 2Mg(s) + O_2(g)$$

at what temperatures would you expect significant changes in $\Delta H_m^{\ominus}$ and/or $\Delta S_m^{\ominus}$ to occur?

22.25 Criticize the following statement: Control of air pollution from blast furnaces is a waste of time and money because it merely improves the appearance of formerly smoky cities.

22.26 Describe the three most common types of steelmaking furnaces and indicate the important differences among them. Which uses least energy? Which is most appropriate for recycling?

22.27 Answer the following questions pertaining to corrosion of iron: (*a*) What is the accepted chemical formula of rust? (*b*) Is rusting more or less likely in a desert climate? (*c*) Does rusting involve a dual electrochemical role for iron? (*d*) Why does an iron nail rust near the head? (*e*) Why does salt used to melt ice on highways cause automobile bodies to rust faster?

22.28 Explain the processes involved in rust prevention by (*a*) galvanizing; (*b*) cathodic protection; (*c*) corrosion inhibitors.

22.29 Define each of the following terms:

a Ligand
b Coordination sphere
c Chelating agent
d Inert
e Labile
f Coordination number

22.30 What ions will be present in solution when each of the following compounds dissolves?

a $[Co(NH_3)_6]Cl_3$
b $[Co(NH_3)_5Cl]Cl_2$
c $[Co(NH_3)_4SO_4]Cl_2$
d $[Co(NH_3)_5Cl]SO_4$
e $[Co(en)_2ClBr]Cl$
f $[Co(NH_3)_6][Cr(CN)_6]$

22.31 If $AgNO_3(aq)$ were added to each of the solutions in Problem 22.30, how many moles of AgCl(*s*) would precipitate immediately per mole of each compound?

22.32 What is the oxidation number of each metal in each of the following compounds?

a $[Co(NH_3)_6]Cl_2$
b $[Co(NH_3)_6]Cl_3$
c $[Pt(NH_3)_2Cl_2]$
d $[Pt(NH_3)_4Cl_2]Cl_2$
e $[Cr(H_2O)_4Cl_2]Cl$
f $Na_2[HgCl_4]$
g $K_4[Pt(CN)_6]$
h $[Pt(NH_3)_4][PtCl_4]$

22.33 Elemental analysis of a compound gives the following results in weight percent: 21.4% Co, 25.4% N, 5.5% H, 12.9% Cl, 11.6% S, 23.2% O. Determine the empirical formula of this compound and, assuming it to be a coordination compound, write the molecular formula. Suggest a further experiment which might confirm that your formula is correct.

22.34 Write the formula for one example of a complex which fits each of the following categories. Do not use the same formula twice.

a Coordination number 2
b Tetrahedral
c Labile
d Octahedral
e Square planar
f Inert
g Neutral complex
h Coordination number 4

22.35 If possible, draw structures for all geometrical isomers of each compound below:

a $[Co(NH_3)_4Cl_2]Cl$
b $K_2[CdCl_2Br_2]$
c $[Pt(en)(NH_3)_2ClBr]SO_4$
d $[Co(NH_3)_3Cl_3]$
e $[Pt(NH_3)_2ClBr]$
f $[Pt(en)(NH_3)_2]Cl_2$

22.36 Choose two chelating agents. Draw the structure of each and indicate each site which is capable of forming a coordinate covalent bond to a metal ion.

22.37 Each of the complex ions listed below contains Cl^- ligands. In which cases would $AgCl(s)$ precipitate immediately if $AgNO_3(aq)$ were added? Give reasons for your choices.

a $[Pt(NH_3)_4Cl_2]^{2+}$
b $[Fe(H_2O)_4Cl_2]^+$
c $[CrCl_6]^{3-}$
d $[CoCl_6]^{4-}$
e $[Pt(NH_3)_3Cl]^+$
f $[Co(NH_3)_5Cl]^{2+}$

22.38 The chelating agent EDTA is used to analyze hard water for calcium ions. Draw the structure of $Ca(EDTA)^{2-}$, a complex ion which is octahedrally coordinated.

22.39 Explain why a molecule like nonactin can transport potassium ions through membranes. Why is nonactin incapable of transporting sodium ions?

APPENDIX 1 Temperature Scales

Suppose that we have measured a temperature as c°C and wish to express it as f°F, or vice versa. To do this we need a general relationship between the number c which occurs in the description of the temperature on the Celsius (formerly centigrade) scale and the number f which occurs in the Fahrenheit description. Such a relationship can be derived by recalling that the freezing and boiling points of pure water at 1 atm pressure are 0 and 100°C on the Celsius scale and 32 and 212°F on the Fahrenheit scale.

As we go from the freezing point of water to the boiling point, the number c increases from 0 to 100, but the number f increases considerably more—from 32 to 212. Thus f is changing $(212 - 32)/(100 - 0) = 180/100 = 9/5$ times as fast as c. This ratio of 9/5 holds for *any* temperature change, not just from the freezing point of water to the boiling point. Suppose, for example, that we go from the freezing point of water to a general temperature described by c°C and f°F. For such a change

$$\frac{\text{Increase in number } f}{\text{Increase in number } c} = \frac{f - 32}{c - 0} = \frac{9}{5}$$

or

$$\frac{f - 32}{c} = \frac{9}{5} \tag{A1.1}$$

This equation may be solved algebraically for either f or c, whichever is wanted.

If a temperature is to be used in any kind of mathematical equation or formula, neither the Celsius nor the Fahrenheit description should be used. Only the thermodynamic (or absolute) temperature (expressed in kelvins) should ever be employed in any kind of calculation. The relationship between the Celsius description of a temperature as c°C and the thermodynamic scale is quite simple because the size of the kelvin and the degree Celsius are the same. This relationship is

$$T = (c + 273.15)\ \text{K} \tag{A1.2}$$

Thus ordinary room temperature (20°C) corresponds to $T = (20 + 273.15)\ \text{K} = 293.15\ \text{K}$.

EXAMPLE A1.1 Express each of the following temperatures on the scale indicated: (a) 38.6°C on the Fahrenheit scale; (b) 125.2°F on the Celsius scale; (c) 62.0°F on the thermodynamic scale.

Solution

a) Solving Eq. (A1.1) for f, $f = \frac{9}{5}c + 32 = \frac{9}{5}(38.6) + 32 = 69.5 + 32 = 101.5$.
Answer: 101.5°F

b) Solving Eq. (A1.1) for c, $c = \frac{5}{9}(f - 32) = \frac{5}{9}(125.2 - 32) = \frac{5}{9}(93.2) = 51.8$.
Answer: 51.8°C

c) Proceeding as in part b, $c = \frac{5}{9}(62.0 - 32) = 16.7$.

Using Eq. (A1.2), $T = (c + 273.15)\ \text{K} = (16.7 + 273.15)\ \text{K} = 289.8\ \text{K}$.

APPENDIX 2 Logarithms and Antilogarithms

The logarithm (base 10) of a number is the power to which 10 must be raised to obtain the number. In mathematical symbols, if

$$\log a = \alpha \tag{A2.1}$$

then

$$10^{\alpha} = a \tag{A2.2}$$

(In these and subsequent equations, ordinary numbers are represented by ordinary letters: a, b, c, etc. The corresponding logarithms are represented by the corresponding Greek letters: α, β, γ, etc.). One of the most convenient properties of logarithms is that the logarithm of a product of two numbers is the sum of the logarithms of each number. Symbolically, if

$$a = b \times c$$

then

$$10^{\alpha} = 10^{\beta} \times 10^{\gamma} = 10^{(\beta+\gamma)}$$

and

$$\alpha = \beta + \gamma$$

or

$$\log a = \log b + \log c$$

This property makes it possible it find the logarithm of any number using a simple table like the one at the end of this appendix. Suppose we want the logarithm of 173. First write the number in scientific notation, that is, as a number between 1 and 10 times a power of 10:

$$173 = 1.73 \times 10^2$$

Then

$$\log 173 = \log 1.73 + \log 10^2$$

The logarithm of 10^2 is exactly 2—10 must be raised to the second power to give 10^2, and the logarithm of 1.73 can be obtained from the table by reading along the row labeled 1.7 until the column headed 0.03 is reached. As indicated by the colored shading in the table, the logarithm of 1.73 is 0.238. Thus

$$\log 173 = 0.238 + 2.000 = 2.238$$

EXAMPLE A2.1 Find the logarithm of each of the following: (a) 3.89×10^7; (b) 6.05×10^{-13}; (c) 1.89×10^{-2}.

Solution

a) $\log(3.89 \times 10^7) = \log 3.89 + \log 10^7 = 0.590 + 7.000 = 7.590$

b) $\log(6.05 \times 10^{-13}) = 0.782 - 13.000 = -12.218$

c) $\log(1.89 \times 10^{-2}) = 0.276 - 2.000 = -1.724$

Quite often we are given a logarithm and asked to find the corresponding number. This process is called taking an antilogarithm and is easily done using Eq. (A2.2). For example, suppose we want the antilogarithm of 4.719. The number a corresponding to this logarithm is

$$a = 10^{\alpha} = 10^{4.719} = 10^{(4+0.719)} = 10^4 \times 10^{0.719}$$

	.00	.01	.02	.03	.04	.05	.06	.07	.08	.09
1.0	.000	.004	.009	.013	.017	.021	.025	.029	.033	.037
1.1	.041	.045	.049	.053	.057	.061	.064	.068	.072	.076
1.2	.079	.083	.086	.090	.093	.097	.100	.104	.107	.111
1.3	.114	.117	.121	.124	.127	.130	.134	.137	.140	.143
1.4	.146	.149	.152	.155	.158	.161	.164	.167	.170	.173
1.5	.176	.179	.182	.185	.188	.190	.193	.196	.199	.201
1.6	.204	.207	.210	.212	.215	.217	.220	.223	.225	.228
1.7	.230	.233	.236	.238	.241	.243	.246	.248	.250	.253
1.8	.255	.258	.260	.262	.265	.267	.270	.272	.274	.276
1.9	.279	.281	.283	.286	.288	.290	.292	.294	.297	.299
2.0	.301	.303	.305	.307	.310	.312	.314	.316	.318	.320
2.1	.322	.324	.326	.328	.330	.332	.334	.336	.338	.340
2.2	.342	.344	.346	.348	.350	.352	.354	.356	.358	.360
2.3	.362	.364	.365	.367	.369	.371	.373	.375	.377	.378
2.4	.380	.382	.384	.386	.387	.389	.391	.393	.394	.396
2.5	.398	.400	.401	.403	.405	.407	.408	.410	.412	.413
2.6	.415	.417	.418	.420	.422	.423	.425	.427	.428	.430
2.7	.431	.433	.435	.436	.438	.439	.441	.442	.444	.446
2.8	.447	.449	.450	.452	.453	.455	.456	.458	.459	.461
2.9	.462	.464	.465	.467	.468	.470	.471	.473	.474	.476
3.0	.477	.479	.480	.481	.483	.484	.486	.487	.489	.490
3.1	.491	.493	.494	.496	.497	.498	.500	.501	.502	.504
3.2	.505	.507	.508	.509	.511	.512	.513	.515	.516	.517
3.3	.519	.520	.521	.522	.524	.525	.526	.528	.529	.530
3.4	.531	.533	.534	.535	.537	.538	.539	.540	.542	.543
3.5	.544	.545	.547	.548	.549	.550	.551	.553	.554	.555
3.6	.556	.558	.559	.560	.561	.562	.563	.565	.566	.567
3.7	.568	.569	.571	.572	.573	.574	.575	.576	.577	.579
3.8	.580	.581	.582	.583	.584	.585	.587	.588	.589	.590
3.9	.591	.592	.593	.594	.595	.597	.598	.599	.600	.601
4.0	.602	.603	.604	.605	.606	.607	.609	.610	.611	.612
4.1	.613	.614	.615	.616	.617	.618	.619	.620	.621	.622
4.2	.623	.624	.625	.626	.627	.628	.629	.630	.631	.632
4.3	.633	.634	.635	.636	.637	.638	.639	.640	.641	.642
4.4	.643	.644	.645	.646	.647	.648	.649	.650	.651	.652
4.5	.653	.654	.655	.656	.657	.658	.659	.660	.661	.662
4.6	.663	.664	.665	.666	.667	.667	.668	.669	.670	.671
4.7	.672	.673	.674	.675	.676	.677	.678	.679	.679	.680
4.8	.681	.682	.683	.684	.685	.686	.687	.688	.688	.689
4.9	.690	.691	.692	.693	.694	.695	.695	.696	.697	.698
5.0	.699	.700	.701	.702	.702	.703	.704	.705	.706	.707
5.1	.708	.708	.709	.710	.711	.712	.713	.713	.714	.715
5.2	.716	.717	.718	.719	.719	.720	.721	.722	.723	.723
5.3	.724	.725	.726	.727	.728	.728	.729	.730	.731	.732
5.4	.732	.733	.734	.735	.736	.736	.737	.738	.739	.740

Table of Common Logarithms (*Continued*)

	.00	.01	.02	.03	.04	.05	.06	.07	.08	.09
5.5	.740	.741	.742	.743	.744	.744	.745	.746	.747	.747
5.6	.748	.749	.750	.751	.751	.752	.753	.754	.754	.755
5.7	.756	.757	.757	.758	.759	.760	.760	.761	.762	.763
5.8	.763	.764	.765	.766	.766	.767	.768	.769	.769	.770
5.9	.771	.772	.772	.773	.774	.775	.775	.776	.777	.777
6.0	.778	.779	.780	.780	.781	.782	.782	.783	.784	.785
6.1	.785	.786	.787	.787	.788	.789	.790	.790	.791	.792
6.2	.792	.793	.794	.794	.795	.796	.797	.797	.798	.799
6.3	.799	.800	.801	.801	.802	.803	.803	.804	.805	.806
6.4	.806	.807	.808	.808	.809	.810	.810	.811	.812	.812
6.5	.813	.814	.814	.815	.816	.816	.817	.818	.818	.819
6.6	.820	.820	.821	.822	.822	.823	.823	.824	.825	.825
6.7	.826	.827	.827	.828	.829	.829	.830	.831	.831	.832
6.8	.833	.833	.834	.834	.835	.836	.836	.837	.838	.838
6.9	.839	.839	.840	.841	.841	.842	.843	.843	.844	.844
7.0	.845	.846	.846	.847	.848	.848	.849	.849	.850	.851
7.1	.851	.852	.852	.853	.854	.854	.855	.856	.856	.857
7.2	.857	.858	.859	.859	.860	.860	.861	.862	.862	.863
7.3	.863	.864	.865	.865	.866	.866	.867	.867	.868	.869
7.4	.869	.870	.870	.871	.872	.872	.873	.873	.874	.874
7.5	.875	.876	.876	.877	.877	.878	.879	.879	.880	.880
7.6	.881	.881	.882	.883	.883	.884	.884	.885	.885	.886
7.7	.886	.887	.888	.888	.889	.889	.890	.890	.891	.892
7.8	.892	.893	.893	.894	.894	.895	.895	.896	.897	.897
7.9	.898	.898	.899	.899	.900	.900	.901	.901	.902	.903
8.0	.903	.904	.904	.905	.905	.906	.906	.907	.907	.908
8.1	.908	.909	.910	.910	.911	.911	.912	.912	.913	.913
8.2	.914	.914	.915	.915	.916	.916	.917	.918	.918	.919
8.3	.919	.920	.920	.921	.921	.922	.922	.923	.923	.924
8.4	.924	.925	.925	.926	.926	.927	.927	.928	.928	.929
8.5	.929	.930	.930	.931	.931	.932	.932	.933	.933	.934
8.6	.934	.935	.936	.936	.937	.937	.938	.938	.939	.939
8.7	.940	.940	.941	.941	.942	.942	.943	.943	.943	.944
8.8	.944	.945	.945	.946	.946	.947	.947	.948	.948	.949
8.9	.949	.950	.950	.951	.951	.952	.952	.953	.953	.954
9.0	.954	.955	.955	.956	.956	.957	.957	.958	.958	.959
9.1	.959	.960	.960	.960	.961	.961	.962	.962	.963	.963
9.2	.964	.964	.965	.965	.966	.966	.967	.967	.968	.968
9.3	.968	.969	.969	.970	.970	.971	.971	.972	.972	.973
9.4	.973	.974	.974	.975	.975	.975	.976	.976	.977	.977
9.5	.978	.978	.979	.979	.980	.980	.980	.981	.981	.982
9.6	.982	.983	.983	.984	.984	.985	.985	.985	.986	.986
9.7	.987	.987	.988	.988	.989	.989	.989	.990	.990	.991
9.8	.991	.992	.992	.993	.993	.993	.994	.994	.995	.995
9.9	.996	.996	.997	.997	.997	.998	.998	.999	.999	.999

We now look for the logarithm 0.719 in the table and read the number (5.2 and 0.04 from the row and column shaded in gray) from the corresponding row and column. Thus the number we want is

$$a = 5.24 \times 10^4$$

(Note that the logarithm 0.719 appears twice in the table. Occasionally the same logarithm may even appear three times. When this happens, you must average the two or three numbers which correspond to the same logarithm and round to three significant figures. In the case of the logarithm 0.719 the two numbers are 5.23 and 5.24. Their average is 5.235, which rounds to 5.24.)

Finding the antilogarithm of a negative number is slightly more complicated because there are no negative logarithms in the table. The safest way to do this is to use the following property of exponents:

$$10^{-\alpha} = \frac{1}{10^{\alpha}}$$

Thus if we want to find the antilogarithm of -4.719, we can write

$$10^{-4.719} = \frac{1}{10^{4.719}} = \frac{1}{10^4 \times 10^{0.719}}$$

The table can now be used to find $10^{0.719}$, and we have

$$10^{-4.719} = \frac{1}{5.24 \times 10^4} = \frac{1}{5.24} \times \frac{1}{10^4} = 0.190 \times 10^{-4} = 1.90 \times 10^{-5}$$

EXAMPLE A2.2 Evaluate each of the following expressions: (a) $10^{5.326}$; (b) antilog(-5.017); (c) $10^{-10.962}$.

Solution

a) $10^{5.326} = 10^5 \times 10^{0.326} = 2.12 \times 10^5$

b) $\text{antilog}(-5.017) = 10^{-5.017} = \frac{1}{10^{5.017}} = \frac{1}{10^5 \times 10^{0.017}} = \frac{1}{1.04 \times 10^5}$

$$= 0.96 \times 10^{-5} = 9.6 \times 10^{-6}$$

c) $10^{-10.962} = \frac{1}{10^{10.962}} = \frac{1}{10^{10} \times 10^{0.962}} = \frac{1}{9.16 \times 10^{10}}$

$$= 0.109 \times 10^{-10} = 1.09 \times 10^{-11}$$

APPENDIX 3 Answers to Selected Questions and Problems

Chapter 1

1.3(a) 0.356 cm; (c) 0.1024 m^2. **1.5**(a) 99°F; (c) 176°F. **1.7**(a) Incorrect, 21 *g*; (c) incorrect—not a unity factor. **1.9**(a) 7.124 73 $\times 10^5$; (c) 3.9268 $\times 10^2$ cm; (e) 4.62 $\times 10^{-10}$ m; (g) 3.00 $\times 10^2$ g. **1.11**(a) 1.96 $\times 10^{10}$; (c) 3.31 $\times 10^2$. **1.13**(a) 78.92; (c) 78.92; (e) 78.92. **1.15**(a) Mm; (c) cm; (e) pm; (g) Eg. **1.17**(a) 1.31 $\times 10^3$ μs; (c) 4.97 $\times 10^{-6}$ g; (e) 236 pm; (g) 1.03 $\times 10^3$ kg m^{-3}. **1.19** 3480 g. **1.21** (for 1.19) $l \rightarrow V \xrightarrow{\rho} m$. **1.23**(a) and (c) Not unity factors; (e) unity factor.

Chapter 2

2.3 Greater mass of Hg atom. **2.5** Ca atoms smaller, packed closer. **2.7** Oxygen in water vapor came from black solid. **2.9**(a) 1.674; (c) 2 to 1. **2.11**(d) No units. **2.13**(a) 2(6.022 $\times 10^{23}$); (c) 0.3333; (e) 2.158 $\times 10^{24}$. **2.15** $N_A = N/n$. **2.17**(a) 137.34 g mol^{-1}; (c) 208.246 g mol^{-1}; (e) 101.96 g mol^{-1}. **2.19**(a) n = 2.051 mol H_2O; (c) N = 1.235 $\times 10^{24}$ molecules H_2O. **2.21**(a) 39.6 g; (c) 78.114 g mol^{-1}; (e) 1.83 $\times 10^{24}$ C atoms. **2.23**(a) 1.853 mol Al; (c) 443.9 g Br; (e) $AlBr_3$. **2.25** Cl_2O_7. **2.27** C_5H_7N; $C_{10}H_{14}N_2$. **2.29** Fe_3O_4, 72.4 g Fe; 1.30 mol Fe. **2.31**(a) $Sb_2S_3 + 3Fe \rightarrow 3FeS + 2Sb$; (c) $3NO_2 + H_2O \rightarrow 2HNO_3 + NO$; (e) $2MnO_2 + 4NaOH + O_2 \rightarrow 2Na_2MnO_4 + 2H_2O$. **2.33**(a) $2H_2 + O_2 \rightarrow 2H_2O$; (c) $H_2O + C \rightarrow CO + H_2$; (e) $CH_4 + 2H_2O \rightarrow CO_2 + 4H_2$.

Chapter 3

3.1(a) 3; (c) 1. **3.3**(a) 3.01; (c) 5.00, 10.00; (e) 55.12. **3.5**(a) 2.64; (c) 35.25; (e) 4.83. **3.7** 517 g H_3BO_3; 50.6 g H_2. **3.9**(a) S; (c) S. **3.11** 906.6 kg Bi_2O_3 produced and consumed (net = 0), 373.9 kg SO_2, 813.1 kg Bi, 163.5 kg CO. **3.13** LiOH. **3.15** 98 percent. **3.17** C_4H_9. **3.19**(a) 600 J; (c) 83.7 J. **3.21** 10.6 kJ mol^{-1}. **3.23** −87.0 kJ mol^{-1}. **3.25**(a) −1411 kJ; (c) 490.7 kJ. **3.27** 0.074 38 mol dm^{-3}. **3.29** 0.074 62 mol dm^{-3}. **3.31**(a) 3.726 mol HCl; (c) 0.002 32 mol HCl. **3.33** 0.0247 dm^3. **3.35** 31 cm^3. **3.37** 955 cm^3. **3.39** 38 Mg. **3.41** 0.1235 dm^3. **3.43** 0.1112 *M*. **3.45** 51.06 percent.

Chapter 4

4.1(a) $2M + 2H_2O \rightarrow 2MOH + H_2$; (c) $2M + H_2 \rightarrow 2MH$; (e) $2M + Cl_2 \rightarrow 2MCl$. **4.3** Period is horizontal; group is vertical; Sn. **4.5** Lengths of periods are 2, 8, 8, 18, 18, 32. **4.9**(a) 3; (c) 1; (e) 4; (g) 4; (i) 5 or 3. **4.13** 9.10 $\times 10^{-28}$ g. **4.17** 1.9 mi. **4.19** Proton, 1.007 g mol^{-1}; neutron, 1.009 g mol^{-1}. **4.21**(a) 51*p*, 51*e*, and 71*n*; (c) 8*p*, 8*c*, and 10*n*; (e) 35*p*, 36*e*, and 45*n*. **4.25** 69.17% ^{63}Cu and 30.83% ^{65}Cu. **4.28** 15.995, 16.999, and 17.999.

Chapter 5

5.1 Atomic numbers are 2 more (Ca) and 2 less (S) than the noble gas Ar. **5.3** Shells 3 to 6 do not fill completely before the next shell begins filling. **5.5** Rb (1), C (4), Al (3), Kr (0), Se (6, 2), Sn (4), Cl (7, 1), Ba (2). **5.7** Both assumed shells; Bohr model more detailed. **5.9** λ is a wave property; *m* is a particle property. **5.11**(b) n = 4; (d) n = 3. **5.13** Uncertainty principle prohibits it. **5.15** Because of the dual wave-particle nature of the electron. **5.17** Probability of finding the electron is highest where the density of dots is greatest. **5.19** Always a finite, but small, possibility. **5.21** Larger n gives a larger area (volume) for the boundary-surface diagram. **5.23**(a) 4 times as great; (c) 9 times. **5.27** Smaller. **5.29** Density of the 2*s* electron is higher near the nucleus; B has one more proton than Be. **5.31** Minimizes repulsion among electrons. **5.33** False (e.g., $4s$ lower than $3d$). **5.35**(a) $1s^22s^22p^63s^23p^2$; (c) $1s^22s^22p^63s^23p^64s^23d^{10}4p^4$; (e) $1s^22s^22p^63s^23p^64s^23d^{10}4p^6$; (g) $1s^22s^22p^63s^23p^64s^23d^{10}4p^65s^1$. **5.37** Representative—outer shell; transition—inner shell. **5.39** Not true for transition elements. **5.41** Possess several subshells which are close in energy. **5.43** Electrons are added ($4f$) two shells below the outermost, hence little effect.

Chapter 6

6.1 Electrons occur in pairs. **6.3** Bond dissociation energy for Cl_2. **6.5** Ignores many other energy terms. **6.7** Nuclei repel when too close. **6.9** Forming Na^{2+} and Cl^{2-} are both energetically unfavorable processes. **6.11** Two of the four $3p$ electrons occupy the same orbital and repel each other strongly, making either one easier to remove. **6.13** Similar size. **6.15**(a) Covalent; (c) no compound; (e) ionic, MgF_2; (g) covalent; (i) ionic, Ba_3N_2. **6.17**(a), (c), (g), and (i) All noble-gas structures; (e) and (k) not noble-gas structures. **6.19**(a) $1s^22s^22p^63s^23p^63d^6$; (c) $1s^22s^22p^63s^23p^63d^3$. **6.21** Similar degrees of attraction for their valence electrons.

6.23(a) 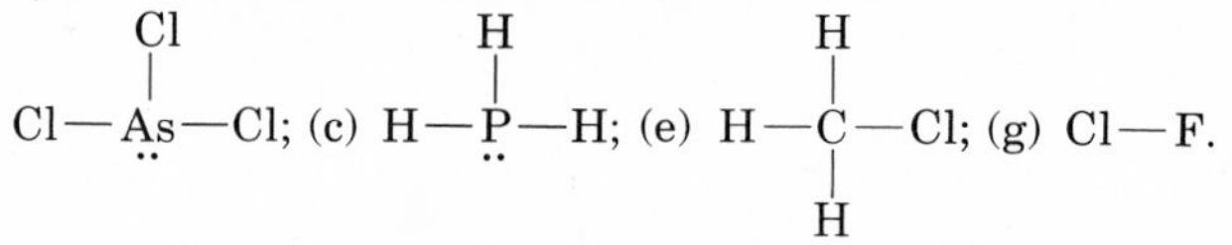

6.25(a) H—C≡N; (c) Cl—N̈=O; (e) 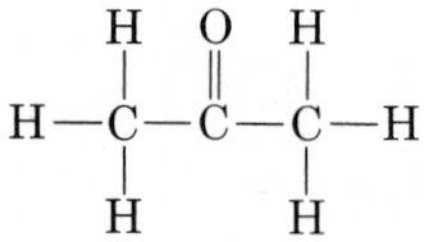

(g) 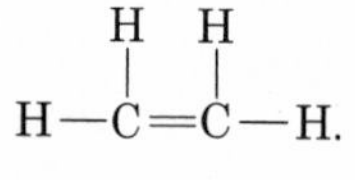 **6.27**(a) 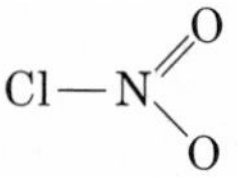 ; (c) 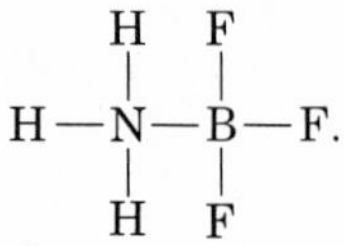

6.29(a) KF ionic; (c) HCl covalent; (e) $K_2Cr_2O_7$ both; (g) MgF_2 ionic. **6.31** $NaNO_3$ (ionic). **6.34**(a) Ca^{2+}, Sr^{2+}, Ba^{2+}, Ra^{2+} (largest); (c) Mg^{2+}, Na^+, Ne, F^- (largest). **6.35**(a) Ar; (c) V^{3+}; (e) Ga^{3+}. **6.37** 143 pm, 286 pm; no.

Chapter 7

7.2(a) and (e) do not; (c) and (g) do. **7.5** BF_4^-, 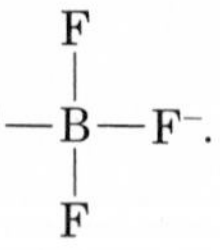 **7.7** No d orbitals available on C. **7.9**(a) Molecular and orbital—both linear; (c) molecular—bent; orbital—tetrahedral; (e) molecular—trigonal pyramid; orbital—tetrahedral. **7.11**(a) iv; (c) ii; (e) iii; (g) iv. **7.13** Yes (AX_2E_0). **7.15**(a) O sp^3; (c) Pb sp^2; (e) S d^2sp^3; (g) Be sp; (i) S sp^3. **7.17** Possible location of the electron at a particular instant. **7.21** 0.168. **7.23** Negative ions are larger, more diffuse; greater polarizability yields more covalent character. **7.25** Power of attraction for shared electrons; F (4.0) highest; Cs (0.7) lowest; increases left to right and bottom to top. **7.27**(a) H—H, H—C, H—N, H—F (most ionic); (c) N—F, C—F, Al—F, K—F (most ionic). **7.29**(a) No; (c) yes; (e) no; (g) yes. **7.31** "Apparent" charge; equals true ionic charge; agrees with direction of polarity. **7.33**(a) $\overset{+1}{K}\overset{+7}{Mn}\overset{-2}{O_4}$; (c) $\overset{+1}{Li}\overset{-1}{H}$; (e) $\overset{+1}{Na_2}\overset{+2}{S_2}\overset{-2}{O_3}$; (g) $\overset{0}{S_8}$; (i) $\overset{+5}{P_2}\overset{-2}{O_7^{4-}}$. **7.35**(a) and (c) Not free radicals; (e) Cl is free radical.

7.37(a) 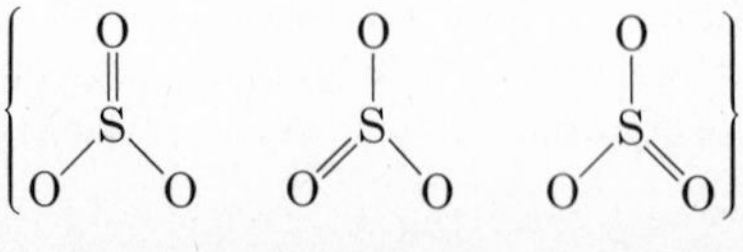 ;

(c) {N=N=O N≡N—O}. **7.39**(a) O=C=O; (c)

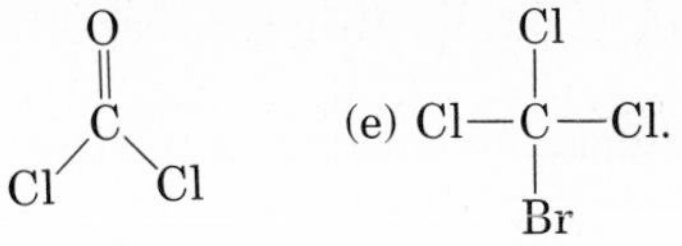

Chapter 8

8.1 They form liquids and solids. **8.3** van der Waals—all intermolecular forces; dipole—intermolecular forces due to polar bonds; London—instantaneous forces in nonpolar molecules. **8.5** $CH_3CH_2CH_2CH_2CH_2CH_3$, $CH_3CH_2CH(CH_3)CH_2CH_3$, $CH_3CH(CH_3)CH_2CH_2CH_3$, $CH_3CH(CH_3)CH(CH_3)CH_3$, $CH_3C(CH_3)_2CH_2CH_3$. **8.8**(a) $2C_8H_{18} + 25O_2 \rightarrow 16CO_2 + 18H_2O$. **8.11**

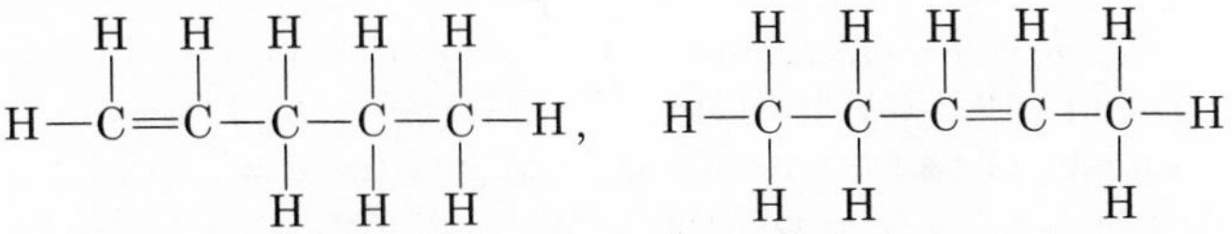

no. **8.13**(a) Alkane; (c) alkene; (e) alkane. **8.15** No double-bond character in cyclohexane. **8.17**(a) Hydrogen bonding; (c) dipole forces; (e) hydrogen bonding. **8.19** $CH_3CH_2OCH_2CH_3$, $CH_3CH_2CH_2OCH_3$, $CH_3CH_2CH_2CH_2OH$, $CH_3CH(OH)CH_2CH_3$.

8.21(a) 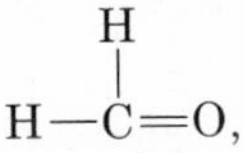, formaldehyde; (c) 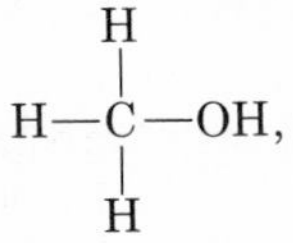, methyl alcohol (methanol).

8.23(a) 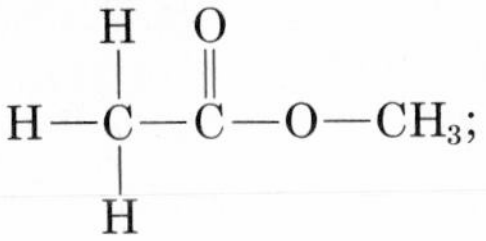; acid has higher boiling point. **8.25**(a)

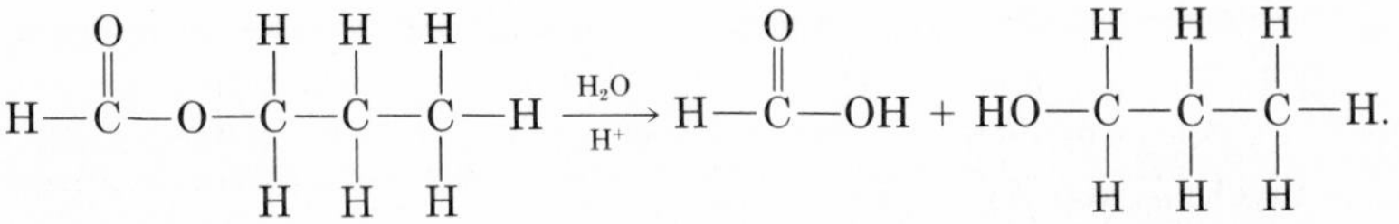

8.27(a) *n*-Butane, straight-chain compound has higher boiling point; (c) acetic acid, two polar groups produce extensive hydrogen bonding; (e) ethylene glycol, extensive hydrogen bonding. **8.29** Carbonyl$-\overset{\overset{\displaystyle O}{\|}}{C}-$; carboxyl$-\overset{\overset{\displaystyle O}{\|}}{C}-OH$. **8.31** $CH_3CH_2CH_2NH_2$, $CH_3CH(NH_2)CH_3$, primary; $CH_3CH_2NH(CH_3)$, secondary; $(CH_3)_3N$, tertiary. **8.33** Partial double-bond character (within layers, not between) in graphite. **8.35** Three full O's for any given pair of Si atoms, plus 1/2 share each of four O's linked to adjacent Si atoms. **8.37**(a)

8.37(a) 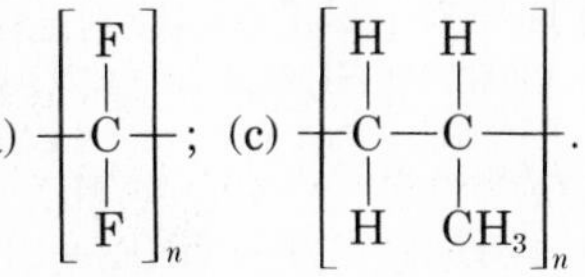

8.40(a) $HO\overset{\overset{\displaystyle O}{\|}}{C}(CH_2)_6\overset{\overset{\displaystyle O}{\|}}{C}OH + H_2N(CH_2)_4NH_2$; (c)  $+ HO(CH_2)_2OH$.

Chapter 9

9.1 55.65 kPa. **9.3** Barometer—single column; manometer—U tube. **9.5** 9.8 kPa; 10.34 m. **9.7** If one increases (P or V), the other (V or P) decreases. **9.9** 122 liters. **9.11** 1213 mmHg. **9.13** 34.2 cm^3. **9.15** Think about the equals signs between the different temperatures (in °C, in °F, in K). **9.17**(a) 0.0110 mol; (c) 0.162 mol. **9.19**(a) 0.0955 g NH_3; (c) 3.11 g CO_2. **9.21** 246 K. **9.23**(a) 867 cm^3; (c) 710 cm^3. **9.25** 0.0506 dm^3. **9.27** 4×10^5 kPa. **9.29** 64.3 kPa. **9.31**(a) 0.001 02 mol; (c) 0.0724 mol. **9.33** 0.761 g Cu. **9.35**(a) 2990 J. **9.37** Root-mean-square velocity is proportional to the square root of T; increase of 10°C does not double the velocity. **9.39** 16 g mol^{-1}. **9.41** Xenon is a larger atom with more electrons. **9.42** The kinetic energy of the gas molecules is smaller; the potential energy of intermolecular attractions is a larger fraction of that kinetic energy; so the effect of intermolecular forces produces a larger deviation in V (and hence PV).

Chapter 10

10.1 Due to strong forces between the constituents (atoms, ions, or molecules) in the solid. **10.5** FCC. **10.7** 185.8 pm. **10.9** BCC (two atoms in unit cell). **10.11** Coordination number counts the number of nearest neighbors, not just those which agree with the formula ratio. **10.13** Long-chain molecules. **10.15** (c), (b), (a), (d). **10.17** Enthalpy of vaporization is proportional to intermolecular forces, not intramolecular bond strengths. **10.20** Boiling point relates to intermolecular forces, not bond strengths. **10.23**(a) Rate of escape increases; reentry rate decreases; (c) both rates decrease. **10.25** Vapor pressure of H_2O causes initial boiling in the evacuated flask until equilibrium vapor pressure is attained. **10.27**(a) 29 kPa; (c) 75°C; (e) pentane. **10.31** 94.77°C. **10.33** 0.1730 g H_2O will evaporate. **10.36** Below its critical T so that it can be liquefied at high P in the can. **10.39**(a) 0.065 66; (c) 0.2643. **10.41** 4.999 percent. **10.43**(a) 0.144. **10.45**(a) CH_3COOH; (c) CH_3CH_2OH; (e) CH_3OCH_3. **10.47** 597.1 mmHg, positive. **10.49** 389.4 mmHg, positive. **10.54** 31.4 mmHg. **10.57** 100.148°C(b.p.). **10.59** 256 g mol^{-1} (S_8 molecules). **10.61** 7780 g mol^{-1}.

Chapter 11

11.1 Extent of conductivity, relative to H_2O. **11.3** Strong acid. **11.5**(a) No reaction; (c) $Ag^+ + Br^- \rightarrow AgBr(s)$; (e) $3Cu^{2+} + 2PO_4^{3-} \rightarrow Cu_3(PO_4)_2(s)$. **11.7** 6.95 g $PbCl_2$. **11.9** 0.1113 M. **11.11** Li^+ and F^- ions are much smaller. **11.13** Minimal attraction among hydrated ions, thus they remain in solution. **11.15** H_3ONO_3; H_3OClO_4 (both contain H_3O^+). **11.17**(a) $NH_3 + H_2O \rightleftharpoons NH_4^+ + OH^-$; (c) $PO_4^{3-} + H_2O \rightleftharpoons HPO_4^{2-} + OH^-$. **11.19**(a) HI acid, H_2O base; (c) HSO_4^- acid, HCO_3^- base; (e) H_2CO_3 acid, NH_3 base. **11.21**(a) Strong acid measured via extensive ionization, not concentration; (c) strong acid readily donates (releases, transfers) protons. **11.23**(a) Strong; (c), (e), (g) and (i) all weak. **11.25**(a) $HCO_3^- + H_3O^+ \rightarrow H_2CO_3 + H_2O$; $HCO_3^- + OH^- \rightarrow CO_3^{2-} + H_2O$; (b) $NH_3CH_2COO + H_3O^+ \rightarrow NH_3CH_2COOH^+ + H_2O$; $NH_3CH_2COO + OH^- \rightarrow NH_2CH_2COO^- + H_2O$. **11.27**(a) NH_4^+; (c) HPO_4^{2-}; (e) H_3O^+. **11.29**(d) Most basic. **11.31**(a) $H_2SO_4 + H_2O \rightarrow H_3O^+ + HSO_4^-$ yes; $HSO_4^- + H_2O \rightarrow H_3O^+ + SO_4^{2-}$ no; (c) $HSO_4^- + CO_3^{2-} \rightarrow HCO_3^- + SO_4^{2-}$. **11.33**(a) SO_3 acid, H_2O base; (c) H_3BO_3 acid, OH^- base; (e) BF_3 acid, $SnCl_2$ base. **11.35**(a) Oxidized; (c) neither; (e) reduced; (g) neither. **11.37**(a) $e^- + 2H^+ + NO_3^- \rightarrow NO_2 + H_2O$; (c) $H_2O + H_3AsO_3 \rightarrow H_3AsO_4 + 2H^+ + 2e^-$; (e) $4OH^- + Cl^- \rightarrow ClO_2 + 2H_2O + 5e^-$. **11.39**(a) $8H^+ + Cr_2O_7^{2-} + 3SO_3^{2-} \rightarrow 2Cr^{3+} + 4H_2O + 3SO_4^{2-}$; (c) $6H^+ + 2MnO_4^- + 5H_2O_2 \rightarrow 2Mn^{2+} + 8H_2O + 5O_2$; (e) $12H^+ + 3VO_3^- + Al \rightarrow 3VO^{2+} + 6H_2O + Al^{3+}$; (g) $2Ce^{4+} + H_2O_2 \rightarrow 2Ce^{3+} + 2H^+ + O_2$. **11.41**(a) and (g) To completion; (c) and (e) scarcely at all. **11.43** 0.2050 M.

Chapter 12

12.1(a) $4s^24p^1$; (c) $3s^1$; (e) $3s^23p^4$; (g) $4s^24p^6$; (i) $2s^22p^1$. **12.3** Lose electrons (ns^1) easily due to their large atomic radii. **12.5**(a) No reaction; (c) $MgCO_3 \xrightarrow{\Delta} MgO + CO_2$; (e) yields mixture of K_2O, K_2O_2, KO_2. **12.7** False; e.g., NH_3 (covalent) gives a basic solution. **12.9**(a) $2Cl^-(aq) + Na^+(aq) + 2H_2O \rightarrow Cl_2 + H_2 + Na^+(aq) + 2OH^-(aq)$. **12.11** High concentration of metal ions (e.g., Ca^{2+}, Mg^{2+}); add Na_2CO_3 to precipitate. **12.13**(a) Small Be^{2+} ion hydrolyzes extensively; (c) bonds are relatively covalent; (e) polymeric structure. **12.15** First element is the smallest, thus forms bonds which are more covalent. **12.17**(a) False; basicity of oxides increases down a family; (c) smaller B^{3+} is a better polarizer and hence more likely to form covalent bonds. **12.19** Behaves as either an acid or a base: $Al_2O_3 + 3H_2O + 2OH^- \rightarrow 2Al(OH)_4^-$; $Al_2O_3 + 6H^+ \rightarrow 2Al^{3+} + 3H_2O$. **12.21**(a) $Ca_{10}(PO_4)_6(OH)_2 + SnF_2 \rightarrow Ca_{10}(PO_4)_6F_2 + Sn(OH)_2$; (c) $2In + 3Br_2 \rightarrow 2InBr_3$; (e) $2Sn^{2+} + O_2 + 4H_3O^+ \rightarrow 2Sn^{4+} + 6H_2O$. **12.23** Resistance of the heavier members of groups III to VIA to be oxidized to their highest state (group number); TlCl; $PbCl_2$. **12.27** NH_4NO_3, $C_3H_5(NO_3)_3$. **12.29**(a) $P_4O_6 + 6H_2O \rightarrow 4H_3PO_3$; (c) $Cl_2 + F^- \rightarrow$ no reaction; (e) $2H_2O_2 \rightarrow 2H_2O + O_2$. **12.31** Toxic clouds of air pollution formed by the action of sunlight on various gases; NO, NO_2, O_3. **12.33**(a) and (e) Possible; (c) very unlikely, Sn(V); (g) very unlikely, six large Br atoms around one S. **12.35**(a) trigonal planar; (c) trigonal bipyramidal; (e) angular. **12.39** 12.6×10^9 kg H_2SO_4 per year via air pollution. **12.41** Strong oxidizing agent; readily forms the hypochlorite (OCl^-) ion. **12.43**(a) IA, IIA; (c) IIIA; (e) IA; (g) VIIA.

Chapter 13

13.3 Equilibrium mixture. **13.4** 1.64. **13.7**(a) $K_c = [NO]^4[H_2O]^6/[NH_3]^4[O_2]^5$; (c) $K_c = [N_2]^2[O_2]/[N_2O]^2$; (e) $K_c = [CO_2]^2/[CO]^2[O_2]$. **13.9**(b) 4.14×10^{-3}. **13.11**(a) 5.1×10^{-4} mol^2 dm^{-6}. **13.13**(a) $K_c = [H_2O]^2$; (c) $K_c = [HCl]^2/[H_2O]$; (e) $K_c = [CO_2][N_2]/[N_2O]^2$. **13.15** 6.39×10^{-8} kPa^{-2}. **13.17** $[CO_2] = [H_2] = 0.67$ mol dm^{-3}; $[CO] = [H_2O] = 0.33$ mol dm^{-3}. **13.19** 1.22 percent. **13.21**(a) 1.93 mol. **13.23**(a) $[CO] = 0.0183$ mol dm^{-3}; $[CO_2] = 0.0158$ mol dm^{-3}. **13.24**(a) (i) Right; (ii) right. (d) (i) Right; (ii) right. (f) (i) Left; (ii) no change. **13.27** Exothermic. **13.29**(a) Less HI; (c) more HI. **13.31** 0.67, same. **13.33** 1.8×10^{-5} mol dm^{-3}.

Chapter 14

14.1(a) Strong acid; (c) weak acid; (g) weak base; (j) strong acid; (l) weak base. **14.3**(a) pH = 1.0; pOH = 13.0. (b) pH × 12.53; pOH × 1.47. **14.5**(a) pH = 12.23; pOH = 1.77. (c) pH = 12.57; pOH = 1.43. **14.7**(a) $HC_2H_3O_2 + H_2O \rightleftharpoons H_3O^+ + C_2H_3O_2^-$; $K_a = [H_3O^+][C_2H_3O_2^-]/[HC_2H_3O_2]$. (c) $H_2CO_3 + H_2O \rightleftharpoons H_3O^+ + HCO_3^-$; $K_{a_1} = [H_3O^+][HCO_3^-]/[H_2CO_3]$. $HCO_3^- + H_2O \rightleftharpoons H_3O^+ + CO_3^{2-}$; $K_{a_2} = [H_3O^+][CO_3^{2-}]/[HCO_3^-]$. (e) $NH_4^+ + H_2O \rightleftharpoons H_3O^+ + NH_3$; $K_a = [H_3O^+][NH_3]/[NH_4^+]$. **14.9**(a) $[H_3O^+] = 7.0 \times 10^{-6}$mol dm^{-3}; $[OH^-] = 1.4 \times 10^{-9}$ mol dm^{-3}. (c) $[H_3O^+] = 1.0 \times 10^{-11}$ mol dm^{-3}; $[OH^-] = 1.0 \times 10^{-3}$ mol dm^{-3}. **14.11**(a) $[H_3O^+] = 1.22 \times 10^{-10}$ mol dm^{-3}; $[OH^-] = 8.22 \times 10^{-5}$ mol dm^{-3}. (c) $[H_3O^+] = 3.61 \times 10^{-9}$ mol dm^{-3}; $[OH^-] = 2.77 \times 10^{-6}$ mol dm^{-3}. **14.13**(b) $[H_3O^+] = 3.6 \times 10^{-3}$ mol dm^{-3}; pH = 2.44. **14.15**(a) 2.55; (c) 1.50. **14.17**(a) 1.83×10^{-5} mol dm^{-3}; (c) 7.31×10^{-6} mol dm^{-3}. **14.19**(b) 11.66. **14.21**(a) $[H_3O^+] = 2.74 \times 10^{-2}$ mol dm^{-3}; pH = 1.56. **14.23**(a) $([H_3O^+][\cancel{F^-}]/[\cancel{HF}]) \times ([\cancel{HF}][OH^-]/[\cancel{F^-}]) = [H_3O^+][OH^-] = K_w$. **14.25**(a) i; (c) iv; (e) v; (g) iv. **14.29**(a) 4.74. **14.31**(a) 8.96. **14.34**(b) 35.1 cm^3. **14.35**(a) Methyl red; (c) alizarin yellow; (e) methyl orange. **14.39** Conductivity, obtain $[OH^-]$ from pH, radioactive tracer. **14.41**(a) 4.8×10^{-2} mol dm^{-3}; (c) 7.8×10^{-5} mol dm^{-3}; (e) 2.2×10^{-10} mol dm^{-3}. **14.43**(a) 5.0×10^{-9} mol; (c) 1.6×10^{-6} mol. **14.45**(d) BaF_2 will precipitate. **14.47**(a) Solubility is lowered.

15.3 Yes. Diatomic gas has larger C_V because energy is required for vibration and rotation. **15.5** 4.00 s. **15.6** 3.44 mol. **15.8** Energy cannot be created or destroyed. $q_V = \Delta U$; $q_P = \Delta U + P\,\Delta V$; $H = U + PV$; $\Delta H^\ominus = \Sigma D$(bonds broken) $- \Sigma D$(bonds formed). **15.11** CF_3CF_3 largest; Xe smallest. **15.13** Electronic energy—see Table 15.3. **15.14** Since $\Delta U = q_V$, macroscopic measurement of heat absorbed at constant volume determines internal energy change. **15.16**(a) $\Delta T = 30$ K; (c) $\Delta T = 6.26$ K; (e) 623.6 J. **15.18** -2981 kJ mol^{-1}. **15.20** -2977 kJ mol^{-1} (for 15.18). **15.23** C = 494 J K^{-1}; $\Delta H_m = 14.91$ kJ mol^{-1}. **15.24**(a) -802.3 kJ mol^{-1}; (c) -518.4 kJ mol^{-1}; (e) -41.2 kJ mol^{-1}. **15.26**(a) -812 kJ mol^{-1}, 1.2 percent; (c) -147 kJ mol^{-1}, 72 percent; (e) -42 kJ mol^{-1}, 1.9 percent. **15.29**(a) Exothermic—stronger bonds in products; (c) exothermic—more bonds in products, weak N=N. **15.32** D_{O-O} (in O_3) = 604.3 kJ mol^{-1}/2 = 302.2 kJ mol^{-1}; value is about the average of $D_{O=O}$ and D_{O-O}. Two resonance structures for O_3. **15.34**(b) Much electricity generated by burning coal, the fuel in greatest supply. **15.38** Milk: 1416 kJ (338 kcal); hamburger: 987 kJ (236 kcal); potato chips: 446 kJ (107 kcal).

Chapter 16

16.1 None violate the first law. The first law does not predict the spontaneous direction. **16.3** 1/9, $1/(3^N)$. **16.5**(a) $K_c = 4$. **16.7**(a) 1; (c) 10. **16.9** 265.6 J K^{-1} mol^{-1}. **16.13**(a) Ga(l)—liquid state more disordered; (c) NaF(s), +1 and −1 ions; (e) C_2H_5OH—greater mass, more complex molecule. **16.15**(a) -173 J K^{-1} mol^{-1}; (c) $+137.2$ J K^{-1} mol^{-1}; (e) -179.8 J K^{-1} mol^{-1}. **16.19** -247.5 J K^{-1} mol^{-1}, cannot tell; $\Delta S_{tot} = -98.8$ J K^{-1} mol^{-1}. **16.21** ΔS_{tot} values (a) $+110$ J K^{-1} mol^{-1}; (c) $+279.9$ J K^{-1} mol^{-1}; (e) $+231.2$ J K^{-1} mol^{-1}. **16.23**(a) -757.0 kJ mol^{-1}; (c) 140.0 kJ mol^{-1}. **16.25**(a) -1482 kJ mol^{-1}; (c) -175.6 kJ mol^{-1}. **16.27**(a) (ii) spontaneous at low T, nonspontaneous at high T; (c) (ii) same as (a); (e) (iv) never spontaneous. **16.29** For Fe: (a) at 800 K, $\Delta G^\ominus = +50.7$ kJ mol^{-1}; no; (b) at 1500 K, $\Delta G^\ominus = -334.3$ kJ mol^{-1}; yes. **16.32**(a) 3.89×10^{-34}; (c) 2×10^{140}. **16.34** $-\Delta H^\ominus/2.303R$ is the slope, and $\Delta S^\ominus/2.303R$ is the intercept of a graph of $\log_{10}K^\ominus$ versus $1/T$. **16.37**(b) $+8.585$ kJ mol^{-1}. **16.39**(b) 7.60 g; (c) 48.04 g.

Chapter 17

17.2 anode, oxidation; cathode, reduction. **17.5**(a) $2H_2O \rightarrow O_2 + 4H^+ + 4e^-$; (c) $2H_2O \rightarrow 2H_2 + O_2$. **17.7** F_2Cl_2, Fe^{3+}, I_2, AgCl, Zn^{2+}, Pb. **17.9**(a) $2H_2O(l) + 2Cl^-(aq) \rightarrow H_2(g) + Cl_2(g) + 2OH^-(aq)$; (c) easier to reduce H_2O than Na^+. **17.11** $Na(Hg)_x$ is a weaker reducing agent than Na(s) because most atoms in $Na(Hg)_x$ are Hg atoms, not Na atoms. **17.13** Anode: $2O^{2-} + C \rightarrow CO_2 + 4e^-$; cathode: $3e^- + Al^{3+} \rightarrow Al$; overall: $4Al^{3+} + 3C + 6O^{2-} \rightarrow 3CO_2 + 4Al$. **17.15** Anode: $Cu(s) \rightarrow Cu^{2+}(aq) + 2e^-$; cathode: $Cu^{2+}(aq) + 2e^- \rightarrow Cu(s)$; overall: no net reaction for Cu(s), $Cu^{2+}(aq)$. **17.17** Fe is oxidized to Fe^{2+} which is a poor oxidizing agent and hence is not readily reduced back to Fe at the cathode. **17.19** Sn and Cu in the anode sludge; Al^{3+} in solution. **17.25** (a) 4.57 A; (c) 4.10 A. **17.27** 0.0925 mol dm^{-3}. **17.29** 168 kg. **17.31** To allow migration of ions and achieve charge balance in both cell compartments; K^+ moves left to right, Cl^- right to left. **17.32**(a) Anode: $Mg(s) \rightarrow Mg^{2+}(aq) + 2e^-$; cathode: $Fe^{3+}(aq) + e^- \rightarrow Fe^{2+}(aq)$; (c) anode; $2I^-(aq) \rightarrow I_2(s) + 2e^-$; cathode; $Cr_2O_7^{2-}(aq) + 14H^+(aq) + 6e^- \rightarrow 2Cr^{3+}(aq) + 7H_2O$; (e) anode: $Sn(s) \rightarrow Sn^{2+}(aq) + 2e^-$; cathode: $Cl_2(g) + 2e^- \rightarrow 2Cl^-(aq)$. **17.35**(a) Anode: $2H_2O + NO \rightarrow NO_3^- + 4H^+ + 3e^-$; cathode: $AuCl_4^- + 3e^- \rightarrow Au + 4Cl^-$; overall: $2H_2O + NO + AuCl_4^- \rightarrow NO_3^- + Au + 4H^+ + 4Cl^-$; (c) anode: $H_2O + V^{3+} \rightarrow VO^{2+} + 2H^+ + 2H^+ + e^-$; cathode:

$Au^+ + e^- \rightarrow Au$; overall: $H_2O + V^{3+} + Au^+ \rightarrow VO^{2+} + 2H^+ + Au$. **17.37**(a) $Mg|Mg^{2+}||Fe^{3+}, Fe^{2+}|Pt$; (c) $Pt|I^-, I_2||Cr_2O_7^{2-}, Cr^{3+}|Pt$; (e) $Sn|Sn^{2+}||Cl^-, Cl_2|Pt$. **17.39**(a) $O_2 + 4H^+ + 4I^- \rightarrow 2H_2O + 2I_2(+0.65\text{ V})$; (c) $2Fe^{3+} + Zn \rightarrow 2Fe^{2+} + Zn^{2+}(+1.53\text{ V})$; (e) $Cl_2 + 2Br^- \rightarrow Br_2 + 2Cl^-(+0.29\text{V})$. **17.41**(a) $Zn + 2MnO_2 + 2NH_4^+ \rightarrow 2MnO(OH) + 2NH_3 + Zn^{2+}$; (c) $Cd + 2Ni(OH)_3 \rightarrow CdO + 2Ni(OH)_2 + H_2O$.

Chapter 18

18.3(a) 0.232 mol dm^{-3} h^{-1}; (c) 0.161 mol dm^{-3} h^{-1}; (e) 0.088 mol dm^{-3} h^{-1}. **18.5** 0.1304 mol dm^{-3} h^{-1} (0 to 5.0 h); 0.119 mol dm^{-3} h^{-1} (2 to 3 h). **18.7**(a) First order in X, in Y, and in Z; third order overall; (c) 1.5 order in X; −1 order in Y; 0.5 order overall. **18.9**(a) First order in NO and H_2; second order overall; (c) 2.4 mol^{-1} dm^3 s^{-1}. **18.11** Number of bimolecular collisions is proportional to the concentration of each of the two reacting species. **18.13**(d) is fastest; (c) is slowest. **18.15** The required termolecular gas-phase collision is very unlikely. **18.17** Rate $= k(c_{NO})^2(c_{Cl_2})$; termolecular, single-step mechanism. **18.19**(a) $CH_3COOCH_3 + H_2O \rightarrow CH_3COOH + HOCH_3$; (c) homogeneous acid catalysis. **18.21**(a), (c), and (d) are compatible with the rate law. **18.23**(a) $NO + O_3 \rightarrow NO_2 + O_2$. **18.25** 101 kJ mol^{-1}. **18.29**(a) 19 s; (c) 21 s. **18.30**(a) 0.16 mol; (c) 120 s. **18.32** $c_0 = 1.00 \times 10^{-3}$ mol dm^{-3}; $t_{1/2} = 259$ s. **18.35**(a), (c), and (d) are true.

Chapter 19

19.1 Chemical reactions involve rearrangements of electrons within atoms; nuclear reactions involve rearrangement of protons and neutrons within nuclei. **19.3**(a) ${}^4_2He$ or ${}^4_2\alpha$; ${}^{\ 0}_{-1}e$ or ${}^{\ 0}_{-1}\beta$; ${}^0_0\gamma$ or γ. (c) Can break bonds or ionize molecules in living tissue, disrupt cell membranes, damage genes, etc.; ${}^4_2\alpha > {}^{\ 0}_{-1}e > \gamma$ in disruptive power; $\gamma > {}^{\ 0}_{-1}e > {}^4_2\alpha$ in penetrating power. **19.5** ${}^{237}_{93}Np \xrightarrow{\alpha} {}^{233}_{91}Pa \xrightarrow{\beta} {}^{233}_{92}U \xrightarrow{\alpha} {}^{229}_{90}Th \xrightarrow{\alpha} {}^{225}_{88}Ra \xrightarrow{\beta} {}^{225}_{89}Ac \xrightarrow{\alpha} {}^{221}_{87}Fr \xrightarrow{\alpha} {}^{217}_{85}At \xrightarrow{\alpha} {}^{213}_{83}Bi \xrightarrow{\beta} {}^{213}_{84}Po \xrightarrow{\alpha} {}^{209}_{82}Pb \xrightarrow{\beta} {}^{209}_{83}Bi$. **19.7** Positive electron; ${}^{\ 0}_{+1}e$. **19.9**(a) 1.40; (c) 1.41; (e) 1.33; (g) 1.56; (i) 1.15, large ratio, β emission; small ratio, positron emission. **19.11**(a) ${}^{237}_{93}Np$, 1.8×10^{14} g cm^{-3}; ${}^{209}_{82}Pb$, 1.8×10^{14} g cm^{-3}. **19.13**(a) ${}^{247}_{96}Cm + {}^{12}_6C \rightarrow {}^{254}_{102}No + 5{}^1_0n$; (c) ${}^{245}_{94}Pu + {}^{12}_6C \rightarrow {}^{253}_{100}Fm + 4{}^1_0n$. **19.15** Bombarding nuclei with α particles gives fast neutrons; moderate by passing through wax to get slow neutrons. **19.17**(a) ${}^{173}_{78}Pt$ odd number of neutrons; proton rich; (c) ${}^{16}_6C$ neutron rich; ${}^{16}_8O$ very stable since the number of each n and p is 8, a magic number. **19.19**(a) 3.13 g; (c) 2.33×10^{-8} g. **19.21** 0.755 mg. **19.23** 1.849 h. **19.25** About 30 000 years. **19.27** Natural radiation from isotopes in our surroundings and from cosmic radiation; cannot escape it; obscures accurate study of the amount of radiation above background. **19.29**(a) Prepare a saturated aqueous solution of $BaSO_4$ labeled with ${}^{131}Ba$; evaluate activity in the liquid phase; (c) use ${}^{18}O$ to label ClO_3^-; check if ${}^{18}O$ is transferred to H_2O or solely to ClO_3^-. **19.31** $\Delta H_m = -216$ GJ mol^{-1}. **19.35** Restrict plasma with a magnetic field; use a laser beam to heat the deuterium and tritium. **19.37** Rate (235/238) = 1.004.

Chapter 20

20.3 Atoms have low mass, small radii, ability to form 1, 2, 3, and 4 covalent bonds, respectively; also, catenation ability of C. **20.5** Stronger bonds are (a) C—C; (c) Si—O; (e) Si—O. **20.7** C—H bonds do not react readily forcing reactions to occur at functional groups. **20.9** Solvent; maintain rigidity of cell walls; thermoregulator. **20.13** $HOCH_2CHOHCH_2OH + C_{17}H_{33}COOH$. **20.17** Saponification of fats gives water-soluble soaps with $—COO^-$ in hydrophilic heads; detergents contain $—SO_3^-$ and do not precipitate with cations. **20.18** $2C_{17}H_{35}COOH(aq) + Ca^{2+}(aq) \rightarrow$

$(C_{17}H_{35}COO)_2Ca(s) + 2H^+(aq)$; soap + hard water forms scummy precipitate. **20.21**

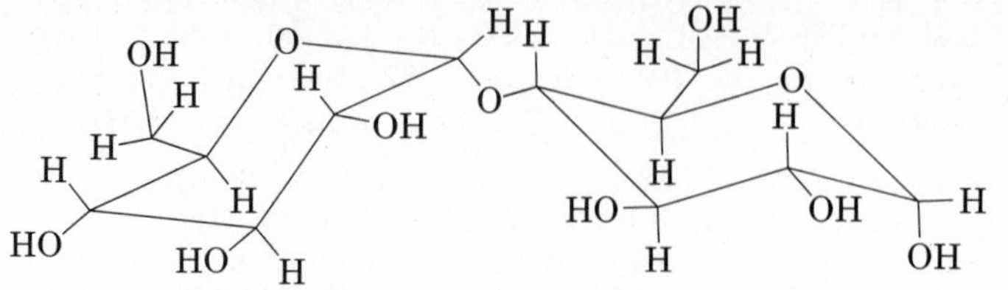

20.24(a) 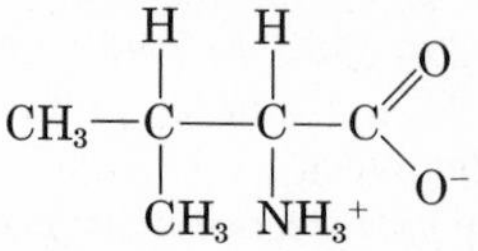 ; (c) 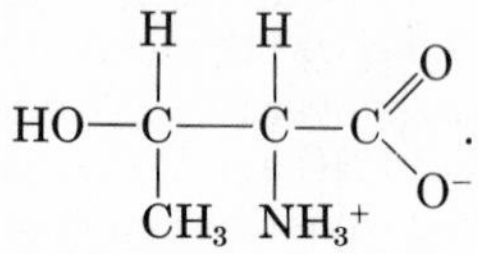 . **20.25** 27 possible tripeptides (3 with only one amino acid, 6 with two, and 18 with three). **20.27**(a) Hydrophobic residues; (c) intermediate. **20.29**(a) Yes ($\Delta G^\ominus = -162.1$ kJ mol^{-1}; emf = +0.42 V). **20.31**(a) 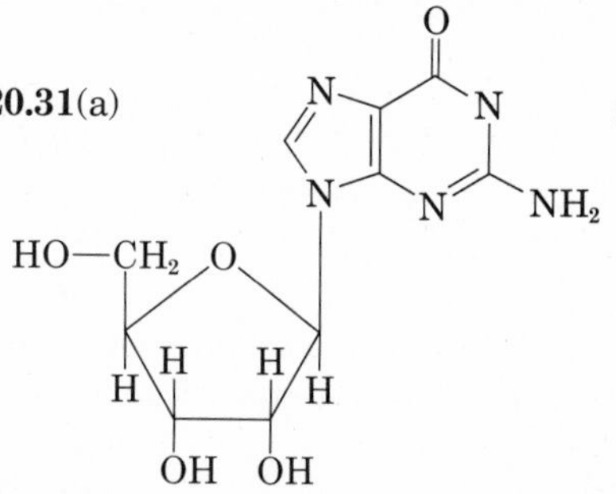 ; (c) 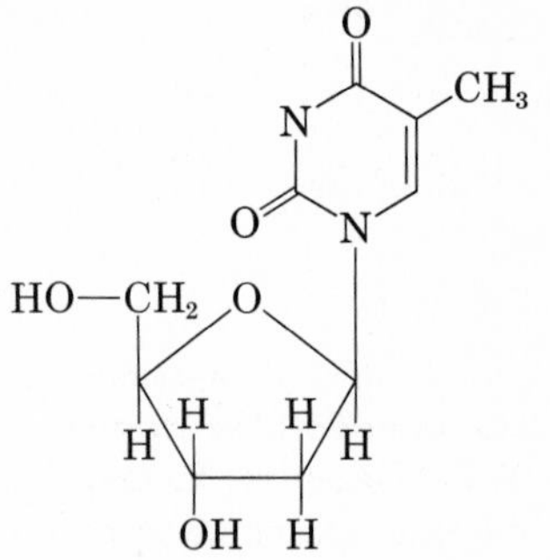

20.34(a) Ser-Cys-STOP-Asn-. **20.35**(c) Contains no termination codon. **20.37**(a) TAACC; (c) TACGT.

Chapter 21

21.1 Visible light is just one form (given range of λ) of electromagnetic radiation. **21.3** $\nu = c/\lambda$. **21.5**(a) Visible (blue); (c) 570 nm; visible (yellow-green); (e) infrared; (g) 39.7 m, radio waves. **21.7** Waves (i.e., points of maximum amplitude). **21.9**(a) $A_1 + 2A_1 + 3A_1 = 6A_1$. **21.11** 4.91×10^{-33} g. **21.13** $E = hc/\lambda$. **21.16** Perfect agreement for (a) 486.2 nm and (c) 410.2 nm. **21.17**(a) 0.4578 aJ, 6.909×10^{14} s^{-1}, 433.9 nm, emitted; (c) 0.1977 aJ, 2.984×10^{14} s^{-1}, 1005 nm, absorbed; (e) 0.4844 aJ, 7.311×10^{14} s^{-1}, 410.1 nm, absorbed. **21.20**(a) 5.450×10^{-19} J; (c) 1.363×10^{-19} J. **21.21**(a) 4; (b) 10. **21.23** Sharp, principal, diffuse, fundamental. **21.25** Molecules absorb IR radiation which does excite them (increase the amplitude with which the molecule vibrates). **21.27** IR absorption peaks correspond to bonds not elements; some elements present may be in bonds which do not absorb in the IR. **21.28**(a) and (d) Polar bonds which produce the necessary change in dipole moment. **21.33** Bond orders: (a) 1; (c) 0; (e) 1; (g) 2.5; (i) 0. (c) and (i) do not exist; others are feasible. **21.35** Paramagnetism—due to unpaired electrons, weakly attracted into a magnetic field; diamagnetism—all e^- paired, repelled. **21.37** Circle is the cloud of delocalized π electron density. **21.39**(a) 6; (c) not identical in shape or energy. **21.41** Black, because light of λ = 589 nm not reflected. **21.43** High-intensity light provides energy to break the H—H or Cl—Cl bonds.

22.3 No difference in electronegativity to account for ionic bonds; not enough valence electrons for localized covalent bonds. **22.5** Reasonable; electron density is diffuse. **22.7**(a) Same as Li except with 3*s*, 3*p* levels; (c) similar to Na except 3*s* band is filled; still a conductor. **22.9** Not true; energy levels close together, but electrons obey the Pauli exclusion principle. **22.11** See Fig. 22.2; metals decrease in conductivity with increasing *T*; semiconductors are the opposite. **22.13**(a) true; (c) false. **22.15** Not true; used only as ores for expensive metals like Zr and Be. **22.18** CO_2 removes OH^- via $CO_2 + OH^- \rightarrow HCO_3^-$. **22.19** At 298 K, only HgO and Ag_2O; at 2000 K, all except Al_2O_3 and MgO. **22.21** MgO, 4210 K; SnO_2, 2155 K. **22.23** 933 to 943 K. **22.25** False; air pollution is a serious health problem, and control of emission conserves iron ore. **22.27**(a) $Fe_2O_3 \cdot \frac{3}{2}H_2O$; (c) yes; selected regions are cathodic, others anodic; (e) the ions from salt improve conduction and speed the electrochemical reaction. **22.31**(a) 3; (c) 2; (e) 1. **22.33** $[Co(NH_3)_5Cl]SO_4$; use $BaCl_2(aq)$ and $AgNO_3(aq)$ to determine whether SO_4^{2-} or Cl^- present in solution. **22.35**(a) Octahedral; one with cis Cl's and the other with trans Cl's; (c) octahedral;

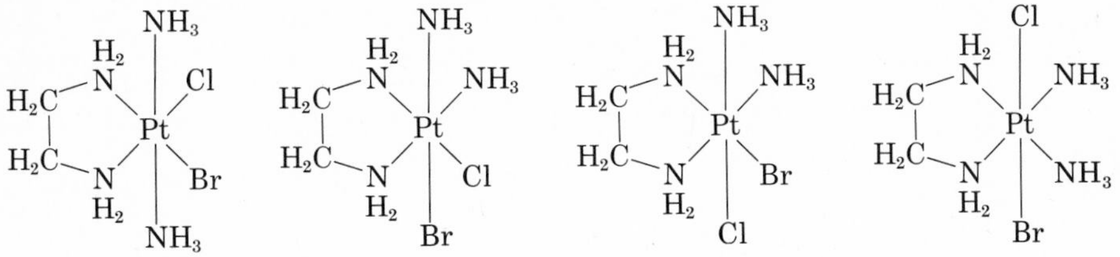

(e) square planar; cis and trans NH_3's. **22.37**(b) and (d)—other complexes are inert. **22.39** Na^+ ion too small to stay bonded inside the template ring.

INDEX

Page references in **boldface** refer to illustrations or tables.